Lecture Notes in Computer Science 7658

Commenced Publication in 1973
Founding and Former Series Editors:
Gerhard Goos, Juris Hartmanis, and Jan van Leeuwen

Xiaoyun Wang   Kazue Sako (Eds.)

Advances in Cryptology – ASIACRYPT 2012

18th International Conference on the Theory
and Application of Cryptology and Information Security
Beijing, China, December 2-6, 2012. Proceedings

 Springer

Volume Editors

Xiaoyun Wang
Tsinghua University
30 Shuangqing Road, 100084 Beijing, China
E-mail: xiaoyunwang@tsinghua.edu.cn

Kazue Sako
NEC Corporation, Central Research Laboratories
1753 Shimonumabe Nakahara, Kawasaki 211-8666, Japan
E-mail: k-sako@ab.jp.nec.com

ISSN 0302-9743　　　　　　　　　　e-ISSN 1611-3349
ISBN 978-3-642-34960-7　　　　　　 e-ISBN 978-3-642-34961-4
DOI 10.1007/978-3-642-34961-4
Springer Heidelberg Dordrecht London New York

Library of Congress Control Number: 2012951486

CR Subject Classification (1998): E.3, D.4.6, F.2, K.6.5, G.2, I.1, J.1

LNCS Sublibrary: SL 4 – Security and Cryptology

Typesetting: Camera-ready by author, data conversion by Scientific Publishing Services, Chennai, India

Printed on acid-free paper

Springer is part of Springer Science+Business Media (www.springer.com)

Preface

ASIACRYPT 2012, the 18th International Conference on Theory and Application of Cryptology and Information Security, was held during December 2–6 in Beijing International Convention Center, Beijing, China. The conference was sponsored by the International Association for Cryptologic Research (IACR) in cooperation with the Chinese Association for Cryptologic Research (CACR). It was also co-sponsored by the National Natural Science Foundation of China, Huawei Technologies Co. Ltd., and Intel Corporation.

From 241 valid submissions, 43 were accepted for publication after a very tough evaluation process. The Program Committee (PC) with the help of 256 external reviewers provided at least three independent reviews for each paper, and five or more for those with PC contributions.

There were also two invited talks. On Monday, Dan Boneh delivered "Pairing-based Cryptography: Past, Present, and Future" as the IACR Distinguished Lecture. On Wednesday, Chuanming Zong spoke on "Some Mathematical Mysteries in Lattices." In addition to the invited talks, the conference also held a Rump Session, full of academic opinions and enjoyment.

We selected a particularly large and broad PC and encouraged members to focus on the positive aspects of submissions. During the one-and-a-half-month-long independent review phase, each PC member had about 28 submissions to review, our PC members and the external reviewers worked very hard and efficiently. In the following one-month daily discussion phase, PC members communicated each other's opinion on the board. We processed the anonymized questions from the PC members to authors, which resulted in a better quality of review.

We would like to thank the authors of all 241 submissions. Their contributions made this conference possible. We are extremely grateful to the PC members for their enormous investment of time and effort in the difficult and delicate process of review and selection, especially given the last decision days were in the midst of summer vacation time. A list of PC members and external reviewers can be found on the succeeding pages of this volume. We would like to thank Xuejia Lai, Zhijun Qiang, Hao Chen, Juan Liu, Dongdai Lin, Bao Li, Meiqin Wang and Jialin Huang for the conference organization. Special thanks go to Shai Halevi for providing and setting up the splendid review software. We are most grateful to Yue Sun, who provided technical support for the entire ASIACRYPT 2012 review process. We are also grateful to Dong Hoon Lee, the ASIACRYPT 2011 Program Chair, for his timely information and replies to the host of questions we posed during the process.

September 2012

Xiaoyun Wang
Kazue Sako

ASIACRYPT 2012

The 18th Annual International Conference on the Theory and Application of Cryptology and Information Security

December 2–6, 2012, Beijing, China

Sponsored by the *International Association for Cryptologic Research (IACR)*

Organized in cooperation with the *Chinese Association for Cryptologic Research (CACR)*

General Chair

Xuejia Lai   Shanghai Jiao Tong University, China

Program Co-chairs

Xiaoyun Wang   Tsinghua University, China
Kazue Sako   NEC, Japan

Program Committee

Feng Bao   I2R, Singapore
Alex Biryukov   University of Luxembourg, Luxembourg
Xavier Boyen   Prime Cryptography, USA
David Cash   IBM T.J. Watson Research Center, USA
Jung Hee Cheon   Seoul National University, Korea
Sherman S.M. Chow   University of Waterloo, Canada
Joan Daemen   STMicroelectronics, Belgium
Jintai Ding   University of Cincinnati, USA
Orr Dunkelman   University of Haifa and Weizmann Institute, Israel
Marc Fischlin   Darmstadt University of Technology, Germany
Vipul Goyal   Microsoft Research, India
Tetsu Iwata   Nagoya University, Japan
Antoine Joux   DGA and Université de Versailles, PRISM, France
Jonathan Katz   University of Maryland, USA
Eike Kiltz   Ruhr University Bochum, Germany
Lars Ramkilde Knudsen   Technical University of Denmark, Denmark
Dong Hoon Lee   Korea University, Korea
Arjen K. Lenstra   EPFL, Switzerland
Dongdai Lin   CAS, China

Mitsuru Matsui   Mitsubishi Electric, Japan
Willi Meier   FHNW, Switzerland
Florian Mendel   KU Leuven, Belgium
Phong Q. Nguyen   INRIA, France and Tsinghua University, China
Tatsuaki Okamoto   NTT, Japan
Bart Preneel   KU Leuven, Belgium
Christian Rechberger   Technical University of Denmark, Denmark
Rei Safavi-Naini   University of Calgary, Canada
Nigel P. Smart   University of Bristol, UK
Ron Steinfeld   Macquarie University, Australia
Hongjun Wu   Nanyang Technological University, Singapore

External Reviewers

Michel Abdalla	Jiazhe Chen	Benedikt Gierlichs
M. Ahmed Abdelraheem	Jie Chen	Serge Gorbunov
Masayuki Abe	Yuanmi Chen	Jens Groth
Shashank Agrawal	Nathan Chenette	Johann Groschädl
Ahmad Ahmadi	Chen-mou Cheng	David Gruenewald
Hadi Ahmadi	Céline Chevalier	Divya Gupta
Mohsen Alimomeni	Seung Geol Choi	Iftach Haitner
Prabhanjan Ananth	Ashish Choudhary	Shai Halevi
Elena Andreeva	Sherman Chow	Nadia Heninger
Kazumaro Aoki	Cheng-Kang Chu	Jens Hermans
Benny Applebaum	Ji Young Chun	Gottfried Herold
Gilles Van Assche	Kai-Min Chung	Shoichi Hirose
Nuttapong Attrapadung	Carlos Cid	Dennis Hofheinz
Jean-Philippe Aumasson	Dana Dachman-Soled	Fumitaka Hoshino
Paul Baecher	Özgür Dagdelen	Lei Hu
Chung Hun Baek	Ivan Damgaard	Zhu Huafei
Aurelie Bauer	Itai Dinur	Tao Huang
Josh Benaloh	Leo Ducas	Jung Yeon Hwang
David Bernhard	Aline Dudeanu	Toshiyuki Isshiki
Guido Bertoni	Pooya Farshim	Mitsugu Iwamoto
Raghav Bhaskar	Xiutao Feng	Tibor Jager
Andrey Bogdanov	Dario Fiore	Dimitar Jetchev
Julia Borghoff	Pierre-Alain Fouque	Mahavir Jhawar
Joppe Bos	Georg Fuchsbauer	Shaoquan Jiang
Charles Bouillaguet	Eiichiro Fujisaki	Saqib Kakvi
Christina Brzuska	Jun Furukawa	Bhavana Kanukurthi
D. Galindo ChacÓn	Tommaso Gagliardoni	Alexandre Karlov
Anne Canteaut	Steven Galbraith	Tomasz Kazana
Angelo De Caro	Nicolas Gama	Qiong Tang
Dario Catalano	Praveen Gauravaram	Aggelos Kiayias
Melissa Chase	Rosario Gennaro	Dongmin Kim

HongTae Kim
Hyoseung Kim
Jihye Kim
Jinsu Kim
Kee Sung Kim
Kitak Kim
Myungsun Kim
Sungwook Kim
Taechan Kim
Mario Kirschbaum
Susumu Kiyoshima
Thorsten Kleinjung
Simon Knellwolf
Yuichi Komano
Woo Kwon Koo
Kaoru Kurosawa
Eyal Kushilevitz
S. Thomas Kutzner
Hidenori Kuwakado
Özgül Küçük
Fabien Laguillaumie
Mario Lamberger
Tanja Lange
Gregor Leander
Hyung Tae Lee
Jooyoung Lee
Kwangsu Lee
Moon Sung Lee
Young-Ran Lee
Younho Lee
Gaëtan Leurent
Allison Bishop Lewko
Pierre-Yvan Liardet
Benoit Libert
Hoon Wei Lim
Yehuda Lindell
Jake Loftus
Jiqiang Lu
Karina M. Magalhäes
Hemanta Maji
Avradip Mandal
Mark Manulis
Giorgia Azzurra Marson
Ben Martin
Takahiro Matsuda

Filippo Melzani
Bart Mennink
Alexander Meurer
Andrea Miele
Kazuhiko Minematsu
Marine Minier
Arno Mittelbach
Payman Mohassel
Ravi Montenegro
Amir Moradi
Nicky Mouha
Tomislav Nad
Michael Naehrig
Jesper Buus Nielsen
Ivica Nikolic
Svetla Nikova
Ryo Nishimaki
Geontae Noh
Ryo Nojima
Adam O'Neill
Cristina Onete
Onur Ozen
Ilya Ozerov
Carles Padro
Dan Page
Omkant Pandey
Jong Hwan Park
Jung Youl Park
Seunghwan Park
Kenny Paterson
Roel Peeters
Chris Peikert
Edoardo Persichetti
Christiane Peters
Duong Hieu Phan
Le Trieu Phong
Josef Pieprzyk
Krzysztof Pietrzak
Thomas Plos
David Pointcheval
Joop van de Pol
Arnab Roy
Hyun Sook Rhee
Alfredo Rial
Vincent Rijmen

Thomas Ristenpart
Alon Rosen
Yannis Rouselakis
Carla Ràfols
Minoru Saeki
Amit Sahai
Bagus Santoso
Santanu Sarkar
Sumanta Sarkar
Yu Sasaki
John Schanck
Martin Schläffer
Jörn-Marc Schmidt
Patrick Schmidt
Michael Schneider
Dominique Schröder
Nicolas Sendrier
Jae Woo Seo
Minjae Seo
Siamak Shahandashti
Kyung-Ah Shim
Ji Sun Shin
Taizo Shirai
Igor Shparlinski
Hervé Sibert
Benjamin Smith
Damien Stehlé
Chunhua Su
Takeshi Sugawara
Ruggero Susella
Daisuke Suzuki
Katsuyuki Takashima
Chengdong Tao
Yannick Teglia
Isamu Teranishi
Stefano Tessaro
Enrico Thomae
Mehdi Tibouchi
Elmar Tischhauser
Deniz Toz
Toyohiro Tsurumaru
Vesselin Velichkov
Vinod Vaikuntanathan
Kerem Varici
Daniele Venturi

Frederik Vercauteren
Vanessa Vitse
Huaxiong Wang
Meiqin Wang
Pengwei Wang
Bogdan Warinschi
Brent Waters
Hoeteck Wee
Lei Wei
Ralf-Philipp Weinmann

Daniel Wichs
Michael Wiener
Chuankun Wu
Keita Xagawa
Xiang Xie
Jing Xu
Bo-yin Yang
Yanjiang Yang
Kazuki Yoneyama
Reo Yoshida

Tsz-Hon Yuen
Aaram Yun
Haibin Zhang
Liangfeng Zhang
Rui Zhang
Yunlei Zhao
Hong-Sheng Zhou
Huafei Zhu
Vassilis Zikas

Sponsoring Institutions

National Natural Science Foundation of China
Huawei Technologies Co. Ltd.
Intel Corporation

Table of Contents

Symmetric Cipher

Security Proof

Public-Key Cryptography II

Lattice-Based Cryptography and Number Theory

Public-Key Cryptography III

Hash Function

Cryptographic Protocol I

Cryptographic Protocol II

Implementation Issues

Erratum

Pairing-Based Cryptography:
Past, Present, and Future

Dan Boneh*

Stanford University
dabo@cs.stanford.edu

Abstract. While pairings were first introduced in cryptography as a tool to attack the discrete-log problem on certain elliptic curves, they have since found numerous applications in the construction of cryptographic systems. To this day many problems can only be solved using pairings. A few examples include collusion-resistant broadcast encryption and traitor tracing with short keys, 3-way Diffie-Hellman, and short signatures.

In this talk we survey some of the existing applications of pairings to cryptography, but mostly focus on open problems that cannot currently be solved using pairings. In particular we explain where the current techniques fail and outline a few potential directions for future progress.

One of the central applications of pairings is identity-based encryption and its generalization to functional encryption. While identity-based encryption can be built using arithmetic modulo composites and using lattices, constructions based on pairings currently provide the most expressive functional encryption systems. Constructing comparable functional encryption systems from lattices and composite arithmetic is a wonderful open problem. Again we survey the state of the art and outline a few potential directions for further progress.

Going beyond pairings (a.k.a bi-linear maps), a central open problem in public-key cryptography is constructing a secure tri-linear or more generally a secure n-linear map. That is, construct groups G and G_T where discrete-log in G is intractable and yet there is an efficiently computable non-degenerate n-linear map $e : G^n \rightarrow G_T$. Such a construct can lead to powerful solutions to the problems mentioned in the first paragraph as well as to new functional encryption and homomorphic encryption systems. Currently, no such construct is known and we hope this talk will encourage further research on this problem.

---

* Supported by NSF, DARPA, AFOSR, Google, and Samsung.

X. Wang and K. Sako (Eds.): ASIACRYPT 2012, LNCS 7658, p. 1, 2012.

Some Mathematical Mysteries in Lattices

Chuanming Zong

Peking University

Lattice, as a basic object in Mathematics, has been studied by many prominent figures, including Gauss, Hermite, Voronio, Minkowski, Davenport, Hlawka,
Rogers and many others still active today. It is one of the most important cor-
nerstones of Geometry of Numbers, a classic branch of Number Theory. During
recent decades, this pure mathematical concept has achieved remarkable applications in Cryptography, in particular its algorithm approaches. The main purpose
of this talk is to demonstrate some basic mathematical problems and results (old
and new) about lattices, which are probably useful in Cryptography in the future. These problems reflect some of the main interests of the mathematicians
about lattices.

Before Minkowski, lattices were mainly studied through positive definitive
quadratic forms. In fact, to determine the minimal value of a positive definitive
quadratic form at integer points is equivalent to determine the length of the
shortest vectors (except $\mathbf{o}$) of a lattice, which is also equivalent to determine the
maximal density of the corresponding lattice ball packings.

It was Minkowski who first studied the density $\delta^*(C)$ of the densest lattice
packings of a given centrally symmetric convex body C. In particular, he obtained the first general lower bound of $\delta^*(C)$ for n-dimensional unit ball B. In
fact, to determine the density $\delta^*(C)$ is to estimate the maximal length of the
shortest vectors of the lattices of determinant 1 with respect to certain metric determined by C. When C is the unit ball, the metric is just the ordinary
Euclidean metric. Therefore, the shortest vector problem is a particular case of
the study about $\delta^*(B)$. There are lower bound and upper bound for $\delta^*(C)$ and
$\delta^*(B)$, however the asymptotic orders of both $\min \delta^*(C)$ and $\delta^*(B)$ are unknown.
For lattice kissing numbers we are facing the similar situation.

The density $\theta^*(C)$ of the thinnest lattice covering of a centrally symmetric
convex body C was first systematically studied by Rogers. In fact, it is equivalent to determine the minimal length of the longest distance from a point to the
lattices of determinant 1 with respect to the metric determined by C. Therefore,
the closest vector problem is a particular case of the study of $\theta^*(B)$. For particular object C, such as a ball in a given dimension, little is known about the
exact value of $\theta^*(C)$.

Let $\gamma^*(C)$ be the smallest number that there is a lattice Λ such that $C + \Lambda$ is
a packing and $\gamma^*(C)C + \Lambda$ is a covering. Equivalently, in every lattice packing
$C + \Lambda$ there is a hole in which one can put a translate of $(\gamma^*(C) - 1)C$. In
1950, Rogers introduced and studied this number, in particular for the unit ball.
In fact, $\gamma^*(C)$ is a bridge connecting $\delta^*(C)$ and $\theta^*(C)$. In other words, it is a

X. Wang and K. Sako (Eds.): ASIACRYPT 2012, LNCS 7658, pp. 2–3, 2012.

bridge connecting the packing radius and the covering radius of a lattice, with respect to the metric determined by C. Some results about $\gamma^*(C)$ and $\gamma^*(B)$ are known. At the same time, a number of fascinating mysteries about $\gamma^*(C)$ and their possible consequences remain unsolved.

Can you imagine that, in every three-dimensional lattice ball packing there is a straight line of infinite length which does not meet any of the balls; when n is large, in every n-dimensional lattice ball packing there is a free hyperplane of dimension more or less $n/\log n$? But, this is true!

Constant-Size Structure-Preserving Signatures: Generic Constructions and Simple Assumptions

Masayuki Abe[1], Melissa Chase[2], Bernardo David[3],
Markulf Kohlweiss[2], Ryo Nishimaki[1], and Miyako Ohkubo[4]

[1] NTT Secure Platform Laboratories
{abe.masayuki,nishimaki.ryo}@lab.ntt.co.jp
[2] Microsoft Research
{melissac,markulf}@microsoft.com
[3] University of Brasilia
bernardo.david@aluno.unb.br
[4] Security Architecture Laboratory, NSRI, NICT
m.ohkubo@nict.go.jp

Abstract. This paper presents efficient structure-preserving signature schemes based on assumptions as simple as Decisional-Linear. We first give two general frameworks for constructing fully secure signature schemes from weaker building blocks such as variations of one-time signatures and random-message secure signatures. They can be seen as refinements of the Even-Goldreich-Micali framework, and preserve many desirable properties of the underlying schemes such as constant signature size *and structure preservation*. We then instantiate them based on simple (i.e., not q-type) assumptions over symmetric and asymmetric bilinear groups. The resulting schemes are structure-preserving and yield constant-size signatures consisting of 11 to 17 group elements, which compares favorably to existing schemes relying on q-type assumptions for their security.

Keywords: Structure-preserving signatures, One-time signatures, Groth-Sahai proof system, Random message attacks.

1 Introduction

A structure-preserving signature (SPS) scheme [1] is a digital signature scheme with two structural properties (i) the verification keys, messages, and signatures are all elements of a bilinear group; and (ii) the verification algorithm checks a conjunction of pairing product equations over the key, the message and the signature. This makes them compatible with the efficient non-interactive proof system for pairing-product equations by Groth and Sahai (GS) [30]. Structure-preserving cryptographic primitives promise to combine the advantages of optimized number theoretic non-blackbox constructions with the modularity and insight of protocols that use only generic cryptographic building blocks.

Indeed the instantiation of known generic constructions with a SPS scheme and the GS proof system has led to many new and more efficient schemes: Groth [29] showed how to construct an efficient simulation-sound zero-knowledge proof system (ss-NIZK)

X. Wang and K. Sako (Eds.): ASIACRYPT 2012, LNCS 7658, pp. 4–24, 2012.

building on generic constructions of [17,39,34]. Abe et al. [4] show how to obtain efficient round-optimal blind signatures by instantiating a framework by Fischlin [20]. SPS are also important building blocks for a wide range of cryptographic functionalities such as anonymous proxy signatures [22], delegatable anonymous credentials [6], transferable e-cash [23] and compact verifiable shuffles [16]. Most recently, [31] show how to construct a structure preserving tree-based signature scheme with a tight security reduction following the approach of [26,18]. This signature scheme is then used to build a ss-NIZK which in turn is used with the Naor-Yung-Sahai [35,38] paradigm to build the first CCA secure public-key encryption scheme with a tight security reduction. Examples for other schemes that benefit from efficient SPS are [7,11,8,32,27,5,37,24,21,28].

Because properties (i) and (ii) are the only dependencies on the SPS scheme made by these constructions, any structure-preserving signature scheme can be used as a drop-in replacement. Unfortunately, all known efficient instantiations of SPS [4,1,2] are based on so-called q-type or interactive assumptions that are primarily justified based on the Generic Group model. An open question since Groth's seminal work [29] (only partially answered by [15]) is to construct a SPS scheme that is both efficient – in particular *constant-size* in the number of signed group elements – and that is based on assumptions that are as weak as those required by the GS proof system itself.

Our contribution. Our first contribution consists of two generic constructions for chosen message attack (CMA) secure signatures that combine variations of one-time signatures and signatures secure against random message attacks (RMA). Both constructions inherit the structure-preserving and constant-size properties from the underlying components. The second contribution consists in the concrete instantiations of these components which result in constant-size structure-preserving signature schemes that produce signatures consisting of only 11 to 17 group elements and that rely only on basic assumptions such as Decisional-Linear (DLIN) for symmetric bilinear groups and analogues of DDH and DLIN for asymmetric bilinear groups. To our knowledge, these are the first constant-size structure-preserving signature schemes that eliminate the use of q-type assumptions while achieving reasonable efficiency.

We instantiate the first generic construction for symmetric (Type-I) and the second for asymmetric (Type-III) pairing groups. See Table 1 in Section 7 for the summary of efficiency of the resulting schemes. We give more details on our generic constructions and their instantiations:

- The first generic construction (SIG1) combines a new variation of one-time signatures which we call *tagged one-time signatures* and signatures secure against *random message attacks* (RMA). A tagged one-time signature scheme, denoted by TOS, is a signature scheme that attaches a fresh tag to a signature. It is unforgeable with respect to tags that are used only once. In our construction, a message is signed with our TOS scheme using a fresh random tag, and then the tag is signed with the second signature scheme, denoted by rSIG. Since the rSIG scheme only signs random tags, RMA-security is sufficient.
- The second generic construction (SIG2) combines *partial one-time signatures* and signatures secure against *extended random message attacks* (XRMA). The latter is a novel notion that we explain below. Partial one-time signatures, denoted by POS, are one-time signatures for which only a part of the one-time key is renewed for

every signing operation. They were first introduced by Bellare and Shoup [9] under the name of two-tier signatures. In our construction, a message is signed with the POS scheme and then the random one-time public-key is certified by the second signature scheme, denoted by xSIG. The difference between a TOS scheme and a POS scheme is that a one-time public-key is associated with a one-time secret-key. Since the one-time secret-key is needed for signing, it must be known to the reduction in the security proof. XRMA-security guarantees that xSIG is unforgeable even if the adversary is given auxiliary information associated with the randomly chosen messages (it is a random coin used for selecting the message). The auxiliary information facilitates access to the one-time secret-key by the reduction.

- To instantiate SIG1, we construct structure-preserving TOS and rSIG signature schemes based on DLIN over Type-I bilinear groups. Our TOS scheme yields constant-size signatures and tags. The resulting SIG1 scheme is structure-preserving, produces signatures consisting of 17 group elements, and relies solely on the DLIN assumption.
- To instantiate SIG2, we construct structure-preserving POS and xSIG signature schemes based on assumptions that are analogues of DDH and DLIN in Type-III bilinear groups. The resulting SIG2 scheme is structure-preserving, produces signatures consisting of 11 group elements for uniliteral messages in a base group or 14 group elements for biliteral messages from both base groups.

The role of partial one-time signatures is to compress a message into a constant number of random group elements. This observation is interesting in light of [3] that implies the impossibility of constructing collision resistant and shrinking structure-preserving hash functions, which could immediately yield constant-size signatures. Our (extended) RMA-secure signature schemes are structure-preserving variants of Waters' dual-signature scheme [41]. In general, the difficulty of constructing CMA-secure SPS arises from the fact that the exponents of the group elements chosen by the adversary as a message are not known to the reduction in the security proof. On the other hand, for RMA security, it is the challenger that chooses the message and therefore the exponents can be known in reductions. This is the crucial advantage for constructing (extended) RMA-secure structure-preserving signature schemes based on Waters' dual-signature scheme.

Finally, we mention a few new applications. Among these is the achievement of a drastic performance improvement when using our partial one-time signatures in the work by Hofheinz and Jager [31] to construct CCA-secure public-key encryption schemes with a proof of security that tightly reduces to DLIN or SXDH.

Related Works. Even, Goldreich and Micali [19] proposed a generic framework (the EGM framework) that combines a one-time signature scheme and a signature scheme that is secure against non-adaptive chosen message attacks (NACMA) to construct a signature scheme that is secure against adaptive chosen message attacks (CMA).

In fact, our generic constructions can be seen as refinements of the EGM framework. There are two reasons why the original framework falls short for our purpose. *The first* is that relaxing to NACMA does not seem a big help in constructing efficient structure-preserving signatures since the messages are still under the control of the adversary and the exponents of the messages are not known to the reduction algorithm in the security proof. As mentioned above, resorting to (extended) RMA is a great help in this regard. In [19], they also showed that CMA-secure signatures exist *iff* RMA-secure signatures exist. The proof, however, does not follow their framework and their impractical construction is mainly a feasibility result. In fact, we argue that RMA-security alone is not sufficient for the original EGM framework. As mentioned above, the necessity of XRMA security arises in the reduction that uses RMA-security to argue security of the ordinary signature scheme, as the reduction not only needs to know the random one-time public-keys, but also their corresponding one-time secret keys in order to generate the one-time signature components of the signatures. The auxiliary information in the XRMA definition facilitates access to these secret keys. Similarly, tagged one-time signatures avoid this problem as tags do not have associated secret values. *The second reason* that the EGM approach is not quite suited to our task is that the EGM framework produces signatures that are linear in the public-key size of the one-time signature scheme. Here, tagged or partial one-time signature schemes come in handy as they allow the signature size to be only linear in the size of the part of the public key that is updated. Thus, to obtain constant-size signatures, we require the one-time part to be constant-size.

Hofheinz and Jager [31] constructed a SPS scheme by following the EGM framework. The resulting scheme allows tight security reduction to DLIN but the size of signatures depends logarithmically to the number of signing operation as their NACMA-secure scheme is tree-based like the Goldwasser-Micali-Rivest signature scheme [26]. Chase and Kohlweiss [15] and Camenisch, Dubovitskaya, and Haralambiev [13] constructed SPS schemes with security based on DLIN that improve the performance of Groth's scheme [29] by several orders of magnitude. The size of the resulting signatures, however, are still linear in the number of signed group elements, and an order of magnitude larger than in our constructions. Camenisch, Dubovitskaya, and Haralambiev constructed a constant-size SPS scheme based on simple assumptions over composite-order groups [12].

Full Version. In this extended abstract, we do not have enough space to write complete proofs, so we omitted them. Please see a full version on Cryptology ePrint Archive (2012/285).

2 Preliminaries

Notation. Appending element y to a sequence $X = (x_1, \ldots, x_n)$ is denoted by (X, y), i.e., $(X, y) = (x_1, \ldots, x_n, y)$. When algorithm A is defined for input x and output y, notation $\boldsymbol{y} \leftarrow A(\boldsymbol{x})$ for $\boldsymbol{x} := \{x_1, \ldots, x_n\}$ means that $y_i \leftarrow A(x_i)$ is executed for $i = 1, \ldots, n$ and $\boldsymbol{y}$ is set as $\boldsymbol{y} := (y_1, \ldots, y_n)$. For set X, notation $a \leftarrow X$ denote a uniform sampling from X. Independent multiple sampling from the same set X is denoted by $a, b, c, .. \leftarrow X$.

Bilinear groups. Let $\mathcal{G}$ be a bilinear group generator that takes security parameter 1^λ and outputs a description of bilinear groups $\Lambda := (p, \mathbb{G}_1, \mathbb{G}_2, \mathbb{G}_T, e)$, where $\mathbb{G}_1$, $\mathbb{G}_2$ and $\mathbb{G}_T$ are groups of prime order p, and e is an efficient and non-degenerating bilinear map $\mathbb{G}_1 \times \mathbb{G}_2 \to \mathbb{G}_T$. Following the terminology in [25] this is a Type-III pairing. In the Type-III setting $\mathbb{G}_1 \neq \mathbb{G}_2$ and there are no efficient mapping between the groups in either direction. In the Type-III setting, we often use twin group elements, $(G^a, \hat{G}^a) \in \mathbb{G}_1 \times \mathbb{G}_2$ for some bases G and $\hat{G}$. For X in $\mathbb{G}_1$, notation $\hat{X}$ denotes for an element in $\mathbb{G}_2$ that $\log X = \log \hat{X}$ where logarithms are with respect to default bases that are uniformly chosen once for all and implicitly associated to Λ. Should their relation be explicitly stated, we write $X \sim \hat{X}$. We count the number of group elements to measure the size of cryptographic objects such as keys, messages, and signatures. For Type-III groups, we denote the size by (x, y) when it consists of x and y elements from $\mathbb{G}_1$ and $\mathbb{G}_2$, respectively. We refer to the Type-I setting when $\mathbb{G}_1 = \mathbb{G}_2$ (i.e., there are efficient mappings in both directions). This is also called the symmetric setting. In this case, we define $\Lambda := (p, \mathbb{G}, \mathbb{G}_T, e)$. When we need to be specific, the group description yielded by $\mathcal{G}$ will be written as Λ_{asym} and Λ_{sym}.

Assumptions. We first define computational and decisional Diffie-Hellman assumptions $(\mathrm{CDH}_1, \mathrm{DDH}_1)$ and decisional linear assumption (DLIN_1) for Type-III bilinear groups. Corresponding more standard assumptions, CDH, DDH, and DLIN, in Type-I groups are obtained by setting $\mathbb{G}_1 = \mathbb{G}_2$ and $G = \hat{G}$ in the respective definitions.

Definition 1 (Computation co-Diffie-Hellman Assumption: CDH_1)
*The CDH_1 assumption holds if, for any p.p.t. algorithm $\mathcal{A}$, the probability $\mathrm{Adv}^{\mathsf{co\text{-}cdh}}_{\mathcal{G},\mathcal{A}}$
$(\lambda) := \Pr[\, Z = G^{xy} \mid \Lambda \leftarrow \mathcal{G}(1^\lambda); x, y \leftarrow \mathbb{Z}_p; Z \leftarrow \mathcal{A}(\Lambda, G, G^x, G^y, \hat{G}, \hat{G}^x, \hat{G}^y)\,]$ is negligible in λ.*

Definition 2 (Decisional Diffie-Hellman Assumption in $\mathbb{G}_1$: DDH_1)
Given $\Lambda \leftarrow \mathcal{G}(1^\lambda)$, $G \leftarrow \mathbb{G}_1^*$, $(G^x, G^y, Z_b) \in \mathbb{G}_1{}^3$ where $Z_1 = G^{x+y}$, $Z_0 \leftarrow \mathbb{G}_1$ for random x and y, any p.p.t. algorithm $\mathcal{A}$ decides whether $b = 1$ or 0 only with advantage $\mathrm{Adv}^{DDH_1}_{\mathcal{G},\mathcal{A}}(\lambda)$ that is negligible in λ.

Definition 3 (Decisional Linear Assumption in $\mathbb{G}_1$: DLIN_1)
Given $\Lambda \leftarrow \mathcal{G}(1^\lambda)$, $(G_1, G_2, G_3) \leftarrow \mathbb{G}_1^{*3}$ and (G_1^x, G_2^y, Z_b) where $Z_1 = G_3^{x+y}$ and $Z_0 = G_3^z$ for random $x, y, z \in \mathbb{Z}_p$, any p.p.t. algorithm $\mathcal{A}$ decides whether $b = 1$ or 0 only with advantage $\mathrm{Adv}^{\mathsf{dlin1}}_{\mathcal{G},\mathcal{A}}(\lambda)$ that is negligible in λ.

For DDH_1 and DLIN_1, we define an analogous assumption in $\mathbb{G}_2$ (DDH_2) by swapping $\mathbb{G}_1$ and $\mathbb{G}_2$ in the respective definitions. In Type-III bilinear groups, it is assumed that both DDH_1 and DDH_2 hold simultaneously. The assumption is called the symmetric external Diffie-Hellman assumption (SXDH), and we define advantage $\mathrm{Adv}^{\mathsf{sxdh}}_{\mathcal{G},\mathcal{C}}$ by $\mathrm{Adv}^{\mathsf{sxdh}}_{\mathcal{G},\mathcal{C}}(\lambda) := \mathrm{Adv}^{\mathsf{ddh1}}_{\mathcal{G},\mathcal{A}}(\lambda) + \mathrm{Adv}^{\mathsf{ddh2}}_{\mathcal{G},\mathcal{B}}(\lambda)$. We extend DLIN in a similar manner as DDH, and SXDH.

Definition 4 (External Decision Linear Assumption in $\mathbb{G}_1$: XDLIN_1)
Given $\Lambda \leftarrow \mathcal{G}(1^\lambda)$, $(G_1, G_2, G_3) \leftarrow \mathbb{G}_1^{*3}$ and $(G_1^x, G_2^y, \hat{G}_1, \hat{G}_2, \hat{G}_3, \hat{G}_1^x, \hat{G}_2^y, Z_b)$ where $(G_1, G_2, G_3) \sim (\hat{G}_1, \hat{G}_2, \hat{G}_3)$, $Z_1 = G_3^{x+y}$, and $Z_0 = G_3^z$ for random $x, y, z \in \mathbb{Z}_p$, any p.p.t. algorithm $\mathcal{A}$ decides whether $b = 1$ or 0 only with advantage $\mathrm{Adv}^{\mathsf{xdlin}}_{\mathcal{G},\mathcal{A}}(\lambda)$ that is negligible in λ.

The XDLIN_1 assumption is equivalent to the DLIN_1 assumption in the generic bilinear group model [40,10] where one can simulate the extra elements, $\hat{G}_1, \hat{G}_2, \hat{G}_3, \hat{G}_1^x, \hat{G}_2^y$, in XDLIN_1 from $G_1, G_2, G_3, G_1^x, G_2^y$ in DLIN_1. We define the XDLIN_2 assumption analogously by giving $\hat{G}_3^{x+y}$ or $\hat{G}_3^z$ as Z_b, to $\mathcal{A}$ instead. Then we define the simultaneous external DLIN assumption, SXDLIN, that assumes that both XDLIN$_1$ and XDLIN$_2$ hold at the same time. By $\mathrm{Adv}_{\mathcal{G},\mathcal{A}}^{\mathrm{xdlin2}}$ ($\mathrm{Adv}_{\mathcal{G},\mathcal{A}}^{\mathrm{sxdlin}}$, resp.), we denote the advantage function for XDLIN$_2$ (and SXDLIN, resp.).

Definition 5 (Double Pairing Assumption in $\mathbb{G}_1$ [4]:DBP$_1$)
Given $\Lambda \leftarrow \mathcal{G}(1^\lambda)$ and $(G_z, G_r) \leftarrow \mathbb{G}_1^{*2}$, any p.p.t. algorithm $\mathcal{A}$ outputs $(Z, R) \in \mathbb{G}_2^{*2}$ that satisfies $1 = e(G_z, Z)\, e(G_r, R)$ only with probability $\mathrm{Adv}_{\mathcal{G},\mathcal{A}}^{\mathrm{dbp1}}(\lambda)$ that is negligible in λ.

The double pairing assumption in $\mathbb{G}_2$ (DBP$_2$) is defined in the same manner by swapping $\mathbb{G}_1$ and $\mathbb{G}_2$. It is known that DBP$_1$ (DBP$_2$, resp.) is implied by DDH$_1$ (DDH$_2$, resp.) and the reduction is tight [4]. Note that the double pairing assumption does not hold in Type-I groups since $Z = G_r$, $R = G_z^{-1}$ is a trivial solution. The following analogous assumption will be useful in Type-I groups.

Definition 6 (Simultaneous Double Pairing Assumption [14]: SDP)
Given $\Lambda \leftarrow \mathcal{G}(1^\lambda)$ and $(G_z, G_r, H_z, H_s) \leftarrow \mathbb{G}^{*4}$, any p.p.t. algorithm $\mathcal{A}$ outputs $(Z, R, S) \in \mathbb{G}^{*3}$ that satisfies $1 = e(G_z, Z)\, e(G_r, R) \wedge 1 = e(H_z, Z)\, e(H_s, S)$ only with probability $\mathrm{Adv}_{\mathcal{G},\mathcal{A}}^{\mathrm{sdp}}(\lambda)$ that is negligible in λ.

As shown in [14] for the Type-I setting, the simultaneous double pairing assumption holds for $\mathcal{G}$ if the decisional linear assumption holds for $\mathcal{G}$.

3   Definitions

Common setup. All building blocks make use of a common setup algorithm Setup that takes the security parameter 1^λ and outputs a global parameters gk that is given to all other algorithms. Usually gk consists of a description Λ of a bilinear group setup and a default generator for each group. In this paper, we include several additional generators in gk for technical reasons. Note that when the resulting signature scheme is used in multi-user applications different additional generators need to be assigned to individual users or one needs to fall back on the common reference string model, whereas Λ and the default generators can be shared. Thus we count the size of gk when we assess the efficiency of concrete instantiations. For ease of notation, we make gk implicit except w.r.t. key generation algorithms.

Signature schemes. We use the following syntax for signature schemes suitable for the multi-user and multi-algorithm setting. The key generation function takes global parameter gk generated by Setup (usually it takes security parameter 1^λ), and the message space $\mathcal{M}$ is determined solely from gk (usually it is determined from a public-key).

Definition 7 (Signature Scheme). *A signature scheme SIG is a tuple of three polynomial-time algorithms* (Key, Sign, Vrf) *that;*

- SIG.Key(gk) *generates a long-term public-key vk and a secret-key sk.*
- SIG.Sign(sk, msg) *takes sk and message msg and outputs signature σ.*
- SIG.Vrf(vk, msg, σ) *outputs 1 for acceptance or 0 for rejection.*

Correctness requires that $1 = $ SIG.Vrf(vk, msg, σ) *holds for any* gk *generated by* Setup, *any keys generated as* $(vk, sk) \leftarrow$ SIG.Key(gk), *any message* $msg \in \mathcal{M}$, *and any signature* $\sigma \leftarrow$ SIG.Sign(sk, msg).

Definition 8 (Attack Game(ATK)). *Let* $\mathcal{O}sig$ *be an oracle and* $\mathcal{A}$ *be an oracle algorithm. We define a meta attack game as a sequence of execution of algorithms as follows:* $ATK(\mathcal{A}, \lambda) =$

$$\left[gk \leftarrow \text{Setup}(1^{\lambda}), pre \leftarrow \mathcal{A}(gk), (vk, sk) \leftarrow \text{SIG.Key}(gk), (\sigma^{\dagger}, msg^{\dagger}) \leftarrow \mathcal{A}^{\mathcal{O}sig}(vk) \right]$$

Adversary $\mathcal{A}$ *commits to* pre, *which is typically a set of messages, in the first run. This formulation is to capture non-adaptive attacks. It is implicit that a state information is passed to the second run of* $\mathcal{A}$. *Let* Q_m *be a set of messages, for which* $\mathcal{A}$ *requests signatures from its oracle before outputting the resulting forgery. The output of ATK is* $(vk, \sigma^{\dagger}, msg^{\dagger}, Q_m)$.

Definition 9 (Adaptive Chosen-Message Attack (CMA)). *Adaptive chosen message attack security is defined by the attack game ATK where* pre *is empty and oracle* $\mathcal{O}sig$ *is the signing oracle that, on receiving a message* msg, *performs* $\sigma \leftarrow$ SIG.Sign(sk, msg), *and returns* σ.

Definition 10 (Random Message Attack (RMA)[19]). *Random message attack security is defined by the attack game ATK where* pre *is empty and oracle* $\mathcal{O}sig$ *is the following: on receiving a request, it chooses* msg *uniformly from* $\mathcal{M}$ *defined by* gk, *computes signature* $\sigma \leftarrow$ SIG.Sign(sk, msg), *and returns* (σ, msg).

Let MSGGen be a uniform message generator. It is a probabilistic algorithm that takes gk and outputs $msg \in \mathcal{M}$ that distributes uniformly over $\mathcal{M}$. Furthermore, MSGGen outputs auxiliary information aux that may give a hint about the random coins used for selecting msg.

Definition 11 (Extended Random Message Attack (XRMA)). *Extended random message attack is attack game ATK where* pre *is empty and oracle* $\mathcal{O}sig$ *is the following. On receiving a request, it runs* $(msg, aux) \leftarrow$ MSGGen(gk), *computes* $\sigma \leftarrow$ SIG.Sign(sk, msg), *and returns* (σ, msg, aux).

Definition 12 (Unforgeability against ATK). *Signature scheme* SIG *is unforgeable against attack ATK (UF-ATK) where* $ATK \in \{CMA, RMA, XRMA\}$, *if for all p.p.t. oracle algorithm* $\mathcal{A}$ *the advantage function* $\text{Adv}_{\text{SIG}, \mathcal{A}}^{\text{uf-atk}} := \Pr\left[msg^{\dagger} \notin Q_m \wedge 1 = \right.$ SIG.Vrf$(vk, \sigma^{\dagger}, msg^{\dagger}) \mid (vk, \sigma^{\dagger}, msg^{\dagger}, Q_m) \leftarrow ATK(\mathcal{A}, \lambda)]$ *is negligibel in* λ.

Fact 1. UF-CMA $\Rightarrow$ UF-XRMA $\Rightarrow$ UF-RMA, i.e., $\text{Adv}_{\text{SIG}, \mathcal{A}}^{\text{uf-cma}}(\lambda) \geq \text{Adv}_{\text{SIG}, \mathcal{A}}^{\text{uf-xrma}}(\lambda) \geq \text{Adv}_{\text{SIG}, \mathcal{A}}^{\text{uf-rma}}(\lambda)$.

Partial one-time and tagged one-time signatures. Partial one-time signatures, also known as two-tier signatures [9], are a variation of one-time signatures where only part of the public-key must be updated for every signing, while the remaining part can be persistent.

Definition 13. *[Partial One-Time Signature Scheme [9]] A partial one-time signatures scheme* POS *is a set of polynomial-time algorithms* POS.{Key, Update, Sign, Vrf}.

- POS.Key(gk) *generates a long-term public-key* pk *and a secret-key* sk. *The message space* $\mathcal{M}_o$ *is associated with* pk. *(Recall that we require that* $\mathcal{M}_o$ *be completely defined by* gk.)
- POS.Update() *takes* gk *as implicit input, and outputs a pair of one-time keys* (opk, osk). *We denote the space for* opk *by* $\mathcal{K}_{opk}$.
- POS.Sign(sk, msg, osk) *outputs a signature* σ *on message* msg *based on* sk *and* osk.
- POS.Vrf(pk, opk, msg, σ) *outputs 1 for acceptance, or 0 for rejection.*

For correctness, it is required that $1 = $ POS.Vrf(pk, opk, msg, σ) *holds except for negligible probability for any* gk, pk, opk, σ, *and* $msg \in \mathcal{M}_o$, *such that* $gk \leftarrow$ Setup(1^λ), (pk, sk) $\leftarrow$ POS.Key(gk), (opk, osk) $\leftarrow$ POS.Update(), $\sigma \leftarrow$ POS.Sign(sk, msg, osk).

A tagged one-time signature scheme is a signature scheme whose signing function in addition to the long-term secret key takes a tag as input. A tag is one-time, i.e., it must be different for every signing.

Definition 14 (Tagged One-Time Signature Scheme). *A tagged one-time signature scheme* TOS *is a set of polynomial-time algorithms* TOS.{Key, Tag, Sign, Vrf}.

- TOS.Key(gk) *generates a long-term public-key* pk *and a secret-key* sk. *The message space* $\mathcal{M}_t$ *is associated with* pk.
- TOS.Tag() *takes* gk *as implicit input and outputs tag. By* $\mathcal{T}$, *we denote the space for tag.*
- TOS.Sign(sk, msg, tag) *outputs signature* σ *for message* msg *based on* sk *and tag.*
- TOS.Vrf(pk, tag, msg, σ) *outputs 1 for acceptance, or 0 for rejection.*

Correctness requires that $1 = $ TOS.Vrf(pk, tag, msg, σ) *holds except for negligible probability for any* gk, pk, tag, σ, *and* $msg \in \mathcal{M}_t$, *such that* $gk \leftarrow$ Setup(1^λ), (pk, sk) $\leftarrow$ TOS.Key(gk), $tag \leftarrow$ TOS.Tag(), $\sigma \leftarrow$ TOS.Sign(sk, msg, tag).

A TOS scheme is POS scheme for which $tag = osk = opk$. We can thus give a security notion for POS schemes that also applies to TOS schemes by reading Update $=$ Tag and $tag = osk = opk$.

Definition 15 (Unforgeability against One-Time Adapative Chosen-Message Attacks). *A partial one-time signature scheme is unforgeable against one-time adaptive chosen message attacks (OT-CMA) if for all p.p.t. oracle algorithm* $\mathcal{A}$ *the advantage function* $\mathrm{Adv}_{\mathsf{POS},\mathcal{A}}^{\mathsf{ot\text{-}cma}}$ *is negligible in* λ, *where* $\mathrm{Adv}_{\mathsf{POS},\mathcal{A}}^{\mathsf{ot\text{-}cma}}(\lambda) :=$

$$\Pr \left[\begin{array}{l} \exists (opk, msg, \sigma) \in Q_m \text{ s.t.} \\ opk^\dagger = opk \ \wedge \ msg^\dagger \neq msg \ \wedge \\ 1 = \mathsf{POS.Vrf}(pk, opk^\dagger, \sigma^\dagger, msg^\dagger) \end{array} \middle| \begin{array}{l} gk \leftarrow \mathsf{Setup}(1^\lambda), \\ (pk, sk) \leftarrow \mathsf{POS.Key}(gk), \\ (opk^\dagger, \sigma^\dagger, msg^\dagger) \leftarrow \mathcal{A}^{\mathcal{O}t, \mathcal{O}sig}(pk) \end{array} \right].$$

Q_m *is initially an empty list.* $\mathcal{O}t$ *is the one-time key generation oracle that on receiving a request invokes a fresh session* j, *performs* (opk_j, osk_j) $\leftarrow$ POS.Update(), *and returns*

opk_j. $\mathcal{O}sig$ is the signing oracle that, on receiving a message msg_j for session j, performs $\sigma_j \leftarrow$ POS.Sign(sk, msg_j, osk_j), returns σ_j to $\mathcal{A}$, and records (opk_j, msg_j, σ_j) to the list Q_m. $\mathcal{O}sig$ works only once for every session. Strong unforgeability is defined as well by replacing condition $msg^\dagger \neq msg$ with $(msg^\dagger, \sigma^\dagger) \neq (msg, \sigma)$.

We define a non-adaptive variant (OT-NACMA) of the above notion by integrating $\mathcal{O}t$ into $\mathcal{O}sig$ so that opk_j and σ_j are returned to $\mathcal{A}$ at the same time. Namely, $\mathcal{A}$ must submit msg_j before seeing opk_j. If a scheme is secure in the sense of OT-CMA, the scheme is also secure in the sense of OT-NACMA. If a scheme is strongly unforgeable, it is unforgeable as well. By $\mathrm{Adv}_{\mathrm{POS},\mathcal{A}}^{\mathrm{ot\text{-}nacma}}(\lambda)$ we denote the advantage of $\mathcal{A}$ in this non-adaptive case. For TOS, we use the same notations, OT-CMA and OT-NACMA, and define advantage functions $\mathrm{Adv}_{\mathrm{TOS},\mathcal{A}}^{\mathrm{ot\text{-}cma}}$ and $\mathrm{Adv}_{\mathrm{TOS},\mathcal{A}}^{\mathrm{ot\text{-}nacma}}$ accordingly. For strong unforge-abiltiy, we use label sot-cma and sot-nacma.

   We define a condition that is relevant for coupling random message secure signature schemes with partial one-time and tagged one-time signature schemes in later sections.

Definition 16 (Tag/One-time Public-Key Uniformity). TOS *is called uniform-tag if* TOS.Tag *outputs tag that uniformly distributes over tag space* $\mathcal{T}$. *Similarly,* POS *is called uniform-key if* POS.Update *outputs opk that uniformly distributes over key space* $\mathcal{K}_{opk}$.

Structure-preserving signatures. A signature scheme is structure-preserving over a bilinear group Λ, if public-keys, signatures, and messages are all base group elements of Λ, and the verification only evaluates pairing product equations. Similarly, POS schemes are structure-preserving if their public-keys, signatures, messages, and tags or one-time public-keys consist of base group elements and the verification only evaluates pairing product equations.

4   Generic Constructions

4.1   SIG1: Combining Tagged One-Time and RMA-Secure Signatures

Let rSIG be a signature scheme with message space $\mathcal{M}_r$, and TOS be a tagged one-time signature scheme with tag space $\mathcal{T}$ such that $\mathcal{M}_r = \mathcal{T}$. We construct a signature scheme SIG1 from rSIG and TOS. Let gk be a global parameter generated by Setup(1^λ).

   - SIG1.Key(gk): Run $(pk_t, sk_t) \leftarrow$ TOS.Key(gk), $(vk_r, sk_r) \leftarrow$ rSIG.Key(gk). Output $vk := (pk_t, vk_r)$ and $sk := (sk_t, sk_r)$.
   - SIG1.Sign(sk, msg): Parse sk into (sk_t, sk_r). Run $tag \leftarrow$ TOS.Tag$()$, $\sigma_t \leftarrow$ TOS.Sign(sk_t, msg, tag), $\sigma_r \leftarrow$ rSIG.Sign(sk_r, tag). Output $\sigma := (tag, \sigma_t, \sigma_r)$.
   - SIG1.Vrf(vk, σ, msg): Parse vk and σ accordingly. Output 1, if $1 =$ TOS.Vrf$(pk_t, tag, \sigma_t, msg)$ and $1 =$ rSIG.Vrf(vk_r, σ_r, tag). Output 0, otherwise.

We prove the above scheme is secure by showing a reduction to the security of each component. As our reductions are efficient in their running time, we only relate success probabilities.

Theorem 17. SIG1 *is UF-CMA if* TOS *is uniform-tag and OT-NACMA, and* rSIG *is UF-RMA. In particular,* $\mathrm{Adv}_{\mathrm{SIG1},\mathcal{A}}^{\mathrm{uf\text{-}cma}}(\lambda) \leq \mathrm{Adv}_{\mathrm{TOS},\mathcal{B}}^{\mathrm{ot\text{-}nacma}}(\lambda) + \mathrm{Adv}_{\mathrm{rSIG},\mathcal{C}}^{\mathrm{uf\text{-}rma}}(\lambda).$

Proof. Any signature that is accepted by the verification algorithm must either reuse an existing tag, or sign a new tag. The success probability $\mathrm{Adv}_{\mathsf{SIG1},\mathcal{A}}^{\mathsf{uf\text{-}cma}}(\lambda)$ of an attacker on SIG1 is bounded by the sum of the success probabilities $\mathrm{Adv}_{\mathsf{TOS},\mathcal{B}}^{\mathsf{ot\text{-}nacma}}(\lambda)$ of an attacker on TOS and the success probability $\mathrm{Adv}_{\mathsf{rSIG},\mathcal{C}}^{\mathsf{uf\text{-}rma}}(\lambda)$ of an attacker on rSIG.

Game 0: The actual Unforgeability game. $\Pr[\textbf{Game 0}] = \mathrm{Adv}_{\mathsf{SIG1},\mathcal{A}}^{\mathsf{uf\text{-}cma}}(\lambda)$.

Game 1: The real security game except that the winning condition is changed to no longer accept repetition of tags.

Lemma 18. $|\Pr[\textbf{Game 0}] - \Pr[\textbf{Game 1}]| \leq \mathrm{Adv}_{\mathsf{TOS},\mathcal{B}}^{\mathsf{ot\text{-}nacma}}(\lambda)$

Game 2: The fully idealized game. The winning condition is changed to reject all signatures.

Lemma 19. $|\Pr[\textbf{Game 1}] - Pr[\textbf{Game 2}]| \leq \mathrm{Adv}_{\mathsf{rSIG},\mathcal{C}}^{\mathsf{uf\text{-}rma}}(\lambda)$

Thus $\mathrm{Adv}_{\mathsf{SIG1},\mathcal{A}}^{\mathsf{uf\text{-}cma}}(\lambda) = \Pr[\textbf{Game 0}] \leq \mathrm{Adv}_{\mathsf{TOS},\mathcal{B}}^{\mathsf{ot\text{-}nacma}}(\lambda) + \mathrm{Adv}_{\mathsf{rSIG},\mathcal{C}}^{\mathsf{uf\text{-}rma}}(\lambda)$ as claimed.

Theorem 20. *If* TOS.Tag *produces constant-size tags and signatures in the size of input messages, the resulting* SIG1 *produces constant-size signatures as well. Furthermore, if* TOS *and* rSIG *are structure-preserving, so is* SIG1.

We omit the proof of Theorem 20 as it is done simply by examining the construction.

4.2 SIG2: Combining Partial One-Time and XRMA-Secure Signatures

Let xSIG be a signature scheme with message space $\mathcal{M}_\mathsf{x}$, and POS be a partial one-time signature scheme with one-time public-key space $\mathcal{K}_{opk}$ such that $\mathcal{M}_\mathsf{x} = \mathcal{K}_{opk}$. We construct a signature scheme SIG2 from xSIG and POS. Let gk be a global parameter generated by $\mathsf{Setup}(1^\lambda)$.

- SIG2.Key(gk): Run $(pk_p, sk_p) \leftarrow$ POS.Key(gk), $(vk_x, sk_x) \leftarrow$ xSIG.Key(gk). Output $vk := (pk_p, vk_x)$ and $sk := (sk_p, sk_x)$.
- SIG2.Sign(sk, msg): Parse sk into (sk_p, sk_x). Run $(opk, osk) \leftarrow$ POS.Update(), $\sigma_p \leftarrow$ POS.Sign(sk_p, msg, osk), $\sigma_x \leftarrow$ xSIG.Sign(sk_x, opk). Output $\sigma := (opk, \sigma_p, \sigma_x)$.
- SIG2.Vrf(vk, σ, msg): Parse vk and σ accordingly. Output 1 if $1 =$ POS.Vrf(pk_p, opk, σ_p, msg), and $1 =$ xSIG.Vrf(vk_x, σ_x, opk). Output 0, otherwise.

Theorem 21. SIG2 *is UF-CMA if* POS *is uniform-key and OT-NACMA, and* xSIG *is UF-XRMA w.r.t.* POS.Update *as the message generator. In particular,* $\mathrm{Adv}_{\mathsf{SIG2},\mathcal{A}}^{\mathsf{uf\text{-}cma}}(\lambda) \leq \mathrm{Adv}_{\mathsf{POS},\mathcal{B}}^{\mathsf{ot\text{-}nacma}}(\lambda) + \mathrm{Adv}_{\mathsf{xSIG},\mathcal{C}}^{\mathsf{uf\text{-}xrma}}(\lambda)$.

Proof. The proof is almost the same as that for Theorem 17. The only difference appears in constructing $\mathcal{C}$ in the second step. Since POS.Update is used as the extended random message generator, the pair (msg, aux) is in fact (opk, osk). Given (opk, osk), adversary $\mathcal{C}$ can run POS.Sign(sk, msg, osk) to yield legitimate signatures.

Theorem 22. *If* POS *produces constant-size one-time public-keys and signatures in the size of input messages, resulting* SIG2 *produces constant-size signatures as well. Furthermore, if* POS *and* xSIG *are structure-preserving, so is* SIG2.

5 Instantiating SIG1

We instantiate the building blocks TOS and rSIG of our first generic construction to obtain our first SPS scheme. We do so in Type-I bilinear group setting. The resulting SIG1 scheme is an efficient structure-preserving signature scheme based only on the DLIN assumption.

Setup for Type-I groups. The following setup procedure is common for all instantiations in this section. The global parameter gk is given to all functions implicitly.

Setup(1^λ): Run $\Lambda = (p, \mathbb{G}, \mathbb{G}_T, e) \leftarrow \mathcal{G}(1^\lambda)$ and pick random generators $(G, C, F, U_1, U_2) \leftarrow \mathbb{G}^{*5}$. Output $gk := (\Lambda, G, C, F, U_1, U_2)$.

The parameters gk fix the message space $\mathcal{M}_r := \{(C^{m_1}, C^{m_2}, F^{m_1}, F^{m_2}, U_1^{m_1}, U_2^{m_2}) \in \mathbb{G}^6 \mid (m_1, m_2) \in \mathbb{Z}_p^2\}$ for the RMA-secure signature scheme defined below. For our generic framework to work, the tagged one-time signature schemes should have the same tag space.

Tagged one-time signature scheme. Basically, a tag in our scheme consists of a pair of elements in $\mathbb{G}$. However, due to a constraint from rSIG we show in the next section, the tags will have to be in an extended form. We therefore parameterize the one-time key generation function Update with a flag $mode \in \{\text{normal}, \text{extended}\}$ so that it outputs a key in the original or extended form. Although $mode$ is given to Update as input, it should be considered as a fixed system-wide parameter that is common for every invocation of Update and the key space is fixed throughout the use of the scheme. Accordingly, this extension does not affect the security model at all.

TOS.Key(gk): Parse $gk = (\Lambda, G, C, F, U_1, U_2)$. Pick random $x_r, y_r, x_s, y_s, x_t, y_t, x_1, y_1, \ldots, x_k, y_k$ in $\mathbb{Z}_p$ such that such that $x_r y_s \neq x_s y_r$ and compute $G_r := G^{x_r}, H_r := G^{y_r}, G_s := G^{x_s}, H_s := G^{y_s}, G_t := G^{x_t}, H_t := G^{y_t}, G_0 := G^{x_0}, H_0 := G^{y_0}, \ldots, G_k := G^{x_k}, H_k := G^{y_k}$. Output $pk := (G_r, G_s, G_t, H_r, H_s, H_t, G_0, \ldots, G_k, H_0, \ldots, H_k)$ and $sk := (x_r, x_s, x_t, y_r, y_s, y_t, x_0, \ldots, x_k, y_0, \ldots, y_k)$

TOS.Tag(): Take generators G, C, F, U_1, U_2 from gk. Choose $w_1, w_2 \leftarrow \mathbb{Z}_p^*$ and compute $tag := (C^{w_1}, C^{w_2}, F^{w_1}, F^{w_2}, U_1^{w_1}, U_2^{w_2})$. Output tag.

TOS.Sign(sk, msg, tag): Parse msg to $(M_1, \ldots, M_k)$ and tag to $(T_1, T_2, \ldots)$. Parse sk accordingly. Choose random $m \leftarrow \mathbb{Z}_p$ and let value $M_0 := G^m \prod_{i=1}^k M_i^{-1}$. (This is uniformly distributed.) Compute $A := G^{-x_t} T_1^{-m} \prod_{i=0}^k M_i^{-x_i}$ and $B := G^{-y_t} T_2^{-m} \prod_{i=0}^k M_i^{-y_i}$. Since $x_r y_s \neq x_s y_r$ we can compute $\left(\begin{smallmatrix} \alpha & \beta \\ \gamma & \delta \end{smallmatrix}\right) = \left(\begin{smallmatrix} x_r & x_s \\ y_r & y_s \end{smallmatrix}\right)^{-1}$. (The determinant is nonzero.) Compute $Z := A^\alpha B^\beta$ and $W := A^\gamma B^\delta$. Output $\sigma := (Z, W, M_0)$.

TOS.Vrf(pk, tag, msg, σ): Accept if the following equalities hold:

$$e(G_r, Z) \cdot e(G_s, W) \cdot e(G_t, G) \prod_{i=0}^k e(G_i T_1, M_i) = 1$$

$$e(H_r, Z) \cdot e(H_s, W) \cdot e(H_t, G) \prod_{i=0}^k e(H_i T_2, M_i) = 1$$

We remark that the correctness of the extended tag $(T_3, \ldots, T_6)$ is not examined within this scheme. (We only need to show that the extended part is simulatable in the security proof.) Since the tag is given to SIGr as a message, it is the verification function of SIGr that verifies the correctness with respect to its message space, which is the same as the tag space. The scheme is obviously structure-preserving and the correctness is easily verified by simple calculation.

Theorem 23. *The above TOS scheme is OT-CMA under the SDP. In particular, for any $\mathcal{A}$ that makes at most q_s signing queries, $\mathrm{Adv}^{\mathsf{ot\text{-}cma}}_{\mathsf{TOS},\mathcal{A}}(\lambda) \le q_s \cdot \mathrm{Adv}^{\mathsf{sdp}}_{\mathcal{G},\mathcal{B}}(\lambda) + 1/p$ holds.*

Proof. We show a reduction algorithm that simulates the one-time adaptive chosen message attack game for the adversary. The reduction gets an instance of the simultaneous double pairing assumption, $\Lambda, G_r, G_s, H_r, H_s$, and proceeds as follows.

Setup and Key Generation. It chooses ξ, η, μ and sets $G_t := G_r^\xi G_s^\eta$, and $H_t := H_r^\xi H_s^\mu$. It chooses $G \in \mathbb{G}$ and random $\omega, \nu, \nu_1, \nu_2$, and computes $gk = (\Lambda, C, F, U_1, U_2) = (\Lambda, G^\omega, G^{\omega\nu}, G^{\omega\nu_1}, G^{\omega\nu_2})$. It chooses random ρ_i, σ_i, τ_i, computes $G_i = G_r^{\rho_i} G_s^{\sigma_i} G_t^{\tau_i} = G_r^{\rho_i+\xi\tau_i} G_s^{\sigma_i+\eta\tau_i}$ and $H_i = H_r^{\rho_i} H_s^{\sigma_i} H_t^{\tau_i} = H_r^{\rho_i+\xi\tau_i} H_s^{\sigma_i+\mu\tau_i}$ for $i = 0 \ldots k$, and sets $pk = (G, G_r, G_s, G_t, H_r, H_s, H_t, G_0, \ldots G_k, H_0, \ldots, H_k)$. (Note that G_i, H_i are correctly distributed and give no information about τ_i.) It sends pk, gk to the adversary. The reduction will pick a random session j^*, and assume that the adversary will try to reuse tag from that session.

Queries to oracle $\mathcal{O}t$. When the adversary makes a query to the tag oracle $\mathcal{O}t$, choose the next new session index j.

- For session $j \ne j^*$: Pick random values $\rho, \sigma, \tau \leftarrow \mathbb{Z}_p$. Compute $(T_1, T_2) = (G_r^\rho G_s^\sigma G_t^\tau, H_r^\rho H_s^\sigma H_t^\tau) = (G_r^{\rho+\xi\tau} G_s^{\sigma+\eta\tau}, H_r^{\rho+\xi\tau} H_s^{\sigma+\mu\tau})$, and set $T = (T_1, T_2, T_1^\nu, T_2^\nu, T_1^{\nu_1}, T_2^{\nu_2})$. Store (j, ρ, σ, τ), and return T to the adversary.
- For session j^*. Pick random values $\rho, \sigma \leftarrow \mathbb{Z}_p$. Compute $(T_1, T_2) = (G_r^\rho G_s^\sigma, H_r^\rho H_s^\sigma)$. Let $T = (T_1, T_2, T_1^\nu, T_2^\nu, T_1^{\nu_1}, T_2^{\nu_2})$. Store (j^*, ρ, σ), and return T to the adversary.

Queries to oracle $\mathcal{O}sig$. When the adversary queries $\mathcal{O}sig$ for message $M = (M_1, \ldots, M_k) \in \mathbb{G}^k$ and session j, proceed as follows.

- If the $\mathcal{O}t$ has not yet produced a tag for session j, or $\mathcal{O}sig$ has already been queried for session j, return $\bot$.
- For session $j \ne j^*$: Look up the stored tuple (j, ρ, σ, τ). Compute $M_0 = (G \prod_{i=1}^k M_i^{\tau_i+\tau})^{-\frac{1}{\tau_0+\tau}}$. Note that for this choice of M_0, it will be the case that $e(G_t, G) \prod_{i=0}^k e(G_t^{\tau_i+\tau}, M_i) = e(G_t, M_0^{\tau_0+\tau} G \prod_{i=1}^k M_i^{\tau_i+\tau}) = 1$ and similarly $e(H_t, G) \prod_{i=0}^k e(H_t^{\tau_i+\tau}, M_i) = e(H_t, M_0^{\tau_0+\tau} G \prod_{i=1}^k M_i^{\tau_i+\tau}) = 1$. Note also that the tag is independent of τ, and since τ is uniformly distributed, then M_0 is independent of $\tau_0, \ldots, \tau_k$ even given tag. (To see this, let $m_0, \ldots, m_k$ be the discrete logarithms of $M_0, \ldots, M_k$ respectively and note that for any choice of $m_1, \ldots, m_k, \tau_0, \ldots, \tau_k$ and for any m_0 such that $m_0 \ne -\sum_{i=1}^k m_i$, there is a $\frac{1}{q}$ chance that we will choose $\tau = \frac{-1-\sum_{i=0}^k m_i \tau_i}{\sum_{i=0}^k m_i}$ which will yield $M_0 = (G \prod_{i=1}^k M_i^{\tau_i+\tau})^{-\frac{1}{\tau_0+\tau}}$.) Now

compute $Z = \prod_{i=0}^{k} M_i^{-\rho_i - \rho}$ and $W = \prod_{i=0}^{k} M_i^{-\sigma_i - \sigma}$ and output the signature (Z, W, M_0).

Note that these are the unique values such that $e(G_r, Z) \cdot e(G_s, W) \cdot e(G_t, G) \prod_{i=0}^{k} e(G_i T_1, M_i) = 1$ and similarly $e(H_r, Z) \cdot e(H_s, W) \cdot e(H_t, G) \prod_{i=0}^{k} e(H_i T_2, M_i) = 1$. Thus, Z, W are uniquely determined by $M_0, M_1, \ldots, M_k, tag$, and pk. $M_1, \ldots, M_k$ are provided by the adversary and, as we have argued, M_0, tag, pk are statistically independent of $\tau_0, \ldots, \tau_k$. We conclude that Z, W reveal no additional information about $\tau_0, \ldots, \tau_k$ even given the rest of the adversary's view.

- For session j^*: Look up the stored tuple (j, ρ, σ). Let $M_0 = (G \prod_{i=1}^{k} M_i^{\tau_i})^{-\frac{1}{\tau_0}}$. Note that for this choice of M_0, it will be the case that $e(G_t, G) \prod_{i=0}^{k} e(G_t^{\tau_i}, M_i) = e(G_t, M_0^{\tau_0} G \prod_{i=1}^{k} M_i^{\tau_i}) = 1$ and similarly $e(H_t, G) \prod_{i=0}^{k} e(H_t^{\tau_i}, M_i) = e(H_t, M_0^{\tau_0} G \prod_{i=1}^{k} M_i^{\tau_i}) = 1$. Note that T_1, T_2 are correctly distributed, that M_0 is statistically close to uniform since $\tau_0, \ldots, \tau_k$ are chosen at random, and furthermore that the only information revealed about $\tau_0, \ldots, \tau_k$ is that $G \prod_{i=0}^{k} M_i^{\tau_i} = 1$. Now, compute $Z = \prod_{i=0}^{k} M_i^{-\rho_i - \rho}$ and $W = \prod_{i=0}^{k} M_i^{-\sigma_i - \sigma}$, and output the signature (Z, W, M_0). Again all values are independent of $\tau_0, \ldots, \tau_k$ with the exception now of M_0, which is chosen so $G \prod_{i=0}^{k} M_i^{\tau_i} = 1$.

Processing the adversary's forgery. Now, suppose that the adversary produces $(M_1^{\dagger}, \ldots M_k^{\dagger})$ and $(Z^{\dagger}, W^{\dagger}, M_0^{\dagger}, T)$ for $T = (T_1, T_2, \ldots)$ used in the j^*th query. Look up the stored tuple (j^*, ρ, σ). Then with non-negligible probability (whenever the adversary succeeds) we have $\mathsf{TOS.Vrf}(pk, T, (M_1^{\dagger}, \ldots, M_k^{\dagger}), (Z^{\dagger}, W^{\dagger}, M_0^{\dagger})) = 1$. This means

$$1 = e(G_r, Z^{\dagger} G^{\xi} \prod_{i=0}^{k} (M_i^{\dagger})^{\rho_i + \rho + \xi \tau_i}) e(G_s, W^{\dagger} G^{\eta} \prod_{i=0}^{k} (M_i^{\dagger})^{\sigma_i + \sigma + \eta \tau_i}), \text{ and}$$

$$1 = e(H_r, Z^{\dagger} G^{\xi} \prod_{i=0}^{k} (M_i^{\dagger})^{\rho_i + \rho + \xi \tau_i}) e(H_s, W^{\dagger} G^{\mu} \prod_{i=0}^{k} (M_i^{\dagger})^{\sigma_i + \sigma + \mu \tau_i}).$$

So if $Z^{\dagger} G^{\xi} \prod_{i=0}^{k} (M_i^{\dagger})^{\rho_i + \rho + \xi \tau_i} \neq 1$, then

$$(Z^{\star}, R^{\star}, S^{\star}) := (Z^{\dagger} G^{\xi} \prod_{i=0}^{k} (M_i^{\dagger})^{\rho_i + \rho + \xi \tau_i}, W^{\dagger} G^{\eta} \prod_{i=0}^{k} (M_i^{\dagger})^{\sigma_i + \sigma + \eta \tau_i}, W^{\dagger} G^{\mu} \prod_{i=0}^{k} (M_i^{\dagger})^{\sigma_i + \sigma + \mu \tau_i})$$

is a valid solution for the simultaneous double pairing assumption.

$Z^{\dagger} G^{\xi} \prod_{i=0}^{k} (M_i^{\dagger})^{\rho_i + \rho + \xi \tau_i} = Z^{\dagger} \prod_{i=0}^{k} (M_i^{\dagger})^{\rho_i + \rho} (G \prod_{i=0}^{k} (M_i^{\dagger})^{\tau_i})^{\xi}$, and a part of $Z^{\dagger} \prod_{i=0}^{k} (M_i^{\dagger})^{\rho_i + \rho}$ is information theoretically hiding. Note that the only information that the adversary has about $\tau_0, \ldots, \tau_1$ is that in the jth session M_0 was chosen so that $G \prod_{i=0}^{k} M_i^{\tau_i} = 1$ (where $M = (M_1, \ldots, M_k)$ is the message signed in the j^*th session). If $M_i^{\dagger} \neq M_i$ for at least one i, then the probability that $G \prod_{i=0}^{k} (M_i^{\dagger})^{\tau_i} = 1$ conditioned on the fact that $G \prod_{i=0}^{k} M_i^{\tau_i} = 1$ is $1/p$. As a result, the probability that $Z^{\dagger} G^{\xi} \prod_{i=0}^{k} (M_i^{\dagger})^{\rho_i + \rho + \xi \tau_i} = 1$ is $1/p$.

Thus, if the guess for j^* is right, we succeed with all but probability $1/p$ whenever $\mathcal{A}$ does. We therefore have $\mathsf{Adv}_{\mathsf{TOS}, \mathcal{A}}^{\mathsf{ot\text{-}cma}}(\lambda) \leq q_s \cdot \mathsf{Adv}_{\mathcal{G}, \mathcal{B}}^{\mathsf{sdp}}(\lambda) + 1/p$.

RMA-secure signature scheme. For our random message signature scheme we will use a construction based on the dual system signature proposed in [41]. While the original scheme is CMA-secure under the DLIN assumption, the security proof makes use of a trapdoor commitment to elements in $\mathbb{Z}_p$ and consequently messages are elements in $\mathbb{Z}_p$ rather than $\mathbb{G}$. Our construction below resorts to RMA-security and removes this commitment to allows messages to be a sequence of random group elements satisfying a particular relation. As mentioned above, the message space $\mathcal{M}_x :=$ $\{(C^{m_1}, C^{m_2}, F^{m_1}, F^{m_2}, U_1^{m_1}, U_2^{m_2}) \in \mathbb{G}^6 \mid (m_1, m_2) \in \mathbb{Z}_p^2\}$ is defined by generators (C, F, U_1, U_2) in gk.

rSIG.Key(gk): Given $gk := (\Lambda, G, C, F, U_1, U_2)$ as input, uniformly select V, V_1, V_2, H from $\mathbb{G}^*$ and $a_1, a_2, b, \alpha,$ and ρ from $\mathbb{Z}_p^*$. Then compute and output $vk := (B, A_1, A_2, B_1, B_2, R_1, R_2, W_1, W_2, V, V_1, V_2, H, X_1, X_2)$ and $sk :=$ (vk, K_1, K_2) where

$$B := G^b, \quad A_1 := G^{a_1}, \quad A_2 := G^{a_2}, \quad B_1 := G^{b \cdot a_1}, \quad B_2 := G^{b \cdot a_2}$$
$$R_1 := VV_1^{a_1}, \quad R_2 := VV_2^{a_2}, \quad W_1 := R_1^b, \quad W_2 := R_2^b,$$
$$X_1 := G^\rho, \quad X_2 := G^{\alpha \cdot a_1 \cdot b / \rho}, \quad K_1 := G^\alpha, \quad K_2 := G^{\alpha \cdot a_1}.$$

rSIG.Sign(sk, msg): Parse msg into $(M_1, M_2, M_3, M_4, M_5, M_6)$. Pick random $r_1, r_2, z_1, z_2 \in \mathbb{Z}_p$. Let $r = r_1 + r_2$. Compute and output signature $\sigma := (S_0, S_1, \dots S_7)$ where

$$S_0 := (M_5 M_6 H)^{r_1}, \quad S_1 := K_2 V^r, \quad S_2 := K_1^{-1} V_1^r G^{z_1}, \quad S_3 := B^{-z_1},$$
$$S_4 := V_2^r G^{z_2}, \quad S_5 := B^{-z_2}, \quad S_6 := B^{r_2}, \quad S_7 := G^{r_1}.$$

rSIG.Vrf(vk, σ, msg): Parse msg into $(M_1, M_2, M_3, M_4, M_5, M_6)$ and σ into $(S_0, S_1, \dots, S_7)$. Also parse vk accordingly. Verify the following pairing product equations:

$$e(S_7, M_5 M_6 H) = e(G, S_0)$$
$$e(S_1, B)\, e(S_2, B_1)\, e(S_3, A_1) = e(S_6, R_1)\, e(S_7, W_1)$$
$$e(S_1, B)\, e(S_4, B_2)\, e(S_5, A_2) = e(S_6, R_2)\, e(S_7, W_2)\, e(X_1, X_2)$$
$$e(F, M_1) = e(C, M_3), \quad e(F, M_2) = e(C, M_4), \quad e(U_1, M_1) = e(C, M_5), \quad e(U_2, M_2) = e(C, M_6)$$

The scheme is structure-preserving by construction and the correctness is easily verified.

Theorem 24. *The above rSIG scheme is UF-RMA under the DLIN assumption. In particular, for any p.p.t. adversary $\mathcal{A}$ against rSIG that makes at most q_s signing queries, there exists p.p.t. algorithm $\mathcal{B}$ for DLIN such that $\mathrm{Adv}_{rSIG, \mathcal{A}}^{uf\text{-}rma}(\lambda) \leq (q_s + 2) \cdot \mathrm{Adv}_{\mathcal{G}, \mathcal{B}}^{dlin}(\lambda)$.*

Proof. We refer to the signatures output by the signing algorithm as a *normal signature*. In the proof we will consider an additional type of signatures to which we refer to as *simulation-type signatures* that are computationally indistinguishable but easier to simulate. For $\gamma \in \mathbb{Z}_p$, simulation-type signatures are of the form $\sigma = (S_0, S_1' = S_1 \cdot G^{-a_1 a_2 \gamma}, S_2' = S_2 \cdot G^{a_2 \gamma}, S_3, S_4' = S_4 \cdot G^{a_1 \gamma}, S_5, \dots, S_7)$. We give the outline of the proof using some lemmas.

Lemma 25. *Any signature that is accepted by the verification algorithm must be formed either as a normal signature, or a simulation-type signature.*

We consider a sequence of games. Let p_i be the probability that the adversary succeeds in **Game i**, and $p_i^{norm}(\lambda)$ and $p_i^{sim}(\lambda)$ that he succeeds with a normal-type respectively simulation-type forgery. Then by Lemma 25, $p_i(\lambda) = p_i^{norm}(\lambda) + p_i^{sim}(\lambda)$ for all i.

Game 0: The actual Unforgeability under Random Message Attacks game.

Lemma 26. *There exists an adversary $\mathcal{B}_1$ such that $p_0^{sim}(\lambda) = \mathrm{Adv}_{\mathcal{G},\mathcal{B}_1}^{dlin}(\lambda)$.*

Game i: The real security game except that the first i signatures that are given by the oracle are simulation-type signatures.

Lemma 27. *There exists an adversary $\mathcal{B}_2$ such that $|p_{i-1}^{norm}(\lambda) - p_i^{norm}(\lambda)| = \mathrm{Adv}_{\mathcal{G},\mathcal{B}_2}^{dlin}(\lambda)$.*

Game q: All sigantures that given by the oracle are simulation-type signatures.

Lemma 28. *There exists an adversary $\mathcal{B}_3$ such that $p_q^{norm}(\lambda) = \mathrm{Adv}_{\mathcal{G},\mathcal{B}_3}^{cdh}(\lambda)$.*

We have shown that in **Game q**, $\mathcal{A}$ can output a normal-type forgery with at most negligible probability. Thus, by Lemma 27 we can conclude that the same is true in **Game 0** and it holds

$$\mathrm{Adv}_{rSIG,\mathcal{A}}^{uf\text{-}rma}(\lambda) = p_0(\lambda) = p_0^{sim}(\lambda) + p_0^{norm}(\lambda) \leq p_0^{sim}(\lambda) + \sum_{i=1}^{q} |p_{i-1}^{norm}(\lambda) - p_i^{norm}(\lambda)| + p_q^{norm}(\lambda)$$

$$\leq \mathrm{Adv}_{\mathcal{G},\mathcal{B}_1}^{dlin}(\lambda) + q\mathrm{Adv}_{\mathcal{G},\mathcal{B}_2}^{dlin}(\lambda) + \mathrm{Adv}_{\mathcal{G},\mathcal{B}_3}^{cdh}(\lambda) \leq (q+2) \cdot \mathrm{Adv}_{\mathcal{G},\mathcal{B}}^{dlin}(\lambda) .$$

Let MSGGen be an extended random message generator that first chooses $aux = (m_1, m_2)$ randomly from $\mathbb{Z}_p^2$ and then computes $msg = (C^{m_1}, C^{m_2}, F^{m_1}, F^{m_2}, U_1^{m_1}, U_2^{m_2})$. Note that this is what the reduction algorithm does in the proof of Theorem 24. Therefore, the same reduction algorithm works for the case of extended random message attacks with respect to message generator MSGGen. We thus have the following.

Corollary 29. *Under the DLIN assumption, rSIG scheme is UF-XRMA w.r.t. the message generator that provides $aux = (m_1, m_2)$ for every message $msg = (C^{m_1}, C^{m_2}, F^{m_1}, F^{m_2}, U_1^{m_1}, U_2^{m_2})$. In particular, for any p.p.t. adversary $\mathcal{A}$ against rSIG that is given at most q_s signatures, there exists p.p.t. algorithm $\mathcal{B}$ such that $\mathrm{Adv}_{rSIG,\mathcal{A}}^{uf\text{-}xrma}(\lambda) \leq (q_s + 2) \cdot \mathrm{Adv}_{\mathcal{G},\mathcal{B}}^{dlin}(\lambda)$.*

Security and efficiency of resulting SIG1. Let SIG1 be the signature scheme obtained from TOS (with *mode* = extended) and rSIG by following the first generic construction in Section 4. From Theorem 17, 20, 23, and 24, the following is immediate.

Theorem 30. *SIG1 is a structure-preserving signature scheme that yields constant-size signatures, and is UF-CMA under the DLIN assumption. In particular, for any p.p.t. adversary $\mathcal{A}$ for SIG1 making at most q_s signing queries, there exists p.p.t. algorithm $\mathcal{B}$ such that $\mathrm{Adv}_{SIG1,\mathcal{A}}^{uf\text{-}cma}(\lambda) \leq (q_s + 3) \cdot \mathrm{Adv}_{\mathcal{G},\mathcal{B}}^{dlin}(\lambda) + 1/p$.*

6 Instantiating SIG2

We instantiate the POS and xSIG building blocks of our second generic construction to obtain our second SPS scheme. Here we choose the Type-III bilinear group setting. The resulting SIG2 scheme is an efficient structure-preserving signature scheme based on SXDH and XDLIN.

Setup for Type-III groups. The following setup procedure is common for all building blocks in this section. The global parameter gk is given to all functions implicitly.

- Setup(1^λ): Run $\Lambda = (p, \mathbb{G}_1, \mathbb{G}_2, \mathbb{G}_T, e) \leftarrow \mathcal{G}(1^\lambda)$ and choose generators $G \in \mathbb{G}_1^*$ and $\hat{G} \in \mathbb{G}_2^*$. Also choose u, f_2, f_3 randomly from $\mathbb{Z}_p^*$, compute $F_2 := G^{f_2}$, $F_3 := G^{f_3}$, $\hat{F}_2 := \hat{G}^{f_2}$, $\hat{F}_3 := \hat{G}^{f_3}$, $U := G^u$, $\hat{U} := \hat{G}^u$, and output $gk :=$ $(\Lambda, G, \hat{G}, F_2, F_3, \hat{F}_2, \hat{F}_3, U, \hat{U})$.

A gk defines a message space $\mathcal{M}_x = \{(\hat{F}_2^m, \hat{F}_3^m, \hat{U}^m) \in \mathbb{G}_2^* \mid m \in \mathbb{Z}_p\}$ for the signature scheme in this section. For our generic construction to work, the partial one-time signature scheme should have the same key space.

Partial one-time signatures for uniliteral messages. We construct a partial one-time signature scheme POSu2 for messages in $\mathbb{G}_2^k$ for $k > 0$. The suffix "u2" indicates that the scheme is uniliteral and messages are taken from $\mathbb{G}_2$. Correspondingly, POSu1 refers to the scheme whose messages belong to $\mathbb{G}_1$, which is obtained by swapping $\mathbb{G}_2$ and $\mathbb{G}_1$ in the following description. Our POSu2 scheme is a minor refinement of the one-time signature scheme introduced in [4]. It comes, however, with a security proof for the new security model.

Basically, a one-time public-key in our scheme consists of one element in the base group $\mathbb{G}_1$ that is the opposite of the group $\mathbb{G}_2$ messages belong to. This property is very useful to construct a POS scheme for signing bilateral messages. As well as tags of TOS in Section 5, the one-time public-keys of POS will have to be in an extended form to meet the constraint from xSIG presented in the sequel. We use $mode \in \{\text{normal}, \text{extended}\}$ for this purpose again.

- POSu2.Key(gk): Take generators U and $\hat{U}$ from gk. Choose w_r randomly from $\mathbb{Z}_p^*$ and compute $G_r := U^{w_r}$. For $i = 1, \ldots, k$, uniformly choose χ_i and γ_i from $\mathbb{Z}_p$ and compute $G_i := U^{\chi_i} G_r^{\gamma_i}$. Output $pk := (G_r, G_1, ..., G_k) \in \mathbb{G}_1^{k+1}$ and $sk := (\chi_1, \gamma_1, ..., \chi_k, \gamma_k, w_r)$.
- POSu2.Update($mode$): Take F_2, F_3, U from gk. Choose $a \leftarrow \mathbb{Z}_p$ and output $opk := U^a \in \mathbb{G}_1$ if $mode = \text{normal}$ or $opk := (F_2^a, F_3^a, U^a) \in \mathbb{G}_1^3$ if $mode = \text{extended}$. Also output $osk := a$.
- POSu2.Sign(sk, msg, osk): Parse msg into $(\hat{M}_1, \ldots, \hat{M}_k) \in \mathbb{G}_2^k$. Take a and w_r from osk and sk, respectively. Choose ρ randomly from $\mathbb{Z}_p$ and compute $\zeta :=$ $a - \rho w_r \bmod p$. Then compute and output $\sigma := (\hat{Z}, \hat{R}) \in \mathbb{G}_2^2$ as the signature, where $\hat{Z} := \hat{U}^\zeta \prod_{i=1}^k \hat{M}_i^{-\chi_i}$ and $\hat{R} := \hat{U}^\rho \prod_{i=1}^k \hat{M}_i^{-\gamma_i}$
- POSu2.Vrf(pk, σ, msg, opk): Parse σ as $(\hat{Z}, \hat{R}) \in \mathbb{G}_2^2$, msg as $(\hat{M}_1, \ldots, \hat{M}_k) \in \mathbb{G}_2^k$, and opk as (A_2, A_3, A) or A depending on $mode$. Return 1, if $e(A, \hat{U}) = e(U, \hat{Z}) e(G_r, \hat{R}) \prod_{i=1}^k e(G_i, \hat{M}_i)$ holds. Return 0, otherwise.

Scheme POS$u2$ is structure-preserving and has uniform one-time public-key property from the construction. We can easily verify that it is correct by simple calculation.

Theorem 31. POS$u2$ *is strongly unforgeable against OT-CMA if DBP$_1$ holds. In particular,* $\mathrm{Adv}^{\mathsf{sot\text{-}cma}}_{\mathsf{POS}u2,\mathcal{A}}(\lambda) \leq \mathrm{Adv}^{\mathsf{dbp1}}_{\mathcal{G},\mathcal{B}}(\lambda) + 1/p.$

Partial one-time signatures for bilateral messages. Using POS$u1$ for $msg \in \mathbb{G}_1^{k_1+1}$ and POS$u2$ for $msg \in \mathbb{G}_2^{k_2}$, we construct a POS$b$ scheme for signing bilateral messages $(msg_1, msg_2) \in \mathbb{G}_1^{k_1} \times \mathbb{G}_2^{k_2}$. The scheme is a simple two-story construction where msg_2 is signed by POS$u2$ with one-time secret-key $osk_2 \in \mathbb{G}_1$ and then the one-time public-key opk_2 is attached to msg_1 and signed by POS$u1$. Public-key opk_2 is included in the signature, and opk_1 is output as a one-time public-key for POSb.

- POSb.Key(gk): Run $(pk_1, sk_1) \leftarrow$ POS$u1$.Key(gk) and $(pk_2, sk_2) \leftarrow$ POS$u2$.Key(gk). Set $pk := (pk_1, pk_2)$ and $sk := (sk_1, sk_2)$, and output (pk, sk).
- POSb.Update($mode$): Run $(opk, osk) \leftarrow$ POS$u1$($mode$) and output (opk, osk).
- POSb.Sign(sk, msg, osk): Parse msg into $(msg_1, msg_2) \in \mathbb{G}_1^{k_1} \times \mathbb{G}_2^{k_2}$, and sk into (sk_1, sk_2). Run $(opk_2, osk_2) \leftarrow$ POS$u2$.Update(normal), and compute $\sigma_2 \leftarrow$ POS$u2$.Sign(sk_2, msg_2, osk_2) and $\sigma_1 \leftarrow$ POS$u1$.Sign($sk_1, (msg_1, opk_2), osk$). Output $\sigma := (\sigma_1, \sigma_2, opk_2)$.
- POSb.Vrf(pk, opk, σ, msg): Parse msg into $(msg_1, msg_2) \in \mathbb{G}_1^{k_1} \times \mathbb{G}_2^{k_2}$, and σ into $(\sigma_1, \sigma_2, opk_2)$. If $1 =$ POS$u1$.Vrf($pk_1, opk, \sigma_1, (msg_1, opk_2)$) $=$ POS$u2$.Vrf($pk_2, opk_2, \sigma_2, msg_2$), output 1. Otherwise, output 0.

For a message in $\mathbb{G}_1^{k_1} \times \mathbb{G}_2^{k_2}$, the above POS$b$ uses a public-key of size $(k+2, k+1)$, yields a one-time public-key of size $(0,1)$ (for $mode = $ normal) or $(0,3)$ (for $mode = $ extended), and a signature of size $(3,2)$. Verification requires 2 pairing product equations. A one-time public-key in extended mode, which is treated as a message to xSIG in this section, is of the form $opk = (\hat{F}_2^a, \hat{F}_3^a, \hat{U}^a) \in \mathbb{G}_2^3$. Structure-preservance and uniform public-key property are taken over from the underlying POS$u1$ and POS$u2$.

Theorem 32. *Scheme POSb is unforgeable against OT-CMA if SXDH holds. In particular,* $\mathrm{Adv}^{\mathsf{ot\text{-}cma}}_{\mathsf{POS}b,\mathcal{A}}(\lambda) \leq \mathrm{Adv}^{\mathsf{sxdh}}_{\mathcal{G},\mathcal{B}}(\lambda) + 2/p.$

XRMA-secure signature scheme. Our construction bases on a variant of Waters' dual system encryption proposed by Ramanna, Chatterjee, and Sarkar [36]. Recall that $gk = (\Lambda, G, \hat{G}, F_2, F_3, \hat{F}_2, \hat{F}_3, U, \hat{U})$ with $\Lambda = (p, \mathbb{G}_1, \mathbb{G}_2, \mathbb{G}_T, e)$ is generated by Setup(1^λ) in advance.

xSIG.Gen(gk): On input gk, select generators $V, V', H \leftarrow \mathbb{G}_1$, $\hat{V}, \hat{V}', \hat{H} \in \mathbb{G}_2$ such that $V \sim \hat{V}, V' \sim \hat{V}', H \sim \hat{H}, F_2 \sim \hat{F}_2, F_3 \sim \hat{F}_3$ and exponent $a, b, \alpha \leftarrow \mathbb{Z}_p$ and $\rho \leftarrow \mathbb{Z}_p^*$, compute $R := V(V')^a$, $\hat{R} := \hat{V}(\hat{V}')^a$, and set $vk := (gk, \hat{G}^b, \hat{G}^a, \hat{G}^{ba}, \hat{R}, \hat{R}^b)$, $sk := (VK, G^\alpha, G^a, G^b)$.

xSIG.Sign(sk, msg): On input message $msg = (\hat{M}_1, \hat{M}_2, \hat{M}_0) = (\hat{F}_2^m, \hat{F}_3^m, \hat{U}^m) \in \mathbb{G}_2^3$ ($m \in \mathbb{Z}_p$), select $r_1, r_2 \leftarrow \mathbb{Z}_p$, set $r := r_1 + r_2$, compute $\sigma_0 := (\hat{M}_0 \hat{H})^{r_1}$, $\sigma_1 := G^\alpha V^r$, $\sigma_2 := (V')^r G^{-z}$, $\sigma_3 := (G^b)^z$, $\sigma_4 := (G^b)^{r_2}$, and $\sigma_5 := G^{r_1}$, and output $\sigma := (\sigma_0, \sigma_1, \ldots, \sigma_5) \in \mathbb{G}_2 \times \mathbb{G}_1^5$.

Table 1. Efficiency of our schemes (SIG1 and SIG2) and comparison to other schemes with constant-size signatures. The top section is for the Type I variant, the middle section is for unilateral messages and the lower section is for bilateral messages. Notation (x, y) represents x elements in $\mathbb{G}_1$ and y in $\mathbb{G}_2$.

Schemes	$|msg|$	$|gk| + |vk|$	$|\sigma|$	#(PPE)	Assumptions
AHO10	k	$2k + 12$	7	2	q-SFP
SIG1	k	$2k + 25$	17	9	DLIN
AHO10	$(k_1, 0)$	$(4, 2k_1 + 8)$	$(5, 2)$	2	q-SFP
AGHO11	$(k_1, 0)$	$(1, k_1 + 4)$	$(3, 1)$	2	q-type
SIG2 : POS$u1$ + xSIG	$(k_1, 0)$	$(7, k_1 + 13)$	$(7, 4)$	5	SXDH, XDLIN$_1$
POSb + AHO10	(k_1, k_2)	$(k_2 + 5, k_1 + 12)$	$(10, 3)$	3	q-SFP
AGHO11	(k_1, k_2)	$(k_2 + 3, k_1 + 4)$	$(3, 3)$	2	q-type
SIG2 : POSb + xSIG	(k_1, k_2)	$(k_2 + 8, k_1 + 14)$	$(8, 6)$	6	SXDH, XDLIN$_1$

xSIG.Vrfy(vk, σ, msg): On input $vk, msg = (\hat{M}_1, \hat{M}_2, \hat{M}_0)$, and signature σ, compute

$$e(F_2, \hat{M}_0) = e(U, \hat{M}_1), \ e(F_3, \hat{M}_0) = e(U, \hat{M}_2), \ e(\sigma_5, \hat{M}_0\hat{H}) = e(G, \sigma_0)$$
$$e(\sigma_1, \hat{G}^b)e(\sigma_2, \hat{G}^{ba})e(\sigma_3, \hat{G}^a) = e(\sigma_4, \hat{R})e(\sigma_5, \hat{R}^b)e(G^\rho, \hat{G}^{ab/\rho}).$$

The scheme is structure-preserving by the construction. We can easily verify the correctness.

Theorem 33. *If the DDH$_2$ and XDLIN$_1$ assumptions hold, then above xSIG scheme is UF-XRMA with respect to the message generator that returns $aux = m$ for every random message $msg = (\hat{F}_2^m, \hat{F}_3^m, \hat{U}^m)$. In particular for any p.p.t. adversary $\mathcal{A}$ for xSIG making at most q signing queries, there exist p.p.t. algorithms $\mathcal{B}_1, \mathcal{B}_2, \mathcal{B}_3$ such that* $\text{Adv}_{\text{xSIG}, \mathcal{A}}^{\text{uf-xrma}}(\lambda) < \text{Adv}_{\mathcal{G}, \mathcal{B}_1}^{\text{ddh2}}(\lambda) + q\text{Adv}_{\mathcal{G}, \mathcal{B}_2}^{\text{xdlin1}}(\lambda) + \text{Adv}_{\mathcal{G}, \mathcal{B}_3}^{\text{co-cdh}}(\lambda)$.

Security and efficiency of resulting SIG2. Let SIG2 be the scheme obtained from POSb (with *mode* = extended) and xSIG. SIG2 is structure-preserving as vk, σ, and msg consist of group elements from $\mathbb{G}_1$ and $\mathbb{G}_2$, and SIG2.Vrf evaluates pairing product equations. From Theorem 21, 32, and 33, we obtain the following theorem.

Theorem 34. SIG2 *is a structure-preserving signature scheme that is unforgeable against adaptive chosen message attacks if SXDH and XDLIN$_1$ hold for $\mathcal{G}$.*

7   Efficiency, Applications and Open Questions

Efficiency. Table 1 summarizes the efficiency of SIG1 and SIG2. For SIG2 we consider both uniliteral and biliteral messages. We count the number of group elements excluding a default generator for each group in gk, and distinguish between $\mathbb{G}_1$ and $\mathbb{G}_2$ and use k_1 and k_2 for the number of message elements in $\mathbb{G}_1$ and $\mathbb{G}_2$, respectively. For comparison, we include the efficiency of the schemes in [4] and [2]. For bilateral messages, AHO10 is combined with POSb from Section 6.

Applications. Structure-preserving signatures (SPS) have become a mainstay in cryptographic protocol design in recent years. From the many applications that benefit from efficient SPS based on simple assumptions, we list only a few recent examples. Using our SIG1 scheme from Section 5 both the construction of a group signature scheme with efficient revocation by Libert, Peters and Yung [33] and the construction of compact verifiable shuffles by Chase et al. [16] can be proven purely under the DLIN assumption. All other building blocks already have efficient instantiations based on DLIN.

Hofheinz and Jager [31] construct a structure-preserving one-time signature scheme and use it to build a tree-based SPS scheme, say tSIG. Instead, we propose to use our partial one-time scheme to construct tSIG. As the resulting tSIG is secure against nonadaptive chosen message attacks, it is secure against extended random message attacks as well. We then combine the POSb scheme and the new tSIG scheme according to our second generic construction. As confirmed with the authors of [31], the resulting signature scheme is significantly more efficient than [31] and is a SPS scheme with a tight security reduction to SXDH. One can do the same in Type-I groups by using the tagged one-time signature scheme in Section 5 whose security tightly reduced to DLIN.

As also shown by [31], SPS schemes allow to implement simulation-sound NIZK proofs based on the Groth-Sahai proof system. Following the Naor-Yung-Sahai [35,38] paradigm, one obtains structure-preserving CCA-secure public-key encryption in a modular fashion.

Open Questions. 1) Can we have (X)RMA-secure schemes with a message space that is a simple Cartesian product of groups without sacrificing on efficiency? 2) The RMA-secure signature schemes developed in this paper are in fact XRMA-secure. Can we have more efficient schemes by resorting to RMA-security? 3) Can we have tagged one-time signature schemes with tight reduction to the underlying simple assumptions? 4) What is the exact lower bound for the size of signatures under simple assumptions? Is it possible to show such a bound?

References

1. Abe, M., Fuchsbauer, G., Groth, J., Haralambiev, K., Ohkubo, M.: Structure-Preserving Signatures and Commitments to Group Elements. In: Rabin, T. (ed.) CRYPTO 2010. LNCS, vol. 6223, pp. 209–236. Springer, Heidelberg (2010)
2. Abe, M., Groth, J., Haralambiev, K., Ohkubo, M.: Optimal Structure-Preserving Signatures in Asymmetric Bilinear Groups. In: Rogaway, P. (ed.) CRYPTO 2011. LNCS, vol. 6841, pp. 649–666. Springer, Heidelberg (2011)
3. Abe, M., Groth, J., Ohkubo, M.: Separating Short Structure-Preserving Signatures from Non-interactive Assumptions. In: Lee, D.H. (ed.) ASIACRYPT 2011. LNCS, vol. 7073, pp. 628–646. Springer, Heidelberg (2011)
4. Abe, M., Haralambiev, K., Ohkubo, M.: Signing on group elements for modular protocol designs. IACR ePrint Archive, Report 2010/133 (2010), http://eprint.iacr.org
5. Abe, M., Ohkubo, M.: A framework for universally composable non-committing blind signatures. IJACT 2(3), 229–249 (2012)
6. Belenkiy, M., Camenisch, J., Chase, M., Kohlweiss, M., Lysyanskaya, A., Shacham, H.: Randomizable Proofs and Delegatable Anonymous Credentials. In: Halevi, S. (ed.) CRYPTO 2009. LNCS, vol. 5677, pp. 108–125. Springer, Heidelberg (2009)

7. Bellare, M., Micciancio, D., Warinschi, B.: Foundations of Group Signatures: Formal Definitions, Simplified Requirements and a Construction based on General Assumptions. In: Biham, E. (ed.) EUROCRYPT 2003. LNCS, vol. 2656, pp. 614–629. Springer, Heidelberg (2003)
8. Bellare, M., Shi, H., Zhang, C.: Foundations of Group Signatures: The Case of Dynamic Groups. In: Menezes, A. (ed.) CT-RSA 2005. LNCS, vol. 3376, pp. 136–153. Springer, Heidelberg (2005)
9. Bellare, M., Shoup, S.: Two-Tier Signatures, Strongly Unforgeable Signatures, and Fiat-Shamir Without Random Oracles. In: Okamoto, T., Wang, X. (eds.) PKC 2007. LNCS, vol. 4450, pp. 201–216. Springer, Heidelberg (2007)
10. Boneh, D., Boyen, X., Shacham, H.: Short Group Signatures. In: Franklin, M. (ed.) CRYPTO 2004. LNCS, vol. 3152, pp. 41–55. Springer, Heidelberg (2004)
11. Boneh, D., Gentry, C., Lynn, B., Shacham, H.: Aggregate and Verifiably Encrypted Signatures from Bilinear Maps. In: Biham, E. (ed.) EUROCRYPT 2003. LNCS, vol. 2656, pp. 416–432. Springer, Heidelberg (2003)
12. Camenisch, J., Dubovitskaya, M., Haralambiev, K.: Efficiently signing group elements under simple assumptions (unpublished manuscript, available from the authors)
13. Camenisch, J., Dubovitskaya, M., Haralambiev, K.: Efficient Structure-Preserving Signature Scheme from Standard Assumptions. In: Visconti, I., De Prisco, R. (eds.) SCN 2012. LNCS, vol. 7485, pp. 76–94. Springer, Heidelberg (2012)
14. Cathalo, J., Libert, B., Yung, M.: Group Encryption: Non-interactive Realization in the Standard Model. In: Matsui, M. (ed.) ASIACRYPT 2009. LNCS, vol. 5912, pp. 179–196. Springer, Heidelberg (2009)
15. Chase, M., Kohlweiss, M.: A New Hash-and-Sign Approach and Structure-Preserving Signatures from DLIN. In: Visconti, I., De Prisco, R. (eds.) SCN 2012. LNCS, vol. 7485, pp. 131–148. Springer, Heidelberg (2012)
16. Chase, M., Kohlweiss, M., Lysyanskaya, A., Meiklejohn, S.: Malleable Proof Systems and Applications. In: Pointcheval, D., Johansson, T. (eds.) EUROCRYPT 2012. LNCS, vol. 7237, pp. 281–300. Springer, Heidelberg (2012)
17. Dolev, D., Dwork, C., Naor, M.: Nonmalleable cryptography. SIAM J. Comput. 30(2), 391–437 (2000)
18. Dwork, C., Naor, M.: An efficient existentially unforgeable signature scheme and its applications. J. Cryptology 11(3), 187–208 (1998)
19. Even, S., Goldreich, O., Micali, S.: On-line/off-line digital signatures. J. Cryptology 9(1), 35–67 (1996)
20. Fischlin, M.: Round-Optimal Composable Blind Signatures in the Common Reference String Model. In: Dwork, C. (ed.) CRYPTO 2006. LNCS, vol. 4117, pp. 60–77. Springer, Heidelberg (2006)
21. Fuchsbauer, G.: Commuting Signatures and Verifiable Encryption. In: Paterson, K.G. (ed.) EUROCRYPT 2011. LNCS, vol. 6632, pp. 224–245. Springer, Heidelberg (2011)
22. Fuchsbauer, G., Pointcheval, D.: Anonymous Proxy Signatures. In: Ostrovsky, R., De Prisco, R., Visconti, I. (eds.) SCN 2008. LNCS, vol. 5229, pp. 201–217. Springer, Heidelberg (2008)
23. Fuchsbauer, G., Pointcheval, D., Vergnaud, D.: Transferable Constant-Size Fair E-Cash. In: Garay, J.A., Miyaji, A., Otsuka, A. (eds.) CANS 2009. LNCS, vol. 5888, pp. 226–247. Springer, Heidelberg (2009)
24. Fuchsbauer, G., Vergnaud, D.: Fair Blind Signatures without Random Oracles. In: Bernstein, D.J., Lange, T. (eds.) AFRICACRYPT 2010. LNCS, vol. 6055, pp. 16–33. Springer, Heidelberg (2010)
25. Galbraith, S.D., Peterson, K.G., Smart, N.P.: Pairings for cryptographers. Discrete Applied Mathematics 156(16), 3113–3121 (2008)

26. Goldwasser, S., Micali, S., Rivest, R.: A digital signature scheme secure against adaptive chosen-message attacks. SIAM Journal on Computing 17(2), 281–308 (1988)
27. Green, M., Hohenberger, S.: Universally Composable Adaptive Oblivious Transfer. In: Pieprzyk, J. (ed.) ASIACRYPT 2008. LNCS, vol. 5350, pp. 179–197. Springer, Heidelberg (2008)
28. Green, M., Hohenberger, S.: Practical Adaptive Oblivious Transfer from Simple Assumptions. In: Ishai, Y. (ed.) TCC 2011. LNCS, vol. 6597, pp. 347–363. Springer, Heidelberg (2011)
29. Groth, J.: Simulation-Sound NIZK Proofs for a Practical Language and Constant Size Group Signatures. In: Lai, X., Chen, K. (eds.) ASIACRYPT 2006. LNCS, vol. 4284, pp. 444–459. Springer, Heidelberg (2006)
30. Groth, J., Sahai, A.: Efficient Non-interactive Proof Systems for Bilinear Groups. In: Smart, N.P. (ed.) EUROCRYPT 2008. LNCS, vol. 4965, pp. 415–432. Springer, Heidelberg (2008)
31. Hofheinz, D., Jager, T.: Tightly Secure Signatures and Public-Key Encryption. In: Safavi-Naini, R. (ed.) CRYPTO 2012. LNCS, vol. 7417, pp. 590–607. Springer, Heidelberg (2012)
32. Kiayias, A., Yung, M.: Group Signatures with Efficient Concurrent Join. In: Cramer, R. (ed.) EUROCRYPT 2005. LNCS, vol. 3494, pp. 198–214. Springer, Heidelberg (2005)
33. Libert, B., Peters, T., Yung, M.: Scalable Group Signatures with Revocation. In: Pointcheval, D., Johansson, T. (eds.) EUROCRYPT 2012. LNCS, vol. 7237, pp. 609–627. Springer, Heidelberg (2012)
34. Lindell, Y.: A simpler construction of CCA2-secure public-key encryption under general assumptions. J. Cryptology 19(3), 359–377 (2006)
35. Naor, M., Yung, M.: Public-key cryptosystems provably secure against chosen ciphertext attacks. In: STOC 1990, pp. 427–437 (1990)
36. Ramanna, S.C., Chatterjee, S., Sarkar, P.: Variants of Waters' Dual System Primitives Using Asymmetric Pairings (Extended Abstract). In: Fischlin, M., Buchmann, J., Manulis, M. (eds.) PKC 2012. LNCS, vol. 7293, pp. 298–315. Springer, Heidelberg (2012)
37. Rückert, M., Schröder, D.: Security of Verifiably Encrypted Signatures and a Construction without Random Oracles. In: Shacham, H., Waters, B. (eds.) Pairing 2009. LNCS, vol. 5671, pp. 17–34. Springer, Heidelberg (2009)
38. Sahai, A.: Non-malleable non-interactive zero-knowledge and chosen-ciphertext security. In: FOCS 1999, pp. 543–553 (1999)
39. De Santis, A., Di Crescenzo, G., Ostrovsky, R., Persiano, G., Sahai, A.: Robust Non-interactive Zero Knowledge. In: Kilian, J. (ed.) CRYPTO 2001. LNCS, vol. 2139, pp. 566–598. Springer, Heidelberg (2001)
40. Shoup, V.: Lower Bounds for Discrete Logarithms and Related Problems. In: Fumy, W. (ed.) EUROCRYPT 1997. LNCS, vol. 1233, pp. 256–266. Springer, Heidelberg (1997)
41. Waters, B.: Dual System Encryption: Realizing Fully Secure IBE and HIBE under Simple Assumptions. In: Halevi, S. (ed.) CRYPTO 2009. LNCS, vol. 5677, pp. 619–636. Springer, Heidelberg (2009)

Dual Form Signatures: An Approach for Proving Security from Static Assumptions

Michael Gerbush[1], Allison Lewko[2], Adam O'Neill[3], and Brent Waters[4,⋆]

[1] The University of Texas at Austin
mgerbush@cs.utexas.edu
[2] Microsoft Research
allew@microsoft.com
[3] Boston University
amoneill@bu.edu
[4] The University of Texas at Austin
bwaters@cs.utexas.edu

Abstract. In this paper, we introduce the abstraction of Dual Form Signatures as a useful framework for proving security (existential unforgeability) from static assumptions for schemes with special structure that are used as a basis of other cryptographic protocols and applications. We demonstrate the power of this framework by proving security under static assumptions for close variants of pre-existing schemes: the LRSW-based Camenisch-Lysyanskaya signature scheme, and the identity-based sequential aggregate signatures of Boldyreva, Gentry, O'Neill, and Yum. The Camenisch-Lysyanskaya signature scheme was previously proven only under the interactive LRSW assumption, and our result can be viewed as a static replacement for the LRSW assumption. The scheme of Boldyreva, Gentry, O'Neill, and Yum was also previously proven only under an interactive assumption that was shown to hold in the generic group model. The structure of the public key signature scheme underlying the BGOY aggregate signatures is quite distinctive, and our work presents the first security analysis of this kind of structure under static assumptions.

1 Introduction

Digital signatures are a fundamental technique for verifying the authenticity of a digital message. The significance of digital signatures in cryptography is also amplified by their use as building blocks for more complex cryptographic

---

⋆ Supported by NSF CNS-0915361 and CNS-0952692, AFOSR Grant No: FA9550-08-1-0352, DARPA through the U.S. Office of Naval Research under Contract N00014-11-1-0382, DARPA N11AP20006, Google Faculty Research award, the Alfred P. Sloan Fellowship, and Microsoft Faculty Fellowship, and Packard Foundation Fellowship. Any opinions, findings, and conclusions or recommendations expressed in this material are those of the author(s) and do not necessarily reflect the views of the Department of Defense or the U.S. Government.

X. Wang and K. Sako (Eds.): ASIACRYPT 2012, LNCS 7658, pp. 25–42, 2012.

protocols. Recently, we have seen several pairing based signature schemes (e.g., [17,13,24,48]) that are both practical and have added structure which has been used to build other primitives ranging from Aggregate Signatures [15,43] to Oblivious Transfer [25,32]. Ideally, for such a fundamental cryptographic primitive we would like to have security proofs from straightforward, static complexity assumptions.

Meeting this goal for certain systems is often challenging. For instance, the Camenisch and Lysyanskaya signature scheme [24][1] has been very influential as it is used as the foundation for a wide variety of advanced cryptographic systems, including anonymous credentials [24,7,6], group signatures [24,5], ecash [22], uncloneable functions [21], batch verification [23], and RFID encryption [4]. While the demonstrated utility of CL signatures has made them desirable, it has been difficult to reduce their security to a static security assumption. Currently, the CL signature scheme is proven secure under the LRSW assumption [44], an interactive complexity assumption that closely mirrors the description of the signature scheme itself. In addition, the interactive assumption transfers to the systems built around these signatures.

The identity-based sequential aggregate signatures of Boldyreva, Gentry, O'Neill, and Yum [9,10] were also proven in the random oracle model under an interactive assumption (justified in the generic bilinear group model), which again closely mirrors the underlying signature scheme itself. (This can be viewed as providing a proof of the scheme only in the generic group model.) Proofs of complicated interactive assumptions in the generic group model have several disadvantages. First, they are themselves complex and prone to error. In fact, the original version of the BGOY identity-based sequential aggregate signature scheme [9] relied on an assumption that was shown to be false, and the scheme was insecure [36]. This scheme and proof were corrected in [10]. Secondly, such proofs do not tend to provide much insight into the security of the scheme. This lack of insight tends to hinder transferring schemes to other settings. For example, many schemes developed in bilinear groups now have lattice-based analogs, and these transformations reused high-level ideas from the original security proofs in the bilinear group setting. Techniques from [48] were used in the lattice setting in [20], techniques from [26] were used in [27], and techniques from [12] were used in [2]. This kind of transference of ideas from the bilinear setting to the lattice setting is unlikely to be achieved through generic group proofs.

In this work, we develop techniques that can be applied to prove security from static assumptions for new signature schemes as well as (slight variants of) pre-existing schemes. Providing new proofs for these existing schemes provides a meaningful sanity check as well as new insight into their security. This kind of sanity check is valuable not only for schemes proven in the generic group model, but also for signatures (CL signatures included) that require extra checks to rule out trivial breaks (e.g. not allowing the message signed to be equal to 0), since these subtleties can easily be missed at first glance. Having new proofs

---

[1] Throughout, we will be discussing the CL signatures based on the LRSW assumption, which should not be confused with those based on the strong RSA assumption.

from static assumptions for variants of schemes like CL signatures and BGOY signatures gives us additional confidence in their security without having to sacrifice the variety of applications built from them. Ultimately, this provides us with a fuller understanding of these kinds of signatures, and is a critical step towards obtaining proofs under the simplest and weakest assumptions.

Dual Form Signatures. Our work is centered around a new abstraction that we call Dual Form Signatures. Dual Form Signatures have similar structure to existing signature schemes, however they have two signing algorithms, Sign_A and Sign_B, that respectively define two forms of signatures that will both verify under the same public key. In addition, the security definition will categorize forgeries into two *disjoint* types, Type I and Type II. Typically, these forgery types will roughly correspond with signatures of form A and B.

In a Dual Form system, we will demand three security properties (stated informally here):

A-I Matching. If an attacker is only given oracle access to Sign_A, then it is hard to create any forgery that is not of Type I.

B-II Matching. If an attacker is only given oracle access to Sign_B, then it is hard to create any forgery that is not of Type II.

Dual-Oracle Invariance. If an attacker is given oracle access to both Sign_A and Sign_B and a "challenge signature" which is either from Sign_A or Sign_B, the attacker's probability of producing a Type I forgery is approximately the same when the challenge signature is from Sign_A as when the challenge signature is from Sign_B.

A Dual Form Signature scheme immediately gives a secure signature scheme if we simply set the signing algorithm $\text{Sign} = \text{Sign}_A$. Unforgeability now follows from a hybrid argument. Consider any EUF-CMA [31] attacker $\mathcal{A}$. By the A-I matching property, we know that it might have a noticeable probability ϵ of producing a Type I forgery, but has only a negligible probability of producing any other kind of forgery. We then show that ϵ must also be negligible. By the dual-oracle invariance property, the probability of producing a Type I forgery will be close to ϵ if we gradually replace the signing algorithm with Sign_B, one signature at a time. Once all of the signatures the attacker receives are from Sign_B, the B-II Matching property implies that the probability of producing a Type I forgery must be negligible in the security parameter.

We demonstrate the usefulness of our framework with two main applications, using significantly different techniques. This illustrates the versatility of our framework and its adaptability to schemes with different underlying structures. In particular, while dual form signatures are related to the dual system encryption methodology introduced by Waters [49] for proving full security of IBE schemes and other advanced encryption functionalities, we demonstrate that our dual form framework can be applied to signature schemes that have no known encryption or IBE analogs. Though all of the applications given here use bilinear groups, the dual form framework can be used in other contexts, including proofs under general assumptions.

Our first application is a slight variant of the Camenisch-Lysyanskaya signature scheme, set in a bilinear group $\mathbb{G}$ of composite order $N = p_1 p_2 p_3$. This application is surprising, since these signatures do not have a known IBE analog. We let $\mathbb{G}_{p_i}$ for each $i = 1, 2, 3$ denote the subgroup of order p_i in the group. The Sign_A algorithm produces signatures which exhibit the CL structure in the $\mathbb{G}_{p_1}$ and $\mathbb{G}_{p_2}$ subgroups and are randomized in the $\mathbb{G}_{p_3}$ subgroup. The Sign_B algorithm produces signatures which exhibit the CL structure in the $\mathbb{G}_{p_1}$ subgroup and are randomized in the $\mathbb{G}_{p_2}$ and $\mathbb{G}_{p_3}$ subgroups. Type I and II forgeries roughly mirror signatures of form A and B. The verification procedure in our scheme will verify that the signature is well formed in the $\mathbb{G}_{p_1}$ subgroup, but not "check" the other subgroups.

We prove security in the dual form framework based on three static subgroup decision-type assumptions, similar to those used in [41]. The most challenging part of the proof is dual-oracle invariance, which we prove by developing a backdoor verification test (performed by the simulator) which acts as an almost-perfect distinguisher between forgery types. Here we face a potential paradox, which is similar to that encountered in dual system encryption [49,41]: we need to create a simulator that does not know whether the challenge signature it produces is distributed as an output of Sign_A or Sign_B, but it also must be able to test the type of the attacker's forgery. To arrange this, we create a "backdoor verification" test, which the simulator can perform to test the form of all but a small space of signatures. Essentially, this backdoor verification test acts an almost-perfect type distinguisher which fails to correctly determine the type of only a very small set of potential forgeries.

The challenge signature of unknown form produced by the simulator will fall within the untestable space; however, with very high probability a forgery by an attacker will not, because some information about this space is information-theoretically hidden from the attacker. This is possible because the elements of the verification key are all in the subgroup $\mathbb{G}_{p_1}$, and the space essentially resides in $\mathbb{G}_{p_2}$. Thus the verification key reveals no information about the hidden space. The only information about the space that the attacker receives is contained in the single signature of unknown type, and we show that this is insufficient for the attacker to be able to construct a forgery that falls inside the space for a *different message*. This is reminiscent of the concept of nominal semi-functionality in dual system encryption (introduced in [41]): in this setting, the simulator produces a key of unknown type which is correlated in its view with the ciphertext it produces, but this correlation is information-theoretically hidden from the attacker. This correlation prevents the simulator from determining the type of the key for itself by testing decryption against a ciphertext.

As a second application of our dual form framework, we prove security from static assumptions for a variant of the BGOY identity-based sequential aggregate signature scheme. Aggregate signatures are useful because they allow signature "compression," meaning that any n individual signatures by n (possibly) different signers on n (possibly) different messages can be transformed into an aggregate signature of the same size as an individual one that nevertheless allows verifying that all these signers signed their messages. However, aggregate signatures do not provide compression of the public keys, which are needed for signature verification. In the identity-based setting, only the identities of the signers are needed – this is a big savings because identities are much shorter than randomly generated keys. However, identity-based aggegate signatures have been notoriously difficult to realize.

We first prove security for a basic public-key version of the scheme, and then show that security for its identity-based sequential aggregate analog reduces to security of the basic scheme (in the random oracle model, as for the original proof). Our techniques here are significantly different, and reflect the different structure of the scheme (it is this structure that allows for aggregation). The core structure of the underlying public key scheme is composed of three group elements of the form $g^{a+bm}g^{r_1 r_2}, g^{r_1}, g^{r_2}$, where m is a message (or a hash of the message), a, b are fixed parameters, and r_1, r_2 are randomly chosen for each signature. There are significant differences between this and the core structure of other notable signatures, like CL and Waters signatures [24,48]. Here, the message term is not multiplicatively randomized, but rather additively randomized by the quadratic term $r_1 r_2$. It is the quadratic nature of this term that allows verification via application of the bilinear map while thwarting attackers who try to combine received signatures by taking linear combinations in the exponents. This unique structure presents a challenge for static security analysis, and we develop new techniques to achieve a proof for a variant of this scheme in our dual form framework.

We still employ composite order subgroups, with the main structure of the scheme reflected in the $\mathbb{G}_{p_1}$ subgroup and the other two subgroups used for differentiating between signature and forgery types. However, to prove dual-oracle invariance, we rely on the fact that the scheme has the basic structure of a one-time signature scheme embedded in it, in addition to the quadratic mechanism to prevent an attacker from forming new signatures by taking combinations of received signatures. We capture the security resulting from this combination of structures through a static assumption for our dual-oracle invariance proof, and we show that this assumption holds in the generic group model. Though we do employ the generic group model as a check on our static assumptions, we believe that our proof provides valuable intuition into the security of the scheme that is not gleaned from a proof based on an interactive assumption or given solely in the generic group model. Also, checking the security of a static assumption in the generic group model is much easier (and less error-prone) than checking the security of an interactive assumption or scheme. We believe that the techniques and insights provided by our proof are an important step toward finding a prime order variant of the scheme that is secure under more standard assumptions, such as the decisional linear assumption.

In the full version, we provide one more application: a signature scheme using the private key structure in the Lewko-Waters IBE system [41]. The LW system itself can be viewed as a composite order extension of the Boneh-Boyen selectively secure IBE scheme [11], although the structure of the proofs of these systems are very different (LW achieves adaptive security). For this reason, we call these "BB-derived" signatures. While the existing LW IBE system can be transformed into a signature scheme using Naor's [2] general transformation, our scheme checks the signature "directly" without going through an IBE encryption. The resulting signature has a constant number of elements in the public key and signatures consist of two group elements.

Further Directions. While we have focused here on applying our techniques for core short signatures, we envision that dual form signatures will be a framework for proving security of many different signature systems that have to this point been difficult to analyze under static assumption Some examples include embed additional structure, such as Attribute-Based signatures [45] and Quoteable signatures [3]. Attribute-based signatures allow a signer to sign a message with a predicate satisfied by his attributes, without revealing any additional information about his attribute set. Our framework could potentially be applied to obtain stronger security proofs for ABS schemes, such as the schemes of [45] proved only in the generic group model. Quoteable signatures enable derivation of signatures from each other under certain conditions, and current constructions are proved only selectively secure [3]. Another future target is signatures that "natively" sign group elements [1].

The primary goal of our work is providing techniques for realizing security under static assumptions, and we leverage composite order groups as a convenient setting for this. A natural future direction is to complement our work by discovering prime order analogs of our techniques. Many previous systems were originally constructed in composite order groups and later transferred into prime order groups [16,34,18,33,19,47,35,38,37,28,29,40,46]. The general techniques presented in [28,39] do not seem directly applicable here, but we emphasize that our dual form framework is not tied to composite order groups and could also be used in the prime order setting. Discussion of additional related works can be found in the full version.

2   Dual Form Signatures

We now define dual form signatures and their security properties. We then show that creating a secure dual form signature system naturally yields an existentially unforgeable signature scheme. We emphasize that the purpose of the dual form signature framework is to provide a template for creating security proofs from static assumptions, but the techniques employed to prove the required properties can be tailored to the structure of the particular scheme.

Definition. We define a dual form signature system to have the following algorithms:

KeyGen(λ)**:** Given a security parameter λ, generate a public key, VK, and a private key, SK.

---

[2] Naor's observation was noted in [14].

Sign$_A$(SK, M): Given a message, M, and the secret key, output a signature, σ.
Sign$_B$(SK, M): Given a message, M, and the secret key, output a signature, σ.
Verify(VK, M, σ): Given a message, the public key, and a signature, output 'true' or 'false'.

We note that a dual form signature scheme is identical to a usual signature scheme, except that it has two different signing algorithms. While only one signing algorithm will be used in the resulting existentially unforgeable scheme, having two different signing algorithms will be useful in our proof of security.

Forgery Classes. In addition to having two signature algorithms, the dual form signature framework also considers two disjoint classes of forgeries. Whether or not a signature verifies depends on the message that it signs as well as the verification key. For a fixed verification key, we consider the set of pairs, $\mathcal{S} \times \mathcal{M}$, over the message space, $\mathcal{M}$, and the signature space, $\mathcal{S}$. Consider the subset of these pairs for which the Verify algorithm outputs 'true': we will denote this set as $\mathcal{V}$. [3] We let $\mathcal{V}_I$ and $\mathcal{V}_{II}$ denote two disjoint subsets of $\mathcal{V}$, and we refer to signatures from these sets as *Type I* and *Type II* forgeries, respectively. In our applications, we will have the property $\mathcal{V} = \mathcal{V}_I \cup \mathcal{V}_{II}$ in addition to $\mathcal{V}_I \cap \mathcal{V}_{II} = \emptyset$, but only the latter property is necessary.

We will use these classes to specify two different types of forgeries received from an adversary in our proof of security. In general, these classes are not the same as the output ranges of our two signing algorithms. However, Type I forgeries will be related to signatures output by the Sign$_A$ algorithm and Type II forgeries will be related to signatures output by the Sign$_B$ algorithm. The precise relationships between the forgery types and the signing algorithms are explicitly defined by the following set of security properties for the dual form system.

Security Properties. We define the following three security properties for a dual form signature scheme. We consider an attacker $\mathcal{A}$ who is initially given the verification key VK produced by running the key generation algorithm. The value SK is also produced, and *not* given to $\mathcal{A}$.

A-I Matching: Let $\mathcal{O}_A$ be an oracle for the algorithm Sign$_A$. More precisely, this oracle takes a message as input, and produces a signature that is identically distributed to an output of the Sign$_A$ algorithm (for the SK produced from the key generation). We say that a dual form signature is *A-I matching* if for all probabilistic polynomial-time (PPT) algorithms, $\mathcal{A}$, there exists a negligible function, $negl(\lambda)$, in the security parameter λ such that:

$$Pr[\mathcal{A}^{\mathcal{O}_A}(\text{VK}) \notin \mathcal{V}_I] = negl(\lambda).$$

This property guarantees that if an attacker is only given oracle access to Sign$_A$, then it is hard to create anything but a Type I forgery.

---

[3] Here we will assume that the Verify algorithm is deterministic. If we consider a nondeterministic Verify algorithm, we could simply take the subset of ordered pairs that are accepted by Verify with non-negligible probability.

B-II Matching: Let $\mathcal{O}_B$ be an oracle for the algorithm Sign_B (which takes in a message and outputs a signature that is identically distributed an output of the Sign_B algorithm). We say that a dual form signature is *B-II matching* if for all PPT algorithms, $\mathcal{A}$:

$$Pr[\mathcal{A}^{\mathcal{O}_B}(\mathrm{VK}) \notin \mathcal{V}_{II}] = negl(\lambda).$$

This property guarantees that if an attacker is only given oracle access to Sign_B, then it is hard to create anything but a Type II forgery.

Dual-Oracle Invariance (DOI): First we define the dual-oracle security game.

1. The key generation algorithm is run, producing a verification key VK and a secret key SK.
2. The adversary, $\mathcal{A}$, is given the verification key VK and oracle access to $\mathcal{O}_0 = \mathrm{Sign}_A(\cdot)$ and $\mathcal{O}_1 = \mathrm{Sign}_B(\cdot)$.
3. $\mathcal{A}$ outputs a challenge message, m.
4. A random bit, $b \leftarrow \{0,1\}$, is chosen, and then a signature $\sigma \leftarrow \mathcal{O}_b(m)$ is computed and given to $\mathcal{A}$. We call σ the challenge signature.
5. $\mathcal{A}$ continues to have oracle access to $\mathcal{O}_0$ and $\mathcal{O}_1$.
6. $\mathcal{A}$ outputs a forgery pair (m^*, σ^*), where $\mathcal{A}$ has not already received a signature for m^*.

We say that a dual form signature scheme has dual-oracle invariance if, for all PPT attackers $\mathcal{A}$, there exists a negligible function, $negl(\lambda)$, in the security parameter λ such that

$$|Pr[(m^*, \sigma^*) \in \mathcal{V}_I | b = 1] - Pr[(m^*, \sigma^*) \in \mathcal{V}_I | b = 0]| = negl(\lambda).$$

We say that a dual form signature scheme is *secure* if it satisfies all three of these security properties.

Secure Signature Scheme. Once we have developed a secure dual form signature system, $(\mathrm{KeyGen}^{DF}, \mathrm{Sign}_A^{DF}, \mathrm{Sign}_B^{DF}, \mathrm{Verify}^{DF})$, this system immediately implies a secure signature scheme. The secure scheme is constructed as follows:

Construction 1. *KeyGen* $= KeyGen^{DF}$, *Sign* $= Sign_A^{DF}$, *Verify* $= Verify^{DF}$.

Our new secure signature scheme is identical to the dual form system except that we have arbitrarily chosen to use Sign_A as our signing algorithm. We could have equivalently elected to use Sign_B. (In which case, we would modify the dual-oracle invariance property to be with respect to Type II forgeries instead of Type I forgeries. Alternatively, we could strengthen the property to address both forgery types.) Now we will prove that this signature scheme is secure.

In the full version, we prove (the argument is rather straightforward):

Theorem 1. *If $\Pi = (KeyGen^{DF}, Sign_A^{DF}, Sign_B^{DF}, Verify^{DF})$ is a secure dual form signature scheme, then Construction 1$= (KeyGen^{DF}, Sign_A^{DF}, Verify^{DF})$ is existentially unforgeable under an adaptive chosen message attack.*

3   Background on Composite Order Bilinear Groups

Composite order bilinear groups were first introduced in [16]. We define a group generator $\mathcal{G}$, an algorithm which takes a security parameter λ as input and outputs a description of a bilinear group $\mathbb{G}$. In our case, we will have $\mathcal{G}$ output $(N = p_1 p_2 p_3, \mathbb{G}, \mathbb{G}_T, e)$ where p_1, p_2, p_3 are distinct primes, $\mathbb{G}$ and $\mathbb{G}_T$ are cyclic groups of order $N = p_1 p_2 p_3$, and $e : \mathbb{G}^2 \to \mathbb{G}_T$ is a map such that:

1. (Bilinear) $\forall g, h \in \mathbb{G}$, $a, b \in \mathbb{Z}_N$, $e(g^a, h^b) = e(g, h)^{ab}$
2. (Non-degenerate) $\exists g \in \mathbb{G}$ such that $e(g, g)$ has order N in $\mathbb{G}_T$.

Computing $e(g, h)$ is also commonly referred to as "pairing" g with h.

We assume that the group operations in $\mathbb{G}$ and $\mathbb{G}_T$ as well as the bilinear map e are computable in polynomial time with respect to λ, and that the group descriptions of $\mathbb{G}$ and $\mathbb{G}_T$ include generators of the respective cyclic groups. We let $\mathbb{G}_{p_1}$, $\mathbb{G}_{p_2}$, and $\mathbb{G}_{p_3}$ denote the subgroups of order p_1, p_2, and p_3 in $\mathbb{G}$ respectively. We note that when $h_i \in \mathbb{G}_{p_i}$ and $h_j \in \mathbb{G}_{p_j}$ for $i \neq j$, $e(h_i, h_j)$ is the identity element in $\mathbb{G}_T$. To see this, suppose we have $h_1 \in \mathbb{G}_{p_1}$ and $h_2 \in \mathbb{G}_{p_2}$. Let g denote a generator of $\mathbb{G}$. Then, $g^{p_1 p_2}$ generates $\mathbb{G}_{p_3}$, $g^{p_1 p_3}$ generates $\mathbb{G}_{p_2}$, and $g^{p_2 p_3}$ generates $\mathbb{G}_{p_1}$. Hence, for some α_1, α_2, $h_1 = (g^{p_2 p_3})^{\alpha_1}$ and $h_2 = (g^{p_1 p_3})^{\alpha_2}$. Then:

$$e(h_1, h_2) = e(g^{p_2 p_3 \alpha_1}, g^{p_1 p_3 \alpha_2}) = e(g^{\alpha_1}, g^{p_3 \alpha_2})^{p_1 p_2 p_3} = 1.$$

This orthogonality property of $\mathbb{G}_{p_1}, \mathbb{G}_{p_2}, \mathbb{G}_{p_3}$ is a useful feature of composite order bilinear groups which we leverage in our constructions and proofs.

If we let g_1, g_2, g_3 denote generators of the subgroups $\mathbb{G}_{p_1}$, $\mathbb{G}_{p_2}$, and $\mathbb{G}_{p_3}$ respectively, then every element h in $\mathbb{G}$ can be expressed as $h = g_1^a g_2^b g_3^c$ for some $a, b, c \in \mathbb{Z}_N$. We refer to g_1^a as the "$\mathbb{G}_{p_1}$ part" or "$\mathbb{G}_{p_1}$ component" of h. If we say that an h has no $\mathbb{G}_{p_2}$ component, for example, we mean that $b \equiv 0 \mod p_2$. Below, we will often use g to denote an element of $\mathbb{G}_{p_1}$ (as opposed to writing g_1).

The original Camenisch-Lysyanskaya scheme and BGOY identity-based sequential aggregate signature scheme both use prime order bilinear groups, i.e. groups $\mathbb{G}$ and $\mathbb{G}_T$ are each of prime order q with an efficiently computable bilinear map $e : \mathbb{G}^2 \to \mathbb{G}_T$.

4   Camenisch-Lysyanskaya Signatures

Now we use the dual form framework to prove security of a signature scheme similar to the one put forward by Camenisch and Lysyanskaya [24]. The Camenisch-Lysyanskaya signature scheme was already shown to be secure under the LRSW assumption. However, the scheme can be naturally adapted to our framework, allowing us to prove security under static, non-interactive assumptions. Our result is not strictly comparable to the result under the LRSW assumption because our signature scheme is not identical to the original. However, this is the first proof of security for a scheme similar to the Camenisch-Lysyanskaya signature scheme from static, non-interactive assumptions.

Our signature scheme will use bilinear groups, $\mathbb{G}$ and $\mathbb{G}_T$, of composite order $N = p_1 p_2 p_3$, where p_1, p_2, and p_3 are all distinct primes. Our construction is identical to the original Camenisch-Lysyanskaya signature scheme in the $\mathbb{G}_{p_1}$ subgroup, but with additional components in the subgroups $\mathbb{G}_{p_2}$ and $\mathbb{G}_{p_3}$. The signatures produced by the Sign_A algorithm will have random components in $\mathbb{G}_{p_3}$ and components in $\mathbb{G}_{p_2}$ which mirror the structure of the scheme in $\mathbb{G}_{p_1}$. The signatures produced by the Sign_B algorithm will have random components in both $\mathbb{G}_{p_2}$ and $\mathbb{G}_{p_3}$. Type I forgeries are those that are distributed exactly like Sign_A signatures in the $\mathbb{G}_{p_2}$ subgroup, while Type II forgeries encompass all other distributions.

To prove dual-oracle invariance, we develop a *backdoor verification* test that the simulator can use to determine the type of the attacker's forgery. We leverage the fact that the simulator will know the discrete logarithms of the public parameters, which will allow it to strip off the components in $\mathbb{G}_{p_1}$ in the forgery and check the distribution of the $\mathbb{G}_{p_2}$ components. This check will fail to determine the type correctly only with negligible probability. In more detail, we create a simulator which must solve a subgroup decision problem and ascertain whether an element T is in $\mathbb{G}_{p_1 p_3}$ or in the full group $\mathbb{G}$. It will use T to create a challenge signature which is either distributed as an output of the Sign_A algorithm or as an output of the Sign_B algorithm, depending on the nature of T. It will be unable to determine the nature of this signature for itself because this will fall into the negligible error space of its backdoor verification test. When the simulator receives a forgery from the attacker, it will perform the backdoor verification test and correctly determine the type of the forgery, unless the attacker manages to produce a forgery for which this test fails. This will occur only with negligible probability, because the attacker will have only limited information about the error space from the challenge signature, and it needs to forge on a *different* message. This is possible because the public parameters are in $\mathbb{G}_{p_1}$, and so reveal no information about the error space of the backdoor test modulo p_2. We use a pairwise independent argument to show that the limited amount of information the attacker can glean from the challenge signature on a message m is insufficient for it to produce a forgery for a different message m^* that causes the backdoor test to err.

4.1   Our Dual Form Scheme

KeyGen(λ): The key generation algorithm chooses two groups, $\mathbb{G} = \langle g \rangle$ and $\mathbb{G}_T$, of order $N = p_1 p_2 p_3$ (where p_1, p_2, and p_3 are all distinct primes of length λ) that have a non-degenerate, efficiently computable bilinear map, $e : \mathbb{G} \times \mathbb{G} \to \mathbb{G}_T$. It then selects uniformly at random $g \in \mathbb{G}_{p_1}$, $g_3 \in \mathbb{G}_{p_3}$, $g_{2,3} \in \mathbb{G}_{p_2 p_3}$, and $x, y, x_e, y_e \in \mathbb{Z}_N$. It sets

$$\text{SK} = (x, y, x_e, y_e, g_3, g_{2,3}),$$

and

$$PK = (N, \mathbb{G}, g, X = g^x, Y = g^y).$$

Sign$_A$(SK, m): Given a secret key $(x, y, x_e, y_e, g_3, g_{2,3})$, a public key $(N, \mathbb{G}, g, X, Y)$, and a message $m \in Z_N^*$, the algorithm chooses a random $r, r' \in \mathbb{Z}_N$, $R_{2,3} \in \mathbb{G}_{p_2 p_3}$, and random R_3, R_3', and $R_3'' \in \mathbb{G}_{p_3}$, and outputs the signature

$$\sigma = (g^r R_{2,3}^{r'} R_3, \ (g^r)^y (R_{2,3}^{r'})^{y_e} R_3', \ (g^r)^{x+mxy} (R_{2,3}^{r'})^{x_e+mx_e y_e} R_3'').$$

Note that the random elements of $\mathbb{G}_{p_3}$ can be obtained by raising g_3 to random exponents modulo N. Likewise, the random elements of $\mathbb{G}_{p_2 p_3}$ can be obtained by raising $g_{2,3}$ to random exponents modulo N. The random exponents modulo N will be uncorrelated modulo p_2 and modulo p_3 by the Chinese Remainder Theorem.

Sign$_B$(SK, m): Given a secret key $(x, y, x_e, y_e, g_3, g_{2,3})$, a public key $(N, \mathbb{G}, g, X, Y)$, and a message $m \in Z_N^*$, the algorithm chooses a random $r \in \mathbb{Z}_N$ and random $R_{2,3}$, $R_{2,3}'$, and $R_{2,3}'' \in \mathbb{G}_{p_2 p_3}$, and outputs the signature

$$\sigma = (g^r R_{2,3}, \ (g^r)^y R_{2,3}', \ (g^r)^{x+mxy} R_{2,3}'').$$

The random elements can be generated in the same way as in Sign$_A$.

Verify(VK, m, σ): Given a public key $pk = (N, \mathbb{G}, g, X, Y)$, message $m \neq 0$, and a signature $\sigma = (\sigma_1, \sigma_2, \sigma_3)$, the verification algorithm checks that:

$$e(\sigma_1, g) \neq 1$$

(which ensures that $\sigma_1 \notin \mathbb{G}_{p_2 p_3}$), and

$$e(\sigma_1, Y) = e(g, \sigma_2) \text{ and } e(X, \sigma_1) \cdot e(X, \sigma_2)^m = e(g, \sigma_3).$$

As in the original CL scheme, messages must be chosen from $\mathbb{Z}_N^*$, so that $m \neq 0$. If we allow $m = 0$, then an adversary can easily forge a valid signature using the public key elements (g, Y, X). Also like the original scheme, the Verify algorithm will not accept a signature where all the elements are the identity in $\mathbb{G}_{p_1}$. It suffices to check that the first element is not the identity in $\mathbb{G}_{p_1}$ and that the other verification equations are satisfied. If σ_1 is the identity in $\mathbb{G}_{p_1}$, then it will be an element of the subgroup $\mathbb{G}_{p_2 p_3}$. To determine if $\sigma_1 \in \mathbb{G}_{p_2 p_3}$, we pair σ_1 with the public key element g under the bilinear map and verify that it does not equal the identity in $\mathbb{G}_T$. Without this check, a signature where all three elements are members of the subgroup $\mathbb{G}_{p_2 p_3}$ would be valid for any message with the randomness $r' = 0 \bmod p_1$.

Notice, until Sign$_A$ is called, no information about the exponents x_e and y_e is given out. Once Sign$_A$ is called, these exponents behave exactly like the secret key exponents x and y, except in the $\mathbb{G}_{p_2}$ subgroup. These exponents will be used to verify that a forgery is of Type I. The additional randomization with the $\mathbb{G}_{p_3}$ elements guarantees that there will be no correlation in the $\mathbb{G}_{p_3}$ subgroup between the three signature elements. Unlike the signatures given out by the Sign$_A$ algorithm, signatures from the Sign$_B$ algorithm will be completely randomized in the $\mathbb{G}_{p_2}$ subgroup as well.

Forgery Classes. We will divide verifiable forgeries according to their correlation in the $\mathbb{G}_{p_2}$ subgroup, similar to the way we have defined the signatures from the Sign_A and Sign_B algorithms. We let z be an exponent in $\mathbb{Z}_N$. By the Chinese Remainder Theorem, we can represent z as an ordered tuple $(z_1, z_2, z_3) \in \mathbb{Z}_{p_1} \times \mathbb{Z}_{p_2} \times \mathbb{Z}_{p_3}$, where $z_1 = z \bmod p_1$, $z_2 = z \bmod p_2$, and $z_3 = z \bmod p_3$. Letting $(z_1, z_2, z_3) = (0 \bmod p_1, 1 \bmod p_2, 0 \bmod p_3)$ and g_2 be a generator of $\mathbb{G}_{p_2}$, we define the forgery classes as follows: Type I forgeries are of the form $\mathcal{V}_I = \{(m^*, \sigma^*) \in \mathcal{V} | (\sigma_1^*)^z = g_2^{r'}, (\sigma_2^*)^z = g_2^{r' y_e}, (\sigma_3^*)^z = g_2^{r'(x_e + m^* x_e y_e)}$ for some $r'\}$, while Type II are of the form $\mathcal{V}_{II} = \{(m^*, \sigma^*) \in \mathcal{V} | (m^*, \sigma^*) \notin \mathcal{V}_I\}$.

Essentially, Type I forgeries will be correlated in the $\mathbb{G}_{p_2}$ subgroup exactly in the same way as they are correlated in the $\mathbb{G}_{p_1}$ subgroup, with the exponents x_e and y_e playing the same role in the $\mathbb{G}_{p_2}$ subgroup that x and y play in the $\mathbb{G}_{p_1}$ subgroup. Type I forgeries will align with the Sign_A algorithm, to guarantee that our scheme is A-I matching. Type II forgeries include any other verifiable signatures, i.e. those not correctly correlated in the $\mathbb{G}_{p_2}$ subgroup. Unlike the signatures produced by the Sign_B algorithm, Type II forgeries need not be completely random in the $\mathbb{G}_{p_2}$ subgroup. However, we will show in our proof of security that this is enough to guarantee B-II matching.

4.2 Complexity Assumptions

We now state our complexity assumptions. We let $\mathbb{G}$ and $\mathbb{G}_T$ denote two cyclic groups of order $N = p_1 p_2 p_3$, where p_1, p_2, and p_3 are distinct primes, and $e : \mathbb{G}^2 \to \mathbb{G}_T$ is an efficient, non-degenerate bilinear map. In addition, we will denote the subgroup of $\mathbb{G}$ of order $p_1 p_2$ as $\mathbb{G}_{p_1 p_2}$, for example.

The first two of these assumptions were introduced in [41], where it is proven that these assumptions hold in the generic group model, assuming it is hard to find a non-trivial factor of the group order, N. These are specific instances of the General Subgroup Decision Assumption described in [8]. The third assumption is new, and in the full version we prove that it also holds in the generic group model, assuming it is hard to find a non-trivial factor of the group order, N.

Assumption 4.1. *Given a group generator $\mathcal{G}$, we define the following distribution:*

$$(N = p_1 p_2 p_3, \mathbb{G}, \mathbb{G}_T, e) \xleftarrow{R} \mathcal{G},$$

$$g, X_1 \xleftarrow{R} \mathbb{G}_{p_1}, X_2 \xleftarrow{R} \mathbb{G}_{p_2}, X_3 \xleftarrow{R} \mathbb{G}_{p_3}$$

$$D = (N, \mathbb{G}, \mathbb{G}_T, e, g, X_1 X_2, X_3)$$

$$T_1 \xleftarrow{R} \mathbb{G}_{p_1 p_2}, T_2 \xleftarrow{R} \mathbb{G}_{p_1}$$

We define the advantage of an algorithm, $\mathcal{A}$, in breaking Assumption 4.1 to be:

$$Adv_{\mathcal{A}}^{4.1}(\lambda) := |Pr[\mathcal{A}(D, T_1) = 1] - Pr[\mathcal{A}(D, T_2) = 1]|.$$

Definition 1. *We say that $\mathcal{G}$ satisfies Assumption 4.1 if for any polynomial time algorithm, $\mathcal{A}$, $Adv_{\mathcal{A}}^{4.1}(\lambda)$ is a negligible function of λ.*

Assumption 4.2. *Given a group generator $\mathcal{G}$, we define the following distribution:*

$$(N = p_1 p_2 p_3, \mathbb{G}, \mathbb{G}_T, e) \xleftarrow{R} \mathcal{G},$$

$$g, X_1 \xleftarrow{R} \mathbb{G}_{p_1}, X_2, Y_2 \xleftarrow{R} \mathbb{G}_{p_2}, X_3, Y_3 \xleftarrow{R} \mathbb{G}_{p_3},$$

$$D = (N, \mathbb{G}, \mathbb{G}_T, e, g, X_1 X_2, X_3, Y_2 Y_3),$$

$$T_1 \xleftarrow{R} \mathbb{G}, T_2 \xleftarrow{R} \mathbb{G}_{p_1 p_3}$$

We define the advantage of an algorithm, $\mathcal{A}$, in breaking Assumption 4.2 to be:

$$Adv_{\mathcal{A}}^{4.2}(\lambda) := |Pr[\mathcal{A}(D, T_1) = 1] - Pr[\mathcal{A}(D, T_2) = 1]|.$$

Definition 2. *We say that $\mathcal{G}$ satisfies Assumption 4.2 if for any polynomial time algorithm, $\mathcal{A}$, $Adv_{\mathcal{A}}^{4.2}(\lambda)$ is a negligible function of λ.*

Assumption 4.3. *Given a group generator $\mathcal{G}$, we define the following distribution:*

$$(N = p_1 p_2 p_3, \mathbb{G}, \mathbb{G}_T, e) \xleftarrow{R} \mathcal{G},$$

$$a, r \xleftarrow{R} \mathbb{Z}_N, g \xleftarrow{R} \mathbb{G}_{p_1}, X_2, X_2', X_2'', Z_2 \xleftarrow{R} \mathbb{G}_{p_2}, X_3 \xleftarrow{R} \mathbb{G}_{p_3},$$

$$D = (N, \mathbb{G}, \mathbb{G}_T, e, g, g^a, g^r X_2, g^{ra} X_2', g^{ra^2} X_2'', Z_2, X_3),$$

We define the advantage of an algorithm, $\mathcal{A}$, in breaking Assumption 4.3 to be:

$$Adv_{\mathcal{A}}^{4.3}(\lambda) := Pr[\mathcal{A}(D) = (g^{r'a^2} R_3, g^{r'} R_3') \text{ and } r' \neq 0 \bmod p_1],$$

where R_3 and R_3' are any values in the subgroup $\mathbb{G}_{p_3}$.

Definition 3. *We say that $\mathcal{G}$ satisfies Assumption 4.3 if for any polynomial time algorithm, $\mathcal{A}$, $Adv_{\mathcal{A}}^{4.3}(\lambda)$ is a negligible function of λ.*

Proof of Security. In the full version, we prove that our signature scheme is secure under these assumptions by proving that it satisfies the three properties of a secure dual form signature scheme.

5   BGOY Signatures

Here we give a public key variant of the BGOY signatures and prove existential unforgeability using our dual form framework. In the full version, we show how this base scheme can be built into an identity-based sequential aggregate signature scheme and reduce the security of the aggregate scheme to the security of this base scheme, in the random oracle model. We will also employ the random oracle model in our proof for the base scheme, although this use of the random oracle can be removed (see below for discussion of this).

Our techniques here are quite different than those employed for the BB-derived and CL signature variants, and they reflect the different structure of this scheme. There are some basic commonalities, however: we again employ a bilinear group of order $N = p_1p_2p_3$, and the main structure of the scheme occurs in the $\mathbb{G}_{p_1}$ subgroup. The signatures produced by the Sign_A algorithm contain group elements which are only in $\mathbb{G}_{p_1}$, while the signatures produced by the Sign_B algorithm additionally have components in $\mathbb{G}_{p_3}$. These components in $\mathbb{G}_{p_3}$ are not fully randomized each time and do not occur on all signature elements: they occur only on three signature elements, and the ratio between two of their exponents is the same for all Sign_B signatures. Our forgery types will be defined in terms of the subgroups present on two of the elements in the forgery.

We design our proof to reflect the structure of the scheme, which essentially combines a one-time signature with a mechanism to prevent an attacker from producing new signatures from linear combinations of old signatures in the exponent. In proving dual-oracle invariance, we leverage these structures by first changing the challenge signature from an output of Sign_A to a signature that has components in $\mathbb{G}_{p_2}$, and then changing it to an output of Sign_B. It is crucial to note that as we proceed through this intermediary step, the challenge signature is the *only* signature which has any non-zero components in $\mathbb{G}_{p_2}$. This allows us to argue that as we make this transition, an attacker cannot change from producing Type I forgeries (which do not have $\mathbb{G}_{p_2}$ components on certain elements) to producing forgeries which do have non-zero $\mathbb{G}_{p_2}$ components in the relevant locations. Intuitively, such an attacker would violate the combination of one-time security and inability to combine signatures, since the attacker has only received one signature with $\mathbb{G}_{p_2}$ elements, and it cannot combine this with any other signatures to produce a forgery on a new message. These aspects seem hard to capture when working directly in a prime order rather than composite order group. (We note, however, that the one-time aspect was also implicit in the security proof of the Gentry-Ramzan scheme [30] on which the BGOY scheme was based; however, differences in the schemes prevent capturing it in the same way for the latter.) The techniques here are also quite different from those used in our proofs for CL and BB-derived signatures: here there is no backdoor verification test or pairwise-independence argument.

5.1   The Dual Form Scheme

KeyGen$(\lambda) \rightarrow \text{VK}, \text{SK}$ The key generation algorithm chooses a bilinear group $\mathbb{G}$ of order $N = p_1p_2p_3$. It chooses two random elements $g, k \in \mathbb{G}_{p_1}$, random elements $g_3, g_3^d \in \mathbb{G}_{p_3}$, and random exponents $a_1, a_2, b_1, b_2, \alpha_1, \alpha_2, \beta_1, \beta_2 \in \mathbb{Z}_N$. It also chooses a function $H : \{0,1\}^* \rightarrow \mathbb{Z}_N$ which will be modeled as a random oracle. It sets the verification key as

$$\text{VK} := \{N, H, \mathbb{G}, g, k, g^{a_1}, g^{a_2}, g^{b_1}, g^{b_2}, g^{\alpha_1}, g^{\alpha_2}, g^{\beta_1}, g^{\beta_2}\}$$

and the secret key as

$$\text{SK} := \{N, H, \mathbb{G}, g, k, g^{a_1 a_2}, g^{b_1 b_2}, g^{\alpha_1 \alpha_2}, g^{\beta_1 \beta_2}, g_3, g_3^d\}.$$

$Sign_A(m, \text{SK}) \to \sigma$ The $Sign_A$ algorithm takes in a message $m \in \{0,1\}^*$. It chooses two random exponents $r_1, r_2 \in \mathbb{Z}_N$, and computes:

$$\sigma_1 := g^{a_1 a_2 + b_1 b_2 H(m)} g^{r_1 r_2}, \quad \sigma_2 := g^{r_1}, \quad \sigma_3 := g^{r_2},$$

$$\sigma_4 := k^{r_2}, \quad \sigma_5 := g^{\alpha_1 \alpha_2 + \beta_1 \beta_2 H(m)} k^{r_1 r_2}.$$

It outputs the signature $\sigma := (\sigma_1, \sigma_2, \sigma_3, \sigma_4, \sigma_5)$.

$Sign_B(m, \text{SK}) \to \sigma$ The $Sign_B$ algorithm takes in a message $m \in \{0,1\}^*$. It chooses two random exponents $r_1, r_2, x, y \in \mathbb{Z}_N$, and computes:

$$\sigma_1 := g^{a_1 a_2 + b_1 b_2 H(m)} g^{r_1 r_2} g_3^x, \quad \sigma_2 := g^{r_1} g_3^y, \quad \sigma_3 := g^{r_2},$$

$$\sigma_4 := k^{r_2}, \quad \sigma_5 := g^{\alpha_1 \alpha_2 + \beta_1 \beta_2 H(m)} k^{r_1 r_2} (g_3^d)^x.$$

It outputs the signature $\sigma := (\sigma_1, \sigma_2, \sigma_3, \sigma_4, \sigma_5)$.

$Verify(m, \sigma, \text{VK}) \to \{True, False\}$ The verification algorithm first checks that:

$$e(\sigma_1, g) = e(g^{a_1}, g^{a_2}) e(g^{b_1}, g^{b_2})^{H(m)} e(\sigma_2, \sigma_3).$$

It also checks that:

$$e(\sigma_5, g) = e(g^{\alpha_1}, g^{\alpha_2}) e(g^{\beta_1}, g^{\beta_2})^{H(m)} e(\sigma_2, \sigma_4).$$

Finally, it checks that:

$$e(g, \sigma_4) = e(k, \sigma_3).$$

If all of these checks pass, it outputs "True." Otherwise, it outputs "False."

We note that the use of the random oracle H to hash messages in $\{0,1\}^*$ to elements in $\mathbb{Z}_N$ in this public key scheme that forms the base of our identity-based sequential aggregate signatures is not necessary, and can be replaced in the following way. Instead of using $g^{a_1 a_2 + H(m) b_1 b_2}$, we can assume our messages are n-bit strings (denoted $m_1 m_2 \ldots m_n$) and use $g^{a_0 b_0} \prod_{i=1}^n g^{m_i a_i b_i}$. Here, $g^{a_0}, \ldots, g^{a_n}, g^{b_0}, \ldots, g^{b_n}$ will be in the public verification key. In the proof, instead of guessing which random oracle query corresponds to the challenge message, the simulator will guess a bit which differs between the challenge message and the message that will be used in the forgery. This guess will be correct with non-negligible probability. However, the use of the random oracle model to prove security for the full identity-based sequential aggregate scheme is still required. Removing the random oracle model altogether remains an open problem.

Forgery Classes. We will divide the forgery types based on whether they have any $\mathbb{G}_{p_2}$ or $\mathbb{G}_{p_3}$ components on σ_1 or σ_5. We let $z_2 \in \mathbb{Z}_N$ denote the exponent represented by the tuple $(0 \bmod p_1, 1 \bmod p_2, 0 \bmod p_3)$, and we let $z_3 \in \mathbb{Z}_N$ denote the exponent represented by the tuple $(0 \bmod p_1, 0 \bmod p_2, 1 \bmod p_3)$. Then we can define the forgery classes as follows. Type I forgeries are of the form $\mathcal{V}_I = \{(m^*, \sigma^*) \in \mathcal{V} | (\sigma_1^*)^{z_2} = 1, (\sigma_1^*)^{z_3} = 1 \text{ and } (\sigma_5^*)^{z_2} = 1, (\sigma_5^*)^{z_3} = 1\}$, while Type II are of the form $\mathcal{V}_{II} = \{(m^*, \sigma^*) \in \mathcal{V} | (\sigma_1^*)^{z_2} \neq 1 \text{ or } (\sigma_5^*)^{z_2} \neq 1 \text{ or } (\sigma_1^*)^{z_3} \neq 1 \text{ or } (\sigma_5^*)^{z_3} \neq 1\}$.

In other words, Type I forgeries have $\sigma_1^*, \sigma_5^* \in \mathbb{G}_{p_1}$, while Type II forgeries have a non-zero component in $\mathbb{G}_{p_3}$ or $\mathbb{G}_{p_2}$ on at least one of these terms. We note that these types are disjoint and exhaustive.

We state our complexity assumptions and prove security of this scheme in the full version. Some the assumptions we employ were previously used in [41,42]. Those that are new are justified in the generic group model.

References

1. Abe, M., Fuchsbauer, G., Groth, J., Haralambiev, K., Ohkubo, M.: Structure-Preserving Signatures and Commitments to Group Elements. In: Rabin, T. (ed.) CRYPTO 2010. LNCS, vol. 6223, pp. 209–236. Springer, Heidelberg (2010)
2. Agrawal, S., Boneh, D., Boyen, X.: Efficient Lattice (H)IBE in the Standard Model. In: Gilbert, H. (ed.) EUROCRYPT 2010. LNCS, vol. 6110, pp. 553–572. Springer, Heidelberg (2010)
3. Ahn, J.H., Boneh, D., Camenisch, J., Hohenberger, S., Shelat, A., Waters, B.: Computing on Authenticated Data, 1–20 (2012)
4. Ateniese, G., Camenisch, J., de Medeiros, B.: Untraceable rfid tags via insubvertible encryption. In: ACM Conference on Computer and Communications Security, pp. 92–101 (2005)
5. Ateniese, G., Camenisch, J., Hohenberger, S., de Medeiros, B.: Practical group signatures without random oracles. Cryptology ePrint Archive, Report 2005/385 (2005), http://eprint.iacr.org/
6. Backes, M., Camenisch, J., Sommer, D.: Anonymous yet accountable access control. In: WPES, pp. 40–46 (2005)
7. Bangerter, E., Camenisch, J., Lysyanskaya, A.: A Cryptographic Framework for the Controlled Release of Certified Data. In: Christianson, B., Crispo, B., Malcolm, J.A., Roe, M. (eds.) Security Protocols 2004. LNCS, vol. 3957, pp. 20–42. Springer, Heidelberg (2006)
8. Bellare, M., Waters, B., Yilek, S.: Identity-Based Encryption Secure against Selective Opening Attack. In: Ishai, Y. (ed.) TCC 2011. LNCS, vol. 6597, pp. 235–252. Springer, Heidelberg (2011)
9. Boldyreva, A., Gentry, C., O'Neill, A., Yum, D.H.: Ordered multisignatures and identity-based sequential aggregate signatures, with applications to secure routing. In: ACM Conference on Computer and Communications Security, pp. 276–285 (2007)
10. Boldyreva, A., Gentry, C., O'Neill, A., Yum, D.H.: Ordered multisignatures and identity-based sequential aggregate signatures, with applications to secure routing. Cryptology ePrint Archive, Report 2007/438 (2007), http://eprint.iacr.org/
11. Boneh, D., Boyen, X.: Efficient Selective-ID Secure Identity-Based Encryption Without Random Oracles. In: Cachin, C., Camenisch, J.L. (eds.) EUROCRYPT 2004. LNCS, vol. 3027, pp. 223–238. Springer, Heidelberg (2004)
12. Boneh, D., Boyen, X.: Secure Identity Based Encryption Without Random Oracles. In: Franklin, M. (ed.) CRYPTO 2004. LNCS, vol. 3152, pp. 443–459. Springer, Heidelberg (2004)
13. Boneh, D., Boyen, X.: Short Signatures Without Random Oracles. In: Cachin, C., Camenisch, J.L. (eds.) EUROCRYPT 2004. LNCS, vol. 3027, pp. 56–73. Springer, Heidelberg (2004)
14. Boneh, D., Franklin, M.K.: Identity-based encryption from the weil pairing. SIAM J. Comput. 32(3), 586–615 (2003)

15. Boneh, D., Gentry, C., Lynn, B., Shacham, H.: Aggregate and Verifiably Encrypted Signatures from Bilinear Maps. In: Biham, E. (ed.) EUROCRYPT 2003. LNCS, vol. 2656, pp. 416–432. Springer, Heidelberg (2003)
16. Boneh, D., Goh, E.-J., Nissim, K.: Evaluating 2-DNF Formulas on Ciphertexts. In: Kilian, J. (ed.) TCC 2005. LNCS, vol. 3378, pp. 325–341. Springer, Heidelberg (2005)
17. Boneh, D., Lynn, B., Shacham, H.: Short signatures from the Weil pairing. Journal of Cryptology 17(4), 297–319 (2004)
18. Boneh, D., Sahai, A., Waters, B.: Fully Collusion Resistant Traitor Tracing with Short Ciphertexts and Private Keys. In: Vaudenay, S. (ed.) EUROCRYPT 2006. LNCS, vol. 4004, pp. 573–592. Springer, Heidelberg (2006)
19. Boneh, D., Waters, B.: Conjunctive, Subset, and Range Queries on Encrypted Data. In: Vadhan, S.P. (ed.) TCC 2007. LNCS, vol. 4392, pp. 535–554. Springer, Heidelberg (2007)
20. Boyen, X.: Lattice Mixing and Vanishing Trapdoors: A Framework for Fully Secure Short Signatures and More. In: Nguyen, P.Q., Pointcheval, D. (eds.) PKC 2010. LNCS, vol. 6056, pp. 499–517. Springer, Heidelberg (2010)
21. Camenisch, J., Hohenberger, S., Kohlweiss, M., Lysyanskaya, A., Meyerovich, M.: How to win the clonewars: efficient periodic n-times anonymous authentication. In: ACM Conference on Computer and Communications Security, pp. 201–210 (2006)
22. Camenisch, J., Hohenberger, S., Lysyanskaya, A.: Compact E-Cash. In: Cramer, R. (ed.) EUROCRYPT 2005. LNCS, vol. 3494, pp. 302–321. Springer, Heidelberg (2005)
23. Camenisch, J., Hohenberger, S., Pedersen, M.Ø.: Batch Verification of Short Signatures. In: Naor, M. (ed.) EUROCRYPT 2007. LNCS, vol. 4515, pp. 246–263. Springer, Heidelberg (2007)
24. Camenisch, J., Lysyanskaya, A.: Signature Schemes and Anonymous Credentials from Bilinear Maps. In: Franklin, M. (ed.) CRYPTO 2004. LNCS, vol. 3152, pp. 56–72. Springer, Heidelberg (2004)
25. Camenisch, J., Neven, G., Shelat, A.: Simulatable Adaptive Oblivious Transfer. In: Naor, M. (ed.) EUROCRYPT 2007. LNCS, vol. 4515, pp. 573–590. Springer, Heidelberg (2007)
26. Canetti, R., Halevi, S., Katz, J.: A Forward-Secure Public-Key Encryption Scheme. In: Biham, E. (ed.) EUROCRYPT 2003. LNCS, vol. 2656, pp. 255–271. Springer, Heidelberg (2003)
27. Cash, D., Hofheinz, D., Kiltz, E., Peikert, C.: Bonsai Trees, or How to Delegate a Lattice Basis. In: Gilbert, H. (ed.) EUROCRYPT 2010. LNCS, vol. 6110, pp. 523–552. Springer, Heidelberg (2010)
28. Freeman, D.M.: Converting Pairing-Based Cryptosystems from Composite-Order Groups to Prime-Order Groups. In: Gilbert, H. (ed.) EUROCRYPT 2010. LNCS, vol. 6110, pp. 44–61. Springer, Heidelberg (2010)
29. Garg, S., Kumarasubramanian, A., Sahai, A., Waters, B.: Building efficient fully collusion-resilient traitor tracing and revocation schemes. In: ACM Conference on Computer and Communications Security, pp. 121–130 (2010)
30. Gentry, C., Ramzan, Z.: Identity-Based Aggregate Signatures. In: Yung, M., Dodis, Y., Kiayias, A., Malkin, T. (eds.) PKC 2006. LNCS, vol. 3958, pp. 257–273. Springer, Heidelberg (2006)
31. Goldwasser, S., Micali, S., Rivest, R.L.: A digital signature scheme secure against adaptive chosen-message attacks. SIAM J. Computing 17(2) (1988)
32. Green, M., Hohenberger, S.: Blind Identity-Based Encryption and Simulatable Oblivious Transfer. In: Kurosawa, K. (ed.) ASIACRYPT 2007. LNCS, vol. 4833, pp. 265–282. Springer, Heidelberg (2007)

33. Groth, J., Ostrovsky, R., Sahai, A.: Non-interactive Zaps and New Techniques for NIZK. In: Dwork, C. (ed.) CRYPTO 2006. LNCS, vol. 4117, pp. 97–111. Springer, Heidelberg (2006)
34. Groth, J., Ostrovsky, R., Sahai, A.: Perfect Non-interactive Zero Knowledge for NP. In: Vaudenay, S. (ed.) EUROCRYPT 2006. LNCS, vol. 4004, pp. 339–358. Springer, Heidelberg (2006)
35. Groth, J., Sahai, A.: Efficient Non-interactive Proof Systems for Bilinear Groups. In: Smart, N.P. (ed.) EUROCRYPT 2008. LNCS, vol. 4965, pp. 415–432. Springer, Heidelberg (2008)
36. Hwang, J.Y., Lee, D.H., Yung, M.: Universal forgery of the identity-based sequential aggregate signature scheme. In: ASIACCS, pp. 157–160 (2009)
37. Iovino, V., Persiano, G.: Hidden-Vector Encryption with Groups of Prime Order. In: Galbraith, S.D., Paterson, K.G. (eds.) Pairing 2008. LNCS, vol. 5209, pp. 75–88. Springer, Heidelberg (2008)
38. Katz, J., Sahai, A., Waters, B.: Predicate Encryption Supporting Disjunctions, Polynomial Equations, and Inner Products. In: Smart, N.P. (ed.) EUROCRYPT 2008. LNCS, vol. 4965, pp. 146–162. Springer, Heidelberg (2008)
39. Lewko, A.: Tools for Simulating Features of Composite Order Bilinear Groups in the Prime Order Setting. In: Pointcheval, D., Johansson, T. (eds.) EUROCRYPT 2012. LNCS, vol. 7237, pp. 318–335. Springer, Heidelberg (2012)
40. Lewko, A., Okamoto, T., Sahai, A., Takashima, K., Waters, B.: Fully Secure Functional Encryption: Attribute-Based Encryption and (Hierarchical) Inner Product Encryption. In: Gilbert, H. (ed.) EUROCRYPT 2010. LNCS, vol. 6110, pp. 62–91. Springer, Heidelberg (2010)
41. Lewko, A., Waters, B.: New Techniques for Dual System Encryption and Fully Secure HIBE with Short Ciphertexts. In: Micciancio, D. (ed.) TCC 2010. LNCS, vol. 5978, pp. 455–479. Springer, Heidelberg (2010)
42. Lewko, A., Waters, B.: Decentralizing Attribute-Based Encryption. In: Paterson, K.G. (ed.) EUROCRYPT 2011. LNCS, vol. 6632, pp. 568–588. Springer, Heidelberg (2011)
43. Lu, S., Ostrovsky, R., Sahai, A., Shacham, H., Waters, B.: Sequential Aggregate Signatures and Multisignatures Without Random Oracles. In: Vaudenay, S. (ed.) EUROCRYPT 2006. LNCS, vol. 4004, pp. 465–485. Springer, Heidelberg (2006)
44. Lysyanskaya, A., Rivest, R.L., Sahai, A., Wolf, S.: Pseudonym Systems (Extended Abstract). In: Heys, H.M., Adams, C.M. (eds.) SAC 1999. LNCS, vol. 1758, pp. 184–199. Springer, Heidelberg (2000)
45. Maji, H.K., Prabhakaran, M., Rosulek, M.: Attribute-Based Signatures. In: Kiayias, A. (ed.) CT-RSA 2011. LNCS, vol. 6558, pp. 376–392. Springer, Heidelberg (2011)
46. Okamoto, T., Takashima, K.: Fully Secure Functional Encryption with General Relations from the Decisional Linear Assumption. In: Rabin, T. (ed.) CRYPTO 2010. LNCS, vol. 6223, pp. 191–208. Springer, Heidelberg (2010)
47. Shi, E., Bethencourt, J., Chan, H.T.-H., Song, D.X., Perrig, A.: Multi-dimensional range query over encrypted data. In: IEEE Symposium on Security and Privacy (2007)
48. Waters, B.: Efficient Identity-Based Encryption Without Random Oracles. In: Cramer, R. (ed.) EUROCRYPT 2005. LNCS, vol. 3494, pp. 114–127. Springer, Heidelberg (2005)
49. Waters, B.: Dual System Encryption: Realizing Fully Secure IBE and HIBE under Simple Assumptions. In: Halevi, S. (ed.) CRYPTO 2009. LNCS, vol. 5677, pp. 619–636. Springer, Heidelberg (2009)

Breaking Pairing-Based Cryptosystems Using η_T Pairing over $GF(3^{97})^\star$

Takuya Hayashi[1,**], Takeshi Shimoyama[2],
Naoyuki Shinohara[3], and Tsuyoshi Takagi[1]

[1] Kyushu University
[2] FUJITSU LABORATORIES Ltd.
[3] National Institute of Information and Communications Technology

Abstract. In this paper, we discuss solving the DLP over $GF(3^{6 \cdot 97})$ by using the function field sieve (FFS) for breaking paring-based cryptosystems using the η_T pairing over $GF(3^{97})$. The extension degree 97 has been intensively used in benchmarking tests for the implementation of the η_T pairing, and the order (923-bit) of $GF(3^{6 \cdot 97})$ is substantially larger than the previous world record (676-bit) of solving the DLP by using the FFS. We implemented the FFS for the medium prime case, and proposed several improvements of the FFS. Finally, we succeeded in solving the DLP over $GF(3^{6 \cdot 97})$. The entire computational time requires about 148.2 days using 252 CPU cores.

Keywords: pairing-based cryptosystems, η_T pairing, discrete logarithm problems, function filed sieve.

1 Introduction

After the advent of the tripartite Diffie-Hellman (DH) key exchange scheme [21] and ID-based encryption using pairing [11], plenty of attractive pairing-based cryptosystems have been proposed, for example, short signature [13], keyword searchable encryption [10], efficient broadcast encryption [12], attribute-based encryption [30], and functional encryption [28]. Pairing-based cryptosystems have become a major research topic in cryptography.

Pairing-based cryptosystems are constructed on the groups G_1, G_1' and G_2 of the same order with a bilinear pairing $G_1 \times G_1' \to G_2$. The security of pairing-based cryptosystems is based on the difficulty in solving several number-theoretic problems such as the computational/decisional bilinear DH problem (CBDH/DBDH), strong DH problem (SDH), decisional linear problem (DLIN), and symmetric external DH problem (SXDH). However, the most important number-theoretic problem in pairing-based cryptosystems is the discrete logarithm problem (DLP) on G_1, G_1', and G_2. All the other number-theoretic problems above are no longer intractable once the DLP on G_1, G_1', or G_2 is broken. Therefore, it is important to investigate the difficulty in solving the DLP.

---

* The full-version of this paper is appeared in [19].
** The author is supported by JSPS KAKENHI Grant Number 10J56502.

X. Wang and K. Sako (Eds.): ASIACRYPT 2012, LNCS 7658, pp. 43–60, 2012.

Table 1. Summary of time data for solving DLP over $GF(3^{6 \cdot 97})$

phase	method	time	machine environment
collecting relations	lattice sieve	53.1 days	212 CPU cores
linear algebra	parallel Lanczos	80.1 days	252 CPU cores
individual logarithm	rationalization and special-Q descent	15.0 days	168 CPU cores
total		148.2 days	252 CPU cores

One of the most efficient algorithms for implementing the pairing is the η_T pairing [5] defined over a supersingular elliptic curve E on the finite field $GF(3^n)$, where n is a positive integer. Since the embedding degree of E is 6, the η_T pairing can reduce a DLP over E on $GF(3^n)$, which is called an ECDLP, to a DLP over $GF(3^{6n})$. Joux proposed the (probably) first cryptographic scheme [21] that uses the pairing over E. Boneh *et al.* then applied the pairing over E to the short signature scheme [13], where a point (x, y) on E for extension degree $n = 97$ can be represented as a signature value, e.g., $x = $ KrpIcVOO9CJ8iyBS8MyVkNrMyE. At CRYPTO 2002, Barreto *et al.* presented algorithms for efficiently computing Tate pairing over E [6]. Many high-speed implementations of pairing over E have subsequently been proposed [3, 7–9, 17, 18, 25]. For many of these implementations, benchmark tests using the extension degree $n = 97$ have been conducted. Therefore, we focus on the DLP over finite field $GF(3^{6 \cdot 97})$ in this paper. The cardinality of the subgroup of the supersingular elliptic curve is 151 bits, and that of $GF(3^{6 \cdot 97})$ is 923 bits. The size of our target DLP is 247 bits larger than the previous world record of solving the DLP over $GF(3^{6 \cdot 71})$, whose cardinality is 676 bits [20]. The current world record for solving an ECDLP is the 112-bit ECDLP [14]. Pollard's ρ method is used for solving the 112-bit ECDLP, and has not reached the ability for solving the 151-bit ECDLP over the subgroup of E.

In this paper, we analyze the difficulty in solving the DLP over $GF(3^{6 \cdot 97})$ by using the function field sieve (FFS), which is known as the asymptotically fastest algorithm [1, 2]. Since the FFS proposed by Joux and Lercier (JL06-FFS) [24] is suitable for solving the DLP over a finite field whose characteristic is small, we use the JL06-FFS and propose several efficient techniques for increasing its speed. Note that the FFS generally consists of four phases: polynomial selection, collecting relations, linear algebra, and individual logarithm, and the time-consuming phases are collecting relations and linear algebra. For the collecting relations phase, we applied several techniques; lattice sieve for the JL06-FFS, lattice sieve with single instruction multiple data (SIMD), and optimization for our parameters. These techniques enable the sieving program to run about 6 times faster. In the linear algebra phase, we applied careful treatments of singleton-clique and merging [15] to the Galois action originating from extension degree 6 of $GF(3^{6 \cdot 97})$, with which the size of the matrix used for the Lanczos method is reduced to approximately 30%. By implementing the JL06-FFS with our improvements, we succeeded in solving the DLP over $GF(3^{6 \cdot 97})$ by using 252 CPU cores (Core2 quad, Xeon, etc) for the target problem discussed in Section 3.1. As shown in Table 1, the computations required 53.1 days for the collecting relations phase, 80.1 days for the linear algebra phase, and 15.0 days for the individual logarithm phase. Thus, a total of 148.2 days were required to solve the DLP over $GF(3^{6\cdot97})$ by using 252 CPU cores. Our computational results contribute to the secure use of pairing-based cryptosystems with the η_T pairing.

2 Pairing-Based Cryptosystems and Discrete Logarithm Problem (DLP)

In this section, we briefly explain the security of pairing-based cryptosystems and give a general overview of the function field sieve (FFS). We also mention its parameters such as the smoothness bound B.

2.1 Pairing-Based Cryptosystems and DLP

Many efficient cryptographic protocols using a bilinear pairing have been proposed (for example [10–13, 21, 28]), and high-speed implementations for the η_T pairing have been reported (for example [3, 6–9, 17, 18, 25]). We discuss the security of pairing-based cryptosystems with the η_T paring over $GF(3^n)$ for an integer n. The security of pairing-based cryptosystems with the η_T paring depends on the difficulty in solving the DLP over the supersingular elliptic curves. Additionally, MOV reduction [27] reduces this problem to a DLP over $GF(3^{6n})^*$ since the embedding degree of the η_T pairing is 6.

In particular, the η_T pairing is a bilinear map such that $\eta_T : G_1 \times G_1 \to G_2$, where G_1 is an additive subgroup of a supersingular elliptic curve over $GF(3^n)$, G_2 is a cyclic subgroup of $GF(3^{6n})^*$, and the cardinalities of G_1, G_2 are the same prime number P. The security of pairing-based cryptosystems with the η_T pairing depends on the difficulty of not only an ECDLP over G_1 but also a DLP over G_2 by MOV reduction. To explain this fact, we take ID-based encryption constructed on pairing-based cryptosystems as an example. The ID-based encryption has a master key $s_{key} \in \mathbb{Z}_P$. Each user ID is deterministically transformed into a point $\mathcal{Q}_{ID} \in G_1$, and the secret key $\mathcal{S}_{ID}$ is defined by $[s_{key}]\mathcal{Q}_{ID}$. Therefore, solving the ECDLP over G_1, namely $\mathcal{S}_{ID} = [s_{key}]\mathcal{Q}_{ID}$, we obtain the master key $s_{key} = \log_{\mathcal{Q}_{ID}} \mathcal{S}_{ID}$. Additionally, for an arbitrary point $\mathcal{R} \in G_1$, we compute $\eta_T(\mathcal{S}_{ID}, \mathcal{R}), \eta_T(\mathcal{Q}_{ID}, \mathcal{R}) \in G_2$, and then have $\eta_T(\mathcal{S}_{ID}, \mathcal{R}) = \eta_T([s_{key}]\mathcal{Q}_{ID}, \mathcal{R}) = \eta_T(\mathcal{Q}_{ID}, \mathcal{R})^{s_{key}} \in G_2$. This implies that $s_{key} = \log_{\eta_T(\mathcal{Q}_{ID}, \mathcal{R})} \eta_T(\mathcal{S}_{ID}, \mathcal{R})$ is also available by solving the DLP over G_2. In this paper, we discuss the DLP over a subgroup of $GF(3^{6n})^*$.

2.2 General Overview of FFS

The FFS is the asymptotically fastest algorithm for solving a DLP over finite fields of small characteristics. Adleman proposed the first FFS in 1994 [1]. After that, several variants of the FFS have been proposed; Adleman and Huang improved the FFS [2], and Joux and Lercier proposed two more practical FFS's,

JL02-FFS [23] and JL06-FFS [24]. The details of JL06-FFS are explained in Sections 3.2.

In this section, we give a general overview of an FFS that consists of four phases: polynomial selection, collecting relations, linear algebra, and individual logarithm. In the overview, we aim at computing $\log_g T$ where $T \in \langle g \rangle \subset GF(3^{6n})^*$.

Polynomial Selection Phase: We select κ from $\kappa = 1, 2, 3, 6$ for the coefficient field of $GF(3^\kappa)[x]$, and a bivariate polynomial $H(x, y) \in GF(3^\kappa)[x, y]$ such that H satisfies the eight conditions proposed by Adleman [1] and $\deg_y H = d_H$ for a given parameter value d_H. We compute a random polynomial $m \in GF(3^\kappa)[x]$ of degree d_m and a monic irreducible polynomial $f \in GF(3^\kappa)[x]$ such that

$$H(x, m) \equiv 0 \pmod{f}, \quad \deg f = 6n/\kappa. \tag{1}$$

We then have $GF(3^{6n}) \cong GF(3^\kappa)[x]/(f)$. Moreover, there is a surjective homomorphism

$$\xi : \begin{cases} GF(3^\kappa)[x, y]/(H) & \to & GF(3^{6n}) \cong GF(3^\kappa)[x]/(f) \\ y & \mapsto & m. \end{cases}$$

We select a positive integer B as a smoothness bound, and define a rational factor base $F_R(B)$ and an algebraic factor base $F_A(B)$ as follows.

$$F_R(B) = \{\mathfrak{p} \in GF(3^\kappa)[x] \mid \deg(\mathfrak{p}) \leq B, \mathfrak{p} \text{ is monic irreducible}\}, \tag{2}$$

$$F_A(B) = \{\langle \mathfrak{p}, y - t \rangle \in \mathrm{Div}(GF(3^\kappa)[x, y]/(H)) \mid \\ \mathfrak{p} \in F_R(B), H(x, t) \equiv 0 \bmod \mathfrak{p}\}, \tag{3}$$

where $\mathrm{Div}(GF(3^\kappa)[x, y]/(H))$ is the divisor group of $GF(3^\kappa)[x, y]/(H)$ and $\langle \mathfrak{p}, y - t \rangle$ is a divisor generated by $\mathfrak{p}$ and $y - t$. Note that $F_R(0) = F_A(0) = \{\emptyset\}$. We simply call the set $F_R(B) \cup F_A(B)$ a factor base and the set $F_R(k) \backslash F_R(k - 1) \cup F_A(k) \backslash F_A(k - 1)$ a factor base of degree k for $k = 1, 2, \ldots, B$.

Collecting Relations Phase: We select positive integers R, S and collect a sufficient amount of pairs $(r, s) \in (GF(3^\kappa)[x])^2$ such that

$$\deg r \leq R, \deg s \leq S, \gcd(r, s) = 1, \tag{4}$$

$$rm + s = \prod_{\mathfrak{p}_i \in F_R(B)} \mathfrak{p}_i^{a_i}, \tag{5}$$

$$\langle ry + s \rangle = \sum_{\langle \mathfrak{p}_j, y - t_j \rangle \in F_A(B)} b_j \langle \mathfrak{p}_j, y - t_j \rangle, \tag{6}$$

for some non-negative integers a_i, b_j by using a sieving algorithm such as the lattice sieve discussed in Section 4.1. To efficiently compute b_j in (6), we use the following equivalent property instead of (6):

$$(-r)^{d_H} H(x, -s/r) = \prod_{\langle \mathfrak{p}_j, y - t_j \rangle \in F_A(B)} \mathfrak{p}_j^{b_j}. \tag{7}$$

The (r, s) satisfying (4), (5), and (7) is called a B-smooth pair. Let h be the class number of the quotient field of $GF(3^\kappa)(x)[y]/(H)$ and assume that h is coprime to $(3^{6n} - 1)/(3^\kappa - 1)$. Then the following congruent holds:

$$\sum_{\mathfrak{p}_i \in F_R(B)} a_i \log_g \mathfrak{p}_i \equiv \sum_{\langle \mathfrak{p}_j, y - t_j \rangle \in F_A(B)} b_j \log_g \mathfrak{s}_j \quad (\text{mod } (3^{6n} - 1)/(3^\kappa - 1)), \quad (8)$$

where $\mathfrak{s}_j = \xi(t_j)^{1/h}$, $\langle t_j \rangle = h \langle \mathfrak{p}_j, y - t_j \rangle$. We call the congruent (8) "relation" in this paper. Moreover, free relation [20] provides additional relations without computation with a sieving algorithm.

Linear Algebra Phase: We generate a system of linear equations described as a large matrix from those collected relations and reduce the rank of the matrix by filtering [15]. The reduced system of linear equations is solved using the parallel Lanczos method [4, 20] or other methods, and the discrete logarithms of elements in the factor base are obtained:

$$\log_g \mathfrak{p}_1, ..., \log_g \mathfrak{p}_{\#F_R(B)}, \log_g \mathfrak{s}_1, ..., \log_g \mathfrak{s}_{\#F_A(B)}.$$

Individual Logarithm Phase: Note that our goal is to compute $\log_g T$. Therefore, we find integers a_i, b_j using the special-Q descent [24] such that,

$$\log_g T \equiv \sum_{\mathfrak{p}_i \in F_R(B)} a_i \log_g \mathfrak{p}_i + \sum_{\langle \mathfrak{p}_j, y - t_j \rangle \in F_A(B)} b_j \log_g \mathfrak{s}_j \quad (\text{mod } (3^{6n} - 1)/(3^\kappa - 1)).$$

The computational time for the individual logarithm phase is smaller than those for the collecting relations and linear algebra phases.

3 Target Problem for $n = 97$ and Setting of Parameters for FFS

We discuss solving the DLP over a subgroup of $GF(3^{6 \cdot 97})^*$, where the cardinality of the subgroup is 151 bits. To estimate the time complexity of solving such a DLP, we unintentionally set a target problem determined from the circular constant π and natural logarithm e. The details are explained in Section 3.1. To solve the target problem effectively, we select the parameter values of the FFS and estimate important numbers, e.g., the number of elements in the factor base, for it. The details are given in Section 3.2.

3.1 Target Problem

For pairing-based cryptosystems, many high-speed implementations of the η_T pairing over supersingular elliptic curves on $GF(3^n)$ have been reported [3, 6–9, 17, 18, 25], and many benchmark tests using the η_T pairing have been conducted for $GF(3^{97})$. In this paper, we deal with a supersingular elliptic curve defined by

$$E := \{(x, y) \in GF(3^{97})^2 \ : \ y^2 = x^3 - x + 1\} \cup \{\mathcal{O}\},$$

where $\mathcal{O}$ is the point at infinity. The order of the E is $3^{97} + 3^{49} + 1 = 7P_{151}$ where P_{151} is a 151-bit prime number as follows:

$$P_{151} = 2726865189058261010774960798134976187171462721.$$

Next, let G_1 be the subgroup of E of order P_{151} and let G_2 be the subgroup of $GF(3^{6\cdot97})^*$ of order P_{151}. Note that, since orders of G_1 and G_2 are prime numbers, every element of $G_1\backslash\{\mathcal{O}\}$ and $G_2\backslash\{1\}$ is a generator of G_1 and G_2, respectively. The η_T pairing for $n = 97$ is a map from $G_1 \times G_1$ to G_2.

Our goal is to solve the ECDLP in G_1. To set our target problem unintentionally, we select two elements $\mathcal{Q}_\pi, \mathcal{Q}_e$ in G_1, which correspond to the circular constant π and natural logarithm e, respectively. We explain how we select $\mathcal{Q}_\pi$ and $\mathcal{Q}_e$ as follows. First, we describe $GF(3^{97})$ as $GF(3)[x]/(x^{97} + x^{16} + 2)$, where the irreducible polynomial $x^{97} + x^{16} + 2 \in GF(3)[x]$ is well used for the fast implementation of field operations. An element in $GF(3^{97})$ is represented by $\sum_{i=0}^{96} d_i x^i$, where $d_i \in GF(3) = \{0, 1, 2\}$. To transform π and e to an element in $GF(3^{97})$ respectively, we define a bijective map $\phi : \sum_{i=0}^{96} d_i x^i \mapsto \sum_{i=0}^{96} d_i 3^i \in \mathbb{Z}$. We then transform π and e to the 3-adic integer of 97 digits by $\lfloor \pi \cdot 3^{95} \rfloor$ and $\lfloor e \cdot 3^{96} \rfloor$, respectively.

From these values, we define $\mathcal{Q}_\pi = (x_\pi, y_\pi)$ and $\mathcal{Q}_e = (x_e, y_e) \in G_1$ as follows. We first find the non-negative smallest 3-adic integers c_π and c_e such that $\phi^{-1}(\lfloor \pi \cdot 3^{95} \rfloor + c_\pi)$ and $\phi^{-1}(\lfloor e \cdot 3^{96} \rfloor + c_e)$ become x-coordinates of the elements $\mathcal{Q}_\pi$ and $\mathcal{Q}_e$ in the subgroup G_1 on the E. In fact we can set $x_\pi = \phi^{-1}(\lfloor \pi \cdot 3^{95} \rfloor + (11)_3)$ and $x_e = \phi^{-1}(\lfloor e \cdot 3^{96} \rfloor + (120)_3)$. There are two points in $G_1\backslash\{\mathcal{O}\}$ of the same x-coordinate. We then set the corresponding y-coordinate by computing $y_\pi = (x_\pi^3 - x_\pi + 1)^{(3^{97}+1)/4}$ and $y_e = (x_e^3 - x_e + 1)^{(3^{97}+1)/4}$ in $GF(3^{97})$, respectively.

Again, our goal is to solve the ECDLP in G_1, i.e., for given $Q_\pi, Q_e \in G_1$ we try to find integer s such that $Q_\pi = [s]Q_e$. On the other hand, the η_T pairing enables us to reduce the ECDLP in G_1 to the DLP over G_2 by the relationship $\eta_T(\mathcal{Q}_\pi, \mathcal{Q}_\pi) = \eta_T(\mathcal{Q}_\pi, \mathcal{Q}_e)^s$. Therefore, we can find s by computing the discrete logarithm

$$s = \log_{\eta_T(\mathcal{Q}_\pi, \mathcal{Q}_e)} \eta_T(\mathcal{Q}_\pi, \mathcal{Q}_\pi) = \log_g \eta_T(\mathcal{Q}_\pi, \mathcal{Q}_\pi)/ \log_g \eta_T(\mathcal{Q}_\pi, \mathcal{Q}_e) \bmod P_{151},$$

for a generator g of G_2.

3.2   Parameter Settings for FFS

In this section, we explain the parameter setting used for our implementations of the FFS. Hayashi et al. [20] reported that, when $n \le 509$, the JL06-FFS [24] is more efficient for solving the DLP over $GF(3^{6n})$ than the JL02-FFS [23]. Thus, we use the JL06-FFS for our computation. In the JL06-FFS, the condition that "r is monic" is introduced into the collecting relations phase in order to compute efficiently. For the remainder of this paper, the FFS refers to the JL06-FFS.

To solve our DLP over $GF(3^{6\cdot97})$, we have to select several parameter values of the FFS, such that its computational time is small enough for a fixed extension degree n. The parameter values for $n = 97$ are listed in [31, Table 3], and we use those parameter values for our computation.

We select the parameter $\kappa \in \{1, 2, 3, 6\}$ as follows. $GF(3^{6\cdot97})$ is described as $GF(3^{\kappa})[x]/(f)$, where $f \in GF(3^{\kappa})[x]$ is an irreducible polynomial of degree $6\cdot97/\kappa$. The appropriate value of κ is given in [31, Table 3], i.e., $\kappa = 6$. However, we select $\kappa = 3$ for the following reasons. In the linear algebra phase, filtering [15] is performed to reduce the size of the matrix. Then it is required that all elements in the factor base correspond to the memory addresses of the PC for efficient computation. The number of elements in the factor base for $\kappa = 6$ is much larger than that for $\kappa = 3$, so $\kappa = 3$ is advantageous on this point. Additionally, [31, Table3] shows that the computational cost of the FFS for $\kappa = 3$ is only about twice as much as that for $\kappa = 6$. We conducted test runs for $\kappa = 3, 6$ in the collecting relations phase, then noticed that our implementation for $\kappa = 3$ was much faster than for $\kappa = 6$, so we set $\kappa = 3$.

Polynomial Selection Phase: We select the bivariate polynomial $H(x, y)$ of the form $x + y^{d_H}$ for a given parameter d_H of the FFS in the same manner as [20]. Then we search an irreducible polynomial $f \in GF(3^{\kappa})[x]$ and a polynomial $m \in GF(3^{\kappa})[x]$ which are satisfying the condition (1), by factoring $H(x, m)$ for a randomly picked polynomial m whose degree is d_m. In fact, we randomly pick up m from $GF(3)[x]$, so that f is also in $GF(3)[x]$ for use of the Galois action. From [31, Table 3], we set d_H and d_m as 6 and 33, respectively.

Next, we select the smoothness bound $B = 6$ by using [31, Table 3] for (2) and (3), i.e., a rational factor base $F_R(B)$ and an algebraic factor base $F_A(B)$. $\#F_R(B)$ is 67576068 and $\#F_A(B)$ is 67572597, thus the number of elements of factor base, i.e., $\#F_R(B) + \#F_A(B)$, is 135148665.

Collecting Relations Phase: In the collecting relations phase, we use the lattice sieve [29] and the free relation [20] and collect many relations (8); $(r, s) \in (GF(3^{\kappa})[x])^2$ satisfying (4), (5), (7), where r is monic. The search range for the lattice sieve depends on the maximum degrees R, S of r, s. We set $R = S = 6$ based on [31, Table 3]. The lattice sieve gives a certain amount of relations for one special-Q, which is defined in Section 4.1. Therefore, we require a sufficient number of special-Q's so that the number of relations obtained in the collecting relations phase is larger than that of all elements in the factor base. The minimum sufficient number of special-Q's is estimated by the following process. We have to select special-Q's from the subset $F_R(6) \backslash F_R(5)$, whose cardinality is 64566684. Let θ_{min} be the minimum sufficient ratio of special-Q's over all elements in $F_R(6) \backslash F_R(5)$. For $n = 97$ and $\kappa = 3$, we can estimate $\theta_{min} = 0.01292$ [31, Table 3]. Therefore, the number of special-Q's must be larger than $\lceil 0.01292 \cdot 64566684 \rceil = 834202$. In our computation, we set 2500000 as the number of special-Q's to obtain more relations than we require since we expect that these excess relations will help us reduce the size of the matrix during filtering, especially in singleton-clique.

4   Implementation

In this section, we propose the following efficient implementation techniques; the lattice sieve for the JL06-FFS and optimization for our parameters in the collecting relations phase, the data structure and the parallel Lanczos method for the Galois action in the linear algebra phase, for reducing the computational cost of the FFS for solving the DLP over $GF(3^{6 \cdot 97})$. Parameters $(\kappa, d_H, d_m, B, R, S)$ are fixed as $(3, 6, 33, 6, 6, 6)$. The reasoning for this is explained in Section 3.2.

4.1   Collecting Relations Phase

In the collecting relations phase, we used the lattice sieve [29] in a similar fashion to factoring a large integer [26] and solving discrete logarithm problems [22, 23]. We give an overview of our implementation of the lattice sieve in the following paragraphs. More details are described in [19].

Lattice Sieve for JL06-FFS: Sieving with the lattice sieve is performed for $(r, s) \in (GF(3^3)[x])^2$ such that the formula (5) given in Section 2.2 is divisible by an element Q chosen from a subset of the rational factor base $F_R(6) \backslash F_R(5)$ (this Q is called a "special-Q"). Recall that $\deg r$ and $\deg s$ are not greater than $R = 6$ and $S = 6$, respectively. Such (r, s) can be represented as $(r, s) = c(r_1, s_1) + d(r_2, s_2)$ for given reduced lattice bases $(r_1, s_1), (r_2, s_2) \in (GF(3^3)[x])^2$ and any $c, d \in GF(3^3)[x]$ such that $\deg(cr_1 + dr_2) \le 6, \deg(cs_1 + ds_2) \le 6$, then sieving is done on the bounded c-d plane. After sieving, we conduct the smoothness test [16] for "candidates" that are evaluated as B-smooth pairs with high probability by using the lattice sieve.

A problem of applying the lattice sieve to the FFS is the condition "r is monic" described in Section 3.2. Since r is represented as $cr_1 + dr_2$, it is difficult to efficiently keep r monic — it might require degree evaluations and branches. Instead of choosing monic r, we introduce the condition $r \equiv 1 \mod x$. To satisfy this condition, we restrict r_1 and r_2 such that $r_1 \equiv 0 \mod x$ and $r_2 \equiv 1 \mod x$. Then sieving is performed on the bounded c-d plane with restriction $d \equiv 1 \mod x$, whose size is reduced to $1/27$ compared with the original bounded c-d plane. This sieving procedure with the restricted condition can be implemented without extra costs such as additional degree evaluations and additional branches.

Lattice Sieve with SIMD: Since operations of $GF(3)$ can be represented using logical instructions [25], operations of $GF(3^3)[x]$ can be performed using a combination of logical and shift instructions. This means SIMD implementation is appropriate for efficient computation of the lattice sieve. We represent $GF(3^3)$ as polynomial basis $GF(3)[\omega]/(\omega^3 - \omega - 1)$, and its element is represented using

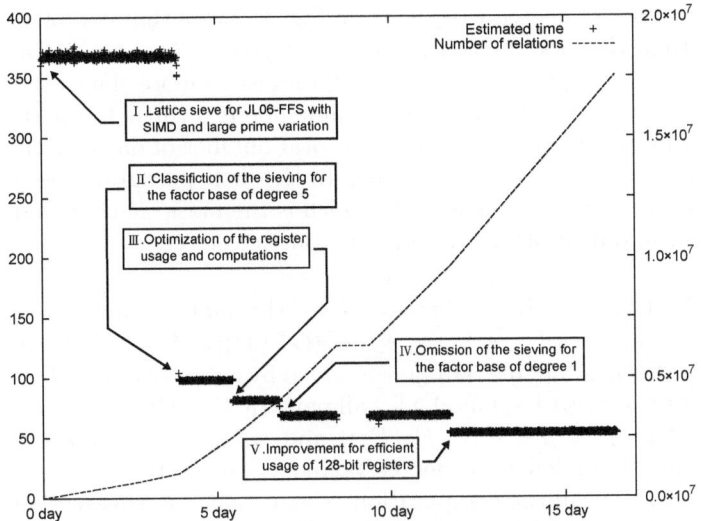

vertical(left)   : estimation days for collecting relations phase
vertical(right) : number of collected relations
horizontal   : first two weeks of computing days for collecting relations phase
(Period with no data between 8-9 days was due to human error in operating PC.)

Fig. 1. Our improvement in collecting relations phase for first two weeks

6-bit $(h_1, \ell_1, h_\omega, \ell_\omega, h_{\omega^2}, \ell_{\omega^2}) \in GF(2)^6$ in our implementation. We then pack 16 elements of $GF(3^3)[x]$ of degree at most 7 into 6 registers of 128 bits, and treat 16 elements with SIMD. Note that the upper bound of the degree of our SIMD data structure is for efficient access to each element in $GF(3^3)[x]$. On the other hand, since we choose B, R, S as all 6, the upper bound of the degrees of $c, d, r_1, s_1, r_2, s_2 \in GF(3^3)[x]$ and $\mathfrak{p}$ in the factor base, which are treated in the lattice sieve, is also 6. Therefore, our SIMD structure can be stored elements treated in the lattice sieve.

History of Our Optimizations: Figure 1 shows the process of our improvements in the collecting relations phase for the first two weeks. We improved our implementation of the lattice sieve four times during this period. We first used large prime variation to omit sieving for the factor base of degree 6 and implemented the lattice sieve for the FFS with SIMD implementation. We then ran the program for the first four days (stage I in Fig. 1). At that point, the estimated total number of days for the collecting relations phase was about 360 days. While the sieving program was running, we found that sieving for the factor base of degree 5 requires heavier computation than sieving for the factor bases of degree 1, 2, 3 and 4. Therefore, we improved sieving for the factor base of degree 5; thus, our sieving program became over 3 times faster than before (stage II in Fig. 1). Next, we optimized register usage for input values and omitted wasteful computations (stage III in Fig. 1). Additionally, we omitted sieving for the factor base of degree 1 (stage IV in Fig. 1), since that computational time was larger than that for the factor bases of degree 2, 3, 4, and 5. Moreover, we improved our sieving program to use 128-bit registers more efficiently (stage V in Fig. 1). Finally, our sieving program became about 6 times faster than the first one (stage I in Fig. 1) and the estimated total number of days for the collecting relations phase became about 53.1 days. In the next paragraph, we explain the details of the improvement in stage II, which is the most effective and important improvement in our implementation of the lattice sieve.

Details of Stage II: In the lattice sieve, the main computation of sieving for given lattice bases (r_1, s_1), $(r_2, s_2) \in (GF(3^3)[x])^2$ is as follows. For fixed $d \in GF(3^3)[x]$, whose degree is upper-bounded by a degree bound D, we compute $c_0 \equiv -d(r_1t+s_1)^{-1}(r_2t+s_2) \bmod \mathfrak{p}$ for all pairs $(\mathfrak{p}, t) \in \{(\mathfrak{p}, t) \mid \mathfrak{p} \in F_R(B), t \equiv m \pmod{\mathfrak{p}}\} \cup \{(\mathfrak{p}, t) \mid \langle \mathfrak{p}, y - t \rangle \in F_A(B)\}$, and compute $c \in GF(3^3)[x]$, whose degree is upper-bounded by a degree bound C, such that $c = c_0 + k\mathfrak{p}$ where $k \in GF(3^3)[x]$. We call the computation "sieving at d" in this section. For given lattice bases, sieving at d is performed for all d of degree not larger than D. Note that c_0 does not need to be computed when $(r_1t + s_1) \equiv 0 \pmod{\mathfrak{p}}$; therefore we assume $(r_1t + s_1) \not\equiv 0 \pmod{\mathfrak{p}}$ in the following description.

In stage I of our implementation, we found that the time of sieving at d for $\deg \mathfrak{p} = 5$ takes over 100 msec, but each sieving time at d for $\deg \mathfrak{p} = 1, 2, 3$ and 4 takes about 10 mesc or less. Therefore, we tried to improve the sieving of degree 5. When we compute c_0 for $\mathfrak{p}$ of degree 5, the degree of c_0 becomes 4 with probability about 26/27. On the other hand, the degree of the lattice bases r_1, s_1, r_2, s_2 is 3 in most cases because the degree of special-Q is 6. On such bases, degree bounds C and D can be chosen as 3 to satisfy condition (4), i.e., $\deg r \leq 6$ and $\deg s \leq 6$. These facts show that about 26/27 of the computation of sieving for $\mathfrak{p}$ of degree 5 are waste computations. Therefore, we discuss how to sieve only with the polynomial c_0, whose degree is not larger than 3, as follows.

Let $\alpha \in GF(3^3)[x]$ be $-(r_1t+s_1)^{-1}(r_2t+s_2) \bmod \mathfrak{p}$, then we have $c_0 = d\alpha \bmod \mathfrak{p}$. Let $\alpha_i \in GF(3^3)$ be the coefficient of the fourth-order term of $x^i\alpha \bmod \mathfrak{p}$ for $i = 0, 1, 2, 3$. Since $\deg d \leq 3$, d is represented as $d_3x^3 + d_2x^2 + d_1x + 1$ for $d_3, d_2, d_1 \in GF(3^3)$. Recall that we restricted $d \equiv 1 \bmod x$ in our implementation of the lattice sieve. Here we know that the degree of c_0 is not larger than 3 if $d_3\alpha_3 + d_2\alpha_2 + d_1\alpha_1 + \alpha_0 = 0$. Therefore, it is sufficient to perform sieving at d for $\mathfrak{p}$ in the factor base of degree 5 for only d satisfying the following property:

$$d_1 = \begin{cases} -K\alpha_1^{-1} & \text{if } \alpha_1 \neq 0 \\ \text{any element in } GF(3^3) & \text{if } \alpha_1 = 0 \text{ and } K = 0 \end{cases} \tag{9}$$

where $K = d_3\alpha_3 + d_2\alpha_2 + \alpha_0$. When $\alpha_1 = 0$ and $K = 0$, we should compute c_0 for d whose d_1 is any element in $GF(3^3)$, and we cannot cut off any d_1; therefore, we assume that $\alpha_1 \neq 0$ in the following description. Suppose that we now fix lattice bases $(r_1, s_1), (r_2, s_2)$ and a pair $(\mathfrak{p}, t)$ where $\deg \mathfrak{p} = 5$, then each α_i for $i = 0, 1, 2, 3$ is also fixed. Therefore, since K depends on d_2 and d_3, the d_1 satisfying (9) is given by d_2 and d_3 and uniquely determined for given

d_2 and d_3. This implies that, since d_1 is in $GF(3^3)$ whose cardinality is 27, we can ignore 26 d_1's not satisfying (9) for given d_2 and d_3. In fact, the time of sieving at d for all pairs $(\mathfrak{p}, t)$ where $\deg \mathfrak{p} = 5$ is reduced to about 1.5 msec by ignoring d_1 not satisfying (9). Note that we need to compute K for given d_2 and d_3 for all pairs $(\mathfrak{p}, t)$. The time of computing K for all $(\mathfrak{p}, t)$ takes about 150 msec in our implementation. Therefore, for all pairs $(\mathfrak{p}, t)$ where $\deg \mathfrak{p} = 5$, the computations of K and sieving at d require about 7.1 msec at stage II, which is over 10 times faster than the computation of sieving at d at stage I. As a result, our implementation of the lattice sieve at stage II becomes over 3 times faster than that at stage I.

4.2 Linear Algebra Phase

After the collecting relations phase, we obtain a system of linear equations modulo P_{151}, which is described in Section 2.1. The Galois action [20, 24] can reduce the number of variables of the system of linear equations to one-third. Additionally, after the Galois action, the numbers of equations and variables of the system of linear equations can be further reduced using filtering [15], i.e., singleton-clique and merging. To solve the system of linear equations defined by this reduced matrix, we use the parallel Lanczos method [4, 20].

Galois Action: The Galois action to $GF(3^{6 \cdot 97})/GF(3^{3 \cdot 97})$ enables us to reduce the number of variables of the system of linear equations to one-third (details of the Galois action are discussed in [20, 24]). However, when we use the Galois action, 151-bit large integers such as $e_0 + e_1\tau + e_2\tau^2$, where $\tau = 3^{97^2} \bmod P_{151}$ and e_i is a small integer of a few bits, are added to elements of the system of linear equations. This unfortunate fact eventually increases the data size of the reduced matrix; therefore, high-capacity memory is required. To allay the increase in the representation size of the elements, we store only a triplet (e_1, e_2, e_3) in the PC memory, not a large 151-bit integer. Since e_i is small enough to be represented by 8 bits, the size of the elements is reduced from 151 to 24 bits on average. We call this representation the "τ-adic structure". Note that the τ-adic structure is used for the Galois action and singleton-clique.

Singleton-Clique and Merging: Filtering consists of two parts, singleton-clique and merging. Singleton-clique deletes unnecessary rows and columns to reduce the size of the matrix. In our implementation of singleton-clique, we performed by maintaining 20000 more rows than columns to prevent accidentally decreasing the rank of the matrix. After that, merging, a weight-controlled Gaussian elimination, is performed. In merging, for small integer k, the column with a weight not larger than k is deleted by row elimination with controlling the pivot selection so that the weight of the matrix is as small as possible. This operation is called k-way merging. In our implementation of merging, we converted the data representation of the matrix from the τ-adic structure to a large 151-bit integer structure, since merging on the τ-adic structure cannot reduce the size of the matrix enough due to the restriction of the pivot selection. More details are described in [19].

Parallel Lanczos Method: By using the parallel Lanczos method [4, 20], we solve the system of linear equations defined by the matrix reduced via the Galois action, singleton-clique, and merging. For parallel computing, the matrix should be split into sub-matrices, i.e., split into $N = N_1 \times N_2$ sub-matrices for N nodes, and nodes communicate among N_1 nodes or N_2 nodes. To reduce the synchronization time before communicating, the matrix is split so that each sub-matrix has almost the same weight. Our machine environment for the parallel Lanczos method consisted of 22 nodes, and each node had 12 CPU cores and 2 NICs. The 2 NICs were connected to a 48-port Gbit HUB, i.e., 44 ports were used for connecting 22 nodes. All 22 nodes could be used, so we had a choice for machine environment; $20 = 5 \times 4$, $21 = 7 \times 3$ or $22 = 11 \times 2$. Using 20 nodes requires the least communication costs but the most computational costs, and using 22 nodes requires the most communication costs but the least computational costs. Using 21 nodes was the best for our implementation; therefore, we used 21 nodes.

For computation in the parallel Lanczos method, many modular multiplications of 151-bit integers × 151-bit integers modulo P_{151} are required due to the Galois action. We implemented Montgomery multiplication optimized to 151-bit integers using assembly language. Our program then becomes several times faster than straightforward modular multiplication using GMP (http://gmplib.org/) for multiple precision arithmetic.

After the computation of the parallel Lanczos method started, we improved our codes of the parallel Lanczos method (for example, efficient register usage, overlapping communications and computations). These improvements are about 15% faster than our initial implementation.

4.3   Individual Logarithm Phase

As mentioned in Section 3.1, $\log_g \eta_T(\mathcal{Q}_\pi, \mathcal{Q}_\pi)$ and $\log_g \eta_T(\mathcal{Q}_\pi, \mathcal{Q}_e)$ are required to solve our target problem. To compute them, rationalization and special-Q descent [24] were used. For simplicity, let T be $\eta_T(\mathcal{Q}_\pi, \mathcal{Q}_\pi)$, or $\eta_T(\mathcal{Q}_\pi, \mathcal{Q}_e)$ in the following paragraphs.

In the rationalization, we randomize T such that the randomized element is M-smooth for a small enough integer $M > B$ by the following process. First, we randomize T by $z \equiv g^\gamma T \pmod{f}$ for a random integer $\gamma \in \mathbb{Z}_{P_{151}}$. We then rationalize z as $z \equiv z_1/z_2 \pmod{f}$ where degrees of z_1 and z_2 are about $\deg f/2$, and check whether both z_1 and z_2 are M-smooth. Then, computing $\log_g T$ is reduced to computing logarithms of irreducible factors of M-smooth elements z_1 and z_2.

M-smooth elements z_i for $i = 1, 2$, contain some irreducible factors of degree larger than B whose logarithms are not computed in the linear algebra phase. To compute these logarithms, the special-Q descent [24] is usually used. In the special-Q descent, the lattice sieve is recursively conducted with an irreducible factor of degree larger than B, which is contained in z_i or in a relation generated during the special-Q descent, as a special-Q.

5   Experimental Results

We succeeded in solving a DLP over $GF(3^{6\cdot97})$ by using the FFS with our efficient implementation techniques discussed in Section 4. In this section, we report our computational results, such as the computational time of each phase of the FFS and the number of relations.

5.1   Polynomial Selection

The FFS has six parameters κ, d_H, d_m, B, R, and S, as defined in Section 2.2, and we set $(\kappa, d_H, d_m, B, R, S) = (3, 6, 33, 6, 6, 6)$ for our target problem, based on the reason given in Section 3.2. In the polynomial selection phase, we can extract appropriate polynomials such as the definition polynomial $H(x, y)$ of a function field described in Section 3.2 in one minute, so the computational cost of the polynomial selection phase is negligibly small.

5.2   Collecting Relations Phase

In the collecting relations phase, we search many relations that are equations of the form (8) to generate a system of linear equations by using the lattice sieve and the free relation. We explain our computational results of the collecting relations phase, e.g., the number of relations obtained in this phase, the computational time of the lattice sieve for one special-Q.

Lattice Sieve. Each special-Q has to be chosen from $F_R(6)\backslash F_R(5)$. The number of elements of $F_R(6)\backslash F_R(5)$ is 64566684, and the size of the table of those elements is about 500 MB. Since our program of the lattice sieve is computed using many nodes, it is not convenient to pick up the element from that 500-MB table as a special-Q. Therefore, we selected a special-Q by randomly generating an irreducible polynomial in $GF(3^3)[x]$ of degree 6, which is in $F_R(6)\backslash F_R(5)$, and iterated the computation of the lattice sieve for the special-Q.

We prepared 47 PCs (in total 212 CPU cores) for the lattice sieve. The computation of the lattice sieve began on May 14, 2011, and we continued optimizing our program of the collecting relations phase. As discussed in Section 4.1, we applied several improvements to our program of the collecting relations phase; the lattice sieve for the JL06-FFS, the lattice sieve with SIMD, and optimization for our parameters. Figure 1 in Section 4.1 shows the process of our improvements in the collecting relations phase for the first two weeks. The total time for the collecting relations phase shortened due to our improvements. Finally, the computation finished on September 9, 2011 and required 118 days. including the loss-time of some programming errors, updating our codes, and power outages. The real computational time of the lattice sieve was equivalent to 53.1 days using 212 CPU cores such as Xeon E5440.

Table 2 summarizes the process of generating relations in the collecting relations phase. It might seem that the number of duplicate relations is very small compared to the integer factorization case using the number field sieve. This

Table 2. Number of collected relations in collecting relations phase

lattice sieve	159032292 relations obtained from 2500000 special-Q's
	(64.91 relations/special-Q, 389 sec/special-Q)
	153815493 unique (non-duplicated) relations
	obtained from 2449991 unique special-Q's
free relation	33786299 relations
total	187602242 relations (consist of 134697663 elements in the factor base)

Table 3. Compressing matrix using Galois action, singleton-clique and merging

method	size of matrix
before compressing	187602242 equations × 134697663 variables
Galois action	159394665 equations × 45049572 variables
singleton-clique	14060794 equations × 14040791 variables
6-way merging	6141443 equations × 6121440 variables

arises from the fact that the size of the sieving space in our parameters is so large compared to that case.

Free Relation. The free relation gives us additional relations not generated by a sieving algorithm such as the lattice sieve. The details of the free relation is given in [20]. As shown in Table 2, the free relation gave us 33786299 relations. Eventually, we obtained a system of linear equations consisting of 187602242 equations and 134697663 variables. Note that there are 451002 elements in the factor base, which does not appear in the 187602242 relations.

5.3   Linear Algebra Phase

In the linear algebra phase, we firstly reduced the size of the matrix by the Galois action and filtering, and then performed the parallel Lanczos method for the reduced matrix. Table 3 shows that the process of the compression of the matrix.

Galois Action. As mentioned in Section 4.2, the Galois action reduced the size of the matrix generated in the collecting relations phase to one-third since $\kappa = 3$. To allay the fact that the size of each element of the matrix increases from a few bits to 151 bits due to the Galois action, we used the τ-adic structure mentioned in Section 4.2.

Singleton-Clique and Merging. After using the Galois action, we additionally reduce the variables and equations of the matrix by singleton-clique and merging [15]. Using a PC, the computation for singleton-clique took about 3 hours, and that for merging took about 10 hours. After 6-way merging, we started the computation of the parallel Lanczos method for the 6-way merged matrix. See [19] for more details about our results of singleton-clique and merging.

Table 4. Computational time of parallel Lanczos method for 6-way merged matrix

calculation time/loop	626.3 msec
synchronization time/loop	46.5 msec
communication time/loop	457.3 msec
total time/loop	1130.1 msec
number of loops	6121438
total time	80.1 days

Parallel Lanczos Method. We used the parallel Lanczos method [4, 20] to solve the system of linear equations defined by the 6-way merged matrix. Note that this matrix is sparse and defined over $\mathbb{Z}_{P_{151}}$, where P_{151} is the 151-bit prime number given in Section 3.1. The computation of the parallel Lanczos method started on January 16, 2012, and was conducted on 21 PCs, which were connected via a 48-port Gbit HUB. As mentioned in Section 4.2, we continued improving our codes of the parallel Lanczos method after computation began. The computational times of our improved codes are listed in Table 4. Finally, computation finished on April 14, 2012. The computation for the parallel Lanczos method took 90 days including time losses similar to our implementation of the lattice sieve. The real computational time is equivalent to 80.1 days using 252 CPU cores such as Xeon X5650.

5.4 Individual Logarithm Phase

Our target is to compute $\log_g \eta_T(\mathcal{Q}_\pi, \mathcal{Q}_e)$ and $\log_g \eta_T(\mathcal{Q}_\pi, \mathcal{Q}_\pi)$ for some $g \in G_2$, as mentioned in Section 3.1.

First, we computed the rationalization described in Section 4.3. Let g be a polynomial $(x+\omega)^{(3^{6\cdot97}-1)/P_{151}} \in G_2$, where ω is a polynomial basis of $GF(3^3) \cong GF(3)[\omega]/(\omega^3 - \omega - 1)$. Note that g is a generator of $G_2 \subset GF(3^{6\cdot97})^*$ and $x+\omega$ is a monic irreducible polynomial in $F_R(B)$ of degree 1. We set $M = 15$ and search a pair (z_1, z_2) (and $(z_1', z_2')) \in (GF(3^3)[x])^2$ such that $\eta_T(\mathcal{Q}_\pi, \mathcal{Q}_e) \cdot g^{\gamma_1} = z_1/z_2$ (and $\eta_T(\mathcal{Q}_\pi, \mathcal{Q}_\pi) \cdot g^{\gamma_2} = z_1'/z_2'$), where z_i (and z_i') are M_i-smooth (where $M_i \leq M$) for some $\gamma_1, \gamma_2 \in \mathbb{Z}_{P_{151}}$ and $i = 1, 2$. We found z_1 and z_2, which are 13- and 15-smooth (and z_1' and z_2' which are 15- and 14-smooth), respectively. These computations were conducted on 168 CPU cores and required 7 days for each computation.

$$\eta_T(\mathcal{Q}_\pi, \mathcal{Q}_e) \cdot g^{\gamma_1} = (13\text{-smooth})/(15\text{-smooth}),$$

$$\gamma_1 = 2514037766787322013334785428291787565870435706,$$

$$\eta_T(\mathcal{Q}_\pi, \mathcal{Q}_\pi) \cdot g^{\gamma_2} = (15\text{-smooth})/(14\text{-smooth}),$$

$$\gamma_2 = 2657516740789758289434702436228062607247517136.$$

Next, we performed special-Q descent for each irreducible factor of smooth elements obtained by the rationalization. These computations were conducted on 168 CPU cores and took about 0.5 days for each $\eta_T(\mathcal{Q}_\pi, \mathcal{Q}_e)$ and $\eta_T(\mathcal{Q}_\pi, \mathcal{Q}_\pi)$.

Thus, the computation of the individual logarithm phase took 15 days; (7 days (for rationalization) + 0.5 days (for special-Q descent)) × 2 elements.

By using the logarithms of the corresponding elements in the factor base obtained from the linear algebra phase, we could compute $\log_g \eta_T(\mathcal{Q}_\pi, \mathcal{Q}_e)$ and $\log_g \eta_T(\mathcal{Q}_\pi, \mathcal{Q}_\pi)$. The logarithm of each element is as follows:

$$\log_g \eta_T(\mathcal{Q}_\pi, \mathcal{Q}_e) = 154096662595700795834782326842395703646965 6370,$$
$$\log_g \eta_T(\mathcal{Q}_\pi, \mathcal{Q}_\pi) = 163028195063550729566380917121783309697044 9894.$$

Finally, we obtained the logarithm of the target element:

$$s = \log_{\eta_T(\mathcal{Q}_\pi, \mathcal{Q}_e)} \eta_T(\mathcal{Q}_\pi, \mathcal{Q}_\pi)$$
$$= 175279958485066813773020730619813142455096 7300.$$

This is the solution of the ECDLP of equation $\mathcal{Q}_\pi = [s]\mathcal{Q}_e$.

6   Concluding Remarks

We evaluated the security of pairing-based cryptosystems using the η_T pairing over supersingular elliptic curves on finite field $GF(3^n)$. We focused on the case of $n = 97$ since many implementers have reported practically relevant high-speed implementations of the η_T pairing with $n = 97$ in both software and hardware. In particular, we examined the difficulty in solving the discrete logarithm problem (DLP) over $GF(3^{6 \cdot 97})$ by our implementation of the function field sieve (FFS).

To reduce the computational cost of the FFS for solving the DLP, we proposed several efficient implementation techniques. In the collecting relations phase, we implemented the lattice sieve for the JL06-FFS with SIMD and introduced improvements by optimizing for factor bases of each degree; therefore, our lattice sieve for the JL06-FFS became about 6 times faster than the first one. The main difference from the number field sieves for integer factorization is the linear algebra phase, namely, we have to deal with a large modulus of 151-bit prime for the computation of the FFS. We thus performed filtering (singleton-clique and merging) by carefully considering the data structure of large integers developing from the Galois action, so that we can efficiently conduct the parallel Lanczos method. From the above improvements, we succeeded in solving the DLP over $GF(3^{6 \cdot 97})$ in 148.2 days by using PCs with 252 CPU cores. Our computational results contribute to the security estimation of pairing-based cryptosystems using the η_T pairing. In particular, they show that the security parameter of such pairing-based cryptosystems must be chosen with $n > 97$.

Finally, we show a very rough estimation of required computational power for solving the DLP over $GF(3^{6n})$ with $n > 97$. Our experiment on the DLP over $GF(3^{6n})$ with $n = 97$ used 252 CPU cores of mainly 2.67 GHz Xeon for 148.2 days, which are equivalent to $2^{62.9}$ clock cycles. From the analysis of [31], the computational complexities of breaking the DLP over $GF(3^{6n})$ with $n = 163$ and 193 become $2^{15.4}$ and $2^{19.1}$ times larger than that with $n = 97$, respectively. Therefore, we could estimate that about $2^{78.3}$ and $2^{82.0}$ clock cycles are required for breaking the DLP over $GF(3^{6n})$ with $n = 163$ and 193, respectively. On the other hand, the currently second fastest supercomputer K has a throughput of about 10.5 petaflop/s from http://www.top500.org/, and it performs about $2^{78.1}$ floating-point operations for one year. If one floating-point operation on the CPU of the K is equivalent to one clock cycle of logical operation on the Xeon core, we might be able to break the DLP over $GF(3^{6\cdot163})$ using our implementation on supercomputer K for one year.

References

1. Adleman, L.M.: The Function Field Sieve. In: Huang, M.-D.A., Adleman, L.M. (eds.) ANTS 1994. LNCS, vol. 877, pp. 108–121. Springer, Heidelberg (1994)
2. Adleman, L.M., Huang, M.-D.A.: Function field sieve method for discrete logarithms over finite fields. Inform. and Comput. 151, 5–16 (1999)
3. Ahmadi, O., Hankerson, D., Menezes, A.: Software Implementation of Arithmetic in F_{3^m}. In: Carlet, C., Sunar, B. (eds.) WAIFI 2007. LNCS, vol. 4547, pp. 85–102. Springer, Heidelberg (2007)
4. Aoki, K., Shimoyama, T., Ueda, H.: Experiments on the Linear Algebra Step in the Number Field Sieve. In: Miyaji, A., Kikuchi, H., Rannenberg, K. (eds.) IWSEC 2007. LNCS, vol. 4752, pp. 58–73. Springer, Heidelberg (2007)
5. Barreto, P.S.L.M., Galbraith, S., ÓhÉigeartaigh, C., Scott, M.: Efficient pairing computation on supersingular Abelian varieties. Des., Codes Cryptogr. 42(3), 239–271 (2007)
6. Barreto, P.S.L.M., Kim, H.Y., Lynn, B., Scott, M.: Efficient Algorithms for Pairing-Based Cryptosystems. In: Yung, M. (ed.) CRYPTO 2002. LNCS, vol. 2442, pp. 354–368. Springer, Heidelberg (2002)
7. Beuchat, J.-L., Brisebarre, N., Detrey, J., Okamoto, E.: Arithmetic Operators for Pairing-Based Cryptography. In: Paillier, P., Verbauwhede, I. (eds.) CHES 2007. LNCS, vol. 4727, pp. 239–255. Springer, Heidelberg (2007)
8. Beuchat, J.-L., Brisebarre, N., Detrey, J., Okamoto, E., Shirase, M., Takagi, T.: Algorithms and arithmetic operators for computing the η_T pairing in characteristic three. IEEE Trans. Comput. 57(11), 1454–1468 (2008)
9. Beuchat, J.-L., Brisebarre, N., Shirase, M., Takagi, T., Okamoto, E.: A Coprocessor for the Final Exponentiation of the η_T Pairing in Characteristic Three. In: Carlet, C., Sunar, B. (eds.) WAIFI 2007. LNCS, vol. 4547, pp. 25–39. Springer, Heidelberg (2007)
10. Boneh, D., Di Crescenzo, G., Ostrovsky, R., Persiano, G.: Public Key Encryption with Keyword Search. In: Cachin, C., Camenisch, J.L. (eds.) EUROCRYPT 2004. LNCS, vol. 3027, pp. 506–522. Springer, Heidelberg (2004)
11. Boneh, D., Franklin, M.: Identity-Based Encryption from the Weil Pairing. In: Kilian, J. (ed.) CRYPTO 2001. LNCS, vol. 2139, pp. 213–229. Springer, Heidelberg (2001)
12. Boneh, D., Gentry, C., Waters, B.: Collusion Resistant Broadcast Encryption with Short Ciphertexts and Private Keys. In: Shoup, V. (ed.) CRYPTO 2005. LNCS, vol. 3621, pp. 258–275. Springer, Heidelberg (2005)
13. Boneh, D., Lynn, B., Shacham, H.: Short Signatures from the Weil Pairing. In: Boyd, C. (ed.) ASIACRYPT 2001. LNCS, vol. 2248, pp. 514–532. Springer, Heidelberg (2001)

14. Bos, J.W., Kaihara, M.E., Kleinjung, T., Lenstra, A.K., Montgomery, P.L.: Solving a 112-bit prime elliptic curve discrete logarithm problem on game consoles using sloppy reduction. International Journal of Applied Cryptography 2(3), 212–228 (2012)
15. Cavallar, S.: Strategies in Filtering in the Number Field Sieve. In: Bosma, W. (ed.) ANTS-IV. LNCS, vol. 1838, pp. 209–231. Springer, Heidelberg (2000)
16. Gordon, D.M., McCurley, K.S.: Massively Parallel Computation of Discrete Logarithms. In: Brickell, E.F. (ed.) CRYPTO 1992. LNCS, vol. 740, pp. 312–323. Springer, Heidelberg (1993)
17. Granger, R., Page, D., Stam, M.: Hardware and software normal basis arithmetic for pairing-based cryptography in characteristic three. IEEE Trans. Comput. 54(7), 852–860 (2005)
18. Hankerson, D., Menezes, A., Scott, M.: Software implementation of pairings. In: Identity-Based Cryptography, pp. 188–206 (2009)
19. Hayashi, T., Shimoyama, T., Shinohara, N., Takagi, T.: Breaking pairing-based cryptosystems using η_T pairing over $GF(3^{97})$. Cryptology ePrint Archive, Report 2012/345 (2012)
20. Hayashi, T., Shinohara, N., Wang, L., Matsuo, S., Shirase, M., Takagi, T.: Solving a 676-bit Discrete Logarithm Problem in $GF(3^{6n})$. In: Nguyen, P.Q., Pointcheval, D. (eds.) PKC 2010. LNCS, vol. 6056, pp. 351–367. Springer, Heidelberg (2010)
21. Joux, A.: A One Round Protocol for Tripartite Diffie-Hellman. In: Bosma, W. (ed.) ANTS-IV. LNCS, vol. 1838, pp. 385–394. Springer, Heidelberg (2000)
22. Joux, A., et al.: Discrete logarithms in $GF(2^{607})$ and $GF(2^{613})$. Posting to the Number Theory List (2005), http://listserv.nodak.edu/cgi-bin/wa.exe?A2=ind0509&L=nmbrthry&T=0&P=3690
23. Joux, A., Lercier, R.: The Function Field Sieve Is Quite Special. In: Fieker, C., Kohel, D.R. (eds.) ANTS 2002. LNCS, vol. 2369, pp. 431–445. Springer, Heidelberg (2002)
24. Joux, A., Lercier, R.: The Function Field Sieve in the Medium Prime Case. In: Vaudenay, S. (ed.) EUROCRYPT 2006. LNCS, vol. 4004, pp. 254–270. Springer, Heidelberg (2006)
25. Kawahara, Y., Aoki, K., Takagi, T.: Faster Implementation of η_T Pairing over $GF(3^m)$ Using Minimum Number of Logical Instructions for $GF(3)$-addition. In: Galbraith, S.D., Paterson, K.G. (eds.) Pairing 2008. LNCS, vol. 5209, pp. 282–296. Springer, Heidelberg (2008)
26. Kleinjung, T., Aoki, K., Franke, J., Lenstra, A.K., Thomé, E., Bos, J.W., Gaudry, P., Kruppa, A., Montgomery, P.L., Osvik, D.A., te Riele, H., Timofeev, A., Zimmermann, P.: Factorization of a 768-Bit RSA Modulus. In: Rabin, T. (ed.) CRYPTO 2010. LNCS, vol. 6223, pp. 333–350. Springer, Heidelberg (2010)
27. Menezes, A., Okamoto, T., Vanstone, S.A.: Reducing elliptic curve logarithms to logarithms in a finite field. IEEE Trans. IT 39(5), 1639–1646 (1993)
28. Okamoto, T., Takashima, K.: Fully Secure Functional Encryption with General Relations from the Decisional Linear Assumption. In: Rabin, T. (ed.) CRYPTO 2010. LNCS, vol. 6223, pp. 191–208. Springer, Heidelberg (2010)
29. Pollard, J.M.: The lattice sieve. In: The development of the number field sieve. LNIM, vol. 1554, pp. 43–49 (1993)
30. Sahai, A., Waters, B.: Fuzzy Identity-Based Encryption. In: Cramer, R. (ed.) EUROCRYPT 2005. LNCS, vol. 3494, pp. 457–473. Springer, Heidelberg (2005)
31. Shinohara, N., Shimoyama, T., Hayashi, T., Takagi, T.: Key Length Estimation of Pairing-Based Cryptosystems using η_T Pairing. In: Ryan, M.D., Smyth, B., Wang, G. (eds.) ISPEC 2012. LNCS, vol. 7232, pp. 228–244. Springer, Heidelberg (2012)

On the (Im)possibility of Projecting Property in Prime-Order Setting

Jae Hong Seo

National Institute of Information and Communications Technology,
4-2-1, Nukui-kitamachi, Koganei, Tokyo, 184-8795, Japan
jaehong@nict.go.jp

Abstract. Projecting bilinear pairings have frequently been used for designing cryptosystems since they were first derived from composite order bilinear groups. There have been only a few studies on the (im)possibility of projecting bilinear pairings. Groth and Sahai showed that projecting bilinear pairings can be achieved in the prime-order group setting. They constructed both projecting *asymmetric* bilinear pairings and projecting *symmetric* bilinear pairings, where a bilinear pairing e is symmetric if it satisfies $e(g, h) = e(h, g)$ for any group elements g and h; otherwise, it is asymmetric.

In this paper, we provide impossibility results on projecting bilinear pairings in a prime-order group setting. More precisely, we specify the lower bounds of

1. the image size of a projecting asymmetric bilinear pairing
2. the image size of a projecting symmetric bilinear pairing
3. the computational cost for a projecting asymmetric bilinear pairing
4. the computational cost for a projecting symmetric bilinear pairing

in a prime-order group setting naturally induced from the k-linear assumption, where the computational cost means the number of generic operations.

Our lower bounds regarding a projecting asymmetric bilinear pairing are tight, i.e., it is impossible to construct a more efficient projecting asymmetric bilinear pairing than the constructions of Groth-Sahai and Freeman. However, our lower bounds regarding a projecting symmetric bilinear pairing differ from Groth and Sahai's results regarding a symmetric bilinear pairing results; We fill these gaps by constructing projecting symmetric bilinear pairings.

In addition, on the basis of the proposed symmetric bilinear pairings, we construct more efficient instantiations of cryptosystems that essentially use the projecting symmetric bilinear pairings in a modular fashion. Example applications include new instantiations of the Boneh-Goh-Nissim cryptosystem, the Groth-Sahai non-interactive proof system, and Seo-Cheon round optimal blind signatures proven secure under the DLIN assumption. These new instantiations are more efficient than the previous ones, which are also provably secure under the DLIN assumption. These applications are of independent interest.

X. Wang and K. Sako (Eds.): ASIACRYPT 2012, LNCS 7658, pp. 61–79, 2012.

1   Introduction

A bilinear group is a tuple of abelian groups with a non-degenerate bilinear pairing. Projecting bilinear pairings, which are bilinear pairings with homomorphisms that satisfy a commutative property, have frequently been used for designing cryptosystems since they were first derived from composite order bilinear groups [10], though Freeman identified and named the projecting property recently [15]. Of special interest is the Groth-Sahai non-interactive proof system [22] and the Boneh-Goh-Nissim cryptosystem [10], both of which essentially use the projecting property and have numerous applications in various fields in cryptography. For example, the Groth-Sahai proofs were used to construct ring signatures [13], group signatures [19], round optimal blind signatures [25], verifiable shuffles [20], a universally composable adaptive oblivious transfer protocol [18], a group encryption scheme [12], anonymous credentials [7,6], and malleable proof systems [14]. For its part, the Boneh-Goh-Nissim cryptosystem was used for designing private searching on streaming data [31], non-interactive zero-knowledge [21], shuffling [5], and privacy-preserving set operations [32].

(Im)possibility of Projecting Bilinear Pairings: Although the projecting bilinear pairings are often used for designing various cryptosystems, there have been only a few studies on the (im)possibility of projecting bilinear pairings. Groth and Sahai [22] demonstrated that projecting bilinear pairings can be achieved in the prime-order group setting. They provided two distinct constructions in prime-order group setting: projecting *asymmetric* bilinear pairings and projecting *symmetric* bilinear pairings, where a bilinear pairing e is symmetric if it satisfies $e(g, h) = e(h, g)$ for any group elements g and h; otherwise, it is asymmetric. On the basis of this idea of projecting bilinear pairings, they developed non-interactive proof systems for quadratic equations over modules that can be instantiated in composite-order bilinear groups, product groups of prime-order bilinear groups with asymmetric bilinear pairings, and product groups of prime-order groups with symmetric bilinear pairings. By extending Groth-Sahai's idea, Freeman [15] generalized Groth-Sahai's projecting asymmetric bilinear pairings.[1] Groth-Sahai and Freeman's constructions of projecting bilinear pairings allow for the simultaneous treatment of subgroup indistinguishability. To use projecting bilinear pairings for designing cryptographic protocols, we need to deal with cryptographic assumptions such as subgroup decision assumption at the same time. Meiklejohn, Shacham, and Freeman [25] have shown some impossibility results for projecting bilinear pairings, e.g., that projecting bilinear pairings cannot simultaneously have a cancelling property if the subgroup indistinguishability is naturally induced from the k-linear assumption [23,36]. Recently, Seo and

---

[1] Freeman identified the other property of bilinear pairings in a composite-order group setting, called *cancelling*, and demonstrated how to achieve the cancelling bilinear pairings in the prime-order group setting.

Cheon [35] proved that bilinear pairings can be simultaneously projecting and cancelling when the subgroup decision assumption holds in the generic group model.[2]

Contribution: In this paper, our contribution is a two-fold. First, we aim to answer the fundamental question how efficient constructions for projecting bilinear pairing can be. Second, we propose a construction of projecting symmetric bilinear pairings that can achieve the efficiency of our lower bounds and then provide several constructions of cryptosystems based on the proposal in a modular fashion.

We focus on constructions only in the prime-order bilinear group setting since this type of group usually supports more efficient (group and bilinear pairing) operations than those in composite-order bilinear groups (see [15] for a detailed comparison of composite and prime-order groups). We present several impossibility results of the projecting bilinear pairings in a prime-order group setting. More precisely, we specify the lower bound of

1. the image size of a projecting asymmetric bilinear pairing
2. the image size of a projecting symmetric bilinear pairing
3. the computational cost for a projecting asymmetric bilinear pairing, and
4. the computational cost for a projecting symmetric bilinear pairing

in a prime-order group setting naturally induced from the decisional Diffie-Hellman (DDH) assumption, the decisional linear (DLIN) assumption, and the k-linear assumption, where the computational cost means the number of generic operations. In this paper, we restrict ourselves to a consideration of a framework in which the subgroup indistinguishability in the framework relies in a natural way on simple assumptions (i.e., the DDH, DLIN, and k-linear assumption). This framework covers all previous constructions by Groth-Sahai and Freeman, and this restriction on the framework has already been used in [25] to show another impossibility result on projecting bilinear pairings. As for the computational cost of projecting bilinear pairings, we consider a slightly restricted computational model since there are typically several ways to perform a given operation, which makes it very difficult to compare all possible (even unknown) ways. We have two basic assumption in our computational model. First, we only count the number of generic operations of the underlying elliptic curve group and the pairings — that is, we assume that one cannot utilize information about the representation of groups and bilinear pairing operations [37,8]. Second, we assume that two inputs of a projecting bilinear pairing are uniformly and independently chosen. In special cases, an additional information about two inputs may lead to an efficient alternative way of computing a pairing operation. For example, when one computes $e(g_1, g_2)$ for the two given inputs g_1 and g_2, where $e : G \times G \to G_t$ is a pairing, if we knows $e(g, g)$, a_1 and a_2 such that $g_1 = g^{a_1}$ and $g_2 = g^{a_2}$ for a generator g of G, then we can perform one field multiplication and one

---

[2] Seo and Cheon's result does not contradict Meiklejohn et al.'s result. Rather, they showed that there is a more general class of bilinear groups than Meiklejohn et al. considered and that some of theses can be both cancelling and projecting.

exponentiation in G_t instead of performing e for $e(g_1, g_2) = e(g, g)^{a_1 a_2}$. Since we want to consider the computational cost of e in general, that is, without any additional information aside from the original two inputs, we assume that two inputs are uniformly and independently distributed in their respective domains: Hence, our computational model rules out special cases like the above example. Although our computational model does not perfectly correspond to the real world, we believe that its lower computational bounds can aid our understanding of the projecting property and enable us to locate efficient constructions for projecting bilinear pairings.

In this study, our lower bounds imply that Freeman's construction of projecting asymmetric bilinear pairings is optimal: that is, it is the most efficient construction for projecting asymmetric bilinear pairings [15]. In contrast, our lower bounds for the projecting symmetric bilinear pairing are different from those of Groth-Sahai [22]. We fill these gaps by constructing projecting symmetric bilinear pairings and demonstrating that our construction can achieve an efficiency coincident with the lower bounds.

The proposed projecting symmetric bilinear pairings can be used to create more efficient instantiations of cryptosystems, which essentially use projecting property and symmetric bilinear pairings, in a modular fashion. To show that the proposed projecting symmetric bilinear pairings can be adapted to various cryptosystems, we apply them to three distinct cryptosystems and create new efficient instantiations of the Groth-Sahai non-interactive proof system [22], the Boneh-Goh-Nissim cryptosystem [10], and Seo-Cheon round optimal blind signatures [35] that are provably secure under the DLIN assumption.[3] The proposed instantiation of the non-interactive proof system has a faster verification than Groth-Sahai's instantiation based on the DLIN assumption, and the proposed instantiation of the Boneh-Goh-Nissim cryptosystem has a smaller ciphertext size and a faster decryption algorithm than Freeman's instantiation based on the DLIN assumption. We can also reduce the verification costs of the Seo-Cheon round optimal blind signatures. These applications are of independent interest. Our new instantiation is based on the DLIN assumption so that we can improve the efficiency of all subsequent protocols using Groth-Sahai's instantiation 3 (based on the DLIN assumption).

We should note here that symmetric bilinear pairings require the use of supersingular elliptic curves and thus the associated bilinear groups are larger than those with asymmetric bilinear pairings using ordinary curves (please see [16] for a detailed comparison). However, some constructions of pairing-based cryptosystems essentially use the symmetric property of bilinear pairings (e.g., Groth-Ostrovsky-Sahai zero-knowledge proofs [21]). Therefore, the proposed projecting symmetric bilinear pairings can be used for designing such cryptosystems.

---

[3] The Seo-Cheon round optimal blind signature scheme can be considered a prime order group version of the Meiklejohn-Shacham-Freeman round optimal blind signature scheme in composite order groups [25]. Since we only consider prime order group settings in this paper, we provide a new instantiation of the Seo-Cheon scheme instead of the Meiklejohn-Shacham-Freeman scheme.

Modular Approach in Cryptography: Generally speaking, a modular approach for cryptosystems leads to a simple design but inefficient constructions in comparison to an ad hoc approach. Recently, we have found a few exceptions for structure preserving cryptography [1,2,11] and mathematical structures [26,27]. Structure preserving schemes enable one to construct modular protocols while preserving conceptual simplicity and yielding reasonable efficiency at the same time. Structure-preserving signatures, commitments [1], and encryptions [11] restrict all components in schemes to group elements, so schemes can easily be combined with Groth-Sahai proofs [22]. In a modular fashion, round optimal blind signatures, group signatures, and anonymous proxy signatures can be derived from structure preserving signatures, and oblivious trusted third parties can be achieved due to the structure preserving encryptions. There has been some impossibility results for structure preserving cryptography [2,3,4]. These save our efforts in terms of impossible goals and widen our understanding regarding modular constructions.

Okamoto and Takashima [26] introduced a mathematical structure called "dual pairing vector spaces" that can be instantiated using a product of bilinear groups or a Jacobian variety of a supersingular curve of genus ≥ 1. On the basis of these dual pairing vector spaces, a homomorphic encryption scheme [26], functional encryption scheme [27,28,30], attribute-based signature scheme [29], and (hierarchical) identity-based encryption scheme [24] have been proposed.

Open Problem: It would be interesting to extend the (im)possibility of the projecting property into a wider framework than ours. Furthermore, finding other applications of projecting pairings is also interesting.

Road Map: In Section 2, we give definitions for bilinear groups, projecting property, and cryptographic assumptions. In Section 3, we explain our impossibility results of projecting bilinear pairings. In Section 4, we show the optimality of Groth-Sahai and Freeman's projecting asymmetric bilinear pairings and give our construction for optimal projecting symmetric bilinear pairings. In Section 5, we apply the proposed projecting symmetric bilinear pairings to three distinct cryptosystems, the Groth-Sahai non-interactive proof system, the Boneh-Goh-Nissim cryptosystem, and the Seo-Cheon round optimal blind signatures.

2   Definition

We use notation $x \xleftarrow{\$} A$ to mean that, if A is a finite group $\mathbb{G}$, an element x is uniformly chosen from $\mathbb{G}$, and, if A is an algorithm, A outputs x by using its own random coins. We use $[i, j]$ to denote a set of integers $\{i, \ldots, j\}$, $\langle g_1, \ldots, g_n \rangle$ to denote a group generated by $g_1, \ldots, g_n$, and $\mathbb{F}_p$ to denote a finite field of prime order p. For a map $\tau : T_D \to T_R$, and any subset S_D of T_D, $\tau(S_D) := \{\tau(s) | s \in S_D\}$. All values in our paper are outputs of some functions taking the security parameter λ and $\approx$ denotes the difference between both sides is a negligible function in λ.

We use two commonly used mathematical notations *internal direct sum*, denoted by $\oplus$, and *tensor product (Kronecker product)*, denoted by $\otimes$. For an abelian group G, if G_1 and G_2 are subgroups of G such that $G = G_1 + G_2 = \{g_1 \cdot g_2 | g_1 \in G_1, g_2 \in G_2\}$ and $G_1 \cap G_2 = \{1_G\}$ for the identity 1_G of G, then we write $G = G_1 \oplus G_2$. If $A = (a_{i,j})$ is a $m_1 \times m_2$ matrix and $B = (b_{i,j})$ is an $\ell_1 \times \ell_2$ matrix, the *tensor product* $A \otimes B$ is the $m_1 \ell_1 \times m_2 \ell_2$ matrix whose (i,j)-th block is $a_{i,j}B$, where we consider $A \otimes B$ as $m_1 \times m_2$ blocks. That is,

$$ A \otimes B = \begin{bmatrix} a_{1,1}B & \cdots & a_{1,m_2}B \\ \vdots & \ddots & \vdots \\ a_{m_1,1}B & \cdots & a_{m_1,m_2}B \end{bmatrix} \in Mat_{m_1 \ell_1 \times m_2 \ell_2}(\mathbb{F}_p). $$

We use several properties of the internal direct sum and tensor product. Every element g in G has a unique representation if $G = G_1 \oplus G_2$. That is, $g \in G$ can be uniquely written as $g = g_1 g_2$ for some $g_1 \in G_1$ and $g_2 \in G_2$. If two matrices A and B are invertible, then $A \otimes B$ is also invertible and the inverse is given by $(A \otimes B)^{-1} = A^{-1} \otimes B^{-1}$. The transposition operation is distributive over the tensor product. That is, $(A \otimes B)^t = A^t \otimes B^t$. We sometimes consider a vector over $\mathbb{F}_p$ as a matrix with one row.

2.1   Bilinear Groups and Projecting Bilinear Pairings

Definition 1. *Let $\mathcal{G}$ be an algorithm that takes as input the security parameter λ. We say that $\mathcal{G}$ is a bilinear group generator if $\mathcal{G}$ outputs a description of five finite abelian groups $(G, G_1, H, H_1,$ and $G_t)$ and a map e such that $G_1 \subset G$, $H_1 \subset H$, and $e : G \times H \to G_t$ is a non-degenerate bilinear pairing; that is, it satisfies*

- *Bilinearity: $e(g_1 g_2, h_1 h_2) = e(g_1, h_1)e(g_1, h_2)e(g_2, h_1)e(g_2, h_2)$ for $g_1, g_2 \in G$ and $h_1, h_2 \in H$,*
- *Non-degeneracy: for $g \in G$, if $e(g, h) = 1 \ \forall h \in H$, then $g = 1$. Similarly, for $h \in H$, if $e(g, h) = 1 \ \forall g \in G$, then $h = 1$.*

In addition, we assume that group operations in each group $(G, H,$ and $G_t)$, bilinear pairing computations, random samplings from each group, and membership-check in each group are efficiently computable (i.e., polynomial time in λ).

If the order of output groups of $\mathcal{G}$ is prime p, we call $\mathcal{G}$ a bilinear group generator of prime order and say $\mathcal{G}_1 \xrightarrow{\$} (p, \mathbb{G}, \mathbb{H}, \mathbb{G}_t, \hat{e})$; that is, $\mathbb{G}, \mathbb{H}$ and $\mathbb{G}_t$ are finite abelian groups of prime order p.

If $G = H$, $G_1 = H_1$, and $e(g, h) = e(h, g)$ for all $g, h \in G$, we say that $\mathcal{G}$ is symmetric. Otherwise, we say that $\mathcal{G}$ is asymmetric.

We define the projecting property of a bilinear pairings.

Definition 2. *Let $\mathcal{G}$ be a bilinear group generator, and $\mathcal{G} \xrightarrow{\$} (G, G_1, H, H_1, G_t, e)$. We say that $\mathcal{G}$ is projecting if there exist a subgroup*

$G'_t \subset G_t$ and three homomorphisms $\pi : G \to G$, $\bar{\pi} : H \to H$, and $\pi_t : G_t \to G_t$ such that

1. $\pi(G) \neq \{1_G\}$, $\bar{\pi}(H) \neq \{1_H\}$, and $\pi_t(e(G, H)) \neq \{1_t\}$, where 1_G, 1_H, and 1_t are identities of G, H, G_t, respectively.
2. $G_1 \subset \ker(\pi)$, $H_1 \subset \ker(\bar{\pi})$, and $G'_t \subset \ker(\pi_t)$.
3. $\pi_t(e(g, h)) = e(\pi(g), \bar{\pi}(h))$ for all $g \in G$ and $h \in H$.

If $\mathcal{G}$ is symmetric, set $\pi = \bar{\pi}$.

Note that in Definition 2 we slightly revised Freeman's original projecting definition to fit our purpose. First, we added a requirement for homomorphisms to be non-trivial (first condition of Definition 2). If we allowed trivial homomorphisms, they would satisfy the projecting property. Since trivial homomorphisms may not be helpful in designing cryptographic protocols, our modification is quite reasonable. Second, our definition requires only the existence of G'_t and homomorphisms while Freeman required them to be output [15]. Since our definition is weaker than Freeman's (if we ignore our first modification), our main results (the lower bounds and optimal construction) are meaningful. Several other researchers [25,24] have used an existence definition like ours instead of Freeman's definition for the projecting property.

2.2 Subgroup Decision Assumption and k-Linear Assumption

Here we define *subgroup decision problem* and *subgroup decision assumption* in the bilinear group setting, which were introduced by Freeman [15].

Definition 3. *Let $\mathcal{G}$ be a bilinear group generator. We define the advantage of an algorithm $\mathcal{A}$ in solving the* subgroup decision problem on the left, *denoted by $Adv_{\mathcal{A},\mathcal{G}}^{SDP_L}(\lambda)$, as*

$$\left| \Pr\left[\mathcal{A}(G, G_1, H, H_1, G_t, e, g) \to 1 \,\middle|\, (G, G_1, H, H_1, G_t, e) \xleftarrow{\$} \mathcal{G}(\lambda), g \xleftarrow{\$} G \right] \right.$$
$$\left. - \Pr\left[\mathcal{A}(G, G_1, H, H_1, G_t, e, g_1) \to 1 \,\middle|\, (G, G_1, H, H_1, G_t, e) \xleftarrow{\$} \mathcal{G}(\lambda), g_1 \xleftarrow{\$} G_1 \right] \right|.$$

We say that $\mathcal{G}$ satisfies the subgroup decision assumption on the left *if, for any PPT algorithm $\mathcal{A}$, its $Adv_{\mathcal{A},\mathcal{G}}^{SDP_L}(\lambda)$ is a negligible function of the security parameter λ.*

We analogously define the *subgroup decision problem on the right*, the advantage $Adv_{\mathcal{A},\mathcal{G}}^{SDP_R}$ of $\mathcal{A}$, and the *subgroup decision assumption on the right* by using H and H_1 instead of G and G_1.

Definition 4. *We say that a bilinear group generator $\mathcal{G}$ satisfies the* subgroup decision assumption *if $\mathcal{G}$ satisfies both the subgroup decision assumptions on the left and subgroup decision assumptions on the right.*

For a subgroup decision assumption in the prime-order group setting, we use the widely-known k-linear assumption which is introduced by Hofheinz and Kiltz and Shacham [23,36], in the bilinear group setting. We give the formal definition of k-linear assumption below.

Definition 5. *Let $\mathcal{G}_1$ be a bilinear group generator of prime order and $k \geq 1$. We define the advantage of an algorithm $\mathcal{A}$ in solving the k-linear problem in $\mathbb{G}$, denoted by $Adv_{\mathcal{A},\mathcal{G}_1}^{k\text{-}Lin_\mathbb{G}}(\lambda)$, to be*

$$\left| \Pr\left[\mathcal{A}(\mathbb{G}, \mathbb{H}, \mathbb{G}_t, e, \mathfrak{g}, \mathfrak{u}_i, \mathfrak{u}_i^{a_i}, \mathfrak{g}^b, \mathfrak{h} \text{ for } i \in [1, k]) \to 1 \right| \right.$$
$$(\mathbb{G}, \mathbb{H}, \mathbb{G}_t, e) \xleftarrow{\$} \mathcal{G}_1(\lambda), \mathfrak{g}, \mathfrak{u}_i \xleftarrow{\$} \mathbb{G}, \mathfrak{h} \xleftarrow{\$} \mathbb{H}, a_i \xleftarrow{\$} \mathbb{F}_p \text{ for } i \in [1, k], b \xleftarrow{\$} \mathbb{F}_p \Big]$$
$$- \Pr\left[\mathcal{A}(\mathbb{G}, \mathbb{H}, \mathbb{G}_t, e, \mathfrak{g}, \mathfrak{u}_i, \mathfrak{u}_i^{a_i}, \mathfrak{g}^b, \mathfrak{h} \text{ for } i \in [1, k]) \to 1 \right|$$
$$\left. (\mathbb{G}, \mathbb{H}, \mathbb{G}_t, e) \xleftarrow{\$} \mathcal{G}_1(\lambda), \mathfrak{g}, \mathfrak{u}_i \xleftarrow{\$} \mathbb{G}, \mathfrak{h} \xleftarrow{\$} \mathbb{H}, a_i \xleftarrow{\$} \mathbb{F}_p \text{ for } i \in [1, k], b = \sum_{i \in [1, k]} a_i \Big] \right|.$$

Then, we say that $\mathcal{G}_1$ satisfies the k-linear assumption in $\mathbb{G}$ if for any PPT algorithm $\mathcal{A}$, $Adv_{\mathcal{A},\mathcal{G}_1}^{k\text{-}Lin_\mathbb{G}}(\lambda)$ is a negligible function of the security parameter.

We can analogously define the k-linear assumption in $\mathbb{H}$. The 1-linear assumption in $\mathbb{G}$ is the DDH assumption in $\mathbb{G}$ and the 2-linear assumption in $\mathbb{G}$ is the decisional linear assumption in $\mathbb{G}$ [9].

3 Impossibility Results of Projecting Bilinear Pairings

In this section, we first formally define natural product groups of prime-order bilinear groups. Next, we derive conditions for projecting bilinear groups, and then provide our impossibility results of projecting bilinear pairings. We begin by defining some notations that will help us to simplify explanations. For group elements $\mathfrak{g}, \mathfrak{g}_1, \ldots, \mathfrak{g}_{k+1} \in \mathbb{G}$, a vector $\overrightarrow{\alpha} = (a_1, \ldots, a_{k+1}) \in \mathbb{F}_p^{k+1}$, and a matrix $M = (m_{i,j}) \in Mat_{(k+1) \times (k+1)}(\mathbb{F}_p)$, we use the notation

$$\mathfrak{g}^{\overrightarrow{\alpha}} := (\mathfrak{g}^{a_1}, \ldots, \mathfrak{g}^{a_{k+1}}) \in \mathbb{G}^{k+1}$$

and

$$(\mathfrak{g}_1, \ldots, \mathfrak{g}_{k+1})^M := \left(\prod_{i \in [1, k+1]} \mathfrak{g}_i^{m_{i,1}}, \ldots, \prod_{i \in [1, k+1]} \mathfrak{g}_i^{m_{i,k+1}} \right).$$

From this notation, we can easily obtain $(\mathfrak{g}^{\overrightarrow{\alpha}})^M = \mathfrak{g}^{(\overrightarrow{\alpha} M)}$.

3.1 Bilinear Groups Naturally Induced from k-linear Assumption

In Figure 1, we provide a generator $\mathcal{G}_k^{\{A_\ell\}_{\ell \in [1,m]}}$ for $A_\ell \in Mat_{(k+1) \times (k+1)}(\mathbb{F}_p)$ and $\ell \in [1, m]$. When we refer to the natural construction of product groups of prime-order bilinear groups such that the subgroup decision assumption "naturally" follows from the k-linear assumption, we mean $\mathcal{G}_k^{\{A_\ell\}_{\ell \in [1,m]}}$.[4] When we

---

[4] Meiklejohn et al. [25] also used the word "natural" to refer to $\mathcal{G}_k^{\{A_\ell\}_{\ell \in [1,m]}}$. They used $\mathcal{G}_k^{\{A_\ell\}_{\ell \in [1,m]}}$ to show the limitation result of both projecting and cancelling: They showed that for any A_ℓ matrices used in $\mathcal{G}_k^{\{A_\ell\}_{\ell \in [1,m]}}$, $\mathcal{G}_k^{\{A_\ell\}_{\ell \in [1,m]}}$ cannot be both projecting and cancelling with overwhelming probability, where the probability goes over the randomness used in $\mathcal{G}_k^{\{A_\ell\}_{\ell \in [1,m]}}$.

1. $\mathcal{G}_k^{\{A_\ell\}_{\ell \in [1,m]}}$ takes the security parameter λ as input.
2. Run $\mathcal{G}_1(\lambda) \rightarrow (p, \mathbb{G}, \mathbb{H}, \mathbb{G}_t, \hat{e})$.
3. Define $G = \mathbb{G}^{k+1}$, $H = \mathbb{H}^{k+1}$, and $G_t = \mathbb{G}_t^m$.
4. Randomly choose $\overrightarrow{x}_1, \ldots, \overrightarrow{x}_k, \overrightarrow{y}_1, \ldots, \overrightarrow{y}_k \in \mathbb{F}_p^{k+1}$ such that the set $\{\overrightarrow{x}_i\}_{i \in [1,k]}$ and $\{\overrightarrow{y}_i\}_{i \in [1,k]}$ are each linearly independent.
5. Randomly choose generators $\mathfrak{g} \in \mathbb{G}$ and $\mathfrak{h} \in \mathbb{H}$, and let $G_1 = \langle \mathfrak{g}^{\overrightarrow{x}_1}, \ldots, \mathfrak{g}^{\overrightarrow{x}_k} \rangle$ and $H_1 = \langle \mathfrak{h}^{\overrightarrow{y}_1}, \ldots, \mathfrak{h}^{\overrightarrow{y}_k} \rangle$.
6. Define a map $e : G \times H \rightarrow G_t$ as an m-tuple of maps $e(\cdot, \cdot)_\ell$ for $\ell \in [1, m]$ as follows:

$$e((\mathfrak{g}_1, \ldots, \mathfrak{g}_{k+1}), (\mathfrak{h}_1, \ldots, \mathfrak{h}_{k+1}))_\ell := \prod_{i,j \in [1,k+1]} \hat{e}(\mathfrak{g}_i, \mathfrak{h}_j)^{a_{i,j}^{(\ell)}},$$

where $A_\ell = (a_{ij}^{(\ell)}) \in Mat_{(k+1) \times (k+1)}(\mathbb{F}_p)$ for $\ell \in [1, m]$.
7. Output description of $(p, G, G_1, H, H_1, G_t, e)$; each group description has its generators only. (e.g., G_1's description has $\mathfrak{g}^{\overrightarrow{x}_1}, \ldots, \mathfrak{g}^{\overrightarrow{x}_k}$, but $\overrightarrow{x}_i$ is not contained in the description of G_1.)

Fig. 1. Description of $\mathcal{G}_k^{\{A_\ell\}_{\ell \in [1,m]}}$

consider the subgroup decision assumption, which is induced from the k-linear assumption, to mean that, given g, it is hard to determine if $g \xleftarrow{\$} G_1$ or $g \xleftarrow{\$} G$, G is a rank-$(k+1)$ $\mathbb{F}_p$-module, and G_1 is a randomly chosen rank-k submodule of G. For any matrices $A_1, \ldots, A_m$ in $Mat_{(k+1) \times (k+1)}(\mathbb{F}_p)$, a group generator $\mathcal{G}_k^{\{A_\ell\}_{\ell \in [1,m]}}$ satisfies the subgroup decision assumption if the underlying prime-order bilinear group generator $\mathcal{G}_1$ satisfies the k-linear assumption.

Theorem 1. *[15, Theorem 2.5] If $\mathcal{G}_1$ satisfies the k-linear assumption in $\mathbb{G}$ and $\mathbb{H}$, $\mathcal{G}_k^{\{A_\ell\}_{\ell \in [1,m]}}$ satisfies the subgroup decision assumption regardless the choice of $\{A_\ell\}_{\ell \in [1,m]}$.*

Note that $\mathcal{G}_k^{\{A_\ell\}_{\ell \in [1,m]}}$ contains Groth-Sahai's constructions based on the DDH assumption ($k = 1$) and the DLIN assumption ($k = 2$).

3.2 Conditions for Symmetric Property

A bilinear pairing e of $\mathcal{G}_k^{\{A_\ell\}_{\ell \in [1,m]}}$ in Figure 1 can be rewritten, using matrix notation, as

$$e(\mathfrak{g}^{\overrightarrow{x}}, \mathfrak{h}^{\overrightarrow{y}})_\ell = \hat{e}(\mathfrak{g}, \mathfrak{h})^{\overrightarrow{x} A_\ell \overrightarrow{y}^t}$$

where $\overrightarrow{x}$ is considered to be a $1 \times (k+1)$ matrix, and $\overrightarrow{y}^t$ is considered to be a $(k+1) \times 1$ matrix.

If $\mathcal{G}_1$ is a symmetric bilinear group generator of prime-order, then one may think that $\mathcal{G}_k^{\{A_\ell\}_{\ell \in [1,m]}}$ is also a symmetric bilinear group generator. However, not all bilinear groups with underlying symmetric bilinear pairings $\hat{e}$ do satisfy symmetric property. The following theorem shows the necessary and sufficient condition of $\{A_\ell\}_{\ell \in [1,m]}$ for $\mathcal{G}_k^{\{A_\ell\}_{\ell \in [1,m]}}$ to be symmetric, that is, $e(g, h) = e(h, g)$ for any group elements g and h.

Theorem 2. $\mathcal{G}_k^{\{A_\ell\}_{\ell \in [1,m]}}$ *is symmetric if and only if* $\mathbb{G} = \mathbb{H}$, $\mathfrak{g} = \mathfrak{h}$, $\overrightarrow{x}_i = \overrightarrow{y}_i$ *for all* $i \in [1, k]$, *and* A_ℓ *is symmetric for all* $\ell \in [1, m]$, *where* $\mathbb{G}, \mathbb{H}, \mathfrak{g}, \mathfrak{h}, \overrightarrow{x}_i$ *and* $\overrightarrow{y}_i$ *are defined in the description of* $\mathcal{G}_k^{\{A_\ell\}_{\ell \in [1,m]}}$.

Because of space constraints, we give the proof of Theorem 2 in the full version of this paper.

3.3 Necessary Condition for Projection Property

Using a tensor product $\otimes$, we can further simplify e computation as follows: Let B be a $(k + 1)^2 \times m$ matrix such that B's $((i - 1)(k + 1) + j, \ell)$ entry is $a_{i,j}^{(\ell)}$, where $A_\ell = (a_{i,j}^{(\ell)})$. Then,

$$e(\mathfrak{g}^{\overrightarrow{x}}, \mathfrak{h}^{\overrightarrow{y}}) = (e(\mathfrak{g}^{\overrightarrow{x}}, \mathfrak{h}^{\overrightarrow{y}})_1, \ldots, e(\mathfrak{g}^{\overrightarrow{x}}, \mathfrak{h}^{\overrightarrow{y}})_m)$$
$$= (\hat{e}(\mathfrak{g}, \mathfrak{h})^{\overrightarrow{x} A_1 \overrightarrow{y}^t}, \ldots, \hat{e}(\mathfrak{g}, \mathfrak{h})^{\overrightarrow{x} A_m \overrightarrow{y}^t}) = \hat{e}(\mathfrak{g}, \mathfrak{h})^{(\overrightarrow{x} \otimes \overrightarrow{y})B}.$$

From now, we use a notation $\mathcal{G}_k^B$ as well as $\mathcal{G}_k^{\{A_\ell\}_{\ell \in [1,m]}}$ to denote a bilinear group generator naturally induced from the k-linear assumption, where B is defined by $\{A_\ell\}_{\ell \in [1,m]}$ as above. This notation is well-defined since there are one-to-one correspondence between B and $\{A_\ell\}_{\ell \in [1,m]}$.

We give a necessary condition of B for $\mathcal{G}_k^B$ to be projecting in Lemma 1. This lemma says that if $G = G_1 \oplus G_2$ and $H = H_1 \oplus H_2$, then $e(G_2, H_2)$ should have at least an element not contained in the subgroup generated by other parts of images.

Lemma 1. 1. *If* $\mathcal{G}_k^B$ *is asymmetric (that is,* $\mathcal{G}_k^B \xrightarrow{\$} (p, G, G_1, H, H_1, G_t, e)$) *and projecting, for decompositions* $G = G_1 \oplus G_2$ *and* $H = H_1 \oplus H_2$ *it satisfies that* $e(G_2, H_2) \not\subset \mathbb{D}$, *where* $\mathbb{D}$ *is the smallest group containing* $e(G_1, H)$ *and* $e(G, H_1)$.

2. *If* $\mathcal{G}_k^B$ *is symmetric (that is,* $\mathcal{G}_k^B \xrightarrow{\$} (p, G, G_1, G_t, e)$) *and projecting, for any decomposition* $G = G_1 \oplus G_2$ *it satisfies that* $e(G_2, G_2) \not\subset \mathbb{D}$, *where* $\mathbb{D}$ *is the smallest group containing* $e(G, G_1)$.

Proof. (1) Suppose that $\mathcal{G}_k^B$ is projecting. Then, there exist three homomorphisms π, $\bar{\pi}$, and π_t. Since π and $\bar{\pi}$ are non-trivial homomorphisms, G_1 and H_1 are proper subgroups of G and H, respectively. Since G_1 and H_1 are proper subgroups, for any decompositions $G = G_1 \oplus G_2$ and $H = H_1 \oplus H_2$, $\{1_G\} \neq G_2 \subset G$ and $\{1_H\} \neq H_2 \subset H$. We show that $G_1, G_2, H_1,$ and H_2 satisfy the condition in the theorem. By definition of $\mathbb{D}$, $\mathbb{D}$ is a group generated by all elements in $e(G_1, H)$ and $e(G, H_1)$ so that every element in $\mathbb{D}$ can be written as a product of elements in $e(G_1, H)$ and $e(G, H_1)$ (though it is not uniquely written). For any $g_1 \in G_1$, $h_1 \in H_1$, $g \in G$, and $h \in H$, $\pi_t(e(g_1, h)e(g, h_1))$ is equal to 1_t since

$$\pi_t(e(g_1, h))\pi_t(e(g, h_1)) = e(\pi(g_1), \bar{\pi}(h))e(\pi(g), \bar{\pi}(h_1)) = e(1_G, \bar{\pi}(h))e(\pi(g), 1_H).$$

We can see that by homomorphic property of π_t, $\pi_t(\mathbb{D}) = 1_t$. If $e(G_2, H_2) \subset \mathbb{D}$, then $e(G, H) \subset \mathbb{D} \subset \ker(\pi_t)$. That is a contradiction of π_t's non-trivial condition.

(2) We can prove similarly as (1). Essential proof idea is same to (1). Thus, we omit it. $\square$

For our impossibility results regarding the image size and computational cost, we will focus on the $(k+1)^2 \times m$ matrix B of $\mathcal{G}_k^B$. All non-zero entries in B imply $\hat{e}$-computations (bilinear pairing $\hat{e}$ of underlying bilinear group generator $\mathcal{G}_1$) and the lower bound of m implies the lower bound of the image size of bilinear pairings. We compute the lower bound of the rank of B of $\mathcal{G}_k^B$, where $\mathcal{G}_k^B$ is asymmetric and projecting, by using the necessary condition of projecting property in Lemma 1. For projecting symmetric bilinear pairings, the overall strategy is similar to those of projecting asymmetric bilinear pairings except that symmetric bilinear pairings have the special form of B as mentioned in Theorem 2. We give the formal statement below.

Lemma 2. *The following statements about $\mathcal{G}_k^B$ are true with overwhelming probability, where the probability goes over the randomness used in the $\mathcal{G}_k^B$.*

1. *If $\mathcal{G}_k^B$ is asymmetric and projecting, then B has $(k+1)^2$ linearly independent rows.*
2. *If $\mathcal{G}_k^B$ is symmetric and projecting, then B has $\frac{(k+1)(k+2)}{2}$ linearly independent rows.*

Proof. (1) Let $\mathcal{G}_k^B$ be a projecting asymmetric bilinear group generator. Let (G, G_1, H, H_1, G_t, e) be the output of $\mathcal{G}_k^B$ and G and H be decomposed by $G = G_1 \oplus G_2$ and $H = H_1 \oplus H_2$, respectively for some subgroups G_2 and H_2. Then, $G_1 = \langle \mathfrak{g}^{\vec{x}_1}, \ldots, \mathfrak{g}^{\vec{x}_k} \rangle$, $H_1 = \langle \mathfrak{h}^{\vec{y}_1}, \ldots, \mathfrak{h}^{\vec{y}_k} \rangle$, $G_2 = \langle \mathfrak{g}^{\vec{x}_{k+1}} \rangle$, and $H_2 = \langle \mathfrak{h}^{\vec{y}_{k+1}} \rangle$ for some sets of linearly independent vectors $\{\vec{x}_i\}_{i \in [1, k+1]}$ and $\{\vec{y}_i\}_{i \in [1, k+1]}$. Let X be a $(k+1) \times (k+1)$ matrix over $\mathbb{F}_p$ with $\vec{x}_i$ as its i-th row, and Y be a $(k+1) \times (k+1)$ matrix over $\mathbb{F}_p$ with $\vec{y}_i$ as its i-th row. Note that X and Y are invertible. Since B is a $(k+1)^2 \times m$ matrix for some m, B can have at most $(k+1)^2$ linear independent rows.

Suppose that B has less than $(k+1)^2$ linearly independent rows. We observe that

$$e(G_2, H_2) = \langle e(\mathfrak{g}^{\vec{x}_{k+1}}, \mathfrak{h}^{\vec{y}_{k+1}}) \rangle = \langle \hat{e}(\mathfrak{g}, \mathfrak{h})^{(\vec{x}_{k+1} \otimes \vec{y}_{k+1})B} \rangle = \langle \hat{e}(\mathfrak{g}, \mathfrak{h})^{\vec{e}_{(k+1)^2}(X \otimes Y)B} \rangle,$$

and similarly

$$\mathbb{D} = \langle \hat{e}(\mathfrak{g}, \mathfrak{h})^{\vec{e}_1(X \otimes Y)B}, \ldots, \hat{e}(\mathfrak{g}, \mathfrak{h})^{\vec{e}_{(k+1)^2 - 1}(X \otimes Y)B} \rangle,$$

where $\vec{e}_i$ is the i-th canonical vector of $\mathbb{F}_p^{(k+1)^2}$. Now, we show that there exists a non-zero vector $\vec{c} \in \mathbb{F}_p^{(k+1)^2}$ with a non-zero in the $(k+1)^2$-th entry such that

$\overrightarrow{c} \cdot (X \otimes Y)B = \overrightarrow{0} \in \mathbb{F}_p^m$. The existence of such a vector $\overrightarrow{c}$ implies that the $(k+1)^2$-th row of $(X \otimes Y)B$ can be represented by the linear combination of upper rows of $(X \otimes Y)B$ so that $e(G_2, H_2) \subset \mathbb{D}$. Then, it would be a contradiction with Lemma 1.

By hypothesis $(rank(B) < (k+1)^2)$, there exists a non-zero vector $\overrightarrow{r} \in \mathbb{F}_p^{(k+1)^2}$ such that $\overrightarrow{r}B = \overrightarrow{0} \in \mathbb{F}_p^m$. For such an $\overrightarrow{r}$, we show that $\overrightarrow{r}(X^{-1} \otimes Y^{-1})$ satisfies conditions for it to be $\overrightarrow{c}$ aforementioned. First, we obtain $\overrightarrow{r}(X^{-1} \otimes Y^{-1}) \cdot (X \otimes Y)B = \overrightarrow{r}B = \overrightarrow{0}$. Next, we argue that $\overrightarrow{r}(X^{-1} \otimes Y^{-1})$'s $(k+1)^2$-th entry is non-zero with overwhelming probability, where the probability goes over the randomness used in $\mathcal{G}_k^B$ (to choose $\overrightarrow{x}_1, \ldots, \overrightarrow{x}_k, \overrightarrow{y}_1, \ldots, \overrightarrow{y}_k$). We consider the $(k+1)$-th column vector $\hat{x}^t$ of X^{-1} such that $\hat{x}$ is orthogonal to all upper k rows of X. Denote the orthogonal complement of $\langle \overrightarrow{x}_1, \ldots, \overrightarrow{x}_k \rangle$ by $\langle \overrightarrow{w} \rangle$. Then, $\hat{x}^t$ is a non-zero vector in $\langle \overrightarrow{w} \rangle$. By definition of $\mathcal{G}_k^B$, $\overrightarrow{x}_1, \ldots, \overrightarrow{x}_k$ are randomly chosen so that $\overrightarrow{w}$ is also uniformly distributed in $\mathbb{F}_p^{k+1}$. Similarly, the $(k+1)$-th column vector $\hat{y}^t$ of Y^{-1} is a non-zero vector in $\langle \overrightarrow{y}_1, \ldots, \overrightarrow{y}_k \rangle^\perp := \langle \overrightarrow{z} \rangle$, and $\overrightarrow{z}$ is uniformly distributed in $\mathbb{F}_p^{k+1}$. The $(k+1)^2$-th entry of $\overrightarrow{r}(X^{-1} \otimes Y^{-1})$ is $\overrightarrow{r}(\hat{x}^t \otimes \hat{y}^t)$, and it is a non-zero constant multiple of $\overrightarrow{r}(\overrightarrow{w} \otimes \overrightarrow{z})^t$. By the first statement of Lemma 3, which is given below, $\overrightarrow{r}(\overrightarrow{w} \otimes \overrightarrow{z})^t$ is non-zero with overwhelming probability. Therefore, we complete the proof of the first statement of theorem.

(2) We can prove the second statement of theorem by using the second statements of Lemma 1 and Lemma 3. The overall strategy is same to the proof of the first statement of theorem. The key observation of the proof of the second statement is that B has a special form due to Theorem 2. We leave the detail of the proof of the second statement in the full version. $\qquad\square$

Lemma 3. *Let V be a subspace of $\mathbb{F}_p^{(k+1)^2}$ generated by $\{\overrightarrow{d}_{i,j}\}_{1 \le i \le j \le k+1}$, where $\overrightarrow{d}_{i,j}$ is a vector with 1 in the $(i-1)(k+1)+j$-th entry, -1 in the $(j-1)(k+1)+i$-th entry, and zeros elsewhere.*

1. *For any non-zero vector $\overrightarrow{r} \in \mathbb{F}_p^{(k+1)^2}$, $\Pr[\overrightarrow{r} \cdot (\overrightarrow{w} \otimes \overrightarrow{z})^t = 0] \le \frac{2}{p}$, where the probability goes over the choice of vectors $\overrightarrow{w}, \overrightarrow{z} \in \mathbb{F}_p^{k+1}$.*

2. *For any vector $\overrightarrow{r} \in \mathbb{F}_p^{(k+1)^2} \setminus V$, $\Pr[\overrightarrow{r} \cdot (\overrightarrow{w} \otimes \overrightarrow{w})^t = 0] \le \frac{2}{p}$, where the probability goes over the choice of a vector $\overrightarrow{w} \in \mathbb{F}_p^{k+1}$.*

We can prove Lemma 3 by using the Schwartz-Zippel lemma [33] and leave a detailed proof in the full version.

3.4 Impossibility of Projecting Property

Basing on Lemma 2, we derive our main theorem on the impossibility results of projecting bilinear pairings. We begin with explaining our computational model for the lower bounds of computational cost of projecting bilinear pairings. In our computational model, we assume two things: First, one who computes projecting bilinear pairings e can not utilize the representation of the underlying bilinear pairing $\hat{e}$ and groups $\mathbb{G}, \mathbb{H}$, and $\mathbb{G}_t$ over which $\hat{e}$ is defined. Note that we rule out techniques for multi-pairings [34,17] in our computational model. This assumption is same to that of the generic group model [37], in particular, generic bilinear group [8]. In [37,8], the generic (bilinear) group model is used to show the computational lower bounds of attacker solving number theoretic problems such as the discrete logarithm problem and q-strong Diffie-Hellman problem. Second, two inputs are uniformly and independently chosen so that any relations with two inputs are unknown. In special cases such that a relation with two inputs are known, there are several alternative way to compute bilinear pairings. For example, one knowing g_1, h_1, $e(g, h)$, and a relation $g_1 = g^2$ and $h_1 = h^3$ can compute $e(g_1, h_1)$ by performing $e(g, h)^6$ instead of performing a bilinear pairing. Since we want to consider the computational cost of e without using any additional information of two inputs, we assume that two inputs are uniformly and independently distributed in their respective domains. We provide our main theorem below.

Theorem 3. *(Lower Bounds) The following statements about $\mathcal{G}_k^B$ are true with overwhelming probability, where the probability goes over the randomness used in the $\mathcal{G}_k^B$.*

1. *The image size of a projecting asymmetric bilinear pairing is at least $(k+1)^2$ elements in $\mathbb{G}_t$.*
2. *The image size of a projecting symmetric bilinear pairing is at least $\frac{(k+1)(k+2)}{2}$ elements in $\mathbb{G}_t$.*
3. *Any construction for a projecting (asymmetric or symmetric) bilinear pairing should perform at least $(k + 1)^2$ computations of $\hat{e}$ in our computational model.*

Proof. (1) Suppose that $\mathcal{G}_k^B$ is asymmetric and projecting. Since a $(k + 1)^2 \times m$ matrix B has at least $(k + 1)^2$ linearly independent rows by Lemma 2, $m \geq (k + 1)^2$. This implies that $G_t = \mathbb{G}_t^m$ consists of m ($\geq (k + 1)^2$) elements in $\mathbb{G}_t$.

(2) If $\mathcal{G}_k^B$ is symmetric and projecting, then $(k+1)^2 \times m$ matrix B has at least $\frac{(k+1)(k+2)}{2}$ linear independent rows by Lemma 2. Thus, $m \geq \frac{(k+1)(k+2)}{2}$; hence, an element in $G_t = \mathbb{G}_t^m$ is m ($\geq \frac{(k+1)(k+2)}{2}$) elements in $\mathbb{G}_t$.

(3) First, we show that for two inputs $g = (g_1, \ldots, g_{k+1}) \in G$ and $h = (h_1, \ldots, h_{k+1}) \in H$, projecting (asymmetric or symmetric) pairings require computing all $\hat{e}(g_i, h_j)$ for all $i, j \in [1, k+1]$. To this end, it is sufficient to show that every row in the matrix B is non-zero. (Recall that $e(g^{\vec{w}}, h^{\vec{z}}) = \hat{e}(g, h)^{(\vec{w} \otimes \vec{z})B}$ and if every row in B is non-zero, then $\hat{e}(g^{w_i}, h^{z_j})$ should be computed at least one time.) If a group generator $\mathcal{G}_k^B$ is projecting and asymmetric, then the rank of B is $(k+1)^2$ by Lemma 1. Since B has $(k+1)^2$ rows, there is no zero rows. If a group generator $\mathcal{G}_k^B$ is projecting and symmetric, then the rank of B is $\frac{(k+1)(k+2)}{2}$ by Lemma 1. We know that the matrix B of symmetric bilinear group generators has the special form by Theorem 2. From Theorem 2, some $\frac{k(k+1)}{2}$ rows in B have respective same rows in B. Since B has $(k+1)^2$ rows and $(k+1)^2 - \frac{k(k+1)}{2}$ is equal to the rank of B, every row in B has at least one non-zero entry.

Next, we show that computing $\hat{e}(\mathfrak{g}_i, \mathfrak{h}_j)$ cannot be generally substitute by a product of other $\hat{e}(\mathfrak{g}_{i'}, \mathfrak{h}_{j'})$ for $i' \in [1, k+1] \setminus \{i\}$ and $j' \in [1, k+1] \setminus \{j\}$ in our computational model. To this end, it is sufficient to show that for any non-zero vector $\vec{r} = (r_1, \ldots, r_{(k+1)^2}) \in \mathbb{F}_p^{(k+1)^2}$,

$$\Pr_{g \xleftarrow{\$} G, h \xleftarrow{\$} H} \left[\prod_{i,j \in [1,k+1]} \hat{e}(\mathfrak{g}_i, \mathfrak{h}_j)^{r_{(i-1)(k+1)+j}} = 1_{\mathbb{G}_t} \right] \approx 0.$$

For two random inputs $\mathfrak{g}^{\vec{w}}$ and $\mathfrak{h}^{\vec{z}}$,

$$\prod_{i,j \in [1,k+1]} \hat{e}(\mathfrak{g}^{w_i}, \mathfrak{h}^{z_j})^{r_{(i-1)(k+1)+j}} = \hat{e}(\mathfrak{g}, \mathfrak{h})^{(\vec{w} \otimes \vec{z})\vec{r}^t},$$

where $\vec{w} = (w_1, \ldots, w_{k+1}) \in \mathbb{F}_p^{k+1}$ and $\vec{z} = (z_1, \ldots, z_{k+1}) \in \mathbb{F}_p^{k+1}$. Since $\vec{r}^t$ is a non-zero vector in $\mathbb{F}_p^{(k+1)^2}$, $(\vec{w} \otimes \vec{z})\vec{r}^t \neq 0$ with overwhelming probability by Lemma 3, and hence we obtain the desired result such that

$$\prod_{i,j \in [1,k+1]} \hat{e}(\mathfrak{g}^{w_i}, \mathfrak{h}^{z_j})^{r_{(i-1)(k+1)+j}} \neq 1_{\mathbb{G}_t}$$

with overwhelming probability.

Therefore, all projecting bilinear pairings require at least $(k+1)^2$ $\hat{e}$-computations. $\square$

4   Optimal Projecting Bilinear Pairings

In this section, we show that our lower bounds are tight; for projecting asymmetric bilinear pairing, we show that Groth-Sahai and Freeman's constructions are optimal (in our computational model), and for projecting symmetric bilinear pairing, we propose a new construction achieving optimal efficiency (in our computational model).

Definition 6. *Let $\mathcal{G}_k^B$ be a projecting asymmetric (symmetric, resp.) bilinear group generator. If the bilinear pairing e consists of $(k+1)^2$ $\hat{e}$-computation in our computational model and $G_t = \mathbb{G}_t^{(k+1)^2}$ ($G_t = \mathbb{G}_t^{\frac{(k+1)(k+2)}{2}}$, resp.), we say that $\mathcal{G}_k^B$ is optimal.*

We can define $\mathcal{G}_k^B$ by defining a $(k+1)^2 \times m$ matrix B, or equivalently a set of $(k+1) \times (k+1)$ matrices $\{A_\ell\}_{\ell \in [1,m]}$. For a projecting asymmetric bilinear group generator, we define B as $I_{(k+1)^2}$, where $I_{(k+1)^2}$ is the identity matrix in $GL_{(k+1)^2}(\mathbb{F}_p)$. Note that $\mathcal{G}_k^{I_{(k+1)^2}}$ is exactly equal to Freeman's projecting asymmetric bilinear group generator [15] (We can easily check that $\mathcal{G}_k^{I_{(k+1)^2}}$ does not satisfy the symmetric property due to Theorem 2). Theorem 3 implies that $\mathcal{G}_k^{I_{(k+1)^2}}$ is optimal. Therefore, we obtain the following theorem.

Theorem 4. $\mathcal{G}_k^{I_{(k+1)^2}}$ *is an optimal projecting asymmetric bilinear group generator.*

$\mathcal{G}_k^{I_{(k+1)^2}}$ covers one of the most interesting cases $k = 1$: $\mathcal{G}_1^{I_4}$ is optimal.[5]

4.1 Optimal Projecting Symmetric Bilinear Pairings

We propose an optimal projecting symmetric bilinear group generator $\mathcal{G}_k^B$ by defining B (equivalently $A_1, \ldots, A_m$). Let a set S be $\{(i, j) \in [1, k+1] \times [1, k+1] | 1 \leq j \leq i \leq k+1\}$. We consider a map $\tau : S \to [1, \frac{(k+1)(k+2)}{2}]$ defined by $(i, j) \mapsto \frac{i(i-1)}{2} + j$.

Lemma 4. τ *is a bijective map.*

We give the proof of Lemma 4 in the full version.

Description of A_ℓ (equivalently B) for optimal projecting symmetric bilinear pairings: Let $\tau^{-1}(\ell) = (i, j)$. For each $\ell \in [1, \frac{(k+1)(k+2)}{2}]$, $A_\ell = (a_{s,t}^{(\ell)})$ is defined as a $(k+1) \times (k+1)$ matrix with

$$\begin{cases} 1 \text{ in the entry } (i, j) \text{ and zeros elsewhere} & \text{if } i = j, \\ 1 \text{ in the entries } (i, j) \text{ and } (j, i), \text{ and zeros elsewhere} & \text{otherwise .} \end{cases}$$

We give an example to easily explain the proposal.

Example 1. For $k = 2$, define

$$A_1 = \begin{pmatrix} 1 & 0 & 0 \\ 0 & 0 & 0 \\ 0 & 0 & 0 \end{pmatrix}, A_2 = \begin{pmatrix} 0 & 1 & 0 \\ 1 & 0 & 0 \\ 0 & 0 & 0 \end{pmatrix}, A_3 = \begin{pmatrix} 0 & 0 & 0 \\ 0 & 1 & 0 \\ 0 & 0 & 0 \end{pmatrix},$$

$$A_4 = \begin{pmatrix} 0 & 0 & 1 \\ 0 & 0 & 0 \\ 1 & 0 & 0 \end{pmatrix}, A_5 = \begin{pmatrix} 0 & 0 & 0 \\ 0 & 0 & 1 \\ 0 & 1 & 0 \end{pmatrix}, A_6 = \begin{pmatrix} 0 & 0 & 0 \\ 0 & 0 & 0 \\ 0 & 0 & 1 \end{pmatrix}.$$

$\square$

Define B as a $(k+1)^2 \times \frac{(k+1)(k+2)}{2}$ matrix such that B's $((s-1)n + t, \ell)$ entry is $a_{s,t}^{(\ell)}$ for $s, t \in [1, k+1]$ and $\ell \in [1, \frac{(k+1)(k+2)}{2}]$. (Then, we implicitly define $G_t = G_t^{\frac{(k+1)(k+2)}{2}}$.) By using the matrix B, we can construct a bilinear group generator $\mathcal{G}_k^B$.

Next, we show that a group generator $\mathcal{G}_k^B$, where B is defined as above, is an optimal projecting symmetric bilinear group generator. The following Theorem 5 provides the desired result.

---

[5] Freeman used the notation $\mathcal{G}_P$, which is equivalent to our notation $\mathcal{G}_1^{I_4}$.

Theorem 5. *Let $\mathcal{G}_k^B$ be a bilinear group generator with restrictions such that $\mathbb{G} = \mathbb{H}$, $\mathfrak{g} = \mathfrak{h}$, $\overrightarrow{x}_i = \overrightarrow{y}_i$ for all $i \in [1, k]$, and B is a $(k+1)^2 \times \frac{(k+1)(k+2)}{2}$ matrix defined as above. Then, $\mathcal{G}_k^B$ is an optimal projecting symmetric bilinear group generator with overwhelming probability, where the probability goes over the randomness used in $\mathcal{G}_k^B$.*

We leave the proof of Theorem 5 in the full version.

Our definition of projecting requires only the existence of homomorphisms satisfying some conditions. However, some applications (ex: Boneh-Goh-Nissim cryptosystem [10,15]) require that such homomorphisms are efficiently computable. We provide the way how to construct efficiently computable homomorphisms (precisely, natural projections) satisfying projecting property in the full version.

Example 2. For $k = 2$, we can construct an optimal projecting symmetric bilinear group generator by using the matrices in example 1. We denote such a bilinear group generator by $\mathcal{G}_2^{B^*}$, where B^* is a 9×6 matrix defined by the $A_1, \ldots, A_6$ matrices in example 1.

$$
B^* = \begin{pmatrix}
\boxed{1} & 0 & 0 & 0 & 0 & 0 \\
0 & \boxed{1} & 0 & 0 & 0 & 0 \\
0 & 0 & 0 & \boxed{1} & 0 & 0 \\
0 & \boxed{1} & 0 & 0 & 0 & 0 \\
0 & 0 & \boxed{1} & 0 & 0 & 0 \\
0 & 0 & 0 & 0 & \boxed{1} & 0 \\
0 & 0 & 0 & \boxed{1} & 0 & 0 \\
0 & 0 & 0 & 0 & \boxed{1} & 0 \\
0 & 0 & 0 & 0 & 0 & \boxed{1}
\end{pmatrix} \quad \text{for } \mathcal{G}_2^{B^*}
$$

By Theorem 5, $\mathcal{G}_2^{B^*}$ is optimal projecting symmetric: Since B^* is a 9×6 matrix, the target group G_t is equal to $\mathbb{G}_t^6$. Moreover, B^* has nine 1's in the entries and zeros elsewhere so that bilinear pairing e requires 9 $\hat{e}$-computations (without any exponentiations).

5 Application

On the basis of our optimal projecting symmetric bilinear pairings, we derive new instantiations of three distinct cryptosystems with improved efficiency. In particular, we apply the projecting symmetric bilinear group generator $\mathcal{G}_2^{B^*}$ in the example 2 for the Groth-Sahai non-interactive proof system, the Boneh-Goh-Nissim Cryptosystem, and the Seo-Cheon round optimal Blind signature scheme. Because of space constraints, we leave details in the full version.

Acknowledgements. We gratefully acknowledge the detailed and helpful comments of anonymous reviewers of ASIACRYPT 2012. We also thank Jung Hee Cheon and Daisuke Moriyama for constructive feedback on an early draft of the paper.

References

1. Abe, M., Fuchsbauer, G., Groth, J., Haralambiev, K., Ohkubo, M.: Structure-Preserving Signatures and Commitments to Group Elements. In: Rabin, T. (ed.) CRYPTO 2010. LNCS, vol. 6223, pp. 209–236. Springer, Heidelberg (2010)
2. Abe, M., Groth, J., Haralambiev, K., Ohkubo, M.: Optimal Structure-Preserving Signatures in Asymmetric Bilinear Groups. In: Rogaway, P. (ed.) CRYPTO 2011. LNCS, vol. 6841, pp. 649–666. Springer, Heidelberg (2011)
3. Abe, M., Groth, J., Ohkubo, M.: Separating Short Structure-Preserving Signatures from Non-interactive Assumptions. In: Lee, D.H. (ed.) ASIACRYPT 2011. LNCS, vol. 7073, pp. 628–646. Springer, Heidelberg (2011)
4. Abe, M., Haralambiev, K., Ohkubo, M.: Group to Group Commitments Do Not Shrink. In: Pointcheval, D., Johansson, T. (eds.) EUROCRYPT 2012. LNCS, vol. 7237, pp. 301–317. Springer, Heidelberg (2012)
5. Adida, B., Wikström, D.: How to Shuffle in Public. In: Vadhan, S.P. (ed.) TCC 2007. LNCS, vol. 4392, pp. 555–574. Springer, Heidelberg (2007)
6. Belenkiy, M., Camenisch, J., Chase, M., Kohlweiss, M., Lysyanskaya, A., Shacham, H.: Randomizable Proofs and Delegatable Anonymous Credentials. In: Halevi, S. (ed.) CRYPTO 2009. LNCS, vol. 5677, pp. 108–125. Springer, Heidelberg (2009)
7. Belenkiy, M., Chase, M., Kohlweiss, M., Lysyanskaya, A.: P-signatures and Noninteractive Anonymous Credentials. In: Canetti, R. (ed.) TCC 2008. LNCS, vol. 4948, pp. 356–374. Springer, Heidelberg (2008)
8. Boneh, D., Boyen, X.: Short Signatures Without Random Oracles. In: Cachin, C., Camenisch, J.L. (eds.) EUROCRYPT 2004. LNCS, vol. 3027, pp. 56–73. Springer, Heidelberg (2004)
9. Boneh, D., Boyen, X., Shacham, H.: Short Group Signatures. In: Franklin, M. (ed.) CRYPTO 2004. LNCS, vol. 3152, pp. 41–55. Springer, Heidelberg (2004)
10. Boneh, D., Goh, E.-J., Nissim, K.: Evaluating 2-DNF Formulas on Ciphertexts. In: Kilian, J. (ed.) TCC 2005. LNCS, vol. 3378, pp. 325–341. Springer, Heidelberg (2005)
11. Camenisch, J., Haralambiev, K., Kohlweiss, M., Lapon, J., Naessens, V.: Structure Preserving CCA Secure Encryption and Applications. In: Lee, D.H. (ed.) ASIACRYPT 2011. LNCS, vol. 7073, pp. 89–106. Springer, Heidelberg (2011)
12. Cathalo, J., Libert, B., Yung, M.: Group Encryption: Non-interactive Realization in the Standard Model. In: Matsui, M. (ed.) ASIACRYPT 2009. LNCS, vol. 5912, pp. 179–196. Springer, Heidelberg (2009)
13. Chandran, N., Groth, J., Sahai, A.: Ring Signatures of Sub-linear Size Without Random Oracles. In: Arge, L., Cachin, C., Jurdziński, T., Tarlecki, A. (eds.) ICALP 2007. LNCS, vol. 4596, pp. 423–434. Springer, Heidelberg (2007)
14. Chase, M., Kohlweiss, M., Lysyanskaya, A., Meiklejohn, S.: Malleable Proof Systems and Applications. In: Pointcheval, D., Johansson, T. (eds.) EUROCRYPT 2012. LNCS, vol. 7237, pp. 281–300. Springer, Heidelberg (2012)
15. Freeman, D.M.: Converting Pairing-Based Cryptosystems from Composite-Order Groups to Prime-Order Groups. In: Gilbert, H. (ed.) EUROCRYPT 2010. LNCS, vol. 6110, pp. 44–61. Springer, Heidelberg (2010),
http://eprint.iacr.org/2009/540

16. Galbraith, S.D., Paterson, K.G., Smart, N.P.: Pairings for cryptographers. Discrete Applied Mathematics 156, 3113–3121 (2008)
17. Granger, R., Smart, N.: On computing products of pairings. Cryptology ePrint Archive, Report 2006/172 (2006)
18. Green, M., Hohenberger, S.: Universally Composable Adaptive Oblivious Transfer. In: Pieprzyk, J. (ed.) ASIACRYPT 2008. LNCS, vol. 5350, pp. 179–197. Springer, Heidelberg (2008)
19. Groth, J.: Fully Anonymous Group Signatures Without Random Oracles. In: Kurosawa, K. (ed.) ASIACRYPT 2007. LNCS, vol. 4833, pp. 164–180. Springer, Heidelberg (2007)
20. Groth, J., Lu, S.: A Non-interactive Shuffle with Pairing Based Verifiability. In: Kurosawa, K. (ed.) ASIACRYPT 2007. LNCS, vol. 4833, pp. 51–67. Springer, Heidelberg (2007)
21. Groth, J., Ostrovsky, R., Sahai, A.: Perfect Non-interactive Zero Knowledge for NP. In: Vaudenay, S. (ed.) EUROCRYPT 2006. LNCS, vol. 4004, pp. 339–358. Springer, Heidelberg (2006)
22. Groth, J., Sahai, A.: Efficient Non-interactive Proof Systems for Bilinear Groups. In: Smart, N.P. (ed.) EUROCRYPT 2008. LNCS, vol. 4965, pp. 415–432. Springer, Heidelberg (2008)
23. Hofheinz, D., Kiltz, E.: Secure Hybrid Encryption from Weakened Key Encapsulation. In: Menezes, A. (ed.) CRYPTO 2007. LNCS, vol. 4622, pp. 553–571. Springer, Heidelberg (2007)
24. Lewko, A.: Tools for Simulating Features of Composite Order Bilinear Groups in the Prime Order Setting. In: Pointcheval, D., Johansson, T. (eds.) EUROCRYPT 2012. LNCS, vol. 7237, pp. 318–335. Springer, Heidelberg (2012)
25. Meiklejohn, S., Shacham, H., Freeman, D.M.: Limitations on Transformations from Composite-Order to Prime-Order Groups: The Case of Round-Optimal Blind Signatures. In: Abe, M. (ed.) ASIACRYPT 2010. LNCS, vol. 6477, pp. 519–538. Springer, Heidelberg (2010)
26. Okamoto, T., Takashima, K.: Homomorphic Encryption and Signatures from Vector Decomposition. In: Galbraith, S.D., Paterson, K.G. (eds.) Pairing 2008. LNCS, vol. 5209, pp. 57–74. Springer, Heidelberg (2008)
27. Okamoto, T., Takashima, K.: Hierarchical Predicate Encryption for Inner-Products. In: Matsui, M. (ed.) ASIACRYPT 2009. LNCS, vol. 5912, pp. 214–231. Springer, Heidelberg (2009)
28. Okamoto, T., Takashima, K.: Fully Secure Functional Encryption with General Relations from the Decisional Linear Assumption. In: Rabin, T. (ed.) CRYPTO 2010. LNCS, vol. 6223, pp. 191–208. Springer, Heidelberg (2010)
29. Okamoto, T., Takashima, K.: Efficient Attribute-Based Signatures for Non-monotone Predicates in the Standard Model. In: Catalano, D., Fazio, N., Gennaro, R., Nicolosi, A. (eds.) PKC 2011. LNCS, vol. 6571, pp. 35–52. Springer, Heidelberg (2011)
30. Okamoto, T., Takashima, K.: Adaptively Attribute-Hiding (Hierarchical) Inner Product Encryption. In: Pointcheval, D., Johansson, T. (eds.) EUROCRYPT 2012. LNCS, vol. 7237, pp. 591–608. Springer, Heidelberg (2012)
31. Ostrovsky, R., Skeith, W.: Private searching on streaming data. Journal of Cryptology 20, 397–430 (2007)

32. Sang, Y., Shen, H.: Efficient and secure protocols for privacy-preserving set oper-
    ations. ACM Transactions on Information and Systems Security 13 (2009)
33. Schwartz, J.: Fast probabilistic algorithms for verification of polynomials identities.
    Journal of the ACM 27, 701–717 (1980)
34. Scott, M.: Computing the Tate Pairing. In: Menezes, A. (ed.) CT-RSA 2005. LNCS,
    vol. 3376, pp. 293–304. Springer, Heidelberg (2005)
35. Seo, J.H., Cheon, J.H.: Beyond the Limitation of Prime-Order Bilinear Groups,
    and Round Optimal Blind Signatures. In: Cramer, R. (ed.) TCC 2012. LNCS,
    vol. 7194, pp. 133–150. Springer, Heidelberg (2012)
36. Shacham, H.: A cramer-shoup encryption scheme from the linear assumption
    and from progressively weaker linear variants. Cryptology ePrint Archive, Report
    2007/074 (2007), http://eprint.iacr.org/2007/074
37. Shoup, V.: Lower Bounds for Discrete Logarithms and Related Problems. In: Fumy,
    W. (ed.) EUROCRYPT 1997. LNCS, vol. 1233, pp. 256–266. Springer, Heidelberg
    (1997)

Optimal Reductions of Some Decisional Problems to the Rank Problem

Jorge Luis Villar[*]

Universitat Politècnica de Catalunya, Spain
jvillar@ma4.upc.edu

Abstract. In the last years the use of large matrices and their algebraic properties proved to be useful to instantiate new cryptographic primitives like Lossy Trapdoor Functions and encryption schemes with improved security, like Key Dependent Message resilience. In these constructions the rank of a matrix is assumed to be hard to guess when the matrix is hidden by elementwise exponentiation. This problem, that we call here the Rank Problem, is known to be related to the Decisional Diffie-Hellman problem, but in the known reductions between both problems there appears a loss-factor in the advantage which grows linearly with the rank of the matrix.

In this paper, we give a new and better reduction between the Rank problem and the Decisional Diffie-Hellman problem, such that the reduction loss-factor depends logarithmically in the rank. This new reduction can be applied to a number of cryptographic constructions, improving their efficiency. The main idea in the reduction is to build from a DDH tuple a matrix which rank shifts from r to $2r$, and then apply a hybrid argument to deal with the general case. In particular this technique widens the range of possible values of the ranks that are tightly related to DDH.

On the other hand, the new reduction is optimal as we show the nonexistence of more efficient reductions in a wide class containing all the "natural" ones (i.e., black-box and algebraic). The result is twofold: there is no (natural) way to build a matrix which rank shifts from r to $2r + \alpha$ for $\alpha > 0$, and no hybrid argument can improve the logarithmic loss-factor obtained in the new reduction.

The techniques used in the paper extend naturally to other "algebraic" problems like the Decisional Linear or the Decisional 3-Party Diffie-Hellman problems, also obtaining reductions of logarithmic complexity.

Keywords: Rank Problem, Decisional Diffie-Hellman Problem, Black-Box Reductions, Algebraic Reductions, Decision Linear Problem.

1 Introduction

Motivation. In the last years the use of large matrices and their algebraic properties proved to be useful to instantiate new cryptographic primitives like

---

[*] Partially supported by the Spanish research project MTM2009-07694, and the European Commission through the ICT programme under contract ICT-2007-216676 ECRYPT II.

X. Wang and K. Sako (Eds.): ASIACRYPT 2012, LNCS 7658, pp. 80–97, 2012.

Lossy Trapdoor Functions [7,8,12,13] and encryption schemes with improved security, like Key Dependent Message [2]. In these constructions the rank of a matrix is assumed to be hard to guess when the matrix is hidden by elementwise exponentiation. This problem, that we call here the Rank Problem, is known to be related to the Decisional Diffie-Hellman (DDH) problem, but in the known reductions between both problems there appears a loss-factor in the adversary's advantage which grows linearly with the rank of the matrix. The Rank Problem first appeared in some papers under the names Matrix-DDH [2] and Matrix d-Linear [10].

In the cryptographic constructions mentioned above, some secret values (messages or keys) are encoded as group element vectors and then hidden by multiplying them by an invertible matrix. The secret value is recovered by inverting the operations: first multiplying by the inverse matrix and then inverting the encoding as group elements. This last step requires to encode a few bits (typically, a single bit) in each group element, forcing the length of the vector and the rank of the matrix to be comparable to the binary length of the secret. Security of these schemes is related to the indistinguishability of full-rank matrices from low-rank (e.g., rank 1) matrices: If the invertible matrix is replaced by a low rank one, the secret value is information-theoretically hidden. Therefore, the security of these schemes is related to the hardness of the Rank problem for matrices of large rank (e.g., 320 or 1024).

Reductions of the DDH problem to the Rank problem are based in the obvious relationship between them in the case of 2×2 matrices. Namely, from a DDH problem tuple (g, g^x, g^y, g^z) one can build a matrix $g^M = \begin{pmatrix} g & g^x \\ g^y & g^z \end{pmatrix}$, which is the elementwise exponentiation of the $\mathbb{Z}_q$ matrix $M = \begin{pmatrix} 1 & x \\ y & z \end{pmatrix}$. For a 0-instance of DDH (i.e., $z = xy$), $\det M = 0$, while for a 1-instance (i.e., $z \neq xy$), $\det M \neq 0$, and therefore, the rank of M shifts from 1 to 2 depending on the DDH instance. This technique can be applied to larger (even non-square) matrices by just padding the previous 2×2 block with some ones in the diagonal and zeroes elsewhere, just increasing the rank from 1 or 2 to $r + 1$ or $r + 2$, where r is the number of ones added to the diagonal.

Now, a general reduction of DDH to any instance of the rank problem (i.e., telling apart hidden matrices of ranks r_1 and r_2) is obtained by applying a hybrid argument, incurring into a loss-factor in the adversary's advantage which grows linearly in the rank difference $r_2 - r_1$.

This loss-factor has an extra impact on the efficiency of the cryptographic schemes based on matrices: For the same security level the size of the group has to be increased, and therefore the sizes of public keys, ciphertexts, etc. increase accordingly.

Until now it was an open problem to find a tighter reduction of DDH to the Rank problem. To face this kind of problems one can choose between building new tighter reductions or showing impossibility results. However, most of the known impossibility results are quite limited because they only claim the nonexistence of reductions of certain type (e.g., black-box, algebraic, etc.). But still these negative results have some value since they capture all possible 'natural' reductions between computational problems, at least in the generic case (e.g., without using specific properties of certain groups and their representation).

Main Results. In this paper, we give a new and better reduction between the Rank and the DDH problems, such that the reduction loss-factor grows logarithmically with the rank of the matrices. This new reduction can be applied to a number of cryptographic constructions improving their efficiency. The main idea in the reduction is to build a matrix from a DDH tuple which rank shifts from r to $2r$, and then apply a hybrid argument to deal with the general case.

On the other hand, the new reduction is optimal: We show the nonexistence of more efficient reductions in a wide class containing all the "natural" ones (i.e., black-box and algebraic). The result is twofold: There is no (natural) way to build a matrix which rank shifts from r to $2r + \alpha$ for $\alpha > 0$, and no hybrid argument can improve the logarithmic loss-factor obtained in the new reduction.

Basically, the new reduction achieves the following result.

(Informal) Theorem 1. *For any ℓ_1, ℓ_2, r_1, r_2 such that $1 \leq r_1 < r_2 \leq \min(\ell_1, \ell_2)$ there is a reduction of the DDH problem to the Rank problem for $\ell_1 \times \ell_2$ matrices of rank either r_1 or r_2, where the advantage of the problem solvers fulfil*

$$\mathbf{AdvRank}(\mathcal{G}, \ell_1, \ell_2, r_1, r_2; t) \leq \left\lceil \log_2 \frac{r_2}{r_1} \right\rceil \mathbf{AdvDDH}(\mathcal{G}; t')$$

and their running times t and t' are essentially equal.

In particular, our reduction relates the DDH Problem to the hardness of telling apart $\ell \times \ell$ full rank matrices from rank 1 matrices with a loss-factor of only $\log_2(\ell)$, instead of the factor ℓ obtained in previous reductions. Moreover, the previous reductions are tight only for ranks r_1 and r_2 such that $r_2 = r_1 + 1$, while our results show that there exists a tight reduction for $r_1 < r_2 \leq 2r_1$.

At this point, it arises the natural question of whether a tight reduction exists for a wider range of the ranks r_1 and r_2. However, we show the optimality of the new reduction by the following negative result.

(Informal) Theorem 2. *For any ℓ_1, ℓ_2, r_1, r_2 such that $1 \leq r_1 < r_2 \leq \min(\ell_1, \ell_2)$ and any 'natural' reduction $\mathcal{R}$ of DDH to the Rank problem, the advantages of the Rank problem solver $\mathcal{A}$ and the DDH solver $\mathcal{R}([\mathcal{A}])$ fulfil*

$$\mathbf{AdvRank}_{\mathcal{R}[\mathcal{A}]}(\mathcal{G}, \ell_1, \ell_2, r_1, r_2; t) \geq \left\lceil \log_2 \frac{r_2}{r_1} \right\rceil \mathbf{AdvDDH}_{\mathcal{A}}(\mathcal{G}; t') - \varepsilon$$

where the running times t, t' are similar and ε is a negligible quantity.

Here, 'natural reduction' basically means a black-box reduction which transforms a DDH tuple into a hidden matrix by performing only (probabilistic) algebraic manipulations, which are essentially linear combinations of the exponents with known integer coefficients, depending on the random coins of the reduction.

All generic reductions from computational problems based on cyclic groups fall into this category. Therefore, this result has to be interpreted as one cannot expect finding a tighter reduction for a large class of groups unless a new (non-black-box or not algebraic) technique is used. Nevertheless, falsifying this negative result would imply an improvement on the efficiency of the cryptosystems based on matrices, or even the discovery of a new reduction technique.

The techniques used in the paper extend naturally to other "algebraic" problems like the Decisional Linear (DLin) or the Decisional 3-Party Diffie-Hellman (D3DH) problems, also obtaining reductions with logarithmic complexity. Actually, these reductions recently appeared in [4] and [5].

(Informal) Theorem 3. *For any ℓ_1, ℓ_2, r_1, r_2 such that $2 \leq r_1 < r_2 \leq \min(\ell_1, \ell_2)$ there is a reduction of the DLin problem to the Rank problem for $\ell_1 \times \ell_2$ matrices of rank either r_1 or r_2, where the advantage of the problem solvers fulfil*

$$\mathbf{AdvRank}(\mathcal{G}, \ell_1, \ell_2, r_1, r_2; t) \leq \left\lceil 1.71 \log_2 \frac{r_2}{r_1 - 1} \right\rceil \mathbf{AdvDLin}(\mathcal{G}; t')$$

and their running times t and t' are essentially equal.

(Informal) Theorem 4. *For any ℓ_1, ℓ_2, r_1, r_2 such that $2 \leq r_1 < r_2 \leq \min(\ell_1, \ell_2)$ there is a reduction of the D3DH problem to the Rank problem for $\ell_1 \times \ell_2$ matrices of rank either r_1 or r_2, where the advantage of the problem solvers fulfil*

$$\mathbf{AdvRank}(\mathcal{G}, \ell_1, \ell_2, r_1, r_2; t) \leq \left\lceil 1.71 \log_2 \frac{r_2}{r_1 - 1} \right\rceil \mathbf{AdvD3DH}(\mathcal{G}; t')$$

and their running times t and t' are essentially equal.

Negative results similar to Theorem 2 are also given, but in these two cases the reductions are shown to be optimal up to a constant factor of 1.71.

Further Research. Some of the ideas and techniques used in the paper suggest that the problem of the optimality of certain type of reductions for a class of decisional assumptions can be studied under the Algebraic Geometric point of view. In particular, this could help to close the gap in the loss-factor between the reduction and the lower bound when reducing DLin or D3DH to Rank, and could made possible to obtain similar results for a broad class of computational problems. A second open problem is how the techniques and results adapt to the case of composite order groups, specially when the factorization of the order, or the order itself is unknown.

Roadmap. The paper starts with some notation and basic lemmas, in Section 2. Then the Rank Problem and the new reduction of DDH is presented in Section 3. The optimality of the reduction is studied in Section 4. In the last section of the paper, the previous results are extended to other "algebraic" decisional problems like DLin or D3DH.

2   Notation and Basic Lemmas

Let $\mathcal{G}$ be a group of prime order q, and let g be a random generator of $\mathcal{G}$. For convenience we will use additive notation for all groups. In particular, $0_{\mathcal{G}}$ denotes the neutral element in $\mathcal{G}$, whereas $1_{\mathcal{G}}$ denotes the generator g. Analogously, $x1_{\mathcal{G}}$, or simply $x_{\mathcal{G}}$, denotes the result of g^x, for any integer $x \in \mathbb{Z}_q$. The additive notation extends to vectors and matrices of elements in $\mathcal{G}$, in the natural way. That is, given a vector $\boldsymbol{x} = (x_1, \ldots, x_\ell) \in \mathbb{Z}_q^\ell$, we will write $\boldsymbol{x}_{\mathcal{G}} = ((x_1)_{\mathcal{G}}, \ldots, (x_\ell)_{\mathcal{G}})$, and the same for matrices. $\mathbb{Z}_q^{\ell_1 \times \ell_2}$ denotes the set of all $\ell_1 \times \ell_2$ matrices, and $\mathbb{Z}_q^{\ell_1 \times \ell_2; r}$ is used for the subset of those matrices with rank r. In the special case of invertible matrices we will write $\mathrm{GL}_\ell(\mathbb{Z}_q) = \mathbb{Z}_q^{\ell \times \ell; \ell}$. The sets of matrices with entries in $\mathcal{G}$, which we write $\mathcal{G}^{\ell_1 \times \ell_2}$, $\mathcal{G}^{\ell_1 \times \ell_2; r}$ and $\mathrm{GL}_\ell(\mathcal{G})$, are defined in the natural way by replacing every matrix M by $M_{\mathcal{G}}$. Notice that the sets are independent of the choice of the group generator $1_{\mathcal{G}}$.

An element $x_{\mathcal{G}} = x1_{\mathcal{G}} \in \mathcal{G}$ and an integer $a \in \mathbb{Z}_q$ can be operated together: $ax_{\mathcal{G}} = (ax \bmod q)1_{\mathcal{G}} = (ax)_{\mathcal{G}} = xa_{\mathcal{G}}$. These operations extend to vectors and matrices in the natural way. Therefore, for any two matrices $A \in \mathbb{Z}_q^{\ell_1 \times \ell_2}$ and $B \in \mathbb{Z}_q^{\ell_2 \times \ell_3}$, we have $A_{\mathcal{G}} B = A B_{\mathcal{G}} = (AB)_{\mathcal{G}}$.

For convenience we will use the notation $A \oplus B$ for block matrix concatenation:

$$A \oplus B = \left(\begin{array}{c|c} A & 0 \\ \hline 0 & B \end{array} \right)$$

In addition, I_ℓ and $0_{\ell_1 \times \ell_2}$ respectively denote the neutral element in $\mathrm{GL}_\ell(\mathbb{Z}_q)$ and the null matrix in $\mathbb{Z}_q^{\ell_1 \times \ell_2}$. The shorthand $0_\ell = 0_{\ell \times \ell}$ is also used. Given a matrix $A \in \mathbb{Z}_q^{\ell_1 \times \ell_2}$, the transpose of A is denoted by $A^\top$, and the vector subspace spanned by the columns of A is denoted by $\mathrm{Span}\, A \subseteq \mathbb{Z}_q^{\ell_1}$, which dimension equals $\mathrm{rank}\, A$.

Uniform sampling of a set S is written as $x \in_{\mathrm{R}} S$. In addition, sampling from a probability distribution D which support is included in S is denoted by $x \leftarrow D$, while $x \leftarrow \mathcal{A}(a)$ denotes that x is the result of running a (probabilistic) algorithm $\mathcal{A}$ on some input a.

As it is usual, a positive function $f : \mathbb{Z}^+ \to \mathbb{R}^+$ is called negligible if $f(\lambda)$ decreases faster than λ^{-c} for any positive constant c. We denote this by $f(\lambda) \in \mathbf{negl}(\lambda)$. Similarly, $f(\lambda) > \mathbf{negl}(\lambda)$ denotes that $f(\lambda)$ is non negligible in λ.

Lemma 1. *The following three natural group actions are transitive:*[1]

1. *the left-action of $\mathrm{GL}_{\ell_1}(\mathbb{Z}_q)$ on $\mathbb{Z}_q^{\ell_1 \times \ell_2; \ell_2}$, for $\ell_1 \geq \ell_2$, defined by $A \mapsto UA$, where $U \in \mathrm{GL}_{\ell_1}(\mathbb{Z}_q)$ and $A \in \mathbb{Z}_q^{\ell_1 \times \ell_2; \ell_2}$,*
2. *the right-action of $\mathrm{GL}_{\ell_2}(\mathbb{Z}_q)$ on $\mathbb{Z}_q^{\ell_1 \times \ell_2; \ell_1}$, for $\ell_1 \leq \ell_2$, defined by $A \mapsto AV$, where $V \in \mathrm{GL}_{\ell_2}(\mathbb{Z}_q)$ and $A \in \mathbb{Z}_q^{\ell_1 \times \ell_2; \ell_1}$,*

---

[1] The action of a group G on a set A is transitive if for any $a, b \in A$ there exists $g \in G$ such that $b = g \cdot a$. As a consequence, if $g \in_{\mathrm{R}} G$ then for any $a \in A$, $g \cdot a$ is uniform in A.

3. *the left-right-action of* $GL_{\ell_1}(\mathbb{Z}_q) \times GL_{\ell_2}(\mathbb{Z}_q)$ *on* $\mathbb{Z}_q^{\ell_1 \times \ell_2;r}$, *defined by* $A \mapsto UAV$, *where* $U \in GL_{\ell_1}(\mathbb{Z}_q)$, $V \in GL_{\ell_2}(\mathbb{Z}_q)$ *and* $A \in \mathbb{Z}_q^{\ell_1 \times \ell_2;r}$.

Lemma 2 (Rank Decomposition). *Given any matrix* $A \in \mathbb{Z}_q^{\ell_1 \times \ell_2;r}$, *there exist matrices* $L \in \mathbb{Z}_q^{\ell_1 \times r;r}$ *and* $R \in \mathbb{Z}_q^{r \times \ell_2;r}$ *such that* $A = LR$.

3 The Rank Problem and the New Reduction of DDH to Rank

We consider an assumption related to matrices, which is weaker than some well-known assumptions like the Decisional Diffie-Hellman, the Decisional Linear [1] and the Decisional 3-Party Diffie-Hellman [3,6,9] assumptions. Given an (additive) cyclic group $\mathcal{G}$ of prime order q of binary length λ, the **Rank**$(\mathcal{G}, \ell_1, \ell_2, r_1, r_2)$ problem informally consists of distinguishing if a given matrix in $\mathbb{Z}_q^{\ell_1 \times \ell_2}$ has either rank r_1 or rank r_2, for given integers $r_1 < r_2$. The problem is formally defined through the following two experiments between a challenger and a distinguisher $\mathcal{A}$.

Experiment **ExpRank**$_{\mathcal{A}}^{b}(\mathcal{G}, \ell_1, \ell_2, r_1, r_2)$ is defined as follows, for $b = 0, 1$.

1. If $b = 0$, the challenger chooses $M \in_R \mathbb{Z}_q^{\ell_1 \times \ell_2;r_1}$ and sends $M_{\mathcal{G}}$ to $\mathcal{A}$.
   If $b = 1$, the challenger chooses $M \in_R \mathbb{Z}_q^{\ell_1 \times \ell_2;r_2}$ and sends $M_{\mathcal{G}}$ to $\mathcal{A}$.
2. The distinguisher $\mathcal{A}$ outputs a bit $b' \in \{0, 1\}$.

Let Ω_b be the event that $\mathcal{A}$ outputs $b' = 1$ in **ExpRank**$_{\mathcal{A}}^{b}(\mathcal{G}, \ell_1, \ell_2, r_1, r_2)$. The advantage of $\mathcal{A}$ is defined as **AdvRank**$_{\mathcal{A}}(\mathcal{G}, \ell_1, \ell_2, r_1, r_2) = |\Pr[\Omega_0] - \Pr[\Omega_1]|$. We can then define

$$\mathbf{AdvRank}(\mathcal{G}, \ell_1, \ell_2, r_1, r_2; t) = \max_{\mathcal{A}} \{\mathbf{AdvRank}_{\mathcal{A}}(\mathcal{G}, \ell_1, \ell_2, r_1, r_2)\}$$

where the maximum is taken over all $\mathcal{A}$ running within time t.

Definition 1. *The* **Rank**$(\mathcal{G}, \ell_1, \ell_2, r_1, r_2)$ *assumption in a group* $\mathcal{G}$ *states that* **AdvRank**$(\mathcal{G}, \ell_1, \ell_2, r_1, r_2; t)$ *is negligible in* $\lambda = \log|\mathcal{G}|$ *for any value of* t *that is polynomial in* λ.

The Rank assumption appeared in recent papers under the names Matrix-DDH [2] and Matrix d-Linear [10]. However, the reduction given in the next proposition substantially improves the reductions previously known. Namely, the loss factor in the new reduction grows no longer linearly but logarithmically in the rank.

Firstly, note that the **Rank**$(\mathcal{G}, \ell_1, \ell_2, r_1, r_2)$ problem is random self-reducible, since by Lemma 1 given $M_0 \in \mathbb{Z}_q^{\ell_1 \times \ell_2;k}$, for random $L \in_R GL_{\ell_1}(\mathbb{Z}_q)$ and $R \in_R GL_{\ell_2}(\mathbb{Z}_q)$ the product LM_0R is uniformly distributed in $\mathbb{Z}_q^{\ell_1 \times \ell_2;k}$.

Lemma 3. *Any distinguisher for* **Rank**$(\mathcal{G}, \ell_1, \ell_2, k - \delta, k)$, $\ell_1, \ell_2 \geq 2$, $k \geq 2$, $1 \leq \delta \leq \lfloor \frac{k}{2} \rfloor$ *can be converted into a distinguisher for the Decisional Diffie-Hellman (DDH) problem, with the same advantage and with essentially the same running time.*

Proof. Given a DDH instance $(1, x, y, z)_{\mathcal{G}}$, the DDH distinguisher builds the $\ell_1 \times \ell_2$ matrix

$$M_{\mathcal{G}} = \underbrace{\begin{pmatrix} 1 & x \\ y & z \end{pmatrix}_{\mathcal{G}} \oplus \cdots \oplus \begin{pmatrix} 1 & x \\ y & z \end{pmatrix}_{\mathcal{G}}}_{\delta \text{ times}} \oplus I_{k-2\delta}{}_{\mathcal{G}} \oplus 0_{(\ell_1-k)\times(\ell_2-k)}{}_{\mathcal{G}}$$

and submits the randomized matrix $LM_{\mathcal{G}}R$ to the $\mathbf{Rank}(\mathcal{G}, \ell_1, \ell_2, k - \delta, k)$ distinguisher, where $L \in_R \mathrm{GL}_{\ell_1}(\mathbb{Z}_q)$ and $R \in_R \mathrm{GL}_{\ell_2}(\mathbb{Z}_q)$. Notice that if $z = xy$ mod q then the resulting matrix is a random matrix in $\mathcal{G}^{\ell_1 \times \ell_2; k-\delta}$. Otherwise, it is a random matrix in $\mathcal{G}^{\ell_1 \times \ell_2; k}$. $\qquad \square$

Theorem 1. *For any ℓ_1, ℓ_2, r_1, r_2 such that $1 \le r_1 < r_2 \le \min(\ell_1, \ell_2)$ we have,*

$$\mathbf{AdvRank}(\mathcal{G}, \ell_1, \ell_2, r_1, r_2; t) \le \left\lceil \log_2 \frac{r_2}{r_1} \right\rceil \mathbf{AdvDDH}(\mathcal{G}; t')$$

where $t' = t + O(\ell_1\ell_2(\ell_1 + \ell_2))$, taking the cost of a scalar multiplication in $\mathcal{G}$ as one time unit.

Proof. We proceed by applying a hybrid argument. Let us consider the sequence of integers $\{n_i\}$ defined by $n_i = r_1 2^i$, and let k be the smallest index such that $n_k \ge r_2$, that is $k = \lceil \log_2 r_2 - \log_2 r_1 \rceil$. Then define a sequence of random matrices $\{M_{i\mathcal{G}}\}$, where $M_i \in_R \mathbb{Z}_q^{\ell_1 \times \ell_2; n_i}$ for $i = 0, \ldots, k - 1$, and $M_k \in_R \mathbb{Z}_q^{\ell_1 \times \ell_2; r_2}$. For any distinguisher $\mathcal{A}_{\mathbf{Rank}}$ with running time upper bounded by t, let $p_i = \Pr[1 \leftarrow \mathcal{A}_{\mathbf{Rank}}(M_{i\mathcal{G}})]$. By Lemma 3,

$$|p_{i+1} - p_i| = \mathbf{AdvRank}_{\mathcal{A}_{\mathbf{Rank}}}(\mathcal{G}, \ell_1, \ell_2, n_i, n_{i+1}) \le \mathbf{AdvDDH}(\mathcal{G}; t')$$

for $i = 0, \ldots, k - 2$, and

$$|p_k - p_{k-1}| = \mathbf{AdvRank}_{\mathcal{A}_{\mathbf{Rank}}}(\mathcal{G}, \ell_1, \ell_2, n_{k-1}, r_2) \le \mathbf{AdvDDH}(\mathcal{G}; t')$$

Therefore,

$$\mathbf{AdvRank}_{\mathcal{A}_{\mathbf{Rank}}}(\mathcal{G}, \ell_1, \ell_2, r_1, r_2) = |p_k - p_0| \le |p_1 - p_0| + \ldots + |p_k - p_{k-1}| \le$$
$$\le k \cdot \mathbf{AdvDDH}(\mathcal{G}; t')$$

which leads to the desired result. $\qquad \square$

4   Optimality of the Reduction

In this section we show that there does not exist any reduction of DDH to the Rank problem that improves the result in Theorem 1, unless it falls out of the class of reductions that we call *black-box algebraic* reductions.

4.1   Black-Box Algebraic Reductions

Formally, a reduction $\mathcal{R}$ of a computational problem $\mathcal{P}_1$ to a problem $\mathcal{P}_2$ efficiently transforms any probabilistic polynomial time algorithm $\mathcal{A}_2$ solving $\mathcal{P}_2$ with a non-negligible advantage ε_2 into another probabilistic polynomial time algorithm $\mathcal{A}_1 = \mathcal{R}[\mathcal{A}_2]$ solving $\mathcal{P}_1$ with a non-negligible advantage ε_1. The reduction $\mathcal{R}$ is called *black-box* if $\mathcal{A}_1$ is just a probabilistic polynomial time algorithm with oracle access to $\mathcal{A}_2$.

In this paper we focus on the optimality of a reduction, measured in terms of the advantages of $\mathcal{A}_1$ and $\mathcal{A}_2$. However, to be meaningful we need to add another requirement to the reduction: The running times of $\mathcal{A}_1$ and $\mathcal{A}_2$ are similar. Otherwise, one can arbitrarily increase the advantage of $\mathcal{A}_1$ by repetition, thus making more than one oracle call to $\mathcal{A}_2$. We must add a qualifier and say that the reduction is then *time-preserving* black-box. However, for simplicity we will omit it and simply refer to black-box reductions.

Following [11], we say that $\mathcal{R}$ is *algebraic* with respect to a group $\mathcal{G}$ if it only performs group operations on the elements of $\mathcal{G}$ (i.e., group operation, inversion and comparison for equality), while there is no limitation in the operations performed on other data types. Although the notion of black-box algebraic reduction is theoretically very limited, it captures all the 'natural' reductions, since all known reductions between problems related to the discrete logarithm in cyclic groups fall into this category. See [11] for a deeper discussion on algebraic reductions and their relation with the generic group model.

In the definition of an algebraic algorithm $\mathcal{R}$ it is assumed that there exists an efficient extractor that, from the inputs of $\mathcal{R}$ (including the random tape) and the code of $\mathcal{R}$, it extracts a representation of every group element in $\mathcal{R}$'s output as a multiexponentiation of the base formed by the group elements in the input of $\mathcal{R}$. However, here we only require that for every value of the random tape of $\mathcal{R}$ there exists such representation, and it is independent of the group elements on the input of $\mathcal{R}$. More precisely, if $g_1, \ldots, g_m \in \mathcal{G}$ are the group elements in the input of $\mathcal{R}$ and $h_1, \ldots, h_n \in \mathcal{G}$ are the group elements in the output, then for any choice of the other inputs and the random tape, there exist coefficients $\alpha_{ij} \in \mathbb{Z}_q$ such that $h_i = \alpha_{i1}g_1 + \ldots + \alpha_{im}g_m$, for $i = 1, \ldots, n$. Notice that this is true as long as $\mathcal{R}$ performs only group operations on the group elements.

We insist in the possible existence of reductions using more intricate operations other than the group operations defined in $\mathcal{G}$. However, there is little hope to be able to control the rank of the manipulated matrices, except for the trivial fact that a random matrix has maximal rank with overwhelming probability.

4.2   Canonical Solvers

In this paper, we consider only reductions $\mathcal{R}$ of some decisional problem (like DDH) to the Rank problem (say $\mathbf{Rank}(\mathcal{G}, \ell_1, \ell_2, r_1, r_2)$). Therefore, in a (time-preserving) black-box reduction, having oracle access to a solver $\mathcal{A}_2$ of Rank exactly means that $\mathcal{R}$ computes some matrix in $\mathcal{G}^{\ell_1 \times \ell_2}$, and uses it as input of

$\mathcal{A}_1$, then obtaining a bit $b' \in \{0, 1\}$ as its output. Therefore, $\mathcal{R}$ is nothing more than a way to obtain a matrix from a DDH instance by an algebraic function.

As Rank problem is random self-reducible, one can consider the notion of a *canonical* solver $\overline{\mathcal{A}}$ for $\mathbf{Rank}(\mathcal{G}, \ell_1, \ell_2, r_1, r_2)$. In a first stage, a canonical solver, on the input of a matrix $M_\mathcal{G} \in \mathcal{G}^{\ell_1 \times \ell_2}$, computes the randomized matrix $M'_\mathcal{G} = L M_\mathcal{G} R$ for randomly chosen $L \in \mathrm{GL}_{\ell_1}(\mathbb{Z}_q)$ and $R \in \mathrm{GL}_{\ell_2}(\mathbb{Z}_q)$, and then uses it as input of the second stage. Observe that $M_\mathcal{G}$ and $M'_\mathcal{G}$ have always the same rank, and they are nearly independent. Indeed $M_\mathcal{G}$ and $M'_\mathcal{G}$ conditioned to any specific value of the rank r are independent random variables, and $M'_\mathcal{G}$ is uniformly distributed in $\mathcal{G}^{\ell_1 \times \ell_2; r}$.

Moreover, for any solver $\mathcal{A}$ of $\mathbf{Rank}(\mathcal{G}, \ell_1, \ell_2, r_1, r_2)$ we build a canonical solver $\overline{\mathcal{A}}$ from $\mathcal{A}$ with the same advantage, by just inserting the initial randomization step. As a consequence, to obtain a negative result about the existence of black-box reductions of some problem to $\mathbf{Rank}(\mathcal{G}, \ell_1, \ell_2, r_1, r_2)$, we only need to consider how the reduction works for canonical solvers of $\mathbf{Rank}(\mathcal{G}, \ell_1, \ell_2, r_1, r_2)$.

Finally, it should be noticed that a canonical solver is completely characterized by a probability vector $\boldsymbol{p}_\mathcal{A} = (p_{\mathcal{A},i})_{i \in \mathbb{Z}^+}$, where $p_{\mathcal{A},i} = \Pr[1 \leftarrow \mathcal{A}(M_\mathcal{G}) : M_\mathcal{G} \in_R \mathcal{G}^{\ell_1 \times \ell_2; i}]$. The advantage of a canonical solver is then $\mathbf{AdvRank}_\mathcal{A} = |p_{\mathcal{A},r_2} - p_{\mathcal{A},r_1}|$. Dealing with all canonical solvers of $\mathbf{Rank}(\mathcal{G}, \ell_1, \ell_2, r_1, r_2)$ means considering all possible probability vectors $\boldsymbol{p}_\mathcal{A}$ such that $|p_{\mathcal{A},r_2} - p_{\mathcal{A},r_1}|$ is non-negligible.

4.3   More Linear Algebra

Let us see the implications of restricting the reductions to be algebraic. Since here we reduce the decisional problem DDH to the $\mathbf{Rank}(\mathcal{G}, \ell_1, \ell_2, r_1, r_2)$ problem, the reduction $\mathcal{R}$ will receive as input either a 0-instance (i.e., $(1_\mathcal{G}, x_\mathcal{G}, y_\mathcal{G}, xy_\mathcal{G})$) or a 1-instance (i.e., $(1_\mathcal{G}, x_\mathcal{G}, y_\mathcal{G}, (xy+s)_\mathcal{G})$) of the decisional problem (where $x, y, s \in_R \mathbb{Z}_q$). In spite of the instance received, $\mathcal{R}$ will compute a matrix $M_\mathcal{G} \in \mathcal{G}^{\ell_1 \times \ell_2}$ that depends 'algebraically' on the input group elements. Therefore, for any value of the random tape of $\mathcal{R}$ there exist matrices $B_1, B_2, B_3, B_4 \in \mathbb{Z}_q^{\ell_1 \times \ell_2}$ such that $M = B_1 + x B_2 + y B_3 + (xy + s) B_4$, where either $s = 0$ or $s \in_R \mathbb{Z}_q$, depending on the type of instance received by $\mathcal{R}$.

Therefore, we need some properties of the sets of matrices that are linear combinations of some fixed matrices with coefficients that are multivariate polynomials. The following lemma informally states that matrices in a linear variety of $\mathbb{Z}_q^{\ell \times \ell}$ (of any dimension) are invertible with either zero or overwhelming probability.

Lemma 4. *Let $\mathcal{M}$ be a coset of a $\mathbb{Z}_q$-vector subspace of $\mathbb{Z}_q^{\ell \times \ell}$, that is, there exist matrices $A, B_1, \ldots, B_k \in \mathbb{Z}_q^{\ell \times \ell}$ for some integer k such that $\mathcal{M} = \{A + x_1 B_1 + \ldots + x_k B_k \mid x_1, \ldots, x_k \in \mathbb{Z}_q\}$. If $\mathrm{GL}_\ell(\mathbb{Z}_q) \cap \mathcal{M} \neq \emptyset$ then,*

$$\nu_\mathcal{M} = \frac{|\mathrm{GL}_\ell(\mathbb{Z}_q) \cap \mathcal{M}|}{|\mathcal{M}|} > 1 - \frac{\ell}{q-1}$$

Proof. [2] Let us choose $A \in \mathrm{GL}_\ell(\mathbb{Z}_q) \cap \mathcal{M}$ and let $\{B_1, \ldots, B_k\}$ be a base of the vector space $\mathcal{M} - A$. In any line $\mathcal{L} \subset \mathcal{M}$ containing A there can be at most ℓ matrices $M \in \mathcal{L}$ such that $\mathrm{rank}\, M < \ell$ (i.e., $\det M = 0$). Indeed, for any line $\mathcal{L}$ there is a nonzero vector $\boldsymbol{x} = (x_1, \ldots, x_k) \in \mathbb{Z}_q^k$ such that $\mathcal{L} = \{A + \mu(x_1 B_1 + \ldots + x_k B_k) \mid \mu \in \mathbb{Z}_q\}$. Therefore the polynomial equation $\det(A + \mu(x_1 B_1 + \ldots + x_k B_k)) = 0$, which is equivalent to $Q_{\boldsymbol{x}}(\mu) = \det(I_\ell + \mu(x_1 B_1 A^{-1} + \ldots + x_k B_k A^{-1})) = 0$, has at most ℓ roots because $Q_{\boldsymbol{x}}(0) = 1$ and $\lambda^{-\ell} Q_{\boldsymbol{x}}(1/\lambda) = \det(\lambda I_\ell + x_1 B_1 A^{-1} + \ldots + x_k B_k A^{-1}) = 0$ if and only if λ is an eigenvalue of $x_1 B_1 A^{-1} + \ldots + x_k B_k A^{-1}$. Finally, since there are exactly $|\mathbb{PZ}_q^{k-1}| = \frac{q^k - 1}{q - 1}$ different lines in $\mathcal{M}$ containing A,

$$\nu_F \geq 1 - \frac{\ell(q^k - 1)/(q - 1)}{q^k} > 1 - \frac{\ell}{q - 1}$$

as k is the dimension of the vector space $\mathcal{M} - A$, and then $|\mathcal{M}| = q^k$. $\square$

This lemma can be easily generalized to parametrical subsets of linear varieties by replacing each variable x_j, $j = 1, \ldots, k$, by a multivariate polynomial $p_j(y_1, \ldots, y_n) \in \mathbb{Z}_q[y_1, \ldots, y_n]$ (or simply, $\mathcal{M}$ is now the range of a multivariate polynomial with matrix coefficients). Here we cannot ensure that the mapping between the parameter vector $\boldsymbol{y} = (y_1, \ldots, y_n)$ and the matrices in $\mathcal{M}$ is one-to-one. Therefore we will define $\nu_\mathcal{M}$ as the probability of obtaining a full-rank matrix when $\boldsymbol{y} \in_R \mathbb{Z}_q^n$ is sampled with the uniform distribution.

Lemma 5. *Let $\mathcal{M}$ be a subset of $\mathbb{Z}_q^{\ell \times \ell}$ defined as $\mathcal{M} = \{p_1(\boldsymbol{y})B_1 + \ldots + p_k(\boldsymbol{y})B_k \mid \boldsymbol{y} \in \mathbb{Z}_q^n\}$, where $p_1(\boldsymbol{y}), \ldots, p_k(\boldsymbol{y}) \in \mathbb{Z}_q[\boldsymbol{y}]$ are multivariate polynomials of total degree at most d, and $B_1, \ldots, B_k \in \mathbb{Z}_q^{\ell \times \ell}$ for some integer k. If $\mathrm{GL}_\ell(\mathbb{Z}_q) \cap \mathcal{M} \neq \emptyset$ then,*

$$\nu_\mathcal{M} = \Pr[M \in GL_\ell(\mathbb{Z}_q) : M = p_1(\boldsymbol{y})B_1 + \ldots + p_k(\boldsymbol{y})B_k, \; \boldsymbol{y} \in_R \mathbb{Z}_q^n] \geq$$
$$\geq 1 - \frac{\ell d}{q - 1} \frac{q^n - 1}{q^n} > 1 - \frac{\ell d}{q - 1}$$

Proof. The proof is similar, but now we choose $A = p_1(\boldsymbol{y}_0)B_1 + \ldots + p_k(\boldsymbol{y}_0)B_k \in \mathrm{GL}_\ell(\mathbb{Z}_q) \cap \mathcal{M}$ and define the new polynomials $q_i(\boldsymbol{z}) = p_i(\boldsymbol{y}_0 + \boldsymbol{z}) - p_i(\boldsymbol{y}_0)$ for $i = 1, \ldots k$. Now, $\mathcal{M} \setminus \{A\}$ is partitioned into subsets $\mathcal{L}^* = \{A + q_1(\mu\boldsymbol{z})B_1 + \ldots + q_k(\mu\boldsymbol{z})B_k) \mid \mu \in \mathbb{Z}_q^\times\}$, where $\boldsymbol{z} \in \mathbb{Z}_q^n \setminus \{\boldsymbol{0}\}$, each one containing at most ℓd singular matrices, since the polynomial $Q_{\boldsymbol{z}}(\mu) = \det(I_\ell + q_1(\mu\boldsymbol{z})B_1 A^{-1} + \ldots + q_k(\mu\boldsymbol{z})B_k A^{-1})$ is nonzero (as $Q_{\boldsymbol{z}}(0) = 1$), and it has degree at most $d\ell$. Finally, the claimed inequality follows from the fact that there are $(q^n - 1)/(q - 1)$ different subsets $\mathcal{L}^*$. $\square$

The above lemmas refer only to invertible matrices but a similar result applies to (even rectangular) matrices with respect to a specific value of the rank.

---

[2] This lemma and the following one can alternatively be proved by using the Schwartz lemma [15] (also referred to as Schwartz-Zippel lemma).

Lemma 6. *Let $\mathcal{M}$ be a subset of $\mathbb{Z}_q^{\ell_1 \times \ell_2}$ defined as $\mathcal{M} = \{p_1(\boldsymbol{y})B_1 + \ldots + p_k(\boldsymbol{y})B_k \mid \boldsymbol{y} \in \mathbb{Z}_q^n\}$, where $p_1(\boldsymbol{y}), \ldots, p_k(\boldsymbol{y}) \in \mathbb{Z}_q[\boldsymbol{y}]$ are multivariate polynomials of total degree at most d, and $B_1, \ldots, B_k \in \mathbb{Z}_q^{\ell_1 \times \ell_2}$ for some integer k. If $r_m = \max_{m \in \mathcal{M}} \operatorname{rank} M$ then,*

$$\nu_{\mathcal{M}} = \Pr[\operatorname{rank} M = r_m \,:\, M = p_1(\boldsymbol{y})B_1 + \ldots + p_k(\boldsymbol{y})B_k, \, \boldsymbol{y} \in_{\mathrm{R}} \mathbb{Z}_q^n] > 1 - \frac{r_m d}{q-1}$$

Proof. We just apply the previous lemma to a projection of the set $\mathcal{M}$. Firstly choose $M_0 \in \mathcal{M}$ such that $\operatorname{rank} M_0 = r_m$ and find matrices $L \in \mathbb{Z}_q^{r_m \times \ell_1; r_m}$ and $R \in \mathbb{Z}_q^{\ell_2 \times r_m; r_m}$ such that $\operatorname{rank} LM_0 R = r_m$, that is $LM_0 R \in \mathrm{GL}_{r_m}(\mathbb{Z}_q)$. This matrices are really easy to build, since by Lemma 2 there exist $L_0 \in \mathbb{Z}_q^{\ell_1 \times r_m; r_m}$ and $R_0 \in \mathbb{Z}_q^{r_m \times \ell_2; r_m}$ such that $M_0 = L_0 R_0$. Therefore, we take any L such that $LL_0 \in \mathrm{GL}_{r_m}(\mathbb{Z}_q)$. For instance, take L as a the all-zero matrix and put r_m ones in its main diagonal, in positions corresponding to r_m linearly independent rows of L_0. We similarly proceed with R_0 and R.

Now, the projected set $\mathcal{M}' = \{LMR \mid M \in \mathcal{M}\}$ fulfils the conditions of Lemma 5 and it contains at least one invertible matrix $LM_0 R$. Thus,

$$\nu_{\mathcal{M}'} = \Pr[M' \in \mathrm{GL}_{r_m}(\mathbb{Z}_q) \,:\, M' = L(p_1(\boldsymbol{y})B_1 + \ldots + p_k(\boldsymbol{y})B_k)R, \, \boldsymbol{y} \in_{\mathrm{R}} \mathbb{Z}_q^n] >$$
$$> 1 - \frac{\ell r_m}{q-1}$$

Moreover, since $\operatorname{rank}(LMR) \leq \operatorname{rank} M \leq r_m$ for all $M \in \mathcal{M}$, then $\operatorname{rank}(LMR) = r_m$ implies $\operatorname{rank} M = r_m$, and

$$\Pr[\operatorname{rank} M = r_m \,:\, M = p_1(\boldsymbol{y})B_1 + \ldots + p_k(\boldsymbol{y})B_k, \, \boldsymbol{y} \in_{\mathrm{R}} \mathbb{Z}_q^n] \geq \nu_{\mathcal{M}'} > 1 - \frac{\ell r_m}{q-1}$$

$\square$

This lemma basically says that in a set $\mathcal{M}$ defined and sampled as above the matrices have a specific rank (the maximal rank in the set) with overwhelming probability, and ranks below the maximal one occur only with negligible probability.

4.4   The Case of DDH

Now let us consider the specific case of the sets $\mathcal{M}_0$ and $\mathcal{M}_1$ generated by a black-box algebraic reduction $\mathcal{R}$ from a DDH 0-tuple or 1-tuple, respectively, for a fixed random tape of $\mathcal{R}$. More precisely, $\mathcal{M}_{\mathrm{DDH}\text{-}0} = \{B_0 + xB_1 + yB_2 + xyB_3 \mid x, y \in \mathbb{Z}_q\}$, while $\mathcal{M}_{\mathrm{DDH}\text{-}1} = \{B_0 + xB_1 + yB_2 + (xy+s)B_3 \mid x, y, s \in \mathbb{Z}_q\}$, for some matrices $B_0, B_1, B_2, B_3 \in \mathbb{Z}_q^{\ell_1 \times \ell_2}$ that could depend on the random tape. Let r_{m0} and r_{m1} be the maximal ranks respectively in $\mathcal{M}_{\mathrm{DDH}\text{-}0}$ and $\mathcal{M}_{\mathrm{DDH}\text{-}1}$. Since the former is a subset of the latter, $r_{m0} \leq r_{m1}$. In addition, it is clear that $\operatorname{rank} B_0 \leq r_{m0}$, but one can also prove that $\operatorname{rank} B_3 \leq r_{m0}$ and therefore $r_{m1} \leq 2r_{m0}$, as claimed in the following lemma.

Lemma 7. *Let r_{m0} and r_{m1} be the maximal ranks respectively in $\mathcal{M}_{DDH\text{-}0}$ and $\mathcal{M}_{DDH\text{-}1}$. Then $r_{m0} \leq r_{m1} \leq 2r_{m0}$.*

Proof. The left inequality is trivial, as mentioned above. To prove the right one we firstly use Lemma 6 to show that $\operatorname{rank} B_3 \leq r_{m0}$. Indeed, the subset $\mathcal{M}_{DDH\text{-}0}^* = \{B_0 + xB_1 + yB_2 + xyB_3 \mid x, y \in \mathbb{Z}_q^\times\}$ differs from $\mathcal{M}_{DDH\text{-}0}$ in that a negligible fraction of it has been removed. Therefore, the probability distributions on both sets (induced by uniformly sampling x and y) are statistically close. Since for all $x, y \in \mathbb{Z}_q^\times$, $\operatorname{rank}(B_0 + xB_1 + yB_2 + xyB_3) = \operatorname{rank}(\frac{1}{xy}B_0 + \frac{1}{y}B_1 + \frac{1}{x}B_2 + B_3)$, and the inversion map $x \mapsto 1/x$ is a bijection in $\mathbb{Z}_q^\times$, the probability distributions of the ranks in $\mathcal{M}_{DDH\text{-}0}^*$ and in $\overline{\mathcal{M}}_{DDH\text{-}0}^* = \{B_3 + xB_2 + yB_1 + xyB_0 \mid x, y \in \mathbb{Z}_q^\times\}$ are identical. Therefore, matrices in $\overline{\mathcal{M}}_{DDH\text{-}0} = \{B_3 + xB_2 + yB_1 + xyB_0 \mid x, y \in \mathbb{Z}_q\}$ have rank r_{m0} with overwhelming probability. Moreover, by Lemma 6, r_{m0} is precisely the maximal rank in $\overline{\mathcal{M}}_{DDH\text{-}0}$ and then, $\operatorname{rank} B_3 \leq r_{m0}$.[3]

Finally, observe that for any $M \in \mathcal{M}_{DDH\text{-}1}$, $M = B_0 + xB_1 + yB_2 + (xy+s)B_3 = (B_0 + xB_1 + yB_2 + xyB_3) + sB_3$ and $\operatorname{rank} M \leq \operatorname{rank}(B_0 + xB_1 + yB_2 + xyB_3) + \operatorname{rank}(sB_3) \leq 2r_{m0}$, because $B_0 + xB_1 + yB_2 + xyB_3 \in \mathcal{M}_{DDH\text{-}0}$. $\qquad \square$

The previous discussion deals with a fixed arbitrary random tape of the reduction $\mathcal{R}$. However, the overall performance of $\mathcal{R}$ depends on the aggregation of the contributions of all possible values of the random tape. Technically, given a particular canonical solver $\mathcal{A}$ of $\mathbf{Rank}(\mathcal{G}, \ell_1, \ell_2, r_1, r_2)$, described by its probability vector $\boldsymbol{p}_{\mathcal{A}}$ as defined in Section 4.2, the advantage of $\mathcal{R}[\mathcal{A}]$ can be computed as

$$\mathbf{AdvDDH}_{\mathcal{R}[\mathcal{A}]}(\mathcal{G}) = \left| \sum_{r=0}^{\min(\ell_1, \ell_2)} (\pi_{0,r} - \pi_{1,r}) p_{\mathcal{A}r} \right| = |(\boldsymbol{\pi}_0 - \boldsymbol{\pi}_1) \cdot \boldsymbol{p}_{\mathcal{A}}|$$

where

$$\pi_{0,r} = \Pr[\operatorname{rank} M = r : M \leftarrow \mathcal{R}(1_{\mathcal{G}}, x_{\mathcal{G}}, y_{\mathcal{G}}, xy_{\mathcal{G}}), x, y \in_{\mathrm{R}} \mathbb{Z}_q]$$

and

$$\pi_{1,r} = \Pr[\operatorname{rank} M = r : M \leftarrow \mathcal{R}(1_{\mathcal{G}}, x_{\mathcal{G}}, y_{\mathcal{G}}, (xy+s)_{\mathcal{G}}), x, y, s \in_{\mathrm{R}} \mathbb{Z}_q]$$

For convenience, we also introduce the cumulative probabilities $\Pi_{b,r} = \sum_{i=0}^{r} \pi_{b,i}$, $b \in \{0, 1\}$.

Since the reduction $\mathcal{R}$ must work for any successful solver $\mathcal{A}$, for every probability vector $\boldsymbol{p}_{\mathcal{A}}$ such that $|p_{\mathcal{A}r_1} - p_{\mathcal{A}r_2}| = \mathbf{AdvRank}_{\mathcal{A}}(\mathcal{G}, \ell_1, \ell_2, r_1, r_2)$ is non-negligible, the advantage $\mathbf{AdvDDH}_{\mathcal{R}[\mathcal{A}]}(\mathcal{G})$ must be also non-negligible. This implies the existence of $\alpha > \mathbf{negl}(\lambda)$ such that[4]

$$|\pi_{0,r} - \pi_{1,r}| \in \mathbf{negl}(\lambda) \qquad \forall r \notin \{r_1, r_2\}$$

---

[3] A very similar trick also shows that $\operatorname{rank} B_1$ and $\operatorname{rank} B_2$ ar at most r_{m0}. However, it is not clear how to extend this argument to arbitrary multivariate polynomials.

[4] To prove it, consider the fact that there cannot exist any probability vector $\boldsymbol{p}_{\mathcal{A}}$ orthogonal to $\boldsymbol{\pi}_0 - \boldsymbol{\pi}_1$ such that $|p_{\mathcal{A}r_1} - p_{\mathcal{A}r_2}| > \mathbf{negl}(\lambda)$.

$$|\pi_{0,r_1} - \pi_{1,r_1}| = \alpha$$
$$|\pi_{0,r_2} - \pi_{1,r_2}| = \alpha \pm \mathbf{negl}(\lambda) \tag{1}$$

Moreover,

$$\mathbf{AdvDDH}_{\mathcal{R}[\mathcal{A}]}(\mathcal{G}) \leq |p_{\mathcal{A}r_1} - p_{\mathcal{A}r_2}|\,\alpha + \mathbf{negl}(\lambda) =$$
$$= \alpha\mathbf{AdvRank}_{\mathcal{A}}(\mathcal{G}, \ell_1, \ell_2, r_1, r_2) + \mathbf{negl}(\lambda)$$

All that remains is to find an upper bound of the reduction loss-factor α.

By Lemma 6 we know that for every value of the random tape, $\Pr[\text{rank}\, M < r_{mb} : M \leftarrow \mathcal{M}_{\text{DDH-b}}] \in \mathbf{negl}\,\lambda$ for $b \in \{0,1\}$, and by definition of r_{mb}, $\Pr[\text{rank}\, M \leq r_{mb} : M \leftarrow \mathcal{M}_{\text{DDH-b}}] = 1$. Therefore, considering all values of the random tape of $\mathcal{R}$,[5]

$$\Pi_{b,i} = \Pr[r_{mb} \leq i] + \mathbf{negl}(\lambda) \qquad b \in \{0,1\} \tag{2}$$

where now r_{m0} and r_{m1} are random variables. By Lemma 7, $r_{m0} \leq r_{m1} \leq 2r_{m0}$, which implies[6] $\Pr[r_{m1} \leq i] \leq \Pr[r_{m0} \leq i] \leq \Pr[r_{m1} \leq 2i]$, for arbitrary i, and by (2),

$$\Pi_{1,i} - \mathbf{negl}(\lambda) \leq \Pi_{0,i} \leq \Pi_{1,2i} + \mathbf{negl}(\lambda) \tag{3}$$

Now, using left hand side of (3) for $i = r_1$ we get $\Pi_{1,r_1} \leq \Pi_{0,r_1} + \mathbf{negl}(\lambda)$, and combined with (1), we obtain $\pi_{0,r_1} = \pi_{1,r_1} + \alpha$ and $\pi_{1,r_2} \leq \pi_{0,r_2} + \alpha + \mathbf{negl}(\lambda)$. In addition, for any i such that $r_1 \leq i < r_2$,

$$\Pi_{0,i} = \Pi_{1,i} + \alpha \pm \mathbf{negl}(\lambda) \tag{4}$$

Let us assume now that $r_2 > 2^k r_1$ for some $k \geq 1$. Then, applying the right hand side of (3) and (4),

$$\Pi_{0,2^k r_1} = \Pi_{1,2^k r_1} + \alpha \pm \mathbf{negl}(\lambda) \geq \Pi_{0,2^{k-1} r_1} + \alpha - \mathbf{negl}(\lambda)$$

and by induction,

$$\Pi_{0,2^k r_1} \geq \Pi_{0,r_1} + k\alpha - \mathbf{negl}(\lambda) \geq (k+1)\alpha - \mathbf{negl}(\lambda)$$

where (4) is used again in the last step.

Finally, since the leftmost sum is upper bounded by 1,

$$\alpha \leq \frac{1 + \mathbf{negl}(\lambda)}{k+1}$$

for any $k < \log_2 r_2 - \log_2 r_1$. Therefore,

$$\alpha \leq \frac{1 + \mathbf{negl}(\lambda)}{\lceil \log_2 r_2 - \log_2 r_1 \rceil}$$

The above discussion proves the following theorem.

---

[5] If $r_{mb} \leq i$ then rank $M \leq i$ with probability 1. Otherwise, rank $M \leq i$ only with negligible probability.

[6] Observe that $r_{m1} \leq i \Rightarrow r_{m0} \leq i \Rightarrow r_{m1} \leq 2i$.

Theorem 2. *For any ℓ_1, ℓ_2, r_1, r_2 such that $1 \le r_1 < r_2 \le \min(\ell_1, \ell_2)$ and any time-preserving black-box algebraic reduction $\mathcal{R}$ of $\mathbf{DDH}(\mathcal{G})$ to the $\mathbf{Rank}(\mathcal{G}, \ell_1, \ell_2, r_1, r_2)$ problem, any canonical Rank solver $\mathcal{A}$ and the corresponding DDH solver $\mathcal{R}([\mathcal{A}])$ fulfil*

$$\mathbf{AdvRank}_{\mathcal{R}[\mathcal{A}]}(\mathcal{G}, \ell_1, \ell_2, r_1, r_2; t) \ge \left\lceil \log_2 \frac{r_2}{r_1} \right\rceil \mathbf{AdvDDH}_{\mathcal{A}}(\mathcal{G}; t') - \mathbf{negl}(\lambda)$$

where the running times t, t' are similar.

$\square$

5   Reductions of Other Decisional Problems

We consider now other well-known computational problems, namely the Decisional Linear (DLin) [1] and the Decisional 3-Party Diffie-Hellman (D3DH) [3,6,9] problems.

The techniques described above can be applied to these problems by defining a suitable basic matrix block M (of suitable size) where the problem instance is embedded, and use as many copies of it as possible. More precisely, we call *algebraic* to any decisional problem (such as DDH, DLin or D3DH) in which the problem instance is defined by a tuple of elements in a (cyclic) group which discrete logarithms fulfil or not a specific algebraic equation. The way the problem instance is embedded into the matrix M is by rewriting the algebraic equation as $\det M = 0$.

5.1   The Decisional Linear Problem

The Decisional Linear problem consists on distinguishing between the distributions $(x_{\mathcal{G}}, y_{\mathcal{G}}, z_{\mathcal{G}}, t_{\mathcal{G}}, (x^{-1}z + y^{-1}t)_{\mathcal{G}}) \in \mathcal{G}^5$ and $(x_{\mathcal{G}}, y_{\mathcal{G}}, z_{\mathcal{G}}, t_{\mathcal{G}}, u_{\mathcal{G}}) \in \mathcal{G}^5$, where $x, y, z, t, u \in_{\mathrm{R}} \mathbb{Z}_q$ are chosen independently and uniformly at random. More formally, we consider the following two experiments between a challenger and a distinguisher $\mathcal{A}$.

Experiment $\mathbf{ExpDLin}_{\mathcal{A}}^b(\mathcal{G})$ is defined as follows, for $b = 0, 1$.

1. The challenger chooses random $x, y, z, t, u \in_{\mathrm{R}} \mathbb{Z}_q$. If $b = 0$, the challenger sends the tuple $(1_{\mathcal{G}}, x_{\mathcal{G}}, y_{\mathcal{G}}, z_{\mathcal{G}}, t_{\mathcal{G}}, (x^{-1}z + y^{-1}t)_{\mathcal{G}}) \in \mathcal{G}^6$ to $\mathcal{A}$. Otherwise, it sends the tuple $(1_{\mathcal{G}}, x_{\mathcal{G}}, y_{\mathcal{G}}, z_{\mathcal{G}}, t_{\mathcal{G}}, u_{\mathcal{G}}) \in \mathcal{G}^6$.
2. The distinguisher $\mathcal{A}$ outputs a bit $b' \in \{0, 1\}$.

Let Ω_b be the event that $\mathcal{A}$ outputs $b' = 1$ in $\mathbf{ExpDLin}_{\mathcal{A}}^b(\mathcal{G})$. The advantage of $\mathcal{A}$ is $\mathbf{AdvDLin}_{\mathcal{A}}(\mathcal{G}) = |\Pr[\Omega_0] - \Pr[\Omega_1]|$. We can then define $\mathbf{AdvDLin}(\mathcal{G}; t) = \max_{\mathcal{A}} \{\mathbf{AdvDLin}_{\mathcal{A}}(\mathcal{G})\}$, where the maximum is taken over all $\mathcal{A}$ running within time t.

Definition 2 (DLin). *The Decisional Linear assumption in a group $\mathcal{G}$ states that $\mathbf{AdvDLin}(\mathcal{G}; t)$ is negligible in $\lambda = \log |\mathcal{G}|$ for any value of t that is polynomial in λ.*

Lemma 8. *Any distinguisher for* $\mathbf{Rank}(\mathcal{G}, \ell_1, \ell_2, k - \delta, k)$, $\ell_1, \ell_2 \geq 3$, $k \geq 3$, $1 \leq \delta \leq \lfloor \frac{k}{3} \rfloor$ *can be converted into a distinguisher for the Decisional Linear (DLin) problem, with the same advantage and running essentially within the same time.*

Proof. Given a DLin instance $(1, x, y, z, t, u)_\mathcal{G}$ the DLin distinguisher builds the $\ell_1 \times \ell_2$ matrix

$$
M_\mathcal{G} = \underbrace{\begin{pmatrix} x & 0 & 1 \\ 0 & y & t \\ z & 1 & u \end{pmatrix}_\mathcal{G} \oplus \cdots \oplus \begin{pmatrix} x & 0 & 1 \\ 0 & y & t \\ z & 1 & u \end{pmatrix}_\mathcal{G}}_{\delta \text{ times}} \oplus I_{k-3\delta\,\mathcal{G}} \oplus 0_{(m-k)\times(n-k)\,\mathcal{G}}
$$

and submits the randomized matrix $LM_\mathcal{G}R$ to the $\mathbf{Rank}(\mathcal{G}, \ell_1, \ell_2, k - \delta, k)$ distinguisher, where $L \in_R \mathrm{GL}_{\ell_1}(\mathbb{Z}_q)$ and $R \in_R \mathrm{GL}_{\ell_2}(\mathbb{Z}_q)$. Notice that if $u = x^{-1}z + y^{-1}t \mod q$ then the resulting matrix is a random matrix in $\mathcal{G}^{\ell_1 \times \ell_2; k - \delta}$. Otherwise, it is a random matrix in $\mathcal{G}^{\ell_1 \times \ell_2; k}$. $\qquad \square$

Theorem 3. *For any* ℓ_1, ℓ_2, r_1, r_2 *such that* $2 \leq r_1 < r_2 \leq \min(\ell_1, \ell_2)$,

$$
\mathbf{AdvRank}(\mathcal{G}, \ell_1, \ell_2, r_1, r_2; t) \leq \left\lceil \frac{\log(3r_2) - \log(3r_1 - 2)}{\log 3 - \log 2} \right\rceil \mathbf{AdvDLin}(\mathcal{G}; t') \leq
$$

$$
\leq \left\lceil 1.71 \log_2 \frac{r_2}{r_1 - 1} \right\rceil \mathbf{AdvDLin}(\mathcal{G}; t')
$$

Proof. We can apply a hybrid argument similar to the one used in Theorem 1. Let us consider the sequence of integers $\{n_i\}$ defined by the recurrence $n_0 = r_1$ and $n_{i+1} = \lfloor \frac{3n_i}{2} \rfloor$, and let k be the smallest index such that $n_k \geq r_2$. Then define a sequence of random matrices $\{M_{i\mathcal{G}}\}$, where $M_i \in_R \mathbb{Z}_q^{\ell_1 \times \ell_2; n_i}$ for $i = 0, \ldots, k-1$, and $M_k \in_R \mathbb{Z}_q^{\ell_1 \times \ell_2; r_2}$. For any distinguisher $\mathcal{A}_{\mathbf{Rank}}$ with running time upper bounded by t, let $p_i = \Pr[1 \leftarrow \mathcal{A}_{\mathbf{Rank}}(M_{i\mathcal{G}})]$. By Lemma 8,

$$
|p_{i+1} - p_i| = \mathbf{AdvRank}_{\mathcal{A}_{\mathbf{Rank}}}(\mathcal{G}, \ell_1, \ell_2, n_i, n_{i+1}) \leq \mathbf{AdvDLin}(\mathcal{G}; t')
$$

for $i = 0, \ldots, k - 2$, and

$$
|p_k - p_{k-1}| = \mathbf{AdvRank}_{\mathcal{A}_{\mathbf{Rank}}}(\mathcal{G}, \ell_1, \ell_2, n_{k-1}, r_2) \leq \mathbf{AdvDLin}(\mathcal{G}; t')
$$

Therefore,

$$
\mathbf{AdvRank}_{\mathcal{A}_{\mathbf{Rank}}}(\mathcal{G}, \ell_1, \ell_2, r_1, r_2) = |p_k - p_0| \leq |p_1 - p_0| + \ldots + |p_k - p_{k-1}| \leq
$$
$$
\leq k \cdot \mathbf{AdvDLin}(\mathcal{G}; t')
$$

On the other hand, as $\lfloor \frac{3x}{2} \rfloor \geq \frac{3x-1}{2}$ then $n_k \geq \left(\frac{3}{2} \right)^k \left(r_1 - \frac{2}{3} \right)$ which implies that $k \leq \frac{\log(3r_2) - \log(3r_1 - 2)}{\log 3 - \log 2}$. $\qquad \square$

The optimality of the reduction presented above can be analyzed with the same tools described in Section 4, but adapting some parts of Subsection 4.4. First of all, we can describe the 0-instances and the 1-instances for the DLin problem in a slightly different way. Namely, $\mathcal{M}_{\text{DLin-0}} = \{B_1 + xB_2 + yB_3 + x\alpha B_4 + y\beta B_5 + (\alpha + \beta)B_6 \mid x, y, \alpha, \beta \in \mathbb{Z}_q\}$, while $\mathcal{M}_{\text{DLin-1}} = \{B_1 + xB_2 + yB_3 + x\alpha B_4 + y\beta B_5 + (\alpha + \beta + s)B_6 \mid x, y, \alpha, \beta, s \in \mathbb{Z}_q\}$, for some matrices $B_1, B_2, B_3, B_4, B_5, B_6 \in \mathbb{Z}_q^{\ell_1 \times \ell_2}$ that could depend on the random tape of the reduction. By a similar trick one can manage to reprove Lemma 7 also for DLin and the rest of the analysis works equally well. The trick in this case is excluding the case $\alpha + \beta = 0$ (which affects to a negligible fraction of the matrices) and then using a more elaborate bijection which transforms $B_1 + xB_2 + yB_3 + x\alpha B_4 + y\beta B_5 + (\alpha + \beta)B_6$ into $\gamma B_1 + x\gamma B_2 + y\gamma B_3 + x\alpha\gamma B_4 + y(1 - \alpha\gamma)B_5 + B_6$, where $\gamma = 1/(\alpha + \beta)$.

However, the logarithmic expression (which is identical to the one in Theorem 2) for the maximal loss-factor in the reduction is different from the loss-factor in the above reduction, leaving a gap that could mean that a better 'natural' reduction is still possible. Nevertheless, the authors think that a more detailed analysis of the maximal ranks r_{m0} and r_{m1} could be possible, which would improve the negative result obtained here.

5.2 The D3DH Problem

The Decisional 3-Party Diffie-Hellman (D3DH) problem [3,6,9] consists in telling apart the two distributions $(x\mathcal{G}, y\mathcal{G}, z\mathcal{G}, (xyz)\mathcal{G}) \in \mathcal{G}^4$ and $(x\mathcal{G}, y\mathcal{G}, z\mathcal{G}, t\mathcal{G}) \in \mathcal{G}^4$, where $x, y, z, t \in_R \mathbb{Z}_q$ are chosen independently at random. The problem is formally defined through the following two experiments between a challenger and a distinguisher $\mathcal{A}$.

Experiment $\mathbf{ExpD3DH}_{\mathcal{A}}^{b}(\mathcal{G})$ is defined as follows, for $b = 0, 1$.

1. The challenger chooses random $x, y, z, t \in_R \mathbb{Z}_q$. If $b = 0$, the challenger sends the tuple $(1_\mathcal{G}, x\mathcal{G}, y\mathcal{G}, z\mathcal{G}, (xyz)\mathcal{G}) \in \mathcal{G}^5$ to $\mathcal{A}$. Otherwise, it sends the tuple $(1_\mathcal{G}, x\mathcal{G}, y\mathcal{G}, z\mathcal{G}, t\mathcal{G}) \in \mathcal{G}^5$.
2. The distinguisher $\mathcal{A}$ outputs a bit $b' \in \{0, 1\}$.

Let Ω_b be the event that $\mathcal{A}$ outputs $b' = 1$ in $\mathbf{ExpD3DH}_{\mathcal{A}}^{b}(\mathcal{G})$. The advantage of $\mathcal{A}$ is $\mathbf{AdvD3DH}_{\mathcal{A}}(\mathcal{G}) = |\Pr[\Omega_0] - \Pr[\Omega_1]|$ and we define $\mathbf{AdvD3DH}(\mathcal{G}, t) = \max_{\mathcal{A}} \{\mathbf{AdvD3DH}_{\mathcal{A}}(\mathcal{G})\}$, where the maximum is taken over all $\mathcal{A}$ running within time t.

Definition 3. *The* Decisional 3-Party Diffie-Hellman assumption *in a group $\mathcal{G}$ states that* $\mathbf{AdvD3DH}(\mathcal{G}, t)$ *is negligible in* $\lambda = \log |\mathcal{G}|$ *for any value of t that is polynomial in* λ.

Similar to the Decisional Linear problem, it turns out that the D3DH problem is easier than the Rank problem.

Theorem 4. *For any* ℓ_1, ℓ_2, r_1, r_2 *such that* $2 \le r_1 < r - 2 \le \min(\ell_1, \ell_2)$,

$$\mathbf{AdvRank}(\mathcal{G}, \ell_1, \ell_2, r_1, r_2; t) \le \left\lceil \frac{\log(3r_2) - \log(3r_1 - 2)}{\log 3 - \log 2} \right\rceil \mathbf{AdvD3DH}(\mathcal{G}; t') \le$$

$$\le \left\lceil 1.71 \log_2 \frac{r_2}{r_1 - 1} \right\rceil \mathbf{AdvD3DH}(\mathcal{G}; t')$$

Proof. The proof only differs from the proof of Proposition 3 in the 3×3 blocks built from a problem instance, in the proof of Lemma 3. Indeed, given the D3DH instance $(1, x, y, z, t)_{\mathcal{G}}$ the matrix

$$\begin{pmatrix} x & -1 & 0 \\ 0 & y & 1 \\ t & 0 & z \end{pmatrix}$$

has rank 2 or 3 depending on whether $t = xyz \mod q$.   □

The analysis of the optimality of this reduction is comparable to the case of the Decisional Linear problem. Here the sets of matrices are $\mathcal{M}_{\text{D3DH-0}} = \{B_1 + xB_2 + yB_3 + zB_4 + xyzB_5 \mid x, y, z \in \mathbb{Z}_q\}$ and $\mathcal{M}_{\text{D3DH-1}} = \{B_1 + xB_2 + yB_3 + zB_4 + (xyz + s)B_5 \mid x, y, z, s \in \mathbb{Z}_q\}$, for some matrices $B_1, B_2, B_3, B_4, B_5 \in \mathbb{Z}_q^{\ell_1 \times \ell_2}$ that could depend on the random tape of the reduction. The same gap between the constructive and negative results is obtained.

5.3   Further Generalizations

The ideas presented before, both the constructive and the negative results for reductions of some decisional problems to the Rank problem seems to be easily applicable to a wide class of decisional problems. On the one hand, the construction of a reduction to the Rank problem only needs a way to encode the difference the 0-instance and the 1-instance of the problem as the determinant of a square matrix M built up from the group elements in the instances. Typically a 0-instance corresponds to $\det M = 0$. Following this approach, it is straightforward to obtain efficient reductions for instance for the family of Decisional r-Linear Problems, with arbitrary r.

On the other hand, the negative results about the existence of efficient reductions also rely on algebraic considerations, mainly related to the sets $\mathcal{M}$ which can be seen as special affine algebraic varieties. It is an open problem to obtain a description of a wide class of algebraic decisional problems for which a general negative result can be derived.

In this paper, only prime order groups are considered. However, it would be interesting to investigate whether the techniques presented here can be applied to composite order groups, where the matrices involved in the analysis are defined over rings, and this can introduce some extra difficulties to deal with notions like the rank and the random self-reducibility.

Acknowledgements. The authors are grateful to Dennis Hofheinz, David Galindo, Javier Herranz, Eike Kiltz, Gottfried Herold and Alexander May for insightful discussions and comments.

References

1. Boneh, D., Boyen, X., Shacham, H.: Short Group Signatures. In: Franklin, M. (ed.) CRYPTO 2004. LNCS, vol. 3152, pp. 41–55. Springer, Heidelberg (2004)
2. Boneh, D., Halevi, S., Hamburg, M., Ostrovsky, R.: Circular-Secure Encryption from Decision Diffie-Hellman. In: Wagner, D. (ed.) CRYPTO 2008. LNCS, vol. 5157, pp. 108–125. Springer, Heidelberg (2008)
3. Boneh, D., Sahai, A., Waters, B.: Fully Collusion Resistant Traitor Tracing with Short Ciphertexts and Private Keys. In: Vaudenay, S. (ed.) EUROCRYPT 2006. LNCS, vol. 4004, pp. 573–592. Springer, Heidelberg (2006)
4. Galindo, D., Herranz, J., Villar, J.L.: Identity-based encryption with master key-dependent message security and applications. IACR Cryptology ePrint Archive, 142 (2012)
5. Galindo, D., Herranz, J., Villar, J.: Identity-Based Encryption with Master Key-Dependent Message Security and Leakage-Resilience. In: Foresti, S., Yung, M., Martinelli, F. (eds.) ESORICS 2012. LNCS, vol. 7459, pp. 627–642. Springer, Heidelberg (2012)
6. Green, M., Hohenberger, S.: Practical Adaptive Oblivious Transfer from Simple Assumptions. In: Ishai, Y. (ed.) TCC 2011. LNCS, vol. 6597, pp. 347–363. Springer, Heidelberg (2011)
7. Hofheinz, D.: All-but-many lossy trapdoor functions. Cryptology ePrint Archive, Report 2011/230 (2011), http://eprint.iacr.org/
8. Hofheinz, D.: All-But-Many Lossy Trapdoor Functions. In: Pointcheval, D., Johansson, T. (eds.) EUROCRYPT 2012. LNCS, vol. 7237, pp. 209–227. Springer, Heidelberg (2012)
9. Laguillaumie, F., Paillier, P., Vergnaud, D.: Universally convertible directed signatures. In: Roy [14], pp. 682–701
10. Naor, M., Segev, G.: Public-Key Cryptosystems Resilient to Key Leakage. In: Halevi, S. (ed.) CRYPTO 2009. LNCS, vol. 5677, pp. 18–35. Springer, Heidelberg (2009)
11. Pascal Paillier and Damien Vergnaud. Discrete-log-based signatures may not be equivalent to discrete log. In: Roy [14], pp. 1–20
12. Peikert, C., Waters, B.: Lossy trapdoor functions and their applications. IACR Cryptology ePrint Archive, 279 (2007)
13. Peikert, C., Waters, B.: Lossy trapdoor functions and their applications. In: Dwork, C. (ed.) STOC, pp. 187–196. ACM (2008)
14. Roy, B. (ed.): ASIACRYPT 2005. LNCS, vol. 3788. Springer, Heidelberg (2005)
15. Schwartz, J.T.: Fast probabilistic algorithms for verification of polynomial identities. J. ACM 27(4), 701–717 (1980)

Signature Schemes Secure against Hard-to-Invert Leakage*

Sebastian Faust[1], Carmit Hazay[2,**], Jesper Buus Nielsen[1],
Peter Sebastian Nordholt[1], and Angela Zottarel[1,***]

[1] Aarhus University, Denmark
[2] Computer Engineering Department, Bar-Ilan University, Israel

Abstract. In the auxiliary input model an adversary is allowed to see
a *computationally hard-to-invert function* of the secret key. The auxiliary input model weakens the bounded leakage assumption commonly
made in leakage resilient cryptography as the hard-to-invert function
may information-theoretically reveal the entire secret key. In this work,
we propose the *first* constructions of digital signature schemes that are
secure in the auxiliary input model. Our main contribution is a digital signature scheme that is secure against *chosen message attacks* when
given an *exponentially hard-to-invert function* of the secret key. As a second contribution, we construct a signature scheme that achieves security
for *random messages* assuming that the adversary is given a *polynomial-
time* hard to invert function. Here, polynomial-hardness is required even
when given the entire public-key – so called *weak* auxiliary input security. We show that such signature schemes readily give us auxiliary input
secure identification schemes.

1 Introduction

Modern cryptography analyzes the security of cryptographic algorithms in the
black-box model. An adversary may view the algorithm's inputs and outputs, but
the secret key as well as all the internal computation remains perfectly hidden.
Unfortunately, the assumption of perfectly hidden keys does not reflect practice where keys frequently get compromised for various reasons. An important
example is side-channel attacks that exploit *information leakage* from the implementation of an algorithm. Side-channel attacks do not only allow the adversary
to gain partial knowledge of the secret key thereby making security proofs less
meaningful, but in many cases may result in complete security breaches.

---

* A full version of this article can be found at http://eprint.iacr.org/2012/045
** This works was done while being affiliated with Aarhus University.
*** The authors acknowledge support from the Danish National Research Foundation
and The National Science Foundation of China (under the grant 61061130540) for
the Sino-Danish Center for the Theory of Interactive Computation, and also from
the CFEM research center (supported by the Danish Strategic Research Council)
within which part of this work was performed.

X. Wang and K. Sako (Eds.): ASIACRYPT 2012, LNCS 7658, pp. 98–115, 2012.

In the last years, significant progress has been made within the theory community to incorporate information leakage into the black-box model (cf. [1, 2, 8, 10, 11, 13, 20, 21] and many more). To this end, these works develop new models to formally describe the information leakage, and design new schemes that can be proven secure therein. The leakage is typically characterized by a *leakage function* h that takes as input the secret key sk and reveals $h(sk)$—the so-called *leakage*—to the adversary. Of course, we cannot allow h to be any function as otherwise it may just reveal the complete secret key. Hence certain restrictions on the class $\mathcal{H}$ of admissible leakage functions are necessary.

With very few exceptions (outlined in the next section) most works assume some form of quantitative restriction on the amount of information leaked to an adversary. More formally, in the *bounded* leakage model, it is assumed that $\mathcal{H}$ is the set of all polynomial-time computable functions $h : \{0,1\}^{|sk|} \to \{0,1\}^{\lambda}$ with $\lambda \ll |sk|$. This restriction can be weakened in many cases. Namely, instead of requiring a concrete bound λ on the amount of leakage, it often suffices that given the leakage $h(sk)$ the secret key still has a "sufficient" amount of min-entropy left [9, 11, 21, 22]. This so-called *noisy leakage* models real-world leakage functions more accurately as now the leakage can be arbitrarily large. Indeed, real-world measurements of physical phenomenons are usually described by several megabytes or even gigabytes of information rather than by a few bits.

While security against bounded or noisy leakage often provides a first good indication for the security of a cryptographic implementation, in practice leakage typically information theoretically determines the entire secret key [25]. The only difficulty of a side-channel adversary lies in extracting the relevant key information efficiently. Formally, this can be modeled by assuming that $\mathcal{H}$ is the set of all polynomial-time computable functions such that given $h(sk)$ it is still computationally "hard" to compute sk. Such *hard-to-invert* leakage are a very natural generalization of both the bounded leakage model and the noisy leakage model, and is the focus of this work. More concretely, we will analyze the security of digital signature schemes in the presence of *hard-to-invert* leakage. We show somewhat surprisingly that simple variants of constructions for the bounded leakage setting [4, 8, 9, 17, 19] also achieve security with respect to the more *general* class of hard-to-invert leakage.

1.1   The Auxiliary Input Model

The *auxiliary input model* of Dodis, Kalai and Lovett [10] introduced the notion of security of cryptographic schemes in the presence of computationally hard-to-invert leakage. They propose constructions for secret key encryption with IND-CPA and IND-CCA security against an adversary who obtains an arbitrary polynomial-time computable hard-to-invert leakage $h(sk)$. Security is shown to hold under a non-standard LPN-related assumption with respect to any *exponentially* hard-to-invert function. We say that h is an exponentially hard-to-invert function of the secret key sk, if there exists a constant $c > 0$ such that, for sufficiently large $k = |sk|$, any PPT adversary $\mathcal{A}$ has probability of at most 2^{-ck} in inverting $h(\mathsf{sk})$. Notice that the result gets stronger, and the class of admissible leakage function gets larger, if c is smaller.

In a follow-up paper, and most relevant for our work, Dodis et al. [7] study the setting of public key encryption. They show that the BHHO encryption scheme [3] based on DDH and variants of the GPV encryption scheme [14] based on LWE are secure with respect to auxiliary input leakage. All their schemes remain secure under *sub-exponentially* hard-to-invert leakage (for a weaker notion that we discuss below [7] achieves security with respect to polynomial hard-to-invert leakages). That is, a function h is sub-exponentially hard-to-invert if there exists a constant $1 > c > 0$ such that $h(\mathsf{sk})$ can be inverted with probability at most 2^{-k^c}.

In the public key setting, some important subtleties arise which are also important for our work.

1. We shall allow the leakage to depend also on the corresponding public key pk. One approach to model this is to let the adversary adaptively choose the leakage function after seeing the public key pk [1]. An alternative that is taken in the work of Dodis et al. [7] assumes admissible leakage functions $h : \{0,1\}^{|\mathsf{sk}|+|\mathsf{pk}|} \to \{0,1\}^*$, where it is hard to compute sk given $h(\mathsf{pk}, \mathsf{sk})$.
2. The public key itself may leak information about the secret key. To illustrate this, consider a contrived scheme, where the public key pk contains the first $k/2$ bits of the secret key in clear. Suppose we want to prove security for leakage functions h with the property that given $h(\mathsf{pk}, \mathsf{sk})$, it is at least $2^{-k/2}$ hard to compute the secret key sk. Given the public key pk and such leakage that reveals the last $k/2$ bits of the secret key, the scheme from above gets completely insecure. To handle this issue, Dodis et al. propose a *weaker* notion of auxiliary input security, which assumes that a function is an admissible leakage if it is hard to compute the secret key *even* when given the public key.

For ease of presentation, we mainly consider in this work this weaker notion of auxiliary input security. As shown in [7], when the public key is short this notion implies security for functions h solely under the assumption that given $h(\mathsf{pk}, \mathsf{sk})$ it is computationally hard to compute sk (i.e., without defining hardness with respect to pk). The underlying idea is that the public key can be guessed within the proof, which implies that the hardness assumption gets stronger when applying this proof technique. Specifically, security is obtained in the presence of exponentially hard-to-invert leakage functions. We further note that this weaker notion already suffices for composition of different cryptographic schemes using the same public key. For instance, consider an encryption and signature scheme sharing the same public key. If the encryption scheme is weakly secure with respect to any polynomially hard-to-invert leakage function,[1] then the scheme remains secure even if the adversary sees arbitrary signatures, as these signatures

---

[1] A function h is *polynomially* hard-to-invert auxiliary information, if any probabilistic polynomial-time adversary computes sk with negligible probability, given the leakage $h(\mathsf{sk}, \mathsf{pk})$.

can be viewed as hard-to-invert leakage. The opposite may not trivially hold for signature schemes that are secure with respect to (sub) exponentially hard-to-invert leakages.

Recently, Brakerski and Goldwasser [5] and Brakerski and Segev [6] proposed further constructions of public key encryptions secure against auxiliary input leakage. In the former, the authors show how to construct a public key encryption scheme secure against sub-exponentially hard-to-invert leakage, based on the QR and DCR hardness assumptions. In the latter, the concept of security against auxiliary input has been introduced in the context of deterministic public key encryption, and several secure constructions were proposed based on DDH and subgroup indistinguishability assumptions.

1.2   Our Contributions

Despite significant progress on constructing encryption schemes in the auxiliary input model, the question of whether digital signature schemes can be built with security against hard-to-invert leakage has remained open so far. This is somewhat surprising as a large number of constructions for the bounded and noisy leakage setting are known [2,4,8,9,17,19]. In this paper, we close this gap and propose the first constructions for digital signature schemes with security in the auxiliary input model. As a first contribution of our work, we propose new security notions that are attainable in the presence of hard-to-invert leakage. We then show that constructions that have been proven to be secure when the amount of leakage is bounded, also achieve security in the presence of hard-to-invert leakage. In a nutshell, our results can be summarized as follows:

1. As shown below, existential unforgeability is unattainable in the presence of polynomially hard-to-invert leakage. We thus weaken the security notion by focusing on the setting where the challenge message is chosen uniformly at random. Our construction uses ideas from [19] to achieve security against polynomially hard-to-invert leakage when prior to the challenge message the adversary only has seen signatures for random messages. Such schemes can straightforwardly be used to construct identification schemes with security against any polynomially hard-to-invert leakage (cf. Sections 3.2).

2. We show that the *generic* constructions proposed in [4, 9, 17] achieve the strongest notion of security, namely *existentially unforgeable under chosen message attacks*, if we restrict the adversary to obtain only *exponentially hard-to-invert* leakage. As basic ingredients these schemes use a family of second preimage resistant hash functions, an IND-CCA secure public key encryption scheme with labels and a reusable CRS-NIZK proof system. For our result to be meaningful, we require both the decryption key and the simulation trapdoor of the underlying encryption scheme to be short when compared to the length of the signing key for the signature scheme (cf. Section 3.3).

3. We show an instantiation of this generic transformation that satisfies our requirements on the length of the keys based on the 2-Linear hardness

assumption in pairing based groups, using the Groth-Sahai proof system [16] (we refer the reader to the full version).

We elaborate on these results in more detail below.

Polynomially Hard-to-Invert Leakage and Random Challenges. Importantly, security with respect to polynomially hard-to-invert leakage is impossible if the message for which the adversary needs to output a forgery, is fixed at the time the leakage function is chosen. This is certainly the case for the standard security notion of existential unforgeability. One potential weakening of the security definition is by requiring the adversary to forge a signature on a random challenge message. In the case when the challenge messages is sampled uniformly at random, even though the leakage may reveal signatures for some messages, it is very unlikely that the adversary hits a forgery for the challenge message.

Specifically, inspired by the work of Malkin et al. [19], we propose a construction that guarantees security in the presence of *any polynomially hard-to-invert* leakage, when the challenge message is chosen uniformly at random. The scheme uses the message as the CRS for a non-interactive zero-knowledge proof of knowledge (NIZKPoK). To sign, we use the CRS to prove knowledge of sk such that vk = $H(\text{sk})$, where H is a second preimage resistant hash function. Therefore, if an adversary forges a signature given vk and the leakage $h(\text{vk}, \text{sk})$ with non-negligible probability, we can use this forgery to extract a preimage of vk which either contradicts the second preimage resistance of H or the assumption that h is polynomially hard-to-invert. An obvious drawback of this scheme is that prior to outputting a forgery for the challenge message the adversary only sees signatures on random messages. Finally, as a natural application of such schemes, we show that auxiliary input security for signatures carries over to auxiliary input security of identification schemes. Hence, our scheme can be readily used to build simple identification schemes with security against any polynomially hard-to-invert leakage function.

Exponentially Hard-to-Invert Leakage and Existential Unforgeability. The standard security notion for signature schemes is existential unforgeability under adaptive chosen-message attacks [15]. Here, one requires that an adversary cannot forge a signature of any message m, even when given access to a signing oracle. We strengthen this notion and additionally give the adversary leakage $h(\text{vk}, \text{sk})$, where h is some admissible function from class $\mathcal{H}$. It is easy to verify that no signature scheme can satisfy this security notion when the only assumption that is made about $h \in \mathcal{H}$, is that it is polynomially hard to compute sk given $h(\text{vk}, \text{sk})$. The reason for this is as follows. Since the secret key must be polynomially hard to compute even given some set of signatures (and the public key), a signature is an admissible leakage function with respect to $\mathcal{H}$. Hence, a forgery is a valid leakage. This observation holds even when we define the hardness of h with respect to the public key as well.

Our first observation towards constructing signatures with auxiliary input security is that the above issues do not necessarily arise when we consider the more restricted class of functions that maintain (sub)-exponentially hardness of inversion. Suppose, for concreteness, that there exists a constant $1 > c > 0$ such that there exists a probabilistic polynomial-time algorithm, taking as input a signature and the public key and outputting sk with probability p. Here, we assume that $\mathsf{negl}(k) \geq p \gg 2^{-k^c}$ for some negligible function $\mathsf{negl}(\cdot)$. Then, if we let $\mathcal{H}$ be the class of functions with hardness at least 2^{-k^c}, the signing algorithm is not in $\mathcal{H}$ and hence the artificial counterexample from above does not work anymore! We instantiate this idea by adding an encryption $C = \mathsf{Enc}_{\mathsf{ek}}(\mathsf{sk})$ of the signing key sk to each signature. The encryption key ek is part of the verification key of the signature scheme, but the decryption key dk associated with ek is not part of the signing key. However, we set up the scheme such that dk can be guessed with probability p. Interestingly, it turns out that recent constructions of leakage resilient signatures [4,9,17], which originally were designed to protect against *bounded* leakage, use as part of the signature an encryption of the secret key. This enables us to prove that these schemes also enjoy security against exponentially hard-to-invert leakages.

One may object that artificially adding an encryption of the secret key to the signature is somewhat counter-intuitive as it seems to reduce the security of the signature scheme. However, all that is needed for this trick is that guessing dk is significantly easier than guessing sk. For a given security level we can therefore pick the length of dk first, as to achieve that security level. After that we can then pick the length of sk as to achieve meaningful leakage bounds. Our concrete security analysis allows to choose these keys as to achieve a given security. Note, also, that adding trapdoors to cryptographic schemes for what superficially only seems to be proof reasons is common in the field – non-interactive zero-knowledge being another prominent example.

For readers familiar with the security proof of the Katz-Vaikuntanathan scheme [17], we note that the crux of our new proof is that in the reduction we cannot generate a CRS together with its simulation trapdoor. Instead, to simulate signatures for chosen messages we will guess the simulation trapdoor. Fortunately, we can show that the loss from guessing the simulation trapdoor only effects the tightness in the reduction to the inversion hardness of the leakage functions. As we use a NIZK proof system with a short simulation trapdoor and only aim for exponential hard-to-invert leakage functions, we can successfully complete the reduction.

Instantiation under the 2-Linear Assumption. As a concrete example, we show in the full version how to instantiate our generic transformation using the Groth-Sahai proofs system based on the 2-linear assumption. This yields security with respect to any $2^{-6k'}$-hard-to-invert leakage. If we do not wish to define the hardness with respect to the public key as well, it is possible to guess it and thus loose an additional factor of $2^{-3k'}$ in the hardness assumption. Here, $k' := \log(p)$ for a prime p that denotes the order of the group for which the 2-linear assumption holds, and the secret key of our scheme has length $k := \ell \cdot k'$ bits for some constant $\ell \in \mathbb{N}$.

1.3 A Road Map

In Section 2 we specify basic security definitions and our modeling for the auxiliary input setting. In Section 3 we present our signature schemes for random messages (Section 3.2) and chosen massage attack security (Section 3.3). In the full version we show how to use signatures on random messages to construct identification schemes with security against any polynomially hard-to-invert leakage. We also show an instantiation of the later signature scheme under the 2-linear hardness assumption.

2 Preliminaries

Basic Notation. We denote the security parameter by k and by PPT probabilistic polynomial-time. For a set S we write $x \leftarrow S$ to denote that x is sampled uniformly from S. We write $y \leftarrow \mathcal{A}(x)$ to indicate that y is the output of an algorithm $\mathcal{A}$ when running on input x. We denote by $\langle a, b \rangle$ the inner product of field elements a and b. We use $\mathsf{negl}(\cdot)$ to denote a negligible function $f : \mathbb{N} \to \mathbb{R}$ and we use the $\approx$ notation to denote computational indistinguishability of families of random variables.

2.1 Public Key Encryption Schemes

We introduce the notion of a labeled public key encryption scheme following the notation used in [9].

Definition 1 (LPKE). *We say that PPT algorithms $\Pi = (\mathsf{KeyGen}, \mathsf{Enc}, \mathsf{Dec})$ is a* labeled public key encryption scheme (LPKE) *with perfect decryption if:*

- KeyGen, *given a security parameter k, outputs keys $(\mathsf{ek}, \mathsf{dk})$, where ek is a public encryption key and dk is a secret decryption key.*
- Enc, *given the public key ek, a label L and a plaintext message m, outputs a ciphertext c encrypting m. We denote this by $c \leftarrow \mathsf{Enc}^L(\mathsf{ek}, m)$.*
- Dec, *given a label L, the secret key dk and a ciphertext c, with $c \leftarrow \mathsf{Enc}^L(\mathsf{ek}, m)$, then with probability 1 outputs m. We denote this by $m \leftarrow \mathsf{Dec}^L(\mathsf{dk}, c)$.*

Definition 2 (IND-LCCA secure encryption scheme). *We say that a labeled public key encryption scheme $\Pi = (\mathsf{KeyGen}, \mathsf{Enc}, \mathsf{Dec})$ is* IND-LCCA *secure encryption scheme if, for every admissible PPT adversary $\mathcal{A} = (\mathcal{A}_1, \mathcal{A}_2)$, there exists a negligible function negl such that the probability IND-LCCA$_{\Pi,\mathcal{A}}(k)$ that $\mathcal{A}$ wins the IND-LCCA game as defined below is at most IND-LCCA$_{\Pi,\mathcal{A}}(k) \leq \frac{1}{2} + \mathsf{negl}(k)$.*

- IND-LCCA game.

$$(\mathsf{ek}, \mathsf{dk}) \leftarrow \mathsf{KeyGen}(1^k)$$
$$(L, m_0, m_1, history) \leftarrow \mathcal{A}_1^{\mathsf{Dec}^{(\cdot)}(\mathsf{dk}, \cdot)}(\mathsf{ek}), \ s.t. \ |m_0| = |m_1|$$
$$c \leftarrow \mathsf{Enc}^L(\mathsf{ek}, m_b), \ where \ b \leftarrow \{0, 1\}$$
$$b' \leftarrow \mathcal{A}_2^{\mathsf{Dec}^{(\cdot)}(\mathsf{dk}, \cdot)}(c, history)$$
$$\mathcal{A} \ wins \ if \ b' = b.$$

An adversary is admissible if it does not query $\mathsf{Dec}^{(\cdot)}(\mathsf{dk}, \cdot)$ with (L, c)

In this work we require a weaker notion, called IND-WLCCA, where the adversary cannot query the decryption oracle with label L. Namely, we change the definition of admissible to mean that the adversary never queries $\mathsf{Dec}^{(\cdot)}(\mathsf{dk}, \cdot)$ with any input of the form $(L, \cdot)$, where L is the label picked to compute the challenge. We discuss further details why this security notion is needed for our construction in Section 3.3.

2.2   Signature Schemes

A signature scheme is a tuple of PPT algorithms $\Sigma = (\mathsf{Gen}, \mathsf{Sig}, \mathsf{Ver})$ defined as follows. The key generation algorithm Gen, on input 1^k outputs a signing and a verification key $(\mathsf{sk}, \mathsf{vk})$. The signing algorithm Sig takes as input a message m and a signing key sk and outputs a signature σ. The verification algorithm Ver, on input (vk, m, σ), outputs either 0 or 1 (respectively rejecting or accepting the signature). A signature scheme has to satisfy the following correctness property: for any message m and keys $(\mathsf{sk}, \mathsf{vk}) \leftarrow \mathsf{Gen}(1^k)$

$$\Pr[\mathsf{Ver}(\mathsf{vk}, m, \mathsf{Sig}(\mathsf{sk}, m)) = 1] = 1$$

The standard security notion for a signature scheme is existentially unforgeability under chosen message attacks. A scheme is said to be secure under this notion if, even after seeing signatures for chosen messages, no adversary can come up with a forgery for a new message. In this article, we extend this security notion and give the adversary additional auxiliary information about the signing key. To this end, we define a set of admissible leakage functions $\mathcal{H}$ and allow the adversary to obtain the value $h(\mathsf{sk}, \mathsf{vk})$ for any $h \in \mathcal{H}$. Notice that by giving vk as input to the leakage function, we capture the fact that the choice of h may depend on vk.

Definition 3 (Existential Unforgeability under Chosen Message and Auxiliary Input Attacks (EU-CMAA)). *We say that a signature scheme* $\Sigma = (\mathsf{Gen}, \mathsf{Sig}, \mathsf{Ver})$ *is* existentially unforgeable against chosen message and auxiliary input attacks (EU-CMAA) *with respect to* $\mathcal{H}$ *if for all PPT adversaries* $\mathcal{A}$ *and any function* $h \in \mathcal{H}$, *the following probability* $\Pr[\mathsf{CMA}_{\Sigma, \mathcal{A}, h}(k) = 1]$ *is negligible in* k, *where* $\mathsf{CMA}_{\Sigma, \mathcal{A}, h}(k)$ *is defined as follows:*

Experiment $\mathsf{CMA}_{\Sigma, \mathcal{A}, h}(k)$	*Oracle* $\mathcal{O}(\mathsf{sk}, m)$
$\quad (\mathsf{vk}, \mathsf{sk}) \leftarrow \mathsf{Gen}(1^k)$	$\quad Return\ (m, \mathsf{Sig}(\mathsf{sk}, m))$
$\quad (m^*, \sigma^*) \leftarrow \mathcal{A}^{\mathcal{O}(\mathsf{sk}, \cdot)}(1^k, h(\mathsf{vk}, \mathsf{sk}), \mathsf{vk})$	
$\quad If\ m^* \notin M\ return\ \mathsf{Ver}(\mathsf{vk}, m^*, \sigma^*),\ else\ return\ 0.$	

Where M *is the set of messages submitted by* $\mathcal{A}$ *to the oracle.*

We note that the leakage may also depend on $\mathcal{A}$'s signature queries as the function h may internally run $\mathcal{A}$, using the access to the secret key in order to emulate the entire security game, including the signature queries made by $\mathcal{A}$.

As outlined in the introduction, we are also interested in a weaker security notion where the adversary is required to output a forgery for a random message after seeing signatures for *random* messages. To this end, we extend the definition from above and let the signing oracle reply with random messages, as well as pick the challenge message at random. This is formally described in the following definition.

Definition 4 (Random Message Unforgeability under Random Message and Auxiliary Input Attacks (RU-RMAA)). *We say that a signature scheme $\Sigma = (\mathsf{Gen}, \mathsf{Sig}, \mathsf{Ver})$ is* random message unforgeable against random message and auxiliary input attacks (RU-RMAA) *with respect to $\mathcal{H}$ if for all PPT adversaries $\mathcal{A}$ and any function $h \in \mathcal{H}$, the probability $\Pr[\mathsf{RMA}_{\Sigma,\mathcal{A},h}(k) = 1]$ is negligible in k, where $\mathsf{RMA}_{\Sigma,\mathcal{A},h}(k)$ is defined as follows:*

Experiment $\mathsf{RMA}_{\Sigma,\mathcal{A},h}(k)$	**Oracle** $\mathcal{O}(\mathsf{sk})$
$(\mathsf{vk}, \mathsf{sk}) \leftarrow \mathsf{Gen}(1^k)$	$m \leftarrow \mathcal{M}$
$m^* \leftarrow \mathcal{M}$, *where $\mathcal{M}$ is the message space*	*Return* $(m, \mathsf{Sig}(\mathsf{sk}, m))$
$\sigma^* \leftarrow \mathcal{A}^{\mathcal{O}(\mathsf{sk})}(1^k, h(\mathsf{vk}, \mathsf{sk}), \mathsf{vk}, m^*)$	
Return $\mathsf{Ver}(\mathsf{vk}, m^*, \sigma^*)$.	

We notice that this notion of security is useful in some settings. For instance, it suffices to construct 2-round identification schemes w.r.t auxiliary inputs. In the full version of this article [12] we propose formal definitions and a simple construction of an identification scheme with security in the presence of auxiliary input leakage.

One way to enhance the security notion obtained by Definition 4 is to allow chosen message attacks, i.e., random message unforgeability under chosen message and auxiliary input attacks (RU-CMAA). In this game the adversary can pick the messages to be signed by itself but still need to forge a signature on a random message; see Section 3.2 for further discussion.

2.3   Classes of Auxiliary Input Functions

The above notions of security require to specify the set of admissible functions $\mathcal{H}$. In the public key setting one can define two different types of classes of leakage functions. In the first class, we require that given the leakage $h(\mathsf{sk}, \mathsf{vk})$ it is computationally hard to compute sk, while in the latter we require hardness of computing sk when additionally given the public key vk. We follow the work of Dodis et al. [7] to formally define this difference. Let in the following $(\mathsf{sk}, \mathsf{vk}) \leftarrow \mathsf{Gen}(1^k)$ be generated randomly.

- Let $\mathcal{H}_{\mathsf{ow}}(\ell(k))$ be the class of polynomial-time computable functions $h : \{0,1\}^{|\mathsf{sk}|+|\mathsf{vk}|} \to \{0,1\}^*$ such that given $h(\mathsf{sk}, \mathsf{vk})$, no PPT adversary can find sk with probability $\ell(k) \geq 2^{-k}$, i.e., for any PPT adversary $\mathcal{A}$: $\Pr_{(\mathsf{sk},\mathsf{vk})\leftarrow\mathsf{Gen}(1^k)}[\mathsf{sk} \leftarrow \mathcal{A}(h(\mathsf{sk},\mathsf{vk}))] < \ell(k)$.
- Let $\mathcal{H}_{\mathsf{vkow}}(\ell(k))$ be the class of polynomial-time computable functions $h : \{0,1\}^{|\mathsf{sk}|+|\mathsf{vk}|} \to \{0,1\}^*$ such that given $(\mathsf{vk}, h(\mathsf{sk}, \mathsf{vk}))$, no PPT adversary

can find sk with probability $\ell(k) \geq 2^{-k}$, i.e., for any PPT adversary $\mathcal{A}$:
$\Pr_{(\mathsf{sk},\mathsf{vk}) \leftarrow \mathsf{Gen}(1^k)}[\mathsf{sk} \leftarrow \mathcal{A}(\mathsf{vk}, h(\mathsf{sk}, \mathsf{vk}))] < \ell(k)$.

Security with respect to auxiliary input gets stronger if $\ell(k)$ is larger. Our goal is typically to make $\ell(k)$ as large as possible while still $\mathsf{negl}(k)$. If a scheme is EU-CMAA for $\mathcal{H}_{\mathsf{vkow}}(\ell(k))$ according to Definition 3, we say for short that it is $\ell(k)$-EU-CMAA. Similarly, if a scheme is RU-RMAA for $\mathcal{H}_{\mathsf{vkow}}(\ell(k))$, then we say that it is an $\ell(k)$-RU-RMAA signature scheme. If the class of admissible leakage functions is $\mathcal{H}_{\mathsf{ow}}(\ell(k))$, we will mention it explicitly.

As outlined in the introduction, we typically prove security with respect to the class $\mathcal{H}_{\mathsf{vkow}}(\ell(k))$. The stronger security notion where hardness is required to hold *only* given the leakage, i.e., for the class of admissible functions $\mathcal{H}_{\mathsf{ow}}(\ell(k))$, can be achieved by a relation between $\mathcal{H}_{\mathsf{ow}}(\cdot)$ and $\mathcal{H}_{\mathsf{vkow}}(\cdot)$ proven by Dodis et al. [7].

Lemma 1 ([7]). *If* $|\mathsf{vk}| = t(k)$ *then for any* $\ell(k)$, *we have*

1. $\mathcal{H}_{\mathsf{vkow}}(\ell(k)) \subseteq \mathcal{H}_{\mathsf{ow}}(\ell(k))$
2. $\mathcal{H}_{\mathsf{ow}}(2^{-t(k)}\ell(k)) \subseteq \mathcal{H}_{\mathsf{vkow}}(\ell(k))$

The first point of Lemma 1 says that if no PPT adversary finds sk given $(\mathsf{vk}, h(\mathsf{sk}, \mathsf{vk}))$ with probability $\ell(k)$ or better, then no PPT adversary finds sk given only $h(\mathsf{sk}, \mathsf{vk})$ with probability $\ell(k)$ or better. Clearly this is the case since knowing vk will not make it harder to guess sk. The second point states that if no PPT adversary finds sk given $h(\mathsf{sk}, \mathsf{vk})$ with probability $2^{-t(k)}\ell(k)$ or better, then any PPT adversary has advantage at most $\ell(k)$ in guessing sk when given additionally vk. To see this consider a PPT adversary $\mathcal{A}$ that finds sk given $(\mathsf{vk}, h(\mathsf{sk}, \mathsf{vk}))$ with probability $\ell'(k) \geq \ell(k)$. $\mathcal{A}$ then implies a PPT adversary $\mathcal{B}$ that given $h(\mathsf{sk}, \mathsf{vk})$ simply tries to guess vk and uses it to run $\mathcal{A}$. Since $\mathcal{B}$ can guess vk with probability at least $2^{-t(k)}$, $\mathcal{B}$ has probability at least $2^{-t(k)}\ell'(k)$ of finding sk. Thus contradicting $h \in \mathcal{H}_{\mathsf{ow}}(2^{-t(k)}\ell(k))$.

3 Signature Schemes with Auxiliary Input Security

3.1 A Warm-Up Construction

In order to illustrate the difficulties encountered in designing cryptographic primitives in the auxiliary input setting we present a warm-up construction of a signature scheme that may seem secure at first glance but, unfortunately, proving its security is impossible. Essentially, the problem arises due to the computational hardness of the leakage and does not occur in other leakage models, where given the leakage the secret key is still information theoretically hidden. For ease of understanding, in this warm-up construction we only aim for the simpler one-time security notion on random messages, where the adversary only views a single signature before it outputs its forgery on a random message. We consider two building blocks for the following scheme:

1. A family H of second preimage resistant hash functions.
2. A non-interactive zero-knowledge proof of knowledge[2] (NIZKPoK) system $\Pi = (\mathsf{CRSGen}, \mathsf{P}, \mathsf{V})$ for proving knowledge of a secret value x so that $y = H_s(x)$ given s and y. We further require that the CRS's of Π are uniformly random strings of some length $p(k)$ for security parameter k and some polynomial $p(\cdot)$. Denote the message space $\mathcal{M}$ by $\{0,1\}^{p(k)}$.

Informally, the signature scheme is built as follows. The signing key sk is a random element x in the domain of the hash function, whereas the verification key vk is $y = H(x)$. The verification key vk also contains a common reference string crs for Π. A signature on a message m is the bit $b = \langle m, \mathsf{sk} \rangle$ together with a non-interactive proof with respect to crs proving that b was computed as the inner product of the preimage of y and the message m. More precisely, define the signature scheme $\Sigma = (\mathsf{Gen}_\Sigma, \mathsf{Sig}_\Sigma, \mathsf{Ver}_\Sigma)$ as follows:

Key Generation, $\mathsf{Gen}_\Sigma(1^k)$: Sample a second preimage resistant hash function H_s from H, a random element x in the domain of H_s and crs $\leftarrow \mathsf{CRSGen}(1^k)$. Output $\mathsf{sk} = x$, $\mathsf{vk} = (H(x), \mathsf{crs})$.

Signing, $\mathsf{Sig}_\Sigma(\mathsf{sk}, m)$: Parse vk as $(H(\mathsf{sk}), \mathsf{crs})$. Compute $b = \langle m, \mathsf{sk} \rangle$. Use the crs to generate a non-interactive zero-knowledge proof of knowledge π, demonstrating that $b = \langle m, \mathsf{sk} \rangle$ and $H(\mathsf{sk}) = y$. Output $\sigma = (b, \pi)$.

Verifying, $\mathsf{Ver}_\Sigma(\mathsf{vk}, m, \sigma)$: Parse vk as $(H(\mathsf{sk}), \mathsf{crs})$ and σ as (b, π). Use crs to verify the proof π. Output 1 if the proof is verified correctly and 0 otherwise.

We continue with an attempt to prove security. Note first that by the properties of Π, the ability to generate a forgery (σ', m') reduces to the ability using the extraction trapdoor to either find a second preimage for the hash function or break the hardness assumption of the leakage function. As the difficulties arise in the reduction to the hardness of the leakage function, we focus in this outline on that part. Assume there is an adversary $\mathcal{A}$ attacking signature scheme Σ given auxiliary input leakage $h(\mathsf{sk}, \mathsf{vk})$ and (y, crs). Then, an attempt to construct $\mathcal{B}$ that breaks the hardness assumption of the leakage function by invoking $\mathcal{A}$ works as follows. $\mathcal{B}$ obtains (y, crs) and the leakage $h(\mathsf{sk}, \mathsf{vk})$ from its challenge oracle. It forwards them to $\mathcal{A}$ who will ask for signature query. Unfortunately, at that point we are not able to answer this query as we cannot simulate a proof without knowing the witness or the trapdoor.

An alternative approach may be to directly prove security with respect to the leakage class $\mathcal{H}_{\mathsf{ow}}(\ell(k))$ and let $\mathcal{B}$ sample the CRS herself using the zero-knowledge simulator to know a trapdoor. Unfortunately, also this approach is deemed to fail as in this case there is no way to learn a $y = H(\mathsf{sk})$ that is consistent with the leakage. Moreover this results into several difficulties in defining the set of admissible leakage functions as they must be different now for $\mathcal{A}$ and $\mathcal{B}$. This can be illustrated as follows. Suppose that the CRS is a public key for an encryption scheme and the trapdoor is the corresponding secret key. As $\mathcal{A}$ only knows the CRS but not the trapdoor a leakage function h that outputs

---

[2] For definition of NIZKPoK we refer to the full version of this article [12].

an encryption of $\mathsf{sk} = x$ is admissible. On the other hand, however, for $\mathcal{B}$ who knows the trapdoor (hence the secret key of the encryption scheme) such leakage cannot be admissible. This shows that we need to consider different approaches when analyzing the security of digital signature schemes in the presence of auxiliary input. In what follows, we demonstrate two different approaches for such constructions, obtaining two different notions of security.

3.2   A RU-RMAA Signature Scheme

In this section we present our construction of a RU-RMAA signature scheme as defined in Definition 4. For this scheme we assume the following building blocks:

1. A family H of second preimage resistant hash functions with input length k_1 and key sampling algorithm Gen_H.
2. A (NIZKPoK) system $\Pi = (\mathsf{CRSGen}, \mathsf{P}, \mathsf{V})$ for proving knowledge of a secret value x so that $y = H_s(x)$ given s and y. We further require that the CRS's of Π are uniformly random strings of some length $p(k)$ for security parameter k and some polynomial $p(\cdot)$. Denote the message space $\mathcal{M}$ by $\{0,1\}^{p(k)}$.

The main idea for the scheme is inspired by the work of Malkin et al. [19] where we view each message m as a common reference string for the proof system Π. Since m is uniformly generated, we are guaranteed that the CRS is generated correctly and knowledge soundness holds. Intuitively since each new message induces a new CRS, each proof is given with respect to an independent CRS. This implies that in the security proof the simulator (playing the role of the signer) *can* use the trapdoor of the CRS that corresponds to the challenge message m^*.
We formally define our scheme $\Sigma = (\mathsf{Gen}, \mathsf{Sig}, \mathsf{Ver})$ as follows.

Key Generation, $\mathsf{Gen}(1^k)$: Sample $s \leftarrow \mathsf{Gen}_H(1^k)$. Sample $x \leftarrow \{0,1\}^{k_1}$ and compute $y = H_s(x)$. Output $\mathsf{sk} = (x, s)$ and $\mathsf{vk} = (y, s)$.
Signing, $\mathsf{Sig}(\mathsf{sk}, m)$: To sign $m \leftarrow \mathcal{M}$, let $\mathsf{crs} = m$ and sample the signature $\sigma \leftarrow \mathsf{P}(\mathsf{crs}, \mathsf{vk}, \mathsf{sk})$ as a proof of knowledge of x such that $y = H_s(x)$.
Verifying, $\mathsf{Ver}(\mathsf{vk}, m, \sigma)$: To verify σ on $m = \mathsf{crs}$, output $\mathsf{V}(\mathsf{crs}, \mathsf{vk}, \sigma)$.

Theorem 1. *Assume that H is a second preimage resistant family of hash functions and $\Pi = (\mathsf{CRSGen}, \mathsf{P}, \mathsf{V})$ is a NIZKPoK system. Then $\Sigma = (\mathsf{Gen}, \mathsf{Sig}, \mathsf{Ver})$ is a $\mathsf{negl}(k)$-RU-RMAA signature scheme.*

The intuition of the proof is that if one can efficiently forge a signature on a random m^* after getting signatures on random messages m, then one can also efficiently compute x, contradicting the assumption that the leakage is hard to efficiently invert. During the simulated attack the signatures on random messages m are simulated by sampling $m = \mathsf{crs}$, where crs is sampled along with the simulation trapdoor. In the end one samples $m^* = \mathsf{crs}$, where crs is sampled along with the extraction trapdoor. Upon getting a forgery on m^*, we can extract x using the extraction trapdoor.

In the standard setting, a simple modification using Chameleon hash functions [18] enables to achieve a stronger notion of security. Recall first that Chameleon hash functions are collision resistance hash functions such that given a trapdoor one can efficiently find collisions for every given preimage and its hashed value. Thereby, instead of signing random messages the scheme can be modified so that the signer signs the hashed value of the message. This achieves chosen message attacks security so that the adversary picks the messages to be signed during the security game, yet the challenge is still picked at random. Nevertheless, when introducing hard-to-invert leakage into the system this approach does not enable to obtain security against polynomially hard-to-invert leakage, because we run into the same problem specified in Section 3.1. Moreover, in Section 3.3 we show how to obtain the strongest security notion of existential unforgeability under chosen message and auxiliary input attacks.

Proof. Let $\mathsf{Exp}_{\Sigma,\mathcal{A},h}$ be as defined in Definition 4 for PPT adversary $\mathcal{A}$ and leakage function $h \in \mathcal{H}_{\mathsf{vkow}}(\mathsf{negl}(k))$. Furthermore let W be the event that $\mathcal{A}$ wins the game. We show that $\Pr[W]$ is negligible. Denote this probability by p_0. Consider the following modification to $\mathsf{Exp}_{\Sigma,\mathcal{A},h}(k)$.

1. Generate $(\mathsf{vk},\mathsf{sk})$ as in $\mathsf{Exp}_{\Sigma,\mathcal{A},h}(k)$.
2. Instead of sampling the challenge m^* as $m^* \leftarrow \mathcal{M}$ sample $(m',\mathsf{td}_e) \leftarrow E_1(1^k)$ and let $m^* = m'$, where $E = (E_1, E_2)$ is the knowledge extractor for Π.
3. Give input to $\mathcal{A}$ as in $\mathsf{Exp}_{\Sigma,\mathcal{A},h}(k)$.
4. To answer the oracle queries of $\mathcal{A}$, sample $(m',\mathsf{td}_s) \leftarrow S_1(1^k)$, let $m = m'$ and return the signature $(m, S_2(m, \mathsf{vk}, \mathsf{td}_s))$, where $S = (S_1, S_2)$ is the simulator for Π.
5. Receive a forgery σ^* from $\mathcal{A}$ as in $\mathsf{Exp}_{\Sigma,\mathcal{A},h}(k)$.
6. Output as in $\mathsf{Exp}_{\Sigma,\mathcal{A},h}(k)$.

Let p_1 be the probability that the modified experiment above outputs 1. Also consider $x' = E_2(m^*, \mathsf{vk}, \mathsf{td}_e, \sigma^*)$. I.e. x' is a signing key extracted from $\mathcal{A}$'s forgery. By Π being a NIZKPoK we have that distributions of messages and signatures in the modified experiment are indistinguishable from the distributions in the original experiment $\mathsf{Exp}_{\Sigma,\mathcal{A},h}(k)$. Thus it follows that p_1 is negligibly close to p_0. Let p_2 be the probability that $H_s(x') = y$. By the knowledge soundness of Π it follows that p_2 is negligibly close to p_0.

Note then that, since S and E are both PPT algorithms, the modified experiment describes a PPT algorithm which computes x' where with probability p_2 it holds that $y = H_s(x')$. Let p_3 be the probability that $y = H_s(x')$ and $x' \neq x$ and let p_4 be the probability that $x' = x$. Note that $p_2 = p_3 + p_4$.

The Event X. Consider the PPT algorithm $\mathcal{B}$ that given vk and leakage $h(\mathsf{sk}, \mathsf{vk})$, where $(\mathsf{sk}, \mathsf{vk}) \leftarrow \mathsf{Gen}(1^k)$, runs steps 2-5 of the modified experiment above and outputs $x^* = E_2(m^*, \mathsf{vk}, \mathsf{td}_e, \sigma^*)$. Denote by X the event in which $\mathcal{B}$ outputs $x^* = x$. Since $(\mathsf{vk}, \mathsf{sk})$ is generated as in $\mathsf{Exp}_{\Sigma,\mathcal{A},h}(k)$ $\Pr[X] \geq p_4$. Thus by definition of $\mathcal{H}_{\mathsf{vkow}}(\mathsf{negl}(k))$, p_4 is negligible.

The Event C. On the other hand, consider the PPT algorithm $\mathcal{B}$ that is given s, x and $y = H_s(x)$. $\mathcal{B}$ lets $\mathsf{vk} = (y, s)$ and runs steps 2-5 of the modified experiment above (notice that $\mathcal{B}$ is given x, so it can compute the leakage h) and outputs $x^* = E_2(m^*, \mathsf{vk}, \mathsf{td}_e, \sigma^*)$. Denote by C the event in which $\mathcal{B}$ outputs $x^* \neq x$ so that $H_s(x^*) = H_s(x)$. Notice again that $((H_s, y), x)$ are generated as in $\mathsf{Exp}_{\Sigma, \mathcal{A}, h}(k)$ and therefore $\Pr[C] \geq p_3$. Thus by the second preimage resistance hardness of the family H, p_3 is negligible.

This implies that p_3 and p_4 are negligible and so is $p_2 = p_3 + p_4$. Since p_0 is negligibly close to p_2, p_0 must also be negligible. By definition $p_0 = \Pr[\mathsf{Exp}_{\Sigma, \mathcal{A}, h}(k) = 1]$ and so by Definition 4, Σ is a $\mathsf{negl}(k)$-RU-RMAA signature scheme. □

Notice that in the above we assume that the CRS of the NIZKPoK Π is a uniformly random bit string. As an example of a NIZKPoK with this property we can use the construction of [23]. In their construction the CRS is a pair (ek, r) where r is a random string and ek is an encryption key for some semantically secure public-key encryption scheme. Thus, we can use the construction of [23] with a public-key encryption scheme where uniformly random bit strings can act as public-keys, like Regev's LWE scheme[24].

3.3   A EU-CMAA Signature Scheme

In this section we build a EU-CMAA signature scheme. We use k to denote the security parameter. We need the following tools:

1. A family of second preimage resistant hash functions H with key sampling algorithm Gen_H, where the input length can be set to be any $k_4 = \mathrm{poly}(k)$ and where the length of the randomness used by $s \leftarrow \mathsf{Gen}_H(1^k)$ is some $l_1 = \mathrm{poly}(k)$ independent of k_4 and where the length of an output $y = H_s(x)$ is some $l_4 = \mathrm{poly}(k)$ independent of k_4. I.e., it is possible to increase the input length of H_s without increasing the randomness used to generate s or the output length.
2. An IND-WLCCA secure labeled public-key encryption scheme $\Gamma = (\mathsf{KeyGen}, \mathsf{Enc}, \mathsf{Dec})$ with perfect decryption (cf. Definition 2), where the length of dk is some $l_2 = \mathrm{poly}(k)$ independent of the length of the messages that Γ can encrypt.
3. A reusable-CRS non-interactive zero-knowledge proof[3] system (NIZK) $\Pi = (\mathsf{CRSGen}, \mathsf{P}, \mathsf{V})$, where the length of the simulation trapdoor td_s at security level k is some $l_3 = \mathrm{poly}(k)$ independent of the size of the proofs that the NIZK can handle.

The IND-WLCCA secure encryption scheme might be replaced by a IND-CPA secure scheme, but at the price of then instead using a simulation sound NIZK: We expect a general proof via true simulation extractability to work along the

---

[3] For definition of reusable-CRS NIZK we refer to the full version of this article [12].

lines of [9]. We chose the above tools as they lean themeselves nicely towards our concrete instantiation.

The reason why we use IND-WLCCA is that our signature scheme requires to encrypt its secret key that is much longer than the decryption key. For that we need to break the secret key into blocks and encrypt each block separately under the *same* label (looking ahead, the label would be the signed message). Note that labeled public-key encryption schemes for arbitrary length massages is not implied by LCCA secure scheme for fixed length messages. This is because the adversary can change the order of the ciphertexts within a specific set of ciphertexts and ask for a decryption. We therefore work with the weaker notion that is sufficient for our purposes to design secure signature schemes, and is easier to instantiate as demonstrated in the full version of this article [12].

Our scheme Σ works as follows:

Key Generation, $\mathsf{Gen}(1^k)$: Sample $s \leftarrow \mathsf{Gen}_H(1^k)$ and $(\mathsf{ek}, \mathsf{dk}) \leftarrow \mathsf{KeyGen}(1^k)$. Furthermore, sample $(\mathsf{crs}, \mathsf{td}_s) \leftarrow S_1(1^k)$ and $x \leftarrow \{0,1\}^{k_4}$, where $S = (S_1, S_2)$ is the simulator for Π.[4] Compute $y = H_s(x)$. Set $(\mathsf{sk}, \mathsf{vk}) = (x, (y, s, \mathsf{ek}, \mathsf{crs}))$.

Signing, $\mathsf{Sig}(\mathsf{sk}, m)$: Compute $C = \mathsf{Enc}^m(\mathsf{ek}, x)$. Using crs and Π, generate a NIZK proof π proving that $\exists x (C = \mathsf{Enc}^m(\mathsf{ek}, x) \wedge y = H_s(x))$. Output $\sigma = (C, \pi)$.

Verifying, $\mathsf{Ver}(\mathsf{vk}, m, \sigma)$: Parse σ as C, π. Use crs and V to verify the NIZK proof π. Output 1 if the proof verifies correctly and 0 otherwise.

As explained in [9], a NIZK proof system together with a CCA-secure encryption scheme are a specific instantiation of *true-simulation extractable (tSE)*. An alternative instantiation would be to compose a simulation-sound NIZK with a CPA-secure encryption scheme. This approach was used in [17]. We note that our proof follows similarly for this instantiation as well.

Theorem 2. *If H, $\Gamma = (\mathsf{KeyGen}, \mathsf{Enc}, \mathsf{Dec})$ and $\Pi = (\mathsf{CRSGen}, \mathsf{P}, \mathsf{V})$ have the properties listed above, then Σ is 2^{-k_5}-EU-CMAA where $k_5 = k + l_2 + l_3$ and where*

- *k is the security parameter of Σ,*
- *l_1 is the length of the randomness used to sample s at security parameter k_1 for H,*
- *l_2 is the length of the decryption key dk at security parameter k_2 for Γ,*
- *l_3 is the length of the simulation trapdoor td_s at security parameter k_3 for Π,*

If we consider the class $\mathcal{H}_{\mathsf{ow}}(\ell(k))$, then our scheme is 2^{-k_6}-EU-CMAA where $k_6 = k + l_1 + l_2 + l_3 + l_4$ and where l_4 is the length of $y = H_s(x)$ at security parameter k_1 for H.

---

[4] It is deliberate that we use a simulated CRS as part of the public key. This makes the set of admissible leakage functions defined relative to a simulated CRS, which we use in the proof. The scheme might be secure for a normal CRS too, but the proof would be more complicated.

Specifically, the best success against Σ in the forging game with 2^{-k_5}-hard leakage by a PPT adversary $\mathcal{A}$ is $2^{-k} + \sum_{i=0}^{3} \varepsilon_i + u\varepsilon_4$, where u is a polynomial and

- *ε_0 and ε_3 are the advantages of some PPT adversaries in the ZK game against Π at security parameter k_3,*
- *ε_1 is the success probability of some PPT adversary in the soundness game against Π at security parameter k_3,*
- *ε_2 is the probability that some PPT adversary wins the second preimage game against H on security parameter k_1 and $x \leftarrow \{0,1\}^{k_4}$,*
- *ε_4 is the advantage of some PPT adversary in the IND-WLCCA game against Γ at security parameter k_2.*

The intuition behind the proof of security is that a forged signature contains an encryption of the secret key x, so forging leads to extracting x using dk, giving a reduction to the assumption that it is hard to compute x given the leakage. In doing this reduction the signing oracle is simulated by encrypting 0^{k_4} and simulating the proofs using the simulation trapdoor td_s. This will clearly still lead to an extraction of x, using reusable-CRS NIZK and IND-WLCCA. The only hurdle is that given $(\mathsf{vk}, h(\mathsf{sk}, \mathsf{vk}))$, we do not know dk or td_s. We can, however, guess these with probability 2^{-l_2} respectively 2^{-l_3}. This is why we only get security $k_W = k + l_2 + l_3$. When we prove security for $\mathcal{H}_{ow}(\ell(k))$ the reduction is not given vk either, so we additionally have to guess s and y, leading to $k_S = k + l_1 + l_2 + l_3 + l_4$.

If we set $k_4 = k + l_2 + l_3 + l_4 + L$, then the min-entropy of x given $y = H_s(x)$ is $k + l_2 + l_3 + L$, so leaking L bits would be an admissible leakage in the 2^{-k_W} security game. Since, by assumption on our primitives, l_2 and l_3 and l_4 does not grow with k_4, it follows that we can set L to be any polynomial and be secure against leaking any fraction $(1 - k^{-O(1)})$ of the secret key. Due to space constraints the complete proof is found in the full version [12].

The following is a corollary to Thm. 2.

Theorem 3. *If H, Γ and Π have the properties listed above, then Σ is 2^{-k_W}-EU-CMAA where $k_W = k + l_2 + l_3$ and l_1 is the length of the randomness used to sample s, l_2 is the length of the decryption key dk for Γ, l_3 is the length of the simulation trapdoor td_s. In particular, Σ is 2^{-k_W}-EU-CMAA for $k_W = \mathrm{poly}(k)$ which do not grow with k_4, i.e, the input length of the hash function.*

If we consider the class $\mathcal{H}_{ow}(\ell(k))$, then Σ is 2^{-k_S}-EU-CMAA where $k_S = k + l_1 + l_2 + l_3 + l_4$ and where l_4 is the length of $y = H_s(x)$.

Our concrete instantiation has all the needed properties, except that s has a length which depends on k_4. This, however, can be handled generically as follows.

Lemma 2. *If there exists an ε-secure family of second preimage resistant hash functions H, with key sampling algorithm Gen_H, and a δ-secure pseudo-random generator prg, then there exists an $(\varepsilon + \delta)$-secure family of second preimage resistant hash function H, with key sampling algorithm Gen'_H, where $s \leftarrow \mathsf{Gen}'_H(1^k)$ can be guessed with probability 2^{-k_0}, where $k_0 = \mathrm{poly}(k)$ is the seed length of prg at security level k.*

Proof. Let $\mathsf{Gen}_H'(1^k; r \in \{0,1\}^{k_0}) = \mathsf{Gen}_H(1^k; \mathsf{prg}(r))$. It is clear that an output of $\mathsf{Gen}_H'(r \in \{0,1\}^{k_0})$ can be guessed with probability 2^{-k_0}, by guessing r. Let

$$\varepsilon = \Pr\nolimits_{x^* \leftarrow \mathcal{A}(s,x) \wedge x \leftarrow \{0,1\}^{k_4} \wedge s \leftarrow \mathsf{Gen}_H}[H_s(x^*) = H_s(x) \wedge x^* \neq x]$$

, and let $\varepsilon' = \Pr\nolimits_{x^* \leftarrow \mathcal{A}(s,x) \wedge x \leftarrow \{0,1\}^{k_4} \wedge s \leftarrow \mathsf{Gen}_H'}[H_s(x^*) = H_s(x) \wedge x^* \neq x]$. Consider the algorithm $\mathcal{B}(s)$ which samples $x \leftarrow \{0,1\}^{k_4}$ and $x^* \leftarrow \mathcal{A}(s,x)$ and outputs 1 iff $H_s(x^*) = H_s(x)$. This algorithm is PPT, and $\varepsilon' = \Pr[\mathcal{B}(\mathsf{Gen}_H(\mathsf{prg}(r \leftarrow \{0,1\}^{k_0}))) = 1]$ and $\varepsilon = \Pr[\mathcal{B}(\mathsf{Gen}_H(r \leftarrow \{0,1\}^*)) = 1]$. By the prg being a δ-pseudo-random generator, it follows that $|\varepsilon' - \varepsilon| \leq \delta$. $\qquad\square$

Remark. We can also prove security in the stronger model, where the leakage function h sees not only sk, but the randomness used by Gen to generate $(\mathsf{vk}, \mathsf{sk})$. In that case we need that the distribution on ek induced by sampling $(\mathsf{ek}, \mathsf{dk})$ with KeyGen_Γ, the distribution of a crs sampled along with a trapdoor and that the distribution on s induced by sampling $s \leftarrow \mathsf{Gen}_H$ can all be sampled with invertible sampling. This is indeed the case for our concrete instantiation. The only problematic point is Lemma 2. Even if $\mathsf{Gen}_H(\{0,1\}^*)$ has invertible sampling, it would be very surprising if $\mathsf{Gen}_H(\mathsf{prg}(\{0,1\}^{k_0}))$ has invertible sampling. So, if the probability of guessing a random $s \leftarrow \mathsf{Gen}_H$ is not independent of the input of H_s, we cannot generically add this property. One can circumvent this problem as in [9] and consider s as a public parameter of the scheme. This is modeled by sampling s in a parameter generation phase prior to the key generation phase and give s as input to all entities. This would in turn make s an input to the reduction (called $\mathcal{B}_7$ in the appendix), circumventing the problem of having to guess s. We would get security when considering the class $\mathcal{H}_{\mathsf{ow}}(\ell(k))$ for $k_S = k + l_2 + l_3 + l_4$.

Acknowledgments. The authors thank Yevgeniy Dodis for discussions at an early stage of this project.

References

1. Akavia, A., Goldwasser, S., Vaikuntanathan, V.: Simultaneous Hardcore Bits and Cryptography against Memory Attacks. In: Reingold, O. (ed.) TCC 2009. LNCS, vol. 5444, pp. 474–495. Springer, Heidelberg (2009)
2. Alwen, J., Dodis, Y., Wichs, D.: Leakage-Resilient Public-Key Cryptography in the Bounded-Retrieval Model. In: Halevi, S. (ed.) CRYPTO 2009. LNCS, vol. 5677, pp. 36–54. Springer, Heidelberg (2009)
3. Boneh, D., Halevi, S., Hamburg, M., Ostrovsky, R.: Circular-Secure Encryption from Decision Diffie-Hellman. In: Wagner, D. (ed.) CRYPTO 2008. LNCS, vol. 5157, pp. 108–125. Springer, Heidelberg (2008)
4. Boyle, E., Segev, G., Wichs, D.: Fully Leakage-Resilient Signatures. In: Paterson, K.G. (ed.) EUROCRYPT 2011. LNCS, vol. 6632, pp. 89–108. Springer, Heidelberg (2011)
5. Brakerski, Z., Goldwasser, S.: Circular and Leakage Resilient Public-Key Encryption under Subgroup Indistinguishability (or: Quadratic Residuosity Strikes Back). In: Rabin, T. (ed.) CRYPTO 2010. LNCS, vol. 6223, pp. 1–20. Springer, Heidelberg (2010)

6. Brakerski, Z., Segev, G.: Better Security for Deterministic Public-Key Encryption: The Auxiliary-Input Setting. In: Rogaway, P. (ed.) CRYPTO 2011. LNCS, vol. 6841, pp. 543–560. Springer, Heidelberg (2011)
7. Dodis, Y., Goldwasser, S., Tauman Kalai, Y., Peikert, C., Vaikuntanathan, V.: Public-Key Encryption Schemes with Auxiliary Inputs. In: Micciancio, D. (ed.) TCC 2010. LNCS, vol. 5978, pp. 361–381. Springer, Heidelberg (2010)
8. Dodis, Y., Haralambiev, K., López-Alt, A., Wichs, D.: Cryptography against continuous memory attacks. In: FOCS, pp. 511–520 (2010)
9. Dodis, Y., Haralambiev, K., López-Alt, A., Wichs, D.: Efficient Public-Key Cryptography in the Presence of Key Leakage. In: Abe, M. (ed.) ASIACRYPT 2010. LNCS, vol. 6477, pp. 613–631. Springer, Heidelberg (2010)
10. Dodis, Y., Kalai, Y.T., Lovett, S.: On cryptography with auxiliary input. In: STOC, pp. 621–630 (2009)
11. Dziembowski, S., Pietrzak, K.: Leakage-resilient cryptography. In: FOCS, pp. 293–302 (2008)
12. Faust, S., Hazay, C., Nielsen, J.B., Nordholt, P.S., Zottarel, A.: Signature schemes secure against hard-to-invert leakage. IACR Cryptology ePrint Archive, 45 (2012)
13. Faust, S., Kiltz, E., Pietrzak, K., Rothblum, G.N.: Leakage-Resilient Signatures. In: Micciancio, D. (ed.) TCC 2010. LNCS, vol. 5978, pp. 343–360. Springer, Heidelberg (2010)
14. Gentry, C., Peikert, C., Vaikuntanathan, V.: Trapdoors for hard lattices and new cryptographic constructions. In: STOC, pp. 197–206 (2008)
15. Goldwasser, S., Micali, S., Rivest, R.L.: A digital signature scheme secure against adaptive chosen-message attacks. SIAM J. Comput. 17(2), 281–308 (1988)
16. Groth, J., Sahai, A.: Efficient Non-interactive Proof Systems for Bilinear Groups. In: Smart, N.P. (ed.) EUROCRYPT 2008. LNCS, vol. 4965, pp. 415–432. Springer, Heidelberg (2008)
17. Katz, J., Vaikuntanathan, V.: Signature Schemes with Bounded Leakage Resilience. In: Matsui, M. (ed.) ASIACRYPT 2009. LNCS, vol. 5912, pp. 703–720. Springer, Heidelberg (2009)
18. Krawczyk, H., Rabin, T.: Chameleon signatures. In: NDSS (2000)
19. Malkin, T., Teranishi, I., Vahlis, Y., Yung, M.: Signatures Resilient to Continual Leakage on Memory and Computation. In: Ishai, Y. (ed.) TCC 2011. LNCS, vol. 6597, pp. 89–106. Springer, Heidelberg (2011)
20. Micali, S., Reyzin, L.: Physically Observable Cryptography (Extended Abstract). In: Naor, M. (ed.) TCC 2004. LNCS, vol. 2951, pp. 278–296. Springer, Heidelberg (2004)
21. Naor, M., Segev, G.: Public-Key Cryptosystems Resilient to Key Leakage. In: Halevi, S. (ed.) CRYPTO 2009. LNCS, vol. 5677, pp. 18–35. Springer, Heidelberg (2009)
22. Pietrzak, K.: A Leakage-Resilient Mode of Operation. In: Joux, A. (ed.) EUROCRYPT 2009. LNCS, vol. 5479, pp. 462–482. Springer, Heidelberg (2009)
23. Rackoff, C., Simon, D.R.: Non-interactive Zero-Knowledge Proof of Knowledge and Chosen Ciphertext Attack. In: Feigenbaum, J. (ed.) CRYPTO 1991. LNCS, vol. 576, pp. 433–444. Springer, Heidelberg (1992)
24. Regev, O.: On lattices, learning with errors, random linear codes, and cryptography. In: Gabow, H.N., Fagin, R. (eds.) STOC, pp. 84–93. ACM (2005)
25. Standaert, F.-X.: Leakage resilient cryptography: a practical overview. Invited Talk at ECRYPT Workshop on Symmetric Encryption, SKEW 2011 (2011)

Completeness for Symmetric Two-Party Functionalities - Revisited[*]

Yehuda Lindell[1], Eran Omri[2], and Hila Zarosim[1]

[1] Bar-Ilan University
lindell@biu.ac.il, zarosih@cs.biu.ac.il
[2] Ariel University Center
omrier@gmail.com

Abstract. Understanding the minimal assumptions required for carrying out cryptographic tasks is one of the fundamental goals of theoretical cryptography. A rich body of work has been dedicated to understanding the complexity of cryptographic tasks in the context of (semi-honest) secure two-party computation. Much of this work has focused on the characterization of trivial and complete functionalities (resp., functionalities that can be securely implemented unconditionally, and functionalities that can be used to securely compute all functionalities).

All previous works define reductions via an ideal implementation of the functionality; i.e., f reduces to g if one can implement f using an ideal box (or oracle) that computes the function g and returns the output to both parties. Such a reduction models the computation of f as an *atomic operation*. However, in the real-world, protocols proceed in rounds, and the output is not learned by the parties simultaneously. In this paper we show that this distinction is significant. Specifically, we show that there exist symmetric functionalities (where both parties receive the same outcome), that are neither trivial nor complete under "ideal-box reductions", and yet the existence of a constant-round protocol for securely computing such a functionality implies infinitely-often oblivious transfer (meaning that it is secure for infinitely-many n's). In light of the above, we propose an alternative definitional infrastructure for studying the triviality and completeness of functionalities.

1 Introduction

Secure Computation and Completeness. In the setting of secure two-party computation, two parties with respective private inputs x and y, wish to compute a function f of their inputs. The computation should preserve a number of security properties, like *privacy* (meaning that nothing but the specified output is learned), *correctness* and more.

In the late 1980s, it was shown that every function can be securely computed in the presence of semi-honest and malicious adversaries, assuming the existence

[*] This work was supported by the ISRAEL SCIENCE FOUNDATION (grant No. 189/11). Hila Zarosim is grateful to the Azrieli Foundation for the award of an Azrieli Fellowship. This work was done while Eran Omri was at Bar-Ilan University.

X. Wang and K. Sako (Eds.): ASIACRYPT 2012, LNCS 7658, pp. 116–133, 2012.

of enhanced trapdoor permutations [18,6]. Soon after, it was shown that any function can be securely computed, given an ideal box for computing the oblivious transfer function [9]. This work demonstrated that there exist "complete" functions for secure computation; that is, functions that can be used to securely compute all other functions. Such functions are of great interest. On the one hand, when attempting to base secure computation on weaker hardness assumptions, it suffices to construct a secure protocol for a complete function based on some weaker assumption, since it will imply that this assumption suffices for securely computing all functions. On the other hand, it is immediate that a complete function is the "hardest" to compute, at least with respect to the minimum hardness assumption. Due to the above, much research has been carried out in an attempt to classify functions as complete or not, and as trivial or not (where triviality means that it can be securely computed without any assumption).

The Complexity of Secure Computation. Currently, we have a good picture regarding the complexity of secure computation, through the aforementioned research of completeness. For example, we know that in the setting of asymmetric functionalities (where only one of the two parties receives output), every two-party (deterministic) asymmetric function is either complete or trivial [1,11]. Thus, no non-trivial asymmetric function can be securely computed under an assumption weaker than that needed for securely computing oblivious transfer.

However, in the setting of symmetric functionalities, where both parties receive the same output, the picture is more complex [10,13,15]. For example, unlike the asymmetric setting, there exist (deterministic) symmetric functions that are neither complete nor trivial; see Figure 1 below for an example of such a function. This begs the following fundamental question:

What hardness assumptions are sufficient and necessary for securely computing functions that are neither complete nor trivial?

The starting point of this work is the above question. We stress that although Kilian [10] separated these functions from all complete functions, hinting that it may be possible to devise secure protocols for such functions relying on assumptions that are strictly weaker than those needed for oblivious transfer, the only known protocols for securely computing non-trivial functions are general protocols that rely on hardness assumptions that can be used to compute *any* function including oblivious transfer.

Black-Box Reductions and Black-Box Separations. As we have mentioned, a large body of work has been dedicated to understanding the complexity of cryptographic tasks in the context of (semi-honest) secure two-party computation (see, e.g., [1,9,10,11,2,7,15]). The idea underlying much of this work is that if the possibility to securely compute a functionality f_1 implies the possibility to securely compute a functionality f_2, then f_1 is at least as hard as f_2. It is then said that f_2 reduces to f_1. A functionality f is called complete if all other functionalities reduce to f. The question of how to define the notion of reduction is of great importance to the implication of these results.

All previous works define a reduction via an ideal implementation of a function; i.e., f_2 reduces to f_1 if a secure protocol for computing f_2 can be constructed given an ideal box (trusted party or oracle) that computes f_1 and gives the outputs to both parties simultaneously.[1] The advantage of (black-box) reductions of the above type is that they provide a constructive way of securely computing one functionality given an implementation of another. However, the disadvantage of black-box reductions is that a *separation* (i.e., a proof that one function does *not* reduce to another) does not necessary imply that one cannot construct a secure protocol for one function given a secure protocol from the other. This is due to the fact that a reduction may be nonblack-box.

Our Contributions. In this work we give substantial evidence that the picture of *computational hardness* of securely computing two-party functionalities in the presence of semi-honest adversaries is different to that drawn by the characterizations of completeness of [10,13]. Specifically, we show that there exist symmetric functionalities f (i.e., where both parties get the same output), that are *not ideal-box-complete* (i.e., OT cannot be implemented using an ideal-box computing f) but may be in some sense as hard to obtain as OT. Specifically, we prove the following:

Theorem 1.1 (informal). *If there exists a constant round protocol π that securely computes a symmetric non-trivial functionality f over a constant-size domain, in the presence of semi-honest adversaries, then there exists an infinitely-often-OT that is secure in the presence of semi-honest uniform adversaries.[2]*

Needless to say, Theorem 1.1 is of interest for functionalities f that are *not* complete; as we have mentioned, such functionalities exist.

Our main observation in proving this result is that in real-world protocols, an ideal-box that simultaneously provides outputs to both parties does not exist. Rather, parties learn their outputs gradually, and hence, in any constant-round protocol, there must be a round in which one party learns substantial information before the other party does. Thus, essentially there is no difference between the symmetric setting (where both parties receive output and there are functions that are neither complete nor trivial) and the asymmetric setting (where only one party receives output and all functions are either trivial or complete).

Alternative Formulation of Completeness – Existential Completeness. In light of the above, we propose a new definition of completeness that is not black box. We define the notion of an "achievable class" of a given functionality f. Informally speaking, the achievable class of a functionality f contains all

---

[1] We stress that the issue of simultaneity has nothing to do with fairness since we consider semi-honest adversaries. Rather, the important point is that both parties receive the same information and it is not possible for one party to learn the output of the function while the other does not. If this were not the case, and only one party receives output then the symmetric setting reduces to the asymmetric setting where all functionalities are either trivial or complete.

[2] Infinitely-often-OT is a protocol for computing OT for which correctness and security hold for infinitely many n's (rather than for all sufficiently large n).

functionalities that can be securely computed, assuming that f can be securely computed. We use this notion in the natural way in order to redefine reductions, and trivial and complete functionalities. Our formulation has the disadvantage of being completely non-constructive. However, it has the advantage of providing a more accurate picture regarding the hardness assumptions required for secure computation.

Related Work. As we have already mentioned, completeness in secure two-party computation was investigated in a large body of work [2,13,10,1,12,14,11,15,16,7]. We discuss a few that are more relevant to our discussion. Kilian [10] and Kushile-vitz [13] consider the symmetric model and give criteria for the existence of uncon-ditionally secure protocols [13] and for completeness [10]. Maji et al. [15] extended the discussion of the symmetric model to the UC-setting. Beimel, Malkin and Mi-cali [1] considered the asymmetric model. They prove a zero-one law for complete-ness vs. triviality in this model. Almost all of these works consider functions with a constant size domain and information-theoretic security. The only exception is [7] who deals with computational security in the asymmetric model.

2 Definitions

2.1 Preliminaries

A function $\mu : \mathbb{N} \to \mathbb{N}$ is negligible if for every positive polynomial $p(\cdot)$ and all sufficiently large n it holds that $\mu(n) < \frac{1}{p(n)}$. We use the abbreviation PPT to denote probabilistic polynomial-time. For an integer ℓ, define $[\ell] = \{1, \ldots, \ell\}$. A *probability ensemble* $X = \{X(a, n)\}_{a \in \{0,1\}^*; n \in \mathbb{N}}$ is an infinite sequence of random variables indexed by a and n. (The value a will represent the parties' inputs and n the security parameter. All polynomials that we will consider will be with respect to the security parameter, unless explicitly stated otherwise; specifically, all polynomial time machines will be polynomial in the security parameter.) We let λ denote the empty word.

Two ensembles $X = \{X(a, n)\}_{a \in \{0,1\}^*; n \in \mathbb{N}}$ and $Y = \{Y(a, n)\}_{a \in \{0,1\}^*; n \in \mathbb{N}}$ are computationally indistinguishable, denoted $X \overset{c}{\equiv} Y$, if for every family $\{C_n\}_{n \in \mathbb{N}}$ of polynomial-size circuits, there exists a negligible function $\mu(\cdot)$ such that for every $a \in \{0, 1\}^*$ and every $n \in \mathbb{N}$,

$$\left| \Pr\left[C_n(X(a, n)) = 1\right] - \Pr\left[C_n(Y(a, n)) = 1\right] \right| < \mu(n).$$

The ensembles X and Y are computationally indistinguishable by uniform machines, denoted $X \overset{c}{\equiv}_U Y$, if the above holds for every PPT distinguisher D.

2.2 Secure Two-Party Computation and Oblivious Transfer

We follow the standard definition of secure two party computation for semi-honest adversaries, as it appears in [5]. In brief, a two-party protocol π is de-fined by two interactive probabilistic polynomial-time Turing machines A and B.

The two Turing machines, called parties, have the security parameter 1^n as their joint input and have private inputs, denoted x and y for A and B, respectively. The computation proceeds in rounds. In each round of the protocol, one of the parties is active and the other party is idle. If party $P \in \{A, B\}$ is active in round i, then in this round P writes some value Out_P^i on its output tape, and sends a message m_i to the other party. Without loss of generality, we assume that A is always active in the odd rounds in π and B in the even rounds. The number of rounds in the protocol is expressed as some function $r(n)$ in the security parameter (where $r(n)$ is bounded by a polynomial).

The view of a party in an execution of the protocol contains its private input, its random string, and the messages it received throughout this execution. The random variable $\mathrm{View}_A^\pi(x, y, 1^n)$ (respectively $\mathrm{View}_B^\pi(x, y, 1^n)$) describes the view of A (resp. B) when executing π on inputs (x, y) (with security parameter n). The output of an execution of π on (x, y) (with security parameter n) is the pair of values written on the output tapes of the parties when the protocol execution terminates. This pair is described by the random variable $\mathrm{Output}^\pi(x, y, 1^n) = (\mathrm{Output}_A^\pi(x, y, 1^n), \mathrm{Output}_B^\pi(x, y, 1^n))$, where $\mathrm{Output}_P^\pi(x, y, 1^n)$ is the output of party $P \in \{A, B\}$ in this execution, and is implicit in the view of P.

In this work, we consider deterministic functionalities over a finite domain. We therefore provide the definition of security only for deterministic functionalities; see [5] for a motivating discussion regarding the definition.

Definition 2.1 (security for deterministic functionalities). *A protocol $\pi = \langle A, B \rangle$ securely computes a deterministic functionality $f = (f_A, f_B)$ in the presence of semi-honest adversaries if the following hold:*

Correctness: *There exists a negligible function $\mu(\cdot)$, such that for every n and every pair of inputs x, y, it holds that*

$$\Pr\left[\mathrm{Output}^\pi(x, y, 1^n) = f(x, y)\right] \geq 1 - \mu(n). \tag{1}$$

where the probability is taken over the random coins of the parties.

Privacy: *There exist two probabilistic polynomial-time (in the security parameter) algorithms $\mathcal{S}_A, \mathcal{S}_B$ (called "simulators"), such that:*

$$\{\mathcal{S}_A(x, f_A(x, y), 1^n)\}_{x,y\in\{0,1\}^*;n\in\mathbb{N}} \overset{c}{\equiv} \{\mathrm{View}_A^\pi(x, y, 1^n)\}_{x,y\in\{0,1\}^*;n\in\mathbb{N}}, \tag{2}$$

$$\{\mathcal{S}_B(y, f_B(x, y), 1^n)\}_{x,y\in\{0,1\}^*;n\in\mathbb{N}} \overset{c}{\equiv} \{\mathrm{View}_B^\pi(x, y, 1^n)\}_{x,y\in\{0,1\}^*;n\in\mathbb{N}}. \tag{3}$$

For most of this paper, we will consider functionalities where both parties receive the same output, meaning that $f_A = f_B$. We call such functions **symmetric** and we denote by $f(x, y)$ the output that both parties receive. We will also only consider the semi-honest model here, and therefore omit this qualification from hereon.

Oblivious Transfer – Naive-OT Variant. The oblivious transfer functionality (OT) is one of the most important cryptographic primitives and is known to be complete for general two-party computation [19,6]. There are several equivalent versions of OT; the most common being Rabin-OT [17] and 1-out-of-2 OT [3]. In this paper we use a slightly different version presented in [7], called Naive-OT, defined by the functionality $OT(b, c) = \begin{cases} (\lambda, \lambda) & \text{if } c = 0 \\ (\lambda, b) & \text{if } c = 1 \end{cases}$, meaning that the sender never learns anything (recall that λ is the empty string), and the receiver learns the sender's bit b if its choice-bit c equals 1, but does not learn anything if $c = 0$. This is the same as Rabin-OT except that the receiver chooses whether or not to receive the sender's bit b. In the semi-honest model it is equivalent to Rabin-OT (and to 1-out-of-2-OT).

2.3   Uniform Infinitely-Often Security

Our main result is a proof that the existence of a constant-round protocol for functionalities that are neither complete nor trivial *almost* implies oblivious transfer. The "almost" in this sentence is due to the fact that we can only prove that it implies oblivious transfer that is secure for *infinitely many n's*, in contrast to all sufficiently large n's. In addition, we can only prove that the oblivious transfer is secure in the presence of uniform distinguishers. We therefore need to define this weaker notion of security.

Definition 2.2 (uniform infinitely-often security). *A protocol π securely computes a deterministic functionality f in the presence of semi-honest adversaries with* uniform infinitely-often security *if there exists an infinite subset $\mathcal{N} \subseteq \mathbb{N}$ such that Equations (1), (2) and (3) hold for every $n \in \mathcal{N}$, and Equations (2) and (3) hold with respect to uniform distinguishers.*

We stress that the correctness and privacy conditions must all hold for every $n \in \mathcal{N}$ (it does not suffice to require infinitely many n's for which each requirement holds since it is possible that they may hold for different n's in which case the function will be trivial).

3   Our Main Technical Result

In this section, we prove Theorem 1.1. In order to formally state the theorem and our result, we first need to define the class of functions that we consider. We therefore begin with preliminaries.

3.1   Preliminaries

Our theorem applies to all non-trivial functionalities, as characterized by Kushilevitz [13]. This characterization uses the notion of "decomposition" of a function. We now define this notion.

Definition 3.1 (equivalence relation $\equiv$ over inputs). *Let $X, Y, Z \subseteq \{0, 1\}^*$, and let $f : X \times Y \to Z$. Two inputs $x_1, x_2 \in X$* existentially coincide, *denoted $x_1 \sim x_2$, if there exists an input $y \in Y$ such that $f(x_1, y) = f(x_2, y)$. We define an equivalence relation $\equiv$ over X to be the transitive closure of the relation $\sim$ over all $x \in X$. The relations $\sim$ and $\equiv$ are defined over Y similarly.*

Definition 3.2 (strongly non-decomposable functions). *A function* g : $X \times Y \to Z$ *is* strongly non-decomposable *if it is not monochromatic, all* $x \in X$ *are equivalent, and all* $y \in Y$ *are equivalent.*

We refer to [13] in order to see why this is called non-decomposable. The binary OR and AND functions are strongly non-decomposable, as is the function $f_{\mathcal{KUSH}}$ defined below:

	y_1	y_2	y_3
x_1	0	0	1
x_2	3	4	1
x_3	3	2	2

Fig. 1. Kushilevitz's function $f_{\mathcal{KUSH}}$

A strongly non-decomposable function has the property that all inputs are equivalent. We now define a non-decomposable function simply to be a function which has a *subfunction* that is strongly non-decomposable.

Definition 3.3 (non-decomposable functions). *A symmetric function* f : $X \times Y \to Z$ *is* non-decomposable *if there exist* $X' \subseteq X$ *and* $Y' \subseteq Y$ *such that* f *restricted to* X' *and* Y' *is strongly non-decomposable; else it is* decomposable.

We remark that Kushilevitz [13] proved that a function is non-trivial if and only if it is non-decomposable. The function $f_{\mathcal{KUSH}}$ is of particular interest since it is neither trivial (as shown by [13]) nor complete (as shown by [10]).

3.2   The Theorem and Proof

Let f be a symmetric non-decomposable functionality with domain of constant size. We show that the existence of a constant-round protocol for computing f implies the existence of a weak variant of oblivious transfer. The idea behind the proof is to run a protocol π for f until the *first* round in which one of the parties learns meaningful information about the input of the other party. Since this is the first round that something is learned and only one party can learn information in any single round, we have that one party has learned something and the other has not. This asymmetry of information suffices for us to construct oblivious transfer.

Our proof proceeds in three stages. First, we prove that a round as described above exists. Intuitively, this is the case since before the protocol execution neither party has any information about the other party's input, but at the end of the execution each party learns significant information about the other party's input. Next, we show that a weak form of oblivious transfer can be constructed from any protocol with such a round (in actuality, we need to prove that such a round exists on a special subset of inputs called a minor, and we demonstrate this in the first step). The OT that we construct is weak in the sense that it is only correct with noticeable probability. Finally, we show how to boost the weak correctness of the OT to fully correct oblivious transfer.

We stress that we do not actually obtain a full oblivious transfer protocol. Rather, our protocol is only secure *infinitely often*; see Definition 2.2. We explain why this is the case at the end of Section 3.3.

Theorem 3.1. *If there exists a constant round protocol π that securely computes a symmetric, deterministic, non-decomposable functionality f (over a constant-size domain), then there exists a uniform infinitely-often OT protocol.*

Proof: Recall that a non-decomposable functionality is a function with a subset of inputs that defines a strongly non-decomposable functionality. Since we consider the semi-honest model and so parties use only their prescribed inputs, it follows that the existence of a secure protocol for a non-decomposable function implies the existence of a secure protocol for the strongly non-decomposable function defined over the appropriate subset. It thus suffices to prove the theorem for a strongly non-decomposable function.

As we have described above, there are three steps in the proof of this theorem. In Section 3.3 we prove the first step. Specifically, in Lemma 3.1 we prove that there exists an "exclusive revelation round" which is a round in which one party has learned while the other has not, and then in Lemma 3.2 we prove that such a round must exist for inputs that form an insecure minor (defined below). We call this an "exclusive revelation minor". Next in Section 3.4 we prove that the existence of an exclusive revelation minor implies the existence of OT with weak correctness, and finally in Section 3.5 we explain how to boost the correctness and thus obtain full OT (with infinitely-often uniform security).   □

3.3   Step 1– The Existence of an Exclusive Revelation Minor

In order to prove our result we exploit the fact that parties obtain information about the output of a computation gradually and that one party learns substantial information before the other party does. We begin with some notation regarding partial protocol executions. For an r-round protocol π and a function $\nu : \mathbb{N} \to \mathbb{N}$ such that $\nu(n) \leq r(n)$ for all $n \in \mathbb{N}$, we denote by π_ν the protocol obtained by halting π after round $\nu(n)$ is completed. Specifically, the random variables $\text{View}_A^{\pi_\nu}(x, y, 1^n)$ and $\text{View}_B^{\pi_\nu}(x, y, 1^n)$ describe the views of A and B (respectively) in a random execution of π_ν on inputs (x, y) with security parameter n.

We next formally define what it means for a party to obtain non-trivial information about the other party's input.

Definition 3.4 (distinguishing between inputs). *Let π be a c-round protocol for computing a functionality f (where c is some function of the security parameter n), and fix $i \in \mathbb{N}$. For a triple x, y, y' of inputs, we say that $A(x)$ distinguishes between y and y' at round i if there exists a polynomial $p(\cdot)$ and a (uniform) PPT machine D such that for infinitely many n's,*

$$\left| \Pr\left[D\left(\text{View}_A^{\pi_i}(x, y, 1^n), 1^n \right) = 1 \right] - \Pr\left[D\left(\text{View}_A^{\pi_i}(x, y', 1^n), 1^n \right) = 1 \right] \right| \geq \frac{1}{p(n)}$$

For a triple x, x', y of inputs we define that $B(y)$ distinguishes between x and x' at round i in an analogous way.

As we will see below, it is crucial that D be a uniform PPT machine, since the parties need to run D in the OT protocol that we construct. For simplicity (and since it suffices for our needs), the above definition considers a fixed round i. This can be easily generalized to any (polynomial time computable) function $i : \mathbb{N} \to \mathbb{N}$ such that $i(n) \le c(n)$ for every n.

We now define the notion of an exclusive revelation round, which is just a round in which one party can distinguish inputs of the other, while the other cannot. Our formulation of this uses Definition 3.4.

Definition 3.5 (exclusive-revelation round). *Let π be a protocol for computing a symmetric functionality f. Then, π has an exclusive revelation at round i if one of the following holds:*

1. *There exists a triplet x, y, y' such that $A(x)$ distinguishes between y and y' at round i, and $B(y)$ does not distinguish between x and x' at round i for any triplet x, x', y (we say that x, y, y' define the revelation round); or*
2. *There exists a triplet x, x', y such that $B(y)$ distinguishes between x and x' at round i and, $A(x)$ does not distinguish between y and y' at round i for any triplet x, y, y' (we say that x, x', y define the revelation round).*

Protocol π has an exclusive-revelation round if there exists $0 \le i \le c$, such that π has an exclusive revelation at round i.

We are now ready to prove that any constant-round protocol for computing a non-constant function (i.e., a function that has at least two different outputs) has an exclusive-revelation round.

Lemma 3.1. *Let f be a symmetric functionality that is not constant (and has domain of constant size). Let π be a constant-round protocol for securely computing f. Then, π has an exclusive-revelation round.*

Proof: For every (round number) $i \le c$, every uniform PPT machine (distinguisher) D, and every triplet x, x', y (recall that there is a constant number of such triplets), we define

$$\varepsilon_{x,x',y}^{i,D}(n) = \left| \Pr\left[D\left(\mathrm{View}_B^{\pi_i}\left(x, y, 1^n\right), 1^n \right) = 1 \right] - \Pr\left[D\left(\mathrm{View}_B^{\pi_i}\left(x', y, 1^n\right), 1^n \right) = 1 \right] \right|$$

and let $r_{x,x',y}^D$ be the minimal round number $0 \le i \le c$ for which there exists a polynomial $p(\cdot)$ such that $\varepsilon_{x,x',y}^{i,D}(n) > \frac{1}{p(n)}$ for infinitely many n's. If no such i exists, we let $r_{x,x',y}^D = c+1$. Note that this means that $r_{x,x',y}^D$ is the first round such that the PPT machine D can distinguish the ensembles $\{\mathrm{View}_B^{\pi_i}(x, y, 1^n)\}_{n \in \mathbb{N}}$ and $\{\mathrm{View}_B^{\pi_i}(x', y, 1^n)\}_{n \in \mathbb{N}}$.

We further define $r_{x,x',y} = \min_D \{r_{x,x',y}^D\}$ (this is well defined, as every $r_{x,x',y}^D \in [c+1]$). Observe that this means that $r_{x,x',y}$ is the minimal round for which there exists *any* uniform PPT machine that can distinguish the two ensembles (equivalently, the minimal round for which $B(y)$ distinguishes between x and x'). For every triplet x, y, y', we define $r_{x,y,y'}$ analogously.

By the correctness of the protocol, for every triplet x, x', y such that $f(x, y) \neq f(x', y)$, the view of both parties after the last round (that is, round c) implies the output and hence there exists a uniform PPT machine D and a negligible function $\mu(\cdot)$ such that for all sufficiently large n's, $\varepsilon_{x,x',y}^{c,D}(n) \geq 1 - \mu(n)$. This in turn implies that for such triplets, there exists a PPT machine D for which $r_{x,x',y}^{D} \leq c$, and hence $r_{x,x',y} \leq c$. Similarly, for every triplet x, y, y' such that $f(x, y) \neq f(x, y')$, it holds that $r_{x,y,y'} \leq c$. Since f is not constant, there either exists a triplet of the former type or of the latter type.

Now, define $i_A^* = \min_{x,y,y'}\{r_{x,y,y'}\}$ and $i_B^* = \min_{x,x',y}\{r_{x,x',y}\}$. Note that i_A^* is the minimal round for which there exists a triplet x, y, y' such that $A(x)$ distinguishes between y and y', and i_B^* is the minimal round for which there exists a triplet x, x', y such that $B(y)$ distinguishes between x and x'. Since f is not constant, it holds that either $i_A^* \leq c$ or $i_B^* \leq c$ (or both). We claim that π has exclusive revelation at either round i_A^* or at round i_B^*.

Assume without loss of generality that $i_A^* \leq i_B^*$; we show that $i_A^* < i_B^*$. It suffices to show that $i_A^* < i_B^*$, since by the definition of i_B^* we know that $B(y)$ does not distinguish between x and x' at any round $i < i_B^*$ and for any triplet x, x', y. A crucial observation is that the view of a party does not change in the round that it is active, and hence, neither does its distinguishing capability. Hence, by the minimality of i_A^*, it must be that B is the one sending a message in round i_A^*, since otherwise A would be able to distinguish already in round $i_A^* - 1$. This means that B's view does not change in round i_A^*, and hence, by the minimality of i_B^* it cannot be that $i_A^* = i_B^*$. The case that $i_B^* \leq i_A^*$ is dealt with analogously. $\qquad\square$

We complete this step of the proof by showing that when a strongly non-decomposable function has a protocol with an exclusive-revelation round, this round is defined by inputs that form an insecure minor. An insecure minor is a tuple of inputs x, x', y, y' such that $f(x, y) = f(x, y')$ and $f(x', y) \neq f(x', y')$ (X-minor), or $f(x, y) = f(x', y)$ and $f(x, y') \neq f(x', y')$ (Y-minor).

Definition 3.6 (exclusive-revelation minor). *Let π be a protocol for computing a symmetric functionality f. If there exists an X-minor x, x', y, y' with respect to f such that x', y, y' define an exclusive revelation round for π, then we say that π has an* exclusive-revelation X-minor; *an* exclusive-revelation Y-minor *is defined analogously. We say that π has an* exclusive-revelation minor *if it has an exclusive revelation X-minor or an exclusive revelation Y-minor.*

The next lemma states that strongly non-decomposable functions have the property that the existence of an exclusive-revelation round implies the existence of an exclusive-revelation minor.

Lemma 3.2. *Let π be a protocol that securely computes a strongly non-decomposable symmetric function f with constant-size domain. If π has an exclusive-revelation round then it has an exclusive-revelation minor.*

Proof: The proof follows by analyzing the general structure of strongly non-decomposable functions. Let f be *any* symmetric strongly non-decomposable function with a constant-size domain. Assume that there exist x_j, y_k, y_ℓ (with $k < \ell$) that define an exclusive revelation at round i; that is, $A(x_j)$ distinguishes between y_k and y_ℓ at round i. We show that this implies that π has an exclusive revelation X-minor. Since f is a strongly non-decomposable function, it holds that $y_k \equiv y_\ell$. Let $y_{i_1}, \dots, y_{i_t}$ be such that $y_k \sim y_{i_1} \sim \dots \sim y_{i_t} \sim y_\ell$ and let $y_{i_0} = y_k$ and $y_{i_{t+1}} = y_\ell$. $A(x_j)$ distinguishes between y_{i_0} and $y_{i_{t+1}}$ at round i, and since t is a constant (recall that f has a constant-size domain), there exists some $h \in [t+1]$ such that $A(x_j)$ distinguishes between $y_{i_{h-1}}$ and y_{i_h} at round i. Now, by definition, since $y_{i_{h-1}} \sim y_{i_h}$, there exists some x such that $f(x, y_{i_{h-1}}) = f(x, y_{i_h})$. Hence, $x, x_j, y_{i_{h-1}}, y_{i_h}$ forms an exclusive-revelation X-minor. The proof for the case that B distinguishes is analogous. □

Infinitely-Often. Observe that the existence of an exclusive revelation minor means that there exists an insecure minor and a round of the protocol such that one party can distinguish the other party's inputs at this round while the other cannot. We stress that a party *distinguishes inputs* if it has polynomial advantage in guessing the input *for infinitely many n's*. It would be preferable to prove this for all sufficiently large n's, since this would enable us to later construct a fully secure oblivious transfer protocol, and not just an infinitely-often secure oblivious transfer protocol. However, we are unable to do this since we need to utilize the existence of a round where one party has learned something and the other has not learned anything. We prove this by taking the first such round, and this guarantees that in any previous round the other party has not learned anything, except possibly for a finite number of n's. This means that it did not learn for infinitely many of the n's in which the other party did learn, as required. In contrast, if we were to take the first round in which one party learns for all sufficiently large n's, then it is possible that the other party has learned for infinitely many of these n's in a previous round, and so security will not be guaranteed.

Constant-Round. We use the assumption that π is constant-round in the proof that π has an exclusive-revelation round (Lemma 3.1). Recall that an exclusive-revelation round is the first round that a party can distinguish between the inputs of the other party. If the number of rounds in π is non-constant, then for every n the concrete number of rounds in the protocol is different and hence we would have to define an "exclusive-revelation function"; that is, a function $\nu : \mathbb{N} \to$ *round number*, that defines the first round (as a function of n) that a party can distinguish between the inputs of the other party. It is not clear how to define such a function, and moreover, how to prove the existence of it.

Constant-Size Domain. We restrict ourselves to functions with constant-size domains (i.e., not dependent on the security parameter) in order to be consistent with previous works studying completeness and triviality of symmetric functions ([10,13]). Extending the study of completeness to functions with non-constant-size domains is beyond the scope of this paper.

3.4   Step 2 – From an Exclusive-Revelation Minor to io-Weak-OT

We now show that if a function has a protocol with an exclusive-revelation minor, then it can be used to obtain a weak version of oblivious transfer. The "weakness" in the OT is with respect to correctness, and not privacy. Formally:

Definition 3.7. *A protocol π is a* infinitely-often uniform *weak oblivious transfer protocol (io-weak-OT) if there exists an infinite set $\mathcal{N} \subseteq \mathbb{N}$ such that Equations (2) and (3) hold for every $n \in \mathcal{N}$ and with respect to uniform distinguishers, and there exists a polynomial $p(\cdot)$ such that Equation (1) holds with probability $\frac{1}{2} + \frac{1}{p(n)}$ for every $n \in \mathcal{N}$.*

We stress that the privacy requirement of the oblivious transfer (Equations (2) and (3)) is identical to uniform infinitely-often security in Definition 2.2. However, the correctness requirement is weaker since it is only required that correctness holds with probability noticeably greater than $1/2$, and not close to 1.

Lemma 3.3. *Let $\pi = \langle A, B \rangle$ be a protocol for securely computing a functionality f. If π has an exclusive-revelation minor, then there exists a PPT protocol $\tilde{\pi}$ that is an infinitely-often uniform weak oblivious transfer.*

Proof: Intuitively, the existence of an exclusive-revelation round in the protocol allows us (in some weak sense) to move to the realm of asymmetric functionalities where one party learns the output, while the other party learns nothing. It is known that an asymmetric functionality containing an insecure minor implies OT. We therefore use the insecure minor guaranteed by the hypothesis of the lemma to construct (a weak form of) OT in a way similar to that used in the world of asymmetric computation. The formal arguments follow.

Let π be a protocol computing a symmetric functionality f. Assume without loss of generality that there exists an X-minor x, x', y, y' with respect to f, such that x', y, y' define an exclusive revelation at round i for π (the case of an exclusive revelation Y-minor is analogous). That is, we have that $A(x')$ distinguishes between y and y' at round i and for *every* triplet $\hat{x}, \hat{x}', \hat{y}$, we have that $B(\hat{y})$ does not distinguish between $\hat{x}$ and $\hat{x}'$ at round i. Let D be the corresponding distinguisher, and assume without loss of generality that it always outputs either 0 or 1. Furthermore, since $f(x, y) = f(x, y')$ (by definition of a minor), by the security of π we also have that $A(x)$ does not distinguish between y and y' at round i (or any round, for that matter). It is without loss of generality (e.g., by interchanging y and y') to assume that for infinitely many n's that

$$\Pr\left[D\left(\text{View}_A^{\pi_i}\left(x', y, 1^n\right), 1^n\right) = 1\right] - \Pr\left[D\left(\text{View}_A^{\pi_i}\left(x', y', 1^n\right), 1^n\right) = 1\right] \geq \frac{1}{p(n)} \tag{4}$$

We now show how to construct an io-weak-OT protocol $\tilde{\pi}$. Before giving the formal description of the protocol, let us give some intuition. The idea is to run the protocol on the inputs of the above minor until round i, and then to halt the execution. By the exclusiveness of the revelation, we are guaranteed that B

learns nothing from the computation, hence the sender $\tilde{S}$ will play the role of B. If the receiver $\tilde{R}$ has input 0, then it will use x as its input and play the role of A, and hence will not learn anything (recall that $f(x,y) = f(x,y')$ and so the output reveals nothing about B's input, meaning that $\tilde{R}$ learns nothing). In case $\tilde{R}$'s input is 1 it will use x' as its input for the computation, and will learn the output by distinguishing as in Equation (4).

Regarding the sender's input, one possibility is to have the sender to use y' as its input for the computation in case $b = 0$ and y in case $b = 1$. The receiver will then output 0 or 1, depending on what the distinguisher outputs. However, it is possible that the distinguisher outputs 0 with probability 3/4 on input (x',y), and with probability $3/4 + 1/p(n)$ on input (x',y'). In such a case, the receiver will output 0 with probability 3/4 even when the output is supposed to be 1, and so weak correctness will not hold (recall that we need correctness with probability greater than 1/2). In order to overcome this, we have the sender use a random input in $\{y,y'\}$ and therefore transfer a random bit r to the receiver (which in turn will try to learn r only if its input is $c = 1$). The sender then sends the receiver the bit $z = r \oplus c$, and the receiver outputs z if the distinguisher output 0 and $z \oplus 1$ otherwise. This has the effect of moving the error to be around 1/2, and so we obtain correctness $1/2 + 1/p(n)$.

Protocol 1 (An io-weak-OT $\tilde{\pi} = \langle \tilde{S}, \tilde{R} \rangle$)

Inputs: *The private input of the sender $\tilde{S}$ is a bit $b \in \{0,1\}$ and the private input of the receiver $\tilde{R}$ is a bit $c \in \{0,1\}$. The common input is 1^n, where n is the security parameter.*

The protocol:

1. *The sender chooses a random bit $r \in \{0,1\}$.*
2. *The parties start an execution of π, where the sender $\tilde{S}$ plays the role of B and the receiver $\tilde{R}$ plays the role of A. The inputs of the parties are set as follows:*
   - *The input of B (played by $\tilde{S}$) is y' if $r = 0$ and y if $r = 1$.*
   - *The input of A (played by $\tilde{R}$) is x if $c = 0$ and x' if $c = 1$.*
   The parties halt after the i-th round of π. Let v_A^i be the partial view of A in this partial execution of π.
3. *The sender $\tilde{S}$ sends $z = r \oplus b$ to the receiver $\tilde{R}$.*
4. *If $c = 0$, the receiver outputs λ. Otherwise (if $c = 1$), the receiver executes D on v_A^i, sets r' to be the output of D, and outputs $z \oplus r'$. The sender always outputs λ.*

Note that the receiver is allowed to use the distinguisher D since D is a uniform Turing machine.

Proving the *Weak-Correctness* of the Protocol. Proving the correctness when $c = 0$ is trivial since both parties will always output λ as required. We consider the case that $c = 1$. We need to show that there exists a polynomial $q(\cdot)$

such that for infinitely many n's, it holds that $\Pr\left[\text{Output}_{\tilde{R}}^{\tilde{\pi}}\left(b, c = 1, 1^n\right) = b\right] \geq \frac{1}{2} + \frac{1}{q(n)}$. We will show that this holds for the polynomial $q(i) = 2p(i)$ and for all n's for which Equation (4) is satisfied. We fix such an n.

Recall that $\tilde{R}$ outputs $z \oplus r'$, where $z = b \oplus r$ and hence the output of $\tilde{R}$ equals b if and only if $r' = r$, where r' denotes the output of D on the partial view v_A^i. Thus, it suffices to give a lower bound on the following term (recall that we consider the case that $\tilde{R}$ uses x' since $c = 1$):

$$\Pr\left[r' = r\right] \tag{5}$$
$$= \Pr\left[r = 0\right] \cdot \Pr\left[r' = 0 \mid r = 0\right] + \Pr\left[r = 1\right] \cdot \Pr\left[r' = 1 \mid r = 1\right]$$
$$= \frac{1}{2} \cdot \Pr\left[D\left(\text{View}_A^{\pi_i}\left(x', y', 1^n\right), 1^n\right) = 0\right] + \frac{1}{2} \cdot \Pr\left[D\left(\text{View}_A^{\pi_i}\left(x', y, 1^n\right), 1^n\right) = 1\right]$$
$$= \frac{1}{2} \cdot \left(1 - \Pr\left[D\left(\text{View}_A^{\pi_i}\left(x', y', 1^n\right), 1^n\right) = 1\right]\right) + \frac{1}{2} \cdot \Pr\left[D\left(\text{View}_A^{\pi_i}\left(x', y, 1^n\right), 1^n\right) = 1\right]$$
$$= \frac{1}{2} + \frac{1}{2} \cdot \left(\Pr\left[D\left(\text{View}_A^{\pi_i}\left(x', y, 1^n\right), 1^n\right) = 1\right] - \Pr\left[D\left(\text{View}_A^{\pi_i}\left(x', y', 1^n\right), 1^n\right) = 1\right]\right).$$

Since Equation (4) is satisfied for n, we have that

$$\Pr\left[D\left(\text{View}_A^{\pi_i}\left(x', y, 1^n\right), 1^n\right) = 1\right] - \Pr\left[D\left(\text{View}_A^{\pi_i}\left(x', y', 1^n\right), 1^n\right) = 1\right] \geq \frac{1}{p(n)}.$$

Hence, we conclude that $\Pr\left[r' = r\right] \geq \frac{1}{2} + \frac{1}{2p(n)}$, and so correctness holds.

Proving the *Privacy* of the Protocol. We now proceed to prove that Equations (2) and (3) in Definition 2.1 hold for all sufficiently large n's (and thus, in particular, for infinitely many n's for which weak correctness holds, as required in Definition 2.2). Due to the lack of space in this extended abstract, we sketch this portion of the proof.

Simulating the View of the Sender. We construct a PPT machine $\mathcal{S}_{\tilde{S}}$ that simulates the sender's view. $\mathcal{S}_{\tilde{S}}$ receives as input the sender's input b and the security parameter 1^n, and works as follows:

1. $\mathcal{S}_{\tilde{S}}$ chooses a random bit $r_{\tilde{S}} \in \{0, 1\}$.
2. $\mathcal{S}_{\tilde{S}}$ then starts an execution of π on the following inputs until the i-th round:
   - If $r_{\tilde{S}} = 0$, the input of B is y' and if $r_{\tilde{S}} = 1$, the input of B is y.
   - The input of A is x.
3. $\mathcal{S}_{\tilde{S}}$ outputs $r_{\tilde{S}}$ and the partial view v_B^i of B.

The difference between the view of the sender in a real execution and in a simulation by $\mathcal{S}_{\tilde{S}}$ is due to the fact that $\mathcal{S}_{\tilde{S}}$ always runs A with x whereas in a real execution A runs with x or x' depending on the receiver's input. Nevertheless, these distributions are computationally indistinguishable since i is an *exclusive revelation round* for A. This means that B learns nothing about A's input up to and including round i, and in particular the view of B when A uses x is computationally indistinguishable from its view when A uses x'. We stress that the fact that i is an exclusive revelation round means that no *uniform distinguisher* given B's view can distinguish (by the notion of distinguishing between inputs; Definition 3.4). This does not necessarily mean that no non-uniform distinguisher can distinguish; thus we only achieve privacy with respect to *uniform* distinguishers.

Simulating the View of the Receiver. In the case that $c = 1$ the simulator receives both the sender's and receiver's inputs c and b and so can perfectly simulate the view of the receiver by just running the protocol on these inputs. We therefore describe the simulator only for the case that $c = 0$. The simulator $S_{\tilde{R}}$ receives as input the bit $c = 0$, the output $OT_R = \lambda$ of the functionality OT to the receiver, and the security parameter 1^n, and works as follows:

1. $S_{\tilde{R}}$ executes π for i rounds, running A with input x and B with input y.
2. $S_{\tilde{R}}$ chooses a random bit $z_S \in \{0, 1\}$.
3. $S_{\tilde{R}}$ outputs z_S appended to the partial view v_A^i of A.

The difference between the simulated view and a real view is that in a real execution, the sender playing B sometimes uses y and sometimes uses y', whereas in the simulated execution it always uses y. In addition, the simulator sends a random z_S that is not correlated to the value r implied by the input used by B in the computation of π. In order to see that this makes no difference, first observe that since x, x', y, y' form an insecure minor, it holds that $f(x, y) = f(x, y')$. Thus, when A has input x in an execution of π, it cannot distinguish the case that B used input y or y'; otherwise, A could learn something that is not revealed by the functionality output. Thus, the view of the receiver (who runs A) in the protocol execution is indistinguishable from its view in the simulation. Given the above, it follows that the distribution of a random bit z_S is indistinguishable from the distribution of $z = r \oplus b$ by the randomness of r. This completes the proof.

$\square$

Uniform Security. As explained above, the privacy of the receiver is preserved by the exclusiveness of the revelation minor (in round i). That is, since the sender in the OT protocol takes the role of the party that cannot distinguish the inputs of the other party (the one active in round i). By Definition 3.4, no *uniform* distinguisher D succeeds with non-negligible probability in distinguishing the two possible inputs of the receiver. It does not, however, rule out the possibility that a *non-uniform* distinguisher has noticeable success probability, yielding the privacy of the receiver vulnerable with respect to non-uniform adversaries.

3.5   From Weak Uniform io-OT to Uniform io-OT

We conclude the proof by arguing that the existence of a uniform infinitely-often weak-OT implies the existence of a uniform infinitely-often OT protocol. Let π be a uniform infinitely-often weak-OT protocol. We construct a uniform infinitely-often OT protocol $\tilde{\pi}$ by having the parties run polynomially many executions of π on their inputs. If $c = 1$, the receiver outputs the majority of the outputs of the receiver in π, and otherwise it outputs λ. It follows from the Chernoff bound that for the infinitely-many n's for which π has weak-correctness, $\tilde{\pi}$ is correct with probability $1 - \mu(n)$, for some negligible function $\mu(\cdot)$. To prove the privacy of $\tilde{\pi}$, we use multiple executions of the simulators of the io-weak-OT. A standard hybrid argument shows that this yields a satisfactory simulation for the io-OT protocol. We stress that a simple hybrid argument works because the parties are semi-honest and hence follow the prescribed protocol (specifically, they select fresh random coins for each execution).

This completes the proof of Theorem 3.1.

4   Ideal-Box and Existential Completeness

Loosely speaking, a functionality is called complete if it can be used to securely compute any functionality. In the standard definitions of completeness used in previous works (cf. [10,13,1]), this is defined via the notion of "reduction". Specifically g reduces to f if it is possible to securely compute g *given access to* f, and a functionality is complete if all functionalities reduce to it. In this section we explore in greater depth how this notion of reduction is defined and what the ramifications of this definition are.

The definition of reduction in all previous works uses the notion of an *ideal black-box* for computing a functionality $f = (f_A, f_B)$. The parties A and B run a protocol for computing g while given access to an incorruptible trusted party who computes f for them throughout the execution (the parties send inputs x and y to the trusted party, who computes $f(x,y) = (f_A(x,y), f_B(x,y))$, and sends them back their respective outputs). A functionality g reduces to a functionality f, if g is securely computable given such a trusted party for computing f. This notion is equivalent to the notion of oracle-aided protocols, defined in [5, Section 7.3.1]. Formally, using the terminology of [5], all previous definitions say that g reduces to f if there exists an oracle-aided protocol π that *information-theoretically* securely computes g when using the oracle functionality f (the only exception is [7] that considers computational security rather than information-theoretic). A functionality f is called complete if all g reduce to it, and it is called trivial if it can be information-theoretically securely computed with no oracle. We call this notion ideal-box completeness since the reduction is *black-box in the functionality*.

The picture of completeness and triviality for the above definition is well known. Specifically, for the case of asymmetric functionalities where only one of the parties receives output, a functionality is complete if it contains an insecure minor, and trivial if not. Furthermore, for the case of symmetric functionalities where the parties receive the same output (i.e., $f_A = f_B$), a functionality is complete if and only if it contains an embedded OR, and is trivial if and only if it is decomposable (see Definition 3.3).

Combining the above with Theorem 3.1, we have the following corollary:

Corollary 4.1. *There exist symmetric deterministic functionalities over a domain of constant-size that are not neither trivial nor ideal-box-complete, such that if there exists a constant round protocol π that securely computes such a function, then there exists a uniform infinitely-often OT protocol.*

We remark that using the results of Kilian [9], one can show that any function-ality can be securely computed with uniform infinitely-often security (Defini-tion 2.2) given a uniform infinitely-often OT protocol. It therefore seems unlikely that such an OT protocol can be constructed under weaker assumption than fully secure OT (at least, infinitely-often secure protocols are not known to be con-structible under weaker assumptions, and the known black-box separations for OT [8,4] hold also for infinitely-often OT).

Existential Completeness – An Alternative Formulation. Corollary 4.1 suggests that there may exist functionalities that are neither trivial nor com-plete, and yet are in some sense complete (albeit, under the caveat of uniform infinitely-often security). This is due to the fact that the definition of ideal-box-completeness relates to the computation of f as *atomic*, whereas in real life, computation is carried out step-by-step, and in particular is *not* black-box in the functionality. We therefore present an alternative notion of completeness which is purely *existential*. Informally, our definition is based on saying that f "implies" g in some sense if the feasibility of securely computing g is implied by the feasibility of securely computing f. Formally:

Definition 4.1. *Let $\mathcal{U}$ denote the set of all polynomial-time computable func-tionalities. The* achievable class *of $f \in \mathcal{U}$, denoted as $\mathcal{C}(f)$, is the set of all $g \in \mathcal{U}$ such that if there exists a computationally secure protocol π_f for computing f, then there exists a computationally secure protocol π_g for computing g.*

Let $f, g \in \mathcal{U}$. We say that g existentially reduces *to f if $g \in \mathcal{C}(f)$. Functionality f is* existentially trivial *if $f \in \mathcal{C}(f_\lambda)$ (where $f_\lambda(\cdot, \cdot) = (\lambda, \lambda)$), and is* existentially complete *if $\mathcal{C}(f) = \mathcal{U}$.*

The above definition follows the intuition that a functionality is trivial if it can be securely computed "with no help", and complete if all functionalities can be securely computed if it can be securely computed. We stress that if (enhanced) trapdoor functions exist, then all functionalities are trivial and complete by this definition. Nevertheless, our definition is helpful since a proof that a functional-ity f is complete (without proving the existence of enhanced trapdoor permu-tations) is essentially a proof that f requires an assumption that implies OT. We remark that this is the same as in the definition of (ideal-box) computa-tional completeness that appears in [7]. We also note that any functionality that is ideal-box-complete, or complete by the computation definition in [7], is also existentially complete.

We conclude by remarking that the definition of existential completeness has the advantage that it can more accurately map the assumptions required for se-curely computing a functionality. In particular, a function that is not complete cannot imply OT, something which *can* happen under the ideal-box definition (as hinted to by Corollary 4.1). However, it is also true that the definition of existential completeness is less helpful due to its non-constructive nature. Specif-ically, it does not enable us to prove or consider a hierarchy of functionalities, and a proof that $g \in \mathcal{C}(f)$ does not necessarily tell us how to securely compute g, even given a protocol for securely computing f.

References

1. Beimel, A., Malkin, T., Micali, S.: The All-or-Nothing Nature of Two-Party Secure Computation. In: Wiener, M. (ed.) CRYPTO 1999. LNCS, vol. 1666, pp. 80–97. Springer, Heidelberg (1999)
2. Chor, B., Kushilevitz, E.: A zero-one law for Boolean privacy. SIAM J. on Discrete Mathematics 4(1), 36–47 (1991)
3. Even, S., Goldreich, O., Lempel, A.: A randomized protocol for signing contracts. CACM 28(6), 637–647 (1985)
4. Gertner, Y., Kannan, S., Malkin, T., Reingold, O., Viswanathan, M.: The relationship between public key encryption and oblivious transfer. In: Proc. of the 41st IEEE Symp. on Foundations of Computer Science, pp. 325–335 (2000)
5. Goldreich, O.: Foundations of Cryptography, vol. II – Basic Applications. Cambridge University Press (2004)
6. Goldreich, O., Micali, S., Wigderson, A.: How to play any mental game. In: Proc. of the 19th ACM Symp. on the Theory of Computing, pp. 218–229 (1987)
7. Harnik, D., Naor, M., Reingold, O., Rosen, A.: Completeness in two-party secure computation: A computational view. In: Proc. of the 36th ACM Symp. on the Theory of Computing, pp. 252–261 (2004)
8. Impagliazzo, R., Rudich, S.: Limits on the provable consequences of one-way permutations. In: Proc. of the 21st ACM Symp. on the Theory of Computing, pp. 44–61 (1989)
9. Kilian, J.: Basing cryptography on oblivious transfer. In: Proc. of the 20th ACM Symp. on the Theory of Computing, pp. 20–31 (1988)
10. Kilian, J.: A general completeness theorem for two-party games. In: Proc. of the 23th ACM Symp. on the Theory of Computing, pp. 553–560 (1991)
11. Kilian, J.: More general completeness theorems for two-party games. In: Proc. of the 32nd ACM Symp. on the Theory of Computing, pp. 316–324 (2000)
12. Kilian, J., Kushilevitz, E., Micali, S., Ostrovsky, R.: Reducibility and completeness in private computations. SIAM J. on Computing 28(4), 1189–1208 (2000); This is the Journal version of [10, 14]
13. Kushilevitz, E.: Privacy and communication complexity. SIAM J. on Discrete Mathematics 5(2), 273–284 (1992)
14. Kushilevitz, E., Micali, S., Ostrovsky, R.: Reducibility and completeness in multi-party private computations. In: Proc. of the 35th IEEE Symp. on Foundations of Computer Science, pp. 478–491 (1994)
15. Maji, H.K., Prabhakaran, M., Rosulek, M.: Complexity of multi-party computation problems: The case of 2-party symmetric secure function evaluation, pp. 256–273 (2009)
16. Mironov, I., Pandey, O., Reingold, O., Vadhan, S.: Computational Differential Privacy. In: Halevi, S. (ed.) CRYPTO 2009. LNCS, vol. 5677, pp. 126–142. Springer, Heidelberg (2009)
17. Rabin, M.O.: How to exchange secrets by oblivious transfer. Technical Report TR-81, Harvard Aiken Computation Laboratory (1981), Cryptology ePrint Archive, Report 2005/187, eprint.iacr.org/2005/187
18. Yao, A.C.: Protocols for secure computations. In: Proc. of the 23th IEEE Symp. on Foundations of Computer Science, pp. 160–164 (1982)
19. Yao, A.C.: How to generate and exchange secrets. In: Proc. of the 27th IEEE Symp. on Foundations of Computer Science, pp. 162–167 (1986)

Adaptively Secure Garbling with Applications to One-Time Programs and Secure Outsourcing

Mihir Bellare[1], Viet Tung Hoang[2], and Phillip Rogaway[2]

[1] Dept. of Computer Science and Eng., University of California, San Diego, USA
[2] Dept. of Computer Science, University of California, Davis, USA

Abstract. Standard constructions of garbled circuits provide only *static* security, meaning the input x is not allowed to depend on the garbled circuit F. But some applications—notably *one-time programs* (Goldwasser, Kalai, and Rothblum 2008) and *secure outsourcing* (Gennaro, Gentry, Parno 2010)—need *adaptive* security, where x may depend on F. We identify gaps in proofs from these papers with regard to adaptive security and suggest the need of a better abstraction boundary. To this end we investigate the adaptive security of *garbling schemes*, an abstraction of Yao's garbled-circuit technique that we recently introduced (Bellare, Hoang, Rogaway 2012). Building on that framework, we give definitions encompassing *privacy, authenticity*, and *obliviousness*, with either *coarse-grained* or *fine-grained* adaptivity. We show how adaptively secure garbling schemes support simple solutions for one-time programs and secure outsourcing, with privacy being the goal in the first case and obliviousness and authenticity the goal in the second. We give transforms that promote static-secure garbling schemes to adaptive-secure ones. Our work advances the thesis that conceptualizing garbling schemes as a first-class cryptographic primitive can simplify, unify, or improve treatments for higher-level protocols.

1 Introduction

OVERVIEW. Yao's garbled-circuit technique [10, 11, 18, 20, 21] has been extremely influential, engendering an enormous number of applications. Yet, at least in its conventional form, the technique provides only *static* security. Some applications, notably one-time programs [13] and secure outsourcing [9], require *adaptive* security.[1] In such cases Yao's technique can be enhanced in *ad hoc* ways, and the enhanced protocol incorporated into the higher-level application.

This paper provides a different approach. We create an abstraction for the goal of *adaptively secure garbling*. Via a single abstraction, we support a variety of applications in a simple and modular way. Let's look at two of the applications that motivate our work.

TWO APPLICATIONS. *One-time programs* are due to Goldwasser, Kalai, and Rothblum (GKR) [13]. The authors aim to compile a program into one that

---

[1] In speaking of adversaries or security, *non-adaptive* and *dynamic* are common synonyms for what we are here calling *static* and *adaptive*.

X. Wang and K. Sako (Eds.): ASIACRYPT 2012, LNCS 7658, pp. 134–153, 2012.

can be executed just once, on an input of the user's choice. Unachievable in any "standard" model of computation, GKR assume what they call *one-time memory*. Their solution makes crucial use of Yao's garbled-circuit technique. Recognizing that this does not support adaptive queries, GKR embellish the method by a technique involving output-masking and n-out-of-n secret sharing.

In a different direction, *secure outsourcing* was formalized and investigated by Gennaro, Gentry, and Parno (GGP) [9]. Here a *client* transforms a function f into a function F that is handed to a *worker*. When, later, the client would like to evaluate f at x, he should be able to quickly map x to a garbled input X and give this to the worker, who will compute and return $Y = F(X)$. The client must be able to quickly reconstruct from this $y = f(x)$. He should be sure that the correct value was computed—the computation is *verifiable*—while the server shouldn't learn anything significant about x, including $f(x)$. GGP again make use of circuit garbling, and they again realize that they need something from it—its authenticity—that is a *novum* for this domain.

ISSUES. Assuming the existence of a one-way function, GKR [13] claim that their construction turns a (statically-secure) garbled circuit into a secure one-time program. We point to a gap in their proof, namely, the absence of a reduction showing that their simulator works based on the one-way function assumption. By presenting an example of a statically-secure garbled circuit that, under their transform, yields a program that is not one-time, we also show that the gap cannot be filled without changing either the construction or the assumption. The problem is that the GKR transform fails to ensure adaptive security of garbled circuits under the stated assumption.

Lindell and Pinkas (LP) [17] prove static security of a version of Yao's protocol assuming a semantically secure encryption scheme satisfying some extra properties (an elusive and efficiently verifiable range). GGP [9] build a one-time outsourcing scheme from the LP protocol, claiming to prove its security based on the same assumption as used in LP. We point to a gap in this proof arising from an implicit assumption of adaptive security of the LP construction.

We do not believe these are major problems for either work. In both cases, alternative ways to establish the the authors' main results already existed. Goyal, Ishai, Sahai, Venkatesan and Wadia [14] present an unconditional one-time compiler (no complexity-theoretic assumption is used at all), while Chung, Kalai and Vadhan [7] present secure outsourcing schemes based solely on FHE (garbled circuits are not employed). Our interpretation of the stated gaps is that they are symptoms of something else—a missing abstraction boundary. As recently argued by Bellare, Hoang and Rogaway (BHR) [4], it is useful and simplifying to see garbling not just as a technique, but as a first-class primitive. To do so, our earlier work defines syntax and security notions for *garbling schemes*, provides proven-correct solutions, then solves some example higher-level problems by employing a garbling scheme that satisfies the appropriate definition. But the security notions of BHR do not go far enough to handle what GKR or GGP need, since BHR deal only with static notions of security. The applications we point to motivate the study of adaptive security for garbling schemes, while the gaps indicate that the issues may be more subtle than recognized.

Of course we communicated our findings to the GKR and GGP authors. GKR responded after a few weeks with an updated manuscript [12]. It modifies the claim from their original paper [13] to now claim that their transform works under the stronger assumption of a sub-exponentially hard one-way function. (This allows "complexity-leveraging," where a static adversary can guess the input that will be used by an adaptive adversary with a probability that, although exponentially-small, is enough under the stronger assumption.) GGP responded to acknowledge the gap and suggest that they would address it by assuming the LP construction, or some related realization of Yao's idea, already provides adaptive security.

DEFINITIONS. We now discuss our contributions in more depth. We start from the abstraction of a garbling scheme—the raw syntax—introduced by BHR [4]. That work gave multiple definitions sitting on top of this syntax, but all were for static adversaries, in the sense that the function f to garble and its input x are selected at the same time. We extend the definitions to adaptive ones, considering two flavors of adaptive security. With *coarse-grained* adaptive security the input x can depend on the garbled function F but x itself is atomic, provided all at once. With *fine-grained* adaptive security not only may x depend on the garbled function F, but individual bits of x can depend on the "tokens" the adversary has so-far learned.[2] We will see that coarse-grained adaptive security is what's needed for GGP's approach to secure outsourcing, while fine-grained adaptive security is what's needed for GKR's approach to one-time programs.

Orthogonal to adaptive security's granularity are the security aims themselves. Following BHR, we consider three different notions: privacy, obliviousness, and authenticity. This gives rise to nine different security notions: {prv, obv, aut} × {static, coarse, fine}. We compactly denote these prv, prv1, prv2, obv, obv1, obv2, aut, aut1, aut2. Informally, when a function f gets transformed into a garbled function F, an encoding function e, and a decoding function d, privacy ensures that F, d, and $X = e(x)$ don't reveal anything beyond $y = f(x)$ that shouldn't be revealed; obliviousness ensures that F and X don't reveal even y; and authenticity ensures that F and X don't enable the computation of a valid $Y \neq F(X)$. Privacy is the classical requirement, while obliviousness and authenticity are motivated by the application to secure outsourcing.

Our primary definitions for adaptive secrecy (prv1, prv2, obv1, obv2) are simulation-based. In the full version of this paper [3] we give indistinguishability-based counterparts as well. For static security this was already done by BHR, but it was not clear how to lift those definitions to the adaptive setting.

RELATIONS. We explore the provable-security relationships among our definitions. As expected, the simulation-based definitions imply indistinguishability-based

---

[2] Fine-grained adaptive security requires the garbling scheme be *projective*: the garbled version of each $x = x_1 \cdots x_n \in \{0,1\}^n$ must be $(X_1^{x_1}, \ldots, X_n^{x_n})$ for some vector of $2n$ strings $(X_1^0, X_1^1, \ldots, X_n^0, X_n^1)$. Typical garbling schemes have this structure.

ones (namely, prv1 $\Rightarrow$ prv1.ind, prv2 $\Rightarrow$ prv2.ind, obv1 $\Rightarrow$ obv1.ind, and obv2 $\Rightarrow$ obv2.ind). But none of the converse statements hold. BHR had earlier shown that, for the static setting, the converse statements *do* hold as long as the associated side-information function[3] is efficiently invertible. In contrast, we show that, for adaptive privacy, this condition still won't guarantee equivalence of simulation-based and indistinguishability-based notions. (For obliviousness, it is true that obv1.ind $\Rightarrow$ obv1 and obv2.ind $\Rightarrow$ obv2 if Φ is efficiently invertible.) The results are our main reason to focus on simulation-based definitions for adaptive privacy. The full version [3] paints a complete picture of the relations among our basic definitions. Apart from the trivial relations (prv2 $\Rightarrow$ prv1 $\Rightarrow$ prv, obv2 $\Rightarrow$ obv1 $\Rightarrow$ obv, and aut2 $\Rightarrow$ aut1 $\Rightarrow$ aut) nothing implies anything else.

ACHIEVING ADAPTIVE SECURITY. Basic garbling-scheme constructions [4, 10, 11, 18] either do not achieve adaptive security or present difficulties in proving adaptive security that we do not know how to overcome. One could give new constructions and directly prove them xxx1 or xxx2 secure, for xxx $\in \{$prv, obv, aut$\}$. An alternative is to provide generic ways to transform statically secure garbling schemes to adaptively secure ones. Combined with results in BHR [4], this would yield adaptively-secure garbling schemes.

The aim of the GKR construction was exactly to add adaptive security to statically-secure garbled circuit constructions. We reformulate it as a transform, OMSS (Output Masking and Secret Sharing), aiming to turn a prv-secure garbling scheme to a prv2-secure one. We show, by counterexample, that OMSS does not achieve this goal.

To give transforms that work we make two steps, first passing from static security to coarse-grained adaptive security, and thence to fine-grained adaptive security. We design these transformations first for privacy (prv-to-prv1, prv1-to-prv2) and then for simultaneously achieving all three goals (all-to-all1 and all1-to-all2). Our prv-to-prv1 transform uses a one-time-padding technique from [14], while our prv1-to-prv2 transform uses the secret-sharing component of OMSS.

APPLICATIONS. We treat the two applications that motivated this work, one-time programs and secure outsourcing. We show that adaptive garbling schemes yield these applications easily and directly. Specifically, we show that a prv2 projective garbling scheme can be turned into a secure one-time program by simply putting the garbled inputs into the one-time memory. We also show how to easily turn an obv1+aut1-secure garbling scheme into a secure one-time outsourcing scheme. (GGP [9] show how to lift one-time outsourcing schemes to many-time ones using FHE.) The simplicity of these transformations underscores our tenet that abstracting garbling schemes and treating adaptive security for them enables modular and rigorous applications of the garbled-circuit technique. Basing the applications on garbling schemes also allows instantiations to inherit efficiency features of future schemes.

---

[3] The side-information function Φ captures that about f one allows to be revealed in its garbled counterpart F.

Transform	Model	Cost	See
prv-to-prv1	standard model	$\|F\| + \|d\| + \|X\|$	Theorem 2
prv1-to-prv2	standard model	$(n+1)\|X\|$	Theorem 3
all-to-all1	standard model	$\|F\| + \|d\| + \|X\| + k$	Theorem 5
all1-to-all2	standard model	$(n+1)\|X\|$	Theorem 6
rom-prv-to-prv1	random-oracle model	$\|X\| + k$	Full paper [3]
rom-prv1-to-prv2	random-oracle model	$\|X\| + nk$	Full paper [3]
rom-all-to-all1	random-oracle model	$\|X\| + 2k$	Full paper [3]
rom-all1-to-all2	random-oracle model	$\|X\| + nk$	Full paper [3]

Fig. 1. Achieving adaptive security. The name of each transform specifies its relevant property. The word all means that prv, obv, and aut are all upgraded. Column "Cost" specifies the length of the garbled input in the constructed scheme in terms of the lengths of the input scheme's garbled function F, decoding function d, garbled input X, number input bits n, and security parameter k.

Applying our prv-to-prv1 and then prv1-to-prv2 transforms to the prv-secure garbling scheme of BHR [4] yields a prv2-secure scheme based on any one-way function. Combining this with the above yields one-time programs based on one-way functions, recovering the claim of GKR [13]. Similarly, applying our all-to-all1 transform to the obv+aut-secure scheme of BHR yields an obv1+aut1-secure garbling scheme based on a one-way function, and combining this with the above yields a secure one-time outsourcing scheme based on one-way functions.

EFFICIENCY. Let us say a garbling scheme has *short* garbled inputs if their length depends only on the security parameter k, the length n of f's input, and the length m of f's output. It does not depend on the length of f. The statically-secure schemes of BHR, as with all classical garbled-circuit constructions, have short garbled inputs. But our prv-to-prv1 and all-to-all1 transforms result in long garbled inputs. In the ROM (random-oracle model) we are able to provide schemes producing short garbled inputs, as illustrated in Fig. 1. Constructing an adaptively secure garbling scheme with short garbled inputs under standard assumptions remains open.

Short garbled inputs are particularly important for the application to secure outsourcing, for in their absence the outsourcing scheme may fail to be non-trivial. (Non-trivial means that the client effort is less than the effort needed to directly compute the function [9].) In particular, the one-time outsourcing scheme we noted above, derived by applying all-to-all1 to BHR, fails to be non-trivial. ROM schemes do not fill the gap because of the use of FHE in upgrading one-time schemes to many-time ones [9]. Thus, a secure and non-trivial instantiation of the GGP method is still lacking. (However, as we have noted before, non-trivial secure outsourcing may be achieved by entirely different means [7].)

FURTHER RELATED WORK. Applebaum, Ishai, and Kushilevitz [1] investigate ideas similar to obliviousness and authenticity. Their approach to obtaining these

ends from privacy can be lifted and formalized in our settings; one could specify transforms prv1-to-all1 and prv2-to-all2, effectively handling the constructive
story "horizontally" instead of "vertically." The line of work on *randomized en-
codings* that the same authors have been at the center of provides an alternative
to garbling schemes [15] but lacks the granularity to speak of adaptive security.

Concurrent work by Kamara and Wei (KW) investigates the garbling what
they call *structured circuits* [16] and, in the process, give definitions somewhat
resembling prv1, obv1, and aut1, although circuit-based, not function-hiding,
and not allowing the adversary to specify the initial function. KW likewise draw
motivation from GKR and GGP, indicating that, in these two setting, the adversary can choose the inputs to the computation as a function of the garbled
circuit, motivating adaptive notions of privacy and unforgeability.

2 Framework

We now review the syntactic framework of garbling schemes from our earlier
work [4]. See the full version for [3] basic notation, including conventions for
randomized algorithms, code-based games, and circuits.

GARBLING SCHEMES. A *garbling scheme* [4] is a five-tuple of algorithms $\mathcal{G} =$
$(\mathsf{Gb}, \mathsf{En}, \mathsf{De}, \mathsf{Ev}, \mathsf{ev})$. The first of these is probabilistic; the rest are deterministic. A
string f, the *original function*, describes the function $\mathsf{ev}(f, \cdot) \colon \{0,1\}^n \to \{0,1\}^m$
that we want to garble. The values $n = f.n$ and $m = f.m$ are efficiently com-
putable from f. On input f and a security parameter $k \in \mathbb{N}$, algorithm Gb
returns a triple of strings $(F, e, d) \leftarrow \mathsf{Gb}(1^k, f)$. String e describes an *encoding function*, $\mathsf{En}(e, \cdot)$, that maps an *initial input* $x \in \{0,1\}^n$ to a *garbled input*
$X = \mathsf{En}(e, x)$. String F describes a *garbled function*, $\mathsf{Ev}(F, \cdot)$, that maps a garbled input X to a *garbled output* $Y = \mathsf{Ev}(F, X)$. String d describes a *decoding
function*, $\mathsf{De}(d, \cdot)$, that maps a garbled output Y to a *final output* $y = \mathsf{De}(d, Y)$.
The correctness requirement is that if $f \in \{0,1\}^*$, $k \in \mathbb{N}$, $x \in \{0,1\}^{f.n}$, and
$(F, e, d) \in [\mathsf{Gb}(1^k, f)]$, then $\mathsf{De}(d, \mathsf{Ev}(F, \mathsf{En}(e, x))) = \mathsf{ev}(f, x)$. We also require
that e and d depend only on k, $f.n$, $f.m$, $|f|$ and the random coins r of Gb. This
non-degeneracy requirement excludes trivial solutions.

A common design in existing garbling schemes is for e to encode a list of
tokens, one pair for each bit in $x \in \{0,1\}^n$. Encoding function $\mathsf{En}(e, \cdot)$ then uses
the bits of $x = x_1 \cdots x_n$ to select from $e = (X_1^0, X_1^1, \ldots, X_n^0, X_n^1)$ the subvector
$X = (X_1^{x_1}, \ldots, X_n^{x_n})$. Formally, we say that garbling scheme $\mathcal{G} = (\mathsf{Gb}, \mathsf{En}, \mathsf{De},$
$\mathsf{Ev}, \mathsf{ev})$ is *projective* if for all f, $x, x' \in \{0,1\}^{f.n}$, $k \in \mathbb{N}$, and $i \in [1..n]$, when
$(F, e, d) \in [\mathsf{Gb}(1^k, f)]$, $X = \mathsf{En}(e, x)$ and $X' = \mathsf{En}(e, x')$, then $X = (X_1, \ldots, X_n)$
and $X' = (X_1', \ldots, X_n')$ are n vectors, $|X_i| = |X_i'|$, and $X_i = X_i'$ if x and x' have
the same ith bit. Let $\mathsf{GS}(\mathsf{proj})$ denote the set of all projective garbling schemes.

Boolean circuits arise often in this work. We say that $\mathcal{G} = (\mathsf{Gb}, \mathsf{En}, \mathsf{De}, \mathsf{Ev}, \mathsf{ev})$
is a circuit-garbling scheme if ev is the canonical circuit evaluation function.

SIDE-INFORMATION FUNCTIONS. A garbled circuit might reveal the size of
the circuit that is being garbled, its topology, the original circuit itself, or

something else. The information that we allow to be revealed is captured by a *side-information function*, Φ, which deterministically maps f to a string $\phi = \Phi(f)$. We parameterize our advantage notions by Φ. We require that $f.n, f.m$ and $|f|$ be easily determined from $\phi = \Phi(f)$. Side-information function Φ_{size} maps a circuit $f = (n, m, q, A, B, G)$ to (n, m, q), while Φ_{topo} maps f to $f^- = \text{Topo}(f) = (n, m, q, A, B)$ and Φ_{circ} is the identity, $\Phi_{\text{circ}}(f) = f$.

SIZES. We say that garbling scheme $\mathcal{G} = (\mathsf{Gb}, \mathsf{En}, \mathsf{De}, \mathsf{Ev}, \mathsf{ev})$ has *short garbled inputs* if there is a polynomial s such that $|\mathsf{En}(e, x)| \leq s(k, f.n, f.m)$ for all $k \in \mathbb{N}$, $f \in \{0, 1\}^*$, $(F, e, d) \in [\mathsf{Gb}(1^k, f)]$, and $x \in \{0, 1\}^{f.n}$. Let T be a transform that maps a garbling scheme $\mathcal{G}$ to a garbling scheme $\mathsf{T}[\mathcal{G}]$. We say that T *preserves short garbled inputs* if $\mathsf{T}[\mathcal{G}]$ has short garbled inputs when $\mathcal{G}$ does.

Typical Yao-style constructions, including Garble1 and Garble2 [4], have short garbled inputs. But they are only statically-secure. Keeping garbled inputs short seems challenging for adaptive security in the standard model.

3   Privacy and One-Time Programs

In this section we define coarse and fine-grained adaptive privacy for garbling schemes. We show that some natural approaches to achieve these aims fail. We provide alternatives that work. In [3], we provide more efficient ones in the ROM. We apply this to get secure one-time programs.

DEFINITIONS FOR ADAPTIVE PRIVACY.   On the top of Fig. 2 we review the defining game for the privacy notion from BHR [4]. The adversary is *static*, in the sense it must commit to its initial function f and its input x at the same time. Thus the latter is independent of the garbled function F (and the decoding function d) derived from f. It is natural to consider stronger privacy notions, ones where the adversary obtains F and *then* selects x. Two formulations for this are specified in Fig. 2. We call these *adaptive* security. The notion in the middle panel, denoted by prv1, this paper, is *coarse-grained* adaptive security. The notion in the bottom panel, denoted by prv2, is *fine-grained* adaptive security. This notion is only applicable for projective garbling schemes.

In detail, let $\mathcal{G} = (\mathsf{Gb}, \mathsf{En}, \mathsf{De}, \mathsf{Ev}, \mathsf{ev})$ be a garbling scheme and let Φ be a side-information function. We define three simulation-based notions of privacy via the games $\mathrm{Prv}_{\mathcal{G}, \Phi, \mathcal{S}}$, $\mathrm{Prv1}_{\mathcal{G}, \Phi, \mathcal{S}}$, and $\mathrm{Prv2}_{\mathcal{G}, \Phi, \mathcal{S}}$ of Fig. 2. Here $\mathcal{S}$, the *simulator*, is an always-terminating algorithm that maintains state across invocations. An adversary $\mathcal{A}$ interacting with any of these games must make exactly one GARBLE query. For game Prv1 it is followed by a single INPUT query. For game Prv2 it is followed by multiple INPUT queries. There, the garbling scheme must be projective. The advantage the adversary gets is defined by

$$\mathbf{Adv}_{\mathcal{G}}^{\mathrm{prv}, \Phi, \mathcal{S}}(\mathcal{A}, k) = 2 \Pr[\mathrm{Prv}_{\mathcal{G}, \Phi, \mathcal{S}}^{\mathcal{A}}(k)] - 1$$

$$\mathbf{Adv}_{\mathcal{G}}^{\mathrm{prv1}, \Phi, \mathcal{S}}(\mathcal{A}, k) = 2 \Pr[\mathrm{Prv1}_{\mathcal{G}, \Phi, \mathcal{S}}^{\mathcal{A}}(k)] - 1$$

$$\mathbf{Adv}_{\mathcal{G}}^{\mathrm{prv2}, \Phi, \mathcal{S}}(\mathcal{A}, k) = 2 \Pr[\mathrm{Prv2}_{\mathcal{G}, \Phi, \mathcal{S}}^{\mathcal{A}}(k)] - 1 .$$

proc GARBLE(f, x) $\qquad\qquad\qquad\qquad\qquad\qquad\qquad\qquad$ Prv$_{\mathcal{G}, \Phi, \mathcal{S}}$ $b \twoheadleftarrow \{0, 1\}$ **if** $x \notin \{0, 1\}^{f.n}$ **then return** $\bot$ **if** $b = 1$ **then** $(F, e, d) \leftarrow \mathsf{Gb}(1^k, f), \ X \leftarrow \mathsf{En}(e, x)$ **else** $y \leftarrow \mathsf{ev}(f, x), \ (F, X, d) \leftarrow \mathcal{S}(1^k, y, \Phi(f))$ **return** (F, X, d)

proc GARBLE(f) $b \twoheadleftarrow \{0, 1\}$ **if** $b = 1$ **then** $(F, e, d) \leftarrow \mathsf{Gb}(1^k, f)$ **else** $(F, d) \leftarrow \mathcal{S}(1^k, \Phi(f), 0)$ **return** (F, d)	**proc** INPUT(x) $\qquad\qquad\quad$ Prv1$_{\mathcal{G}, \Phi, \mathcal{S}}$ **if** $x \notin \{0, 1\}^{f.n}$ **then return** $\bot$ **if** $b = 1$ **then** $X \leftarrow \mathsf{En}(e, x)$ **else** $y \leftarrow \mathsf{ev}(f, x), \ X \leftarrow \mathcal{S}(y, 1)$ **return** X

proc GARBLE(f)
 $b \twoheadleftarrow \{0, 1\}; \ n \leftarrow f.n; \ Q \leftarrow \emptyset; \ \tau \leftarrow \varepsilon$
 if $b = 1$ **then**
 $\quad (F, (X_1^0, X_1^1, \ldots, X_n^0, X_n^1), d) \leftarrow \mathsf{Gb}(1^k, f)$
 else
 $\quad (F, d) \leftarrow \mathcal{S}(1^k, \Phi(f), 0)$
 return (F, d)
	proc INPUT(i, c) $\qquad\quad$ Prv2$_{\mathcal{G}, \Phi, \mathcal{S}}$
	 if $i \notin \{1, \ldots, n\} \setminus Q$ **then return** $\bot$
	 $x_i \leftarrow c; \ Q \leftarrow Q \cup \{i\}$
	 if $|Q| = n$ **then**
	 $\quad x \leftarrow x_1 \cdots x_n; \ y \leftarrow \mathsf{ev}(f, x); \ \tau \leftarrow y$
	 if $b = 1$ **then** $X_i \leftarrow X_i^{x_i}$
	 else $X_i \leftarrow \mathcal{S}(\tau, i, |Q|)$
	 return X_i

Fig. 2. Three kinds of privacy: prv, prv1, prv2. Games to define the static, coarse-grained, and fine-grained privacy of $\mathcal{G} = (\mathsf{Gb}, \mathsf{En}, \mathsf{De}, \mathsf{Ev}, \mathsf{ev})$. FINALIZE$(b')$ returns the predicate $(b = b')$. Notation $s \twoheadleftarrow S$ denotes uniform sampling from a finite set.

For xxx $\in \{\mathrm{prv}, \mathrm{prv1}, \mathrm{prv2}\}$ we say that $\mathcal{G}$ is xxx-secure with respect to (or over) Φ if for every PT adversary $\mathcal{A}$ there exists a PT simulator $\mathcal{S}$ such that $\mathbf{Adv}_{\mathcal{G}}^{\mathrm{xxx}, \Phi, \mathcal{S}}(\mathcal{A}, \cdot)$ is negligible. We let $\mathsf{GS}(\mathrm{xxx}, \Phi)$ be the set of all garbling schemes that are xxx-secure over Φ.

Let us now explain the three games, beginning with static privacy. Here we let the adversary select f and x and we do one of two things: garble f to make (F, e, d) and encode x to make X, giving the adversary (F, X, d); or, alternatively, we ask the simulator produce a "fake" (F, X, d) based only on the security parameter k, the partial information $\Phi(f)$ about f, and the output $y = \mathsf{ev}(f, x)$. The adversary will have to guess if the garbling was real or fake.

For coarse-grained adaptive privacy, we begin by letting the adversary pick f. Either we garble it to $(F, e, d) \leftarrow \mathsf{Gb}(1^k, f)$ and give the adversary (F, d); or else we ask the simulator to devise a fake (F, d) based solely on k and $\phi = \Phi(f)$. Only after the adversary has received (F, d) do we ask it to provide an input x. Corresponding to the two choices we either encode x to $X = \mathsf{En}(e, x)$ or ask the simulator to produce a fake X, assisting it only by providing $\mathsf{ev}(f, x)$.

Coarse-grained adaptive privacy is arguably not all *that* adaptive, as the adversary specifies its input x all in one shot. This is unavoidable as long as the encoding function e operates on x atomically. But if the encoding function e is projective, then we can dole out the garbled input component-by-component. Only after the adversary specifies all n bits, one by one, is the input fully determined. At that point the simulator is handed y, which might be needed for constructing the final token $X_i^{x_i}$.

THE OMSS TRANSFORM. In the process of constructing one-time programs from garbled circuits, GKR [13] recognize the need for adaptive privacy of the garbled circuits. Their construction incorporates a technique to provide it. This technique is easily abstracted to provide, in our terminology, a transform that aims to convert a projective, prv garbling scheme into a projective, prv2 garbling scheme. Instead of garbling f we pick $r \twoheadleftarrow \{0,1\}^m$ and garble the circuit g defined by $g(x) = f(x) \oplus r$ for every $x \in \{0,1\}^n$ where $n = f.n$ and $m = f.m$. Then we secret share r as $r = r_1 \oplus \cdots \oplus r_n$ and include r_i in the i-th token, so that evaluation reconstructs r and it can be xored back at decoding time to recover $\mathsf{ev}(f, x)$ as $\mathsf{ev}(g, x) \oplus r$. Intuitively, this should work because the simulator can garble a dummy constant function with random output s and does not have to commit to r until it gets the target output value y of f and needs to provide the last token, at which point it can pick $r = s \oplus y$ so that y as desired [13]. Just the same, we show by counterexample that the OMSS does not in work, in general, to convert a prv-secure scheme to a prv2-secure one: we present a prv secure $\mathcal{G}$ such that $\mathsf{OMSS}[\mathcal{G}]$ is not prv2 secure. While this does not show that OMSS fails in the context in which GMR use it, our counterexample extends to that setting as well; see the full paper [3].

Now proceeding formally, we associate to circuit-garbling scheme $\mathcal{G} = (\mathsf{Gb}, \mathsf{En}, \mathsf{De}, \mathsf{Ev}, \mathsf{ev}) \in \mathsf{GS}(\mathsf{proj})$ the circuit-garbling scheme $\mathsf{OMSS}[\mathcal{G}] = (\mathsf{Gb}_2, \mathsf{En}_2, \mathsf{De}_2, \mathsf{Ev}_2, \mathsf{ev}) \in \mathsf{GS}(\mathsf{proj})$ defined at the top of Fig. 3. For simplicity we are assuming that the decoding rule d in $\mathcal{G}$ is always vacuous, meaning $d = \varepsilon$. (We do not need non-trivial d to achieve privacy [4], and this lets us stay closer to GKR [13], whose garbled circuits have no analogue of our decoding rule.) In the code, $g(\cdot) \leftarrow f(\cdot) \oplus r$ means that we construct from f, r a circuit g such that $\mathsf{ev}(g, x) = \mathsf{ev}(f, x) \oplus r$ for all $x \in \{0,1\}^{f.n}$. (Note we can do this in such a way that $\Phi_{\mathrm{topo}}(g) = \Phi_{\mathrm{topo}}(f)$.)

The claim under consideration is that if $\mathcal{G}$ is prv-secure relative to $\Phi = \Phi_{\mathrm{topo}}$ then $\mathcal{G}_2$ is prv2-secure relative to $\Phi = \Phi_{\mathrm{topo}}$. To prove this, we would need to let $\mathcal{A}_2$ be an arbitrary PT adversary and build a PT simulator $\mathcal{S}_2$ such that $\mathbf{Adv}_{\mathcal{G}_2}^{\mathrm{prv2}, \Phi, \mathcal{S}_2}(\mathcal{A}_2, \cdot)$ is negligible. GKR suggest a plausible strategy for the simulator that, in particular, explains the intuition for the transform. We present here our understanding of this strategy adapted to our setting. In its first phase the simulator $\mathcal{S}_2$ has input $1^k, \phi, 0$ where $\phi = \Phi(f)$, with f being the query made by the adversary to GARBLE. Simulator $\mathcal{S}_2$ picks $s \twoheadleftarrow \{0,1\}^n$ and lets f_s be the circuit that has output s on all inputs and $\Phi_{\mathrm{topo}}(f_s) = \phi$. It also picks random m-bit strings $s_1, \ldots, s_n$ and a random input $w \twoheadleftarrow \{0,1\}^n$. It lets $(G, (X_1^0, X_1^1, \ldots, X_n^0, X_n^1), \varepsilon) \twoheadleftarrow \mathsf{Gb}(1^k, f_s)$ and returns G to the adversary, saving $\sigma = (s, s_1, \ldots, s_n)$ as state information. In the second phase, when given input τ, i, j, for $j \leq n - 1$, the simulator lets $T_i \leftarrow (X_i^{w_i}, s_i)$ and returns T_i to the adversary as the token for bit i of the input. In the case that $j = n$, the simulator obtains (from τ as per our game) the output $y = \mathsf{ev}(f, x)$ of the function on input x, the latter defined by the adversary's queries to INPUT. It now resets

proc $\mathsf{Gb}_2(1^k, f)$	**proc** $\mathsf{En}_2((T_1^0, T_1^1, \ldots, T_n^0, T_n^1), x)$
$n \leftarrow f.n, \; r_1, \ldots, r_n \twoheadleftarrow \{0,1\}^{f.m}$	$x_1 \cdots x_n \leftarrow x$
$r \leftarrow r_1 \oplus \cdots \oplus r_n, \; g(\cdot) \leftarrow f(\cdot) \oplus r$	**return** $(T_1^{x_1}, \ldots, T_n^{x_n})$
$(G, (X_1^0, X_1^1, \ldots, X_n^0, X_n^1), \varepsilon) \twoheadleftarrow \mathsf{Gb}(1^k, g)$	
for $i \in \{1, \ldots, n\}$ **do**	
$\quad T_i^0 \leftarrow (X_i^0, r_i), \; T_i^1 \leftarrow (X_i^1, r_i)$	
return $(G, (T_1^0, T_1^1, \ldots, T_n^0, T_n^1), \varepsilon)$	
proc $\mathsf{Ev}_2(G, (T_1, \ldots, T_n))$	**proc** $\mathsf{De}_2(\varepsilon, (Y, r))$
for $i \in \{1, \ldots, n\}$ **do** $(X_i, r_i) \leftarrow T_i$	**return** $\mathsf{De}(\varepsilon, Y) \oplus r$
$Y \leftarrow \mathsf{Ev}(G, (X_1, \ldots, X_n))$	
$r \leftarrow r_1 \oplus \cdots \oplus r_n$	
return (Y, r)	

proc $\mathsf{Gb}(1^k, g)$	**proc** $\mathsf{Ev}(G, (X_1, \ldots, X_n))$
$(n, m) \leftarrow (g.n, g.m)$	**for** $i \in \{1, \ldots, n\}$ **do** $(Z_i, V_i) \leftarrow X_i$
$(G', (Z_1^0, Z_1^1, \ldots, Z_n^0, Z_n^1), \varepsilon) \twoheadleftarrow \mathsf{Gb}'(1^k, g)$	$(G', v, V) \leftarrow G$
for $i \in \{1, \ldots, n\}$ **do** $V_i^0, V_i^1 \twoheadleftarrow \{0,1\}^m$	**return** $\mathsf{Ev}'(G', (Z_1, \ldots, Z_n))$
$v_1 \cdots v_n \leftarrow v \twoheadleftarrow \{0,1\}^n, \quad V \twoheadleftarrow \{0,1\}^m$	
if $n \geq k$ **then**	
$\quad V \leftarrow \mathsf{ev}(g, \overline{v}) \oplus V_1^{v_1} \oplus \cdots \oplus V_n^{v_n}$	**proc** $\mathsf{En}((X_1^0, X_1^1, \ldots, X_n^0, X_n^1), x)$
for $i \in \{1, \ldots, n\}$ **do**	$x_1 \cdots x_n \leftarrow x$
$\quad X_i^0 \leftarrow (Z_i^0, V_i^0), \quad X_i^1 \leftarrow (Z_i^1, V_i^1)$	**return** $(X_1^{x_1}, \ldots, X_n^{x_n})$
$G \leftarrow (G', v, V)$	
return $(G, (X_1^0, X_1^1, \ldots, X_n^0, X_n^1), \varepsilon)$	

Fig. 3. **OMSS definition (top).** Scheme $\mathsf{OMSS}[\mathcal{G}] = (\mathsf{Gb}_2, \mathsf{En}_2, \mathsf{De}_2, \mathsf{Ev}_2, \mathsf{ev})$ where $\mathcal{G} = (\mathsf{Gb}, \mathsf{En}, \mathsf{De}, \mathsf{Ev}, \mathsf{ev})$. **OMSS counterexample (bottom).** The garbling scheme $\mathcal{G} = (\mathsf{Gb}, \mathsf{En}, \mathsf{De}, \mathsf{Ev}, \mathsf{ev})$ obtained from $\mathcal{G}' = (\mathsf{Gb}', \mathsf{En}', \mathsf{De}, \mathsf{Ev}', \mathsf{ev})$ is prv secure when $\mathcal{G}'$ is, but $\mathsf{OMSS}[\mathcal{G}]$ is not prv2 secure.

$s_i = y \oplus s \oplus s_i \oplus s_1 \oplus \cdots \oplus s_n$ and returns (X_i, s_i), so that evaluation of the garbled function indeed results in output y.

This simulation strategy is intuitive, but trying to prove it correct runs into problems. We have to show that $\mathbf{Adv}_{\mathcal{G}_2}^{\mathrm{prv2}, \Phi, \mathcal{S}_2}(\mathcal{A}_2, \cdot)$ is negligible. We must utilize the assumption of prv security to do this, which means we must perform a reduction. The only plausible path towards this is to construct from $\mathcal{A}_2$ an adversary $\mathcal{A}$ against the prv-security of $\mathcal{G}$ and then exploit the existence of a simulator $\mathcal{S}$ such that $\mathbf{Adv}_{\mathcal{G}}^{\mathrm{prv}, \Phi, \mathcal{S}}(\mathcal{A}, \cdot)$ is negligible. However, it is not clear how to construct $\mathcal{A}$, let alone how its simulator comes into play. (As we will see when proving our transforms, the proof template that works is different, not trying first to build $\mathcal{S}_2$, but instead building $\mathcal{A}$ from $\mathcal{A}_2$ and then $\mathcal{S}_2$ from $\mathcal{S}$.)

The problem turns out to be more than technical, for we will see that the transform itself does not work in general. By this we mean that we can exhibit a (projective) circuit-garbling scheme $\mathcal{G} = (\mathsf{Gb}, \mathsf{En}, \mathsf{De}, \mathsf{Ev}, \mathsf{ev})$ that is prv-secure relative to $\Phi = \Phi_{\mathrm{topo}}$ but the transformed scheme $\mathcal{G}_2 = \mathsf{OMSS}[\mathcal{G}]$ is subject to

proc $\mathsf{Gb}_1(1^k, f)$	**proc** $\mathsf{En}_1(e_1, x)$
$(F, e, d) \leftarrow \mathsf{Gb}(1^k, f)$	$(e, d', F') \leftarrow e_1, \quad X \leftarrow \mathsf{En}(e, x)$
$F' \leftarrow \{0, 1\}^{\lvert F \rvert}, \quad d' \leftarrow \{0, 1\}^{\lvert d \rvert}$	**return** (X, d', F')
$F_1 \leftarrow F \oplus F', \quad d_1 \leftarrow d \oplus d'$	
$e_1 \leftarrow (e, d', F')$	
return (F_1, e_1, d_1)	
proc $\mathsf{Ev}_1(F_1, X_1)$	**proc** $\mathsf{De}_1(d_1, Y_1)$
$(X, d', F') \leftarrow X_1, \quad F \leftarrow F_1 \oplus F'$	$(Y, d') \leftarrow Y_1, \quad d \leftarrow d_1 \oplus d'$
$Y \leftarrow \mathsf{Ev}(F, X)$	**return** $\mathsf{De}(d, Y)$
return (Y, d')	

proc $\mathsf{Gb}_2(1^k, f)$	**proc** $\mathsf{Ev}_2(F, X_2)$
$(F, e, d) \leftarrow \mathsf{Gb}_1(1^k, f)$	$((U_1, S_1), \ldots, (U_n, S_n)) \leftarrow X_2$
$(X_1^0, X_1^1, \ldots, X_n^0, X_n^1) \leftarrow e$	$Z \leftarrow S_1 \oplus \cdots \oplus S_n$
$N \leftarrow \lvert \mathsf{En}_1(e, 0^n) \rvert$	$(Z_1, \ldots, Z_n) \leftarrow Z$
for $i \in \{1, \ldots, n\}$ **do**	$X \leftarrow (U_1 \oplus Z_1, \ldots, U_n \oplus Z_n)$
$\quad Z_i \leftarrow \{0, 1\}^{\lvert X_i^0 \rvert}, \quad S_i \leftarrow \{0, 1\}^N$	**return** $\mathsf{Ev}_1(F, X)$
$Z \leftarrow (Z_1, \ldots, Z_n)$	
$S_n \leftarrow Z \oplus S_1 \oplus \cdots \oplus S_{n-1}$	**proc** $\mathsf{En}_2(e_2, x)$
for $i \in \{1, \ldots, n\}$ **do**	$(T_1^0, X_1^1, \ldots, T_n^0, X_n^1) \leftarrow e_2$
$\quad T_i^0 \leftarrow (X_i^0 \oplus Z_i, S_i), \quad T_i^1 \leftarrow (X_i^1 \oplus Z_i, S_i)$	$x_1 \cdots x_n \leftarrow x$
return $(F, (T_1^0, T_1^1, \ldots, T_n^0, T_n^1), d)$	**return** $(T_1^{x_1}, \ldots, T_n^{x_n})$

Fig. 4. Transform prv-to-prv1 (top): Scheme $\mathcal{G}_1 = (\mathsf{Gb}_1, \mathsf{En}_1, \mathsf{De}_1, \mathsf{Ev}_1, \mathsf{ev}) \in$ $\mathsf{GS}(\mathsf{prv1}, \varPhi)$ obtained by applying the prv-to-prv1 transform to $\mathcal{G} = (\mathsf{Gb}, \mathsf{En}, \mathsf{De}, \mathsf{Ev}, \mathsf{ev}) \in \mathsf{GS}(\mathsf{prv}, \varPhi)$. **Transform prv1-to-prv2 (bottom):** Projective garbling scheme $\mathcal{G}_2 = (\mathsf{Gb}_2, \mathsf{En}_2, \mathsf{De}, \mathsf{Ev}_2, \mathsf{ev}) \in \mathsf{GS}(\mathsf{prv2}, \varPhi)$ obtained by applying the prv1-to-prv2 transform to projective garbling scheme $\mathcal{G}_1 = (\mathsf{Gb}_1, \mathsf{En}_1, \mathsf{De}_1, \mathsf{Ev}_1, \mathsf{ev}) \in \mathsf{GS}(\mathsf{prv1}, \varPhi)$

an attack showing that it is not prv2 secure. This means, in particular, that the above simulation strategy does not in general work.

To carry this out, we start with an arbitrary projective circuit-garbling scheme $\mathcal{G}' = (\mathsf{Gb}', \mathsf{En}', \mathsf{De}, \mathsf{Ev}', \mathsf{ev})$ assumed to be prv-secure relative to $\varPhi = \varPhi_{\mathrm{topo}}$. We then transform it into the projective circuit-garbling scheme $\mathcal{G} = (\mathsf{Gb}, \mathsf{En}, \mathsf{De}, \mathsf{Ev}, \mathsf{ev})$ shown at the bottom of Fig. 3. (We assume the decoding rule of $\mathcal{G}'$ is vacuous, a feature inherited by $\mathcal{G}$. We are letting $\overline{v}$ denote the bitwise complement of a string v.) The following proposition, whose proof is in the full paper [3], says that $\mathcal{G}$ continues to be prv-secure but an attack shows that $\mathsf{OMSS}[\mathcal{G}]$ is not prv2-secure. (The proof shows it is in fact not even prv1 secure.)

Proposition 1. Let ev be the canonical circuit-evaluation function. Assume $\mathcal{G}' = (\mathsf{Gb}', \mathsf{En}', \mathsf{De}, \mathsf{Ev}', \mathsf{ev}) \in \mathsf{GS}(\mathsf{prv}, \varPhi_{\mathrm{topo}}) \cap \mathsf{GS}(\mathsf{proj})$ and let $\mathcal{G} = (\mathsf{Gb}, \mathsf{En}, \mathsf{De}, \mathsf{Ev}, \mathsf{ev}) \in \mathsf{GS}(\mathsf{proj})$ be the garbling scheme shown at the bottom of Fig. 3. Then **(1)** $\mathcal{G} \in \mathsf{GS}(\mathsf{prv}, \varPhi_{\mathrm{topo}}) \cap \mathsf{GS}(\mathsf{proj})$, but **(2)** $\mathsf{OMSS}[\mathcal{G}] \notin \mathsf{GS}(\mathsf{prv2}, \varPhi_{\mathrm{topo}})$.

ACHIEVING PRV1 SECURITY. We now describe a transform prv-to-prv1 that successfully turns a prv secure circuit garbling scheme into a prv1 secure one.

Combined with established results [4], this yields prv1-secure schemes based on standard assumptions. The idea is to use one-time pads to mask F and d, and then append the pads to X. This will ensure that the adversary learns nothing about F and d until it fully specifies function f and x. Given a (not necessarily projective) garbling scheme $\mathcal{G} = (\mathsf{Gb}, \mathsf{En}, \mathsf{De}, \mathsf{Ev}, \mathsf{ev})$, the prv-to-prv1 transform returns the garbling scheme prv-to-prv1$[\mathcal{G}] = (\mathsf{Gb}_1, \mathsf{En}_1, \mathsf{De}_1, \mathsf{Ev}_1, \mathsf{ev})$ at the top of Fig. 4. We claim:

Theorem 2. For any Φ, if $\mathcal{G} \in \mathsf{GS}(\mathrm{prv}, \Phi)$ then prv-to-prv1$[\mathcal{G}] \in \mathsf{GS}(\mathrm{prv1}, \Phi)$.

The intuition behind the prv-to-prv1 transform (outlined above) is simple, but the proof template is instructive in indicating how to move from the intuition to a formal proof. Given any PT adversary $\mathcal{A}_1$ against the prv1-security of $\mathcal{G}_1$ we build a PT adversary $\mathcal{A}$ against the prv-security of $\mathcal{G}$. Now the assumption of prv-security yields a PT simulator $\mathcal{S}$ for $\mathcal{A}$ such that $\mathbf{Adv}_{\mathcal{G}}^{\mathrm{prv}, \Phi, \mathcal{S}}(\mathcal{A}, \cdot)$ is negligible. Now we build from $\mathcal{S}$ a PT simulator $\mathcal{S}_1$ such that for all $k \in \mathbb{N}$ we have $\mathbf{Adv}_{\mathcal{G}_1}^{\mathrm{prv1}, \Phi, \mathcal{S}_1}(\mathcal{A}_1, k) \leq \mathbf{Adv}_{\mathcal{G}}^{\mathrm{prv}, \Phi, \mathcal{S}}(\mathcal{A}, k)$. This yields the theorem. In the full paper [3] we provide a full proof that shows how to build $\mathcal{A}$ and $\mathcal{S}_1$.

ACHIEVING PRV2 SECURITY.   Next we show how to transform a prv1 scheme into a prv2 one. Formally, given a projective garbling scheme $\mathcal{G} = (\mathsf{Gb}, \mathsf{En}, \mathsf{De}, \mathsf{Ev}, \mathsf{ev}) \in \mathsf{GS}(\mathrm{prv1}, \Phi)$, the prv1-to-prv2 transform returns the projective garbling scheme prv1-to-prv2$[\mathcal{G}] = (\mathsf{Gb}_2, \mathsf{En}_2, \mathsf{De}, \mathsf{Ev}_2, \mathsf{ev})$ shown at the bottom of Fig. 4. The idea is to mask the garbled input and then use the second part of GKR's idea as represented by OMSS, namely secret-share the mask, putting a piece in each token, so that unless one has all tokens, one learns nothing about the garbled input. The formal proof of the following is in the full paper [3].

Theorem 3. For any Φ, if $\mathcal{G}_1 \in \mathsf{GS}(\mathrm{prv1}, \Phi) \cap \mathsf{GS}(\mathrm{proj})$ then prv1-to-prv2$[\mathcal{G}_1] \in \mathsf{GS}(\mathrm{prv2}, \Phi) \cap \mathsf{GS}(\mathrm{proj})$.

ONE-TIME COMPILERS.   Starting from garbling schemes with prv2 security, we give simple designs, and proofs, for one-time programs. We begin with the definitions. Following GKR [13], the intent is that possession of a one-time program P for a function f should enable one to evaluate f at any single value x; but, beyond that, the one-time program should be useless. Unachievable in any standard model of computation (where possession of P would enable its repeated evaluation at multiple point), GKR suggest achieving one-time programs in a model of computation that provides *one-time memory*—tamper-resistant hardware whose read-once i-th location returns, on query $(i, b) \in \mathbb{N} \times \{0, 1\}$, the string T_i^b, immediately thereafter expunging T_i^{1-b}. A *one-time compiler* probabilistically transforms the description of a function f into a one-time program P and its associated one-time memory T.

For a formal treatment, we begin by specifying two stateful oracles; see Fig. 5. The first, OTP_f, formalizes the desired behavior of a one-time program for f. Here f will now be regarded as a string, not a function, but this string represents a circuit computing a function $\mathsf{ev}(f) : \{0, 1\}^{f.n} \to \{0, 1\}^{f.m}$; we write ev for the canonical circuit-evaluation function [4]. The agent calling out to OTP_f

proc $\mathrm{OTP}_f(x)$	**proc** $\mathrm{OTM}_T(i, b)$
if $x \notin \{0,1\}^{f.n}$ **then ret** $\bot$	$(T_1^0, T_1^1, \ldots, T_\ell^0, T_\ell^1) \leftarrow T$
if *called* **then ret** $\bot$	**if** $i \notin [1..\ell]$ **or** *used$_i$* **or** $b \notin \{0,1\}$ **then ret** $\bot$
called $\leftarrow$ true	*used$_i$* $\leftarrow$ true
ret $\mathrm{ev}(f, x)$	**ret** T_i^b

Fig. 5. Oracles model one-time programs and one-time memory. Oracle OTP depends on a string f representing a boolean circuit. Oracle OTM depends on a list of strings T.

provides x and, on the first query, it gets $\mathrm{ev}(f, x)$. Subsequent queries return nothing. On the right-hand side of Fig. 5 we similarly define an oracle OTM_T, this to model possession of a one-time-memory system. Given a list of ℓ pairs of strings (establish some convention so that every string T is regarded as denoting a list of ℓ pairs of strings, for some $\ell \in \mathbb{N}$) the oracle returns at most one string from each pair, otherwise satisfying each request.

Elaborating on GKR, we now define a *one-time compiler* as a pair of probabilistic algorithms $\varPi = (\mathrm{Co}, \mathrm{Ex})$ (for *compile* and *execute*). Algorithm Co, on input 1^k and a string f, produces a pair $(P, T) \leftarrow \mathrm{Co}(1^k, f)$ where P (the one-time program) is a string and T (the one-time-memory) encodes a list of 2ℓ strings, for some ℓ. Algorithm Ex, on input of strings P and x, and given access to an oracle $\mathcal{O}$, returns a string $y \leftarrow \mathrm{Ex}^{\mathcal{O}}(P, x)$. We require the following *correctness* condition of $\varPi = (\mathrm{Co}, \mathrm{Ex})$: if $(P, T) \leftarrow \mathrm{Co}(1^k, f)$ and $x \in \{0,1\}^{f.n}$ then $\mathrm{Ex}^{\mathrm{OTM}_T(\cdot, \cdot)}(P, x) = \mathrm{ev}(f, x)$.

The security of $\varPi = (\mathrm{Co}, \mathrm{Ex})$ will be relative to a side-information function $\varPhi$; the value $\phi = \varPhi(f)$ captures the information about f that P is allowed to reveal. So fix a one-time compiler $\varPi = (\mathrm{Co}, \mathrm{Ex})$, an adversary $\mathcal{A}$, a security parameter k, and a string f. (1) Consider the distribution $\mathrm{Real}_{\varPi, \mathcal{A}, f}(k)$ determined by the following experiment: first, sample $(P, T) \leftarrow \mathrm{Co}(1^k, f)$; then, run $\mathcal{A}^{\mathrm{OTM}_T(\cdot)}(1^k, P)$ and output whatever $\mathcal{A}$ outputs. (2) Alternatively, fix a one-time compiler $\varPi = (\mathrm{Co}, \mathrm{Ex})$, an information function $\varPhi$, a simulator $\mathcal{S}$, a security parameter k, and a string f. Consider the distribution $\mathrm{Fake}_{\varPi, \varPhi, \mathcal{S}, f}(k)$ determined by the following experiment: run $\mathcal{S}^{\mathrm{OTP}_f(\cdot)}(1^k, \varPhi(f))$ and output whatever $\mathcal{S}$ outputs. For $\mathcal{D}$ an algorithm and $\varPi$, $\varPhi$, $\mathcal{A}$, $\mathcal{S}$, and k as above, let

$$\mathbf{Adv}_{\varPi, \varPhi, \mathcal{A}, \mathcal{S}, \mathcal{D}}^{\mathrm{otc}}(k) = \Pr[(f, \sigma) \leftarrow \mathcal{D}(1^k); v \leftarrow \mathrm{Real}_{\varPi, \mathcal{A}, f}(k): \ \mathcal{D}(\sigma, v) \Rightarrow 1] -$$
$$\Pr[(f, \sigma) \leftarrow \mathcal{D}(1^k); v \leftarrow \mathrm{Fake}_{\varPi, \varPhi, \mathcal{S}, f}(k): \ \mathcal{D}(\sigma, v) \Rightarrow 1]$$

One-time compiler $\varPi$ is said to be (OTC-) *secure* with respect to side-information function $\varPhi$ if for any PPT adversary $\mathcal{A}$ there is a PPT simulator $\mathcal{S}$ such that for all PPT distinguishers $\mathcal{D}$, function $\mathbf{Adv}_{\varPi, \varPhi, \mathcal{A}, \mathcal{S}, \mathcal{D}}^{\mathrm{otc}}(k)$ is negligible.

CONSTRUCTING AN OTC FROM A GARBLING SCHEME. A circuit-garbling scheme $\mathcal{G} = (\mathrm{Gb}, \mathrm{En}, \mathrm{De}, \mathrm{Ev}, \mathrm{ev})$ can be turned into a one-time compiler $\varPi = (\mathrm{Co}, \mathrm{Ex})$ in a natural way: let $\mathrm{OTC}[\mathcal{G}] = (\mathrm{Co}, \mathrm{Ex})$ be defined as follows. (1) $\mathrm{Co}(1^k, f)$: let $(F, e, d) \leftarrow \mathrm{Gb}(f)$ and return (P, T) where $P = (F, d)$ and $T = e$.

(2) $\mathsf{Ex}^{\mathcal{O}}(P, x)$: Let $(F, d) \leftarrow P$, let $x_1 \cdots x_n \leftarrow x$, query oracle $\mathcal{O}$ on $(1, x_1)$, $\ldots, (n, x_n)$ to obtain $X_1, \ldots, X_n$, respectively, and return $\mathsf{De}(d, \mathsf{Ev}(F, X))$ with $X = (X_1, \ldots, X_n)$. The proof of the following is in the full paper [3].

Theorem 4. If $\mathcal{G}$ is a prv2-secure garbling scheme over side-information function Φ then $\mathrm{OTC}[\mathcal{G}]$ is OTC-secure with respect to side-information Φ.

The straightforwardness of the construction and its trivial proof are, we believe, points in our favor, evidence of our claim that the garbling scheme abstraction and appropriate security notions for it engender applications in direct, simple and less error-prone ways.

SEPARATION. In the full paper [3], we elaborate on how Proposition 1 gives an example of a garbling scheme $\mathcal{G}$ such that $\mathrm{OTC}[\mathrm{OMSS}[\mathcal{G}]]$ is not otc-secure. We explain why this refutes GKR's claim [13] that their construction provides a secure one-time compiler assuming one-way functions.

4   Obliviousness, Authenticity and Secure Outsourcing

We define obliviousness and authenticity, both with either the coarse-grained or fine-grained adaptivity. We show how to achieve these goals, in combination with adaptive privacy, via generic transforms and in the standard model. In the full paper [3] we provide more efficient transforms in the ROM. Finally we apply this to obtain extremely simple and modular designs, and security proofs, for verifiable outsourcing schemes based on the paradigm of GGP [9].

OBLIVIOUSNESS. Intuitively, a garbling scheme is *oblivious* if garbled function F and garbled input X, these corresponding to f and x, reveal nothing of f or x beyond side-information $\Phi(f)$. In particular, possession F and X will not allow the calculation of $y = \mathsf{ev}(f, x)$.

The formal definition for *static* obliviousness is from BHR [4]. See the top of Fig. 6. We add to this two new definitions, to incorporate either coarse-grained or fine-grained adaptive security. See the rest of Fig. 6. Fine-grained adaptive security continues to require that $\mathcal{G}$ be projective. The games used for defining obliviousness closely mirror their privacy counterparts. The first important difference is that the adversary does not get the decoding function d. The second important difference is that the simulator must do without $y = \mathsf{ev}(f, x)$. For a garbling scheme $\mathcal{G}$, side-information Φ, simulator $\mathcal{S}$, adversary $\mathcal{A}$, and security parameter $k \in \mathbb{N}$, we let $\mathbf{Adv}_{\mathcal{G}}^{\mathrm{obv}, \Phi, \mathcal{S}}(\mathcal{A}, k) = 2 \Pr[\mathrm{Obv}_{\mathcal{G}, \Phi, \mathcal{S}}^{\mathcal{A}}(k)] - 1$, $\mathbf{Adv}_{\mathcal{G}}^{\mathrm{obv1}, \Phi, \mathcal{S}}(\mathcal{A}, k) = 2 \Pr[\mathrm{Obv1}_{\mathcal{G}, \Phi, \mathcal{S}}^{\mathcal{A}}(k)] - 1$, and finally $\mathbf{Adv}_{\mathcal{G}}^{\mathrm{obv2}, \Phi, \mathcal{S}}(\mathcal{A}, k) = 2 \Pr[\mathrm{Obv2}_{\mathcal{G}, \Phi, \mathcal{S}}^{\mathcal{A}}(k)] - 1$. Garbling scheme $\mathcal{G}$ is obv-secure with respect to Φ if for every PPT $\mathcal{A}$ there exists a simulator $\mathcal{S}$ such that $\mathbf{Adv}_{\mathcal{G}}^{\mathrm{obv}, \Phi, \mathcal{S}}(\mathcal{A}, k)$ is negligible. We similarly define obv1 and obv2 security. For xxx $\in \{\mathrm{obv}, \mathrm{obv1}, \mathrm{obv2}\}$ we let $\mathrm{GS}(\mathrm{xxx}, \Phi)$ denote the set of all garbling schemes that are xxx-secure over Φ.

Fig. 6 also formalizes the games underlying three definitions of authenticity, capturing an adversary's inability to create from F and X a garbled output

proc $\text{GARBLE}(f, x)$	$\text{Obv}_{\mathcal{G}, \Phi, \mathcal{S}}$

$b \leftarrow \{0, 1\}$
if $x \notin \{0, 1\}^{f.n}$ **then return** $\perp$
if $b = 1$ **then** $(F, e, d) \leftarrow \text{Gb}(1^k, f), \ X \leftarrow \text{En}(e, x)$
else $(F, X) \leftarrow \mathcal{S}(1^k, \Phi(f))$
return (F, X)

proc $\text{GARBLE}(f)$	**proc** $\text{INPUT}(x)$	$\text{Obv1}_{\mathcal{G}, \Phi, \mathcal{S}}$

$b \leftarrow \{0, 1\}$ **if** $x \notin \{0, 1\}^{f.n}$ **then return** $\perp$
if $b = 1$ **then** $(F, e, d) \leftarrow \text{Gb}(1^k, f)$ **if** $b = 1$ **then** $X \leftarrow \text{En}(e, x)$
else $F \leftarrow \mathcal{S}(1^k, \Phi(f), 0)$ **else** $X \leftarrow \mathcal{S}(1)$
return F **return** X

proc $\text{GARBLE}(f)$	**proc** $\text{INPUT}(i, c)$	$\text{Obv2}_{\mathcal{G}, \Phi, \mathcal{S}}$

$b \leftarrow \{0, 1\}; \ n \leftarrow f.n; \ Q \leftarrow \emptyset; \ \sigma \leftarrow \varepsilon$ **if** $i \notin \{1, \ldots, n\} \setminus Q$ **then return** $\perp$
if $b = 1$ **then** $x_i \leftarrow c; \ Q \leftarrow Q \cup \{i\}$
     $(F, (X_1^0, X_1^1, \ldots, X_n^0, X_n^1), d) \leftarrow \text{Gb}(1^k, f)$ **if** $b = 1$ **then** $X_i \leftarrow X_i^{x_i}$
else $F \leftarrow \mathcal{S}(1^k, \Phi(f), 0)$ **else** $X_i \leftarrow \mathcal{S}(i, |Q|)$
return F **return** X_i

proc $\text{GARBLE}(f, x)$	$\text{Aut}_{\mathcal{G}}$

if $x \notin \{0, 1\}^{f.n}$ **then return** $\perp$
$(F, e, d) \leftarrow \text{Gb}(1^k, f), \ X \leftarrow \text{En}(e, x)$
return (F, X)

proc $\text{GARBLE}(f)$	**proc** $\text{INPUT}(x)$	$\text{Aut1}_{\mathcal{G}}$

$(F, e, d) \leftarrow \text{Gb}(1^k, f)$ **if** $x \notin \{0, 1\}^{f.n}$ **then return** $\perp$
return F $X \leftarrow \text{En}(e, x)$
     return X

proc $\text{GARBLE}(f)$	**proc** $\text{INPUT}(i, c)$	$\text{Aut2}_{\mathcal{G}}$

$n \leftarrow f.n; \ Q \leftarrow \emptyset; \ \sigma \leftarrow \varepsilon$ **if** $i \notin \{1, \ldots, n\} \setminus Q$ **then return** $\perp$
$(F, (X_1^0, X_1^1, \ldots, X_n^0, X_n^1), d) \leftarrow \text{Gb}(1^k, f)$ $x_i \leftarrow c; \ Q \leftarrow Q \cup \{i\}, \ X_i \leftarrow X_i^{x_i}$
return F **if** $|Q| = n$ **then** $X \leftarrow (X_1, \ldots, X_n)$
   return X_i

Fig. 6. Obliviousness (top). Games for defining the obv, obv1, and obv2 security of $\mathcal{G} = (\text{Gb}, \text{En}, \text{De}, \text{Ev}, \text{ev})$. For each game, $\text{FINALIZE}(b')$ returns $(b = b')$. **Authenticity (bottom).** Games for defining the aut, aut1, and aut2 security of $\mathcal{G} = (\text{Gb}, \text{En}, \text{De}, \text{Ev}, \text{ev})$. Procedure $\text{FINALIZE}(Y)$ of each game returns $(\text{De}(d, Y) \neq \perp$ **and** $Y \neq \text{Ev}(F, X))$.

$Y \neq F(X)$ that will be deemed authentic. The static definition of BHR [4] is strengthened either to allow the adversary to specify x subsequent to obtaining F, or, stronger, the bits of x are provided one-by-one, each corresponding token then issued. For the second case, game Aut2, the garbling scheme must once again be projective. For a garbling scheme $\mathcal{G}$, adversary $\mathcal{A}$, and security parameter $k \in \mathbb{N}$, we let $\mathbf{Adv}_{\mathcal{G}}^{\text{aut}}(\mathcal{A}, k) = 2 \Pr[\text{Aut}_{\mathcal{G}}^{\mathcal{A}}(k)] - 1$, $\mathbf{Adv}_{\mathcal{G}}^{\text{aut1}}(\mathcal{A}, k) =$

proc $\mathsf{Gb}_1(1^k, f)$	**proc** $\mathsf{En}_1(e_1, x)$
$(F, e, d) \leftarrow \mathsf{Gb}(1^k, f)$	$(e, d', F', \mathrm{tag}) \leftarrow e_1$
$F' \twoheadleftarrow \{0,1\}^{|F|}, \quad d' \twoheadleftarrow \{0,1\}^{|d|}$	**return** $(\mathsf{En}(e, x), d', F', \mathrm{tag})$
$F_1 \leftarrow F \oplus F', \quad K \twoheadleftarrow \{0,1\}^k, \quad d_1 \leftarrow (d \oplus d', K)$	
$\mathrm{tag} \leftarrow \mathsf{F}_K(d'), \quad e_1 \leftarrow (e, d', F', \mathrm{tag})$	
return (F_1, e_1, d_1)	
proc $\mathsf{Ev}_1(F_1, X_1)$	**proc** $\mathsf{De}_1(d_1, Y_1)$
$(X, d', F', \mathrm{tag}) \leftarrow X_1, \quad F \leftarrow F_1 \oplus F'$	$(Y, d', \mathrm{tag}) \leftarrow Y_1$
$Y \leftarrow \mathsf{Ev}(F, X)$	$(D, K) \leftarrow d_1, \quad d \leftarrow D \oplus d'$
return (Y, d', tag)	**if** $\mathrm{tag} \neq \mathsf{F}_K(d')$ **then return** $\perp$
	return $\mathsf{De}(d, Y)$

Fig. 7. Scheme $\mathsf{all\text{-}to\text{-}all1}[\mathcal{G}] = (\mathsf{Gb}_1, \mathsf{En}_1, \mathsf{De}_1, \mathsf{Ev}_1, \mathsf{ev}) \in \mathsf{GS}(\mathrm{prv1}, \varPhi) \cap \mathsf{GS}(\mathrm{obv1}, \varPhi) \cap \mathsf{GS}(\mathrm{aut1})$ obtained from scheme $\mathcal{G} = (\mathsf{Gb}, \mathsf{En}, \mathsf{De}, \mathsf{Ev}, \mathsf{ev}) \in \mathsf{GS}(\mathrm{prv}, \varPhi) \cap \mathsf{GS}(\mathrm{obv}, \varPhi) \cap \mathsf{GS}(\mathrm{aut})$. The transform uses a PRF $\mathsf{F} : \{0,1\}^k \times \{0,1\}^* \to \{0,1\}^k$.

$2 \Pr[\mathsf{Aut1}_{\mathcal{G}}^{\mathcal{A}}(k)] - 1$, and $\mathbf{Adv}_{\mathcal{G}}^{\mathrm{aut2}}(\mathcal{A}, k) = 2 \Pr[\mathsf{Aut2}_{\mathcal{G}}^{\mathcal{A}}(k)] - 1$. Garbling scheme $\mathcal{G}$ is aut-secure with respect to $\varPhi$ if for every PPT $\mathcal{A}$ $\mathbf{Adv}_{\mathcal{G}}^{\mathrm{aut}}(\mathcal{A}, k)$ is negligible. We similarly define aut1 and aut2 security. For $\mathrm{xxx} \in \{\mathrm{aut}, \mathrm{aut1}, \mathrm{aut2}\}$ we let $\mathsf{GS}(\mathrm{xxx})$ denote the set of all garbling schemes that are xxx-secure.

ACHIEVING OBV1 AND AUT1 SECURITY. It is tempting to think that the prv-to-prv1 operator in Fig. 4 also promotes xxx-security, with $\mathrm{xxx} \in \{\mathrm{obv}, \mathrm{aut}\}$, to xxx1-security, but it does not. We now show how to change prv-to-prv1 to an operator all-to-all1 that promotes any $\mathrm{xxx} \in \{\mathrm{prv}, \mathrm{obv}, \mathrm{aut}\}$ to being xxx1 secure. See Fig. 7. The proof of the following is in the full paper [3].

Theorem 5. (1) For any $\varPhi$ and any $\mathrm{xxx} \in \{\mathrm{prv}, \mathrm{obv}\}$, if $\mathcal{G} \in \mathsf{GS}(\mathrm{xxx}, \varPhi)$ then $\mathsf{all\text{-}to\text{-}all1}[\mathcal{G}] \in \mathsf{GS}(\mathrm{xxx1}, \varPhi)$ (2) If $\mathcal{G} \in \mathsf{GS}(\mathrm{aut})$ then $\mathsf{all\text{-}to\text{-}all1}[\mathcal{G}] \in \mathsf{GS}(\mathrm{aut1})$ (3) If $\mathcal{G} \in \mathsf{GS}(\mathrm{proj})$ then $\mathsf{all\text{-}to\text{-}all1}[\mathcal{G}] \in \mathsf{GS}(\mathrm{proj})$.

ACHIEVING OBV2 AND AUT2 SECURITY. The transform to promote coarse-grained to fine-grained security is unchanged. We let $\mathsf{all1\text{-}to\text{-}all2} = \mathsf{prv1\text{-}to\text{-}prv2}$ be the transform at the bottom of Fig. 4. We claim it has additional features captured by the following, whose proof is in the full paper [3].

Theorem 6. (1) For any $\varPhi$ and any $\mathrm{xxx} \in \{\mathrm{prv}, \mathrm{obv}\}$ if $\mathcal{G}_1 \in \mathsf{GS}(\mathrm{xxx1}, \varPhi) \cap \mathsf{GS}(\mathrm{proj})$ then $\mathsf{all1\text{-}to\text{-}all2}[\mathcal{G}_1] \in \mathsf{GS}(\mathrm{xxx2}, \varPhi) \cap \mathsf{GS}(\mathrm{proj})$ (2) If $\mathcal{G}_1 \in \mathsf{GS}(\mathrm{aut1}) \cap \mathsf{GS}(\mathrm{proj})$ then $\mathsf{all1\text{-}to\text{-}all2}[\mathcal{G}_1] \in \mathsf{GS}(\mathrm{aut2}) \cap \mathsf{GS}(\mathrm{proj})$.

OUTSOURCING DEFINITIONS. Towards the application to secure outsourcing, we begin with the definitions, following GGP [9]. An outsourcing scheme $\varPi = (\mathsf{Gen}, \mathsf{Inp}, \mathsf{Out}, \mathsf{Comp}, \mathsf{ev})$ is a tuple of PT algorithms that, intuitively, will be run partly on a *client* and partly on a *server*. Generation algorithm Gen is run by the client on input of the unary encoding 1^k and a string f describing the function $\mathsf{ev}(f, \cdot) : \{0,1\}^{f.n} \to \{0,1\}^{f.m}$ to be evaluated (so that ev, like in a garbling scheme, is a deterministic evaluation algorithm) to get back a public key pk that

is sent to the server and a secret key sk that is kept by the client. Algorithm Inp
is run by the client on input pk, sk and $x \in \{0,1\}^{f.n}$ to return a garbled input
X that is sent to the server. Associated state information St is preserved by
the client. Algorithm Comp is run by the server on input pk, X to get a garbled
output Y that is returned to the client. The latter runs deterministic algorithm
Out on pk, sk, Y, St to get back $y \in \{0,1\}^{f.n} \cup \{\bot\}$. Correctness requires that
for all $k \in \mathbb{N}$, all $f \in \{0,1\}^*$, and all $x \in \{0,1\}^{f.n}$, if $(pk, sk) \leftarrow \mathsf{Gen}(1^k, f)$,
$(X, St) \leftarrow \mathsf{Inp}(pk, sk, x)$, $Y \leftarrow \mathsf{Comp}(pk, X)$, and $y \leftarrow \mathsf{Out}(pk, sk, Y, St)$, then
$y = \mathsf{ev}(f, x)$. Our syntax is the same as that of GGP [9] except for distinguishing
between functions and their descriptions, as represented the addition of ev to
the list.

The games OSVF_Π and $\mathrm{OSPR}_{\Pi,\Phi,\mathcal{S}_{os}}$ of Fig. 8 are used to define *verifiability*
and *privacy* of an outsourcing scheme $\Pi = (\mathsf{Gen}, \mathsf{Inp}, \mathsf{Out}, \mathsf{Comp}, \mathsf{ev})$, where Φ
is a side-information function and $\mathcal{S}_{os}$ is a simulator. In both games, the adver-
sary is allowed only one GETPK query, and this must be its first oracle query.
For adversaries $\mathcal{A}_{os}$ and $\mathcal{B}_{os}$, we let $\mathbf{Adv}_\Pi^{osvf}(\mathcal{A}_{os}, k) = \Pr[\mathrm{OSVF}_\Pi^{\mathcal{A}_{os}}(k)]$ and
$\mathbf{Adv}_\Pi^{ospr,\Phi,\mathcal{S}_{os}}(\mathcal{B}_{os}, k) = 2\Pr[\mathrm{OSPR}_{\Pi,\Phi,\mathcal{S}_{os}}^{\mathcal{B}_{os}}(k)] - 1$. We say that Π is verifiable if
$\mathbf{Adv}_\Pi^{osvf}(\mathcal{A}_{os}, \cdot)$ is negligible for all PT adversaries $\mathcal{A}_{os}$. We say that Π is pri-
vate over Φ if for all PT adversaries $\mathcal{B}_{os}$ there is a PT simulator $\mathcal{S}_{os}$ such that
$\mathbf{Adv}_\Pi^{ospr,\Phi,\mathcal{S}_{os}}(\mathcal{A}_{os}, \cdot)$ is negligible. An adversary is said to be one-time if it makes
only one INPUT query. We say that Π is one-time verifiable if $\mathbf{Adv}_\Pi^{osvf}(\mathcal{A}_{os}, \cdot)$ is
negligible for all PT one-time adversaries $\mathcal{A}_{os}$. We say that Π is one-time private
over Φ if for all PT one-time adversaries $\mathcal{B}_{os}$ there is a PT simulator $\mathcal{S}_{os}$ such
that $\mathbf{Adv}_\Pi^{ospr,\Phi,\mathcal{S}_{os}}(\mathcal{A}_{os}, \cdot)$ is negligible.

Our verifiability definition coincides with that of GGP [9] but our privacy
definition is stronger: it requires not just "input privacy" (concealing each input x) but, also, privacy of the function f (relative to Φ). (As in our garbling
definitions this is subject to $\Phi(f)$ being revealed). Also, while GGP use an
indistinguishability-style formalization, we use a simulation-style one, as this
is stronger for some side-information functions.

To be "interesting" the work of the client in an outsourcing scheme should
be less than the work required to compute the function directly, for otherwise
outsourcing is not buying anything. An outsourcing scheme is said to be non-
trivial if this condition is met.

FROM GARBLING TO OUTSOURCING.   GGP show how to use FHE to turn
any one-time verifiable and private outsourcing scheme into a fully verifiable
and private one. This allows us to focus on designing the former. We show how
a garbling scheme that is both aut1 and obv1 secure immediately implies a
one-time verifiable and private outsourcing scheme. The construction, given in
Fig. 8, is very direct, and the proof of the following, given in the full paper [3],
is trivial, points which reinforce our claim that the garbling scheme abstraction
and adaptive security may be easily used in applications:

Theorem 7. If $\mathcal{G} \in \mathsf{GS}(\mathsf{obv1}, \Phi) \cap \mathsf{GS}(\mathsf{aut1})$ then outsourcing scheme $\Pi[\mathcal{G}]$ is
one-time verifiable and also one-time private over Φ.

proc GETPK(f) OSVF$_\Pi$	proc GETPK(f) OSPR$_{\Pi,\Phi,\mathcal{S}_{os}}$
$(pk, sk) \leftarrow \mathsf{Gen}(1^k, f),\ \ i \leftarrow 0$	$c \leftarrow \{0, 1\}$
return pk	if $c = 1$ then $(pk, sk) \leftarrow \mathsf{Gen}(1^k, f)$
	else $(pk, \sigma) \leftarrow \mathcal{S}_{os}(1^k, \Phi(f))$
proc INPUT(x)	return pk
if $x \notin \{0, 1\}^{f.n}$ then return $\bot$	
$i \leftarrow i + 1,\ \ x_i \leftarrow x$	proc INPUT(x)
$(X_i, St_i) \leftarrow \mathsf{Inp}(pk, sk, x)$	if $x \notin \{0, 1\}^{f.n}$ then return $\bot$
return X_i	if $c = 1$ then $(X, St) \leftarrow \mathsf{Inp}(pk, sk, x)$
	else $(X, \sigma) \leftarrow \mathcal{S}_{os}(\sigma)$
proc FINALIZE(Y, j)	return X
if $j \notin \{1, \ldots, i\}$ then return false	
$y \leftarrow \mathsf{Out}(pk, sk, Y, St_j)$	proc FINALIZE(c')
return $(y \notin \{\mathsf{ev}(f, x_j), \bot\})$	return $(c = c')$

$\mathsf{Gen}(1^k, f)$	$\mathsf{Inp}(F, (e, d), x)$	$\mathsf{Comp}(F, X)$	$\mathsf{Out}(F, (e, d), Y, St)$
$(F, e, d) \leftarrow \mathsf{Gb}(1^k, f)$	$X \leftarrow \mathsf{En}(e, x)$	$Y \leftarrow \mathsf{Ev}(F, x)$	$y \leftarrow \mathsf{De}(d, Y)$
return $(F, (e, d))$	return (X, ε)	return Y	return y

Fig. 8. Games to define the **verifiability** (OSVF) and **privacy** (OSPR) of outsourcing scheme $\Pi = (\mathsf{Gen}, \mathsf{Inp}, \mathsf{Out}, \mathsf{Comp}, \mathsf{ev})$. **Bottom:** constructing the outsourcing scheme $\Pi[\mathcal{G}] = (\mathsf{Gen}, \mathsf{Inp}, \mathsf{Out}, \mathsf{Comp}, \mathsf{ev})$ from garbling scheme $\mathcal{G} = (\mathsf{Gb}, \mathsf{En}, \mathsf{De}, \mathsf{Ev}, \mathsf{ev})$.

A benefit of our modular approach is that we may use any obv1 + aut1 garbling scheme as a starting point while GGP were tied to the scheme of [17]. However, the latter scheme is not adaptively secure, which brings us to our next point.

DISCUSSION. GGP give a proof that their outsourcing scheme is one-time verifiable assuming the encryption scheme underlying the garbled-circuit construction of [17] meets the condition called Yao-secure in [17]. However, their proof has a gap. Quoting [9, p. 12 of Aug 2010 ePrint version]: "For any two values x, x' with $f(x) = f(x')$, the security of Yao's protocol implies that no efficient player P_2 can distinguish if x or x' was used." This claim is correct if both x and x' are chosen independently of the randomness in the garbled circuit. But in their setting, the string x is chosen *after* the adversary sees the garbled circuit, and the security proof given by [17] no longer applies.

One may try to give a new proof that the LP garbling scheme satisfies aut1 security. However, this seems to be difficult. Intuitively, an adaptive attack on the garbling scheme allows the adversary to mount a key-revealing selective-opening (SOA-K) attack on the underlying encryption scheme. But SOA-K secure encryption is notoriously hard to achieve [2]. The only known way to achieve it is via non-committing encryption [5, 6, 8], which is only possible with keys as long as the total number of bits of message ever encrypted [19], so the outsourcing scheme may fail to be non-trivial.

This brings us to a more full discussion of non-triviality. The obv1 + aut1 secure scheme obtained via our all-to-all1 transform has long garbled inputs, so the one-time verifiable outsourcing scheme yielded by Theorem 7, while secure, is not non-trivial. Our ROM transforms coupled with Theorem 7 yield a non-trivial one-time outsourcing scheme in the ROM but the FHE-based method of GGP of lifting to a many-time scheme fails in the ROM. Finding a obv1 + aut1 garbling scheme with short garbled inputs in the standard model under standard assumptions is an open problem. We think Theorem 7 is still useful because it can be used at any point such a scheme emerges. All this again is an indication of the subtleties and hidden challenges underlying adaptive security of garbled circuits that seem to have been overlooked in the literature.

Acknowledgments. Thanks to the ASIACRPYT reviewers for their helpful comments, and thanks to the NSF for their continuing support: Bellare was supported in part by NSF grants CNS-1116800, CNS 0904380 and CCF-0915675, while Hoang and Rogaway were supported in part by NSF grant CNS 0904380.

References

1. Applebaum, B., Ishai, Y., Kushilevitz, E.: From Secrecy to Soundness: Efficient Verification via Secure Computation. In: Abramsky, S., Gavoille, C., Kirchner, C., Meyer auf der Heide, F., Spirakis, P.G. (eds.) ICALP 2010, Part I. LNCS, vol. 6198, pp. 152–163. Springer, Heidelberg (2010)
2. Bellare, M., Dowsley, R., Waters, B., Yilek, S.: Standard Security Does Not Imply Security against Selective-Opening. In: Pointcheval, D., Johansson, T. (eds.) EUROCRYPT 2012. LNCS, vol. 7237, pp. 645–662. Springer, Heidelberg (2012)
3. Bellare, M., Hoang, V., Rogaway, P.: Adaptively secure garbling with applications to one-time programs and secure outsourcing. Cryptology ePrint Archive (2012)
4. Bellare, M., Hoang, V., Rogaway, P.: Foundations of garbled circuits. In: ACM Computer and Communications Security (CCS 2012). Association for Computing Machinery. ACM (2012); Full version as ePrint Archive, Report 2012/265 (May 2012)
5. Canetti, R., Feige, U., Goldreich, O., Naor, M.: Adaptively secure multi-party computation. In: 28th ACM STOC, pp. 639–648. ACM Press (May 1996)
6. Choi, S.G., Dachman-Soled, D., Malkin, T., Wee, H.: Improved Non-committing Encryption with Applications to Adaptively Secure Protocols. In: Matsui, M. (ed.) ASIACRYPT 2009. LNCS, vol. 5912, pp. 287–302. Springer, Heidelberg (2009)
7. Chung, K.-M., Kalai, Y., Vadhan, S.: Improved Delegation of Computation Using Fully Homomorphic Encryption. In: Rabin, T. (ed.) CRYPTO 2010. LNCS, vol. 6223, pp. 483–501. Springer, Heidelberg (2010)
8. Damgård, I., Nielsen, J.B.: Improved Non-committing Encryption Schemes Based on a General Complexity Assumption. In: Bellare, M. (ed.) CRYPTO 2000. LNCS, vol. 1880, pp. 432–450. Springer, Heidelberg (2000)
9. Gennaro, R., Gentry, C., Parno, B.: Non-interactive Verifiable Computing: Outsourcing Computation to Untrusted Workers. In: Rabin, T. (ed.) CRYPTO 2010. LNCS, vol. 6223, pp. 465–482. Springer, Heidelberg (2010)
10. Goldreich, O.: Foundations of Cryptography: Basic Applications, vol. 2. Cambridge University Press, Cambridge (2004)
11. Goldreich, O., Micali, S., Wigderson, A.: How to play any mental game or a completeness theorem for protocols with honest majority. In: Aho, A. (ed.) STOC, pp. 218–229. ACM (1987)

12. Goldwasser, S., Kalai, Y.T., Rothblum, G.N.: One-time programs. Manuscript, full version of [13] (July 2012)
13. Goldwasser, S., Kalai, Y.T., Rothblum, G.N.: One-Time Programs. In: Wagner, D. (ed.) CRYPTO 2008. LNCS, vol. 5157, pp. 39–56. Springer, Heidelberg (2008)
14. Goyal, V., Ishai, Y., Sahai, A., Venkatesan, R., Wadia, A.: Founding Cryptography on Tamper-Proof Hardware Tokens. In: Micciancio, D. (ed.) TCC 2010. LNCS, vol. 5978, pp. 308–326. Springer, Heidelberg (2010)
15. Ishai, Y., Kushilevitz, E.: Randomizing polynomials: A new representation with applications to round-efficient secure computation. In: 41st FOCS, pp. 294–304. IEEE Computer Society Press (November 2000)
16. Kamara, S., Wei, L.: Special-purpose garbled circuits. (manuscript, 2012)
17. Lindell, Y., Pinkas, B.: A proof of security of Yao's protocol for two-party computation. Journal of Cryptology 22(2), 161–188 (2009)
18. Naor, M., Pinkas, B., Sumner, R.: Privacy preserving auctions and mechanism design. In: Proceedings of the 1st ACM Conference on Electronic Commerce, pp. 129–139. ACM (1999)
19. Nielsen, J.B.: Separating Random Oracle Proofs from Complexity Theoretic Proofs: The Non-committing Encryption Case. In: Yung, M. (ed.) CRYPTO 2002. LNCS, vol. 2442, pp. 111–126. Springer, Heidelberg (2002)
20. Yao, A.: Protocols for secure computations (extended abstract). In: FOCS, pp. 160–164. IEEE Computer Society (1982)
21. Yao, A.: How to generate and exchange secrets. In: 27th Annual Symposium on Foundations of Computer Science, pp. 162–167. IEEE (1986)

The Generalized Randomized Iterate and Its Application to New Efficient Constructions of UOWHFs from Regular One-Way Functions

Scott Ames[1], Rosario Gennaro[2],
and Muthuramakrishnan Venkitasubramaniam[1]

[1] University of Rochester, Rochester, NY 14611, USA
[2] IBM T.J.Watson Research Center, Hawthore, NY 10532, USA

Abstract. This paper presents the *Generalized Randomized Iterate* of a (regular) one-way function f and show that it can be used to build Universal One-Way Hash Function (UOWHF) families with $O(n^2)$ key length.

We then show that Shoup's technique for UOWHF domain extension can be used to improve the efficiency of the previous construction. We present the *Reusable Generalized Randomized Iterate* which consists of $k \geq n + 1$ iterations of a regular one-way function composed at each iteration with a pairwise independent hash function, where we only use $\log k$ such hash functions, and we "schedule" them according to the same scheduling of Shoup's domain extension technique. The end result is a UOWHF construction from regular one-way functions with an $O(n \log n)$ key. These are the first such efficient constructions of UOWHF from regular one-way functions of unknown regularity.

Finally we show that the Shoup's domain extension technique can also be used in lieu of derandomization techniques to improve the efficiency of PRGs and of hardness amplification constructions for regular one-way functions.

1 Introduction

One of the central results in Modern Cryptography is that one-way functions imply digital signatures (as defined in [6]). This result was first established by Naor and Yung in [12] for one-way permutations via the notion of Universal One-Way Hash Functions (UOWHF). Later Rompel in [13] proved that UOWHFs can be built from any one-way function. The notion of UOWHF is interesting on its own, apart from its connection to digital signatures. UOWHFs are *compressing* functions (i.e. the output is shorter than the input) which enjoy a *target collision resistance* property: a function family $\mathcal{G}$ is a UOWHF if no efficient adversary A succeeds in the following game with non-negligible probability:

- A chooses a target input z;
- a randomly chosen function $g \in \mathcal{G}$ is selected;
- A finds a *collision* for $g(z)$, i.e. an input $z' \neq z$ such that $g(z) = g(z')$.

X. Wang and K. Sako (Eds.): ASIACRYPT 2012, LNCS 7658, pp. 154–171, 2012.

A seemingly weaker notion is *second preimage resistance* where the target input z is randomly chosen (rather than by A). It is however well known how to convert a second preimage resistant function family into a UOWHF.

The security of these constructions is proven by *reductions*: given an adversary A that wins the above UOWHF game, we build an "inverter" I that is able to solve a computationally hard problem, e.g. invert a one-way function. A crucial feature of these reductions is their *efficiency*, i.e. the relationship between the running time of D (or A) and I, and the resulting degradation in the security parameters. For the case of UOWHFs one of the most important efficiency measures is the size of the key needed to run the algorithm.

Unfortunately the construction of UOWHFs based on general one-way functions do not fare very well on that front. If n is the security parameter, the original Rompel construction yielded a key of size $\tilde{O}(n^{12})$ which was later improved to $\tilde{O}(n^7)$ by Haitner *et al.* in [8]. Conversely under the much stronger assumption of one-way *permutations* Naor and Yung in [12] achieve linear key size. Apart from the above works, we are aware of only one work by De Santis and Yung [2] that constructs UOWHFs from *regular* one-way functions (i.e. functions that have constant size preimages). Their construction achieves $O(n \log n)$ key size but is very complicated and more importantly requires knowledge of the regularity parameter.

We go back to investigating the construction of UOWHFs from regular one-way functions. We obtain a very simple construction with $O(n \log n)$ key size, which does *not* require knowledge of the regularity parameter. These are the first such efficient constructions of UOWHF from regular one-way functions of unknown regularity.

Somewhat surprisingly our UOWHF construction is obtained via a simple "tweak" on a well-known algorithm for pseudo-random number generation from regular one-way functions: the *Randomized Iterate* [4,7]. Another surprising connection established by this paper is that Shoup's domain extension technique [15] can be used to improve the seed size in *both* the PRG and UOWHF.

MOTIVATION. Collision resistant hashing is an ubiquitous tool in Cryptography and in practice a stronger notion of collision resistance is used where the adversary is given as input just $H \in \mathcal{H}$ and must find z, z' that collide (we will refer to this notion as *full collision resistance* as opposed to the target collision-resistance property enjoyed by UOWHFs).

This is problematic because there is strong evidence that this stronger notion cannot be achieved by assuming just OWFs. Simon [16] proves that there is no black-box construction[1] of a fully collision resistant hash function from one-way permutations. While a non black-box construction based on OWFs remains theoretically possible, such construction would probably be very inefficient, since efficient constructions based on general assumptions seem to be black-box ones.

---

[1] Informally, a black-box construction accesses the underlying OWF only via input queries, without any knowledge of its internal structure.

Furthermore the cryptanalysis of practical and widely adopted supposedly collision-resistant functions have reminded us of the importance of construct-ing *efficient* candidates for collision-resistant functions which are also *provably secure*, i.e. have a security reduction to a well established computational hard problem. The above explains why researchers and practitioners alike are looking at UOWHFs to replace full collision resistant hashing in practical applications (such as certifications – see for example the work of Halevi and Krawczyk on randomized hashing [10]).

Current efficient candidates for UOWHFs have either no proof of security or make stronger assumptions than the existence of OWPs[2]. Achieving a truly efficient UOWHF construction based on OWFs would offer practitioners a target collision-resistant function which can be used in practice and gives the peace of mind of a strong security guarantee.

In order to achieve this goal, our construction slightly relaxes the assumption to *regular* OWFs, yielding a dramatic improvement to a $O(n \log n)$ key size. We are following the same approach as [7] for pseudo-random generators: looking at the more limited case of regular OWFs not only to improve the efficiency, but also to explore techniques that might benefit constructions in the general case (which is what happened in the PRG case).

1.1   Our Contribution

We present a new algorithm (we call it the *Generalized Randomized Iterate GRI*) which depending on its parameters can be used to build *either* PRGs or UOWHFs starting from *regular* one-way functions.

First proposed in [4] the original Randomized Iterate construction involves composing the regular one-way function with different n-wise (later improved to simply pair-wise independent in [7]) universal hash functions at each iteration. More specifically if f is a regular one-way function, and $h_1, \ldots, h_m$ are pairwise independent hash functions all from $\{0,1\}^n$ to $\{0,1\}^n$, the mth randomized iter-ate of f using the h_i is defined as $f^k = f \circ h_k \circ f \circ h_{k-1} \circ \ldots f \circ h_1 \circ f$. In [4,7] it is shown that this function is hard to invert at each stage and therefore can be used to construct PRGs in conjunction with a generic hard-core predicate (such as the Goldreich-Levin bit [5]).

We generalize the Randomized Iterate to use compressing pair-wise indepen-dent hash functions h_i at each stage. Somewhat surprisingly we then show that the resulting family (see Definition 8) is second-preimage resistant.

Notice that in the above applications the universal hash functions h_i are part of the secret key of the resulting algorithm (the seed for the PRG, the index key for the UOWHF). Therefore it is desirable to have constructions in which the number of functions can be minimized.

---

[2] For example Halevi and Krawczyk in [10] propose a mode of operation for typical hash function such as SHA-1 that creates a UOWHF under an assumption on the compression function which is seemingly stronger than OWF, but somewhat weaker than full collision resistance.

The Randomized Iterate PRG construction in [7] has an $O(n^2)$ seed, but it was also shown how an $O(n \log n)$ seed could be achieved by using generic de-randomization techniques. First we point out that this approach does not immediately work in the UOWHF case, as in order to reduce the key size, the de-randomization procedure requires an additional property[3].

We then explore another fascinating and somewhat unexpected connection. We observe that instead of using de-randomization techniques, the structure of the Generalized Randomized Iterate can be improved by using Shoup's domain extension technique for UOWHFs [15]. We define the *Reusable Generalized Randomized Iterate* RGRI : Using Shoup's approach we prove that it is possible to "recycle" some of the hash functions in the Generalized Randomized Iterate, to $O(\log m)$ for m iterations (instead of m). The net result is that we achieve a UOWHF with $O(n \log n)$ key size.

Finally we point out that the RGRI also yields an $O(n \log n)$-seed PRG from regular one-way function, and can be also used for hardness amplification of regular one-way functions, obtaining alternative proofs of results already appearing in [7].

1.2   Comparison with Previous Work

We already mention the previous works on UOWHFs based on general assumptions [12,13,8,2] and how they compare to our work.

As discussed above our UOWHF construction uses in a crucial way tools that were developed for the task of pseudo-random generation. In this sense our work follows the path of recent papers on *inaccessible entropy* [9,8]. Those beautiful works elegantly show that the known constructions of PRGs and UOWHFs can be interpreted as similar manipulation techniques on different forms of computational entropy (pseudo-entropy for PRGs and inaccessible entropy for UOWHFs). While less general, our work shows a more direct and specific connection: a single algorithm (the Generalized Randomized Iterate) which is sufficiently "flexible" to be used either as a PRG or as a UOWHF.

1.3   Paper Organization

We briefly recall the relevant definitions in Section 2. In Section 3 we introduce the Generalized Randomized Iterate and its Reusable variant; we also prove a main technical Lemma that is at the heart of the efficiency claim for our UOWHF construction which appears in Section 4. We present our alternative constructions of a $O(n \log n)$-seed PRG, and the hardness amplification result in Section 5 (the proofs of these constructions will appear in the full-version). We conclude with some discussions and open problems in Section 6.

---

[3] The actual de-randomization algorithm (the Nisan-Zuckerman PRG for space-bounded computations) used in [7] has this property, but a generic PRG for space-bounded computation might not.

2   Preliminaries and Definitions

2.1   One-Way Functions

Definition 1. *Let $f : \{0,1\}^* \to \{0,1\}^*$ be a polynomial-time computable function. f is* one-way *if for every* PPT *machine A, there exists a negligible function $\nu(\cdot)$ such that*

$$\Pr[x \leftarrow \{0,1\}^n; y = f(x) : A(1^n, y) \in f^{-1}(f(x))] \leq \nu(n)$$

Definition 2 (Regular One-Way Functions). *Let $f : \{0,1\}^* \to \{0,1\}^*$ be a one-way function. f is regular if there exists a function $\alpha : N \to N$ such that for every $n \in N$ and every $x \in \{0,1\}^n$ we have:*

$$|f^{-1}(f(x))| = \alpha(n)$$

We assume that the regularity $\alpha(\cdot)$ of a function f is not known (i.e. not polynomial time computable). Without loss of generality, we assume the one-way function is length preserving i.e. $f(\{0,1\}^n) \subseteq \{0,1\}^n$.

2.2   Hardcore Predicates

Definition 3. *Let $f : \{0,1\}^n \to \{0,1\}^*$ and $b : \{0,1\}^n \to \{0,1\}$ be polynomial-time computable functions. We say b is a hardcore predicate of f, if for every* PPT *machine A, there exists a negligible function $\nu(\cdot)$ such that*

$$\Pr[x \leftarrow \{0,1\}^n; y = f(x) : A(1^n, y) = b(x)] \leq \frac{1}{2} + \nu(n)$$

If f is a one-way function over $\{0,1\}^n$ then Goldreich and Levin in [5] prove that the one-way function f' over $\{0,1\}^{2n}$ defined as $f'(x, r) = (f(x), r)$ admits the following hard-core predicate $b(x, r) = <x, r> = \Sigma x_i r_i \bmod 2$ where x_i, r_i is the ith bit of x, r respectively. In the following we refer to this predicate as the GL bit of f.

2.3   Pseudorandom Generators

Definition 4. *Let $G : \{0,1\}^n \to \{0,1\}^{l(n)}$ be a polynomial time computable function where $l(n) > n$. We say G is a pseudorandom generator, if for every* PPT *machine A, there exists a negligible function $\nu(n)$ such that*

$$\left| \Pr[x \leftarrow \{0,1\}^n; y \leftarrow G(x) : A(1^n, y) = 1] \right.$$
$$\left. - \Pr[x \leftarrow \{0,1\}^{l(n)} : A(1^n, y) = 1] \right| \leq \nu(n)$$

2.4   Universal One-Way Hash Function Families

Definition 5. *Let $\mathcal{G} = \{g_k\}_{k \in \mathcal{K}}$ be a family of functions where each function g_k goes from $\{0,1\}^{n+\ell}$ to $\{0,1\}^n$. We say that $\mathcal{G}$ is a a* Universal One-Way Hash Function Family *if (i) the functions g_k are efficiently computable and (ii) for every efficient adversary A , the probability that A succeeds in the following game is negligible in n:*

- *Let $(x, \sigma) \leftarrow A(1^n)$*
- *Choose $k \leftarrow \mathcal{K}$*
- *Let $x' \leftarrow A(\sigma, k)$*
- *A succeeds if $x \neq x'$ and $g_k(x) \neq g_k(x')$*

Universal One-Way Hash Function Families [12] as defined above enjoy the property of target collision-resistance. Next, we define the seemingly weaker notion of *Second Preimage Resistance* where the adversary cannot find a collision for randomly chosen input and key. It is well-known how to construct UOWHFs from second preimage resistant families.

Definition 6 (Second Preimage Resistance). *Let $\mathcal{G} = \{g_k\}_{k \in \mathcal{K}}$ be a family of functions where each function g_k goes from $\{0,1\}^{n+\ell}$ to $\{0,1\}^n$. We say that $\mathcal{G}$ is a a* Second Preimage Resistant Hash Function Family *if (i) the functions g_k are efficiently computable and (ii) for every efficient adversary A , then the following probability*

$$Pr[z \leftarrow \{0,1\}^{n+\ell} \; ; \; k \leftarrow \mathcal{K} \; ; \; A(z, k) = z' \; : \; z \neq z' \text{ and } g_k(z) = g_k(z')]$$

is negligible in n.

2.5   Universal Hash Function Families

Definition 7. *Let $\mathcal{H}$ be a family of functions where each function $h \in \mathcal{H}$ goes from $\{0,1\}^{n+\ell}$ to $\{0,1\}^n$. We say that $\mathcal{H}$ is a an efficient family of pairwise independent hash functions if (i) the functions $h \in \mathcal{H}$ can be described with a polynomial (in n) number of bits; (ii) there is a polynomial (in n) time algorithm to compute $h \in \mathcal{H}$; (iii) for all $x \neq x' \in \{0,1\}^{n+\ell}$ and for all $y, y' \in \{0,1\}^n$*

$$Pr_{h \in \mathcal{H}}[h(x) = y \text{ and } h(x') = y'] = 2^{-2n}$$

3   The Generalized Randomized Iterate

A well known fact about one-way functions is that if you iterate them, you may not end up with a function that is difficult to invert. Indeed while a permutation f, when iterated $f^{(i)} = f \circ \ldots \circ f$ (i.e. f composed with itself i times) remains one-way, this is not true for general one-way functions as a single application could concentrate the outputs on a very small fraction of the inputs of f, where f might even be easy to invert.

Goldreich, Krawczyk and Luby in [4] introduced the *Randomized Iterate* construction where a randomization step is added between two application of f, in its iteration. As shown in [7] when using pair-wise independent hashing to implement this randomization step, the Randomized Iterate is hard to invert.

We introduce the Generalized Randomized Iterate (GRI) and we show how it can be used to construct *both* pseudo-random generators *and* target collision-resistant hashing. We then show a randomness efficient form of the (generalized) Randomized Iterate, where some of the hash functions are "recycled" during the iteration. This *Reusable Generalized Randomized Iterate* is the core of our efficient construction of UOWHFs.

Definition 8. *Let* $f : \{0,1\}^n \to \{0,1\}^n$ *and let* $\mathcal{H}$ *be an efficient family of pairwise-independent hash functions from* $\{0,1\}^{n+\ell}$ *to* $\{0,1\}^n$. *For input* $x \in \{0,1\}^n$, $z \in \{0,1\}^{\ell k}$, $h_1, \ldots, h_m \in \mathcal{H}$ *and* $m \geq k$, *define the* k^{th} *Generalized Randomized Iterate* $g^k : \{0,1\}^n \times \{0,1\}^{\ell k} \times \mathcal{H}^m \to \{0,1\}^n$ *recursively as:*

$$g^k(x, z, h_1, \ldots, h_m) = h_k(f(g^{k-1}(x, z, h_1, \ldots, h_m))||z_{[(k-1)\ell+1\ldots k\ell]})$$

where $g^0(x, z, h_1, \ldots, h_m) = x$, $||$ *denotes concatenation and* $z_{[a\ldots b]}$ *is the substring of* z *from position* a *to position* b.

In other words at each iteration of the Generalized Randomized Iterate, first f is applied to the output of the previous iteration, then a block of ℓ bits from z are appended to the output, and then a pair-wise independent hash function is applied. Note that at each iteration a new hash function is used.

While we are defining GRI for any value of ℓ, we are going to be interested to two cases:

- $\ell = 0$ in which case z is the empty string, and the pair-wise independent hash functions map n bits to n bits. This case is equivalent to the Randomized Iterate from [4,7] and as shown there it can be used to build PRGs;
- $\ell = 1$ in which case z is k-bits long, and the hash functions compress one bit. We will show in Section 4 that this function is a second preimage resistant function (from which a UOWHF can be easily built).

3.1   The Reusable Generalized Randomized Iterate

We now introduce the *Reusable Generalized Randomized Iterate* (RGRI) which is a version of the Randomized Generalized Iterate that uses fewer hash functions. While the GRI described in the previous Section use new distinct hash functions at each iteration, we "recycle" some of this hash functions during the process. More specifically we sample m hash functions $h_1, \ldots, h_m$ from $\mathcal{H}$ and then in the ith iteration of the RGRI we use the function $h_{\phi(i)}$ where $\phi(i)$ is the function that on input i, outputs the highest power of 2 that divides i. It is not hard to see that if we have k iterations it is sufficient to set $m = \lceil \log k \rceil + 1$. This "scheduling" of the hash functions is identical to the way Shoup recycles random masks in his construction of a domain extender for TCR functions [15].

Definition 9. *Let $f : \{0,1\}^n \to \{0,1\}^n$ and let $\mathcal{H}$ be an efficient family of pairwise-independent hash functions from $\{0,1\}^{n+\ell}$ to $\{0,1\}^n$. For input $x \in \{0,1\}^n$, $z \in \{0,1\}^{\ell k}$, $h_1, \ldots, h_m \in \mathcal{H}$ and $m \geq \lceil \log k \rceil + 1$, define the kth Reusable Generalized Randomized Iterate $\widetilde{g^k} : \{0,1\}^n \times \{0,1\}^{\ell k} \times \mathcal{H}^m \to \{0,1\}^n$ recursively as:*

$$\widetilde{g^k}(x, z, h_1, \ldots, h_m) = \begin{cases} h_{\phi(k)}(f(\widetilde{g^{k-1}}(x, z, h_1, \ldots, h_m))\|z_{[(k-1)\ell+1\ldots k\ell]}) & k > 0 \\ x & otherwise \end{cases}$$

where $\phi(n)$ is one greater than the highest power of 2 that divides n.

3.2   A Technical Lemma

We now prove a preliminary Lemma which is crucial in allowing us to achieve logarithmic key size for our UOWHF construction. This Lemma abstracts the property of the "Shoup domain extension" technique we use to construct the RGRI : intuitively the Lemma proves a preliminary result that will allows us later to claim that the distribution induced by the RGRI is not that far from the distribution induced by the GRI with distinct (i.e. non-reused hash functions).

The goal of the Lemma is to count how many input pairs lead to two specific values a_0, a_1 as outputs of the RGRI.

Lemma 1. *Fix two arbitrary values $a_0, a_1 \in \{0,1\}^n$ and an integer i. The number of pairs $[(x_0, z_0, h_1, \ldots, h_m), (x_1, z_1, h_1, \ldots, h_m)]$ such that*

$$\widetilde{g^i}(x_0, z_0, h_1, \ldots, h_m) = a_0 \quad and \quad \widetilde{g^i}(x_1, z_1, h_1, \ldots, h_m) = a_1$$

is bounded by $2^{2\ell k} \cdot |\mathcal{H}|^m$.

Note that in the Lemma we are counting the pairs with possibly distinct inputs x, z but same hash functions h_i.

Proof: To prove the Lemma we use a "key-reconstruction" strategy introduced by Shoup in [15]. The algorithm in Figure 1 on input $i \in [0..k]$, $z_0, z_1 \in \{0,1\}^{\ell k}$ and $a_0, a_1 \in \{0,1\}^n$ generates a pair of inputs $(x_0, \overline{h})$ and $(x_1, \overline{h})$ such that the output of the ith iterate is a_0 and a_1, i.e.

$$\widetilde{g^i}(x_0, z_0, h_1, \ldots, h_m) = a_0 \quad and \quad \widetilde{g^i}(x_1, z_1, h_1, \ldots, h_m) = a_1$$

We prove that this algorithm outputs all possible input pairs $(x_0, \overline{h})$ and $(x_1, \overline{h})$ with some probability. To complete the proof of the claim we show that the total number of distinct outputs by the algorithm is $|\mathcal{H}|^m$ (the Lemma follows since there are $2^{2\ell k}$ possible values of z_0, z_1).

The high-level idea of the Shoup reconstruction strategy described in Figure 1 is the following. Consider the simple case of the randomized iterate function g^k (where a different hash function is used after each iterate). Since, we use different hash functions at every iterate, we choose all the hash functions

$h_1, \ldots, h_{i-1}, h_{i+1}, \ldots, h_m$ arbitrarily, except the one in the ith iterate (i.e. h_i). Using $x_0, z_0, h_1, \ldots, h_{i-1}$ and $x_1, z_1, h_1, \ldots, h_{i-1}$ we compute y_0, y_1 as the outputs of $f \circ g^{i-1}$. We then choose h_i so that $h_i(y_0 || z_{0,[(i-1)\ell+1\ldots i\ell]}) = a_0$ and $h_i(y_1 || z_{1,[(i-1)\ell+1\ldots i\ell]}) = a_1$ simultaneously holds. This is possible since $\mathcal{H}$ is a pairwise-independent family. Furthermore, the number of such functions h_i is equal to $|\mathcal{H}|/2^{2n}$. Observe that, every input pair satisfying the conditions is output by the strategy for some random choices and every random choice yields different outputs satisfying the conditions. Therefore, the total number of pairs satisfying the conditions equals the total number of random choices made by the strategy and that is $2^{2n} 2^{2\ell i} |\mathcal{H}|^{m-1} \times |\mathcal{H}|/2^{2n} = 2^{2\ell i} |\mathcal{H}|^m$.

However this procedure does not work for the reusable randomized iterate since the hash functions are recycled. Instead, we consider segments and perform a "right to left" sweep from the ith iterate to the first iterate, ensuring that each segment is locally consistent. More precisely, in each segment, for a particular a, the algorithm selects hash functions and string x such that if x is fed as input to the jth iterate, then the output of the computation at the ith iterate ($i > j$) is a. For the segments to compose, we need to ensure that the hash functions selected by different segments do no conflict with each other and that is the technical part of the proof. To extend the algorithm to achieve consistency for two inputs it suffices to observe that for all $x_0 \neq x_1$ and arbitrary values a_0, a_1, there exists an h such that $h(x_0) = a_0$ and $h(x_1) = a_1$. The formal description of the algorithm is presented in Figure 1.

First, we prove correctness and then compute the number of colliding pairs.

Sub-Claim 1. *If the algorithm in Figure 1 outputs $(x_0, \overline{h}), (x_1, \overline{h})$, then it holds that $\widetilde{g}^i(x_0, z_0, \overline{h}) = a_0$ and $\widetilde{g}^i(x_1, z_1, \overline{h}) = a_1$.*

Proof: Every iteration of the algorithm, considers the segment from the jth iterate to the ith iterate and achieves the following: if x_0^j (and x_1^j) is fixed as the partial input to the jth iterate then a_0 (and a_1) is the output of the ith iterate. This follows from the fact that, $h_{\phi(i)}$ is assigned a value at step 2(d) after knowing what the output of the $i - 1^{st}$ iterate is computed. It only remains to show that two iterations do not assign values to the same hash function. The algorithm assigns value to a hash function in steps 2(b), 2(d) and 4. By construction step 2(b) and 4 only assign values to hash functions that have not been defined yet (indicated by the flag being false). It suffices to ensure that there are no conflicts in the assignment made at step 2(d). This is ensured by maintaining the invariant that $h_{\phi(i)}$ is undefined before executing 2(d) in any iteration. Observe that, in every iteration, $\phi(j) > \phi(i)$ and for all c such that $j < c < i$, $\phi(c) < \phi(i)$. Hence, before step 2(d) is reached in any iteration, the only hash-functions that are defined are those with indices c such that $\phi(c) < \phi(j)$. $\square$

Sub-Claim 2. *The number of distinct pairs output by the Shoup Reconstruction algorithm is bounded by $|\mathcal{H}|^m$.*

Proof: From Sub-Claim 1, we know that every pair output of the algorithm satisfies the condition that a_0 and a_1 are the output of the ith iterate. Furthermore, every pair that satisfies the condition occurs as an output for some choice

Input: i, z_0, z_1, a_0, a_1

1. Set Flags $F_0, \dots, F_{m-1}$ to *false* // Flags indicate which hash-functions are assigned
2. while $i \neq 0$
   (a) $j \longleftarrow (i - 2^{\phi(i)})$ // The new condition will be at position j
   (b) Randomly choose x_0^j, x_1^j from $\{0, 1\}^n$. For all $j < c < i$, if $F_{\phi(c)} = false$, randomly choose $h_{\phi(c)}$ from $\mathcal{H}$ and set $F_{\phi(c)} \longleftarrow true$.
   (c) Compute

$$x_0^j \xrightarrow{f} \overset{||z_{0,[j\ell+1\dots(j+1)\ell]}}{\longrightarrow} \xrightarrow{h_{\phi(j+1)}} \xrightarrow{f} \cdots \xrightarrow{h_{\phi(i-1)}} \xrightarrow{f} y_0$$

$$x_1^j \xrightarrow{f} \overset{||z_{1,[j\ell+1\dots(j+1)\ell]}}{\longrightarrow} \xrightarrow{h_{\phi(j+1)}} \xrightarrow{f} \cdots \xrightarrow{h_{\phi(i-1)}} \xrightarrow{f} y_1$$

   (d) Randomly choose $h \in \mathcal{H}$ conditioned on

$$h(y_0 || z_{0,[(i-1)\ell+1\dots i\ell]}) = a_0 \quad \text{and} \quad h(y_1 || z_{1,[(i-1)\ell+1\dots i\ell]}) = a_1$$

   Set $h_{\phi(i)} \longleftarrow h, F_{\phi(i)} \longleftarrow true$.
   (e) $i \longleftarrow j, a_0 \longleftarrow x_0^j, a_1 \longleftarrow x_1^j$
3. endwhile
4. For all c, if $F_{\phi(c)} = false$, pick $h_{\phi(c)}$ uniformly from $\mathcal{H}$ and set $F_{\phi(c)}$ to *true*.
5. output $(x_0 = x_0^1, h_1, \dots, h_m), (x_1 = x_1^1, h_1, \dots, h_m)$

Fig. 1. Shoup Reconstruction Algorithm

made by the algorithm and each choice made by the algorithm yields distinct outputs. Therefore, it suffices to compute the total number of choices made by the algorithm. To compute the number of pairs, observe that, for every choice made for x_0^j and x_1^j (such that $x_0^j \neq x_1^j$) in step (b), the number of hash functions h such that

$$h(y_0 || z_{0,[(i-1)\ell+1\dots i\ell]}) = a_0 \quad \text{and} \quad h(y_1 || z_{1,[(i-1)\ell+1\dots i\ell]}) = a_1$$

is $\frac{|\mathcal{H}|}{2^{2n}}$, by the pairwise independence property. We treat the choices made for x_0^j, x_1^j as a choice made for $h_{\phi(i)}$ set in step 2(d). Thus, the number of choices for the hash function in step 2(d) is at most $2^{2n} \times \frac{|\mathcal{H}|}{2^{2n}} = |\mathcal{H}|$. The only other choices are the hash functions picked in step 2(b) and 4. Since they can take any value, they have $|\mathcal{H}|$ many choices. Hence, corresponding to every hash function the algorithm makes $|\mathcal{H}|$ many choices. Thus, the total number of pairs is bounded by $|\mathcal{H}|^m$. $\quad\square$

This concludes the proof of Lemma 1. $\quad\square$

The following Corollary is proven by using the same counting argument and the same "reconstruction strategy" of Lemma 1 (intuitively, the bound results from the fact that you can choose x in 2^n ways, z in 2^{lk} ways, $m-1$ hash functions uniformly at random in $\mathcal{H}$, and the hash function h_i via pairwise independence among $|\mathcal{H}|/2^{2n}$ possible candidates).

Corollary 1. *Fix arbitrary values $a_0, a_1 \in \{0,1\}^n$, $y \in \{0,1\}^{n+\ell}$ and an integer i. The number of inputs $(x, z, h_1, \ldots, h_m)$ such that*

$$\widetilde{g^i}(x, z, h_1, \ldots, h_m) = a_0 \quad \text{and} \quad h_i(y) = a_1$$

is bounded by $2^{\ell k - n} \cdot |\mathcal{H}|^m$. Moreover there exists a polynomial time algorithm that samples such an input uniformly at random.

Remark: We point out that the "reconstruction" property outlined in Lemma 1 is exactly what is needed in order to prove the security of our UOWHF with $O(n \log n)$ key based on the RGRI.

This is in contrast to the case of PRG [7] where any PRG for space-bounded computation would work to "de-randomize" the seed from n^2 to $n \log n$. We can show that the particular space-bounded PRG used in [7] satisfies a Lemma similar to Lemma 1, and therefore could be used to reduce the size of the key of our UOWHF. For simplicity we just show the construction based on Shoup's technique.

4   Constructions of Universal One-Way Hash Functions

In this section, we show how to construct second preimage resistant functions from regular one-way functions. We start with a simple construction (that already improves the efficiency from previous work) of quadratic key size. We then provide a more efficient and essentially optimal solution with $O(n \log n)$ key size. Note that our functions compress a single bit (higher compression can be achieved by standard modes of iteration). Note also that UOWHFs can be easily built from second preimage resistant families.

4.1   A Construction with Linear Key Size

Definition 10. *Let $f : \{0,1\}^n \to \{0,1\}^n$ and let $\mathcal{K} = \{0,1\}^n \times \mathcal{H}^{n+1}$ where $\mathcal{H}$ is an efficient family of pairwise-independent hash functions from $\{0,1\}^{n+1}$ to $\{0,1\}^n$. Define the function $g(z, k)$ with input space $z \in \{0,1\}^{n+1}$ and key-space $k = (x, h_1, \ldots, h_{n+1}) \in \mathcal{K}$ as follows:*

$$g(z, (x, h_1, \ldots, h_{n+1})) = g^{n+1}(x, z, h_1, \ldots, h_{n+1})$$

where g^i is the Generalized Randomized Iterate with $\ell = 1$.

Theorem 1. *Suppose f is a 2^r-regular one-way function. Then g defined according to Definition 10 is a second preimage resistant function family.*

Proof Overview: To understand how our construction works, let us assume (as a simplifying assumption) that we can uniformly sample pairs (a_1, a_2) such that $f(a_1) = f(a_2)$. Let us refer to such pairs as *siblings* for f.

Given such a pair it is possible to set up the hash functions in the above construction so that if the adversary finds a collision, then we invert the one-way function on a point y. Intuitively this is done as follows: given a random input z for the UOWHF, we choose the hash functions (i.e. the key k) so that $g^i(z, k) = a_1$ and $h_i(y||b) = a_2$ for a random index i and a random bit b. We then run the adversary on z, k and if the adversary finds a collision z', with non-negligible probability the collision "goes through" a_2 at index i, i.e. $g^i(z', k) = a_2$ allowing us to find a preimage of y.

The intuition here is that given any input z, and key k, at each iterate the input going into the one-way function has most 2^r collisions w.r.t f. For a collision to occur at a particular iterate, it must be the case that some range element y of the one-way function f must occur at the previous iterate and the hash function takes y and an input bit into one of the 2^r collisions in the next iterate. Since there are at most 2^{n-r} range elements, in expectation over hash functions, the number of possible inputs at the previous iterate that are mapped into the 2^r collisions are small, in fact $O(1)$. Thus the hash functions selected above will succeed with high probability.

But how do we get to sample a_1, a_2, i.e. siblings for f in the first place? For this we use the adversary again. Indeed when an adversary finds a collision to input z (say z'), it must be that at some iterate, the inputs into the intermediate hash functions are different and the outputs to the next iterate are strings a_1 and a_2 such that $f(a_1) = f(a_2)$, i.e. siblings for f. It remains to argue that sampling a_1 and a_2 by first querying the adversary is good enough, and this is established using a collision-probability-type analysis. We now proceed to a formal proof.

Proof: Assume for contradiction, there exists an adversary A and polynomial $p(\cdot)$ such that for infinitely many lengths n, the probability with which A finds a collision on a random input $z \in \{0, 1\}^n$ and key $k = (x, \overline{h}) \in \mathcal{K}$ is at least $\epsilon \geq \frac{1}{p(n)}$. We assume for simplicity that A is deterministic. Fix a particular n for which this happens. Using A, we construct a machine M that inverts f with probability that is polynomially related to ϵ and thus arrive at a contradiction.

The machine M on input $y \in \{0, 1\}^n$ internally incorporates the code of A and proceeds as follows:

1. Sample a random input z and key $k = (x, \overline{h})$. Internally run A on input (z, k). If A fails to return a collision, halt outputting $\perp$. Otherwise, let z' be the output of A.

2. Let i be the smallest index such that $f(g^{i-1}(z, k))||z_i \neq f(g^{i-1}(z', k))||z_i'$ and $f(g^i(z, k)) = f(g^i(z', k))$ (since $g(z, k) = g(z', k)$ such an i must exists). Let $a_1 = g^i(z, k)$ and $a_2 = g^i(z', k)$. It follows now that $f(a_1) = f(a_2)$. For any two colliding inputs such as z and z' with key k, we call this i the colliding-index.

3. Choose $z^*, k^* = (x^*, h_1^*, \ldots, h_{n+1}^*)$ and a random bit b such that $g^i(z^*, k^*) = a_1$ and $h_i^*(y||b) = a_2$. This can be done using the pairwise independence

property of $\mathcal{H}$. More precisely, choose z^*, x^* and all the hash functions except h_i^* at random and set h_i^* so that both the conditions hold. Run A on input (z^*, k^*). If A fails to return a collision or such a hash function h_i can not be sampled,[4] halt outputting $\perp$. Otherwise, let z'' be the output of A.

4. If $f(g^{i-1}(z'', k^*)) \neq y$, halt outputting $\perp$. Otherwise, output $g^{i-1}(z'', k^*)$.

It follows from the construction that if M outputs w, then $f(w) = y$. We now proceed to compute the success probability of M. But first, we require the following definition. Define sets $N(i, a_1, a_2)$ to contain all input-key pairs (z, k) such that the following hold true: $f(a_1) = f(a_2)$ and $g^i(z, k) = a_1$, and A on input (z, k) returns z' such that $g^i(z', k) = a_2$ and i is the colliding-index. We first express the success probability of M using these sets.

Claim 1. *The probability with which M succeeds in inverting f is*

$$2^{n+r-1} \sum_{i, a_1, a_2} \frac{|N(i, a_1, a_2)|^2}{\left(2^{2n+1}|\mathcal{H}|^{n+1}\right)^2}$$

Proof: Given a tuple (z^*, k^*, i, a_1, a_2) such that $(z^*, k^*) \in N(i, a_1, a_2)$, define the following events:

Event $E1$: The randomly chosen input-key pair (z, k) by M in Step 1 is in $N(i, a_1, a_2)$.

Since the input and key are chosen uniformly at random, it holds that $\Pr[E_1] = 1/2^{n+1} \times 1/2^n \times 1/|\mathcal{H}|^{n+1} \times |N(i, a_1, a_2)| = |N(i, a_1, a_2)|/2^{2n+1}|\mathcal{H}|^{n+1}$

Event $E2$: If A on input (z^*, k^*) returns z'—where $k^* = (x, h_1, \ldots, h_n)$— this event denotes that M's random choice $b = z_i'$ and M's input is y such that $g^{i-1}(z', k^*) = y$. Therefore, $h_i(y||b) = h_i(y||z_i') = a_2$.

The probability that $b = z_i'$ is $1/2$. Therefore, since f is a 2^r-regular OWF, $\Pr[E_2] = 1/2 \cdot 2^r/2^n = 2^{r-1}/2^n$.

Event $E3$: M chooses z^*, k^* in Step 3.

From the pairwise-independence property of $\mathcal{H}$, it follows that[5] $\Pr[E_3] = 1/(2^{2n+1}\frac{|\mathcal{H}|^{n+1}}{2^{2n}}) = 1/2|\mathcal{H}|^{n+1}$

It follows from the description that for any tuple (z^*, k^*, i, a_1, a_2) such that $(z^*, k^*) \in N(i, a_1, a_2)$, if E_1, E_2 and E_3 occurs, M inverts y. Note that E_1, E_2 and E_3 are independent. Therefore, for a fixed tuple (z^*, k^*, i, a_1, a_2) such that $(z^*, k^*) \in N(i, a_1, a_2)$ the probability that E_1, E_2 and E_3 occurs is

$$\frac{|N(i, a_1, a_2)|}{2^{2n+1}|\mathcal{H}|^{n+1}} \times \frac{2^{r-1}}{2^n} \times \frac{1}{2|\mathcal{H}|^{n+1}}$$

---

[4] This occurs when $a_1 \neq a_2$ and $f(g^{i-1}(z^*, k^*)) = y$ and $z_i = b$.

[5] z, x and all the hash functions except h_i are randomly chosen. There are $2^{2n+1}|\mathcal{H}|^{m-1}$ such tuples. h_i is chosen so that two of its values are fixed. Since $\mathcal{H}$ is a pairwise-independent family of hash functions, there are exactly $\frac{|\mathcal{H}|}{2^{2n}}$ such functions. Finally, one of these tuples are chosen uniformly at random.

It follows from the definition of the sets $N(\cdot, \cdot, \cdot)$, that for every (z, k) there exists at most one tuple (i, a_1, a_2) such that $(z, k) \in N(i, a_1, a_2)$. Therefore, the success probability of M can be expressed as the sum of the success probability of M on each tuple (z^*, k^*, i, a_1, a_2) such that $(z^*, k^*) \in N(i, a_1, a_2)$. More precisely, the success probability of M is,

$$\sum_{i, a_1, a_2} \sum_{(z^*, k^*) \in N(i, a_1, a_2)} \frac{|N(i, a_1, a_2)|}{2^{2n+1} |\mathcal{H}|^{n+1}} \times \frac{2^{r-1}}{2^n} \times \frac{1}{2|\mathcal{H}|^{n+1}}$$

$$= \sum_{i, a_1, a_2} \frac{|N(i, a_1, a_2)|^2}{2^{2n+1} |\mathcal{H}|^{n+1}} \times \frac{2^{r-1}}{2^n} \times \frac{1}{2|\mathcal{H}|^{n+1}}$$

$$= 2^{n+r-1} \sum_{i, a_1, a_2} \frac{|N(i, a_1, a_2)|^2}{(2^{2n+1} |\mathcal{H}|^{n+1})^2}$$

□

We now relate this expression to the success probability of A.

Claim 2. *If A succeeds with probability ϵ then* $\displaystyle\sum_{(i, a_1, a_2)} \frac{|N(i, a_1, a_2)|^2}{(2^{2n+1} |\mathcal{H}|^{n+1})^2} \geq \frac{\epsilon^2}{n 2^{n+r}}$

Proof: Since for every pair (z, k), there exists at most one tuple (i, a_1, a_2) such that $(z, k) \in N(i, a_1, a_2)$ and by definition if $(z, k) \in N(i, a_1, a_2)$ then A succeeds on input (z, k), we have that the success probability of A is

$$1/ \left(2^{2n+1} |\mathcal{H}|^{n+1}\right) \times \sum_{(i, a_1, a_2)} |N(i, a_1, a_2)| = \epsilon$$

Let us consider the sum in the left-hand side and use the Cauchy-Schwartz inequality to obtain a bound on the sum of the squares of each term. It suffices to consider the sum over all tuples (i, a_1, a_2) such that $N(i, a_1, a_2)$ is not empty. In particular, they are not empty only if $f(a_1) = f(a_2)$. Therefore, the total number of such tuples is at most $n 2^{n+r}$. Using the Cauchy-Schwartz inequality, we have that

$$\sum_{(i, a_1, a_2)} \frac{|N(i, a_1, a_2)|^2}{(2^{2n+1} |\mathcal{H}|^{n+1})^2} \geq \frac{\epsilon^2}{n 2^{n+r}}$$

□

Now, we conclude the proof of the theorem. Applying Claim 2 to Claim 1, we obtain that the success probability of M is at least $2^{n+r-1} \times \frac{\epsilon^2}{n 2^{n+r}} = \frac{\epsilon^2}{2n}$ which is non-negligible. Therefore, M inverts f with non-negligible probability and we arrive at a contradiction. □

4.2 A Construction with Logarithmic Key Size

We now show how to construct a more efficient second preimage resistant family from regular one-way functions, by showing that if f is a regular OWF then the Reusable Generalized Randomized Iterate is second preimage resistant.

Definition 11. *Let $f : \{0,1\}^n \to \{0,1\}^n$ and let $\mathcal{K} = \{0,1\}^n \times \mathcal{H}^m$ where $\mathcal{H}$ is an efficient family of pairwise-independent hash functions from $\{0,1\}^{n+1}$ to $\{0,1\}^n$ and $m = O(\log n)$. Define the function $g(z,k)$ with input space $z \in \{0,1\}^{n+1}$ and key-space $k = (x, h_1, \ldots, h_m) \in \mathcal{K}$ as follows:*

$$g(z, (x, h_1, \ldots, h_m)) = \widetilde{g^{n+1}}(x, z, h_1, \ldots, h_m)$$

where $\widetilde{g^i}$ is the Reusable Generalized Randomized Iterate with $\ell = 1$.

Theorem 2. *Suppose f is a 2^r-regular one-way function. Then g defined according to Definition 11 is a second preimage resistant function family.*

Proof: Assume for contradiction, there exists an adversary A and polynomial $p(\cdot)$ such that for infinitely many lengths n, the probability with which A finds a collision on a random input $z \in \{0,1\}^n$ and key $k = (x, \overline{h}) \in \mathcal{K}$ is $\epsilon \geq \frac{1}{p(n)}$. As before, we assume for simplicity that A is deterministic.

Fix a particular n for which this happens. Using A, we construct a machine M that inverts f with probability that is polynomially related to ϵ and thus arrive at a contradiction. The machine M on input $y \in \{0,1\}^n$ internally incorporates the code of A and proceeds as follows:

1. Sample a random input z and key $k = (x, \overline{h})$. Internally run A on input (z, k). If A fails to return a collision, halt outputting $\bot$. Otherwise, let z' be the output of A.
2. Let i be the colliding-index. Let $a_1 = g^i(z, k)$ and $a_2 = g^i(z', k)$
3. Choose $z^*, k^* = (z^*, h_1^*, \ldots, h_m^*)$ and a random bit b such that $g^i(z^*, k^*) = a_1$ and $h_{\phi(i)}^*(y||b) = a_2$. This can be done in polynomial time following Corollary 1. Internally run A on input (z^*, k^*). If A fails to return a collision, halt outputting $\bot$. Otherwise, let z'' be A's output.
4. If $f(g^{i-1}(z'', k^*)) \neq y$, halt outputting $\bot$. Otherwise, output $g^{i-1}(z'', k^*)$.

As before, we define sets $N(i, a_1, a_2)$ that satisfy the same condition with the exception that we rely on $\widetilde{g^i}$ instead of g^i. The next claim relates these sets to the success probability of M.

Claim 3. *The probability with which M succeeds in inverting the one-way function f is $2^{n+r-1} \sum\limits_{i, a_1, a_2} |N(i, a_1, a_2)|^2 / \left(2^{2n+1} |\mathcal{H}|^m\right)^2$*

Proof: Consider the events E1,E2 and E3 exactly as before. We now have that given a tuple (z, k, i, a_1, a_2),

- Probability that E_1 occurs is $1/2^{n+1} \times 1/2^n \times 1/|\mathcal{H}|^m \times |N(i, a_1, a_2)| = |N(i, a_1, a_2)|/2^{2n+1}|\mathcal{H}|^m$.
- Probability that E_2 occurs is $2^{r-1}/2^n$ as before.
- Probability that E_3 occurs given E_1 and E_2 occurs is $1/(2^{2n+1} \frac{|\mathcal{H}|^m}{2^{2n}}) = 1/2|\mathcal{H}|^m$. This follows from Corollary 1 for $\ell = 1$ and $k = n + 1$ (which are the parameters used in this construction).

Again, we have that every (z, k) belongs to at most one set $N(i, a_1, a_2)$. Therefore, the success probability of M is

$$\sum_{i, a_1, a_2} \sum_{(z,k) \in N(i, a_1, a_2)} \frac{|N(i, a_1, a_2)|}{2^{2n+1}|\mathcal{H}|^m} \times \frac{2^{r-1}}{2^n} \times \frac{1}{2|\mathcal{H}|^m} = 2^{n+r-1} \sum_{i, a_1, a_2} \frac{|N(i, a_1, a_2)|^2}{(2^{2n+1}|\mathcal{H}|^m)^2}$$

$\square$

The next claim follows identically to Claim 2.

Claim 4. *If A succeeds with probability ϵ then* $\sum_{(i, a_1, a_2)} \frac{|N(i, a_1, a_2)|^2}{(2^{2n+1}|\mathcal{H}|^m)^2} \geq \frac{\epsilon^2}{n 2^{n+r}}$

As before applying Claim 3 to Claim 4, we obtain that the success probability of M is at least $\frac{\epsilon^2}{2n}$ and thus we arrive at a contradiction. $\square$

5   PRG Construction and Hardness Amplification

The idea of iterating a one-way permutation f on itself to obtain a PRG originates from the work of Blum, Micali and Yao [1,17]. Since f is a permutation, the function $f^{(i)} = f \circ \ldots \circ f$ (f iterated on itself i times) is also one-way. This means that the hardcore bit of every intermediate step is unpredictable. Iterating $n + 1$ times on a random input of length n and outputting all the hardcore bits would then yield a PRG that stretches by 1 bit. We refer to this as the BMY construction[6].

This approach, unfortunately does not work for general one-way functions. For the special case of regular one-way functions, Goldreich, Krawczyk and Luby [4], showed how to extend the BMY construction by adding a randomization step using an n-wise independent hash-function between every two applications of f. Haitner, et. al [7] simplified the construction to use just pair-wise hashing and further derandomized the construction by showing how to generate the n hash-functions required at the randomization steps using just $n \log n$ bits thus obtaining a PRG of seed length $O(n \log n)$.

In [7], Haitner et. al, showed that the same randomized iterate can also be used for hardness amplification to obtain strong one-way function from any regular weakly one-way function with unknown regularity. They also showed that similar derandomization yielded corresponding efficiency gains.

Using the Reusable Generalized Randomized Iterate, we obtain analogous PRG constructions and hardness-amplification with same efficiency. More precisely, we obtain the following results.

Theorem 3. *Let $f : \{0, 1\}^n \to \{0, 1\}^n$ be a regular one-way function and $\mathcal{H}$ be an efficient family of pairwise-independent length preserving hash functions. Define $G : \{0, 1\}^{2n} \times \mathcal{H}^m \to \{0, 1\}^{2n+1} \times \mathcal{H}^m$ as*

$$G(x, r, \overline{h}) = (b(\widetilde{f}^0(x, \overline{h}), r), \ldots, b(\widetilde{f}^n(x, \overline{h}), r), r, \overline{h})$$

---

[6] If f is a permutation over n-bit strings a more efficient construction is to set the generator G as $G(x) = f(x).b(x)$. However this uses in a crucial way the property that f is a permutation (since if x is uniform then $f(x)$ is also uniform).

where $\widetilde{f^k}(x, h_1, \ldots, h_m) = f(\widetilde{g^k}(x, h_1, \ldots, h_m))$ and $\widetilde{g^k}$ is the RGRI defined by $x, h_1, \ldots, h_m$ with $\ell = 0$ and b is the Goldreich-Levin hardcore predicate. Then G is a pseudorandom generator.

Theorem 4. *Let f be a $\frac{1}{p(n)}$-weak one-way function for some polynomial $p(\cdot)$.*[7] *Let $k = 4np(n)$ and $m = \lceil \log k \rceil$. For input $x \in \{0, 1\}^n$, $\overline{h} = [h_1, \ldots, h_m] \in \mathcal{H}^m$, define $g(x, \overline{h}) = (\widetilde{f^k}(x, \overline{h}), \overline{h})$ where $\widetilde{f^k}$ is the Reusable Randomized Iterate of f. Then, g is a (strong) one-way function.*

The proofs of both these theorems appear in the full-version of the paper and on a high-level follow the proofs presented in [7].

6   Discussion and Conclusions

This paper presented the Reusable Generalized Randomized Iterate, and its application to new efficient constructions of Universal One-Way Hash Functions based on regular one-way functions. These are the first such efficient constructions of UOWHF from regular one-way functions of unknown regularity.

We also showed that the Reusable Generalized Randomized Iterate can be used to construct PRGs based on regular-one way functions, obtaining an alternative proof of a result by [7].

An interesting question raised by our work is the following: can we replace Shoup's technique for TCR domain extension with any appropriate log-space derandomizer? This is not immediately clear, since the reconstruction algorithm of Lemma 1 plays a crucial role in our construction and such a property does not follow from the definition of derandomizers (although, the current derandomizers indeed have that property).

A more conceptual contribution of this paper is to show that by combining techniques from the collision-resistant hashing and PRG toolboxes we can improve efficiency in both areas. Following [9,8] we believe that exploring the interplay between the two fields, and the possibility to apply techniques from one field to the other can lead to new and interesting discoveries.

The works in [9,8] highlight a inherent "black-box duality" between PRGs and UOWHFs. Starting from the PRG constructions based on OWPs [1] and OWFs [11], one can obtain the UOWHF constructions based on OWPs [12] and OWFs [13,8] using the following "parallelism". If there is "unpredictable" entropy in an input to an application of the one-way function in the PRG construction from which pseudo-entropy can be extracted, then there exists a symmetric TCR construction with the same structure where the output of the application of the one-way function has "inaccessible" entropy and can be compressed.

Our Generalized Randomized Iterate justifies this observation for the case of regular one-way functions in a more direct way, by showing a *single* algorithm that yields either a PRG or a UOWHF depending on the parameters.

---

[7] A function f is an ϵ-weak one-way function, if no adversary can succeed in inverting f with probability better than $1 - \epsilon$.

The approaches in [9,8] and in this paper can hopefully help in addressing the following interesting open problem. Is there a transformation that takes any PRG construction from a primitive $\mathcal{P}$ with ℓ-bit-expansion to a TCR construction from the same primitive $\mathcal{P}$ with ℓ-bit-compression and vice-versa. For example, given a OWP with a large ℓ-bit hard-core function, we know how to build a PRG that expands by ℓ bits per invocation of the OWP: is it possible to obtain a TCR which compresses by ℓ bits per invocation of the OWP? Conversely, an answer to the above general question would allow us to achieve more efficient PRG constructions from stronger primitives such as collision-resistant hash-functions.

References

1. Blum, M., Micali, S.: How to generate cryptographically strong sequences of pseudo random bits. SIAM Journal of Computing, 112–117 (1982)
2. De Santis, A., Yung, M.: On the Design of Provably-Secure Cryptographic Hash Functions. In: Damgård, I.B. (ed.) EUROCRYPT 1990. LNCS, vol. 473, pp. 412–431. Springer, Heidelberg (1991)
3. Gennaro, R., Trevisan, L.: Lower Bounds on the Efficiency of Generic Cryptographic Constructions. In: FOCS 2000, pp. 305–313 (2000)
4. Goldreich, O., Krawczyk, H., Luby, M.: On the existence of pseudorandom generators. SIAM Journal of Computing 22(6), 1163–1175 (1993)
5. Goldreich, O., Levin, L.A.: A hard-core predicate for all one-way functions. In: STOC 1989, pp. 25–32 (1989)
6. Goldwasser, S., Micali, S., Rivest, R.: A digital signature scheme secure against adaptive chosen-message attacks. SIAM Journal on Computing 17, 281–308 (1988)
7. Haitner, I., Harnik, D., Reingold, O.: On the Power of the Randomized Iterate. In: Dwork, C. (ed.) CRYPTO 2006. LNCS, vol. 4117, pp. 22–40. Springer, Heidelberg (2006)
8. Haitner, I., Holenstein, T., Reingold, O., Vadhan, S., Wee, H.: Universal One-Way Hash Functions via Inaccessible Entropy. In: Gilbert, H. (ed.) EUROCRYPT 2010. LNCS, vol. 6110, pp. 616–637. Springer, Heidelberg (2010)
9. Haitner, I., Reingold, O., Vadhan, S., Wee, H.: Inaccessible Entropy. In: STOC 2009, pp. 611–620 (2009)
10. Halevi, S., Krawczyk, H.: Strengthening Digital Signatures Via Randomized Hashing. In: Dwork, C. (ed.) CRYPTO 2006. LNCS, vol. 4117, pp. 41–59. Springer, Heidelberg (2006)
11. Hastad, J., Impagliazzo, R., Levin, L.A., Luby, M.: A pseudorandom generator from any one-way function. SIAM Journal of Computing 28(4), 1364–1396 (1989)
12. Naor, M., Yung, M.: Universal One-Way Hash Functions and their Cryptographic Applications. In: STOC 1989, pp. 33–43 (1989)
13. Rompel, J.: One-Way Functions are Necessary and Sufficient for Secure Signatures. In: STOC 1990, pp. 387–394 (1990)
14. Sarkar, P.: Masking-based domain extenders for UOWHFs: bounds and constructions. IEEE Transactions on Information Theory 51(12), 4299–4311 (2005)
15. Shoup, V.: A Composition Theorem for Universal One-Way Hash Functions. In: Preneel, B. (ed.) EUROCRYPT 2000. LNCS, vol. 1807, pp. 445–452. Springer, Heidelberg (2000)
16. Simon, D.R.: Findings Collisions on a One-Way Street: Can Secure Hash Functions Be Based on General Assumptions? In: Nyberg, K. (ed.) EUROCRYPT 1998. LNCS, vol. 1403, pp. 334–345. Springer, Heidelberg (1998)
17. Yao, A.: Theory and applications of trapdoor functions. In: FOCS, pp. 80–91 (1982)

Perfect Algebraic Immune Functions*

Meicheng Liu, Yin Zhang, and Dongdai Lin

SKLOIS, Institute of Information Engineering, CAS, Beijing 100195, P.R. China
meicheng.liu@gmail.com, zhangy@is.iscas.ac.cn, ddlin@iie.ac.cn

Abstract. A perfect algebraic immune function is a Boolean function with perfect immunity against algebraic and fast algebraic attacks. The main results are that for a perfect algebraic immune balanced function the number of input variables is one more than a power of two; for a perfect algebraic immune unbalanced function the number of input variables is a power of two. Also, for n equal to a power of two, the Carlet-Feng functions on $n+1$ variables and the modified Carlet-Feng functions on n variables are shown to be perfect algebraic immune functions.

Keywords: Boolean functions, Algebraic immunity, Fast algebraic attacks.

1 Introduction

The study of the cryptanalysis of the filter and combination generators of stream ciphers based on linear feedback shift registers (LFSRs) has resulted in a wealth of cryptographic criteria for Boolean functions, such as balancedness, high algebraic degree, high nonlinearity, high correlation immunity and so on. An overview of cryptographic criteria for Boolean functions with extensive bibliography is given in [3].

In recent years, algebraic and fast algebraic attacks [1,5,6] have been regarded as the most successful attacks on LFSR-based stream ciphers. These attacks cleverly use overdefined systems of multivariable nonlinear equations to recover the secret key. Algebraic attacks make use of the equations by multiplying a nonzero function of low degree, while fast algebraic attacks make use of the equations by linear combination.

Thus the algebraic immunity $(\mathcal{AI})$, the minimum algebraic degree of nonzero annihilators of f or $f + 1$, was introduced by W. Meier et al. [20] to measure the ability of Boolean functions to resist algebraic attacks. It was shown by N. Courtois and W. Meier [5] that maximum $\mathcal{AI}$ of n-variable Boolean functions is $\lceil \frac{n}{2} \rceil$. The properties and constructions of Boolean functions with maximum $\mathcal{AI}$ were researched in a large number of papers, e.g., [8,15,16,18,4,24,25].

* Supported by the National 973 Program of China under Grant 2011CB302400, the National Natural Science Foundation of China under Grants 10971246, 60970152, and 61173134, the Strategic Priority Research Program of the Chinese Academy of Sciences under Grant XDA06010701, and the CAS Special Grant for Postgraduate Research, Innovation and Practice.

X. Wang and K. Sako (Eds.): ASIACRYPT 2012, LNCS 7658, pp. 172–189, 2012.

The resistance against fast algebraic attacks is not covered by algebraic immunity [7,2,17]. At Eurocrypt 2006, F. Armknecht et al. [2] introduced an effective algorithm for determining the immunity against fast algebraic attacks, and showed that a class of symmetric Boolean functions (the majority functions) have poor resistance against fast algebraic attacks despite their resistance against algebraic attacks. Later M. Liu et al. [17] stated that almost all the symmetric functions including these functions with good algebraic immunity behave badly against fast algebraic attacks. In [22] P. Rizomiliotis introduced a method to evaluate the behavior of Boolean functions against fast algebraic attacks using univariate polynomial representation. However, it is unclear what is maximum immunity to fast algebraic attacks.

A preprocessing of fast algebraic attacks on LFSR-based stream ciphers, which use a Boolean function $f : GF(2)^n \to GF(2)$ as the filter or combination generator, is to find a nonzero function g of small algebraic degree such that the multiple gf has algebraic degree not too large [6]. N. Courtois [6] proved that for any pair of positive integers (e, d) such that $e + d \geq n$, there is a nonzero function g of degree at most e such that gf has degree at most d. This result reveals an upper bound on maximum immunity to fast algebraic attacks. It implies that the function f has maximum possible resistance against fast algebraic attacks, if for any pair of positive integers (e, d) such that $e + d < n$ and $e < n/2$, there is no nonzero function g of degree at most e such that gf has degree at most d. Such functions are said to be perfect algebraic immune ($\mathcal{PAI}$). Note that one can use the fast general attack [6, Theorem 7.1.1] by splitting the function into two $f = h + l$ with l being the linear part of f. In this case, $h = f + l$ rather than $h = gf$ is used, then e equals 1, i.e., the degree of the linear function l, and d equals the degree of the function h, i.e., the degree of f. Thus $\mathcal{PAI}$ functions have algebraic degree at least $n - 1$.

A $\mathcal{PAI}$ function also achieves maximum $\mathcal{AI}$. As a consequence, a $\mathcal{PAI}$ function has perfect immunity against classical and fast algebraic attacks. Although preventing classical and fast algebraic attacks is not sufficient for resisting algebraic attacks on the augmented function [12], the resistance against these attacks depends on the update function and tap positions used in a stream cipher and in actual fact it is not a property of the Boolean function. Thus the use of $\mathcal{PAI}$ functions does not guarantee that a stream cipher is not vulnerable to algebraic attacks since the attacker can also exploit suitable relations for the augmented functions as suggested in [6,12].

It is an open question whether there are $\mathcal{PAI}$ functions for arbitrary number of input variables. This problem was also noticed in [4] at Asiacrypt 2008. It seems that $\mathcal{PAI}$ functions are quite rare. In [4] C. Carlet and K. Feng observed that the Carlet-Feng functions on 9 variables are $\mathcal{PAI}$. One can check that the Carlet-Feng functions on 5 variables are also $\mathcal{PAI}$ (see also [10]). However, no function is shown to be $\mathcal{PAI}$ for arbitrary number of variables. On the contrary, M. Liu et al. [17] proved that no symmetric functions are $\mathcal{PAI}$, and in [26] the authors proved that no rotation symmetric functions are $\mathcal{PAI}$ for even number (except a power of two) of variables.

In this paper, we study the upper bounds on the immunity to fast algebraic attacks, and solve the above question. The immunity against fast algebraic attacks is related to a matrix thanks to Theorem 1 of [2]. By a simple transformation on this matrix we obtain a symmetric matrix whose elements are the coefficients of the algebraic normal form of a given Boolean function. We improve the upper bounds on the immunity to fast algebraic attacks by proving that the symmetric matrix is singular in some cases. The results are that for an n-variable function, we have: (1) if n is a power of 2 then a $\mathcal{PAI}$ function has algebraic degree n (showing that the function is unbalanced); (2) if n is one more than a power of 2 then a $\mathcal{PAI}$ function has algebraic degree $n - 1$ (which is also balanced); (3) otherwise, the function is not $\mathcal{PAI}$. We then prove that the Carlet-Feng functions, which have algebraic degree $n - 1$, are $\mathcal{PAI}$ for n equal to one more than a power of 2, and are almost $\mathcal{PAI}$ for the other cases. Also we prove that the modified Carlet-Feng functions, which have algebraic degree n, are $\mathcal{PAI}$ for n equal to a power of 2, and are almost $\mathcal{PAI}$ for the other cases. The results show that our bounds on the immunity to fast algebraic attacks are tight, and that the Carlet-Feng functions are optimal against fast algebraic attacks as well as classical algebraic attacks. Our results explain the experimental observations of C. Carlet and K. Feng [4] and also prove their conjecture.

The remainder of this paper is organized as follows. In Section 2 some basic concepts are provided. Section 3 presents the improved upper bounds on the immunity of Boolean functions against fast algebraic attacks while Section 4 shows that the Carlet-Feng functions and their modifications achieve these bounds. Section 5 concludes the paper.

2   Preliminary

Let $\mathbb{F}_2$ denote the binary field $GF(2)$ and $\mathbb{F}_2^n$ the n-dimensional vector space over $\mathbb{F}_2$. An n-variable Boolean function is a mapping from $\mathbb{F}_2^n$ into $\mathbb{F}_2$. Denote by $\mathbf{B}_n$ the set of all n-variable Boolean functions. An n-variable Boolean function f can be uniquely represented as its truth table, i.e., a binary string of length 2^n,

$$f = [f(0, 0, \cdots, 0), f(1, 0, \cdots, 0), \cdots, f(1, 1, \cdots, 1)].$$

The support of f is given by $\mathrm{supp}(f) = \{x \in \mathbb{F}_2^n \mid f(x) = 1\}$. The Hamming weight of f, denoted by $\mathrm{wt}(f)$, is the number of ones in the truth table of f. An n-variable function f is said to be balanced if its truth table contains equal number of zeros and ones, that is, $\mathrm{wt}(f) = 2^{n-1}$.

An n-variable Boolean function f can also be uniquely represented as a multivariate polynomial over $\mathbb{F}_2$,

$$f(x) = \sum_{c \in \mathbb{F}_2^n} a_c x^c, \ a_c \in \mathbb{F}_2, \ x^c = x_1^{c_1} x_2^{c_2} \cdots x_n^{c_n}, \ c = (c_1, c_2, \cdots, c_n),$$

called the algebraic normal form (ANF). The algebraic degree of f, denoted by $\deg(f)$, is defined as $\max\{\mathrm{wt}(c) \mid a_c \neq 0\}$.

Let $\mathbb{F}_{2^n}$ denote the finite field $GF(2^n)$. The Boolean function f considered as a mapping from $\mathbb{F}_{2^n}$ into $\mathbb{F}_2$ can be uniquely represented as

$$f(x) = \sum_{i=0}^{2^n-1} a_i x^i, \ a_i \in \mathbb{F}_{2^n}, \tag{1}$$

where $f^2(x) \equiv f(x) (\bmod\, x^{2^n} - x)$. Expression (1) is called the univariate polynomial representation of the function f. It is well known that $f^2(x) \equiv f(x) (\bmod\, x^{2^n} - x)$ if and only if $a_0, a_{2^n-1} \in \mathbb{F}_2$ and for $1 \leq i \leq 2^n - 2$, $a_{2i \bmod (2^n-1)} = a_i^2$. The algebraic degree of the function f equals $\max_{a_i \neq 0} \mathrm{wt}(i)$, where $i = \sum_{k=1}^{n} i_k 2^{k-1}$ is considered as $(i_1, i_2, \cdots, i_n) \in \mathbb{F}_2^n$.

Let α be a primitive element of $\mathbb{F}_{2^n}$. The a_i's of Expression (1) are given by $a_0 = f(0), a_{2^n-1} = f(0) + \sum_{j=0}^{2^n-2} f(\alpha^j)$ and

$$a_i = \sum_{j=0}^{2^n-2} f(\alpha^j) \alpha^{-ij}, \text{ for } 1 \leq i \leq 2^n - 2. \tag{2}$$

For more details with regard to the representation of Boolean functions, we refer to [3].

The algebraic immunity of Boolean functions is defined as follows. Maximum algebraic immunity of n-variable Boolean functions is $\lceil \frac{n}{2} \rceil$ [5].

Definition 1. *[20] The algebraic immunity of a function $f \in \mathbf{B}_n$, denoted by $\mathcal{AI}(f)$, is defined as*

$$\mathcal{AI}(f) = \min\{\deg(g) \mid gf = 0 \text{ or } g(f+1) = 0, 0 \neq g \in \mathbf{B}_n\}.$$

The immunity of f against fast algebraic attacks is related to the algebraic degree e of a function g and the algebraic degree d of gf with $e \leq d$. For an n-variable function f and any positive integer e with $e < n/2$, there is a nonzero function g of degree at most e such that gf has degree at most $n - e$ [6]. There are several notions about the immunity of Boolean functions against fast algebraic attacks in previous literatures, such as [13,21]. The perfect algebraic immune function we define below is actually a Boolean function which is algebraic attack resistant (see [21]) and has degree at least $n-1$. The latter is necessary for perfect algebraic immune function since a function of degree less than $n - 1$ admits $e = 1$ and $d = \deg(f) < n - 1 = n - e$ (taking g being a nonzero constant).

Definition 2. *Let f be an n-variable Boolean function. The function f is said to be perfect algebraic immune if for any positive integers $e < n/2$, the product gf has degree at least $n - e$ for any nonzero function g of degree at most e.*

A perfect algebraic immune ($\mathcal{PAI}$) function achieves maximum $\mathcal{AI}$ and is therefore a Boolean function perfectly resistant to classical and fast algebraic attacks. As a matter of fact, if a function does not achieve maximum $\mathcal{AI}$, then it admits a nonzero function g of degree less than $n/2$ such that $gf = 0$ or $gf = g$, which means that it is not $\mathcal{PAI}$.

3   The Immunity of Boolean Functions against Fast Algebraic Attacks

In this section, we present the upper bounds on the immunity of Boolean functions against fast algebraic attacks. We first recall the previous results for determining the immunity against fast algebraic attacks, then state our bounds.

Denote by $\mathcal{W}_i$ the ordered set $\{x \in \mathbb{F}_2^n \mid \mathrm{wt}(x) \leq i\}$ in lexicographic order and by $\overline{\mathcal{W}}_i$ the ordered set $\{x \in \mathbb{F}_2^n \mid \mathrm{wt}(x) \geq i+1\}$ in the reverse of lexicographic order. According to the definitions of $\mathcal{W}_i$ and $\overline{\mathcal{W}}_i$, it follows that if x is the j-th element in $\mathcal{W}_e$, then $\bar{x}$ is the j-th element in $\overline{\mathcal{W}}_{n-e-1}$, where $\bar{x} = (x_1 + 1, \ldots, x_n + 1)$. Here are some additional notational conventions: for $y, z \in \mathbb{F}_2^n$, let $z \subset y$ be an abbreviation for $\mathrm{supp}(z) \subset \mathrm{supp}(y)$, where $\mathrm{supp}(x) = \{i \mid x_i = 1\}$, and let $y \cap z = (y_1 \wedge z_1, \ldots, y_n \wedge z_n)$, $y \cup z = (y_1 \vee z_1, \ldots, y_n \vee z_n)$, where $\wedge$ and $\vee$ are the AND and OR operations respectively. We can see that $z \subset y$ if and only if $y^z = y_1^{z_1} y_2^{z_2} \cdots y_n^{z_n} = 1$.

Let g be a function of algebraic degree at most e $(e < n/2)$ such that $h = gf$ has algebraic degree at most d $(e \leq d)$. Let

$$f(x) = \sum_{c \in \mathbb{F}_2^n} f_c x^c, \ f_c \in \mathbb{F}_2,$$

$$g(x) = \sum_{z \in \mathcal{W}_e} g_z x^z, \ g_z \in \mathbb{F}_2,$$

and

$$h(x) = \sum_{y \in \mathcal{W}_d} h_y x^y, \ h_y \in \mathbb{F}_2$$

be the ANFs of f, g and h respectively. For $y \in \overline{\mathcal{W}}_d$, we have $h_y = 0$ and therefore

$$0 = h_y = \sum_{c \in \mathbb{F}_2^n} \sum_{\substack{c \cup z = y \\ z \in \mathcal{W}_e}} f_c g_z = \sum_{z \in \mathcal{W}_e} g_z \sum_{\substack{c \cup z = y \\ c \in \mathbb{F}_2^n}} f_c. \tag{3}$$

The above equations on g_z's are homogeneous linear. Denote by $V(f; e, d)$ the coefficient matrix of the equations, which is a $\sum_{i=d+1}^{n} \binom{n}{i} \times \sum_{i=0}^{e} \binom{n}{i}$ matrix with the (i, j)-th element equal to

$$v_{yz} = \sum_{\substack{c \cup z = y \\ c \in \mathbb{F}_2^n}} f_c = \sum_{\substack{y \cap \bar{z} \subset c \subset y \\ z \subset y}} f_c = y^z \sum_{y \cap \bar{z} \subset c \subset y} f_c, \tag{4}$$

where y is the i-th element in $\overline{\mathcal{W}}_d$ and z is the j-th element in $\mathcal{W}_e$. Then f admits no nonzero function g of algebraic degree at most e such that $h = gf$ has algebraic degree at most d if and only if the rank of the matrix $V(f; e, d)$ equals the number of g_z's which is $\sum_{i=0}^{e} \binom{n}{i}$, i.e., $V(f; e, d)$ has full column rank (see also [2,10]).

Theorem 1. *[2,10] Let $f \in \mathbf{B}_n$ and $\sum_{c \in \mathbb{F}_2^n} f_c x^c$ be the ANF of f. Let $V(f; e, d)$ be the matrix whose (i, j)-th element equals $\sum_{c \cup z = y} f_c$, where y is the i-th element in $\overline{\mathcal{W}}_d$ and z is the j-th element in $\mathcal{W}_e$.*

Then there exists no nonzero function g of degree at most e such that the product gf has degree at most d if and only if the matrix $V(f; e, d)$ has full column rank.

Remark 1. The theorem shows that $\mathcal{AI}(f) > e$ if and only if the matrix $V(f; e, e)$ has full column rank (since $\mathcal{AI}(f) > e$ if and only if there exists no nonzero function g of degree at most e such that $h = gf$ has degree at most e). Then $\mathcal{AI}(f) = \lceil \frac{n}{2} \rceil$ if and only if the matrix $V(f; \lceil \frac{n}{2} \rceil - 1, \lceil \frac{n}{2} \rceil - 1)$ has full column rank.

Now we show that performing some column operations on the matrix $V(f; e, d)$ creates a matrix with f_c's as its elements.

Lemma 2. $\sum_{z^* \subset z} v_{yz^*} = f_{y \cap \bar{z}}$.

Proof. Note that $c \cup z = y$ if and only if $c \subset y, z \subset y$ and $y \subset c \cup z$, that is, $y^c = 1, y^z = 1$ and $(c \cup z)^y = 1$. By (4) we have

$$\sum_{z^* \subset z} v_{yz^*} = \sum_{z^* \subset z} \sum_{c \cup z^* = y} f_c$$

$$= \sum_{z^* \subset z} \sum_{c \in \mathbb{F}_2^n} y^c y^{z^*} (c \cup z^*)^y f_c$$

$$= \sum_{c \in \mathbb{F}_2^n} y^c f_c \sum_{z^* \subset z} y^{z^*} (c \cup z^*)^y$$

$$= \sum_{c \subset y} f_c \sum_{\substack{z^* \subset y \cap z \\ y \subset c \cup z^*}} 1$$

$$= \sum_{c \subset y} f_c \sum_{y \cap \bar{c} \subset z^* \subset y \cap z} 1$$

$$= \sum_{c \subset y, y \cap \bar{c} = y \cap z} f_c$$

$$= f_{y \cap \bar{z}}.$$

$\square$

Lemma 2 shows that the matrix $V(f; e, d)$ can be transformed into a matrix, denoted by $W(f; e, d)$, with the (i, j)-th element equal to

$$w_{yz} = f_{y \cap \bar{z}}, \tag{5}$$

where y is the i-th element in $\overline{\mathcal{W}}_d$ and z is the j-th element in $\mathcal{W}_e$.

The (j, i)-th element of $W(f; e, d)$ is equal to

$$w_{\bar{z}\bar{y}} = f_{\bar{z} \cap \bar{\bar{y}}} = f_{y \cap \bar{z}} = w_{yz},$$

since $\bar{z}$ is the j-th element in $\overline{\mathcal{W}}_d$ and $\bar{y}$ is the i-th element in $\mathcal{W}_e$ by the definitions of $\overline{\mathcal{W}}_d$ and $\mathcal{W}_e$. Recall that $V(f; e, d)$ and $W(f; e, d)$ are $\sum_{i=d+1}^{n} \binom{n}{i} \times \sum_{i=0}^{e} \binom{n}{i}$ matrices. Therefore the matrix $W(f; e, n - e - 1)$ is a symmetric $\sum_{i=0}^{e} \binom{n}{i} \times \sum_{i=0}^{e} \binom{n}{i}$ matrix, denoted by $W(f; e)$.

Theorem 3. *Let $f \in \mathbf{B}_n$ and $\sum_{c \in \mathbb{F}_2^n} f_c x^c$ be the ANF of f. Let $W(f; e, d)$ be the matrix whose (i, j)-th element equals $f_{y \cap \bar{z}}$, where y is the i-th element in $\overline{\mathcal{W}}_d$ and z is the j-th element in $\mathcal{W}_e$.*

Then there exists no nonzero function g of degree at most e such that gf has degree at most d if and only if $W(f; e, d)$ has full column rank.

Proof. Lemma 2 shows that $V(f; e, d)$ and $W(f; e, d)$ have the same rank. Then the theorem follows from Theorem 1.   ☐

Remark 2. The theorem shows that $\mathcal{AI}(f) > e$ if and only if the matrix $W(f; e, e)$ has full column rank. Then $\mathcal{AI}(f) = \lceil \frac{n}{2} \rceil$ if and only if the matrix $W(f; \lceil \frac{n}{2} \rceil - 1, \lceil \frac{n}{2} \rceil - 1)$ has full column rank.

Next we concentrate on the upper bounds on the immunity of Boolean functions against fast algebraic attacks. As mentioned in Section 2, for an n-variable function f and any positive integer e with $e < n/2$, there is a nonzero function g of degree at most e such that gf has degree at most $n - e$. This can also be explained by Theorem 1 or Theorem 3: the matrices $V(f; e, n - e)$ and $W(f; e, n - e)$ do not have full column rank since they are $\sum_{i=0}^{e-1} \binom{n}{i} \times \sum_{i=0}^{e} \binom{n}{i}$ matrices. From Theorem 3 the bounds on the immunity to fast algebraic attacks are related to the question whether the symmetric matrix $W(f; e)$ is invertible.

Before stating our main results, we list a useful lemma about the determinant of a symmetric matrix over a field with characteristic 2.

Lemma 4. *Let $A = (a_{ij})_{m \times m}$ be a symmetric $m \times m$ matrix over a field with characteristic 2, and $a_{ii} = a_{1i}^2$ for $2 \le i \le m$, that is,*

$$A = \begin{pmatrix} a_{11} & a_{12} & a_{13} & \cdots & a_{1m} \\ a_{12} & a_{12}^2 & a_{23} & \cdots & a_{2m} \\ a_{13} & a_{23} & a_{13}^2 & \cdots & a_{3m} \\ \vdots & \vdots & \vdots & \ddots & \vdots \\ a_{1m} & a_{2m} & a_{3m} & \cdots & a_{1m}^2 \end{pmatrix}. \tag{6}$$

If $a_{11} = (m + 1) \bmod 2$, then $\det(A) = 0$.

Proof. Let S_m be the symmetric group of degree m. Then

$$\det(A) = \sum_{\sigma \in S_m} \prod_{i=1}^{m} a_{i, \sigma(i)}$$

$$= \sum_{\sigma \in S_m, \sigma^2 = 1} \prod_{i=1}^{m} a_{i, \sigma(i)} + \sum_{\sigma \in S_m, \sigma^2 \ne 1} \prod_{i=1}^{m} a_{i, \sigma(i)}$$

$$\left(\text{since } \prod_{i=1}^{m} a_{i,\sigma(i)} = \prod_{i=1}^{m} a_{\sigma(i),i} = \prod_{i=1}^{m} a_{\sigma(i),\sigma^{-1}(\sigma(i))} = \prod_{i=1}^{m} a_{i,\sigma^{-1}(i)}\right)$$

$$= \sum_{\sigma \in S_m, \sigma^2=1} \prod_{i=1}^{m} a_{i,\sigma(i)}.$$

If m is odd, then $a_{11} = 0$ and therefore

$$\det(A) = \sum_{\substack{j=2 \\ \sigma^2=1 \\ \sigma(1)=j}}^{m} \sum a_{1j} \prod_{i=2}^{m} a_{i,\sigma(i)}$$

$$= \sum_{\substack{j=2 \\ \sigma^2=1 \\ \sigma(1)=j}}^{m} \sum a_{1j}^2 \prod_{\substack{2 \le i \le m \\ i \ne j}} a_{i,\sigma(i)}$$

(for odd m and $\sigma^2 = 1$, there is j' such that $j' \ne j$ and $\sigma(j') = j'$)

$$= \sum_{j=2}^{m} \sum_{\substack{\sigma^2=1 \\ \sigma(1)=j, \sigma(j')=j'}} a_{1j}^2 a_{1j'}^2 \prod_{\substack{2 \le i \le m \\ i \ne j,j'}} a_{i,\sigma(i)}$$

(there is unique σ' such that $\sigma'(1) = j'$, $\sigma'(j') = 1$, $\sigma'(j) = j$, and $\sigma'(i) = \sigma(i)$ for $i \notin \{1, j, j'\}$)

$$= 0.$$

If m is even, then $a_{11} = 1$ and therefore

$$\det(A) = \sum_{\substack{\sigma^2=1 \\ \sigma(1)=1}} \prod_{i=2}^{m} a_{i,\sigma(i)} + \sum_{\substack{j=2 \\ \sigma(1)=j}}^{m} \sum_{\sigma^2=1} a_{1j}^2 \prod_{\substack{2 \le i \le m \\ i \ne j}} a_{i,\sigma(i)}$$

$$= \sum_{j=2}^{m} \sum_{\substack{\sigma^2=1 \\ \sigma(1)=1, \sigma(j)=j}} a_{1j}^2 \prod_{\substack{2 \le i \le m \\ i \ne j}} a_{i,\sigma(i)} + \sum_{\substack{j=2 \\ \sigma(1)=j}}^{m} \sum_{\sigma^2=1} a_{1j}^2 \prod_{\substack{2 \le i \le m \\ i \ne j}} a_{i,\sigma(i)}$$

$$= 0.$$

$\square$

Remark 3. For the matrix A of Lemma 4 it holds that $\det(A) = \det(A^{(1,1)})$ if $a_{11} = m \bmod 2$, where $A^{(i,j)}$ is the $(m-1) \times (m-1)$ matrix that results from A by removing the i-th row and the j-th column.

Theorem 5. *Let $f \in \mathbf{B}_n$ and f_{2^n-1} be the coefficient of the monomial $x_1 x_2 \cdots x_n$ in the ANF of f. Let e be a positive integer less than $n/2$. If $f_{2^n-1} = \binom{n-1}{e} + 1 \bmod 2$, then there exists a nonzero function g with degree at most e such that gf has degree at most $n - e - 1$.*

Proof. According to Theorem 3 we need to prove that the square matrix $W(f;e)$ is singular when $f_{2^n-1} = \binom{n-1}{e} + 1 \bmod 2$. Let W_{ij} be the (i,j)-th element of $W(f;e)$. Since $\mathbf{1} = (1,1,\cdots,1)$ and $\mathbf{0} = (0,0,\cdots,0)$ are the first elements in $\overline{\mathcal{W}}_{n-e-1}$ and $\mathcal{W}_e$ respectively, by (5) we have $W_{11} = w_{\mathbf{1,0}} = f_{2^n-1}$. Because $\sum_{i=0}^{e}\binom{n}{i} = \sum_{i=1}^{e}\binom{n-1}{i} + \sum_{i=1}^{e}\binom{n-1}{i-1} + 1 \equiv \binom{n-1}{e}(\bmod 2)$, we know $W_{11} = \sum_{i=0}^{e}\binom{n}{i} + 1 \bmod 2$ when $f_{2^n-1} = \binom{n-1}{e} + 1 \bmod 2$. As mentioned previously, $W(f;e)$ is a symmetric $\sum_{i=0}^{e}\binom{n}{i} \times \sum_{i=0}^{e}\binom{n}{i}$ matrix over $\mathbb{F}_2$. We wish to show that $W(f;e)$ has the form of (6). By (5) we have $W_{1i}^2 = W_{1i} = w_{1z} = f_{1\cap\bar{z}} = f_{\bar{z}} = f_{\bar{z}\cap\bar{z}} = w_{\bar{z}\bar{z}} = W_{ii}$ where $\bar{z}$ is the i-th element in $\overline{\mathcal{W}}_{n-e-1}$ and z is the i-th element in $\mathcal{W}_e$. It follows from Lemma 4 that the matrix $W(f;e)$ is singular. $\square$

Corollary 6. *Let n be an even number and $f \in \mathbf{B}_n$. If f is balanced, then there exists a nonzero function g with degree at most 1 such that the product gf has degree at most $n-2$.*

Proof. If f is balanced, then $f_{2^n-1} = 0$. For even n, it holds that $\binom{n-1}{1} + 1 \equiv 0(\bmod 2)$. Therefore the result follows from Theorem 5. $\square$

From Corollary 6 it seems that for the number n of input variables, odd numbers are better than even ones from a cryptographic point of view (since cryptographic functions must be balanced).

Lucas' theorem states that for positive integers m and i, the following congruence relation holds:

$$\binom{m}{i} \equiv \prod_{k=1}^{s} \binom{m_k}{i_k}(\bmod 2),$$

where $m = \sum_{k=1}^{s} m_k 2^{k-1}$ and $i = \sum_{k=1}^{s} i_k 2^{k-1}$ are the binary expansion of m and i respectively. It means that $\binom{m}{i} \bmod 2 = 1$ if and only if $i \subset m$.

Note that $f_{2^n-1} = 1$ if and only if $\deg(f) = n$. Theorem 5 shows that for an n-variable function f of degree n and $e \not\subset n-1$, there is a nonzero function g of degree at most e such that gf has degree at most $n-e-1$, and that for an n-variable function f of degree less than n and $e \subset n-1$, there is a nonzero function g of degree at most e such that gf has degree at most $n-e-1$.

For the case $n-1 \notin \{2^s, 2^s-1\}$, there are integers e, e^* with $0 < e, e^* < n/2$ such that $e \subset n-1$ and $e^* \not\subset n-1$, and thus an n-variable function is not $\mathcal{PAI}$. This shows that for a $\mathcal{PAI}$ function the number n of input variables is 2^s+1 or 2^s. For $n = 2^s+1$ (resp. 2^s), it holds that $e \not\subset n-1$ (resp. $e \subset n-1$) for positive integer $e < n/2$, and thus an n-variable function with degree equal to n (resp. less than n) is not $\mathcal{PAI}$. Recall that a function on odd number of variables with maximum $\mathcal{AI}$ is always balanced [9]. For $n = 2^s+1$, a $\mathcal{PAI}$ function has degree $n-1$ and is balanced since it has maximum $\mathcal{AI}$. For $n = 2^s$, a $\mathcal{PAI}$ function has degree n and is then unbalanced, since a function has an odd Hamming weight if and only if it has degree n. Consequently the following theorem is obtained.

Theorem 7. *Let $f \in \mathbf{B}_n$ be a perfect algebraic immune function. Then n is one more than or equal to a power of 2. Further, if f is balanced, then n is one more than a power of 2; if f is unbalanced, then n is a power of 2.*

4 The Immunity of Boolean Functions against Fast Algebraic Attacks Using Univariate Polynomial Representation

In this section we focus on the immunity of Boolean functions against fast algebraic attacks using univariate polynomial representation and show that the bounds presented in Section 3 can be achieved.

Recall that $\mathcal{W}_e$ is the ordered set $\{x \in \mathbb{F}_2^n \mid \mathrm{wt}(x) \leq e\}$ in lexicographic order and $\overline{\mathcal{W}}_d$ is the ordered set $\{x \in \mathbb{F}_2^n \mid \mathrm{wt}(x) \geq d+1\}$ in the reverse of lexicographic order. Hereinafter, an element $x = (x_1, x_2, \cdots, x_n)$ in $\mathcal{W}_e$ or $\overline{\mathcal{W}}_d$ is considered as an integer $x_1 + x_2 2 + \cdots + x_n 2^{n-1}$ from 0 to $2^n - 1$, and the operations "+" and "$-$" may be considered as addition and subtraction operations modulo $2^n - 1$ respectively if there is no ambiguity.

Let f, g and h be n-variable Boolean functions, and let g be a function of algebraic degree at most e $(e < n/2)$ satisfying that $h = gf$ has algebraic degree at most d $(e \leq d)$. Let

$$f(x) = \sum_{i=0}^{2^n-1} f_i x^i, \ f_i \in \mathbb{F}_{2^n},$$

$$g(x) = \sum_{z \in \mathcal{W}_e} g_z x^z, \ g_z \in \mathbb{F}_{2^n},$$

and

$$h(x) = \sum_{y \in \mathcal{W}_d} h_y x^y, \ h_y \in \mathbb{F}_{2^n},$$

be the univariate polynomial representations of f, g and h respectively. For $y \in \overline{\mathcal{W}}_d$, we have $h_y = 0$ and thus

$$0 = h_y = \sum_{\substack{i+z=y \\ z \in \mathcal{W}_e}} f_i g_z = \sum_{z \in \mathcal{W}_e} f_{y-z} g_z. \tag{7}$$

The above equations on g_z's are homogeneous linear. Denote by $U(f; e, d)$ the coefficient matrix of the equations, which is a $\sum_{i=d+1}^{n} \binom{n}{i} \times \sum_{i=0}^{e} \binom{n}{i}$ matrix with the (i, j)-th element equal to

$$u_{yz} = f_{y-z}, \tag{8}$$

where y is the i-th element in $\overline{\mathcal{W}}_d$ and z is the j-th element in $\mathcal{W}_e$. More precisely, for $(i, j) = (1, 1)$ we have $(y, z) = (2^n - 1, 0)$ and $u_{yz} = f_{2^n-1}$; for $(i, j) \neq (1, 1)$ we have $y - z \notin \{0, 2^n - 1\}$ and $u_{yz} = f_{(y-z) \bmod (2^n-1)}$ when $e \leq d$.

If the matrix $U(f; e, d)$ has full column rank, i.e., the rank of $U(f; e, d)$ equals the number of g_z's, then f admits no nonzero function g of algebraic degree at most e such that $h = gf$ has algebraic degree at most d.

If the matrix $U(f; e, d)$ does not have full column rank, then there always exists a nonzero Boolean function satisfying Equations (7). More precisely, if $g(x) = \sum_{z \in \mathcal{W}_e} g_z x^z$ $(g_z \in \mathbb{F}_{2^n})$ satisfies (7), then

$$0 = h_y^2 = \sum_{z \in \mathcal{W}_e} f_{y-z}^2 g_z^2 = \sum_{z \in \mathcal{W}_e} f_{2y-2z} g_z^2, \ y \in \overline{\mathcal{W}}_d, \tag{9}$$

where $f_{2(2^n-1)} = f_{2^n-1}$ and f_{2i} is considered as $f_{2i \bmod (2^n-1)}$ for $i \neq 2^n - 1$, and thus $g^2(x) = \sum_{z \in \mathcal{W}_e} g_z^2 x^{2z} \bmod (x^{2^n} - x)$ satisfies (9). Note that the system of (7) and the system of (9) are actually the same. Therefore, if $g(x)$ satisfies Equations (7) then $\mathrm{Tr}(g(x))$ satisfies Equations (7), where $\mathrm{Tr}(x) = x + x^2 + \cdots + x^{2^{n-1}}$. Also it follows that if $g(x)$ satisfies Equations (7) then $\beta g(x)$ and $\mathrm{Tr}(\beta g(x))$ satisfy Equations (7) for any $\beta \in \mathbb{F}_{2^n}$. If $g(x) \neq 0$, then there is $c \in \mathbb{F}_{2^n}$ such that $g(c) \neq 0$, and there is $\beta \in \mathbb{F}_{2^n}$ such that $\mathrm{Tr}(\beta g(c)) \neq 0$ and thus $\mathrm{Tr}(\beta g(x)) \neq 0$. Now we can see that $\mathrm{Tr}(\beta g(x))$ is a nonzero Boolean function and satisfies (7). Hence if there is a nonzero solution for (7), then there always exists a nonzero Boolean function g satisfying (7).

Thus the following theorem is obtained.

Theorem 8. *Let $f \in \mathbf{B}_n$ and $\sum_{i=0}^{2^n-1} f_i x^i$ be the univariate polynomial representation of f. Let $U(f; e, d)$ be the matrix whose (i, j)-th element equals f_{y-z}, where y is the i-th element in $\overline{\mathcal{W}}_d$ and z is the j-th element in $\mathcal{W}_e$.*

Then there exists no nonzero function g of algebraic degree at most e such that the product gf has algebraic degree at most d if and only if the matrix $U(f; e, d)$ has full column rank.

Remark 4. As described at the beginning of this section, the sets $\mathcal{W}_e$ and $\overline{\mathcal{W}}_d$ of Theorem 8 are subsets of $\{0, 1, \cdots, 2^n - 1\}$, while the sets $\mathcal{W}_e$ and $\overline{\mathcal{W}}_d$ of Theorem 1 and Theorem 3 are subsets of $\mathbb{F}_2^n$.

Remark 5. The theorem gives a method using one matrix to evaluate the immunity of Boolean functions against fast algebraic attacks based on univariate polynomial representation while in [22] P. Rizomiliotis used three matrices.

Remark 6. The theorem shows that $\mathcal{AI}(f) > e$ if and only if the matrix $U(f; e, e)$ has full column rank. Then $\mathcal{AI}(f) = \lceil \frac{n}{2} \rceil$ if and only if the matrix $U(f; \lceil \frac{n}{2} \rceil - 1, \lceil \frac{n}{2} \rceil - 1)$ has full column rank.

Remark 7. The matrix $U(f; e, n - e - 1)$, denoted by $U(f; e)$, is symmetric since

$$u_{\bar{z}\bar{y}} = f_{\bar{z}-\bar{y}} = f_{(2^n-1-z)-(2^n-1-y)} = f_{y-z} = u_{yz}.$$

Further, we have

$$u_{y\bar{y}} = f_{y-\bar{y}} = f_{y-(2^n-1-y)} = f_{2y} = f_y^2 = u_{y,0}^2,$$

and therefore $U(f; e)$ has the form of (6). Hence Theorem 5 can also be derived from Theorem 8 and Lemma 4.

4.1   Carlet-Feng Functions

The class of the Carlet-Feng functions were first presented in [11] and further studied by C. Carlet and K. Feng [4]. Such functions have maximum algebraic immunity and good nonlinearity. It was observed through computer experiments by Armknecht's algorithm [2] that the functions also have good behavior against fast algebraic attacks. In [23], P. Rizomiliotis determined the immunity of the Carlet-Feng functions against fast algebraic attacks by computing the linear complexity of a sequence, which is more efficient than Armknecht's algorithm but is not yet feasible for large n. In this section, we further discuss the immunity of the Carlet-Feng functions against fast algebraic attacks and prove that the functions achieve the bounds of Theorem 5.

Let n be an integer and α a primitive element of $\mathbb{F}_{2^n}$. Let $f \in \mathbf{B}_n$ and

$$\operatorname{supp}(f) = \{\alpha^l, \alpha^{l+1}, \alpha^{l+2}, \cdots, \alpha^{l+2^{n-1}-1}\}, 0 \le l \le 2^n - 2. \tag{10}$$

Then $\mathcal{AI}(f) = \lceil \frac{n}{2} \rceil$ according to [11,4]. As a matter of fact, the support of the function $f(\alpha^{l+2^{n-1}}x) + 1$ is $\{0, 1, \alpha, \cdots, \alpha^{2^{n-1}-2}\}$, which is a Carlet-Feng function. It means that these functions are affine equivalent.

A similar proof of [4, Theorem 2] applies to the following result. Here we give a proof for self-completeness.

Proposition 9. *Let $\sum_{i=0}^{2^n-1} f_i x^i$ be the univariate polynomial representation of the function f of (10). Then $f_0 = 0$, $f_{2^n-1} = 0$, and for $1 \le i \le 2^n - 2$,*

$$f_i = \frac{\alpha^{-il}}{1 + \alpha^{-i/2}}.$$

Hence the algebraic degree of f is equal to $n - 1$.

Proof. We have $f_0 = f(0) = 0$ and $f_{2^n-1} = 0$ since f has even Hamming weight and thus algebraic degree less than n. For $1 \le i \le 2^n - 2$, by (2) we have

$$f_i = \sum_{j=0}^{2^n-2} f(\alpha^j)\alpha^{-ij} = \sum_{j=l}^{l+2^{n-1}-1} \alpha^{-ij} = \alpha^{-il} \sum_{j=0}^{2^{n-1}-1} \alpha^{-ij}$$

$$= \alpha^{-il} \frac{1 + \alpha^{-i2^{n-1}}}{1 + \alpha^{-i}} = \alpha^{-il} \frac{1 + \alpha^{-i/2}}{1 + \alpha^{-i}} = \frac{\alpha^{-il}}{1 + \alpha^{-i/2}}.$$

We can see that $f_{2^n-2} \neq 0$ and therefore f has algebraic degree $n - 1$. $\square$

Remark 8. For the function f of (10), the (i, j)-th element of the matrix $U(f; e, d)$ with $e \le d$ is equal to

$$u_{yz} = f_{y-z} = \frac{\alpha^{-yl}\alpha^{zl}}{1 + \alpha^{-y/2}\alpha^{z/2}}, \text{ for } (i, j) \neq (1, 1),$$

where y is the i-th element in $\overline{\mathcal{W}}_d$ and z is the j-th element in $\mathcal{W}_e$.

Lemma 10. *Let K be a field of characteristic 2. Let $A = (a_{ij})_{m \times m}$ be an $m \times m$ matrix over K and $a_{ij} = (1 + \beta_i \gamma_j)^{-1}$, $\beta_i, \gamma_j \in K$ and $\beta_i \gamma_j \neq 1$, $1 \leq i, j \leq m$. Then the determinant of A is equal to*

$$\prod_{1 \leq i < j \leq m} (\beta_i + \beta_j)(\gamma_i + \gamma_j) \prod_{1 \leq i, j \leq m} a_{ij}.$$

Furthermore, the determinant of A is nonzero if and only if $\beta_i \neq \beta_j$ and $\gamma_i \neq \gamma_j$ for $i \neq j$.

Proof. The second half part of this lemma is derived from the first half part. The proof of the first half part is given by induction on m. First we can check that the statement is certainly true for $m = 1$. Now we verify the induction step. Suppose that it holds for $m - 1$. Thus we suppose that

$$\det(A^{(1,1)}) = \prod_{2 \leq i < j \leq m} (\beta_i + \beta_j)(\gamma_i + \gamma_j) \prod_{2 \leq i, j \leq m} a_{ij},$$

where $A^{(i,j)}$ is the $(m-1) \times (m-1)$ matrix that results from A by removing the i-th row and the j-th column.

We wish to show that it also holds for m. Let $B = (b_{ij})_{m \times m}$ with $b_{1j} = a_{1j}$ and for $i > 1$,

$$\begin{aligned}
b_{ij} &= a_{ij} + a_{11}^{-1} a_{i1} a_{1j} \\
&= \frac{1}{1 + \beta_i \gamma_j} + \left(\frac{1}{1 + \beta_1 \gamma_1}\right)^{-1} \cdot \frac{1}{1 + \beta_i \gamma_1} \cdot \frac{1}{1 + \beta_1 \gamma_j} \\
&= \frac{(1 + \beta_i \gamma_1)(1 + \beta_1 \gamma_j) + (1 + \beta_1 \gamma_1)(1 + \beta_i \gamma_j)}{(1 + \beta_i \gamma_j)(1 + \beta_i \gamma_1)(1 + \beta_1 \gamma_j)} \\
&= \frac{\beta_i \gamma_1 + \beta_1 \gamma_j + \beta_1 \gamma_1 + \beta_i \gamma_j}{(1 + \beta_i \gamma_j)(1 + \beta_i \gamma_1)(1 + \beta_1 \gamma_j)} \\
&= \frac{(\beta_1 + \beta_i)(\gamma_1 + \gamma_j)}{(1 + \beta_i \gamma_j)(1 + \beta_i \gamma_1)(1 + \beta_1 \gamma_j)} \\
&= a_{ij} \cdot (\beta_1 + \beta_i) a_{i1} \cdot (\gamma_1 + \gamma_j) a_{1j}.
\end{aligned}$$

Let

$$P = \text{diag}(1, (\beta_1 + \beta_2) a_{21}, \cdots, (\beta_1 + \beta_m) a_{m1})$$

and

$$Q = \text{diag}(1, (\gamma_1 + \gamma_2) a_{12}, \cdots, (\gamma_1 + \gamma_m) a_{1m})$$

where $\text{diag}(x_1, \cdots, x_m)$ denotes a diagonal matrix whose diagonal entries starting in the upper left corner are $x_1, \cdots, x_m$. Then

$$B = P \begin{pmatrix} a_{11} & * \\ 0 & A^{(1,1)} \end{pmatrix} Q.$$

Hence

$$\det(A) = \det(B)$$

$$= \det(P) \cdot a_{11} \det(A^{(1,1)}) \cdot \det(Q)$$

$$= \left(\prod_{i=2}^{m}(\beta_1 + \beta_i)a_{i1}\right) \cdot a_{11} \det(A^{(1,1)}) \cdot \left(\prod_{j=2}^{m}(\gamma_1 + \gamma_j)a_{1j}\right)$$

$$= \prod_{1 \le i < j \le m}(\beta_i + \beta_j)(\gamma_i + \gamma_j) \prod_{1 \le i,j \le m} a_{ij}.$$

It has now been proved by mathematical induction that the first half part of this lemma holds for all positive integers m. □

Lemma 11. *Let $A = (a_{ij})_{m \times m}$ and $B = (b_{ij})_{m \times m}$ be $m \times m$ matrices with $a_{ij} = \beta_i \gamma_j b_{ij}$ and $\beta_i \ne 0, \gamma_j \ne 0$ for $1 \le i, j \le m$. Then $\det(A) \ne 0$ if and only if $\det(B) \ne 0$.*

Proof. Let $P = \mathrm{diag}(\beta_1, \beta_2, \cdots, \beta_m)$ and $Q = \mathrm{diag}(\gamma_1, \gamma_2, \cdots, \gamma_m)$. Then $A = PBQ$ and hence $\det(A) = \det(B) \prod_{i=1}^{m} \beta_i \gamma_i$, which proves this lemma. □

Proposition 12. *Let e be a positive integer less than $n/2$ and f be the function of (10). Then $U(f; e)$ is invertible if $\binom{n-1}{e} \equiv 0 (\mathrm{mod}\, 2)$, and $U(f; e, n - e - 2)$ has full column rank if $\binom{n-1}{e} \equiv 1 (\mathrm{mod}\, 2)$.*

Proof. Let $U = U(f; e)$ and U_{ij} be the (i, j)-th element of U. We have $U_{11} = f_{2^n - 1} = 0$. By Remark 7 we know that U is a symmetric matrix of order $\sum_{i=0}^{e} \binom{n}{i}$ in the form of (6). For the case $\binom{n-1}{e} \bmod 2 = 0$, we have $\sum_{i=0}^{e} \binom{n}{i} \bmod 2 = 0 = U_{11}$. By Remark 3 it holds that $\det(U) = \det(U^{(1,1)})$. Remark 8 shows that the (i, j)-th element of $U^{(1,1)}$ is

$$U_{ij}^{(1,1)} = \frac{\alpha^{-yl}\alpha^{zl}}{1 + \alpha^{-y/2}\alpha^{z/2}},$$

where y is the i-th element in $\overline{\mathcal{W}}_{n-e-1} \setminus \{2^n - 1\}$ and z is the j-th element in $\mathcal{W}_e \setminus \{0\}$, since $e \le n - e - 1$ for $e < n/2$. Let U^* be a $(\sum_{i=0}^{e} \binom{n}{i} - 1) \times (\sum_{i=0}^{e} \binom{n}{i} - 1)$ matrix with the (i, j)-th element equal to

$$U_{ij}^* = \frac{1}{1 + \alpha^{-y/2}\alpha^{z/2}}.$$

Since $\alpha^{-y/2} \ne \alpha^{-y'/2}$ for $y \ne y'$ ($y, y' \in \overline{\mathcal{W}}_{n-e-1} \setminus \{2^n - 1\}$) and $\alpha^{z/2} \ne \alpha^{z'/2}$ for $z \ne z'$ ($z, z' \in \mathcal{W}_e \setminus \{0\}$), from Lemma 10 we have $\det(U^*) \ne 0$. Then by Lemma 11 it holds that $\det(U^{(1,1)}) \ne 0$. Hence, U is invertible.

For the case $\binom{n-1}{e} \bmod 2 = 1$, we consider the $\sum_{i=0}^{e+1} \binom{n}{i} \times \sum_{i=0}^{e} \binom{n}{i}$ matrix $U(f; e, n - e - 2)$. For even n, we always have $e \le n - e - 2$ for $e < n/2$. For odd n, we always have $e \le n - e - 2$ for $e \le (n - 3)/2$ and $\binom{n-1}{e} \bmod 2 = 0$ for $e = \frac{n-1}{2}$. Thus for $\binom{n-1}{e} \bmod 2 = 1$ and $e < n/2$, we always have $e \le n - e - 2$. Let U^{**} be the $\sum_{i=0}^{e} \binom{n}{i} \times \sum_{i=0}^{e} \binom{n}{i}$ matrix that results from $U(f; e, n - e - 2)$ by removing the first $\binom{n}{e+1}$ rows. A similar proof of $\det(U^{(1,1)}) \ne 0$ also applies to $\det(U^{**}) \ne 0$. Then $U(f; e, n - e - 2)$ has full column rank. □

The proof of the proposition shows that for the function f of (10) the rank of the matrix $U(f;e)$ is at least $\sum_{i=0}^{e} \binom{n}{i} - 1$ (since the matrix $U^{(1,1)}$ is invertible). Then, by Theorem 5, f admits a unique nonzero function g with algebraic degree e such that gf has algebraic degree at most $n - e - 1$ when $\binom{n-1}{e} \equiv 1 \pmod 2$.

Theorem 13. *Let e be a positive integer less than $n/2$ and f be the function of (10). Then f admits no nonzero function g with algebraic degree at most e such that gf has algebraic degree at most $n - e - 1$ if $\binom{n-1}{e} \equiv 0 \pmod 2$, and admits no nonzero function g with algebraic degree at most e such that gf has algebraic degree at most $n - e - 2$ if $\binom{n-1}{e} \equiv 1 \pmod 2$.*

Proof. It is derived from Theorem 8 and Proposition 12. □

Corollary 14. *Let $n = 2^s + 1$ and $f \in \mathbf{B}_n$ be the function of (10). Then f is $\mathcal{PAI}$.*

Proof. It is obtained from Theorem 13 since $\binom{n-1}{e} = \binom{2^s}{e} \equiv 0 \pmod 2$ for $1 \le e < n/2$. □

Theorem 13 states that the Carlet-Feng functions achieve the bounds of Theorem 5 and thus the bounds of Theorem 5 are tight for the functions with algebraic degree less than n, while Corollary 14 states that the Carlet-Feng functions on $2^s + 1$ variables are $\mathcal{PAI}$. The theorem explains the experimental results of [4,10] on the immunity of the Carlet-Feng functions against fast algebraic attacks, and implies the conjecture of C. Carlet and K. Feng [4, Section 5].

Next we consider the Boolean functions with algebraic degree equal to n.

Let n be an integer and α a primitive element of $\mathbb{F}_{2^n}$. Let $f \in \mathbf{B}_n$ and

$$\operatorname{supp}(f) = \{0, \alpha^l, \alpha^{l+1}, \cdots, \alpha^{l+2^{n-1}-1}\}, 0 \le l \le 2^n - 2. \tag{11}$$

The function of (11) is a function that results from the function of (10) by flipping the output at $x = 0$.

A similar proof of Proposition 9 applies to the following result.

Proposition 15. *Let $\sum_{i=0}^{2^n-1} f_i x^i$ be the univariate polynomial representation of the function f of (11). Then $f_0 = 1$, $f_{2^n-1} = 1$, and for $1 \le i \le 2^n - 2$,*

$$f_i = \frac{\alpha^{-il}}{1 + \alpha^{-i/2}}.$$

Hence the algebraic degree of f is equal to n.

A similar proof of Proposition 12 also applies to the following result.

Proposition 16. *Let e be a positive integer less than $\frac{n-1}{2}$ and f be the function of (11). Then $U(f;e)$ is invertible if $\binom{n-1}{e} \equiv 1 \pmod 2$, and $U(f;e,n-e-2)$ has full column rank if $\binom{n-1}{e} \equiv 0 \pmod 2$.*

Theorem 17. *Let e be a positive integer less than $\frac{n-1}{2}$ and f be the function of (11). Then f admits no nonzero function g with algebraic degree at most e such that gf has algebraic degree at most $n - e - 1$ if $\binom{n-1}{e} \equiv 1 \pmod 2$, and admits no nonzero function g with algebraic degree at most e such that gf has algebraic degree at most $n - e - 2$ if $\binom{n-1}{e} \equiv 0 \pmod 2$.*

Proof. It is confirmed by Theorem 8 and Proposition 16.   □

Similarly to the function of (10), the function of (11) admits a unique nonzero function g with algebraic degree e such that gf has algebraic degree at most $n - e - 1$ when $\binom{n-1}{e} \equiv 0 \pmod 2$.

In Theorem 17 we do not consider the case $e = \frac{n-1}{2}$ for odd n, since Theorem 5 shows that for odd n, an n-variable function f with algebraic degree n admits a nonzero function g with algebraic degree at most $\frac{n-1}{2}$ such that gf has algebraic degree at most $\frac{n-1}{2}$ (noting that $\binom{n-1}{\frac{n-1}{2}} \bmod 2 = 0$).

Corollary 18. *Let $n = 2^s$ and $f \in \mathbf{B}_n$ be the function of (11). Then f is $\mathcal{PAI}$.*

Proof. It is obtained from Theorem 17 since $\binom{n-1}{e} = \binom{2^s-1}{e} \equiv 1 \pmod 2$ for $1 \le e < n/2$.   □

Theorem 17 states that the modified Carlet-Feng functions achieve the bounds of Theorem 5 and thus the bounds of Theorem 5 are tight for the functions with algebraic degree equal to n, while Corollary 18 states that the modified Carlet-Feng functions on 2^s variables are $\mathcal{PAI}$.

Consequently, as mentioned above, the bounds of Theorem 5 are tight and there exist $\mathcal{PAI}$ functions on 2^s and $2^s + 1$ variables. More precisely, there exist n-variable $\mathcal{PAI}$ functions with degree $n - 1$ (balanced functions) if and only if $n = 2^s + 1$; there exist n-variable $\mathcal{PAI}$ functions with degree n (unbalanced functions) if and only if $n = 2^s$.

5   Conclusion

In this paper, several open problems about the immunity of Boolean functions against fast algebraic attacks have been solved. We proved the maximum immunity to fast algebraic attacks, and identified the immunity of the Carlet-Feng functions against fast algebraic attacks. It seems that for a balanced function, in terms of the immunity to fast algebraic attacks, the optimal value of the number n of input variables is one more than a power of two. The Carlet-Feng functions previously shown to have maximum algebraic immunity and good nonlinearity are proved to be optimal against fast algebraic attacks among the balanced functions. To the best of our knowledge this is the first time that a class of Boolean functions are shown to have such cryptographic property.

Acknowledgement. The authors thank the anonymous referees for their valuable comments on this paper. The authors are also grateful to Tianze Wang for his careful reading of the manuscript, and to Shaoyu Du, Lin Jiao, Yao Lu, Wenlun Pan, Tao Shi, and Wenhao Wang for their participation in FAA seminar at SKLOIS in December 2011. The first author would especially like to thank Dingyi Pei for his enlightening conversations on the resistance of Boolean functions against algebraic attacks.

References

1. Armknecht, F.: Improving Fast Algebraic Attacks. In: Roy, B., Meier, W. (eds.) FSE 2004. LNCS, vol. 3017, pp. 65–82. Springer, Heidelberg (2004)
2. Armknecht, F., Carlet, C., Gaborit, P., Künzli, S., Meier, W., Ruatta, O.: Efficient Computation of Algebraic Immunity for Algebraic and Fast Algebraic Attacks. In: Vaudenay, S. (ed.) EUROCRYPT 2006. LNCS, vol. 4004, pp. 147–164. Springer, Heidelberg (2006)
3. Carlet, C.: Boolean functions for cryptography and error correcting codes. In: Crama, Y., Hammer, P. (eds.) Boolean Methods and Models in Mathematics, Computer Science, and Engineering, pp. 257–397. Cambridge University Press, Cambridge (2010)
4. Carlet, C., Feng, K.: An Infinite Class of Balanced Functions with Optimal Algebraic Immunity, Good Immunity to Fast Algebraic Attacks and Good Nonlinearity. In: Pieprzyk, J. (ed.) ASIACRYPT 2008. LNCS, vol. 5350, pp. 425–440. Springer, Heidelberg (2008)
5. Courtois, N.T., Meier, W.: Algebraic Attacks on Stream Ciphers with Linear Feedback. In: Biham, E. (ed.) EUROCRYPT 2003. LNCS, vol. 2656, pp. 345–359. Springer, Heidelberg (2003)
6. Courtois, N.T.: Fast Algebraic Attacks on Stream Ciphers with Linear Feedback. In: Boneh, D. (ed.) CRYPTO 2003. LNCS, vol. 2729, pp. 176–194. Springer, Heidelberg (2003)
7. Courtois, N.T.: Cryptanalysis of Sfinks. In: Won, D.H., Kim, S. (eds.) ICISC 2005. LNCS, vol. 3935, pp. 261–269. Springer, Heidelberg (2006)
8. Dalai, D.K., Maitra, S., Sarkar, S.: Basic theory in construction of Boolean functions with maximum possible annihilator immunity. Designs, Codes and Cryptography 40(1), 41–58 (2006)
9. Dalai, D.K., Gupta, K.C., Maitra, S.: Results on Algebraic Immunity for Cryptographically Significant Boolean Functions. In: Canteaut, A., Viswanathan, K. (eds.) INDOCRYPT 2004. LNCS, vol. 3348, pp. 92–106. Springer, Heidelberg (2004)
10. Du, Y., Zhang, F., Liu, M.: On the Resistance of Boolean Functions against Fast Algebraic Attacks. In: Kim, H. (ed.) ICISC 2011. LNCS, vol. 7259, pp. 261–274. Springer, Heidelberg (2012)
11. Feng, K., Liao, Q., Yang, J.: Maximal values of generalized algebraic immunity. Designs, Codes and Cryptography 50(2), 243–252 (2009)
12. Fischer, S., Meier, W.: Algebraic Immunity of S-Boxes and Augmented Functions. In: Biryukov, A. (ed.) FSE 2007. LNCS, vol. 4593, pp. 366–381. Springer, Heidelberg (2007)
13. Gong, G.: Sequences, DFT and Resistance against Fast Algebraic Attacks. In: Golomb, S.W., Parker, M.G., Pott, A., Winterhof, A. (eds.) SETA 2008. LNCS, vol. 5203, pp. 197–218. Springer, Heidelberg (2008)
</caption>

14. Hawkes, P., Rose, G.G.: Rewriting Variables: The Complexity of Fast Algebraic Attacks on Stream Ciphers. In: Franklin, M. (ed.) CRYPTO 2004. LNCS, vol. 3152, pp. 390–406. Springer, Heidelberg (2004)

15. Li, N., Qu, L., Qi, W., Feng, G., Li, C., Xie, D.: On the construction of Boolean Functions with optimal algebraic immunity. IEEE Transactions on Information Theory 54(3), 1330–1334 (2008)

16. Li, N., Qi, W.-F.: Construction and Analysis of Boolean Functions of $2t+1$ Variables with Maximum Algebraic Immunity. In: Lai, X., Chen, K. (eds.) ASIACRYPT 2006. LNCS, vol. 4284, pp. 84–98. Springer, Heidelberg (2006)

17. Liu, M., Lin, D., Pei, D.: Fast algebraic attacks and decomposition of symmetric Boolean functions. IEEE Transactions on Information Theory 57(7), 4817–4821 (2011)

18. Liu, M., Pei, D., Du, Y.: Identification and construction of Boolean functions with maximum algebraic immunity. Science China Information Sciences 53(7), 1379–1396 (2010)

19. MacWilliams, F.J., Sloane, N.J.A.: The theory of error correcting codes. North-Holland, New York (1977)

20. Meier, W., Pasalic, E., Carlet, C.: Algebraic Attacks and Decomposition of Boolean Functions. In: Cachin, C., Camenisch, J.L. (eds.) EUROCRYPT 2004. LNCS, vol. 3027, pp. 474–491. Springer, Heidelberg (2004)

21. Pasalic, E.: Almost Fully Optimized Infinite Classes of Boolean Functions Resistant to (Fast) Algebraic Cryptanalysis. In: Lee, P.J., Cheon, J.H. (eds.) ICISC 2008. LNCS, vol. 5461, pp. 399–414. Springer, Heidelberg (2009)

22. Rizomiliotis, P.: On the resistance of Boolean functions against algebraic attacks using univariate polynomial representation. IEEE Transactions on Information Theory 56(8), 4014–4024 (2010)

23. Rizomiliotis, P.: On the security of the Feng-Liao-Yang Boolean functions with optimal algebraic immunity against fast algebraic attacks. Designs, Codes and Cryptography 57(3), 283–292 (2010)

24. Tu, Z., Deng, Y.: A conjecture about binary strings and its applications on constructing Boolean functions with optimal algebraic immunity. Designs, Codes and Cryptography 60(1), 1–14 (2011)

25. Zeng, X., Carlet, C., Shan, J., Hu, L.: More balanced Boolean functions with optimal algebraic immunity and good nonlinearity and resistance to fast algebraic attacks. IEEE Transactions on Information Theory 57(9), 6310–6320 (2011)

26. Zhang, Y., Liu, M., Lin, D.: On the immunity of rotation symmetric Boolean functions against fast algebraic attacks. Cryptology ePrint Archive, Report 2012/111, http://eprint.iacr.org/

Differential Analysis of the LED Block Cipher[*]

Florian Mendel, Vincent Rijmen, Deniz Toz, and Kerem Varıcı

KU Leuven, ESAT/COSIC and IBBT, Belgium
{florian.mendel,vincent.rijmen,deniz.toz,kerem.varici}@esat.kuleuven.be

Abstract. In this paper, we present a security analysis of the lightweight block cipher LED proposed by Guo et al. at CHES 2011. Since the design of LED is very similar to the Even-Mansour scheme, we first review existing attacks on this scheme and extend them to related-key and related-key-cipher settings before we apply them to LED. We obtain results for 12 and 16 rounds (out of 32) for LED-64 and 16 and 24 rounds (out of 48) for LED-128. Furthermore, we present an observation on full LED in the related-key-cipher setting[1]. For all these attacks we need to find good differentials for one step (4 rounds) of LED. Therefore, we extend the study of plateau characteristics for AES-like structures from two rounds to four rounds when the key addition is replaced with a constant addition. We introduce an algorithm that can be used to find good differentials and right pairs for one step of LED. To be more precise, we can find more than 2^{10} right pairs for one step of LED with complexity of 2^{16} and memory requirement of 5×2^{17}. Moreover, a similar algorithm can also be used to find iterative characteristics for the LED.

1 Introduction

Security in embedded systems, such as RFID and sensor networks, where the area is restricted is getting more and more important since people started interacting with them in daily life more often. Improving the efficiency while preserving the security is one of the main challenges in this area and it has been an ongoing research problem. Recently, many algorithms have been developed to address this problem: hash functions like QUARK [1], PHOTON [13], SPONGENT [3] as well as block ciphers like Piccolo [23], LED [14], TWINE [24] and Klein [12]. Each of them uses the advantage of the improved knowledge on the design and analysis of symmetric key components.

LED [14] is a lightweight block cipher proposed by Guo et al. at CHES 2011. While being dedicated to compact hardware implementation with one of the smallest area consumptions (among block ciphers with comparable parameters),

---

[*] This work was sponsored he Research Fund KU Leuven, OT/08/027, by the IAP Programme P6/26 BCRYPT of the Belgian State (Belgian Science Policy) and by the European Commission through the ICT Programme under Contract ICT-2007-216676 (ECRYPT II).

[1] Due to the page limitations, details of this part is given in the full version of the paper in [19].

X. Wang and K. Sako (Eds.): ASIACRYPT 2012, LNCS 7658, pp. 190–207, 2012.

LED also offers reasonable performance in software. The design bears some resemblance with the (generalized) Even-Mansour construction [4] with the difference that the same key is used in each step for LED-64 or every second step in the case of the larger variant LED-128. The step function is based on AES-like design principles that provide good bounds against large classes of attacks including differential and linear cryptanalysis. Additionally, LED offers strong security arguments against attacks even in the related-key model.

To the best of our knowledge, no external analysis of LED with respect to differential cryptanalysis has been published so far. The best existing differential attacks are distinguishers for 15 (out of 32) rounds of LED-64 and 27 (out of 48) rounds of LED-128 in a hash setting, where the key is known to (or even chosen by) the attacker, described by the designers. Moreover, the security of LED against meet-in-the-middle attack has been investigated recently by Isobe et al. [16]. They describe attacks for 8 (out of 32) and 16 (out of 48) rounds for LED-64 and LED-128, respectively.

Our Contribution. In this paper, we present the first external cryptanalysis of LED with respect to differential cryptanalysis. First, we show attacks for LED-64 reduced to 12 and 16 rounds. Furthermore, we present an observation on full LED in the related-cipher setting [25]. All our attacks are based on the attack of Daemen [5] on Even-Mansour construction [11] that is extended in a straightforward way to the related-key setting.

Secondly, we show how to improve the bound for the maximum expected differential probability (MEDP) for four rounds (one STEP) of LED from 2^{-32} to $2^{-41.75}$ using mega-boxes and the result of Park et al. [20].

Furthermore, we present algorithms to find differential characteristics with high probability that can be used in our attacks. By using the ideas of plateau characteristics [9] and extending the work with mega boxes [6], we are able to obtain characteristics for four rounds of LED. We find more than 2^{10} right pairs for a differential with a complexity less than 2^{16} time and 5×2^{17} memory and an iterative characteristic with six right pairs with the same complexities. We emphasize that our method is not specific to the block cipher LED and it can be used in the analysis of any AES-like construction where the key addition is replaced with a constant addition.

Outline. This paper is organized as follows. In Section 2 we give a brief description of LED and introduce the required definitions for our analysis. In Section 3, we describe the attacks on Even-Mansour construction and show how they can be extended to attack LED. We continue with differential analysis and give an algorithm to find the number of right pairs in a plateau characteristic in Section 4. We generalize this algorithm find characteristics for four rounds of LED in Section 5 and we provide the results for characteristics with high probability and iterative characteristics that can be used in our attacks in Section 6.

2   Description of LED

LED [14] is a conservative lightweight block cipher whose design can be seen as a special case of the generalized Even and Mansour construction [11] depicted in Figure 1.

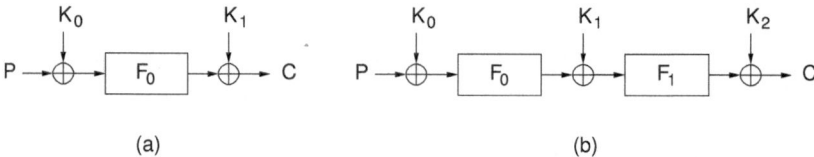

Fig. 1. Even-Mansour Construction with (a) $t = 1$ and (b) $t = 2$

LED accepts a 64-bit plaintext P, represented by a 4×4 array, and a 64-bit (or 128-bit) user key as inputs, and is composed of 8 (or 12) STEP functions preceded by a key addition. The STEP function is an AES-like design composed of four rounds. Each round is combination of Constant addition, S-boxes, ShiftRows, and (a variant of) MixColumns. LED uses the PRESENT S-box. In MixColumnsSerial, each column vector is multiplied by a matrix and replaced with the resulting vector. Note that the round constants for the second column are obtained from a linear shift register while the round constants for the remaining three columns do not change.

Key Schedule: LED has a simple key schedule where the 64-bit user key K is used as it is in each round whereas the 128-bit key is divided into two parts, $K = K_0 || K_1$, and used alternately. For the remainder of this paper, we refer to these two versions as LED-64 and LED-128. For more detail, please check the specification of LED [14].

One observation is that the S-boxes and linear transformations in the round function of the cipher can be described by structure of a *super box*:

Definition 1 (Super box [9]). *A super box maps an array a of m elements a_i to an array e of m elements e_i. Each of the elements has size n. A super box takes a key k of size $m \times n = n_b$ where n_b is the block size. It consists of the sequence of four transformations (layers): Substitution, Mixing, Round Key Addition, Substitution.*

Similar to AES [7], two rounds of LED can also be described alternatively as four parallel instances of the LED *super box* where the key addition is replaced by the constant addition. So, instead of dealing with the classical 4-bit S-boxes, one can consider 16-bit super boxes each composed of two S-box layers surrounding one MixColumnsSerial (MC) and one AddConstants (AC) function.

Four rounds of LED can be described as a mega-box, where the elements are 16-bit words and the LED super boxes defined above are seen as S-boxes. The linear transformation in the middle is a combination of ShiftRows, MixColumnsSerial and ShiftRows respectively. We will refer to this linear transformation as SMS.

3 Attacks on the Even-Mansour Construction and Application to LED

The Even-Mansour construction is a simple and yet provably secure block cipher construction. The designers have shown that the number of queries needed to break the scheme is bounded by $2^{n/2}$, where n is the blocklength ($n = 64$ for LED). A generic key recovery attack with chosen plaintexts showing that this bound is tight was introduced by Daemen [5]. Twenty years later, the construction was revisited. It was shown that the same bound applies to the known plaintext setting by using the slidex attack, an extended version of the slide attack [10].

Simultaneously, Bogdanov et al. generalized the construction in [4] to more steps and discussed its security. They even provided a security proof for the construction in the single-key setting. However, as pointed out by the authors, the scheme is insecure in the related-key setting. In this section, we focus on the attack of Daemen on the Even-Mansour construction, since it is the basis for all our attacks on LED. First we show how it can be extended to a related key attack on the generalized Even-Mansour construction. Then, we will use it to attack reduced versions of the LED block cipher.

3.1 Daemen's Attack

At Asiacrypt 1991 Daemen presented a generic key-recovery attack with complexity of $2^{n/2}$ [5]. It can be summarized as follows.

1. Choose a difference Δ.
2. For ℓ values of a compute $\Delta F_0 = F_0(a) \oplus F_0(a \oplus \Delta)$ and save the pair $(\Delta F_0, a)$ in a list L.
3. Choose an arbitrary plaintext P with $P' = P \oplus \Delta$ and ask for the ciphertexts C and C'
4. Compute $\Delta C = C \oplus C'$ and check if ΔC is in the list L to get a.
   - If ΔC is in the list L then a candidate for the key is found. Compute $K_0 = a \oplus P$ and $K_1 = F_0(a) \oplus C$.
   - Else go back to Step 3.

After repeating steps $3 - 4$ about $2^n/\ell$ times one expects to find the correct key with complexity of about $2^n/\ell + \ell$. Obviously the attack has the best complexity by choosing $\ell = 2^{n/2}$ resulting in a final attack complexity of about $2^{n/2}$ and similar memory requirements.

Note that, the attack can be applied in an iterative way to attack the Even-Mansour construction with $t > 1$ with complexity of $2^{t \cdot n/2}$ and similar memory requirements. For instance, if $t = 2$ then we get a complexity of 2^n.

3.2 Using Daemen's Attack in a Related-Key Setting

In certain scenarios one considers also related-key attacks where the adversary is allowed to get encryptions under several related keys. In this setting Daemen's attack can be adapted to attack t steps of the Even-Mansour construction with complexity of $t \cdot 2^{n/2}$ and similar memory requirements. For the sake of simplicity we first describe the attack for $t = 2$.

Related Key Attack with $t = 2$. Let K, K' be two related keys, where $K = K_0 \| K_1 \| K_2$ and $K' = K_0 \oplus \Delta_0 \| K_1 \oplus \Delta_1 \| K_2 \oplus \Delta_2$, with arbitrary (but known) $\Delta_0, \Delta_1, \Delta_2$ and $\Delta_1 \neq 0$. Then we can do a key recovery attack on the Even-Mansour construction with $t = 2$ with complexity of roughly $2^{n/2}$ and similar memory requirements using the attack of Daemen [5]. It can be summarized as follows.

1. For ℓ values of a compute $\Delta F_1 = F_1(a) \oplus F_1(a \oplus \Delta_1)$ and save the pair $(\Delta F_1, a)$ in a list L.
2. Choose an arbitrary P and $P' = P \oplus \Delta_0$ and ask for the ciphertexts C and C'
3. Compute $\Delta C = C \oplus (C' \oplus \Delta_2)$ and check if ΔC is in the list L to get a.
   - If ΔC is in the list L then a candidate for K_2 is found, $K_2 = F_1(a) \oplus C$.
   - Else go back to Step 2.

After repeating steps $2 - 3$ about $2^n/\ell$ times, the expected number of matches in the list L (i.e., candidates for K_2) is at least one. Note that, if we have more than one candidate for K_2 then we have to repeat the attack to get new candidates for K_2. The intersection of both sets of candidates gives us the correct key. Note that it is very unlikely that this intersection will have more than one solution.

Once K_2 is known one can apply the attack of Daemen to find K_0 and K_1. This results in a final attack complexity of about $2 \cdot 2^n/\ell + 2\ell$ and memory requirements of ℓ. Again, the attack has the best complexity by choosing $\ell \approx 2^{n/2}$ resulting in a final attack complexity of about $2 \cdot 2^{n/2}$ and memory requirements of $2^{n/2}$.

Related Key Attack with $t > 2$. The related key attack can be extended to more steps by applying the attack for $t = 2$ iteratively using more related keys with certain properties. Assume $t = 3$ and there are two related keys $K = K_0 \| K_1 \| K_2 \| K_3$ and $K' = K_0 \oplus \Delta_0 \| K_1 \| K_2 \oplus \Delta_2 \| K_3 \oplus \Delta_3$, with arbitrary (but known) $\Delta_0, \Delta_2, \Delta_3$ and $\Delta_2 \neq 0$. Then one can find K_3 similar as in the attack on the Even-Mansour construction with $t = 2$ with a complexity of roughly $2^{n/2}$. Once K_3 is found one can apply the attack for $t = 2$ with another pair of related keys to recover K_0, K_1 and K_2. In general, one can find the key for $t = i$ using i related keys with certain properties.

3.3   Attacks on Reduced LED

In this section, we will discuss the application of the attacks described in the previous section to the LED block cipher. Due to the fact that in LED the same key is used more than once the number of steps that can be attacked is significantly reduced. However, the attack can still be used in a straightforward way to break one and two steps of LED-64 in a single-key and related-key setting, respectively.

Both attacks have a complexity of about $2^{n/2}$ and similar memory requirements. Note that a similar related-attack was described recently in [4].

However, both attacks can be extended to more steps in the case of LED-128. In more detail, we can attack four and six steps of LED-128 in the single-key and related-key setting, respectively. First, we describe an attack on four steps of LED-128 based on Daemen's attack. It is based on the following simple observation (cf. Figure 2).

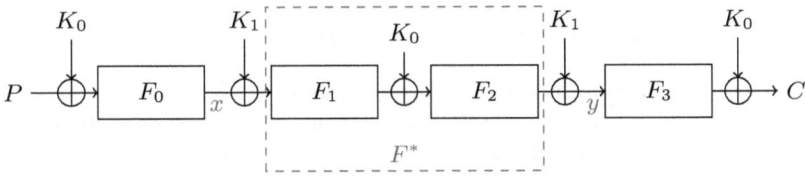

Fig. 2. Structure of LED-128 with $t = 4$

Assume K_0 is known, then one can peel off the first and last key addition. Thus, the attacker can remove one iteration at each side of the cipher with a complexity of about 2^{64} tries on K_0. Moreover, assuming that K_0 is known two steps of LED-128 can be viewed as one big iteration using only K_1. In other words, we get a 'new' Even-Mansour construction with $t = 1$ and one key K_1 where we can apply Daemen's attack to recover the key. Using this, one can find K_0 and K_1 for four steps of LED-128 with a complexity of about $2^{3n/2}$. It can be summarized as follows.

1. Guess the key K_0.
2. For $2^{n/2}$ values a and a fixed Δ compute $\Delta F^* = F^*(a) \oplus F^*(a \oplus \Delta)$ with $F^*(a) = F_2(F_1(a) \oplus K_0)$ and save the pair $(\Delta F^*, a)$ in a list L.
3. Choose an arbitrary P and compute $P' = F_0^{-1}(x \oplus \Delta) \oplus K_0$ with $x = F_0(P \oplus K_0)$. Ask for the ciphertexts C and C'.
4. Compute $\Delta y = y \oplus y'$ with $y = F_3^{-1}(C \oplus K_0)$ and $y' = F_3^{-1}(C' \oplus K_0)$. Check if Δy is in the list L to get a.
   - If Δy is in the list L then a candidate for the key is found. Compute $K_1 = a \oplus x$.
   - Else go back to Step 3.
5. Once K_1 is found check if the key $K = K_0 \| K_1$ is correct.

Since the expected number of K_0 guesses that we need to make to find the correct key is 2^n, we need to repeat the attack 2^n times. Since for each guess of K_0 we need about $2^{n/2}$ computations to find K_1, the complexity of the attack is roughly $2^{3n/2}$. Note that the above attack needs the whole codebook. However, at the cost of a higher attack complexity, the data complexity of the attack can be reduced. To be more precise, in step 3 of the attack we can always choose P from a predefined subset and when computing P' we check if it is also in this subset, if not then we repeat this step. Thus, the data complexity of the attack can be reduced by simultaneously increasing the time complexity.

The attack can be extended to six steps of LED-128 using related keys as in the attack on the Even-Mansour construction with $t = 2$. The attack is very similar as the attack on four steps. Basically only steps $2 - 4$ (Daemen's attack) are replaced by the related key attack described in the previous section. The result is a key-recovery attack on six steps (24 rounds) of LED-128 with complexity of about $2^{3n/2}$. Again, as in the attack on 4 steps the data complexity of the attack can be reduced on the cost of a higher attack complexity.

3.4   Extending the Attack to More Steps

In this section, we discuss how the attacks can be extended to more steps of LED. First, we show that by exploiting differential properties of the STEP-function F, it might be possible to extend the attacks on LED-64 by one or two steps. Moreover, the attack on 4 steps can also be used in related-cipher attack [25] with related key setting on full LED-128. We represent our observation in the full version of our paper [19].

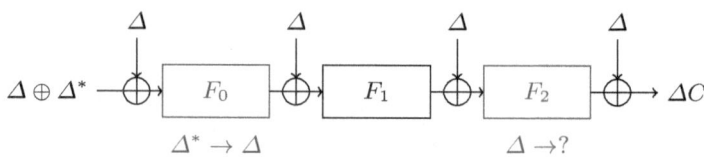

Fig. 3. Attack on LED-64 with $t = 3$

In the following, we show how the attack can be extended to t steps of LED-64. The attack is based on the assumption that one can find a good related-key differential for the first $t - 2$ steps such that one gets a zero difference after the key addition of step $t - 2$. Then one can use Daemen's attack on the last 2 steps to recover the key. For In the attack on 3 steps we a differential with good probability in F_0 is used, see Figure 3. The attack can be summarized as follows.

1. Assume we have given two related keys K_0 and $K_0' = K_0 \oplus \Delta$ and furthermore the differential $\Delta^* \to \Delta$ for F_0 holds with probability $p \gg 2^{-64}$.
2. For $2^{(n+\frac{1}{p})/2}$ values a compute $\Delta F_2 = F_2(a) \oplus F_2(a \oplus \Delta)$ and save the pair $(\Delta F_2, a)$ in a list L.
3. Choose an arbitrary P and $P' = P \oplus \Delta^* \oplus \Delta$ and ask for the ciphertexts C and C'
4. Compute $\Delta C = C \oplus (C' \oplus \Delta)$ and check if ΔC is in the list L to get a.
   - If ΔC is in the list L then a candidate for K_0 is found, $K_0 = F_2(a) \oplus C$.
   - Else go back to step 3.

After repeating steps $3-4$ about $2^{(n+\frac{1}{p})/2}$ times, the expected number of matches in the list L (and hence candidates for the key K_0) is $1/p$. Since the differential in F_0 will hold with probability p, for one of these matches we will have $\Delta F_1 = 0$.

Table 1. Summary of the attacks on LED

algorithm	# STEP functions	time complexity	memory complexity	attack type	reference
LED-64	3	$2^{(n+\frac{1}{p})/2}$	$2^{(n+\frac{1}{p})/2}$	related-key	Section 3.4
	4	$2^{(n+\frac{1}{p})/2}$	$2^{(n+\frac{1}{p})/2}$	related-key	Section 3.4
LED-128	4	$2^{3n/2}$	$2^{n/2}$	single-key	Section 3.3
	6	$2^{3n/2}$	$2^{n/2}$	related-key	Section 3.3
	12	$2^{(n+\frac{1}{p})/2}$	$2^{(n+\frac{1}{p})/2}$	related-key-cipher	[19]

Hence, one will find the right key after testing all candidates for K_0 resulting from the $1/p$ matches in the list L. The complexity and memory requirements of the attack depends on p, i.e. $2^{(n+\frac{1}{p})/2}$.

The attack on three steps can be extended to four steps of LED-64. Assume we can find a good iterative differential for F_1 that holds with probability p. Then this differential can be easily extended to a differential for the first 2 steps with the same probability (see Figure 4), resulting in an attack on 4 steps of LED-64 with complexity of $2^{(n+\frac{1}{p})/2}$ and similar memory requirements.

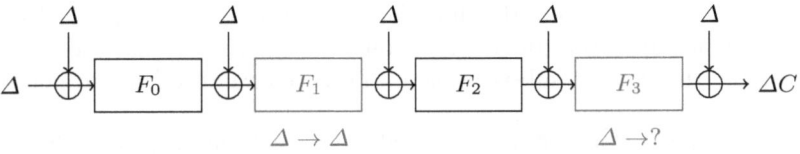

Fig. 4. Attack on LED-64 with $t = 4$

In the Table 1, we summarize the attacks on LED that are given in Section 3. We will discuss in the following sections how to find good (iterative) differential characteristics for one step of LED that can be used in the attacks on three and four steps.

4   Differential Analysis and Plateau Characteristics

In this section, we start with some definitions that will be helpful to understand the rest of the paper. We then give an introduction of the previous work on AES [9] and describe how we can use this method to find two/four round characteristics efficiently (and the corresponding right pairs).

4.1   Characteristics and Differentials

Differential cryptanalysis [2] is one of the most powerful techniques used in analysis of block ciphers, hash functions, stream ciphers, etc. It investigates how an input difference (generally XOR) propagates through the target function. The concept of differential cryptanalysis starts with analyzing the components of the function, mostly focusing on S-boxes since they are the smallest nonlinear building block. In the analysis, we call an S-box *active* if it has a non-zero input difference, otherwise we call it *passive*.

A *differential characteristic* $Q = (\Delta_0, \Delta_1, \cdots, \Delta_m)$ is a sequence of differences through various stages of the encryption. The sequence consists of an input difference Δ_0, followed by the output differences of all the steps $(\Delta_1, \Delta_2, \cdots, \Delta_m)$.

A *differential* [17] over a map is denoted by (Δ_0, Δ_m) where Δ_0 is the input difference and Δ_m is the output difference. The *differential probability* $DP(\Delta_0, \Delta_m)$ of a differential over a map f is the fraction of pairs with input difference Δ_0 that have output difference Δ_m.

For a keyed map, we can define differential probabilities $DP[k](\Delta_0, \Delta_m)$ and $DP[k](Q)$ for each value k of the key. Then, the *expected differential probability* (EDP) is the average of the differential probability over all keys. The *weight* of a differential or a characteristic is minus the binary logarithm of their EDP. Moreover, we define the *height* of a possible differential or a characteristic as the binary logarithm of the number of their right pairs satisfying (Δ_0, Δ_m) for a fixed key.

A differential characteristic through the AES-like (including LED) super boxes consist of a sequence of four differences: the input difference a, the difference after the first substitution b, the difference after the mixing step which is equal to the difference after the round (key) constant addition d, and the output difference after the second substitution e. These characteristics are denoted by $Q = (a, b, d, e)$.

It can be shown that SMS is a map whose branch number is 5. Therefore, a characteristic over a mega-box consists of 5 to 8 sub-characteristics, each over an LED super box. We denote the characteristics over the first and the second layer of super boxes by (a, b, d, e) and (f, g, i, j), respectively.

4.2   The Maximum Expected Differential Probability of LED

Differential cryptanalysis plays a crucial role in the analysis of symmetric key components since most of the cryptanalysis techniques are based on it. Therefore, giving bounds for resistance against differential cryptanalysis is one of the first steps in the evaluation of a design. In LED, the AES-like structure in the STEP function makes it possible to apply the previous work of [20] to bound the MEDP. By a straightforward computation of the formula stated in [20, Theorem 4], the designers compute the bound for the MEDP as 2^{-32}. This bound can be improved by considering the STEP function as a mega-box and then using [20, Theorem 1] to bound the MEDP of LED as

$$\max\left\{ \max_{\substack{1\leq i\leq 8 \\ 1\leq x\leq 2^{16}-1}} \sum_{y=1}^{2^{16}-1} \left\{DP^{sb_i}(x,y)\right\}^5, \quad \max_{\substack{1\leq i\leq 8 \\ 1\leq x\leq 2^{16}-1}} \sum_{y=1}^{2^{16}-1} \left\{DP^{sb_i}(y,x)\right\}^5 \right\} = 2^{-41.75}.$$

Here $\mathrm{DP}^{\mathrm{sb}_i}(x, y)$ is the probability of the characteristic (x, y) for the i-th super box obtained from the Difference Distribution Table (DDT). This result improves the approximations used in [14, Table 1]. We provided the bound for the first STEP function; the results for the other super boxes are similar.

4.3   Planar Differentials

Let γ be a map and let $F_{(a,b)}$, $G_{(a,b)}$ be the sets that contain the input values, respectively the output values, for the right pairs of the differential (a, b). i.e., $F_{(a,b)} = \{x | \gamma(x) + \gamma(x + a) = b\}$ and $G_{(a,b)} = \gamma(F_{(a,b)})$. A differential (a, b) is called a *planar differential*, if $F_{(a,b)}$ and $G_{(a,b)}$ form affine subspaces [9]. In that case, we can write:

$$F_{(a,b)} = p + U_{(a,b)}$$
$$G_{(a,b)} = q + V_{(a,b)},$$

where $U_{(a,b)}$ and $V_{(a,b)}$ are uniquely defined vector spaces, p any element in $F_{(a,b)}$ and q any element in $G_{(a,b)}$. Note that, if a differential (a, b) has exactly two or four right pairs, then it is always planar [9].

Plateau characteristics [9] are a special type of characteristics whose probability for each value k of the key, $DP[k](Q)$, depends on the key and can have only two values. For a fraction $2^{n_b - (weight(Q) + height(Q))}$ of the keys $DP[k](Q) = 2^{height(Q) - n_b}$ and for all other keys the it is zero. Note that the height is independent of the key.

Two-Round Plateau Characteristic Theorem states that a characteristic $Q = (a, b, c)$ over a map consisting of two steps with a key addition in between, in which the differentials (a, b) and (b, c) are planar, is a plateau characteristic with $height(Q) = dim(V_{(a,b)} \cap U_{(b,c)})$.

4.4   Algorithm for Number of Right Pairs in a Plateau Characteristic

Here, we describe the algorithm to find the number of right pairs of a given characteristic $Q = (a, b, d, e)$ through a super box. If the sub-characteristics (a, d) and (d, e) are planar then we can use the *Two-Round Plateau Characteristic Theorem* to compute the right pairs. Our aim in the algorithm is to build the matrix B containing the basis vectors of $(M(V_{(a,b)}))$ and $U_{(d,e)}$ where M is the mixing operation and $M(V) = \{M(v) | v \in V\}$. We denote vectors by *rows* of n_b bits.

The first step of our algorithm is to determine $V_{(a,b)}$ and $U_{(d,e)}$. Since, the super box is a set of m parallel maps, $V_{(a,b)}$ and $U_{(d,e)}$ can be written as:

$$V_{(a,b)} = V_{(a_1,b_1)} \times V_{(a_2,b_2)} \times \cdots \times V_{(a_m,b_m)}$$
$$U_{(d,e)} = U_{(d_1,e_1)} \times U_{(d_2,e_2)} \times \cdots \times U_{(d_m,e_m)}$$

by using the Lemma 4 in [9]. Now, if $|G_{(a_i,b_i)}| > 0$, we are interested in the output values of the right pairs.

- If $|G_{(a_i,b_i)}| = 2$, then the right pairs have input values in the set $\{q + \{0, b_i\}\}$ for some q in $G_{(a_i,b_i)}$, the basis vector for $V_{(a_i,b_i)}$ being b_i.
- If $|G_{(a_i,b_i)}| = 2^k$ where $2 \leq k < n$, then $V_{(a_i,b_i)} = < b_i, \beta_i^1, \ldots, \beta_i^k >$ and hence $V_{(a_i,b_i)}$ is said to be spanned by b_i and β_i^j's.
- If $(a_i, b_i) = (0,0)$ then $G_{(a_i,b_i)}$ covers the whole space and $V_{(a_i,b_i)} = < w_0, w_1, \cdots, w_{n-1} >$ where w_j is a coordinate vector (i.e. a vector with 1 at position j and zero at all other positions) and V is the standard basis.

Similarly, if $|F_{(d_i,e_i)}| > 0$, we are interested in the input values of the right pairs. When we find the right pairs for each parallel map we can compute the height by using Algorithm 1. The number of dependent rows in B gives $dim(M(V_{(a,b)}) \cap U_{(d,e)})$ which is equal to the height.

Algorithm 1 calls the following subroutines. $\texttt{Add}(v)$ adds the vector v as a new row to the matrix B. $\texttt{RowReduce}$ is the Gaussian Elimination and $\texttt{RowCount}$ gives the number of nonzero rows of a matrix.

Algorithm 1 Algorithm to compute the height of a given plateau characteristic

---

Input: Characteristic $Q = (a, b, d, e)$ with $EDP(Q) > 0$
Output: height(Q)

1: **procedure** PRECOMPUTE
2: **for** $i = 1 \rightarrow m$ **do**
3:    Compute $V_{(a_i,b_i)} = < b_i, \beta_i^1, \ldots, \beta_i^{k_i^v} >$ and $U_{(d_i,e_i)} = < d_i, \delta_i^1, \ldots, \delta_i^{k_i^u} >$
4: **end for**
5: **end procedure**

6: **procedure** HEIGHT
7: //at the input of Mixing
8: **for** $i = 0 \rightarrow m$ **do**
9:    **if** $b_i = 0$ **then**
10:      **for** $j = 0 \rightarrow n$ **do**
11:        Add($M(w_{4i+j})$)
12:      **end for**
13:    **else if** $b_i > 0$ **then**
14:      Add($M(b_i)$)
15:      **if** $|V_{(a_i,b_i)}| > 2$ **then**
16:        **for** $j = 1 \rightarrow k_i^v$ **do**
17:          Add($M(\beta_i^j)$)
18:        **end for**
19:      **end if**
20:    **end if**
21: **end for**

22: //at the output of Mixing
23: **for** $i = 0 \rightarrow m$ **do**
24:    **if** $d_i = 0$ **then**
25:      **for** $j = 0 \rightarrow n$ **do**
26:        Add(w_{4i+j})
27:      **end for**
28:    **else if** $d_i > 0$ **then**
29:      Add(d_i)
30:      **if** $|U_{(d_i,e_i)}| > 2$ **then**
31:        **for** $j = 1 \rightarrow k_i^u$ **do**
32:          Add(δ_i^j)
33:        **end for**
34:      **end if**
35:    **end if**
36: **end for**

37: B' = RowReduce (B).
38: **return** height(Q) = RowCount(B) - RowCount(B')
39: **end procedure**

The algorithm also gives us an insight on how to find the right pairs which can be determined by intersecting the affine spaces $F_{(a,b)} \cap (G_{(b,c)} \oplus k)$. This can be efficiently done by preparing the set of linear equations to solve. We would like to emphasize that, for a fraction of the keys the right pairs exists and their values differ depending on the key. On the other hand, if the constant operation is used instead of the key addition operation in the cipher, then it is not guaranteed always to have a solution.

5   Non-plateau Characteristics: LED Mega-Box

As we mentioned in Section 2, two rounds of LED can be considered as a super box and four rounds is defined as a mega-box. Let (a, b, d, e) and (f, g, i, j) denote the characteristic through the super boxes at the input and the output of SMS respectively. Since the super boxes are key independent, we consider them as 16-bit S-boxes. This allows us to omit the middle values (b, d) and (g, i) and use the differentials (a, e) and (f, j) in our analysis.

In order to use the two-round plateau characteristic theorem, it is required that the set of output values $G_{(a,e)}$ and the set of input values $F_{(f,j)}$ for the right pairs must be affine spaces/planar. However, this is not always guaranteed when the number of right pairs is greater than 4. Although the difference between the values of each pair is known and constant, some extra conditions between the pairs are also required for a set to become affine/planar. Therefore, we have to work with a union of affine spaces in order to compute the number of right pairs of a given characteristic. In the following, we will denote by height* the binary logarithm of the maximum number of right pairs of a given characteristic, over all values of the key. For a plateau characteristic, height* equals the height.

The details of our algorithm are given below. An algorithmic description can be found in Algorithm 2.

Precomputation: The first step of our algorithm is finding $G_{(a,e)}$ and $F_{(f,j)}$ for the given path, and the next step is obtaining the subspace decompositions of $V_{(a,e)}$ and $U_{(f,j)}$. If $V_{(a_i,e_i)}$ is affine then $V_{(a_i,e_i)} = < e, \varepsilon_1, \ldots, \varepsilon_n >$, otherwise it is a union of smaller vector spaces, i.e. $V_{(a_i,e_i)} = V^1_{(a_i,e_i)} \cup V^2_{(a_i,e_i)} \cup \ldots V^m_{(a_i,e_i)}$ where $m \geq 2$. Therefore, we have to find the corresponding basis vectors (ε_i's) for each subspace. The results are then stored in a list, L_i, for each active super box.

Analysis: We then use the Two-Round Plateau Characteristic Theorem to compute the height using the basis vectors obtained in the precomputation phase. Since the solution exist only for a fraction of the constant values, we check whether the given round constant is in the solution set or not. This step can also be done by solving a system of linear equations as in two-rounds, but this time the equations are obtained from the SMS layer and the basis vectors of the super boxes.

Here, we would like to emphasize that the solution does not always exist for the round constant of LED. Denote by K_q, the set of values, k, such that

---

Algorithm 2 Algorithm to compute the height* of a four-round characteristic

---

Input: Characteristic $Q = (a, e, f, j)$
Output: Upper bound for height*(Q)

1: **procedure** PRECOMPUTE
2: $L_0 = L_1 = \ldots = L_7 = \emptyset$
3: **for** $i = 0 \to 3$ **do**
4:     Compute $G_{(a_i, e_i)}$ and $F_{(f_i, j_i)}$

$$\bigcup_m V_{(a_i, e_i)}^m = \text{Decompose}(V_{(a_i, e_i)}) \text{ and } < \varepsilon_1^m, \varepsilon_2^m, \ldots \varepsilon_{d_m}^m >= V_{(a_i, e_i)}^m, \ d_m = |V_{(a_i, e_i)}^m|$$

$$\bigcup_n U_{(f_i, j_i)}^n = \text{Decompose}(U_{(f_i, j_i)}) \text{ and } < \varepsilon_1^n, \varepsilon_2^n, \ldots \varepsilon_{d_n}^n >= V_{(f_i, j_i)}^n, \ d_n = |V_{(f_i, j_i)}^n|$$

5:     Store$(L_i, \{(a_i, e_i), \varepsilon_1^m, \varepsilon_2^m, \ldots \varepsilon_{d_m}^m\})$ and Store$(L_{4+i}, \{(f_i, j_i), \varepsilon_1^n, \varepsilon_2^n, \ldots \varepsilon_{d_n}^n\})$
6: **end for**
7: **end procedure**

8: **procedure** ANALYZE
9: count $= 0$
10: **for all** $q \in L_0 \times L_1 \times \ldots \times L_7$ **do**
11:     $h = \text{HEIGHT}(q);$
12:     count $=$ count $+ 2^h$
13: **end for**
14: **return** $\log_2(\text{count})$
15: **end procedure**

DP$[k](q) > 0$. Since constants are used in the round function of LED, it is not guaranteed that the round constant, $c_r \in K_q$ for all q. Therefore, the algorithm gives an upper bound for height*(Q). If the key addition was used in the round function rather than constant addition, it could be possible to find a key value $k \in K_q$ for all q satisfying the upper bound.

On the other hand, if the key addition was used, the Algorithm 2 could not be applied immediately, since the lists L_i would depend on the key values and would not be unique. This would require recomputation of the lists for each key value increasing the complexity of the algorithm.

Note that, since height* for four rounds is the summation over all possible decompositions $q \in L_0 \times L_1 \times \ldots \times L_7$ of the characteristic Q, height*(Q) is not guaranteed to be an integer, although height(q) is integer for all q.

In Algorithm 2, Store adds input/output differences and the basis vectors $\{\varepsilon_1, \varepsilon_2, \ldots\}$ to the list L. HEIGHT is given in Algorithm 1 used with parameters $m = 4$ and $n = 16$.

6   Application of the Algorithms 1 and 2

In this section, we give two examples to demonstrate how Algorithm 1 and Algorithm 2 work. These examples can directly be used with attacks described

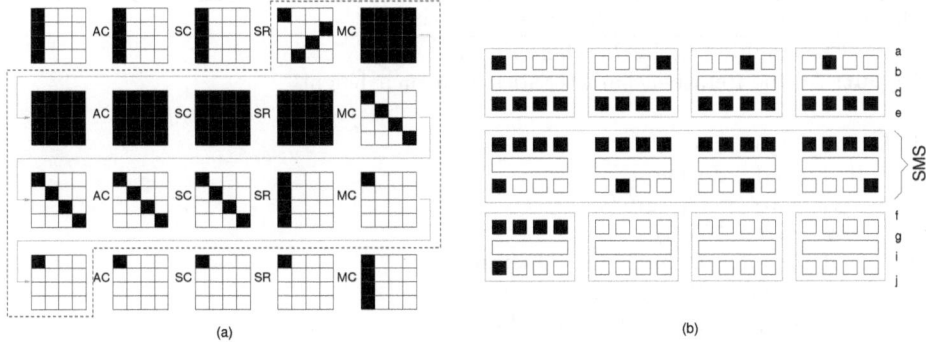

Fig. 5. (a)Path for iterative characteristics of the LED cipher (b)Mega-box representation of the same path

in Section 3.4. We do not claim that these are the best characteristics in terms of probability for the STEP function of LED that one can find. For both examples, we fix the number of active S-boxes to 25 for four rounds of LED. Since, we know from previous work [9] that all the characteristics with high probability are expected to have a low weight and a low number of active S-boxes. This also allows us to reduce the time and memory complexities of our algorithm and make the computation feasible.

6.1   Iterative Characteristics

Our aim is to find iterative characteristics (i.e., characteristics that have the same input and output difference) for the STEP function of the LED block cipher. We show that it is possible to obtain multiple iterative characteristics by using the 16-bit boxes and the two round plateau characteristic theorem in 2^{16} time and around 5×2^{17} memory. In terms of efficiency, this computation can be compared with the inbound technique of the rebound attack [18]. The main advantage of our computation is that many characteristics can be found whereas with the rebound attack, the expected number of characteristics that we find, equals one, using the same time complexity and slightly less data complexity.

In our analysis we used the differential path given in Figure 5. It is possible to adopt the algorithm for the other possible differential paths. The algorithm is summarized as follows:

Precomputation: For each of the active super boxes, obtain the differentials (a_i, e_i) (or (f_i, j_i)) for the given path and find the corresponding right pairs $G_{(a_i, e_i)}$ (or $F_{(f_i, j_i)}$). Then compute their affine subspace decomposition and the corresponding basis vectors. Store the input/output differences together with the basis vectors in a list. We denote these lists as L_0, L_1, L_2, L_3 for the super boxes at the input and L_4 for the super box at the output of the SMS layer. Note that this calculation is done for all possible differentials. Each list has around 2^{17} elements, therefore the total memory requirement of this step is 5×2^{17}.

Algorithm 3 Compute iterative characteristics

Input: Precomputed tables L_i where $i \in \{1, \cdots, 5\}$
Output: All the iterative characteristics with their height*

  1: **for all** $(e, f) \in$ S **do**
  2:    **if** $(f_0, j_0) \in L_4$ **then**
  3:       $\Delta = MC \circ SR \circ AC(j)$
  4:       $a = SR \circ AC(\Delta)$
  5:       **if** $(a_i, e_i) \in L_i$ for $1 \leq i \leq 4$ **then**
  6:          $h =$HEIGHT$^*(Q)$
  7:          **Output** $Q = (a, e, f, j)$ and h
  8:       **end if**
  9:    **end if**
10: **end for**

Analysis: We start from the four `MixColumnsSerial` operations in the SMS layer. Each of them has only one 4-bit word active at the output, hence we have $15^4 \approx 2^{16}$ possibilities for the differences at f (call the set of possibilities S). For each of these differences, we obtain the possible differences at j by using the precomputed list L_4. Then, we compute $(MC \circ SR \circ AC)(j) = \Delta$ which is the output difference after four rounds of the STEP function and is also equal to the input difference of the STEP function since we are interested in iterative characteristics. We make one more computation $(SR \circ AC)(\Delta)$ to obtain the difference at a. Note that by choosing a difference for f, we have already fixed the difference at e. We then check whether (a_i, e_i) is in the list L_i for $0 \leq i \leq 3$. If it does for all i, we use the Algorithm 2 to compute the height* and find the right pairs.

Results: In our analysis we found 240 iterative characteristics for the pattern given in Figure 5 but not all of them have a solution for the round constants of the LED block cipher. One of these characteristics is given below. It has 6 right pairs and the corresponding right pairs are given in [19].

a	0x6000	0x0003	0x0070	0x0C00
e	0x6962	0x5848	0x46A3	0x5CBF

f	0x943C	0x0000	0x0000	0x0000
j	0x8000	0x0000	0x0000	0x0000

6.2   Characteristics with High Height*

In this section, our aim is to find characteristics with high height* for the STEP function of the LED block cipher. We show that it is possible to obtain such characteristics by using a similar algorithm to Algorithm 3 with 2^{16} time complexity and 5×2^{17} memory complexity. In our analysis we focused on the differential path given in Figure 6 and searched for characteristics whose height* is greater than 5.

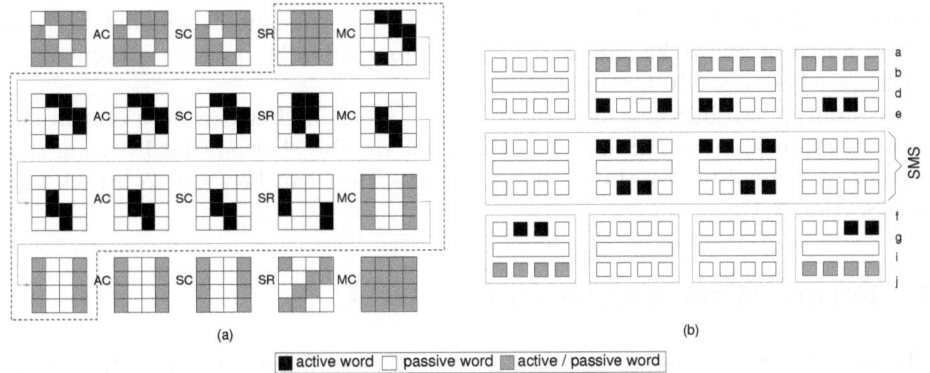

(a)   (b)

■ active word □ passive word ▨ active / passive word

Fig. 6. (a)Characteristics of the LED cipher with high height* (b)Mega-box representation of the same characteristics

Precomputation: All possible differentials together with the basis vectors of their affine space decomposition are stored in the lists L_1, L_2, L_3 for each of the super boxes at the input and in the lists L_4, L_7 for the super boxes at the output of the SMS layer. Again, each list has around 2^{17} elements, and the total the memory requirement of this step is 5×2^{17}.

Analysis: We start from the two active `MixColumnsSerial` operations in the SMS layer. Each of them has two 4-bit words active at the output, hence we have $(15^2)^2 \approx 2^{16}$ possibilities for the differences at f. For each of these possibilities, we obtain the possible differences at a by using the precomputed lists L_1, L_2 and L_3. Similarly, the possible differences at j are obtained by using the lists L_4 and L_7. We then use Algorithm 2 to compute the height and find the right pairs.

Results: Assume that $\dim(V_{(a,e)}) > 0$ and $\dim(U_{(f,j)}) > 0$, then we can write $V_{(a,e)} = \bigcup_m V_{(a,e)}^m$ and $U_{(f,j)} = \bigcup_n U_{(f,j)}^n$. We define a partition by Q_{mn} where $Q_{mn} = SMS(V_{(a,e)}^m) \cap U_{(f,j)}^n$. Then we know that $\text{height}^*(Q) \leq log_2(\sum_{m,n} 2^{\dim(Q_{mn})})$ (see Algorithm 2). In our analysis we observed that it is not easy to find a partition whose height is greater than six, but by combining all partitions, we were able to find characteristics which have height* greater than eleven or twelve. One example of such characteristics is provided below.

a	0x0000	0x0F91	0x2F0B	0x2803
e	0x0000	0xC00D	0x8F00	0x0F50
f	0x0CD0	0x0000	0x0000	0x00C8
j	0x8C07	0x0000	0x0000	0x50BF

The upper bound for height* is computed as 12.16 by using the formula. However, not all partitions have a solution for the given round constant, and we obtain only 1026 right pairs for the round constants used in LED. We also computed the number of right pairs by changing the round constant used in round three of the STEP function. The number of right pairs is computed as $1024 \pm \epsilon$ where $\epsilon \leq 116$ for all constants.

To sum up, we introduced not only a new method that can be useful in the security evaluation of AES-like structures but we also showed that by using this method it is possible to obtain characteristics that can be used to attack LED (see Section 3.4).

7 Future Work and Open Problems

The analysis of super boxes and mega-boxes play an important role in the cryptanalysis of AES-like ciphers. In this paper, we focused on characteristics for the block cipher LED with 25 active S-boxes. Since it is not feasible to compute the whole distribution of the characteristics for four rounds of LED, we focus only on characteristics that may have many right pairs. Therefore, our results cover characteristics with high height* and iterative characteristics with a fixed pattern. The examples given in this paper are the best ones that we computed. But still, it is possible to cover other patterns with 25 active S-boxes and they might give better results and at the same time result in improvements of our attacks.

We want to note that the algorithms given in this paper can also be used to compute the differentials for constructions using four rounds of AES as internal building block such as Pelican [8] giving new insights on these designs. Moreover, these algorithms might also be used in the computation of the inbound phase of the rebound attack.

References

1. Aumasson, J.-P., Henzen, L., Meier, W., Naya-Plasencia, M.: QUARK: A Lightweight Hash. In: Mangard, S., Standaert, F.-X. (eds.) CHES 2010. LNCS, vol. 6225, pp. 1–15. Springer, Heidelberg (2010)
2. Biham, E., Shamir, A.: Differential Cryptanalysis of DES-like Cryptosystems. In: Menezes, A., Vanstone, S.A. (eds.) CRYPTO 1990. LNCS, vol. 537, pp. 2–21. Springer, Heidelberg (1991)
3. Bogdanov, A., Knezevic, M., Leander, G., Toz, D., Varıcı, K., Verbauwhede, I.: SPONGENT: A Lightweight Hash Function. In: Preneel, Takagi [22], pp. 312–325
4. Bogdanov, A., Knudsen, L.R., Leander, G., Standaert, F.X., Steinberger, J.P., Tischhauser, E.: Key-Alternating Ciphers in a Provable Setting: Encryption Using a Small Number of Public Permutations - (Extended Abstract). In: Pointcheval, Johansson [21], pp. 45–62
5. Daemen, J.: Limitations of the Even-Mansour Construction. In: Imai, et al. [15], pp. 495–498
6. Daemen, J., Lamberger, M., Pramstaller, N., Rijmen, V., Vercauteren, F.: Computational aspects of the expected differential probability of 4-round AES and AES-like ciphers. Computing 85(1-2), 85–104 (2009)
7. Daemen, J., Rijmen, V.: The Design of Rijndael: AES - The Advanced Encryption Standard. Springer (2002)

8. Daemen, J., Rijmen, V.: The Pelican MAC Function. IACR Cryptology ePrint Archive 2005, 88 (2005)
9. Daemen, J., Rijmen, V.: Plateau characteristics. IET Information Security 1(1), 11–17 (2007)
10. Dunkelman, O., Keller, N., Shamir, A.: Minimalism in Cryptography: The Even-Mansour Scheme Revisited. In: Pointcheval, Johansson [21], pp. 336–354
11. Even, S., Mansour, Y.: A Construction of a Cipher From a Single Pseudorandom Permutation. In: Imai, et al. [15], pp. 210–224
12. Gong, Z., Nikova, S., Law, Y.W.: KLEIN: A New Family of Lightweight Block Ciphers. In: Juels, A., Paar, C. (eds.) RFIDSec 2011. LNCS, vol. 7055, pp. 1–18. Springer, Heidelberg (2012)
13. Guo, J., Peyrin, T., Poschmann, A.: The PHOTON Family of Lightweight Hash Functions. In: Rogaway, P. (ed.) CRYPTO 2011. LNCS, vol. 6841, pp. 222–239. Springer, Heidelberg (2011)
14. Guo, J., Peyrin, T., Poschmann, A., Robshaw, M.J.B.: The LED Block Cipher. In: Preneel, Takagi [22], pp. 326–341
15. Imai, H., Rivest, R.L., Matsumoto, T.: ASIACRYPT 1991. LNCS, vol. 739. Springer, Heidelberg (1993)
16. Isobe, T., Shibutani, K.: Security Analysis of the Lightweight Block Ciphers XTEA, LED and Piccolo. In: Susilo, W., Mu, Y., Seberry, J. (eds.) ACISP 2012. LNCS, vol. 7372, pp. 71–86. Springer, Heidelberg (2012)
17. Lai, X., Massey, J.L., Murphy, S.: Markov Ciphers and Differential Cryptanalysis. In: Davies, D.W. (ed.) EUROCRYPT 1991. LNCS, vol. 547, pp. 17–38. Springer, Heidelberg (1991)
18. Mendel, F., Rechberger, C., Schläffer, M., Thomsen, S.S.: The Rebound Attack: Cryptanalysis of Reduced Whirlpool and Grøstl. In: Dunkelman, O. (ed.) FSE 2009. LNCS, vol. 5665, pp. 260–276. Springer, Heidelberg (2009)
19. Mendel, F., Rijmen, V., Toz, D., Varıcı, K.: Differential Analysis of the LED Block Cipher. Cryptology ePrint Archive, Report 2012/544 (2012), http://eprint.iacr.org/
20. Park, S., Sung, S.H., Lee, S., Lim, J.: Improving the Upper Bound on the Maximum Differential and the Maximum Linear Hull Probability for SPN Structures and AES. In: Johansson, T. (ed.) FSE 2003. LNCS, vol. 2887, pp. 247–260. Springer, Heidelberg (2003)
21. Pointcheval, D., Johansson, T. (eds.): EUROCRYPT 2012. LNCS, vol. 7237, pp. 2012–2031. Springer, Heidelberg (2012)
22. Preneel, B., Takagi, T. (eds.): CHES 2011. LNCS, vol. 6917, pp. 2011–2013. Springer, Heidelberg (2011)
23. Shibutani, K., Isobe, T., Hiwatari, H., Mitsuda, A., Akishita, T., Shirai, T.: Piccolo: An Ultra-Lightweight Blockcipher. In: Preneel, Takagi [22], pp. 342–357
24. Suzaki, T., Minematsu, K., Morioka, S., Kobayashi, E.: Twine: A Lightweight, Versatile Blockcipher. In: ECRYPT Workshop on Lightweight Cryptography (2011), http://www.uclouvain.be/crypto/ecrypt_lc11/static/post_proceedings.pdf
25. Wu, H.: Related-Cipher Attacks. In: Deng, R.H., Qing, S., Bao, F., Zhou, J. (eds.) ICICS 2002. LNCS, vol. 2513, pp. 447–455. Springer, Heidelberg (2002)

PRINCE – A Low-Latency Block Cipher for Pervasive Computing Applications

Extended Abstract*

Julia Borghoff[1,**], Anne Canteaut[1,2,***], Tim Güneysu[3], Elif Bilge Kavun[3], Miroslav Knezevic[4], Lars R. Knudsen[1], Gregor Leander[1,†], Ventzislav Nikov[4], Christof Paar[3], Christian Rechberger[1], Peter Rombouts[4], Søren S. Thomsen[1], and Tolga Yalçın[3]

[1] Technical University of Denmark
[2] INRIA, Paris-Rocquencourt, France
[3] Ruhr-University Bochum, Germany
[4] NXP Semiconductors, Leuven, Belgium

Abstract. This paper presents a block cipher that is optimized with respect to latency when implemented in hardware. Such ciphers are desirable for many future pervasive applications with real-time security needs. Our cipher, named PRINCE, allows encryption of data within one clock cycle with a very competitive chip area compared to known solutions. The fully unrolled fashion in which such algorithms need to be implemented calls for innovative design choices. The number of rounds must be moderate and rounds must have short delays in hardware. At the same time, the traditional need that a cipher has to be iterative with very similar round functions disappears, an observation that increases the design space for the algorithm. An important further requirement is that realizing decryption and encryption results in minimum additional costs. PRINCE is designed in such a way that the overhead for decryption on top of encryption is negligible. More precisely for our cipher it holds that decryption for one key corresponds to encryption with a related key. This property we refer to as α-reflection is of independent interest and we prove its soundness against generic attacks.

1 Introduction

The area of lightweight cryptography, i.e., ciphers with particularly low implementation costs, has drawn considerable attention over the last years. Among the

---

* Due to page limitations, several details are omitted in this proceedings version. A full version is available at [10].
** Supported by the Danish research council for Technology and Production Sciences, Grant No.10-093667.
*** Partially supported by DGA/DS under Contract 2011.60.055.
† Partially supported by the Danish-Chinese Center for Applications of Algebraic Geometry in Coding Theory and Cryptography (Danish National Research Foundation and the National Science Foundation of China, Grant No.11061130539).

X. Wang and K. Sako (Eds.): ASIACRYPT 2012, LNCS 7658, pp. 208–225, 2012.

best studied algorithms are the block ciphers CLEFIA, Hight, KATAN, KTAN-TAN, Klein, mCrypton, LED, Piccolo and PRESENT [33,23,15,20,29,21,32,9], as well as the stream ciphers Grain, Mickey, and Trivium [22,2,16]. Particular interest in lightweight symmetric ciphers is coming from industry, as becoming evident in the adoption of CLEFIA and PRESENT in the ISO/IEC Standard 29192-2. The dominant metric according to which the majority of lightweight ciphers have been optimized is chip area, typically measured in gate equivalences (GE), i.e., the cipher area normalized to the area of a 2-input NAND gate in a given standard cell library. This is certainly a valid optimization objective in cases where there are extremely tight power or cost constraints, in particular passive RFID tags. However, depending on the application, there are several other implementation parameters according to which a cipher should have lightweight characteristics. There are several important applications for which a low-latency encryption and instant response time is highly desirable, such as instant authentication or block-wise read/write access to memory devices, e.g., solid-state hard disks. There are also embedded applications where current block ciphers in multiple-clock architectures could be sufficiently fast, but the needed high clock rates are not supported by the system. For instance, in many FPGA designs clock rates above 200 MHz are often difficult to realize. It can also be anticipated that given the ongoing growth of pervasive computing, there will be many more future embedded systems that require low-latency encryption, especially applications with real-time requirements, e.g., in the automotive domain. Moreover, [27] as well as [24] show that low-latency goes hand in hand with energy efficiency, another crucial criterion in many (other) applications.

For all these cases, we like to have symmetric ciphers that can *instantaneously* encrypt a given plaintext, i.e., the entire encryption *and* decryption should take place within the shortest possible delay. This seemingly simple problem poses a considerable challenge with today's cryptosystems — in particular if encryption and decryption should both be available on a given platform. Software implementations of virtually all strong ciphers take hundreds or thousands of clock cycles, making them ill suited for a designer aiming for low-latency cryptography. In the case of stream ciphers implemented in hardware, the high number of clock cycles for the initialization phase makes them not suitable for this task, especially when secret keys need to be regularly changed. Moreover, if we want to encrypt small blocks selected at random (e.g., encryption of sectors on solid-state disks), stream ciphers are not suited[1]. This leaves block ciphers as the remaining viable solution. However, the round-based, i.e., iterative, nature of virtually all existing block ciphers, as shown for the case of AES, makes low-latency implementation a non-trivial task. A round-based hardware architecture of the AES-128 requires ten clock cycles to output a ciphertext which we do not consider instantaneous as it is still too long for some applications. As a remedy, the ten rounds can be loop-unrolled, i.e., the circuit that realizes the single round is repeated ten times. Now, the cipher returns a ciphertext within a single clock cycle — but at the cost of a very long critical path. This yields a very slow absolute response time

---

[1] A possible exception are random-access stream ciphers such as Salsa [5].

and clock frequencies, e.g., in the range of a few MHz. Furthermore, the unrolled architecture has a high gate count in the range of several tens of thousand GE, implying a high power consumption and costs. Both features are undesirable, especially if one considers that many of the applications for instantaneous ciphers are in the embedded domain. Following the same motivation and reasoning as above [27] compares several lightweight ciphers with respect to latency and as a conclusion calls for new designs that are optimized for low-latency.

Our Contribution. Based on the above discussion our goal is to design a new block cipher which is optimized with respect to the following criteria if implemented in hardware:

1. The cipher can perform instantaneous encryption, a ciphertext is computed within a single clock cycle. There is no warm-up phase.
2. If implemented in modern chip technology, low delays resulting in moderately high clock rates can be achieved.
3. The hardware costs are moderate (i.e., considerably lower than fully unrolled versions of AES or PRESENT).
4. Encryption and decryption should both be possible with low costs and overhead.

We would like to remark that existing lightweight ciphers such as PRESENT do *not* fulfill Criteria 2 and 3 (low delay, small area) due to their large number of rounds. In order to fulfill Criterion 4, one needs to design a cipher for which decryption and encryption use (almost) identical pieces of hardware. This is an important requirement since the unrolled nature of instantaneous ciphers leads to circuits which are large and it is thus clearly advantageous if large parts of the implementation can be used both for encryption and decryption.

Besides designing a new lightweight cipher that is for the first time optimized with respect to the goals above, PRINCE has several innovative features that we like to highlight.

First, a fully unrolled design increases the possible design choices enormously. With a fully unrolled cipher, the traditional need that a cipher has to be iterative with very similar round functions disappears. This in turn allows us to efficiently implement a cipher where decryption with one key corresponds to encryption with a related key. This property we refer to as α-reflection is of independent interest and we prove its soundness against generic attacks. As a consequence, the overhead of implementing decryption over encryption becomes negligible. Note that previous approaches to minimizing the overhead of decryption over encryption, for example in the ciphers NOEKEON and ICEBERG usually require multiplexer in each round. While for a round-based implementation this does not make a difference, our approach is clearly preferable for a fully unrolled implementation, as we require multiplexer only once at the beginning of the circuit.

Another difference to known lightweight ciphers like PRESENT is that we balance the cost of an Sbox-layer and the linear layer. As it turns out optimizing the cost of the Sbox chosen has a major influence on the overall cost of the cipher. As an Sbox that performs well in one technology does not necessarily perform well in another technology, we propose the PRINCE-family of ciphers that allows to freely choose the Sbox within a (large) set of Sboxes fulfilling certain criteria. Our choice for the linear layer can be seen as being inbetween a bit-permutation layer PRESENT (implemented with wires only) and AES (implemented with considerable combinatorial logic). With the expense of only 2 additional XOR-gates per bit over a simple bit-permutation layer, we achieve an almost-MDS property that helps to prove much better bounds against various classes of attacks and in turn allows to significantly reduce the number of rounds and hence latency.

As a result, PRINCE compares very favorable to existing ciphers. For the same time constraints and technologies, PRINCE uses 6-7 times less area than PRESENT-80 and 14-15 times less area than AES-128. In addition to this, our design uses about 4-5 times less area than other ciphers in the literature (see Section 5 and in particular Tables 1 and 2 for a detailed comparison and technology details). To facilitate further study and fairer comparisons, we also report synthesis results using the open-source standard-cell library NANGATE [30]. We also like to mention that, although this is not the main objective of the cipher, PRINCE compares reasonably well to other lightweight ciphers when implemented in a round-based fashion (see [10]).

We believe that our consideration can be of major value for industry and can at the same time stimulate the scientific community to pursue research on lightweight ciphers with different optimization goals.

Organization of the Paper. We introduce an instance of PRINCE-family of ciphers and state our security claims in Section 2. Design decisions are discussed in Section 3 where we also describe the entire PRINCE-family. We provide security proofs and evaluations considering cryptanalytical attacks in Section 4. In Section 5 we finally present implementation results and comparisons with other lightweight ciphers for a range of hardware technologies. For further details, including a detailed security analysis against standard attacks as well as test vectors, we refer to [10].

2   Cipher Description

PRINCE is a 64-bit block cipher with a 128-bit key. The key is split into two parts of 64 bits each,

$$k = k_0 || k_1$$

and extended to 192 bits by the mapping

$$(k_0 || k_1) \rightarrow (k_0 || k_0' || k_1) := (k_0 || (k_0 \ggg 1) \oplus (k_0 \gg 63) || k_1).$$

PRINCE is based on the so-called FX construction [7,25]: the first two subkeys k_0 and k'_0 are used as whitening keys, while the key k_1 is the 64-bit key for a 12-round block cipher we refer to as PRINCE_{core}. We provide test vectors in the full version of the paper [10].

Specification of PRINCE_{core}.

The whole encryption process of PRINCE_{core} is depicted below.

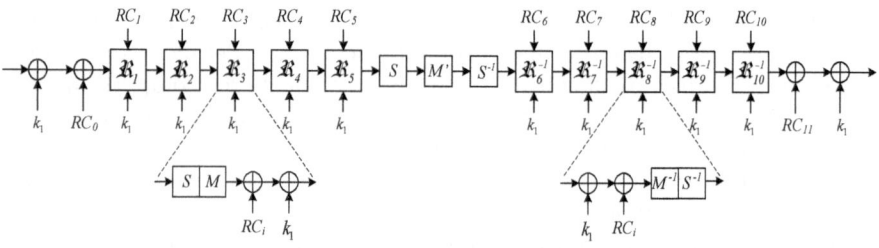

Each round of PRINCE_{core} consist of a key addition, an Sbox-layer, a linear layer, and the addition of a round constant.

k_i-add. Here the 64-bit state is xored with the 64-bit subkey.

S-Layer. The cipher uses one 4-bit Sbox. The action of the Sbox in hexadecimal notation is given by the following table.

x	0	1	2	3	4	5	6	7	8	9	A	B	C	D	E	F
$S[x]$	B	F	3	2	A	C	9	1	6	7	8	0	E	5	D	4

The Matrices: M/M'-layer. In the M and M'-layer the 64-bit state is multiplied with a 64×64 matrix M (resp. M') defined in Section 3.3.

RC_i-add. In the RC_i-add step a 64-bit round constant is xored with the state. We define the constants used below (in hex notation)

RC_0	0000000000000000
RC_1	13198a2e03707344
RC_2	a4093822299f31d0
RC_3	082efa98ec4e6c89
RC_4	452821e638d01377
RC_5	be5466cf34e90c6c
RC_6	7ef84f78fd955cb1
RC_7	85840851f1ac43aa
RC_8	c882d32f25323c54
RC_9	64a51195e0e3610d
RC_{10}	d3b5a399ca0c2399
RC_{11}	c0ac29b7c97c50dd

Note that, for all $0 \le i \le 11$, $RC_i \oplus RC_{11-i}$ is the constant $\alpha = \text{c0ac29b7c97c50dd}$, $RC_0 = 0$ and that $RC_1, \ldots, RC_5$ and α are derived from the fraction part of $\pi = 3.141\ldots$.

From the fact that the round constants satisfy $RC_i \oplus RC_{11-i} = \alpha$ and that M' is an involution, we deduce that the core cipher is such that the inverse of PRINCE$_{core}$ parametrized with k is equal to PRINCE$_{core}$ parametrized with $(k \oplus \alpha)$. We call this property of PRINCE$_{core}$ the α-*reflection property*. It follows that, for any expanded key $(k_0||k_0'||k_1)$,

$$D_{(k_0||k_0'||k_1)}(\cdot) = E_{(k_0'||k_0||k_1 \oplus \alpha)}(\cdot)$$

where α is the 64-bit constant $\alpha = \text{c0ac29b7c97c50dd}$. Thus, for decryption one only has to do a very cheap change to the master key and afterwards reuse the exact same circuit.

Security Claims. For an adversary that is able to acquire 2^n plaintext/ciphertext pairs in a model with a single fixed unknown key k, we claim that the effort to find the key is not significantly less expensive than 2^{127-n} calls to the encryption or decryption function. In Section 4.1 we give a bound matching this claim in the ideal cipher model that does consider the special relation between the encryption and decryption operations. One way to interpret this is that any attack violating our security claim will have to use more properties of the cipher than the relation between the encryption and decryption operations.

We explicitly state that we do not have claims in related-key or known- and chosen-key models as we do not consider them to be relevant for the intended use cases. In particular, as for any cipher based on the FX construction or on the Even-Mansour scheme [18], there exists a trivial distinguisher for PRINCE in the related-key model: for any difference Δ, the ciphertexts corresponding to m and $(m \oplus \Delta)$ encrypted under keys $(k_0||k_1)$ and $((k_0 \oplus \Delta)||k_1)$ respectively, differ from $((\Delta \ggg 1) \oplus (\Delta \gg 63))$ with probability 1.

Reduced Versions. Many classes of cryptanalytic attacks become more difficult with an increased number of rounds. In order to facilitate third-party cryptanalysis and estimate the security margin, reduced-round variants need to be considered. We encourage to study round-reduced variants of PRINCE where the symmetry around the middle is kept, and rounds are added in an inside-out fashion, i.e. for every additional round $\mathfrak{R}_i$ its inverse is also added. Another natural way to reduce PRINCE is to consider the cipher without the key whitening layer, PRINCE$_{core}$.

3   Design Decisions

In this section we explain our design decisions. First note that an SP-network is preferable over a Feistel-cipher, since a Feistel-cipher operates only on half the state resulting often in a higher number of rounds. In order to minimize the number of rounds and still achieve security against linear and differential attacks, we adopted the wide-trail strategy [11]. As not all round functions have to be identical for a cipher aiming for a fully unrolled implementation as PRINCE, it is very tempting to directly use the concept of code-concatenation [13] to achieve a high number of active Sboxes over 4 rounds of the cipher. However, not only a serial implementation benefits from similar round functions. It is also very helpful for ensuring a minimum number of active Sboxes. Assume that, using the code-concatenation approach, one can ensure that rounds R_i to R_{i+3} have at least 16 active Sboxes. While this is nice, the problem is that it does not ensure that rounds R_{i-1} to R_{i+2} or R_{i+1} to R_{i+4} have 16 active Sboxes as well if the individual rounds are very different in nature. We therefore decided to follow a design that on one hand allows to use the freedom given by a fully enrolled design and on the other hand still keeps the round functions similar enough to prove some bounds on the resistance against linear and differential attacks.

In this context, one of the main features of the design is that decryption can be implemented on top of encryption with a minimal overhead. This is achieved by designing a cipher which is symmetric around the middle round, a very simple key scheduling, and a special choice of round constants.

3.1   Aligning Encryption with Decryption

The use of a core cipher having the α-reflection property, with two additional whitening keys, offers a nice alternative to the usual design strategy which consists in using involutional components — Noekeon [12], Khazad [4], Anubis [3], Iceberg [35] or SEA [34] are some examples of such ciphers with involutional components. Actually, the general construction used in PRINCE has the following advantages:

- It allows a much larger choice of Sboxes, which may lead to a lower implementation cost, since the Sbox is not required to be an involution. It is worth noticing that the fact that both the Sbox and its inverse are involved in the encryption function does not affect the cost of the fully-unrolled implementations we consider;

- In ciphers with involutional components, the overhead due to the implementation of the inverse key scheduling can be reduced by adding some symmetry in the subkey sequence. But this may introduce weak keys or potential slide attacks. The fact that all components are involutions may also introduce some regularities in the cyclic structure of the cipher which can be exploited in some attacks [6]. The resistance of PRINCE to this type of attacks will be extensively discussed in Section 4.2.
- It is an open problem to prove the security of ciphers with ideal, involutional components against generic attacks. We show in Section 4.1 that ciphers with the α-reflection property (for $\alpha \neq 0$) has a proof of security similar to that of the FX construction.
- Previous approaches to minimizing the overhead of decryption over encryption usually require multiplexer in each round while our approach requires multiplexer only once at the beginning of the circuit.

3.2   The PRINCE-Family: Choosing the Sbox

As discussed in more detail in Section 5, the cost of the Sbox, i.e., its area and critical path, is a substantial part of the overall cost. Thus, choosing an Sbox which minimizes those costs is crucial for obtaining competitive results. As the cost of an Sbox depends on various parameters, such as the technology, the synthesis tool, and the library used, one cannot expect that there is one optimal Sbox for all environments. In fact, in order to achieve optimal results it is preferable to choose your favorite Sbox. In order to ensure the security of the resulting design, an Sbox $S : \mathbb{F}_2^4 \to \mathbb{F}_2^4$ for the PRINCE-Family has to fulfill the following criteria.

1. The maximal probability of a differential is $1/4$
2. There are exactly 15 differentials with probability $1/4$.
3. The maximal absolute bias of a linear approximation is $1/4$.
4. There are exactly 30 linear approximations with absolute bias $1/4$.
5. Each of the 15 non-zero component functions has algebraic degree 3.

As it can be deduced for example from [28] up to affine equivalence there are only 8 Sboxes fulfilling those criteria. Thus, another way of defining an Sbox for the PRINCE-Family is to say that it has to be affine equivalent to one of the eight Sboxes S_i given in the full version of this paper [10].

3.3   The Linear Layer

In the M and M'-layer the 64-bit state is multiplied with a 64×64 matrix M (resp. M') defined below. We have different requirements for the two different linear layers. The M'-layer is only used in the middle round, thus M' has to be an involution to ensure the α-reflection property. This requirement does not apply for the M-layer used in the round functions. Here we want to ensure full diffusion after two rounds. To achieve this we combine the M'-mapping with an application of matrix SR which behaves like the AES shift rows and permutes the 16 nibbles in the following way.

$$\boxed{0\,1\,2\,3\,4\,5\,6\,7\,8\,9\,10\,11\,12\,13\,14\,15} \longrightarrow \boxed{0\,5\,10\,15\,4\,9\,14\,3\,8\,13\,2\,7\,12\,1\,6\,11}$$

Thus $M = SR \circ M'$.

Additionally the implementation costs should be minimized, meaning that the number of ones in the matrices M' and M should be minimal, while at the same time it should be guaranteed that at least 16 Sboxes are active in 4 consecutive rounds (see full version [10] for details). Thus, trivially each output bit of an Sbox has to influence 3 Sboxes in the next round and therefore the minimum number of ones per row and column is 3. Thus we can use the following four 4×4 matrices as building blocks for the M'-layer.

$$M_0 = \begin{pmatrix} 0000 \\ 0100 \\ 0010 \\ 0001 \end{pmatrix}, \ M_1 = \begin{pmatrix} 1000 \\ 0000 \\ 0010 \\ 0001 \end{pmatrix}, \ M_2 = \begin{pmatrix} 1000 \\ 0100 \\ 0000 \\ 0001 \end{pmatrix}, \ M_3 = \begin{pmatrix} 1000 \\ 0100 \\ 0010 \\ 0000 \end{pmatrix}$$

In the next step we generate a 4×4 block matrix $\hat{M}$ where each row and column is a permutation of the four 4×4 matrices $M_0, \ldots, M_3$. The row permutations are chosen such that we obtain a symmetric block matrix. The choice of the building blocks and the symmetric structure ensures that the resulting 16×16 matrix is an involution. We define

$$\hat{M}^{(0)} = \begin{pmatrix} M_0 \ M_1 \ M_2 \ M_3 \\ M_1 \ M_2 \ M_3 \ M_0 \\ M_2 \ M_3 \ M_0 \ M_1 \\ M_3 \ M_0 \ M_1 \ M_2 \end{pmatrix} \quad \hat{M}^{(1)} = \begin{pmatrix} M_1 \ M_2 \ M_3 \ M_0 \\ M_2 \ M_3 \ M_0 \ M_1 \\ M_3 \ M_0 \ M_1 \ M_2 \\ M_0 \ M_1 \ M_2 \ M_3 \end{pmatrix}.$$

In order to obtain a permutation for the full 64-bit state we construct a 64×64 block diagonal matrix M' with $(\hat{M}^{(0)}, \hat{M}^{(1)}, \hat{M}^{(1)}, \hat{M}^{(0)})$ as diagonal blocks. The matrix M' is an involution with 2^{32} fixed points, which is average for a randomly chosen involution [19, Page 596]. The linear layer M is not an involution anymore due to the composition of M' and shift rows, which is not an involution.

3.4   The Key Expansion

The 128-bit key $(k_0 \| k_1)$ is extended to a 192-bit key $(k_0 \| k_0' \| k_1)$ by a linear mapping of the form

$$(k_0 \| k_1) \mapsto (k_0 \| P(k_0) \| k_1) .$$

This expansion should be such that it makes peeling of rounds (both at the beginning and at the end) by partial key guessing difficult for the attacker. In particular, we would like that each pair of subkeys among k_1 and the quantities $(k_0 \oplus k_1)$ and $(k_0' \oplus k_1)$ takes all the 2^{128} possible values when $(k_0 \| k_1)$ varies in the set of 128-bit words. In other words, the set of all triples $(k_0 \| P(k_0) \| k_1)$ should correspond to an MDS code of length 3 and size 2^{128} over $\mathbf{F}_2^{64}$. This equivalently means that both $x \mapsto P(x)$ and $x \mapsto x \oplus P(x)$ should be permutations of $\mathbf{F}_2^{64}$. Note that no bit-permutation P satisfies this condition. Indeed, both the all-zero vector and the all-one vector satisfy $P(x) \oplus x = 0$.

Thus, a hardware-optimal choice for P such that both P and $P \oplus \mathrm{Id}$ are permutations is

$$P(x) = (x \ggg 1) \oplus (x \gg 63) ,$$

i.e., $P(x_{63}, \ldots, x_0) = (x_0, x_{63}, \ldots, x_2, x_1 \oplus x_{63})$. Then, we can easily check that $P(x) = 0$ (resp. $P(x) = x$) has a unique solution.

4   Security Analysis

This section investigates the security of the general construction of PRINCE. In particular, we show that the α-reflection property of the core cipher does not introduce any generic attack with complexity significantly lower than the known generic attacks against the FX construction. However, in the particular case of PRINCE$_{core}$, the α-reflection property comes from some symmetries in the construction, including the use of an involution as middle round. Thus, we investigate in Section 4.2 whether weaknesses similar to those identified for involutional ciphers could also appear in the case of PRINCE. An evaluation of the security of PRINCE regarding more classical attacks, including linear, differential and algebraic but also to the recently introduced biclique attacks is provided in the full version [10].

4.1   On Generic Attacks: Security Proof

The FX construction, introduced by Rivest for increasing the resistance of DES to exhaustive key-search [7], consists in deriving a block cipher E with $(2n + \kappa)$-bit key and n-bit block from a block cipher F with κ-bit key and n-bit block by xoring the input and output of F with a pre-whitening key and a post-whitening key:

$$E_{k_0,k_1,k_2}(x) = F_{k_1}(x \oplus k_0) \oplus k_2 .$$

Kilian and Rogaway [25,26] proved that, if the core cipher F is ideal, then this construction achieves $(\kappa + n - 1 - \log T)$-bit security where T is the number of pairs of inputs and outputs for F known by the attacker. This result obviously does not apply in the case of PRINCE since the core cipher F in PRINCE can be easily distinguished from a family of random permutations due to the α-reflection property, i.e., $F_k^{-1} = F_{k \oplus \alpha}$ for any k. Here, we want to quantify the impact of this property on the generic attacks against the FX construction. For instance, it appears that a decryption oracle also gives a related-key oracle with the fixed-key relation $(k_0, k_2, k_1) \rightarrow (k_2, k_0, k_1 \oplus \alpha)$ and it is important to determine whether an adversary can profit from this relation.

A similar question was investigated by Kilian and Rogaway for showing that the complementation property of DES decreases the security level by a single bit [25, Section 4]. In the case of the α-reflection property, we like to model the core cipher F as an ideal cipher, that is as a set of random permutations, with the (only!) additional relation that $F_{k \oplus \alpha}(x) = F_k^{-1}(x)$. Informally, this can be seen as picking only half of the 2^κ permutations independently at random, while the second half is defined by the encryption vs decryption relation above.

More precisely, we consider for F a keyed permutation with a $(\kappa - 1)$-bit key, operating on n-bit blocks. Let α be a nonzero element in $\mathbf{F}_2^\kappa$. We decompose the set of κ-bit words into two subsets as $\mathbf{F}_2^\kappa = H \cup (\alpha \oplus H)$ where H is some linear subspace of dimension $(\kappa - 1)$ which does not contain α, e.g., if $\mathrm{lsb}(\alpha) = 1$, H is the set of all n-bit words x with $\mathrm{lsb}(x) = 0$. In the following, H is identified with the set of $(\kappa - 1)$-bit words. It is worth noticing that such a decomposition does not exist when $\alpha = 0$, i.e., when F is an involution. Therefore, the following construction is defined for $\alpha \neq 0$ only. Now, we derive from F a block cipher with $(2n + \kappa)$ key bits and n-bit blocks:

$$E_{k_0,k_1,k_2}(m) = \begin{cases} F_{k_1}(m \oplus k_0) \oplus k_2 & \text{if } k_1 \in H \\ F_{k_1 \oplus \alpha}^{-1}(m \oplus k_0) \oplus k_2 & \text{if } k_1 \in (\alpha \oplus H) \end{cases}$$

This construction, we refer to as $\tilde{F}X$-construction, corresponds to the FX construction applied to $\tilde{F}$ where $\tilde{F}$ is the family of 2^κ permutations defined by

$$\tilde{F}_k(x) = \begin{cases} F_k(x) & \text{if } k \in H \\ F_{k \oplus \alpha}^{-1}(x) & \text{if } k \in (\alpha \oplus H) \end{cases}$$

The only difference with the construction considered in the case of the complementation property is that F is extended by using the inverse permutations $F_k, k \in H$, instead of the permutations themselves. But, we can obtain a similar result.

More precisely, when analyzing the original FX construction, Kilian and Rogaway [25] consider the following problem. Let $\mathcal{A}$ be an adversary with access to three oracles: E, F and F^{-1}. During the game, the adversary may make queries to E, to F and F^{-1}. Any query to the F/F^{-1} oracle consists of a pair (k, x) in $\mathbf{F}_2^\kappa \times \mathbf{F}_2^n$ and the oracle returns an element in $\mathbf{F}_2^n$. A query to the E oracle consists of an n-bit element, and an n-bit value is returned. The aim of this adversary is then to guess whether the E oracle computes FX_k for some random key k, or if it computes π for a random permutation of $\mathbf{F}_2^n$. Then, a game-hoping argument leads to the following upper-bound on the advantage of any such adversary.

Theorem 1. *[25] The advantage of any adversary who makes D queries to the E oracle and T queries to the F/F^{-1} oracle satisfies*

$$\mathsf{Adv}_{FX}^{\mathsf{CPA}}(\mathcal{A}) = \Big| \Pr[k \xleftarrow{\$} \mathbf{F}_2^{\kappa+2n}, F \xleftarrow{\$} (\mathcal{P}_n)^{2^\kappa} \; : \; \mathcal{A}^{FX_k, F, F^{-1}} = 1]$$

$$- \Pr[\pi \xleftarrow{\$} \mathcal{P}_n, F \xleftarrow{\$} (\mathcal{P}_n)^{2^\kappa} \; : \; \mathcal{A}^{\pi, F, F^{-1}} = 1] \Big| \leq DT2^{-(n+\kappa-1)},$$

where $x \xleftarrow{\$} S$ means that x is uniformly chosen at random from a set S, $\mathcal{P}_n$ denotes the set of permutations of $\mathbf{F}_2^n$ and $F \xleftarrow{\$} (\mathcal{P}_n)^{2^\kappa}$ means that F is a family of 2^κ independently chosen random permutations.

We deduce a similar result for the $\tilde{F}X$ construction.

Corollary 1. *The advantage of any adversary who makes D queries to the E oracle and T queries to the F/F^{-1} oracle satisfies*

$$\mathsf{Adv}^{\mathsf{CPA}}_{\widetilde{F}X}(\mathcal{A}) = \Big| \Pr[k \xleftarrow{\$} \mathbf{F}_2^{\kappa+2n}, F \xleftarrow{\$} (\mathcal{P}_n)^{2^{\kappa-1}} : \mathcal{A}^{\widetilde{F}X_k,F,F^{-1}} = 1]$$
$$- \Pr[\pi \xleftarrow{\$} \mathcal{P}_n, F \xleftarrow{\$} (\mathcal{P}_n)^{2^{\kappa-1}} : \mathcal{A}^{\pi,F,F^{-1}} = 1]\Big| \le DT2^{-(n+\kappa-2)}$$

Proof. We decompose

$$P_c = \Pr[k \xleftarrow{\$} \mathbf{F}_2^{\kappa+2n}, F \xleftarrow{\$} (\mathcal{P}_n)^{2^{\kappa-1}} : \mathcal{A}^{\widetilde{F}X_k,F,F^{-1}} = 1]$$
$$= \Pr[k_0, k_2 \xleftarrow{\$} \mathbf{F}_2^n, k_1 \xleftarrow{\$} H, F \xleftarrow{\$} (\mathcal{P}_n)^{2^{\kappa-1}} : \mathcal{A}^{\widetilde{F}X_{k_0,k_1,k_2},F,F^{-1}} = 1]$$
$$\times \Pr[k_1 \in H]$$
$$+ \Pr[k_0, k_2 \xleftarrow{\$} \mathbf{F}_2^n, k_1 \xleftarrow{\$} \alpha \oplus H, F \xleftarrow{\$} (\mathcal{P}_n)^{2^{\kappa-1}} : \mathcal{A}^{\widetilde{F}X_{k_0,k_1,k_2},F,F^{-1}} = 1]$$
$$\times \Pr[k_1 \in \alpha \oplus H]$$
$$= \frac{1}{2}\Pr[k_0, k_2 \xleftarrow{\$} \mathbf{F}_2^n, k_1 \xleftarrow{\$} H, F \xleftarrow{\$} (\mathcal{P}_n)^{2^{\kappa-1}} : \mathcal{A}^{FX_{k_0,k_1,k_2},F,F^{-1}} = 1]$$
$$+ \frac{1}{2}\Pr[k_0, k_2 \xleftarrow{\$} \mathbf{F}_2^n, k_1 \xleftarrow{\$} H, F \xleftarrow{\$} (\mathcal{P}_n)^{2^{\kappa-1}} : \mathcal{A}^{F^{-1}X_{k_0,k_1,k_2},F,F^{-1}} = 1] ,$$

since

$$\widetilde{F}X_{k_0,k_1,k_2}(x) = \begin{cases} FX_{k_0,k_1,k_2}(x) & \text{if } k_1 \in H \\ F^{-1}X_{k_0,k_1\oplus\alpha,k_2}(x) & \text{if } k_1 \in \alpha \oplus H . \end{cases}$$

Obviously,

$$\Pr[\mathcal{A}^{F^{-1}X_{k_0,k_1,k_2},F,F^{-1}} = 1] = \Pr[\mathcal{A}^{FX_{k_0,k_1,k_2},F,F^{-1}} = 1]$$

leading to

$$P_c = \Pr[k_0, k_2 \xleftarrow{\$} \mathbf{F}_2^n, k_1 \xleftarrow{\$} H, F \xleftarrow{\$} (\mathcal{P}_n^{2^{\kappa-1}}) : \mathcal{A}^{FX_{k_0,k_1,k_2},F,F^{-1}} = 1] .$$

It directly follows from Theorem 1 that

$$\mathsf{Adv}^{\mathsf{CPA}}_{\widetilde{F}X}(\mathcal{A}) = \mathsf{Adv}^{\mathsf{CPA}}_{FX}(\mathcal{A}) \le DT2^{-(n+\kappa-2)} .$$

$\square$

As noticed in [25], this bound is still valid in a chosen-ciphertext scenario; it can also be extended to the case where the whitening keys are related, for instance if $k_2 = k_0$ or $k_2 = P(k_0)$ as in PRINCE. Both generalizations apply to the $\widetilde{F}X$ construction as well.

The bound obtained for the FX construction is achieved, for instance by the slide attack due to Biryukov and Wagner [8] and by its recent generalization named slidex [17]. A chosen-plaintext variant of this attack allows to exploit the α-reflection property for reducing the security level by one bit, compared to the original FX construction. This attack, detailed in the full version, has an average time complexity corresponding to $2^{\kappa+n-\log_2 D}$ computations of the core cipher F for any number D of pairs of chosen plaintexts-ciphertexts.

4.2   Impact of the Construction Implementing the α-Reflection Property

As mentioned earlier, one particular feature of PRINCE is the α-reflection property of PRINCE$_{core}$. But, not surprisingly, the construction we used for obtaining this feature also has structural properties, including an involutional middle round, and care has to be taken when designing a cipher with such a structure. In this section we analyse the influence of this construction on the security of the cipher. In particular, we are interested in the so-called profile of the core cipher, *i.e.*, in the sequence of the lengths of all cycles in the decomposition of PRINCE$_{core}$.

A first strategy for exploiting some information on the profile of the core cipher is the following. If the decomposition of the core cipher is independent from the key, then this decomposition can be used as a distinguishing property for recovering some information on the whitening keys. The simplest illustration of this type of attack is when the core cipher is an involution, *i.e.* when $\alpha = 0$ which is the only case where Corollary 1 does not apply. Indeed, the attack presented by Dunkelman *et al.* [17, Section 5.2] allows to recover the sum of the two whitening keys ($k_0 \oplus k_2$) in the FX construction when F is an involution. This attack uses the fact that for two plaintext-ciphertext pairs (m, c) and (m', c') related by $m' = E^{-1}_{k_0, k_1, k_2}(m \oplus k_0 \oplus k_2)$ it holds that $m \oplus c = m' \oplus c'$. Indeed,

$$
\begin{aligned}
m' \oplus c' &= E^{-1}_{k_0, k_1, k_2}(m \oplus k_0 \oplus k_2) \oplus m \oplus k_0 \oplus k_2 \\
&= k_0 \oplus F^{-1}_{k_1}(m \oplus k_0) \oplus m \oplus k_0 \oplus k_2 = F_{k_1}(m \oplus k_0) \oplus m \oplus k_2 \\
&= m \oplus c
\end{aligned}
$$

where the last-but-one equality uses that F_{k_1} is an involution. Thus, plaintext-ciphertext pairs (m, c) and (m', c') such that $c' = m \oplus k_0 \oplus k_2$ can be easily detected. Such a collision can be found if the attacker has access to $2^{\frac{n+1}{2}}$ known plaintext-ciphertext pairs, and it provides the value of $(k_0 \oplus k_2)$. Moreover, in the particular case of PRINCE, k_2 is related to k_0 by $k_2 = P(k_0)$ where $x \mapsto x \oplus P(x)$ is a permutation (see Section 3.4). Therefore, the whitening key k_0 can be deduced from $(k_0 \oplus k_2)$ in this case. It follows that, when the core cipher is an involution, the whole key can then be recovered with time complexity 2^κ (corresponding to an exhaustive search for k_1) and data complexity $2^{\frac{n+1}{2}}$. This confirms that Corollary 1 does not hold for $\alpha = 0$.

This type of attack can be generalized to the case where the profile of the core cipher does not depend on k_1: since PRINCE$_{core}$ has a reasonable block size, its cycle structure could be precomputed and then used as a distinguishing property for $(k_0 \oplus k_2)$. Indeed, the profile of $E_{k_0, k_1, k_2} : m \mapsto k_2 \oplus F_{k_1}(m \oplus k_0)$ depends on $(k_0 \oplus k_2)$ only. It follows that, for each n-bit word δ, we could compute one or a few cycles of $x \mapsto F_{k_1}(x \oplus k_0 \oplus k_2 \oplus \delta)$ in a chosen-plaintext scenario where the attacker knows a sequence of plaintext-ciphertext pairs (m_i, c_i) with $m_{i+1} = c_i \oplus \delta$. A valid candidate for $(k_0 \oplus k_2)$ is a value δ which leads to a cycle having a length which appears in the precomputed profile of F_{k_1}.

We checked whether the cycle structure of PRINCE_{core} has some peculiarities which do not depend on its key. Based on the technique used by Biryukov for analyzing involutional ciphers [6], we can observe the profile of the reduced version of PRINCE_{core} with 4 Sbox layers where we keep the symmetry around the middle does not depend on the key. Actually, this reduced version can be written as

$$G = \left(R_5^{-1} \circ \text{Add}_{k_1 \oplus \alpha}\right) \circ \left(S^{-1} \circ M' \circ S\right) \circ \left(\text{Add}_{k_1} \circ R_5\right)$$

where R_5 corresponds to $\Re_5$ without the key addition. Since $S^{-1} \circ M' \circ S$ is an involution, the cycle structure of $\text{Add}_{k_1 \oplus \alpha} \circ \left(S^{-1} \circ M' \circ S\right) \circ \text{Add}_{k_1}$ depends on α only and not on k_1. Its profile then remains unchanged after a right composition with R_5 and a left composition with its inverse. However, this property does not hold anymore when an additional round is included since the next key addition $\text{Add}_{k_1 \oplus \alpha} \circ G \circ \text{Add}_{k_1}$ modifies the cycle structure of G in a way which depends on the values G, and not only on its profile. Therefore, it appears that the previously mentioned attack strategy does not apply if PRINCE_{core} contains more than 6 Sbox layers.

In the light of the previous analysis, a more relevant attack method consists in using the fact that the core cipher may have a peculiar cycle decomposition for some weak keys. For instance, if there exists some weak keys k_1 for which PRINCE_{core} is an involution, then this class of keys can be detected from the knowledge of $2^{\frac{n+1}{2}}$ pairs of plaintext-ciphertext by counting the number of collisions for $m \oplus c$. And the technique from [17] that we have previously described also recovers the whitening key. It is worth noticing that this attack applies to DESX and allows to detect the use of the four weak keys of DES [14] for which DES is an involution. A similar weakness would appear if, in PRINCE_{core}, we have used two subkeys k_1 and k_1' in turn as round keys. Keeping the remaining structure of PRINCE_{core} results in the following relation

$$F_{(k_1||k_1')}^{-1} = F_{(k_1' \oplus \alpha || k_1 \oplus \alpha)} \ .$$

However, this has serious – and interesting – consequences for the security of the resulting cipher. For the class of keys such that $k_1' = k_1 \oplus \alpha$, it holds that

$$F_{(k_1||k_1')}^{-1} = F_{(k_1||k_1')},$$

that is, the core cipher is an involution. This class of weak keys can then be easily detected. It then appears that some particular related-key distinguishers for the core cipher may be exploited for detecting the corresponding class of keys. To be very clear, we do not consider related key-attacks here in the classical sense of enlarging the power of an adversary. But without a careful choice, the construction we used for implementing the α-reflection property might result in key-recovery attacks for certain weak-key classes, as soon as the core cipher is vulnerable to related key-attacks.

5   Implementation

Besides the main target low-latency, low-cost hardware implementation is one of the design objectives of PRINCE. To achieve low-latency, a fully unrolled design should be considered for implementation. During the design process of PRINCE the cost of each function was investigated and each component was carefully designed in order to get the lowest possible gate count without compromising security. One of the most critical and expensive operations of the cipher is the substitution, where we use the same Sbox 16 times (rather than having 16 different Sboxes). Therefore, the implementation of PRINCE started with a search for the most suitable Sbox for the target design specifications. In order to achieve an implementation with low delay and gate count, we analyzed many Sbox instances to identify one with optimal combinational logic and propagation paths. Then, the targeted unrolled design was implemented with the resulting *optimal* Sbox.

In the implementation process, *Cadence NCVerilog 06.20-p001* is used for simulation and *Cadence Encounter RTL Compiler v10.1* for synthesis. Since gate count and delay parameters are heavily technology dependent, the implementations have been synthesized for three different technology libraries: 130 nm and 90 nm low-leakage Faraday libraries from UMC, and 45 nm generic NANGATE Open Cell Library. In all syntheses, typical operating conditions were assumed.

The unrolled version of PRINCE is a direct mapping to hardware of the cipher defined in Section 2. Multiplexers select encryption and decryption keys accordingly. The only costs associated with the key whitening stages are XOR gates and multiplexers used for whitening key selection. However, in practice, due to the unrolled nature of the implementation, these additions reduce to XOR operations with constants, which in turn reduce to inverters or no additional gates at all. Furthermore, these inverters are combined with the preceding or following matrix multiplications, which are implemented with cascaded XOR gates. In cases where an XOR is sourced by the output from an inverter, or is sourcing input of an inverter, it is simply replaced by an XNOR gate and the sourced/sourcing inverter is removed. Since both XOR and XNOR have the same gate count, the overall effect of the round constant addition on area reduces to zero.

The unrolled implementation of PRINCE results are listed in Table 1 for different technologies with respect to different timing constraints. In this table, a unit delay (UD) parameter is used to enable a fair comparison between different technologies. It is the average delay of a single inverter gate (with lowest drive - X1) within a ring oscillator under zero wireload conditions in the target technology (6.7 ps, 31.9 ps, and 43.6 ps for 45 nm, 90 nm, and 130 nm, respectively). We also implemented PRESENT-80, PRESENT-128, LED-128 and AES-128 and applied the same metrics to adequately evaluate the achievements of our new cipher (note that in some cases the key size – and also our security claim – is different: PRINCE does not claim to offer 128-bit security and security against related key-attacks). In order to achieve both encryption and decryption capability in PRESENT and LED, we had to implement both true and inverse Sboxes and select their output by a multiplexer, which doubled the Sbox area with respect to an encryption-only implementation. For AES, we just had to implement the inverse affine transform since the finite field inversion module could be shared between encryption and decryption. In addition to this comparison, Table 2 shows the extrapolated results (which are calculated by removing register and control logic area from the total gate count, and multiplying the rest by the number of rounds) for other unfolded cipher instances obtained from round-based cipher implementations provided by previous works. Note that all ciphers in the table include encryption and decryption functionality with 128-bit key size, however the comparison is difficult as the block size is different in some cases (also note that the ciphers having 128-bit block size are obviously much bigger and more power consuming than a 64-bit block cipher).

We also measured maximum frequencies achievable by unrolled versions of PRINCE under two different conditions: The frequency where the area of synthesized design starts to deviate from the unconstrained area – 158.9, 38.4 and 35.5 MHz, and the frequencey where the timing slack becomes zero – 212.8, 71.8 and 54.3 MHz. Both figures are given for 45 nm, 90 nm, and 130 nm, respectively.

Table 1. Area/power comparison of unrolled versions of PRINCE and other ciphers

	Tech.	Nangate 45nm Generic			UMC 90nm Faraday			UMC 130nm Faraday		
	Constr.(UD)	1000	3162	10000	1000	3162	10000	1000	3162	10000
PRINCE˜	Area(GE)	8260	8263	8263	7996	7996	7996	8679	8679	8679
	Power(mW)	38.5	17.9	8.3	26.3	10.9	3.9	29.8	11.8	4.1
PRESENT-80	Area(GE)	63942	51631	50429	113062	49723	49698	119196	51790	51790
	Power(mW)	1304.6	320.9	98.0	1436.9	144.9	45.5	1578.4	134.9	42.7
PRESENT-128	Area(GE)	68908	56668	55467	120271	54576	54525	126351	56732	56722
	Power(mW)	1327.1	330.4	99.1	1491.1	149.9	47.8	1638.7	137.4	43.6
LED-128	Area(GE)	109811	109958	109697	281240	286779	98100	236770	235106	111496
	Power(mW)	2470.7	835.7	252.3	5405.0	1076.3	133.7	5274.8	1133.9	163.6
AES-128	Area (GE)	135051	135093	118440	421997	130835	118522	347860	141060	130764
	Power (mW)	3265.8	1165.7	301.6	8903.2	587.4	186.8	8911.2	876.8	229.1

Table 2. Extrapolated area of unrolled versions of other ciphers against PRINCE

Technology	Area* (GE)
CLEFIA-128 [1]	28035 (18 rounds unfolded, 130nm CMOS)
HIGHT-128 [23]	42688 (32 rounds unfolded, 250nm CMOS)
mCrypton-128 [29]	37635 (13 rounds unfolded, 130nm CMOS)
Piccolo-128 [32]	25668 (31 rounds unfolded, 130nm CMOS)

* Area requirements extrapolated from round-based implementations.

References

1. Akishita, T., Hiwatari, H.: Very Compact Hardware Implementations of the Block Cipher CLEFIA. In: Miri, A., Vaudenay, S. (eds.) SAC 2011. LNCS, vol. 7118, pp. 278–292. Springer, Heidelberg (2012)
2. Babbage, S., Dodd, M.: The MICKEY Stream Ciphers. In: Robshaw, Billet [31], pp. 191–209

3. Barreto, P.S.L.M., Rijmen, V.: The ANUBIS Block Cipher. Submission to the NESSIE project (2000), http://www.larc.usp.br/~pbarreto/AnubisPage.html
4. Barreto, P.S.L.M., Rijmen, V.: The Khazad Legacy-level Block Cipher. Submission to the NESSIE project (2000), http://www.larc.usp.br/~pbarreto/KhazadPage.html
5. Bernstein, D.J.: The Salsa20 Family of Stream Ciphers. In: Robshaw, Billet [31], pp. 84–97
6. Biryukov, A.: Analysis of Involutional Ciphers: Khazad and Anubis. In: Johansson, T. (ed.) FSE 2003. LNCS, vol. 2887, pp. 45–53. Springer, Heidelberg (2003)
7. Biryukov, A.: DES-X (or DESX). In: Encyclopedia of Cryptography and Security, 2nd edn., p. 331. Springer (2011)
8. Biryukov, A., Wagner, D.: Advanced Slide Attacks. In: Preneel, B. (ed.) EURO-CRYPT 2000. LNCS, vol. 1807, pp. 589–606. Springer, Heidelberg (2000)
9. Bogdanov, A., Knudsen, L.R., Leander, G., Paar, C., Poschmann, A., Robshaw, M., Seurin, Y., Vikkelsoe, C.: PRESENT: An Ultra-Lightweight Block Cipher. In: Paillier, P., Verbauwhede, I. (eds.) CHES 2007. LNCS, vol. 4727, pp. 450–466. Springer, Heidelberg (2007)
10. Borghoff, J., Canteaut, A., Güneysu, T., Kavun, E.B., Knezevic, M., Knudsen, L.R., Leander, G., Nikov, V., Paar, C., Rechberger, C., Rombouts, P., Thomsen, S.S., Yalçın, T.: PRINCE – A Low-latency Block Cipher for Pervasive Computing Applications. IACR Cryptology ePrint Archive, 529 (2012)
11. Daemen, J.: Cipher and Hash Function Design, Strategies Based on Linear and Differential Cryptanalysis. PhD thesis, Katholieke Universiteit Leuven (1995)
12. Daemen, J., Peeters, M., Van Assche, G., Rijmen, V.: The NOEKEON Block Cipher. Submission to the NESSIE project (2000), http://gro.noekeon.org/
13. Daemen, J., Rijmen, V.: Codes and Provable Security of Ciphers. In: Enhancing Cryptographic Primitives with Techniques from Error Correcting Codes. NATO Science for Peace and Security Series D - Information and Communication Security 23, vol. 1807, pp. 60–80. IOS Press (2009)
14. Davies, D.W.: Some Regular Properties of the 'Data Encryption Standard' Algorithm. In: Advances in Cryptology, CRYPTO 1982, pp. 89–96. Plenum Press, New York (1982)
15. De Cannière, C., Dunkelman, O., Knežević, M.: KATAN and KTANTAN — A Family of Small and Efficient Hardware-Oriented Block Ciphers. In: Clavier, C., Gaj, K. (eds.) CHES 2009. LNCS, vol. 5747, pp. 272–288. Springer, Heidelberg (2009)
16. De Cannière, C., Preneel, B.: Trivium Specifications. eSTREAM, ECRYPT Stream Cipher Project (2006)
17. Dunkelman, O., Keller, N., Shamir, A.: Minimalism in Cryptography: The Even-Mansour Scheme Revisited. In: Pointcheval, D., Johansson, T. (eds.) EURO-CRYPT 2012. LNCS, vol. 7237, pp. 336–354. Springer, Heidelberg (2012)
18. Even, S., Mansour, Y.: A Construction of a Cipher From a Single Pseudorandom Permutation. In: Matsumoto, T., Imai, H., Rivest, R.L. (eds.) ASIACRYPT 1991. LNCS, vol. 739, pp. 210–224. Springer, Heidelberg (1993)
19. Flajolet, P., Sedgewick, R.: Analytic Combinatorics. Cambridge University Press (2009)
20. Gong, Z., Nikova, S., Law, Y.W.: KLEIN: A New Family of Lightweight Block Ciphers. In: Juels, A., Paar, C. (eds.) RFIDSec 2011. LNCS, vol. 7055, pp. 1–18. Springer, Heidelberg (2012)

21. Guo, J., Peyrin, T., Poschmann, A., Robshaw, M.: The LED Block Cipher. In: Preneel, B., Takagi, T. (eds.) CHES 2011. LNCS, vol. 6917, pp. 326–341. Springer, Heidelberg (2011)

22. Hell, M., Johansson, T., Meier, W.: Grain: A Stream Cipher for Constrained Environments. International Journal of Wireless and Mobile Computing 2(1), 86–93 (2007)

23. Hong, D., Sung, J., Hong, S., Lim, J., Lee, S., Koo, B.-S., Lee, C., Chang, D., Lee, J., Jeong, K., Kim, H., Kim, J.-S., Chee, S.: HIGHT: A New Block Cipher Suitable for Low-Resource Device. In: Goubin, L., Matsui, M. (eds.) CHES 2006. LNCS, vol. 4249, pp. 46–59. Springer, Heidelberg (2006)

24. Kerckhof, S., Durvaux, F., Hocquet, C., Bol, D., Standaert, F.-X.: Towards Green Cryptography: A Comparison of Lightweight Ciphers from the Energy Viewpoint. In: Prouff, E., Schaumont, P. (eds.) CHES 2012. LNCS, vol. 7428, pp. 390–407. Springer, Heidelberg (2012)

25. Kilian, J., Rogaway, P.: How to Protect DES against Exhaustive Key Search. In: Koblitz, N. (ed.) CRYPTO 1996. LNCS, vol. 1109, pp. 252–267. Springer, Heidelberg (1996)

26. Kilian, J., Rogaway, P.: How to Protect DES Against Exhaustive Key Search (An Analysis of DESX). J. Cryptology 14(1), 17–35 (2001)

27. Knežević, M., Nikov, V., Rombouts, P.: Low-Latency Encryption – Is "Lightweight = Light + Wait"? In: Prouff, E., Schaumont, P. (eds.) CHES 2012. LNCS, vol. 7428, pp. 426–446. Springer, Heidelberg (2012)

28. Leander, G., Poschmann, A.: On the Classification of 4 Bit S-Boxes. In: Carlet, C., Sunar, B. (eds.) WAIFI 2007. LNCS, vol. 4547, pp. 159–176. Springer, Heidelberg (2007)

29. Lim, C.H., Korkishko, T.: mCrypton – A Lightweight Block Cipher for Security of Low-Cost RFID Tags and Sensors. In: Song, J.-S., Kwon, T., Yung, M. (eds.) WISA 2005. LNCS, vol. 3786, pp. 243–258. Springer, Heidelberg (2006)

30. NANGATE. The NanGate 45nm Opencell Library, http://www.nangate.com

31. Robshaw, M., Billet, O. (eds.): New Stream Cipher Designs. LNCS, vol. 4986. Springer, Heidelberg (2008)

32. Shibutani, K., Isobe, T., Hiwatari, H., Mitsuda, A., Akishita, T., Shirai, T.: *Piccolo*: An Ultra-Lightweight Blockcipher. In: Preneel, B., Takagi, T. (eds.) CHES 2011. LNCS, vol. 6917, pp. 342–357. Springer, Heidelberg (2011)

33. Shirai, T., Shibutani, K., Akishita, T., Moriai, S., Iwata, T.: The 128-Bit Blockcipher CLEFIA (Extended Abstract). In: Biryukov, A. (ed.) FSE 2007. LNCS, vol. 4593, pp. 181–195. Springer, Heidelberg (2007)

34. Standaert, F.-X., Piret, G., Gershenfeld, N., Quisquater, J.-J.: SEA: A Scalable Encryption Algorithm for Small Embedded Applications. In: Domingo-Ferrer, J., Posegga, J., Schreckling, D. (eds.) CARDIS 2006. LNCS, vol. 3928, pp. 222–236. Springer, Heidelberg (2006)

35. Standaert, F.-X., Piret, G., Rouvroy, G., Quisquater, J.-J., Legat, J.-D.: ICEBERG: An Involutional Cipher Efficient for Block Encryption in Reconfigurable Hardware. In: Roy, B., Meier, W. (eds.) FSE 2004. LNCS, vol. 3017, pp. 279–299. Springer, Heidelberg (2004)

Analysis of Differential Attacks
in ARX Constructions

Gaëtan Leurent

University of Luxembourg – LACS

Abstract. In this paper, we study differential attacks against ARX
schemes. We build upon the generalized characteristics of de Cannière
and Rechberger; we introduce new multi-bit constraints to describe differential characteristics in ARX designs more accurately, and quartet
constraints to analyze boomerang attacks. We also describe how to propagate those constraints; this can be used either to assist manual construction of a differential characteristic, or to extract more information
from an already built characteristic. We show that our new constraints
are more precise than what was used in previous works, and can detect
more cases of incompatibility.

In particular, we show that several published attacks are in fact fact invalid because the differential characteristics cannot be satisfied. This highlights the importance of verifying differential attacks more thoroughly.

Keywords: Symmetric ciphers, Hash functions, ARX, Generalized characteristics, Differential attacks, Boomerang attacks.

1 Introduction

A popular way to construct cryptographic primitives is the so-called ARX design,
where the construction only uses Additions ($a \boxplus b$), Rotations ($a \ggg i$), and Xors
($a \oplus b$). These operations are very simple and can be implemented efficiently in
software or in hardware, but when mixed together, they interact in complex and
non-linear ways. In particular, two of the SHA-3 finalists, Blake and Skein, follow
this design strategy. More generally, functions of the MD/SHA family are built
using Additions, Rotations, Xors, but also bitwise Boolean functions, and logical
shifts; they are sometimes also referred to as ARX. This stategy as also been
used for stream ciphers such as Salsa20 and ChaCha, and block ciphers, such
as TEA, XTEA, HIGHT, or SHACAL (RC5 uses additions and data-dependant
rotations, but we only consider construction with fixed rotations).

The ARX design philosophy is opposed to S-Box based designs such as the
AES. Analysis of S-Box based designs usually happen at the word-level, and
differential characteristic are relatively easy to build, but efficient attacks often
need novel techniques, such as the rebound attack against hash functions [17].
For ARX designs, the analysis is done on a bit-level; finding good differential
characteristics remains an important challenge. In particular, the seminal attacks on the MD/SHA-familiy by the team of X. Wang are based on differential
characteristics built by hand [28,27,29], and an important effort has been devoted

X. Wang and K. Sako (Eds.): ASIACRYPT 2012, LNCS 7658, pp. 226–243, 2012.

to building tools to construct automatically such characteristics [6,23,8,15,24]. This effort has been quite successful for functions of the MD/SHA family, and it has allowed new attacks based on specially designed characteristics: attacks against HMAC [9], the construction of a rogue MD5 CA certificate [25], and attacks against combiners [16].

Another important problem is that the components of an ARX design can interact in complex and unexpected ways. Differential characteristics are usually built by looking at each operation individually, and multiplying the probabilities of each non-linear operation, but this approach can lead to very misleading results. For SHA-0 and SHA-1 differential characteristics, it has been shown that the hypothesis of independence between the local collisions is flawed, and some patterns of local collisions lead to impossible characteristics [4,21,14]. Problems have also been identified for differential attacks on SHACAL [26]. More recently, Mendel, Nad, and Schläffer have tackled the problem of building differential characteristics for SHA-2, and found that many of them are in fact incompatible [15].

A similar problem has been discussed in the context of boomerang attacks by Murphy [20]: the assumption that the differential characteristics are independent does not necessarily hold. Several recent works have found characteristics that turned out to be incompatible when analyzing ARX hash functions such as HAVAL [22], SHA-256 [2], or Skein [11].

Our Results. In this paper, we try to provide a framework to study these problems for ARX designs. In pure ARX functions, the modular addition is the only source of non-linearity (with respect to the xor difference). Therefore it is important to capture its behaviour as accurately as possible.

We extend the generalized characteristics of de Cannière and Rechberger [6] by introducing constraints involving several consecutive bits of a variable (*i.e.* $x^{[i]}$ and $x^{[i-1]}$), instead of considering bits one by one. We show that constraints on 2 consecutive bits can completely capture the modular difference, and we introduce reduced sets of constraints on 1.5 and 2.5 consecutive bits. This is motivated by the analysis of modular addition, but since these constraints are still local, they interact well with bitwise Boolean operations and rotations, and we can use them to study pure ARX as well as SHA-like constructions. We show that they capture more information than the single bit constraints of [6]. In particular, we describe cases of incompatibility in ARX characteristics due to interactions between consecutive bits, and we show that a proposed path for Skein is invalid [29]. This is detected automatically by our new constraints.

We also study boomerang attacks, and introduce constraints on *quartets* of variables, instead of considering each characteristic separately with constraints on *pairs* of variables. This allows to capture some extra information in the middle of the attack, when the top characteristic and the bottom characteristic meet. In particular, we can automatically detect incompatibilities in previously published attacks against Skein [1,5] and Blake [3].

As opposed to [15], our work is focused on local conditions, and we try to extract as much information as possible from a single operation. If needed, it can be combined with more computing intensive techniques considering several operations simultaneously.

Additionally, we give a complete description of how to compute the probability of a characteristic using these constraints, and how to do constraints propagation. All our code will be available from our webpage[1] so that these tools can be used by the community to build or verify differential attacks. Our tools are quite generic and we hope that they can be used to study more primitives. We don't provide a complete solution to automatically find differential characteristics in ARX schemes, but we believe our work is an important step in this direction.

This paper is organized as follows: first, we explain the theory of S-systems and how to solve them efficiently in Section 2, and we show how to use S-systems to study differential attacks using the generalized characteristic of de Cannière and Rechberger in Section 3. In Section 4, we introduce multi-bit constraints and show how they improve over previous results. Finally, in Section 5, we describe quartet constraint to study boomerang attacks, and show that they can detect incompatibilities in several attacks.

2   Analysis of S-systems

Since ARX systems in general are hard to analyze, we first study systems without rotations. An important remark is that a system of Additions and Xors, can be seen as a T-function [10], or more precisely, as an S-function [19]. We use the following definitions:

T-function. A T-function on n-bit words with k inputs and l outputs is a function from $(\{0,1\}^n)^k$ to $(\{0,1\}^n)^l$ with the following property:
> *For all t, the t least significant bits of the outputs can be computed from the t least significant bits of the inputs.*

S-function. An S-function on n-bit words is a function from $(\{0,1\}^n)^k$ to $(\{0,1\}^n)^l$, for which we can define a small set of *states* $\mathcal{S}$, and an initial state $S[-1] \in \mathcal{S}$ with the following property:
> *For all t, bit t of the outputs and the state $S[t] \in \mathcal{S}$ can be computed from bit t of the inputs, and the state $S[t-1]$.*

In practice, our analysis will be linear in the number of states, and the number of states can be exponential in the size of the system. We can only study systems with a limited number of states.

For instance, the modular addition is an S-function, with a 1-bit state corresponding to the carry. An S-function can also include bitwise functions, shifts to the left by a fixed number of bits, or multiplications by constants. However, a shift to the left by i bits, or multiplication by constant of i bits, leads to an increase of the state by a factor of 2^i, so the analysis will only be practical for small values of i. Note that the multiplication of two variables, or a data-dependant shift to the left, are T-functions, but are not S-functions because the size of the state has to grow with n.

---

[1] http://www.di.ens.fr/~leurent/arxtools.html

In this work, we consider systems of the form $f(P, x) = 0$ where f is an S-function, P is a vector of p parameters, and x is a vector of v unknown variables. This defines a family of systems, and we are interested in properties of the set of solutions of the unknown x for a given P. We call such a system an S-system.

A simple and yet important example is the system

$$x \oplus \Delta = x \boxplus \delta \tag{1}$$

where the parameter are Δ, δ. Solving this system is equivalent to finding a pair of variables with a given modular difference and a given xor difference, and was an important part of a recent attack on BMW [12].

It is well-known that those systems are T-functions, and can be solved from the least significant bit to the most significant bit. However, the naive approach to solve such a system uses backtracking, and can lead to an exponential complexity in the worst case.[2]

2.1   Representation of S-systems Using Finite State Machines

A more efficient strategy is to use an approach based on Finite State Machines, or automata: any system of such equations can be represented by an automaton, and solving a particular instance take time proportional to the word length. This kind of approach has been used to study differential properties of S-functions in [19].

The first step to apply this technique is to build an automaton corresponding to the system of equations. The states of this automaton correspond to the states of the S-function in $\mathcal{S}$, *i. e.* the carry bits: a system with s modular additions gives an automaton with 2^s states. The alphabet of the automaton is $\{0, 1\}^{p+v}$; each transition reads one bit from each parameter and each variable, starting from the least significant bit. The automaton just accepts (P, x) if and only if $f(P, x) = 0$.

We can then count the number of solutions to the system by counting paths in the graph corresponding to the automaton. In this work we mainly use this technique to decide whether a system is solvable, but we can also compute a random solution, or enumerate the set of solutions.

If the S-system is given as an expression with additions and bitwise Boolean operations, the transition table of the automaton can easily be constructed by evaluating the expression for every possible state, every possible 1-bit parameters, and every possible 1-bit variable.

Decision Automaton. When we remove the information about the variables from the edges, we obtain a non-deterministic automaton which can decide whether a system is solvable or not, *i. e.* whether there *exists* a choice of the variable x so that $f(P, x) = 0$ for a given P. We can then optionally build an equivalent deterministic automaton using the powerset construction.

---

[2] *e.g.* to solve the system $x \oplus \text{0x80000000} = x$, the backtracking algorithm will try all possible values for the 31 lower bits of x before concluding that there is no solution.

Implementation. We have automated the construction of the FSM from a simple description of the S-system. Our tool can deal with any system of additions, and bitwise Boolean functions. For instance System (1) will be written as `V0^P0==V0+P1`, and System (2) will be written as `P0|V0|V1; P1|V0|~V1; P2|~V0|V1; P3|~V0|~V1`. The variables are denoted by Vi and the parameters by Pi, and the operations are written naturally with a C-like syntax. The tool outputs the transition table of the automaton, and we have a collection of function to compute properties of the system from this table. From the FSM representation of an S-system, we can automatically derive:

- Whether a given set of parameter leads to a compatible system
- A random solution when the system is compatible
- The number of solutions (and the probability that a random x is a solution)
- A description of the solution set, from which we can efficiently iterate over the solutions

3   Study of Differential Characteristics

The most basic approach to describe a differential characteristic is to choose a difference operation (usually the modular difference $\boxminus$ or the xor difference $\oplus$), and to specify the difference $x' - x$ for every internal variable of a cipher. One can compute the probability of reaching the specified output difference for each operation, and the probability of the full characteristic is computed by multiplying the probabilities of each operation, under the assumption that the probabilities are independent.

However, this approach is not very successful for ARX designs, because the assumption of independence is very often false. To overcome this, Wang *et al.* introduced the notion of a signed difference. For each bit, we now consider three different possibilities:

- $x^{[i]} = x'^{[i]}$, this is denoted as 0;
- $x^{[i]} = 0$, $x'^{[i]} = 1$, this is denoted as $+1$;
- $x^{[i]} = 1$, $x'^{[i]} = 0$, this is denoted as -1.

This gives much better results in the presence of modular addition, because it combines both the modular difference and the xor difference.

3.1   Generalized Constraints

More generally, de Cannière and Rechberger noted that we can define a difference characteristic by allowing certain subsets of the values of (x, x') for each bit of the cipher [6].

Table 1 shows the symbol they use to denote all the possible subsets of $\mathcal{P}(\{0,1\}^2)$. For a given internal state variable x, and a constraint Δ, we write $\delta(x, x') = \Delta$ — or $\delta = \Delta$ if there is no ambiguity — to means that (x, x') is restricted to the subset defined by Δ.

Table 1. Constraints used in [6]   **Table 2.** Trivial encoding

(x, x'):		$(0,0)$	$(0,1)$	$(1,0)$	$(1,1)$		P_0	P_1	P_2	P_3
?	*anything*	✓	✓	✓	✓	?	1	1	1	1
-	$x = x'$	✓	-	-	✓	-	1	0	0	1
x	$x \neq x'$	-	✓	✓	-	x	0	1	1	0
0	$x = x' = 0$	✓	-	-	-	0	1	0	0	0
u	$(x, x') = (0,1)$	-	✓	-	-	u	0	1	0	0
n	$(x, x') = (1,0)$	-	-	✓	-	n	0	0	1	0
1	$x = x' = 1$	-	-	-	✓	1	0	0	0	1
#	*incompatible*	-	-	-	-	#	0	0	0	0
3	$x = 0$	✓	✓	-	-	3	1	1	0	0
5	$x' = 0$	✓	-	✓	-	5	1	0	1	0
7		✓	✓	✓	-	0	1	1	1	0
A	$x' = 1$	-	✓	-	✓	A	0	1	0	1
B		✓	✓	-	✓	B	1	1	0	1
C	$x = 1$	-	-	✓	✓	C	0	0	1	1
D		✓	-	✓	✓	D	1	0	1	1
E		-	✓	✓	✓	E	0	1	1	1

Since the definition of δ only involves bitwise operation, we can write it as an S-system, if we encode Δ as shown in Table 2:

$$P_0 = 0 \Rightarrow (x, x') \neq (0,0) \quad P_1 = 0 \Rightarrow (x, x') \neq (0,1)$$
$$P_2 = 0 \Rightarrow (x, x') \neq (1,0) \quad P_3 = 0 \Rightarrow (x, x') \neq (1,1)$$

or equivalently:

$$P_0 \vee x \vee x' \quad P_1 \vee x \vee \bar{x}' \quad P_2 \vee \bar{x} \vee x' \quad P_3 \vee \bar{x} \vee \bar{x}'. \qquad (2)$$

3.2 Differential Characteristics

In order to describe a differential characteristics with this framework, we specify a difference for each internal variable of a cipher, and we consider the operations that connect the variables. For each operation $\odot$, we can write an S-system[3]:

$$\delta x = \Delta_x \qquad \delta y = \Delta_y \qquad \delta z = \Delta_z \qquad z = x \odot y \qquad z' = x' \odot y', \qquad (3)$$

where x, y, z, x', y', z' are unknowns, and $\Delta_x, \Delta_y, \Delta_z$ are parameters. Using this S-system, we can verify if the differences specified input and output patterns for each operation are compatible. Moreover, we can compute the probability to reach the specified output pattern by counting the number of solutions. Assuming that the probabilities of each operations are independent, we can compute the probability of the full characteristic by multiplying the probabilities of each operations. We deal with the rotations $y = x \ggg i$ by just rotating the constraint pattern: if $\delta x = \Delta_x$ then we use $\delta y = \Delta_x \ggg i$.

---

[3] We assume that all the operations except the rotations are S-function, as is the case in ARX designs.

3.3 Propagation of Constraints

This approach can also be used to propagate the constraints associated with a differential characteristic. The main idea is to consider each bit constraint, and to split it into two disjoint subsets; if one of the subsets result in an incompatible system, we known that we can restrict the constraint to the other subset without reducing the number of solutions. More precisely, we use the following splits for the 1-bit constraints of [6]:

$$? \rightarrow -/x,\, 3/C,\, 5/A,\, 0/E,\, 1/7,\, u/D,\, n/B \quad - \rightarrow 0/1 \qquad\qquad x \rightarrow u/n$$
$$3 \rightarrow 0/u \qquad\qquad C \rightarrow 1/n \qquad\qquad 5 \rightarrow 0/n \qquad\qquad A \rightarrow 1/u$$
$$7 \rightarrow 0/x,\, u/5,\, n/3 \quad B \rightarrow 0/A,\, u/-,\, 1/3 \quad D \rightarrow 0/C,\, n/-,\, 1/5 \quad E \rightarrow u/C,\, n/A,\, 1/x$$

For instance, if a bit is specified as ?, we test whether the system is still compatible when it is restricted to - and to x, respectively. If one of the systems becomes incompatible, we can turn the ? constraint into x or -, accordingly. If both are still compatible, we then try to restrict the ? bit to 3 and C, and try all the available splits.

This will be repeated with the S-systems corresponding to each operation in the cipher. We can not apply this strategy to bigger chunks because the resulting system would be too large. Still, the constraints found in one system will be given as input to other systems involving the same variable, and can generate new constraints. The technique will discover *necessary* constraints, and output a characteristic more precise than the input characteristic.

This can also be combined with more global techniques such as Section 2.3 of [7]. or the "Complete Condition Check" of [15]. When we a constraint is split into two subsets, we can look for contradictions by running the propagation algorithm on the full path, instead of running it on a single operation. However, this becomes very expensive for large systems and it can take hours to try to split each constraint. In this work, we focus on discovering local conditions efficiently, and we leave the analysis of less local techniques for future work.

All this can be implemented quite efficiently using automata to solve S-systems. If we build deterministic decision automata, we can test whether a system is compatible with only n table access. This approach is very similar to the technique used in [6], and explained in more details in [21] and [18]. The main difference is that we iterate over all possible choices for the variables only when building the automaton, not when using it. In previous work, a similar result is achieved by caching the results of the computations.

4 Multi-bit Constraints

In this work, we extend this framework by considering constraints on several consecutive bits, instead of strictly bitwise constraints. This allows to express some conditions that occur naturally when considering carry extension, such as $x^{[i]} = x^{[i-1]}$. Two-bit conditions have already been proposed in [15], but they are treated separately from the main characteristic. In particular, two-bit conditions

Table 3. New 1.5-bit constraints

$(x \oplus x', x \oplus 2x, x)$:		$(0,0,0)$	$(0,0,1)$	$(0,1,0)$	$(0,1,1)$	$(1,0,0)$	$(1,0,1)$	$(1,1,0)$	$(1,1,1)$
=	$x' = x = 2x$	✓	✓	-	-	-	-	-	-
!	$x' = x \neq 2x$	-	-	✓	✓	-	-	-	-
<	$x' \neq x = 2x$	-	-	-	-	✓	✓	-	-
>	$x' \neq x \neq 2x$	-	-	-	-	-	-	✓	✓

are not used to deduce further constraints through the propagation algorithm. In our work, multi-bit constraints can only deal with consecutive bits of a variable, but they are part of the characteristic, and they can be propagated efficiently.

1.5-bit Constraints. First, we consider constraints on pairs of consecutive bits. Intuitively, this is used to capture the fact that in a carry chain, even if we don't known the sign of the modular difference, we know that the active bits all have the same sign, except the final one. For instance, if we have ---x$\boxplus$----→ -xxx, we know that output difference must be either -nuu (if the input difference is ---n) or -unn (if the input difference is ---u). We can capture this behaviour using constraints that link the sign of an active bit to the sign of the previous bit. In our implementation, we introduce a set of 16 constraints described in Table 3 and 1: ?, -, x, 0, u, n, 1, #, 3, C, 5, A, =, !, <, >. For instance, the symbol < means that the current bit is active, and that bit i of x is equal to bit i of $2x$, i. e. to bit $i-1$ of x — this can be written as $x'^{[i]} \neq x^{[i]} = x^{[i-1]}$, and it appears in the middle of carry chain. The situation of a carry extension with an unknown sign as in ---x$\boxplus$----→ -xxx can now be written more accurately as ->< x.

The constraints of Table 3 are written as subsets of $(x^{[i]}, x'^{[i]}, x^{[i-1]})$; we call them 1.5-bit constraints because we use $x^{[i-1]}$ but we do not use $x'^{[i-1]}$.

2-bit Constraints. The 1.5-bit constraints are quite efficient to capture information about the carries when the xor difference is known. However, when the xor difference is not known *a priori*, we still loose a lot of information. To overcome this problem, we considered the full set of 2^{16} possible constraints on $(x^{[i]}, x'^{[i]}, x^{[i-1]}, x'^{[i-1]})$, and we discovered an important property: they can restrict the pair (x, x') to *exactly* the set of values with any given modular difference. More precisely, this is achieved using the 10 constraints of Table 4. We found this set of constraints experimentally, by testing all 8-bit differences.

This is an important result because it allows to express the modular difference using only local constraints. Local constraints can easily go through rotations, and can be expressed as S-functions. Therefore we can compute the probability of a differential characteristic expressed in this way, and we can propagate these constraints automatically.

We denote the first four constraints as U, V, N and M; the remaining six can be obtained by combining previous constraints. The most important constraints are

Table 4. 2-bit constraints sufficient to describe exactly the modular difference

$(2x, 2x', x \oplus x')$:	(0,0,0) -0	(0,0,1) x0	(0,1,0) -u	(0,1,1) xu	(1,0,0) -n	(1,0,1) xn	(1,1,0) -1	(1,1,1) x1
U ≡ {--,-u,xn}	✓	-	-	✓	✓	-	✓	-
V not U	-	✓	✓	-	-	✓	-	✓
N ≡ {--,-n,xu}	✓	-	✓	-	-	✓	✓	-
M not N	-	✓	-	✓	✓	-	-	✓
≡ x0	-	✓	-	-	-	-	-	-
≡ -0	✓	-	-	-	-	-	-	-
≡ x-	-	✓	-	-	-	-	-	✓
≡ --	✓	-	-	-	-	-	✓	-
≡ Ux	-	-	-	✓	✓	-	-	-
≡ Nx	-	-	✓	-	-	✓	-	-

Table 5. New 2.5-bit constraints

$(x \oplus x', x \oplus 2x, x)$: $2x \oplus 2x', 4x \oplus 2x$:	(0,0,0)	(0,0,1)	(0,1,0)	(0,1,1)	(1,0,0) (0,0)/(0,1)/(1,0)/(1,1)	(1,0,1)	(1,1,0)	(1,1,1)
X carry chain	✓	✓	✓	✓	- /✓	- /✓	- /✓	- /✓
U u carry	✓ /✓	✓ / -	✓ / -	✓ /✓	- / -	- /✓	- /✓	- / -
N n carry	✓ / -	✓ /✓	✓ /✓	✓ / -	- /✓	- / -	- / -	- /✓
/ x carry	✓/✓/-/✓	✓/✓/-/✓	✓/✓/-/✓	✓/✓/-/✓	-/-/✓/-	-/-/✓/-	-/-/✓/-	-/-/✓/-
\ X minus /	-/-/✓/-	-/-/✓/-	-/-/✓/-	-/-/✓/-	-/-/-/✓	-/-/-/✓	-/-/-/✓	-/-/-/✓

the one denoted as U and N: they can capture the carry extension of a positive (resp. negative) modular difference. For instance a modular difference of $+1$ can be realized with 4-bit words as ---u, --un, -unn, unnn or nnnn, depending on the carry extension. For each of the potential carry bits (1–3), we can see that the difference pattern of bit i and bit $i-1$ is always one of --, -u, un, or nn. Reciprocally, if bits 1–3 follow these patterns, then the full difference has to be one of the previous patterns, and the modular difference will be $+1$. The U constraint correspond to these patterns.

In our implementation, we only use the U and N constraints, which are sufficient to express sparse modular differences.

2.5-bit Constraints. To obtain an efficient technique to study differential characteristics in ARX constructions, we want to combine the results of the 1.5-bit constraints, and the 2-bit constraints. On the one hand, the 1.5-bit constraints are constructed from the 1-bit constraints in order to capture information about the carry when the sign of the difference is not known. On the other hand, the 2-bit constraints can capture exactly the modular difference, but we need to know the sign of the difference. We now introduce constraints to capture the modular difference when the sign is not known.

Table 6. Comparison of the constraint sets. We show how simple difference sets can be encoded with our constraints, and the number of pairs allowed by each constraint.

Diff, carry	1-bit cstr.	1.5-bit cstr.	2-bit cstr.	2.5-bit cstr.
$+1$, k-bit (2^{n-k})	-unnn (2^{n-k})	-unnn (2^{n-k})	-unnn (2^{n-k})	-unnn (2^{n-k})
± 1, k-bit (2^{n-k+1})	-xxxx (2^n)	-><<x (2^{n-k+1})	-><<x (2^{n-k+1})	-><<x (2^{n-k+1})
$+1$, any (2^n)	????x (2^{2n-1})	????x (2^{2n-1})	UUUUx (2^n)	UUUUx (2^n)
± 1, any (2^{n+1})	????x (2^{2n-1})	????x (2^{2n-1})	XXXXx $(2^n \times n)$	///Xx (2^{n+1})

Following the analysis of the 2-bit constraints, we study the patterns created by a carry extension with an unknown sign. Using the 1.5-bit constraints, we can see that the constraints of bits i and $i-1$ are either --, ->, or x<. Reciprocally, if all the bits follow these patterns, this result in a valid carry extension. We denote the corresponding set of possibilities for $(x^{[i]}, x'^{[i]}, x^{[i-1]}, x'^{[i-1]}, x^{[i-2]})$ as /.

As shown in Table 5, we introduce the following new constraints: X $\equiv$ $\{--, -x, xx\}$, U $\equiv$ $\{--, -u, xn\}$, N $\equiv$ $\{--, -n, xu\}$, / $\equiv$ $\{--, ->, x<\}$, \ $\equiv$ $\{-<, x>\}$. For efficiency reasons, we keep a set of only 16 constraints by removing the less useful ones: ?, -, x, 0, u, n, 1, =, !, <, >, X, U, N, /, \.

4.1   Comparison

To compare the sets of constraints, we show how they can be used in simple situations in Table 6. We consider 4 situations, were we describe a set of pairs with a modular difference of ± 1:

- First, we assume that we know the sign of the difference, and the length of the carry (*e.g.* -----u $\boxplus$ ------ $\rightarrow$ -xxxxx). In this case all the constraints systems give an optimal characterization of the set of allowed pairs.
- Second, we assume that we don't known the sign of the difference, but we know the length of he carry (*e.g.* -----x $\boxplus$ ------ $\rightarrow$ -xxxxx). In the case, we need constraints on 1.5 bits to optimally capture the relations in the carry-extended bits.
- Third, we assume that we know the sign of the difference, but we don't know the length of the carry (*e.g.* -----u $\boxplus$ ------ $\rightarrow$??????). In this situation, the 2-bit constraints can express precisely the modular difference.
- Finally, we assume that we don't know the sign of the difference, nor the length of the carry (*e.g.* -----x $\boxplus$ ------ $\rightarrow$??????). Here, we need constraints on 2.5 bits to restrict the set of pairs optimally using relations between the bits.

4.2   Use as S-sytems

We also denote the new sets of constraint by δ. Since the definition of δ only involves bitwise operation and left-shift by a few bits, $\delta x = \Delta$ can by written as a S-system, similar to System (2). We can use the tools of Section 2 to compute the probability of a characteristic specified with the new constraints, and to propagate the new constraints, by build the automata associated with the systems of each operation, as given in (3). These automata are quite large, because the state of the automaton has to include the values of $x^{[i-1]}$, $x'^{[i-1]}$, and $x^{[i-2]}$.

In practice, we implemented the 1.5-bit constraints and the 2.5-bit constraints. With the 1.5-bit constraints, we have 5 bits of state for the S-system of the addition, but the transition automaton only reaches 16 different states. When using the powerset construction to build a deterministic decision, we obtain 12929 states, and the full table takes 102MB. With the 2.5-bit constraints, we have 11 bits of state, and the transition automaton reaches 160 different states (we cannot build a deterministic decision automaton in this case).

We could easily include more constraints in our framework, but this set of symbol is quite expressive, and a larger set of constraints would result in larger tables. We will see that those constraints give good results in practice. Moreover, we note that many cases can be expressed using the constraints of two consecutive bits. For instance, the constraint $x^{[i]} = x'^{[i]} = x^{[i-1]} = 0$ cannot be expressed in Table 3, but it will be coded with constraint = for bit i, and constraint 3 for bit $i-1$ (if some more information is known for bit $i-1$, it will become 0 or u).

When we deal with a rotation, we have to relax the constraints slightly if the multi-bit constraints are broken by the rotation. For a rotation of i bits to the right, if $\Delta_x^{[i]}$ is one of =, !, < or >, it will be relaxed to -, -, x and x, respectively.

4.3   Propagation of Constraints

To propagate the new constraints, we need to define how to split the new constraints. We use the following splits for the 1.5-bit constraints:

$$? \rightarrow -/x, 3/C, 5/A \qquad - \rightarrow 0/1, =/! \qquad x \rightarrow u/n, </>$$
$$3 \rightarrow 0/u \qquad\qquad C \rightarrow 1/n \qquad\quad 5 \rightarrow 0/n \qquad\quad A \rightarrow 1/u$$
$$= \rightarrow 0/1 \qquad\qquad ! \rightarrow 0/1 \qquad\quad > \rightarrow u/n \qquad\quad < \rightarrow u/n$$

For the 2.5-bit constraints, some useful subsets are not included in the 16 constraints, but can be obtained by restricting both $\Delta_x^{[i]}$ and $\Delta_x^{[i-1]}$. We use the following splits:

$$? \rightarrow -/x, X/x- \qquad X \rightarrow U/Nx, N/Ux, -/xx, //\backslash \qquad N \rightarrow -/xu$$
$$\backslash \rightarrow -</x> \qquad\quad / \rightarrow U/Nx, N/Ux, -/x< \qquad\qquad U \rightarrow -/xn$$

This approach is quite efficient. As an example, let us consider this system:

$$\delta x = \text{x--x} \qquad \delta y = \text{----} \qquad z = x \boxplus y$$
$$\delta u = \text{---x} \qquad \delta v = \text{----} \qquad z = u \boxplus v \qquad \delta z = \text{-???}.$$

It is easy to see that this system is incompatible when considering modular differences: the difference in $x \boxplus y$ is $\pm 8 \pm 1$, while the difference in $u \boxplus v$ is ± 1. However, when using only the xor difference, or the constraints of [6], this system seems to be compatible, and constraint propagation gives $\delta z = \text{-xxx}$. Using our new constraints, the algorithm can further deduces $\delta z = \text{-<<x}$ from the first

Table 7. Experiments with a few rounds of a 4-bit Skein. We give the number of input/output differences accepted by each technique, and the ratio of false positive.

Method	4 rounds (total: 2^{32})		6 rounds (sparse[1])	
	Accepted	Fp.	Accepted	Fp.
Exhaustive search	35960536 $(2^{25.1})$	0	427667 $(2^{18.7})$	0
2.5-bit constraints	40820032 $(2^{25.3})$	0.13	746742 $(2^{19.5})$	0.7
1.5-bit constraints	40820032 $(2^{25.3})$	0.13	1372774 $(2^{20.4})$	2.2
1-bit constraints	43564288 $(2^{25.4})$	0.21	1762857 $(2^{20.7})$	3.1
Checking additions independently	56484732 $(2^{25.8})$	0.57		

[1] Weight 4 differences. The total number of input/output differences is $\binom{24}{4}\binom{24}{4} \approx 2^{26.75}$

addition and δz = -><x from the second addition, and the incompatibility is detected. Moreover, the incompatibility can be detected without specifying the difference in z beforehand using the 2.5 bit constraints.

4.4 Comparison with Previous Works

To compare the efficiency of the constraints, we did some experiments with reduced versions of Skein. We test a set of input and output xor differences, and we compare several methods to detect if the differences are compatible. We use small versions so that we can find exact results with exhaustive search. We verify that no false-negative are found, and we compare how many false-positive are found with each technique.

First we use a reduced Skein with two rounds and 4 words of 4 bits each. We note that for a two-round Skein, all the intermediary xor difference can be computed from the input and output xor differences; therefore we have a full xor differential characteristic. As a reference point, we can check whether each non-linear operation has a non-zero probability. Our result in Table 7 show that the assumption of independence of the operations can be quite flawed: we found many paths where each operation has a non-zero probability, but no pair can satisfy the differential. This motivates the use of more advanced constraints in order to extract information from one operation and combine it with another operation. We also see that our 1.5-bit constraints can detect more problems that the 1-bit constraints of [6]. In this setting the 2.5-bit constraints are no better than the 1.5-bit constraints because the xor differences are all known.

We also did experiments with a reduced Skein with three rounds and 4 words of 6 bits each. We only use sparse differences (less than 4 active bits in the input and output), because the full space is too large to be exhausted in practice. Our results are given in Table 7, and show that in this setting, the 2.5-bit constraints reduces the number of false positives threefold over the 1.5-bit constraints. The 2.5-bit constraints provide much better results than previous works when the xor difference is not known beforehand.

4.5   Description of Some Case of Incompatibility

We have developed a graphical tool that can display such a characteristic, and allows the user to easily modify the characteristic by adding and removing constraints. The tool can automatically propagate the new constraints, and show incompatibilities if there are some.

We have studied published differential trails with this tool and we found problems in several of them. It seems that many characteristics following a natural construction, and seemingly valid when verified manually, are in fact incompatible. We will now describe some of the patterns that can lead to unexpected problem.

Problems with Modular Addition. A simple class of problems is related to the modular additions when using xor differences. Techniques to check the validity of these operations are well known [13,19], but in some cases the results are somewhat unexpected. In particular, the valid differences are quite constrained in the least significant bit, because the incoming carry is fixed to zero. For instance the following path is built with a simple linearization, but it is in fact incompatible:

$$\delta a = \texttt{---x} \qquad \delta b = \texttt{---x} \qquad \delta c = \texttt{---x}$$
$$x = a \boxplus b \boxplus c \qquad \delta x = \texttt{---x}.$$

More generally, some pattern which seem valid when studied with a signed difference are in fact incompatible. The characteristic used in a recent near-collision attack against Skein [29] contains a pattern similar to this one[4]:

$$\delta a = \texttt{--xxxxx-} \qquad \delta b = \texttt{---xx---}$$
$$x = a \boxplus b \qquad \delta x = \texttt{-xxxx-x-}.$$

This seems valid when considering signed differences: the difference should be ± 2 in a, ± 8 in b, and $\pm 2 \pm 8$ in x. In fact, this does not have any solution, and it does not seem easy to modify the characteristic of [29] to obtain a valid attack.

Problems with Carry Extensions. Carry extensions in modular additions generate constraints between consecutive bits which can be detected with our framework. For instance, let us consider the following simple path:

$$\delta a = \texttt{-xx---} \qquad c = a \boxplus b \qquad c' = c \ggg 2 \qquad u = c' \boxplus d$$
$$\delta b = \texttt{xxx---} \qquad \delta c = \texttt{------} \qquad \delta d = \texttt{---xx-} \qquad \delta u = \texttt{---xx-}.$$

The first addition generate a constraint $c^{[4]} \neq c^{[3]}$ (*i. e.* $\delta c = \texttt{-!----}$), and the second addition generate a constraints $c'^{[2]} = c'^{[1]}$ (*i. e.* $\delta c' = \texttt{----=--}$). Obviously these constraints are contradictory through the rotation. In this example the problem will be detected by our new constraints, but not when looking at each operation individually, or using the single-bit constraints of [6].

---

[4] This can be found at round 20, in the addition $c_{20} = c_{19} \boxplus d_{19}$, with the following xor-differences: $\Delta c_{19} = \texttt{0x020030a0000f80a0}$, $\Delta d_{19} = \texttt{0xf8f87ca007f7c7a7}$, $\Delta c_{20} = \texttt{0x7ef8f50001104501}$.

5   Constraints for the Analysis of Boomerang Attacks

We also study differential characteristics in the context of boomerang attacks. The traditional approach is to specify each characteristic separately, and to assume that they are all independent. In this work, we consider a boomerang characteristic mostly as collection of constraints for the top characteristics and the bottom characteristics.

Let x be some internal state variable, and $x^{(0)}$, $x^{(1)}$, $x^{(2)}$, $x^{(3)}$ be the corresponding variables in a boomerang quartet. A boomerang property is built by specifying a top trail for $(x^{(0)}, x^{(2)})$ and $(x^{(1)}, x^{(3)})$, and a bottom trail for $(x^{(0)}, x^{(1)})$ and $(x^{(2)}, x^{(3)})$. For more generality, we allow the two characteristics to be different in each case (*e.g.*, the signs might be different). The top trail will be mostly unconstrained for the bottom part of the cipher, while the bottom trail will be mostly unconstrained for the top part.

Unfortunately, the hypothesis of independence might be wrong in practice, and we can find paths that are impossible to satisfy simultaneously, as shown by Murphy [20]. In fact, this kind of problem seem to be quite common with ARX designs, as shown in the case of HAVAL [22], SHA-256 [2], or Skein [11]. To capture this kind of dependency, we use constraints on *quartets* of variables, instead of constraints on pairs of variables. We can not use the full set of 2^{16} constraints, because the resulting system is too large, but we use a set of 81 constraints given in Table 8 to specify the xor difference in each of the four sides of the quartet. For (i, j) in $\{(0, 1), (2, 3), (0, 2), (1, 3)\}$, we restrict $x^{(i)} \oplus x^{(j)}$ to 0 or 1, or leave it unrestricted. Note that some constraints are actually contradictory[5] or redundant[6], but this uniform set is much easier to work with than a reduced set without the extra constraints.

We use three different kinds of S-systems to propagate constraints in a boomerang characteristic:

1. systems with multi-bit constraints and non-linear operations in each individual path, following System (3) (for (i, j) in $\{(0, 1), (2, 3), (0, 2), (1, 3)\}$):

$$\delta(x^{(i)}, x^{(j)}) = \Delta_x^{i,j} \quad \delta(y^{(i)}, y^{(j)}) = \Delta_y^{i,j} \quad \delta(z^{(i)}, z^{(j)}) = \Delta_z^{i,j}$$
$$x^{(i)} \boxplus y^{(i)} = z^{(i)} \quad x^{(j)} \boxplus y^{(j)} = z^{(j)};$$

2. systems with quartet constraints and non-linear operations:

$$\delta(u^{(0)}, u^{(1)}, u^{(2)}, u^{(3)}) = \Delta_u^{0,1,2,3}, \qquad \text{for all } u \text{ in } \{x, y, z\}$$
$$x^{(i)} \boxplus y^{(i)} = z^{(i)}, \qquad \text{for all } i \text{ in } \{0, 1, 2, 3\}$$

3. systems with multi-bit constraints linking the four variables of a quartet:

$$\delta(x^{(0)}, x^{(2)}) = \Delta_x^{0,2} \qquad \delta(x^{(1)}, x^{(3)}) = \Delta_x^{1,3} \qquad \textit{(Top path)}$$
$$\delta(x^{(0)}, x^{(1)}) = \Delta_x^{0,1} \qquad \delta(x^{(2)}, x^{(3)}) = \Delta_x^{2,3}. \qquad \textit{(Bottom path)}$$

---

[5] *e.g.* ---x means $x^{(0)} \oplus x^{(1)} = 0$, $x^{(2)} \oplus x^{(3)} = 0$, $x^{(0)} \oplus x^{(2)} = 0$, $x^{(1)} \oplus x^{(3)} = 1$, which is impossible.

[6] *e.g.* ---? and ---- allow the same values for the $x^{(i)}$'s.

Table 8. New boomerang constraints

$(x^{(0)} \oplus x^{(1)}, x^{(2)} \oplus x^{(3)}, x^{(0)} \oplus x^{(2)}, x^{(1)} \oplus x^{(3)})$:

	0000	0001	0010	0011	0100	0101	0110	0111	1000	1001	1010	1011	1100	1101	1110	1111
????	✓	✓	✓	✓	✓	✓	✓	✓	✓	✓	✓	✓	✓	✓	✓	✓
x???	-	-	-	-	-	-	-	-	✓	✓	✓	✓	✓	✓	✓	✓
-???	✓	✓	✓	✓	✓	✓	✓	✓	-	-	-	-	-	-	-	-
?x??	-	-	-	-	✓	✓	✓	✓	-	-	-	-	✓	✓	✓	✓
xx??	-	-	-	-	-	-	-	-	-	-	-	-	✓	✓	✓	✓
----	✓	-	-	-	-	-	-	-	-	-	-	-	-	-	-	-

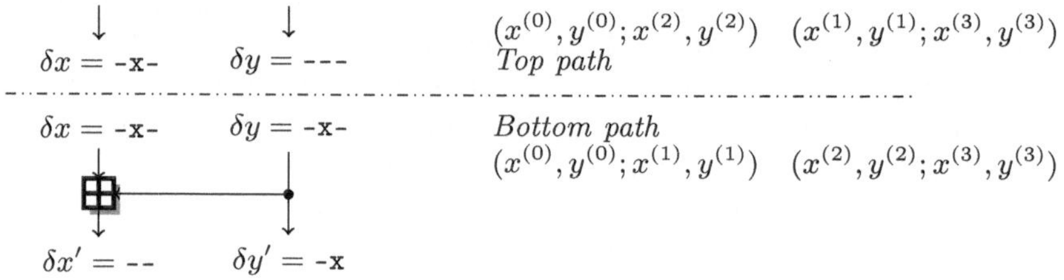

$\delta x = \texttt{-x-} \qquad \delta y = \texttt{---}$

$\delta x = \texttt{-x-} \qquad \delta y = \texttt{-x-}$

$\delta x' = \texttt{--} \qquad \delta y' = \texttt{-x}$

$(x^{(0)}, y^{(0)}; x^{(2)}, y^{(2)}) \quad (x^{(1)}, y^{(1)}; x^{(3)}, y^{(3)})$
Top path

Bottom path
$(x^{(0)}, y^{(0)}; x^{(1)}, y^{(1)}) \quad (x^{(2)}, y^{(2)}; x^{(3)}, y^{(3)})$

Fig. 1. Example of incompatible characteristics

5.1 Incompatibility in Boomerang Characteristics

We found that some very simple patterns can lead to incompatibilities. Figure 1 gives an example of a pattern that results in incompatible characteristics. If a quartet follows these characteristics, the middle bit of the variables has to satisfy:

$$x^{(0)} \oplus x^{(2)} = x^{(1)} \oplus x^{(3)} = 1 \quad y^{(0)} \oplus y^{(2)} = y^{(1)} \oplus y^{(3)} = 0 \quad \textit{(Top path)}$$
$$x^{(0)} \oplus x^{(1)} = x^{(2)} \oplus x^{(3)} = 1 \quad y^{(0)} \oplus y^{(1)} = y^{(2)} \oplus y^{(3)} = 1 \quad \textit{(Bottom path)}$$
$$x^{(0)} \boxplus y^{(0)} = x^{(1)} \boxplus y^{(1)} \qquad x^{(2)} \boxplus y^{(2)} = x^{(3)} \boxplus y^{(3)}$$

We can assume that $x^{(0)} = 0$, and deduce $x^{(1)} = 1$, $x^{(2)} = 1$, $x^{(3)} = 0$. Since the difference in $(y^{(0)}, y^{(1)})$ must cancel the difference in $(x^{(0)}, x^{(1)})$, we have $y^{(0)} = 1$, $y^{(1)} = 0$, and we can deduce $y^{(2)} = 1$, $y^{(3)} = 0$. But the difference in $(y^{(2)}, y^{(2)})$ can not cancel the difference in $(x^{(2)}, x^{(3)})$. A more detailed analysis shows that this pattern can lead to incompatibilities even if we allow some incoming carries.

This pattern seem to appear very frequently when using linearized characteristics in ARX designs.

5.2 Application

We used our tools to verify several boomerang attacks in the literature, and found some attack using incompatible paths.

Blake-256. First, we studied the boomerang attacks on Blake from Biryukov *et.al* in [3]. When looking at the paths used for the attacks on 7 and 8 round of the keyed permutation, our tool detects an incompatibility. More precisely, when starting from a middle quartet with the specified differences, and going backward through G_3, it is impossible to get the specified difference simultaneously in both paths. We verified experimentally that we could not find such quartets, even with significantly more trials than predicted under the assumption that the paths are independent.

With the help of the authors of [3], we found out an alternative path that give a valid boomerang attack. More precisely we modify the top path by using a difference on bit 25 instead of 31, and rotating all the difference patterns. We verified experimentally that this leads to a valid attack, but the cost of the attack becomes higher than reported in [3].

Similarly, for the compression function attacks, our tool detects that the path used for the 6.5 and 7-round attacks is invalid. We found that this can corrected by modifying the top path to use differences on bits 4 and 20 instead of 15 and 31.

Skein-512. We also used our tool to study the boomerang attacks on Skein. We start with only the linearized (or almost linearized) xor differential characteristics for rounds 12–16 and 16–20, with the key addition in between to provide extra freedom, and we use our tool to propagate the constraints. We found that the following paths lead to contradictions:

- The paths for the 32-round attack of [5];
- The paths for the 33- and 34-round attack of [5];
- The paths for the attack of[1], based on the old rotation constants, and inverse permutations; as well as a modified version using the correct permutations.

In each case, our tool detect the contradiction automatically. More recently, a new path has been proposed [30], and a middle quartet was given to show that the paths are compatible.

6   Conclusion

In this paper, we study differential characteristics in ARX constructions. We extend the framework of de Cannière and Rechberger with new constraints. First we introduce multi-bit constraints that can be propagated more accurately through modular addition. We show that a set of 2-bit constraints can express exactly the modular difference of a pair of variables, and describe a reduced set of 2.5-bit constraints that can express the modular difference in simple cases and can also capture the carry extensions of an unsigned difference. Second, we introduce new quartet constraints to work with boomerang attacks.

We provide experimental results showing that our constraints can automatically detect several cases of incompatibility in differential characteristics undetected by previous techniques; and we point out several published attacks that turn out to be invalid. We show that some paths can in fact be incompatibile; this shows the importance of verifying differential attacks.

We hope that the tools will be useful to other cryptanalysts, and they are available at `http://www.di.ens.fr/~leurent/arxtools.html`.

Acknowledgement. We would like to thank the anonymous reviewers of Asiacrypt for their insightful comments. We would also like to thank the authors of [3] for helping us verify the problem with their attack, and finding an alternative path.

Gaëtan Leurent is supported by the AFR grant PDR-10-022 of the FNR Luxembourg.

References

1. Aumasson, J.-P., Çalık, Ç., Meier, W., Özen, O., Phan, R.C.-W., Varıcı, K.: Improved Cryptanalysis of Skein. In: Matsui, M. (ed.) ASIACRYPT 2009. LNCS, vol. 5912, pp. 542–559. Springer, Heidelberg (2009)
2. Biryukov, A., Lamberger, M., Mendel, F., Nikolić, I.: Second-Order Differential Collisions for Reduced SHA-256. In: Lee, D.H., Wang, X. (eds.) ASIACRYPT 2011. LNCS, vol. 7073, pp. 270–287. Springer, Heidelberg (2011)
3. Biryukov, A., Nikolić, I., Roy, A.: Boomerang Attacks on BLAKE-32. In: Joux, A. (ed.) FSE 2011. LNCS, vol. 6733, pp. 218–237. Springer, Heidelberg (2011)
4. Chabaud, F., Joux, A.: Differential Collisions in SHA-0. In: Krawczyk, H. (ed.) CRYPTO 1998. LNCS, vol. 1462, pp. 56–71. Springer, Heidelberg (1998)
5. Chen, J., Jia, K.: Improved Related-Key Boomerang Attacks on Round-Reduced Threefish-512. In: Kwak, J., Deng, R.H., Won, Y., Wang, G. (eds.) ISPEC 2010. LNCS, vol. 6047, pp. 1–18. Springer, Heidelberg (2010)
6. De Cannière, C., Rechberger, C.: Finding SHA-1 Characteristics: General Results and Applications. In: Lai, X., Chen, K. (eds.) ASIACRYPT 2006. LNCS, vol. 4284, pp. 1–20. Springer, Heidelberg (2006)
7. Grechnikov, E.A.: Collisions for 72-step and 73-step sha-1: Improvements in the method of characteristics. Cryptology ePrint Archive, Report 2010/413 (2010), `http://eprint.iacr.org/`
8. Fouque, P.A., Leurent, G., Nguyen, P.: Automatic Search of Differential Path in MD4. ECRYPT Hash Worshop – Cryptology ePrint Archive, Report 2007/206 (2007), `http://eprint.iacr.org/`
9. Fouque, P.-A., Leurent, G., Nguyen, P.Q.: Full Key-Recovery Attacks on HMAC/NMAC-MD4 and NMAC-MD5. In: Menezes, A. (ed.) CRYPTO 2007. LNCS, vol. 4622, pp. 13–30. Springer, Heidelberg (2007)
10. Klimov, A., Shamir, A.: A New Class of Invertible Mappings. In: Kaliski Jr., B.S., Koç, Ç.K., Paar, C. (eds.) CHES 2002. LNCS, vol. 2523, pp. 470–483. Springer, Heidelberg (2003)
11. Leurent, G., Roy, A.: Boomerang Attacks on Hash Function Using Auxiliary Differentials. In: Dunkelman, O. (ed.) CT-RSA 2012. LNCS, vol. 7178, pp. 215–230. Springer, Heidelberg (2012)
12. Leurent, G., Thomsen, S.S.: Practical Near-Collisions on the Compression Function of BMW. In: Joux, A. (ed.) FSE 2011. LNCS, vol. 6733, pp. 238–251. Springer, Heidelberg (2011)

13. Lipmaa, H., Moriai, S.: Efficient Algorithms for Computing Differential Properties of Addition. In: Matsui, M. (ed.) FSE 2001. LNCS, vol. 2355, pp. 336–350. Springer, Heidelberg (2002)

14. Manuel, S.: Classification and generation of disturbance vectors for collision attacks against SHA-1. Des. Codes Cryptography 59(1-3), 247–263 (2011)

15. Mendel, F., Nad, T., Schläffer, M.: Finding SHA-2 Characteristics: Searching through a Minefield of Contradictions. In: Lee, D.H., Wang, X. (eds.) ASIACRYPT 2011. LNCS, vol. 7073, pp. 288–307. Springer, Heidelberg (2011)

16. Mendel, F., Rechberger, C., Schläffer, M.: MD5 Is Weaker Than Weak: Attacks on Concatenated Combiners. In: Matsui, M. (ed.) ASIACRYPT 2009. LNCS, vol. 5912, pp. 144–161. Springer, Heidelberg (2009)

17. Mendel, F., Rechberger, C., Schläffer, M., Thomsen, S.S.: The Rebound Attack: Cryptanalysis of Reduced Whirlpool and Grøstl. In: Dunkelman, O. (ed.) FSE 2009. LNCS, vol. 5665, pp. 260–276. Springer, Heidelberg (2009)

18. Mouha, N., De Cannière, C., Indesteege, S., Preneel, B.: Finding Collisions for a 45-Step Simplified HAS-V. In: Youm, H.Y., Yung, M. (eds.) WISA 2009. LNCS, vol. 5932, pp. 206–225. Springer, Heidelberg (2009)

19. Mouha, N., Velichkov, V., De Cannière, C., Preneel, B.: The Differential Analysis of S-Functions. In: Biryukov, A., Gong, G., Stinson, D.R. (eds.) SAC 2010. LNCS, vol. 6544, pp. 36–56. Springer, Heidelberg (2011)

20. Murphy, S.: The Return of the Cryptographic Boomerang. IEEE Transactions on Information Theory 57(4), 2517–2521 (2011)

21. Peyrin, T.: Analyse de fonctions de hachage cryptographiques. PhD thesis, University of Versailles (2008)

22. Sasaki, Y.: Boomerang Distinguishers on MD4-Family: First Practical Results on Full 5-Pass HAVAL. In: Miri, A., Vaudenay, S. (eds.) SAC 2011. LNCS, vol. 7118, pp. 1–18. Springer, Heidelberg (2012)

23. Schläffer, M., Oswald, E.: Searching for Differential Paths in MD4. In: Robshaw, M. (ed.) FSE 2006. LNCS, vol. 4047, pp. 242–261. Springer, Heidelberg (2006)

24. Stevens, M., Lenstra, A.K., de Weger, B.: Chosen-Prefix Collisions for MD5 and Colliding X.509 Certificates for Different Identities. In: Naor, M. (ed.) EUROCRYPT 2007. LNCS, vol. 4515, pp. 1–22. Springer, Heidelberg (2007)

25. Stevens, M., Sotirov, A., Appelbaum, J., Lenstra, A., Molnar, D., Osvik, D.A., de Weger, B.: Short Chosen-Prefix Collisions for MD5 and the Creation of a Rogue CA Certificate. In: Halevi, S. (ed.) CRYPTO 2009. LNCS, vol. 5677, pp. 55–69. Springer, Heidelberg (2009)

26. Wang, G., Keller, N., Dunkelman, O.: The Delicate Issues of Addition with Respect to XOR Differences. In: Adams, C., Miri, A., Wiener, M. (eds.) SAC 2007. LNCS, vol. 4876, pp. 212–231. Springer, Heidelberg (2007)

27. Wang, X., Yin, Y.L., Yu, H.: Finding Collisions in the Full SHA-1. In: Shoup, V. (ed.) CRYPTO 2005. LNCS, vol. 3621, pp. 17–36. Springer, Heidelberg (2005)

28. Wang, X., Yu, H.: How to Break MD5 and Other Hash Functions. In: Cramer, R. (ed.) EUROCRYPT 2005. LNCS, vol. 3494, pp. 19–35. Springer, Heidelberg (2005)

29. Yu, H., Chen, J., Ketingjia, W.X.: Near-Collision Attack on the Step-Reduced Compression Function of Skein-256. Cryptology ePrint Archive, Report 2011/148 (2011), http://eprint.iacr.org/ (last revised March 31, 2011)

30. Yu, H., Chen, J., Wang, X.: The Boomerang Attacks on the Round-Reduced Skein-512. In: SAC (2012)

Integral and Multidimensional Linear Distinguishers with Correlation Zero

Andrey Bogdanov[1,*], Gregor Leander[2,*], Kaisa Nyberg[3,*], Meiqin Wang[4,*]

[1] KU Leuven, ESAT/SCD/COSIC and IBBT, Belgium
[2] Technical University of Denmark, Denmark
[3] Aalto University, Finland
[4] Shandong University, Key Laboratory of Cryptologic Technology and Information Security, Ministry of Education, Shandong University, Jinan 250100, China

Abstract. Zero-correlation cryptanalysis uses linear approximations holding with probability exactly 1/2. In this paper, we reveal fundamental links of zero-correlation distinguishers to integral distinguishers and multidimensional linear distinguishers. We show that an integral implies zero-correlation linear approximations and that a zero-correlation linear distinguisher is actually a special case of multidimensional linear distinguishers. These observations provide new insight into zero-correlation cryptanalysis which is illustrated by attacking a Skipjack variant and round-reduced CAST-256 without weak key assumptions.

Keywords: zero-correlation cryptanalysis, integral distinguishers, multidimensional linear distinguishers, Skipjack, CAST-256.

1 Introduction

1.1 Zero-Correlation

Zero-correlation cryptanalysis [7, 8] is a novel promising attack technique for block ciphers. The distinguishing property used in zero-correlation cryptanalysis is the existence of *zero-correlation linear approximations* over (a part of) the cipher. Those are linear approximations that hold true with a probability p of exactly 1/2, that is, strictly unbiased approximations having a *correlation* $c = 2p - 1$ equal to 0.

The original work [7] provides a simple and efficient technique to find zero-correlation approximation but the distinguisher was rather weak. Recently, the work [8] has proposed a more powerful distinguisher by exploiting the fact that zero-correlation approximations are numerous in susceptible ciphers. Though working fine in practice and being useful in cryptanalysis, the distinguisher of [8] has some constraints that we would like to overcome: (1) If there are ℓ zero-correlation linear approximations for an n-bit block cipher, the distinguisher of [8] has to make $\mathcal{O}(2^n/\sqrt{\ell})$ queries. So the data complexity does not go down as fast as ℓ grows. (2) The distinguisher of [8] relies on the assumption that all linear approximations with

---

* All authors are corresponding authors.

X. Wang and K. Sako (Eds.): ASIACRYPT 2012, LNCS 7658, pp. 244–261, 2012.

correlation zero are independent. In most cases, including the attacks of [8] in fact, this assumption is formally not met, since all classes of zero-correlation approximations known so far are actually truncated, building linear spaces of dimension $\log_2 \ell$. That is, almost all ℓ approximations used will be linearly dependent, formally jeopardizing the assumption and another theory is needed to support the zero-correlation.

1.2   Our Contributions

Zero-Correlation and Integrals. Integral distinguishers were originally proposed by Knudsen as a dedicated attack against the Rijndael-predecessor Square [12]. Integral distinguishers [21] are also known as square distinguishers for this reason, especially when applied to Square-type ciphers such as AES. Variants of integral distinguishers include saturation [23] and multiset distinguishers [5]. Integral distinguishers mainly make use of the observation that it is possible to fix some parts of the plaintext such that specific parts of the ciphertext are balanced, i.e. each possible partial value occurs the exact same number of times in the output.

In this paper, we demonstrate that an integral implies zero-correlation linear approximations, see Fig. 1. In the other direction, a zero-correlation distinguisher implies an integral distinguisher only if input and output linear masks in zero-correlation approximations are independent of each other. Note that the condition for the input and output masks to be detached from each other implies that, for instance, the 5-round zero-correlation property of balanced Feistel ciphers of [7] is not directly described by an integral.

In this sense, the fact the integrals imply zero-correlation distinguishers is especially intriguing as not only the ways the distinguishers are constructed are different but also the ways the resulting attacks work seem inherently different. In particular, this link allows using ℓ input masks and one output mask with correlation zero in a distinguisher with a data complexity of $2^n/\ell$. Thus, in these settings the above outlined link allows to reduce the data complexity of zero-correlation distinguishers by a factor of $\sqrt{\ell}$ (at the price of transforming the attack into a chosen-plaintext attack) compared to previous works.

Zero-Correlation and Multidimensional Linear Distinguishers. The basic idea of multidimensional cryptanalysis [1,4,13,15,17,18] is that, given correlations of all linear approximations with non-zero correlation on a linear space formed by some cipher data, the probability distribution of the cipher data can be determined. Then, instead of the statistical behavior of a large set of mutually dependent linear approximations, one can examine the data distribution. Indeed, statistical behavior of multiple linear approximations has been analyzed only under the assumption of statistical independence [4]. The main advantage of the multidimensional approach is that it allows rigorous statistical analysis of linear approximations without the independence assumption. In traditional linear cryptanalysis, the focus is on linear approximations with correlations of large magnitude. The larger are the magnitudes of correlations, the more non-uniform is the distribution of the cipher data under consideration. The linear distinguisher is then based on distinguishing the nonuniform cipher data distribution from an uniform distribution. For a more comprehensive recent survey on multidimensional linear distinguishers, the reader is referred to e.g. [16].

In this paper, we consider linear spaces of cipher data where correlations of all linear approximations are equal to zero. Our starting observation here is that in fact, being truncated, *zero-correlation approximations constitute a special case of multidimensional linear approximations*. However, unlike traditional multidimensional linear distinguishers where the cipher data behaves non-uniformly, the cipher data for zero-correlation is uniformly distributed. This requires the development of a statistical theory to distinguish a sample of such cipher data from a sample of random data drawn from an uniform distribution.

In contrast to [8], the new distinguisher does not need the assumption of the statistical independence for multiple zero-correlation linear approximations. While still requiring about $\mathcal{O}(2^n/\sqrt{\ell})$ cipher queries, it allows taking full advantage of all zero-correlation linear approximations available, independent or not. The distribution of the cipher data is accurately modeled as sampling from a multivariate hypergeometric distribution, while the random data is drawn from a multinomial distribution. This establishes an inherent link of zero-correlation to multidimensional linear distinguishers. In their essence, zero-correlation distinguishers constitute a special case of multidimensional linear-correlation distinguishers, see Fig. 1. We expect this technique to be useful in the cryptanalysis of many ciphers.

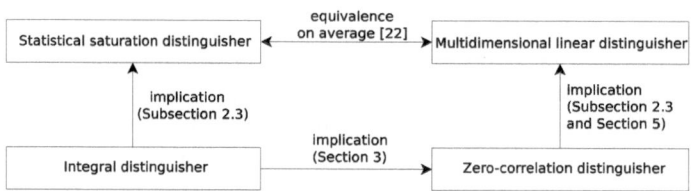

Fig. 1. Relations among distinguishers: zero-correlation, integral, statistical saturation, and multidimensional linear

Applications: Attacks on Skipjack Variant and CAST-256. To emphasize the practical meaningfulness of our findings, we apply the new distinguishers to mount key recovery attacks on block ciphers.

Skipjack is the only block cipher known to be designed by NSA. It is a 32-round 4-line unbalanced Feistel-type network based on interleaving two types of round functions – Rule A and Rule B. The best known cryptanalytic result for Skipjack is the impossible differential cryptanalysis for 31 rounds given by Biham et al. [2] based on a 24-round impossible differential. We change the order of Rules A and B in Skipjack such that the longest impossible differential identified is over 21 rounds and show that it has a 30-round zero-correlation property. We can recover its key for 31 rounds with practical complexity using an integral zero-correlation attack.

CAST-256 was proposed as an AES candidate. It has 48 rounds. The best cryptanalysis so far in the classical single-key model without the weak-key assumption has been a linear attack on 24 rounds. We find 24-round zero-correlation linear approximations for CAST-256 and attack 28 rounds of CAST-256 using multidimensional zero-correlation cryptanalysis. At the same time, the longest impossible differential we are aware of is over 18 rounds (though there is an unspecified impossible differential for 20 rounds mentioned in the literature). Our multidimensional zero-correlation attack is the first attack on more than half of the full-round AES-candidate CAST-256 without the weak key assumption.

The remainder of the paper is organized as follows. In Section 2, we introduce some basic concepts and notions which will be useful throughout the paper. Section 3 establishes a strong link between the properties of integrals and zero-correlation approximations. Using an integral zero-correlation distinguisher, Section 4 cryptanalyzes a Skipjack variant resistant to impossible differential attack. Section 5 describes a link of zero-correlation approximations to multidimensional linear approximations and introduces a novel zero-correlation multidimensional linear distinguisher. Section 6 uses it to recover the key of 28 rounds of CAST-256. We conclude in Section 7.

2   Preliminaries

2.1   Linear Approximations and Balanced Functions

$\mathbb{F}_2$ denotes the binary field of two elements and $\mathbb{F}_2^n$ is its extension of dimension n. Let x and $a \in \mathbb{F}_2^n$. Then $\langle a, x \rangle$ denotes their cannonical inner product on $\mathbb{F}_2^n$. Given a function $H : \mathbb{F}_2^n \to \mathbb{F}_2^k$ the *correlation* c of the *linear approximation*

$$\langle b, H(x) \rangle + \langle a, x \rangle$$

for a k-bit output mask b and an n-bit input mask a is defined by

$$\Pr(\langle b, H(x) \rangle + \langle a, x \rangle = 0) = \frac{1 + c}{2}$$

where the probability is taken over all choices of inputs x. A related measure for this correlation is *the Walsh- or Fourier-transformation*, defined as

$$\widehat{H}(a, b) = \sum_x (-1)^{\langle b, H(x) \rangle + \langle a, x \rangle}.$$

The fundamental relation between the Fourier transformation of H and the correlation of the linear approximation is given by

$$c = \frac{\widehat{H}(a, b)}{2^n}$$

and, thus, studying the correlation and studying the Fourier transformation are, up to scaling, equivalent.

We say a function $F : \mathbb{F}_2^n \to \mathbb{F}_2^k$ is *balanced* if all preimages have identical size, i.e. if the size of the set

$$F^{-1}(y) := \{x \in \mathbb{F}_2^n \mid F(x) = y\}$$

is independent of y. Note that F being balanced implies $k \leq n$. We recall the following well-known characterization of balanced functions, see for example [10, Proposition 2]: A function $F : \mathbb{F}_2^n \to \mathbb{F}_2^k$ is balanced if and only if all its component functions are balanced, that is, if and only if for any non-zero $b \in \mathbb{F}_2^k$ it holds that $\widehat{F}(0, b) = 0$.

2.2   Decomposition of the Target Cipher

Assume that $H : \mathbb{F}_2^n \to \mathbb{F}_2^n$ is a (part of) cipher. To simplify notation and without loss of generality we split the inputs and outputs into two parts each.

$$H : \mathbb{F}_2^r \times \mathbb{F}_2^s \to \mathbb{F}_2^t \times F_2^u, H(x, y) = \begin{pmatrix} H_1(x, y) \\ H_2(x, y) \end{pmatrix}$$

Furthermore, the function T_λ defined by

$$T_\lambda : \mathbb{F}_2^s \to \mathbb{F}_2^t, T_\lambda(y) = H_1(\lambda, y)$$

will play a key role. The function T_λ is the function H when the first r bits of its input are fixed to λ and only the first t bits of the output are taken into account.

Table 1. Defining properties of some important distinguishers

Distinguisher	Defining property
multidimensional linear	$\sum_{a_1,b_1} \widehat{H}(a,b)^2$ non-random
statistical saturation	$\forall \lambda : \sum_{b_1} \widehat{T}_\lambda(0,b_1)^2$ non-random
integral	$\forall \lambda, b_1 : \widehat{T}_\lambda(0,b_1) = 0$
zero-correlation	$\forall a_1, b_1 : \widehat{H}(a,b) = 0$

2.3   Distinguishers and Relations

Here we briefly outline the concepts behind four types of relevant distinguishers that we will be dealing with in this paper, which are also summarized in Table 1: *Zero-correlation distinguisher* uses the property that, for all input and output masks $a = (a_1, 0)$ and $b = (b_1, 0)$, the Fourier transformation of the cipher yields zero, $\widehat{H}(a, b) = 0$. *Integral distinguisher* is based on the property that, for all partial input fixations λ, the partial function of the cipher with this fixation is balanced in parts of its output. *Multidimensional linear distinguisher* relies upon the property that multiple Fourier coefficients of the cipher behave in a non-random way, i.e. $\sum_{a_1,b_1} \widehat{H}(a,b)^2$ is non-random. *Statistical saturation distinguisher* builds upon the property that, for all partial input fixations λ, the partial function of the cipher with this fixation is non-random under Fourier transformation, i.e. $\sum_{b_1} \widehat{T}_\lambda(0,b_1)^2$ is non-random. While statistical saturation and multidimensional linear distinguishers concentrate on the cumulative properties holding for the partial Fourier spectra, integral and zero-correlation distinguishers deal with a set of individual properties of Fourier coefficients.

3   Zero-Correlation and Integral Distinguishers

3.1   Conditional Equivalence Result

We start by stating the main result of this section, which is summarized in the following statement:

Proposition 1. *If the input and output linear masks a and b are independent, the approximation $\langle b, H(x)\rangle + \langle a, x\rangle$ has correlation zero for any $a = (a_1, 0)$ and any $b = (b_1, 0) \neq 0$ (zero-correlation) if and only if the function T_λ is balanced for any λ (integral).*

This basically means that, at least in terms of their defining properties, integral distinguishers imply zero-correlation distinguishers. The proof of Proposition 1 follows directly from the two lemmata below whose proofs are provided in the full version of this paper [6]. The tools used in the proofs mainly originate from results in the area of Boolean functions [22]. For instance, Lemma 2 is stated in different notation e.g. in [11, Proposition 9]).

The main technical tool is the next lemma linking the correlation of T_λ to the correlation of H.

Lemma 1. *With the notation from above, the following holds for any λ, b_1:*

$$2^s \widehat{T}_\lambda(0, b_1) = \sum_{a_1} (-1)^{\langle a_1, \lambda \rangle} \widehat{H}((a_1, 0), (b_1, 0)) \tag{1}$$

Lemma 1 already proves one direction of Proposition 1, namely, that zero-correlation approximations imply an integral under the condition that b_1 remains the same with the change of a_1. Lemma 1 is also especially useful for defining an integral distinguisher that is based on zero-correlation properties: Given a number of zero-correlation linear approximations (on the right-hand side of (1)), one checks if the corresponding partial function of the cipher is balanced (the left-hand side of (1)). This can be done for each partial input fixation λ separately.

The following direct corollary of Lemma 1 is even more telling and is the key in exhibiting the close link between zero-correlation distinguishers and integral distinguishers:

Lemma 2. *The following holds for any b_1:*

$$2^s \sum_\lambda \widehat{T}_\lambda(0, b_1)^2 = \sum_{a_1} \widehat{H}_1((a_1, 0), b_1)^2$$

This lemma proves both directions of Proposition 1, including the fact that an integral implies zero-correlation distinguishers. In the sequel, we provide a more detailed description of the link and an example.

3.2   From Zero-Correlation to Integral Distinguishers (Conditional)...

First, assume that $H : \mathbb{F}_2^n \to \mathbb{F}_2^n$ is a (part of) cipher vulnerable to zero-correlation attacks. More precisely, assume that for any $a = (a_1, 0)$ and any $b = (b_1, 0) \neq 0$ the relation $\langle b, H(x) \rangle + \langle a, x \rangle$ has correlation zero. We'd like to highlight two points here. The restriction to masks of the form $a = (a_1, 0)$ and $b = (b_1, 0)$, that is, to the masks where the last bits are fixed to zero, is solely for the simplicity of notations. However, the zero-correlation distinguishers considered here are of a special case: We assume not only that the used input and output masks form subspaces but also that this space of input and output masks is actually the direct product of the space of input masks and the space of output masks. Informally, the masks must not be coupled as they are for example in the attack on CAST-256 described in Section 6. We call such uncoupled input-output masks, for our equivalence result applies, *detached masks*.

Under those conditions, it follows from Lemma 2 above that $\widehat{T}_\lambda(0, b_1)$ equals zero for all $b_1 \neq 0$ and all λ. This yields that, for any λ the function T_λ mapping s bits to t bits is balanced. In other words, H exhibits the following integral distinguisher: Fixing the first s bits of H arbitrarily and encrypting all remaining 2^r possible plaintext, each possible t bits string occurs equally often in the first t bits of the output of H. In the particular case of $s = t$, the function T_λ is a permutation and, thus, each possible t-bit string should occur exactly once.

3.3   ...And Back Again (Unconditional)

On the other hand, let us consider the case of a cipher that is vulnerable to an integral distinguisher in the following sense. Assume that, by fixing some (without loss of generality, the first s) bits in the input and encrypting all possible remaining plaintexts, one can identify a subset of t bits (again without loss of generality, the first t bits), each possible t-bit string occurs equally often. Then H is also vulnerable to a zero-correlation attack. More precisely, $\widehat{H}((a_1, 0), (b_1, 0)) = 0$ for all $a_1 \in \mathbb{F}_2^s$ and $b_1 \in \mathbb{F}_2^t$. Again, this follows directly from Lemma 2. In fact, an integral unconditionally implies zero-correlation.

3.4   Discussion of the Link

As pointed out, this relation is intriguing as zero-correlation distinguishers and integral distinguishers are constructed quite differently. Moreover, not only the ways the distinguishers are constructed are different but also the ways the resulting attacks work seem inherently different.

The first difference is that zero-correlation attacks are usually known plaintext attacks (or using known distinct plaintexts, while integral attacks are usually chosen plaintext attacks. Moreover, for zero-correlation attacks, appending rounds before the distinguisher normally does not increase the data complexity. On the other hand, appending rounds before an integral distinguisher often results in an increased data complexity as, for each (partial) key guess, one has to ensure that some values are fixed according to the distinguisher. Finally, integral distinguishers have the advantage that it is often possible to extend the distinguisher by relaxing the balanced property to a zero-sum property (or equivalently to the fact that a certain subfunction does not have maximal algebraic degree). For zero correlation attacks, such an extension is not known so far.

Thus, besides being interesting from a theoretical perspective, the above mentioned link clearly calls for further work on combining the specific advantages offered by both attacks.

Before discussing an application of this relation to mount an integral attack on a variant of Skipjack, we'd like to illustrate the above with AES as an example.

3.5   Example with AES

Fig.2 depicts the well-known 3-round integral distinguisher for AES. Starting with one active byte and fixing all other bytes results in all bytes being active after ShiftRows in the third round. In terms of zero-correlation distinguisher, the above discussion implies that for any non-zero input mask with (at least) one zero byte and any non-zero output mask which is zero in all but one byte the corresponding linear approximation is unbiased.

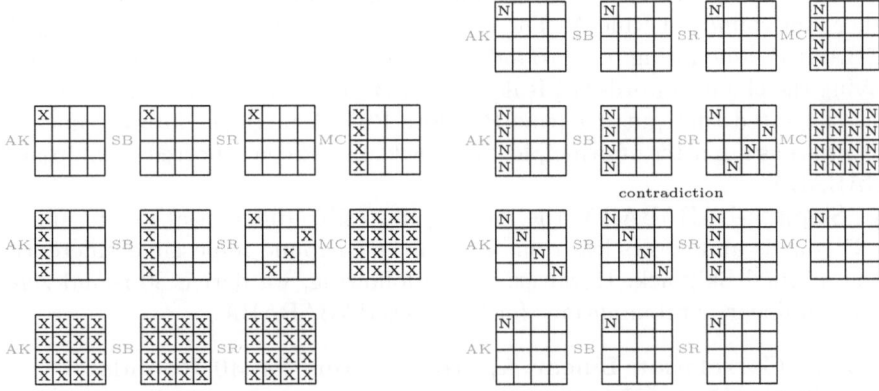

Fig. 2. The integral distinguisher on 3 rounds of AES. The X denotes an active byte.

Fig. 3. Zero correlation distinguisher on 4 rounds of AES. The N denotes a non-zero byte in the mask.

Reciprocally, Fig.3 shows the 4 round zero-correlation distinguisher from [7]. For any non-zero mask which is zero in all-but-one bytes and any output mask with the same condition, the corresponding linear approximation is unbiased.

Now, again using the above discussion, this implies the following integral distinguisher on 4 rounds of AES. Fix any byte in the plaintext and encrypt all remaining 2^{120} possible plaintexts. Check if the output restricted to any byte results is a balanced function, that is, of each out of the possible 256 values is obtained exactly 2^{112} times. Note that this distinguisher was implicitly used for example in [14].

4   Integral Zero-Correlation for a Skipjack Variant

4.1   Skipjack-BABABABA vs the Original Skipjack-AABBAABB

Skipjack [25] is the only block cipher known to be designed by NSA. Skipjack is a 64-bit block cipher with an 80-bit key. It is an unbalanced Feistel network with 32 rounds of two types, called Rule A and Rule B. Each round is described in the form of a linear feedback shift register with additional non-linear keyed G permutation. Rule B is basically the inverse of Rule A with minor positioning differences. Skipjack applies eight rounds of Rule A, followed by eight rounds of Rule B, followed by another eight rounds of Rule A, followed by another eight rounds of Rule B. We refer to this original Skipjack algorithm as Skipjack-AABBAABB – A denoting four rounds of Rule A and B standing for four rounds of Rule B. The best known cryptanalytic result for the original Skipjack-AABBAABB is the impossible differential cryptanalysis for 31 rounds given by Biham et.al. [2] based on a 24-round impossible differential.

In Skipjack-BABABABA, four rounds of Rule B are applied first, followed by four rounds of Rule A, followed by another four rounds of Rule B, followed by another four rounds of Rule A. The rest of the cipher is exactly as in Skipjack-AABBAABB, amounting to 32 rounds in total. See the Fig.4a. Skipjack variants involving the change of order of Rules A and B were studied in [19,20]. Though it was suggested that putting Rule B before Rule A might facilitate truncated differentials as a matter of principle, no attacks have been reported on Skipjack-BABABABA.

For Skipjack-BABABABA, the longest impossible differential we can find is over 21 rounds and covers less rounds than the 24-round impossible differential for the original Skipjack. However, in the following, we derive 30-round zero-correlation linear approximations for Skipjack-BABABABA.

4.2   Zero-Correlation Linear Approximations for 30 Rounds of Skipjack-BABABABA

Let the input masks for the first round be $(L_1, L_1, 0, 0)$ and the output mask for the last round be $(L_2, L_2, 0, 0)$ for any non-zero L_1 and L_2. Fig.4b depicts the evolution of both masks from the top and from the bottom towards the middle of the cipher. In the figure, M_i denotes an undetermined non-zero mask and R_i denotes an undetermined mask (zero or non-zero). From the input mask $(L_1, L_1, 0, 0)$ at the first round, the output mask of the 19-th round is (M_4, R_2, R_1, M_5). From the output mask $(L_2, L_2, 0, 0)$ at the 30-th round, the input mask of the 20-th

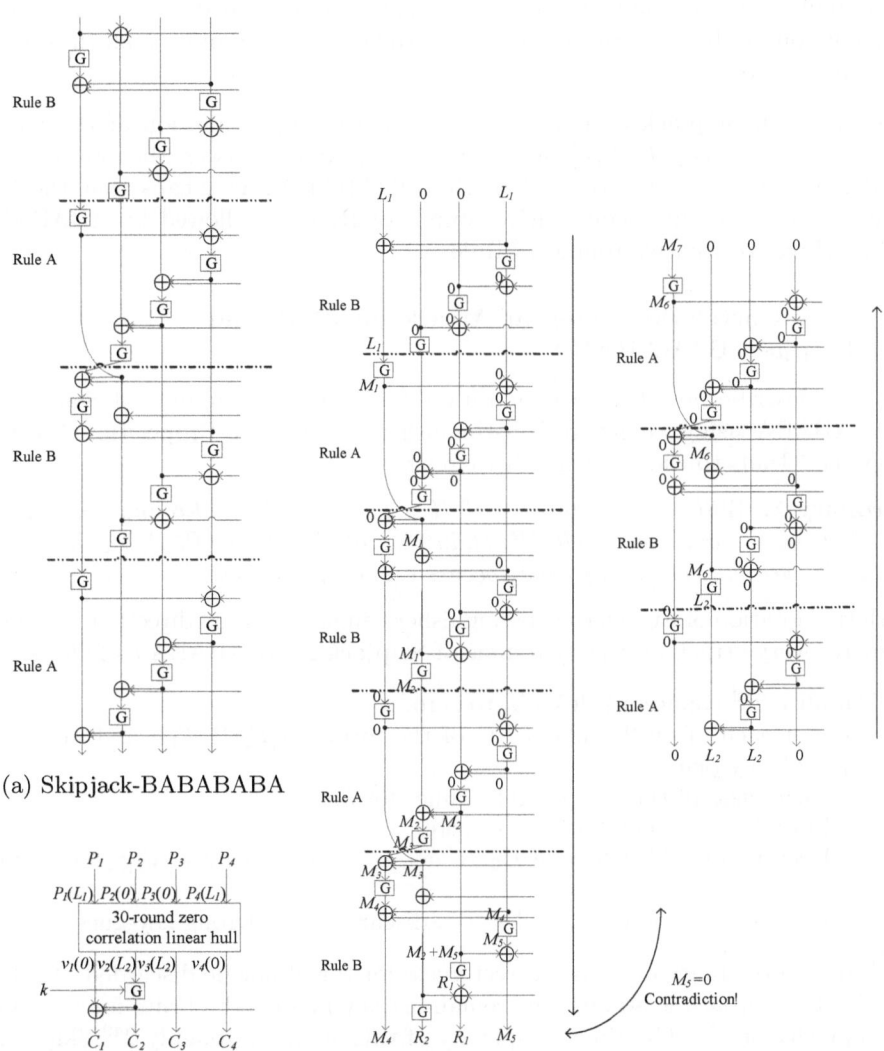

(a) Skipjack-BABABABA

(c) Integral zero-correlation attack on 31-round Skipjack-BABABABA (values in brackets are masks, plain values are actual data processed)

(b) 30-round zero-correlation linear approximations for Skipjack-BABABABA (M_i denotes an undetermined non-zero mask and R_i denotes an undetermined mask – zero or non-zero)

Fig. 4. Integral zero-correlation cryptanalysis of 31-round Skipjack-BABABABA

round is $(M_7, 0, 0, 0)$. Here we conclude that $(M_4, R_2, R_1, M_5) \neq (M_7, 0, 0, 0)$ as equality would imply that $M_5 = 0$ contradicting that $M_5 \neq 0$. Therefore, the linear hull of the 30-round linear approximation $(L_1, L_1, 0, 0) \rightarrow (L_2, L_2, 0, 0)$ does not contain linear trails of non-zero correlation contribution and, thus, has correlation zero.

Property 1. In Skipjack-BABABABA, each linear approximation of the form $(L_1, L_1, 0, 0) \rightarrow (L_2, L_2, 0, 0)$ for non-zero L_1 and L_2 over the 30 rounds $B^3 ABABABA^3$ has zero correlation. Here $B^3 ABABABA^3$ means that the 30 rounds start from three consecutive rounds of Rule B, followed by ABABAB and by three consecutive rounds of Rule A.

4.3   Zero-Correlation Integral Attack on 31-Round Skipjack-BABABABA

Here we describe how to use Proposition 1 to attack 31 rounds of Skipjack-$B^3 ABABABA$ using an integral distinguisher. Combining Proposition 1 with Property 1 leads to the following distinguisher.

Corollary 1. *With the notation of Fig.4c, for the 30-round Skipjack-$B^3 ABABABA^3$, encrypting all 2^{48} plaintexts of the form $(P_1|P_2|P_3|P_1)$ each of the 2^{16} possible values of $v_2 \oplus v_3$ occurs exactly 2^{32} times.*

With the notation of Fig.4c, this distinguisher can now be used directly to mount a key-recovery attack on the 31 rounds of Skipjack-$B^3 ABABABA$ as follows.

- Initialize 2^{32} counters $V_1[C_2|C_3]$ to zero.
- Encrypt each of all 2^{48} plaintexts of the form $(P_1|P_2|P_3|P_1)$, and increase $V_1[C_2|C_3]$ by one.
- For each guess of the 2^{32} possible values for k:
  - Initialize 2^{16} counters $V_2[v]$ to zero.
  - Decrypt all 2^{16} values of C_2 to get $v_2|v_3$ and increase $V_2[v_2 \oplus v_3]$ by $V_1[C_2|C_3]$.
  - If one of the counters $V_2[v] \neq 2^{16}$, discard k as a wrong key-guess.

With high probability only the correct guess for k will not be discarded. As the key size for Skipjack is 80 bits, the remaining key bits can be brute-forced with a complexity of 2^{48}. The time complexity of this attack is roughly 2^{49} Skipjack encryptions and we have to store roughly 2^{32} counters. The data complexity is 2^{48} chosen plaintexts. Thus, this attack has practical complexities.

5   Zero-Correlation and Multidimensional Linear Distinguishers

5.1   Multidimensional Linear Setting

Given m linear approximations

$$\langle u_i, x \rangle + \langle w_i, y \rangle, \quad i = 1, \ldots, m,$$

where $x \in \mathbb{F}_2^n$ is plaintext and $y \in \mathbb{F}_2^t$ is some part of data in the encryption process, one obtains an m-tuple of bits by evaluating those for a plaintext-ciphertext pair. Instead of considering each such bit and its distribution independently as x varies, multidimensional linear cryptanalysis focuses on the analysis of the distribution of the m-tuples

$$z = (z_1, \ldots, z_m), \quad z_i = \langle u_i, x \rangle + \langle w_i, y \rangle.$$

Then we have the following relationship between the probability distribution of z and the correlations c_γ of all linear approximations $\gamma \in \mathbb{F}_2^m$:

$$\Pr[z] = 2^{-m} \sum_{\gamma \in \mathbb{F}_2^m} (-1)^{\langle \gamma, z \rangle} c_\gamma. \tag{2}$$

Note that this is actually the key in proving that for a balanced function all component functions have zero-correlation.

We denote by U and W the $m \times n$ and $m \times t$ matrices with rows u_i and w_i, respectively. Then we have $z = Ux + Wy$ and can write

$$\langle \gamma, z \rangle = \langle \gamma, Ux + Wy \rangle = \langle U^T \gamma, x \rangle + \langle W^T \gamma, y \rangle, \tag{3}$$

where $U^T \gamma$ and $W^T \gamma$ are linear combinations of the linear masks u_i and w_i, $i = 0, \ldots, m$, respectively.

5.2   How to Make Zero-Correlation Multidimensional

Now we are ready to formulate the zero-correlation distinguishing property as a special case of the multidimensional distinguishing property.

Zero-correlation distinguisher assumes that the correlations of all linear approximations $\langle u_i, x \rangle + \langle w_i, y \rangle$, $i = 1, \ldots, m$, and their nonzero linear combinations are equal to zero. (Note that this means, in particular, that these m linear approximations are statistically independent.) By (3), it follows that $c_\gamma = 0$, for all $\gamma \neq 0$. When substituting this information in the formula of $\Pr[z]$ in (2), we obtain that z has a uniform distribution in $\mathbb{F}_2^m$.

Let the adversary be given N distinct plaintexts for an n-bit block cipher and m linear approximations such that all their nonzero linear combinations have correlation zero. Then he can construct, as shown above, a function from $\mathbb{F}_2^n$ to $\mathbb{F}_2^m$ whose outputs z computed for all plaintexts are uniformly distributed m-tuples of bits in $\mathbb{F}_2^m$.

Such a completely uniform distribution is very unlikely to have been obtained from selecting the values at random in $\mathbb{F}_2^m$, even if the probability of each value is equal, spanning a linear space of $\ell = 2^m$ zero-correlation approximations of dimension m. But as we will see, it is possible to distinguish the non-random behavior of the cipher data already with much less data than the full codebook. The distribution of the cipher data follows *multivariate hypergeometric distribution*, while the data drawn at random from a uniform distribution on $\mathbb{F}_2^m$ follows *multinomial distribution*. These distributions have essentially different parameters for large sample sizes N and can be distinguished from each other. The distinguisher can be obtained as follows.

5.3   Multidimensional Distinguisher for Correlation Zero

For each of the 2^m data values $z \in \mathbb{F}_2^m$, the attacker initializes a counter $V[z]$, $z = 0, 1, 2, \ldots, 2^m - 1$, to zero value. Then, for each distinct plaintext, the attacker computes the corresponding data value in $\mathbb{F}_2^m$ (by evaluating the m basis linear approximations) and increments the counter $V[z]$ of this data value by one. Then the attacker computes the statistic T for this distribution as

$$T = \sum_{i=0}^{2^m - 1} \frac{(V[z] - N2^{-m})^2}{N2^{-m}(1 - 2^{-m})}. \tag{4}$$

The statistic T will have two distinct distributions for the cipher exhibiting zero-correlation and a randomly drawn permutation which is our wrong-key hypothesis assumption:

Proposition 2. *For sufficiently large sample size N and number ℓ of zero-correlation linear approximations given for the cipher, the statistic T follows a χ^2-distribution for the cipher approximately with mean and variance*

$$\mu_0 = \mathrm{Exp}(T_{cipher}) = (\ell - 1)\frac{2^n - N}{2^n - 1} \quad and \quad \sigma_0^2 = \mathrm{Var}(T_{cipher}) = 2(\ell - 1)\left(\frac{2^n - N}{2^n - 1}\right)^2$$

and for a randomly drawn permutation with mean and variance

$$\mu_1 = \mathrm{Exp}(T_{random}) = \ell - 1 \quad and \quad \sigma_1^2 = \mathrm{Var}(T_{random}) = 2(\ell - 1).$$

The proof of this proposition is available in the full version of this paper [6].

5.4   Distinguishing Complexity

Applying the standard normal approximation of χ^2 to the two different distributions of the statistic T in Proposition 2, one can compute data complexities N of the distinguisher, given error probabilities. As a rule of thumb, we can conclude that it is sufficient to have $N \approx 2^{n+2-\frac{m}{2}}$ distinct plaintexts and their corresponding ciphertexts to distinguish the cipher distribution from randomly drawn permutation. A more precise distinguishing complexity is given by the following statement.

Corollary 2. *Under the assumptions of Proposition 2, for type-I error probability α_0 (the probability to wrongfully discard the cipher), type-II error probability α_1 (the probability to wrongfully accept a randomly chosen permutation as the cipher), for an n-bit block cipher exhibiting ℓ zero-correlation linear approximations forming an $\log_2 \ell$-dimensional linear space, the distinguishing complexity N can be approximated as*

$$N = \frac{2^n(q_{1-\alpha_0} + q_{1-\alpha_1})}{\sqrt{\ell/2} - q_{1-\alpha_1}},$$

where $q_{1-\alpha_0}$ and $q_{1-\alpha_1}$ are the respective quantiles of the standard normal distribution.

Note that this statistical test is based on the decision threshold of $\tau = \mu_0 + \sigma_0 q_{1-\alpha_0} = \mu_1 - \sigma_1 q_{1-\alpha_1}$: If the statistic $T \leq \tau$, the test outputs 'cipher'. Otherwise, if the statistic $T > \tau$, the test returns 'random'.

6   Multidimensional Zero-Correlation for 28-Round CAST-256

6.1   Description of CAST-256

As a first-round AES candidate, CAST-256 is designed based on CAST-128. The block size is 128 bits, and the key size can be 128, 192 or 256 bits. CAST-256 has 48 rounds for all key sizes. The design of CAST-256 is a generalized Feistel network with 4 lines as illustrated in Fig.5a.

We denote the 128-bit block of CAST-256 as $\beta = (A|B|C|D)$, where A, B, C and D are 32 bits each. Two types of round function, the *forward quad-round* $Q(\cdot)$ and the *reverse quad-round* $\bar{Q}(\cdot)$ are used in CAST-256.

The forward quad-round $\beta := Q_i(\beta)$ is defined as consecutive application of 4 rounds as follows:

$$C = C \oplus F_1(D, K_{R_1}{}^{(i)}, K_{M_1}{}^{(i)}), \quad B = B \oplus F_2(C, K_{R_2}{}^{(i)}, K_{M_2}{}^{(i)}),$$
$$A = A \oplus F_3(B, K_{R_3}{}^{(i)}, K_{M_3}{}^{(i)}), \quad D = D \oplus F_1(A, K_{R_4}{}^{(i)}, K_{M_4}{}^{(i)}).$$

Similarly, the reverse quad-round $\beta := \bar{Q}_i(\beta)$ is defined as:

$$D = D \oplus F_1(A, K_{R_4}{}^{(i)}, K_{M_4}{}^{(i)}), \quad A = A \oplus F_3(B, K_{R_3}{}^{(i)}, K_{M_3}{}^{(i)}),$$
$$B = B \oplus F_2(C, K_{R_2}{}^{(i)}, K_{M_2}{}^{(i)}), \quad C = C \oplus F_1(D, K_{R_1}{}^{(i)}, K_{M_1}{}^{(i)}),$$

where $K_R{}^{(i)} = \{K_{R_1}{}^{(i)}, K_{R_2}{}^{(i)}, K_{R_3}{}^{(i)}, K_{R_4}{}^{(i)}\}$ is the set of rotation keys for the i-th quad-round, and $K_M{}^{(i)} = \{K_{M_1}{}^{(i)}, K_{M_2}{}^{(i)}, K_{M_3}{}^{(i)}, K_{M_4}{}^{(i)}\}$ is the set of masking keys for the i-th quad-round.

The encryption procedure for CAST-256 consists of 6 forward quad-rounds followed by 6 reverse quad-rounds, counting 48 rounds in total. Decryption is identical to encryption except that the sets of quad-round keys $K_R{}^{(i)}$ and $K_M{}^{(i)}$ are applied in the reverse order. The keys are obtained from an up to 256-bit master key by encrypting it with a CAST-256-type cipher (acting on on eight 32-bit words) with known constants as subkeys.

The functions F_1, F_2 and F_3 are exactly those of CAST-128. They use four 8x32-bit S-boxes based on bent functions, modular addition, modular subtraction, XOR and key-dependent rotation. See Fig. 5a.

6.2   24-Round Zero-Correlation Linear Approximations for CAST-256

Property 2. For 24-round CAST-256 (3 forward quad-rounds followed by 3 reverse quad-rounds, or rounds 13-36), if the input mask is $(0|0|0|L_1)$ and the output mask is $(0|0|0|L_2)$, the correlation of the linear approximation for the 24-round CAST-256 is zero, where $L_1 \neq L_2, L_1 \neq 0$, and $L_2 \neq 0$.

The proof of this property is available in the full version of this paper [6].

As compared to this 24-round property, the longest impossible differential for CAST-256 we are aware of covers 18 rounds [28]. The work [3] claims unspecified 20-round impossible differentials. Thus, the zero-correlation property for CAST-256 is at least 4 rounds longer than the one of impossible differential.

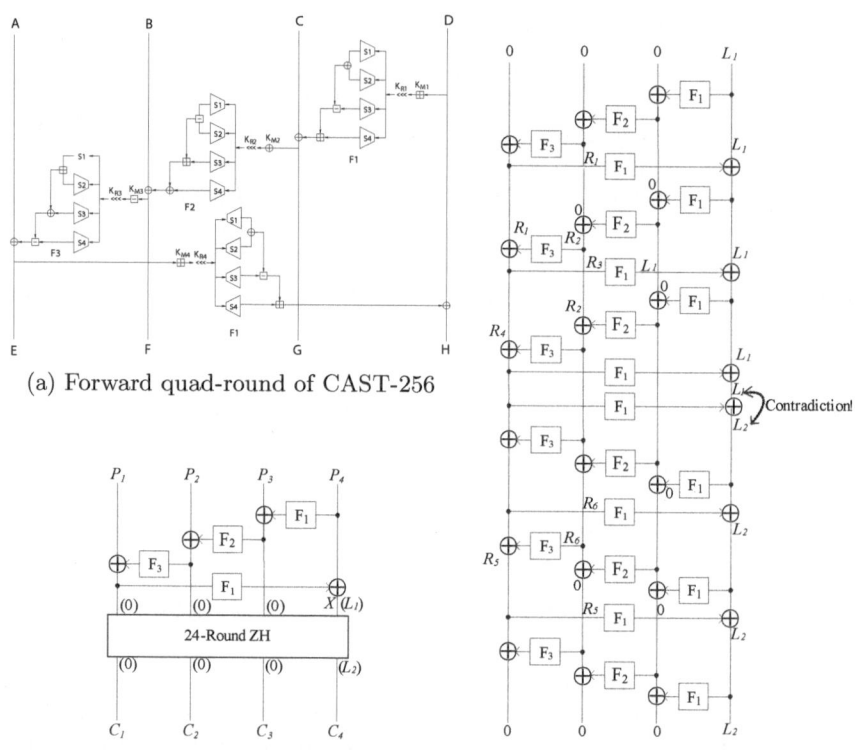

(a) Forward quad-round of CAST-256

(c) Multidimensional zero-correlation cryptanalysis of 28-round CAST-256: 4 forward quad-rounds and 3 reverse quad-rounds (values in brackets denote masks, plain values are actual data processed)

(b) 24-round zero-correlation linear approximations of CAST-256 (3 forward and 3 reverse quad-rounds)

Fig. 5. Multidimensional zero-correlation cryptanalysis of 28-round CAST-256

6.3   Key Recovery for 28-Round CAST-256

We use the 24-round zero-correlation linear approximations of Property 2 to attack 28 rounds of CAST-256. Fig. 5c illustrates the recovery of the subkey values from the first round to the fourth round. The attack works as follows.

For each possible 148-bit subkey value $\kappa = K_R^{(1)}|K_M^{(1)}$:

1. Allocate a 64-bit global counter $V[z]$ for each of 2^{64} possible values of the 64-bit vector z and set it to 0. $V[z]$ will contain the number of times the vector value z occurs for the current key guess κ. The vector z is the concatenation of evaluations of 64 basis zero-correlation masks.
2. For each of N distinct plaintext-ciphertext pairs:
   (a) Partially encrypt 4 rounds and get 64-bit value for $X|C_4$.
   (b) Evaluate all 64 basis zero-correlation masks on $X|C_4$ and put the evaluations to the vector z.
   (c) Increment $V[z]$.
3. Compute the χ^2 statistic $T = N2^{64} \sum_{z=0}^{2^{64}-1} \left(\frac{V[z]}{N} - \frac{1}{2^{64}} \right)^2$.
4. If $T < \tau$, then the subkey guess κ is a possible subkey candidate and all master keys it is compatible with are tested exhaustively against a maximum of 3 plaintext-ciphertext pairs.

Table 2. Summary of attacks on CAST-256: KP = Known Plaintexts, CP = Chosen Plaintexts

Rounds	Key size	Attack	Data	Time	Memory (bytes)	Ratio of weak keys	Ref.
16	128, 192, 256	boomerang	$2^{49.3}$CP	–	–	1	[26]
24	192 or 256	linear	$2^{124.1}$KP	$2^{156.52}$	–	1	[27]
36	256	differential	2^{123}CP	2^{182}	–	2^{-35}	[24]
28	256	multidim. ZC	$2^{98.8}$KP	$2^{246.9}$	2^{68}	1	Here

In this attack, using Corollary 2, we set the type-I error probability (the probability to miss the right key) to $\alpha_0 = 2^{-2.7}$ and the type-II error probability (the probability to accept a wrong key) to $\alpha_1 = 2^{-14}$. Thus, we get $q_{1-\alpha_0} = 1$ and $q_{1-\alpha_1} = 3.84$. Here, $\tau = \sigma_1 \cdot q_{\alpha_1} + \mu_1 \approx 2^{64}$.

Corollary 2 suggests that the data complexity is $N = 2^{98.8}$ distinct plaintext-ciphertexts with those parameters. The success probability of the entire attack is $1 - \alpha_0 \approx 0.846$.

The time complexity is $2^{246.8}$ times of one-round encryption and $2^{246.8}$ memory accesses to a memory of size 2^{64}. Under the assumption that one memory access with size 2^{64} is equivalent to one 28-round CAST-256 encryption, the total time complexity would be about $2^{246.9}$ 28-round CAST-256 encryptions. Due to $\alpha_1 = 2^{-14}$ and the total number of recovered bits is 148, the number of the remaining subkey values is $2^{-14} \cdot 2^{148} = 2^{134}$. Then we exhaustively search other $256 - 148 = 108$ subkey bits, the time complexity will be $2^{134+108} = 2^{242}$ times of 28-round encryptions.

The memory requirements are 2^{64} 128-bit words needed for $V[z]$, or 2^{68} bytes.

In all, the data complexity is about $2^{98.8}$ known plaintexts, the time complexity is about $2^{246.9}$ 28-round CAST-256 encryptions and the memory requirements are 2^{64} blocks. This is the first attack on more than half of the full-round AES-candidate CAST-256 without the weak key assumption. See Table 2 for a summary and a comparison of attacks.

7   Conclusions

In this paper, we establish fundamental links between zero-correlation distinguishers on the one hand and integral and multidimensional linear distinguishers on the other. In particular, an integral implies a zero-correlation property and zero-correlation distinguishers can be seen as a special case of multidimensional linear distinguishers. These findings result in two novel distinguishers for zero-correlation based on integral and multidimensional linear distinguishers. To obtain the latter, we refine the theory of multidimensional linear distinguishers. We illustrate these new distinguishers by mounting attacks on a Skipjack variant and CAST-256.

Acknowledgements. Andrey Bogdanov is postdoctoral fellow of the Fund for Scientific Research - Flanders (FWO). This work has been supported in part by the IAP Programme P6/26 BCRYPT of the Belgian State, by the European Commission under contract number ICT-2007-216676 ECRYPT NoE phase II, by KU Leuven-BOF (OT/08/027), by the Research Council KU Leuven (GOA TENSE), by NSFC Projects (No.61133013 and No.61070244), by 973 project (2013CB834205) as well as Interdisciplinary Research Foundation of Shandong University (No.2012JC018).

References

1. Baignères, T., Junod, P., Vaudenay, S.: How Far Can We Go Beyond Linear Cryptanalysis? In: Lee, P.J. (ed.) ASIACRYPT 2004. LNCS, vol. 3329, pp. 432–450. Springer, Heidelberg (2004)
2. Biham, E., Biryukov, A., Shamir, A.: Cryptanalysis of Skipjack Reduced to 31 Rounds Using Impossible Differentials. In: Stern, J. (ed.) EUROCRYPT 1999. LNCS, vol. 1592, pp. 12–23. Springer, Heidelberg (1999)
3. Biham, E., Biryukov, A., Shamir, A.: Miss in the Middle Attacks on IDEA and Khufu. In: Knudsen, L.R. (ed.) FSE 1999. LNCS, vol. 1636, pp. 124–138. Springer, Heidelberg (1999)
4. Biryukov, A., De Cannière, C., Quisquater, M.: On Multiple Linear Approximations. In: Franklin, M. (ed.) CRYPTO 2004. LNCS, vol. 3152, pp. 1–22. Springer, Heidelberg (2004)
5. Biryukov, A., Shamir, A.: Structural Cryptanalysis of SASAS. In: Pfitzmann, B. (ed.) EUROCRYPT 2001. LNCS, vol. 2045, pp. 394–405. Springer, Heidelberg (2001)
6. Bogdanov, A., Leander, G., Nyberg, K., Wang, M.: Integral and Multidimensional Linear Distinguishers with Correlation Zero. IACR ePrint Archive report (2012)
7. Bogdanov, A., Rijmen, V.: Linear Hulls with Correlation Zero and Linear Cryptanalysis of Block Ciphers. Designs, Codes and Cryptography. Springer (to appear, 2012); preprint available as Cryptology ePrint Archive: Report 2011/123, http://eprint.iacr.org/2011/123
8. Bogdanov, A., Wang, M.: Zero Correlation Linear Cryptanalysis with Reduced Data Complexity. In: Canteaut, A. (ed.) FSE 2012. LNCS, vol. 7549, pp. 29–48. Springer, Heidelberg (2012)

9. Borst, J., Knudsen, L.R., Rijmen, V.: Two Attacks on Reduced IDEA. In: Fumy, W. (ed.) EUROCRYPT 1997. LNCS, vol. 1233, pp. 1–13. Springer, Heidelberg (1997)
10. Carlet, C.: Vectorial (multi-output) Boolean Functions for Cryptography. Cambridge University Press (to appear)
11. Carlet, C.: Boolean Functions for Cryptography and Error Correcting Codes. Cambridge University Press (to appear), preliminary version,
http://www-rocq.inria.fr/codes/Claude.Carlet/chap-fcts-Bool.pdf
12. Daemen, J., Knudsen, L.R., Rijmen, V.: The Block Cipher SQUARE. In: Biham, E. (ed.) FSE 1997. LNCS, vol. 1267, pp. 149–165. Springer, Heidelberg (1997)
13. Englund, H., Maximov, A.: Attack the Dragon. In: Maitra, S., Veni Madhavan, C.E., Venkatesan, R. (eds.) INDOCRYPT 2005. LNCS, vol. 3797, pp. 130–142. Springer, Heidelberg (2005)
14. Ferguson, N., Kelsey, J., Lucks, S., Schneier, B., Stay, M., Wagner, D., Whiting, D.: Improved Cryptanalysis of Rijndael. In: Schneier, B. (ed.) FSE 2000. LNCS, vol. 1978, pp. 213–230. Springer, Heidelberg (2001)
15. Hermelin, M., Cho, J.Y., Nyberg, K.: Multidimensional Extension of Matsui's Algorithm 2. In: Dunkelman, O. (ed.) FSE 2009. LNCS, vol. 5665, pp. 209–227. Springer, Heidelberg (2009)
16. Hermelin, M., Nyberg, K.: Linear cryptanalysis using multiple linear approximations. In: Junod, P., Canteaut, A. (eds.) Advanced Linear Cryptanalysis of Block and Stream Ciphers. IOS Press (2011)
17. Hermelin, M., Cho, J.Y., Nyberg, K.: Multidimensional Linear Cryptanalysis of Reduced Round Serpent. In: Mu, Y., Susilo, W., Seberry, J. (eds.) ACISP 2008. LNCS, vol. 5107, pp. 203–215. Springer, Heidelberg (2008)
18. Kaliski Jr., B.S., Robshaw, M.J.B.: Linear Cryptanalysis Using Multiple Approximations. In: Desmedt, Y.G. (ed.) CRYPTO 1994. LNCS, vol. 839, pp. 26–39. Springer, Heidelberg (1994)
19. Knudsen, L.R., Robshaw, M.J.B., Wagner, D.: Truncated Differentials and Skipjack. In: Wiener, M. (ed.) CRYPTO 1999. LNCS, vol. 1666, pp. 165–180. Springer, Heidelberg (1999)
20. Knudsen, L.R., Wagner, D.: On the structure of Skipjack. Discrete Applied Mathematics 111(1-2), 103–116 (2001)
21. Knudsen, L.R., Wagner, D.: Integral Cryptanalysis. In: Daemen, J., Rijmen, V. (eds.) FSE 2002. LNCS, vol. 2365, pp. 112–127. Springer, Heidelberg (2002)
22. Leander, G.: On Linear Hulls, Statistical Saturation Attacks, PRESENT and a Cryptanalysis of PUFFIN. In: Paterson, K.G. (ed.) EUROCRYPT 2011. LNCS, vol. 6632, pp. 303–322. Springer, Heidelberg (2011)
23. Lucks, S.: The Saturation Attack - A Bait for Twofish. In: Matsui, M. (ed.) FSE 2001. LNCS, vol. 2355, pp. 1–15. Springer, Heidelberg (2002)
24. Seki, H., Kaneko, T.: Differential Cryptanalysis of CAST-256 Reduced to Nine Quad-rounds. IEICE Transactions on Fundamentals of Electronics Communications and Computer Sciences E84A(4), 913–918 (2001)
25. Skipjack and KEA Algorithm Specifications, Version 2.0, (May 29, 1998); The National Institute of Standards and Technology's web page,
http://csrc.nist.gov/groups/ST/toolkit/documents/skipjack/skipjack.pdf
26. Wagner, D.: The Boomerang Attack. In: Knudsen, L.R. (ed.) FSE 1999. LNCS, vol. 1636, pp. 156–170. Springer, Heidelberg (1999)
27. Wang, M., Wang, X., Hu, C.: New Linear Cryptanalytic Results of Reduced-Round of CAST-128 and CAST-256. In: Avanzi, R.M., Keliher, L., Sica, F. (eds.) SAC 2008. LNCS, vol. 5381, pp. 429–441. Springer, Heidelberg (2009)
28. Wang, Q., Chen, J.: 18-Round Impossible Differential for CAST-256 (2012)

Differential Attacks against Stream Cipher ZUC[*]

Hongjun Wu, Tao Huang, Phuong Ha Nguyen, Huaxiong Wang, and San Ling

Division of Mathematical Sciences,
School of Physical and Mathematical Sciences
Nanyang Technological University, Singapore
{wuhj,huangtao,ng007ha,hxwang,lingsan}@ntu.edu.sg

Abstract. Stream cipher ZUC is the core component in the 3GPP confidentiality and integrity algorithms 128-EEA3 and 128-EIA3. In this paper, we present the details of our differential attacks against ZUC 1.4. The vulnerability in ZUC 1.4 is due to the non-injective property in the initialization, which results in the difference in the initialization vector being cancelled. In the first attack, difference is injected into the first byte of the initialization vector, and one out of $2^{15.4}$ random keys result in two identical keystreams after testing $2^{13.3}$ IV pairs for each key. The identical keystreams pose a serious threat to the use of ZUC 1.4 in applications since it is similar to reusing a key in one-time pad. Once identical keystreams are detected, the key can be recovered with average complexity $2^{99.4}$. In the second attack, difference is injected into the second byte of the initialization vector, and every key can result in two identical keystreams with about 2^{54} IVs. Once identical keystreams are detected, the key can be recovered with complexity 2^{67}. We have presented a method to fix the flaw by updating the LFSR in an injective way in the initialization. Our suggested method is used in the later versions of ZUC. The latest ZUC 1.6 is secure against our attacks.

1 Introduction

Comparing to block ciphers, dedicated stream ciphers normally require less computation for achieving the same security level. Stream ciphers are widely used in applications. For example, RC4 [10] is used in SSL and WEP, and A5/1 [8] is used in GSM (the Global System for Mobile Communications). But the use of RC4 in WEP is insecure [7], and A5/1 is very weak [4]. ECRYPT (2004–2008) has organised the eSTREAM competition, which stimulated the study on stream ciphers, and a number of new stream ciphers were proposed [1–3, 5, 6, 9, 15].

The 3rd Generation Partnership Project (3GPP) was set up for making globally applicable 3G mobile phone system specifications based on the GSM specifications. Stream cipher ZUC was designed by the Data Assurance and Communication Security Research Center of the Chinese Academy of Sciences.

---

[*] This research is supported by the National Research Foundation Singapore under its Competitive Research Programme (CRP Award No. NRF-CRP2-2007-03) and Nanyang Technological University NAP startup grant (M4080529.110).

X. Wang and K. Sako (Eds.): ASIACRYPT 2012, LNCS 7658, pp. 262–277, 2012.

It is the core component of the 3GPP Confidentiality and Integrity Algorithms 128-EEA3 & 128-EIA3 which were proposed for inclusion in the "4G" mobile standard LTE (Long Term Evolution). In July 2010, the ZUC 1.4 [11] was made public for evaluation. We developed two key recovery attacks against ZUC 1.4 [16], and our attacks directly led to the tweak of ZUC 1.4 into ZUC 1.5 [12] in Jan 2011. (Note that it was reported independently in [14] that the non-injective initialization of ZUC 1.4 may result in identical keystreams.) The latest version, ZUC 1.6 [13], was released in June 2011 (ZUC 1.6 and ZUC 1.5 have almost the same specifications).

In this paper, we present the details of our differential attacks against ZUC 1.4. Our attacks against ZUC is similar to the differential attacks against Py, Py6 and Pypy [17], in which different IVs result in identical keystreams. In the first attack against ZUC 1.4, the difference is at the first byte of the IV, and one in $2^{15.4}$ keys results in identical keystreams after testing $2^{13.3}$ IV pairs for each key. Once identical keystreams are detected, the key can be recovered with complexity $2^{99.4}$. In the second attack against ZUC 1.4, the difference is at the second byte of the IV, and identical keystreams can be obtained after testing 2^{54} IVs. The key can be recovered with complexity 2^{67}.

This paper is organized as follows. The notations and the description of ZUC 1.4 are give in Sect. 2. The overview of the attack is is given in Sect. 3. In Section 4 and 5, we present the key recovery attack with difference at the first byte and the second byte of IV, respectively. We suggest the tweak to fix the flaw in Sect. 6. Section 7 concludes the paper.

2   Preliminaries

2.1   The Notations

In this paper, we follow the notations used in the ZUC specifications [11].

$+$	The addition of two integers
$\oplus$	The bit-wise exclusive-or operation of integers
$\boxplus$	The modulo 2^{32} addition
ab	The product of integers a and b
$a\|\|b$	The concatenation of a and b
$a <\!<\!< k$	The k-bit cyclic shift of a to the left
$a >\!>\!> k$	The k-bit cyclic shift of a to the right
$a >\!> k$	The k-bit right shift of integer a
a_H	The most significant 16 bits of integer a
a_L	The least significant 16 bits of integer a
$(a_1, a_2, \ldots, a_n) \rightarrow (b_1, b_2, \ldots, b_n)$	It assigns the values of a_i to b_i in parallel

$$0_n \qquad \text{The sequence of } n \text{ bits } 0$$
$$1_n \qquad \text{The sequence of } n \text{ bits } 1$$
$$\bar{y} \qquad \text{The bitwise complement of } y$$

An integer a can be written in different formats. For example,

$$a = 25 \qquad \text{decimal representation}$$
$$= 0x19 \qquad \text{hexadecimal representation}$$
$$= 00011001_2 \quad \text{binary representation}$$

We number the least significant bit with 1 and use $A[i]$ to denote the ith bit of a A. And use $B[i..j]$ to denote the bit i to bit j of B.

2.2   The General Structure of ZUC 1.4

ZUC is a word-oriented stream cipher with 128-bit secret key and a 128-bit initial vector. It consists of three main components: the linear feedback shift register (LFSR), the bit-reorganization (BR) and a nonlinear function F. The general structure of the algorithm is illustrated in Fig. 1.

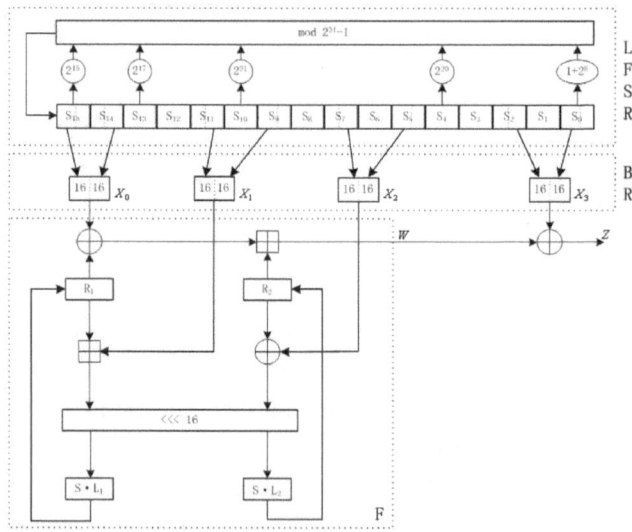

Fig. 1. General structure of ZUC

Linear Feedback Shift Register(LFSR). It consists of sixteen 31-bit registers $s_0, s_1, \ldots, s_{15}$, and each register is an integer in the range $\{1, 2, \ldots, 2^{31} - 1\}$. During the keystream generation stage, the LFSR is updated as follows:

LFSRUpdate():

1. $s_{16} = (2^{15}s_{15} + 2^{17}s_{13} + 2^{21}s_{10} + 2^{20}s_4 + (1 + 2^8)s_0)\mod(2^{31} - 1)$;
2. If $s_{16} = 0$ then set $s_{16} = 2^{31} - 1$;
3. $(s_1, s_2, \ldots, s_{15}, s_{16}) \to (s_0, s_1, \ldots, s_{14}, s_{15})$.

Bit-Reorganization Function. It extracts 128 bits from the state of the LFSR and forms four 32-bit words X_0, X_1 X_2 and X_3 as follows:

Bitreorganization():

1. $X_0 = s_{15H} \| s_{14L}$;
2. $X_1 = s_{11L} \| s_{9H}$;
3. $X_2 = s_{7L} \| s_{5H}$;
4. $X_3 = s_{2L} \| s_{0H}$;

Nonlinear Function F. It contains two 32-bit memory words R_1 and R_2. The description of F is given below. In function F, S is the Sbox layer and L_1 and L_2 are linear transformations as defined in [11]. The output of function F is a 32-bit word W. The keystream word Z is given as $Z = W \oplus X_3$.

$F(X_0, X_1, X_2)$:

1. $W = (X_0 \oplus R_1) \boxplus R_2$;
2. $W_1 = R_1 \boxplus X_1$;
3. $W_2 = R_2 \oplus X_2$;
4. $R_1 = S(L_1(W_{1L} \| W_{2H}))$;
5. $R_2 = S(L_2(W_{2L} \| W_{1H}))$;

2.3 The Initialization of ZUC 1.4

The initialization of ZUC 1.4 consists of two steps: loading the key and IV into the register, and running the cipher for 32 steps with the keystream word being used to update the state.

Key and IV Loading. Denote the 16 key bytes as k_i ($0 \le i \le 15$), the 16 IV bytes as iv_i ($0 \le i \le 15$). We load the key and IV into the register as: $s_i = (k_i \| d_i \| iv_i)$. The values of the constants d_i are given in [11]. The two memory words R_1 and R_2 in function F are set as 0.

Running the Cipher for 32 Steps. At the initialization stage, the keystream word Z is used to update the LFSR as follows:

LFSRWithInitialisationMode(u):

1. $v = (2^{15}s_{15} + 2^{17}s_{13} + 2^{21}s_{10} + 2^{20}s_4 + (1 + 2^8)s_0)\mod(2^{31} - 1)$;
2. If $v = 0$ then set $v = 2^{31} - 1$;
3. $s_{16} = v \oplus u$;
4. If $s_{16} = 0$ then set $s_{16} = 2^{31} - 1$;
5. $(s_1, s_2, \ldots, s_{15}, s_{16}) \to (s_0, s_1, \ldots, s_{14}, s_{15})$.

The cipher runs for 32 steps at the initialization stage as follows:
   InitializationStage():
        for $i = 0$ **to** 31 {

        1. Bitreorganization();
        2. $Z = F(X_0, X_1, X_2) \oplus X_3$;
        3. LFSRWithInitialisationMode($Z \gg 1$).

   }

3   Overview of the Attacks

We notice that the LFSR in ZUC is defined over $GF(2^{31}-1)$, with the element 0 being replaced with $2^{31}-1$. To the best of our knowledge, it is the first time that $GF(2^{31}-1)$ is used in the design of stream cipher. In the initialization of ZUC 1.4, we notice that XOR is involved in the update of LFSR ($s_{16} = v \oplus u$). When XOR is applied to the elements in $GF(2^{31}-1)$, we obtain the following undesirable property:

Property 1. Suppose that a and a' are two elements in $GF(2^{31}-1)$, $a \neq a'$, and $\bar{a} = a'$. If $b = a$ or $b = \bar{a}$, then $a \oplus b \bmod (2^{31}-1) = a' \oplus b \bmod (2^{31}-1) = 0$.

The above property shows that the difference between a and a' can get eliminated with an XOR operation! In the rest of this paper, we exploit this property to attack ZUC 1.4 by eliminating the difference in the state.

   In our attacks, we try to eliminate the difference in the state without the difference in the state being injected into the nonlinear function F. The reason is that if a difference is injected into F, then Sboxes would be involved, and the difference would remain in F until additional difference being injected into F, thus the probability that the difference in the state being eliminated would get significantly reduced.

   We now investigate what are the IV differences that would result in the difference in the state being eliminated with high probability. The IV differences are classified into the following three types:

Type 1. $\Delta iv_i \neq 0$ for at least one value of i ($7 \leq i \leq 15$).

After loading this type of IVs into LFSR, the difference would appear at the least significant byte of at least one of the LFSR elements $s_7, s_8, \cdots, s_{15}$. Note that the least significant byte of s_7 is part of X_2 in the Bit-reorganization function since $X_2 = s_{7L} || s_{5H}$, and X_2 is an input to function F. Due to the shift of LFSR, the difference at the least significant byte of $s_7, s_8, \cdots, s_{15}$ would be injected into F. Thus we would not use this type of IV difference in our attacks.

Type 2. $\Delta iv_i = 0$ for $7 \leq i \leq 15$, $\Delta iv_i \neq 0$ for at least one value of i ($2 \leq i \leq 6$).
After loading this type of IVs into LFSR, the difference would appear at the least

significant byte of at least one of the LFSR elements s_2, s_3, $\cdots$, s_6. Note that the least significant byte of s_2 is part of X_3 in the Bit-reorganization function since $X_3 = s_{2L}||s_{0H}$, X_3 is XORed with the output of F to generate keystream word Z, and Z is used to update the LFSR. Two steps later, the difference in iv_2 would appear in the feedback function to update LFSR. It means that if there is difference in iv_2, the difference in s_2 would be used to update the LFSR twice, and the probability that the difference would be eliminated is very small. Due to the shift of LFSR, the difference at s_2, s_3, $\cdots$, s_7 would be eliminated with very small probability. Thus we did not use this type of IV difference in our attacks.

Type 3. $\Delta iv_i = 0$ for $2 \leq i \leq 15$, $\Delta iv_0 \neq 0$ or $\Delta iv_1 \neq 0$.
The focus of our attacks is on this type of IV differences. In order to increase the chance of success, we consider the difference at only one byte of the IV. We discuss below how the difference in the state can be eliminated when there is difference in s_0 (the analysis for the difference in s_1 is similar). At the first step in the initialization,

$$s_0 = (k_0||d_0||iv_0)\,, \tag{1}$$
$$v = 2^{15}s_{15} + 2^{17}s_{13} + 2^{21}s_{10} + 2^{20}s_4 + (1 + 2^8)s_0 \quad \mathrm{mod}\ (2^{31} - 1)\,, \tag{2}$$
$$s_{16} = v \oplus u\,. \tag{3}$$

Suppose that the difference is only at iv_0, and $iv_0 - iv_0' = \Delta iv_0 > 0$. From (1) and (2) we know that

$$v - v' = (1 + 2^8)(iv_0 - iv_0') \quad \mathrm{mod}\ (2^{31} - 1)$$
$$= \Delta iv_0 \parallel \Delta iv_0\,. \tag{4}$$

If we need to eliminate the difference in s_{16}, from Property 1 and (3), the following condition should be satisfied:

$$v \oplus v' = 1_{31} \tag{5}$$
$$u = v \quad \text{or} \quad u = v' \tag{6}$$

According to (5), v and v' have XOR difference in the left-most 15 bits (i.e.$v[17..31]$ and $v'[17..31]$), while according to (4), the subtraction difference of those bits are 0. The only possible reason is that the 15 bits, $v[17..31]$, are all affected by the carries from the addition of Δiv_0 to v'. After testing all the one-byte differences, we found that v must be in one of the following four forms (the values of v and v' can be swapped):

$$v = 111111111111111_2 \parallel y \parallel 1_2 \parallel y$$
$$\text{or} \quad v = 011111111111111_2 \parallel y \parallel 0_2 \parallel y$$
$$\text{or} \quad v = 000000000000000_2 \parallel \bar{y} \parallel 0_2 \parallel \bar{y} \tag{7}$$
$$\text{or} \quad v = 100000000000000_2 \parallel \bar{y} \parallel 1_2 \parallel \bar{y}$$
$$(y \text{ is a 7-bit integer.})$$

There are 510 possible values of v ($v = 1_{31}$ and $v = 0_{31}$ are excluded since one of v and $\bar{v}$ cannot be 0). All the (v, v') pairs and their differences are given in Table 1 in Appendix A. Notice that we ignored the order of v and v' as they are exchangeable. We have obtained all the possible values of v and u for generating identical keystreams.

We highlight the following property in the table: the difference between v and v' uniquely determines the value of pair (v, v') in the table. As a result, if we know the difference of IVs that results in the collision of the state, we can determine the value of (v, v') immediately.

By eliminating the difference in the state as illustrated above, we developed two attacks against ZUC 1.4. The first attack is to exploit the difference at iv_0, and the second attack is to exploit the difference at iv_1. The details are given in the following two sections.

4   Attack ZUC 1.4 with Difference at iv_0

In this section, we present our first differential attack on the initialization by using IV difference at iv_0 and generating identical keystream. The keys that generate the same keystream are called weak keys in this attack. We will show that a weak key exists with probability $2^{-15.4}$, and a weak key can be detected with about $2^{13.3}$ chosen IVs. Once a weak key is detected, its effective key size is reduced from 128 bits to around 100 bits.

4.1   The Weak Keys for Δiv_0

We will show that when there is difference at iv_0, about one in $2^{15.4}$ keys would result in identical keystream. For a random key, we will check whether there exists a pair of IVs such that (5), (6) and (7) can be satisfied.

We start with analyzing how keys and IVs are involved in the expression of u and v in the first step of initialization. From the specifications of the initialization, we have

$$
\begin{aligned}
u &= Z \gg 1 = (X_0 \oplus X_3) \gg 1 = ((s_{15H}||s_{14L}) \oplus (s_{2L}||s_{0H})) \gg 1 \\
&= ((k_{15} \parallel iv_2 \parallel k_0 \parallel iv_{14}) \oplus \text{0x6b8f9a89}) \gg 1
\end{aligned} \tag{8}
$$

In (2) and (8), there are 5 bytes of key, $\{k_0, k_4, k_{10}, k_{13}, k_{15}\}$, and 7 bytes of IV, $\{iv_0, iv_2, iv_4, iv_{10}, iv_{13}, iv_{14}, iv_{15}\}$ being involved in the computation of u and v. The complexity would be very high if we directly try all possible combinations of the keys and IVs. However, with analysis on the expressions of u and v, we can reduce the search space from 2^{96} to around $2^{26.3}$.

Solve (5), (6), (7) and (8), we obtain the following four groups of solutions:

Group 1.

$$u = v = 1111111111111111_2 \parallel y \parallel 1_2 \parallel y$$
$$k_{15} = \text{0x94}$$
$$iv_2 = \text{0x70} \tag{9}$$
$$k_0 = \text{0x9a} \oplus (y \parallel 1_2)$$
$$iv_{14} >> 1 = \text{0x44} \oplus y$$

Group 2.

$$u = v = 0111111111111111_2 \parallel y \parallel 0_2 \parallel y$$
$$k_{15} = \text{0x14}$$
$$iv_2 = \text{0x70} \tag{10}$$
$$k_0 = \text{0x9a} \oplus (y \parallel 0_2)$$
$$iv_{14} >> 1 = \text{0x44} \oplus y$$

Group 3.

$$u = v = 0000000000000000_2 \parallel \bar{y} \parallel 0_2 \parallel \bar{y}$$
$$k_{15} = \text{0x6b}$$
$$iv_2 = \text{0x8f} \tag{11}$$
$$k_0 = \text{0x9a} \oplus (\bar{y} \parallel 0_2)$$
$$iv_{14} >> 1 = \text{0xbb} \oplus \bar{y}$$

Group 4.

$$u = v = 1000000000000000_2 \parallel \bar{y} \parallel 1_2 \parallel \bar{y}$$
$$k_{15} = \text{0xeb}$$
$$iv_2 = \text{0x8f} \tag{12}$$
$$k_0 = \text{0x9a} \oplus (\bar{y} \parallel 1_2)$$
$$iv_{14} >> 1 = \text{0xbb} \oplus \bar{y}$$

Furthermore, from (2) we compute v as follows (note that the property $2^k s_i$ mod $(2^{31} - 1) = s_i <<< k$):

$$v = (1 + 2^{23})k_0 + 2^7 k_{15} + 2^9(k_{13} + 2^3 k_4 + 2^4 k_{10}) + (1 + 2^8)iv_0$$
$$+ 2^{15}(iv_{15} + 2^2 iv_{13} + 2^5 iv_4 + 2^6 iv_{10}) + \text{0x451bfe1b} \mod (2^{31} - 1) \tag{13}$$

Let $sum_1 = k_{13} + 2^3 k_4 + 2^4 k_{10}$, $sum_2 = iv_{15} + 2^2 iv_{13} + 2^5 iv_4 + 2^6 iv_{10}$. The value of sum_1 ranges from 0 to 6375, and the value of sum_2 ranges from 0 to 25755. We developed Algorithm 1 to search for weak keys.

Algorithm 1. Find weak keys for Δiv_0

---

for (k_{15}, iv_2) in each of the 4 groups of solutions (9), (10), (11), (12) **do**

    for $y = 0$ to 127 **do**

        determine $iv_{14} \gg 1$ and k_0

        for $sum_1 = 0$ to 6375 **do**

            for $iv_0 = 0$ to 255 **do**

                $keySum \leftarrow 2^7 k_{15} + (2^{23} + 1)k_0 + 2^9 sum_1 \mod (2^{31} - 1)$

                $sum_2 \leftarrow (u - keySum - (1 + 2^8)iv_0 - 0\text{x}451\text{bfe1b})/2^{15} \mod (2^{31} - 1)$

                if sum_2 is less than 25756 **then**

                    $v = u;\ v' = u \oplus 1_{32};$

                    if $(v - v') \mod (2^{31} - 1)$ is a multiple of $1 + 2^8$ **then**

                        $\Delta iv_0 = (v - v') \mod (2^{31} - 1)/(1 + 2^8);$

                        $iv_0' = iv_0 - \Delta iv_0;$

                  else

                      $\Delta iv_0 = (v' - v) \mod (2^{31} - 1)/(1 + 2^8);$

                      $iv_0' = iv_0 + \Delta iv_0;$

                  end if

                  output $u, k_0, k_{15}, sum_1, iv_0, iv_0', iv_2, iv_{14} \gg 1, sum_2$

                end if

            end for

        end for

    end for

end for

Each output from Algorithm 1 gives the value of $(k_{15}, k_0, sum_1, iv_0, iv_0', iv_2, iv_{14}, sum_2)$ that results in identical keystreams. Running Algorithm 1, we found $9934 = 2^{13.28}$ different outputs. We note that on average, each sum_1 from the output of the algorithm represents $2^{24}/6376 = 2^{11.36}$ possible choices of (k_4, k_{10}, k_{13}). Thus there are $2^{13.3} \times 2^{11.4} = 2^{24.7}$ weak values of $(k_0, k_4, k_{10}, k_{13}, k_{15})$. Hence, there are $2^{24.7}$ weak keys out of 2^{40} possible values of the 5 key bytes. The probability that a random key is weak for IV difference at iv_0 is $2^{-15.4}$. The complexity of Algorithm 1 is $4 \times 128 \times 6376 \times 256 = 2^{26.3}$.

Identical Keystreams. We give below a weak key and an IV pair with difference at iv_0 that result in identical keystreams.

$$key = 87,4,95,13,161,32,199,61,20,147,56,84,126,205,165,148$$
$$IV = 166,166,112,38,192,214,34,211,170,25,18,71,4,135,68,5$$
$$IV' = 116,166,112,38,192,214,34,211,170,25,18,71,4,135,68,5$$

For both IV and IV', the identical keystreams are: 0xbfe800d5 0360a22b 6c4554c8 67f00672 2ce94f3f f94d12ba 11c382b3 cbaf4b31. . . ..

4.2 Detecting Weak Keys for Δiv_0

We have shown above that a random key is weak with probability $2^{-15.4}$. In the attack against ZUC, we will first detect a weak key, then recover it. To detect a weak key, our approach is to use the IV pairs generated from Algorithm 1 to test whether identical keystreams are generated. Note that for a particular value of sum_2, we can always find a combination of $(iv_4, iv_{10}, iv_{13}, iv_{15})$ that satisfies $sum_2 = iv_{15} + 2^2 iv_{13} + 2^5 iv_4 + 2^6 iv_{10}$. Thus a pair of IVs $(iv_0, iv_2, iv_4, iv_{10}, iv_{13}, iv_{14}, iv_{15})$ and $(iv'_0, iv_2, iv_4, iv_{10}, iv_{13}, iv_{14}, iv_{15})$ can be determined by each output of Algorithm 1. Using this result, we developed Algorithm 2 to detect weak keys for Δiv_0.

Algorithm 2. Detecting weak keys for Δiv_0

---

1. Choose one of the $2^{13.28}$ outputs of Algorithm 1.
2. Find the pair of IVs determined by this output (if iv_j does not appear in the first initialization step, set it as some fixed constant).
3. Use the IV pair to generate two key steams.
4. If the keystreams are identical, output the IVs and conclude the key is weak.
5. If all outputs of Algorithm 1 have been checked, and there are no identical keystreams, we conclude that the key is not weak.

In Algorithm 2, we need to test at most $2^{13.3}$ pairs of IVs to determine if a key is weak for difference at iv_0.

4.3   Recovering Weak Keys for Δiv_0

After detecting a weak key, we proceed to recover the weak key. Once a key is detected as weak (as given from Algorithm 2), from the IV pair being used to generate identical keystreams, we immediately know the value of k_0, k_{15} and sum_1. Note that $sum_1 = (k_{13} + 2^3 k_4 + 2^4 k_{10})$. In the best situations, the sum is 0 or 25755, then we can uniquely determine k_4, k_{10} and k_{13}. In the worst situation, there are 2^{12} possible choices for k_4, k_{10} and k_{13}, and therefore, we need 2^{12} tests to determine the correct values for k_4, k_{10} and k_{13}. On average, for each value of sum_1, we need to test $2^{11.4}$ combinations of (k_4, k_{10}, k_{13}).

Since there are only five key bytes being recovered in our attack, the remaining 11 key bytes should be recovered with exhaustive search. Hence, the complexity to recover all key bits is $2^{88} \times 2^{11.4} = 2^{99.4}$. From the analysis above, we also know that the best complexity is 2^{88} and the worst complexity is 2^{100}.

5   Attack ZUC 1.4 with Difference at iv_1

In this section, we present the differential attack on ZUC 1.4 for IV difference at iv_1. Different from the attack in Section 4, we need to consider the computation of u and v in the second step of the initialization. For this type of IV difference, for every key, there are some IV pairs that result in identical keystreams since more IV bytes are involved. Once we found such an IV pair, we can recover the key with complexity around 2^{67}.

5.1   Identical Keystreams for Δiv_1

The computation of u and v in the second initialization step involves more key and IV bytes. The v in the second initialization step is computed as:

$$v = (2^{15}s_{16} + 2^{17}s_{14} + 2^{21}s_{11} + 2^{20}s_5 + (1 + 2^8)s_1) \bmod (2^{31} - 1),$$
$$s_{16} = ((2^{15}s_{15} + 2^{17}s_{13} + 2^{21}s_{10} + 2^{20}s_4 + (1 + 2^8)s_0) \bmod (2^{31} - 1)) \quad (14)$$
$$\oplus (((k_{15} \parallel iv_2 \parallel k_0 \parallel iv_{14}) \oplus \text{0x6b8f9a89}) \gg 1)$$

And u is given as:

$$u = (((X_0 \oplus R_1) + R_2) \oplus X_3) \gg 1$$
$$X_0 = (s_{16H} \parallel 10101100_2 \parallel iv_{15})$$
$$X_3 = (01011110_2 \parallel iv_3 \parallel k_1 \parallel 01001101_2) \quad (15)$$
$$R_1 = S(L_1(s_{9H} \parallel s_{7L})) = f_1(iv_7, k_9)$$
$$R_2 = S(L2(s_{5H} \parallel s_{11L})) = f_2(iv_{11}, k_5)$$

where f_1 and f_2 are some deterministic non-linear functions.

There are 10 IV bytes involved in the expression of v, i.e. (iv_0, iv_1, iv_2, iv_4, iv_5, iv_{10}, iv_{11}, iv_{13}, iv_{14}, iv_{15}) and 8 IV bytes involved in the expression of u, i.e. (iv_0, iv_3, iv_4, iv_7, iv_{10}, iv_{11}, iv_{13}, iv_{15}). In total, there are 12 IV bytes being involved in the computation of u and v, and every bit of u and v can be affected by IV. We conjecture that for every key, the conditions (5) and (6) can be satisfied, and identical keystreams can be generated. To verify it, we tested 1000 random keys. Our experimental results show that there is always an IV pair for each key that results in identical keystreams.

In the attack, a random key and a random iv pair with difference at iv_1, the probability that v and u satisfy the conditions (5) and (6) is $2^{-31} \times 2^{-31} \times 2 = 2^{-61}$. Choosing 2^8 ivs with difference at iv_1, we have around 2^{15} pairs. The identical keystream pair appears with probability $2^{-61+15} = 2^{-46}$ with 2^8 IVs. We thus need about $2^{46} \times 2^8 = 2^{54}$ IVs to obtain identical keystreams.

Identical Keystreams. We give below a key and an IV pair with difference at iv_1 that result in identical keystreams. The algorithm being used to find the IV pair is given in Appendix B. The algorithm is a bit complicated since a number of optimization tricks are involved. The explanation of the optimization details is omitted here since our focus is to develop a key recovery attack.

$$key = 123, 149, 193, 87, 42, 150, 117, 4, 209, 101, 85, 57, 46, 117, 49, 243$$
$$IV = 92, 80, 241, 10, 0, 217, 47, 224, 48, 203, 0, 45, 204, 0, 0, 17$$
$$IV' = 92, 182, 241, 10, 0, 217, 47, 224, 48, 203, 0, 45, 204, 0, 0, 17$$

The identical keystreams are: 0xf09cc17d 41f12d3f 453ac0c3 cadcef9f f98fb964 ca6e576e b48b813 6c43da22 ....

5.2   Key Recovery for Δiv_1

After identical keystreams are generated from an IV pair with difference at iv_1, we proceed to recover the secret key. From Table 1 in Appendix A, we know the value of (v, v') since we know the difference at iv_1 of the chosen IV pair, and we also know the value of u since $u = v$ or $u = v'$. In the following, we illustrate a key recovery attack after identical keystreams have been detected.

1. In the expression of u in (15), (k_1, k_5, k_9, s_{16H}) is involved. Note that there are only two possible values of the 31-bit u. We try all the possible values of (k_1, k_5, k_9, s_{16H}), then there would be $2^{8\times3+16} \times 2^{-31} \times 2 = 2^{10}$ possible values of (k_1, k_5, k_9, s_{16H}) that generate the two possible values of u. The complexity of this step is 2^{40}.
2. Next we use the expression of s_{16} in (14). For each of the 2^{10} possible values of (k_1, k_5, k_9, s_{16H}), we try all the possible values of $(k_0, k_4, k_{10}, k_{13}, k_{15})$ and check whether the values of s_{16H} is computed correctly or not. There would be $2^{8\times5} \times 2^{-16} = 2^{24}$ possible values of $(k_0, k_4, k_{10}, k_{13}, k_{15})$ left. Considering that there are 2^{10} possible values of (k_1, k_5, k_9, s_{16H}), about $2^{10} \times 2^{24} = 2^{34}$ possible values of $(k_0, k_1, k_4, k_5, k_9, k_{10}, k_{13}, k_{15}, s_{16H})$ remain. The complexity of this step is $2^{8\times5} \times 2^{10} = 2^{50}$.
3. Then we use the expression of v in (14). For each of the 2^{34} possible values of $(k_0, k_1, k_4, k_5, k_9, k_{10}, k_{13}, k_{15}, s_{16H})$, we try all the possible values of (k_{11}, k_{14}) and check whether the value of v is correct or not. A random value of (k_{11}, k_{14}) would pass the test with probability $2^{8\times2} \times 2^{-31} = 2^{-15}$ Considering that there are 2^{34} possible values of $(k_0, k_1, k_4, k_5, k_9, k_{10}, k_{13}, k_{15}, s_{16H})$, about $2^{34} \times 2^{-15} = 2^{19}$ possible values of $(k_0, k_1, k_4, k_5, k_9, k_{10}, k_{11}, k_{13}, k_{14}, k_{15})$ remain. The complexity of this step is $2^{8\times2} \times 2^{34} = 2^{50}$.
4. For each of the 2^{19} possible values of $(k_0, k_1, k_4, k_5, k_9, k_{10}, k_{11}, k_{13}, k_{14}, k_{15})$, we recover the remaining 6 key bytes $(k_2, k_3, k_6, k_7, k_8, k_{12})$ by exhaustive search. The complexity of this step is $2^{19} \times 2^{8\times6} = 2^{67}$.

The overall computational complexity to recover a key is $2^{40} + 2^{50} + 2^{50} + 2^{67} \approx 2^{67}$. And we need about 2^{54} IVs in the attack. Note that the complexity in the first, second and third steps can be significantly reduced with optimization since we are dealing with simple functions. For example, meet-in-the-middle attack can be used in the first step, and the sum of a few key bytes can be considered in the second and third steps. However, the complexity of those three steps has little effect on the overall complexity of the attack, so we do not present the details of the optimization here.

6   Improving ZUC 1.4

From the analysis in Sect. 3, the weakness of the initialization comes from the non-injective update of the LFSR. To fix the flaw, we proposed the tweak in the rump session of Asiacrypt 2010. Instead of using the XOR operation, it is better to use addition modulo operation over $GF(2^{31} - 1)$. More specifically, the operation $s_{16} = v \oplus u$ is changed to $s_{16} = v + u \mod (2^{31} - 1)$. With this tweak, the difference in v would always result in the difference in s_{16} if there is no difference in u, and the attack against ZUC 1.4 can no longer be applied. In the later versions ZUC 1.5 and 1.6 (ZUC 1.5 and 1.6 have almost the same specifications), the computation of s_{16} is modified using our suggested method.

7   Conclusion

In this paper, we developed two chosen IV attacks against the initialization of ZUC 1.4. In our attacks, identical keystreams are generated from different IVs, then key recovery attacks are applied. Our attacks are independent of the number of steps in initialization. The lesson from this paper is that when non-injective functions are used in cipher design, we should pay special attention to ensure that the difference cannot be eliminated with high probability.

References

1. Babbage, S., Dodd, M.: The MICKEY Stream Ciphers. In: Robshaw, M., Billet, O. (eds.) New Stream Cipher Designs. LNCS, vol. 4986, pp. 191–209. Springer, Heidelberg (2008)
2. Berbain, C., Billet, O., Canteaut, A., Courtois, N.T., Gilbert, H., Goubin, L., Gouget, A., Granboulan, L., Lauradoux, C., Minier, M., Pornin, T., Sibert, H.: SOSEMANUK, a Fast Software-Oriented Stream Cipher. In: Robshaw, M., Billet, O. (eds.) New Stream Cipher Designs. LNCS, vol. 4986, pp. 98–118. Springer, Heidelberg (2008)
3. Bernstein, D.J.: The Salsa20 Family of Stream Ciphers. In: Robshaw, M., Billet, O. (eds.) New Stream Cipher Designs. LNCS, vol. 4986, pp. 84–97. Springer, Heidelberg (2008)
4. Biryukov, A., Shamir, A., Wagner, D.: Real Time Cryptanalysis of A5/1 on a PC. In: Schneier, B. (ed.) FSE 2000. LNCS, vol. 1978, pp. 1–44. Springer, Heidelberg (2001)
5. Boesgaard, M., Vesterager, M., Zenner, E.: The Rabbit Stream Cipher. In: Robshaw, M., Billet, O. (eds.) New Stream Cipher Designs. LNCS, vol. 4986, pp. 69–83. Springer, Heidelberg (2008)
6. De Cannière, C., Preneel, B.: TRIVIUM. In: Robshaw, M., Billet, O. (eds.) New Stream Cipher Designs. LNCS, vol. 4986, pp. 244–266. Springer, Heidelberg (2008)
7. Fluhrer, S., Mantin, I., Shamir, A.: Weaknesses in the Key Scheduling Algorithm of RC4. In: Vaudenay, S., Youssef, A.M. (eds.) SAC 2001. LNCS, vol. 2259, pp. 1–24. Springer, Heidelberg (2001)
8. Golić, J.D.: Cryptanalysis of Alleged A5 Stream Cipher. In: Fumy, W. (ed.) EUROCRYPT 1997. LNCS, vol. 1233, pp. 239–255. Springer, Heidelberg (1997)
9. Hell, M., Johansson, T., Maximov, A., Meier, W.: The Grain Family of Stream Ciphers. In: Robshaw, M., Billet, O. (eds.) New Stream Cipher Designs. LNCS, vol. 4986, pp. 179–190. Springer, Heidelberg (2008)
10. Rivest, R.L.: The RC4 Encryption Algorithm. RSA Data Security, Inc. (March 1992)

11. ETSI/SAGE Specification. Specification of the 3GPP Confidentiality and Integrity Algorithms 128-EEA3 & 128-EIA3. Document 2: ZUC Specification; Version: 1.4 (July 30, 2010)
12. ETSI/SAGE Specification. Specification of the 3GPP Confidentiality and Integrity Algorithms 128-EEA3 & 128-EIA3. Document 2: ZUC Specification; Version: 1.5 (January 4, 2011)
13. ETSI/SAGE Specification. Specification of the 3GPP Confidentiality and Integrity Algorithms 128-EEA3 & 128-EIA3. Document 2: ZUC Specification; Version: 1.6 (June 28, 2011)
14. Sun, B., Tang, X., Li, C.: Preliminary Cryptanalysis Results of ZUC. In: First International Workshop on ZUC Algorithm, vol. 12 (2010)
15. Wu, H.: The Stream Cipher HC-128. In: Robshaw, M., Billet, O. (eds.) New Stream Cipher Designs. LNCS, vol. 4986, pp. 39–47. Springer, Heidelberg (2008)
16. Wu, H., Nguyen, P.H., Wang, H., Ling, S.: Cryptanalysis of the Stream Cipher ZUC in the 3GPP Confidentiality & Integrity Algorithms 128-EEA3 & 128-EIA3. In: Rump Session of Asiacrypt 2010 (2008)
17. Wu, H., Preneel, B.: Differential Cryptanalysis of the Stream Ciphers Py, Py6 and Pypy. In: Naor, M. (ed.) EUROCRYPT 2007. LNCS, vol. 4515, pp. 276–290. Springer, Heidelberg (2007)

A   The List of Possible v and v' for Collision

Table 1. The list of possible v, v'

Index	v	v'	Δiv	Index	v	v'	Δiv	Index	v	v'	Δiv
1	0x3fff8000	0x40007fff	0xff	86	0x3fffd555	0x40002aaa	0x55	171	0x7fffaaaa	0x5555	0xaa
2	0x3fff8101	0x40007efe	0xfd	87	0x3fffd656	0x400029a9	0x53	172	0x7fffabab	0x5454	0xa8
3	0x3fff8202	0x40007dfd	0xfb	88	0x3fffd757	0x400028a8	0x51	173	0x7fffacac	0x5353	0xa6
4	0x3fff8303	0x40007cfc	0xf9	89	0x3fffd858	0x400027a7	0x4f	174	0x7fffadad	0x5252	0xa4
5	0x3fff8404	0x40007bfb	0xf7	90	0x3fffd959	0x400026a6	0x4d	175	0x7fffaeae	0x5151	0xa2
6	0x3fff8505	0x40007afa	0xf5	91	0x3fffda5a	0x400025a5	0x4b	176	0x7fffafaf	0x5050	0xa0
7	0x3fff8606	0x400079f9	0xf3	92	0x3fffdb5b	0x400024a4	0x49	177	0x7fffb0b0	0x4f4f	0x9e
8	0x3fff8707	0x400078f8	0xf1	93	0x3fffdc5c	0x400023a3	0x47	178	0x7fffb1b1	0x4e4e	0x9c
9	0x3fff8808	0x400077f7	0xef	94	0x3fffdd5d	0x400022a2	0x45	179	0x7fffb2b2	0x4d4d	0x9a
10	0x3fff8909	0x400076f6	0xed	95	0x3fffde5e	0x400021a1	0x43	180	0x7fffb3b3	0x4c4c	0x98
11	0x3fff8a0a	0x400075f5	0xeb	96	0x3fffdf5f	0x400020a0	0x41	181	0x7fffb4b4	0x4b4b	0x96
12	0x3fff8b0b	0x400074f4	0xe9	97	0x3fffe060	0x40001f9f	0x3f	182	0x7fffb5b5	0x4a4a	0x94
13	0x3fff8c0c	0x400073f3	0xe7	98	0x3fffe161	0x40001e9e	0x3d	183	0x7fffb6b6	0x4949	0x92
14	0x3fff8d0d	0x400072f2	0xe5	99	0x3fffe262	0x40001d9d	0x3b	184	0x7fffb7b7	0x4848	0x90
15	0x3fff8e0e	0x400071f1	0xe3	100	0x3fffe363	0x40001c9c	0x39	185	0x7fffb8b8	0x4747	0x8e
16	0x3fff8f0f	0x400070f0	0xe1	101	0x3fffe464	0x40001b9b	0x37	186	0x7fffb9b9	0x4646	0x8c
17	0x3fff9010	0x40006fef	0xdf	102	0x3fffe565	0x40001a9a	0x35	187	0x7fffbaba	0x4545	0x8a
18	0x3fff9111	0x40006eee	0xdd	103	0x3fffe666	0x40001999	0x33	188	0x7fffbbbb	0x4444	0x88
19	0x3fff9212	0x40006ded	0xdb	104	0x3fffe767	0x40001898	0x31	189	0x7fffbcbc	0x4343	0x86
20	0x3fff9313	0x40006cec	0xd9	105	0x3fffe868	0x40001797	0x2f	190	0x7fffbdbd	0x4242	0x84
21	0x3fff9414	0x40006beb	0xd7	106	0x3fffe969	0x40001696	0x2d	191	0x7fffbebe	0x4141	0x82
22	0x3fff9515	0x40006aea	0xd5	107	0x3fffea6a	0x40001595	0x2b	192	0x7fffbfbf	0x4040	0x80
23	0x3fff9616	0x400069e9	0xd3	108	0x3fffeb6b	0x40001494	0x29	193	0x7fffc0c0	0x3f3f	0x7e
24	0x3fff9717	0x400068e8	0xd1	109	0x3fffec6c	0x40001393	0x27	194	0x7fffc1c1	0x3e3e	0x7c
25	0x3fff9818	0x400067e7	0xcf	110	0x3fffed6d	0x40001292	0x25	195	0x7fffc2c2	0x3d3d	0x7a
26	0x3fff9919	0x400066e6	0xcd	111	0x3fffee6e	0x40001191	0x23	196	0x7fffc3c3	0x3c3c	0x78
27	0x3fff9a1a	0x400065e5	0xcb	112	0x3fffef6f	0x40001090	0x21	197	0x7fffc4c4	0x3b3b	0x76
28	0x3fff9b1b	0x400064e4	0xc9	113	0x3ffff070	0x40000f8f	0x1f	198	0x7fffc5c5	0x3a3a	0x74
29	0x3fff9c1c	0x400063e3	0xc7	114	0x3ffff171	0x40000e8e	0x1d	199	0x7fffc6c6	0x3939	0x72
30	0x3fff9d1d	0x400062e2	0xc5	115	0x3ffff272	0x40000d8d	0x1b	200	0x7fffc7c7	0x3838	0x70
31	0x3fff9e1e	0x400061e1	0xc3	116	0x3ffff373	0x40000c8c	0x19	201	0x7fffc8c8	0x3737	0x6e
32	0x3fff9f1f	0x400060e0	0xc1	117	0x3ffff474	0x40000b8b	0x17	202	0x7fffc9c9	0x3636	0x6c
33	0x3fffa020	0x40005fdf	0xbf	118	0x3ffff575	0x40000a8a	0x15	203	0x7fffcaca	0x3535	0x6a
34	0x3fffa121	0x40005ede	0xbd	119	0x3ffff676	0x40000989	0x13	204	0x7fffcbcb	0x3434	0x68
35	0x3fffa222	0x40005ddd	0xbb	120	0x3ffff777	0x40000888	0x11	205	0x7fffcccc	0x3333	0x66
36	0x3fffa323	0x40005cdc	0xb9	121	0x3ffff878	0x40000787	0xf	206	0x7fffcdcd	0x3232	0x64
37	0x3fffa424	0x40005bdb	0xb7	122	0x3ffff979	0x40000686	0xd	207	0x7fffcece	0x3131	0x62
38	0x3fffa525	0x40005ada	0xb5	123	0x3ffffa7a	0x40000585	0xb	208	0x7fffcfcf	0x3030	0x60
39	0x3fffa626	0x400059d9	0xb3	124	0x3ffffb7b	0x40000484	0x9	209	0x7fffd0d0	0x2f2f	0x5e
40	0x3fffa727	0x400058d8	0xb1	125	0x3ffffc7c	0x40000383	0x7	210	0x7fffd1d1	0x2e2e	0x5c
41	0x3fffa828	0x400057d7	0xaf	126	0x3ffffd7d	0x40000282	0x5	211	0x7fffd2d2	0x2d2d	0x5a
42	0x3fffa929	0x400056d6	0xad	127	0x3ffffe7e	0x40000181	0x3	212	0x7fffd3d3	0x2c2c	0x58
43	0x3fffaa2a	0x400055d5	0xab	128	0x3fffff7f	0x40000080	0x1	213	0x7fffd4d4	0x2b2b	0x56
44	0x3fffab2b	0x400054d4	0xa9	129	0x7fff8080	0x7f7f	0xfe	214	0x7fffd5d5	0x2a2a	0x54
45	0x3fffac2c	0x400053d3	0xa7	130	0x7fff8181	0x7e7e	0xfc	215	0x7fffd6d6	0x2929	0x52
46	0x3fffad2d	0x400052d2	0xa5	131	0x7fff8282	0x7d7d	0xfa	216	0x7fffd7d7	0x2828	0x50
47	0x3fffae2e	0x400051d1	0xa3	132	0x7fff8383	0x7c7c	0xf8	217	0x7fffd8d8	0x2727	0x4e
48	0x3fffaf2f	0x400050d0	0xa1	133	0x7fff8484	0x7b7b	0xf6	218	0x7fffd9d9	0x2626	0x4c
49	0x3fffb030	0x40004fcf	0x9f	134	0x7fff8585	0x7a7a	0xf4	219	0x7fffdada	0x2525	0x4a
50	0x3fffb131	0x40004ece	0x9d	135	0x7fff8686	0x7979	0xf2	220	0x7fffdbdb	0x2424	0x48
51	0x3fffb232	0x40004dcd	0x9b	136	0x7fff8787	0x7878	0xf0	221	0x7fffdcdc	0x2323	0x46
52	0x3fffb333	0x40004ccc	0x99	137	0x7fff8888	0x7777	0xee	222	0x7fffdddd	0x2222	0x44
53	0x3fffb434	0x40004bcb	0x97	138	0x7fff8989	0x7676	0xec	223	0x7fffdede	0x2121	0x42
54	0x3fffb535	0x40004aca	0x95	139	0x7fff8a8a	0x7575	0xea	224	0x7fffdfdf	0x2020	0x40
55	0x3fffb636	0x400049c9	0x93	140	0x7fff8b8b	0x7474	0xe8	225	0x7fffe0e0	0x1f1f	0x3e
56	0x3fffb737	0x400048c8	0x91	141	0x7fff8c8c	0x7373	0xe6	226	0x7fffe1e1	0x1e1e	0x3c
57	0x3fffb838	0x400047c7	0x8f	142	0x7fff8d8d	0x7272	0xe4	227	0x7fffe2e2	0x1d1d	0x3a
58	0x3fffb939	0x400046c6	0x8d	143	0x7fff8e8e	0x7171	0xe2	228	0x7fffe3e3	0x1c1c	0x38
59	0x3fffba3a	0x400045c5	0x8b	144	0x7fff8f8f	0x7070	0xe0	229	0x7fffe4e4	0x1b1b	0x36
60	0x3fffbb3b	0x400044c4	0x89	145	0x7fff9090	0x6f6f	0xde	230	0x7fffe5e5	0x1a1a	0x34
61	0x3fffbc3c	0x400043c3	0x87	146	0x7fff9191	0x6e6e	0xdc	231	0x7fffe6e6	0x1919	0x32
62	0x3fffbd3d	0x400042c2	0x85	147	0x7fff9292	0x6d6d	0xda	232	0x7fffe7e7	0x1818	0x30
63	0x3fffbe3e	0x400041c1	0x83	148	0x7fff9393	0x6c6c	0xd8	233	0x7fffe8e8	0x1717	0x2e
64	0x3fffbf3f	0x400040c0	0x81	149	0x7fff9494	0x6b6b	0xd6	234	0x7fffe9e9	0x1616	0x2c
65	0x3fffc040	0x40003fbf	0x7f	150	0x7fff9595	0x6a6a	0xd4	235	0x7fffeaea	0x1515	0x2a
66	0x3fffc141	0x40003ebe	0x7d	151	0x7fff9696	0x6969	0xd2	236	0x7fffebeb	0x1414	0x28
67	0x3fffc242	0x40003dbd	0x7b	152	0x7fff9797	0x6868	0xd0	237	0x7fffecec	0x1313	0x26
68	0x3fffc343	0x40003cbc	0x79	153	0x7fff9898	0x6767	0xce	238	0x7fffeded	0x1212	0x24
69	0x3fffc444	0x40003bbb	0x77	154	0x7fff9999	0x6666	0xcc	239	0x7fffeeee	0x1111	0x22
70	0x3fffc545	0x40003aba	0x75	155	0x7fff9a9a	0x6565	0xca	240	0x7fffefef	0x1010	0x20
71	0x3fffc646	0x400039b9	0x73	156	0x7fff9b9b	0x6464	0xc8	241	0x7ffff0f0	0xf0f	0x1e
72	0x3fffc747	0x400038b8	0x71	157	0x7fff9c9c	0x6363	0xc6	242	0x7ffff1f1	0xe0e	0x1c
73	0x3fffc848	0x400037b7	0x6f	158	0x7fff9d9d	0x6262	0xc4	243	0x7ffff2f2	0xd0d	0x1a
74	0x3fffc949	0x400036b6	0x6d	159	0x7fff9e9e	0x6161	0xc2	244	0x7ffff3f3	0xc0c	0x18
75	0x3fffca4a	0x400035b5	0x6b	160	0x7fff9f9f	0x6060	0xc0	245	0x7ffff4f4	0xb0b	0x16
76	0x3fffcb4b	0x400034b4	0x69	161	0x7fffa0a0	0x5f5f	0xbe	246	0x7ffff5f5	0xa0a	0x14
77	0x3fffcc4c	0x400033b3	0x67	162	0x7fffa1a1	0x5e5e	0xbc	247	0x7ffff6f6	0x909	0x12
78	0x3fffcd4d	0x400032b2	0x65	163	0x7fffa2a2	0x5d5d	0xba	248	0x7ffff7f7	0x808	0x10
79	0x3fffce4e	0x400031b1	0x63	164	0x7fffa3a3	0x5c5c	0xb8	249	0x7ffff8f8	0x707	0xe
80	0x3fffcf4f	0x400030b0	0x61	165	0x7fffa4a4	0x5b5b	0xb6	250	0x7ffff9f9	0x606	0xc
81	0x3fffd050	0x40002faf	0x5f	166	0x7fffa5a5	0x5a5a	0xb4	251	0x7ffffafa	0x505	0xa
82	0x3fffd151	0x40002eae	0x5d	167	0x7fffa6a6	0x5959	0xb2	252	0x7ffffbfb	0x404	0x8
83	0x3fffd252	0x40002dad	0x5b	168	0x7fffa7a7	0x5858	0xb0	253	0x7ffffcfc	0x303	0x6
84	0x3fffd353	0x40002cac	0x59	169	0x7fffa8a8	0x5757	0xae	254	0x7ffffdfd	0x202	0x4
85	0x3fffd454	0x40002bab	0x57	170	0x7fffa9a9	0x5656	0xac	255	0x7ffffefe	0x101	0x2

B Generating Identical Keystreams for Δiv_1

Here we describe more details of an algorithm that is used to generate identical keystreams for the IV difference at iv_1:

1. Initialize $iv_0, iv_1, \ldots, iv_{15}$ with 0. Set $iv_{13} = 64$.
2. Denote $(iv_4 + 8iv_{13} + 16iv_{10})$ as sum_1 and guess sum_1 with 1 of the 6376 possible values.
3. Guess $iv_2[1, 2]$, and compute v, until the condition $v[1..7] - (v \gg 8)[1..7] \leq 1$ is satisfied. If not possible, go to (2) .
4. Guess iv_7 and iv_{11}, and compute u, until $u[24..31] = \mathtt{0xff}$ is satisfied. We store the intermediate state s_{16}. If not possible, go to (3).
5. Guess iv_{15} and re-compute u, until $u[1..7] = u[9..15]$ and $u[8] = 0$ are satisfied. If not possible, go to (4).
6. Now we compare the current s_{16} with stored s_{16} to capture the change. By properly changing iv_2 and iv_{13}(this is the reason iv_{13} is initialized as 64), we can always change the current s_{16} back to the saved value. Hence, $u[24..31]$ will remain.
7. Determine iv_1 as follows:
   - If $v[8] \neq v[16]$, then if $u[1..16] < v[1..16]$ is satisfied, $iv_1 = 256 + u[1..16] - v[1..16]$ and update v, otherwise, go to (5).
   - If $v[8] = v[16]$, then if $u[1..16] >= v[1..16]$ is satisfied, $iv_1 = u[1..16] - v[1..16]$ and update v, otherwise, go to (5).
8. Guess iv_0, iv_5 and iv_{14}, compute v, until $v[16..31] = \mathtt{0xffff}$. If not possible, go to (5).
9. If $(u \oplus v)[1] = 1$, let $iv_2 = iv_2 \oplus 2$. Choose iv_3 properly to ensure $u[16..23] = \mathtt{0xff}$. Check if we indeed have $v = u$, then output $iv_0, iv_1, \ldots, iv_{15}$. Otherwise, go to (8).

In this algorithm, we restrict the forms of v and u to those starting with $\mathtt{0x7fff}$ to reduce the search space.

An Asymptotically Tight Security Analysis of the Iterated Even-Mansour Cipher

Rodolphe Lampe[1,*], Jacques Patarin[1], and Yannick Seurin[2,**]

[1] University of Versailles, France
[2] ANSSI, Paris, France
rodolphe.lampe@gmail.com,
jacques.patarin@uvsq.fr,yannick.seurin@m4x.org

Abstract. We analyze the security of the iterated Even-Mansour cipher (*a.k.a.* key-alternating cipher), a very simple and natural construction of a blockcipher in the random permutation model. This construction, first considered by Even and Mansour (J. Cryptology, 1997) with a single permutation, was recently generalized to use t permutations in the work of Bogdanov *et al.* (EUROCRYPT 2012). They proved that the construction is secure up to $\mathcal{O}(N^{2/3})$ queries (where N is the domain size of the permutations), as soon as the number t of rounds is 2 or more. This is tight for $t = 2$, however in the general case the best known attack requires $\Omega(N^{t/(t+1)})$ queries. In this paper, we give asymptotically tight security proofs for two types of adversaries:

1. for non-adaptive chosen-plaintext adversaries, we prove that the construction achieves an optimal security bound of $\mathcal{O}(N^{t/(t+1)})$ queries;
2. for adaptive chosen-plaintext and ciphertext adversaries, we prove that the construction achieves security up to $\mathcal{O}(N^{t/(t+2)})$ queries (for t even). This improves previous results for $t \geq 6$.

Our proof crucially relies on the use of a *coupling* to upper-bound the statistical distance of the outputs of the iterated Even-Mansour cipher to the uniform distribution.

Keywords: blockcipher, Even-Mansour cipher, key-alternating cipher, random permutation model, coupling, provable security.

1 Introduction

The Even-Mansour Cipher. Even and Mansour [9] proposed the following "minimal" construction of a blockcipher on message space $\{0,1\}^n$: given a public permutation P on $\{0,1\}^n$ (*e.g.* AES-128 with a fixed, publicly known key), encrypt x by computing $y = k_1 \oplus P(k_0 \oplus x)$, where k_0, k_1 are two n-bit keys. Their work was motivated by the DESX construction proposed by Rivest (1984, unpublished) and later formally analyzed by Kilian and Rogaway [13], in which

---

* This author is partially supported by the French Department of Defense (DGA).
** This author is partially supported by the French National Agency of Research: ANR-11-INS-011.

X. Wang and K. Sako (Eds.): ASIACRYPT 2012, LNCS 7658, pp. 278–295, 2012.

Rivest suggested to strengthen DES against exhaustive key search by using two independent pre-whitening and post-whitening keys xored respectively to the input and the output of DES (thereby augmenting the key size of the resulting cipher from 56 to 184 bits). Even and Mansour analyzed their proposal in the random permutation model, where P is replaced by an oracle implementing a random (invertible) permutation, publicly accessible to all parties including the adversary. They showed that an adversary with black-box access to both P and the cipher with a random unknown key (as well as their inverse), has only a negligible probability to correctly inverse the cipher on an un-queried ciphertext of its choice (or to compute the ciphertext corresponding to some un-queried plaintext). In fact, the Even-Mansour cipher yields a (strong) pseudorandom permutation (in the random permutation model) in the sense that the system $(P, \text{EM}_{P,(k_0,k_1)})$, where $\text{EM}_{P,(k_0,k_1)}$ is the Even-Mansour cipher built from P with two uniformly random keys k_0 and k_1, is indistinguishable from an ideal system (P, Q), where Q is an independent random permutation. More precisely, any distinguisher has to make $\Omega(2^{n/2})$ queries to distinguish these two systems with non-negligible advantage.

The Iterated Even-Mansour Cipher. The Even-Mansour cipher was recently generalized in a very natural way by Bogdanov et al. [4] as follows: given t public permutations $P_1, \ldots, P_t$ on $\{0,1\}^n$, encrypt x by computing:

$$y = k_t \oplus P_t(k_{t-1} \oplus P_{t-1}(\cdots P_1(k_0 \oplus x) \cdots)) \ ,$$

where $k_0, \ldots, k_t$ are $t+1$ keys of n bits. They used the moniker (first coined in [7]) *key alternating cipher* for this construction, but we will prefer the name *iterated Even-Mansour cipher* in this paper to emphasize that we work in the random permutation model. We will refer to t as the number of *rounds* of the construction.

The main result of [4] is a proof (again, in the random permutation model for $P_1, \ldots P_t$) that the iterated Even-Mansour cipher with $t \geq 2$ rounds is secure (*i.e.*, indistinguishable from an independent random permutation) up to $\mathcal{O}(N^{2/3})$ queries (where $N = 2^n$). They also gave a distinguishing attack (in fact a key-recovery attack) requiring $\Omega(N^{t/(t+1)})$ queries. Hence, their analysis is tight for $t = 2$, but they left the security gap for $t > 2$ as an interesting open problem.

Our Contribution. In this work, we strengthen the security bounds of [4]. We obtain two distinct results depending on which type of adversaries we consider. For non-adaptive chosen-plaintext (NCPA for short) adversaries, we prove that the iterated Even-Mansour cipher with t rounds is secure up to $\mathcal{O}(N^{t/(t+1)})$ queries. Given that the attack described by [4] falls into this category of adversaries, this is tight up to constant factors. Tough this type of adversaries was not explicitly considered by [4], we note that this improves their general bound as soon as $t \geq 3$.

For adaptive chosen-plaintext and ciphertext (CCA for short) adversaries (*i.e.* the most powerful ones in terms of how queries may be issued to the system), we prove that the iterated Even-Mansour cipher with t rounds is secure up to $\mathcal{O}(N^{t/(t+2)})$ queries when t is even. When t is odd, we get the same bound as for $t - 1$ (since it is clear that adding a round to the construction cannot improve the advantage of a distinguisher). Our bound becomes better than $\mathcal{O}(N^{2/3})$, therefore improving [4]'s result, for $t \geq 6$. In particular, for $t = 6$, we obtain an improved security bound of $\mathcal{O}(N^{3/4})$ queries. Our findings are summarized in Table 1.

Our Techniques. Our proof strategy is very different and much simpler than the one of [4] (the counterpart of which is that for the interesting case of CCA adversaries, we improve their results only for t "large", where large means at least 6). One of the main ingredient of our proof is a well-known tool of the theory of Markov chains, namely the *coupling* technique. Indeed, a crucial step of our proof is to upper-bound, for any possible tuple of plaintext queries $(x^1, \ldots, x^{q_e})$ to the iterated Even-Mansour cipher, the statistical distance of the outputs of the cipher to the uniform distribution, conditioned on some partial information about the inner permutations $P_1, \ldots, P_t$ (namely equations of the form $P_i(a) = b$) that was gathered from the queries to these permutations. The outputs of permutations P_i, $i = 1, \ldots, t$, when computing the ciphertexts for inputs $(x^1, \ldots, x^{q_e})$, can be seen as the state of a Markov chain, so that we can reformulate the problem as studying how quick the distribution of this Markov chain converges to the uniform (as a function of the number of rounds). The coupling technique is one of the most efficient way to analyze this convergence rate (often named the *mixing time* of the Markov chain), and this is exactly the technique we adopt. Couplings were previously used in cryptography by Mironov [16] to analyze the RC4 stream cipher, and more recently by Morris *et al.* [17] to study maximally unbalanced Feistel networks and by Hoang and Rogaway [12] who generalized the results of [17] to many variants of the Feistel construction. In fact, our analysis was strongly inspired by the works of [17,12].

However, the coupling technique only enables to treat adversaries choosing their queries to the cipher non-adaptively. To leverage the result from NCPA-security to CCA-security, we use a composition strategy which is very similar to what is often referred to as the "two weak make one strong" technique [14,15]. For "classical" pseudorandom permutations (*i.e.* not build from ideal primitives as the Even-Mansour cipher), this strategy enables to prove the following: if $\{F_k\}$ and $\{G_{k'}\}$ are two permutation families secure against NCPA attacks (with upper-bounds resp. ε_F and ε_G on the advantage of any NCPA-distinguisher), then the composition $\{G_{k'}^{-1} \circ F_k\}$ is secure against CCA attacks (with advantage upper-bounded by $\varepsilon_F + \varepsilon_G$). This was proved by Maurer and Pietrzak [14] up to logarithmic factors and then refined by Maurer *et al.* [15], in the formalism of random systems. However, subtle complications appear when trying to use these results directly because of the additional inner permutation oracles $P_1, \ldots, P_t$, so that we prefer a more direct approach, very similar to the "H coefficients" technique of Patarin [18].

A Caveat. We warn that the value of our results is similar to security proofs in the random oracle model [2], meaning that they offer no guarantee once the inner permutations are instantiated with real, standard permutations [5]. They show however that any attack beating our bounds cannot use the inner permutations as black-boxes.

Table 1. Summary of our results. The NCPA (resp. CCA) column gives the constant c such that the iterated Even-Mansour cipher is secure up to N^c queries against NCPA-distinguishers (resp. CCA-distinguishers). Gray cells indicate when we improve the $N^{2/3}$ bound of [4]. The last column gives, for $n = 128$, the log in base 2 of the minimal number of queries a CCA-distinguisher has to make to have advantage at least $1/2$ in distinguishing the cipher from random (we only give this number when our bound improves the one of [4]).

t	NCPA	CCA	CCA ($n = 128$)
2	2/3	1/2	—
3	3/4	1/2	—
4	4/5	2/3	—
5	5/6	2/3	—
6	6/7	3/4	93
7	7/8	3/4	93
8	8/9	4/5	100

Related Work. We focus on security proofs in this work, but we stress that quite a few papers explored attacks (mainly key-recovery ones) against the Even-Mansour cipher. Daemen [6] gave a differential-style attack requiring q_p (direct) chosen queries to P and q_e chosen plaintext queries to the cipher, with $q_p q_e = \Omega(2^n)$ (hence the total query complexity is minimized for $q_p = q_e = \Omega(2^{n/2})$). Later, Biryukov and Wagner [3] gave an attack requiring $\Omega(2^{n/2})$ queries to both P and the cipher, but allowing to use known plaintexts rather than chosen ones. However, their method does not allow any trade-off between queries to P and the cipher as is possible in Daemen's attack. Recently, Dunkelman *et al.* [8] refined the work of [3] by giving a known-plaintext attack where such a trade-off is possible, thereby providing an optimal attack on the Even-Mansour cipher.

On the provable-security side, Gentry and Ramzan [10] showed that the Even-Mansour cipher remains secure when the random permutation oracle P is replaced by a Feistel construction with four rounds, where the round functions are public random function oracles.

Open Problems. Our work settles the case of non-adaptive chosen-plaintext adversaries; there remains however a gap for adaptive chosen-plaintext and cipher-text attacks between the proven bound of $\mathcal{O}(N^{t/(t+2)})$ queries and the best attack requiring $\Omega(N^{t/(t+1)})$ queries. The two practically appealing cases where all keys are identical (as was for example recently proposed in the blockcipher LED [11]), and where all inner permutations are identical, also remain interesting directions of research. It may even be possible that using both identical keys *and* a single inner permutation provides some level of security greater than $2^{n/2}$.[1]

Organization. In Section 2, we introduce the general notation, formally define the adversarial model, and give the necessary background on couplings. In Section 3, we prove our main result on the statistical distance of the outputs of the iterated Even-Mansour cipher to the uniform distribution using a coupling, which enables us to treat NCPA-adversaries. In Section 4, we deal with CCA-adversaries.

2   Preliminaries

2.1   General Notation

In all the following, we fix an integer $n \geq 1$. We denote $\mathcal{I}_n = \{0, 1\}^n$ the set of binary strings of length n and $N = 2^n$. Given an integer $q \geq 1$, we denote $(\mathcal{I}_n)^{*q}$ the set of all sequences of pairwise distinct elements of $\mathcal{I}_n$ of length q. Given integers $q_1, \ldots, q_t$ we denote $(\mathcal{I}_n)^{*q_1, \ldots, q_t} = (\mathcal{I}_n)^{*q_1} \times \cdots \times (\mathcal{I}_n)^{*q_t}$. We denote $(N)_q = N(N-1) \cdots (N-q+1)$ the falling factorial. Note that $|(\mathcal{I}_n)^{*q}| = (N)_q$. We denote $[i; j]$ the set of integers k such that $i \leq k \leq j$.

The set of permutations on $\mathcal{I}_n$ will be denoted $\mathcal{P}_n$. Given $P \in \mathcal{P}_n$ and two sequences $x = (x^1, \ldots, x^q)$ and $y = (y^1, \ldots, y^q)$ of $(\mathcal{I}_n)^{*q}$, we will write $P(x) = y$ to mean that $P(x^i) = y^i$ for $i = 1, \ldots, q$. Given a tuple of permutations $\boldsymbol{P} = (P_1, \ldots, P_t) \in (\mathcal{P}_n)^t$ and two sequences $a = (a_1, \ldots, a_t)$ and $b = (b_1, \ldots, b_t)$ of $(\mathcal{I}_n)^{*q_1, \ldots, q_t}$, with $a_i = (a_i^1, \ldots, a_i^{q_i})$ and $b_i = (b_i^1, \ldots, b_i^{q_i})$, we will write $\boldsymbol{P}(a) = b$ to mean that $P_i(a_i) = b_i$ for $i = 1, \ldots, t$ (*i.e.* $P_i(a_i^j) = b_i^j$ for $j = 1, \ldots, q_i$).

Given a value $k \in \{0, 1\}^n$, $\oplus_k$ denotes the mapping $x \mapsto x \oplus k$ from $\{0, 1\}^n$ to itself. Fix an integer $t \geq 1$. Let $\boldsymbol{P} = (P_1, \ldots, P_t)$ be a tuple of permutations on $\{0, 1\}^n$. Then the *iterated Even-Mansour cipher* associated with $\boldsymbol{P}$ is the cipher with message space $\{0, 1\}^n$ and key space $(\{0, 1\}^n)^{t+1}$ where the permutation associated with key $k = (k_0, \ldots, k_t)$ is defined as (see Fig. 1):

$$\mathrm{EM}_{\boldsymbol{P}, k} = \oplus_{k_t} \circ P_t \circ \oplus_{k_{t-1}} \circ \cdots \circ \oplus_{k_1} \circ P_1 \circ \oplus_{k_0} .$$

We denote $\Omega_t = (\mathcal{P}_n)^t \times (\mathcal{I}_n)^{t+1}$. An element $(\boldsymbol{P}, k)$ of Ω_t names a tuple of permutations and a key for the resulting Even-Mansour cipher.

2.2   Distinguishers

We consider distinguishers interacting with systems constituted of $t + 1$ permutations. A query to such a system is a triplet (i, b, z) where $i \in [1; t+1]$ names which permutation is being queried, b is a bit indicating whether the

---

[1] Note however that, as observed by [4], using P and P^{-1} for the construction with $t = 2$ rounds causes the security to drop to $2^{n/2}$, even with three independent keys.

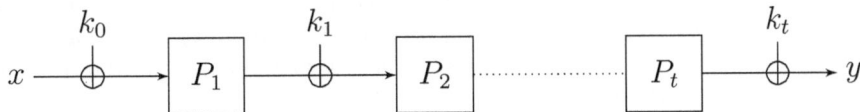

Fig. 1. The iterated Even-Mansour cipher

query is forward or backward, and $z \in \{0,1\}^n$ is the actual query to the permutation. The goal of the distinguisher is to tell whether it is interacting with a tuple of $t+1$ uniformly random and independent (URI for short) permutations $(P_1, \ldots, P_t, Q)$, or with $(P_1, \ldots, P_t, \mathrm{EM}_{\boldsymbol{P},k})$ where $(P_1, \ldots, P_t)$ are URI and $\mathrm{EM}_{\boldsymbol{P},k}$ is the Even-Mansour cipher associated with $\boldsymbol{P} = (P_1, \ldots, P_t)$ with a uniformly random key $k = (k_0, \ldots, k_t)$. In the following we will refer to the first t permutations of the system as the *inner permutations*, by opposition to the last permutation of the system (which may be an independent random permutation Q or the Even-Mansour cipher $\mathrm{EM}_{\boldsymbol{P},k}$) to which we will refer to as the *outer permutation*. A $(q_1, \ldots, q_t, q_e)$-distinguisher is a distinguisher that makes at most q_i queries to inner permutation P_i for $i = 1, \ldots, t$ and q_e queries to the outer permutation. We will consider only computationally unbounded distinguishers. As usual we restrict ourself *wlog* to deterministic distinguishers that never make redundant queries and always make the maximal number of allowed queries to each permutation of the system.

The way we define chosen-plaintext/-ciphertext and adaptive/non-adaptive distinguishers is very specific to the context of our work. The qualifier chosen-plaintext/-ciphertext will only refer to the queries the distinguisher is allowed to make to the *outer permutation* of the system (it will always be allowed to make both forward and backward queries to the inner permutations). As well, adaptivity will only refer to how the distinguisher is allowed to choose its queries to the outer permutation (it will always be allowed to choose its queries to the inner permutations adaptively), and also to whether the distinguisher is allowed to query the inner permutations as a function of the answers received from the outer permutation. We now give a precise definition of the two types of distinguishers we consider: non-adaptive chosen-plaintext (NCPA) distinguishers and adaptive chosen-plaintext and ciphertext (CCA) distinguishers.

Definition 1. *A $(q_1, \ldots, q_t, q_e)$-NCPA-distinguisher runs in two phases:*

1. *in a first phase, it can only query the inner permutations $(P_1, \ldots, P_t)$. These queries can be adaptive, and both forward and backward queries are allowed. During this phase it makes exactly q_i queries to P_i for $i = 1, \ldots, t$;*
2. *in a second phase, it chooses a tuple of q_e non-adaptive[2] forward queries $x = (x^1, \ldots, x^{q_e})$ to the outer permutation of the system, and receives the corresponding answers.*

---

[2] By non-adaptive we mean that all queries have to be chosen before receiving any corresponding answer from the outer permutation. However the choice of x may depend on the answers received from the inner permutations during the first phase.

A $(q_1, \ldots, q_t, q_e)$-*CCA-distinguisher is the most general one: it is allowed to make both forward and backward queries to all permutations of the system, in any order it wishes (in particular it may interleave queries to the outer permutation and to the inner permutations).*

In all the following, the probability of an event E when $\mathcal{D}$ interacts with $t + 1$ URI permutations $(P_1, \ldots, P_t, Q)$ will simply be denoted $\mathrm{Pr}^*[E]$, whereas the probability of an event E when $\mathcal{D}$ interacts with $(P_1, \ldots, P_t, \mathrm{EM}_{\boldsymbol{P},k})$, where $\boldsymbol{P} = (P_1, \ldots, P_t)$ are URI permutations and the key k is uniformly random, will simply be denoted $\mathrm{Pr}[E]$. With these notations, the advantage of a distinguisher $\mathcal{D}$ is defined as $|\mathrm{Pr}[\mathcal{D}(1^n) = 1] - \mathrm{Pr}^*[\mathcal{D}(1^n) = 1]|$ (we omit the oracles in this notation since they can be deduced from the notation $\mathrm{Pr}[\cdot]$ or $\mathrm{Pr}^*[\cdot]$). The maximum advantage of a $(q_1, \ldots, q_t, q_e)$-ATK-distinguisher against the iterated Even-Mansour cipher with t rounds (where ATK is NCPA or CCA) will be denoted $\mathbf{Adv}_{\mathcal{EM}[t]}^{\mathrm{atk}}(q_1, \ldots, q_t, q_e)$. When considering distinguishers making at most q queries in total, we simply denote $\mathbf{Adv}_{\mathcal{EM}[t]}^{\mathrm{atk}}(q)$.

Remark 1. We warn that our NCPA-security notion should not be considered as interesting in itself, but rather as a preliminary step towards proving CCA-security. The reason why it is rather artificial is that once the distinguisher has received the answers to its queries to the outer permutation, it is not allowed to query the inner permutations any more. This is not satisfying since these permutations are public primitives, and hence adversaries should be allowed to query them in their entire discretion.

2.3 Total Variation Distance and Coupling

Given a finite event space Ω and two probability distributions μ and ν defined on Ω, the *total variation distance* (or statistical distance) between μ and ν, denoted $\|\mu - \nu\|$ is defined as:

$$\|\mu - \nu\| = \frac{1}{2} \sum_{x \in \Omega} |\mu(x) - \nu(x)| \ .$$

The following definitions can easily be seen equivalent:

$$\|\mu - \nu\| = \max_{S \subset \Omega} \{\mu(S) - \nu(S)\} = \max_{S \subset \Omega} \{\nu(S) - \mu(S)\} = \max_{S \subset \Omega} \{|\mu(S) - \nu(S)|\} \ .$$

A *coupling* of μ and ν is a distribution λ on $\Omega \times \Omega$ such that for all $x \in \Omega$, $\sum_{y \in \Omega} \lambda(x, y) = \mu(x)$ and for all $y \in \Omega$, $\sum_{x \in \Omega} \lambda(x, y) = \nu(y)$. In other words, λ is a joint distribution whose marginal distributions are resp. μ and ν. The fundamental result of the coupling technique is the following one. For completeness, we provide the proof in Appendix A.

Lemma 1 (Coupling Lemma). *Let μ and ν be probability distributions on a finite event space Ω, let λ be a coupling of μ and ν, and let $(X, Y) \sim \lambda$ (i.e. (X, Y) is a random variable sampled according to distribution λ). Then $\|\mu - \nu\| \leq \mathrm{Pr}[X \neq Y]$.*

For the analysis of CCA attacks, we will rely on the following observation.

Lemma 2. *Let Ω be some finite event space and ν be the uniform probability distribution on Ω. Let μ be a probability distribution on Ω such that $\|\mu - \nu\| \leq \varepsilon$. Then there is a set $S \subset \Omega$ such that:*

- $|S| \geq (1 - \sqrt{\varepsilon})|\Omega|$
- $\forall x \in S, \mu(x) \geq (1 - \sqrt{\varepsilon})\nu(x)$

Proof. Define $S = \{x \in \Omega : \mu(x) \geq (1 - \sqrt{\varepsilon})\nu(x)\}$. We will show that $|S| \geq (1 - \sqrt{\varepsilon})|\Omega|$. Assume for contradiction that $|S| < (1 - \sqrt{\varepsilon})|\Omega|$, or equivalently $|\bar{S}| > \sqrt{\varepsilon}|\Omega|$, i.e. $\nu(\bar{S}) > \sqrt{\varepsilon}$. By definition, for any $x \in \bar{S}$, $\nu(x) - \mu(x) > \sqrt{\varepsilon}\nu(x)$. Consequently,

$$\nu(\bar{S}) - \mu(\bar{S}) > \sqrt{\varepsilon}\nu(\bar{S}) > (\sqrt{\varepsilon})^2 = \varepsilon \ ,$$

a contradiction with $\|\mu - \nu\| \leq \varepsilon$. $\qquad\square$

3 Security against Non-adaptive Distinguishers

In this section, we start with dealing with NCPA-distinguishers. The crucial point will be to upper bound the statistical distance between the outputs of the iterated Even-Mansour cipher *conditioned on partial information on the inner permutations* (namely $\boldsymbol{P}(a) = b$ for some tuples $a, b \in (\mathcal{I}_n)^{*q_1, \ldots, q_t}$) and the uniform distribution on $(\mathcal{I}_n)^{*q_e}$. We introduce the following important definitions and notations.

Definition 2. *Let $q_1, \ldots, q_t, q_e$ be positive integers. Fix tuples $a, b \in (\mathcal{I}_n)^{*q_1, \ldots, q_t}$ and $x \in (\mathcal{I}_n)^{*q_e}$. We denote $\mu_x(\cdot|\boldsymbol{P}(a) = b)$ the distribution of $\mathrm{EM}_{\boldsymbol{P},k}(x)$ conditioned on the event $\boldsymbol{P}(a) = b$ (i.e. when the key $k = (k_0, \ldots, k_t)$ is uniformly random and the permutations $\boldsymbol{P} = (P_1, \ldots, P_t)$ are uniformly random among permutations satisfying $\boldsymbol{P}(a) = b$). We also denote $\mu_{q_e}^* = 1/(N)_{q_e}$ the uniform distribution on $(\mathcal{I}_n)^{*q_e}$.*

We have the following expression for $\mu_x(\cdot|\boldsymbol{P}(a) = b)$.

Lemma 3. *Let $a, b \in (\mathcal{I}_n)^{*q_1, \ldots, q_t}$ and $x \in (\mathcal{I}_n)^{*q_e}$. Then for any $y \in (\mathcal{I}_n)^{*q_e}$ one has:*

$$\mu_x(y|\boldsymbol{P}(a) = b) = \frac{\#\{(\boldsymbol{P}, k) \in \Omega_t : \boldsymbol{P}(a) = b \wedge \mathrm{EM}_{\boldsymbol{P},k}(x) = y\}}{|\Omega_t| / \prod_{i=1}^{t}(N)_{q_i}} \ .$$

Proof. This follows easily from the observation that the number of $(\boldsymbol{P}, k) \in \Omega_t$ such that $\boldsymbol{P}(a) = b$ is $|\Omega_t| / \prod_{i=1}^{t}(N)_{q_i}$. $\qquad\square$

The following lemma states that the advantage of a NCPA-distinguisher is upper-bounded by the total variation distance between $\mu_x(\cdot|\boldsymbol{P}(a) = b)$ and $\mu_{q_e}^*$. This is a classical result regarding the advantage of the best NCPA-distinguisher for a pseudorandom permutation, however we need to adapt it here to fit the random permutation model.

Lemma 4. *Let $q_1, \ldots, q_t, q_e$ be positive integers. Assume that there exists α such that for any tuples $a, b \in (\mathcal{I}_n)^{*q_1, \ldots, q_t}$ and $x \in (\mathcal{I}_n)^{*q_e}$, one has*

$$\|\mu_x(\cdot | \boldsymbol{P}(a) = b) - \mu_{q_e}^*\| \leq \alpha \ .$$

Then $\mathbf{Adv}_{\mathcal{EM}[t]}^{\mathrm{ncpa}}(q_1, \ldots, q_t, q_e) \leq \alpha$.

Proof. Fix a $(q_1, \ldots, q_t, q_e)$-NCPA-distinguisher $\mathcal{D}$. Such a distinguisher first queries the inner permutations $(P_1, \ldots, P_t)$. Let τ be the resulting transcript, *i.e.* the ordered sequence of $q_1 + \ldots + q_t$ queries with the corresponding answer (i, b, z, z'), where $i \in [1; t]$ names which permutation is being queried, b is a bit indicating whether the query is forward or backward, $z \in \{0, 1\}^n$ is the actual query and z' the answer. Let also Φ be the function that maps a tuple of permutations $\boldsymbol{P} = (P_1, \ldots, P_t)$ to the transcript of the first phase of the attack when $\mathcal{D}$ interacts with $(P_1, \ldots, P_t, *)$, where $*$ is either an independent random permutation Q or $\mathrm{EM}_{\boldsymbol{P}, k}$ (this is clearly irrelevant since $\mathcal{D}$ does not query the outer permutation during the first phase of the attack). We say that a transcript τ is *consistent* if there exists a tuple of permutations $\boldsymbol{P}$ such that $\Phi(\boldsymbol{P}) = \tau$, and we denote Γ the set of consistent transcripts. Finally, from a consistent transcript τ, we build the sequences $a(\tau), b(\tau) \in (\mathcal{I}_n)^{*q_1, \ldots, q_t}$ as follows: let (i, b, z, z') be the j-th query and corresponding answer to P_i in the transcript. If this is a forward query ($b = 0$), then we define $a_i^j = z$ and $b_i^j = z'$; else, when this is a backward query ($b = 1$), we define $a_i^j = z'$ and $b_i^j = z$. Note that for a consistent transcript τ, $\Phi(\boldsymbol{P}) = \tau$ iff $\boldsymbol{P}(a(\tau)) = b(\tau)$. The number of consistent transcripts can be exactly determined:

$$|\Gamma| = \prod_{i=1}^{t} (N)_{q_i} \ . \tag{1}$$

This can be easily seen as follows. The first query of $\mathcal{D}$ is fixed in all executions. Assume *wlog* that this is a query to P_1. There are exactly N possible answer. The next query is determined by the answer received to the first query. If this is again a query to P_1, there are now $N - 1$ possible answers, whereas if this a query to P_i, $i \neq 1$, there are N possible answers. This can be easily extended by induction to obtain the above claim.

The tuple of non-adaptive plaintext queries $x = (x^1, \ldots, x^{q_e}) \in (\mathcal{I}_n)^{*q_e}$ of $\mathcal{D}$ to the outer permutation is a deterministic function of the transcript τ of the first phase of the attack. Let Ψ denote the function which maps a consistent transcript τ to the corresponding tuple of queries. The output of $\mathcal{D}$ is then a deterministic function of τ and the answers $y = (y^1, \ldots, y^{q_e})$ received from the outer permutation to the tuple of queries $\Psi(\tau)$. For any consistent transcript τ, we denote Σ_τ the set of tuples y such that $\mathcal{D}$ outputs 1 when receiving answers y to the queries $\Psi(\tau)$. Then, by definition we have:

$$\Pr{}^*[\mathcal{D}(1^n) = 1] = \sum_{\tau \in \Gamma} \sum_{y \in \Sigma_\tau} \frac{\#\{(\boldsymbol{P}, Q) \in (\mathcal{P}_n)^{t+1} : \Phi(\boldsymbol{P}) = \tau \wedge Q(\Psi(\tau)) = y\}}{|\mathcal{P}_n|^{t+1}}$$

$$= \sum_{\tau \in \Gamma} \sum_{y \in \Sigma_\tau} \frac{\#\{(\boldsymbol{P}, Q) \in (\mathcal{P}_n)^{t+1} : \boldsymbol{P}(a(\tau)) = b(\tau) \wedge Q(\Psi(\tau)) = y\}}{|\mathcal{P}_n|^{t+1}}$$

$$= \sum_{\tau \in \Gamma} \sum_{y \in \Sigma_\tau} \frac{1}{(N)_{q_e} \prod_{i=1}^{t} (N)_{q_i}} . \tag{2}$$

Also, we have:

$$\Pr[\mathcal{D}(1^n) = 1] =$$

$$\sum_{\tau \in \Gamma} \sum_{y \in \Sigma_\tau} \frac{\#\{(\boldsymbol{P}, k) \in \Omega_t : \Phi(\boldsymbol{P}) = \tau \wedge \mathrm{EM}_{\boldsymbol{P},k}(\Psi(\tau)) = y\}}{|\Omega_t|} . \tag{3}$$

We now use the assumption that, for all tuples $a, b \in (\mathcal{I}_n)^{*q_1, \ldots, q_t}$ and $x \in (\mathcal{I}_n)^{*q_e}$, one has $\|\mu_x(\cdot | \boldsymbol{P}(a) = b) - \mu_{q_e}^*\| \leq \alpha$. By Lemma 3, this exactly means that for all tuples a, b, x and any subset $S \subset (\mathcal{I}_n)^{*q_e}$, one has:

$$\left| \sum_{y \in S} \frac{\#\{(\boldsymbol{P}, k) \in \Omega_t : \boldsymbol{P}(a) = b \wedge \mathrm{EM}_{\boldsymbol{P},k}(x) = y\}}{|\Omega_t| / \prod_{i=1}^{t}(N)_{q_i}} - \sum_{y \in S} \frac{1}{(N)_{q_e}} \right| \leq \alpha .$$

For any $\tau \in \Gamma$ we can apply the above inequality with $(a, b) = (a(\tau), b(\tau))$, $x = \Psi(\tau)$, and $S = \Sigma_\tau$ to get:

$$\left| \sum_{y \in \Sigma_\tau} \frac{\#\{(\boldsymbol{P}, k) \in \Omega_t : \boldsymbol{P}(a(\tau)) = b(\tau) \wedge \mathrm{EM}_{\boldsymbol{P},k}(\Psi(\tau)) = y\}}{|\Omega_t|} - \right.$$

$$\left. \sum_{y \in \Sigma_\tau} \frac{1}{(N)_{q_e} \prod_{i=1}^{t}(N)_{q_i}} \right| \leq \frac{\alpha}{\prod_{i=1}^{t}(N)_{q_i}} . \tag{4}$$

Combining Eqs. (2-3-4), and using that for a consistent transcript τ, $\Phi(\boldsymbol{P}) = \tau$ iff $\boldsymbol{P}(a(\tau)) = b(\tau)$, we obtain:

$$|\Pr[\mathcal{D}(1^n) = 1] - \Pr{}^*[\mathcal{D}(1^n) = 1]| \leq \sum_{\tau \in \Gamma} \frac{\alpha}{\prod_{i=1}^{t}(N)_{q_i}} .$$

Finally, we deduce using Eq. (1) that the advantage of $\mathcal{D}$ is less than α, which concludes the proof. $\qquad\square$

The rest of this section is devoted to establishing an appropriate upper bound α for $\|\mu_x(\cdot | \boldsymbol{P}(a) = b) - \mu_{q_e}^*\|$ as required to apply Lemma 4. The following lemma can be regarded as the main contribution of this work.

Lemma 5. *Let $q_1, \ldots, q_t, q_e$ be positive integers. Fix tuples $a, b \in (\mathcal{I}_n)^{*q_1, \ldots, q_t}$ and $x \in (\mathcal{I}_n)^{*q_e}$. Then:*

$$\|\mu_x(\cdot | \boldsymbol{P}(a) = b) - \mu_{q_e}^*\| \leq 2^t \frac{q_e \prod_{i=1}^{t} q_i}{N^t} \ .$$

Proof. Fix tuples $a, b \in (\mathcal{I}_n)^{*q_1, \ldots, q_t}$ and $x \in (\mathcal{I}_n)^{*q_e}$, with $x = (x^1, \ldots, x^{q_e})$. For each $\ell \in [0; q_e]$, let $(z^1, \ldots, z^{q_e})$ be a tuple of queries such that $z^i = x^i$ for $i \leq \ell$, and z^i is uniformly random in $\{0,1\}^n \setminus \{z^1, \ldots, z^{i-1}\}$ for $i > \ell$. Denote ν_ℓ the distribution of the tuple of q_e outputs when $\text{EM}_{\boldsymbol{P}, k}$ receives inputs $(z^1, \ldots, z^{q_e})$, conditioned on $\boldsymbol{P}(a) = b$. Note that $\nu_0 = \mu_{q_e}^*$ since for $\ell = 0$ the tuple of inputs is uniformly random in $(\mathcal{I}_n)^{*q_e}$, and $\nu_{q_e} = \mu_x(\cdot | \boldsymbol{P}(a) = b)$. Hence we have:

$$\|\mu_x(\cdot | \boldsymbol{P}(a) = b) - \mu_{q_e}^*\| = \|\nu_{q_e} - \nu_0\| \leq \sum_{l=0}^{q_e - 1} \|\nu_{\ell+1} - \nu_\ell\| \ . \tag{5}$$

It remains to upper bound the total variation distance between $\nu_{\ell+1}$ and ν_ℓ, for each $\ell \in [0; q_e - 1]$. For this, we will construct a suitable coupling of the two distributions. Note that we only have to consider the first $\ell + 1$ elements of the two tuples of outputs since for both distributions, the i-th inputs for $i > \ell + 1$ are sampled at random. In other words, $\|\nu_{\ell+1} - \nu_\ell\| = \|\nu'_{\ell+1} - \nu'_\ell\|$, where $\nu'_{\ell+1}$ and ν'_ℓ are the respective distributions of the $\ell + 1$ first outputs of the cipher. To define the coupling of $\nu'_{\ell+1}$ and ν'_ℓ, we consider the iterated Even-Mansour cipher $\text{EM}_{\boldsymbol{P}, k}$, where $\boldsymbol{P}$ satisfies $\boldsymbol{P}(a) = b$, that receives inputs $x' = (x^1, \ldots, x^{\ell+1})$, so that $\text{EM}_{\boldsymbol{P}, k}(x')$ is distributed according to $\nu'_{\ell+1}$. We will construct a second Even-Mansour cipher $\text{EM}_{\boldsymbol{P}', k'}$, with inputs $u = (u^1, \ldots, u^{\ell+1})$, satisfying the following properties:

1) $u^i = x^i$ for $i = 1, \ldots, \ell$, and $u^{\ell+1}$ is uniformly random in $\{0,1\}^n \setminus \{u^1, \ldots, u^\ell\}$;
2) for $i = 1, \ldots, \ell + 1$, if the outputs of the j-th inner permutation in the computations of $\text{EM}_{\boldsymbol{P}, k}(x^i)$ and $\text{EM}_{\boldsymbol{P}', k'}(u^i)$ are equal, then this also holds for any subsequent inner permutation;
3) $\boldsymbol{P}'$ is uniformly random among permutation tuples satisfying $\boldsymbol{P}'(a) = b$ and k' is uniformly random in $(\mathcal{I}_n)^{t+1}$.

Note that properties 1) and 3) will ensure that $\text{EM}_{\boldsymbol{P}', k'}(u)$ is distributed according to ν'_ℓ. We warn that $(\boldsymbol{P}', k')$ will not be *independent* from $(\boldsymbol{P}, k)$, however this is not required for the Coupling Lemma to apply. The only requirement is that both $(\boldsymbol{P}, k)$ and $(\boldsymbol{P}', k')$ have the correct marginal distribution.

We now describe how the second iterated Even-Mansour cipher is constructed. First, it uses exactly the same keys as the original one, namely $k' = (k_0, \ldots, k_t)$. In order to construct permutations $\boldsymbol{P}'$ (on points encountered when computing $\text{EM}_{\boldsymbol{P}', k'}(u)$), we compare the computations of $\text{EM}_{\boldsymbol{P}, k}(x^i)$ and $\text{EM}_{\boldsymbol{P}', k'}(u^i)$ for $i = 1, \ldots, \ell + 1$. For $j = 1, \ldots, t$, we define x_j^i as the output of P_j when computing $\text{EM}_{\boldsymbol{P}, k}(x^i)$, and similarly u_j^i as the output of P_j' when computing $\text{EM}_{\boldsymbol{P}', k'}(u^i)$, i.e.

$$x_j^i = P_j(k_{j-1} \oplus P_{j-1}(\cdots P_1(x^i \oplus k_0) \cdots))$$
$$\text{and } u_j^i = P_j'(k_{j-1} \oplus P_{j-1}'(\cdots P_1'(u^i \oplus k_0) \cdots)) \ .$$

We also let $x_0^i = x^i$ and $u_0^i = u^i$. For $j = 0, \ldots, t-1$ we use the following rules:

i) if $u_j^i \oplus k_j \in a_{j+1}$, then $u_{j+1}^i = P_{j+1}'(u_j^i \oplus k_j)$ is determined by the constraint $\boldsymbol{P}'(a) = b$;

ii) if $u_j^i \oplus k_j \notin a_{j+1}$ and $x_j^i \oplus k_j \in a_{j+1}$, then we choose $u_{j+1}^i = P_{j+1}'(u_j^i \oplus k_j)$ uniformly at random in $\{0,1\}^n \setminus (b_{j+1} \cup \{u_{j+1}^1, \ldots, u_{j+1}^{i-1}\})$;

iii) if $u_j^i \oplus k_j \notin a_{j+1}$ and $x_j^i \oplus k_j \notin a_{j+1}$, then we define $u_{j+1}^i = x_{j+1}^i$, that is $P_{j+1}'(u_j^i \oplus k_j) = P_{j+1}(x_j^i \oplus k_j)$.

Property 2) can easily be seen to follow from these rules and the fact that the keys are the same in both ciphers. Since $\boldsymbol{P}$ is uniformly random among permutation tuples satisfying $\boldsymbol{P}(a) = b$, so is $\boldsymbol{P}'$. This follows from the fact that when using rule iii), $x_j^i \oplus k_j \notin a_{j+1}$ implies that x_{j+1}^i is uniformly random in $\{0,1\}^n \setminus (b_{j+1} \cup \{x_{j+1}^1, \ldots, x_{j+1}^{i-1}\})$, and hence u_{j+1}^i is uniformly random in $\{0,1\}^n \setminus (b_{j+1} \cup \{u_{j+1}^1, \ldots, u_{j+1}^{i-1}\})$ as well. This justifies Property 3). Hence, the joint distribution probability we created for the random variable $(\mathrm{EM}_{\boldsymbol{P},k}(x'), \mathrm{EM}_{\boldsymbol{P}',k'}(u))$ is such that the marginal distributions of $\mathrm{EM}_{\boldsymbol{P},k}(x')$ and $\mathrm{EM}_{\boldsymbol{P}',k'}(u)$ are respectively $\nu_{\ell+1}'$ and ν_ℓ'. We can now apply Lemma 1 to obtain:

$$\|\nu_{\ell+1} - \nu_\ell\| = \|\nu_{\ell+1}' - \nu_\ell'\| \leq \Pr\left[(x_t^1, \ldots, x_t^{\ell+1}) \neq (u_t^1, \ldots, u_t^{\ell+1})\right]$$

where we used $\mathrm{EM}_{\boldsymbol{P},k}(x^i) = x_t^i \oplus k_{t+1}$ and $\mathrm{EM}_{\boldsymbol{P}',k'}(u^i) = u_t^i \oplus k_{t+1}$. Clearly, the rules (combined with the fact that $u^i = x^i$ for $i = 1, \ldots, \ell$) imply that $u_j^i = x_j^i$ for $i = 1, \ldots, \ell$ and $j = 0, \ldots, t$, so that the above expression simplifies to $\|\nu_{\ell+1} - \nu_\ell\| \leq \Pr[x_t^{\ell+1} \neq u_t^{\ell+1}]$. Hence, we are left with the task of upper-bounding the probability not to equate $x_j^{\ell+1}$ and $u_j^{\ell+1}$ in any of the t rounds.

Consider the first round. Unless we have $u_0^{\ell+1} \oplus k_0 \in a_1$ or $x_0^{\ell+1} \oplus k_0 \in a_1$, we will use rule iii) so that we will have $u_1^{\ell+1} = x_1^{\ell+1}$. Since the size of a_1 is q_1, and k_0 is uniformly random, we see that $\Pr[x_1^{\ell+1} \neq u_1^{\ell+1}] \leq 2q_1/N$. Assume now that $x_j^{\ell+1} \neq u_j^{\ell+1}$ for some $j \in [1; t-1]$. As in the preceding case, unless $u_j^{\ell+1} \oplus k_j \in a_{j+1}$ or $x_j^{\ell+1} \oplus k_j \in a_{j+1}$, we will have $u_{j+1}^{\ell+1} = x_{j+1}^{\ell+1}$, so that $\Pr[x_{j+1}^{\ell+1} \neq u_{j+1}^{\ell+1} | x_j^{\ell+1} \neq u_j^{\ell+1}] \leq 2q_{j+1}/N$. Using a chain of conditional probabilities, we get:

$$\|\nu_{\ell+1} - \nu_\ell\| \leq \Pr[x_t^{\ell+1} \neq u_t^{\ell+1}] \leq \frac{2q_1}{N} \cdot \frac{2q_2}{N} \cdots \frac{2q_t}{N} = 2^t \frac{\prod_{i=1}^t q_i}{N^t} .$$

Finally, using Eq. (5), we see that

$$\|\mu_x(\cdot|\boldsymbol{P}(a) = b) - \mu_{q_e}^*\| = \|\nu_{q_e} - \nu_0\| \leq 2^t \frac{q_e \prod_{i=1}^t q_i}{N^t} ,$$

as claimed. $\square$

Remark 2. It can easily be checked that the final key k_t does not play any role in the proof of Lemma 5. Hence it also holds for iterated Even-Mansour cipher without the last key.

Remark 3. The proof of Lemma 5 can be straightforwardly extended to handle distinguishers that are allowed to make both forward *and* backward queries to the outer permutation, in a non-adaptive way (such adversaries could be named NCCA). However, notations become quite cumbersome, so that we omit the details.

Combining Lemmata 4 and 5, we obtain the following theorem.

Theorem 1. *Let $q_1, \ldots, q_t, q_e$ be positive integers. Then:*

$$\mathbf{Adv}^{\mathrm{ncpa}}_{\mathcal{EM}[t]}(q_1, \ldots, q_t, q_e) \leq 2^t \frac{q_e \prod_{i=1}^{t} q_i}{N^t} \ .$$

In particular, for any positive integer q:

$$\mathbf{Adv}^{\mathrm{ncpa}}_{\mathcal{EM}[t]}(q) \leq 2^t \frac{q^{t+1}}{N^t} \ .$$

This remains true for the iterated Even-Mansour cipher where the last key k_t is omitted.

More concretely, the iterated Even-Mansour cipher with t rounds achieves NCPA-security up to $N^{\frac{t}{t+1}}$ queries. This is optimal (neglecting constant factors) considering the attack described in [4].

4   From Non-adaptive to Adaptive Distinguishers

In this section, we turn to the case of CCA-distinguishers. For this, we will need the following refinement to Lemma 4, which relies on a stronger assumption on the distribution of the outputs of the iterated Even-Mansour cipher.

Lemma 6. *Let $q_1, \ldots, q_t, q_e$ be positive integers. Assume that there exists β such that for any tuples $a, b \in (\mathcal{I}_n)^{*q_1, \ldots, q_t}$ and $x, y \in (\mathcal{I}_n)^{*q_e}$, one has*

$$\Pr[\boldsymbol{P}(a) = b \wedge \mathrm{EM}_{\boldsymbol{P},k}(x) = y] \geq \frac{1 - \beta}{(N)_{q_e} \prod_{i=1}^{t}(N)_{q_i}} \ .$$

Then $\mathbf{Adv}^{\mathrm{cca}}_{\mathcal{EM}[t]}(q_1, \ldots, q_t, q_e) \leq \beta$.

Proof. The proof is very similar to the one of Lemma 4. Fix a $(q_1, \ldots, q_t, q_e)$-CCA-distinguisher $\mathcal{D}$. Let τ be the transcript of the interaction of $\mathcal{D}$ with the system of $t + 1$ permutations, *i.e.* the ordered sequence of $q_1 + \ldots + q_t + q_e$ queries with the corresponding answer (i, b, z, z'), where $i \in [1; t+1]$ names which permutation is being queried, b is a bit indicating whether the query is forward or backward, $z \in \{0,1\}^n$ is the actual query and z' the answer. Let also Φ be the function that maps a tuple of permutations $(\boldsymbol{P}, P_{t+1}) \in (\mathcal{P}_n)^{t+1}$ to the transcript of the attack when $\mathcal{D}$ interacts with $(\boldsymbol{P}, P_{t+1})$. We say that a transcript is *consistent* if there exists a tuple of permutations $(\boldsymbol{P}, P_{t+1})$ such that $\Phi(\boldsymbol{P}, P_{t+1}) = \tau$, and we denote Γ the set of consistent transcripts. Finally, from a consistent transcript τ, we build the sequences $a(\tau), b(\tau) \in (\mathcal{I}_n)^{*q_1, \ldots, q_t}$ and $x(\tau), y(\tau) \in (\mathcal{I}_n)^{*q_e}$ as follows. For $i = 1, \ldots, t$, let (i, b, z, z') be the j-th query and corresponding answer to P_i in the transcript. If this is a forward query ($b = 0$), then we define $a_i^j = z$ and $b_i^j = z'$; else, when this is a backward query ($b = 1$), we define $a_i^j = z'$ and $b_i^j = z$. Similarly, let $(t+1, b, z, z')$ be the j-th query and corresponding answer to the outer permutation P_{t+1} in the transcript. If this is a forward query ($b = 0$), then we define $x^j = z$ and $y^j = z'$; else, when this is a backward query ($b = 1$), we define $x^j = z'$ and $y^j = z$. Note that for a consistent transcript τ, $\Phi(\boldsymbol{P}, P_{t+1}) = \tau$ iff $\boldsymbol{P}(a(\tau)) = b(\tau)$ and $P_{t+1}(x(\tau)) = y(\tau)$.

The output of $\mathcal{D}$ is a deterministic function of the transcript. We let Σ denote the set of consistent transcripts τ such that $\mathcal{D}$ outputs 1 when the transcript is τ. Then, by definition we have:

$$
\begin{aligned}
\Pr^*[\mathcal{D}(1^n) = 1] &= \sum_{\tau \in \Sigma} \frac{\#\{(\boldsymbol{P}, Q) \in (\mathcal{P}_n)^{t+1} : \Phi(\boldsymbol{P}, Q) = \tau\}}{|\mathcal{P}_n|^{t+1}} \\
&= \sum_{\tau \in \Sigma} \frac{\#\{(\boldsymbol{P}, Q) \in (\mathcal{P}_n)^{t+1} : \boldsymbol{P}(a(\tau)) = b(\tau) \wedge Q(x(\tau)) = y(\tau)\}}{|\mathcal{P}_n|^{t+1}} \\
&= \sum_{\tau \in \Sigma} \frac{1}{(N)_{q_e} \prod_{i=1}^{t}(N)_{q_i}} .
\end{aligned}
\tag{6}
$$

Also, we have:

$$
\begin{aligned}
\Pr[\mathcal{D}(1^n) = 1] &= \sum_{\tau \in \Sigma} \frac{\#\{(\boldsymbol{P}, k) \in \Omega_t : \Phi(\boldsymbol{P}, \mathrm{EM}_{\boldsymbol{P},k}) = \tau\}}{|\Omega_t|} \\
&= \sum_{\tau \in \Sigma} \Pr\left[\boldsymbol{P}(a(\tau)) = b(\tau) \wedge \mathrm{EM}_{\boldsymbol{P},k}(x(\tau)) = y(\tau)\right] .
\end{aligned}
\tag{7}
$$

Using the assumption and Eq. (6), we see that:

$$
\Pr[\mathcal{D}(1^n) = 1] \geq \sum_{\tau \in \Sigma} \frac{1 - \beta}{(N)_{q_e} \prod_{i=1}^{t}(N)_{q_i}} = (1 - \beta)\Pr^*[\mathcal{D}(1^n) = 1] ,
$$

so that $\Pr^*[\mathcal{D}(1^n) = 1] - \Pr[\mathcal{D}(1^n) = 1] \leq \beta$. Applying the same reasoning to the distinguisher $\mathcal{D}'$ which outputs the negation of $\mathcal{D}$'s output, we obtain

$$
(1 - \Pr^*[\mathcal{D}(1^n) = 1]) - (1 - \Pr[\mathcal{D}(1^n) = 1]) \leq \beta ,
$$

which implies that the advantage of $\mathcal{D}$ is at most β. This concludes the proof. $\quad\square$

We will now derive an appropriate bound β refining Lemma 5 by doubling the number of rounds of the construction and using Lemma 2.

Lemma 7. *Let t be an even integer and $t' = t/2$. Let $q_1, \ldots, q_t, q_e$ be positive integers. We denote:*

$$\alpha_1 = 2^{t'} \frac{q_e \prod_{i=1}^{t'} q_i}{N^{t'}} \quad and \quad \alpha_2 = 2^{t'} \frac{q_e \prod_{i=t'+1}^{t} q_i}{N^{t'}} .$$

Then for any tuples $a, b \in (\mathcal{I}_n)^{*q_1, \ldots, q_t}$ and $x, y \in (\mathcal{I}_n)^{*q_e}$, one has

$$\Pr[\boldsymbol{P}(a) = b \wedge \mathrm{EM}_{\boldsymbol{P}, k}(x) = y] \geq \frac{1 - \beta}{(N)_{q_e} \prod_{i=1}^{t} (N)_{q_i}} ,$$

where $\beta = 2(\sqrt{\alpha_1} + \sqrt{\alpha_2})$.

Proof. First, we slightly modify how the Even-Mansour cipher with $2t'$ rounds is defined in order to write it as the composition of two Even-Mansour ciphers with t' rounds. For this, we write the middle key $k_{t'}$ between permutations $P_{t'}$ and $P_{t'+1}$ as the xor of two independent keys $k_{t'}^1$ and $k_{t'}^2$, and we redefine $\mathrm{EM}_{\boldsymbol{P}, k}$ where $\boldsymbol{P} = (P_1, \ldots, P_{2t'}) \in (\mathcal{P}_n)^{2t'}$ and $k = (k_0, \ldots, k_{t'-1}, k_{t'}^1, k_{t'}^2, k_{t'+1}, \ldots, k_{2t'}) \in (\mathcal{I}_n)^{2t'+2}$, as:

$$\mathrm{EM}_{\boldsymbol{P}, k} = \underbrace{\oplus_{k_{2t'}} \circ P_{2t'} \circ \oplus_{k_{2t'-1}} \circ \cdots \circ P_{t'+1} \circ \oplus_{k_{t'}^2} \circ}_{\mathrm{EM}_{\boldsymbol{P}_2, \tilde{k}_2}}$$
$$\underbrace{\oplus_{k_{t'}^1} \circ P_{t'} \circ \cdots \circ \oplus_{k_1} \circ P_1 \circ \oplus_{k_0}}_{\mathrm{EM}_{\boldsymbol{P}_1, \tilde{k}_1}} .$$

Clearly, this does not change the quantity $\Pr[\boldsymbol{P}(a) = b \wedge \mathrm{EM}_{\boldsymbol{P}, k}(x) = y]$ since $k_{t'}^1 \oplus k_{t'}^2$ is uniformly distributed when $k_{t'}^1$ and $k_{t'}^2$ are. This enables to write $\mathrm{EM}_{\boldsymbol{P}, k} = \mathrm{EM}_{\boldsymbol{P}_2, \tilde{k}_2} \circ \mathrm{EM}_{\boldsymbol{P}_1, \tilde{k}_1}$, where $\boldsymbol{P}_1 = (P_1, \ldots, P_{t'})$, $\boldsymbol{P}_2 = (P_{t'+1}, \ldots, P_{2t'})$, $\tilde{k}_1 = (k_0, \ldots, k_{t'-1}, k_{t'}^1)$, $\tilde{k}_2 = (k_{t'}^2, k_{t'+1}, \ldots, k_{2t'})$. In the following we denote $\widehat{\Omega}_{2t'} = (\mathcal{P}_n)^{2t'} \times (\mathcal{I}_n)^{2t'+2}$. Note that $|\widehat{\Omega}_{2t'}| = |\Omega_{t'}|^2$.

Fix tuples $a, b \in (\mathcal{I}_n)^{*q_1, \ldots, q_t}$ and $x, y \in (\mathcal{I}_n)^{*q_e}$. We denote $\tilde{a}_1 = (a_1, \ldots, a_{t'})$, $\tilde{a}_2 = (a_{t'+1}, \ldots, a_{2t'})$, $\tilde{b}_1 = (b_1, \ldots, b_{t'})$, and $\tilde{b}_2 = (b_{t'+1}, \ldots, b_{2t'})$. We will apply Lemma 2 independently to each half of the cipher $\mathrm{EM}_{\boldsymbol{P}_1, \tilde{k}_1}$ and $\mathrm{EM}_{\boldsymbol{P}_2, \tilde{k}_2}$. Consider the first half $\mathrm{EM}_{\boldsymbol{P}_1, \tilde{k}_1}$. By Lemma 5, we have $\|\mu_x^1(\cdot | \boldsymbol{P}_1(\tilde{a}_1) = \tilde{b}_1) - \mu_x^*\| \leq \alpha_1$, where $\mu_x^1(\cdot | \boldsymbol{P}_1(\tilde{a}_1) = \tilde{b}_1)$ is the distribution of $\mathrm{EM}_{\boldsymbol{P}_1, \tilde{k}_1}(x)$ conditioned on $\boldsymbol{P}_1(\tilde{a}_1) = \tilde{b}_1$. Hence Lemma 2 ensures that there is a subset $S_x \subset (\mathcal{I}_n)^{*q_e}$ of size at least $(1 - \sqrt{\alpha_1})(N)_{q_e}$ such that for all $z \in S_x$:

$$\mu_x^1(z | \boldsymbol{P}_1(\tilde{a}_1) = \tilde{b}_1) = \frac{\#\{(\boldsymbol{P}_1, \tilde{k}_1) \in \Omega_{t'} : \boldsymbol{P}_1(\tilde{a}_1) = \tilde{b}_1 \wedge \mathrm{EM}_{\boldsymbol{P}_1, \tilde{k}_1}(x) = z\}}{|\Omega_{t'}| / \prod_{i=1}^{t'} (N)_{q_i}}$$
$$\geq (1 - \sqrt{\alpha_1}) \frac{1}{(N)_{q_e}} .$$

Applying a similar reasoning to the distribution $\mu_y^2(\cdot | \boldsymbol{P}_2(\tilde{a}_2) = \tilde{b}_2)$ of $\mathrm{EM}^{-1}_{\boldsymbol{P}_2, \tilde{k}_2}(y)$ conditioned on $\boldsymbol{P}_2(\tilde{a}_2) = \tilde{b}_2$, we see that there exits a subset $S_y \subset (\mathcal{I}_n)^{*q_e}$ of size at least $(1 - \sqrt{\alpha_2})(N)_{q_e}$ such that for all $z \in S_y$:

$$\mu_y^2(z | \boldsymbol{P}_2(\tilde{a}_2) = \tilde{b}_2) = \frac{\#\{(\boldsymbol{P}_2, \tilde{k}_2) \in \Omega_{t'} : \boldsymbol{P}_2(\tilde{a}_2) = \tilde{b}_2 \wedge \mathrm{EM}^{-1}_{\boldsymbol{P}_2, \tilde{k}_2}(y) = z\}}{|\Omega_{t'}| / \prod_{i=t'+1}^{t}(N)_{q_i}}$$

$$\geq (1 - \sqrt{\alpha_2}) \frac{1}{(N)_{q_e}} \ .$$

We can now lower-bound the number of $(\boldsymbol{P}, k) \in \tilde{\Omega}_{2t'}$ satisfying $\boldsymbol{P}(a) = b$ and $\mathrm{EM}_{\boldsymbol{P}, k}(x) = y$ by summing, over all intermediate values $z \in S_x \cap S_y$, the product of the number of $(\boldsymbol{P}_1, \tilde{k}_1) \in \Omega_{t'}$ satisfying $\boldsymbol{P}_1(\tilde{a}_1) = \tilde{b}_1$ and $\mathrm{EM}_{\boldsymbol{P}_1, \tilde{k}_1}(x) = z$ times the number of $(\boldsymbol{P}_2, \tilde{k}_2) \in \Omega_{t'}$ satisfying $\boldsymbol{P}_2(\tilde{a}_2) = \tilde{b}_2$ and $\mathrm{EM}_{\boldsymbol{P}_2, \tilde{k}_2}(z) = y$. Combining the two above equations yields:

$$\#\{(\boldsymbol{P}, k) \in \tilde{\Omega}_{2t'} : \boldsymbol{P}(a) = b \wedge \mathrm{EM}_{\boldsymbol{P}, k}(x) = y\} \geq$$
$$\frac{|S_x \cap S_y| (1 - \sqrt{\alpha_1})(1 - \sqrt{\alpha_2}) |\Omega_{t'}|^2}{((N)_{q_e})^2 \prod_{i=1}^{t}(N)_{q_i}} \ .$$

Finally, noting that $|S_x \cap S_y| \geq (1 - \sqrt{\alpha_1} - \sqrt{\alpha_2})(N)_{q_e}$, dividing both terms by $|\Omega_{t'}|^2 = |\tilde{\Omega}_{2t'}|$, and using

$$(1 - \sqrt{\alpha_1} - \sqrt{\alpha_2})(1 - \sqrt{\alpha_1})(1 - \sqrt{\alpha_2}) \geq 1 - 2(\sqrt{\alpha_1} + \sqrt{\alpha_2}) \ ,$$

we obtain:

$$\Pr[\boldsymbol{P}(a) = b \wedge \mathrm{EM}_{\boldsymbol{P}, k}(x) = y] \geq \frac{1 - \beta}{(N)_{q_e} \prod_{i=1}^{t}(N)_{q_i}} \ ,$$

with $\beta = 2(\sqrt{\alpha_1} + \sqrt{\alpha_2})$, which concludes the proof. $\square$

Combining Lemmata 6 and 7, we finally obtain our main theorem.

Theorem 2. *Let t be an even integer and $t' = t/2$. Let $q_1, \ldots, q_t, q_e$ be positive integers. Then:*

$$\mathbf{Adv}^{\mathrm{cca}}_{\mathcal{EM}[t]}(q_1, \ldots, q_t, q_e) \leq \left(\frac{2^{t'+2} q_e \prod_{i=1}^{t'} q_i}{N^{t'}} \right)^{1/2} + \left(\frac{2^{t'+2} q_e \prod_{i=t'+1}^{t} q_i}{N^{t'}} \right)^{1/2} \ .$$

In particular, for any positive integer q:

$$\mathbf{Adv}^{\mathrm{cca}}_{\mathcal{EM}[t]}(q) \leq 2^{t/4+3} \frac{q^{(t+2)/4}}{N^{t/4}} \ .$$

For odd t, we have $\mathbf{Adv}^{\mathrm{cca}}_{\mathcal{EM}[t]} \leq \mathbf{Adv}^{\mathrm{cca}}_{\mathcal{EM}[t-1]}$, so that we can use the above bounds with $t - 1$.

More concretely, the iterated Even-Mansour cipher with t rounds achieves CCA-security up to $N^{\frac{t}{t+2}}$ queries.

References

1. Aldous, D.J.: Random walks on finite groups and rapidly mixing Markov chains. In: Séminaire de Probabilités XVII. Lecture Notes in Mathematics, vol. 986, pp. 243–297. Springer (1983)
2. Bellare, M., Rogaway, P.: Random Oracles are Practical: A Paradigm for Designing Efficient Protocols. In: ACM Conference on Computer and Communications Security, pp. 62–73 (1993)
3. Biryukov, A., Wagner, D.: Advanced Slide Attacks. In: Preneel, B. (ed.) EURO-CRYPT 2000. LNCS, vol. 1807, pp. 589–606. Springer, Heidelberg (2000)
4. Bogdanov, A., Knudsen, L.R., Leander, G., Standaert, F.-X., Steinberger, J., Tischhauser, E.: Key-Alternating Ciphers in a Provable Setting: Encryption Using a Small Number of Public Permutations (Extended Abstract). In: Pointcheval, D., Johansson, T. (eds.) EUROCRYPT 2012. LNCS, vol. 7237, pp. 45–62. Springer, Heidelberg (2012)
5. Canetti, R., Goldreich, O., Halevi, S.: The Random Oracle Methodology, Revisited (Preliminary Version). In: Symposium on Theory of Computing, STOC 1998, pp. 209–218. ACM (1998); Full version available at http://arxiv.org/abs/cs.CR/0010019.
6. Daemen, J.: Limitations of the Even-Mansour Construction. In: Imai, H., Rivest, R.L., Matsumoto, T. (eds.) ASIACRYPT 1991. LNCS, vol. 739, pp. 495–498. Springer, Heidelberg (1993)
7. Daemen, J., Rijmen, V.: Probability Distributions of Correlations and Differentials in Block Ciphers. ePrint Archive Report 2005/212 (2005), http://eprint.iacr.org/2005/212.pdf
8. Dunkelman, O., Keller, N., Shamir, A.: Minimalism in Cryptography: The Even-Mansour Scheme Revisited. In: Pointcheval, D., Johansson, T. (eds.) EURO-CRYPT 2012. LNCS, vol. 7237, pp. 336–354. Springer, Heidelberg (2012)
9. Even, S., Mansour, Y.: A Construction of a Cipher from a Single Pseudorandom Permutation. Journal of Cryptology 10(3), 151–162 (1997)
10. Gentry, C., Ramzan, Z.: Eliminating Random Permutation Oracles in the Even-Mansour Cipher. In: Lee, P.J. (ed.) ASIACRYPT 2004. LNCS, vol. 3329, pp. 32–47. Springer, Heidelberg (2004)
11. Guo, J., Peyrin, T., Poschmann, A., Robshaw, M.: The LED Block Cipher. In: Preneel, B., Takagi, T. (eds.) CHES 2011. LNCS, vol. 6917, pp. 326–341. Springer, Heidelberg (2011)
12. Hoang, V.T., Rogaway, P.: On Generalized Feistel Networks. In: Rabin, T. (ed.) CRYPTO 2010. LNCS, vol. 6223, pp. 613–630. Springer, Heidelberg (2010)
13. Kilian, J., Rogaway, P.: How to Protect DES Against Exhaustive Key Search (an Analysis of DESX). Journal of Cryptology 14(1), 17–35 (2001)
14. Maurer, U.M., Pietrzak, K.: Composition of Random Systems: When Two Weak Make One Strong. In: Naor, M. (ed.) TCC 2004. LNCS, vol. 2951, pp. 410–427. Springer, Heidelberg (2004)
15. Maurer, U.M., Pietrzak, K., Renner, R.: Indistinguishability Amplification. In: Menezes, A. (ed.) CRYPTO 2007. LNCS, vol. 4622, pp. 130–149. Springer, Heidelberg (2007)
16. Mironov, I.: (Not So) Random Shuffles of RC4. In: Yung, M. (ed.) CRYPTO 2002. LNCS, vol. 2442, pp. 304–319. Springer, Heidelberg (2002)

17. Morris, B., Rogaway, P., Stegers, T.: How to Encipher Messages on a Small Domain. In: Halevi, S. (ed.) CRYPTO 2009. LNCS, vol. 5677, pp. 286–302. Springer, Heidelberg (2009)
18. Patarin, J.: New Results on Pseudorandom Permutation Generators Based on the DES Scheme. In: Feigenbaum, J. (ed.) CRYPTO 1991. LNCS, vol. 576, pp. 301–312. Springer, Heidelberg (1992)

A   Proof of the Coupling Lemma

The original statement and proof of the Coupling Lemma is due to Aldous [1]. Here we follow closely a proof by Vigoda.[3]

Let λ be a coupling of μ and ν, and $(X, Y) \sim \lambda$. By definition, we have that for any $z \in \omega$, $\lambda(z, z) \leq \min\{\mu(z), \nu(z)\}$. Moreover, $\Pr[X = Y] = \sum_{z \in \Omega} \lambda(z, z)$. Hence we have:

$$\Pr[X = Y] \leq \sum_{z \in \Omega} \min\{\mu(z), \nu(z)\} \ .$$

Therefore:

$$\Pr[X \neq Y] \geq 1 - \sum_{z \in \Omega} \min\{\mu(z), \nu(z)\}$$

$$= \sum_{z \in \Omega} (\mu(z) - \min\{\mu(z), \nu(z)\})$$

$$= \sum_{\substack{z \in \Omega \\ \mu(z) \geq \nu(z)}} (\mu(z) - \nu(z))$$

$$= \max_{S \subset \Omega} \{\mu(S) - \nu(S)\}$$

$$= \|\mu - \nu\| \ .$$

---

[3] Available from www.cc.gatech.edu/~vigoda/MCMC_Course/MC-basics.pdf

3kf9: Enhancing 3GPP-MAC
beyond the Birthday Bound

Liting Zhang[2], Wenling Wu[1], Han Sui[2], and Peng Wang[2]

[1] Institute of Software, Chinese Academy of Sciences
State Key Laboratory of Information Security
[2] Institute of Information Engineering, Chinese Academy of Sciences
{zhangliting,wwl,suihan}@is.iscas.ac.cn, wp@is.ac.cn

Abstract. Among various cryptographic schemes, CBC-based MACs belong to the few ones most widely used in practice. Such MACs iterate a blockcipher E_K in the so called Cipher-Block-Chaining way, i.e. $C_i = E_K(M_i \oplus C_{i-1})$, offering high efficiency in practical applications. In the paper, we propose a new deterministic variant of CBC-based MACs that is provably secure beyond the birthday bound. The new MAC 3kf9 is obtained by combining $f9$ (3GPP-MAC) and EMAC sharing the same internal structure, and so it is almost as efficient as the original CBC MAC. 3kf9 offers $O(\frac{l^3 q^3}{2^{2n}} + \frac{lq}{2^n})$ PRF-security when its underlying n-bit blockcipher is pseudorandom with three independent keys. This makes it more secure than traditional CBC-based MACs, especially when they are applied with lightweight blockciphers. Therefore, 3kf9 is expected to be a possible candidate MAC in resource-restricted environments.

Keywords: MAC, Birthday Bound, CBC, Mode of Operation.

1 Introduction

1.1 Background

BIRTHDAY BOUND. In cryptography, birthday attack is a generic attack that exploits no specific properties within cryptographic schemes, but just takes the advantage of birthday paradox in probability theory. This paradox says, approximately $2^{n/2}$ independently random n-bit points will collide with a probability close-to-1, where $2^{n/2}$ is called the birthday bound [28,20]. The birthday attack itself is not fatal to the practical security of cryptographic schemes, because people can choose long-enough security parameters to defend, e.g. by restricting the output length of hash functions to be no shorter than 224 bits [3], or by preventing attackers from getting sufficient number of input-output pairs, to make this attack infeasible in recent years.

However, being constrained by some particular software/hardware environments, there still exist many actual applications using short security parameters. For example, the 64-bit blockcipher KASUMI is currently a standard algorithm in mobile communication systems [7]. With the rapid developments of Internet

X. Wang and K. Sako (Eds.): ASIACRYPT 2012, LNCS 7658, pp. 296–312, 2012.

of Things, several lightweight primitives have been proposed in recent years, e.g. PRESENT and PHOTON [11,14]. These algorithms take small-size internal states and output values, usually are much easier to be realized in software and require smaller area in hardware, offering better performance than normal-size ones. Unfortunately, their small sizes imply vulnerability when they are used with traditional modes of operation, most of which are only secure within the birthday bound [19,2]. To ensure practical security in such cases, those modes have to be combined with stateful or random values, or to limit the lengths of their input messages, or to update secret keys frequently, resulting in inconveniences and security risks if misused.

MAC. Message Authentication Code is a widely-used cryptographic scheme for data integrity protection and data origin authentication. Practical applications usually require them to be not only secure (outputting unpredictable tags for new messages) but also efficient. A common way to design a MAC algorithm is to iterate a blockcipher $E : \mathcal{K}_E \times \{0,1\}^n \to \{0,1\}^n$ in the Cipher-Block-Chaining (CBC) manner. That is, in each step, a new chaining value C_i is obtained by encrypting the XOR result of the current message block M_i and the previous chaining value C_{i-1}, i.e. $C_i = E_K(M_i \oplus C_{i-1})$. The CBC structure is so common in the design of many cryptographic schemes that it has been considerably studied for many years [8,27,9,16,24].

Up to now, many excellent CBC-based MACs have been proposed, e.g. EMAC, XCBC, OMAC, CMAC and GCBC [27,9,16,4,24]. Besides, PMAC takes a fully parallelizable construction and can offer extremely high speed in parallel environments [10]. All of the above MAC algorithms are deterministic (needing no stateful or random values), and provably secure when their underlying blockcipher is assumed to be a pseudorandom permutation (PRP). However, their security bounds all fall within the birthday bound, and can not be further improved because there exist birthday attacks on them, i.e. the birthday bound is tight for them [19,2].

There are also a few CBC-based MACs with provable security beyond the birthday bound. For example, RMAC replaces the second key in EMAC by XORing its first key and a random value [18,2], and MAC-R1 and MAC-R2 inject n-bit randomness into the internal states of CBC-based MACs [23]. Obviously, their high security relies on not only the PRP security of blockciphers but also the randomness of the injected values.

In fact, all the deterministic blockcipher-based MACs fall within the birthday bound until Yasuda shows algorithm 6 in the ISO standard is an exception, conditioned on some restrictions on messages [1,30]. In the same paper, Yasuda also introduces SUM-ECBC to reduce the key size in algorithm 6, by XORing the results from two CBC-based MACs, providing half of the efficiency that normal CBC-based MACs offer in serial implementations (rate 2 [1]). On the other hand, Dodis and Steinberger build a MAC from unpredictable blockciphers, with

---

[1] For each message of l blocks long, it has to call the underlying blockcipher roughly $2l$ times.

security beyond the birthday bound, but pay by very high efficiency cost [12]. Very recently, Yasuda proposes PMAC_Plus that improves PMAC beyond the birthday bound [31]. By pre-calculating sufficiently large number (as many as the number of message blocks) of masks, this MAC would provide high efficiency due to the fully parallelizable structure in PMAC and rate-1 design.

3GPP-MAC. To promote the global system for mobile communications, the 3rd Generation Partnership Project (3GPP) proposes f9 as its first MAC algorithm, which is based on blockcipher KASUMI and produces 32-bit tags [6]. $f9$ inherits the structure of original CBC MAC, but in the end encrypts the sum of all chaining values, other than the last chaining value, to obtain the tag. The analysis for $f9$ tends to be tough due to this particular feature [17]. Knudsen and Mitchell are the first to give birthday attacks on $f9$, which need $2^{(n+1)/2}$ known (Message, MAC) pairs and $2^{n/2+1}$ chosen (Message, MAC) pairs to make a forgery against $f9$ without truncations [20]. Then, Iwata and Kohno proved that when KASUMI is secure against a special kind of related-key attacks (RK-PRP), a generalized version of $f9$ (named with $f9'$) is PRF-secure within the birthday bound [15]. This implies the previous birthday attack is the best one without knowledge of internal information.

Despite the fact that the birthday attacks on MACs need on-line invocations, making it much more harder than those on hash functions (needing only off-line computations), people still take several countermeasures for large enough security margin. For example, in the practical applications of $f9$, it has been demanded that each message should be prepended with a fresh value, the length of messages should be no longer than 20000 bits, the secret key should be changed after each invocation, and the outputs should be truncated [5,6].

1.2 Our Work

In this paper, we attempt to design a rate-1 CBC-based MAC with provable security beyond the birthday bound. A direct application of such a scheme is to enforce the security level of current CBC-based MACs, especially in the situations where small-size (lightweight) blockciphers are used, e.g. 3GPP and smart cards. Another application is to make it serve as a highly-secure pseudorandom number generator for various protocols, which therefore would improve the security level of the latter.

To do this, stateful or random values (e.g. counter, fresh) can help, but we would not consider them for practical convenience. Another possible way is to enlarge the size of internal states but still output normal-size tags. As for CBC-based MACs, however, their internal states have the same size as their underlying blockcipher, so one may want to use a large-size blockcipher in CBC-based MACs and truncate their outputs. Unfortunately, the efficiency of such a solution will not be satisfying, because a large-size blockcipher usually runs no faster than a small-size one, not to mention many other costs, e.g. memory and area requirements.

Our starting point is $f9$, in favor of its double-blocksize internal states providing a possible chance to resist the birthday attacks. Inspired by the design of SUM-ECBC and PMAC_Plus, we append one more blockcipher invocation to the end of the $f9$ structure, as illustrated in Fig. 1. The resulting MAC is named with 3kf9, for it enhances $f9$ and needs three independent keys. From another point of view, it is also an extension of EMAC [27], ignoring E_{K_3} and the last XOR operation.

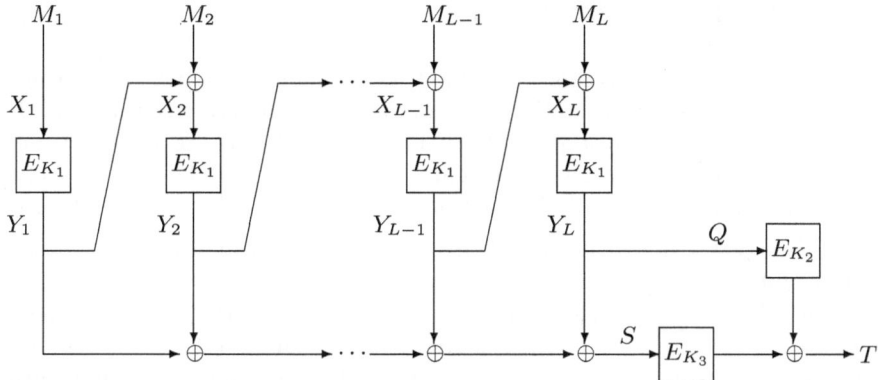

Fig. 1. Illustration of 3kf9

When authenticating messages, 3kf9 can start to work without stateful values or message length information (on-line), requires no pre-computation and only two block-size memory for internal states, besides those for its underlying blockcipher. Specially, it needs no multiplications, comparing with PMAC_Plus. Therefore, 3kf9 will provide high efficiency in serial implementations.

A more detailed comparison with related MACs is given in Table 1.

Table 1. Comparison among 3kf9 and its related deterministic MACs

	key size	rate	structure	multi.	upper bounds	bBB. [a]	Ref.
Alg. 6 in ISO std.[b] SUM-ECBC	$6k$ $4k$	2	CBC	none	$O(\frac{l^4 q^3}{2^{2n}})$ or restricted $O(\frac{l^3 q^3}{2^{2n}})$	conditional	[1] [30]
PMAC_Plus 3kf9	$3k$	1	parallel CBC	$4l-1$ none	$O(\frac{l^3 q^3}{2^{2n}} + \frac{lq}{2^n})$	yes	[31] This Work
$f9$ EMAC	k [c] $2k$	1	CBC	none	$O(\frac{l^2 q^2}{2^n})$	no	[15] [27]

[a] bBB stands for "beyond the Birthday Bound".
[b] It has been removed from the latest version ISO/IEC 9797-1:2011.
[c] Its second key is obtained by $K_2 = K_1 \oplus \text{KM}$, where KM is a non-zero k-bit value.

1.3 Organization

The rest of this paper is organized as follows. Section 2 introduces necessary symbols and 3kf9 specification. Section 3 gives our provable security analysis for 3kf9, including security definitions, the main result and its proof. The proof will be completed in Section 4. In Section 5, we give some suggestions for practical usages of 3kf9. Finally, we conclude this work in Section 6.

2 Symbols and Specification

$\{0,1\}^n$ is the set of all n-bit strings and $\{0,1\}^*$ is the set of all strings. For strings $a, b \in \{0,1\}^*$, $a\|b$ is a concatenation of a and b, and $|a|$ is its length in bits. If a, b have equal lengths then $a \oplus b$ is their bitwise XOR. Denote $\mathrm{Perm}(n)$ and $\mathrm{Rand}(n, n)$ as the sets of all permutations and functions over $\{0,1\}^n$ respectively. $\mathrm{Rand}(*, n)$ stands for the set of all functions whose range belongs to $\{0,1\}^n$. If A is a set, then $\#A$ denotes the size of set A, and $x \xleftarrow{\$} A$ means that x is chosen from set A uniformly at random.

A message M can be alternatively seen as a bit string $M \in \{0,1\}^*$. Then, by $M \leftarrow M\|10^{n-1-|M| \bmod n}$ we mean we append a single bit "1" to the end of M, followed by as many as $n - 1 - |M| \bmod n$ bit "0"s such that the length of the padded string is a multiple of n. For any such string M ($|M| = nL$), $M_1 M_2 \cdots M_L \leftarrow \textbf{Partition}(M)$ means we break M into L successive n-bit blocks such that $M_1\|M_2\|\cdots\|M_L = M$.

MAC Algorithm 3kf9[E]

Input: $K_1, K_2, K_3 \xleftarrow{\$} \mathcal{K}$, $M \in \{0,1\}^*$
Output: $T \in \{0,1\}^n$
01. $M \leftarrow M\|10^{n-1-|M| \bmod n}$
02. $M_1 M_2 \cdots M_L \leftarrow \textbf{Partition}(M)$
03. $S \leftarrow 0^n$
04. $Y_0 \leftarrow 0^n$
05.   **for** $l \leftarrow 1$ **to** L **do**
06. $X_l \leftarrow Y_{l-1} \oplus M_l$
07. $Y_l \leftarrow E_{K_1}(X_l)$
08. $S \leftarrow S \oplus Y_l$
09.   **end for**
10. $T \leftarrow E_{K_2}(Y_L) \oplus E_{K_3}(S)$
11.   **return** T

Fig. 2. Specification of 3kf9

For any message $M \in \{0,1\}^*$, 3kf9 takes a blockcipher $E : \mathcal{K}_E \times \{0,1\}^n \to \{0,1\}^n$ as its underlying primitive, calling it iteratively as specified in Fig. 2 to deal with M, and finally outputs $T \in \{0,1\}^n$ as a tag. If necessary, T can be truncated to be of some particular length less than n.

3kf9 needs three keys K_1, K_2 and K_3, each of which should be independently selected from $\mathcal{K} = \mathcal{K}_E$ uniformly at random. We use $3kf9[E_{K_1}, E_{K_2}, E_{K_3}]$ to stand for this MAC algorithm and we also write it as $3kf9[E]$ for short.

3 Security Proof

3.1 Security Definitions

W need to introduce PRP/PRF definitions here, which are frequently used in the analysis of modes of operation for blockciphers [8,27,9,16,24].

These two definitions focus on the randomness of a keyed function f_K, which is selected from a function family $f : \mathcal{K}_f \times \{0,1\}^* \to \{0,1\}^n$ by selecting a random key K. To measure its randomness, f_K is compared with a random function $R \overset{\$}{\leftarrow} \mathrm{Rand}(*, n)$ (or a random permutation $P \overset{\$}{\leftarrow} \mathrm{Perm}(n)$ if f consists of only permutations).

The comparison is done as, informally, allowing adversaries (without knowing K) to query an oracle, which is either f_K or R with equal probability. The oracle will answer with the corresponding outputs. After some number of queries, the adversaries are asked to tell what the oracle is. The precise definition is given by

$$
\begin{cases}
\mathbf{Adv}_f^{\mathrm{prf}}(\mathcal{A}) \overset{\text{def}}{=} |\Pr[K \overset{\$}{\leftarrow} \mathcal{K}_f : \mathcal{A}^{f_K(\cdot)} = 1] - \Pr[R \overset{\$}{\leftarrow} \mathrm{Rand}(*, n) : \mathcal{A}^{R(\cdot)} = 1]|, \\
\mathbf{Adv}_f^{\mathrm{prf}}(t, q, \mu) \overset{\text{def}}{=} \max_{\mathcal{A}} \{\mathbf{Adv}_f^{\mathrm{prf}}(\mathcal{A})\},
\end{cases}
$$

$$
\begin{cases}
\mathbf{Adv}_f^{\mathrm{prp}}(\mathcal{A}) \overset{\text{def}}{=} |\Pr[K \overset{\$}{\leftarrow} \mathcal{K}_f : \mathcal{A}^{f_K(\cdot)} = 1] - \Pr[P \overset{\$}{\leftarrow} \mathrm{Perm}(n) : \mathcal{A}^{P(\cdot)} = 1]|, \\
\mathbf{Adv}_f^{\mathrm{prp}}(t, q, \mu) \overset{\text{def}}{=} \max_{\mathcal{A}} \{\mathbf{Adv}_f^{\mathrm{prp}}(\mathcal{A})\},
\end{cases}
$$

and the maximum is over all adversaries taking time at most t, making oracle queries at most q, whose total length is at most μ bits. If $\mathbf{Adv}_f^{\mathrm{prf}}(t, q, \mu)$ (or $\mathbf{Adv}_f^{\mathrm{prp}}(t, q, \mu)$) is sufficiently small, we say function family f is a pseudorandom function (PRF) (or a pseudorandom permutation (PRP)).

It has been proved that a PRF is a secure MAC [8].

3.2 Main Results

Let $3kf9[P_1, P_2, P_3]$ stand for $3kf9[E_{K_1}, E_{K_2}, E_{K_3}]$ when blockcipher E with three independent keys are replaced by three independently random permutations P_1, P_2 and P_3, and we further write it as $3kf9[P]$ for simplicity. Then, the following theorem says that $3kf9[P]$ is a PRF with an upper bound beyond the birthday bound.

Theorem 1 (Main Theorem). *For any computationally unbounded adversary* $\mathcal{A}$, *after querying the oracle* q *times, with each query no longer than* $l_{\max}$ *blocks, its advantage to distinguish* $3\text{kf9}[P]$ *from a random function* $R \overset{\$}{\leftarrow} \text{Rand}(*, n)$ *is upper bounded by*

$$|\Pr[\mathcal{A}^{3\text{kf9}[P]} = 1] - \Pr[\mathcal{A}^R = 1]| \leq \frac{ql_{\max}+q}{2^{n-2}} + \frac{2q^3 l_{\max}^3 + q^3 l_{\max}^2 + 2q^3 l_{\max} + 2q^3}{2^{2n-1}}.$$

We conclude this theorem by the "coefficient H technique" initially proposed by Patarin [25,26]. This method is a useful tool for proving pseudorandom properties of blockcipher structures and modes of operation, and it has been frequently used before [25,13,16,24].

To simplify our proof, we also adopt the framework used in the proofs for SUM-ECBC and PMAC_Plus [30,31], which separates the inputs to P_2 and P_3 into four cases. Taking advantage of some known results for CBC structure, $f9$ and sum of PRPs [9,15,22], the first three cases can be easily upper bounded. For the last case, we prove it by Lemma 1 in the next section.

Proof. Since $\mathcal{A}$ is computationally unbounded, w.l.o.g. we assume $\mathcal{A}$ is a deterministic algorithm, otherwise we can maximize $\mathcal{A}$ by running it over all possible cases and choose the most powerful one. Based on this, the i-th query $M^i \notin \{M^1, M^2, \cdots, M^{i-1}\}$ $\mathcal{A}$ would make is fully determined by the previous $i-1$ input-output pairs (M^1, T^1), (M^2, T^2), $\cdots$, (M^{i-1}, T^{i-1}). Then, if we fix a q-tuple $\overrightarrow{T} = (T^1, T^2, \cdots, T^q)$, we know
  - all $\mathcal{A}$'s queries are uniquely determined,
  - the number of queries q is uniquely determined, and
  - the output of $\mathcal{A}$ (0 or 1) is uniquely determined.

Denote $\text{Tset}_1 = \{(T^1, T^2, \cdots, T^q)\}$ is the set that contains all q-tuple $\overrightarrow{T} = (T^1, T^2, \cdots, T^q)$ such that $\mathcal{A}$ outputs 1, and $N = \#\text{Tset}_1$. Then we have

Evaluation for random function R.

$$\Pr[\mathcal{A}^R = 1] = \sum\nolimits_{\overrightarrow{T} \in \text{Tset}_1} \Pr[R(M^i) = T^i, i = 1, 2, \cdots, q] = \frac{N}{2^{qn}}.$$

Evaluation for $3\text{kf9}[P]$.

$$\Pr[\mathcal{A}^{3\text{kf9}[P]} = 1]$$
$$= \sum_{\overrightarrow{T} \in \text{Tset}_1} \Pr[3\text{kf9}[P](M^i) = T^i, i = 1, 2, \cdots, q]$$
$$\geq \sum_{\overrightarrow{T} \in \text{Tset}_1} \left(\Pr[3\text{kf9}[P] \text{ outputs } q \text{ random values}] \times (\frac{1}{2^n})^q\right)$$
$$= \frac{N}{2^{qn}} \times \Pr[3\text{kf9}[P] \text{ outputs } q \text{ random values}]. \tag{1}$$

Denote $\underline{\text{CBC}}[P_1]$ as the internal structure of $3kf9[P]$, i.e. $(Q, S) \leftarrow \underline{\text{CBC}}[P_1](M)$, and $3kf9[P](M) = P_2(Q) \oplus P_3(S) = T$, as in Fig. 1. In the following analysis, we do step by step for each $i = 1, 2, \cdots, q$. Suppose in the previous $i-1$ queries, the $i-1$ outputs $T^1, T^2, \cdots, T^{i-1}$ are independently random values. Let $\text{Domain}[P_2] = \{Q^1, Q^2, \cdots, Q^{i-1}\}$ and $\text{Domain}[P_3] = \{S^1, S^2, \cdots, S^{i-1}\}$. Then, for the i-th query M^i, its corresponding $(Q^i, S^i) \leftarrow \underline{\text{CBC}}[P_1](M^i)$ will definitely fall into one of the following four cases,

Case A: $Q^i \in \text{Domain}[P_2]$ and $S^i \notin \text{Domain}[P_3]$,
Case B: $Q^i \notin \text{Domain}[P_2]$ and $S^i \in \text{Domain}[P_3]$,
Case C: $Q^i \notin \text{Domain}[P_2]$ and $S^i \notin \text{Domain}[P_3]$,
Case D: $Q^i \in \text{Domain}[P_2]$ and $S^i \in \text{Domain}[P_3]$.

For Case A, Black and Rogaway have shown that the probability for any two messages to collide in CBC structure (with an independent ending blockcipher invocation, e.g. EMAC, ECBC) is upper bounded by the birthday bound, i.e. $\Pr[Q^j = Q^i] \leq \frac{4(l_{\max}+1)^2}{2^n}$ (See Lemma 3 in [9]). In such a case, we still have randomness for $T^i = P_2(Q^i) \oplus P_3(S^i)$ because $S^i \notin \text{Domain}[P_3]$ and we can do lazy sampling $P_3(S^i)$. Since at this moment $\#\text{Domain}[P_3] \leq i-1$, the advantage to distinguish $P_3(S^i)$ from a random value $r \xleftarrow{\$} \{0,1\}^n$ is no more than $\frac{i-1}{2^n}$. Then, the advantage to distinguish T^i from r is upper bounded by $\binom{i-1}{1}\frac{4(l_{\max}+1)^2}{2^n} \times \frac{i-1}{2^n}$.

For Case B, Iwata and Kohno have pointed out that the probability for any two messages to collide in $f9$ (with an independent ending block cipher invocation) is also upper bounded by the birthday bound, i.e. $\Pr[S^j = S^i] \leq \frac{(2l_{\max}+2)^2+2^2}{2^{n+1}} = \frac{2l_{\max}^2+4l_{\max}+4}{2^n}$ (See Lemma B.1 in [15], and note that we apply $\sigma \leq 2l_{\max} + 2$ and $q = 2$ here). Then, by lazy sampling for $P_2(Q^i)$, we know the advantage to distinguish T^i from r is upper bounded by $\binom{i-1}{1}\frac{2l_{\max}^2+4l_{\max}+4}{2^n} \times \frac{i-1}{2^n}$.

For Case C, Lucks has proved that the advantage to distinguish $T^i = P_2(Q^i)\oplus P_3(S^i)$ from r is upper bounded by $\frac{(i-1)^2}{(2^n-(i-1))^2} \leq \frac{4(i-1)^2}{2^{2n}}$ (See the proof for Theorem 5 in [22]).

As for Case D, we will show by Lemma 1 in the next section that $\Pr[\exists i \in [1, q] : \text{Case D occurs}] \leq \frac{ql_{\max}+q}{2^{n-2}} + \frac{q^3 l_{\max}^3}{2^{2n-2}}$.

Denote $[T^i \not\sim r]$ as the event that T^i is not an independently random value. Then, based on the none occurrence of Case D, we get

$$\Pr[T^i \not\sim r]$$
$$= \Pr[\text{Case A}]\Pr[T^i \not\sim r | \text{Case A}] + \Pr[\text{Case B}]\Pr[T^i \not\sim r | \text{Case B}] +$$
$$\Pr[\text{Case C}]\Pr[T^i \not\sim r | \text{Case C}]$$
$$\leq \binom{i-1}{1}\frac{4(l_{\max}+1)^2}{2^n} \times \frac{i-1}{2^n} + \binom{i-1}{1}\frac{2l_{\max}^2+4l_{\max}+4}{2^n} \times \frac{i-1}{2^n} + 1 \times \frac{4(i-1)^2}{2^{2n}}$$
$$= \frac{(i-1)^2(3l_{\max}^2+5l_{\max}+6)}{2^{2n-1}}.$$

This allows us to have

$$\Pr[3kf9[P] \text{ doesn't output } q \text{ random values}]$$

$$\leq \Pr[\text{Case D}] + \sum_{i=1}^{q} \Pr[T^i \not\sim r]$$

$$\leq \frac{q l_{\max} + q}{2^{n-2}} + \frac{q^3 l_{\max}^3}{2^{2n-2}} + \sum_{i=1}^{q} \frac{(i-1)^2 (3 l_{\max}^2 + 5 l_{\max} + 6)}{2^{2n-1}}$$

$$\leq \frac{q l_{\max} + q}{2^{n-2}} + \frac{2 q^3 l_{\max}^3 + q^3 l_{\max}^2 + 2 q^3 l_{\max} + 2 q^3}{2^{2n-1}}$$

$$= \epsilon,$$

which implies $\Pr[\mathcal{A}^{3kf9[P]} = 1] \geq \frac{N}{2^{qn}} \times (1 - \epsilon)$ by applying it to inequality (1).

Comparison

By the above analysis, we can get

$$\Pr[\mathcal{A}^R = 1] - \Pr[\mathcal{A}^{3kf9[P]} = 1] \leq \frac{N}{2^{qn}} - \frac{N}{2^{qn}} \times (1 - \epsilon) \leq \frac{N}{2^{qn}} \times \epsilon \leq \epsilon.$$

On the other side, if we define Tset_0 and by similar analysis we can get

$$\Pr[\mathcal{A}^R = 0] - \Pr[\mathcal{A}^{3kf9[P]} = 0] \leq \epsilon,$$

which implies $(1 - \Pr[\mathcal{A}^R = 1]) - (1 - \Pr[\mathcal{A}^{3kf9[P]} = 1]) \leq \epsilon$. Thus we get $\Pr[\mathcal{A}^{3kf9[P]} = 1] - \Pr[\mathcal{A}^R = 1] \leq \epsilon$.

Finally, we conclude

$$|\Pr[\mathcal{A}^{3kf9[P]} = 1] - \Pr[\mathcal{A}^R = 1]| \leq \frac{q l_{\max} + q}{2^{n-2}} + \frac{2 q^3 l_{\max}^3 + q^3 l_{\max}^2 + 2 q^3 l_{\max} + 2 q^3}{2^{2n-1}}.$$

$\square$

Based on the main theorem, we can say that $3kf9[E]$ is a PRF if blockcipher E is secure. More precisely, we have

Theorem 2. *If blockcipher $E : \mathcal{K}_E \times \{0,1\}^n \rightarrow \{0,1\}^n$ is a PRP, then $3kf9[E]$ is a PRF for all adversaries, who make at most q queries, each of which is no longer than $l_{\max}$ blocks. That is,*

$$\text{Adv}^{\text{prf}}_{3kf9[E]}(t, q, \mu) \leq \frac{q l_{\max} + q}{2^{n-2}} + \frac{2 q^3 l_{\max}^3 + q^3 l_{\max}^2 + 2 q^3 l_{\max} + 2 q^3}{2^{2n-1}} + 3 \text{Adv}^{\text{prp}}_E(t', q', \mu'),$$

where $t' = t + O(t)$, $q' \leq q(l_{\max} + 1)$, and $\mu' \leq \mu + qn$.

4   Key Lemma

The none occurrence of Case D implies the q pairs (Q^i, S^i) $(i = 1, 2, \cdots, q)$ are *free*. By "free", we mean for each $i \in [1, q]$, either Q^i is unique in its corresponding sequence $Q^1, Q^2, \cdots, Q^q$ or S^i is unique in its corresponding sequence

$S^1, S^2, \cdots, S^q$. This property is closely related to the newly appeared Cover Free notion [12], which says the q outputs $(N_1^i, N_2^i, \cdots, N_w^i)$ $(1 \leq i \leq q)$ from a cover-free function should satisfy the following property. For each i, there exists at least one $j \in [1, w]$ such that N_j^i is unique in its own subsequence $N_j^1, N_j^2, \cdots, N_j^q$. Unfortunately, the internal structure $\underline{CBC}[P_1]$ can not satisfy the cover free property, when its outputs are made public. However, if adversaries can not get its internal states, $\underline{CBC}[P_1]$ holds a similar property, as the following lemma says.

Lemma 1. *If P_1, P_2 and P_3 are independently random permutations from $\mathrm{Perm}(n)$, then for all computationally unbounded adversaries, who querying $3kf9[P]$ no more than q times, with each query no longer than $l_{\max}$ blocks, the probability for internal states (Q^i, S^i) $(i = 1, 2, \cdots, q)$ to satisfy Case D is upper bounded by*

$$\Pr[\exists i \in [1, q] : \text{ Case D occurs}] \leq \frac{q l_{\max} + q}{2^{n-2}} + \frac{q^3 l_{\max}^3}{2^{2n-2}}.$$

In the following proof, we will prove an even stronger result. That is, all the pairs (Y_l^i, S_l^i) for $l = 1, 2, \cdots, L^i$ and $i = 1, 2, \cdots, q$ are free with this probability, excluding the trivial case that $(Y_l^i, S_l^i) = (Y_l^j, S_l^j)$ with $l \leq d$ for two different messages M^i and M^j, which after being padded are written as $M_1^i||M_2^i||\cdots||M_{L^i}^i$ and $M_1^j||M_2^j||\cdots||M_{L^j}^j$ and having common prefix $M_1^i||M_2^i||\cdots||M_d^i = M_1^j||M_2^j||\cdots||M_d^j$ for some $d \leq \min\{L^i, L^j\}$. To do this, we check the process detail of $\underline{CBC}[P_1]$ in dealing with the querying messages $M^1, M^2, \cdots, M^q$ step by step, and record every Y_l^i and S_l^i for $l = 1, 2, \cdots, L^i$ and $i = 1, 2, \cdots, q$ with two sets YRange and SRange. By lazy sampling for P_1, we upper bound the probability for the events $Y_l^i \in$ YRange and $S_l^i \in$ SRange to occur at the same time, and in the end we sum up all these probabilities to get the final result.

Proof. For any q pairwise distinct queries $M^1, M^2, \cdots, M^q$, we use a program to show the process of $\underline{CBC}[P_1]$ in dealing with them, as in Fig. 3. To better analyze the target probability, we do lazy sampling for P_1. Furthermore, we denote three flags `Zero`, `Cover` and `Bad`. `Zero` is used to identify whether there exists $Y_l^i = 0^n$, which may be easily used to undermine the freeness consistence of (Y_l^i, S_l^i) for $l = 1, 2, \cdots, L^i$ and $i = 1, 2, \cdots, q$. `Cover` is used directly to identify the freeness of (Y_l^i, S_l^i). Either [`Zero` = `True`] or [`Cover` = `True`] implies [`Bad` = `True`], so $\Pr[\exists i \in [1, q] : \text{Case D occurs}] = \Pr[\texttt{Bad} = \texttt{True}] \leq \Pr[\texttt{Zero} = \texttt{True}] + \Pr[\texttt{Cover} = \texttt{True}]$.

Then, it is easy to get that $\Pr[\texttt{Zero} = \texttt{True}] \leq \sum_{j=1}^{q(l_{\max}+1)} \frac{1}{2^n - (j-1)} \leq \frac{q(l_{\max}+1)}{2^{n-1}}$, because for the q messages whose length is no more than $l_{\max}+1$ blocks after being padded, we do no more than $q(l_{\max}+1)$ lazy sampling for P_1, and in the j-th sampling for a new output Y, $\Pr[Y = 0^n] \leq \frac{1}{2^n - (j-1)}$. Here we use $q(l_{\max}+1) < 2^{n-1}$ to get the final bound.

To upper bound $\Pr[\texttt{Cover} = \texttt{True}]$ for all (Y_l^i, S_l^i), we will upper bound the probability for each lazy sampling that may result in the occurrence of $[Y_l^i \in$ YRange $\wedge$ $S_l^i \in$ SRange] with $l = 1, 2, \cdots, L^i$ and $i = 1, 2, \cdots, q$, and then sum up them. For better understanding the following analysis, we work on a simple case first (see Fig. 4 for an illustration), and then generalize it step by step.

```
00.   Domain[P₁], Range[P₁], YRange, SRange ← φ; Zero, Cover, Bad ← False;
for A's i-th query Mⁱ ∈ {0,1}*, do
01.   Mⁱ ← Mⁱ||10ⁿ⁻¹⁻|Mⁱ| mod n; M₁ⁱM₂ⁱ ··· M_{Lⁱ}ⁱ ← Partition(Mⁱ);
02.   S₀ⁱ ← 0ⁿ; Y₀ⁱ ← 0ⁿ;
03.   for l ← 1 to Lⁱ do
04.       Xₗⁱ ← Y_{l-1}ⁱ ⊕ Mₗⁱ;
05.       if Xₗⁱ ∈ Domain[P₁] then Yₗⁱ ← P₁(Xₗⁱ);
06.       else Yₗⁱ ←$ {0,1}ⁿ \ Range[P₁];
07.           if Yₗⁱ = 0ⁿ then Zero ← True; Bad ← True; end if
08.           Range[P₁] ← Range[P₁] ∪ {Yₗⁱ};
09.           Domain[P₁] ← Domain[P₁] ∪ {Xₗⁱ};
10.       end if
11.       Sₗⁱ ← S_{l-1}ⁱ ⊕ Yₗⁱ;
12.       if Yₗⁱ ∈ YRange and Sₗⁱ ∈ SRange and
13.           ∄j < i s.t. M₁ⁱ||M₂ⁱ|| ··· ||Mₗⁱ = M₁ʲ||M₂ʲ|| ··· ||Mₗʲ
14.       then Cover ← True; Bad ← True;
15.       else YRange ← YRange ∪ {Yₗⁱ}; SRange ← SRange ∪ {Sₗⁱ};
16.       end if
17.   end for
```

Fig. 3. A program showing the process of CBC[P₁]

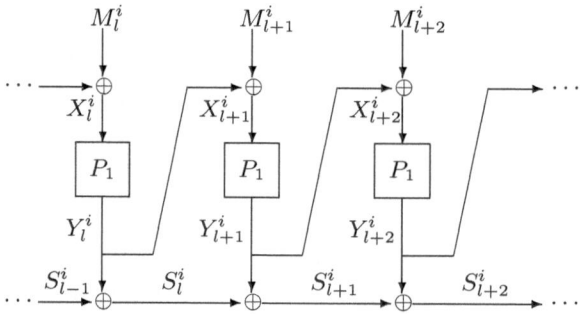

Fig. 4. An insight view on the internal structure of CBC[P₁]

4.1 The Most Common Case

For a new input $X_l^i \notin \mathrm{Domain}[P_1]$, we will choose a value $Y_l^i \xleftarrow{\$} \{0,1\}^n \backslash \mathrm{Range}[P_1]$ by lazy sampling. Since Y_l^i is a new output, it is definite that (Y_l^i, S_l^i) is consistent with the previous pairs for freeness. However, if it happens that $X_{l+1}^i = Y_l^i \oplus M_{l+1}^i \in \mathrm{Domain}[P_1]$, then event $[Y_{l+1}^i \in \mathrm{YRange}]$ would occur, and the freeness consistence of pairs will rely only on the none occurrence of the event $[S_{l+1}^i \in \mathrm{SRange}]$. Consider the following two subcases:

1. $X_{l+1}^i = X_l^i$. This implies $Y_{l+1}^i = Y_l^i$ and $S_{l+1}^i = S_{l-1}^i$, and thus undermining the freeness consistence. The probability for this event to occur is no more

than $\Pr[X_{l+1}^i = X_l^i] = \Pr[Y_l^i = X_l^i \oplus M_{l+1}^i] \le \frac{1}{2^n - \#\text{Range}[P_1]} \le \frac{1}{2^{n-1}}$, where we assume $\#\text{Range}[P_1] < 2^{n-1}$.

2. $X_{l+1}^i \in \text{Domain}[P_1] \setminus \{X_l^i\}$. This implies $Y_l^i \oplus M_{l+1}^i \in \text{Domain}[P_1] \setminus \{X_l^i\}$, and so Y_l^i has no more than $\#\text{Domain}[P_1] \setminus \{X_l^i\}$ choices. Choose any one such choice and fix Y_l^i, then $Y_{l+1}^i = P_1(X_{l+1}^i) = P_1(Y_l^i \oplus M_{l+1}^i)$ would be fixed, so is $S_{l+1}^i = \sum_{c=1}^{l+1} Y_c^i$. On the other hand, the elements in SRange are $\sum_{c=1}^d Y_c^j$ $(1 \le d \le L^j, 1 \le j \le i-1)$ and $\sum_{c=1}^d Y_c^i$ $(1 \le d \le l)$. Then, event $[S_{l+1}^i \in \text{SRange}]$ implies no more than $\#\text{SRange}$ equations, all of which can be written as linear combination of Y_b^a equals to linear combination of message blocks (i.e. $M_{l+1}^i \oplus M_{b+1}^a$ or 0^n) with $0 \le b \le L^a$, $1 \le a \le i-1$ or $0 \le b \le l-1$, $a = i$. Specially, note that Y_l^i is not included here because $X_{l+1}^i \in \text{Domain}[P_1] \setminus \{X_l^i\}$ implies Y_l^i can be written as $M_{l+1}^i \oplus X_{l+1}^i = M_{l+1}^i \oplus \overline{Y} \oplus \overline{M}$, where $\overline{Y}$ and $\overline{M}$ appear in the previous (Y, S) pairs and queries respectively ($\overline{Y}$ may be 0^n if $b = 0$). Furthermore, notice that we have upper bounded $\Pr[Y = 0^n]$ by analyzing $[\text{Zero} = \text{True}]$, so we can assume all Y_b^a $(b \ge 1)$ are non-zero values. Then, excluding the trivial case that two different messages would collide in their common prefix part, the possibility for each of these equations to hold is no more than $1/2^{n-1}$, because all Y_b^a $(b \ge 1)$ are chosen by the previous lazy samplings, from a space with roughly $2^n - \#\text{Domain}[P_1] - \#\text{Range}[P_1] - 1 \le 2^{n-1}$ size. $2^n - \#\text{Range}[P_1]$ is naturally understood, "1" is respect to 0^n, and "$\#\text{Domain}[P_1]$" is respect to the number of bad points that may result in $Y_b^a \oplus M_{b+1}^a \in \text{Domain}[P_1]$. So the linear combinations of Y_b^a has at least 2^{n-1} possible values, and their real values are hidden in the internal structure $\underline{\text{CBC}}[P_1]$, not known by adversaries. So, in this subcase,

$$\Pr[Y_{l+1}^i \in \text{Range}[P_1] \setminus \{Y_l^i\} \wedge S_{l+1}^i \in \text{SRange}]$$
$$= \Pr[X_{l+1}^i \in \text{Domain}[P_1] \setminus \{X_l^i\} \wedge S_{l+1}^i \in \text{SRange}]$$
$$= \Pr[Y_l^i \oplus M_{l+1}^i \in \text{Domain}[P_1] \setminus \{X_l^i\} \wedge S_{l+1}^i \in \text{SRange}] \qquad (2)$$
$$\le \Pr[Y_l^i \oplus M_{l+1}^i \in \text{Domain}[P_1] \setminus \{X_l^i\}] \times \Pr[S_{l+1}^i \in \text{SRange}] \qquad (3)$$
$$\le \frac{\#\text{Domain}[P_1] \setminus \{X_l^i\}}{2^n - \#\text{Range}[P_1]} \times \frac{\#\text{SRange}}{2^{n-1}}$$
$$\le \frac{(\#\text{Domain}[P_1])^2}{2^{2n-2}},$$

Where we apply $\#\text{Range}[P_1] < 2^{n-1}$. Notice that $P_1[X_l^i] = Y_l^i \xleftarrow{\$} \{0,1\}^n \setminus \text{Range}[P_1]$ is a new lazy sampling, and $S_{l+1}^i \in \text{SRange}$ is only related with previous lazy samplings ($X_{l+1}^i \in \text{Domain}[P_1] \setminus \{X_l^i\}$ implying $Y_l^i = M_{l+1}^i \oplus \overline{Y} \oplus \overline{M}$ can be calculated by the previous pairs and queries), so the probability in (2) can be separated, thus we obtain inequality (3).

In this most common case, the probability for lazy sampling $P_1[X_l^i] = Y_l^i \xleftarrow{\$} \{0,1\}^n \setminus \text{Range}[P_1]$ to undermine the freeness consistence is at most $\frac{1}{2^{n-1}} + \frac{(\#\text{Domain}[P_1])^2}{2^{2n-2}}$.

4.2   Generalized Case 1

The above lazy sampling may further induce the occurrence of event $[X_{l+2}^i \in \text{Domain}[P_1]]$, so the previous analysis is not complete, and here we generalize it in this direction.

Suppose $P_1[X_l^i] = Y_l^i \xleftarrow{\$} \{0,1\}^n \setminus \text{Range}[P_1]$ induces series of occurrences, i.e. $[X_{l+1}^i \in \text{Domain}[P_1]]$, $[X_{l+2}^i \in \text{Domain}[P_1]]$, $\cdots$, $[X_{l+u-1}^i \in \text{Domain}[P_1]]$, with $u \leq L^i - l + 1$, let us consider the probability to undermine the freeness consistence. First, we have $\Pr[X_{l+1}^i = X_l^i] \leq \frac{1}{2^n-1}$ as before. Then, conditioned on $X_{l+1}^i \neq X_l^i$, those $u-1$ events imply $Y_l^i \oplus M_{l+1}^i \in \text{Domain}[P_1] \setminus \{X_{l+1}^i\}$ and $Y_{l+a}^i \oplus M_{l+a+1}^i \in \text{Domain}[P_1]$ for $1 \leq a \leq u-2$, and so Y_l^i has at most $\#\text{Domain}[P_1] \setminus \{X_{l+1}^i\}$ choices. Choose any one such choice and fix Y_l^i, then S_{l+a}^i $(0 \leq a \leq u-1)$ are also fixed. To keep freeness consistence, none of the events $[S_{l+1+a} \in \text{SRange} \cup \{S_l^i, S_{l+1}^i, \cdots, S_{l+a}^i\}]$ $(0 \leq a \leq u-2)$ should occur. These events imply no more than $(u-1)\#\text{SRange} + \frac{(u-1)(u-2)}{2}$ equations, and each has a probability of $1/2^{n-1}$ to occur, with similar reasons given in the most common case. So, here the probability for this lazy sampling to keep freeness consistence is upper bounded by $\frac{1}{2^{n-1}} + \frac{\#\text{Domain}[P_1] \setminus \{X_{l+1}^i\}}{2^n - \#\text{Range}[P_1]} \times \frac{(u-1)\#\text{SRange} + \frac{(u-1)(u-2)}{2}}{2^{n-1}} \leq \sum_{a=1}^{u} \left(\frac{1}{2^{n-1}} + \frac{(\#\text{Domain}[P_1]+a-1)^2}{2^{2n-2}}\right)$. Notice that u is the number of invocations to P_1 related to lazy sampling $P_1[X_l^i] = Y_l^i \xleftarrow{\$} \{0,1\}^n \setminus \text{Range}[P_1]$.

4.3   Generalized Case 2

Since we assume adversaries can make any q pairwise distinct queries M^1, M^2, $\cdots$, M^q, it is possible that some queries share a common prefix. Here we generalize the probability for lazy sampling $P_1[X_l^i] = Y_l^i \xleftarrow{\$} \{0,1\}^n \setminus \text{Range}[P_1]$ to undermine the freeness consistence in this direction.

Without loss of generality, we assume $M^i, M^{i+1}, \cdots, M^{i+v-1}$ share a common prefix (This can be reached by sorting the queries), and M_l^i is the last block in their prefix. If $X_{l+1}^{i+b} = Y_l^{i+b} \oplus M_{l+1}^{i+b} \notin \text{Domain}[P_1]$ for all $b \in [0, v-1]$, then Y_{l+1}^{i+b} can keep freeness consistence. However, if $\exists b \in [0, v-1]$ s.t. $X_{l+1}^{i+b} = Y_l^{i+b} \oplus M_{l+1}^{i+b} = X_l^{i+b} = X_l^i$, then the events $[Y_{l+1}^{i+b} = Y_l^{i+b}]$ and $[S_{l+1}^{i+b} = S_{l-1}^{i+b}]$ will occur, and thus undermine the freeness consistence. This probability is no more than $\Pr[\exists b \in [0, v-1], X_{l+1}^{i+b} = X_l^{i+b}] \leq \frac{v}{2^{n-1}}$. Based on its none occurrence, we focus on the probability of $[\exists b \in [0, v-1], X_{l+1}^{i+b} \in \text{Domain}[P_1] \setminus \{X_l^i\}]$. Note that some particular choices of Y_l^i may result in several $[X_{l+1}^{i+b} \in \text{Domain}[P_1] \setminus \{X_l^i\}]$ to occur at the same time, and the number of Y_l^i that induces v' such events is no more than $\#\text{Domain}[P_1]v/v'$. W.l.o.g. we assume $X_{l+1}^i, X_{l+1}^{i+1}, \cdots, X_{l+1}^{i+v'-1} \in \text{Domain}[P_1] \setminus \{X_l^i\}$ for some $v' \in [1, v]$. Choose any one such Y_l^i and fix it, then $Y_{l+1}^i, Y_{l+1}^{i+1}, \cdots, Y_{l+1}^{i+v'-1}$ would be fixed, so are $S_{l+1}^i, S_{l+1}^{i+1}, \cdots, S_{l+1}^{i+v'-1}$. The events $[S_{l+1}^{i+j} \in \text{SRange} \cup \{S_{l+1}^i, S_{l+1}^{i+1}, \cdots, S_{l+1}^{i+j-1}\}]$ $(0 \leq j \leq v'-1)$ imply no more than $v'\#\text{SRange} + \frac{v'(v'-1)}{2}$ equations, with probability $1/2^{n-1}$ to occur each. Then it is not hard to get the probability to keep freeness consistence in this

case is no more than $\frac{v}{2^{n-1}} + \frac{\#\mathrm{Domain}[P_1]v/v'}{2^n-\#\mathrm{Range}[P_1]} \times \frac{v'\#\mathrm{SRange}+\frac{v'(v'-1)}{2}}{2^{n-1}} \leq \sum_{b=1}^{v}(\frac{1}{2^{n-1}} +$
$\frac{(\#\mathrm{Domain}[P_1]+b-1)^2}{2^{2n-2}})$. Notice that v is the number of invocations to P_1 related to
lazy sampling $P_1[X_l^i] = Y_l^i \overset{\$}{\leftarrow} \{0,1\}^n \setminus \mathrm{Range}[P_1]$.

4.4   The Most General Case

Based on the above, we generalize the most common case in two directions, as
in Generalized case 1 and 2.

The analysis here is the same as that in Generalized case 2, until Y_l^i is fixed.
and w.l.o.g. we assume $X_{l+1}^i, X_{l+1}^{i+1}, \cdots, X_{l+1}^{i+v'-1} \in \mathrm{Domain}[P_1] \setminus \{X_l^i\}$ for some
$v' \in [1,v]$ occurs. Then we take Generalized case 1 into account.

Suppose for X_{l+1}^{i+b} ($0 \leq b \leq v'-1$), its following calls to P_1 $X_{l+2}^{i+b}, X_{l+3}^{i+b}, \cdots,$
$X_{l+u[b]-1}^{i+b} \in \mathrm{Domain}[P_1]$, with $u[b] \leq L^{i+b} - l + 1$. Then $S_l^{i+b}, S_{l+1}^{i+b}, \cdots, S_{l+u[b]-1}^{i+b}$
can be fixed by Y_l^i. The events $[S_{l+a+1}^{i+b} \in \mathrm{SRange} \cup \{S_l^{i+b}, S_{l+1}^{i+b}, \cdots, S_{l+a}^{i+b}\}]$ with
$0 \leq a \leq u[b]-2$ and $0 \leq b \leq v'-1$ imply no more than $\sum_{w=1}^{s}(\#\mathrm{SRange}+w-1)$
equations ($s = \sum_{b=0}^{v'-1} u[b]$), with probability $1/2^{n-1}$ to occur each. Then we can get
the probability for lazy sampling $P_1[X_l^i] = Y_l^i \overset{\$}{\leftarrow} \{0,1\}^n \setminus \mathrm{Range}[P_1]$ to undermine
the freeness consistence is at most $\frac{v}{2^{n-1}} + \frac{\#\mathrm{Domain}[P_1]v/v'}{2^n-\#\mathrm{Range}[P_1]} \times \frac{\sum_{w=1}^{s}(\#\mathrm{SRange}+w-1)}{2^{n-1}} \leq$
$\sum_{w=1}^{s}(\frac{1}{2^{n-1}} + \frac{(\#\mathrm{Domain}[P_1]+w-1)^2}{2^{2n-2}})$. Notice that $s = \sum_{b=0}^{v'-1} u[b]$ is the number of
invocations to P_1 related to lazy sampling $P_1[X_l^i] = Y_l^i \overset{\$}{\leftarrow} \{0,1\}^n \setminus \mathrm{Range}[P_1]$.

4.5   Summing Up

From the most common case to the most general case, we have observed that for
every lazy sampling $P_1[X_l^i] = Y_l^i \overset{\$}{\leftarrow} \{0,1\}^n \setminus \mathrm{Range}[P_1]$, its probability to under-
mine the freeness consistence is no more than $\sum_{w=1}^{s}(\frac{1}{2^{n-1}} + \frac{(\#\mathrm{Domain}[P_1]+w-1)^2}{2^{2n-2}})$,
where s is the number of invocations to P_1 related to this lazy sampling. Suppose
in dealing with $M^1, M^2, \cdots, M^q$, we do z times lazy sampling in total, and the
invocations to P_1 related to them are $s_1, s_2, \cdots, s_z$ respectively. Thus,

$$\Pr[\mathtt{Cover} = \mathtt{True}] \leq \sum_{j=1}^{z} \Pr[\mathtt{Cover} = \mathtt{True} \text{ in lazy sampling } j]$$

$$\leq \sum_{j=1}^{z} \sum_{w=1}^{s_j} (\frac{1}{2^{n-1}} + \frac{(\#\mathrm{Domain}[P_1]+w-1)^2}{2^{2n-2}})$$

$$\leq \frac{q(l_{\max}+1)}{2^{n-1}} + \sum_{w=1}^{q(l_{\max}+1)} \frac{(w-1)^2}{2^{2n-2}}$$

$$\leq \frac{ql_{\max}+q}{2^{n-1}} + \frac{q^3 l_{\max}^3}{2^{2n-2}},$$

where we apply $\sum_{j=1}^{z} s_j \le q(l_{\max} + 1)$ and note that $\#\text{Domain}[P_1]$ is a variable growing from 0 to some value no larger than $q(l_{\max} + 1)$, with lazy samplings.

At last, we get $\Pr[\exists i \in [1, q] : \text{Case D occurs}] = \Pr[\text{Bad} = \text{True}] \le \Pr[\text{Zero} = \text{True}] + \Pr[\text{Cover} = \text{True}] \le \frac{q(l_{\max}+1)}{2^{n-1}} + \frac{ql_{\max}+q}{2^{n-1}} + \frac{q^3 l_{\max}^3}{2^{2n-2}} = \frac{ql_{\max}+q}{2^{n-2}} + \frac{q^3 l_{\max}^3}{2^{2n-2}}$.   □

5   Some Suggestions

The key size in 3kf9 is three times of that for its underlying blockcipher, and this may be too large to be stored securely in some resource-restricted environments. For such cases, we give the following solutions:

1. Derive a master key $K \xleftarrow{\$} \{0,1\}^k$, and generate $K_i = E_K(\text{Cst}_i)$ $(i = 1, 2, 3)$ with three different constants Cst_i. Then we need only to store the master key K securely. The security of the resulting scheme is still guaranteed by the PRP assumption on blockcipher E.

2. Derive $K_1 \xleftarrow{\$} \{0,1\}^k$, and generate $K_i = K_1 \oplus \text{Cst}_i$ for $i = 2, 3$, with two non-zero constants $\text{Cst}_2, \text{Cst}_3$. Then we need only to store K_1 securely. However, this solution requires blockcipher E should be a RK-PRP (pseudorandom against a kind of related-key attacks) [15].
   We warn that generating $K_2 = E_{K_1}(\text{Cst}_2)$ and $K_3 = E_{K_1}(\text{Cst}_3)$ may result in security flaws in 3kf9, because $E_K(K \oplus \cdot)$ may not reach pseudorandomness given E is a PRP [29].

3. Adopt a beyond-birthday-bound tweakable blockcipher TBC as the underlying primitive in 3kf9. Then, we can replace E_{K_1}, E_{K_2} and E_{K_3} by $\text{TBC}_K^{T_1}$, $\text{TBC}_K^{T_2}$ and $\text{TBC}_K^{T_3}$, where T_1, T_2, T_3 are three public tweaks. Such a TBC has recently been introduced by Landecker, Shrimpton and Terashima [21], but the current TBC scheme still needs key size reducing.

Since CMAC has been widely used in practical applications [4], someone may want to use $\text{CMAC}_{K_1}(\cdot) \oplus \text{CMAC}_{K_2}(\cdot)$ to get a highly secure MAC. We note that the precise security of this proposal is still unclear [30], and it is rate-2, implying more power consumption and lower efficiency in serial implementations.

6   Conclusion

We propose a rate-1 CBC-based MAC 3kf9 with provable security beyond the birthday bound in this paper. 3kf9 is efficient for its rate-1 design, and highly-secure for its $O(\frac{l^3 q^3}{2^{2n}} + \frac{lq}{2^n})$ PRF bound. Moreover, 3kf9 is light in the sense that it needs only XOR operations besides blockcipher invocations, and thus it immediately turns into a lightweight MAC when equipped with a lightwight blockcipher. However, its key size seems to be too large in some particular environments, requiring further improvements therefore.

Acknowledgments. The authors would like to thank the anonymous referees at both FSE 2012 and Asiacrypt 2012 and the attendees at ASK 2011 for their valuable comments. Special thanks to Lei Wang for pointing out a flaw in an earlier proof, to Tetsu Iwata for some technical comments, and to Yuefei Sui for some editorial comments. Furthermore, this work is supported by the National Natural Science Foundation of China (No. 61272476, 91118006, 60903219 and 61202422), and the National Grand Fundamental Research 973 Program of China .

References

1. ISO/IEC 9797-1:1999. Information technology – Security Techniques – Message Authentication Codes (MACs) – Part 1: Mechanisms Using a Block Cipher. Revised by ISO/IEC 9797-1:2011
2. Public Commnets,
   http://csrc.nist.gov/groups/ST/toolkit/BCM/comments.html
3. Requirements for SHA-3 by NIST, Federal Register vol. 72(212),
   http://csrc.nist.gov/groups/ST/hash/sha-3/index.html
4. Special Publication 800-38B. Recommendation for Block Cipher Modes of Operation: The CMAC Mode for Authentication. National Institute of Standards and Technology,
   http://csrc.nist.gov/groups/ST/toolkit/BCM/current_modes.html
5. TS 33.105. 3G Security: Cryptographic Algorithm Requirements,
   http://www.3gpp.org/ftp/Specs/html-info/33-series.htm
6. TS 35.201. 3G Security: Specification of the 3GPP Confidentiality and Integrity Algorithms; Document 1: f8 and f9 Specifications,
   http://www.3gpp.org/ftp/Specs/html-info/35-series.htm
7. TS 35.202. 3G Security: Specification of the 3GPP Confidentiality and Integrity Algorithms; Document 2: Kasumi Specification,
   http://www.3gpp.org/ftp/Specs/html-info/35-series.htm
8. Bellare, M., Kilian, J., Rogaway, P.: The Security of Cipher Block Chaining. In: Desmedt, Y.G. (ed.) CRYPTO 1994. LNCS, vol. 839, pp. 341–358. Springer, Heidelberg (1994)
9. Black, J., Rogaway, P.: CBC MACs for Arbitrary-Length Messages:The Three-Key Constructions. In: Bellare, M. (ed.) CRYPTO 2000. LNCS, vol. 1880, pp. 197–215. Springer, Heidelberg (2000)
10. Black, J., Rogaway, P.: A Block-Cipher Mode of Operation for Parallelizable Message Authentication. In: Knudsen, L.R. (ed.) EUROCRYPT 2002. LNCS, vol. 2332, pp. 384–397. Springer, Heidelberg (2002)
11. Bogdanov, A., Knudsen, L.R., Leander, G., Paar, C., Poschmann, A., Robshaw, M., Seurin, Y., Vikkelsoe, C.: PRESENT: An Ultra-Lightweight Block Cipher. In: Paillier, P., Verbauwhede, I. (eds.) CHES 2007. LNCS, vol. 4727, pp. 450–466. Springer, Heidelberg (2007)
12. Dodis, Y., Steinberger, J.: Domain Extension for MACs Beyond the Birthday Barrier. In: Paterson, K.G. (ed.) EUROCRYPT 2011. LNCS, vol. 6632, pp. 323–342. Springer, Heidelberg (2011)
13. Gilbert, H., Minier, M.: New Results on the Pseudorandomness of Some Blockcipher Constructions. In: Matsui, M. (ed.) FSE 2001. LNCS, vol. 2355, pp. 248–266. Springer, Heidelberg (2002)

14. Guo, J., Peyrin, T., Poschmann, A.: The PHOTON Family of Lightweight Hash Functions. In: Rogaway, P. (ed.) CRYPTO 2011. LNCS, vol. 6841, pp. 222–239. Springer, Heidelberg (2011)

15. Iwata, T., Kohno, T.: New Security Proofs for the 3GPP Confidentiality and Integrity Algorithms. In: Roy, B., Meier, W. (eds.) FSE 2004. LNCS, vol. 3017, pp. 427–445. Springer, Heidelberg (2004)

16. Iwata, T., Kurosawa, K.: OMAC: One-Key CBC MAC. In: Johansson, T. (ed.) FSE 2003. LNCS, vol. 2887, pp. 129–153. Springer, Heidelberg (2003)

17. Iwata, T., Kurosawa, K.: On the Correctness of Security Proofs for the 3GPP Confidentiality and Integrity Algorithms. In: Paterson, K.G. (ed.) Cryptography and Coding 2003. LNCS, vol. 2898, pp. 306–318. Springer, Heidelberg (2003)

18. Jaulmes, É., Joux, A., Valette, F.: On the Security of Randomized CBC-MAC Beyond the Birthday Paradox Limit: A New Construction. In: Daemen, J., Rijmen, V. (eds.) FSE 2002. LNCS, vol. 2365, pp. 237–251. Springer, Heidelberg (2002)

19. Joux, A., Poupard, G., Stern, J.: New Attacks against Standardized MACs. In: Johansson, T. (ed.) FSE 2003. LNCS, vol. 2887, pp. 170–181. Springer, Heidelberg (2003)

20. Knudsen, L.R., Mitchell, C.J.: Analysis of 3gpp-MAC and Two-key 3gpp-MAC. Discrete Applied Mathematics 128(1), 181–191 (2003)

21. Landecker, W., Shrimpton, T., Terashima, R.S.: Tweakable Blockciphers with Beyond Birthday-Bound Security. In: Safavi-Naini, R., Canetti, R. (eds.) CRYPTO 2012. LNCS, vol. 7417, pp. 14–30. Springer, Heidelberg (2012)

22. Lucks, S.: The Sum of PRPs Is a Secure PRF. In: Preneel, B. (ed.) EUROCRYPT 2000. LNCS, vol. 1807, pp. 470–484. Springer, Heidelberg (2000)

23. Minematsu, K.: How to Thwart Birthday Attacks against MACs via Small Randomness. In: Hong, S., Iwata, T. (eds.) FSE 2010. LNCS, vol. 6147, pp. 230–249. Springer, Heidelberg (2010)

24. Nandi, M.: Fast and Secure CBC-Type MAC Algorithms. In: Dunkelman, O. (ed.) FSE 2009. LNCS, vol. 5665, pp. 375–393. Springer, Heidelberg (2009)

25. Patarin, J.: Pseudorandom Permutations Based on the DES Scheme. In: Cohen, G.D., Charpin, P. (eds.) EUROCODE 1990. LNCS, vol. 514, pp. 193–204. Springer, Heidelberg (1991)

26. Patarin, J.: The "Coefficients H" Technique. In: Avanzi, R.M., Keliher, L., Sica, F. (eds.) SAC 2008. LNCS, vol. 5381, pp. 328–345. Springer, Heidelberg (2009)

27. Petrank, E., Rackoff, C.: CBC MAC for Real-Time Data Sources. J. Cryptology 13(3), 315–338 (2000)

28. Preneel, B., van Oorschot, P.C.: MDx-MAC and Building Fast MACs from Hash Functions. In: Coppersmith, D. (ed.) CRYPTO 1995. LNCS, vol. 963, pp. 1–14. Springer, Heidelberg (1995)

29. Wang, P., Feng, D., Wu, W., Zhang, L.: On the Unprovable Security of 2-Key XCBC. In: Mu, Y., Susilo, W., Seberry, J. (eds.) ACISP 2008. LNCS, vol. 5107, pp. 230–238. Springer, Heidelberg (2008)

30. Yasuda, K.: The Sum of CBC MACs Is a Secure PRF. In: Pieprzyk, J. (ed.) CT-RSA 2010. LNCS, vol. 5985, pp. 366–381. Springer, Heidelberg (2010)

31. Yasuda, K.: A New Variant of PMAC: Beyond the Birthday Bound. In: Rogaway, P. (ed.) CRYPTO 2011. LNCS, vol. 6841, pp. 596–609. Springer, Heidelberg (2011)

Understanding Adaptivity: Random Systems Revisited

Dimitar Jetchev[1], Onur Özen[1], and Martijn Stam[2]

[1] EPFL IC IIF LACAL, Station 14, CH-1015 Lausanne, Switzerland
dimitar.jetchev@epfl.ch, oezen.onur@gmail.com
[2] Department of Computer Science, University of Bristol, Merchant Venturers Building,
Woodland Road, Bristol BS8 1UB, UK
stam@compsci.bristol.ac.uk

Abstract. We develop a conceptual approach for probabilistic analysis of adaptive adversaries via Maurer's methodology of random systems (Eurocrypt'02). We first consider a well-known comparison theorem of Maurer according to which, under certain hypotheses, adaptivity does not help for achieving a certain event. This theorem has subsequently been misinterpreted, leading to a misrepresentation with one of Maurer's hypotheses being omitted in various applications. In particular, the only proof of (a misrepresentation of) the theorem available in the literature contained a flaw. We clarify the theorem by pointing out a simple example illustrating why the hypothesis of Maurer is necessary for the comparison statement to hold and provide a correct proof. Furthermore, we prove several technical statements applicable in more general settings where adaptivity might be helpful, which can be seen as the random system analogue of the game-playing arguments recently proved by Jetchev, Özen and Stam (TCC'12).

1 Introduction

One of the key concepts in cryptographic security definitions and proofs is the notion of indistinguishability [3]. In the information-theoretic setting, the simplest example is how easy it is for a computationally unbounded adversary to distinguish two random variables X and Y based on a single sample from either of the two variables. It is not hard to see that the success probability of the optimal distinguishing algorithm (the distinguisher's advantage) is simply the statistical distance of the two probability distributions for X and Y. Yet, the analysis of current cryptographic systems typically requires much more than distinguishing two random variables. For instance, the related cryptographic primitive of a pseudo-random function allows an adversary to make multiple queries and hence, obtain multiple related samples in order to distinguish between either a truly random function or a pseudo-random one. Moreover, the distinguisher can interact with the system by choosing the queries *adaptively*, i.e., based on the previous queries and corresponding responses. Adversarial adaptivity is notoriously difficult to deal with, not only in the context of pseudorandomness, but across the cryptologic landscape.

With the increasing number of sophisticated cryptographic schemes appearing in the literature (e.g., authenticated encryption, compression functions, message authentication codes), the level of complexity of proving even relatively straightforward security notions such as pseudorandomness or collision resistance becomes ever more

X. Wang and K. Sako (Eds.): ASIACRYPT 2012, LNCS 7658, pp. 313–330, 2012.

involved and complicated. Even though the building blocks of the proofs rarely extend beyond basic notions such as conditional probabilities, Bayes' rule or basic concepts from stochastic processes, combining these building blocks into a rigorous proof poses a challenge in many cases. Consequently, developing a more conceptual approach towards rigorous security analyses of adaptive adversaries is an important challenge in theoretical cryptology.

Games and Random Systems. One of the general methods for security proofs is based on "game-playing" [2,8,16]. A common technique involves the introduction to the game of a flag bad (initially set to `false`). The *fundamental lemma of game playing* [2, §3.4] states that for games that are identical until bad, distinguishing between these games is at most as hard as setting bad to `true`. Several common and a few new techniques employed to prove preimage and collision security of compression functions based on ideal primitives were recently abstracted using game playing by Jetchev, Özen and Stam [7].

A different approach to indistinguishability and probabilistic analysis of adaptive adversaries is through the concept of *random systems*, as introduced by Maurer [11]. This abstraction unifies many existing security proofs and it allows for proving new indistinguishability results. Intuitively, a random system takes a generally unbounded sequence of inputs (queries) and produces an output (response) for each input using a specific source of randomness. Random systems are rigorously modeled in such a way that they exploit the input-output behavior via specifying (abstractly) a set of conditional probability distributions (see Definition 1 for more details).

A *distinguisher* (see Definition 4) can be thought of as another random system that is allowed to query either one of the two random systems and that outputs a binary decision bit at the end. Estimating the advantage in the case of non-adaptive adversaries is often much simpler than estimating the advantage for adaptive ones. Maurer gave a two step approach to deal with adaptive distinguishers effectively.

First, in analogy with the fundamental lemma of game playing, it is always possible to rephrase the problem of upper bounding the advantage of any adversary in distinguishing two arbitrary random systems into one where an adversary has to provoke an event instead [11, 14, Thm.1]. Most of the indistinguishability proofs indeed follow along these lines.

Next, Maurer [11, Thm.2] presented a result stating that, under certain hypotheses, adaptivity does not help to cause an event. Throughout the paper, we often refer to this statement as the adaptive–non-adaptive (ANA) switching lemma (see Section 4.1). It can also be used in the context of events that are meaningful in their own right, such as finding collisions for a hash function.

Our Contribution. In this paper, we revisit and refine the currently existing techniques based on random systems for bounding the advantage of an adaptive adversary for provoking a certain event. Our contribution is twofold. On the one hand, we show that Maurer's phrasing of the ANA switching lemma has been been misinterpreted, in the sense that an essential hypothesis has been omitted in subsequent applications.

This applies to the only proof given in the literature (by Pietrzak [15, §3.2]) which consequently contains an incorrect step. We restate and prove a corrected version that luckily works for most uses of the lemma in the literature. We explain why the original hypothesis is indeed necessary by providing a simple example where adaptivity does help, yet, where the remaining hypotheses have been satisfied. On the other hand, we examine existing techniques to bound the advantage of adaptive adversaries directly in the context of random systems. This can be seen as a generalization of the earlier work by Jetchev, Özen and Stam [7].

The example is rather simple and intuitive: finding a fixed point in a uniformly random permutation. Here, one can easily see that adaptivity is helpful after the first query/response pair is obtained since (assuming that the first query has not produced a fixed point) an adaptive adversary can choose its second query based on the response to the first query and the condition that there is no fixed point yet (see Section 4.1). Indeed, an adaptive adversary can already eliminate one choice for the second query (two for the third and so on), as opposed to a non-adaptive adversary who commits all of its queries in advance. Thus the best adaptive adversary will have a significantly better advantage than any non-adaptive one. Nevertheless, as we demonstrate, the hypotheses of Pietrzak's (mis)interpretation of the ANA switching lemma *are* satisfied, thus completing our counterexample.

We proceed to examine Pietrzak's proof of the lemma to determine what underlies the mistake and whether the proof can be fixed. To some extent, the problem originates from the elliptical notation that the theory of random systems occasionally suffers from. We propose a restatement of the lemma (Theorem 12) together with a correct proof. We then perform the important (if somewhat tedious) task of investigating known examples in the literature where an incorrect version of the ANA switching lemma has been exploited (see the full version). Fortunately, to the best of our knowledge, the flaw uncovered by us does not lead to a violation of any security claim based on the incorrect ANA switching lemma (as the modified hypotheses are still satisfied).

Our second contribution is a string of technical statements, all phrased in the language of random systems, that are applicable in the more general setting where adaptivity might be helpful in triggering an event. The first result (Proposition 9) is the random system interpretation of a well-known technique, where a union bound is computed over the subevent that an adversary provokes the event at the jth step, where the required "stepwise" probabilities (for the subevents) are maximized in a greedy-type manner. This is a standard and often-used argument from security proofs that has not been previously linked to random systems. It makes derivation of the overall bound relatively easy. Yet, in many cases the overall upper bound is not tight enough due to the maximal probabilities occuring for rather unlikely query/response histories or due to overcounting.

Several proofs in the literature tackle the problem of "bad" query/response history by the introduction of an auxiliary event explicitly bounding such a bad history occuring (e.g., [10,17]). Subsequent bounding of the probabilities of on the one hand the auxiliary event and on the other of the actual event conditioned on the auxiliary bad event not occurring, leads to a tighter bound. Proposition 13 generalizes this method in the context of random systems.

Lee et al. [9] recently introduced "wish lists" to the analysis of adaptive adversaries to limit the effect of overcounting. The idea is to cut up the analysis in two parts. First, one upper bounds the maximum size W of the wish list, i.e., the total number of query-response pairs that could ever lead to an adversarial win (to get useful bounds, one typically needs to introduce an auxiliary flag as in the discussion above). Next, one upper bounds the probability p of any particular wish to be granted, i.e., the probability that a query on the wish list gets to the wished for response when actually being asked by the adversary. Finally, one observes that in order to win, at some point an adversary needs to have some wish granted. Intuitively, a union bound over all wishes in the list means the advantage of an adaptive adversary is then at most pW. We formalize this approach in Proposition 14, which assumes as a hypothesis an upper bound on the sum of the stepwise probabilities of success for each query/response history and thus avoids the greedy-type argument. We refine this in Proposition 15 by adding an auxiliary flag event.

Yet, the most subtle and useful (in terms of applications) bounds are provided in Proposition 16. Here, an adaptive adversary is trying to achieve a certain event more than once. A simple example is an adversary trying to obtain more than κ fixed points in a random permutation, but it could also relate to a scenario where an adversary needs to see multiple wishes being granted. The techniques we develop here are very similar to those used for the analysis of a recent incidence-based compression function construction [7]. We illustrate the usefulness of our result by revisiting the analysis of an auxiliary collinearity event needed for the security proof of that construction (see the full version). The strong emphasis on conditional probabilities in the random systems methodology makes it very natural to express the various bounds on an adaptive adversary's advantage, providing a different and arguably clearer perspective on the original proof.

Related Work. Modification of the adversary is an important technique, orthogonal to our work, that is often used to bound the advantage of an adaptive adversary. In particular free queries have been used to great effect in the analysis of double length hash functions [1,6,9]. A typical proof will first modify the adversary—adding the free queries with the somewhat paradoxic effect of taking away some of the adaptivity of the adversary by making it more it more powerful—followed by an analysis of the advantage of this modified adversary. For bounding the advantage of the modified adversary our work comes into play.

Very recently, during their analysis of key-alternating ciphers, Bogdanov et al. [4] uncovered an interesting scenario where a distinguisher surprisingly benefits from adaptivity. While it would be straightforward to describe their problem (and the supporting counterexample) in the random systems framework and subsequently applying the first step of Maurer's two step approach to move it from distinguishing to causing an event, the resulting event cannot be expressed as a predicate, ruling out direct application of many of our theorems. It is an interesting open problem to see if our approach can be extended to improve upon the bounds already obtained by Steinberger [18] and Bogdanov et al.

2   Preliminaries

Notation. Following the terminology and notation of [11, 15], we denote random variables by capital letters (e.g., X), their values by lower-case letters (e.g., x) and their finite[1] sampling spaces by calligraphic letters (e.g., $\mathcal{X}$). For a fixed sample space $\mathcal{X}$, let $\mathcal{X}^k$ be k-fold Cartesian product of $\mathcal{X}$. The corresponding random variables and their values are denoted analogously (i.e., X^k and x^k, respectively). For brevity, we use $\mathsf{P}_A[a]$ to denote the probability $\Pr[A = a]$ and similarly, $\mathsf{P}_{A|BC}[a; b, c]$ for $\Pr[A = a | B = b \wedge C = c]$. If it is clear from the context, we sometimes omit the specific values and simply use, e.g., $\mathsf{P}_{A|BC}$ to denote $\Pr[A = a | B = b \wedge C = c]$.

Random Systems. Various cryptographic systems can be seen as random systems [11] that are modeled as the mathematical abstraction of interactive systems: an $(\mathcal{X}, \mathcal{Y})$-random system takes the inputs $X_1, X_2, \ldots \in \mathcal{X}$ and for each input X_i it generates an output $Y_i \in \mathcal{Y}$ depending probabilistically on $X^i = (X_1, \ldots, X_i)$ and $Y^{i-1} = (Y_1, \ldots, Y_{i-1})$. Random systems have been used in the literature (see e.g., [11–14]) to unify, simplify, generalize, and in some cases strengthen security proofs.

Definition 1 (Random System). *An $(\mathcal{X}, \mathcal{Y})$-random system $\mathbf{F}$ is a (possibly infinite) sequence of conditional probability distributions $\mathsf{P}^{\mathbf{F}}_{Y_i | X^i Y^{i-1}}$ for $i \geq 1$; specifically, the distribution of the outputs Y_i conditioned on $X^i = x^i$ (i.e., the ith query x_i and all previous queries $x^{i-1} = (x_1, \ldots, x_{i-1})$) and $Y^{i-1} = y^{i-1}$ (i.e., all previous outputs $y^{i-1} = (y_1, \ldots, y_{i-1})$). Define*

$$\mathsf{P}^{\mathbf{F}}_{Y^i | X^i} := \prod_{j=1}^{i} \mathsf{P}^{\mathbf{F}}_{Y_j | X^j Y^{j-1}},$$

where, for completeness, $\mathsf{P}^{\mathbf{F}}_{Y_1 | X^1 Y^0} := \mathsf{P}^{\mathbf{F}}_{Y_1 | X^1} = \mathsf{P}^{\mathbf{F}}_{Y_1 | X_1}$. Two $(\mathcal{X}, \mathcal{Y})$-random systems $\mathbf{F}$ and $\mathbf{G}$ are said to be equivalent (denoted by $\mathbf{F} \equiv \mathbf{G}$) if $\mathsf{P}^{\mathbf{F}}_{Y_i | X^i Y^{i-1}} = \mathsf{P}^{\mathbf{G}}_{Y_i | X^i Y^{i-1}}$ for all $i \geq 1$ and all arguments $(x^i, y^i) \in \mathcal{X}^i \times \mathcal{Y}^i$.

Example 2 (Random system). Random functions and random permutations are special cases of random systems. If $(\mathcal{X}, \mathcal{Y})$ is any pair of sets, a *random function* $\mathcal{X} \to \mathcal{Y}$ is a random variable whose values are functions $\mathcal{X} \to \mathcal{Y}$. For any finite set $\mathcal{X}$, a random permutation is a random variable taking values in the set of permutations of $\mathcal{X}$. A *uniformly random function* $\mathbf{R}$ is a random function with uniform distribution over all functions $\mathcal{X} \to \mathcal{Y}$. Using random systems, we have the following:

$$\mathsf{P}^{\mathbf{R}}_{Y_i | X^i Y^{i-1}}[y_i; x^i, y^{i-1}] = \begin{cases} 1 & \text{if } x_i = x_j \text{ for some } j < i \text{ and } y_i = y_j, \\ 0 & \text{if } x_i = x_j \text{ for some } j < i \text{ and } y_i \neq y_j, \\ 1/|\mathcal{Y}| & \text{else.} \end{cases} \quad (1)$$

A uniformly random permutation is defined analogously.

---

[1] Most of the results and arguments in this paper generalize to infinite sampling spaces; for simplicity, we restrict to finite spaces as the latter are the ones relevant for cryptographic applications.

Distinguishing Random Systems. In order to distinguish two $(\mathcal{X}, \mathcal{Y})$-random systems $\mathbf{F}$ and $\mathbf{G}$, we use the notion of a distinguisher that can be regarded as a random system itself. A distinguisher interacts with random systems by making queries to either $\mathbf{F}$ or $\mathbf{G}$ and outputs a binary decision bit after a certain number of queries. In the sequel, we consider information-theoretic distinguishers only; they are computationally unbounded and the only measure of complexity is the number of queries made by them.

In the literature, distinguishers are classified based on how they interact with the random systems. For instance, adaptive distinguishers choose their ith query X_i depending on the history (i.e., all previous query-response pairs), whereas non-adaptive distinguishers commit all their queries in advance. Throughout, we let Ad and NAd be the classes of all adaptive and non-adaptive distinguishers, respectively. Definition 4 formally introduces the concept of a distinguisher as well as its interaction with random systems via probability theory.

Definition 3 (Distinguisher). *An $(\mathcal{X}, \mathcal{Y})$-distinguisher $\mathbf{D}$ is a $(\mathcal{Y}, \mathcal{X})$-random system defined by a sequence of conditional probability distributions $\mathsf{P}^{\mathbf{D}}_{X_i \mid Y^{i-1} X^{i-1}}$. That is, it is a $(\mathcal{Y}, \mathcal{X})$-random system that is one query ahead. A $(\mathcal{X}, \mathcal{Y})$-distinguisher $\mathbf{D}$ and an $(\mathcal{X}', \mathcal{Y}')$-random system $\mathbf{F}$ are said to be compatible if $\mathcal{X}' = \mathcal{X}$ and $\mathcal{Y}' = \mathcal{Y}$.*

One models the interaction of a distinguisher with a random system via a random experiment that is a sequence of conditional probability distributions. This is denoted by $\mathsf{P}^{\mathbf{D} \lozenge \mathbf{F}}_{X_i Y_i \mid X^{i-1} Y^{i-1}}$ and defined simply as

$$\mathsf{P}^{\mathbf{D} \lozenge \mathbf{F}}_{X_i Y_i \mid X^{i-1} Y^{i-1}} = \mathsf{P}^{\mathbf{F}}_{Y_i \mid X^i Y^{i-1}} \mathsf{P}^{\mathbf{D}}_{X_i \mid X^{i-1} Y^{i-1}} \, .$$

Intuitively, this models the probabilities of the distinguisher choosing a given query x_i at the ith step and the random system returning a given response y_i conditioned on the history. Moreover, we define

$$\mathsf{P}^{\mathbf{D} \lozenge \mathbf{F}}_{X^i Y^i} = \prod_{j=1}^{i} \mathsf{P}^{\mathbf{D} \lozenge \mathbf{F}}_{X_j Y_j \mid X^{j-1} Y^{j-1}} \, .$$

We are now interested in distinguishing two random systems $\mathbf{F}$ and $\mathbf{G}$ where we assume that both systems are compatible with the distinguisher $\mathbf{D}$. The performance of $\mathbf{D}$ (known as the *advantage* of $\mathbf{D}$ in distinguishing $\mathbf{F}$ from $\mathbf{G}$) is generally measured as follows:

Definition 4. *Let $\mathbf{F}$ and $\mathbf{G}$ be two $(\mathcal{X}, \mathcal{Y})$-random systems that are compatible with a distinguisher $\mathbf{D}$. Given an integer $i > 0$, the advantage of $\mathbf{D}$ in distinguishing $\mathbf{F}$ from $\mathbf{G}$ in i queries is defined to be*

$$\Delta_i^{\mathbf{D}}(\mathbf{F}, \mathbf{G}) := \frac{1}{2} \sum_{(x^i, y^i) \in \mathcal{X}^i \times \mathcal{Y}^i} |\mathsf{P}^{\mathbf{D} \lozenge \mathbf{F}}_{X^i Y^i} - \mathsf{P}^{\mathbf{D} \lozenge \mathbf{G}}_{X^i Y^i}| \, .$$

Let $\mathcal{C}$ be a class of distinguishers trying to distinguish $\mathbf{F}$ from $\mathbf{G}$. We define the advantage of the best $\mathcal{C}$-distinguisher making i queries to $\mathbf{F}$ and $\mathbf{G}$ as

$$\Delta_i^{\mathcal{C}}(\mathbf{F}, \mathbf{G}) := \max_{\mathbf{D} \in \mathcal{C}} \left\{ \Delta_i^{\mathbf{D}}(\mathbf{F}, \mathbf{G}) \right\} \, .$$

Random Systems with Monotone Conditions. One of the similarities between random systems and game-playing is a notion known as *monotone condition* or *monotone event*. Intuitively, it represents an event that once set, it cannot be "reset" by additional queries. The notion of *monotone event/condition* is more general and should not be confused with *monotone predicate* (or *monotone binary output* as discussed in [5, §2.3]). To explain the difference, let $\mathcal{A} = \{\mathbf{a}_i\}$ be the sequence of events $\mathbf{a}_1, \mathbf{a}_2, \ldots$.

Monotone predicates (or binary outputs) are simpler and less general since the query/response pairs (x^i, y^i) at step i uniquely determine whether the corresponding event $\mathbf{a}_i$ holds or not, whereas the former could be more complex (e.g., $\mathbf{a}_i$ could be the (monotone) event that a certain flag is set in at most 10 steps). In other words, in the case of monotone predicates, the conditional probability of $\mathbf{a}_i$ occurring conditioned on $X^i = x^i \wedge Y^i = y^i$ is binary, whereas monotone events could be more general. For simplicity, we assume that our monotone events are monotone predicates and consider a sequence of boolean predicates a_i indicating whether $\mathbf{a}_i$ holds (i.e., $a_i \Leftrightarrow \mathbf{a}_i$ holds; equivalently, $\neg a_i \Leftrightarrow \mathbf{a}_i$ does not hold) with the property that $\neg a_i \Rightarrow \neg a_{i+1}$ (the latter guarantees monotonicity).

As an example, consider the monotone event $\mathbf{a}_i$ that after the ith query to a uniformly random function, all distinct inputs result in distinct outputs (i.e., there exists no output collisions). It is not difficult to see that $\mathcal{A} = \{\mathbf{a}_i\}$ is a monotone binary output as $\neg a_i \Rightarrow \neg a_{i+1}$ and a_i is completely determined from (x^i, y^i). Equivalently, if there is an output collision for the ith step, there is also an output collision for all the subsequent steps. The monotonicity condition gives rise to a sequence of binary probabilities $\mathsf{P}^{\mathbf{F}}_{a_i|X^iY^i} \in \{0, 1\}$ with the property that

$$\forall i \geq 1, \qquad \mathsf{P}^{\mathbf{F}}_{a_i|X^iY^i} = 1 \Rightarrow \mathsf{P}^{\mathbf{F}}_{a_{i-1}|X^{i-1}Y^{i-1}} = 1 . \qquad (2)$$

Associated to a random system with a monotone binary output, we have the following data:

- **D.0 (data defining F):** these are simply the probability distributions $\mathsf{P}^{\mathbf{F}}_{Y_i|X^iY^{i-1}}$,
- **D.1 (binary probabilities for $\mathcal{A}$):** these are the binary probabilities $\mathsf{P}^{\mathbf{F}}_{a_i|X^iY^i}$ (describing the predicates a_i and $\neg a_i$) satisfying (2).

Remark 5. In the case of monotone conditions, the defining probabilities $\mathsf{P}^{\mathbf{F}}_{a_i|X^iY^i}$ can be arbitrary real numbers in the interval $[0, 1]$.

We can derive various other probabilities using conditional probabilities/Bayes' rule as well as **D.0** and **D.1**:

Event Probabilities for $\mathcal{A}$: These are the probabilities denoted by $\mathsf{P}^{\mathbf{F}}_{a_i|a_{i-1}X^iY^{i-1}}$. Intuitively, $\mathsf{P}^{\mathbf{F}}_{a_i|a_{i-1}X^iY^{i-1}}$ models the probability of the predicate a_i conditioned on the query/response history, as well as on the predicate a_{i-1}. We derive it from **D.0** and **D.1** as follows:

$$\mathsf{P}^{\mathbf{F}}_{a_i|a_{i-1}X^iY^{i-1}} = \sum_{y_i} \mathsf{P}^{\mathbf{F}}_{a_i|X^iY^i} \mathsf{P}^{\mathbf{F}}_{Y_i|X^iY^{i-1}} .$$

Here, one can also derive the probability distributions $P^{\mathbf{F}}_{\neg a_i | a_{i-1} X^i Y^{i-1}}$ simply as $1 - P^{\mathbf{F}}_{a_i | a_{i-1} X^i Y^{i-1}}$. It is important to note that if the condition $a_{i-1} X^i Y^{i-1}$ evaluates to false for all y_i for a given (x^i, y^{i-1}), this probability is set to zero (for reasons that will become clear later). We remark that in a similar manner, one can adjoin yet another monotone condition $\mathcal{B}$ to a random system with a monotone condition $\mathcal{A}$.

A Random System Conditioned on $\mathcal{A}$ not Failing (denoted by $\mathbf{F}|\mathcal{A}$): These are probability distributions $P^{\mathbf{F}}_{Y_i | a_i X^i Y^{i-1}}$ and can be derived from Bayes' rule as follows:

$$P^{\mathbf{F}}_{Y_i | X^i Y^{i-1}} P^{\mathbf{F}}_{a_i | X^i Y^i} = P^{\mathbf{F}}_{a_i Y_i | X^i Y^{i-1}} = P^{\mathbf{F}}_{Y_i | a_i X^i Y^{i-1}} P^{\mathbf{F}}_{a_i | X^i Y^{i-1}} , \qquad (3)$$

where the middle term (which has not been defined yet) is a formal symbol for the corresponding probability. Assuming that $P^{\mathbf{F}}_{a_{i-1} | X^{i-1} Y^{i-1}} = 1$ together with the monotonicity of $\mathcal{A}$, we see that $P^{\mathbf{F}}_{a_i | X^i Y^{i-1}} = P^{\mathbf{F}}_{a_i | a_{i-1} X^i Y^{i-1}} \neq 0$. One can thus derive the conditional probabilities

$$P^{\mathbf{F}}_{Y_i | a_i X^i Y^{i-1}} = \frac{P^{\mathbf{F}}_{Y_i | X^i Y^{i-1}} P^{\mathbf{F}}_{a_i | X^i Y^i}}{P^{\mathbf{F}}_{a_i | a_{i-1} X^i Y^{i-1}}}.$$

Intuitively, this looks like a random system except that we have conditioned on the predicate a_i. Note that this need not be a probability distribution: for instance, consider the example of a random function $\mathbf{R} \colon \{0,1\}^n \to \{0,1\}^n$ and define the $\mathbf{a}_i$ as the event of having a collision between an input and an output. It might occur that $x_2 = y_1$ in which case $a_2 X^2 Y^1$ will always evaluate to false and thus, the probability $P^{\mathbf{R}}_{Y_2 | a_2 X^2 Y^1} = 0$ for all y_2, so it will not represent a well-defined distribution on the variable Y_2. In cases when this degeneracy does not occur, we can consider $\mathbf{F}|\mathcal{A}$ as a true random system $\mathbf{G}$ (see Hypothesis 8), denoted $\mathbf{F}|\mathcal{A} \equiv \mathbf{G}$. Note that this particular notion of equivalence of a random system and a random system with a monotone condition can be extended slightly in the case of degeneracies too. As described in [11, Defn.6], we say that $\mathbf{F}|\mathcal{A}$ is *equivalent* to $\mathbf{G}$ if $P^{\mathbf{F}}_{Y_i | a_i X^i Y^{i-1}} = P^{\mathbf{G}}_{Y_i | X^i Y^{i-1}}$ for any i and any values of the parameters for which $P^{\mathbf{F}}_{Y_i | a_i X^i Y^{i-1}}$ is not identically zero (i.e.., is a distribution). Finally, we note that $\mathbf{F}|\mathcal{A}$ appeared in, e.g., [15, Defn.7].

A Random System with a Condition $\mathcal{A}$ (denoted by $\mathbf{F}^{\mathcal{A}}$): This is the random system corresponding to [15, Defn.6]) and can be derived by

$$P^{\mathbf{F}}_{a_i Y_i | a_{i-1} X^i Y^{i-1}} := P^{\mathbf{F}}_{Y_i | a_i X^i Y^{i-1}} P^{\mathbf{F}}_{a_i | a_{i-1} X^i Y^{i-1}} .$$

We also define

$$P^{\mathbf{F}}_{a_i Y^i | X^i} := \prod_{j=1}^{i} P^{\mathbf{F}}_{a_j Y_j | a_{j-1} X^j Y^{j-1}} .$$

Moreover, we consider distinguishers trying to provoke the negated event $\neg a_i$ again via a sequence of probability distributions. To indicate the link with $\mathbf{a}_i$, we denote these distributions by $P^{\mathbf{D}}_{X_i | a_{i-1} X^{i-1} Y^{i-1}}$. As in the case of true random systems, this models the probability distribution of an adversary choosing the ith query based on the previous responses and the predicate a_{i-1} (meaning that the desired event $\neg a_{i-1}$ has not occurred after the $(i-1)$st query/response pair).

Using this data, we can derive various probabilities and distributions by imposing Bayes' rule. We define the probabilities for the random experiment $\mathbf{D}\Diamond\mathbf{F}$ by

$$\mathsf{P}^{\mathbf{D}\Diamond\mathbf{F}}_{a_i X_i Y_i | a_{i-1} X^{i-1} Y^{i-1}} := \mathsf{P}^{\mathbf{F}}_{a_i Y_i | a_{i-1} X^i Y^{i-1}} \mathsf{P}^{\mathbf{D}}_{X_i | a_{i-1} X^{i-1} Y^{i-1}}.$$

Intuitively, this models the probability of choosing a particular query, obtaining a particular response and the predicate a_i (resp., $\neg a_i$) conditioned on the history and the predicate a_{i-1}. Finally, let

$$\mathsf{P}^{\mathbf{D}\Diamond\mathbf{F}}_{a_i X^i Y^i} := \prod_{j=1}^{i} \mathsf{P}^{\mathbf{D}\Diamond\mathbf{F}}_{a_j X_j Y_j | a_{j-1} X^{j-1} Y^{j-1}}.$$

Similarly, we define an expression for $\neg a_i$. We are now ready to define the advantage of the distinguisher (adversary) $\mathbf{D}$ in provoking the desired event $\neg a_i$:

Definition 6. *Let $\mathcal{C}$ be a class of distinguishers $\mathbf{D}$ that are trying to provoke $\neg a_i$. Given $i > 0$, define $\nu^{\mathbf{D}}(\mathbf{F}, \neg a_i)$ to be the advantage of the distinguisher $\mathbf{D}$ in provoking the event $\neg a_i$ in the random experiment $\mathbf{D}\Diamond\mathbf{F}$. That is*

$$\nu^{\mathbf{D}}(\mathbf{F}, \neg a_i) = \sum_{(x^i, y^i) \in \mathcal{X}^i \times \mathcal{Y}^i} \mathsf{P}^{\mathbf{D}\Diamond\mathbf{F}}_{\neg a_i X^i Y^i}.$$

Furthermore, for all $i \geq 1$, define $\nu^{\mathcal{C}}(\mathbf{F}, \neg a_i) := \max_{\mathbf{D} \in \mathcal{C}} \nu^{\mathbf{D}}(\mathbf{F}, \neg a_i)$ to be the maximum advantage over all distinguishers in the class $\mathcal{C}$ trying to provoke $\neg a_i$.

Finally, we explain the analogue (in the context of random systems) of the fundamental lemma of game-playing and comment on why the random-system statement is more general. Suppose that $\mathbf{F}$ is a random system with a monotone condition $\mathcal{A}$ and let $\mathbf{G}$ be another random system. The analogue of the hypothesis of the fundamental lemma of game-playing (that two games are equivalent up to statements that are evaluated only if a_i is set to $\texttt{true}$) is simply $\mathbf{F}|\mathcal{A} \equiv \mathbf{G}$. Under that hypothesis, we expect that one can bound the distinguishing advantage $\mathbf{\Delta}^{\mathbf{D}}_i(\mathbf{F}, \mathbf{G})$ via the advantage $\nu^{\mathbf{D}}(\mathbf{F}, \neg a_i)$ of an adversary to provoke $\neg a_i$. Interestingly enough, one can deduce the latter from a weaker hypothesis, namely the hypothesis that

$$\mathsf{P}^{\mathbf{F}}_{a_j Y^j | X^j} \leq \mathsf{P}^{\mathbf{G}}_{Y^j | X^j}, \ \forall j \leq i.$$

The following lemma is proven in [15, Lem.6] (see also [11, Thm.1]):

Lemma 7. *Assume that $\mathsf{P}^{\mathbf{F}}_{a_j Y^j | X^j} \leq \mathsf{P}^{\mathbf{G}}_{Y^j | X^j}$ holds for all $j \leq i$. Then for any distinguisher $\mathbf{D}$,*

$$\mathbf{\Delta}^{\mathbf{D}}_i(\mathbf{F}, \mathbf{G}) \leq \nu^{\mathbf{D}}(\mathbf{F}, \neg a_i).$$

In the following sections, we develop techniques to upper bound $\nu^{\mathbf{D}}(\mathbf{F}, \neg a_i)$.

3   A Standard Method for Probabilistic Analysis of Adaptive Adversaries

Let $\mathcal{A}$ be a monotone condition and let $\mathbf{F}$ be an $(\mathcal{X}, \mathcal{Y})$-random system. Our goal is to compute an upper bound for $\nu^{\mathsf{Ad}}(\mathbf{F}, \neg a_i)$. The standard way to deal with the overall probability of setting $\neg a_i$ is to bound it by a sum (over $j \leq i$) of the maximum (over all adversaries) probability of winning at the jth step, where these "stepwise" probabilities are only taken over the probability distributions describing $\mathbf{F}$. In other words, for each j, we maximize individually the probability of winning at the jth step assuming that we have not won at step $j - 1$. This greedy-type approach for producing an upper bound can be formalized in Proposition 9 (see Appendix of the full version for its proof). We first state an hypothesis that is commonly used throughout the paper.

Hypothesis 8. Let $\mathbf{F}$ be an $(\mathcal{X}, \mathcal{Y})$-random system and let $\mathcal{A}$ be a monotone condition on $\mathbf{F}$. There exists an $(\mathcal{X}, \mathcal{Y})$-random system $\mathbf{G}$ such that $\mathbf{F}|\mathcal{A} \equiv \mathbf{G}$, i.e., for all $i \geq 1$ and all $(x^i, y^i) \in \mathcal{X}^i \times \mathcal{Y}^i$,

$$\mathsf{P}^{\mathbf{F}}_{Y_i|a_i X^i Y^{i-1}} = \mathsf{P}^{\mathbf{G}}_{Y_i|X^i Y^{i-1}}.$$

Proposition 9. *Let $\mathbf{F}$ be an $(\mathcal{X}, \mathcal{Y})$-random system and let $\mathcal{A}$ be a monotone condition on $\mathbf{F}$. Assuming that $\sum_{j=1}^{i} \max_{(x^j, y^{j-1})} \left\{ \mathsf{P}^{\mathbf{F}}_{\neg a_j|a_{j-1} X^j Y^{j-1}} \right\} < 1$, we have*

$$\nu^{\mathsf{Ad}}(\mathbf{F}, \neg a_i) \leq \sum_{j=1}^{i} \max_{(x^j, y^{j-1})} \left\{ \mathsf{P}^{\mathbf{F}}_{\neg a_j|a_{j-1} X^j Y^{j-1}} \right\}.$$

4   When Adaptivity Does Not Help

4.1   Revisiting the Result of Maurer and Pietrzak

Maurer [11] and Pietrzak [15] provide a general method for proving that under certain hypotheses, adaptive strategies are no better than non-adaptive ones in forcing a condition to fail. In other words, if these hypotheses are satisfied, the advantage of the best Ad- and NAd-distinguisher are equal. Here, we show that the hypothesis (Hypothesis 8) used by Pietrzak is not sufficient for the comparison result of [11, 15] (ANA switching lemma) to hold by providing a particular counterexample in Proposition 10 where the hypothesis is clearly satisfied and where adaptivity does help. We then explain the problem in the ANA switching lemma in detail and suggest different ways to remedy it in Section 4.2. The following statement appears in [15, Lem.6]:

Adaptive–Non-Adaptive (ANA) Switching Lemma. Let $\mathcal{A}$ be a monotone condition and let $\mathbf{F}$ be an $(\mathcal{X}, \mathcal{Y})$-random system. If Hypothesis 8 holds for $\mathbf{F}$, $\mathcal{A}$ and an $(\mathcal{X}, \mathcal{Y})$-random system $\mathbf{G}$, then adaptivity does not help in provoking $\neg a_i$. More precisely,

$$\nu^{\mathsf{Ad}}(\mathbf{F}, \neg a_i) = \nu^{\mathsf{NAd}}(\mathbf{F}, \neg a_i).$$

Now, we present an example of a random system where adaptive adversaries have *better* advantage than the adaptive ones in provoking a welldefined monotone event.

Proposition 10. *Let $\mathcal{X} = \{0,1\}^n$ and let $\mathbf{P} \colon \mathcal{X} \to \mathcal{X}$ be a uniformly random permutation. Let $\mathbf{a}_i$ be the event that $y_j \neq x_j$ for all $j \leq i$ where $y_j = \mathbf{P}(x_j)$. Then $\mathcal{A}$ is monotone and*

$$\nu^{\mathsf{Ad}}(\mathbf{P}, \neg a_i) > \nu^{\mathsf{NAd}}(\mathbf{P}, \neg a_i) .$$

Proof. The sequence of predicates $\{a_i\}$ is monotone by definition. We calculate the probability of obtaining a fixed point for $\mathbf{P}$ after at most two queries; the case for general i follows by inspection. After querying $\mathbf{P}$ with any $X_1 = x_1 \in \{0,1\}^n$, the response $y_1 \in \{0,1\}^n$ is uniformly random. Thus, with probability $1/2^n$ a fixed point is found after the first query. Hence,

$$\mathsf{P}^{\mathbf{P}}[Y_2 = x_2 \vee Y_1 = x_1] = \mathsf{P}^{\mathbf{P}}[Y_2 = x_2 \wedge Y_1 \neq x_1] + \mathsf{P}^{\mathbf{P}}[Y_1 = x_1] =$$
$$= \mathsf{P}^{\mathbf{P}}[Y_2 = x_2 \wedge Y_1 \neq x_1] + 1/2^n .$$

The distinction between an adaptive and a non-adaptive strategy shows up after the second query: the latter commits the second query in advance whereas the former chooses it adaptively based on the first query and its response.

Case 1: Non-adaptive adversary. If the adversary were non-adaptive, she would have fixed $x_2 \neq x_1$ prior to obtaining the response y_1 and since $\mathbf{P}(x_2) \neq \mathbf{P}(x_1)$ and $\mathbf{P}$ is a uniformly random permutation, $\mathbf{P}(x_2) \in \{0,1\}^n - \{y_1\}$. Note however that if $x_2 = y_1$, no y_2 could lead to a fixed point. Hence (by Bayes' rule),

$$\mathsf{P}^{\mathbf{P}}[Y_2 = x_2 \wedge Y_1 \neq x_1] = \mathsf{P}^{\mathbf{P}}[Y_2 = x_2 \mid Y_1 \neq x_1, x_2]\mathsf{P}^{\mathbf{P}}[Y_1 \neq x_1, x_2] .$$

Clearly, $\mathsf{P}^{\mathbf{P}}[Y_1 \neq x_1, x_2] = (2^n - 2)/2^n$. Moreover, y_2 is uniformly random among $\{0,1\}^n - \{y_1\}$, so

$$\mathsf{P}^{\mathbf{P}}[Y_2 = x_2 \wedge Y_1 \neq x_1] = \frac{1}{2^n - 1} \cdot \frac{2^n - 2}{2^n} \Rightarrow \mathsf{P}^{\mathbf{P}}[Y_2 = x_2 \vee Y_1 = x_1] =$$
$$= \frac{1}{2^n - 1} \cdot \frac{2^n - 2}{2^n} + \frac{1}{2^n} < \frac{1}{2^{n-1}} .$$

Since the above analysis holds for any non-adaptive adversary, we conclude that $\nu^{\mathsf{NAd}}(\mathbf{P}, \neg a_2) < 1/2^{n-1}$.

Case 2: Adaptive adversary. Knowing x_1, y_1 and $y_1 \neq x_1$ from the first query, an adaptive adversary can eliminate one choice for the second query x_2 different from x_1, namely $x_2 = y_1$. Thus, a clever adversary will choose $x_2 \in \{0,1\}^n - \{x_1, y_1\}$ so that the chance of finding a fixed point after the second step is $1/(2^n - 1)$. Thus,

$$\mathsf{P}^{\mathbf{P}}[Y_2 = x_2 \wedge Y_1 \neq x_1] = \mathsf{P}^{\mathbf{P}}[Y_2 = x_2 \mid Y_1 \neq x_1 \wedge Y_1 \neq x_2]\mathsf{P}^{\mathbf{P}}[Y_1 \neq x_1 \wedge Y_1 \neq x_2] =$$
$$= \frac{1}{2^n - 1} \cdot \frac{2^n - 1}{2^n} ,$$

and we conclude that

$$\nu^{\text{Ad}}(\mathbf{P}, \neg a_2) \geq \frac{1}{2^n} + \frac{1}{2^n - 1} \cdot \frac{2^n - 1}{2^n} = \frac{1}{2^{n-1}} > \nu^{\text{NAd}}(\mathbf{P}, \neg a_2) \,.$$

$\square$

We now explain why Hypothesis 8 holds for the monotone event sequence $\mathcal{A}$ and the random system $\mathbf{P}$.

Proposition 11. *Let $\mathbf{P}$ and $\mathcal{A}$ be as in Proposition 10. Then, Hypothesis 8 holds using the monotone condition $\mathcal{A}$, along with taking $\mathbf{P}$ as $\mathbf{F}$.*

Proof. Let $i = 2$. We simply need to define the distributions (i) $\mathsf{P}^{\mathbf{G}}_{Y_1 | X^1}$ for all $y_1 \in \mathcal{Y}$ and $x^1 \in \mathcal{X}^1$, and (ii) $\mathsf{P}^{\mathbf{G}}_{Y_2 | X^2 Y^1}$ for all $y_2 \in \mathcal{Y}$, $x^2 \in \mathcal{X}^2$ and $y^1 \in \mathcal{Y}^1$. For (i), define

$$\mathsf{P}^{\mathbf{G}}_{Y_1 | X^1} = \begin{cases} \frac{1}{2^n - 1} & \text{if } y_1 \neq x_1, \\ 0 & \text{otherwise.} \end{cases}$$

Clearly, $\mathsf{P}^{\mathbf{F}}_{Y_1 | a_1 X^1} = \mathsf{P}^{\mathbf{G}}_{Y_1 | X^1}$. For (ii), assuming $y_1 \neq x_1$ and $x_1 \neq x_2$, define the distribution in the following two cases:

Case 1: $x_2 = y_1$. There are $2^n - 1$ possible values for $y_2 = \mathbf{P}(x_2)$ occurring with equal probabilities and none of these values can lead to a fixed point, so we have

$$\mathsf{P}^{\mathbf{G}}_{Y_2 | X^2 Y^1} = \begin{cases} 0 & \text{if } y_2 = y_1 = x_2, \\ \frac{1}{2^n - 1} & \text{otherwise}. \end{cases}$$

Case 2: $x_2 \neq y_1$. Here, the case of $y_2 = x_2 \neq y_1$ causes $\neg a_2$, so one can define:

$$\mathsf{P}^{\mathbf{G}}_{Y_2 | X^2 Y^1} = \begin{cases} 0 & \text{if } y_2 = y_1 \text{ or } y_2 = x_2 \neq y_1 \\ \frac{1}{2^n - 2} & \text{otherwise}. \end{cases}$$

We easily verify that in all cases, $\mathsf{P}^{\mathbf{F}}_{Y_2 | a_2 X^2 Y^1} = \mathsf{P}^{\mathbf{G}}_{Y_2 | X^2 Y^1}$. $\square$

4.2   Another Look at the Comparison of Adaptive vs. Non-adaptive Adversaries

Propositions 10 and 11 show that the ANA switching lemma cannot hold as stated in [15, Lem.6]. We now analyze in detail the proof of the ANA switching lemma given in [15], identify the step that causes the discrepancy and propose a fix.

The Mistake in the Original Proof [15]. The ANA switching lemma first appears in [11, Thm.2] with the correct hypothesis (see (1) of loc. cit.), but without a proof. A slightly different version referring to the original claim is given in [12, Prop.2] (again without a proof). The only proof, to the best of our knowledge, appears in [15, Lem.6] and is based on a chain of equalities and inequalities starting with

$$1 - \nu^{\mathsf{Ad}}(\mathbf{F}, \neg a_i) = \min_{D \in \mathsf{Ad}} \left\{ \sum_{(x^i, y^i)} \left(\prod_{j=1}^{i} \mathsf{P}^{\mathbf{F}}_{a_j Y_j | X^j Y^{j-1}} \mathsf{P}^{\mathbf{D}}_{X^j | Y^{j-1}} \right) \right\}.$$

Similarly to Proposition 9, the proof is based on applying Bayes' rule to $\mathsf{P}^{\mathbf{F}}_{a_j Y_j | X^j Y^{j-1}}$. The application of the Bayes' rule in [15, Lem.6] is, however, incorrect[2]. The correct application yields (assuming that the conditional distributions are well-defined)

$$\mathsf{P}^{\mathbf{F}}_{Y_j a_j | X^j Y^{j-1}} = \mathsf{P}^{\mathbf{F}}_{Y_j | a_j X^j Y^{j-1}} \mathsf{P}^{\mathbf{F}}_{a_j | X^j Y^{j-1}}.$$

The problem is that the term $\mathsf{P}^{\mathbf{F}}_{a_i | X^i} = \prod_{j=1}^{i} \mathsf{P}^{\mathbf{F}}_{a_j | X^j Y^{j-1}}$ is assumed to be independent of Y^{i-1} (see the top line of [15, p.30]) - step (2.26)). There is no reason why (for a fixed x^i) this term should be independent of Y^{i-1}; yet, this is used implicitly in the argument. We have seen in Proposition 10 that the probability $\mathsf{P}^{\mathbf{F}}_{a_2 | X^2 Y^1}$ depends on Y^1, so the ANA switching lemma does not apply.

Strengthening the Hypotheses. We now propose a simple fix to the ANA switching lemma by adding an extra hypothesis, essentially stating that the probability of achieving a success on the jth query is independent of the *answers* to all the previous queries. This statement (albeit in a different formulation) already appears as (1) in Maurer's original [11, Thm.2], as well as a rephrased reproduction [12, Prop.2]. Neither of these statements comes with a proof and both omit mention of Hypothesis 8, although in [11, Thm.2] an alternative condition (2) is given such that (2) is claimed to imply both (1) and Hypothesis 8.

Our proof of Theorem 12 follows largely along the lines of the (incorrect) proof of Pietrzak, but obviously with fixes applied where necessary. Here, Hypothesis 8 is needed to guarantee that all conditional probabilities $\mathsf{P}^{\mathbf{F}}_{Y_j | a_j X^j Y^{j-1}}$ are well-defined and are also distributions when considered as functions on $y_j \in \mathcal{Y}$. The second hypothesis simply says that if there is no dependency of the conditionals $\mathsf{P}^{\mathbf{F}}[a_j | a_{j-1} \wedge X^j = x^j \wedge Y^{j-1} = y^{j-1}]$ on the previous outputs then adaptivity should not help at all.

Theorem 12. *Let $\mathbf{F}$ be an $(\mathcal{X}, \mathcal{Y})$-random system and let $\mathcal{A}$ be a monotone condition on $\mathbf{F}$. Let $i > 0$ be an integer. Suppose that Hypothesis 8 holds for $\mathbf{F}$ and $\mathcal{A}$. If, in addition, for every $j \leq i$ and $x^j \in \mathcal{X}^j$, $\mathsf{P}^{\mathbf{F}}[a_j | a_{j-1} \wedge X^j = x^j \wedge Y^{j-1} = y^{j-1}]$ is independent of $y^{j-1} \in \mathcal{Y}^{j-1}$, then adaptivity does not help in provoking $\neg a_i$, i.e.,*

$$\nu^{\mathsf{Ad}}(\mathbf{F}, \neg a_i) = \nu^{\mathsf{NAd}}(\mathbf{F}, \neg a_i).$$

Proof of Theorem 12. We first note that $\nu^{\mathsf{Ad}}(\mathbf{F}, \neg a_i) \geq \nu^{\mathsf{NAd}}(\mathbf{F}, \neg a_i)$ holds. The rest of the proof follows by showing the other direction of the inequality; we have that $1 - \nu^{\mathsf{Ad}}(\mathbf{F}, \neg a_i)$ equals

---

[2] Furthermore, the argument in [15, Lem.6] does not state whether the conditional probabilities $\mathsf{P}^{\mathbf{F}}_{Y_j | a_j X^j Y^{j-1}}$ are well-defined, for all $j \leq i$.

$$\min_{D \in \mathsf{Ad}} \left\{ \sum_{(x^i, y^i)} \left(\prod_{j=1}^{i} \mathsf{P}^{\mathbf{F}}_{Y_j | a_j X^j Y^{j-1}} \mathsf{P}^{\mathbf{F}}_{a_j | a_{j-1} X^j Y^{j-1}} \mathsf{P}^{\mathbf{D}}_{X_j | a_{j-1} X^{j-1} Y^{j-1}} \right) \right\}$$

$$\stackrel{(*)}{=} \min_{D \in \mathsf{Ad}} \left\{ \sum_{(x^i, y^i)} \left(\prod_{j=1}^{i} \mathsf{P}^{\mathbf{G}}_{Y_j | X^j Y^{j-1}} \mathsf{P}^{\mathbf{F}}_{a_j | a_{j-1} X^j} \mathsf{P}^{\mathbf{D}}_{X_j | a_{j-1} X^{j-1} Y^{j-1}} \right) \right\}$$

$$= \min_{D \in \mathsf{Ad}} \left\{ \sum_{x^i} \left(\prod_{j=1}^{i} \mathsf{P}^{\mathbf{F}}_{a_j | a_{j-1} X^j} \right) \sum_{y^i} \left(\prod_{j=1}^{i} \mathsf{P}^{\mathbf{G}}_{Y_j | X^j Y^{j-1}} \mathsf{P}^{\mathbf{D}}_{X_j | a_{j-1} X^{j-1} Y^{j-1}} \right) \right\}$$

$$= \min_{D \in \mathsf{Ad}} \left\{ \sum_{x^i} \left(\prod_{j=1}^{i} \mathsf{P}^{\mathbf{F}}_{a_j | a_{j-1} X^j} \right) \right\}$$

$$\geq \min_{D \in \mathsf{Ad}} \left\{ \sum_{x^i} \left(\prod_{j=1}^{i} \mathsf{P}^{\mathbf{F}}_{a_j | a_{j-1} X^j} \mathsf{P}^{\mathbf{D}}_{X_j | a_{j-1}} \right) \right\} = \min_{D \in \mathsf{Ad}} \left\{ \sum_{x^i} \mathsf{P}^{\mathbf{D} \Diamond \mathbf{F}}_{a_i X^i} \right\}$$

$$\geq = (1 - \nu^{\mathsf{NAd}}(\mathbf{F}, \neg a_i)) .$$

Here, (*) uses Hypothesis 8, as well as the extra hypothesis that $\mathsf{P}^{\mathbf{F}}_{a_j | X^j Y^{j-1}}$ is independent of y^{j-1}. Hence, $\nu^{\mathsf{NAd}}(\mathbf{F}, \neg a_i)) \geq \nu^{\mathsf{Ad}}(\mathbf{F}, \neg a_i))$ and the claim follows.   □

5   Towards Obtaining Better Bounds

5.1   Using an Auxiliary Flag

The standard approach given in Section 3 has the disadvantage that for more complex constructions, the maximal probabilities can get too large. This is often due to the fact that the maximum is achieved for rather degenerate values of (x^i, y^i) that occur with very low probability. Assuming that one can bound the probability of the degeneracy, one way to refine the analysis of the adaptive adversary is to introduce an auxiliary event (flag) that is set only for non-degenerate pairs (x^i, y^i). More precisely, if $\mathbf{a}_i$ is the monotone event to be studied, we introduce a flag event $\mathbf{b}_i$ (together with a corresponding predicate b_i indicating whether $\mathbf{b}_i$ has occurred or not) and we use the fact that

$$\neg \mathbf{a}_i \Leftrightarrow (\neg \mathbf{a}_i \wedge \mathbf{b}_i) \vee (\neg \mathbf{a}_i \wedge \neg \mathbf{b}_i) \Rightarrow (\neg \mathbf{a}_i \wedge \mathbf{b}_i) \vee \neg \mathbf{b}_i.$$

Now, bounding the advantage of achieving $\neg \mathbf{a}_i$ amounts to bounding the advantage of achieving $\neg \mathbf{a}_i \wedge \mathbf{b}_i$ together with bounding the probability of degeneracy (or, of $\neg \mathbf{b}_i$). The latter can be done via Proposition 9; yet for the former we need to introduce new definitions.

All this can be rigorously modeled using random systems as follows: suppose that $\mathbf{F}$ is a random system with a monotone condition $\mathcal{B}$ (here, $\mathcal{B}$ represents the flag event). Suppose further that $\mathbf{F}|\mathcal{B}$ is equivalent to another random system $\mathbf{G}$ (i.e., $\mathbf{F}|\mathcal{B} \equiv \mathbf{G}$). Now, we simply impose a monotone condition $\mathcal{A}$ on $\mathbf{G}$. Equivalently, we need to specify the corresponding probabilities and distributions from Section 2. Suppose that we are given the following data:

- Event probabilities $\mathsf{P}^{\mathbf{G}}_{a_i|a_{i-1}X^iY^{i-1}}$ also denoted by $\mathsf{P}^{\mathbf{F}}_{a_i|a_{i-1}b_iX^iY^{i-1}}$ (to indicate better what they are supposed to model),
- The random system $\mathbf{G}|\mathcal{A}$, namely, probabilities $\mathsf{P}^{\mathbf{G}}_{Y_i|a_iX^iY^{i-1}}$ that we also denote by $\mathsf{P}^{\mathbf{F}}_{Y_i|a_ib_iX^iY^{i-1}}$,
- Distinguisher relative to $\mathcal{A}$, namely, probability distributions denoted by $\mathsf{P}^{\mathbf{D}}_{X_i|a_{i-1}b_{i-1}X^{i-1}Y^{i-1}}$.

This data allows us to upper bound the advantage $\nu^{\mathsf{Ad}}(\mathbf{F}, \neg a_i \wedge b_i)$ (by defining $\mathsf{P}^{\mathbf{D}\Diamond\mathbf{F}}_{\neg a_i|b_i} = \mathsf{P}^{\mathbf{D}\Diamond\mathbf{G}}_{\neg a_i}$) following exactly the same steps as in Section 2 (for the random system $\mathbf{G}$ and the monotone event $\mathcal{A}$). Moreover, we assume all the corresponding notation. The following proposition provides an upper bound on the adaptive advantage (see Appendix of the full version for its proof):

Proposition 13. *Let $\mathbf{F}$ be a random system with a monotone condition $\mathcal{B}$ with the property that there exists a random system $\mathbf{G}$ such that $\mathbf{F}|\mathcal{B} \equiv \mathbf{G}$. Let $\mathcal{A}$ be a monotone condition on $\mathbf{G}$. Assuming that*

$$\sum_{j=1}^{i} \max_{(x^j, y^{j-1})} \left\{ \mathsf{P}^{\mathbf{F}}_{\neg a_j|a_{j-1}b_jX^jY^{j-1}} \right\} < 1 ,$$

we have

$$\nu^{\mathsf{Ad}}(\mathbf{F}, \neg a_i \wedge b_i) \leq \sum_{j=1}^{i} \max_{(x^j, y^{j-1})} \left\{ \mathsf{P}^{\mathbf{F}}_{\neg a_j|a_{j-1}b_jX^jY^{j-1}} \right\} .$$

5.2 Improving the Bounds Obtained from Step-Specific Maximization

The greedy approach based on step-specific maximization often has limitations in the sense that the produced bounds are not tight enough. One can obtain better bounds via the simple observation that the advantage of an adversary in provoking $\neg a_i$ for a monotone event $\mathcal{A}$ can be bounded by the sum of the event probabilities for the negated events $\neg a_j$ for $j \leq i$ that are part of the data defining the monotone condition $\mathcal{A}$. Consequently, if one is able to provide upper bounds on these sums, one would automatically obtain an upper bound on the adaptive advantage.

In order to carry out this idea rigorously, we consider two methods that are formally stated in Propositions 14 and 15 (see Appendix of the full version for the proof of the former; the proof of the latter follows from the proof of Propositions 13 and 14). We first give ourselves an upper bound B_{Σ} on the sum of the event probabilities and then show that the same B_{Σ} bounds the adaptive advantage as well. The second method is a variation of the first where one uses an auxiliary event. These two techniques are important whenever the bounds given in Propositions 9 and 13 are not sufficiently tight). A good example of that is the analysis an adaptive adversary trying to achieve a collision in the compression function of [7] (see Appendix of the full version for the details).

Proposition 14. *Let* **F** *be a random system with a monotone event* $\mathcal{A}$*. If there exists a value* $B_\Sigma \in (0,1)$ *such that for all* $(x^i, y^i) \in \mathcal{X}^i \times \mathcal{Y}^i$

$$\sum_{j=1}^{i} \mathsf{P}^{\mathbf{F}}_{\neg a_j | a_{j-1} X^j Y^{j-1}} \leq B_\Sigma \,,$$

then $\nu^{\mathsf{Ad}}(\mathbf{F}, \neg a_i) \leq B_\Sigma$.

The following proposition shows the natural generalization of the above proposition to the case of auxiliary events (its proof follows from the proof of Propositions 13 and 14):

Proposition 15. *Let* **F** *be a random system with a monotone condition* $\mathcal{B}$ *with the property that there exists a random system* **G** *such that* $\mathbf{F}|\mathcal{B} \equiv \mathbf{G}$*. Let* $\mathcal{A}$ *be a monotone condition on* **G***. Suppose that there exists a value* $B_\Sigma \in (0,1)$ *such that for all* $(x^i, y^i) \in \mathcal{X}^i \times \mathcal{Y}^i$

$$\sum_{j=1}^{i} \mathsf{P}^{\mathbf{F}}_{\neg a_j | a_{j-1} b_j X^j Y^{j-1}} = \sum_{j=1}^{i} \mathsf{P}^{\mathbf{G}}_{\neg a_j | a_{j-1} X^j Y^{j-1}} \leq B_\Sigma \,.$$

Then $\nu^{\mathsf{Ad}}(\mathbf{F}, \neg a_i \wedge b_i) \leq B_\Sigma$.

Counting Successes. In Proposition 14, we are mainly interested in estimating the maximal probability of the event (success) occurring once. Nevertheless, in some cases the major monotone event $\mathcal{A}$ might depend on an auxiliary condition that intrinsically requires an event (success) to occur more than once. As a simple example, consider a generalization of the case studied in Proposition 10: let **P** be a uniformly random permutation $\mathbf{P} \colon \mathcal{X} \to \mathcal{X}$ for $\mathcal{X} = \{0,1\}^n$ and let $\neg a_i$ be the event that $y_j = x_j$ for more than κ values of $j \leq i$ where $y_j = \mathbf{P}(x_j)$ and κ is a positive integer. More precisely, a_i is the predicate that there exist at most κ fixed points after the ith query.

Such a general problem can be modeled and studied using random systems as follows: suppose that **F** is an $(\mathcal{X}, \mathcal{Y})$-random system. We then attach an event called hit_i to the random system **F** - this is the success event at step i. Note that hit_i is not monotone. Moreover, we introduce a random variable ctr_i to indicate the number of successes up to step i. In other words, $\mathsf{ctr}_0 = 0$ and for every $j \geq 1$, $\mathsf{ctr}_j = \mathsf{ctr}_{j-1} + 1$ if hit_j occurs and $\mathsf{ctr}_j = \mathsf{ctr}_{j-1}$ otherwise. Finally, we can associate monotone events $\mathcal{A}_\kappa = \{a_{\kappa,i}\}$ for every integer $\kappa \geq 0$, so that $a_{\kappa,i}$ is event that there are at most κ successes after the ith query. In other words, $a_{\kappa,i}$ is the event that $\mathsf{ctr}_i \leq \kappa$.

In order to attach the success event to the random system, we provide the following additional data to **D.0**:

H.1: Binary probabilities $\mathsf{P}^{\mathbf{F}}_{\mathsf{hit}_i | X^i Y^i}$ for every $x^i \in \mathcal{X}^i$ and $y^i \in \mathcal{Y}^i$.

We can derive the following probabilities from **D.0** and **H.1** via Bayes' rule:

- Probabilities $\mathsf{P}^{\mathbf{F}}_{\mathsf{hit}_i | X^i Y^{i-1}}$ for every $x^i \in \mathcal{X}^i$ and $y^{i-1} \in \mathcal{Y}^{i-1}$ defined by

$$\mathsf{P}^{\mathbf{F}}_{\mathsf{hit}_i | X^i Y^{i-1}} = \sum_{y_i} \mathsf{P}^{\mathbf{F}}_{\mathsf{hit}_i | X^i Y^i} \mathsf{P}^{\mathbf{F}}_{Y_i | X^i Y^{i-1}}.$$

- The data for each of the monotone events $\mathcal{A}_\kappa$.

Proposition 16 sets an upper bound on $\nu^{\mathsf{Ad}}(\mathbf{F}, \neg a_{i,\kappa})$ (see Appendix of the full version for its proof).

Proposition 16. *Let κ be a non-negative integer and suppose that there exists a value $B_\Sigma \in (0,1)$ such that for all $(x^i, y^i) \in \mathcal{X}^i \times \mathcal{Y}^i$,*

$$\sum_{j=0}^{i} \mathsf{P}^{\mathbf{F}}_{\mathsf{hit}_j | X^j Y^{j-1}} \leq B_\Sigma \quad \text{and} \quad \mathsf{P}^{\mathbf{F}}_{\mathsf{hit}_i | X^i Y^{i-1}} > 0.$$

Then $\nu^{\mathsf{Ad}}(\mathbf{F}, \neg a_{i,\kappa}) \leq B_\Sigma^{\kappa+1}$.

Remark 17. We should indicate the analogy between Proposition 14 and Proposition 16 with [7, Prop.7] and [7, Prop.9], respectively. We believe that having such statements and techniques developed in the general context of random systems could serve as a guiding tool for more conceptual security proofs for other constructions in the future.

Acknowledgement. The authors gratefully acknowledge Krzysztof Pietrzak for insightful discussions and the Crypto'12 and Asiacrypt'12 program committees for their useful feedback. This work has been supported in part by the European Commission through the ICT programme under contract ICT-2007-216676 ECRYPT II. Jetchev and Özen were supported by a grant of the Swiss National Science Foundation, 200021-122162. The work was initiated while Stam was at EPFL.

References

1. Armknecht, F., Fleischmann, E., Krause, M., Lee, J., Stam, M., Steinberger, J.: The Preimage Security of Double-Block-Length Compression Functions. In: Lee, D.H., Wang, X. (eds.) ASIACRYPT 2011. LNCS, vol. 7073, pp. 233–251. Springer, Heidelberg (2011)
2. Bellare, M., Rogaway, P.: The Security of Triple Encryption and a Framework for Code-Based Game-Playing Proofs. In: Vaudenay, S. (ed.) EUROCRYPT 2006. LNCS, vol. 4004, pp. 409–426. Springer, Heidelberg (2006)
3. Blum, M., Micali, S.: How to generate cryptographically strong sequences of pseudo random bits. In: FOCS, pp. 112–117. IEEE Computer Society (1982)
4. Bogdanov, A., Knudsen, L.R., Leander, G., Standaert, F.-X., Steinberger, J., Tischhauser, E.: Key-Alternating Ciphers in a Provable Setting: Encryption Using a Small Number of Public Permutations (Extended Abstract). In: Pointcheval, D., Johansson, T. (eds.) EUROCRYPT 2012. LNCS, vol. 7237, pp. 45–62. Springer, Heidelberg (2012)
5. Gaži, P., Maurer, U.: Free-Start Distinguishing: Combining Two Types of Indistinguishability Amplification. In: Kurosawa, K. (ed.) ICITS 2009. LNCS, vol. 5973, pp. 28–44. Springer, Heidelberg (2010)
6. Hirose, S.: Some Plausible Constructions of Double-Block-Length Hash Functions. In: Robshaw, M. (ed.) FSE 2006. LNCS, vol. 4047, pp. 210–225. Springer, Heidelberg (2006)
7. Jetchev, D., Özen, O., Stam, M.: Collisions Are Not Incidental: A Compression Function Exploiting Discrete Geometry. In: Cramer, R. (ed.) TCC 2012. LNCS, vol. 7194, pp. 303–320. Springer, Heidelberg (2012)
8. Kilian, J., Rogaway, P.: How to Protect DES Against Exhaustive Key Search (an Analysis of DESX). J. Cryptology 14(1), 17–35 (2001)

9. Lee, J., Stam, M., Steinberger, J.: The Collision Security of Tandem-DM in the Ideal Cipher Model. In: Rogaway, P. (ed.) CRYPTO 2011. LNCS, vol. 6841, pp. 561–577. Springer, Heidelberg (2011)

10. Lucks, S.: A collision-resistant rate-1 double-block-length hash function. In: Biham, E., Handschuh, H., Lucks, S., Rijmen, V. (eds.) Symmetric Cryptography. No. 07021 in Dagstuhl Seminar Proceedings, Internationales Begegnungs- und Forschungszentrum für Informatik (IBFI), Schloss Dagstuhl, Germany, Dagstuhl, Germany (2007),
    http://drops.dagstuhl.de/opus/volltexte/2007/1017

11. Maurer, U.M.: Indistinguishability of Random Systems. In: Knudsen, L.R. (ed.) EURO-CRYPT 2002. LNCS, vol. 2332, pp. 110–132. Springer, Heidelberg (2002)

12. Maurer, U., Pietrzak, K.: The Security of Many-Round Luby–Rackoff Pseudo-Random Permutations. In: Biham, E. (ed.) EUROCRYPT 2003. LNCS, vol. 2656, pp. 544–561. Springer, Heidelberg (2003)

13. Maurer, U.M., Pietrzak, K.: Composition of Random Systems: When Two Weak Make One Strong. In: Naor, M. (ed.) TCC 2004. LNCS, vol. 2951, pp. 410–427. Springer, Heidelberg (2004)

14. Maurer, U.M., Pietrzak, K., Renner, R.S.: Indistinguishability Amplification. In: Menezes, A. (ed.) CRYPTO 2007. LNCS, vol. 4622, pp. 130–149. Springer, Heidelberg (2007)

15. Pietrzak, K.: Indistinguishability and Composition of Random Systems. ETH Zurich, Ph.D. thesis (2005),
    http://homepages.cwi.nl/%7Epietrzak/publications/thesis05.ps

16. Shoup, V.: Sequences of Games: A Tool for Taming Complexity in Security Proofs. Cryptology ePrint Archive, Report 2004/332 (2004), http://eprint.iacr.org/

17. Steinberger, J.P.: The Collision Intractability of MDC-2 in the Ideal-Cipher Model. In: Naor, M. (ed.) EUROCRYPT 2007. LNCS, vol. 4515, pp. 34–51. Springer, Heidelberg (2007)

18. Steinberger, J.P.: Improved security bounds for key-alternating ciphers via hellinger distance. Cryptology ePrint Archive, Report 2012/481 (2012), http://eprint.iacr.org/

RKA Security beyond the Linear Barrier: IBE, Encryption and Signatures

Mihir Bellare[1], Kenneth G. Paterson[2], and Susan Thomson[3]

[1] Department of Computer Science & Engineering,
University of California San Diego
mihir@eng.ucsd.edu
cseweb.ucsd.edu/~mihir/
[2] Information Security Group, Royal Holloway, University of London
kenny.paterson@rhul.ac.uk
www.isg.rhul.ac.uk/~kp
[3] Information Security Group, Royal Holloway, University of London
s.thomson@rhul.ac.uk

Abstract. We provide a framework enabling the construction of IBE schemes that are secure under related-key attacks (RKAs). Specific instantiations of the framework yield RKA-secure IBE schemes for sets of related key derivation functions that are non-linear, thus overcoming a current barrier in RKA security. In particular, we obtain IBE schemes that are RKA secure for sets consisting of all affine functions and all polynomial functions of bounded degree. Based on this we obtain the first constructions of RKA-secure schemes for the same sets for the following primitives: CCA-secure public-key encryption, CCA-secure symmetric encryption and Signatures. All our results are in the standard model and hold under reasonable hardness assumptions.

1 Introduction

Related-key attacks (RKAs) were first conceived as tools for the cryptanalysis of blockciphers [22,9]. However, the ability of attackers to modify keys stored in memory via tampering [13,10] raises concerns that RKAs can actually be mounted in practice. The key could be an IBE master key, a signing key of a certificate authority, or a decryption key, making RKA security important for a wide variety of primitives.

Provably achieving security against RKAs, however, has proven extremely challenging. This paper aims to advance the theory with new feasibility results showing achievability of security under richer classes of attacks than previously known across a variety of primitives.

CONTRIBUTIONS IN BRIEF. The primitive we target in this paper is IBE. RKA security for this primitive was defined by Bellare, Cash, and Miller [4]. As per the founding theoretical treatment of RKAs by Bellare and Kohno [5], the definition is parameterized by the class Φ of functions that the adversary is allowed to apply to the target key. (With no restrictions, security is unachievable.) For future reference we define a few relevant classes of functions over the space $\mathcal{S}$ of master

X. Wang and K. Sako (Eds.): ASIACRYPT 2012, LNCS 7658, pp. 331–348, 2012.

keys. The set $\Phi^c = \{\phi_c\}_{c \in \mathcal{S}}$ with $\phi_c(s) = c$ is the set of constant functions. If $\mathcal{S}$ is a group under an operation $*$ then $\Phi^{\mathrm{lin}} = \{\phi_a\}_{a \in \mathcal{S}}$ with $\phi_a(s) = a * s$ is the class of linear functions. (Here $*$ could be multiplication or addition.) If $\mathcal{S}$ is a field we let $\Phi^{\mathrm{aff}} = \{\phi_{a,b}\}_{a,b \in \mathcal{S}}$ with $\phi_{a,b}(s) = as + b$ be the class of affine functions and $\Phi^{\mathrm{poly}(d)} = \{\phi_q\}_{q \in \mathcal{S}_d[x]}$ with $\phi_q(s) = q(s)$ the class of polynomial functions, where q ranges over the set $\mathcal{S}_d[x]$ of polynomials over $\mathcal{S}$ of degree at most d. RKA security increases and is a more ambitious target as we move from Φ^{lin} to Φ^{aff} to $\Phi^{\mathrm{poly}(d)}$.

The choice of IBE as a primitive is not arbitrary. First, IBE is seeing a lot of deployment, and compromise of the master secret key would cause widespread damage, so we are well motivated to protect it against side-channel attacks. Second, IBE was shown in [4] to be an enabling primitive in the RKA domain: achieving RKA-secure IBE for any class Φ immediately yields Φ-RKA-secure CCA-PKE (CCA-secure public-key encryption) and Sig (signature) schemes. These results were obtained by noting that the CHK [12] IBE-to-CCA-PKE transform and the Naor IBE-to-Sig transform both preserve RKA security. Thus, results for IBE would immediately have wide impact.

We begin by presenting attacks showing that existing IBE schemes such as those of Boneh-Franklin [14] and Waters [25] are not RKA secure, even for Φ^{lin}. This means we must seek new designs.

We present a framework for constructing RKA-secure IBE schemes. It is an adaptation of the framework of Bellare and Cash [3] that builds RKA-secure PRFs based on key-malleable PRFs and fingerprinting. Our framework has two corresponding components. First, we require a starting IBE scheme that has a key-malleability property relative to our target class Φ of related-key deriving functions. Second, we require the IBE scheme to support what we call collision-resistant identity renaming. We provide a simple and efficient way to transform any IBE scheme with these properties into one that is Φ-RKA secure.

To exploit the framework, we must find key-malleable IBE schemes. Somewhat paradoxically, we show that the very attack strategies that broke the RKA security of existing IBE schemes can be used to show that these schemes are Φ-key-malleable, not just for $\Phi = \Phi^{\mathrm{lin}}$ but even for $\Phi = \Phi^{\mathrm{aff}}$. We additionally show that these schemes support efficient collision-resistant identity renaming. As a consequence we obtain Φ^{aff}-RKA-secure IBE schemes based on the same assumptions used to prove standard IBE security of the base IBE schemes.

From the practical perspective, the attraction of these results is that our schemes modify the known ones in a very small and local way limited only to the way identities are hashed. They thus not only preserve the efficiency of the base schemes, but implementing them would require minimal and modular software changes, so that non-trivial RKA security may be added without much increase in cost. From the theoretical perspective, the step of importance here is to be able to achieve RKA security for non-linear functions, and this without extra computational assumptions. As we will see below, linear RKAs, meaning Φ^{lin}-RKA security, has so far been a barrier for most primitives.

However, we can go further, providing a $\Phi^{\mathrm{poly}(d)}$-RKA-secure IBE scheme. Our scheme is an extension of Waters' scheme [25]. The proof is under a q-type hardness assumption that we show holds in the generic group model. The

Primitive	Linear	Affine	Polynomial
IBE	[4]+[3]	✓	✓
Sig	[4]+[3]	✓	✓
CCA-PKE	[26], [4]+[3]	✓	✓
CPA-SE	[2], [4]+[3]	[21]	[21]
CCA-SE	[4]+[3]	✓	✓*
PRF	[3]	–	–

Fig. 1. Rows are indexed by primitives. Columns are indexed by the class Φ of related-key derivation functions, $\Phi^{\mathrm{lin}}, \Phi^{\mathrm{aff}}$ and $\Phi^{\mathrm{poly}(d)}$ respectively. Entries indicate work achieving Φ-RKA security for the primitive in question. Checkmarks indicate results from this paper that bring many primitives all the way to security under polynomial RKAs in one step. The table only considers achieving the strong, adaptive notions of security from [4]; non-adaptively secure signature schemes for non-linear RKAs were provided in [21]. Note that symmetric key primitives cannot be RKA secure against constant RKD functions, so affine and polynomial RKA security for the last three rows is with respect to the RKD sets $\Phi^{\mathrm{aff}} \setminus \Phi^c$ and $\Phi^{\mathrm{poly}(d)} \setminus \Phi^c$. The "*" in the CCA-SE row is because our CCA-SE construction is insecure against RKD functions where the linear coefficient is zero, so does not achieve RKA security against the full set $\Phi^{\mathrm{poly}(d)} \setminus \Phi^c$. See the full version for details.

significance of this result is to show that for IBE we can go well beyond linear RKAs, something not known for PRFs.

As indicated above, we immediately get Φ-RKA-secure CCA-PKE and Sig schemes for any class Φ for which we obtained Φ-RKA-secure IBE schemes, and under the same assumptions. When the base IBE scheme has a further malleability property, the CCA-PKE scheme so obtained can be converted into a Φ-RKA-secure CCA-SE (CCA-secure symmetric encryption) scheme. This yields the first RKA secure schemes for the primitives Sig, CCA-PKE, and CCA-SE for non-linear RKAs, meaning beyond Φ^{lin}.

BACKGROUND AND CONTEXT. The theoretical foundations of RKA security were laid by Bellare and Kohno [5], who treated the case of PRFs and PRPs. Research then expanded to consider other primitives [20,2,21,4]. In particular, Bellare, Cash and Miller [4] provide a comprehensive treatment including strong definitions for many primitives and ways to transfer Φ-RKA security from one primitive to another.

RKA-security is finding applications beyond providing protection against tampering-based sidechannel attacks [19], including instantiating random oracles in higher-level protocols and improving efficiency [2,1].

With regard to achieving security, early efforts were able to find PRFs with proven RKA security only for limited Φ or under very strong assumptions. Eventually, using new techniques, Bellare and Cash [3] were able to present DDH-based PRFs secure against linear RKAs ($\Phi = \Phi^{\mathrm{lin}}$). But it is not clear how to take their techniques further to handle larger RKA sets Φ.

Fig. 1 summarizes the broad position. Primitives for which efforts have now been made to achieve RKA security include CPA-SE (CPA secure symmetric encryption), CCA-SE (CCA secure symmetric encryption), CCA-PKE (CCA secure public-key encryption[1]) Sig (Signatures), and IBE (CPA secure identity-based encryption). Schemes proven secure under a variety of assumptions have been provided. But the salient fact that stands out is that prior to our work, results were all for linear RKAs with the one exception of CPA-SE where a scheme secure against polynomial (and thus affine) RKAs was provided by [21].

In more detail, Bellare, Cash and Miller [4] show how to transfer RKA security from PRF to any other primitive, assuming an existing standard-secure instance of the primitive. Combining this with [3] yields DDH-based schemes secure against linear RKAs for all the primitives, indicated by a "[4]+[3]" table entry. Applebaum, Harnik and Ishai [2] present LPN and LWE-based CPA-SE schemes secure against linear RKAs. Wee [26] presents CCA-PKE secure schemes for linear RKAs. Goyal, O'Neill and Rao [21] gave a CPA-SE scheme secure against polynomial RKAs. (We note that their result statement should be amended to exclude constant RKD functions, for no symmetric primitive can be secure under these.) Wee [26] (based on a communication of Wichs) remarks that AMD codes [18] may be used to achieve RKA security for CCA-PKE, a method that extends to other primitives including IBE (but not PRF), but with current constructions of these codes [18], the results continue to be restricted to linear RKAs. We note that we are interested in the stronger, adaptive versions of the definitions as given in [4], but non-adaptively secure signature schemes for non-linear RKAs were provided in [21].

In summary, a basic theoretical question that emerges is how to go beyond linear RKAs. A concrete target here is to bring other primitives to parity with CPA-SE by achieving security for affine and polynomial RKAs. Ideally, we would like approaches that are general, meaning each primitive does not have to be treated separately. As discussed above, we are able to reach these goals with IBE as a starting point.

A CLOSER LOOK. Informally, key-malleability means that user-level private keys obtained by running the IBE scheme's key derivation algorithm $\mathcal{K}$ using a modified master secret key $\phi(s)$ (where $\phi \in \Phi$ and $s \in \mathcal{S}$, the space of master secret keys) can alternatively be computed by running $\mathcal{K}$ using the original master secret key s, followed by a suitable transformation. A collision-resistant identity renaming transform maps identities from the to-be-constructed RKA-secure IBE scheme back into identities in the starting IBE scheme in such a way as to "separate" the sets of identities coming from different values of $\phi(s)$. By modifying the starting IBE scheme to use renamed identities instead of the original ones, we obtain a means to handle otherwise difficult key extraction queries in the RKA setting.

---

[1] RKAs are interesting for symmetric encryption already in the CPA case because encryption depends on the secret key, but for public-key encryption they are only interesting for the CCA case because encryption does not depend on the secret key.

To show that the framework is applicable to the Boneh-Franklin [14] and Waters [25] IBE schemes with $\Phi = \Phi^{\text{aff}}$ (the space of master keys here is $\mathbb{Z}_p$), we exploit specific algebraic properties of the starting IBE schemes. In the Waters case, we obtain an efficient, Φ^{aff}-RKA-secure IBE scheme in the standard model, under the Decisional Bilinear Diffie-Hellman (DBDH) assumption. In the Boneh-Franklin case, we obtain an efficient, Φ^{aff}-RKA-secure IBE scheme under the Bilinear Diffie-Hellman (BDH) assumption with more compact public keys at the expense of working in the Random Oracle Model. Going further, we exhibit a simple modification of the Waters scheme which allows us to handle related key attacks for $\Phi^{\text{poly}(d)}$, this being the set of polynomial functions of bounded degree d. This requires the inclusion of an extra $2d - 2$ elements in the master public key, and a modified, q-type hardness assumption. We show that this assumption holds in the generic group model.

Applying the results of [4] to these IBE schemes, we obtain the first constructions of RKA-secure CCA-PKE and signature schemes for Φ^{aff} and $\Phi^{\text{poly}(d)}$. Again, our schemes are efficient and our results hold in the standard model under reasonable hardness assumptions. The CCA-PKE schemes, being derived via the CHK transform [12], just involve the addition of a one-time signature and verification key to the IBE ciphertexts and so incur little additional overhead for RKA security. As an auxiliary result that improves on the corresponding result of [4], we show in the full version [6] that the more efficient MAC-based transform of [15,12] can be used in place of the CHK transform. The signature schemes arise from the Naor trick, wherein identities are mapped to messages, IBE user private keys are used as signatures, and a trial encryption and decryption on a random plaintext are used to verify the correctness of a signature. This generic construction can often be improved by tweaking the verification procedure, and the same is true here: for example, for the Waters-based signature scheme, we can base security on the CDH assumption instead of DBDH, and can achieve more efficient verification. We stress that our signature schemes are provably unforgeable in a fully adaptive related-key setting, in contrast to the recently proposed signatures in [21].

Note that RKA-secure PRFs for sets Φ^{aff} and $\Phi^{\text{poly}(d)}$ cannot exist, since these sets contain constant functions, and we know that no PRF can be RKA-secure in this case [5]. Thus we are able to show stronger results for IBE, CCA-PKE and Sig than are possible for PRF. Also, although Bellare, Cash and Miller [4] showed that Φ-RKA security for PRF implies Φ-RKA security for Sig and CCA-PKE, the observation just made means we cannot use this result to get Φ^{aff} or $\Phi^{\text{poly}(d)}$ RKA-secure IBE, CCA-PKE or Sig schemes. This provides further motivation for starting from RKA-secure IBE as we do, rather than from RKA-secure PRF.

Finally we note that even for linear RKAs where IBE schemes were known via [4]+[3], our schemes are significantly more efficient.

FURTHER CONTRIBUTIONS. In the full version [6], as a combination of the results of [4] and [24], we provide definitions for RKA security in the joint security setting, where the same key pair is used for both signature and encryption functions, and show that a Φ-RKA-secure IBE scheme can be used to build a Φ-RKA and jointly secure combined signature and encryption scheme. This construction can be instantiated using any of our specific IBE schemes, by which we obtain the first concrete jointly secure combined signature and encryption schemes for the RKA setting.

We also show in [6] how to adapt the KEM-DEM (or hybrid encryption) paradigm to the RKA setting, and describe a highly efficient, Φ^{aff}-RKA-secure CCA-KEM that is inspired by our IBE framework and is based on the scheme of Boyen, Mei and Waters [17]. Our CCA-KEM's security rests on the hardness of the DBDH problem for asymmetric pairings $e : \mathbb{G}_1 \times \mathbb{G}_2 \to \mathbb{G}_T$; its ciphertexts consist of 2 group elements (one in $\mathbb{G}_1$ and one in $\mathbb{G}_2$), public keys are 3 group elements (two in $\mathbb{G}_2$ and one in $\mathbb{G}_T$), encryption is pairing-free, and the decryption cost is dominated by 3 pairing operations.

The final contribution (also in [6]) is an extension of our framework that lets us build an RKA-secure CCA-SE scheme from any IBE scheme satisfying an additional master public key malleability property. Such an IBE scheme, when subjected to our transformation, meets a notion of strong Φ-RKA security [4] where the challenge encryption is also subject to RKA. Applying the CHK transform gives a strong Φ-RKA-secure CCA-PKE scheme which can be converted into a Φ-RKA-secure CCA-SE scheme in the natural way.

PAPER ORGANIZATION. Section 2 contains preliminaries, Section 3 describes some IBE schemes and RKA attacks on them, while Section 4 presents our framework for constructing RKA-secure IBE schemes. Section 5 applies the framework to specific schemes, and sketches the CCA-PKE and signature schemes that result from applying the techniques of [4].

2   Preliminaries

NOTATION. For sets X, Y let $\mathsf{Fun}(X, Y)$ be the set of all functions mapping X to Y. If S is a set then $|S|$ denotes its size and $s \leftarrow_{\$} S$ the operation of picking a random element of S and denoting it by s. Unless otherwise indicated, an algorithm may be randomized. An adversary is an algorithm. By $y \leftarrow_{\$} A(x_1, x_2, \dots)$ we denote the operation of running A on inputs $x_1, x_2, \dots$ and letting y denote the outcome. We denote by $[A(x_1, x_2, \dots, x_n)]$ the set of all possible outputs of A on inputs $x_1, x_2, \dots, x_n$.

GAMES. Some of our definitions and proofs are expressed through code-based games [8]. Recall that such a game consists of an INITIALIZE procedure, procedures to respond to adversary oracle queries, and a FINALIZE procedure. A game G is executed with an adversary A as follows. First, INITIALIZE executes and its output is the input to A. Then A executes, its oracle queries being answered by the corresponding procedures of G. When A terminates, its output becomes the input to the FINALIZE procedure. The output of the latter is called the output of the game. We let G^A denote the event that this game output takes value true. The running time of an adversary, by convention, is the worst case time for the execution of the adversary with any of the games defining its security, so that the time of the called game procedures is included.

RKD FUNCTIONS AND CLASSES. We say that ϕ is a related-key deriving (RKD) function over a set $\mathcal{S}$ if $\phi \in \mathsf{Fun}(\mathcal{S}, \mathcal{S})$. We say that Φ is a class of RKD functions over $\mathcal{S}$ if $\Phi \subseteq \mathsf{Fun}(\mathcal{S}, \mathcal{S})$ and $\mathsf{id} \in \Phi$ where id is the identity function on $\mathcal{S}$. In our constructs, $\mathcal{S}$ will have an algebraic structure, such as being a group, ring or field. In the last case, for $a, b \in \mathcal{S}$ we define $\phi_b^+, \phi_a^*, \phi_{a,b}^{\mathrm{aff}} \in \mathsf{Fun}(\mathcal{S}, \mathcal{S})$ via $\phi_b^+(s) = s + b$, $\phi_a^*(s) = as$, and $\phi_{a,b}^{\mathrm{aff}}(s) = as + b$ for all $s \in \mathcal{S}$. For a polynomial q over field $\mathcal{S}$, we define $\phi_q^{\mathrm{poly}}(s) = q(s)$ for all $s \in \mathcal{S}$. We let $\Phi^+ = \{\, \phi_b^+ \,:\, b \in \mathcal{S} \,\}$ be the class of additive RKD functions, $\Phi^* = \{\, \phi_a^* \,:\, a \in \mathcal{S} \,\}$ be the class of multiplicative RKD functions, $\Phi^{\mathrm{aff}} = \{\, \phi_{a,b}^{\mathrm{aff}} \,:\, a, b \in \mathcal{S} \,\}$ the class of affine RKD functions, and for any fixed positive integer d, we let $\Phi^{\mathrm{poly}(d)} = \{\, \phi_q^{\mathrm{poly}} \,:\, \deg q \leq d \,\}$ be the set of polynomial RKD functions of bounded degree d.

If $\phi \neq \phi'$ are distinct functions in a class Φ there is of course by definition an s such that $\phi(s) \neq \phi'(s)$, but there could also be keys s on which $\phi(s) = \phi'(s)$. We say that a class Φ is claw-free if the latter does not happen, meaning for all distinct $\phi \neq \phi'$ in Φ we have $\phi(s) \neq \phi'(s)$ for *all* $s \in \mathcal{S}$. With the exception of [21], all previous constructions of Φ-RKA-secure primitives with proofs of security have been for claw-free classes [5,23,20,3,4,26]. In particular, key fingerprints are defined in [3] in such a way that their assumption of a Φ-key fingerprint automatically implies that Φ is claw-free.

IBE SYNTAX. We specify an IBE scheme $\mathcal{IBE} = (\mathcal{S}, \mathcal{P}, \mathcal{K}, \mathcal{E}, \mathcal{D})$ by first specifying a non-empty set $\mathcal{S}$ called the *master-key space* from which the master secret key s is drawn at random. The master public key $\pi \leftarrow \mathcal{P}(s)$ is then produced by applying to s a deterministic master public key generation algorithm $\mathcal{P}$. A decryption key for an identity u is produced via $dk_u \leftarrow_{\$} \mathcal{K}(s, u)$. A ciphertext C encrypting a message M for u is generated via $C \leftarrow_{\$} \mathcal{E}(\pi, u, M)$. A ciphertext C is deterministically decrypted via $M \leftarrow \mathcal{D}(dk, C)$. Correctness requires that $\mathcal{D}(\mathcal{K}(s, u), \mathcal{E}(\pi, u, M)) = M$ with probability one for all $M \in \mathsf{MSp}$ and all $u \in \mathsf{USp}$ where $\mathsf{MSp}, \mathsf{USp}$ are, respectively, the message and identity spaces associated to $\mathcal{IBE}$.

The usual IBE syntax specifies a single parameter generation algorithm that produces s, π together, and although there is of course a space from which the master secret key is drawn, it is not explicitly named. But RKD functions will have domain the space of master keys of the IBE scheme, which is why it is convenient in our context to make it explicit in the syntax. Saying the master public key is a deterministic function of the master secret key is not strictly necessary for us, but it helps make some things a little simpler and is true in all known schemes, so we assume it.

We make an important distinction between parameters and the master public key, namely that the former may not depend on s while the latter might. Parameters will be groups, group generators, pairings and the like. They will be fixed and available to all algorithms without being named as explicit inputs.

proc INITIALIZE
$s \leftarrow_{\$} \mathcal{S}; \pi \leftarrow \mathcal{P}(s)$
$b \leftarrow_{\$} \{0, 1\}$
$u^{*} \leftarrow \perp; I \leftarrow \emptyset$
Ret π

proc FINALIZE(b')
Ret $(b = b')$

proc KD(ϕ, u)
$s' \leftarrow \phi(s)$
If $(s' = s)$ $I \leftarrow I \cup \{u\}$
If $(u^{*} \in I)$ Ret $\perp$
Ret $dk \leftarrow_{\$} \mathcal{K}(s', u)$

proc LR(u, M_0, M_1)
If $(|M_0| \neq |M_1|)$ Ret $\perp$
$u^{*} \leftarrow u$
If $(u^{*} \in I)$ Ret $\perp$
Ret $C \leftarrow_{\$} \mathcal{E}(\pi, u^{*}, M_b)$

Fig. 2. Game IBE defining Φ-RKA-security of IBE scheme $\mathcal{IBE} = (\mathcal{S}, \mathcal{P}, \mathcal{K}, \mathcal{E}, \mathcal{D})$

$\mathcal{P}(s)$:
$\pi \leftarrow g^{s}$
Ret π

$\mathcal{K}(s, u)$:
$dk \leftarrow H_1(u)^{s}$
Ret dk

$\mathcal{E}(\pi, u, M)$:
$t \leftarrow_{\$} \mathbb{Z}_p$
$C_1 \leftarrow g^{t}$
$C_2 \leftarrow H_2(e(\pi, H_1(u))^{t}) \oplus M$
Ret (C_1, C_2)

$\mathcal{D}(dk, C)$:
$M \leftarrow C_2 \oplus H_2(e(dk, C_1))$
Ret M

$\mathcal{P}(s)$:
$\pi \leftarrow g^{s}$
Ret π

$\mathcal{K}(s, u)$:
$r \leftarrow_{\$} \mathbb{Z}_p$
$dk_1 \leftarrow g_1^{s} \cdot H(u)^{r}$
$dk_2 \leftarrow g^{r}$
Ret (dk_1, dk_2)

$\mathcal{E}(\pi, u, M)$:
$t \leftarrow_{\$} \mathbb{Z}_p$
$C_1 \leftarrow g^{t}$
$C_2 \leftarrow H(u)^{t}$
$C_3 \leftarrow e(\pi, g_1)^{t} \cdot M$
Ret (C_1, C_2, C_3)

$\mathcal{D}(dk, C)$:
$M \leftarrow C_3 \cdot \frac{e(dk_2, C_2)}{e(dk_1, C_1)}$
Ret M

Fig. 3. Boneh-Franklin IBE scheme on the left, Waters IBE scheme on the right

RKA-SECURE IBE. We define Φ-RKA security of IBE schemes following [4]. Game IBE of Fig. 2 is associated to $\mathcal{IBE} = (\mathcal{S}, \mathcal{P}, \mathcal{K}, \mathcal{E}, \mathcal{D})$ and a class Φ of RKD functions over $\mathcal{S}$. An adversary is allowed only one query to LR. Let $\mathbf{Adv}_{\mathcal{IBE}, \Phi}^{\mathrm{ibe-rka}}(A)$ equal $2 \Pr[\mathrm{IBE}^{A}] - 1$. A feature of the definition we draw attention to is that the key derivation oracle KD refuses to act only when the identity it is given matches the challenge one *and* the derived key equals the real one. This not only creates a strong security requirement but one that is challenging to achieve because a simulator, not knowing s, cannot check whether or not the IBE adversary succeeded. This difficulty is easily resolved if Φ is claw-free but not otherwise. We consider this particular RKA security definition as, in addition to its strength, it is the level of RKA security required of an IBE scheme so that application of the CHK and Naor transforms results in RKA-secure CCA-PKE and signature schemes.

3   Existing IBE Schemes and RKA Attacks on Them

The algorithms of the Boneh-Franklin BasicIdent IBE scheme [14] are given in Figure 3. The parameters of the scheme are groups $\mathbb{G}_1, \mathbb{G}_T$ of prime order p, a symmetric pairing $e : \mathbb{G}_1 \times \mathbb{G}_1 \to \mathbb{G}_T$, a generator g of $\mathbb{G}_1$ and hash functions $H_1 : \{0, 1\}^{*} \to \mathbb{G}_1$, $H_2 : \mathbb{G}_T \to \{0, 1\}^{n}$ which are modeled as random oracles in the security analysis. Formally, these are output by a pairing parameter generator on input 1^k. This scheme is IND-CPA secure in the usual model for IBE security, under the Bilinear Diffie-Hellman (BDH) assumption.

The algorithms of the Waters IBE scheme [25] are also given in Figure 3. The parameters of the scheme are groups $\mathbb{G}_1, \mathbb{G}_T$ of prime order p, a symmetric pairing $e : \mathbb{G}_1 \times \mathbb{G}_1 \to \mathbb{G}_T$, generators g, g_1 of $\mathbb{G}_1$ and group elements $h_0, \ldots, h_n \in \mathbb{G}_1$ specifying the hash function $H(u) = h_0 \prod_{i \in u} h_i$. The Waters IBE scheme is also IND-CPA secure in the usual model for IBE security, under the DBDH assumption.

The Waters IBE scheme is not RKA secure if Φ includes a function $\phi_a^*(s) = as$. A call to the key derivation oracle with any such ϕ yields a user secret key $(dk_1, dk_2) = (g_1^{as} \cdot H(u)^r, g^r)$. Raising this to a^{-1} gives $(dk_1', dk_2') = (g_1^s \cdot H(u)^{ra^{-1}}, g^{ra^{-1}})$, so that (dk_1', dk_2') is a user secret key for identity u under the original master secret key with randomness $r' = ra^{-1}$. An RKA adversary can thus obtain the user secret key for any identity of his choosing and hence break the RKA security of the Waters scheme. A similar attack applies to the Boneh-Franklin scheme.

4   Framework for Deriving RKA-Secure IBE Schemes

In the previous section we saw that the Boneh-Franklin and Waters schemes are not RKA secure. Here we will show how to modify these and other schemes to be RKA secure by taking advantage, in part, of the very algebra that leads to the attacks. We describe a general framework for creating RKA-secure IBE schemes and then apply it obtain several such schemes.

We target a very particular type of framework, one that allows us to reduce RKA security of a modified IBE scheme directly to the normal IBE security of a base IBE scheme. This will allow us to exploit known results on IBE in a blackbox way and avoid re-entering the often complex security proofs of the base IBE schemes.

KEY-MALLEABILITY. We say that an IBE scheme $I\mathcal{BE} = (\mathcal{S}, \mathcal{P}, \mathcal{K}, \mathcal{E}, \mathcal{D})$ is Φ-key-malleable if there is an algorithm T, called the key simulator, which, given π, an identity u, a decryption key $dk' \leftarrow_\$ \mathcal{K}(s, u)$ for u under s and an RKD function $\phi \in \Phi$, outputs a decryption key dk for u under master secret key $\phi(s)$ that is distributed identically to the output of $\mathcal{K}(\phi(s), u)$. The formalization takes a little more care for in talking about two objects being identically distributed one needs to be precise about relative to what other known information this is true. A simple and rigorous definition here can be made using games. We ask that

$$\Pr[\text{KMReal}_{I\mathcal{BE}, \Phi}^M] = \Pr[\text{KMSim}_{I\mathcal{BE}, \Phi, T}^M]$$

for all (not necessarily computationally bounded) adversaries M, where the games are as follows. The INITIALIZE procedure of both picks s at random from $\mathcal{S}$ and returns $\pi \leftarrow \mathcal{P}(s)$ to the adversary. In game KMReal$_{I\mathcal{BE}, \Phi}$, oracle KD(ϕ, u) returns $dk \leftarrow_\$ \mathcal{K}(\phi(s), u)$ but in game KMSim$_{I\mathcal{BE}, \Phi, T}$ it lets $dk' \leftarrow_\$ \mathcal{K}(s, u)$ and returns $T(\pi, u, dk', \phi)$. There are no other oracles, and FINALIZE(b') returns ($b' = 1$).

USING KM. Intuitively, key-malleability allows us to simulate a Φ-RKA adversary via a normal adversary and would thus seem to be enough to prove Φ-RKA security of IBE based on its normal security. Let us see how this argument goes and then see the catches that motivate a transformation of the scheme via collision-resistant identity renaming. Letting $\overline{A}$ be an adversary attacking the Φ-RKA security of IBE, we aim to build an adversary A such that

$$\mathbf{Adv}^{ibe-rka}_{IBE,\Phi}(\overline{A}) \leq \mathbf{Adv}^{ibe}_{IBE}(A) . \tag{1}$$

On input π, adversary A runs $\overline{A}(\pi)$. When the latter makes a KD(ϕ, u) query, A lets $dk \leftarrow$ KD(id, u), where KD is A's own key derivation oracle. It then lets $\overline{dk} \leftarrow T(\pi, u, dk, \phi)$ and returns $\overline{dk}$ to $\overline{A}$. Key-malleability tells us that $\overline{dk}$ is distributed identically to an output of KD(ϕ, u), so the response provided by A is perfectly correct. When $\overline{A}$ makes a LR(u, M_0, M_1) query, A lets $C \leftarrow$ LR(u, M_0, M_1) and returns C to A. Finally when $\overline{A}$ halts with output a bit b', adversary A does the same.

The simulation seems perfect, so we appear to have established Equation (1). What's the catch? The problem is avoiding *challenge key derivation*. Suppose $\overline{A}$ made a KD(ϕ, u) query for a ϕ such that $\phi(s) \neq s$; then made a LR(u, M_0, M_1) query; and finally, given C, correctly computed b. It would win its game, because the condition $\phi(s) \neq s$ means that identity u may legitimately be used both in a key derivation query and in the challenge LR query. But our constructed adversary A, in the simulation, would make query KD(id, u) to answer $\overline{A}$'s KD(ϕ, u) query, and then make query LR(u, M_0, M_1). A would thus have queried the challenge identity u to the key-extraction oracle and would not win.

This issue is dealt with by transforming the base scheme via what we call identity renaming, so that Φ-RKA security of the transformed scheme can be proved based on the Φ-key-malleability of the base scheme.

IDENTITY RENAMING. Renaming is a way to map identities in the new scheme back to identities of the given, base scheme. Let us now say how renaming works more precisely and then define the modified scheme.

Let $IBE = (\mathcal{S}, \mathcal{P}, \mathcal{K}, \mathcal{E}, \mathcal{D})$ denote the given, base IBE scheme, and let USp be its identity space. A renaming scheme is a pair (SI, PI) of functions where SI: $\mathcal{S} \times \overline{USp} \rightarrow$ USp and PI: $[\mathcal{P}(\mathcal{S})] \times \overline{USp} \times \Phi \rightarrow$ USp where $\overline{USp}$, implicitly specified by the renaming scheme, will be the identity space of the new scheme we will soon define. The first function SI, called the secret renaming function, uses the master secret key, while its counterpart public renaming function PI uses the master public key. We require that SI($\phi(s), \overline{u}$) = PI($\pi, \overline{u}, \phi$) for all $s \in \mathcal{S}$, all $\pi \in [\mathcal{P}(s)]$, all $\overline{u} \in \overline{USp}$ and all $\phi \in \Phi$. This *compatibility condition* says that the two functions arrive, in different ways, at the same outcome.

THE TRANSFORM. The above is all we need to specify our Identity Renaming Transform **IRT** that maps a base IBE scheme $IBE = (\mathcal{S}, \mathcal{P}, \mathcal{K}, \mathcal{E}, \mathcal{D})$ to a new IBE scheme $\overline{IBE} = (\mathcal{S}, \mathcal{P}, \overline{\mathcal{K}}, \overline{\mathcal{E}}, \mathcal{D})$. As the notation indicates, the master key space, master public key generation algorithm and decryption algorithm are unchanged. The other algorithms are defined by

$$\overline{\mathcal{K}}(s, \overline{u}) = \mathcal{K}(s, \mathrm{SI}(s, \overline{u})) \quad \text{and} \quad \overline{\mathcal{E}}(\pi, \overline{u}, M) = \mathcal{E}(\pi, \mathrm{PI}(\pi, \overline{u}, \mathsf{id}), M) .$$

We clarify that algorithms of the new IBE scheme do not, and cannot, have as input the RKD functions ϕ used by the attacker. We are defining an IBE scheme, and algorithm inputs must follow the syntax of IBE schemes. When the new encryption algorithm invokes PI, it sets ϕ to the identity function id. (Looking ahead, the simulation will call the renaming functions with ϕ emanating from the adversary attacking the new IBE scheme.) The key derivation algorithm has s but not π (recall we cannot give it π because otherwise it becomes subject to the RKA) and thus uses the secret renaming function. On the other hand the encryption algorithm has π but obviously not s and thus uses the public renaming function. This explains why we need two, compatible renaming functions. The new scheme has the same message space as the old one. Its identity space is inherited from the renaming scheme, being the space $\overline{\mathsf{USp}}$ from which the renaming functions draw their identity inputs.

The above compatibility requirement implies that $\mathrm{SI}(s, \overline{u}) = \mathrm{PI}(\pi, \overline{u}, \mathsf{id})$. From this it follows that $\overline{\mathcal{IBE}}$ preserves the correctness of $\mathcal{IBE}$. We now go on to specifying properties of the base IBE scheme and the renaming functions that suffice to prove Φ-RKA security of the new scheme.

A trivial renaming scheme is obtained by setting $\mathrm{SI}(s, \overline{u}) = \overline{u} = \mathrm{PI}(\pi, \overline{u}, \phi)$. This satisfies the compatibility condition. However, the transformed IBE scheme $\overline{\mathcal{IBE}}$ ends up identical to the base $\mathcal{IBE}$ and thus this trivial renaming cannot aid in getting security. We now turn to putting a non-trivial condition on the renaming scheme that we will show suffices.

COLLISION-RESISTANCE. The renaming scheme $(\mathrm{SI}, \mathrm{PI})$ will be required to have a collision-resistance property. In its simplest and strongest form the requirement is that

$$(\phi(s), \overline{u}_1) \neq (s, \overline{u}_2) \quad \Rightarrow \quad \mathrm{SI}(\phi(s), \overline{u}_1) \neq \mathrm{SI}(s, \overline{u}_2)$$

for all $s \in \mathcal{S}$, all $\overline{u}_1, \overline{u}_2 \in \overline{\mathsf{USp}}$ and all $\phi \in \Phi$. This *statistical collision-resistance* will be enough to prove that $\overline{\mathcal{IBE}}$ is Φ-RKA secure if $\mathcal{IBE}$ is Φ-key-malleable (cf. Theorem 1). We will now see how this goes. Then we will instantiate these ideas to get concrete Φ-RKA-secure schemes for many interesting classes Φ including Φ^{aff} and $\Phi^{\mathrm{poly}(d)}$.

Theorem 1. *Let $\mathcal{IBE} = (\mathcal{S}, \mathcal{P}, \mathcal{K}, \mathcal{E}, \mathcal{D})$ be a Φ-key-malleable IBE scheme with key simulator T. Let $\overline{\mathcal{IBE}} = (\mathcal{S}, \mathcal{P}, \overline{\mathcal{K}}, \overline{\mathcal{E}}, \mathcal{D})$ be obtained from $\mathcal{IBE}$ and renaming scheme $(\mathrm{SI}, \mathrm{PI})$ via the transform **IRT** described above. Assume the renaming scheme is statistically collision-resistant. Let $\overline{A}$ be a Φ-RKA adversary against $\overline{\mathcal{IBE}}$ that makes q key derivation queries. Then there is an adversary A making q key derivation queries such that*

$$\mathbf{Adv}^{\mathrm{ibe-rka}}_{\overline{\mathcal{IBE}}, \Phi}(\overline{A}) \leq \mathbf{Adv}^{\mathrm{ibe}}_{\mathcal{IBE}}(A) . \tag{2}$$

proc INITIALIZE $/\!\!/ \mathrm{G}_0$
000 $s \leftarrow_\$ \mathcal{S}$; $\pi \leftarrow \mathcal{P}(s)$
001 $b \leftarrow_\$ \{0,1\}$; $\overline{u}^* \leftarrow \perp$
002 $\overline{I} \leftarrow \emptyset$
003 Ret π

proc INITIALIZE $/\!\!/ \mathrm{G}_1, \mathrm{G}_2, \mathrm{G}_3$
100 $s \leftarrow_\$ \mathcal{S}$; $\pi \leftarrow \mathcal{P}(s)$
101 $b \leftarrow_\$ \{0,1\}$; $u^* \leftarrow \perp$
102 $I \leftarrow \emptyset$
103 Ret π

proc KD$(\phi, \overline{u})$ $/\!\!/ \mathrm{G}_0$
010 $s' \leftarrow \phi(s)$
011 If $(s' = s)$ $\overline{I} \leftarrow \overline{I} \cup \{\overline{u}\}$
012 If $(\overline{u}^* \in \overline{I})$ Ret $\perp$
013 $u \leftarrow \mathrm{SI}(s', \overline{u})$
014 Ret $\overline{dk} \leftarrow_\$ \mathcal{K}(s', u)$

proc KD$(\phi, \overline{u})$ $/\!\!/ \mathrm{G}_1$
110 $s' \leftarrow \phi(s)$
111 $u \leftarrow \mathrm{SI}(s', \overline{u})$
112 $I \leftarrow I \cup \{u\}$
113 If $(u^* \in I)$ Ret $\perp$
114 Ret $\overline{dk} \leftarrow_\$ \mathcal{K}(s', u)$

proc KD$(\phi, \overline{u})$ $/\!\!/ \mathrm{G}_2$
210 $u \leftarrow \mathrm{PI}(\pi, \overline{u}, \phi)$
211 $I \leftarrow I \cup \{u\}$
212 If $(u^* \in I)$ Ret $\perp$
213 Ret $\overline{dk} \leftarrow_\$ \mathcal{K}(\phi(s), u)$

proc KD$(\phi, \overline{u})$ $/\!\!/ \mathrm{G}_3$
310 $u \leftarrow \mathrm{PI}(\pi, \overline{u}, \phi)$
311 $I \leftarrow I \cup \{u\}$
312 If $(u^* \in I)$ Ret $\perp$
313 $dk \leftarrow_\$ \mathcal{K}(s, u)$
314 Ret $\overline{dk} \leftarrow T(\pi, u, dk, \phi)$

proc LR$(\overline{u}, M_0, M_1)$ $/\!\!/ \mathrm{G}_0$
020 If $(|M_0| \neq |M_1|)$ Ret $\perp$
021 $\overline{u}^* \leftarrow \overline{u}$
022 If $(\overline{u}^* \in \overline{I})$ Ret $\perp$
023 $u^* \leftarrow \mathrm{SI}(s, \overline{u}^*)$
024 Ret $C \leftarrow_\$ \mathcal{E}(\pi, u^*, M_b)$

proc LR$(\overline{u}, M_0, M_1)$ $/\!\!/ \mathrm{G}_1$
120 If $(|M_0| \neq |M_1|)$ Ret $\perp$
121 $u^* \leftarrow \mathrm{SI}(s, \overline{u})$
122 If $(u^* \in I)$ Ret $\perp$
123 Ret $C \leftarrow_\$ \mathcal{E}(\pi, u^*, M_b)$

proc LR$(\overline{u}, M_0, M_1)$ $/\!\!/ \mathrm{G}_2, \mathrm{G}_3$
220 If $(|M_0| \neq |M_1|)$ Ret $\perp$
221 $u^* \leftarrow \mathrm{PI}(\pi, \overline{u}, \mathrm{id})$
222 If $(u^* \in I)$ Ret $\perp$
223 Ret $C \leftarrow_\$ \mathcal{E}(\pi, u^*, M_b)$

proc FINALIZE(b') $/\!\!/ \mathrm{All}$
030 Ret $(b = b')$

Fig. 4. Games for proof of Theorem 1

Furthermore, the running time of A is that of $\overline{A}$ plus the time for q executions of T and $q + 1$ executions of PI.

Proof (Theorem 1). Consider the games of Fig. 4. Game G_0 is written to be equivalent to game $\mathrm{IBE}_{\overline{\mathcal{IBE}}}$, so that

$$\mathbf{Adv}_{\overline{\mathcal{IBE}}, \Phi}^{\mathrm{ibe-rka}}(\overline{A}) = 2 \Pr[\mathrm{G}_0^{\overline{A}}] - 1 . \tag{3}$$

In answering a KD$(\phi, \overline{u})$ query, G_0 must use the key-generation algorithm $\overline{\mathcal{K}}$ of the new scheme $\overline{\mathcal{IBE}}$ but with master secret key $s' = \phi(s)$. From the definition of $\overline{\mathcal{K}}$, it follows that not only is the key-generation at line 014 done under s', but also the identity renaming at line 013. LR, correspondingly, should use $\overline{\mathcal{E}}$, and thus the public renaming function PI. The compatibility property however allows us at line 023 to use SI instead. This will be useful in exploiting statistical collision-resistance in the next step, after which we will revert back to PI.

The adversary A we aim to construct will not know s. A central difficulty in the simulation is thus lines 011, 012 of G_0 where the response provided to $\overline{A}$ depends on the result of a test involving s, a test that A cannot perform. Before we can design A we must get rid of this test. Statistical collision-resistance is what will allow us to do so. KD of game G_1 moves the identity renaming up before the list of queried identities is updated to line 111 and then, at line 112, adds the transformed identity to the list. LR is likewise modified so its test now involves the transformed (rather than original) identities. We claim this makes no difference, meaning

$$\Pr[\mathrm{G}_0^{\overline{A}}] = \Pr[\mathrm{G}_1^{\overline{A}}] . \tag{4}$$

Indeed, statistical collision-resistance tell us that $(s', \overline{u}) = (s, \overline{u}^*)$ iff $\mathrm{SI}(s', \overline{u}) = \mathrm{SI}(s, \overline{u}^*)$. This means that lines 011, 012 and lines 112, 113 are equivalent.

Compatibility is invoked to use PI in place of SI in both KD and in LR in G_2, so that

$$\Pr[G_1^{\overline{A}}] \;=\; \Pr[G_2^{\overline{A}}] \,. \tag{5}$$

Rather than use s' for key generation as at 213, G_3 uses s at 313 and then applies the key simulator T. We claim the key-malleability implies

$$\Pr[G_2^{\overline{A}}] \;=\; \Pr[G_3^{\overline{A}}] \,. \tag{6}$$

To justify this we show that there is an adversary M such that

$$\Pr[\mathrm{KMReal}_{\mathcal{IBE}, \Phi}^M] \;=\; \Pr[G_2^{\overline{A}}] \quad \text{and} \quad \Pr[\mathrm{KMSim}_{\mathcal{IBE}, \Phi, T}^M] \;=\; \Pr[G_3^{\overline{A}}] \,.$$

Adversary M, on input π, begins with the initializations $u^* \leftarrow \perp$; $I \leftarrow \emptyset$; $b \leftarrow\!\!{}_\$ \{0, 1\}$ and then runs $\overline{A}$ on input π. When $\overline{A}$ makes a $\mathrm{KD}(\phi, \overline{u})$ query, M does the following:

$$u \leftarrow \mathrm{PI}(\pi, \overline{u}, \phi) \,;\; I \leftarrow I \cup \{u\} \,;\; \text{If } (u^* \in I) \text{ Ret } \perp \,;\; \overline{dk} \leftarrow \mathrm{KD}(\phi, u).$$

If M is playing game KMReal then its KD oracle will behave as line 213 in game G_2, while if M is playing game KMSim its KD oracle will behave as lines 313, 314 in game G_3. When $\overline{A}$ makes its $\mathrm{LR}(\overline{u}, M_0, M_1)$ query M sets $u^* \leftarrow \mathrm{PI}(\pi, \overline{u}, \mathrm{id})$ and checks if $u^* \in I$, returning $\perp$ if so. M then computes $C \leftarrow\!\!{}_\$ \mathcal{E}(\pi, u^*, M_b)$ which it returns to $\overline{A}$. When $\overline{A}$ halts with output b', M returns the result of $(b' = b)$. If M is playing game KMReal then game G_2 is perfectly simulated, while if M is playing KMSim then game G_3 is perfectly simulated, so M returns 1 with the same probability that $\overline{A}$ wins in each case and by the key-malleability of $\mathcal{IBE}$ Equation (6) holds.

Finally, we design A so that

$$\mathbf{Adv}_{\mathcal{IBE}}^{\mathrm{ibe}}(A) \;=\; 2\Pr[G_3^{\overline{A}}] - 1 \,. \tag{7}$$

On input π, adversary A runs $\overline{A}(\pi)$. When the latter makes a $\mathrm{KD}(\phi, \overline{u})$ query, A does the following:

$$u \leftarrow \mathrm{PI}(\pi, \overline{u}, \phi) \,;\; dk \leftarrow \mathrm{KD}(\mathrm{id}, u) \,;\; \overline{dk} \leftarrow T(\pi, u, dk, \phi).$$

It then returns $\overline{dk}$ to $\overline{A}$. The KD invoked in this code is A's own oracle. Compatibility tells us that $u = \mathrm{SI}(\phi(s), \overline{u})$ and thus from the definition of $\overline{\mathcal{IBE}}$, the response to $\overline{A}$'s query is distributed according to $\mathcal{K}(\phi(s), u)$. But key-malleability then tells us that $\overline{dk}$ is distributed identically to this, so the response provided by A is perfectly correct. When $\overline{A}$ makes a $\mathrm{LR}(\overline{u}, M_0, M_1)$ query, A does the following:

$$u \leftarrow \mathrm{PI}(\pi, \overline{u}, \mathrm{id}) \,;\; C \leftarrow \mathrm{LR}(u, M_0, M_1).$$

It then returns C to $\overline{A}$. The LR invoked in this code is A's own oracle. The definition of $\overline{\mathcal{IBE}}$ implies that the response provided by A is again perfectly correct. Finally when $\overline{A}$ halts with output a bit b', adversary A does the same.

5   Applying the Framework

AFFINE RKD FUNCTIONS FOR BONEH-FRANKLIN AND WATERS. We show how
the framework can be instantiated with the IBE schemes of Boneh-Franklin and
Waters to achieve IBE schemes secure against affine related-key attacks. First
we look at key-malleability. Keys in the Boneh-Franklin IBE scheme are of the
form $dk' = H_1(u)^s$, so the algorithm T is as follows:

$$T(\pi, u, dk', \phi_{a,b}): \ dk \leftarrow dk'^a \cdot H_1(u)^b; \ \text{Ret} \ dk$$

The output of T is a valid key for user u under master secret key $\phi_{a,b}(s)$, since:
$dk'^a \cdot H_1(u)^b = H_1(u)^{sa} \cdot H_1(u)^b = H_1(u)^{as+b}$. Since the key derivation algorithm
is deterministic, the keys output by T are distributed identically to the keys
output by $\mathcal{K}(\phi(s), u)$, and so the Boneh-Franklin IBE scheme is key-malleable.

Keys in the Waters IBE scheme are of the form $(dk'_1, dk'_2) = (g_1^s \cdot H(u)^r, g^r)$
for some r in $\mathbb{Z}_p$, so the algorithm T is as follows:

$$T(\pi, u, dk', \phi_{a,b}):$$
If $(a = 0)$ then $r \leftarrow_{\$} \mathbb{Z}_p$; $dk_1 \leftarrow g_1^b \cdot H(u)^r$; $dk_2 \leftarrow g^r$
Else $dk_1 \leftarrow dk'^a_1 \cdot g_1^b$; $dk_2 \leftarrow dk'^a_2$
Ret (dk_1, dk_2)

When the RKD function is a constant function, T behaves exactly as the key
derivation algorithm under master secret key b, so its output is valid and correctly
distributed. Otherwise, the output of T is still a valid key for user u under master
secret key $\phi_{a,b}(s)$, now under randomness ra, since:

$$dk'^a_1 \cdot g_1^b = (g_1^s \cdot H(u)^r)^a \cdot g_1^b = g_1^{as+b} H(u)^{ra} \qquad dk'^a_2 = g^{ra} \ .$$

Since r is uniformly distributed in $\mathbb{Z}_p$, ra is also uniformly distributed in $\mathbb{Z}_p$ and
so the keys output by T are distributed identically to those output by $\mathcal{K}(\phi(s), u)$.
Hence the Waters IBE scheme is key-malleable.

The same identity renaming scheme can be used for both IBE schemes.
Namely, $\text{SI}(s, \overline{u})$ returns $\overline{u}||g^s$ and $\text{PI}(\pi, \overline{u}, \phi_{a,b})$ returns $\overline{u}||\pi^a \cdot g^b$. The com-
patibility requirement is satisfied and the renaming scheme is clearly collision-
resistant since $\overline{u}_1||g^{\phi(s)} = \overline{u}_2||g^s \Rightarrow \overline{u}_1 = \overline{u}_2 \wedge \phi(s) = s$. Thus the IBE schemes
of Boneh-Franklin and Waters are key-malleable and admit a suitable identity
renaming scheme, and so satisfy the requirements of Theorem 1. Notice that in
the Waters case, we must increase the parameter n by the bit length of elements
of $\mathbb{G}_1$ (and hence increase the size of the description of the scheme parameters)
to allow identities of the form $\overline{u}||g^s$ to be used in the renaming scheme.

The following theorem is obtained by combining Theorem 1 with [14], and
the running time of B below may be obtained in the same way.

Theorem 2. *Let $\overline{\mathcal{IBE}} = (\mathcal{S}, \mathcal{P}, \overline{\mathcal{K}}, \overline{\mathcal{E}}, \mathcal{D})$ be the Boneh-Franklin IBE scheme
shown in Fig. 3 under the above identity renaming transform. Let $\overline{A}$ be a Φ^{aff}-
RKA adversary against $\overline{\mathcal{IBE}}$ making q_{KD} key derivation queries and q_{H_2} queries
to random oracle H_2. Then there is an algorithm B solving the Decision Bilinear
Diffie-Hellman problem such that*

$$\mathbf{Adv}^{\text{ibe-rka}}_{\overline{\mathcal{IBE}}, \Phi^{\text{aff}}}(\overline{A}) \leq \frac{e(1 + q_{KD})q_{H_2}}{2} \cdot \mathbf{Adv}^{\text{dbdh}}(B) \ . \tag{8}$$

The following theorem is obtained by combining Theorem 1 with [25], and the running time of B below may be obtained in the same way. Concrete-security improvements would be obtained by using instead the analysis of Waters' scheme from [7].

Theorem 3. *Let $\overline{IBE} = (\mathcal{S}, \mathcal{P}, \overline{\mathcal{K}}, \overline{\mathcal{E}}, \mathcal{D})$ be the Waters scheme shown in Fig. 3 under the above identity renaming transform. Let $\overline{A}$ be a $\varPhi^{\text{aff}}$-RKA adversary against $\overline{IBE}$ making q_{KD} key derivation queries. Then there is an algorithm B solving the Decision Bilinear Diffie-Hellman problem such that*

$$\mathbf{Adv}_{\overline{IBE}, \varPhi^{\text{aff}}}^{\text{ibe-rka}}(\overline{A}) \leq 32(n+1) \cdot q_{KD} \cdot \mathbf{Adv}^{\text{dbdh}}(B) . \tag{9}$$

We recall from [4] that, given a $\varPhi$-RKA-secure IBE scheme, the CHK transform [12] yields a $\varPhi$-RKA-secure CCA-PKE scheme at the cost of adding a strongly unforgeable one-time secure signature and its verification key to the IBE ciphertexts. In the full version [6] we show that the more efficient Boneh-Katz transform [12] can also be used to the same effect. We omit the details of the $\varPhi^{\text{aff}}$-RKA-secure CCA-PKE schemes that result from applying these transforms to the above IBE schemes. We simply note that the resulting CCA-PKE schemes are as efficient as the pairing-based schemes of Wee [26], which are only $\varPhi^{\text{lin}}$-RKA-secure. Similarly, using a result of [4], we may apply the Naor transform to these IBE schemes to obtain $\varPhi^{\text{aff}}$-RKA-secure signature schemes that are closely related to (and as efficient as) the Boneh-Lynn-Shacham [16] and Waters [25] signature schemes. The verification algorithms of these signature schemes can be improved by replacing Naor's trial encryption and decryption procedure by bespoke algorithms, exactly as in [16,25].

AN IBE SCHEME HANDLING RKAS FOR BOUNDED DEGREE POLYNOMIALS. We show how to construct an IBE scheme that is RKA secure when the RKD function set equals $\varPhi^{\text{poly}(d)}$, the set of all polynomials of degree at most d, for an arbitrary d chosen at the time of master key generation. The scheme is obtained through a simple extension of the IBE scheme of Waters combined with the identity renaming transform used above. The only change we make to the Waters scheme is in the master public key, where we add the extra elements $g^{s^2}, \ldots, g^{s^d}, g_1^{s^2}, \ldots, g_1^{s^d}$ alongside g^s. These elements assist in achieving key-malleability for the set $\varPhi^{\text{poly}(d)}$. The master public-key generation algorithm $\mathcal{P}$ of the extended Waters scheme, on input s, returns $\pi \leftarrow (g^s, g^{s^2}, \ldots, g^{s^d}, (g_1)^{s^2}, \ldots, (g_1)^{s^d})$. The other algorithms and keys remain unchanged; in particular, key derivation does not make use of these new elements. This extended Waters IBE scheme is secure (in the usual IND-CPA sense for IBE) under the q-type extension of the standard DBDH assumption captured by the game in Fig. 5. We define the advantage of an adversary A against the problem as $\mathbf{Adv}^{q\text{-edbdh}}(A) = 2\Pr[q\text{-EDBDH}^A] - 1$.

Theorem 4. *Let $IBE = (\mathcal{S}, \mathcal{P}, \mathcal{K}, \mathcal{E}, \mathcal{D})$ be the extended Waters scheme. Let A be an adversary against IBE making q_{KD} key derivation queries. Then there is*

proc INITIALIZE

$g \leftarrow_\$ \mathbb{G}_1 \; ; \; x, y, z \leftarrow_\$ \mathbb{Z}_p \; ; \; b \leftarrow_\$ \{0,1\}$
If $(b = 1)$ $T \leftarrow e(g,g)^{xyz}$
Else $T \leftarrow_\$ \mathbb{G}_T^*$
Ret $g, g^x, g^{x^2}, \ldots, g^{x^q}, g^y, g^{(x^2)y}, g^{(x^3)y}, \ldots, g^{(x^q)y}, g^z, T$

proc FINALIZE(b')
Ret $(b = b')$

Fig. 5. q-Extended Decision Bilinear Diffie-Hellman (q-EDBDH) game

an algorithm B solving the q-Extended Decision Bilinear Diffie-Hellman problem for $q = d$ such that

$$\mathbf{Adv}_{\overline{IBE}}^{ibe}(A) \leq 32(n+1) \cdot q_{KD} \cdot \mathbf{Adv}^{q\text{-edbdh}}(B) \; . \qquad (10)$$

To see this, observe that the original proof of security for Waters' scheme [25,7] also goes through for the extended scheme, using the elements g, g^x, g^y, T from the q-EDBDH problem to run the simulation as in the original proof and using the additional elements from the q-EDBDH problem to set up the master public key in the extended scheme.

We give evidence for the validity of the q-EDBDH assumption by examining the difficulty of the problem in the generic group model. The problem falls within the framework of the generic group model "master theorem" of Boneh, Boyen and Goh [11]. In their notation, we have $P = \{1, x, x^2, \ldots, x^q, y, x^2y, \ldots, x^qy, z\}, Q = 1$, and $f = xyz$. It is clear by inspection that P, Q and f meet the independence requirement of the master theorem, and it gives a lower bound on an adversary's advantage of solving the q-EDBDH problem in a generic group of the form $(q+1)(q_\xi + 4q + 6)^2/p$ where q_ξ is a bound on the number of queries made by the adversary to the oracles computing the group operations in $\mathbb{G}, \mathbb{G}_T$. While a lower bound in the generic group model does not rule out an efficient algorithm when the group is instantiated, it lends heuristic support to our assumption.

The extended Waters IBE scheme is $\Phi^{\mathrm{poly}(d)}$-key malleable with algorithm T as follows:

$T(\pi, u, dk', \phi_{a_0, a_1, \ldots, a_d})$:
If $(a_0 = 0)$ then $r \leftarrow_\$ \mathbb{Z}_p \; ; \; dk_1 \leftarrow g_1^{a_0} \cdot H(u)^r \cdot (g_1^{s^2})^{a_2} \cdots (g_1^{s^d})^{a_d} \; ; \; dk_2 \leftarrow g^r$
Else $dk_1 \leftarrow g_1^{a_0} \cdot dk_1'^{a_1} \cdot (g_1^{s^2})^{a_2} \cdots (g_1^{s^d})^{a_d} \; ; \; dk_2 \leftarrow dk_2'^{a_1}$
Ret (dk_1, dk_2)

The identity renaming scheme is then defined via

$$\mathrm{SI}(s, \overline{u}) = \overline{u} \| g^s \quad \text{and} \quad \mathrm{PI}(\pi, \overline{u}, \phi_{a_0, a_1, \ldots, a_d}) = \overline{u} \| g^{a_0} \cdot \pi^{a_1} \cdot (g^{s^2})^{a_2} \cdots (g^{s^d})^{a_d}$$

which clearly meets the compatibility and collision-resistance requirements. Combining Theorem 1 with Theorem 4 gives the following theorem.

Theorem 5. *Let $\overline{IBE} = (\mathcal{S}, \mathcal{P}, \overline{\mathcal{K}}, \overline{\mathcal{E}}, \mathcal{D})$ be the extended Waters scheme under the above identity renaming transform. Let $\overline{A}$ be a $\Phi^{\mathrm{poly}(d)}$-RKA adversary against $\overline{IBE}$ making q_{KD} key derivation queries. Then there is an algorithm B*

solving the q-Extended Decision Bilinear Diffie-Hellman problem for $q = d$ such that

$$\mathbf{Adv}_{\overline{IBE}, \Phi^{\mathrm{poly}(d)}}^{\mathrm{ibe-rka}}(\overline{A}) \leq 32(n+1) \cdot q_{KD} \cdot \mathbf{Adv}^{q\text{-edbdh}}(B) . \tag{11}$$

As in the affine case, we may apply results of [4] to obtain a $\Phi^{\mathrm{poly}(d)}$-RKA-secure CCA-PKE scheme and a $\Phi^{\mathrm{poly}(d)}$-RKA-secure signature scheme. We omit the detailed but obvious description of these schemes, noting merely that they are efficient and secure in the standard model under the q-EDBDH assumption.

Acknowledgments. Bellare was supported in part by NSF grants CNS-1116800, CNS 0904380 and CCF-0915675. Paterson and Thomson were supported by EP-SRC Leadership Fellowship EP/H005455/1.

References

1. Applebaum, B.: Garbling XOR gates "for free" in the standard model. Cryptology ePrint Archive, Report 2012/516 (2012), http://eprint.iacr.org/
2. Applebaum, B., Harnik, D., Ishai, Y.: Semantic security under related-key attacks and applications. In: Yao, A.C.-C. (ed.) ICS 2011, Tsinghua University Press (2011)
3. Bellare, M., Cash, D.: Pseudorandom Functions and Permutations Provably Secure against Related-Key Attacks. In: Rabin, T. (ed.) CRYPTO 2010. LNCS, vol. 6223, pp. 666–684. Springer, Heidelberg (2010)
4. Bellare, M., Cash, D., Miller, R.: Cryptography Secure against Related-Key Attacks and Tampering. In: Lee, D.H., Wang, X. (eds.) ASIACRYPT 2011. LNCS, vol. 7073, pp. 486–503. Springer, Heidelberg (2011)
5. Bellare, M., Kohno, T.: A Theoretical Treatment of Related-Key Attacks. In: Biham, E. (ed.) EUROCRYPT 2003. LNCS, vol. 2656, pp. 491–506. Springer, Heidelberg (2003)
6. Bellare, M., Paterson, K.G., Thomson, S.: RKA security beyond the linear barrier: IBE, encryption and signatures. Cryptology ePrint Archive, Report 2012/514 (2012), Full version of this abstract, http://eprint.iacr.org/
7. Bellare, M., Ristenpart, T.: Simulation without the Artificial Abort: Simplified Proof and Improved Concrete Security for Waters' IBE Scheme. In: Joux, A. (ed.) EUROCRYPT 2009. LNCS, vol. 5479, pp. 407–424. Springer, Heidelberg (2009)
8. Bellare, M., Rogaway, P.: The Security of Triple Encryption and a Framework for Code-Based Game-Playing Proofs. In: Vaudenay, S. (ed.) EUROCRYPT 2006. LNCS, vol. 4004, pp. 409–426. Springer, Heidelberg (2006)
9. Biham, E.: New Types of Cryptanalytic Attacks Using Related Keys (Extended Abstract). In: Helleseth, T. (ed.) EUROCRYPT 1993. LNCS, vol. 765, pp. 398–409. Springer, Heidelberg (1994)
10. Biham, E., Shamir, A.: Differential Fault Analysis of Secret Key Cryptosystems. In: Kaliski Jr., B.S. (ed.) CRYPTO 1997. LNCS, vol. 1294, pp. 513–525. Springer, Heidelberg (1997)
11. Boneh, D., Boyen, X., Goh, E.-J.: Hierarchical Identity Based Encryption with Constant Size Ciphertext. In: Cramer, R. (ed.) EUROCRYPT 2005. LNCS, vol. 3494, pp. 440–456. Springer, Heidelberg (2005)
12. Boneh, D., Canetti, R., Halevi, S., Katz, J.: Chosen-ciphertext security from identity-based encryption. SIAM J. Comput. 36(5), 1301–1328 (2007)

13. Boneh, D., DeMillo, R.A., Lipton, R.J.: On the Importance of Checking Cryptographic Protocols for Faults (Extended Abstract). In: Fumy, W. (ed.) EUROCRYPT 1997. LNCS, vol. 1233, pp. 37–51. Springer, Heidelberg (1997)
14. Boneh, D., Franklin, M.: Identity-based encryption from the Weil pairing. SIAM J. Comput. 32(3), 586–615 (2003)
15. Boneh, D., Katz, J.: Improved Efficiency for CCA-Secure Cryptosystems Built Using Identity-Based Encryption. In: Menezes, A. (ed.) CT-RSA 2005. LNCS, vol. 3376, pp. 87–103. Springer, Heidelberg (2005)
16. Boneh, D., Lynn, B., Shacham, H.: Short Signatures from the Weil Pairing. In: Boyd, C. (ed.) ASIACRYPT 2001. LNCS, vol. 2248, pp. 514–532. Springer, Heidelberg (2001)
17. Boyen, X., Mei, Q., Waters, B.: Direct chosen ciphertext security from identity-based techniques. In: Atluri, V., Meadows, C., Juels, A. (eds.) ACM CCS 2005, pp. 320–329. ACM Press (November 2005)
18. Cramer, R., Dodis, Y., Fehr, S., Padró, C., Wichs, D.: Detection of Algebraic Manipulation with Applications to Robust Secret Sharing and Fuzzy Extractors. In: Smart, N.P. (ed.) EUROCRYPT 2008. LNCS, vol. 4965, pp. 471–488. Springer, Heidelberg (2008)
19. Gennaro, R., Lysyanskaya, A., Malkin, T., Micali, S., Rabin, T.: Algorithmic Tamper-Proof (ATP) Security: Theoretical Foundations for Security against Hardware Tampering. In: Naor, M. (ed.) TCC 2004. LNCS, vol. 2951, pp. 258–277. Springer, Heidelberg (2004)
20. Goldenberg, D., Liskov, M.: On Related-Secret Pseudorandomness. In: Micciancio, D. (ed.) TCC 2010. LNCS, vol. 5978, pp. 255–272. Springer, Heidelberg (2010)
21. Goyal, V., O'Neill, A., Rao, V.: Correlated-Input Secure Hash Functions. In: Ishai, Y. (ed.) TCC 2011. LNCS, vol. 6597, pp. 182–200. Springer, Heidelberg (2011)
22. Knudsen, L.R.: Cryptanalysis of LOKI91. In: Seberry, J., Zheng, Y. (eds.) AUSCRYPT 1992. LNCS, vol. 718, pp. 196–208. Springer, Heidelberg (1993)
23. Lucks, S.: Ciphers Secure against Related-Key Attacks. In: Roy, B., Meier, W. (eds.) FSE 2004. LNCS, vol. 3017, pp. 359–370. Springer, Heidelberg (2004)
24. Paterson, K.G., Schuldt, J.C.N., Stam, M., Thomson, S.: On the Joint Security of Encryption and Signature, Revisited. In: Lee, D.H., Wang, X. (eds.) ASIACRYPT 2011. LNCS, vol. 7073, pp. 161–178. Springer, Heidelberg (2011)
25. Waters, B.: Efficient Identity-Based Encryption Without Random Oracles. In: Cramer, R. (ed.) EUROCRYPT 2005. LNCS, vol. 3494, pp. 114–127. Springer, Heidelberg (2005)
26. Wee, H.: Public Key Encryption against Related Key Attacks. In: Fischlin, M., Buchmann, J., Manulis, M. (eds.) PKC 2012. LNCS, vol. 7293, pp. 262–279. Springer, Heidelberg (2012)

Fully Secure Unbounded Inner-Product and Attribute-Based Encryption

Tatsuaki Okamoto[1] and Katsuyuki Takashima[2]

[1] NTT
okamoto.tatsuaki@lab.ntt.co.jp
[2] Mitsubishi Electric
Takashima.Katsuyuki@aj.MitsubishiElectric.co.jp

Abstract. In this paper, we present the first inner-product encryption (IPE) schemes that are *unbounded* in the sense that the public parameters do not impose additional limitations on the predicates and attributes used for encryption and decryption keys. All previous IPE schemes were *bounded*, or have a bound on the size of predicates and attributes given public parameters fixed at setup. The proposed unbounded IPE schemes are *fully (adaptively) secure and fully attribute-hiding* in the standard model under a standard assumption, the decisional linear (DLIN) assumption. In our unbounded IPE schemes, the inner-product relation is generalized, where the two vectors of inner-product can be different sizes and it provides a great improvement of efficiency in many applications. We also present the first *fully secure unbounded* attribute-based encryption (ABE) schemes, and the security is proven under the DLIN assumption in the standard model. To achieve these results, we develop novel techniques, *indexing* and *consistent randomness amplification*, on the (extended) dual system encryption technique and the dual pairing vector spaces (DPVS).

1 Introduction

1.1 Background

IPE and ABE. The notions of *inner-product encryption* (IPE) and *attribute-based encryption* (ABE) introduced by Katz, Sahai and Waters [6] and Sahai and Waters [18] constitute an advanced class of encryption, *functional encryption* (FE), and provide more flexible and fine-grained functionalities in sharing and distributing sensitive data than traditional symmetric and public-key encryption as well as identity-based encryption (IBE).

In FE, there is a relation $R(v, x)$, that determines whether a secret key associated with a parameter v can decrypt a ciphertext encrypted under another parameter x. The parameters for IPE are expressed as vectors $\vec{x}$ (for encryption) and $\vec{v}$ (for a secret key), where $R(\vec{v}, \vec{x})$ holds, i.e., a secret key with $\vec{v}$ can decrypt a ciphertext with $\vec{x}$, iff $\vec{v} \cdot \vec{x} = 0$. (Here, $\vec{v} \cdot \vec{x}$ denotes the standard inner-product.) In ABE systems, either one of the parameters for encryption and secret key is a set of attributes, and the other is an access policy (structure) or (monotone)

X. Wang and K. Sako (Eds.): ASIACRYPT 2012, LNCS 7658, pp. 349–366, 2012.

span program over a universe of attributes, e.g., a secret key for a user is associated with an access policy and a ciphertext is associated with a set of attributes, where a secret key can decrypt a ciphertext, iff the attribute set satisfies the policy. If the access policy is for a secret key, it is called key-policy ABE (KP-ABE), and if the access policy is for encryption, it is ciphertext-policy ABE (CP-ABE).

For some applications, the parameters for encryption are required to be hidden from ciphertexts. To capture the security requirement, Katz, Sahai and Waters [6] introduced *attribute-hiding* (based on the same notion for hidden vector encryption (HVE) by Boneh and Waters [4]), a security notion for FE that is stronger than the basic security requirement, *payload-hiding*. Roughly speaking, attribute-hiding requires that a ciphertext conceal the associated parameter as well as the plaintext, while payload-hiding only requires that a ciphertext conceal the plaintext. A weaker notion of attribute-hiding than the original one [6] was given by [7]. The weaker notion is called *weakly attribute-hiding*, and the original one is *fully attribute-hiding*. Informally, in the fully attribute-hiding, the secrecy of attribute x is ensured even against an adversary having a secret key with v such that $R(v, x)$ holds (i.e., no information is released on x except $R(v, x)$ holds), while it is ensured only when $R(v, x)$ does not hold in the weakly attribute-hiding (see Definition 4 for the definition of the fully attribute-hiding).

To the best of our knowledge, the widest class of attribute-hiding FE is IPE [6, 7, 12, 14] (KSW08, LOS+10, OT10 and OT12 schemes). Inner-products for IPE represent a fairly wide class of relations including equality tests as the simplest case (i.e., anonymous IBE and HVE are very special classes of attribute-hiding IPE), disjunctions or conjunctions of equality tests, and, more generally, CNF or DNF formulas. We note, however, that inner-product relations are less expressive than a class of relations (on span programs) for ABE, while existing ABE schemes for such a wider class of relations are not attribute-hiding but only payload-hiding.

Among the existing IPE schemes, only the OT12 IPE scheme [14] achieves the *full (adaptive)* security and *fully attribute-hiding* simultaneously, whereas other attribute-hiding IPE schemes [6, 11, 7, 12] are selectively secure or weakly attribute-hiding, and some IPE schemes [1, 13] only achieve payload-hiding. As for ABE, Lewko et.al. and Okamoto-Takashima ABE schemes [7, 12] are fully secure in the standard model, while ABE schemes [18, 5, 16, 20] before [7, 12] were *selectively* secure.

Unbounded IPE and ABE. All previous constructions of IPE and ABE except the Lewko-Waters ABE scheme [9] have restriction, or are *bounded*, in the choice of the parameters for secret key and encryption once the public parameters have been set. The only *unbounded* ABE scheme [9], however, is *selectively* secure, while they presented an *unbounded* hierarchical identity-based encryption (HIBE) that is *fully secure* in the standard model. No *unbounded* IPE scheme has been presented. Therefore, no *fully secure* and *unbounded* scheme for an advanced class of encryption like IPE or ABE has been presented.

In practice, it is highly desirable that the parameters for secret key and encryption should be flexible or *unbounded* by the public parameters fixed at setup, since if we set the public parameters for a possible maximum size (e.g., the maximum dimension of predicate and attribute vectors for IPE), the size of the public parameters should be huge.

Removing the restrictions for fully secure IPE and ABE, however, is quite challenging. As mentioned above, no *fully secure* and *unbounded* scheme for an advanced class of encryption like IPE or ABE has been presented. The difficulty resides in the existing techniques for proving the *full (or adaptive) security* of such an advanced class of encryption.

The only known technique to prove the full security of an (attribute-hiding) IPE or ABE system is the dual system encryption by Waters [19] and its extension [14]. In the techniques, information theoretical arguments (e.g., conceptual change due to the same distribution and the independent randomness of two distributions etc.) over some (hidden) parts of a secret-key and challenge ciphertext play a key role in the security proof, provided that the adversary follows the secret-key-query condition in the security games. To execute a security proof based on the information theoretical arguments, an appropriate distribution of randomness consistent with the key-query condition should be supplied in the proof games transformed from the original proof game.

As for *bounded* IPE and ABE schemes, the public parameters can supply immanent randomness enough for the arguments, since the size of parameters for secret-keys and encryption is bounded by the public parameters. For example, when the dimension of vectors for IPE is required to be n, the public parameters whose size is $O(n)$ with respect to n should be given in *bounded* IPE, and the size of secret randomness to generate the public parameter is $O(n^2)$. Such an amount of randomness can be enough for the arguments over n-dimensional vectors.

In contrast, for *unbounded* IPE and ABE schemes, some (unbounded amount of) randomness whose distribution is consistent with the key-query condition should be supplied in addition to the randomness provided by the public parameters. For example, even when the dimension of vectors for IPE is required to be n, the size of the public parameters is $O(1)$ in *unbounded* IPE, i.e., the size of secret randomness to generate the public parameters is $O(1)$. Clearly, such a size of randomness is not sufficient for the information theoretical arguments over n-dimensional vectors. Therefore, any additional source of randomness should be provided, and the distribution of the randomness should be specific (i.e., consistent with the key-query condition). For the unbounded HIBE scheme [9], where the equality (un-)matching is the key-query condition, a simple compression technique works well to create such randomness since equality can be simply compressed with preserving the property. The key-query condition for IPE and ABE, however, is in general much more complicated than just the equality matching for (H)IBE, and no technique was known to create randomness consistent with such a complicated condition in some security proofs. This is a reason why [9] succeeds in realizing a fully secure unbounded HIBE but not for ABE (and not for IPE).

Restriction on IPE. The existing IPE schemes have another restriction on the parameters (i.e., vectors) for secret key and encryption that the dimensions of $\vec{x}$ (for encryption) and $\vec{v}$ (for a secret key) should be equivalent. Such a restriction may be considered to be inevitable for the inner-product relation on $\vec{v} \cdot \vec{x}$, but it is required to be relaxed in various applications to improve the efficiency, especially in *unbounded* IPE systems where the setup (public) parameters give no restriction on the dimensions of vectors.

Let us consider an example on a genetic profile data of an individual. It is desirable that such a sensitive data be treated as encrypted data even for data processing and retrievals. Although a genetic profile may include a large amount of information, only a part of the profile is examined in many applications. For example, let $X_1, \ldots, X_{100}$ be variables of 100 genetic properties and $x_1, \ldots, x_{100}$ be Alice's values of these variables. To evaluate if $f(x_1, \ldots, x_{100}) = 0$ for any examination (multivariate) polynomial f with degree 3, or the truth value of the corresponding predicate $\phi_f(x_1, \ldots, x_{100})$, the attribute vector $\vec{x}$ of Alice should be a monomial vector of Alice's values with degree 3, $\vec{x} := (1, x_1, \ldots, x_{100}, x_1^2, x_1 x_2, \ldots, x_{100}^2, x_1^3, x_1^2 x_2, \ldots, x_{100}^3)$, whose dimension is around 10^6. A predicate vector $\vec{v}$ for a secret key can be associated with predicate ϕ_f.

To ensure the private data processing of $\vec{x}$, it should be encrypted (say c for a ciphertext of $\vec{x}$) by a *fully attribute-hiding* IPE scheme, since whether $\phi_f(x_1, \ldots, x_{100})$ holds can be examined with releasing no other information by checking whether c can be decrypted by a secret key with $\vec{v}$ (i.e., $R(\vec{v}, \vec{x})$ holds). Here, if c is encrypted by *fully* attribute-hiding IPE, it releases no information on $\vec{x}$ except that $R(\vec{v}, \vec{x})$ holds, or $\phi_f(x_1, \ldots, x_{100})$ holds, however, if it is encrypted by *weakly* attribute-hiding IPE, such desirable security cannot be ensured.

Let a predicate for $\vec{v}$ be $((X_5 = a) \vee (X_{16} = b)) \wedge (X_{57} = c)$, which focuses only three factors, X_5, X_{16}, X_{57}, among the 100 genetic properties. It can be represented by a polynomial equation, $r_1(X_5 - a)(X_{16} - b) + r_2(X_{57} - c) = 0$ (where $r_1, r_2 \xleftarrow{\mathsf{U}} \mathbb{F}_q$), i.e., $(r_1 ab - r_2 c) - r_1 b X_5 - r_1 a X_{16} + r_2 X_{57} + r_1 X_5 X_{16} = 0$. In order that $r_1(x_5 - a)(x_{16} - b) + r_2(x_{57} - c) = 0$ iff $\vec{v} \cdot \vec{x} = 0$, vector $\vec{v}$ should be $((r_1 ab - r_2 c), 0, \ldots, 0, -r_1 b, 0, \ldots, 0, -r_1 a, 0, \ldots, 0, r_2, 0, \ldots, 0, r_1, 0, \ldots, 0)$, whose dimension is equivalent to that of $\vec{x}$, i.e., around 10^6, although the effective dimension of $\vec{v}$ is just 5. This is due to the above-mentioned restriction on the inner-product relation of the existing IPE schemes. The size of secret key for $\vec{v}$ then should be in proportion to the dimension of $\vec{v}$ (and $\vec{x}$), around 10^6. This example shows us a strong practical motivation, especially for *unbounded* IPE schemes, to relax this restriction on the inner-product relation and to shorten the length of the secret key to that in proportion to the effective dimension, e.g., 5, instead of around 10^6.

1.2   Our Results

1. This paper introduces a new concept of IPE, generalized IPE, which relaxes the above-mentioned restriction of IPE and consists of three types of IPE, Types 0, 1 and 2. Here the notion of Types 1 and 2 is introduced in this paper, and Type 0 is the traditional one (see Remark below).

Table 1. Comparison of *attribute-hiding IPE schemes*, where $|\mathbb{G}|$ and $|\mathbb{G}_T|$ represent size of an element of $\mathbb{G}$ and that of $\mathbb{G}_T$, respectively. AH, IP, PK, SK, CT, GSD and eDDH stand for attribute-hiding, inner-product, master public key (public parameters), secret key, ciphertext, general subgroup decision [3] and extended decisional Diffie-Hellman [7], respectively.

	KSW08 [6]	LOS+10 [7]	OT10 [12]	OT12 [14] (basic)	OT12 [14] (variant)	Proposed IPE (type 1 or 2) Section 4.1	Proposed IPE (type 0) Section 4.2
Bounded or Unbounded	bounded	bounded	bounded	bounded	bounded	unbounded	unbounded
Restriction on IP relation	restricted*	restricted	restricted	restricted	restricted	relaxed	restricted
Security	selective & fully-AH	adaptive & weakly-AH	adaptive & weakly-AH	adaptive & fully-AH	adaptive & fully-AH	adaptive & fully-AH	adaptive & fully-AH
Order of $\mathbb{G}$	composite	prime	prime	prime	prime	prime	prime
Assump.	2 variants of GSD	n-eDDH	DLIN	DLIN	DLIN	DLIN	DLIN
PK size	$O(n)|\mathbb{G}|$	$O(n^2)|\mathbb{G}|$	$O(n^2)|\mathbb{G}|$	$O(n^2)|\mathbb{G}|$	$O(n)|\mathbb{G}|$	$O(1)|\mathbb{G}|$	$O(1)|\mathbb{G}|$
SK size	$(2n+1)|\mathbb{G}|$	$(2n+3)|\mathbb{G}|$	$(3n+2)|\mathbb{G}|$	$(4n+2)|\mathbb{G}|$	$11|\mathbb{G}|$	$(15n+5)|\mathbb{G}|$	$(21n+9)|\mathbb{G}|$
CT size	$(2n+1)|\mathbb{G}|$ $+ |\mathbb{G}_T|$	$(2n+3)|\mathbb{G}|$ $+ |\mathbb{G}_T|$	$(3n+2)|\mathbb{G}|$ $+ |\mathbb{G}_T|$	$(4n+2)|\mathbb{G}|$ $+ |\mathbb{G}_T|$	$(5n+1)|\mathbb{G}|$ $+ |\mathbb{G}_T|$	$(15n'+5)|\mathbb{G}|$ $+ |\mathbb{G}_T|$	$(21n'+9)|\mathbb{G}|$ $+ |\mathbb{G}_T|$

* It can be easily relaxed.

Remark: We now roughly explain the three types of inner-product relations. To relax the above-mentioned restriction on the inner-product relation, we introduce a new type of inner-product (generalized inner-product) for $\vec{v}$ and $\vec{x}$, where their dimensions can be different (say n and n' for the dimensions of $\vec{v}$ and $\vec{x}$). In this notion, vector $\vec{v}$ and $\vec{x}$ are expressed by $\{(t, v_t) \mid t \in I_{\vec{v}}, \sharp I_{\vec{v}} = n\}$ and $\{(t, x_t) \mid t \in I_{\vec{x}}, \sharp I_{\vec{x}} = n'\}$, respectively, where $t \in \mathbb{N}$ is an index for vectors, whose semantics is given by each application. Here note that we abuse the same vector notation, $\vec{v}$, for the new expression as well as for the conventional one, $(v_1, \ldots, v_n)$. In the above-mentioned example, $\vec{x} := \{(1, 1), (2, x_1), \ldots, (101, x_{100}), (102, x_1^2), (103, x_1 x_2), \ldots, (n', x_{100}^3)\}$ where $I_{\vec{x}} := \{1, 2, \ldots, n'\}$, and $\vec{v} := \{(1, r_1 a b - r_2 c), (6, -r_1 b), (17, -r_1 a), (58, r_2), (517, r_1)\}$ where $I_{\vec{v}} := \{1, 6, 17, 58, 517\}$. The generalized inner-product of $\vec{v}$ over $\vec{x}$ is defined by $\sum_{t \in I_{\vec{v}}} v_t x_t$ if $I_{\vec{v}} \subseteq I_{\vec{x}}$. Otherwise, it is undefined. By using the generalized inner-product notion, the secret key size can be in proportion to the effective dimension (e.g., 5 instead of around 10^6).

We then introduce three types of IPE schemes. For Type 1, relation $R(\vec{v}, \vec{x})$ holds iff the generalized inner-product of $\vec{v}$ over $\vec{x}$ is 0, while for Type 2 it holds iff the generalized inner-product of $\vec{x}$ over $\vec{v}$ is 0. We call Type 0 for the conventional inner-products, i.e., relation $R(\vec{v}, \vec{x})$ is defined by the standard inner-product of $\vec{v}$ and $\vec{x}$, where $\vec{v}$ and $\vec{x}$ have the same dimension

Table 2. Comparison of *KP-ABE Schemes*, where $|\mathbb{G}|$ represents the size of an element of $\mathbb{G}$, and PK, SK, CT and GSD stand for master public key (public parameters), secret key, ciphertext and general subgroup decision [3], respectively. And, d, n, n_{max}, ℓ and k_{max} are the number of sub-universes of attributes, the number of attributes for a CT, the maximum number of attributes for a CT, the row size of an access policy matrix for a SK and the maximum value of the degree of access policies, respectively.

	LW11 [9]	LOS+10 [7]		OT10 [12]		Proposed KP-ABE	
		(basic)	(modified)	(basic)	(modified)	(basic) Section 5	(modified) in full ver.
Bounded or Unbounded	unbounded	bounded	bounded	bounded	bounded	unbounded	unbounded
Security	selective	full	full	full	full	full	full
Order of $\mathbb{G}$	composite	composite	composite	prime	prime	prime	prime
Assump.	GSD	GSD	GSD	DLIN	DLIN	DLIN	DLIN
Degree of access policies	arbitrary	1	arbitrary	1	arbitrary	1	arbitrary
PK size	$O(1)|\mathbb{G}|$	$O(n_{max})|\mathbb{G}|$	$O(n_{max})|\mathbb{G}|$	$O(d)|\mathbb{G}|$	$O(d)|\mathbb{G}|$	$O(1)|\mathbb{G}|$	$O(1)|\mathbb{G}|$
SK size	$O(\ell)|\mathbb{G}|$	$O(\ell)|\mathbb{G}|$	$O(\ell)|\mathbb{G}|$	$O(\ell)|\mathbb{G}|$	$O(\ell)|\mathbb{G}|$	$O(\ell)|\mathbb{G}|$	$O(\ell)|\mathbb{G}|$
CT size	$O(n)|\mathbb{G}|$	$O(n)|\mathbb{G}|$	$O(k_{max}n)|\mathbb{G}|$	$O(n)|\mathbb{G}|$	$O(k_{max}n)|\mathbb{G}|$	$O(n)|\mathbb{G}|$	$O(k_{max}n)|\mathbb{G}|$

(in other words, the inner-product for Type 0 is defined iff these dimensions are equivalent.)

2. We present the first *unbounded* inner-product encryption (IPE) schemes. The proposed unbounded IPE schemes are *fully (adaptively) secure and fully attribute-hiding* in the standard model under a standard assumption, the decisional linear (DLIN) assumption. The proposed unbounded IPE schemes consist of the above-mentioned types of generalized IPE, Types 0, 1 and 2, For comparison of attribute-hiding IPE schemes, see Table 1.

3. We present the first *unbounded* KP- and CP-ABE schemes that are *fully secure* (adaptively payload-hiding) in the standard model. The proposed unbounded ABE schemes are fully secure under the DLIN assumption, and are for a wide class of relations, non-monotone access structures (see the full version for the proposed CP-ABE scheme). See Table 2 for comparison of KP-ABE schemes.

Remark: Similarly to the existing fully secure ABE schemes in the standard model [7, 12, 8] except [10], our basic ABE scheme (Section 5) has a restriction that the degree of access policies is 1^1. A modified KP-ABE scheme is shown in the full version of this paper to relax the restriction or to achieve an arbitrary degree k of access policies with preserving the fully

---

[1] Informally, the degree may imply the number of appearance of a variable in a formula, e.g., formula $((x = a) \vee (x = b)) \wedge (y = c)$ has degree 2 for variable x. For the definition of the degree of access policies in our schemes, see the full version. The degree should be a bit differently defined in [18, 5, 16, 20, 7, 8], where degree 1 is called *one-use*.

secure and unbounded property. It, however, shares a shortcoming of the existing fully secure (modified) ABE schemes [7, 12, 8] that the ciphertext size grows linearly with k. Here, a (maximum) value of k can be determined in each application of our ABE scheme, while the public parameters are fixed and commonly shared by all applications and users.

1.3   Key Techniques

As mentioned above, the difficulty of realizing a fully secure unbounded IPE or ABE scheme arises from the hardness of supplying an *unbounded amount of* randomness *consistent* with the complicated key-query condition for the (dual system encryption) security arguments on IPE or ABE. To overcome this difficulty, we develop novel techniques, *indexing* and *consistent randomness amplification*, on the dual system encryption and the dual pairing vector spaces (DPVS). Roughly speaking, the *indexing* technique is for supplying a source of unbounded amount of randomness and the *consistent randomness amplification* technique is for amplifying the randomness of the source through a computational assumption (e.g., the DLIN assumption in our case) and the randomness of hidden bases as well as for adjusting the distribution of the amplified randomness to be consistent with a condition. This methodology could provide a general framework for proving the security in unbounded situations.

In DPVS, a pair of dual (or orthonormal) bases for N-dimensional linear spaces, $\mathbb{B} := (\boldsymbol{b}_1, \ldots, \boldsymbol{b}_N)$ and $\mathbb{B}^* := (\boldsymbol{b}_1^*, \ldots, \boldsymbol{b}_N^*)$, are randomly generated using a secret random linear transformation X (random $N \times N$ matrix) (see Section 2). In a typical application of DPVS to cryptography, a part of $\mathbb{B}$ (say $\hat{\mathbb{B}}$) is used as a public key (public parameters), and $\mathbb{B}^*$ as a secret key, where X is the top level secret key and the source of randomness.

In a typical construction of *bounded* IPE schemes [7, 12, 14] which are based on DPVS, once a basis of DPVS, a part of the basis of a N-dimensional space is published as public parameters, the dimension n of predicate and attribute vectors for secret key and encryption is bounded or fixed, e.g., $n \le N/4$ (i.e., $N = O(n)$). The full security is proven through the information theoretical arguments, and the randomness of secret matrix X (e.g., the amount of the randomness is $O(n^2)$) supplies enough randomness for the arguments.

In contrast, the dimension, n, of the predicate and attribute vectors is not bounded by the public parameters in *unbounded* IPE. For example, in one of the proposed IPE schemes (Section 4), the public parameters consist of a constant number of elements, 9 elements of bases (or 105 pairing group elements), $\hat{\mathbb{B}}_0 := (\boldsymbol{b}_{0,1}, \boldsymbol{b}_{0,3}, \boldsymbol{b}_{0,5})$ and $\hat{\mathbb{B}} := (\boldsymbol{b}_1, \ldots, \boldsymbol{b}_4, \boldsymbol{b}_{14}, \boldsymbol{b}_{15})$, where random matrices of constant sizes, $X_0 \xleftarrow{\mathsf{U}} \mathbb{F}_q^{5 \times 5}$ and $X_1 \xleftarrow{\mathsf{U}} \mathbb{F}_q^{15 \times 15}$, are employed to generate the public parameters. The randomness of the public parameters, just a constant amount with respect to n, is clearly insufficient for the (dual system encryption) arguments on the proof of full security.

To supply additional randomness for the purpose, in our IPE schemes, we introduce a technique called *indexing*, where two-dimensional index vectors,

$\sigma_t(1, t)$ and $\mu_t(t, -1)$ are embedded into ciphertext $\boldsymbol{c}_t$ and secret key $\boldsymbol{k}_t^*$, respectively, where σ_t and μ_t are freshly random for each t. In our IPE scheme (Section 4) where $n = n'$ for simplicity, for example, secret key $(\boldsymbol{k}_1^*, \ldots, \boldsymbol{k}_n^*)$ for $\vec{v} := (v_1, \ldots, v_n)$ can be expressed by a coefficient vector, $(\mu_t(t, -1), \delta v_t, \ldots)$, for $t = 1, \ldots, n$, over basis $\mathbb{B}^*$, i.e., $\boldsymbol{k}_t^* := (\mu_t(t, -1), \delta v_t, \ldots)_{\mathbb{B}^*}$ and ciphertext $(\boldsymbol{c}_1, \ldots, \boldsymbol{c}_n)$ for $\vec{x} := (x_1, \ldots, x_n)$ can be expressed by $\boldsymbol{c}_t := (\sigma_t(1, t), \omega x_t, \ldots)_{\mathbb{B}}$ for $t = 1, \ldots, n$, where δ, ω are randomly selected. While the size of the public parameters or its randomness is constant in n, an unbounded amount of randomness, $\{\mu_t\}_{t=1,\ldots,n}, \{\sigma_t\}_{t=1,\ldots,n}$, can be supplied to secret key and ciphertext. This is a key idea of the *indexing* technique.

Although the technique supplies an unbounded amount of randomness, i.e., $O(n)$-size of randomness, it is not enough for our purpose. We need more and a specific distribution of randomness. This is because: in the proof of full security on dual system encryption and the extension, such a *real* randomness provided by the indexing technique should be expanded into a *hidden* part in spaces over bases $\mathbb{B}$ and $\mathbb{B}^*$, and the distribution should be also adjusted to (or consistent with) the key-query condition for IPE or ABE. For this purpose, i.e., in order to amplify the randomness to a hidden subspace and to adjust it to a specific distribution, we develop another technique, *consistent randomness amplification*.

For a bit more detailed explanation of the consistent randomness amplification technique, we will briefly review a hidden part (subspace) of DPVS. As mentioned above, in a typical application of DPVS to cryptography, a part of $\mathbb{B}$ (say $\hat{\mathbb{B}}$) is used as a public key (public parameters). Therefore, the basis, $\mathbb{B} - \hat{\mathbb{B}}$, is information theoretically concealed against an adversary, i.e., even an infinite power adversary has no idea on which basis is selected as $\mathbb{B} - \hat{\mathbb{B}}$ when $\hat{\mathbb{B}}$ is published. The underlying dual vector spaces, $\mathsf{span}\langle \mathbb{B} \rangle$ and $\mathsf{span}\langle \mathbb{B}^* \rangle$, are 15-dimensional for our IPE scheme (Type 1 or 2) and 14-dimensional for our ABE scheme. The subspaces employed for public parameters are just 6-dimensional and other 2 dimensional basis can be public. Hence, the basis for the remaining 7 or 6-dimensional subspace is information theoretically concealed (uncertain). The consistent randomness amplification technique is executed over these 7 or 6-dimensional hidden subspaces. For example, as mentioned above, a real secret key $\{\boldsymbol{k}_t^*\}$ and ciphertext $\{\boldsymbol{c}_t\}$ are expressed by $\boldsymbol{k}_t^* := (\mu_t(t, -1), \delta v_t, s_t, \boxed{0^7}, \ldots)_{\mathbb{B}^*}$ and $\boldsymbol{c}_t := (\sigma_t(1, t), \omega x_t, \widetilde{\omega}, \boxed{0^7}, \ldots)_{\mathbb{B}}$. This technique provides a transformation (for the dual system encryption technique and the extension) to the following forms: $\boldsymbol{k}_t^* := (\mu_t(t, -1), \delta v_t, s_t, \boxed{0^4, (\pi v_t, a_t) \cdot U_t, 0}, \ldots)_{\mathbb{B}^*}$ and $\boldsymbol{c}_t := (\sigma_t(1, t), \omega x_t, \widetilde{\omega}, \boxed{\ldots, (\tau x_t, \widetilde{\tau}) \cdot Z_t, 0}, \ldots)_{\mathbb{B}}$, where Z_t is an independently random 2×2 matrix for each t and $U_t := (Z_t^{\mathrm{T}})^{-1}$, and other new variables are random. Here, the box-framed parts are the information theoretically hidden subspaces, the randomness of the hidden parts is amplified and the distribution of $(\pi v_t, a_t) \cdot U_t$ and $(\tau x_t, \widetilde{\tau}) \cdot Z_t$ is consistent with the key-query condition.

The consistent randomness amplification technique is composed of several computational and conceptual (information theoretical) transformations. One of the key tricks of the transformations is to amplify a source of randomness to a hidden part by applying a computational assumption, the DLIN assumption. Another computational trick is to swap two vectors in different positions under DLIN. Information theoretical key tricks are inter-subspace and intra-subspace types of conceptual transformations (see the full version for more details).

The security proofs of our IPE and ABE schemes are hierarchically constructed in a modular manner. The very top level of the security proof is based on the dual system encryption and its extension. Several problems in the middle level support the top level arguments. Our key techniques, the indexing and consistent randomness amplification techniques, which are also constructed in a hierarchical manner, are employed in the lowest level to reduce the hardness of the middle level problems to the DLIN assumption.

1.4   Notations

When A is a random variable or distribution, $y \xleftarrow{\mathsf{R}} A$ denotes that y is randomly selected from A according to its distribution. When A is a set, $y \xleftarrow{\mathsf{U}} A$ denotes that y is uniformly selected from A. $y := z$ denotes that y is set, defined or substituted by z. We denote the finite field of order q by $\mathbb{F}_q$, $\mathbb{F}_q \setminus \{0\}$ by $\mathbb{F}_q^\times$, and the set of positive integers by $\mathbb{N}$. The vector $\vec{0}$ is abused as the zero vector in $\mathbb{F}_q^n$ for any n. X^{T} denotes the transpose of matrix X. A bold face letter denotes an element of vector space $\mathbb{V}$, e.g., $\boldsymbol{x} \in \mathbb{V}$. When $\boldsymbol{b}_i \in \mathbb{V}$ $(i = 1, \ldots, n)$, $\mathrm{span}\langle \boldsymbol{b}_1, \ldots, \boldsymbol{b}_n \rangle \subseteq \mathbb{V}$ (resp. $\mathrm{span}\langle \vec{x}_1, \ldots, \vec{x}_n \rangle$) denotes the subspace generated by $\boldsymbol{b}_1, \ldots, \boldsymbol{b}_n$ (resp. $\vec{x}_1, \ldots, \vec{x}_n$). For bases $\mathbb{B} := (\boldsymbol{b}_1, \ldots, \boldsymbol{b}_N)$ and $\mathbb{B}^* := (\boldsymbol{b}_1^*, \ldots, \boldsymbol{b}_N^*)$, $(x_1, \ldots, x_N)_\mathbb{B} := \sum_{i=1}^N x_i \boldsymbol{b}_i$ and $(y_1, \ldots, y_N)_{\mathbb{B}^*} := \sum_{i=1}^N y_i \boldsymbol{b}_i^*$. $\vec{e}_1$ and $\vec{e}_2$ denote the canonical basis vectors in $\mathbb{F}_q^2$, i.e., $\vec{e}_1 := (1, 0)$ and $\vec{e}_2 := (0, 1)$. $GL(n, \mathbb{F}_q)$ denotes the general linear group of degree n over $\mathbb{F}_q$.

2   Dual Pairing Vector Spaces by Direct Product of Symmetric Pairing Groups

Definition 1. *"Symmetric bilinear pairing groups" $(q, \mathbb{G}, \mathbb{G}_T, G, e)$ are a tuple of a prime q, cyclic additive group $\mathbb{G}$ and multiplicative group $\mathbb{G}_T$ of order q, $G \neq 0 \in \mathbb{G}$, and a polynomial-time computable nondegenerate bilinear pairing $e : \mathbb{G} \times \mathbb{G} \to \mathbb{G}_T$ i.e., $e(sG, tG) = e(G, G)^{st}$ and $e(G, G) \neq 1$. Let $\mathcal{G}_{\mathsf{bpg}}$ be an algorithm that takes input 1^λ and outputs a description of bilinear pairing groups $(q, \mathbb{G}, \mathbb{G}_T, G, e)$ with security parameter λ.*

Definition 2. *"Dual pairing vector spaces (DPVS)" $(q, \mathbb{V}, \mathbb{G}_T, \mathbb{A}, e)$ by a direct product of symmetric pairing groups $(q, \mathbb{G}, \mathbb{G}_T, G, e)$ are a tuple of prime q, N-dimensional vector space $\mathbb{V} := \overbrace{\mathbb{G} \times \cdots \times \mathbb{G}}^{N}$ over $\mathbb{F}_q$, cyclic group $\mathbb{G}_T$ of order q, canonical basis $\mathbb{A} := (\boldsymbol{a}_1, \ldots, \boldsymbol{a}_N)$ of $\mathbb{V}$, where $\boldsymbol{a}_i := (\overbrace{0, \ldots, 0}^{i-1}, G, \overbrace{0, \ldots, 0}^{N-i})$, and pairing $e : \mathbb{V} \times \mathbb{V} \to \mathbb{G}_T$. The pairing is defined by $e(\boldsymbol{x}, \boldsymbol{y}) := \prod_{i=1}^N e(G_i, H_i) \in \mathbb{G}_T$*

where $\boldsymbol{x} := (G_1, \ldots, G_N) \in \mathbb{V}$ and $\boldsymbol{y} := (H_1, \ldots, H_N) \in \mathbb{V}$. This is nondegenerate bilinear i.e., $e(s\boldsymbol{x}, t\boldsymbol{y}) = e(\boldsymbol{x}, \boldsymbol{y})^{st}$ and if $e(\boldsymbol{x}, \boldsymbol{y}) = 1$ for all $\boldsymbol{y} \in \mathbb{V}$, then $\boldsymbol{x} = \boldsymbol{0}$. For all i and j, $e(\boldsymbol{a}_i, \boldsymbol{a}_j) = e(G, G)^{\delta_{i,j}}$ where $\delta_{i,j} = 1$ if $i = j$, and 0 otherwise, and $e(G, G) \neq 1 \in \mathbb{G}_T$. DPVS generation algorithm $\mathcal{G}_{\mathsf{dpvs}}$ takes input 1^λ ($\lambda \in \mathbb{N}$) and $N \in \mathbb{N}$, and outputs a description of $\mathsf{param}_{\mathbb{V}} := (q, \mathbb{V}, \mathbb{G}_T, \mathbb{A}, e)$ with security parameter λ and N-dimensional $\mathbb{V}$. It can be constructed by using $\mathcal{G}_{\mathsf{bpg}}$.

For the asymmetric version of DPVS, see Appendix A.2 in [12]. We describe random dual orthonormal basis generator $\mathcal{G}_{\mathsf{ob}}$, which is used as a subroutine in our IPE and ABE schemes.

$$\mathcal{G}_{\mathsf{ob}}(1^\lambda, (N_t)_{t=0,1}) : \mathsf{param}_{\mathbb{G}} := (q, \mathbb{G}, \mathbb{G}_T, G, e) \xleftarrow{\mathsf{R}} \mathcal{G}_{\mathsf{bpg}}(1^\lambda), \quad \psi \xleftarrow{\mathsf{U}} \mathbb{F}_q^\times,$$

$$\text{for } t = 0, 1, \quad \mathsf{param}_{\mathbb{V}_t} := (q, \mathbb{V}_t, \mathbb{G}_T, \mathbb{A}_t, e) := \mathcal{G}_{\mathsf{dpvs}}(1^\lambda, N_t, \mathsf{param}_{\mathbb{G}}),$$

$$X_t := (\chi_{t,i,j})_{i,j=1,\ldots,N_t} \xleftarrow{\mathsf{U}} GL(N_t, \mathbb{F}_q),$$

$$X_t^* := (\vartheta_{t,i,j})_{i,j=1,\ldots,N_t} := \psi \cdot (X_t^{\mathsf{T}})^{-1}, \quad \text{hereafter, } \vec{\chi}_{t,i} \text{ and } \vec{\vartheta}_{t,i}$$

denote the i-th rows of X_t and X_t^* for $i = 1, \ldots, N_t$, respectively,

$$\boldsymbol{b}_{t,i} := (\vec{\chi}_{t,i})_{\mathbb{A}_t} = \sum_{j=1}^{N_t} \chi_{t,i,j} \boldsymbol{a}_{t,j} \text{ for } i = 1, \ldots, N_t, \quad \mathbb{B}_t := (\boldsymbol{b}_{t,1}, \ldots, \boldsymbol{b}_{t,N_t}),$$

$$\boldsymbol{b}_{t,i}^* := (\vec{\vartheta}_{t,i})_{\mathbb{A}_t} = \sum_{j=1}^{N_t} \vartheta_{t,i,j} \boldsymbol{a}_{t,j} \text{ for } i = 1, \ldots, N_t, \quad \mathbb{B}_t^* := (\boldsymbol{b}_{t,1}^*, \ldots, \boldsymbol{b}_{t,N_t}^*),$$

$$g_T := e(G, G)^\psi, \quad \mathsf{param} := (\{\mathsf{param}_{\mathbb{V}_t}\}_{t=0,1}, g_T), \quad \text{return } (\mathsf{param}, \mathbb{B}, \mathbb{B}^*).$$

We note that $g_T = e(\boldsymbol{b}_{t,i}, \boldsymbol{b}_{t,i}^*)$ for $t = 0, 1; i = 1, \ldots, N_t$. Hereafter, for simplicity, we denote $N := N_1, \mathbb{V} := \mathbb{V}_1, \mathbb{A} := \mathbb{A}_1, \mathbb{B} := \mathbb{B}_1$ and $\mathbb{B}^* := \mathbb{B}_1^*$ for variables with $t = 1$.

3   Definitions of Generalized Inner-Product Encryption (IPE) and Attribute-Based Encryption (ABE)

3.1   Generalized Inner-Product Encryption

This section defines generalized inner product encryption (IPE) and its security.

The parameters of generalized inner-product predicates are expressed as a vector $\vec{x} := \{(t, x_t) \mid t \in I_{\vec{x}}, \ x_t \in \mathbb{F}_q\} \setminus \{\vec{0}\}$ with finite index set $I_{\vec{x}} \subset \mathbb{N}$ for encryption and a vector $\vec{v} := \{(t, v_t) \mid t \in I_{\vec{v}}, \ v_t \in \mathbb{F}_q\} \setminus \{\vec{0}\}$ with finite index set $I_{\vec{v}} \subset \mathbb{N}$ for a secret key, respectively. Here there are three types of unbounded IPE with respect to the decryption condition. For Type 1, $R(\vec{v}, \vec{x}) = 1$ iff $I_{\vec{v}} \subseteq I_{\vec{x}}$ and $\sum_{t \in I_{\vec{v}}} v_t x_t = 0$. For Type 2, $R(\vec{v}, \vec{x}) = 1$ iff $I_{\vec{v}} \supseteq I_{\vec{x}}$ and $\sum_{t \in I_{\vec{x}}} v_t x_t = 0$.

We will consider Type 0 inner-product predicate only for conventional prefix type vectors $\vec{v} := (v_1, \ldots, v_n)$ and $\vec{x} := (x_1, \ldots, x_{n'})$. For Type 0, $R(\vec{v}, \vec{x}) = 1$ iff $n = n'$ and $\vec{v} \cdot \vec{x} := \sum_{t=1}^{n} v_t x_t = 0$.

Definition 3. *An inner product encryption scheme (for generalized inner-product relation $R(\vec{v}, \vec{x})$) consists of probabilistic polynomial-time algorithms* $\mathsf{Setup}, \mathsf{KeyGen}, \mathsf{Enc}$ *and* Dec. *They are given as follows:*

Setup *takes as input security parameter* 1^λ. *It outputs public parameters* pk *and (master) secret key* sk.

KeyGen *takes as input public parameters* pk, *secret key* sk, *and vector* $\vec{v}$. *It outputs a corresponding secret key* sk$_{\vec{v}}$.

Enc *takes as input public parameters* pk, *message* m *in some associated message space,* msg, *and vector* $\vec{x}$. *It returns ciphertext* ct$_{\vec{x}}$.

Dec *takes as input the master public key* pk, *secret key* sk$_{\vec{v}}$ *and ciphertext* ct$_{\vec{x}}$. *It outputs either* $m' \in$ msg *or the distinguished symbol* $\perp$.

A generalized IPE scheme should have the following correctness property: for all $(\text{pk}, \text{sk}) \xleftarrow{\text{R}} \text{Setup}(1^\lambda)$, all vectors $\vec{v}$ and $\vec{x}$, all secret keys sk$_{\vec{v}} \xleftarrow{\text{R}}$ KeyGen(pk, sk, $\vec{v}$), all messages m, all ciphertext ct$_{\vec{x}} \xleftarrow{\text{R}}$ Enc(pk, $m, \vec{x}$), it holds that $m =$ Dec(pk, sk$_{\vec{v}}$, ct$_{\vec{x}}$) if $R(\vec{v}, \vec{x}) = 1$. Otherwise, it holds with negligible probability.

Definition 4. *The model for defining the adaptively fully-attribute-hiding security of IPE against adversary* $\mathcal{A}$ *(under chosen plaintext attacks) is given by the following game:*

Setup. *The challenger runs the setup algorithm,* $(\text{pk}, \text{sk}) \xleftarrow{\text{R}} \text{Setup}(1^\lambda)$, *and gives public parameters* pk *to* $\mathcal{A}$.

Phase 1. $\mathcal{A}$ *may adaptively make a polynomial number of key queries for vectors,* $\vec{v}$, *to the challenger. In response, the challenger gives the corresponding key* sk$_{\vec{v}} \xleftarrow{\text{R}}$ KeyGen(pk, sk, $\vec{v}$) *to* $\mathcal{A}$.

Challenge. $\mathcal{A}$ *submits challenge vectors* $(\vec{x}^{(0)}, \vec{x}^{(1)})$ *with the same index set* $I_{\vec{x}^{(0)}} = I_{\vec{x}^{(1)}}$ *(or* $n'^{(0)} = n'^{(1)}$ *for Type 0) and challenge messages* $(m^{(0)}, m^{(1)})$, *subject to the following restrictions:*

- *Any key query* $\vec{v}$ *in Phase 1 satisfies* $R(\vec{v}, \vec{x}^{(0)}) = R(\vec{v}, \vec{x}^{(1)}) = 0$, *or*
- *Two challenge messages are equal, i.e.,* $m^{(0)} = m^{(1)}$, *and any key query* $\vec{v}$ *in Phase 1 satisfies* $R(\vec{v}, \vec{x}^{(0)}) = R(\vec{v}, \vec{x}^{(1)})$.

The challenger flips a coin $b \xleftarrow{\text{U}} \{0, 1\}$, *and gives* ct$_{\vec{x}^{(b)}} \xleftarrow{\text{R}}$ Enc(pk, $m^{(b)}, \vec{x}^{(b)}$) *to* $\mathcal{A}$.

Phase 2. *Phase 1 is repeated with the above restriction for key query* $\vec{v}$ *and challenge,* $(\vec{x}^{(0)}, \vec{x}^{(1)})$ *and* $(m^{(0)}, m^{(1)})$.

Guess. $\mathcal{A}$ *outputs a bit* b', *and wins if* $b' = b$.

The advantage of $\mathcal{A}$ *in the above game is defined as* $\text{Adv}_{\mathcal{A}}^{\text{IPE,AH}}(\lambda) := \Pr[\mathcal{A} \text{ wins }] - 1/2$ *for any security parameter* λ. *An IPE scheme is* adaptively fully-attribute-hiding (AH) against chosen plaintext attacks *if all probabilistic polynomial-time adversaries* $\mathcal{A}$ *have at most negligible advantage in the above game. For each run of the game, the variable* s *is defined as* $s := 0$ *if* $m^{(0)} \neq m^{(1)}$ *for challenge messages* $m^{(0)}$ *and* $m^{(1)}$, *and* $s := 1$ *otherwise.*

3.2   Attribute-Based Encryption with Non-monotone Access Structures

Span Programs and Non-Monotone Access Structures

Definition 5 (Span Programs [2]). *Let* $\{p_1, \ldots, p_n\}$ *be a set of variables. A span program over* $\mathbb{F}_q$ *is a labeled matrix* $\hat{M} := (M, \rho)$ *where* M *is a* $(\ell \times r)$ *matrix* over $\mathbb{F}_q$ and ρ is a labeling of the rows of M by literals from $\{p_1, \ldots, p_n, \neg p_1, \ldots, \neg p_n\}$ (every row is labeled by one literal), i.e., $\rho : \{1, \ldots, \ell\} \rightarrow \{p_1, \ldots, p_n, \neg p_1, \ldots, \neg p_n\}$.

A span program accepts or rejects an input by the following criterion. For every input sequence $\delta \in \{0, 1\}^n$ define the submatrix M_δ of M consisting of those rows whose labels are set to 1 by the input δ, i.e., either rows labeled by some p_i such that $\delta_i = 1$ or rows labeled by some $\neg p_i$ such that $\delta_i = 0$. (i.e., $\gamma : \{1, \ldots, \ell\} \rightarrow \{0, 1\}$ is defined by $\gamma(j) = 1$ if $[\rho(j) = p_i] \wedge [\delta_i = 1]$ or $[\rho(j) = \neg p_i] \wedge [\delta_i = 0]$, and $\gamma(j) = 0$ otherwise. $M_\delta := (M_j)_{\gamma(j)=1}$, where M_j is the j-th row of M.)

The span program $\hat{M}$ accepts δ if and only if $\vec{1} \in \mathsf{span}\langle M_\delta \rangle$, i.e., some linear combination of the rows of M_δ gives the all one vector $\vec{1}$. (The row vector has the value 1 in each coordinate.) A span program computes a Boolean function f if it accepts exactly those inputs δ where $f(\delta) = 1$.

A span program is called monotone if the labels of the rows are only the positive literals $\{p_1, \ldots, p_n\}$. Monotone span programs compute monotone functions. (So, a span program in general is "non"-monotone.)

We assume that no row M_i $(i = 1, \ldots, \ell)$ of the matrix M is $\vec{0}$. We now introduce a non-monotone access structure with evaluating map γ that is employed in the proposed attribute-based encryption schemes.

Definition 6 (Access Structures). $\mathcal{U}_t$ $(t = 1, \ldots, d$ and $\mathcal{U}_t \subset \{0, 1\}^*)$ is a sub-universe, a set of attributes, each of which is expressed by a pair of sub-universe id and value of attribute, i.e., (t, v), where $t \in \{1, \ldots, d\}$ and $v \in \mathbb{F}_q$.

We now define such an attribute to be a variable p of a span program $\hat{M} := (M, \rho)$, i.e., $p := (t, v)$. An access structure $\mathbb{S}$ is span program $\hat{M} := (M, \rho)$ along with variables $p := (t, v), p' := (t', v'), \ldots$, i.e., $\mathbb{S} := (M, \rho)$ such that $\rho : \{1, \ldots, \ell\} \rightarrow \{(t, v), (t', v'), \ldots, \neg(t, v), \neg(t', v'), \ldots\}$.

Let Γ be a set of attributes, i.e., $\Gamma := \{(t, x_t) \mid x_t \in \mathbb{F}_q, 1 \le t \le d\}$, where $1 \le t \le d$ means that t is an element of some subset of $\{1, \ldots, d\}$.

When Γ is given to access structure $\mathbb{S}$, map $\gamma : \{1, \ldots, \ell\} \rightarrow \{0, 1\}$ for span program $\hat{M} := (M, \rho)$ is defined as follows: For $i = 1, \ldots, \ell$, set $\gamma(i) = 1$ if $[\rho(i) = (t, v_i)] \wedge [(t, x_t) \in \Gamma] \wedge [v_i = x_t]$ or $[\rho(i) = \neg(t, v_i)] \wedge [(t, x_t) \in \Gamma] \wedge [v_i \ne x_t]$. Set $\gamma(i) = 0$ otherwise.

Access structure $\mathbb{S} := (M, \rho)$ accepts Γ iff $\vec{1} \in \mathsf{span}\langle (M_i)_{\gamma(i)=1} \rangle$.

We now construct a secret-sharing scheme for a non-monotone access structure or span program.

Definition 7. A secret-sharing scheme for span program $\hat{M} := (M, \rho)$ is:

1. Let M be $\ell \times r$ matrix. Let column vector $\vec{f}^{\mathrm{T}} := (f_1, \ldots, f_r)^{\mathrm{T}} \xleftarrow{\mathsf{U}} \mathbb{F}_q^r$. Then, $s_0 := \vec{1} \cdot \vec{f}^{\mathrm{T}} = \sum_{k=1}^r f_k$ is the secret to be shared, and $\vec{s}^{\mathrm{T}} := (s_1, \ldots, s_\ell)^{\mathrm{T}} := M \cdot \vec{f}^{\mathrm{T}}$ is the vector of ℓ shares of the secret s_0 and the share s_i belongs to $\rho(i)$.

2. *If span program $\hat{M} := (M, \rho)$ accept δ, or access structure $\mathbb{S} := (M, \rho)$ accepts Γ, i.e., $\vec{1} \in \text{span}\langle (M_i)_{\gamma(i)=1} \rangle$ with $\gamma : \{1, \ldots, \ell\} \to \{0, 1\}$, then there exist constants $\{\alpha_i \in \mathbb{F}_q \mid i \in I\}$ such that $I \subseteq \{i \in \{1, \ldots, \ell\} \mid \gamma(i) = 1\}$ and $\sum_{i \in I} \alpha_i s_i = s_0$. Furthermore, these constants $\{\alpha_i\}$ can be computed in time polynomial in the size of matrix M.*

Key-Policy Attribute-Based Encryption. In key-policy attribute-based encryption (KP-ABE), encryption (resp. a secret key) is associated with attributes Γ (resp. access structure $\mathbb{S}$). Relation R for KP-ABE is defined as $R(\mathbb{S}, \Gamma) = 1$ iff access structure $\mathbb{S}$ accepts Γ.

Definition 8 (Key-Policy Attribute-Based Encryption: KP-ABE). *A key-policy attribute-based encryption scheme consists of probabilistic polynomial-time algorithms* Setup, KeyGen, Enc *and* Dec. *They are given as follows:*

Setup *takes as input security parameter 1^λ. It outputs public parameters* pk *and master secret key* sk.

KeyGen *takes as input public parameters* pk, *master secret key* sk, *and access structure $\mathbb{S} := (M, \rho)$. It outputs a corresponding secret key* $\text{sk}_\mathbb{S}$.

Enc *takes as input public parameters* pk, *message m in some associated message space* msg, *and a set of attributes, $\Gamma := \{(t, x_t) | x_t \in \mathbb{F}_q, 1 \leq t \leq d\}$. It outputs a ciphertext* ct_Γ.

Dec *takes as input public parameters* pk, *secret key $\text{sk}_\mathbb{S}$ for access structure $\mathbb{S}$, and ciphertext ct_Γ that was encrypted under a set of attributes Γ. It outputs either $m' \in$* msg *or the distinguished symbol $\perp$.*

A KP-ABE scheme should have the following correctness property: for all $(\text{pk}, \text{sk}) \xleftarrow{\text{R}} \text{Setup}(1^\lambda)$, all access structures $\mathbb{S}$, all secret keys $\text{sk}_\mathbb{S} \xleftarrow{\text{R}} \text{KeyGen}(\text{pk}, \text{sk}, \mathbb{S})$, all messages m, all attribute sets Γ, all ciphertexts $\text{ct}_\Gamma \xleftarrow{\text{R}} \text{Enc}(\text{pk}, m, \Gamma)$, it holds that $m = \text{Dec}(\text{pk}, \text{sk}_\mathbb{S}, \text{ct}_\Gamma)$ if $\mathbb{S}$ accepts Γ. Otherwise, it holds with negligible probability.

Definition 9. *The model for defining the adaptively payload-hiding security of KP-ABE under chosen plaintext attack is given by the following game:*

Setup. *The challenger runs the setup algorithm, $(\text{pk}, \text{sk}) \xleftarrow{\text{R}} \text{Setup}(1^\lambda)$, and gives public parameters* pk *to the adversary.*

Phase 1. *The adversary is allowed to adaptively issue a polynomial number of key queries, $\mathbb{S}$, to the challenger. The challenger gives $\text{sk}_\mathbb{S} \xleftarrow{\text{R}} \text{KeyGen}(\text{pk}, \text{sk}, \mathbb{S})$ to the adversary.*

Challenge. *The adversary submits two messages $m^{(0)}, m^{(1)}$ and a set of attributes, Γ, provided that no $\mathbb{S}$ queried to the challenger in Phase 1 accepts Γ. The challenger flips a coin $b \xleftarrow{\text{U}} \{0, 1\}$, and computes $\text{ct}_\Gamma^{(b)} \xleftarrow{\text{R}} \text{Enc}(\text{pk}, m^{(b)}, \Gamma)$. It gives $\text{ct}_\Gamma^{(b)}$ to the adversary.*

Phase 2. *Phase 1 is repeated with the restriction that no queried $\mathbb{S}$ accepts challenge Γ.*

Guess. *The adversary outputs a guess b' of b, and wins if $b' = b$.*

The advantage of adversary $\mathcal{A}$ in the above game is defined as $\mathsf{Adv}_{\mathcal{A}}^{\mathsf{KP\text{-}ABE,PH}}(\lambda) := \Pr[\mathcal{A} \text{ wins }] - 1/2$ *for any security parameter λ. A KP-ABE scheme is adaptively payload-hiding secure if all polynomial time adversaries have at most a negligible advantage in the above game.*

4   Proposed IPE Schemes

4.1   Type 1 IPE Scheme

Construction Idea for Our Type 1 and 2 IPE Schemes. In the existing constructions [11, 7, 12–15] of IPE on DPVS, around cn ($c \geq 1$) dimensional vector spaces are used for n-dimensional attribute and predicate vectors. Here, the vectors are encoded in an n-dimensional subspace. Although this is a typical strategy of constructing IPE on DPVS, we cannot employ this idea in the *unbounded* setting, where we can use only constant dimensional spaces. In our construction, each component x_t of $\vec{x}$ (resp. v_t of $\vec{v}$) is encoded in a constant dimensional space. In order to meet the decryption condition, we employ the *indexing* technique and n-out-of-n secret sharing trick. For example, in Type 1 construction, 4-dimensional vector $(\mu_t(t, -1), \delta v_t, s_t)$ is encoded in key $\boldsymbol{k}_t^*$, and $(\sigma_t(1, t), \omega x_t, \widetilde{\omega})$ is encoded in ciphertext $\boldsymbol{c}_t$. The first 2-dimension is used for indexes, and s_t in the fourth component of $\boldsymbol{k}_t^*$ is for the secret sharing. Informally, a ciphertext can be decrypted if all n pieces of shares s_t are recovered. A Type 2 IPE scheme can be constructed from our Type 1 scheme by setting the secret-sharing mechanism in the ciphertext side instead of the secret key side.

Construction of Type 1 IPE

$\mathsf{Setup}(1^\lambda):$ $(\mathsf{param}, (\mathbb{B}_0, \mathbb{B}_0^*), (\mathbb{B}, \mathbb{B}^*)) \xleftarrow{\mathsf{R}} \mathcal{G}_{\mathsf{ob}}(1^\lambda, (N_0 := 5, N := 15)),$

$\quad \widehat{\mathbb{B}}_0 := (\boldsymbol{b}_{0,1}, \boldsymbol{b}_{0,3}, \boldsymbol{b}_{0,5}),\ \widehat{\mathbb{B}} := (\boldsymbol{b}_1, .., \boldsymbol{b}_4, \boldsymbol{b}_{14}, \boldsymbol{b}_{15}),$

$\quad \widehat{\mathbb{B}}_0^* := (\boldsymbol{b}_{0,1}^*, \boldsymbol{b}_{0,3}^*, \boldsymbol{b}_{0,4}^*),\ \widehat{\mathbb{B}}^* := (\boldsymbol{b}_1^*, .., \boldsymbol{b}_4^*, \boldsymbol{b}_{12}^*, \boldsymbol{b}_{13}^*),$

$\quad \text{return}\ \ \mathsf{pk} := (1^\lambda, \mathsf{param}, \widehat{\mathbb{B}}_0, \widehat{\mathbb{B}}),\ \mathsf{sk} := (\widehat{\mathbb{B}}_0^*, \widehat{\mathbb{B}}^*).$

$\mathsf{KeyGen}(\mathsf{pk}, \mathsf{sk}, \vec{v} := \{(t, v_t) \mid t \in I_{\vec{v}}\}):\ s_t, \delta, \eta_0 \xleftarrow{\mathsf{U}} \mathbb{F}_q \text{ for } t \in I_{\vec{v}},$

$\quad s_0 := \sum_{(t, v_t) \in \vec{v}} s_t,\quad \boldsymbol{k}_0^* := (\ -s_0,\ 0,\ 1,\ \eta_0,\ 0\ )_{\mathbb{B}_0^*},$

$\quad \text{for } t \in I_{\vec{v}},\quad \mu_t, \eta_{t,1}, \eta_{t,2} \xleftarrow{\mathsf{U}} \mathbb{F}_q,$

$$\boldsymbol{k}_t^* := (\ \overbrace{\mu_t(t,\ -1),\ \delta v_t,\ s_t}^{4}\ \ \overbrace{0^7,}^{7}\ \overbrace{\eta_{t,1}, \eta_{t,2},}^{2}\ \overbrace{0^2}^{2}\ )_{\mathbb{B}^*},$$

$\quad \text{return}\ \ \mathsf{sk}_{\vec{v}} := (I_{\vec{v}}, \boldsymbol{k}_0^*, \{\boldsymbol{k}_t^*\}_{t \in I_{\vec{v}}}).$

$\mathsf{Enc}(\mathsf{pk}, m, \vec{x} := \{(t, x_t) \mid t \in I_{\vec{x}}\}):\ \omega, \widetilde{\omega}, \zeta, \varphi_0 \xleftarrow{\mathsf{U}} \mathbb{F}_q,$

$\quad \boldsymbol{c}_0 := (\ \widetilde{\omega},\ 0,\ \zeta,\ 0,\ \varphi_0\ )_{\mathbb{B}_0},\ c_T := g_T^\zeta,$

for $t \in I_{\vec{x}}, \quad \sigma_t, \varphi_{t,1}, \varphi_{t,2} \xleftarrow{\mathsf{U}} \mathbb{F}_q,$

$$c_t := (\overbrace{\sigma_t(1, \ t), \ \omega x_t, \ \widetilde{\omega}}^{4} \quad \overbrace{0^7,}^{7} \quad \overbrace{0^2,}^{2} \quad \overbrace{\varphi_{t,1}, \varphi_{t,2}}^{2})_{\mathbb{B}},$$

return $\mathsf{ct}_{\vec{x}} := (I_{\vec{x}}, c_0, \{c_t\}_{t \in I_{\vec{x}}}, c_T).$

$\mathsf{Dec}(\mathsf{pk}, \ \mathsf{sk}_{\vec{v}} := (I_{\vec{v}}, k_0^*, \{k_t^*\}_{t \in I_{\vec{v}}}), \ \mathsf{ct}_{\vec{x}} := (I_{\vec{x}}, c_0, \{c_t\}_{t \in I_{\vec{x}}}, c_T)) :$

if $I_{\vec{v}} \subseteq I_{\vec{x}}, \quad K := e(c_0, k_0^*) \cdot \prod_{t \in I_{\vec{v}}} e(c_t, k_t^*), \quad$ return $m' := c_T / K,$

else return $\perp$.

[Correctness] If $I_{\vec{v}} \subseteq I_{\vec{x}}$ and $\sum_{t \in I_{\vec{v}}} v_t x_t = 0, \quad e(c_0, k_0^*) \cdot \prod_{t \in I_{\vec{v}}} e(c_t, k_t^*) =$

$g_T^{-\widetilde{\omega}s_0 + \zeta} \cdot \prod_{t \in I_{\vec{v}}} g_T^{\delta \omega v_t x_t + \widetilde{\omega}s_t} = g_T^{-\widetilde{\omega}s_0 + \zeta} \cdot g_T^{\delta\omega(\sum_{t \in I_{\vec{v}}} v_t x_t) + \widetilde{\omega}(\sum_{t \in I_{\vec{v}}} s_t)} = g_T^{-\widetilde{\omega}s_0 + \zeta + \widetilde{\omega}s_0}$

$= g_T^{\zeta}.$

Theorem 1. *The proposed Type 1 IPE scheme is adaptively fully-attribute-hiding against chosen plaintext attacks under the DLIN assumption.*

The proof of Theorem 1 is given in the full version of this paper.

4.2   Type 0 IPE Scheme

Construction Idea for Our Type 0 IPE Scheme. In Type 1 construction, 4-dimensional vector $(\mu_t(t, -1), \delta v_t, s_t)$ is encoded in key k_t^*, and $(\sigma_t(1, t), \omega x_t, \widetilde{\omega})$ is encoded in ciphertext c_t. Here, secret-sharing system, s_t for $t \in I_{\vec{v}}$, in k_t^* are used to assure one of the decryption conditions, $I_{\vec{v}} \subseteq I_{\vec{x}}$. In Type 0 scheme, to achieve its decryption condition $I_{\vec{v}} = I_{\vec{x}}$ for $\vec{v} := (v_1, \ldots, v_n), \vec{x} := (x_1, \ldots, x_{n'})$ i.e., that is equivalent to $n = n'$, we use the above mechanism also to ciphertext side. Then, in our Type 0 scheme, we encode 5-dimensional $(\mu_t(t, -1), \delta v_t, s_t, \widetilde{\delta})$ in the first part of k_t^*, and $(\sigma_t(1, t), \omega x_t, \widetilde{\omega}, f_t)$ in the first part of c_t with random $\mu_t, \sigma_t, \omega, \widetilde{\omega}, \delta, \widetilde{\delta}, s_t, f_t \xleftarrow{\mathsf{U}} \mathbb{F}_q.$

Construction of Type 0 IPE

$\mathsf{Setup}(1^\lambda) : \quad (\mathsf{param}, (\mathbb{B}_0, \mathbb{B}_0^*), (\mathbb{B}, \mathbb{B}^*)) \xleftarrow{\mathsf{R}} \mathcal{G}_{\mathsf{ob}}(1^\lambda, (N_0 := 9, N := 21)),$

$\widehat{\mathbb{B}}_0 := (b_{0,1}, b_{0,2}, b_{0,5}, b_{0,8}, b_{0,9}), \quad \widehat{\mathbb{B}} := (b_1, \ldots, b_5, b_{19}, \ldots, b_{21}),$

$\widehat{\mathbb{B}}_0^* := (b_{0,1}^*, b_{0,2}^*, b_{0,5}^*, \ldots, b_{0,7}^*), \quad \widehat{\mathbb{B}}^* := (b_1^*, \ldots, b_5^*, b_{16}^*, \ldots, b_{18}^*),$

return $\mathsf{pk} := (1^\lambda, \mathsf{param}, \widehat{\mathbb{B}}_0, \widehat{\mathbb{B}}), \quad \mathsf{sk} := (\widehat{\mathbb{B}}_0^*, \widehat{\mathbb{B}}^*).$

$\mathsf{KeyGen}(\mathsf{pk}, \mathsf{sk}, \vec{v} := (v_1, \ldots, v_n)) : \quad s_t, \delta, \widetilde{\delta}, \eta_{0,1}, \eta_{0,2} \xleftarrow{\mathsf{U}} \mathbb{F}_q$ for $t = 1, \ldots, n,$

$s_0 := \sum_{t=1}^n s_t, \quad k_0^* := (-s_0, \ \widetilde{\delta}, \ 0^2, \ 1, \ \eta_{0,1}, \eta_{0,2}, \ 0^2)_{\mathbb{B}_0^*},$

for $t = 1, \ldots, n, \quad \mu_t, \eta_{t,1}, .., \eta_{t,3} \xleftarrow{\mathsf{U}} \mathbb{F}_q,$

$$k_t^* := (\overbrace{\mu_t(t, \ -1), \ \delta v_t, \ s_t, \ \widetilde{\delta},}^{5} \quad \overbrace{0^{10},}^{10} \quad \overbrace{\eta_{t,1}, .., \eta_{t,3},}^{3} \quad \overbrace{0^3}^{3})_{\mathbb{B}^*},$$

return $\mathsf{sk}_{\vec{v}} := \{k_t^*\}_{t=0,\ldots,n}.$

$\text{Enc}(\text{pk}, m, \vec{x} := (x_1, \ldots, x_{n'})): \quad f_t, \omega, \widetilde{\omega}, \zeta, \varphi_{0,1}, \varphi_{0,2} \xleftarrow{\mathsf{U}} \mathbb{F}_q \text{ for } t = 1, \ldots, n',$

$\quad f_0 := \sum_{t=1}^{n'} f_t, \quad c_0 := (\widetilde{\omega}, -f_0, 0^2, \zeta, 0^2, \varphi_{0,1}, \varphi_{0,2})_{\mathbb{B}_0}, \quad c_T := g_T^\zeta,$

$\quad \text{for } t = 1, \ldots, n', \quad \sigma_t, \varphi_{t,1}, \ldots, \varphi_{t,3} \xleftarrow{\mathsf{U}} \mathbb{F}_q,$

$$c_t := (\overbrace{\sigma_t(1, t), \omega x_t, \widetilde{\omega}, f_t,}^{5} \overbrace{0^{10},}^{10} \overbrace{0^3,}^{3} \overbrace{\varphi_{t,1}, \ldots, \varphi_{t,3}}^{3})_\mathbb{B},$$

$\quad \text{return } \text{ct}_{\vec{x}} := (\{c_t\}_{t=0,\ldots,n'}, c_T).$

$\text{Dec}(\text{pk}, \text{sk}_{\vec{v}} := \{k_t^*\}_{t=0,\ldots,n}, \text{ct}_{\vec{x}} := (\{c_t\}_{t=0,\ldots,n'}, c_T)):$

$\quad \text{if } n = n', K := \prod_{t=0}^n e(c_t, k_t^*), \quad \text{return } m' := c_T/K, \quad \text{else return } \bot.$

Correctness of the scheme can be shown in a similar manner to that of our Type 1 IPE scheme.

Theorem 2. *The proposed Type 0 IPE scheme is adaptively fully-attribute-hiding against chosen plaintext attacks under the DLIN assumption.*

The proof of Theorem 2 is given in the full version of this paper.

5   Proposed KP-ABE Scheme (Basic)

We define function $\widetilde{\rho} : \{1, .., \ell\} \to \{1, .., d\}$ by $\widetilde{\rho}(i) := t$ if $\rho(i) = (t, v)$ or $\rho(i) = \neg(t, v)$, where ρ is given in access structure $\mathbb{S} := (M, \rho)$. In the proposed scheme, we assume that $\widetilde{\rho}$ is injective for $\mathbb{S} := (M, \rho)$ in $\text{sk}_\mathbb{S}$. For the modified scheme without such a restriction, see the full version. Let $d := poly(\lambda)$, where $poly(\cdot)$ is a polynomial.

$\text{Setup}(1^\lambda): \quad (\text{param}, (\mathbb{B}_0, \mathbb{B}_0^*), (\mathbb{B}, \mathbb{B}^*)) \xleftarrow{\mathsf{R}} \mathcal{G}_{\text{ob}}(1^\lambda, (N_0 := 5, N := 14)),$

$\quad \widehat{\mathbb{B}}_0 := (b_{0,1}, b_{0,3}, b_{0,5}), \quad \widehat{\mathbb{B}} := (b_1, .., b_4, b_{13}, b_{14}),$

$\quad \widehat{\mathbb{B}}_0^* := (b_{0,1}^*, b_{0,3}^*, b_{0,4}^*), \quad \widehat{\mathbb{B}}^* := (b_1^*, .., b_4^*, b_{11}^*, b_{12}^*),$

$\quad \text{return } \text{pk} := (1^\lambda, \text{param}, \widehat{\mathbb{B}}_0, \widehat{\mathbb{B}}), \quad \text{sk} := (\widehat{\mathbb{B}}_0^*, \widehat{\mathbb{B}}^*).$

$\text{KeyGen}(\text{pk}, \text{sk}, \mathbb{S} := (M, \rho)): \quad \vec{f} \xleftarrow{\mathsf{U}} \mathbb{F}_q^r, \quad s_0 := \vec{1} \cdot \vec{f}^\mathsf{T},$

$\quad \vec{s}^\mathsf{T} := (s_1, \ldots, s_\ell)^\mathsf{T} := M \cdot \vec{f}^\mathsf{T}, \quad \eta_0 \xleftarrow{\mathsf{U}} \mathbb{F}_q, \quad k_0^* := (-s_0, 0, 1, \eta_0, 0)_{\mathbb{B}_0^*},$

$\quad \text{for } i = 1, \ldots, \ell, \quad \mu_i, \theta_i, \eta_{i,1}, \eta_{i,2} \xleftarrow{\mathsf{U}} \mathbb{F}_q,$

$\quad \text{if } \rho(i) = (t, v_i),$

$$k_i^* := (\overbrace{\mu_i(t, -1), s_i + \theta_i v_i, -\theta_i}^{4} \overbrace{0^6,}^{6} \overbrace{\eta_{i,1}, \eta_{i,2},}^{2} \overbrace{0^2}^{2})_{\mathbb{B}^*},$$

$\quad \text{if } \rho(i) = \neg(t, v_i),$

$$k_i^* := (\overbrace{\mu_i(t, -1), s_i(v_i, -1),}^{4} \overbrace{0^6,}^{6} \overbrace{\eta_{i,1}, \eta_{i,2},}^{2} \overbrace{0^2}^{2})_{\mathbb{B}^*},$$

$\quad \text{return } \text{sk}_\mathbb{S} := (\mathbb{S}, \{k_i^*\}_{i=0,\ldots,\ell}).$

$\mathsf{Enc}(\mathsf{pk},\ m,\ \Gamma := \{(t, x_t) \mid 1 \le t \le d\}) :\ \omega, \zeta, \varphi_0 \xleftarrow{\mathsf{U}} \mathbb{F}_q,$

$\quad \boldsymbol{c}_0 := (\omega,\ 0,\ \zeta,\ 0,\ \varphi_0)_{\mathbb{B}_0},\quad c_{d+1} := g_T^{\zeta} m,$

$\quad \text{for } (t, x_t) \in \Gamma,\quad \sigma_t, \varphi_{t,1}, \varphi_{t,2} \xleftarrow{\mathsf{U}} \mathbb{F}_q,$

$$\boldsymbol{c}_t := (\ \overbrace{\sigma_t(1,\ t),\ \omega(1,\ x_t),}^{4}\quad \overbrace{0^6,}^{6}\quad \overbrace{0^2,}^{2}\quad \overbrace{\varphi_{t,1},\ \varphi_{t,2}}^{2}\ )_{\mathbb{B}},$$

$\quad \text{return } \mathsf{ct}_\Gamma := (\Gamma, \boldsymbol{c}_0, \{\boldsymbol{c}_t\}_{(t,x_t)\in\Gamma}, c_{d+1}).$

$\mathsf{Dec}(\mathsf{pk},\ \mathsf{sk}_\mathbb{S} := (\mathbb{S}, \{\boldsymbol{k}_i^*\}_{i=0,\dots,\ell}),\ \mathsf{ct}_\Gamma := (\Gamma, \boldsymbol{c}_0, \{\boldsymbol{c}_t\}_{(t,x_t)\in\Gamma}, c_{d+1})) :$

$\quad$ If $\mathbb{S} := (M, \rho)$ accepts $\Gamma := \{(t, x_t)\}$, then compute I and $\{\alpha_i\}_{i\in I}$ such that

$\quad \vec{1} = \sum_{i\in I} \alpha_i M_i$, where M_i is the i-th row of M, and

$\quad I \subseteq \{i \in \{1, \dots, \ell\} \mid [\rho(i) = (t, v_i) \land (t, v_i) \in \Gamma]$

$$\lor\, [\rho(i) = \neg(t, v_i) \land (t, x_t) \in \Gamma \land v_i \ne x_t]\, \},$$

$\quad K := e(\boldsymbol{c}_0, \boldsymbol{k}_0^*) \prod_{i\in I\, \land\, \rho(i)=(t,v_i)} e(\boldsymbol{c}_t, \boldsymbol{k}_i^*)^{\alpha_i} \prod_{i\in I\, \land\, \rho(i)=\neg(t,v_i)} e(\boldsymbol{c}_t, \boldsymbol{k}_i^*)^{\alpha_i/(v_i - x_t)},$

$\quad$ return $m' := c_{d+1}/K$, else return $\perp$.

[Correctness] If $\mathbb{S} := (M, \rho)$ accepts $\Gamma := \{(t, x_t)\}$,

$K = g_T^{-\omega s_0 + \zeta} \prod_{i\in I\, \land\, \rho(i)=(t,v_i)} g_T^{\omega \alpha_i s_i} \prod_{i\in I\, \land\, \rho(i)=\neg(t,v_i)} g_T^{\omega \alpha_i s_i (v_i - x_t)/(v_i - x_t)} = $

$g_T^{\omega(-s_0 + \sum_{i\in I} \alpha_i s_i) + \zeta} = g_T^{\zeta}.$

Theorem 3. *The proposed KP-ABE scheme is adaptively payload-hiding against chosen plaintext attacks under the DLIN assumption.*

The proof of Theorem 3 is given in the full version of this paper.

References

1. Attrapadung, N., Libert, B.: Functional Encryption for Inner Product: Achieving Constant-Size Ciphertexts with Adaptive Security or Support for Negation. In: Nguyen, P.Q., Pointcheval, D. (eds.) PKC 2010. LNCS, vol. 6056, pp. 384–402. Springer, Heidelberg (2010)
2. Beimel, A.: Secure schemes for secret sharing and key distribution. PhD Thesis, Israel Institute of Technology, Technion, Haifa (1996)
3. Bellare, M., Waters, B., Yilek, S.: Identity-Based Encryption Secure against Selective Opening Attack. In: Ishai, Y. (ed.) TCC 2011. LNCS, vol. 6597, pp. 235–252. Springer, Heidelberg (2011)
4. Boneh, D., Waters, B.: Conjunctive, Subset, and Range Queries on Encrypted Data. In: Vadhan, S.P. (ed.) TCC 2007. LNCS, vol. 4392, pp. 535–554. Springer, Heidelberg (2007)
5. Goyal, V., Pandey, O., Sahai, A., Waters, B.: Attribute-based encryption for fine-grained access control of encrypted data. In: Juels, A., Wright, R.N., di Vimercati, S.D.C. (eds.) ACM CCS 2006, pp. 89–98. ACM (2006)
6. Katz, J., Sahai, A., Waters, B.: Predicate Encryption Supporting Disjunctions, Polynomial Equations, and Inner Products. In: Smart, N.P. (ed.) EUROCRYPT 2008. LNCS, vol. 4965, pp. 146–162. Springer, Heidelberg (2008)

7. Lewko, A., Okamoto, T., Sahai, A., Takashima, K., Waters, B.: Fully Secure Functional Encryption: Attribute-Based Encryption and (Hierarchical) Inner Product Encryption. In: Gilbert, H. (ed.) EUROCRYPT 2010. LNCS, vol. 6110, pp. 62–91. Springer, Heidelberg (2010), full version available at http://eprint.iacr.org/2010/110

8. Lewko, A., Waters, B.: Decentralizing attribute-based encryption. In: Paterson [17], pp. 568–588

9. Lewko, A., Waters, B.: Unbounded HIBE and attribute-based encryption. In: Paterson [17], pp. 547–567

10. Lewko, A., Waters, B.: New Proof Methods for Attribute-Based Encryption: Achieving Full Security through Selective Techniques. In: Safavi-Naini, R., Canetti, R. (eds.) CRYPTO 2012. LNCS, vol. 7417, pp. 180–198. Springer, Heidelberg (2012)

11. Okamoto, T., Takashima, K.: Hierarchical Predicate Encryption for Inner-Products. In: Matsui, M. (ed.) ASIACRYPT 2009. LNCS, vol. 5912, pp. 214–231. Springer, Heidelberg (2009)

12. Okamoto, T., Takashima, K.: Fully Secure Functional Encryption with General Relations from the Decisional Linear Assumption. In: Rabin, T. (ed.) CRYPTO 2010. LNCS, vol. 6223, pp. 191–208. Springer, Heidelberg (2010), full version is available at http://eprint.iacr.org/2010/563

13. Okamoto, T., Takashima, K.: Achieving Short Ciphertexts or Short Secret-Keys for Adaptively Secure General Inner-Product Encryption. In: Lin, D., Tsudik, G., Wang, X. (eds.) CANS 2011. LNCS, vol. 7092, pp. 138–159. Springer, Heidelberg (2011), full version is available at http://eprint.iacr.org/2011/648

14. Waters, B.: Ciphertext-Policy Attribute-Based Encryption: An Expressive, Efficient, and Provably Secure Realization. In: Catalano, D., Fazio, N., Gennaro, R., Nicolosi, A. (eds.) PKC 2011. LNCS, vol. 6571, pp. 53–70. Springer, Heidelberg (2011), full version is available at http://eprint.iacr.org/2011/543

15. Okamoto, T., Takashima, K.: Efficient (hierarchical) inner product encryption tightly reduced from the decisional linear assumption. In: IEICE Trans. Fundamentals E96-A(1) (to appear)

16. Ostrovsky, R., Sahai, A., Waters, B.: Attribute-based encryption with non-monotonic access structures. In: Ning, P., di Vimercati, S.D.C., Syverson, P.F. (eds.) ACM CCS 2007, pp. 195–203. ACM (2007)

17. Paterson, K.G. (ed.): EUROCRYPT 2011. LNCS, vol. 6632. Springer, Heidelberg (2011)

18. Sahai, A., Waters, B.: Fuzzy Identity-Based Encryption. In: Cramer, R. (ed.) EUROCRYPT 2005. LNCS, vol. 3494, pp. 457–473. Springer, Heidelberg (2005)

19. Waters, B.: Dual System Encryption: Realizing Fully Secure IBE and HIBE under Simple Assumptions. In: Halevi, S. (ed.) CRYPTO 2009. LNCS, vol. 5677, pp. 619–636. Springer, Heidelberg (2009)

20. Waters, B.: Ciphertext-Policy Attribute-Based Encryption: An Expressive, Efficient, and Provably Secure Realization. In: Catalano, D., Fazio, N., Gennaro, R., Nicolosi, A. (eds.) PKC 2011. LNCS, vol. 6571, pp. 53–70. Springer, Heidelberg (2011)

Computing on Authenticated Data: New Privacy Definitions and Constructions

Nuttapong Attrapadung[1,*], Benoît Libert[2,**], and Thomas Peters[2,***]

[1] Research Institute for Secure Systems, AIST (Japan)
[2] Université Catholique de Louvain, ICTEAM Institute (Belgium)

Abstract. Homomorphic signatures are primitives that allow for public computations on authenticated data. At TCC 2012, Ahn *et al.* defined a framework and security notions for such systems. For a predicate P, their notion of P-homomorphic signature makes it possible, given signatures on a message set M, to publicly derive a signature on any message m' such that $P(M, m') = 1$. Beyond unforgeability, Ahn *et al.* considered a strong notion of privacy – called strong context hiding – requiring that derived signatures be perfectly indistinguishable from signatures newly generated by the signer. In this paper, we first note that the definition of strong context hiding may not imply unlinkability properties that can be expected from homomorphic signatures in certain situations. We then suggest other definitions of privacy and discuss the relations among them. Our strongest definition, called *complete* context hiding security, is shown to imply previous ones. In the case of linearly homomorphic signatures, we only attain a slightly weaker level of privacy which is nevertheless stronger than in previous realizations in the standard model. For subset predicates, we prove that our strongest notion of privacy is satisfiable and describe a completely context hiding system with constant-size public keys. In the standard model, this construction is the first one that allows signing messages of arbitrary length. The scheme builds on techniques that are very different from those of Ahn *et al.*

Keywords: Homomorphic signatures, provable security, privacy, unlinkability, standard model.

1 Introduction

With the advent of fully homomorphic encryption [24], much attention has been paid to the problem of computing on encrypted data (see, e.g., [24,37]) in the recent years. This also revived the interest of the research community in homomorphic signatures, which allow for computations on authenticated data.

---

* This author is supported by KAKENHI (Grant-in-Aid for Young Scientists B) No. 22700020. This work was done while the author visited ENS Paris.
** This author was supported by the Belgian Fund for Scientific Research (F.R.S.-F.N.R.S.) via a "Collaborateur scientifique" fellowship.
*** Supported by the Walloon Region Camus Project.

X. Wang and K. Sako (Eds.): ASIACRYPT 2012, LNCS 7658, pp. 367–385, 2012.

Informally, a signer has a set of messages $\{m_i\}_{i=1}^k$ and generates a corresponding set of signatures $\{\sigma_i\}_{i=1}^k$ with $\sigma_i = \mathsf{Sign}(\mathsf{sk}, m_i)$ for each i. The signed dataset $\{(m_i, \sigma_i)\}_{i=1}^k$ is then archived on a remote server. Later on, the server can publicly compute $(m, \sigma) = \mathsf{Evaluate}(\mathsf{pk}, \{(m_i, \sigma_i)\}_{i=1}^k, f)$ such that $\mathsf{Verify}(\mathsf{pk}, m, \sigma) = 1$, where $m = f(m_1, \ldots, m_k)$ for some function f.

In the last decade, the area was investigated by several lines of research: examples include homomorphic signatures for arithmetic functions [10,22,11,12] but also redactable signatures [34,15,16,14] and various other forms of algebraic signatures [33,7,26,27].

Recently, Ahn et al. [3] defined a framework for computing on signed data. For a predicate P, their notion of P-homomorphic signature allows anyone who observes signatures on a message m to publicly derive signatures on messages m' such that $P(m, m') = 1$. This framework is geared towards capturing homomorphic signatures supporting quoting and redacting, arithmetic functions and more. Ahn et al. [3] gave thorough definitions for the unforgeability of P-homomorphic signatures. Besides, they introduced a strong notion of privacy, called *strong context hiding*, that captures the infeasibility of linking a derived signature to the signature it was derived from. A scheme is said strongly context hiding when a derived signature is statistically indistinguishable from a freshly generated signature, *even* when the original signature is available.

1.1 Related Work

Homomorphic signatures were first considered by Johnson, Molnar, Song and Wagner [32]. Boneh, Freeman, Katz and Waters [10] used them to sign vector spaces in order to prevent pollution attacks in network coding. They adapted the definitions of [32] to the network coding setting and designed a linearly homomorphic scheme in the random oracle model using bilinear maps. Gennaro, Katz, Krawczyk and Rabin subsequently described a homomorphic signature [22] over the integers based on the RSA assumption in the random oracle model. Later on, Boneh and Freeman [11] gave a linearly homomorphic construction over binary fields. They also formalized a notion, called *weak privacy*, which requires derived signatures to hide the original dataset they were derived from.

In the network coding scenario, constructions in the standard model were given by Attrapadung and Libert [4] and Catalano, Fiore and Warinschi [17,18]. Recently, Freeman [20] defined a framework for constructing linearly homomorphic signatures satisfying enhanced security properties. In the standard model, the framework of [20] notably provides constructions based on the RSA, Diffie-Hellman and Strong Diffie-Hellman assumptions. In the meantime, Boneh and Freeman [12] used lattices to move beyond linear functions and described homomorphic signatures (in the random oracle model) supporting the evaluation of multivariate polynomials over signed data.

Recently, Ahn et al. [3] realized strongly context hiding P-homomorphic signatures for quoting and subset predicates: a signed message allows deriving signatures on substrings or arbitrary subsets of that message, respectively. They also showed that linearly homomorphic signatures [10,11,17,20] give P-homomorphic signatures allowing for the computation of weighted averages and Fourier transforms on signed data. The construction of [10] was notably shown strongly context hiding thanks to its uniqueness of signatures property.

1.2 Our Contributions

NEW DEFINITIONS OF PRIVACY. In this paper, we first reconsider the definition of strong context hiding security in [3] and point out a subtlety that arises in the context of randomizable signatures. While the definition of Ahn *et al.* [3] aims at perfect indistinguishability, it only considers honestly generated original signatures. In specific schemes, signatures may satisfy the verification algorithm without being produced by the legitimate signing algorithm. Signatures [30,4,23] derived from Waters' dual system encryption technique [39] – which is currently the only known way to prove the standard unforgeability property for certain predicates – are typical examples. For these constructions, the definition of [3] does not guarantee the unlinkability when the original signature is adversarially chosen (e.g., by re-randomizing original signatures). This may be a concern in certain applications. In network coding, suppose that we want to hide the path taken by specific packets. If a curious target node colludes with some intermediate nodes that maliciously re-randomize signatures on the road, they may infer information on the rest of the path downstream.

To address this issue, we suggest other definitions of unlinkability and discuss the relations among them. We first define a security property, called *adaptive context hiding*, that allows for adversarially-generated original signatures. Since this definition only asks for computational security, it does not imply strong context hiding security [3]: we show examples of schemes that are context hiding according to one definition and fall short of satisfying the other one. In order to unify these definitions, we thus define a notion of *completely context hiding* homomorphic signature, which requires statistical unlinkability and implies *both* strong and adaptive context hiding properties.

NEW LINEARLY HOMOMORPHIC SIGNATURES. Using the dual system technique [39,30], we describe a new linearly homomorphic signature and prove it (in the standard model) both strongly context hiding and context hiding on adversarially-chosen signatures with private key exposure. To our knowledge, all previous such schemes fail to simultaneously satisfy both security notions. The scheme of [4] is actually the only strongly context hiding realization in the standard model but, as we shall see, it is provably not adaptively context hiding. Since the new construction is only adaptively context hiding for computationally bounded distinguishers, it does not meet our strongest definition. This shortcoming seems inherent to all signature schemes [4,23] based on the dual system paradigm. We leave it as an open problem to achieve information-theoretic unlinkability in that sense without resorting to the random oracle model.

If we settle for weak context hiding security[1] (as in most linearly homomorphic signatures [11,20]), a variant of our scheme provides the shortest linearly homomorphic signature based on a simple assumption in the standard model. At the expense of being context hiding in a weaker sense than [10], the scheme can be proved unforgeable under the standard computational Diffie-Hellman (CDH) assumption. Each signature consists of two group elements and one scalar, which shortens Freeman's CDH-based signatures [20] by about 25%.

HANDLING SUBSET PREDICATES FOR MESSAGES OF ARBITRARY LENGTH. Finally, the paper puts forward a new method for dealing with subset predicates. Ahn *et al.* [3] showed how to obtain such signatures from a certain class of ciphertext-policy attribute-based encryption (CP-ABE) systems, by applying a Naor-like transformation [9]. With currently available fully secure CP-ABE schemes [29,35], this technique is limited to support messages of bounded length: the maximal length n_{max} of original messages must be fixed at key generation time and public keys comprise at least $O(n_{max})$ group elements. This limitation could be avoided using a fully secure unbounded [31] CP-ABE scheme. However, no such system is currently available: the only known [31,28] unbounded ABE constructions to date are selectively secure key-policy ABE schemes.

To fill this gap, we suggest an alternative design principle which yields constant-size public keys and allows signing messages of arbitrary length. Our construction departs from the ABE-based approach of [3] and rather uses the randomizability properties of Groth-Sahai proofs [25]. In a nutshell, when original signatures are computed for a set of words $\{m_1, \ldots, m_n\}$, the signer generates a fresh public key pk', which is certified using the long-term secret key of the system, and uses sk' to compute $\sigma_i = \mathsf{Sign}(sk', m_i)$ for each i. This construction is made unlinkable by letting pk' and all signatures $\{\sigma_i\}_{i=1}^n$ appear in committed form, accompanied with non-interactive witness indistinguishable proofs of their validity. The general idea is instantiated by combining the structure-preserving signature of [1] with Waters signatures [38] – which are both partially randomizable – in such a way that we only need to manipulate linear pairing product equations (in the terminology of [25]). This makes it easy to re-randomize Groth-Sahai proofs when deriving signatures. As a result, the system provably satisfies our strongest definition of unlinkability.

We believe this approach to be of interest in its own right for the design of P-homomorphic signatures. Indeed, if we compare it with the dual system technique [39], it allows us to more easily obtain completely context hiding schemes.

1.3   Organization

We first review previous security definitions for P-homomorphic signatures and introduce new definitions of privacy in Section 2.1. Section 3 discusses the relations among these privacy definitions. In Section 4, we describe a new linearly

---

[1] This property relaxes strong context hiding security by only requiring the indistinguishability when the original signatures are not given.

homomorphic constructions, for which a CDH-based weakly context-hiding variant is described in the full version of the paper. Section 5 finally presents our completely context hiding system for subset predicates.

2 Background

2.1 Definitions for Homomorphic Signatures

Definition 1 ([3]). *Let $\mathcal{M}$ be a message space and $2^{\mathcal{M}}$ be its powerset. Let $P : 2^{\mathcal{M}} \times \mathcal{M} \to \{0,1\}$ be a predicate. A message m' is said* **derivable** *from $M \subset \mathcal{M}$ if $P(M, m') = 1$. As in [3], $P^i(M)$ is the set of messages derivable from $P^{i-1}(M)$, where $P^0(M) := \{m' \in \mathcal{M} \mid P(M, m') = 1\}$. Finally, $P^*(M) := \cup_{i=0}^{\infty} P^i(M)$ denotes the set of messages derivable from M by iterated derivation.*

Definition 2 ([3]). *A P-homomorphic signature for a predicate $P : 2^{\mathcal{M}} \times \mathcal{M} \to \{0,1\}$ is a triple of algorithms* (Keygen, SignDerive, Verify) *such that:*

Keygen(λ): *takes as input a security parameter $\lambda \in \mathbb{N}$ and outputs a key pair* (sk, pk). *As in [3], the private key* sk *is seen as a signature on the empty tuple $\varepsilon \in \mathcal{M}$.*

SignDerive$(\mathsf{pk}, (\{\sigma_m\}_{m \in M}, M), m')$: *is a possibly randomized algorithm that takes as input a public key* pk, *a set of messages $M \subset \mathcal{M}$, a corresponding set of signatures $\{\sigma_m\}_{m \in M}$ and a derived message $m' \in \mathcal{M}$. If $P(M, m') = 0$, it returns $\perp$. Otherwise, it outputs a derived signature σ'*

Verify(pk, σ, m): *is a deterministic algorithm that takes as input a public key* pk, *a signature σ and a message m. It outputs 0 or 1.*

Note that the empty tuple $\varepsilon \in \mathcal{M}$ satisfies $P(\varepsilon, m) = 1$ for each $m \in \mathcal{M}$. Like [3], we define the algorithm Sign(pk, sk, m) that runs SignDerive(pk, (sk, ε), m) and returns the resulting output. For any set $M = \{m_1, \ldots, m_k\} \subset \mathcal{M}$, we define Sign(sk, M) := $\{$Sign(sk, m_1), ..., Sign(sk, m_k)$\}$. Also, Verify(pk, M, $\{\sigma_m\}_{m \in M}$) = 1 means that Verify(pk, m, σ_m) = 1 for each $m \in M$.

CORRECTNESS. It is mandated that, for all pairs (pk, sk) $\leftarrow$ Keygen(λ), for any set $M \subset \mathcal{M}$, any message $m' \in \mathcal{M}$ such that $P(M, m') = 1$, then, we have

- SignDerive(pk, (Sign(sk, M), M), m') $\neq \perp$.
- Verify$\big($pk, m', SignDerive(pk, (Sign(sk, M), M), m')$\big)$ = 1.

Definition 3 ([3]). *A P-homomorphic signature* (Keygen, SignDerive, Verify) *is said* **unforgeable** *if no probabilistic polynomial-time (PPT) adversary has non-negligible advantage in this game:*

1. *The challenger generates* (pk, sk) $\leftarrow$ Keygen(λ) *and gives* pk *to the adversary $\mathcal{A}$. It initializes two initially empty tables T and Q.*
2. *$\mathcal{A}$ adaptively interleaves the following queries.*
   - *Signing queries: $\mathcal{A}$ chooses a message $m \in \mathcal{M}$. The challenger replies by choosing a handle h, runs $\sigma \leftarrow$ Sign(sk, m) and stores (h, m, σ) in a table T. The handle h is returned to $\mathcal{A}$.*

- *Derivation queries: $\mathcal{A}$ chooses a vector of handles $\vec{h} = (h_1, \ldots, h_k)$ and a message $m' \in \mathcal{M}$. The challenger retrieves the tuples $\{(h_i, m_i, \sigma_i)\}_{i=1}^k$ from T and returns $\perp$ if one of these does not exist. Otherwise, it defines $M := (m_1, \ldots, m_k)$ and $\{\sigma_m\}_{m \in M} = \{\sigma_1, \ldots, \sigma_k\}$. If $P(M, m') = 1$, the challenger runs $\sigma' \leftarrow \mathsf{SignDerive}(\mathsf{pk}, (\{\sigma_m\}_{m \in M}, M), m')$, chooses a handle h', stores (h', m', σ') in T and returns h' to $\mathcal{A}$.*
- *Reveal queries: $\mathcal{A}$ chooses a handle h. If no tuple of the form (h, m', σ') exists in T, the challenger returns $\perp$. Otherwise, it returns σ' to $\mathcal{A}$ and adds (m', σ') to the set Q.*
3. *$\mathcal{A}$ outputs a pair (σ', m') and wins if the following conditions hold.*
   - *$\mathsf{Verify}(\mathsf{pk}, m', \sigma') = 1$.*
   - *If $M \subset \mathcal{M}$ is the set of messages in Q, then $m' \notin P^*(M)$.*

Definition 4 ([3]). *A homomorphic signature $(\mathsf{Keygen}, \mathsf{Sign}, \mathsf{SignDerive}, \mathsf{Verify})$ is **strongly context hiding** for the predicate P if, for all key pairs $(\mathsf{pk}, \mathsf{sk}) \leftarrow \mathsf{Keygen}(\lambda)$, for all messages $M \subset \mathcal{M}^*$ and $m' \in \mathcal{M}$ such that $P(M, m') = 1$, the following two distributions are statistically close:*

$$\{(\mathsf{sk}, \{\sigma_m\}_{m \in M} \leftarrow \mathsf{Sign}(\mathsf{sk}, M), \ \mathsf{Sign}(sk, m'))\}_{sk, M, m'},$$

$$\left\{\left(\mathsf{sk}, \{\sigma_m\}_{m \in M} \leftarrow \mathsf{Sign}(\mathsf{sk}, M), \ \mathsf{SignDerive}(\mathsf{pk}, (\{\sigma_m\}_{m \in M}, M), m')\right)\right\}_{sk, M, m'}.$$

In [3] Ahn *et al.* showed that, if a scheme is strongly context hiding, then Definition 3 can be simplified by removing the SignDerive and Reveal oracles and only providing the adversary with an ordinary signing oracle.

As we will see, specific constructions leave a gap between signatures accepted by the verification algorithm and those generated by the original signing procedure. For these schemes, a stronger definition than Definition 4 may be necessary in some situations.

To illustrate this, we first give an alternative definition which is almost identical to the computational security definition of [3][Appendix A]: the only difference is that, in the challenge phase, one of the signatures is supplied by the adversary instead of being honestly generated by the challenger. This modification is motivated by re-randomizable signatures. It allows for adversaries who attempt to re-randomize one of the signatures obtained from the oracle in order to embed some subliminal information that would help them win the game.

Definition 5. *A P-homomorphic signature $(\mathsf{Keygen}, \mathsf{Sign}, \mathsf{SignDerive}, \mathsf{Verify})$ is **weakly adaptively context hiding** if no PPT adversary has non-negligible advantage in the following game:*

1. *The challenger runs $(\mathsf{sk}, \mathsf{pk}) \leftarrow \mathsf{Keygen}(\lambda)$ and gives pk to the adversary.*
2. *The adversary $\mathcal{A}$ adaptively interleaves queries exactly as in Definition 3.*
3. *The adversary $\mathcal{A}$ chooses a message set $M \subset \mathcal{M}$ together with a set of signatures $\{\sigma_m\}_{m \in M}$ as well as another message $m' \in \mathcal{M}$. If $P(M, m') = 0$ or $\mathsf{Verify}(\mathsf{pk}, M, \{\sigma_m\}_{m \in M}) = 0$, return $\perp$. Otherwise, the challenger flips a fair binary coin $\beta \xleftarrow{R} \{0, 1\}$. If $\beta = 0$, it computes a derived signature $\sigma^* = \mathsf{SignDerive}(\mathsf{pk}, (\{\sigma_m\}_{m \in M}, M), m')$. If $\beta = 1$, it computes $\sigma^* = \mathsf{Sign}(sk, m')$. In either case, σ^* is sent as a challenge to $\mathcal{A}$.*

4. $\mathcal{A}$ is allowed to make another series of queries as in stage 2.
5. Eventually, $\mathcal{A}$ outputs a bit $\beta' \in \{0, 1\}$ and wins if $\beta' = \beta$. As usual, $\mathcal{A}$'s advantage is defined to be $\mathbf{Adv}(\mathcal{A}) = |\Pr[\beta' = \beta] - 1/2|$.

The latter definition can be seen as an analogue of a definition of unlinkability given by Prabhakaran and Rosulek [36] for homomorphic encryption: both models account for adversarially-chosen original signatures or ciphertexts.

We will see that Definitions 4 and 5 do not imply each other. While incomparable, we believe that they both make sense in practice. For example, when it comes to conceal the path followed by packets in network coding signatures, Definition 5 ensures that each node only learns the last node visited by incoming packets, even if it colludes with another node far upstream.

Towards unifying previous definitions, we now simplify Definition 5 as follows. Instead of providing the adversary $\mathcal{A}$ with a signing oracle, $\mathcal{A}$ is directly given the private key at the beginning.

Definition 6. *A P-homomorphic signature is* **adaptively context hiding** *if no PPT adversary has non-negligible advantage in the following game:*

1. *The challenger runs* $(\mathsf{sk}, \mathsf{pk}) \leftarrow \mathsf{Keygen}(\lambda)$ *and hands* $(\mathsf{sk}, \mathsf{pk})$ *to* $\mathcal{A}$.
2. *The adversary* $\mathcal{A}$ *chooses a message set* $M \subset \mathcal{M}$ *together with a set of signatures* $\{\sigma_m\}_{m \in M}$ *as well as another message* $m' \in \mathcal{M}$. *If* $P(M, m') = 0$ *or* $\mathsf{Verify}(\mathsf{pk}, M, \{\sigma_m\}_{m \in M}) = 0$, *return* $\perp$. *Otherwise, the challenger flips a fair binary coin* $\beta \xleftarrow{R} \{0, 1\}$. *If* $\beta = 0$, *it computes a derived signature* $\sigma^* = \mathsf{SignDerive}(\mathsf{pk}, (\{\sigma_m\}_{m \in M}, M), m')$. *If* $\beta = 1$, *it computes* $\sigma^* = \mathsf{Sign}(\mathsf{sk}, m')$. *In either case,* σ^* *is sent as a challenge to* $\mathcal{A}$.
3. *Eventually,* $\mathcal{A}$ *outputs a bit* $\beta' \in \{0, 1\}$ *and wins if* $\beta' = \beta$. *As usual,* $\mathcal{A}$'s *advantage is defined to be* $\mathbf{Adv}(\mathcal{A}) = |\Pr[\beta' = \beta] - 1/2|$.

While the latter definition seems sufficient for many applications, it still does not imply Definition 4 and we may want signatures to be unlinkable in the statistical sense. The resulting stronger definition implies both Definition 6 and Definition 4 and goes as follows.

Definition 7. *A P-homomorphic signature* $(\mathsf{Keygen}, \mathsf{Sign}, \mathsf{SignDerive}, \mathsf{Verify})$ *is* **completely context hiding** *if, for all pairs* $(\mathsf{pk}, \mathsf{sk}) \leftarrow \mathsf{Keygen}(\lambda)$, *all messages* $M \subset \mathcal{M}^*$ *and* $m' \in \mathcal{M}$ *such that* $P(M, m') = 1$, *for all* $\{\sigma_m\}_{m \in M}$ *such that* $\mathsf{Verify}(\mathsf{pk}, M, \{\sigma_m\}_{m \in M}) = 1$, *the distribution* $\{(\mathsf{sk}, \mathsf{Sign}(\mathsf{sk}, m'))\}_{\mathsf{sk}, M, m'}$ *is statistically close to* $\{(\mathsf{sk}, \mathsf{SignDerive}(\mathsf{pk}, (\{\sigma_m\}_{m \in M}, M), m'))\}_{\mathsf{sk}, M, m'}$.

In all schemes based on the dual system approach [4,23], the existence of an alternative distribution of acceptable signatures makes it seemingly impossible to satisfy the above definition. In these schemes, the combination of strong (*i.e.*, Definition 4) and adaptive context hiding security thus appears as the best we can hope for. For this reason, we chose to present Definition 6 first instead of directly working with Definition 7.

Definition 7 assumes honestly generated keys $(\mathsf{sk}, \mathsf{pk})$. It can be strengthened by allowing the adversary to generate a pair $(\mathsf{sk}, \mathsf{pk})$ of its own. In the random oracle model, the construction of [10] is easily seen to satisfy such a stronger definition (if we assume that all public keys live in a cyclic group which is part of common public parameters) because it has unique signatures. In the standard model, we do not know of any scheme that would be secure in that sense.

In the following, we can satisfy Definition 7 with our homomorphic signature for subset predicates. In the case of linearly homomorphic signatures, we are only able to meet Definition 6.

2.2   Complexity Assumptions

We consider groups $(\mathbb{G}, \mathbb{G}_T)$ of composite order $N = p_1 p_2 p_3$, for which a bilinear map $e : \mathbb{G} \times \mathbb{G} \to \mathbb{G}_T$ is computable. For each $i \in \{1, 2, 3\}$, we denote by $\mathbb{G}_{p_i}$ the subgroup of order p_i. Also, for all distinct i, j, we call $\mathbb{G}_{p_i p_j}$ the subgroup of order $p_i p_j$. An important property of composite order groups is that pairing two elements of order p_i and p_j, with $i \neq j$, always gives the identity element $1_{\mathbb{G}_T}$.

In these groups, we rely on the following assumptions introduced in [30].

Assumption 1. Given $g \xleftarrow{R} \mathbb{G}_{p_1}, X_3 \xleftarrow{R} \mathbb{G}_{p_3}$, and T, it is infeasible to efficiently decide if $T \in_R \mathbb{G}_{p_1 p_2}$ or $T \in_R \mathbb{G}_{p_1}$.

Assumption 2. Let $g, X_1 \xleftarrow{R} \mathbb{G}_{p_1}, X_2, Y_2 \xleftarrow{R} \mathbb{G}_{p_2}, Y_3, Z_3 \xleftarrow{R} \mathbb{G}_{p_3}$. Given a tuple $(g, X_1 X_2, Z_3, Y_2 Y_3)$ and T, it is hard to decide if $T \in_R \mathbb{G}$ or $T \in_R \mathbb{G}_{p_1 p_3}$.

Assumption 3. Let elements $g, w, g^t, X_1 \xleftarrow{R} \mathbb{G}_{p_1}$ with $t \xleftarrow{R} \mathbb{Z}_N, X_2, Y_2, Z_2 \xleftarrow{R} \mathbb{G}_{p_2}, X_3, Y_3, Z_3 \xleftarrow{R} \mathbb{G}_{p_3}$. Given $(g, w, g^t, X_1 X_2, X_3, Y_2 Y_3)$, and $T \in \mathbb{G}$, decide if $T = w^t Z_3$ or $T = w^t Z_2 Z_3$.

Assumption 4. Let $g \xleftarrow{R} \mathbb{G}_{p_1}, X_2, Y_2, Z_2 \xleftarrow{R} \mathbb{G}_{p_2}, X_3 \xleftarrow{R} \mathbb{G}_{p_3}$ and $a, b, c \xleftarrow{R} \mathbb{Z}_N$. Given $(g, g^a, g^b, g^{ab} X_2, X_3, g^c Y_2, Z_2)$, it is infeasible to compute $e(g, g)^{abc}$.

We also use bilinear maps $e : \mathbb{G} \times \mathbb{G} \to \mathbb{G}_T$ over groups of prime order p. In these groups, we rely on the following hardness assumptions.

Definition 8 ([8]). *The **Decision Linear Problem** (DLIN) in $\mathbb{G}$, is to distinguish the distributions $(g^a, g^b, g^{ac}, g^{bd}, g^{c+d})$ and $(g^a, g^b, g^{ac}, g^{bd}, g^z)$, where $a, b, c, d \xleftarrow{R} \mathbb{Z}_p^*, z \xleftarrow{R} \mathbb{Z}_p^*$. The **Decision Linear Assumption** is the intractability of DLIN for any PPT distinguisher $\mathcal{D}$.*

Definition 9 ([1]). *In a group $\mathbb{G}$, the q-**Simultaneous Flexible Pairing Problem** (q-SFP) is, given $(g_z, h_z, g_r, h_r, a, \tilde{a}, b, \tilde{b} \in \mathbb{G})$ and q tuples $(z_j, r_j, s_j, t_j, u_j, v_j, w_j) \in \mathbb{G}^7$ such that*

$$e(a, \tilde{a}) = e(g_z, z_j) \cdot e(g_r, r_j) \cdot e(s_j, t_j), \quad e(b, \tilde{b}) = e(h_z, z_j) \cdot e(h_r, u_j) \cdot e(v_j, w_j), \tag{1}$$

to find a new tuple $(z^\star, r^\star, s^\star, t^\star, u^\star, v^\star, w^\star) \in \mathbb{G}^7$ satisfying (1) and such that $z^\star \notin \{1_{\mathbb{G}}, z_1, \ldots, z_q\}$.

2.3   Structure-Preserving Signatures

Privacy-preserving protocols often require to sign elements of bilinear groups as if they were ordinary messages. Abe, Haralambiev and Ohkubo [1,2] (AHO) described such an efficient structure-preserving signature. The description hereunder assumes public parameters $\mathsf{pp} = ((\mathbb{G}, \mathbb{G}_T), g)$ consisting of bilinear groups $(\mathbb{G}, \mathbb{G}_T)$ of prime order $p > 2^\lambda$, where $\lambda \in \mathbb{N}$ and a generator $g \in \mathbb{G}$.

Keygen(pp, n): given an upper bound $n \in \mathbb{N}$ on the number of group elements per signed message, choose generators $G_r, H_r \xleftarrow{R} \mathbb{G}$. Pick $\gamma_z, \delta_z \xleftarrow{R} \mathbb{Z}_p$ and $\gamma_i, \delta_i \xleftarrow{R} \mathbb{Z}_p$, for $i = 1$ to n. Then, compute $G_z = G_r^{\gamma_z}$, $H_z = H_r^{\delta_z}$ and $G_i = G_r^{\gamma_i}$, $H_i = H_r^{\delta_i}$ for each $i \in \{1, \ldots, n\}$. Finally, choose $\alpha_a, \alpha_b \xleftarrow{R} \mathbb{Z}_p$ and define $A = e(G_r, g^{\alpha_a})$ and $B = e(H_r, g^{\alpha_b})$. The public key is defined to be

$$pk = \big(G_r, \ H_r, \ G_z, \ H_z, \ \{G_i, H_i\}_{i=1}^n, \ A, \ B\big) \in \mathbb{G}^{2n+4} \times \mathbb{G}_T^2$$

while the private key is $sk = \big(\alpha_a, \alpha_b, \gamma_z, \delta_z, \{\gamma_i, \delta_i\}_{i=1}^n\big)$.

Sign$(sk, (M_1, \ldots, M_n))$: to sign a vector $(M_1, \ldots, M_n) \in \mathbb{G}^n$ using sk, choose $\zeta, \rho_a, \rho_b, \omega_a, \omega_b \xleftarrow{R} \mathbb{Z}_p$ and compute $\theta_1 = g^\zeta$ as well as

$$\theta_2 = g^{\rho_a - \gamma_z \zeta} \cdot \prod_{i=1}^n M_i^{-\gamma_i}, \qquad \theta_3 = G_r^{\omega_a}, \qquad \theta_4 = g^{(\alpha_a - \rho_a)/\omega_a},$$

$$\theta_5 = g^{\rho_b - \delta_z \zeta} \cdot \prod_{i=1}^n M_i^{-\delta_i}, \qquad \theta_6 = H_r^{\omega_b}, \qquad \theta_7 = g^{(\alpha_b - \rho_b)/\omega_b},$$

The signature consists of $\sigma = (\theta_1, \theta_2, \theta_3, \theta_4, \theta_5, \theta_6, \theta_7) \in \mathbb{G}^7$.

Verify$(pk, \sigma, (M_1, \ldots, M_n))$: given $\sigma = (\theta_1, \theta_2, \theta_3, \theta_4, \theta_5, \theta_6, \theta_7)$, return 1 iff these equalities hold:

$$A = e(G_z, \theta_1) \cdot e(G_r, \theta_2) \cdot e(\theta_3, \theta_4) \cdot \prod_{i=1}^n e(G_i, M_i),$$

$$B = e(H_z, \theta_1) \cdot e(H_r, \theta_5) \cdot e(\theta_6, \theta_7) \cdot \prod_{i=1}^n e(H_i, M_i).$$

The scheme was proved [1,2] existentially unforgeable under chosen-message attacks under the q-SFP assumption, where q is the number of signing queries.

As showed in [1,2], signature components $\{\theta_i\}_{i=2}^7$ can be publicly randomized to obtain a different signature $\{\theta_i'\}_{i=1}^7 \leftarrow \mathsf{ReRand}(pk, \sigma)$ on $(M_1, \ldots, M_n)$. After randomization, we have $\theta_1' = \theta_1$ while $\{\theta_i'\}_{i=2}^7$ are uniformly distributed among the values $(\theta_2, \ldots, \theta_7)$ such that the equalities $e(G_r, \theta_2') \cdot e(\theta_3', \theta_4') = e(G_r, \theta_2) \cdot e(\theta_3, \theta_4)$ and $e(H_r, \theta_5') \cdot e(\theta_6', \theta_7') = e(H_r, \theta_5) \cdot e(\theta_6, \theta_7)$ hold. This re-randomization is performed by choosing $\varrho_2, \varrho_5, \mu, \nu \xleftarrow{R} \mathbb{Z}_p$ and computing

$$\theta_2' = \theta_2 \cdot \theta_4^{\varrho_2}, \qquad\qquad \theta_3' = (\theta_3 \cdot G_r^{-\varrho_2})^{1/\mu}, \qquad\qquad \theta_4' = \theta_4^{\mu} \qquad (2)$$

$$\theta_5' = \theta_5 \cdot \theta_7^{\varrho_5}, \qquad\qquad \theta_6' = (\theta_6 \cdot H_r^{-\varrho_5})^{1/\nu}, \qquad\qquad \theta_7' = \theta_7^{\nu}.$$

As a result, $\{\theta'_i\}_{i\in\{3,4,6,7\}}$ are statistically independent of the message and other signature components. This implies that, in privacy-preserving protocols, re-randomized $\{\theta'_i\}_{i\in\{3,4,6,7\}}$ can be safely given in the clear as long as $(M_1, \ldots, M_n)$ and $\{\theta'_i\}_{i\in\{1,2,5\}}$ are given in committed form.

3   Separation Results

SEPARATING DEFINITIONS 4 AND 5. Let us consider the following variant[2] of the construction in [4], which relies on the Lewko-Waters signatures [30] and bilinear groups whose order is a product $N = p_1 p_2 p_3$ of three primes. If n denotes the dimension of signed vectors, the public key is $\mathsf{pk} = (g, e(g,g)^\alpha, u, v, \{h_i\}_{i=1}^n, X_3)$, where $\alpha \in_R \mathbb{Z}_N$, $g, u, v, h_1, \ldots, h_n \in \mathbb{G}_{p_1}$, $X_3 \in \mathbb{G}_{p_3}$ and the private key consists of $\mathsf{sk} = (g^\alpha, \kappa)$, where κ is the seed of a pseudorandom function. The latter is used to de-randomize the scheme and make sure that all vectors of the same file will be signed using partially identical random coins.

To sign a vector $\vec{v} = (v_1, \ldots, v_n) \in \mathbb{Z}_N^n$ using the file identifier τ, the signer computes a pseudorandom $r = \Psi(\kappa, \tau) \in \mathbb{Z}_N$ which is used to compute

$$(\sigma_1, \sigma_2, \sigma_3) = \left(g^\alpha \cdot (u^\tau \cdot v)^r \cdot R_3, \ g^r \cdot R'_3, \ (\prod_{i=1}^n h_i^{v_i})^r \cdot R''_3 \right),$$

with $R_3, R'_3, R''_3 \xleftarrow{R} \mathbb{G}_{p_3}$. The homomorphic property follows from the fact that all vectors of the same dataset are signed using the same $r \in \mathbb{Z}_N$. The homomorphic evaluation algorithm proceeds in the obvious way and combines signatures $\{(\sigma_{i,1}, \sigma_{i,2}, \sigma_{i,3})\}_{i=1}^\ell$ by linearly combining the $\{\sigma_{i,3}\}_{i=1}^\ell$ and re-randomizing the $\mathbb{G}_{p_3}$ components. Note that the underlying exponent r is not re-randomized, so that all $\{(\sigma_{i,1}, \sigma_{i,2})\}_{i=1}^\ell$ share the same $\mathbb{G}_{p_1}$ components.

It is easy to see that the construction is strongly context hiding in the sense of Definition 4. Indeed, the signing algorithm is honestly run in the first distribution of Definition 4. This implies that, for any message set $M = \{(\tau, \vec{v}_1), \ldots, (\tau, \vec{v}_k)\} \subset \mathcal{M}$, the underlying $\log_g(\sigma_2)$ will have the same value no matter if the second signature (σ_1, σ_2) is produced by Sign or SignDerive.

However, the scheme does not satisfy Definition 5. Indeed, in step 2, the adversary can first invoke the signing oracle on k occasions to obtain signatures for some set $M = \{(\tau, \vec{v}_1), \ldots, (\tau, \vec{v}_k)\}$ of its choice. If we denote by $\{\sigma_m\}_{m\in M}$ the resulting signatures, the adversary re-randomizes $\{\sigma_m\}_{m\in M}$ in such a way that each randomized σ_m is of the form $\left(g^\alpha \cdot (u^\tau \cdot v)^{r'} \cdot \tilde{R}_3, \ g^{r'} \cdot \tilde{R}'_3, \ (\prod_{i=1}^n h_i^{v_i})^{r'} \cdot \tilde{R}''_3 \right)$, for some fresh $r' \in_R \mathbb{Z}_N$. The adversary $\mathcal{A}$ can then choose a random message $m' \in \mathcal{M}$ such that $P(M, m') = 1$ and send $\left((M, \{\sigma'_m\}_{m\in M}), m'\right)$ to the challenger. The latter returns a challenge signature $\sigma^\star = (\sigma_1^\star, \sigma_2^\star, \sigma_3^\star)$ on m' and $\mathcal{A}$ can immediately figure out if $\sigma^\star$ is fresh or derived, by testing if $e(\sigma_2^\star, g) = e(\sigma_{m,2}, g)$. With overwhelming probability, the latter equality only holds if $\beta = 0$.

---

[2] This variant is obtained by applying Freeman's framework [20] to Lewko-Waters signatures [31], which guarantees its unforgeability.

SEPARATING DEFINITIONS 5 AND 6. The original construction of [4] works exactly like the scheme outlined in the previous paragraph with the difference that it prevents public randomizations of the $\mathbb{G}_{p_1}$ components of signatures $(\sigma_1, \sigma_2, \sigma_3)$. More precisely, the scheme makes use of an additional collision-resistant hash function $H : \{0,1\}^* \to \mathbb{Z}_N$. If the file identifier is τ, a vector $\vec{v} = (v_1, \ldots, v_n)$ is signed by computing $r = \Psi(\kappa, \tau) \in \mathbb{Z}_N$, $\tau' = H(\tau, e(g,g)^r)$ and returning

$$(\sigma_1, \sigma_2, \sigma_3) = \left(g^\alpha \cdot (u^{\tau'} \cdot v)^r \cdot R_3, \ g^r \cdot R_3', \ (\prod_{i=1}^n h_i^{v_i})^r \cdot R_3''\right),$$

with $R_3, R_3', R_3'' \xleftarrow{R} \mathbb{G}_{p_3}$. The security proof of [4] implies that, if the adversary is given signatures $\{(\sigma_{i,1}, \sigma_{i,2}, \sigma_{i,3})\}_{i=1}^\ell$ on messages $(\tau, \vec{v}_1), \ldots, (\tau, \vec{v}_\ell)$, the adversary cannot generate a signature $(\sigma_1, \sigma_2, \sigma_3)$ on $(\tau, \vec{y})$ such that $e(\sigma_2, g) \neq e(\sigma_{i,2}, g)$ for each i. Essentially, since $(\sigma_{i,1}, \sigma_{i,2})$ can be seen as a Lewko-Waters signature on the message $H(\tau, e(g,g)^r)$, any valid signature $(\sigma_1, \sigma_2, \sigma_3)$ for which $e(\sigma_{i,2}, g) \neq e(\sigma_2, g)$ implies either an attack against the signature scheme of [30] or a breach in the collision-resistance of H.

Let us consider an adversary in the sense of Definition 5. Since signatures cannot be publicly randomized, when the adversary enters the challenge phase in step 3, it can only choose a message set $M = \{(\tau, \vec{v}_1), \ldots, (\tau, \vec{v}_\ell)\}$ and signatures $\{(\sigma_{m,1}, \sigma_{m,2}, \sigma_{m,3})\}_{m \in M}$ for which $\{e(\sigma_{m,2}, g)\}_{m \in M}$ has the same value as in signatures obtained from the signing oracle at step 2. Therefore, the only way for $\mathcal{A}$ to have non-negligible advantage in the game of Definition 5 is to choose $(M, \{\sigma_m\}_{m \in M})$ where $\{\sigma_m\}_{m \in M}$ is obtained by introducing a $\mathbb{G}_{p_2}$ component in a signature obtained from the signing oracle. Otherwise, the distribution of the challenge signature $(\sigma_1^\star, \sigma_2^\star, \sigma_3^\star)$ does not depend on $\beta \in \{0,1\}$ in step 3. Using the same arguments as in the proof of Theorem 1, we can prove that Assumption 1 can be broken if $\mathcal{A}$ can output a set $\{\sigma_m\}_{m \in M}$ where one of the signatures contains a $\mathbb{G}_{p_2}$ component. If H is collision-resistant and under the assumptions used in [4], the scheme is thus weakly adaptively context hiding.

Now, we easily observe that the original scheme of [4] is not adaptively context hiding in the sense of Definition 6. Recall that the adversary is given the private key $sk = (g^\alpha, \kappa)$ at the beginning of the game. In the challenge phase, it can thus choose a message set $M \subset \mathcal{M}$ and signatures $\{\sigma_m\}_{m \in M}$ for which each σ_m is of the form $(\sigma_{m,1}, \sigma_{m,2}, \sigma_{m,3}) = (g^\alpha \cdot (u^{\tau'} \cdot v)^{r'} \cdot R_3, \ g^{r'} \cdot R_3', \ (\prod_{i=1}^n h_i^{v_i})^{r'} \cdot R_3'')$, with $R_3, R_3', R_3'' \in_R \mathbb{G}_{p_3}$, and for some random $r' \in_R \mathbb{Z}_N \setminus \{\Psi(\kappa, \tau)\}$. When receiving $(M, \{\sigma_m\}_{m \in M})$ and m' such that $P(M, m') = 1$, the challenger runs SignDerive on $\{\sigma_m\}_{m \in M}$ if $\beta = 0$. If $\beta = 1$, it ignores $\{\sigma_m\}_{m \in M}$ and simply generates a fresh signature on m'. In the letter case, the challenge signature $(\sigma_1^\star, \sigma_2^\star, \sigma_3^\star)$ is such that $\log_g(\sigma_2^\star) = \Psi(\kappa, \tau) \bmod p_1$ and, since the adversary knows κ, it can easily test whether $e(\sigma_2^\star, g) = e(g,g)^{\Psi(\kappa,\tau)}$ and, if so, return $\beta' = 1$.

Later on, we will see an example of scheme that satisfies Definition 6 but fails to be secure as per Definition 4. The two definitions are thus incomparable.

4   An Adaptively Context Hiding Linearly Homomorphic Scheme in the Standard Model

So far, the scheme of [4] is seemingly the only linearly homomorphic signature in the standard model to satisfy Definition 4. This section presents a linearly homomorphic signature satisfying both Definition 4 and the adaptive context hiding property captured by Definition 6.

The scheme works over groups whose order is a product $N = p_1 p_2 p_3$ of three primes. Like [4], it builds on Lewko-Waters signatures, where public keys contain $(g, e(g,g)^\alpha, u, v)$, with $g, u, v \in \mathbb{G}_{p_1}$ and $\alpha \in \mathbb{Z}_N$, and a signature on m consists of $(g^\alpha \cdot (u^m \cdot v)^r \cdot R_3, g^r \cdot R_3')$, for some $R_3, R_3' \in \mathbb{G}_{p_3}$. A difference with [4] is that $e(g,g)^\alpha$ is replaced by g^α in the public key and signatures are obtained by aggregating a Lewko-Waters signature on the file identifier τ and a signed vector hash $(\prod_{i=1}^n g_i^{v_i})^\alpha$ of the vector $\vec{v} = (v_1, \dots, v_n)$, where $(g_1, \dots, g_n) \in \mathbb{G}_{p_1}^n$ is part of the public key. We note that $(\prod_{i=1}^n g_i^{v_i})^\alpha$ is not a secure homomorphic signature in general: it can actually be seen as a one-time linearly homomorphic signature where only one message set $M = \{(\tau, \vec{v}_1), \dots, (\tau, \vec{v}_k)\}$ can be signed. Nevertheless, we will show that aggregating the two components actually provides unforgeability. Moreover, beyond providing a stronger flavor of privacy than [4], it also shortens signatures by 33%.

For simplicity, the scheme is described in terms of composite order groups. It is very plausible that Lewko's techniques [28] apply to translate the scheme in the prime order setting.

4.1   Construction

Keygen(λ, n): given $\lambda \in \mathbb{N}$ and an integer $n \in \mathsf{poly}(\lambda)$, choose bilinear groups $(\mathbb{G}, \mathbb{G}_T)$ of order $N = p_1 p_2 p_3$, where $p_i > 2^\lambda$ for each $i \in \{1, 2, 3\}$. Choose $\alpha \xleftarrow{R} \mathbb{Z}_N$, $g, u, v \xleftarrow{R} \mathbb{G}_{p_1}$, $X_{p_3} \xleftarrow{R} \mathbb{G}_{p_3}$, $g_i \xleftarrow{R} \mathbb{G}_{p_1}$ for $i = 1$ to n. Then, select an identifier space $\mathcal{T}$. The private key is $\mathsf{sk} := \alpha$ while the public key is

$$\mathsf{pk} := \Big((\mathbb{G}, \mathbb{G}_T), \ N, \ g, \ g^\alpha, \ u, \ v, \ \{g_i\}_{i=1,\dots,n}, \ X_{p_3} \Big).$$

Sign$(\mathsf{sk}, \tau, \vec{v})$: on input of a vector $\vec{v} = (v_1, \dots, v_n) \in \mathbb{Z}_N^n$, a file identifier $\tau \in \mathcal{T}$ and the private key $\mathsf{sk} = \alpha \in \mathbb{Z}_N$, return $\perp$ if[3] $\vec{v} = \vec{0}$. Otherwise, conduct the following steps. First, choose $r \xleftarrow{R} \mathbb{Z}_N$ and $R_3, R_3' \xleftarrow{R} \mathbb{G}_{p_3}$. Then, compute a signature $\sigma = (\sigma_1, \sigma_2)$ as

$$\sigma_1 = (g_1^{v_1} \cdots g_n^{v_n})^\alpha \cdot (u^\tau \cdot v)^r \cdot R_3, \qquad \sigma_2 = g^r \cdot R_3',$$

SignDerive$(\mathsf{pk}, \tau, \{(\beta_i, \sigma_i)\}_{i=1}^\ell)$: given pk, a file identifier τ and ℓ tuples (β_i, σ_i), parse σ_i as $\sigma_i = (\sigma_{i,1}, \sigma_{i,2})$ for $i = 1$ to ℓ. Then, choose $\tilde{r} \xleftarrow{R} \mathbb{Z}_N$, $\tilde{R}_3, \tilde{R}_3' \xleftarrow{R} \mathbb{Z}_N$ and compute $\sigma_1 = \prod_{i=1}^\ell \sigma_{i,1}^{\beta_i} \cdot (u^\tau \cdot v)^{\tilde{r}} \cdot \tilde{R}_3$ and $\sigma_2 = \prod_{i=1}^\ell \sigma_{i,2}^{\beta_i} \cdot g^{\tilde{r}} \cdot \tilde{R}_3'$ and output (σ_1, σ_2).

---

[3] In the construction, we disallow signatures on the all-zeroes vector $\vec{0}$. This is not a restriction since, in all applications of linearly homomorphic signatures, a unit vector $(0, \dots, 1, \dots, 0)$ of appropriate length is appended to signed vectors.

Verify$(\mathsf{pk}, \tau, \vec{y}, \sigma)$: given a public key pk, a signature $\sigma = (\sigma_1, \sigma_2)$ and a message $(\tau, \vec{y})$, where $\tau \in \mathbb{Z}_N$ and $\vec{y} = (y_1, \ldots, y_n) \in (\mathbb{Z}_N)^n$, return $\perp$ if $\vec{y} = \vec{0}$. Otherwise, return 1 if and only if $e(\sigma_1, g) = e(g_1^{y_1} \cdots g_n^{y_n}, g^\alpha) \cdot e(u^\tau \cdot v, \sigma_2)$.

Verifying the correctness of the scheme is straightforward since pairing an element of $\mathbb{G}_{p_1}$ with an element of $\mathbb{G}_{p_3}$ always gives the identity element in $\mathbb{G}_T$.

4.2   Security

Theorem 1. *The scheme is adaptively context hiding if Assumption 1 holds.* (The proof is given in the full version of the paper).

As already mentioned, computational adaptive context hiding security does not imply statistical strong context hiding security (cf. Definition 4) in general. Let us consider a simple modification of the scheme. The public key includes $e(g, g)^\varphi$, for some $\varphi \in_R \mathbb{Z}_N$ which is *not* part of sk. Original signatures are augmented with $\sigma_3 = e(g, g)^{\varphi \cdot r}$, which is ignored by the verification algorithm. Also, $\mathsf{SignDerive}$ replaces σ_3 by a random element of $\mathbb{G}_T$. Although this artificial scheme can be proved adaptively context hiding under Assumptions 1 and 4, it does not meet the requirements of Definition 4.

   Yet, it is immediate that the system of Section 4.1 is also secure in the sense of Definition 4.

Theorem 2. *The scheme is unforgeable assuming that Assumptions 1, 2, 3 and 4 hold.* (The proof is given in the full version of the paper).

In the full version of the paper, we show that the same scheme can be safely instantiated in prime order groups if we settle for the weaker privacy definition used in [11,12,20]. The unforgeability of this modified scheme can be proved under the standard Diffie-Hellman assumption. To date, this construction turns out to be the shortest linearly homomorphic signature based on a simple assumption.

5   A Construction with Short Keys for Subset Predicates

In this section, we use the malleability properties of Groth-Sahai proofs (already exploited in, e.g., [6,21,19]) to construct a homomorphic signature for subset predicates. The main advantage over the approach of [3] is that we obtain constant-size[4] public keys in the standard model. In the standard model, the CP-ABE approach of [3] is currently limited to provide linear-size public keys in the maximal length of signed messages.

   This limitation could be avoided using a ciphertext-policy adaption of the unbounded key-policy ABE system of [31]. However, the ABE construction of [31] is only known to be selectively secure and, for the time being, no fully secure unbounded CP-ABE system is available. Conceivably, such a scheme can

---

[4] By "constant", we mean that it only depends on the security parameter and not on the length of messages to be signed.

be obtained by extending the techniques of [31]. Still, the resulting system would probably encounter the same difficulties as in Section 4 when it comes to obtain complete context hiding security. In contrast, our scheme is proved completely context hiding and fully (as opposed to selective-message) secure. It also allows for messages of unbounded (but polynomial) length.

In homomorphic signatures for subset predicates, the message space $\mathcal{M}$ can be defined as the set of tuples $\mathcal{M} := \Sigma^*$, where Σ is a set of words. The predicate P is defined in such a way that, for any polynomials $\{n_i\}_i$ and n', we have

$$P\Big(\{m_1, \ldots, m_n\}, \{m'_1, \ldots, m'_{n'}\}\Big) = 1$$
$$\iff \quad (n' \leq n) \wedge (m'_j \in \{m_1, \ldots, m_n\} \text{ for } j = 1 \text{ to } n').$$

The intuition of the scheme begins with the following naive construction, based on any digital signature, that only works when privacy is not a concern. The public key of the scheme is a standard digital signature key pair (sk, pk). When a message $\mathsf{Msg} = \{m_1, \ldots, m_n\}$ must be signed, the signer generates a fresh public key (sk', pk'), certifies pk' by computing $\sigma_{pk'} \leftarrow \mathsf{Sign}(sk, pk')$ and returning $(pk', \sigma_{pk'}, \{\sigma_i = \mathsf{Sign}(sk', m_i)\}_{i=1}^n)$. This simple construction immediately allows signature derivations for subset predicates. Moreover, since each signed set of words Msg involves a different public key pk', there is no way to generate a signature on a message $\mathsf{Msg}^\star$ that mixes words from two distinct signed messages Msg_1, Msg_2. However, the latter construction is trivially not context hiding. To achieve the latter property, instead of leaving pk' and $\{\sigma_i\}_{i=1}^{n'}$ appear in the clear within signatures, we let them appear in committed form and appeal to non-interactive witness indistinguishable (NIWI) arguments of knowledge of these signatures and keys. Then, the randomizability properties of Groth-Sahai proofs come in handy to obtain the desired privacy properties.

To realize the above idea, we work with Waters signatures [38] and the structure-preserving signature of Abe $et\ al.$ [1,2] because they make it possible to work with $linear$ pairing product equations. As observed in [21], these equations have proofs that only depend on the randomness of Groth-Sahai commitments and not on the committed witnesses or on the right-hand-side member of the equation. In the $\mathsf{SignDerive}$ algorithm, this allows updating some of the witnesses in such a way that the old proof remains valid.

In the following notations, we define a coordinate-wise pairing $E : \mathbb{G} \times \mathbb{G}^3 \to \mathbb{G}_T^3$ such that, for any element $h \in \mathbb{G}$ and any vector $\vec{g} = (g_1, g_2, g_3)$, we have $E(h, \vec{g}) = (e(h, g_1), e(h, g_2), e(h, g_3))$. In the following, when $X \in \mathbb{G}$ (resp. $Y \in \mathbb{G}_T$), the notation $\iota_\mathbb{G}(X)$ (resp. $\iota_{\mathbb{G}_T}(Y)$) will be used to denote the vector $(1_\mathbb{G}, 1_\mathbb{G}, X) \in \mathbb{G}^3$ (resp. the vector $(1_{\mathbb{G}_T}, 1_{\mathbb{G}_T}, Y) \in \mathbb{G}_T^3$).

Keygen(λ): given a security parameter $\lambda \in \mathbb{N}$, choose bilinear groups $(\mathbb{G}, \mathbb{G}_T)$ of prime order $p > 2^\lambda$. Then, do the following.

1. Generate a Groth-Sahai CRS $\mathbf{f} = (\vec{f}_1, \vec{f}_2, \vec{f}_3)$ for the perfect witness indistinguishability setting. Namely, choose $\vec{f}_1 = (f_1, 1, g)$, $\vec{f}_2 = (1, f_2, g)$, and $\vec{f}_3 = \vec{f}_1^{\,\xi_1} \cdot \vec{f}_2^{\,\xi_2} \cdot (1, 1, g)^{-1}$, with $f_1, f_2 \xleftarrow{R} \mathbb{G}$, $\xi_1, \xi_2 \xleftarrow{R} \mathbb{Z}_p$.

2. Generate a key pair $(sk_{\mathsf{AHO}}, pk_{\mathsf{AHO}})$ for the AHO signature in order to sign messages consisting of a single group element. This key pair are

$$pk_{\mathsf{AHO}} = \Big(G_r, \ H_r, \ G_z = G_r^{\gamma_z}, \ H_z = H_r^{\delta_z}, \ G_1 = G_r^{\gamma_1}, H_1 = H_r^{\delta_1}, \ A, \ B\Big)$$

and $sk_{\mathsf{AHO}} = \big(\alpha_a, \alpha_b, \gamma_z, \delta_z, \gamma_1, \delta_1\big)$.

3. Generate parameters for the Waters signature. Namely, choose group elements $h \xleftarrow{R} \mathbb{G}$, and $(u_0, u_1, \ldots, u_L) \xleftarrow{R} \mathbb{G}^{L+1}$. These are used to implement a hash function $H_{\mathbb{G}} : \{0,1\}^L \to \mathbb{G}$ such that, for any string $m = m[1] \ldots m[L] \in \{0,1\}^L$, $H_{\mathbb{G}}(m) = u_0 \cdot \prod_{i=1}^{L} u_i^{m[i]}$.

The public key is defined to be $\mathsf{pk} := \Big((\mathbb{G}, \mathbb{G}_T), g, \ \mathbf{f}, \ pk_{\mathsf{AHO}}, \ h, \ \{u_i\}_{i=0}^{L}\Big)$ and the private key is $\mathsf{sk} = sk_{\mathsf{AHO}}$. The public key defines $\Sigma = \{0,1\}^L$.

Sign$(\mathsf{sk}, \mathsf{Msg})$: on input of a message $\mathsf{Msg} = \{m_i\}_{i=1}^{n}$, where $m_i \in \{0,1\}^L$ for each i, and the private key $\mathsf{sk} = sk_{\mathsf{AHO}}$, do the following.

1. Choose a new public key $X = g^x$ for Waters signatures, with $x \xleftarrow{R} \mathbb{Z}_p$. Generate a Groth-Sahai commitment $\vec{C}_X = \iota_{\mathbb{G}}(X) \cdot \vec{f_1}^{\,r_X} \cdot \vec{f_2}^{\,s_X} \cdot \vec{f_3}^{\,t_X}$, with $r_X, s_X, t_X \xleftarrow{R} \mathbb{Z}_p$.

2. Generate an AHO signature $(\theta_1, \ldots, \theta_7) \in \mathbb{G}^7$ on the group element $X \in \mathbb{G}$. Then, for each $j \in \{1,2,5\}$, generate Groth-Sahai commitments $\vec{C}_{\theta_j} = \iota_{\mathbb{G}}(\theta_j) \cdot \vec{f_1}^{\,r_{\theta_j}} \cdot \vec{f_2}^{\,s_{\theta_j}} \cdot \vec{f_3}^{\,t_{\theta_j}}$. Finally, generate NIWI proofs $\vec{\pi}_{\mathsf{AHO},1}, \vec{\pi}_{\mathsf{AHO},2} \in \mathbb{G}^3$ that committed variables $(X, \theta_1, \theta_2, \theta_5)$ satisfy

$$A \cdot e(\theta_3, \theta_4)^{-1} = e(G_z, \theta_1) \cdot e(G_r, \theta_2) \cdot e(G_1, X) \tag{3}$$
$$B \cdot e(\theta_6, \theta_7)^{-1} = e(H_z, \theta_1) \cdot e(H_r, \theta_5) \cdot e(H_1, X)$$

These proofs are obtained as

$$\vec{\pi}_{\mathsf{AHO},1} = \big(G_z^{-r_{\theta_1}} G_r^{-r_{\theta_2}} G_1^{-r_X}, \ G_z^{-s_{\theta_1}} G_r^{-s_{\theta_2}} G_1^{-s_X}, \ G_z^{-t_{\theta_1}} G_r^{-t_{\theta_2}} G_1^{-t_X}\big)$$
$$\vec{\pi}_{\mathsf{AHO},2} = \big(H_z^{-r_{\theta_1}} H_r^{-r_{\theta_5}} H_1^{-r_X}, \ H_z^{-s_{\theta_1}} H_r^{-s_{\theta_5}} H_1^{-s_X}, \ H_z^{-t_{\theta_1}} H_r^{-t_{\theta_5}} H_1^{-t_X}\big)$$

3. For each $i \in \{1, \ldots, n\}$, generate a Waters signature $(\sigma_{i,1}, \sigma_{i,2})$ on the word $m_i \in \{0,1\}^L$ by computing $(\sigma_{i,1}, \sigma_{i,2}) = \big(h^x \cdot H_{\mathbb{G}}(m_i)^{\chi_i}, \ g^{\chi_i}\big)$ for a randomly chosen $\chi_i \xleftarrow{R} \mathbb{Z}_p$. Then, generate a Groth-Sahai commitment $\vec{C}_{\sigma_{i,1}} = \iota_{\mathbb{G}}(\sigma_{i,1}) \cdot \vec{f_1}^{\,r_{i,1}} \cdot \vec{f_2}^{\,s_{i,1}} \cdot \vec{f_3}^{\,t_{i,1}}$, with $r_{i,1}, s_{i,1}, t_{i,1} \xleftarrow{R} \mathbb{Z}_p$, and a NIWI proof $\pi_{W,i}$ that $(X, \sigma_{i,1})$ satisfy

$$e(H_{\mathbb{G}}(m_i), \sigma_{i,2}) = e(X, h)^{-1} \cdot e(\sigma_{i,1}, g). \tag{4}$$

This proof is obtained as $\pi_{W,i} = \big(h^{r_X} \cdot g^{-r_{i,1}}, \ h^{s_X} \cdot g^{-s_{i,1}}, \ h^{t_X} \cdot g^{-t_{i,1}}\big)$.

4. Return the signature

$$\sigma = \Big(\vec{C}_X, \{\vec{C}_{\theta_j}\}_{j \in \{1,2,5\}}, \{\theta_j\}_{j \in \{3,4,6,7\}}, \vec{\pi}_{\mathsf{AHO},1}, \vec{\pi}_{\mathsf{AHO},2}, \{\vec{C}_{\sigma_{i,1}}, \sigma_{i,2}, \vec{\pi}_{W,i}\}_{i=1}^{n}\Big). \tag{5}$$

Note that proofs $\vec{\pi}_{\mathsf{AHO},1}, \vec{\pi}_{\mathsf{AHO},2}$ and $\{\vec{\pi}_{W,i}\}_i$ only depend on the randomness used in commitments and not on the committed values or on the left-hand-side members of pairing-product equations (3) and (4).

SignDerive$(\mathsf{pk}, \mathsf{Msg}, \mathsf{Msg}', \sigma)$**:** given pk, $\mathsf{Msg} = \{m_i\}_{i=1}^n$ and $\mathsf{Msg}' = \{m_i'\}_{i=1}^{n'}$, return $\bot$ if there exists $i \in \{1, \dots, n'\}$ such that $m_i' \notin \{m_i\}_{i=1}^n$. Otherwise, parse σ as in (5). For each $i \in \{1, \dots, n'\}$, let $\rho(i) \in \{1, \dots, n\}$ be the index such that $m_i' = m_{\rho(i)}$. Then, for each $i \in \{1, \dots, n'\}$, do the following.

1. Re-randomize the commitment $\vec{C}_X$ and the proofs $\vec{\pi}_{\mathsf{AHO},1}, \vec{\pi}_{\mathsf{AHO},2}, \{\vec{\pi}_{W,i}\}_i$ accordingly. Let $\vec{C}_X', \vec{\pi}_{\mathsf{AHO},1}', \vec{\pi}_{\mathsf{AHO},2}'$, and $\{\vec{\pi}_{W,i}'\}_i$ be the randomized commitment and proofs. Note that, in all of these commitments and proofs (r_X, s_X, t_X) have been updated consistently.

2. Re-randomize $\{\vec{C}_{\theta_j}\}_{j \in \{2,5\}}$ and $\{\theta_j\}_{j \in \{3,4,6,7\}}$ by choosing $\varrho_2, \varrho_5, \mu, \nu$ and computing

$$\vec{C}_{\theta_2}' = \vec{C}_{\theta_2} \cdot \iota_{\mathbb{G}}(\theta_4)^{\varrho_2} \qquad \theta_3' = \left(\theta_3 \cdot G_r^{-\varrho_2}\right)^{1/\mu} \qquad \theta_4' = \theta_4^\mu,$$
$$\vec{C}_{\theta_5}' = \vec{C}_{\theta_5} \cdot \iota_{\mathbb{G}}(\theta_7)^{\varrho_5} \qquad \theta_6' = \left(\theta_6 \cdot H_r^{-\varrho_5}\right)^{1/\nu} \qquad \theta_7' = \theta_7^\nu.$$

We note that, although the committed values inside $\vec{C}_{\theta_2}', \vec{C}_{\theta_5}'$ have changed. The proofs $\pi_{\mathsf{AHO},1}', \pi_{\mathsf{AHO},2}'$ are still valid for the new committed values. Then, compute $\{\vec{C}_{\theta_j}''\}_{j \in \{1,2,5\}}$ by re-randomizing the commitments $\vec{C}_{\theta_1}, \{\vec{C}_{\theta_j}'\}_{j \in \{2,5\}}$ and re-randomize the proofs $\pi_{\mathsf{AHO},1}', \pi_{\mathsf{AHO},2}'$ again. Let $\vec{\pi}_{\mathsf{AHO},1}'', \vec{\pi}_{\mathsf{AHO},2}''$ be the re-randomized proofs.

3. For each $i \in \{1, \dots, n'\}$, choose $\chi_i' \xleftarrow{R} \mathbb{Z}_p$ and compute

$$\vec{C}_{\sigma_{\rho(i),1}}' = \vec{C}_{\sigma_{\rho(i),1}} \cdot \iota_{\mathbb{G}}\left(H_{\mathbb{G}}(m_{\rho(i)})^{\chi_i'}\right), \qquad \sigma_{\rho(i),2}' = \sigma_{\rho(i),2} \cdot g^{\chi_i'}.$$

Even though the committed value inside $\vec{C}_{\sigma_{\rho(i),1}}'$ has changed, $\vec{\pi}_{W,\rho(i)}'$ remains a valid proof that the updated committed value $\sigma_{\rho(i),1}'$ satisfies $e(X, h) \cdot e(H_{\mathbb{G}}(m_{\rho(i)}), \sigma_{\rho(i),2}') = e(\sigma_{\rho(i),1}', g)$. The commitment $\vec{C}_{\sigma_{\rho(i),1}}'$ is then re-randomized and the proof $\vec{\pi}_{W,\rho(i)}'$ is re-randomized accordingly. Let $\vec{C}_{\sigma_{\rho(i),1}}''$ and $\vec{\pi}_{W,\rho(i)}''$ denote the new commitment and proof.

4. Return the signature

$$\sigma' = \Big(\vec{C}_X', \{\vec{C}_{\theta_j}''\}_{j \in \{1,2,5\}}, \{\theta_j'\}_{j \in \{3,4,6,7\}},$$
$$\vec{\pi}_{\mathsf{AHO},1}'', \vec{\pi}_{\mathsf{AHO},2}'', \{\vec{C}_{\sigma_{\rho(i),1}}'', \sigma_{\rho(i),2}', \vec{\pi}_{W,\rho(i)}''\}_{i=1}^{n'}\Big). \quad (6)$$

Verify$(\mathsf{pk}, \mathsf{Msg}, \sigma)$**:** given pk, σ and $\mathsf{Msg} = \{m_i\}_{i=1}^n$, parse σ as per (5).

1. Return 0 if $\vec{\pi}_{\mathsf{AHO},1} = (\pi_1, \pi_2, \pi_3)$ and $\vec{\pi}_{\mathsf{AHO},2} = (\pi_4, \pi_5, \pi_6)$ do not satisfy.

$$\iota_{\mathbb{G}_T}(A) \cdot E\left(\theta_3, \iota_{\mathbb{G}}(\theta_4)\right)^{-1} = E(G_z, \vec{C}_{\theta_1}) \cdot E(G_r, \vec{C}_{\theta_2}) \cdot E(G_1, \vec{C}_X) \cdot \prod_{j=1}^{3} E(\pi_j, \vec{f}_j)$$

$$\iota_{\mathbb{G}_T}(B) \cdot E\left(\theta_6, \iota_{\mathbb{G}}(\theta_7)\right)^{-1} = E(H_z, \vec{C}_{\theta_1}) \cdot E(H_r, \vec{C}_{\theta_5}) \cdot E(H_1, \vec{C}_X) \cdot \prod_{j=1}^{3} E(\pi_{j+3}, \vec{f}_j).$$

2. Return 1 if and only if, for each i, $\vec{\pi}_{W,i} = (\pi_{W,i,1}, \pi_{W,i,2}, \pi_{W,i,3})$ satisfies

$$E(h, \vec{C}_X) \cdot E(H_{\mathbb{G}}(m_i), (1, 1, \sigma_{i,2})) = E(g, \vec{C}_{\sigma_{i,1}}) \cdot \prod_{j=1}^{3} E(\pi_{W,i,j}, \vec{f}_j).$$

In the full version of the paper, we prove that the scheme is unforgeable under the DLIN and q-SFP assumptions and completely context hiding.

References

1. Abe, M., Haralambiev, K., Ohkubo, M.: Signing on Elements in Bilinear Groups for Modular Protocol Design. Cryptology ePrint Archive: Report 2010/133 (2010)
2. Abe, M., Fuchsbauer, G., Groth, J., Haralambiev, K., Ohkubo, M.: Structure-Preserving Signatures and Commitments to Group Elements. In: Rabin, T. (ed.) CRYPTO 2010. LNCS, vol. 6223, pp. 209–236. Springer, Heidelberg (2010)
3. Ahn, J.H., Boneh, D., Camenisch, J., Hohenberger, S., Shelat, A., Waters, B.: Computing on Authenticated Data. In: Cramer, R. (ed.) TCC 2012. LNCS, vol. 7194, pp. 1–20. Springer, Heidelberg (2012)
4. Attrapadung, N., Libert, B.: Homomorphic Network Coding Signatures in the Standard Model. In: Catalano, D., Fazio, N., Gennaro, R., Nicolosi, A. (eds.) PKC 2011. LNCS, vol. 6571, pp. 17–34. Springer, Heidelberg (2011)
5. Barreto, P.S.L.M., Naehrig, M.: Pairing-Friendly Elliptic Curves of Prime Order. In: Preneel, B., Tavares, S. (eds.) SAC 2005. LNCS, vol. 3897, pp. 319–331. Springer, Heidelberg (2006)
6. Belenkiy, M., Camenisch, J., Chase, M., Kohlweiss, M., Lysyanskaya, A., Shacham, H.: Randomizable Proofs and Delegatable Anonymous Credentials. In: Halevi, S. (ed.) CRYPTO 2009. LNCS, vol. 5677, pp. 108–125. Springer, Heidelberg (2009)
7. Bellare, M., Neven, G.: Transitive Signatures Based on Factoring and RSA. In: Zheng, Y. (ed.) ASIACRYPT 2002. LNCS, vol. 2501, pp. 397–414. Springer, Heidelberg (2002)
8. Boneh, D., Boyen, X., Shacham, H.: Short Group Signatures. In: Franklin, M. (ed.) CRYPTO 2004. LNCS, vol. 3152, pp. 41–55. Springer, Heidelberg (2004)
9. Boneh, D., Franklin, M.: Identity-Based Encryption from the Weil Pairing. SIAM Journal of Computing 32(3), 586–615 (2003); In: Kilian, J. (ed.) CRYPTO 2001. LNCS, vol. 2139, pp. 213–229. Springer, Heidelberg (2001)
10. Boneh, D., Freeman, D.M., Katz, J., Waters, B.: Signing a Linear Subspace: Signature Schemes for Network Coding. In: Jarecki, S., Tsudik, G. (eds.) PKC 2009. LNCS, vol. 5443, pp. 68–87. Springer, Heidelberg (2009)
11. Boneh, D., Freeman, D.M.: Linearly Homomorphic Signatures over Binary Fields and New Tools for Lattice-Based Signatures. In: Catalano, D., Fazio, N., Gennaro, R., Nicolosi, A. (eds.) PKC 2011. LNCS, vol. 6571, pp. 1–16. Springer, Heidelberg (2011)
12. Boneh, D., Freeman, D.M.: Homomorphic Signatures for Polynomial Functions. In: Paterson, K.G. (ed.) EUROCRYPT 2011. LNCS, vol. 6632, pp. 149–168. Springer, Heidelberg (2011)
13. Boneh, D., Shen, E., Waters, B.: Strongly Unforgeable Signatures Based on Computational Diffie-Hellman. In: Yung, M., Dodis, Y., Kiayias, A., Malkin, T. (eds.) PKC 2006. LNCS, vol. 3958, pp. 229–240. Springer, Heidelberg (2006)

14. Brzuska, C., Busch, H., Dagdelen, O., Fischlin, M., Franz, M., Katzenbeisser, S., Manulis, M., Onete, C., Peter, A., Poettering, B., Schröder, D.: Redactable Signatures for Tree-Structured Data: Definitions and Constructions. In: Zhou, J., Yung, M. (eds.) ACNS 2010. LNCS, vol. 6123, pp. 87–104. Springer, Heidelberg (2010)

15. Brzuska, C., Fischlin, M., Freudenreich, T., Lehmann, A., Page, M., Schelbert, J., Schröder, D., Volk, F.: Security of Sanitizable Signatures Revisited. In: Jarecki, S., Tsudik, G. (eds.) PKC 2009. LNCS, vol. 5443, pp. 317–336. Springer, Heidelberg (2009)

16. Brzuska, C., Fischlin, M., Lehmann, A., Schröder, D.: Unlinkability of Sanitizable Signatures. In: Nguyen, P.Q., Pointcheval, D. (eds.) PKC 2010. LNCS, vol. 6056, pp. 444–461. Springer, Heidelberg (2010)

17. Catalano, D., Fiore, D., Warinschi, B.: Adaptive Pseudo-free Groups and Applications. In: Paterson, K.G. (ed.) EUROCRYPT 2011. LNCS, vol. 6632, pp. 207–223. Springer, Heidelberg (2011)

18. Catalano, D., Fiore, D., Warinschi, B.: Efficient Network Coding Signatures in the Standard Model. In: Fischlin, M., Buchmann, J., Manulis, M. (eds.) PKC 2012. LNCS, vol. 7293, pp. 680–696. Springer, Heidelberg (2012)

19. Chase, M., Kohlweiss, M., Lysyanskaya, A., Meiklejohn, S.: Malleable Proof Systems and Applications. In: Pointcheval, D., Johansson, T. (eds.) EUROCRYPT 2012. LNCS, vol. 7237, pp. 281–300. Springer, Heidelberg (2012)

20. Freeman, D.M.: Improved Security for Linearly Homomorphic Signatures: A Generic Framework. In: Fischlin, M., Buchmann, J., Manulis, M. (eds.) PKC 2012. LNCS, vol. 7293, pp. 697–714. Springer, Heidelberg (2012)

21. Fuchsbauer, G.: Commuting Signatures and Verifiable Encryption. In: Paterson, K.G. (ed.) EUROCRYPT 2011. LNCS, vol. 6632, pp. 224–245. Springer, Heidelberg (2011)

22. Gennaro, R., Katz, J., Krawczyk, H., Rabin, T.: Secure Network Coding over the Integers. In: Nguyen, P.Q., Pointcheval, D. (eds.) PKC 2010. LNCS, vol. 6056, pp. 142–160. Springer, Heidelberg (2010)

23. Gerbush, M., Lewko, A., O'Neill, A., Waters, B.: Dual Form Signatures: An Approach for Proving Security from Static Assumptions. In: Wang, X., Sako, K. (eds.) ASIACRYPT 2012. LNCS, vol. 7658, pp. 25–42. Springer, Heidelberg (2012)

24. Gentry, C.: Fully homomorphic encryption using ideal lattices. In: STOC 2009, pp. 169–178 (2009)

25. Groth, J., Sahai, A.: Efficient Non-interactive Proof Systems for Bilinear Groups. In: Smart, N.P. (ed.) EUROCRYPT 2008. LNCS, vol. 4965, pp. 415–432. Springer, Heidelberg (2008)

26. Hevia, A., Micciancio, D.: The Provable Security of Graph-Based One-Time Signatures and Extensions to Algebraic Signature Schemes. In: Zheng, Y. (ed.) ASIACRYPT 2002. LNCS, vol. 2501, pp. 379–396. Springer, Heidelberg (2002)

27. Kiltz, E., Mityagin, A., Panjwani, S., Raghavan, B.: Append-Only Signatures. In: Caires, L., Italiano, G.F., Monteiro, L., Palamidessi, C., Yung, M. (eds.) ICALP 2005. LNCS, vol. 3580, pp. 434–445. Springer, Heidelberg (2005)

28. Lewko, A.: Tools for Simulating Features of Composite Order Bilinear Groups in the Prime Order Setting. In: Pointcheval, D., Johansson, T. (eds.) EUROCRYPT 2012. LNCS, vol. 7237, pp. 318–335. Springer, Heidelberg (2012)

29. Lewko, A., Okamoto, T., Sahai, A., Takashima, K., Waters, B.: Fully Secure Functional Encryption: Attribute-Based Encryption and (Hierarchical) Inner Product Encryption. In: Gilbert, H. (ed.) EUROCRYPT 2010. LNCS, vol. 6110, pp. 62–91. Springer, Heidelberg (2010)

30. Lewko, A., Waters, B.: New Techniques for Dual System Encryption and Fully Secure HIBE with Short Ciphertexts. In: Micciancio, D. (ed.) TCC 2010. LNCS, vol. 5978, pp. 455–479. Springer, Heidelberg (2010)

31. Lewko, A., Waters, B.: Unbounded HIBE and Attribute-Based Encryption. In: Paterson, K.G. (ed.) EUROCRYPT 2011. LNCS, vol. 6632, pp. 547–567. Springer, Heidelberg (2011)

32. Johnson, R., Molnar, D., Song, D., Wagner, D.: Homomorphic Signature Schemes. In: Preneel, B. (ed.) CT-RSA 2002. LNCS, vol. 2271, pp. 244–262. Springer, Heidelberg (2002)

33. Micali, S., Rivest, R.L.: Transitive Signature Schemes. In: Preneel, B. (ed.) CT-RSA 2002. LNCS, vol. 2271, pp. 236–243. Springer, Heidelberg (2002)

34. Miyazaki, K., Hanaoka, G., Imai, H.: Digitally signed document sanitizing scheme based on bilinear maps. In: AsiaCCS 2006, pp. 343–354 (2006)

35. Okamoto, T., Takashima, K.: Fully Secure Functional Encryption with General Relations from the Decisional Linear Assumption. In: Rabin, T. (ed.) CRYPTO 2010. LNCS, vol. 6223, pp. 191–208. Springer, Heidelberg (2010)

36. Prabhakaran, M., Rosulek, M.: Homomorphic Encryption with CCA Security. In: Aceto, L., Damgård, I., Goldberg, L.A., Halldórsson, M.M., Ingólfsdóttir, A., Walukiewicz, I. (eds.) ICALP 2008, Part II. LNCS, vol. 5126, pp. 667–678. Springer, Heidelberg (2008)

37. van Dijk, M., Gentry, C., Halevi, S., Vaikuntanathan, V.: Fully Homomorphic Encryption over the Integers. In: Gilbert, H. (ed.) EUROCRYPT 2010. LNCS, vol. 6110, pp. 24–43. Springer, Heidelberg (2010)

38. Waters, B.: Efficient Identity-Based Encryption Without Random Oracles. In: Cramer, R. (ed.) EUROCRYPT 2005. LNCS, vol. 3494, pp. 114–127. Springer, Heidelberg (2005)

39. Waters, B.: Dual System Encryption: Realizing Fully Secure IBE and HIBE under Simple Assumptions. In: Halevi, S. (ed.) CRYPTO 2009. LNCS, vol. 5677, pp. 619–636. Springer, Heidelberg (2009)

A Coding-Theoretic Approach to Recovering Noisy RSA Keys

Kenneth G. Paterson, Antigoni Polychroniadou, and Dale L. Sibborn

Information Security Group, Royal Holloway, University of London

Abstract. Inspired by cold boot attacks, Heninger and Shacham (Crypto 2009) initiated the study of the problem of how to recover an RSA private key from a noisy version of that key. They gave an algorithm for the case where some bits of the private key are known with certainty. Their ideas were extended by Henecka, May and Meurer (Crypto 2010) to produce an algorithm that works when all the key bits are subject to error. In this paper, we bring a coding-theoretic viewpoint to bear on the problem of noisy RSA key recovery. This viewpoint allows us to cast the previous work as part of a more general framework. In turn, this enables us to explain why the previous algorithms do not solve the motivating cold boot problem, and to design a new algorithm that does (and more). In addition, we are able to use concepts and tools from coding theory – channel capacity, list decoding algorithms, and random coding techniques – to derive bounds on the performance of the previous and our new algorithm.

1 Introduction

Cold boot attacks [6, 7] are a class of attacks wherein memory remanence effects are exploited to extract data from a computer's memory. The idea is that modern computer memories retain data for periods of time after power is removed, so an attacker with physical access to a machine may be able to recover, for example, cryptographic key information. The time during which data is retained can be increased by cooling the memory chips. However, because the memory gradually degrades over time once power is removed, only a noisy version of the data may be recoverable. The question then naturally arises: given a noisy version of a cryptographic key, is it possible to reconstruct the original key?

This question was addressed for broad classes of cryptosystems, both symmetric and asymmetric, by Halderman *et al.* in [6,7] and specifically for RSA private keys in [8,9]. Similar problems arise in the context of side-channel analysis of cryptographic implementations, where noisy key information may leak through power consumption [11] or timing [2]. The question is also linked to the classical cryptanalysis problem of recovering an RSA private key when some bits of the key are known, for example the most or least significant bits, or contiguous bits spread over a number of blocks (see, for example, the surveys in [1,12] and [10]).

X. Wang and K. Sako (Eds.): ASIACRYPT 2012, LNCS 7658, pp. 386–403, 2012.

Heninger and Shacham (HS) [9] considered the setting where a random fraction of the RSA private key bits is known with certainty. Their approach exploits the fact that the individual bits of an RSA private key of the form sk $= (p, q, d, d_p, d_q)$ must satisfy certain algebraic relations. This enables the recovery of the private key in a bit-by-bit fashion, starting with the least significant bits, by growing a search tree. It is easy to prune the search tree to remove partial solutions which do not match with the known key bits. The resulting algorithm will always succeed in recovering the private key, since the pruning process will never remove a partial version of the correct solution. On the other hand, when only few bits are known, the search tree may grow very large, and the HS algorithm will blow up. It was proved in [9] that, under reasonable assumptions concerning randomness of incorrect solutions, the HS algorithm will efficiently recover an n-bit RSA private key in time $O(n^2)$ with probability $1 - 1/n^2$ when a random fraction of at least 0.27 of the private key bits are known with certainty. These theoretical results are well-matched by experiments reported in [9]. These experiments also confirm that the HS algorithm has good performance when the known fraction is as small as 0.24, and the analysis of [9] extends to cases where the RSA private key sk is of the form (p, q, d) or (p, q).

Henecka, May and Meurer (HMM) [8] took the ideas of [9] and developed them further to address the situation where no key bits are known with certainty. They consider the *symmetric* case where the two possible bit flips $0 \to 1$, $1 \to 0$ have equal probability δ. Their main idea was to consider t bit-slices at a time of possible solutions to the equations relating the bits of sk, instead of single bits at a time as in the HS algorithm. In the formulation where sk $= (p, q, d, d_p, d_q)$, this yields 2^t candidate solutions on $5t$ new private key bits for each starting candidate at each stage of the algorithm. The HMM algorithm then computes the Hamming distance between the candidate solutions and the noisy key, keeping all candidates for which this metric is less than some carefully chosen threshold C. This replaces the procedure of looking for exact matches used in the HS algorithm. Of course, now the correct solution may fail this statistical test and be rejected; moreover the number of candidate solutions retained may explode if C is set too loosely. Nevertheless, it was shown in [8] that the HMM algorithm is efficient and has reasonable success in outputting the correct solution provided that $\delta < 0.237$. Again, the analysis depends on assumptions concerning the random behaviour of wrong solutions. To support the analysis, [8] reports the results of experiments for different noise levels and algorithmic parameters. For example, the algorithm can cope with $\delta = 0.20$.

In recent work independent of ours, Sarkar and Maitra [13] revisited the work of [8], applying the HMM algorithm to break Chinese Remainder implementations of RSA with low weight decryption exponents and giving *ad hoc* heuristics to improve the algorithm.

Limitations of Previous Work and Open Questions: Although inspired by cold boot attacks, it transpires that neither the HS algorithm nor the HMM algorithm actually solve the motivating cold boot problem. Let us see why.

One observation made in [6,7] is that for a given region of memory, the decay of memory bits is overwhelmingly either $0 \to 1$ or $1 \to 0$, while the decay direction in a given region can be inferred by comparing the number of 0s and 1s (since for an uncorrupted private key, we expect these to be roughly equal). Thus, in a $1 \to 0$ region, a 1 bit in the noisy version of the key is known (with high probability) to correspond to a 1 bit in the original key.

In the case of [9], the assumption is made that a certain fraction of the RSA private key bits – both 0s and 1s – is known with certainty. But, in the cold boot scenario, only 1 (or 0) bits are known, and not a mixture of both. Fortunately, the authors of [9] have informed us that their algorithm does still work when only 0 or only 1 bits are known, but this is not the case it was designed for, and, formally, the performance guarantees obtained in [9] do not apply in this case. Furthermore, in a real cold boot attack, bits are never known with *absolute* certainty, because even in a $1 \to 0$ region, say, bit flips in the reverse direction can occur. Halderman *et al.* report rates of 0.05% to 0.1% for this event. Such an event will completely derail the HS algorithm, as it will result in the correct solution being eliminated from the search tree. Based on an occurrence rate of 0.1%, this kind of fatal event can be expected to arise around 2.5 to 5 times in a real key recovery attack for 1024-bit RSA moduli with $\mathsf{sk} = (p, q, d, d_p, d_q)$. Thus, the HS algorithm really only applies to an "idealised" cold boot setting, where some bits are known for sure.

The HMM algorithm is designed to work for the symmetric case where the two possible bit flips have equal probability δ. Yet, in a cold boot attack, in a $1 \to 0$ region say, $\alpha := \Pr(0 \to 1)$ will be very small (though non-zero), while $\beta := \Pr(1 \to 0)$ may be relatively large, and perhaps even greater than 0.5 in a very degraded case. The use of Hamming distance as a metric for comparison and the setting of the threshold C are closely tied to the symmetric case, and it is not immediately clear how one can generalise the HMM approach to handle the type of errors occurring in real cold boot attacks. So it does not solve the cold boot problem for RSA keys.

Intriguing features of the work in [8,9] are the constants 0.27 and 0.237, which bound the fraction of known bits/noise rate the HS and HMM algorithms can handle. One can trace through the relevant analysis to see how these numbers emerge, but it would be more satisfying to have a deeper, unifying explanation. One might also wonder if these bounds are best possible or whether significant improvements might yet be forthcoming. Is there any ultimate limit to the noise level that these kinds of algorithms can deal with? And can we design an algorithm that works in the true cold boot setting, or for fully general noise models that might be expected to occur in other types of side channel attack?

Our contributions: We show how to recast the problem of noisy RSA key recovery as a problem in coding theory. That such a connection exists should be no surprise: after all, we are in a situation where bits are only known with certain probabilities and we wish to recover the true bits. However, this connection opens up the opportunity to apply to our problem the full gamut of sophisticated tools that have been developed by coding theorists over the last 60 years. We sketch this connection and its main consequences next.

Recall that in the HMM algorithm, we generate from each solution so far a set of 2^t candidate solutions on $5t$ new bits. We now view the set of 2^t candidates as being a *code*, with one codeword s (representing bits of the true private key) being selected and transmitted over a noisy channel, resulting in a received word r (representing $5t$ bits of the noisy version of the key). In the HMM case, the noise is realised via bit flipping with probability δ. The HS algorithm can be seen as arising from the special case $t = 1$, where the noise now corresponds to erasing a fraction of key bits instead of flipping them. Alternatively, we can consider a generalisation of the HS algorithm which considers $5t$ bits at a time, generated just as in the HMM algorithm, and which then filters the resulting 2^t candidates based on matching with known key bits. Because filtering is based on exact matching, this algorithm has the same output as the original HS algorithm. This brings the two algorithms under a single umbrella.

In general, in coding theory, the way in which s is transformed into r depends on the *channel model*, which in its full generality defines the probabilities $\Pr(r|s)$ over all possible pairs (s, r). In the case of [9], the assumption is that particular bits are known with certainty and others are not known at all, with the bits all being treated independently. The appropriate channel model is then an *erasure* channel, meaning that bits are independently either erased or transmitted correctly over the channel, with the receiver knowing the positions of the erasures. In the case of [8], the appropriate channel model is the binary symmetric channel with cross-over probability δ. It also emerges that the appropriate channel model for the true cold boot setting is a binary *non-symmetric* channel with cross-over probabilities (α, β). In general, the problem we are faced with is to decode r, with the aim being to reproduce s with high probability.

When couched in this language, it becomes obvious that the HS and HMM algorithms do not solve the original cold boot problem – simply put these algorithms use inappropriate channel models for that specific problem. We can also use this viewpoint to derive limits on the performance of *any* procedure for selecting which candidate solutions to keep in an HMM-style algorithm. To see why, we recall that the converse to Shannon's noisy-channel coding theorem [14] states that *no* combination of code and decoding procedure can jointly achieve arbitrarily reliable decoding when the code rate exceeds the (Shannon) capacity of the channel. Moreover, there are analogues of the converse of Shannon's theorem for so-called *list decoding* that essentially show that channel capacity is also the barrier to any efficient algorithm outputting lists of candidates, as the HS and HMM algorithms do.

When sk is of the form (p, q, d, d_p, d_q), for example, the code rate is fixed at $1/5$ (we have 2^t codewords and length $5t$). The channel capacity can be calculated as a function of the channel model and its parameters. For example, for the erasure channel with erasure probability ρ (meaning that a fraction $1 - \rho$ of the bits are known with certainty), the capacity is simply $1 - \rho$. Then we see that the limiting value is $\rho = 0.8$, meaning that the fraction of known bits must be *at least* 0.2 to achieve arbitrarily reliable, efficient decoding. The analysis in [9] needs that fraction to be at least 0.27, though a fraction as low as 0.24 could be handled in practice. Thus a capacity analysis suggests that there should be room to improve the HS algorithm further, but capacity shows that it is impossible go below a fraction 0.2 of known bits with an efficient algorithm. See Section 3 for further details on list decoding and its application to the analysis of the HS and HMM algorithms.

Informed by our coding-theoretic viewpoint, we derive a new key recovery algorithm that works for any (memoryless) binary channel and therefore *is* applicable to the cold boot setting (and more). In essence, we modify the HMM algorithm to use a likelihood statistic in place of the Hamming metric when selecting from the candidate codewords. We keep the L codewords having the highest values of this likelihood statistic and reject the others. An important consequence of this algorithmic choice is that our algorithm has *deterministic* running time $O(L2^t n/t)$ and, when implemented using a stack, *deterministic* memory consumption $O(L+t)$. This stands in contrast to the running time and memory usage of the HS and HMM algorithms, which may blow up when the erasure/error rates are high. We note that private RSA keys are big enough that they may cross regions when stored in memory. We can handle this by changing the likelihood statistic used in our algorithm at the appropriate transition points, requiring only a simple modification to our approach. In the full version, we give an analysis of the success probability of our new algorithm, under different randomness hypotheses, using coding-theoretic tools. Essentially, we are able to show that, as $t \to \infty$, its success probability tends to 1 provided the code rate (1/5 when $\mathsf{sk} = (p, q, d, d_p, d_q)$) remains below the channel capacity. Moreover, from the converse to Shannon's theorem, we are unlikely to be able to improve this result if reliable key recovery is required.

We include the results of extensive experiments using our new algorithm. These demonstrate that our approach matches or outperforms the HS and HMM algorithms in the cases they are designed for, and achieves results close to the limits imposed by our capacity analysis more generally. For example, in the symmetric case with $\delta = 0.20$, we can achieve a 20% success rate in recovering keys for $t = 18$ and $L = 32$. This is comparable to the results of [8]. Furthermore, for the same t and L we achieve a 4% success rate for $\delta = 0.22$, whilst [8] does not report any experiments for an error rate this high. As another example, our algorithm can handle the idealised cold boot scenario by setting $\alpha = 0$ (in which case all the 1 bits in r are known with certainty, i.e. we are in a $1 \to 0$ region). Here, our capacity analysis puts a bound of 0.666 on β for reliable key recovery. Using our algorithm, we can recover keys for $\beta = 0.6$ with a 13% success rate using $t = 18$ and $L = 32$, whereas the HS algorithm can only reach $\beta = 0.52$ (and this under the assumption that the experimental results reported in [9] for a mixture of known 0 and 1 bits do translate to the same performance for the case where only 1 bits are known). In the same setting, we can even recover keys up to $\beta = 0.63$ with a non-zero success rate. We also have similar experimental results for the 'true' cold boot setting where both α and β are non-zero, and for the situation where sk is of the form (p, q, d) or (p, q).

Paper Organisation: The remainder of this paper is organised as follows. In the next section, we give further background on the algorithms of [8,9]. In Section 3, we develop the connection with coding theory and explain how to use it to derive limits on the performance of noisy RSA key recovery algorithms. Section 4 describes our new maximum likelihood list decoding algorithm. Section 5 presents our experimental results. Finally, Section 6 contains some closing remarks and open problems.

2   The HS and HMM Algorithms

Let (N, e) be the RSA public key, where $N = pq$ is an n-bit RSA modulus, and p, q are balanced primes. As with [8,9], we assume throughout that e is small, say $e = 3$ or $e = 2^{16} + 1$; for empirical justification of this assumption, see [15]. We start by assuming that private keys sk follow the PKCS#1 standard and so are of the form $(N, p, q, e, d, d_p, d_q, q_p^{-1})$, where d is the decryption key, $d_p = d \mod p - 1$, $d_q = d \mod q - 1$ and $q_p = q^{-1} \mod p$. However, neither the algorithms of [8,9] nor ours make use of q_p^{-1}, so we henceforth omit this information. Furthermore, we assume N and e are publicly known, so we work only with the tuple sk $= (p, q, d, d_p, d_q)$. We will also consider attacks where the private key contains less information – either sk $= (p, q, d)$ or sk $= (p, q)$.

Now assume we are given a degraded version of the key $\widetilde{\text{sk}} = (\tilde{p}, \tilde{q}, \tilde{d}, \tilde{d}_p, \tilde{d}_q)$. We start with the four RSA equations:

$$N = pq \tag{1}$$

$$ed = k(N - p - q + 1) + 1 \tag{2}$$

$$ed_p = k_p(p - 1) + 1 \tag{3}$$

$$ed_q = k_q(q - 1) + 1. \tag{4}$$

where k, k_p and k_q are integers to be determined. A method for doing so is given in [9]: first it is shown that $0 < k < e$; then, since e is small, we may enumerate

$$d(k') := \left\lfloor \frac{k'(N + 1) + 1}{e} \right\rfloor$$

for all $0 < k' < e$. We then find the k' such that $d(k')$ is "closest" to $\tilde{d}$ in the most significant half of the bits. Simple procedures for doing this are given in [8,9]. In the more general setting where bit flips can occur in both directions and with different probabilities, we proceed as follows. First, we estimate parameters $\alpha = \Pr(0 \to 1)$ and $\beta = \Pr(1 \to 0)$ from known bits, e.g. from a noisy version of N that is adjacent in memory to the private key. Second, we compute for each k' an approximate log-likelihood using the expression

$$n_{01} \log \alpha + n_{00} \log(1 - \alpha) + n_{10} \log \beta + n_{11} \log(1 - \beta)$$

where n_{01} is the number of positions in the most significant half where a 0 appears in $d(k')$ and a 1 appears in $\tilde{d}$, etc. Finally, we select the k' that provides the highest log-likelihood.

At the end of this procedure, with high probability we will have $k' = k$ and we will have recovered the most significant half of the bits of d. Now we wish to find k_p and k_q. By manipulating the above equations we see that

$$k_p^2 - (k(N-1)+1)k_p - k \equiv 0 \mod e$$

If e is prime (as in the most common case $e = 2^{16} + 1$) there will only be two solutions to this equation. One will be k_p and the other k_q. If e is not prime we will have to try all possible pairs of solutions in the remainder of the algorithm.

Now, for integers x, we define $\tau(x) := \max\{i \in \mathbb{N} : 2^i \mid x\}$. Then it is easy to see that $2^{\tau(k_p)+1}$ divides $k_p(p-1)$, $2^{\tau(k_q)+1}$ divides $k_q(q-1)$ and $2^{\tau(k)+2}$ divides $k\phi(N)$. These facts, along with relations (2) – (4), allow us to see that

$$d_p \equiv e^{-1} \mod 2^{\tau(k_p)+1}$$
$$d_q \equiv e^{-1} \mod 2^{\tau(k_q)+1}$$
$$d \equiv e^{-1} \mod 2^{\tau(k)+2}.$$

This allows us to correct the least significant bits of d, d_p and d_q. Furthermore we can calculate slice(0), where we define

$$\mathsf{slice}(i) := (p[i], q[i], d[i + \tau(k)], d_p[i + \tau(k_p)], d_q[i + \tau(k_q)]).$$

with $x[i]$ denoting the i-th bit of the string x.

Now we are ready to explain the main idea behind the algorithm of [9]. Suppose we have a solution (p', q', d', d_p', d_q') from slice(0) to slice($i-1$). Then [9] uses a multivariate version of Hensel's Lemma to show that the bits involved in slice(i) must satisfy the following congruences:

$$p[i] + q[i] = (N - p'q')[i] \mod 2$$
$$d[i + \tau(k)] + p[i] + q[i] = (k(N+1) + 1 - k(p'+q') - ed')[i + \tau(k)] \mod 2$$
$$d_p[i + \tau(k_p)] + p[i] = (k_p(p'-1) + 1 - ed_p')[i + \tau(k_p)] \mod 2$$
$$d_q[i + \tau(k_q)] + q[i] = (k_q(q'-1) + 1 - ed_q')[i + \tau(k_q)] \mod 2.$$

Because we have 4 constraints on 5 unknowns, there are exactly 2 possible solutions for slice(i), rather than 32. This is then used in [9] as the basis of building a search tree for the unknown private key bits. At each node in the tree, representing a partial solution up to slice($i-1$), at most two successor nodes are added by the above procedure. Moreover, since a random fraction of the bits is assumed to be known with certainty, the tree can be pruned of any partial solutions that are not consistent with these known bits. Clearly, if the fraction of known bits is large enough, then the tree will be highly pruned and the number of nodes in the tree will be small. The analysis of [9] shows that if the fraction of known bits is at least 0.27, then the tree's size remains close to linear in n, the size of the RSA modulus, meaning that an efficient algorithm results. A similar algorithm and analysis can be given for the case where sk is of the form (p, q, d) or (p, q); in each case, there are exactly 2 possible solutions for each slice(i).

Instead of doing Hensel lifting bit-by-bit and pruning on each bit, the HMM algorithm performs t Hensel lifts for some parameter t, yielding, for each surviving candidate solution on slice(0) to slice($i - 1$), a tree of depth t whose 2^t leaf nodes represent candidate solutions on slices slice(0) to slice($i + t - 1$), involving $5t$ new bits (in slice(i) to slice($i + t - 1$)). A solution is kept for the next iteration if the Hamming distance between the $5t$ new bits and the corresponding vector of noisy bits is less than some threshold C. Clearly the HS algorithm could also be modified in this way, lifting t times and then doing pruning based on matching known key bits. Alternatively, one can view the HS algorithm as being the special case $t = 1$ of the HMM algorithm (with a different pruning procedure). The HMM algorithm can also be adapted to work with sk of the form (p, q, d) or (p, q). Henecka et al. [8] showed how to select C and t so as to guarantee that their algorithm is efficient and produces the correct solution with a reasonable success rate. In particular, they were able to show that this is the case provided the probability of a bit flip δ is at most 0.237.

At each stage in the HMM algorithm, candidate solutions on t new slices are constructed. Then roughly $n/2t$ iterations or stages of the algorithm are needed, since all the quantities being recovered contain at most $n/2$ bits. As pointed out in [8], only half this number of stages is required since once we have the least significant half of the bits of the private key, the entire private key can be recovered using a result of Coppersmith [3]. At their conclusion, the HS and HMM algorithms outputs lists of candidate solutions rather than a single solution. But it is easy to verify the correctness of each candidate by using a trial encryption and decryption, say. Thus the success rate of the algorithms is defined to be the probability that the correct solution is on the output list. We adopt the same measure of success in the remainder of the paper.

3   The Coding-Theoretic Viewpoint

In this section, we develop our coding-theoretic viewpoint on the HS and HMM algorithms, using it to derive limits on the performance of these and similar algorithms. In particular, we will explain how channel capacity plays a crucial role in setting these limits.

We begin by defining the parameter m. We set $m = 5$ when sk $= (p, q, d, d_p, d_q)$, $m = 3$ when sk $= (p, q, d)$, and $m = 2$ when sk $= (p, q)$. Consider a stage of the HMM algorithm, commencing with M partial solutions that have survived the previous stage's pruning step. The HMM algorithm produces a total of $M2^t$ candidate solutions on mt bits, prior to pruning. We label these $s_1, \ldots, s_{M2^t}$, let $\mathcal{C}$ denote the set of all $M2^t$ candidates, and use r to denote the corresponding vector of mt noisy bits in sk.

Now we think of $\mathcal{C}$ as being a code. This code has rate $R \geq 1/m$, but its other standard parameters such as its minimum distance are unknown (and immaterial to our analysis). The problem of recovering the correct candidate s_j given r is clearly just the problem of decoding this code. Now both the HS and HMM algorithms have pruning steps that output lists of candidates for the correct solution, with the list size being dynamic in both cases and depending on the number of candidates surviving the relevant filtering process (based either on exact matches for the HS algorithm or on Hamming distance for the HMM algorithm). In this sense, the HS and HMM algorithms are performing types of *list decoding*, an alternative to the usual unique decoding of codes that was originally proposed by Elias [4].

To complete the picture, we need to discuss what error and channel models are used in [8,9], and what models are appropriate to the cold boot setting. As noted in the introduction, [9] assumes that some bits of r are known exactly, while no information at all is known about the other bits. This corresponds to an *erasure* model for errors, and an *erasure* channel. Usually, this is defined in terms of a parameter ρ representing the fraction of erasures. So $1 - \rho$ represents the fraction of known bits, a parameter denoted δ in [9]. On the other hand, [8] assumes that all bits of r are obtained from the correct s_j by independent bit flipping with probability δ. In standard coding terminology, we have a (memoryless) binary symmetric channel with crossover probability δ. From the experimental data reported in [6,7], an appropriate model for the cold boot setting would be a binary non-symmetric channel with crossover probabilities (α, β), with α being small and β being significantly larger in a $1 \rightarrow 0$ region (and vice-versa in a $0 \rightarrow 1$ region). In an idealised cold boot case, we could assume $\alpha = 0$, meaning that a $0 \rightarrow 1$ bit flip can never occur, so that all 1 bits in r are known with certainty. This is better known as a Z-channel in the coding-theoretic literature.

This viewpoint highlights the exact differences between the settings considered in [8,9] and the cold boot setting. It also reveals that, while the HS algorithm can be applied for the Z-channel seen in the idealised cold boot setting, there is no guarantee that the performance proven for it in [9] for the erasure channel will transfer to the Z-channel. Moreover, one might hope for substantial improvements to the HS algorithm if one could somehow take into account the (partial) information known about 0 bits as well as the exact information known about 1 bits.

3.1   The Link to Channel Capacity

We can use this coding viewpoint to derive limits on the performance of *any* procedure for selecting which candidate solutions to keep in the HS and HMM algorithms. To see why, we recall that the converse to Shannon's noisy-channel coding theorem [14] states that *no* combination of code and decoding procedure can jointly achieve arbitrarily reliable decoding when the code rate exceeds the capacity of the channel. Our code rate is at least $1/m$ where $m = 2, 3$ or 5 and the channel capacity can be calculated as a function of the channel model and its parameters.

Two *caveats* must be made here. Firstly, capacity only puts limits on *reliable* decoding, and even decoding with low success probability is of interest in cryptanalysis. Secondly, Shannon's result applies only to decoding algorithms that output a single codeword s, while both the HS and HMM algorithms are permitted to output many candidates at each stage, with the final output list only being required to contain the correct private key. Perhaps such list-outputting algorithms can surpass the bounds imposed by Shannon's theorem? Indeed, the HS algorithm is guaranteed to output the correct key provided the algorithm terminates. Similarly, the threshold C in the HMM algorithm can always be set to a value that ensures that every candidate passes the test and is kept for the next stage, thus guaranteeing that the algorithm is always successful. However, neither of these variants would be *efficient* and in fact there are analogues of the converse of Shannon's noisy-channel coding theorem that essentially show that capacity is the barrier for efficient list decoding too.

For the binary symmetric channel, it is shown in [5, Theorem 3.4] that if $\mathcal{C}$ is *any* code of length n and rate $1 - H_2(\delta) + \epsilon$ for some $\epsilon > 0$, then some word r is such that the Hamming sphere of radius δn around r contains at least $2^{\epsilon n/2}$ codewords. Here $H_2(\cdot)$ is the binary entropy function:

$$H_2(x) = -x \log_2(x) - (1 - x) \log_2(1 - x)$$

and $1 - H_2(\delta)$ is just the capacity of the channel. The proof also shows that, over a random choice of r, the average number of codewords in a sphere of radius δn around r is $2^{\epsilon n/2}$. Since the expected number of errors in r is δn, we expect the correct codeword to be in this sphere, along with $2^{\epsilon n/2}$ other codewords. This implies that, if the rate of the code exceeds the channel capacity $1 - H_2(\delta)$ by a constant amount ϵ, then $\mathcal{C}$ cannot be list decoded using a polynomial-sized list, either in the worst case or on average, as $n \to \infty$.

An analogue result can be proved for the erasure channel, based on a similarly simple counting argument as was used in the proof of [5, Theorem 3.4]: if ρ is the erasure probability and $\mathcal{C}$ is any code of rate $1 - \rho + \epsilon$ (i.e. ϵ above the erasure channel's capacity), then it can be shown that on average there will be $2^{\epsilon n}$ codewords that differ from r in its erasure positions, assuming r contains ρn erasure symbols. Hence reliable list decoding for $\mathcal{C}$ cannot be achieved using a polynomial-sized list.

In the next sub-section, we will examine in more detail the implications of these results on list decoding for the HS and HMM algorithms.

3.2   Implications of the Capacity Analysis

The Binary Symmetric Channel and the HMM Algorithm. If the HMM algorithm is to have reasonable success probability in recovering the key, then at each stage, it must set the threshold C in such a way that all words $s_i \in \mathcal{C}$ with $d_H(s_i, r) \approx \delta m t$ are accepted by the algorithm. This is because $\delta m t$ is the expected number of errors occurring in r, and if the threshold is set below this value, then the correct codeword is highly likely to be rejected by the algorithm. (In fact, the HMM algorithm sets C to be slightly higher than this, which makes good sense given that there is an even chance of there being more than $\delta m t$

errors.) Recall that we have rate $R \geq 1/m$. Now suppose δ is such that $R = 1 - H_2(\delta) + \epsilon$ for some $\epsilon > 0$, i.e. δ is chosen so that that the code rate is just above capacity. Then the argument above shows that there will be on average at least $2^{\epsilon m t/2}$ codewords on the output list at each stage. Thus, as soon as δ is such that R exceeds capacity by a constant amount ϵ, then there must be a blow-up in the algorithm's output size at each stage, and the algorithm will be inefficient asymptotically.

We write $C_{\mathrm{BSC}}(\delta) = 1 - H_2(\delta)$ for the capacity of the binary symmetric channel. Table 1 shows that $C_{\mathrm{BSC}}(\delta) = 0.2$ when $\delta = 0.243$. Thus what our capacity analysis shows is that the best error rate one could hope to deal with in the HMM algorithm when $m = 5$ is $\delta = 0.243$. Notice that this value is rather close to, but slightly higher than, the corresponding value of 0.237 arising from the analysis in [8]. The same is true for the other entries in this table. This means that significantly improving the theoretical performance of the HMM algorithm (or indeed any HMM-style algorithm) whilst keeping the algorithm efficient will not be possible. The experimental work in [8] gives results up to a maximum δ of 0.20; compared to the capacity bound of 0.243, it appears that there is some room for practical improvement in the symmetric case.

The Erasure Channel and the HS Algorithm. As noted above, for the erasure channel, the capacity is $1 - \rho$, where ρ is the fraction of bits erased by the channel. Note that the list output by the HS algorithm is independent of whether pruning is done after each lift or in one pass at the end (but obviously doing so on a lift-by-lift basis is more efficient in terms of the total number of candidates examined). Then considering the HS algorithm in its entirety (i.e. over $n/2$ Hensel lifts), we see that it acts as nothing more than a list decoder for the erasure channel, with the code $\mathcal{C}$ being the set of all $2^{n/2}$ words on $mn/2$ bits generated by doing $n/2$ Hensel lifts without any pruning, and the received word r being the noisy version of the entire private key sk.

Then our analysis above applies to show that the HS algorithm will produce an exponentially large output list, and will therefore be inefficient, when the rate (which in this case is exactly $1/m$) exceeds the capacity $1 - \rho$. For $m = 5$, we have rate 0.2 and so our analysis shows that the HS algorithm will produce an exponentially large output list whenever ρ exceeds 0.8. Now [9] reports good results (in the sense of having a reasonable running time) for ρ as high as 0.76 (corresponding to Heninger and Shacham's parameter δ being equal to 0.24),

Table 1. Private key-type, equivalent rate R, and maximum crossover probability δ allowing reliable key recovery, symmetric channel case

sk	R	δ
(p, q, d, d_p, d_q)	1/5	0.243
(p, q, d)	1/3	0.174
(p, q)	1/2	0.110

Table 2. Private key-type, equivalent rate R, and maximum error probability ρ allowing reliable key recovery, Z-channel case

sk	R	β
(p, q, d, d_p, d_q)	1/5	0.666
(p, q, d)	1/3	0.486
(p, q)	1/2	0.304

leaving a gap between the experimental performance and the theoretical bound. Similar remarks apply for the cases $m = 2, 3$: for $m = 2$, the HS algorithm should be successful for $\rho = 0.43$ ($\delta = 0.57$), while the bound from capacity is 0.50; for $m = 3$, we have $\rho = 0.58$ ($\delta = 0.42$) and the capacity bound is 0.67. Hence, further improvements for $m = 2, 3$ are not ruled out by the capacity analysis.

The Z-channel. We may also apply the above capacity analysis to the idealised cold boot setting, where the crossover probabilities are of the form $(0, \beta)$. Here we have a Z-channel, whose capacity can be written as:

$$C_Z(\beta) = \log_2(1 + (1 - \beta)\beta^{\frac{\beta}{1-\beta}}).$$

Solving the equation $C_Z(\beta) = R$ for $R = 1/5$, $1/3$, $1/2$ gives us the entries in Table 2. We point out the large gap between these figures and what we would expect to obtain both theoretically and experimentally if we were to directly apply the HS algorithm to the idealised cold boot setting. For example, when $m = 5$, the analysis of [9] suggests that key recovery should be successful provided that β does not exceed 0.46 (the value of $\delta = 0.27$ translates into a β value of 0.46 using the formula $\delta = (1 - \beta)/2$ given in [9]), whereas the capacity analysis suggests a maximum β value of 0.666. This illustrates that the HS algorithm is not well-matched to the Z-channel. Our new algorithm will close this gap substantially.

The True Cold Boot Setting. For the true cold boot setting, we must consider the general case of a memoryless, binary channel with crossover probabilities (α, β). We can calculate the capacity $C(\alpha, \beta)$ of this channel and obtain the regions for which $C(\alpha, \beta) > R$ for $R = 1/5$, $1/3$, $1/2$. The results are shown in Figure 1. Notice that these plots include as special cases the data from Tables 1 and 2. If we set $\alpha = 0.001$, say, we see that the maximum achievable β is quite close to that in the idealised cold boot setting. Note also that the plots are symmetric about the lines $y = x$ and $y = 1 - x$, reflecting the fact that capacity is preserved under the transformations $(\alpha, \beta) \rightarrow (\beta, \alpha)$ and $(\alpha, \beta) \rightarrow (1 - \alpha, 1 - \beta)$.

However, we must caution that capacity-based bounds for list decoding for the general binary non-symmetric channel (including the Z-channel) are not known in the coding-theoretic literature. Strictly speaking, then, our capacity analysis for this case does not bound the performance of key recovery algorithms that are allowed to output many key candidates, but only the limited class of algorithms that output a *single* key candidate. This said, our capacity analysis sets a target for our new algorithm, which follows.

4   The New Algorithm and Its Analysis

In this section, we give our new algorithm for noisy RSA key recovery that works for any memoryless, binary channel, as characterised by the cross-over probabilities (α, β). Our algorithm has the same basic structure as the HMM

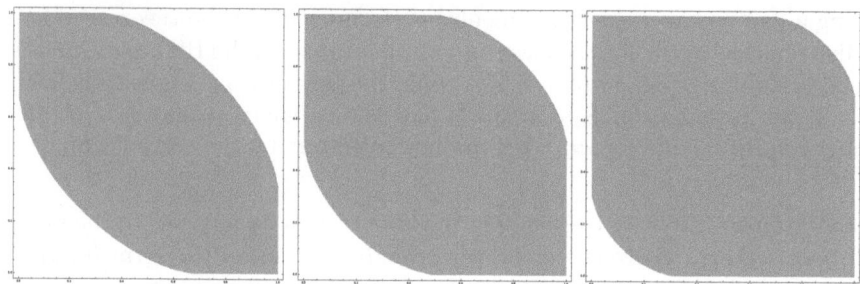

Fig. 1. Plots showing achievable (α, β) pairs for private keys containing 5, 3 and 2 components, respectively. The vertical axis is β, the horizontal axis is α. The shaded area in each case represents the unachievable region.

algorithm but uses a different procedure to decide which candidate solutions to retain and which to reject. Specifically, we use a likelihood measure in place of Hamming distance.

Recall that we label the $M2^t$ candidate solutions on mt bits arising at some stage in the HMM algorithm $s_1, \ldots, s_{M2^t}$ and let us name the corresponding vector of mt noisy bits in the RSA private key r. Then the Maximum Likelihood (ML) estimate for the correct candidate solution is simply:

$$\arg \max_{1 \leq i \leq M2^t} \Pr(s_i | r).$$

that is, the choice of i that maximises the conditional probability $\Pr(s_i | r)$. Using Bayes' theorem, this can be rewritten as:

$$\arg \max_{1 \leq i \leq M2^t} \frac{\Pr(r | s_i) \Pr(s_i)}{\Pr(r)}.$$

Here, $\Pr(r)$ is a constant for a given set of bits r. Let us make the further mild assumption that $\Pr(s_i)$ is also a constant, independent of i. Then the ML estimate is obtained from

$$\arg \max_{1 \leq i \leq M2^t} (\Pr(r | s_i)) = \arg \max_{1 \leq i \leq M2^t} \left((1 - \alpha)^{n_{00}^i} \alpha^{n_{01}^i} (1 - \beta)^{n_{11}^i} \beta^{n_{10}^i} \right)$$

where $\alpha = \Pr(0 \rightarrow 1)$ and $\beta = \Pr(1 \rightarrow 0)$ are the crossover probabilities, n_{00}^i denotes the number of positions where s_i and r both have 0 bits, n_{01}^i denotes the number of positions where s_i has a 0 and r has a 1, and so on.

Equivalently, we may maximise the log of these probabilities, and so we seek:

$$\arg \max_{1 \leq i \leq M2^t} (\log \Pr(r | s_i))$$

$$= \arg \max_{1 \leq i \leq M2^t} \left(n_{00}^i \log(1 - \alpha) + n_{01}^i \log \alpha + n_{11}^i \log(1 - \beta) + n_{10}^i \log \beta \right)$$

which provides us with a simpler form for computational purposes.

Algorithm 1. Pseudo-code for the maximum likelihood list decoding algorithm for reconstructing RSA private keys.

---

list ← slice(0);
for stage = 1 **to** $n/2t$ **do**
> Replace each entry in list with a set of 2^t candidate solutions obtained by Hensel lifting;
> Calculate the log-likelihood $\log \Pr(r|s_i)$ for each entry s_i on list;
> Keep the L entries in list having the highest log-likelihoods and delete the remainder;

Output list;

Then our proposed algorithm is simply this: select at each stage from the candidates generated by Hensel lifting those L candidates s_i which produce the highest values of the log-likelihood $\log \Pr(r|s_i)$ as in the equation above. These candidates are then passed to the next stage. So at each stage except the first we will generate a total of $L2^t$ candidates and keep the best L. We may then test each entry in the final list by trial encryption and decryption to recover a single candidate for the private key. Pseudo-code for this algorithm is shown in Algorithm 1. Note that here we assume there are $n/2t$ stages; this number can be halved as in the HS and HMM algorithms.

Our algorithm has fixed running time $O(L2^t)$ for each of the $n/2t$ stages, and fixed memory consumption $O(L2^t)$. This is a consequence of choosing to keep the L best candidates at each stage in place of all candidates surpassing some threshold as in the HMM algorithm. The memory consumption can be reduced to $O(L+t)$ by using a depth-first approach to generating and filtering the candidates. The main overhead is then the Hensel lifting to generate candidate solutions; the subsequent computation of log-likelihoods for each candidate is relatively cheap. Notice that if $\alpha = 0$ (as in the Z-channel for an idealised cold boot setting), then any instance of a $0 \rightarrow 1$ bit flip is very heavily penalised by the log-likelihood statistic – it adds a $-\infty$ term to $\log \Pr(r|s_i)$. In practice, for $\alpha = 0$, we just reject any solution containing a $0 \rightarrow 1$ transition. For the erasure channel, we reject any candidate solution that does not match r in the known bits.

A special case of our algorithm arises when $L = 1$ and corresponds to just keeping the single ML candidate at each stage. This algorithm then corresponds to Maximum Likelihood (ML) decoding. However, at a given stage, it is likely that the correct solution will be rejected because a wrong solution happens to have the highest likelihood. This is especially so in view of how similar some candidates will be to the correct solution. Therefore, ML decoding is likely to go awry at some stage of the algorithm.

4.1   Remarks on the Asymptotic Analysis of Our Algorithm

In the full version, we give two analyses of our algorithm, using tools from coding theory to assist us. The first analysis uses a strong randomness assumption, that

Table 3. Success probabilities for the symmetric case $((\alpha, \beta) = (\delta, \delta))$. Experiments with $\delta \leq 0.16$ are based on 500 trials. Capacity bound on δ is 0.243.

δ	0.08	0.10	0.12	0.14	0.16	0.18	0.19	0.2	0.21	0.22
t	6	8	10	12	16	18	18	18	18	18
L	4	4	8	32	32	32	32	32	32	64
Success rate	1	0.921	0.932	0.963	0.84	0.60	0.38	0.20	0.08	0.04
Time per trial (ms)	113	98	474	4323	85662	395069	399451	380139	377342	722341

the $L2^t$ candidates s_i generated at each stage of Algorithm 1 are independent and uniformly random mt-bit vectors. It shows that, asymptotically, our algorithm will be successful in recovering the RSA private key provided $1/m$ is less than the capacity of the memoryless, binary channel with crossover probabilities (α, β). In fact, this result follows as a simple application of Shannon's noisy-channel coding theorem [14], which states that, asymptotically, the use of random codes in combination with Maximum Likelihood (ML) decoding achieves arbitrarily small decoding error probability, provided that the code rate stays below the capacity of the channel. Unfortunately, it is easy to see that our strong randomness assumption is in fact *not* true for the codes C generated in our algorithm, because of the iterative nature of the Hensel lifting. The second analysis proves a similar result for the symmetric case under weaker randomness assumptions for which we have good experimental evidence. Details can be found in the full version.

5   Experimental Results

For our experiments, we used a multi-threaded implementation based on Java code kindly supplied by the authors of [8]. We ran our code on an 8x virtual CPU hosted on a 2x Intel Xeon X5650, clocked at 2.67 GHz (IBM BladeCenter HS22V). Except where noted below, our experiments were run for 100 trials using a randomly-generated RSA key for each trial. Except where noted, our results refer to private keys of the form $\mathsf{sk} = (p, q, d, d_p, d_q)$ and are all for 1024-bit RSA moduli.

We have conducted extensive experiments for the symmetric case considered in [8]. Our results are shown in Table 3. For small values of δ, we achieve a success rate of 1 or very close to 1 using only moderate amounts of computation. By contrast the HMM algorithm does not achieve such high success rate for small δ. This cannot be solved by increasing t in the HMM algorithm because this leads to a blow-up in running time. For larger δ, the success rate of our algorithm is comparable to that of [8] for similar values of t. We were able to obtain a non-zero success rate for $\delta = 0.22$, while [8] only reached $\delta = 0.20$. The bound from capacity is 0.243.

For the idealised cold boot setting where $\alpha = 0$, our experimental results are shown in Table 4. Recall that the HS algorithm can also be applied to this case. Translating the fraction of known bits $(1 - \rho)$ to the idealised cold boot setting,

Table 4. Success probabilities for the idealised cold boot case ($\alpha = 0$). Capacity bound on β is 0.666.

ρ	0.1	0.2	0.3	0.4	0.46	0.5	0.55	0.6	0.62	0.63
t	6	6	8	12	16	18	18	18	18	18
L	4	4	8	8	8	16	16	16	64	64
Success rate	1	1	1	0.98	0.87	0.81	0.43	0.13	0.07	0.03
Time per trial (ms)	69	88	147	1518	22349	292834	282235	290254	692532	683421

Table 5. Success probabilities for the true cold-boot case with $\alpha = 0.001$. Capacity bound on β is 0.658.

β	0.1	0.2	0.3	0.4	0.5	0.55	0.6	0.61
t	6	6	8	12	16	18	18	18
L	4	4	8	8	16	32	64	64
Success rate	1	1	0.97	0.97	0.66	0.31	0.09	0.04
Time per trial (ms)	80	80	273	4268	42732	384262	740244	735169

and assuming the HS algorithm works just as well when only 1 bits are known (instead of a mixture of 0 and 1 bits), the maximum value of β that could be handled by the HS algorithm theoretically would be 0.46 (though results reported in [9] would allow β as high as 0.52). Our algorithm still has a reasonable success rate for β as high as 0.6 and non-zero success rate even for $\beta = 0.63$, beating the HS algorithm by some margin. Our capacity analysis for this case suggests that the maximum value of β will be 0.666. Thus our algorithm is operating within 5% of capacity here.

Table 6. Success probabilities for the true cold-boot case with $\alpha = 0.001$ and sk = (p, q, d). Capacity bound on β is 0.479.

β	0.1	0.15	0.20	0.25	0.30	0.35	0.40	0.43
t	6	10	14	16	18	18	18	18
L	4	16	16	16	16	16	32	64
Success rate	0.99	0.99	0.98	0.96	0.63	0.55	0.12	0.04
Time per trial (ms)	46	371	4441	19906	117502	108523	165418	301457

Table 7. Success probabilities for the true cold-boot case with $\alpha = 0.001$ and sk = (p, q). Capacity bound on β is 0.298.

β	0.05	0.1	0.15	0.20	0.26
t	10	12	16	18	18
L	8	8	16	32	64
Success rate	0.95	0.83	0.68	0.29	0.06
Time per trial (ms)	404	904	9492	87273	217214

We present experimental results for the true cold boot setting in Table 5. Given $\alpha = 0.001$, it follows from our asymptotic analysis that the theoretical maximum value of β which can be handled by our algorithms is 0.658. Our algorithm still has a non-zero success rate for β as high as 0.61. We reiterate that this true cold boot setting is not handled by any of the algorithms previously reported in the literature.

Furthermore, for private keys of the form $\mathsf{sk} = (p, q, d)$ and $\mathsf{sk} = (p, q)$, our algorithm performs very well in the true cold boot setting. For $\mathsf{sk} = (p, q, d)$, the maximum value of β suggested by our capacity analysis is 0.479. With $\beta = 0.4$, $t = 20$ and $L = 16$ our success rate is 0.12 and we have non-zero success rate even with $\beta = 0.43$. Similarly, when $\mathsf{sk} = (p, q)$ our capacity analysis shows that the maximum β is 0.298. When $\beta = 0.2$, $t = 18$ and $L = 16$ we still have a success rate of 0.29, but we can even recover keys with non-zero success rate for β as high as 0.26. Tables 6 and 7 show our results for these cases.

In the full version, we report further results for the erasure channel that improve on the results of [9] and nearly close the gap to our capacity bound. For example, when $m = 5$, we can achieve reliable key recovery up to an erasure rate of 0.79 for this channel, where the bound from capacity is 0.80. By contrast, the best result reported in [9] is for erasure rate 0.76. These and other improvements are obtained using an optimised 'C' implementation of a depth-first search.

6   Conclusions

We have introduced an coding-theoretic viewpoint to the problem of recovering an RSA private key from a noisy version of the key. This provides new insights on the HS and HMM algorithms and leads to a new algorithm which is efficiently implementable and enjoys good performance at high error rates. In particular, ours is the first algorithm that works for the true cold boot case, where both $\Pr(0 \to 1)$ and $\Pr(1 \to 0)$ are non-zero. Our algorithm is amenable to asymptotic analysis, and our experimental results indicate that this analysis provides a good guide to what is actually achievable with reasonable computing resources. Open problems include:

1. Developing a rigorous asymptotic analysis of our algorithm in the general case. However, in view of the state-of-the-art in list decoding, this seems to be hard to obtain.
2. Generalising our approach to the situation where soft information is available about the private key bits, for example reliability estimates of the bits. In general, and by analogy with the situation in the coding theory literature, one would expect to achieve better performance by exploiting such information.

Acknowledgements. The first and third authors were supported by EPSRC Leadership Fellowship, EP/H005455/1. The second author was supported by the Lilian Voudouri Foundation. We thank Mihir Bellare and the referees of Crypto 2012 for thought-provoking comments on an earlier version of this paper.

References

1. Boneh, D.: Twenty years of attacks on the RSA cryptosystem. Notices of the American Mathematical Society 46(2), 203–313 (1999)
2. Brumley, D., Boneh, D.: Remote timing attacks are practical. Computer Networks 48(5), 701–716 (2005)
3. Coppersmith, D.: Small solutions to polynomial equations, and low exponent RSA vulnerabilities. J. Cryptology 10(4), 233–260 (1997)
4. Elias, P.: List decoding for noisy channels. Technical Report 335, Research Laboratory of Electronics. MIT (1957)
5. Guruswami, V.: Algorithmic results in list decoding. Foundations and Trends in Theoretical Computer Science 2(2) (2006)
6. Alex Halderman, J., Schoen, S.D., Heninger, N., Clarkson, W., Paul, W., Calandrino, J.A., Feldman, A.J., Appelbaum, J., Felten, E.W.: Lest we remember: Cold boot attacks on encryption keys. In: van Oorschot, P.C. (ed.) USENIX Security Symposium, pp. 45–60. USENIX Association (2008)
7. Alex Halderman, J., Schoen, S.D., Heninger, N., Clarkson, W., Paul, W., Calandrino, J.A., Feldman, A.J., Appelbaum, J., Felten, E.W.: Lest we remember: cold-boot attacks on encryption keys. Commun. ACM 52(5), 91–98 (2009)
8. Henecka, W., May, A., Meurer, A.: Correcting Errors in RSA Private Keys. In: Rabin, T. (ed.) CRYPTO 2010. LNCS, vol. 6223, pp. 351–369. Springer, Heidelberg (2010)
9. Heninger, N., Shacham, H.: Reconstructing RSA Private Keys from Random Key Bits. In: Halevi, S. (ed.) CRYPTO 2009. LNCS, vol. 5677, pp. 1–17. Springer, Heidelberg (2009)
10. Herrmann, M., May, A.: Solving Linear Equations Modulo Divisors: On Factoring Given Any Bits. In: Pieprzyk, J. (ed.) ASIACRYPT 2008. LNCS, vol. 5350, pp. 406–424. Springer, Heidelberg (2008)
11. Kocher, P.C.: Timing Attacks on Implementations of Diffie-Hellman, RSA, DSS, and Other Systems. In: Koblitz, N. (ed.) CRYPTO 1996. LNCS, vol. 1109, pp. 104–113. Springer, Heidelberg (1996)
12. May, A.: Using LLL-reduction for solving RSA and factorization problems: A survey. In: Nguyen, P. (ed.) Proceedings of LLL+25, p. 3 (June 2007)
13. Sarkar, S., Maitra, S.: More on correcting errors in RSA private keys: Breaking CRT-RSA with low weight decryption exponents. Cryptology ePrint Archive, Report 2012/106 (2012); To appear at CHES (2012)
14. Shannon, C.E.: A mathematical theory of communication. Bell System Technical Journal 27, 379–423, 623–656 (1948)
15. Yilek, S., Rescorla, E., Shacham, H., Enright, B., Savage, S.: When private keys are public: results from the 2008 Debian OpenSSL vulnerability. In: Feldmann, A., Mathy, L. (eds.) Internet Measurement Conference, pp. 15–27. ACM (2009)

Certifying RSA

Saqib A. Kakvi, Eike Kiltz, and Alexander May

Faculty of Mathematics
Horst-Görtz Institute for IT Security
Ruhr-University Bochum, Germany
{saqib.kakvi,eike.kiltz,alex.may}@rub.de

Abstract. We propose an algorithm that, given an arbitrary N of unknown factorization and prime $e \geq N^{\frac{1}{4}+\varepsilon}$, certifies whether the RSA function $\mathsf{RSA}_{N,e}(x) := x^e \bmod N$ defines a permutation over $\mathbb{Z}_N^*$ or not. The algorithm uses Coppersmith's method to find small solutions of polynomial equations and runs in time $O(\varepsilon^{-8} \log^2 N)$. Previous certification techniques required $e > N$.

Keywords: RSA, certified trapdoor permutations, Coppersmith.

1 Introduction

One of the most well known cryptographic primitives is the RSA function [25]. Given a public modulus N (which is usually the product of two primes) and an exponent e, it is defined as $\mathsf{RSA}_{N,e} : \mathbb{Z}_N^* \to \mathbb{Z}_N^*$, $x \mapsto x^e \bmod N$. It is well known that the RSA function defines a permutation over the domain $\mathbb{Z}_N^*$ iff $\gcd(e, \varphi(N)) = 1$. Furthermore, with the right choice of parameters, the RSA function even defines a *trapdoor* permutation since the prime factorization of N allows to efficiently invert $\mathsf{RSA}_{N,e}$.

Trapdoor permutations have many applications to public-key cryptosystems and serve as a building block for (often quite complex) cryptographic protocols. In a large number of applications of trapdoor functions, the fact that the function is a permutation is required to be publicly verifiable. The importance of trapdoor permutations with an efficient permutation checking procedure was first noted by Bellare and Yung [2,3], who called them *certified trapdoor permutations*. Certified trapdoor permutations are in particular important in scenarios where one party (for example, the prover) sends a description of a trapdoor permutation to another party (for example, the verifier). A dishonest prover may send a malicious description of a trapdoor function which is not a permutation. If this remains unnoticed by the verifier, it may allow the prover to cheat in the protocol. See Section 1.2 for a list of applications of certified trapdoor permutations.

RSA AS A CERTIFIED TRAPDOOR PERMUTATION. The question whether the RSA function is a certified trapdoor permutation was first addressed by Bellare and Yung who wrote in [2,3]:

X. Wang and K. Sako (Eds.): ASIACRYPT 2012, LNCS 7658, pp. 404–414, 2012.

> *In particular, RSA is (probably) not certified [...]. This is because [...]*
> *the (description of) the trapdoor permutation f includes a number which*
> *is a product of two primes, and there is (probably) no polynomial time*
> *procedure to test whether or not a number is a product of two primes.*

To overcome this problem, Bellare and Yung showed that every trapdoor permutation can be *transformed* into a certified trapdoor permutation by presenting pre-images (under the function) of random elements specified in a common reference string (CRS), hence certifying that the function is (almost) a permutation. While this result is certainly interesting at a theoretical level, the Bellare-Yung transformation has two main disadvantages. First, it comes with an additional computational overhead (consisting of a number of evaluations of the function) and is therefore relatively inefficient. Second, in order to keep the same data structures one would rather prefer that the initial trapdoor function (e.g., RSA) can be certified directly, without any additional overhead such as a CRS or pre-images. Related transformations for RSA were proposed in [14,6,7].

Subsequently, two results were obtained about the direct certifiability of RSA, i.e., without using a CRS and expanding the public description. First, [5,20] observed that if $e > N$ and e is prime, then the RSA function $\mathsf{RSA}_{N,e}$ is a certified permutation. (This is, since if e is a prime, then it can never divide $\varphi(N) < N$ and hence $\gcd(e, \varphi(N)) = 1$.) However, choosing $e > N$ is usually avoided in practice due to the costs for modular exponentiation. Second, Kiltz et al. [19] noted that if $e < N^{1/4}$, then $\mathsf{RSA}_{N,e}$ is a *lossy trapdoor permutation* [24] (under the phi-Hiding Assumption [5]) and hence it cannot be certified. This is because a lossy trapdoor permutation is in some sense the opposite of a certified trapdoor permutation: a honestly generated (N, e) with $N = pq$ and $\gcd(e, \varphi(N)) = 1$ cannot be efficiently distinguished from (N, e) for which $\mathsf{RSA}_{N,e}$ is many-to-1 and hence not a permutation.

To summarize, if $e < N^{1/4}$, then the RSA function is lossy and cannot be certified (unless the phi-Hiding assumption is wrong); if $e > N$, then it is certified [5,20]; if $N^{1/4} < e < N$, nothing is known and therefore generic Bellare-Yung NIZK proofs [3] have to be added to certify RSA.

1.1 Our Results

In this work we close the above gap by showing an efficient certification procedure that works for any prime exponent $e > N^{1/4}$. Concretely, we construct an algorithm that, given an arbitrary modulus N (with unknown factorization) and a prime $e \geq N^{1/4+\varepsilon}$, returns 1 iff $\mathsf{RSA}_{N,e}$ defines a permutation over $\mathbb{Z}_N^*$. The running time of the algorithm is $O(\varepsilon^{-8} \log^2(N))$ bit operations plus additional $O(\log^4 N)$ if e needs to be checked for primality.

OUR CERTIFICATION ALGORITHM. The idea of our new certification algorithm is as follows. The $\mathsf{RSA}_{N,e}$ function defines a permutation over $\mathbb{Z}_N^*$ iff e does not divide $\varphi(N)$. Hence given N, e, our goal is to identify if $\gcd(e, \varphi(N)) = 1$ or not. First, we use Coppersmith's algorithm [8,21] to find prime divisors p of N in a specific range. Concretely, our algorithm FindFactor run with parameter β

successfully identifies if a given prime $e > N^{1/4+\varepsilon}$ divides $p-1$ iff there exists a divisor p of N in the range $[N^\beta, N^{\beta^2+1/4+\epsilon}]$. If we could assume that $N = pq$ is the product of two primes, both of size roughly $N^{1/2}$, then we could run FindFactor with parameter $\beta = 1/2$ to identify whether e divides $\varphi(N)$ or not. However, the certification algorithm has to view N as an arbitrary integer with unknown factorization. If $N = pq$ with $p \approx N^{2/3}$ and $q \approx N^{1/3}$, then FindFactor run with parameter $\beta = 1/2$ does not work any more. To get around this, we run the FindFactor algorithm multiple times (with different parameters β) to check for various ranges of the prime factors of N. Our main technical contribution is to show that the number of invocations of FindFactor in our certification algorithm is $poly(\varepsilon)$ if $e \geq N^{1/4+\varepsilon}$.

EXTENSIONS. Our certification algorithm works only for prime e but it can be extended to the case where the factorization of $e = \prod e_i^{z_i}$ is known. In that case we can give an efficient certification procedure if $e_i \geq N^{1/4+\varepsilon}$, for all i. If, for one i, we have $e_i < N^{1/4}$, then $\mathsf{RSA}_{N,e}$ is (at least) e_i-to-1 (lossy) under the phi-hiding assumption. Extending our methods to work with arbitrary integers e of unknown factorization remains an open problem.

1.2 Certified Trapdoor Permutations and Applications

The only known candidate trapdoor permutations are the (factoring-based) Blum-Blum-Shub permutation [4], the RSA permutation [25], and Paillier [23]. Since the Blum-Blum-Shub function is lossy assuming one cannot distinguish $N = pq$ from $N = pqr$ [22,12], the RSA trapdoor function is the most efficient certified trapdoor permutation currently known. Our results show that one can use RSA with prime $e = N^{1/4+\varepsilon}$ (rather than $e > N$) as a certified trapdoor permutation.

We now mention a number of cryptographic protocols that are using certified (rather than standard) trapdoor permutations as a building block. Most importantly, NIZK protocols for any NP-statement can be built from (doubly-enhanced) certified trapdoor permutations [11,17,15,16]. Since the RSA trapdoor permutation is doubly-enhanced [17] we obtain simplified and more efficient NIZK protocols from the RSA assumption (with $e > N^{1/4}$), that do not suffer from the Bellare-Yung certification overhead. Apart from that, [10] used certified trapdoor permutations to construct ZAPS and verifiable PRFs; [13] to construct round-optimal blind signatures; [20,1] to build sequential aggregate signatures. We stress that requiring the trapdoor permutation to be certified is not only an artifact of the security proofs. In almost all cases the use of a *lossy trapdoor permutation* leads to a concrete attack on the scheme. For example, the security of the RSA-based aggregate signatures scheme of [20] can be broken (assuming the Phi-Hiding Assumption) when instantiated with $e < N^{1/4}$ (e.g., using the common choices $e = 3$ or $e = 2^{16} + 1$). The same holds for the NIZK protocols for any NP statement [17]. Recently, [18] showed that a full-domain hash impossibility result by Coron [9] only holds if the trapdoor function is certified.

2   Definitions

2.1   Notation

We denote our security parameter as k. For all $n \in \mathbb{N}$, we denote by 1^n the n-bit string of all ones. For any element x in a set S, we use $x \in_R S$ to indicate that we choose x uniformly at random from S. We denote the set of prime numbers by $\mathbb{P}$ and the set of n-bit prime numbers by $\mathbb{P}_n$. We denote by $\mathbb{Z}_N^* = \{x \in \mathbb{Z}_N : \gcd(x, N) = 1\}$ the multiplicative group modulo an integer N. All logarithms are base 2 unless otherwise stated.

2.2   Families of Permutations

Definition 1. *A family of permutations* $\mathsf{P} = (\mathsf{Gen}, \mathsf{Eval})$ *consists of the following two polynomial-time algorithms.*

1. *A probabilistic algorithm* Gen, *which on input* 1^k *outputs a public description* pub *which includes an efficiently sampleable domain* Dom_{pub}.
2. *A deterministic algorithm* Eval, *which on input pub and* $x \in \mathsf{Dom}_{pub}$, *outputs* $y \in \mathsf{Dom}_{pub}$. *We write* $f(x) = \mathsf{Eval}(pub, x)$.

We require that for all $k \in \mathbb{N}$ *and all pub output by* $\mathsf{Gen}(1^k)$, $\mathsf{Eval}(pub, \cdot)$ *defines a permutation over* Dom_{pub}.

Definition 1 extends to families of trapdoor permutations, where Gen additionally outputs a trapdoor *trap* which can be used by a deterministic polynomial-time algorithm Invert to compute $f^{-1}(y)$, for any $y \in \mathsf{Dom}_{pub}$.

We want to point out that $\mathsf{Eval}(pub, \cdot)$ is only required to be a permutation for correctly generated *pub* but not every bit-string *pub* yields a permutation. A family of permutations Π is said to be *certified* [3] if the fact that it is a permutation can be verified in polynomial time given *pub*.

Definition 2. $\mathsf{CP} = (\mathsf{Gen}, \mathsf{Eval}, \mathsf{Certify})$ *is called a family of certified permutations if* $(\mathsf{Gen}, \mathsf{Eval})$ *is a family of permutations and* $\mathsf{Certify}$ *is a deterministic polynomial-time algorithm that, on input of* 1^k *and an arbitrary pub (potentially not generated by* Gen), *returns 1 iff* $\mathsf{Eval}(pub, \cdot)$ *defines a permutation over* Dom_{pub}.

Definition 2 also extends to families of certified trapdoor permutations.

We remark that Definition 2 follows [20] and is slightly weaker than that of Bellare and Yung [3], where, for all inputs, the $\mathsf{Certify}$ algorithm is required to return 1 iff *pub* was generated by $\mathsf{Gen}(1^k)$, with some constant error probability (in the sense of a BPP algorithm).[1] In fact, it seems that the certification

---

[1] The difference between the two definitions can be explained for the case of RSA. Suppose the original Gen algorithm outputs $pub = (N = pq, e)$ with $\gcd(e, \varphi(N)) = 1$. This cannot define a certified permutation with respect to the Bellare-Yung definition since if $pub' = (N' = pqr, e')$ with $\gcd(e', \varphi(N')) = 1$ then $pub \approx pub'$ under the 2-vs-3 prime assumption but pub' is never output by Gen. However, since $\gcd(N', e') = 1$, $\mathsf{RSA}_{N', e'}$ defines a permutation so there is some hope that it still meets Definition 2.

constructions by Bellare and Yung [3, Section 3] only meet our weaker definition which is, in particular, sufficient for their applications to NIZK for all NP
languages.

2.3   RSA Trapdoor Permutation

In Figure 1 we give a description of a family of trapdoor permutations $\mathsf{RSA}_\gamma =$
$(\mathsf{RSAGen}_\gamma, \mathsf{RSAEval}, \mathsf{RSAInvert})$, parametrized by some function $\gamma > 0$ (which
controls the size of the exponent $e \approx N^\gamma$). The domain is defined as $\mathsf{Dom}_{pub} =$
$\mathbb{Z}_N^*$.

algorithm $\mathsf{RSAGen}_\gamma(1^k)$	algorithm $\mathsf{RSAEval}(pk, x)$	algorithm $\mathsf{RSAInvert}(td, y)$
$p, q \in_R \mathbb{P}_{k/2}$	return $x^e \mod N$	return $y^d \mod N$
$N = pq$		
repeat		
$\quad e \in_R \mathbb{P}_{\gamma k}$		
until $(gcd(e, \varphi(N)) = 1)$		
$d = e^{-1} \mod \varphi(N)$		
return $(pk = (N, e), td = d)$		

Fig. 1. RSA permutation algorithms

3   RSA Certification Algorithm

In this section we will give a certification algorithm for the RSA trapdoor permutation RSA_γ from Section 2.3. Our algorithm can be derived from the following
main theorem.

Theorem 3. *Let N be an integer of unknown factorization and $e < N$ be a
prime integer such that $\gamma = \log_N e = \frac{1}{4} + \varepsilon$ and $gcd(e, N) = 1$. We can decide if
$gcd(e, \varphi(N)) = 1$ or $gcd(e, \varphi(N)) = e$ in time $O(\varepsilon^{-8} \log^2 N)$.*

Proof. Let us write $N = \prod_{i=1}^n p_i^{z_i}$, with prime p_i. Therefore,

$$\varphi(N) = \prod_{i=1}^n p_i^{z_i - 1}(p_i - 1).$$

Since e is prime, we can only have $gcd(e, \varphi(N)) = 1$ or $gcd(e, \varphi(N)) = e$. In the
last case, we must have $e | \varphi(N)$. If $e > N$ then we know that $gcd(e, \varphi(N)) = 1$
[20]. When $e < N$, then we need to perform some further checks.

Let us look at the case $e | \varphi(N)$. If $e | p_i^{z_i - 1}$ then $gcd(e, N) = e$, which contradicts
the prerequisite that e and N are coprime. Hence we must have $e | (p_i - 1)$ for
some i. Let us denote $p = p_i$. There exists an $x_0 \in \mathbb{N}$ s.t.

$$ex_0 + 1 = p.$$

Our goal is to recover x_0 and thus to find p. Notice that x_0 is a small root of the polynomial equation $f(x) = ex + 1$ modulo p.

This allows us to use Coppersmith's algorithm for finding small roots of modular polynomial equations.

Theorem 4 (Coppersmith). *Let N be an integer of unknown factorization, which has a divisor $p \geq N^\beta, 0 < \beta \leq 1$. Let $0 < \mu \leq \frac{1}{7}\beta$. Furthermore, let $f(x)$ be a univariate monic polynomial of degree δ. Then we can find all solutions x_0 for the equation:*

$$f(x_0) = 0 \bmod p \quad \text{with } |x_0| \leq \frac{1}{2} N^{\frac{\beta^2}{\delta} - \mu}$$

This can be achieved in time $O(\mu^{-7} \delta^5 \log^2 N)$. The number of solutions x_0 is bounded by $O(\mu^{-1}\delta)$.

A proof can be found in [21].

We use Coppersmith's algorithm to find prime divisors p of N in a specific range as specified in the following lemma.

Lemma 5. *Let N be an integer of unknown factorization with divisor $p \geq N^\beta$ for some $\beta \in (0, 1]$. Let $\mu \in (0, \frac{\beta}{7}]$. Further, let $e = N^\gamma$ with $e|p-1$. Then there is an algorithm* FindFactor *that on input N, e, β, μ outputs p in time $O(\mu^{-7} \log^2 N)$ provided that*

$$p \leq N^{\beta^2 + \gamma - \mu}.$$

If FindFactor *cannot find a non-trivial factor of N, it outputs $\perp$.*

Proof. Since $e|p-1$, we have $ex_0 = p - 1$ for some $x_0 \in \mathbb{N}$. Thus the polynomial $f(x) = ex + 1$ has the root x_0 modulo p. Multiplication of $f(x)$ by e^{-1} modulo N gives us a monic polynomial with the same root modulo p. Let us bound the size of our desired root x_0. We have

$$x_0 = \frac{p-1}{e} < \frac{N^{\beta^2 + \gamma - \mu}}{N^\gamma} = N^{\beta^2 - \mu}.$$

Thus we can recover x_0 by Theorem 4 in time $O(\mu^{-7} \log^2 N)$. Also by Theorem 4, the number of candidates for x_0 is bounded by $O(\mu^{-1})$. For every candidate we check whether $\gcd(ex_0 + 1, N)$ gives us the divisor p. This can be done in time $O(\mu^{-1} \log^2 N)$, which concludes the proof.

Lemma 5 can be used to check whether $e|p - 1$ for some prime divisor p in the range $[N^\beta, N^{\beta^2 - \mu + \gamma}]$. Our goal is to check whether $e|p - 1$ for some p in the entire range $[e, N]$, which we will call the target range.

Obviously $p \leq N$. Thus, we can set the upper bound to $\beta^2 + \gamma - \mu = 1$. This in turn implies a lower bound of $\beta = \sqrt{1 - (\gamma - \mu)}$. Hence, we can first search for a divisor p in the interval $[N^{\sqrt{1-(\gamma-\mu)}}, N]$. If we do not find a divisor p in this interval, then we know that any divisor p must satisfy $p \leq N^{\sqrt{1-(\gamma-\mu)}}$. This defines a new upper bound, and in turn a new lower bound.

In total, we cover the target range by a sequence of intervals $[N^{\beta_1}, N^{\beta_0}], \ldots,$ $[N^{\beta_n}, N^{\beta_{n-1}}]$ where the β_i are defined by the recurrence relation

$$\beta_{i+1} = \max\{\sqrt{\beta_i - (\gamma - \mu)}, \gamma\} \text{ with } \beta_0 = 1.$$

Two examples of such an interval sequence are illustrated in Figure 2.

The following lemma shows that our recurrence reaches γ and thus covers the target range $[e, N]$ after a certain number of steps.

Lemma 6. *Let* $\frac{1}{4} < \gamma - \mu < \gamma < 1$. *Then the recurrence relation*

$$\beta_{i+1} = \max\{\sqrt{\beta_i - (\gamma - \mu)}, \gamma\} \text{ with } \beta_0 = 1$$

satifies $\beta_k = \gamma$ *for some* $k \leq \left\lceil \frac{1-\gamma}{\gamma-\mu-\frac{1}{4}} \right\rceil + 1.$

Proof. Since by definition $\gamma \leq \beta_i \leq 1$ for all i and $\mu > 0$, we have $\beta_i - (\gamma - \mu) > 0$ and therefore $\sqrt{\beta_i - (\gamma - \mu)}$ is defined in $\mathbb{R}$.

We now show by induction that the sequence of the β_i is monotone decreasing. Let us start with $\beta_1 < \beta_0$. Since $\beta_0 - (\gamma - \mu) < 1$, we have $\max\{\sqrt{\beta_0 - (\gamma - \mu)}, \gamma\} < 1$ and therefore $\beta_1 < \beta_0$.

Our inductive hypothesis is $\beta_i \leq \beta_{i-1}$ for all $i \leq n$. Now $\beta_n \leq \beta_{n-1}$ implies

$$\beta_n - (\gamma - \mu) \leq \beta_{n-1} - (\gamma - \mu)$$

and therefore by monotonicity of the square root function

$$\sqrt{\beta_n - (\gamma - \mu)} \leq \sqrt{\beta_{n-1} - (\gamma - \mu)}.$$

This yields $\max\{\sqrt{\beta_n - (\gamma - \mu)}, \gamma\} \leq \max\{\sqrt{\beta_{n-1} - (\gamma - \mu)}, \gamma\}$. Thus, $\beta_{n+1} \leq \beta_n$.

Since the sequence of the β_i is monotone decreasing and bounded below by γ, it converges. Now we show that we can upper bound the number $k-1$ of intervals $[\beta_i, \beta_{i-1}], 1 \leq i < k$ for which $\beta_i > \gamma$. This implies that our sequence stabilizes after k steps at the point $\beta_k = \gamma$.

Let us define a function $\Delta(\beta_{i-1}) = \beta_{i-1} - \beta_i \geq 0$, which gives us the length of the ith interval. For $\beta_i > \gamma$ we obtain $\Delta(\beta_{i-1}) = \beta_{i-1} - \sqrt{\beta_{i-1} - (\gamma - \mu)}$. Since the first two derivatives of $\Delta(\beta)$ satisfy

$$\Delta'(\beta) = 1 - \frac{1}{2}(\beta - (\gamma - \mu))^{-\frac{1}{2}} \text{ and } \Delta''(\beta) = \frac{1}{4}(\beta - (\gamma - \mu))^{-\frac{3}{2}} > 0,$$

an easy computation shows that $\Delta(\beta)$ achieves its minimum at the point $\beta^{(0)} = \frac{1}{4} + \gamma - \mu$. Therefore, each interval length is of size at least

$$\Delta(\beta^{(0)}) = \gamma - \mu - \frac{1}{4}.$$

This in turn means that the number $k - 1$ of intervals with $\beta_i > \gamma$ is at most

$$k - 1 \leq \left\lceil \frac{1-\gamma}{\gamma-\mu-\frac{1}{4}} \right\rceil,$$

which concludes the proof.

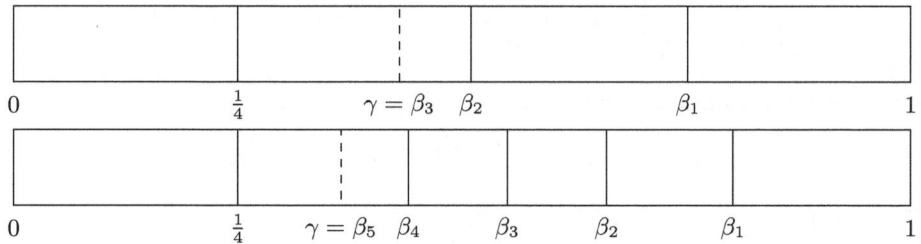

Fig. 2. Values Obtained for $(\gamma = 0.43, \mu = 0.06)$ and $(\gamma = 0.365, \mu = 0.05)$

We continue the proof of Theorem 3. We can use the algorithm FindFactor from Lemma 5 with the parameters (N, e, β_i, μ) to test if there is any factor p such that $e|p-1$ in the sub-range $[N^{\beta_i}, N^{\beta_i^2 - \mu + \gamma}]$. If we run FindFactor multiple times with the β_i values computed using the relation in Lemma 6, we can test the entire range as required.

We now discuss the choice of the parameter μ. Lemma 5 gives us the condition $\mu \le \beta_i/7$ for all values of i. We know from the proof of Lemma 6 that $\gamma \le \beta_i$ for all values of i. Hence it is sufficient to pick μ such that $\mu \le \gamma/7$.

Furthermore, from Lemma 6 we have the condition $\mu < \gamma - \frac{1}{4}$. It is easy to verify that both conditions

$$\mu \le \gamma/7 \text{ and } \mu < \gamma - \tfrac{1}{4}$$

are satisfied by the choice $\mu := \frac{1}{7}(\gamma - \frac{1}{4}) = \frac{1}{7}\varepsilon$ for all $\gamma > \frac{1}{4}$.

We give the whole algorithm GCDDecide for deciding whether $\gcd(e, \phi(N)) = 1$ in Figure 3.

It remains to determine the running time $t_{\text{GCDDecide}}$ of GCDDecide. We know from Lemma 6 that we need at most $\lceil (1 - \gamma)/(\gamma - \mu - \frac{1}{4}) \rceil + 1$ iterations of FindFactor, which can be bounded as

algorithm GDCDecide(N, e)	**algorithm** RSACertify(N, e)
if $(e > N)$ then return 1	if $(!\text{PRIME}(e))$ then return $\bot$
$\gamma = \log_N e,\ \varepsilon = \gamma - \frac{1}{4}$	if $(gcd(e, N)! = 1)$ then return $\bot$
if $(\varepsilon \le 0)$ then return $\bot$	if $(\text{GCDDecide}(N, e)! = 1)$
$\mu = \frac{1}{7}\varepsilon$	$\qquad$ then return false
$\beta_0 = 1, i = 0$	else return true
while$(\beta_i >= \gamma)$	
$\qquad$ if $(\text{FindFactor}(N, e, \beta_i, \mu) \ne \bot)$ return e	
$\qquad i++$	
$\qquad \beta_i = \max\{\sqrt{\beta_{i-1} - (\gamma - \mu)}, \gamma\}$	
wend	
return 1	

Fig. 3. GCD Decision and RSA Certification algorithms

$$\left\lceil \frac{1-\gamma}{\gamma - \mu - \frac{1}{4}} \right\rceil + 1 \leq \left\lceil \frac{1}{6\mu - \frac{1}{4}} \right\rceil + 1 = O(\mu^{-1}) = O(\varepsilon^{-1}).$$

Since each iteration takes time $O(\mu^{-7} \log^2 N)$, we obtain

$$t_{\mathsf{GCDDecide}} = O(\mu^{-8} \log^2 N) = O(\varepsilon^{-8} \log^2 N).$$

This concludes the proof of Theorem 3.

We now describe our full certification algorithm $\mathsf{RSACertify}$ that certifies the RSA trapdoor permutation RSA_γ from Section 2.3, for $\gamma = 1/4 + \varepsilon$. Note that we assume in Theorem 3 that e is prime and that $gcd(e, N) = 1$. If we want to check these prerequisites, we have an additional overhead of $O(\log^4 N)$ for the primality test on e and $O(\log^2 N)$ for the GCD computation. The complete certification algorithm $\mathsf{RSACertify}$ is described in Figure 3. The total running time of $\mathsf{RSACertify}$, denoted by $t_{\mathsf{RSACertify}}$, is given by the expression

$$t_{\mathsf{RSACertify}} = O(\log^4 N) + O(\log^2 N) + t_{\mathsf{GCDDecide}}$$
$$= O(\max\{\log^4 N, \varepsilon^{-8} \log^2 N\}).$$

Let $\mathsf{CRSA}_\gamma = (\mathsf{RSAGen}_\gamma, \mathsf{RSAEval}, \mathsf{RSAInvert}, \mathsf{RSACertify})$, as described in Figures 1 and 3, where γ controls the size of $e \approx N^\gamma$. By Theorem 3 we can see that, for any $\gamma = 1/4 + 1/poly(k)$, CRSA_γ defines a family of certified trapdoor permutations with respect to Definition 2.

Acknowledgements. Saqib Kakvi and Eike Kiltz were funded by a Sofja Kovalevskaja Award of the Alexander von Humboldt Foundation and the German Federal Ministry for Education and Research.

References

1. Bellare, M., Namprempre, C., Neven, G.: Unrestricted Aggregate Signatures. In: Arge, L., Cachin, C., Jurdziński, T., Tarlecki, A. (eds.) ICALP 2007. LNCS, vol. 4596, pp. 411–422. Springer, Heidelberg (2007)
2. Bellare, M., Yung, M.: Certifying Cryptographic Tools: The Case of Trapdoor Permutations. In: Brickell, E.F. (ed.) CRYPTO 1992. LNCS, vol. 740, pp. 442–460. Springer, Heidelberg (1993)
3. Bellare, M., Yung, M.: Certifying permutations: Noninteractive zero-knowledge based on any trapdoor permutation. Journal of Cryptology 9(3), 149–166 (1996)
4. Blum, L., Blum, M., Shub, M.: A simple unpredictable pseudo-random number generator. SIAM Journal on Computing 15(2), 364–383 (1986)
5. Cachin, C., Micali, S., Stadler, M.A.: Computationally Private Information Retrieval with Polylogarithmic Communication. In: Stern, J. (ed.) EUROCRYPT 1999. LNCS, vol. 1592, pp. 402–414. Springer, Heidelberg (1999)

6. Camenisch, J.L., Michels, M.: Separability and Efficiency for Generic Group Signature Schemes (Extended Abstract). In: Wiener, M. (ed.) CRYPTO 1999. LNCS, vol. 1666, pp. 413–430. Springer, Heidelberg (1999)

7. Catalano, D., Pointcheval, D., Pornin, T.: Trapdoor hard-to-invert group isomorphisms and their application to password-based authentication. Journal of Cryptology 20(1), 115–149 (2007)

8. Coppersmith, D.: Finding a Small Root of a Univariate Modular Equation. In: Maurer, U.M. (ed.) EUROCRYPT 1996. LNCS, vol. 1070, pp. 155–165. Springer, Heidelberg (1996)

9. Coron, J.-S.: On the Exact Security of Full Domain Hash. In: Bellare, M. (ed.) CRYPTO 2000. LNCS, vol. 1880, pp. 229–235. Springer, Heidelberg (2000)

10. Dwork, C., Naor, M.: Zaps and their applications. In: 41st Annual Symposium on Foundations of Computer Science, pp. 283–293. IEEE Computer Society Press (November 2000)

11. Feige, U., Lapidot, D., Shamir, A.: Multiple non-interactive zero knowledge proofs based on a single random string (extended abstract). In: FOCS, pp. 308–317 (1990)

12. Freeman, D.M., Goldreich, O., Kiltz, E., Rosen, A., Segev, G.: More Constructions of Lossy and Correlation-Secure Trapdoor Functions. In: Nguyen, P.Q., Pointcheval, D. (eds.) PKC 2010. LNCS, vol. 6056, pp. 279–295. Springer, Heidelberg (2010)

13. Garg, S., Rao, V., Sahai, A., Schröder, D., Unruh, D.: Round Optimal Blind Signatures. In: Rogaway, P. (ed.) CRYPTO 2011. LNCS, vol. 6841, pp. 630–648. Springer, Heidelberg (2011)

14. Gennaro, R., Micciancio, D., Rabin, T.: An efficient non-interactive statistical zero-knowledge proof system for quasi-safe prime products. In: ACM CCS 1998: 5th Conference on Computer and Communications Security, pp. 67–72. ACM Press (November 1998)

15. Goldreich, O.: Foundations of Cryptography: Basic Tools, vol. 1. Cambridge University Press, Cambridge (2001)

16. Goldreich, O.: Foundations of Cryptography: Basic Applications, vol. 2. Cambridge University Press, Cambridge (2004)

17. Goldreich, O.: Basing Non-Interactive Zero-Knowledge on (Enhanced) Trapdoor Permutations: The State of the Art. In: Goldreich, O. (ed.) Studies in Complexity and Cryptography. LNCS, vol. 6650, pp. 406–421. Springer, Heidelberg (2011)

18. Kakvi, S.A., Kiltz, E.: Optimal Security Proofs for Full Domain Hash, Revisited. In: Pointcheval, D., Johansson, T. (eds.) EUROCRYPT 2012. LNCS, vol. 7237, pp. 537–553. Springer, Heidelberg (2012)

19. Kiltz, E., O'Neill, A., Smith, A.: Instantiability of RSA-OAEP under Chosen-Plaintext Attack. In: Rabin, T. (ed.) CRYPTO 2010. LNCS, vol. 6223, pp. 295–313. Springer, Heidelberg (2010)

20. Lysyanskaya, A., Micali, S., Reyzin, L., Shacham, H.: Sequential Aggregate Signatures from Trapdoor Permutations. In: Cachin, C., Camenisch, J. (eds.) EUROCRYPT 2004. LNCS, vol. 3027, pp. 74–90. Springer, Heidelberg (2004)

21. May, A.: Using LLL-reduction for solving RSA and factorization problems. In: The LLL Algorithm, Information Security and Cryptography, pp. 315–348. Springer (2010)

22. Mol, P., Yilek, S.: Chosen-Ciphertext Security from Slightly Lossy Trapdoor Functions. In: Nguyen, P.Q., Pointcheval, D. (eds.) PKC 2010. LNCS, vol. 6056, pp. 296–311. Springer, Heidelberg (2010)
23. Paillier, P.: Public-Key Cryptosystems Based on Composite Degree Residuosity Classes. In: Stern, J. (ed.) EUROCRYPT 1999. LNCS, vol. 1592, pp. 223–238. Springer, Heidelberg (1999)
24. Peikert, C., Waters, B.: Lossy trapdoor functions and their applications. In: Ladner, R.E., Dwork, C. (eds.) 40th ACM STOC Annual ACM Symposium on Theory of Computing, pp. 187–196. ACM Press (May 2008)
25. Rivest, R.L., Shamir, A., Adleman, L.M.: A method for obtaining digital signature and public-key cryptosystems. Communications of the Association for Computing Machinery 21(2), 120–126 (1978)

Faster Gaussian Lattice Sampling Using Lazy Floating-Point Arithmetic

Léo Ducas[1] and Phong Q. Nguyen[2]

[1] ENS, Dept. Informatique, 45 rue d'Ulm, 75005 Paris, France
http://www.di.ens.fr/~ducas/
[2] INRIA, France and Tsinghua University, Institute for Advanced Study, China
http://www.di.ens.fr/~pnguyen/

Abstract. Many lattice cryptographic primitives require an efficient algorithm to sample lattice points according to some Gaussian distribution. All algorithms known for this task require long-integer arithmetic at some point, which may be problematic in practice. We study how much lattice sampling can be sped up using floating-point arithmetic. First, we show that a direct floating-point implementation of these algorithms does not give any asymptotic speedup: the floating-point precision needs to be greater than the security parameter, leading to an overall complexity $\tilde{\mathcal{O}}(n^3)$ where n is the lattice dimension. However, we introduce a laziness technique that can significantly speed up these algorithms. Namely, in certain cases such as NTRUSIGN lattices, laziness can decrease the complexity to $\tilde{\mathcal{O}}(n^2)$ or even $\tilde{\mathcal{O}}(n)$. Furthermore, our analysis is practical: for typical parameters, most of the floating-point operations only require the double-precision IEEE standard.

1 Introduction

Lattice-based cryptography has been attracting considerable interest in the past few years (see the survey [22]), due to unique features such as security based on worst-case assumptions [3] or more recently fully-homomorphic encryption [11]. But it has several differences compared to classical public-key cryptography based on factoring and discrete logarithms: in particular, the description of many lattice schemes (such as the seminal Ajtai-Dwork cryptosystem [4] and its LWE variants [27], or schemes using lattice sampling [12]) involves real numbers at some point. Although the descriptions usually mention that one can replace these real numbers by approximations with sufficiently high precision, which guarantees efficiency in an asymptotical sense, the practical impact is unclear: no article seems to specify exactly which precision one should take, and how all the operations will be performed exactly. This was not an issue when lattice-based cryptography was considered to be mostly of theoretical interest, but recent works [22,26,17,28,18,20] suggest that the time has come to assess the practicality of lattice-based constructions.

There is another reason to study carefully the use of floating-point arithmetic in lattice-based cryptography. Many recent lattice schemes (*e.g.* trapdoor signatures [12,6] and ID-based encryption [12,7,1,2]) require a *Gaussian sampler*,

X. Wang and K. Sako (Eds.): ASIACRYPT 2012, LNCS 7658, pp. 415–432, 2012.

that is an efficient algorithm to sample lattice points according to a Gaussian-like distribution, given a (short secret) basis and a target vector. There are two approaches for this task: Klein's randomized variant [15] (as analyzed by Gentry *et al.* [12]) of Babai's nearest plane algorithm [5], and algorithms [26,20] based on convolution for the so-called q-ary lattices.

The cost of Klein's algorithm is the same as Babai's algorithm, namely $\tilde{\mathcal{O}}(n^3 \log B)$ (or $\mathcal{O}(n^4 \log^2 B)$ without fast integer arithmetic), where n is the lattice dimension, and B is the maximal norm of the input basis vectors: since B is polynomial in n for trapdoor bases used in lattice cryptography, the usual cost is $\tilde{\mathcal{O}}(n^3)$ (or $\tilde{\mathcal{O}}(n^4)$ without fast integer arithmetic). The main reason behind the cost of Klein's algorithm is the use of long-integer arithmetic: it relies on Gram-Schmidt orthogonalization, which involves rational numbers of bit-length $\mathcal{O}(n \log B)$. A natural way to improve the efficiency is to use floating-point arithmetic (FPA) to replace exact Gram-Schmidt by suitable approximations. Indeed, Klein's algorithm is a variant of Babai's nearest plane algorithm, which itself is simply the size-reduction subroutine used extensively in the LLL algorithm [16]; and floating-point arithmetic is classically used to speed up LLL (see [29,25,23]). But the use of FPA is not straightforward, and it is unclear at first sight how much speed up can be gained, if any.

On the other hand, the convolution algorithms [26,20] based on Peikert's work [26] have two phases: an offline phase (depending on the secret basis only) and an online phase (depending on the target vector). The online phase costs $\tilde{\mathcal{O}}(n^2)$ for q-ary lattices (which are widespread in lattice cryptography), or even $\tilde{\mathcal{O}}(n)$ in the so-called ring setting (*i.e.* special lattices such as NTRU lattices); but the offline phase is the generation of a noise following some discrete Gaussian distribution, which seems to have the same cost $\tilde{\mathcal{O}}(n^3)$ as Klein's algorithm, and involves floating-point arithmetic whose exact cost is not analyzed in [26,20]. Both algorithms [26,20] can use the same offline phase, which will later be referred to as Peikert's offline Algorithm.

It should be stressed that the offline phase is not a precomputation: this phase must be repeated before each sampling, which is reminiscent of DSA one-time pairs (k, k^{-1}), which can be precomputed as coupons or generated online; but unlike a precomputation it should not be re-used. In some scenario, this computational cost might be acceptable, but it is clearly valuable to analyze and improve the offline phase.

Our results. We develop techniques to improve all three samplers, obtaining the first algorithms with quasi-optimal complexity to sample the discrete Gaussian distribution over lattices: their running time is quasi-linear in the size of the input basis. More precisely, our optimized variant of Klein's algorithm runs in $\tilde{\mathcal{O}}(n^2)$ (for certain bases) and our variant of Peikert's offline algorithm runs in average time $\tilde{\mathcal{O}}(n)$ in some ring setting (where n is the lattice dimension). In both cases, our improvements do not introduce any loss of quality.

To do so, we study how much lattice sampling can be sped up using FPA. As a starting point, we present FPA variants of Klein's algorithm with statistically close output. Surprisingly, the basic FPA variant has the same asymptotical complexity $\tilde{\mathcal{O}}(n^3)$ as Klein's algorithm, because the precision needs to be greater than the security parameter. However, we also present an optimized algorithm with an improved complexity $\tilde{\mathcal{O}}(n^2)$: it is based on a so-called laziness technique which combines high and low precision FPA. But this optimized complexity only applies to a special class of bases which include NTRUSIGN bases [13], namely the inverse basis must be small.

Next, we show that the same optimization can be used to speed up Peikert's offline algorithm, improving the total complexity, to bring its offline complexity down to that of its online complexity for both sampling algorithms of [26,20]. More precisely, we apply our laziness technique to reduce the offline complexity to $\tilde{\mathcal{O}}(n^2)$. And for certain ring settings (precisely when the ring is $\mathcal{R} = X^b \pm 1$), we show that the offline phase can also be sped up to average quasi-linear time. This is achieved by using two additional tricks: a structured square-root algorithm and an improved rejection sampler for Gaussians over $\mathbb{Z}$.

As a direct application of this last result, one can strengthen the security of NTRUSIGN [13] by replacing their heuristic perturbation technique with our optimized sampler, without any loss of efficiency asymptotically. This prevents learning attacks [24,10] on NTRUSIGN as the signature scheme is now provably secure in the random-oracle model (see [12]), under the (reasonable) assumption that finding close vectors in NTRUSIGN lattices is hard.

While numerical analysis has often be used [29,25,23] to speed up lattice reduction algorithms in a rigorous way, our work might be its first application to provable security.

Practical impact of laziness. The precision used for floating-point arithmetic has non-negligible practical impact, because fp-operations become much more expensive when the precision goes over the hardware precision. For instance, modern processors typically provide floating-point arithmetic following the double IEEE standard (53-bit precision), but quad-float FPA (113-bit precision simulated by software libraries) is usually about 10-20 times slower for basic operations, and the overhead is much more for multiprecision FPA.

Our complexity results are stated in an asymptotical manner, but our analysis can give concrete bounds (which are provided in the full version [9]). It turns out that in typical cryptographic settings, the *double-precision* (53-bit) IEEE standard can be selected as the "low precision" of our lazy algorithm, which means that most of our fp-operations are hardware fp-operations, even though the security level is not limited to 53 bits.

Roadmap. We start in Sect. 2 with background and notation on lattices, sampling and FPA. In Sect. 3, we present our basic FPA variant of Klein's algorithm, which we optimize using laziness in Sect. 4. In Sect. 5, we apply laziness to speedup Peikert's Offline Algorithm. Eventually, in Sect. 6, we explain how to reach quasi-linear time complexity in the ring setting. Missing proofs and additional details, such as non-asymptotic bounds can be found in the full version [9].

2   Preliminaries

Throughout the paper, we use row representations of matrices (to match lattice software), and use bold fonts to denote vectors: if $B = (\mathbf{b}_1, \ldots, \mathbf{b}_n)$ is a matrix, then its row vectors are the $\mathbf{b}_i$'s. Notation $\mathcal{M}_n$ and $\mathcal{S}_n^+$ denote respectively the square matrices, and the square symmetric definite positive matrices of dimension n over $\mathbb{R}$.

2.1   Notation

Lattices. Lattices are discrete subgroups of $\mathbb{R}^m$. A lattice L is represented by a *basis*, that is, a set of linearly independent vectors $\mathbf{b}_1, \ldots, \mathbf{b}_n$ in $\mathbb{R}^m$ such that L is equal to the set $L(\mathbf{b}_1, \ldots, \mathbf{b}_n) = \{\sum_{i=1}^n x_i \mathbf{b}_i, x_i \in \mathbb{Z}\}$ of all integer linear combinations of the $\mathbf{b}_i$'s. The integer n is the *dimension* of the lattice L. The *volume* $\mathrm{vol}(L)$ is the n-dimensional volume of the parallelepiped generated by any basis of L. In lattice-based cryptography, one mainly uses the so-called q-ary lattices, which include NTRU lattices [14,13] and Ajtai's worst-case/average-case lattices [3]. A q-*ary* lattice is simply a full-rank integer lattice $L \subseteq \mathbb{Z}^n$ such that $q\mathbb{Z}^n \subseteq L$, where q is a somewhat small integer. For such a lattice, $\mathrm{vol}(L)$ divides q^n.

Norms. For a vector $\mathbf{x} \in \mathbb{R}^n$, $\|\mathbf{x}\| = \sqrt{\langle \mathbf{x}, \mathbf{x} \rangle}$ will denote its Euclidean norm. The norm of a matrix B is the maximal norm of its rows: $\|B\| = \max_{i=1}^n \|\mathbf{b}_i\|$. The spectral norm of a square $n \times n$ matrix M is: $\|M\|_s = \max_{\mathbf{x} \in \mathbb{R}^n/\{0\}} \frac{\|\mathbf{x} \cdot M\|}{\|\mathbf{x}\|}$.

Orthogonalization. An $n \times m$ basis $B = (\mathbf{b}_1, \ldots, \mathbf{b}_n)$ can be written uniquely as $B = \mu \cdot D \cdot Q$ where $\mu = (\mu_{i,j})$ is an $n \times n$ lower-triangular matrix with unit diagonal, D an n-dimensional positive diagonal matrix and Q an $n \times m$ matrix with orthonormal row vectors. Then μD is a lower triangular representation of B (with respect to Q), $B^\star = DQ = (\mathbf{b}_1^\star, \ldots, \mathbf{b}_n^\star)$ is the Gram-Schmidt orthogonalization of the basis, and D is the diagonal matrix formed by the $\|\mathbf{b}_i^\star\|$'s. With those notations, we have $\mu_{i,j} = \langle \mathbf{b}_i, \mathbf{b}_j^\star \rangle / \|\mathbf{b}_j^\star\|^2$.

For any $\sigma > 0$, we let $\sigma_i = \sigma/\|\mathbf{b}_i^\star\|$ and $\hat{\sigma} = \max_{i=1}^n \sigma_i$. Since the $\mathbf{b}_i^\star$'s are orthogonal, we have $\hat{\sigma} = \sigma/(\min_{i=1}^n \|\mathbf{b}_i^\star\|) = \sigma \|B^{\star-1}\|_s \leq \sigma \|B^{-1}\|_s \|\mu\|_s \leq \sigma \|B^{-1}\|_s n\hat{\mu}$ where $\hat{\mu} \geq 1$ upper bounds the coefficients of μ.

Gaussian Distribution. The (unnormalized) weight of Gaussian distribution of parameter $\sigma \in \mathbb{R}$ and center $c \in \mathbb{R}$ at $x \in \mathbb{R}$ is defined by $\rho_{\sigma,c}(x) = \exp\left(-\pi\frac{(x-c)^2}{\sigma^2}\right)$, and more generally by $\rho_{\sigma,\mathbf{c}}(\mathbf{x}) = \exp\left(-\pi\frac{\|\mathbf{x}-\mathbf{c}\|^2}{\sigma^2}\right)$ for $\mathbf{c}, \mathbf{x} \in \mathbb{R}^n$. The discrete Gaussian distribution over $\mathbb{Z}$ is defined by $D_{\mathbb{Z},\sigma,c}(x) = \rho_{\sigma,c}(x)/\rho_{\sigma,c}(\mathbb{Z})$, and more generally, over a lattice L by $D_{L,\sigma,\mathbf{c}}(\mathbf{x}) = \rho_{\sigma,\mathbf{c}}(\mathbf{x})/\rho_{\sigma,\mathbf{c}}(L)$. Peikert [26] generalized the discrete Gaussian distribution over a lattice L using a positive definite matrix $\Sigma > 0$ (which generalizes $\sigma \in \mathbb{R}$) as follows: the density $D_{L,\sqrt{\Sigma},\mathbf{c}}(\mathbf{x})$ is proportional to $\rho_{1,0}((\mathbf{x}-\mathbf{c})B^{-1})$ where $\Sigma = B^t B$, for $\mathbf{x} \in L$.

2.2   Gaussian Lattice Sampling

The goal of Gaussian lattice sampling is to efficiently sample lattice points according to a distribution statistically close to $D_{L,\sigma,c}$. All lattice samplers known [15,12,26,20] have constraints on the parameter σ and the statistical distance, which are related to the so-called smoothing parameter. The sampling parameter σ determines the average distance of the sampled lattice point to the target point: the smaller σ, the better for cryptographic applications. For instance, σ impacts the verification threshold of lattice-based signatures [12] and therefore the security of the scheme; a lower quality forces to increase lattice parameters. And for a security level of λ bits, we need a statistical distance less than $2^{-\lambda}$.

Smoothing Parameter. For any n-dimensional lattice L and any real $\iota > 0$, the *smoothing parameter* $\eta_\iota(L)$ (see [21]) is the smallest real $s > 0$ such that $\rho_{1/s}(L^* \backslash \{0\}) \leq \iota$, where L^* is the dual lattice of L. For details on the importance of this parameter, please refer to [21,12].

Klein's sampling. Gentry *et al.* showed in [12] that given as input a lattice basis B of an n-dimensional lattice L such that $\sigma \geq \|B^*\|\omega(\sqrt{\log n})$, Klein's algorithm [15] outputs lattice points with a distribution statistically close to $D_{L,\sigma,c}(\mathbf{x})$. For applications, it is more convenient to have a concrete bound on the statistical distance, and to separate this bound from the lattice dimension n. We therefore use the following concrete analysis of Klein's algorithm:

Theorem 1 (Concrete version of [12, Th. 4.1]). *Let $n, \lambda \in \mathbb{N}$ be any positive integers, and $\iota = 2^{-\lambda}/(2n)$. For any n-dimensional lattice L generated by a basis $B \in \mathbb{Z}^{n \times n}$, and for any target vector $\mathbf{c} \in \mathbb{Z}^{1 \times n}$, Alg. 2 is such that the statistical distance $\Delta(D_{L,\sigma,c}, \mathbf{SampleLattice}_\infty(B, \sigma, \mathbf{c}))$ is less than $2^{-\lambda}$, under the condition:*

$$\sigma \geq \|B^*\| \eta_\iota(\mathbb{Z}) \qquad where \quad \eta_\iota(\mathbb{Z}) \lessapprox \sqrt{(\lambda \ln 2 + \ln n)/\pi} \ .$$

Tailcut. We will also use a tailcut parameter τ, chosen such that (informally) a sample from a normal distribution of parameter σ is at distance at most $\tau\sigma$ from the center with overwhelming probability:

Corollary 1 (Tailcut error, Corollary of [21, Lemma 2.10]). *Let L be an n-dimensional lattice, $\iota \leq 1/2$, $\sigma \geq \eta_\iota(L)$, $\tau > 1$ $\delta_\tau \in (0,1)$ and $\mathbf{c} \in \mathbb{R}^n$. For $x \leftarrow D_{L,\sigma,c}$ we have: $\Pr\left[\|\mathbf{x} - \mathbf{c}\| \geq (1 - \delta_\tau)\tau\sigma\right] \leq 3E_{\mathbf{tailcut}}(\tau, \delta_\tau)^n$ where $E_{\mathbf{tailcut}}(\tau, \delta_\tau) \stackrel{def}{=} \tau\sqrt{2\pi e} \cdot e^{-\pi(1-\delta_\tau)^2\tau^2}$.*

2.3   Floating-Point Arithmetic

We consider floating-point arithmetic (FPA) with m bits of mantissa, which we denote by $\mathbb{FP}_m$: the precision is $\epsilon = 2^{-m+1}$. A floating-point number $\bar{f} \in \mathbb{FP}_m$ is a triplet $\bar{f} = (s, e, v)$ where $s \in \{0, 1\}$, $e \in \mathbb{Z}$ and $v \in \mathbb{N}_{2^m - 1}$, which represents the real number $R(\bar{f}) = (-1)^s \cdot 2^{e-m} \cdot v \in \mathbb{R}$. Every FPA-operation $\bar{\circ} \in \{\bar{+}, \bar{-}, \bar{\times}, \bar{/}\}$ and its respective arithmetic operation on $\mathbb{R}$, $\circ \in \{+, -, \cdot, /\}$ verify:

$$\forall \bar{f}_1, \bar{f}_2 \in \mathbb{FP}_m, \left| R(\bar{f}_1 \bar{\circ} \bar{f}_2) - (R(\bar{f}_1) \circ R(\bar{f}_2)) \right| \le (R(\bar{f}_1) \circ R(\bar{f}_2))\epsilon \qquad (1)$$

We require a floating-point implementation of the exponentiation function $\bar{\exp}(\cdot)$ and we assume that it verifies a similar error bound: for any $\bar{f} \in \mathbb{FP}_m$, $\left| R(\bar{\exp}(\bar{f})) - \exp(R(\bar{f})) \right| \le \epsilon$. Finally, we note that if an integer $x \in \mathbb{Z}$ verifies $|x| \le 2^m$, it can be converted to a float $\bar{f} \in \mathbb{FP}_m$ with no error, i.e. $R(\bar{f}) = x$. For the rest of the article, we omit the function R and consider $\mathbb{FP}_m$ as a subset of $\mathbb{R}$.

2.4 Pseudo-code

Types. Variables are typed, and the type is given at each initialization and assignment, as follows: $variable \leftarrow value : type$. We use a simpler syntax for the definition of local functions: $\{variable \mapsto value\}$. Functional types are denoted by $(t_1 \to t_2)$.

Primitives. We use the basic arithmetic operations $\{+, -, \cdot, /\}$, as well as squaring $\square^2$ and exponentiation $\exp$; the arguments are either integers in $\mathbb{Z}$, or floating-point numbers in $\mathbb{FP}_m$. We extend these notations to vectors and matrices. We also use the following additional primitives:

RandInt$(a, b) : \mathbb{Z} \times \mathbb{Z} \to \mathbb{Z}$: return a random uniform integer in the range $[a, b]$.
RandFloat$_m() : \text{void} \to \mathbb{FP}_m$: return a random uniform float in the range $[0, 1)$.
ExtRandFloat$_{m',m}(r) : \mathbb{FP}_{m'} \to \mathbb{FP}_m$: return a random uniform floating-point number in the range $[r, r+2^{-m'})$. For a random $r \leftarrow$ **RandFloat**$_{m'}()$, the output follows the same distribution as **RandFloat**$_m()$.

3 A Basic Floating-Point Variant of Klein's Algorithm

3.1 Description

Algorithm 2 describes both Klein's algorithm [15] and our basic floating-point variant: given a basis B of a lattice L, a target $\mathbf{c}$ and a parameter σ, the algorithm outputs a vector with distribution statistically close to $D_{L,\sigma,\mathbf{c}}$. It uses two subroutines: **DecomposeGS**$_m$ (Alg. 3) to compute the coordinates t_i's of the target vector $\mathbf{c}$ with respect to the Gram-Schmidt basis $B^\star$, and **SampleZ**$_m$ (Alg. 1) to sample according to the Gaussian distribution over $\mathbb{Z}$. Algorithm 2 comes in two flavors:

- **SampleLattice**$_\infty$ is the exact version, which corresponds to Klein's original algorithm [15]. The $\mu_{i,j}$'s and the t_i's are represented exactly by rational numbers, and all the computations use exact integer arithmetic. Assuming $\sigma \in \mathbb{Q}$, we can only ensure that $\sigma_i \in \sqrt{\mathbb{Q}}$, thus we can represent them exactly by their square. We also assume that this version has access to a perfect primitive (or an oracle) **SampleZ$_\infty$**$(\sigma_i, t_i, \tau = \infty)$ that given $t_i, \sigma_i^2 \in \mathbb{Q}$ answers an integer $x : \mathbb{Z}$ exactly according to the distribution $D_{\mathbb{Z},\sigma_i,t_i}$. It does not matter how to sample such a perfect distribution, as the purpose of this perfect algorithm is to be a reference for inexact ones.

– **SampleLattice$_m$** is our basic floating-point version, using $\mathbb{FP}_m$. The matrices μ and $B^\star$ and values σ_i may have been pre-computed exactly, but only approximations are stored.

Algorithm 1. SampleZ$_m$: Rejection Sampling for Discrete Gaussian on $\mathbb{Z}$

input: A center $t : \mathbb{FP}_m$, and a parameter $\sigma : \mathbb{FP}_m$, and a tailcut parameter $\tau : \mathbb{FP}_m$

output: output $x : \mathbb{Z}$, with distribution statistically close to $D_{\mathbb{Z},t,\sigma}$

1: $h \leftarrow -\pi/\sigma^2 : \mathbb{FP}_m$; $x_{\max} \leftarrow \lceil t + \tau\sigma \rceil : \mathbb{Z}$; $x_{\min} \leftarrow \lfloor t - \tau\sigma \rfloor : \mathbb{Z}$
2: $x \leftarrow$ **RandInt**$(x_{\min}, x_{\max}) : \mathbb{Z}$; $p \leftarrow \exp(h \cdot (x - t)^2) : \mathbb{FP}_m$
3: $r \leftarrow$ **RandFloat**$_m() : \mathbb{FP}_m$;   **if** $r < p$ **then** return x
4: **Goto** Step 2.

Algorithm 2. SampleLattice$_m$: Gaussian Sampling over a lattice

input: a (short) lattice basis $B = (\mathbf{b}_1, \ldots, \mathbf{b}_n) : \mathbb{Z}^{n \times n}$, parameter $\sigma : \mathbb{FP}_m$, A target vector $\mathbf{c} : \mathbb{Z}^{1 \times n}$, and a tailcut parameter $\tau : \mathbb{FP}_m$ **Precomputation:** The GS decomposition $(B^\star = (\mathbf{b}_1^\star, \ldots, \mathbf{b}_n^\star), (\mu_{i,j}) = (\boldsymbol{\mu}_1, \ldots, \boldsymbol{\mu}_n))$, norms $r_i = \|\mathbf{b}_i^\star\| : \mathbb{FP}_m$ and $\sigma_i = \sigma/r_i : \mathbb{FP}_m$

output: a vector $\mathbf{v} : \mathbb{Z}^{1 \times n}$ drawn approximately from $D_{L,c,\sigma}$ where $L = L(B)$

1: $\mathbf{v}, \mathbf{z} \leftarrow \mathbf{0} : \mathbb{Z}^n$; $\mathbf{t} \leftarrow$ **DecomposeGS**$_m(\mathbf{c}, B^\star) : \mathbb{FP}_m$
2: **for** $i = n$ downto 1 **do**
3:    $z_i \leftarrow$ **SampleZ**$_m(\sigma_i, t_i, \tau) : \mathbb{Z}$
4:    $\mathbf{v} \leftarrow \mathbf{v} + z_i \cdot \mathbf{b}_i : \mathbb{Z}^n$; $\mathbf{t} \leftarrow \mathbf{t} - z_i \cdot \boldsymbol{\mu}_i : \mathbb{FP}_m^n$
5: **end for**
6: **return** $\mathbf{v}$

Algorithm 3. DecomposeGS$_m$: Decompose a vector c over the GS Basis

input: A vector $\mathbf{c} : \mathbb{Z}^{1 \times n}$, an orthogonal basis $B^\star = (\mathbf{b}_1^\star, \ldots, \mathbf{b}_n^\star) : \mathbb{Q}^{n \times n}$, and $r_i^2 = \|\mathbf{b}_i^\star\|^2 \in \mathbb{FP}_m$

output: output $\mathbf{t} : \mathbb{Q}^n$ such that $c = t_1 \mathbf{b}_1^\star + \cdots + t_n \mathbf{b}_n^\star$
1: $\mathbf{y} \leftarrow \mathbf{c} \cdot B^{\star t}$ $: \mathbb{Z}^{1 \times n}$
2: **return** $(y_1/r_1^2, \ldots, y_n/r_n^2)$

The description of **SampleLattice$_\infty$** differs from the original description [15,12] only in the way we compute and update the coordinates t_i's. In our version, the final value of t_i before it is used is $t_i = \langle \mathbf{c}, \mathbf{b}_i^\star \rangle / r_i^2 - \sum_{j>i}^n z_j \mu_{j,i}$, which matches with the original value :

$$t_i' = \left\langle \mathbf{c} - \sum_{j>i}^n z_j \mathbf{b}_j , \mathbf{b}_i^\star \right\rangle / r_i^2 = \left(\langle \mathbf{c}, \mathbf{b}_i^\star \rangle - \sum_{j>i}^n z_j \langle \mathbf{b}_j, \mathbf{b}_i^\star \rangle \right) / r_i^2 = t_i$$

We unroll this computation and update the sum after each value z_i is known. This allows a parallelization up to n processors without the usual $\log n$ factor required for summing up all terms.

Since we use the matrix μ in the main loop, we might want to get rid of $B^\star$ for the **DecomposeGS** algorithm, to save some precomputation and storage, by computing $\mathbf{c}' \leftarrow \mathbf{c} \cdot B^t$ and then solving the triangular system $\mathbf{y} \, \mu^t = \mathbf{c}'$. Solving this system also requires n^2 operations, however when using FPA, it would produce a relative error exponential in the dimension n, because we recursively use previous results.

Our main loop may also be seen as solving a triangular system, where we apply Gaussian rounding at each step. It is worth noting that this additional rounding prevents such relative exponential error, as our proof will show.

Efficiency of **SampleLattice**$_\infty$. The algorithm **SampleLattice**$_\infty$ performs $\mathcal{O}(n^2)$ arithmetic operations on rational numbers of size $\mathcal{O}(n \log B)$, which leads to a complexity of $\tilde{\mathcal{O}}(n^4)$ for cryptographic use. Here, we ignored the calls to the oracle **SampleZ**$_\infty(\cdot, \cdot, \tau = \infty)$.

Termination of **SampleZ**$_\infty(\cdot, \cdot, \tau < \infty)$. We upper bound the number of trials of Rejection Sampling, ignoring issues related to the transcendental function exp:

Fact 2. *If $\sigma \geq 4$ and $\tau \geq 1$, and uniforms $x \leftarrow \mathbb{Z} \cap [x_{min}, x_{max}]$ and $r \leftarrow [0, 1)$, we have* $\Pr\left[r < \rho_{\sigma,t}(x)\right] > 1/(6\tau)$ *where $x_{min} = \lceil t - \tau\sigma \rceil$ and $x_{max} = \lfloor t + \tau\sigma \rfloor$.*

Thus **SampleZ**$_\infty(\cdot, \cdot, \tau)$ performs less than 6τ trials on average.

3.2   Correctness

We give the list of assumptions needed for our correctness results (Theorems 3 and 5), and which we refer to as conditions $\mathcal{A}$.

Assumption on Gram-Schmidt Precomputation. We assume that the Gram-Schmidt values are (possibly approximately) precomputed, and that the computed values $\bar{\mu}_{i,j}$, $\bar{b}^\star_{i,j}$ and $\bar{\sigma}_i$ verify:

$$|\Delta\mu_{i,j}| = |\mu_{i,j} - \bar{\mu}_{i,j}| \leq \hat{\mu}\epsilon, \, |\Delta b^\star_{i,j}| = |b^\star_{i,j} - \bar{b}^\star_{i,j}| \leq \|\mathbf{b}^\star_i\| \, \epsilon,$$
$$|\Delta\sigma_i| = |\sigma_i - \bar{\sigma}_i| \leq \sigma_i\epsilon,$$

where $\hat{\mu}$ denotes the maximal absolute value of the sub-diagonal coefficient of μ. Those condition can be achieved by running the precomputation exactly, then convert the result to floating points of mantissa size m.

Assumption on the Target Vector. We assume that the components c_i of the input target vector $\mathbf{c}$ satisfy: $|c_i| \leq q$ for a parameter q. This holds in all known cryptographic applications of lattice sampling, for which the lattice is q-ary. But we do not require that the lattice is q-ary.

Assumption on the Parameters

$$\mathcal{A} \begin{cases} \epsilon \leq 0.01, \quad K^n = (1+\epsilon)^n \leq 1.1, 1 + nK^n\epsilon \leq 1.01 \\ n\iota \leq 0.01, \forall i, \sigma_i \geq \eta_\iota(\mathbb{Z}), \qquad \forall i, \sigma_i \geq 4 \\ n \geq 10 \qquad \tau \geq 4 \end{cases}$$

The assumptions on ϵ are easily achievable for a mantissa size m at least logarithmic in the dimension n. The condition on ι is not restrictive as it needs to be negligible. Similarly, conditions on σ_i's are not restrictive since the security requires all $\sigma_i \geq \eta_\iota(\mathbb{Z}) > 4$ for security parameters $\lambda \geq 80$.

For the rest of the analysis, we assume that all parameters B, $\mathbf{c}$ and σ are fixed. Our main result states that with enough precision, the outputs of the exact sampler **SampleLattice**$_\infty$ and the floating-point sampler **SampleLattice**$_m$ are statistically close:

Theorem 3. *There exist constants C_λ, C_τ, C_m, such that for any security parameter $\lambda \geq C_\lambda$, and under conditions $\mathcal{A}$, the statistical distance between* **SampleLattice**$_m$ *and* **SampleLattice**$_\infty$ *is less than $2^{-\lambda}$ on the same input if the following conditions are satisfied:*

$$\tau \geq C_\tau\sqrt{\lambda \log n} \quad m \geq C_m + \lambda + 2\log_2(\||B^{-1}\||_s) + \log_2\left(\hat{\mu}^2 n^4(q+\sigma^2)\tau^3\right)$$

Furthermore, under those conditions, the integers manipulated by **SampleLattice**$_m$ *can be represented by floating-point numbers without errors.*

3.3   Efficiency

We deduce the efficiency of the basic floating-point sampler from Theorem 3. We first analyze **SampleZ**$_m$:

Fact 4. *There is a constant C_m such that for any $m \geq C_m$, and any $\tau \geq 1$,* **SampleZ**$_m(\cdot, \cdot, \tau)$ *performs less than 6τ trials on the average.*

This can be easily derived from Fact 2 and appropriate error bound (see full version). This ensures that **SampleLattice**$_m$ performs $\sim 6n^2$ $\mathbb{FP}_m$-operations as long as $\tau = o(n)$.

Arbitrary bases. To minimize the FPA-precision m in Theorem 3, we need to evaluate $\log(\||B^{-1}\||_s)$: this is always less than $\approx n\log(B)$ by Cramer's rule. This leads to the constraint $m \geq \lambda + n\ell$ where ℓ is logarithmic in n and B, yielding a $\tilde{\mathcal{O}}(n^3)$ bit-complexity as long as $\lambda = \mathcal{O}(n)$, or $\tilde{\mathcal{O}}(n^4)$ without fast integer arithmetic.

The exact algorithm **SampleLattice**$_\infty$ also has complexity $\tilde{\mathcal{O}}(n^3)$. However, the constants are likely to be smaller for the FPA sampler. Indeed, the exact algorithm must handle integers of size $\log(\max_{1\leq i\leq n}\text{vol}(\mathbf{b}_1,\ldots,\mathbf{b}_i))$, whereas the quantity $\log(\||B^{-1}\||_s)$ is typically smaller, though they have similar worst-case asymptotical bounds. And the constants of the FPA sampler can be improved by processing the basis, for instance using LLL reduction.

Furthermore, in cryptographic applications, we may focus on bases B of a particular shape. More precisely, we will consider the following type of basis:

Small-inverse bases. A sequence $\mathcal{C} = (\mathcal{C}_n)$ of square matrices generating q_n-ary lattices of dimension n is a class of *small-inverse bases* if there exists a polynomial function f such that for any basis $B \in \mathcal{C}_n$, $\|B\|_s \leq f(n)$ and $\|B^{-1}\|_s \leq f(n)$.

In particular, the bases used by the NTRUSIGN signature scheme [13] form a small-inverse class (see [13]). For such bases, we only need $m \geq \lambda + \ell$ for ℓ logarithmic in λ. This still gives a $\tilde{\mathcal{O}}(n^3)$ complexity for cryptographic use (when $\lambda \sim n$), but with much better constants.

4   A Lazy Floating-Point Variant of Klein's Algorithm

Overview. We now describe our optimized sampler, which is more efficient than the basic sampler, due to a better use of FPA. The analysis of the basic sampler showed that it was sufficient to compute t_i up to $\approx \lambda$ bits below the unity to get an error below $2^{-\lambda}$ on the output distribution. However, a careful analysis of the rejection sampling algorithm (Alg. 1) shows that most of the time, many of those bits are not used: the precision of t_i impacts the precision of $p = \rho_{\sigma,t}(x)$, which is only used to make a comparison with a uniform random real $r \in [0,1)$. For all $j > 1$, such a comparison is determined by the first j bits, except with probability 2^{-j} (exactly when the j first bits of r and p match); and on average only the first two bits contribute to the decision.

However, we still need to decide properly this comparison even when the first $j \leq \lambda$ bits match, to output a proper distribution. This suggests a new strategy: compute lazily the bits of t_i and p. We first only compute most significant bits and backtrack for additional bits until the comparison can be determined. We choose a simple lazyness control, using only two levels of precision (for simplicity, but also for practical efficiency). Informally, we choose $k \leq \lambda$, and compute t_i up to a precision m' that only guarantees the first k bits of p, draw the first k bits of the random real r. If the comparison is decided with those k bits, continue normally. Otherwise (which happens with probability less than 2^{-k}), recompute t_i and p at a precision m to ensure λ correct bits.

4.1   Description

Our optimized sampler **LazySampleLattice**$_{m',m}$ (Alg 4) works with two floating-point types, $\mathbb{FP}_m$ (high precision) and $\mathbb{FP}_{m'}$ (low precision), where $m > m'$. The algorithm works similarly to the original one, except it now works most of the time at low precision m'. The subroutine for sampling over $\mathbb{Z}$ is replaced by **LazySample$\mathbb{Z}$**$_{m',m}$, which takes the usual arguments at low precision, plus an error bound, and access to high-precision arguments: σ is precomputed thus requiring no special care, however, the access to high precision value of t is given through a function that takes no argument.

This new subroutine **LazySample$\mathbb{Z}$**$_{m',m}$ (Alg. 5) works identically to the original **Sample$\mathbb{Z}$**$_{m'}$ as long as the decisive comparison is trusted, i.e. as long as the difference $|r' - p'|$ is higher than the error bound δ_p. Otherwise, the high precision is triggered, and high-precision inputs are requested through the function F. Then all sample trials are computed with high precision.

Algorithm 4. LazySampleLattice$_{m',m}$: Lazy Gaussian Sampling over a lattice

---

input: Same as **SampleLattice** plus low precision versions of $\mu, B^\star$ and σ_i's values:
$\mu', B^{\star'} : \mathbb{FP}_{m'}^{n \times n}, \sigma_i' : \mathbb{FP}_{m'}$, and an error bound δ_p
output: Same as **SampleLattice**
1: $\mathbf{v}, \mathbf{z} \leftarrow \mathbf{0} : \mathbb{Z}^n$; $\mathbf{t}' \leftarrow \mathbf{DecomposeGS}_{m'}(\mathbf{c}, B^{\star'}) : \mathbb{FP}_{m'}^n$
2: **for** $i = n$ downto 1 **do**
3:     $F_i \leftarrow \{() \mapsto \langle \mathbf{c}, \mathbf{b}_i^\star \rangle - \langle \mathbf{z}, [\mu^t]_i \rangle\}$: (void $\to \mathbb{FP}_m$)
4:     $z_i \leftarrow \mathbf{LazySampleZ}_{m',m}(\sigma_i', \tau, t_i', \delta_p, \sigma_i, F_i) : \mathbb{Z}$
5:     $\mathbf{v} \leftarrow \mathbf{v} + z_i \cdot \mathbf{b}_i : \mathbb{Z}^n$; $\mathbf{t}' \leftarrow \mathbf{t}' - z_i \cdot \mu_i' : \mathbb{FP}_{m'}^n$
6: **end for**
7: **return** $\mathbf{v}$

Algorithm 5. LazySampleZ$_{m',m}(\sigma', \tau, t', \delta_p : \mathbb{FP}_{m'}, \sigma : \mathbb{FP}_m, F : (\text{void} \to \mathbb{FP}_m))$

1: $h' \leftarrow -\pi/\sigma^2 : \mathbb{FP}_{m'}$; $x_{\max} \leftarrow \lceil t' + \tau\sigma' \rceil : \mathbb{Z}$; $x_{\min} \leftarrow \lfloor t' - \tau\sigma' \rfloor : \mathbb{Z}$; highprec $\leftarrow$ false : bool
2: $x \leftarrow \mathbf{RandInt}(x_{\min}, x_{\max}) : \mathbb{Z}$; $r' \leftarrow \mathbf{RandFloat}_{m'}() : \mathbb{FP}_{m'}$
3: **if** not(highprec) **then**
4:     $p' \leftarrow \exp(h' \cdot (x - t')^2) : \mathbb{FP}_{m'}$
5:     **if** $|r' - p'| \leq \delta_p$ **then** $\{t \leftarrow F() : \mathbb{FP}_m; h \leftarrow -\pi/\sigma^2 : \mathbb{FP}_m$; highprec $\leftarrow$ true $\}$
6:     **else if** $r' < p'$ **then** return x
7: **end if**
8: **if** highprec **then**
9:     $r \leftarrow \mathbf{ExtRandFloat}_{m',m}(r') : \mathbb{FP}_m$; $p \leftarrow \exp(h \cdot (x - t)^2) : \mathbb{FP}_m$
10:     **if** $r < p$ **then** return x
11: **end if**
12: **Goto** Step 2.

4.2 Correctness

We need to determine a proper value for the error bound δ_p in terms of the basis and m' (the size of the low precision), to ensure correctness. For this parameter, the lower the better, since it determines the probability to trigger the re-computation of t at high precision, as detailed in the next section. The behavior of the new subroutine is analyzed by the following:

Lemma 1 (Informal, see [9] for a formal statement). *The behaviour of* **LazySampleZ$_{m,m'}$** *given approximate inputs $\sigma \pm \delta_\sigma$ and $t \pm \delta_t$ and δ_p, is similar to* **SampleZ$_m$** *on input σ, t under the condition:*

$$\delta_p \geq 4\sigma^2 \epsilon' + 1.7\sigma\delta_\sigma + (1.7/\sigma)\delta_t \quad \text{where } \epsilon' = 2^{1-m'}$$

From this lemma, we prove the correctness of **LazySampleLattice$_{m',m}$**, summarized by the following result.

Theorem 5. *There exist constants $C_\lambda, C_\tau, C_m, C_m', C_{\delta_p}$, such that for any security parameter $\lambda \geq C_\lambda$, and under Conditions $\mathcal{A}$, the statistical distance between* **LazySampleLattice$_{m,m'}$** *and* **SampleLattice$_\infty$** *is less than $2^{-\lambda}$ on the same input if the following conditions are satisfied:*

$$\tau \geq C_\tau \sqrt{\lambda \log n}$$
$$m \geq C_m + \lambda + 2\log_2(\|B^{-1}\|_s) + \log_2\left(\hat{\mu}^2 n^4 q\sigma^2 \tau^3\right)$$
$$m' \geq C_{m'} + 2\log_2(\|B^{-1}\|_s) + \log_2\left(\hat{\mu}^2 n^4 (Q + \sigma^2)\tau^3\right)$$
$$\delta_p \geq 2^{-k} \text{ where } k = m' - \left(C_{\delta_p} + 2\log_2(\|B^{-1}\|_s) + \log_2\left(\hat{\mu}^2 n^3 \tau\sigma^2 q\right)\right)$$

Furthermore, under those conditions, the integers manipulated by the algorithm can be represented by low-precision floating-point numbers (FP$_{m'}$) without errors.

4.3   Efficiency

The error bound δ_p impacts the efficiency of the optimized sampler as follows:

Lemma 2. *Under the conditions of Theorem 5, each call to* **LazySample$\mathbb{Z}_{m,m'}$** *triggers high precision with probability less than $12\tau\delta_p$. On the average, the algorithm* **LazySampleLattice$_{m,m'}$** *performs less than $\mathcal{O}(n^2\tau\delta_p)$ high-precision floating-point operations.*

Proof. At each trial performed by **LazySample$\mathbb{Z}_{m,m'}$**, the probability to trigger high precision is less than $2\delta_p$: indeed it happens only if the randomness $r' \leftarrow [0,1)$ falls in the interval $[p' - \delta_p, p' + \delta_p]$. It remains to bound the average number of trials performed by **LazySample$\mathbb{Z}_{m,m'}$**. The condition of Theorem 5 ensures that it behaves similarly to **Sample$\mathbb{Z}_m$**. Thus, for a large enough m, Fact 4 ensures that the average number of trials is less than 6τ.

Triggering high precision during **LazySample$\mathbb{Z}_{m,m'}$** requires $\mathcal{O}(n)$ high-precision FPA operations. This subroutine is called n times, thus on the average less than $\mathcal{O}(n^2\tau\delta_p)$ high-precision FPA operations.   □

This leads to our main result: with Small-Inverse bases, the discrete Gaussian distribution can be sampled in quasi-quadratic time, with an exponentially small statistical distance, and no sacrifice on the quality compared to the analysis of [12].

Theorem 6 (Gaussian sampling in quasi-quadratic time). *Let $(\mathcal{C}_n)$ be a Small-Inverse class of bases. For any implicit function λ, such that $\lambda \sim n$, and σ polynomial in n, there exist implicit functions m, m', τ, δ_p of n such that, for any basis $B \in \mathcal{C}_n$ generating a lattice L:*

- **LazySampleLattice$_{m,m'}(B, \sigma, \mathbf{c}, \tau, \delta_p)$** *runs in expected time $\tilde{\mathcal{O}}(n^2)$ without fast integer arithmetic.*
- $\Delta(D_{L,\sigma,\mathbf{c}}, \textbf{LazySampleLattice}_{m,m'}(B, \sigma, \mathbf{c}, \tau, \delta_p)) \leq 2^{-\lambda}$ *whenever σ verifies $\sigma \geq \|B^\star\| \eta_\iota(\mathbb{Z})$ with $\iota = 2^{-\lambda}/(4n)$.*

Proof. For a small-inverse class of bases, the conditions of Theorem 5 can be satisfied with functions verifying:

$$\tau = \mathcal{O}(\sqrt{n}), m = \mathcal{O}(n), m' = \mathcal{O}(\log n), \delta_p = \mathcal{O}(1/n^{5/2}).$$

Lemma 2 states that on the average, less than $\mathcal{O}(n^2\tau\delta_p)$ high-precision operations are performed, which in our case is a $\mathcal{O}(1)$. Without fast integer arithmetic, the total complexity is thus less than $\mathcal{O}(n^2)\mathcal{O}(m'^2) + \mathcal{O}(1)\mathcal{O}(m^2) \leq \tilde{\mathcal{O}}(n^2)$.   □

5   Speeding Up Peikert's Offline Algorithm

Peikert [26] recently proposed a different sampling algorithm based on convolution, which was inspired by NTRUSign's perturbation countermeasure [13]. This algorithm offers a different trade-off than Klein's algorithm, with slightly worse constraints on sampling parameters (see [26] for details). The discrete Gaussian distribution is obtained by adding two points, one generated by an offline phase, the other generated by a (cheaper) online phase. The online phase is essentially a randomized variant of Babai's round-off algorithm [5], which only involves small-integer arithmetic when the input is a q-ary lattice, and thus runs in $\tilde{O}(n^2)$ time, and even $\tilde{O}(n)$ in ring settings. This offline phase is itself essentially the generation of some discrete Gaussian distribution, which requires long-integer arithmetic, and is not fully analyzed in [26], but seems to be $\tilde{O}(n^3)$ (even $\tilde{O}(n^4)$ without fast integer arithmetic) like Klein's algorithm. In the follow-up work of Micciancio and Peikert [20], a new kind of lattice trapdoor is introduced to optimize efficiency and geometric quality, which allows an even faster online phase, but the same kind of offline computations is required. We refer to this common offline phase as Peikert's offline algorithm.

5.1   Peikert's Offline Algorithm

Let B be the input basis of the lattice for which one wants to generate the discrete Gaussian distribution. In both [26,20], the offline phase consists of generating a (centered) discrete Gaussian noise over $\mathbb{Z}^n$ of parameter $\Sigma \in \mathcal{S}_n^+$ such that $B^t B + \Sigma = sI_n$ where s is some appropriate real number: this implies certain constraints on B which are discussed in [26]. Letting $\Sigma = C^t C$, this distribution $D_{\mathbb{Z}^n, \sqrt{\Sigma}}$ has support $\mathbb{Z}^n$ and density at $\mathbf{x}$ proportional to $\rho_{1,0}(\mathbf{x}C^{-1})$: in other words, this is "essentially' the discrete Gaussian distribution $\mathcal{D}$ over the lattice spanned by C^{-1}, since the density of $\mathbf{x} \in \mathbb{Z}^n$ is proportional to the density of the lattice point $\mathbf{x}C^{-1}$ in $\mathcal{D}$. The offline-phase algorithm is described in Alg. 6 (from [26]): it generates this discrete Gaussian distribution by convolution (see [26]), which is a different strategy than Klein's algorithm, and has different constraints. The main idea is to consider a "shift" $\Sigma' = \Sigma - \eta^2 I_n$ of Σ such that $\Sigma' \in \mathcal{S}_n^+$ (which implies that $\Sigma \geq \eta^2 I_n$) and $\eta \geq \eta_\iota(\mathbb{Z}^n)$, and to compute a square-root L of Σ', i.e. $\Sigma' = L^t L$. To implement this, it is suggested in [26] to use a Cholesky decomposition. The parameters selected to reach security λ are $\eta = \tau = \eta_\iota(\mathbb{Z})$

Algorithm 6. Peikert's Offline Algorithm

input: $\Sigma \in \mathcal{S}_n^+$, a real $\eta \geq \eta_\iota(\mathbb{Z}^n)$ such that $\Sigma' = \Sigma - \eta^2 I_n \in \mathcal{S}_n^+$ and ι is negligible, and a square-root L of Σ' i.e. $\Sigma' = L^t L$.

output: An integer vector $\mathbf{z} \in \mathbb{Z}^n$ following the distribution $D_{\mathbb{Z}^n, \sqrt{\Sigma}}$
1: Choose $\mathbf{x} : \mathbb{R}^n$ according to the continuous Gaussian distribution of covariance I_n
2: $\mathbf{y} = \mathbf{x} \cdot L$
3: **for** $i = 1$ to n **do** $z_i \leftarrow$ **SampleZ**$_m(\eta, y_i, \tau)$
4: return $\mathbf{z}$

$= \tilde{\mathcal{O}}(\sqrt{\lambda})$. The choice of the floating-point precision is not discussed in [26,20], however a quick analysis shows that one should take $m = \lambda + \ell$ where ℓ is logarithmic in n, s and τ. Thus, a naive implementation would have a running-time of $\tilde{\mathcal{O}}(n^2\lambda^2)$, the main cost being a (non-structured) matrix-vector product: that is n^2 floating-point operations, at precision $\tilde{\mathcal{O}}(\lambda)$.

5.2 Using Laziness in Peikert's Offline Algorithm

Like in Klein's sampling algorithm, the offline phase of Peikert's algorithm [26] only uses non-integer values to compute the input of the $\mathbf{SampleZ}_m(\eta, \cdot, \tau)$ subroutine. High-precision bits of this input are useless except with small probability: one may apply the laziness technique to improve efficiency to $\tilde{\mathcal{O}}(n^2)$, by replacing the subroutine by $\mathbf{LazySampleZ}_{m',m}$. We sketch a proof.

The floating-point computation $y_j = \sum_{i=1}^{n} x_i L_{j,i}$ with m bits of precision produces an error less than $\tilde{\mathcal{O}}(n^2 \|x\|_\infty \|L\|_\infty \epsilon)$ where $\epsilon = 2^{1-m}$. For $\tau = \tilde{\mathcal{O}}(\sqrt{n})$ we have that $\|x\|_\infty \leq \tau$ with overwhelming probability, and $\|L\|_\infty \leq \|L\|_S \leq s$ since $L^t L = C' \leq \sigma^2 \mathrm{Id}$. The error propagation is thus polynomial in n, and Lemma 1 ensures correction with the following parameters:

$$\tau = \mathcal{O}(\sqrt{n}), m = \mathcal{O}(n), m' = \mathcal{O}(\log n), \delta_p = \mathcal{O}(1/n^{5/2}).$$

Similarly to Lemma 2, one easily proves that, on average, less than $\mathcal{O}(n^2\tau\delta_p)$ high-precision operations are performed, which in our case is $\mathcal{O}(1)$. Without fast integer arithmetic, the total complexity is thus less than $\mathcal{O}(n^2)\mathcal{O}(m'^2) + \mathcal{O}(1)\mathcal{O}(m^2) \leq \tilde{\mathcal{O}}(n^2)$.

6   Quasi-Linear Complexity in Ring Settings $\mathcal{R} = \mathbb{Z}_q[X]/(X^b \pm 1)$

For efficiency purposes, lattice cryptography often uses a special class of "algebraic" lattices arising from polynomial rings *i.e.* $\mathcal{R} = \mathbb{Z}[X]/(P(X))$ for some polynomial P of degree b. More precisely, the lattices are generated by an $\mathcal{R}$-basis, and can also be viewed as an integer lattice of dimension ℓb for some $\ell \geq 1$.

In this section, we show that for the ring settings $\mathcal{R} = \mathbb{Z}_q[X]/(X^b \pm 1)$, it is possible to achieve quasi-linear complexity using two improvement on top of our lazy variant of Peikert's offline phase [26,20]. The first improvement is to use special square-root algorithms (*e.g.* Babylonian Method or the Denman-Beavers iteration [8]) to preserve matrix structures, unlike Cholesky decomposition. In our case, we use block-circulant or block-skew-circulant structures, which are stable under transposition and multiplication, which implies that $\Sigma' = \Sigma - \eta^2 I_n = (s - \eta^2)I_n - B^t B$ has the same structure. The second improvement targets $\mathbf{SampleZ}$.

6.1   Structured Square-Root for $\mathcal{R} = \mathbb{Z}_q[X]/(X^b \pm 1)$

Consider the special ring setting $\mathcal{R} = \mathbb{Z}_q[X]/(X^b \pm 1)$, which includes $\mathbb{Z}_q[X]/(X^b - 1)$ for the class of NTRU lattices [13], and some cyclotomic lattices $\mathbb{Z}_q[X]/(\Phi_m)$ the m-th cyclotomic ring, when m is a power of two, made popular by the hardness results of [19].

When $P(X) = X^b - 1$ (resp. $P(X) = X^b + 1$) the integer representation $B \in \mathcal{M}^{bk \times bl}(\mathbb{Z})$ of any $\mathcal{R}$-basis is a b-block circulant, (resp. b-block skew-circulant) matrix, *i.e.* a matrix composed with $(b \times b)$-blocks of the form :

$$\begin{bmatrix} a_1 & a_2 & \cdots & a_b \\ a_b & a_1 & \cdots & a_{b-1} \\ \vdots & \ddots & \ddots & \vdots \\ a_2 & \cdots & a_b & a_1 \end{bmatrix}, \quad \text{resp.} \quad \begin{bmatrix} a_1 & a_2 & \cdots & a_b \\ -a_b & a_1 & \cdots & a_{b-1} \\ \vdots & \ddots & \ddots & \vdots \\ -a_2 & \cdots & -a_b & a_1 \end{bmatrix}.$$

We denote these families by $\mathcal{C}_b$ (resp. $\mathcal{C}_b^\top$). These families are stable under ring operations (addition, product and inverse, when defined) because of the ring isomorphism with matrices over $\mathcal{R}$. Such isomorphisms also exist for other polynomials P, defining other b-block structures. However, circulant and skew-circulant structures have a key property for our improvement:

Fact 7. *Matrix families $\mathcal{C}_b$ and $\mathcal{C}_b^\top$ are stable under transposition.*

From this, we deduce that $\Sigma' = \Sigma - \eta^2 I_n = (s - \eta^2) I_n - B^t B \in \mathcal{C}_b$ (or $\mathcal{C}_b^\top$) when working in this ring setting. At this point, one would want to find a square root of Σ that is still structured. Interestingly, the solution can be found in algorithms that were designed to extract another notion of square root; namely, the Babylonian Method, or the Denman-Beavers iteration [8]. Indeed, those algorithms are searching for an Y such that $Y \cdot Y = X$, without symmetry requirement on X, and no guarantee of convergence in general. Lemma 3 proves that given as input $X \in \mathcal{S}_n^+$, such methods (quickly) converge to some $Y \in \mathcal{S}_n^+$ such that $Y^t \cdot Y = X$.

Definition. The Babylonian Method approximates the limit of the sequence:

$$Y_0(X) = I_n; \quad Y_{k+1}(X) = (Y_k(X) + X \cdot Y_k(X)^{-1})/2 \tag{2}$$

and if this sequence converges to an invertible limit $Y(X)$, it must verify $Y(X) = \frac{1}{2}(Y(X) + X \cdot Y(X)^{-1})$, which is equivalent to $Y(X) \cdot Y(X) = X$. The Denman-Beavers iteration is similar, using the sequences:

$$\begin{cases} Y_0(X) = X \\ Z_0(X) = \text{Id} \end{cases} \quad \begin{cases} Y_{k+1}(X) = \left(Y_k(X) + Z_k(X)^{-1}\right)/2 \\ Z_{k+1}(X) = \left(Z_k(X) + Y_k(X)^{-1}\right)/2 \end{cases} \tag{3}$$

it verifies the invariant $Y_k \cdot Z_k^{-1} = Z_k^{-1} \cdot Y_k = X$, and if it converges, the limit Y of Y_k verifies $Y \cdot Y = X$.

Lemma 3. *Let* $X \in \mathcal{S}_n^+$ *be a symmetric positive definite matrix, then the Babylonian Method, as defined by the sequence* $Y_k(X)$ *in (2) converges quadratically[3] to some* $Y(X) \in \mathcal{S}_n^+$. *Furthermore, if* $X \in \mathcal{C}_b$ *(resp.* $\mathcal{C}_b^\top$*) then* $Y(X) \in \mathcal{C}_b$ *(resp.* $\mathcal{C}_b^\top$*), which implies that* $Y(X)^t Y(X) = X$. *Similar results also hold for the Denman-Beavers iteration (3).*

Proof (sketch). By induction, write $Y_i(X)$ as QD_iQ^t for a fixed orthogonal matrix Q and diagonal matrices D_i. Each diagonal entry of (D_i) follows the Babylonian Square-Root sequence over $\mathbb{R}$, which allows to prove convergence. Structure preservation follows from ring and topological closure of $\mathcal{C}_b$ and $\mathcal{C}_b^\top$.

6.2 Improved Efficiency

Assuming the square root L of Σ was precomputed using one of the structure-preserving algorithms described below, each computation of $\mathbf{y} = \mathbf{x} \cdot L$ at precision m' can now be done in time $\tilde{\mathcal{O}}(nm'^2)$, but some coordinate may need to be recomputed at precision m. Using a similar analysis as in Sect. 5.2 with:

$$\tau = \mathcal{O}(\sqrt{n}), m = \mathcal{O}(n), m' = \mathcal{O}(\log n), \delta_p = \mathcal{O}(1/n^{7/2}).$$

we show that the "average" time[4] spent on the computation of $\mathbf{y} = x \cdot L$ is indeed $\tilde{\mathcal{O}}(n)$.

By combining Laziness and Structured-Square-Root, we move the complexity bottleneck to the **LazySample$\mathbb{Z}$** subroutine, which is called n times and requires $\tilde{\mathcal{O}}(\tau) = \tilde{\mathcal{O}}(\sqrt{\lambda})$ trials in average. For $\lambda \sim n$, this leads to an overall average complexity of $\tilde{\mathcal{O}}(n^{1.5})$.

To reach quasi-linear complexity we need a third trick, detailed in the full version [9]. There, we improve the rejection sampling algorithm **Sample$\mathbb{Z}$** so that it only needs a constant number of trials on average. This is done by sampling from a distribution before rejection which is much closer to the target distribution than the uniform distribution used in **Sample$\mathbb{Z}$**.

By combining the three techniques, we eventually obtain an implementation of Peikert's offline phase which runs in average[4] quasi-linear time. These results also apply to the recent variant of Micciancio and Peikert [20].

---

[3] The number of correct bits grows quadratically with the number k of iterations: $|s_k - s_\infty| \le c\, 2^{-c'k^2}$ for some $c, c' > 0$.

[4] We explain what we mean by average. As high-precision is triggered independently with small probability over n trials, the running times of the optimized Klein's Sampler and optimized Peikert's Offline Phase are bounded by some function $\tilde{\mathcal{O}}(n^2)$, except with negligible probability. However, when applying laziness in the ring setting, triggering high-precision once in the whole algorithm raises this instance's running time to $\tilde{\mathcal{O}}(n\lambda^2)$: only the average cost is below that bound. And dealing with average running times is less problematic in an offline phase, than in an online phase which is more subject to timing attacks.

Acknowledgements. We thank T. Lepoint, C. Peikert, O. Regev and D. Stehlé for useful discussions. We also thanks anonymous reviewers for their comments. Part of this work is supported by the Commission of the European Communities through the ICT program under contract ICT-2007-216676 ECRYPT II, and by China's 973 Program (Grant 2013CB834205).

References

1. Agrawal, S., Boneh, D., Boyen, X.: Efficient Lattice (H)IBE in the Standard Model. In: Gilbert, H. (ed.) EUROCRYPT 2010. LNCS, vol. 6110, pp. 553–572. Springer, Heidelberg (2010)
2. Agrawal, S., Boneh, D., Boyen, X.: Lattice Basis Delegation in Fixed Dimension and Shorter-Ciphertext Hierarchical IBE. In: Rabin, T. (ed.) CRYPTO 2010. LNCS, vol. 6223, pp. 98–115. Springer, Heidelberg (2010)
3. Ajtai, M.: Generating hard instances of lattice problems. In: Proc. STOC 1996, pp. 99–108. ACM (1996)
4. Ajtai, M., Dwork, C.: A public-key cryptosystem with worst-case/average-case equivalence. In: Proc. ACM STOC 1997, pp. 284–293 (1997)
5. Babai, L.: On Lovász lattice reduction and the nearest lattice point problem. Combinatorica 6(1), 1–13 (1986)
6. Boyen, X.: Lattice Mixing and Vanishing Trapdoors: A Framework for Fully Secure Short Signatures and More. In: Nguyen, P.Q., Pointcheval, D. (eds.) PKC 2010. LNCS, vol. 6056, pp. 499–517. Springer, Heidelberg (2010)
7. Cash, D., Hofheinz, D., Kiltz, E., Peikert, C.: Bonsai Trees, or How to Delegate a Lattice Basis. In: Gilbert, H. (ed.) EUROCRYPT 2010. LNCS, vol. 6110, pp. 523–552. Springer, Heidelberg (2010)
8. Denman, E., Beavers, A.: The matrix sign function and computations in systems. American Elsevier (1976)
9. Ducas, L., Nguyen, P.Q.: Faster Gaussian lattice sampling using lazy floating-point arithmetic. Full version of the ASIACRYPT 2012 article (2012)
10. Ducas, L., Nguyen, P.Q.: Learning a Zonotope and More: Cryptanalysis of NTRUSign Countermeasures. In: Wang, X., Sako, K. (eds.) ASIACRYPT 2012. LNCS, vol. 7658, pp. 433–450. Springer, Heidelberg (2012)
11. Gentry, C.: Fully homomorphic encryption using ideal lattices. In: Proc. STOC 2009, pp. 169–178. ACM (2009)
12. Gentry, C., Peikert, C., Vaikuntanathan, V.: Trapdoors for hard lattices and new cryptographic constructions. In: Proc. STOC 2008, pp. 197–206. ACM (2008)
13. Hoffstein, J., Howgrave-Graham, N., Pipher, J., Silverman, J.H., Whyte, W.: NTRUSIGN: Digital Signatures Using the NTRU Lattice. In: Joye, M. (ed.) CT-RSA 2003. LNCS, vol. 2612, pp. 122–140. Springer, Heidelberg (2003)
14. Hoffstein, J., Pipher, J., Silverman, J.H.: NTRU: A Ring-Based Public Key Cryptosystem. In: Buhler, J.P. (ed.) ANTS 1998. LNCS, vol. 1423, pp. 267–288. Springer, Heidelberg (1998)
15. Klein, P.N.: Finding the closest lattice vector when it's unusually close. In: Proc. ACM SODA, pp. 937–941 (2000)
16. Lenstra, A.K., Lenstra Jr., H.W., Lovász, L.: Factoring polynomials with rational coefficients. Mathematische Ann. 261, 513–534 (1982)
17. Lindner, R., Peikert, C.: Better Key Sizes (and Attacks) for LWE-Based Encryption. In: Kiayias, A. (ed.) CT-RSA 2011. LNCS, vol. 6558, pp. 319–339. Springer, Heidelberg (2011)

18. Lyubashevsky, V.: Lattice Signatures without Trapdoors. IACR Cryptology ePrint Archive, 2011:537 (2011); In: Pointcheval, D., Johansson, T. (eds.) EUROCRYPT 2012. LNCS, vol. 7237, pp. 738–755. Springer, Heidelberg (2012)
19. Lyubashevsky, V., Peikert, C., Regev, O.: On Ideal Lattices and Learning with Errors over Rings. Cryptology ePrint Archive, Report 2012/230 (2010); In: Gilbert, H. (ed.) EUROCRYPT 2010. LNCS, vol. 6110, pp. 1–23. Springer, Heidelberg (2010)
20. Micciancio, D., Peikert, C.: Trapdoors for Lattices: Simpler, Tighter, Faster, Smaller. IACR Cryptology ePrint Archive, 2011:501 (2011); In: Pointcheval, D., Johansson, T. (eds.) EUROCRYPT 2012. LNCS, vol. 7237, pp. 700–718. Springer, Heidelberg (2012)
21. Micciancio, D., Regev, O.: Worst-case to average-case reductions based on Gaussian measures. In: Annual IEEE Symposium on Foundations of Computer Science, pp. 372–381 (2004)
22. Micciancio, D., Regev, O.: Lattice-based cryptography. In: Post-Quantum Cryptography, pp. 147–191. Springer, Berlin (2009)
23. Morel, I., Stehlé, D., Villard, G.: H-LLL: using Householder inside LLL. In: Proc. ISSAC 2009, pp. 271–278. ACM (2009)
24. Nguyen, P.Q., Regev, O.: Learning a parallelepiped: Cryptanalysis of GGH and NTRU signatures. J. Cryptology 22(2), 139–160 (2009)
25. Nguyen, P.Q., Stehlé, D.: An LLL algorithm with quadratic complexity. SIAM J. Comput. 39(3), 874–903 (2009)
26. Peikert, C.: An Efficient and Parallel Gaussian Sampler for Lattices. In: Rabin, T. (ed.) CRYPTO 2010. LNCS, vol. 6223, pp. 80–97. Springer, Heidelberg (2010)
27. Regev, O.: The learning with errors problem (invited survey). In: Proc. IEEE Conference on Computational Complexity, pp. 191–204. IEEE Computer Society (2010)
28. Rückert, M., Schneider, M.: Estimating the security of lattice-based cryptosystems. Cryptology ePrint Archive, Report 2010/137 (2010), http://eprint.iacr.org/
29. Schnorr, C.-P.: A more efficient algorithm for lattice basis reduction. J. Algorithms 9(1), 47–62 (1988)

Learning a Zonotope and More: Cryptanalysis of NTRUSign Countermeasures

Léo Ducas[1] and Phong Q. Nguyen[2]

[1] ENS, Dept. Informatique, 45 rue d'Ulm, 75005 Paris, France
http://www.di.ens.fr/~ducas/
[2] INRIA, France and Tsinghua University, Institute for Advanced Study, China
http://www.di.ens.fr/~pnguyen/

Abstract. NTRUSIGN is the most practical lattice signature scheme. Its basic version was broken by Nguyen and Regev in 2006: one can efficiently recover the secret key from about 400 signatures. However, countermeasures have been proposed to repair the scheme, such as the perturbation used in NTRUSIGN standardization proposals, and the deformation proposed by Hu *et al.* at IEEE Trans. Inform. Theory in 2008. These two countermeasures were claimed to prevent the NR attack. Surprisingly, we show that these two claims are incorrect by revisiting the NR gradient-descent attack: the attack is more powerful than previously expected, and actually breaks both countermeasures in practice, *e.g.* 8,000 signatures suffice to break NTRUSIGN-251 with one perturbation as submitted to IEEE P1363 in 2003. More precisely, we explain why the Nguyen-Regev algorithm for learning a parallelepiped is heuristically able to learn more complex objects, such as zonotopes and deformed parallelepipeds.

1 Introduction

There is growing interest in cryptography based on hard lattice problems (see the survey [22]). The field started with the seminal work of Ajtai [2] back in 1996, and recently got a second wind with Gentry's breakthrough work [7] on fully-homomorphic encryption. It offers asymptotical efficiency, potential resistance to quantum computers and new functionalities. There has been significant progress in provably-secure lattice cryptography in the past few years, but from a practical point of view, very few lattice schemes can compete with standardized schemes for now. This is especially true in the case of signature schemes, for which there is arguably only one realistic lattice alternative: NTRUSIGN [11], which is an optimized instantiation of the Goldreich-Goldwasser-Halevi (GGH) signature scheme [9] using the compact lattices introduced in NTRU encryption [14] and whose performances are comparable with ECDSA. By comparison, signatures have size beyond 10,000 bits (at 80-bit security level) for the most efficient provably-secure lattice signature scheme known, namely the recent scheme of Lyubashevsky [19].

X. Wang and K. Sako (Eds.): ASIACRYPT 2012, LNCS 7658, pp. 433–450, 2012.

However, NTRUSIGN has no provable-security guarantee. In fact, the GGH signature scheme and its simplest NTRUSIGN instantiation were broken at EUROCRYPT '06 by Nguyen and Regev [23], who presented a polynomial-time key-recovery attack using a polynomial number of signatures: in the case of NTRUSIGN, 400 signatures suffice in practice to disclose the secret key within a few hours. In the GGH design, a signature is a lattice point which is relatively close to the (hashed) message. Clearly, many lattice points could be valid signatures, but GGH selects one which is closely related to the secret key: each message–signature pair actually discloses a sample almost uniformly distributed in a secret high-dimensional parallelepiped. The NR attack works by learning such a parallelepiped: given a polynomial number of samples of the form $\sum_{i=1}^{n} x_i \mathbf{b}_i$ where the x_i's are picked uniformly at random from $[-1, 1]$ and the secret vectors $\mathbf{b}_1, \ldots, \mathbf{b}_n \in \mathbb{R}^n$ are linearly independent, the attack recovers the parallelepiped basis $(\mathbf{b}_1, \ldots, \mathbf{b}_n)$, by finding minima of a certain multivariate function, thanks to a well-chosen gradient descent. The NR attack motivated the search of countermeasures to repair NTRUSIGN:

- The very first countermeasure already appeared in half of the parameter choices of NTRU's IEEE P1363.1 standardization proposal [17], the other half being broken by NR. It consists of applying the signature generation process twice, using two different NTRU lattices, the first one being kept secret: here, the secret parallelepiped becomes the Minkowski sum of two secret parallelepipeds, which is a special case of zonotopes. This slows down signature generation, and forces to increase parameters because the signature obtained is less close to the message. However, no provable security guarantee was known or even expected. In fact, heuristic attacks have been claimed by both the designers of NTRUSIGN [10] and more recently by Malkin *et al.* [20], but both are impractical: the most optimistic estimates [10,20] state that they both require at least 2^{60} signatures, and none have been fully implemented. Yet, as a safety precaution, the designers of NTRUSIGN [10] only claim the security of NTRUSIGN with perturbation up to 1 million signatures in [11]. Still, breaking this countermeasure was left as an open problem in [23].
- In 2008, Hu, Wang and He [16] proposed a simpler and faster countermeasure in IEEE Trans. Inform. Theory, which we call IEEE-IT, where the secret parallelepiped is deformed. Again, the actual security was unknown.
- Gentry, Peikert and Vaikuntanathan [8] proposed the first provably secure countermeasure for GGH signatures, by using a randomized variant [18] of Babai's nearest plane algorithm. However, this slows down signature generation significantly, and forces to increase parameters because the signatures obtained are much less close to the message. As a result, the resulting signature for NTRUSIGN does not seem competitive with classical signatures: no concrete parameter choice has been proposed.

OUR RESULTS. We revisit the Nguyen-Regev gradient-descent attack to show that it is much more powerful than previously expected: in particular, an optimized NR attack can surprisingly break in practice both NTRU's perturbation

technique [11] as recommended in standardization proposals [17,13], and the
IEEE-IT countermeasure [16]. For instance, we can recover the NTRUSIGN secret key in a few hours, using 8,000 signatures for the original NTRUSIGN-251
scheme with one perturbation submitted to IEEE P1363 standardization in 2003,
or only 5,000 signatures for the latest 80-bit-security parameter set [13] proposed
in 2010. These are the first successful experiments fully breaking NTRUSIGN
with countermeasures. Note that in the perturbation case, we have to slightly
modify the original NR attack. The warning is clear: our work strongly suggests
to dismiss all GGH/NTRUSIGN countermeasures which are not supported by
some provable security guarantee.

Our work sheds new light on the NR attack. The original analysis of Nguyen
and Regev does not apply to any of the two NTRUSIGN countermeasures, and
it seemed *a priori* that the NR attack would not work in these cases. We show
that the NR attack is much more robust than anticipated, by extending the
original analysis of the Nguyen-Regev algorithm for learning a parallelepiped,
to tackle more general objects such as zonotopes (to break the NTRUSIGN
countermeasure with a constant number of perturbations) or deformed parallelepipeds (to break the IEEE-IT countermeasure). For instance, in the zonotope
case, the parallelepiped distribution $\sum_{i=1}^{n} x_i \mathbf{b}_i$ is replaced by $\sum_{i=1}^{m} x_i \mathbf{v}_i$ where
$\mathbf{v}_1, \dots, \mathbf{v}_m \in \mathbb{R}^n$ are secret vectors with $m \geq n$. The key point of the NR attack
is that all the local minima of a certain multivariate function are connected to
the directions $\mathbf{b}_i$'s of the secret parallelepiped. We show that there is somewhat
a similar (albeit more complex) phenomenon when the parallelepiped is replaced
by zonotopes or deformed parallelepipeds: there, we establish the existence of
local minima connected to the secret vectors spanning the object, but we cannot rule out the existence of other minima. Yet, the attack works very well in
practice, as if there were no other minima.

ROADMAP. In Sect. 2, we recall background on NTRUSIGN and the NR attack.
In Sect. 3, we attack NTRU's perturbation countermeasure, by learning a zonotope. In Sect. 4, we attack the IEEE-IT countermeasure, by learning a deformed
parallelepiped. More information is provided in the full version [5].

2   Background and Notation

2.1   Notation

Sets. $\mathbb{Z}_q$ is the ring of integers modulo q. $\mathbb{N}$ and $\mathbb{Z}$ denote the usual sets. $[n]$
denotes $\{1, \cdots, n\}$. $\mathbb{S}_n$ is the unit sphere of $\mathbb{R}^n$ for the Euclidean norm $\|.\|$,
whose inner product is $\langle, \rangle$.

Linear Algebra. Vectors of $\mathbb{R}^n$ will be row vectors denoted by bold lowercase
letters. A (row) matrix is denoted by $[\mathbf{b}_1, \dots, \mathbf{b}_n]$. We denote by $\mathcal{M}_{m,n}(\mathcal{R})$ the
set of $m \times n$ matrices over a ring $\mathcal{R}$. The group of $n \times n$ invertible matrices with
real coefficients will be denoted by $\mathcal{GL}_n(\mathbb{R})$ and $\mathcal{O}_n(\mathbb{R})$ will denote the subgroup
of orthogonal matrices. The transpose of a matrix M will be denoted by M^t,

and M^{-t} will mean the inverse of the transpose. For a set $\mathcal{S}$ of vectors in $\mathbb{R}^n$ and $M \in \mathcal{M}_{n,m}(\mathbb{R})$, $\mathcal{S} \cdot M$ denotes the set $\{\mathbf{s} \cdot M : \mathbf{s} \in \mathcal{S}\}$. We denote by I_n the $n \times n$ identity matrix.

Rounding. We denote by $\lceil x \rfloor$ the closest integer to x. Naturally, $\lceil \mathbf{b} \rfloor$ denotes the operation applied to all the coordinates of $\mathbf{b}$.

Distributions. If X is a random variable, we denote by $\mathbb{E}[X]$ its expectation. For any set S, we denote by $\mathcal{U}(S)$ the uniform distribution over S, when applicable. If $\mathcal{D}$ is a distribution over $\mathbb{R}^n$, its *covariance* is the $n \times n$ symmetric positive matrix $\mathrm{Cov}(\mathcal{D}) = \mathbb{E}_{\mathbf{x} \leftarrow \mathcal{D}}[\mathbf{x}^t \mathbf{x}]$. The notation $\mathcal{D} \oplus \mathcal{D}'$ denotes the convolution of two distributions, that is the distribution of $\mathbf{x} + \mathbf{y}$ where $\mathbf{x} \leftarrow \mathcal{D}$ and $\mathbf{y} \leftarrow \mathcal{D}'$ are sampled independently. Furthermore, we denote by $\mathcal{D} \cdot B$ the distribution of $\mathbf{x}B$ where $\mathbf{x} \leftarrow \mathcal{D}$.

Zonotopes and Parallelepipeds. A zonotope is the Minkowski sum of finitely many segments. Here, we use centered zonotopes: the *zonotope* spanned by an $m \times n$ row matrix $V = [\mathbf{v}_1, \ldots, \mathbf{v}_m]$ is the set $\mathcal{Z}(V) = \{\sum_{i=1}^m x_i \mathbf{v}_i, -1 \le x_i \le 1\}$. We denote by $\mathcal{D}_{\mathcal{Z}(V)}$ the convolution distribution over $\mathcal{Z}(V)$ obtained by picking independently each x_i uniformly at random from $[-1,1]^n$: in other words, $\mathcal{D}_{\mathcal{Z}(V)} = \mathcal{U}([-1,1]^n) \cdot V$, which in general is not the uniform distribution over $\mathcal{Z}(V)$. However, in the particular case $V \in \mathcal{GL}_n(\mathbb{R})$, $\mathcal{Z}(V)$ is simply the *parallelepiped* $\mathcal{P}(V)$ spanned by V, and $\mathcal{D}_{\mathcal{P}(V)}$ is equal to the uniform distribution over $\mathcal{P}(V)$.

Differentials. Let f be a function from $\mathbb{R}^n$ to $\mathbb{R}$. The *gradient* of f at $\mathbf{w} \in \mathbb{R}^n$ is denoted by $\nabla f(\mathbf{w}) = (\frac{\partial f}{\partial x_1}(\mathbf{w}), \ldots, \frac{\partial f}{\partial x_n}(\mathbf{w}))$. The *Hessian matrix* of f at $\mathbf{w} \in \mathbb{R}^n$ is denoted by $\mathrm{H} f(\mathbf{w}) = (\frac{\partial^2 f}{\partial x_i \partial x_j}(\mathbf{w}))_{1 \le i,j \le n}$.

Running Times. All given running times were measured using a 2.27-GHz Intel Xeon E5520 core.

Lattices. We refer to the survey [24] for a bibliography on lattices. In this paper, by the term lattice, we mean a full-rank discrete subgroup of $\mathbb{R}^n$. A non-empty set $L \subseteq \mathbb{R}^n$ is a lattice if and only if there exists $B = [\mathbf{b}_1, \ldots, \mathbf{b}_n] \in \mathcal{GL}_n(\mathbb{R})$ such that $L = \{\sum_{i=1}^n n_i \mathbf{b}_i \mid n_i \in \mathbb{Z}\}$. Any such B is called a basis of L, and the absolute value of its determinant is the lattice volume $\mathrm{vol}(L)$ of the lattice L. The *closest vector problem* (CVP) is the following: given a basis of $L \subseteq \mathbb{Z}^n$ and a target $\mathbf{t} \in \mathbb{Q}^n$, find a lattice vector $\mathbf{v} \in L$ minimizing the distance $\|\mathbf{v} - \mathbf{t}\|$. If d is the minimal distance, then approximating CVP to a factor k means finding $\mathbf{v} \in L$ such that $\|\mathbf{v} - \mathbf{t}\| \le kd$. *Bounded Distance Decoding* (BDD) is a special case of CVP where the distance to the lattice is known to be small.

2.2 The GGH Signature Scheme

The GGH scheme [9] works with a lattice L in $\mathbb{Z}^n$. The secret key is a non-singular matrix $R \in \mathcal{M}_n(\mathbb{Z})$, with very short row vectors. Following [21], the public key is the Hermite normal form (HNF) of L. The messages are hashed onto a "large enough" subset of $\mathbb{Z}^n$, for instance a large hypercube. Let $\mathbf{m} \in \mathbb{Z}^n$ be the hash of the message to be signed. The signer applies Babai's round-off CVP approximation algorithm [3] to get a lattice vector close to $\mathbf{m}$:

$$\mathbf{s} = \lfloor \mathbf{m} R^{-1} \rceil R, \tag{1}$$

so that $\mathbf{s} - \mathbf{m} \in \frac{1}{2}\mathcal{P}(R)$. To verify the signature $\mathbf{s}$ of $\mathbf{m}$, one checks that $\mathbf{s} \in L$ using the public basis B, and that the distance $\|\mathbf{s} - \mathbf{m}\|$ is sufficiently small.

2.3   NTRUSign

Basic Scheme. NTRUSIGN [11] is an instantiation of GGH using the compact lattices from NTRU encryption [14], which we briefly recall: we refer to [11,4] for more details. In the former NTRU standards [4] proposed to IEEE P1363.1 [17], $N = 251$ and $q = 128$. Let $\mathcal{R}$ be the ring $\mathbb{Z}[X]/(X^N - 1)$ whose multiplication is denoted by $*$. One computes $(f, g, F, G) \in \mathcal{R}^4$ such that $f * G - g * F = q$ in $\mathcal{R}$ and f is invertible mod q, where f and g have 0–1 coefficients (with a prescribed number of 1), while F and G have slightly larger coefficients, yet much smaller than q. This quadruplet is the NTRU secret key. Then the secret basis is the following $(2N) \times (2N)$ block-wise circulant matrix:

$$R = \begin{bmatrix} \mathcal{C}(f) & \mathcal{C}(g) \\ \mathcal{C}(F) & \mathcal{C}(G) \end{bmatrix} \text{ where } \mathcal{C}(a) \text{ denotes } \begin{bmatrix} a_0 & a_1 & \cdots & a_{N-1} \\ a_{N-1} & a_0 & \cdots & a_{N-2} \\ \vdots & & \ddots & \vdots \\ a_1 & \cdots & a_{N-1} & a_0 \end{bmatrix},$$

and f_i denotes the coefficient of X^i of the polynomial f. Thus, the lattice dimension is $n = 2N$. Due to the special structure of R, a single row of R is sufficient to recover the whole secret key. Because f is chosen invertible mod q, the polynomial $h = g/f \bmod q$ is well-defined in $\mathcal{R}$: this is the NTRU public key. Its fundamental property is that $f * h \equiv g \bmod q$ in $\mathcal{R}$. The polynomial h defines the following (natural) public basis of the lattice: $\begin{bmatrix} I_n & \mathcal{C}(h) \\ 0 & qI_n \end{bmatrix}$, which implies that the lattice volume is q^N.

The messages are assumed to be hashed in $\{0, \ldots, q-1\}^{2N}$. Let $\mathbf{m}$ be such a hash. We write $\mathbf{m} = (\mathbf{m}_1, \mathbf{m}_2)$ with $\mathbf{m}_i \in \{0, \ldots, q-1\}^N$. The signature is the vector $(\mathbf{s}, \mathbf{t}) \in \mathbb{Z}^{2N}$ which would have been obtained by applying Babai's round-off CVP approximation algorithm to $\mathbf{m}$, except that it is computed more efficiently using convolution products and can even be compressed (see [11]). We described the basic NTRUSIGN scheme [11], as used in half of the parameter choices of the former NTRU standards [4].

Perturbations. The second half of parameter choices of NTRU standards [4] use perturbation techniques [10,4,12] to strengthen security, which are described in Sect. 2.5. But there is a second change: instead of the standard NTRU secret key, one uses the so-called *transpose basis*, which is simply R^t, then the public basis remains the same, except that one defines the public key as $h = F/f = G/g \bmod q$ rather than $h = g/f \bmod q$.

New Parameters. In the latest NTRU article [13], new parameters for NTRUSIGN have been proposed. These include different values of (N, q) and a different shape for f and g: the coefficients of f and g are now in $\{0, \pm 1\}$, rather than $\{0, 1\}$ like in [11]. But the scheme itself has not changed.

2.4 The Nguyen-Regev Attack

We briefly recall the Nguyen-Regev attack [23], using a slightly different presentation. The NR attack solves the following idealized problem:

Problem 1 (The Hidden Parallelepiped Problem or HPP). *Let $V = [\mathbf{v}_1, \ldots, \mathbf{v}_n] \in \mathcal{GL}_n(\mathbb{R})$ and let $\mathcal{P}(V) = \{\sum_{i=1}^n x_i \mathbf{v}_i : x_i \in [-1, 1]\}$ be the parallelepiped spanned by V. The input to the HPP is a sequence of $\mathrm{poly}(n)$ independent samples from the uniform distribution $\mathcal{D}_{\mathcal{P}(V)}$. The goal is to find a good approximation of the rows of $\pm V$.*

In practice, instead of samples from $\mathcal{D}_{\mathcal{P}(V)}$, the attack uses $2(\mathbf{s} - \mathbf{m})$ for all given message-signature pairs $(\mathbf{m}, \mathbf{s})$: this distribution is heuristically close to $\mathcal{D}_{\mathcal{P}(V)}$ where R is the secret basis. To recover rows of R, the attack simply rounds the approximations found to integer vectors. The NR attack has two stages: morphing and minimization.

Morphing the Parallelepiped into a Hypercube. The first stage of the NR attack is to transform the hidden parallelepiped into a hidden hypercube (see Alg. 1), using a suitable linear transformation L. It is based on the following elementary lemma [23, Lemmas 1 and 2]:

Lemma 1. *Let $V \in \mathcal{GL}_n(\mathbb{R})$ and denote by $G \in \mathcal{GL}_n(\mathbb{R})$ the symmetric positive definite matrix $V^t V$. Then:*

- $\mathrm{Cov}(\mathcal{D}_{\mathcal{P}(V)}) = G/3$.
- *If $L \in \mathcal{GL}_n(\mathbb{R})$ satisfies $LL^t = G^{-1}$ and we let $C = VL$, then $C \in \mathcal{O}_n(\mathbb{R})$ and $\mathcal{D}_{\mathcal{P}(V)} \cdot L = \mathcal{D}_{\mathcal{P}(C)}$.*

Algorithm 1. Morphing($\mathcal{X}$): Morphing a Parallelepiped into a Hybercube

Input: A set $\mathcal{X}$ of vectors $\mathbf{x} \in \mathbb{R}^n$ sampled from the uniform distribution $\mathcal{D}_{\mathcal{P}(V)}$ over a parallelepiped.
Output: A matrix L such that $\mathcal{D}_{\mathcal{P}(V)} \cdot L$ is close to $\mathcal{D}_{\mathcal{P}(C)}$ for some $C \in \mathcal{O}_n(R)$.
1: Compute an approximation G of $V^t V$ using the set $\mathcal{X}$, using $\mathrm{Cov}(\mathcal{D}_{\mathcal{P}(V)}) = V^t V/3$ (see Lemma 1).
2: Return L such that $LL^t = G^{-1}$

This stage is exactly (up to scaling) the classical preprocessing used in independent component analysis to make covariance equal to the identity matrix:

Lemma 2. *Let G be the covariance matrix of a distribution $\mathcal{D}$ over $\mathbb{R}^n$. If $L \in \mathcal{GL}_n(\mathbb{R})$ satisfies $LL^t = G^{-1}$, then $\mathrm{Cov}(\mathcal{D} \cdot L) = I_n$.*

Learning a Hypercube. The second stage of the NR attack is to solve the hidden hypercube problem, using minimization with a gradient descent (see Alg. 2). Nguyen and Regev [23] showed that for any $V \in \mathcal{O}_n(\mathbb{R})$, if $\mathcal{D}$ denotes the distribution $\mathcal{D}_{\mathcal{P}(V)}$:

- The function $\text{mom}_{\mathcal{D},4}(\mathbf{w}) = \mathbb{E}_{\mathbf{x} \leftarrow \mathcal{D}}[\langle \mathbf{x}, \mathbf{w} \rangle^4]$ has exactly $2n$ local minima over the unit sphere $\mathbb{S}_n$, which are located at $\pm\mathbf{v}_1, \cdots, \pm\mathbf{v}_n$, and are global minima.
- It is possible to find all minima of $\text{mom}_{\mathcal{D},4}(\cdot)$ over $\mathbb{S}_n$ in random polynomial time, using Alg. 2 with parameter $\delta = 3/4$, thanks to the nice shape of $\text{mom}_{\mathcal{D},4}(\cdot)$. Alg. 2 is denoted by $\mathbf{Descent}(\mathcal{X}, \mathbf{w}, \delta)$ which, given a point $\mathbf{w} \in \mathbb{S}_n$, performs a suitable gradient descent using the sample set $\mathcal{X}$, and returns an approximation of some $\pm\mathbf{v}_i$.

Algorithm 2. $\mathbf{Descent}(\mathcal{X}, \mathbf{w}, \delta)$: Solving the Hidden Hypercube Problem by Gradient Descent

Input: A set $\mathcal{X}$ of samples from the distribution $\mathcal{D}_{\mathcal{P}(V)}$ where $V \in \mathcal{O}_n(\mathbb{R})$, a vector $\mathbf{w}$ chosen uniformly at random from $\mathbb{S}_n$ and a descent parameter δ.
Output: An approximation of some row of $\pm V$.
1: Compute an approximation $\mathbf{g}$ of the gradient $\nabla \text{mom}_{V,4}(\mathbf{w})$ using $\mathcal{X}$.
2: Let $\mathbf{w}_{new} = \mathbf{w} - \delta \mathbf{g}$.
3: Divide $\mathbf{w}_{new}$ by its Euclidean norm $\|\mathbf{w}_{new}\|$.
4: **if** $\text{mom}_{V,4}(\mathbf{w}_{new}) \geq \text{mom}_{V,4}(\mathbf{w})$ where the moments are approximated using $\mathcal{X}$
   then
5:     **return** the vector $\mathbf{w}$.
6: **else**
7:     Replace $\mathbf{w}$ by $\mathbf{w}_{new}$ and go back to Step 1.
8: **end if**

The whole NR attack is summarized by Alg. 3.

Algorithm 3. $\mathbf{SolveHPP}(\mathcal{X})$: Learning a Parallelepiped [23]

Input: A set $\mathcal{X}$ of vectors $\mathbf{x} \in \mathbb{R}^n$ sampled from $\mathcal{D}_{\mathcal{P}(V)}$, where $V \in \mathcal{GL}_n(\mathbb{R})$
Output: An approximation of a random row vector of $\pm V$
1: $L := \mathbf{Morphing}(\mathcal{X})$ using Alg. 1
2: $\mathcal{X} := \mathcal{X} \cdot L$
3: Pick $\mathbf{w}$ uniformly at random from $\mathbb{S}_n$
4: Compute $\mathbf{r} := \mathbf{Descent}(\mathcal{X}, \mathbf{w}, \delta) \in \mathbb{S}^n$ using Alg. 2: use $\delta = 3/4$ in theory and $\delta = 0.7$ in practice.
5: Return $\mathbf{r}L^{-1}$

Shrinking the number of NTRUSIGN-*signatures.* In practice, the NR attack requires a polynomial number of signatures, but it is possible to experimentally decrease this amount by a linear factor [23], using a well-known symmetry of NTRU lattices. We define the NTRUSIGN symmetry group $\mathfrak{S}_N^{\text{NTRU}}$ as the group spanned by $\sigma \in \mathcal{O}_n(\mathbb{R}) : (x_1, \ldots x_N | y_1, \cdots y_N) \mapsto (x_2, \ldots x_N, x_1 | y_2, \cdots y_N, y_1)$.

If L is the NTRU lattice, then $\sigma(L) = L$. Furthermore, $(\sigma(\mathbf{m}), \sigma(\mathbf{s}))$ follows the same distribution as uniformly random $(\mathbf{m}, \mathbf{s})$. So, any pair $(\mathbf{m}, \mathbf{s})$ gives rise to N parallelepiped samples. This technique also allows a N-factor speedup for covariance computation, which is the most time consuming part of the attack.

2.5   Countermeasures

NTRUSIGN *perturbation: Summing Parallelepipeds.* Roughly speaking, these techniques perturbates the hashed message $\mathbf{m}$ before signing it with the NTRU secret basis. More precisely, the hashed message $\mathbf{m}$ is first signed using a second NTRU secret basis (of another NTRU lattice, which is kept secret), and the resulting signature is then signed as before. Heuristically, the effect on the sample distribution of the transcript is as follows: if R and R' are the two secret bases, the distribution of $\mathbf{s} - \mathbf{m}$ becomes the convolution $\mathcal{P}(R) \oplus \mathcal{P}(R')$, *i.e.* a natural distribution over the Minkowski sum of the two parallelepipeds obtained by adding the uniform distributions of both parallelepipeds.

IEEE-IT perturbation: Parallelepiped Deformation. Hu et al. [16] suggested another approach to secure NTRUSIGN in the journal IEEE Trans. IT. Their definition are specific to NTRUSIGN-bases, but it can be generalized to GGH, and we call this technique "Parallelepiped deformation". Let $\delta : [\text{-}1/2, 1/2)^n \to \mathbb{Z}^n$ be a function, possibly secret-key dependent. The signature generation (1) is replaced by:

$$\mathbf{s} = \left(\lceil \mathbf{m}R^{-1} \rfloor + \delta(\mathbf{m}R^{-1} - \lceil \mathbf{m}R^{-1} \rfloor) \right) R \qquad (2)$$

If δ outputs small integer vectors, then the signature $\mathbf{s}$ is still valid. The associated deformation function is $d_\delta(\mathbf{x}) = \mathbf{x} + \delta(\mathbf{x})$. The sample distribution of $\mathbf{s} - \mathbf{m}$ is deformed in the following way : $d_\delta(\mathcal{U}^n) \cdot R$ where $d_\delta(\mathcal{U}^n)$ denotes the distribution of $\mathbf{x} + \delta(\mathbf{x})$ with $\mathbf{x} \leftarrow \mathcal{U}^n$. In [16], the deformation δ_{IEEE} for a NTRUSIGN secret key (f, g, F, G) is as follows:

- Let $U \subset [N]$ be the set of indexes u such that the u-th entry of $f + g + F + G$ is 1 modulo 2, and let $A = \#U$. On the average, $A \approx N/2$, and it is assumed that $A \geq 25$, otherwise a new secret key must be generated.
- Let $1 \leq u_1 < u_2 < \cdots < u_A \leq N$ be the elements of U. For $i \notin [A]$, u_i denotes $u_{(i \bmod A)}$.
- Let the input of δ_{IEEE} be the concatenation of two vectors $\mathbf{x}, \mathbf{y} \in [\text{-}1/2, 1/2)^N$. Then the i-th entry of $\delta_{\text{IEEE}}(\mathbf{x}|\mathbf{y})$ is:

$$\left[\delta_{\text{IEEE}}(\mathbf{x}|\mathbf{y}) \right]_i = \begin{cases} 0 & \text{if } i \notin U \\ s(x_{u_j}, y_{u_j}, y_{u_{j+1}}, y_{u_{j+3}}, y_{u_{j+7}}, y_{u_{j+12}}) & \text{if } i = u_j \end{cases}$$

$$\text{where } s(a_0, \dots, a_5) = \begin{cases} 1 & \text{if } a_i < 0 \text{ for all } i \\ -1 & \text{if } a_i > 0 \text{ for all } i \\ 0 & \text{otherwise} \end{cases}$$

Gaussian Sampling. Gentry *et al.* [8] described the first provably secure countermeasure: Gaussian sampling. In previous schemes, the distribution of $\mathbf{s} - \mathbf{m}$ was related to the secret key. In [8], the distribution becomes independent of the secret key: it is some discrete Gaussian distribution, which gives rise a to a security proof in the random-oracle model, under the assumption that finding close vectors is hard in the NTRU lattice. Unfortunately, this countermeasure is not very competitive in practice: the sampling algorithm [18] is much less efficient than NTRUSIGN generation, and the new signature is less close to the message, which forces to increase parameters. But its efficiency has recently been improved, see [26,6].

3   Learning a Zonotope: Breaking NTRUSign with Perturbations

In Sect. 3.1, we introduce the hidden zonotope problem (HZP), which is a natural generalization of the hidden parallelepiped problem (HPP), required to break NTRUSIGN with perturbations. In Sect. 3.2, we explain why the Nguyen-Regev HPP algorithm (Alg. 3) can heuristically solve the HZP, in cases that include NTRUSIGN, provided that Step 5 is slightly modified.   Yet, the approximations obtained by the algorithm are expected to be worse than in the non-perturbed case, so we use a folklore meet-in-the-middle algorithm for BDD in NTRU lattices, which is described in [5]. Finally, in Sect. 3.3, we present experimental results with our optimized NR attack which show that NTRUSIGN with one (or slightly more) perturbation(s) is completely insecure, independently of the type of basis. In particular, we completely break the original NTRUSIGN proposed to IEEE P1363 standardization [4]: only one half of the parameter sets was previously broken in [23].

3.1   The Hidden Zonotope Problem

Assume that one applies $k - 1$ NTRUSIGN perturbations as a countermeasure, which corresponds to k NTRUSIGN lattices $L_1, \ldots, L_k$ (with secret bases $R_1, \ldots, R_k$ where only L_k is public. One signs a hashed message $\mathbf{m} \in \mathbb{Z}^n$ by computing $\mathbf{s}_1 \in L_1$ such that $\mathbf{s}_1 - \mathbf{m} \in \frac{1}{2}\mathcal{P}(R_1)$, then $\mathbf{s}_2 \in L_2$ such that $\mathbf{s}_2 - \mathbf{s}_1 \in \frac{1}{2}\mathcal{P}(R_2)$, ..., and finally $\mathbf{s}_k \in L_k$ such that $\mathbf{s}_k - \mathbf{s}_{k-1} \in \frac{1}{2}\mathcal{P}(R_k)$. It follows that $\mathbf{s}_k$ is somewhat close to $\mathbf{m}$, because $\mathbf{s}_k - \mathbf{m}$ is in the Minkowski sum $\frac{1}{2}\mathcal{P}(R_1) + \frac{1}{2}\mathcal{P}(R_2) + \cdots + \frac{1}{2}\mathcal{P}(R_k)$, which is a zonotope spanned by $\frac{1}{2}R_1, \ldots, \frac{1}{2}R_k$. And heuristically, the distribution of $2(\mathbf{s}_k - \mathbf{m})$ is the convolution of all the k uniform distributions $\mathcal{D}_{\mathcal{P}(R_i)}$. In other words, similarly to the perturbation-free case, an attacker wishing to recover the secret key of a GGH-type signature scheme using perturbations using a polynomial number of signatures is faced with the following problem with $m = kn$:

Problem 2 (The Hidden Zonotope Problem or HZP). *Let $m \geq n$ be integers, and $V = [\mathbf{v}_1, \ldots, \mathbf{v}_m]$ be an $m \times n$ row matrix of rank n. The input to the HZP is a sequence of $\mathrm{poly}(n, m)$ independent samples from $\mathcal{D} =$*

$\mathcal{D}_{\mathcal{Z}(V)}$ over $\mathbb{R}^n$, which is the convolution distribution over the zonotope $\mathcal{Z}(V) = \{\sum_{i=1}^m x_i \mathbf{v}_i, -1 \leq x_i \leq 1\}$ spanned by V. The goal is to find a good approximation of the rows of $\pm V$.

Here, we assume V to have rank n, because this is the setting of NTRUSIGN with perturbation, and because the HPP is simply the HZP with $m = n$.

3.2   Extending the Nguyen-Regev Analysis to Zonotopes

Here, we study the behavior of the original Nguyen-Regev algorithm for learning a parallelepiped (**SolveHPP**$(\mathcal{X})$, Alg. 3) on a HZP instance, that is, when the secret matrix V is not necessarily square, but is an arbitrary $m \times n$ matrix of rank n with $m \geq n$. To do this, we need to change the analysis of Nguyen and Regev [23], and we will have to slightly change Alg. 3 to make the attack still work: Alg. 4 is the new algorithm. Recall that the input distribution $\mathcal{D}_{\mathcal{Z}(V)}$ is formed by $\sum_{i=1}^m x_i \mathbf{v}_i$ where the x_i's are uniformly chosen in $[-1, 1]$. We study how the two stages of the NR attack behave for $\mathcal{D}_{\mathcal{Z}(V)}$.

Morphing Zonotopes. We start with a trivial adaptation of Lemma 1 to zonotopes:

Lemma 3. *Let V be an $m \times n$ matrix over $\mathbb{R}$ of rank n. Let G be the symmetric definite positive matrix $V^t V$. Then:*

- $\mathrm{Cov}(\mathcal{D}_{\mathcal{Z}(V)}) = G/3$.
- *If $L \in \mathcal{GL}_n(\mathbb{R})$ satisfies $LL^t = G^{-1}$ and we let $C = VL$, then $C^t C = I_n$ and $\mathcal{D}_{\mathcal{Z}(V)} \cdot L = \mathcal{D}_{\mathcal{Z}(C)}$.*

Lemma 3 shows that if we apply **Morphing**$(\mathcal{X})$ (Alg. 1) to samples from $\mathcal{D}_{\mathcal{Z}(V)}$ (rather than $\mathcal{D}_{\mathcal{P}(V)}$), the output transformation L will be such that $\mathcal{D}_{\mathcal{Z}(V)} \cdot L$ is close to $\mathcal{D}_{\mathcal{Z}(C)}$ for some $m \times n$ matrix C such that $C^t C = I_n$.

In other words, the effect of Step. 2 in **SolveHPP**$(\mathcal{X})$ (Alg. 3) is to make the zonotope matrix V have orthonormal columns: $V^t V = I_n$. The following lemma gives elementary properties of such matrices, which will be useful for our analysis:

Lemma 4. *Let V be an $m \times n$ row matrix $[\mathbf{v}_1, \ldots, \mathbf{v}_m]$ such that $V^t V = I_n$. Then:*

- $\|\mathbf{w}\|^2 = \sum_{i=1}^m \langle \mathbf{w}, \mathbf{v}_i \rangle^2$ *for all $\mathbf{w} \in \mathbb{R}^n$.*
- $\|\mathbf{v}_i\| \leq 1$ *for all $1 \leq i \leq m$.*
- $\sum_{i=1}^m \|\mathbf{v}_i\|^2 = n$ *and $\mathrm{Exp}_{\mathbf{x} \leftarrow \mathcal{U}(\mathbb{S}_n)}(\|\mathbf{x}VV^t\|^2) = n/m$.*

Learning an "Orthogonal" Zonotope. Nguyen and Regev [23] used the target function $\mathrm{mom}_{\mathcal{D},4}(\mathbf{w}) = \mathbb{E}_{\mathbf{x} \leftarrow \mathcal{D}}[\langle \mathbf{x}, \mathbf{w} \rangle^4]$ for $\mathbf{w} \in \mathbb{S}_n$, $\mathcal{D} = \mathcal{D}_{\mathcal{P}(V)}$ and $V \in \mathcal{O}_n(\mathbb{R})$ to recover the hidden hypercube. We need to study this function when $\mathcal{D}$ is the zonotope distribution $\mathcal{D} = \mathcal{D}_{\mathcal{Z}(V)}$ to recover the hidden zonotope. Nguyen and Regev [23] gave elementary formulas for $\mathrm{mom}_{\mathcal{D},4}$ and $\nabla \mathrm{mom}_{\mathcal{D},4}$ when $\mathcal{D} = \mathcal{D}_{\mathcal{P}(V)}$ and $V \in \mathcal{O}_n(\mathbb{R})$, which can easily be adapted to the zonotope distribution $\mathcal{D}_{\mathcal{Z}(V)}$ if $V^t V = I_n$, as follows:

Lemma 5. *Let V be a $m \times n$ matrix over $\mathbb{R}$ such that $V^t V = I_n$, and $\mathcal{D}$ be the convolution distribution $\mathcal{D}_{\mathcal{Z}(V)}$ over the zonotope spanned by V. Then, for any $\mathbf{w} \in \mathbb{R}^n$:*

$$\mathrm{mom}_{\mathcal{D},4}(\mathbf{w}) = \frac{1}{3} \|\mathbf{w}\|^4 - \frac{2}{15} \sum_{i=1}^m \langle \mathbf{v}_i, \mathbf{w} \rangle^4$$

$$\nabla \mathrm{mom}_{\mathcal{D},4}(\mathbf{w}) = \frac{4}{3} \mathbf{w} - \frac{8}{15} \sum_{i=1}^m \langle \mathbf{v}_i, \mathbf{w} \rangle^3 \mathbf{v}_i \quad \textit{if } \mathbf{w} \in \mathbb{S}_n$$

Corollary 1. *Under the same hypotheses as Lemma 5, the minima over $\mathbb{S}_n$ of the function $\mathrm{mom}_{\mathcal{D},4}(\mathbf{w})$ are the maxima (over $\mathbb{S}_n$) of $f(\mathbf{w}) = \sum_{i=1}^m f_{\mathbf{v}_i}(\mathbf{w})$ where $f_{\mathbf{v}}(\mathbf{w}) = \langle \mathbf{v}, \mathbf{w} \rangle^4$ is defined over $\mathbb{R}^n$.*

In [23, Lemma 3], Nguyen and Regev used Lagrange multipliers to show that when $V \in \mathcal{O}_n(\mathbb{R})$, the local minima of $\mathrm{mom}_{\mathcal{D}_{\mathcal{P}(V)},4}$ were located at $\pm\mathbf{v}_1, \ldots, \mathbf{v}_n$, and these minima are clearly global minima. However, this argument breaks down when V is a rectangular $m \times n$ matrix of rank n such that $V^t V = I_n$. To tackle the zonotope case, we use a different argument, which requires to study each function $f_{\mathbf{v}_i}(\mathbf{w}) = \langle \mathbf{v}_i, \mathbf{w} \rangle^4$ individually:

Lemma 6. *Let $\mathbf{v} \in \mathbb{R}^n$ and $f_{\mathbf{v}}(\mathbf{w}) = \langle \mathbf{v}, \mathbf{w} \rangle^4$ for $\mathbf{w} \in \mathbb{R}^n$. Then:*

1. *The gradient and Hessian matrix of $f_{\mathbf{v}}$ are $\nabla f_{\mathbf{v}}(\mathbf{w}) = 4 \langle \mathbf{w}, \mathbf{v} \rangle^3 \cdot \mathbf{v}$ and $\mathrm{H} f_{\mathbf{v}}(\mathbf{w}) = 12 \langle \mathbf{w}, \mathbf{v} \rangle^2 \cdot \mathbf{v}^t \mathbf{v}$.*
2. *There are only two local maxima of $f_{\mathbf{v}}$ over $\mathbb{S}_n$, which are located at $\pm\mathbf{v}/\|\mathbf{v}\|$, and their value is $\|\mathbf{v}\|^4$.*
3. *The local minima of $f_{\mathbf{v}}$ over $\mathbb{S}_n$ are located on the hyperplane orthogonal to $\mathbf{v}$, and their value is 0.*
4. *The mean value of $f_{\mathbf{v}}$ over $\mathbb{S}_n$ is $3\|\mathbf{v}\|^4/(n(n+2))$.*

This already gives a different point of view from Nguyen and Regev in the special case where $V \in \mathcal{O}_n(\mathbb{R})$: for all $1 \leq j \leq n$, $\mathbf{v}_j$ is a local maximum of $f_{\mathbf{v}_j}$ and a local minimum of $f_{\mathbf{v}_i}$ for all $i \neq j$ because $\mathbf{v}_i \perp \mathbf{v}_j$; and therefore $\pm\mathbf{v}_1, \ldots, \mathbf{v}_n$ are local extrema of $\mathrm{mom}_{\mathcal{U} \cdot V,4}$.

In the general case where V is an $m \times n$ matrix such that $V^t V = I_n$, let $\mathbf{d}_i = \mathbf{v}_i/\|\mathbf{v}_i\| \in \mathbb{S}_n$ for $1 \leq i \leq m$. The direction $\mathbf{d}_j$ is a local maximum of $f_{\mathbf{v}_j}$ over $\mathbb{S}_n$. On the other hand, $f_{\mathbf{v}_i}(\mathbf{d}_j)$ is likely to be small for $i \neq j$. This suggests that $\mathbf{d}_j$ should be very close to a local maximum of the whole sum $\sum_{i=1}^m f_{\mathbf{v}_i}(\mathbf{d}_j)$, provided that the local maximum $\|\mathbf{v}_j\|^4$ of $f_{\mathbf{v}_j}$ is somewhat larger than $\sum_{i \neq j} f_{\mathbf{v}_i}(\mathbf{d}_j)$. In fact, this local maximum $\mathbf{d}_j$ is intuitively shifted by $\mathbf{g}/(2\|\mathbf{v}_j\|^4)$ where $\mathbf{g}$ is the gradient of $\sum_{i=1}^m f_{\mathbf{v}_i}(\mathbf{d}_j)$ at $\mathbf{d}_j$, because this is exactly what happens for its second-order Taylor approximation. This is formalized by our main result, which provides a sufficient condition on V guaranteeing that a given direction $\mathbf{v}_j/\|\mathbf{v}_j\|$ is close to a local minimum of $\mathrm{mom}_{\mathcal{D}_{\mathcal{Z}(V)},4}$:

Theorem 3 (Local Minima for Zonotopes). *Let V be a $m \times n$ matrix over $\mathbb{R}$ such that $V^t V = I_n$. Assume that there is $\alpha \geq 1$ such that V is α-weakly-orthogonal, that is, its m rows satisfy for all $i \neq j$: $|\langle \mathbf{v}_i, \mathbf{v}_j \rangle| \leq \alpha \|\mathbf{v}_i\| \|\mathbf{v}_j\| / \sqrt{n}$. Let $1 \leq j \leq m$ and $0 < \varepsilon < 1/\sqrt{2}$ such that:*

$$\varepsilon \|\mathbf{v}_j\|^4 > 6 \left(\frac{\alpha}{\sqrt{n}} + \varepsilon \right)^2 \varepsilon + \frac{4}{\|\mathbf{v}_j\|^3} \| \sum_{i \neq j} \langle \mathbf{v}_j, \mathbf{v}_i \rangle^3 \mathbf{v}_i \| \tag{3}$$

which holds in particular if $\|\mathbf{v}_j\| \geq \dfrac{2\sqrt{\alpha}}{n^{1/12}}$ and $\varepsilon = \dfrac{5\alpha^3}{\sqrt{n} \|\mathbf{v}_j\|^4} < 1/\sqrt{2}$. Then, over the unit sphere, the function $\mathrm{mom}_{\mathcal{D}_{\mathcal{Z}(V)}, 4}$ has a local minimum at some point $\mathbf{m}_j \in \mathbb{S}_n$ such that $\mathbf{m}_j$ is close to the direction of $\mathbf{v}_j$, namely:

$$\left\langle \mathbf{m}_j, \frac{\mathbf{v}_j}{\|\mathbf{v}_j\|} \right\rangle > 1 - \frac{\epsilon^2}{2} \quad and \quad \left\| \mathbf{m}_j - \frac{\mathbf{v}_j}{\|\mathbf{v}_j\|} \right\| \leq \epsilon.$$

And the local minimum $\mathrm{mom}_{\mathcal{D}_{\mathcal{Z}(V)}, 4}(\mathbf{m}_j)$ discloses an approximation of $\|\mathbf{v}_j\|$, namely:

$$\left| \mathrm{mom}_{\mathcal{D}_{\mathcal{Z}(V)}, 4}(\mathbf{m}_j) - \left(\frac{1}{3} - \frac{2\|\mathbf{v}_j\|^4}{15} \right) \right| \leq \frac{2}{15} \left(5\varepsilon^3 + 6\varepsilon^2 + 4\varepsilon + m \left(\varepsilon + \frac{\alpha}{\sqrt{n}} \right)^4 \right).$$

Proof. (Sketch of the proof in [5]) Let $\mathcal{B} = \{ \mathbf{w} \in \mathbb{S}_n : \|\mathbf{w} - \mathbf{d}_j\| < \varepsilon \}$ be the open ball of $\mathbb{S}_n$ of radius ε, where $\mathbf{d}_j = \mathbf{v}_j / \|\mathbf{v}_j\| \in \mathbb{S}_n$ Notice that for all $\mathbf{w} \in \mathbb{S}_n$:

$$\|\mathbf{w} - \mathbf{d}_j\|^2 = \|\mathbf{w}\|^2 + \|\mathbf{d}_j\|^2 - 2 \langle \mathbf{w}, \mathbf{d}_j \rangle = 2(1 - \langle \mathbf{w}, \mathbf{d}_j \rangle).$$

Therefore $\mathcal{B} = \{ \mathbf{w} \in \mathbb{S}_n : \langle \mathbf{d}_j, \mathbf{w} \rangle > 1 - \varepsilon^2/2 \}$, whose closure and boundary are denoted respectively by $\bar{\mathcal{B}}$ and $\partial \mathcal{B}$. Recall that $f = \sum_{i=1}^m f_{\mathbf{v}_i}$. We will prove the following property:

$$\forall \mathbf{w} \in \partial \mathcal{B}, f(\mathbf{w}) < f(\mathbf{d}_j), \tag{4}$$

which allows to conclude the proof of Th. 3. Indeed, by continuity, the restriction of f to $\bar{\mathcal{B}}$ has a global maximum at some point $\mathbf{m}_j \in \bar{\mathcal{B}}$. And (4) implies that $\mathbf{m}_j \notin \partial \mathcal{B}$, therefore $\mathbf{m}_j \in \mathcal{B}$. Thus, $\mathbf{m}$ is a global maximum of f over the open set $\mathcal{B}$: in other words, $\mathbf{m}_j$ is a local maximum of f, and therefore a local minimum of $\mathrm{mom}_{\mathcal{D}, 4}$. Furthermore, by definition of $\mathcal{B}$, we have: $\|\mathbf{m}_j - \mathbf{d}_j\| < \varepsilon$ and $\langle \mathbf{d}_j, \mathbf{m}_j \rangle > 1 - \varepsilon^2/2$. And the final inequality follows from:

$$\mathrm{mom}_{\mathcal{D}, 4}(\mathbf{m}_j) - \left(\frac{1}{3} - \frac{2\|\mathbf{v}_j\|^4}{15} \right) = \frac{2}{15} \left(\langle \mathbf{v}_j, \mathbf{d}_j \rangle^4 - \langle \mathbf{v}_j, \mathbf{m}_j \rangle^4 - \sum_{i \neq j} \langle \mathbf{v}_i, \mathbf{m}_j \rangle^4 \right).$$

We now prove (4). Let $\mathbf{w} \in \partial \mathcal{B}$. To show $f(\mathbf{d}_j) - f(\mathbf{w}) > 0$, we decompose it as:

$$\left(f_{\mathbf{v}_j}(\mathbf{d}_j) - f_{\mathbf{v}_j}(\mathbf{w}) \right) + \sum_{i \neq j} \left(f_{\mathbf{v}_i}(\mathbf{d}_j) - f_{\mathbf{v}_i}(\mathbf{w}) \right) \tag{5}$$

On the one hand, the left-hand term of (5) is:

$$f_{\mathbf{v}_j}(\mathbf{d}_j) - f_{\mathbf{v}_j}(\mathbf{w}) = \|\mathbf{v}_j\|^4 - (1 - \frac{\varepsilon^2}{2})^4 \|\mathbf{v}_j\|^4 \geq \varepsilon^2 \|\mathbf{v}_j\|^4 \tag{6}$$

because $\varepsilon < 1/\sqrt{2}$. On the other hand, we upper bound the right-hand term of (5) by the Taylor-Lagrange formula, which states that there exists $\theta \in (0,1)$ such that $\sum_{i \neq j} (f_{\mathbf{v}_i}(\mathbf{w}) - f_{\mathbf{v}_i}(\mathbf{d}_j))$ is equal to:

$$\left\langle \sum_{i \neq j} \nabla f_{\mathbf{v}_i}(\mathbf{d}_j), \mathbf{w} - \mathbf{d}_j \right\rangle + \frac{1}{2}(\mathbf{w} - \mathbf{d}_j) \sum_{i \neq j} H f_{\mathbf{v}_i}(\mathbf{d}_j + \theta(\mathbf{w} - \mathbf{d}_j))(\mathbf{w} - \mathbf{d}_j)^t \tag{7}$$

Let $\mathbf{g} = \sum_{i \neq j} \nabla f_{\mathbf{v}_i}(\mathbf{d}_j) = 4 \sum_{i \neq j} \langle \mathbf{d}_j, \mathbf{v}_i \rangle^3 \mathbf{v}_i$ by Lemma 6. The left-hand term of (7) is bounded as:

$$\left| \left\langle \sum_{i \neq j} \nabla f_{\mathbf{v}_i}(\mathbf{d}_j), \mathbf{w} - \mathbf{d}_j \right\rangle \right| \leq \varepsilon \|\mathbf{g}\|. \tag{8}$$

Using Lemma 6, the right-hand term of (7) can be bounded as:

$$\left| (\mathbf{w} - \mathbf{d}_j) \sum_{i \neq j} H f_{\mathbf{v}_i}(\mathbf{d}_j + \theta(\mathbf{w} - \mathbf{d}_j))(\mathbf{w} - \mathbf{d}_j)^t \right| \leq 12(\alpha/\sqrt{n} + \varepsilon)^2 \varepsilon^2. \tag{9}$$

Collecting (6), (7), (8) and (9), we obtain:

$$f(\mathbf{d}_j) - f(\mathbf{w}) \geq \left(\varepsilon \|\mathbf{v}_j\|^4 - \|\mathbf{g}\| - 6(\alpha/\sqrt{n} + \varepsilon)^2 \varepsilon \right) \varepsilon,$$

which is > 0 by (3). To conclude, it remains to prove that (3) is satisfied when $\|\mathbf{v}_j\| \geq \frac{2\sqrt{\alpha}}{n^{1/12}}$ and $\varepsilon = \frac{5\alpha^3}{\sqrt{n}\|\mathbf{v}_j\|^4} < 1/\sqrt{2}$. This is shown by tedious computations, using weak-orthogonality and Lemma 4. □

Th. 3 states that under suitable assumptions on V (which we will discuss shortly), if $\|\mathbf{v}_j\|$ is not too small, then the secret direction $\mathbf{v}_j/\|\mathbf{v}_j\|$ is very close to a local minimum of $\text{mom}_{\mathcal{D}_{\mathcal{Z}(V)},4}$, whose value discloses an approximation of $\|\mathbf{v}_j\|$, because it is $\approx \frac{1}{3} - \frac{2}{15}\|\mathbf{v}_j\|^4$. This suggests **SolveHZP**$(\mathcal{X})$ (Alg. 4) for learning a zonotope: **SolveHZP**$(\mathcal{X})$ is exactly **SolveHPP**$(\mathcal{X})$ (Alg. 3), except that Step 5 of **SolveHPP**$(\mathcal{X})$ has been modified, to take into account that $\|\mathbf{v}_j\|$ is no longer necessarily equal to 1, but can fortunately be approximated by the value of the local minimum.

First, we discuss the value of α in Th. 3 . Note that weak-orthogonality is a natural property, as shown by the following basic result:

Lemma 7. *Let $\mathbf{v} \in \mathbb{S}^n$ and denote by X the random variable $X = \langle \mathbf{v}, \mathbf{w} \rangle^2$ where $\mathbf{w}$ has uniform distribution over $\mathbb{S}_n$. Then X has distribution $Beta(1/2, (n-1)/2)$, $\text{Exp}(X) = \frac{1}{n}$, $\text{Exp}(X^2) = \frac{3}{n(n+2)}$, $\text{Exp}(X^3) = \frac{15}{n(n+2)(n+4)}$ and more generally: $\text{Exp}(X^k) = \frac{k-1/2}{n/2+k-1} \text{Exp}(X^{k-1})$.*

Algorithm 4. SolveHZP($\mathcal{X}$): Learning a Zonotope

Input: A set $\mathcal{X}$ of vectors $\mathbf{x} \in \mathbb{R}^n$ sampled from $\mathcal{D}_{\mathcal{Z}(V)}$, where V is an $m \times n$ matrix of rank n.
Output: An approximation of some row vector of $\pm V$.
1: $L := \mathbf{Morphing}(\mathcal{X})$ using Alg. 1
2: $\mathcal{X} := \mathcal{X} \cdot L$
3: Pick $\mathbf{w}$ uniformly at random from $\mathbb{S}_n$
4: Compute $\mathbf{r} := \mathbf{Descent}(\mathcal{X}, \mathbf{w}, \delta) \in \mathbb{S}^n$ using Alg. 2: use $\delta = 3/4$ in theory and $\delta = 0.7$ in practice.
5: Return $\lambda \mathbf{r} L^{-1}$ where $\lambda = ((\frac{1}{3} - \mathrm{mom}_{\mathcal{X},4}(\mathbf{r}))\frac{15}{2})^{1/4}$

By studying more carefully the Beta distribution, it is possible to obtain strong bounds. For instance, Ajtai [1, Lemma 47] showed that for all sufficiently large n, if $\mathbf{v} \in \mathbb{S}^n$ is fixed and $\mathbf{w}$ has uniform distribution over $\mathbb{S}_n$, then $|\langle \mathbf{v}, \mathbf{w} \rangle| \leq (\log n)/\sqrt{n}$ with probability $\geq 1 - \frac{1}{n^{(\log n)/2 - 1}}$. Since the probability is subexponentially close to 1, this implies that if $m = n^{O(1)}$ and we assume that all the directions $\mathbf{v}_i/\|\mathbf{v}_i\|$ are random, then V is $(\log n)$-weakly orthogonal with probability asymptotically close to 1.

This gives strong evidence that, if $m = n^{O(1)}$, the assumption on V in Th. 3 will be satisfied for $\alpha = \log n$. We can now discuss the remaining assumptions. If $\alpha = \log n$, we may take any index j such that $\|\mathbf{v}_j\| \geq \Omega(1/n^{13})$: in particular, if $\|\mathbf{v}_j\| = \Omega(1)$, we may take $\varepsilon = O(\log^3 n)/\sqrt{n}$. And higher values of α can be tolerated, as while as $\alpha = o(n^{1/6})$. Now recall that $\sum_{i=1}^m \|\mathbf{v}_i\|^2 = n$, thus $\max_i \|\mathbf{v}_i\| \geq \sqrt{n/m}$ and $\|\mathbf{v}_i\|$ is on average $\sqrt{n/m}$. In particular, if the number of perturbations is constant, then $m = O(n)$ and $\max_i \|\mathbf{v}_i\| \geq \Omega(1)$, therefore Th. 3 applies to at least one index j, provided that $\alpha = o(n^{1/6})$. In fact, one can see that the result can even tolerate slightly bigger values of m than $\Theta(n)$, such as $m = o(n^{7/6}/\log n)$.

While Th. 3 explains why **SolveHZP($\mathcal{X}$)** (Alg. 4) can heuristically solve the HZP, it is not a full proof, as opposed to the simpler parallelepiped case. The obstructions are the following:

- First, we would need to prove that the distance is sufficiently small to enable the recovery of the original zonotope vectors, using an appropriate BDD solver. Any error on $\mathbf{v}_j/\|\mathbf{v}_j\|$ is multiplied by $L^{-1}\|\mathbf{v}_j\|$. In [23], the error on $\mathbf{v}_j$ could be made polynomially small for any polynomial, provided that the number of samples was (polynomially) large enough. But ε cannot be chosen polynomially small for any arbitrary polynomial in Th. 3.
- Second, we would need to prove that **Descent($\mathcal{X}, \mathbf{w}, \delta$)** (Alg. 2) finds a random local minimum of $\mathrm{mom}_{\mathcal{D}_{\mathcal{Z}(V)},4}$ in polynomial time, even in the presence of noise to compute $\mathrm{mom}_{\mathcal{D}_{\mathcal{Z}(V)},4}$. Intuitively, this is not unreasonable since the function $\mathrm{mom}_{\mathcal{D}_{\mathcal{Z}(V)},4}$ is very regular, but it remains to be proved.
- Finally, we would need to prove that there are no other local minima, or at least, not too many of them.

Regarding the third obstruction, it is easy to prove the following weaker statement, which implies that global minima of $\text{mom}_{\mathcal{D}_{\mathcal{Z}(V)},4}$ over the unit sphere are close to some direction $\mathbf{v}_j/\|\mathbf{v}_j\|$:

Lemma 8. *Let V be a $m \times n$ matrix over $\mathbb{R}$ such that $V^t V = I_n$, and $\mathcal{D}$ be the distribution $\mathcal{D}_{\mathcal{Z}(V)}$. Let $\mathbf{w}$ be a global maximum of $f(\mathbf{w}) = \sum_{i=1}^{m} f_{\mathbf{v}_i}(\mathbf{w})$ over $\mathbb{S}_n$. Then there exists $j \in \{1, \ldots, m\}$ such that:* $\frac{1}{m^{1/4}} < \frac{|\langle \mathbf{v}_j, \mathbf{w} \rangle|}{\|\mathbf{v}_j\|} \leq 1$.

3.3   Experiments

We now report on experiments with the attack performed on NTRUSIGN, with n up to 502. Our experiments are real-world experiments using signatures of uniformly distributed messages.

Conditions of Th. 3. Our discussion following Th. 3 suggested that the matrix V should be heuristically weakly-orthogonal for $\alpha = \log n$. In practice, we may in fact take $\alpha \approx 5$ for both types of NTRUSIGN secret bases.

Regarding the norms $\|\mathbf{v}_i\|$ after morphing, we experimentally verified that $\|\mathbf{v}_i\| \approx \sqrt{1/k}$ where k is the number of perturbations for NTRUSIGN transposed bases (see [5]), as expected by $\sum_{i=1}^{m} \|\mathbf{v}_i\|^2 = n$. But for the so-called standard bases, the situation is a bit different: half of the $\|\mathbf{v}_i\|$'s are very small, and the remaining half are close to $\sqrt{2/k}$. This can be explained by the fact that standard bases are unbalanced: half of the vectors are much shorter than the other vectors.

For a number of perturbations ≤ 8, we experimentally verified that the "gradient" $\mathbf{g} = \frac{4}{\|\mathbf{v}_j\|^3} \| \sum_{i \neq j} \langle \mathbf{v}_j, \mathbf{v}_i \rangle^3 \mathbf{v}_i \|$ appearing in the conditions of Th. 3 satisfies $\|\mathbf{g}\| = O(1/n)$ with a small constant ≤ 4 (see [5]).

To summarize, the conditions of Th. 3 are experimentally verified for a number of perturbations ≤ 8: for all vectors $\mathbf{v}_j$'s in the case of transposed bases, and for half of the vectors $\mathbf{v}_j$'s in the case of standard bases.

Modifications to the original NR attack. We already explained that the original NR algorithm **SolveHPP**$(\mathcal{X})$ (Alg. 3) had to be slightly modified into **SolveHZP**$(\mathcal{X})$ (Alg. 4): more precisely, Step 5 is modified.

However, because Th. 3 states that the secret direction might be perturbed by some small ε, we also implemented an additional modification: instead of the elementary BDD algorithm by rounding, we used in the final stage a special BDD algorithm tailored for NTRU lattices, which is a tweaked version of Odlyzko's meet-in-the-middle attack on NTRU described in [15]. Details are given in [5].

Practical cryptanalysis. We first applied successfully the optimized NR-attack on the original NTRUSIGN-251 scheme with one perturbation (which corresponds to a lattice dimension of 502), as initially submitted to the IEEE P1363 standard: about 8,000 signatures were sufficient to recover the secret key, which should be compared with the 400 signatures of the original attack [23] when there was no perturbation. This means that the original NTRUSIGN-251 scheme [10] is now completely broken.

Furthermore, we performed additional experiments for varying dimension and number of perturbations, for the parameters proposed in the latest NTRU article [13], where transposed bases are used. Table 1 summarizes the results obtained: each successful attack took less than a day, and the MiM error recovery algorithm ran with less than 8Gb of memory.

Table 1. Experiments with the generalized NR-attack on the latest NTRUSign parameters [13]

Security level : dimension n	Toy : 94	80-bit : 314	112-bit : 394	128-bit : 446
0 perturbation	300:(0,1)	400:(0,1)	400:(0,1)	600:(0,1)
1 perturbation	1000:(1,2)	5000:(0,1)	4000:(0,1)	4000:(0,0)
2 perturbations	10000:(5,3)	12000:(0,2)		
3 perturbations	12000:(5,4)			
4 perturbations	100000:(0,1)			

In this table, each non-empty cell represents a successful attack for a given transposed basis (the column indicates the security level and the dimension) and number of perturbations (row). These cells have the form $s : (e = \|\epsilon_F\|_1, w = \|\epsilon_G\|_\infty)$ where s is the number of signatures used by the learning algorithm, and where $(\epsilon_F|\epsilon_G)$ is the error vector of the best approximation given by a descent. The running time of our MiM-Algorithm is about $(n/2)^{\lceil e/2 \rceil + 1}$ for such small w.

Our experiments confirm our theoretical analysis: NTRUSign with a constant number of perturbations is insecure, but we see that the number of signatures required increases with the number of perturbations.

4   Learning a Deformed Parallelepiped: Breaking the IEEE-IT Countermeasure

In this section, we show that the deformation suggested in [16] is unlikely to prevent the NR attack [23]. More generally, we show that the NR attack heuristically still works if the deformation is only *partial*, which means that it preserves at least one of the canonical axes, namely there exists at least one index i such that:

– for all $\mathbf{x} \in [\text{-}1/2, 1/2)^n$, $[\delta(\mathbf{x})]_i = 0$
– $\delta(\mathbf{x})$ is independent of $x_i : (\forall j \neq i, x_j = y_j) \Rightarrow \delta(\mathbf{x}) = \delta(\mathbf{y})$

Such an index i is said to be ignored by the deformation δ. And it is clear that δ_{IEEE} is partial by definition (see Sect. 2.5), because it ignores exactly all index $i \notin U$. Our main result is the following, whose proof is given in [5].

Theorem 4. *Let δ be a partial deformation, and i be an index ignored by δ. Let $\mathcal{D} = 2 \cdot d_\delta(\mathcal{U}^n)$ and $M \in \mathcal{GL}_n(\mathbb{R})$ be an invertible matrix and $G = \text{Cov}(\mathcal{D} \cdot M)$. Let L be such that $LL^t = G^{-1}$. Then $\mathbf{r} = \frac{1}{\sqrt{3}} \cdot \mathbf{m}_i L$ is a local minimum of $\text{mom}_{4,\mathcal{D}'}(\cdot)$ over the unit sphere, where $\mathcal{D}' = \mathcal{D} \cdot M \cdot L$.*

While this is a strong theoretical argument supporting why the NR attack still works, it is not a full proof, for reasons similar to the zonotope case (see the previous section): there may be other minima, and we did not prove that the gradient descent efficiently finds minima.

Experimental results. The attack was run, using 300,000 signatures, to recover the secret key in 80-bit, 112-bit and 128-bit NTRUSIGN security level settings, and each run led to a secret key recovery, in about two days. No other local minimum was found. Though the samples no longer belong to a set stable by NTRU symmetry group $\mathfrak{S}_N^{\mathrm{NTRU}}$, we may still try to apply the symmetry trick, to multiply the number of samples by N, like in [23]. This modifies the distribution of the sample to the average of its orbit : $\mathfrak{S}_N^{\mathrm{NTRU}}(\mathcal{D}) = \sigma(\mathbf{x}) : \mathbf{x} \leftarrow \mathcal{D}, \sigma \leftarrow \mathcal{U}(\mathfrak{S}_N^{\mathrm{NTRU}})$. It turns out that applying the attack on such an averaged distribution leads once again to descents converging to some basis vectors: in fact, by symmetry, all of them are equally likely. The attack used 2,000 signatures, and ran in less than an hour, on the same basis. Intuitively, this averaging strongly reduces the co-dependence between the coordinates of $\mathbf{x} \leftarrow \mathcal{D}_\sigma$, making the resulting distribution much closer to a parallelepiped than $\mathcal{D}$.

Acknowledgements. Part of this work is supported by the Commission of the European Communities through the ICT program under contract ICT-2007-216676 ECRYPT II, the European Research Council, and by China's 973 Program (Grant 2013CB834205).

References

1. Ajtai, M.: Generating random lattices according to the invariant distribution (Draft of March 2006)
2. Ajtai, M.: Generating hard instances of lattice problems. In: Complexity of Computations and Proofs. Quad. Mat, vol. 13, Dept. Math., Seconda Univ. Napoli, Caserta, pp. 1–32 (2004)
3. Babai, L.: On Lovász lattice reduction and the nearest lattice point problem. Combinatorica 6, 1–13 (1986)
4. Consortium for Efficient Embedded Security. Efficient embedded security standards #1: Implementation aspects of NTRUEncrypt and NTRUSign. Version 2.0 [17] (June 2003)
5. Ducas, L., Nguyen, P.Q.: Learning a Zonotope and More: Cryptanalysis of NTRUSign Countermeasures. In: Wang, X., Sako, K. (eds.) ASIACRYPT 2012. LNCS, vol. 7658, pp. 433–450. Springer, Heidelberg (2012)
6. Ducas, L., Nguyen, P.Q.: Faster Gaussian Lattice Sampling Using Lazy Floating-Point Arithmetic. In: Wang, X., Sako, K. (eds.) ASIACRYPT 2012. LNCS, vol. 7658, pp. 415–432. Springer, Heidelberg (2012)
7. Gentry, C.: Fully homomorphic encryption using ideal lattices. In: Proc. STOC 2009, pp. 169–178. ACM (2009)
8. Gentry, C., Peikert, C., Vaikuntanathan, V.: Trapdoors for hard lattices and new cryptographic constructions. In: Proc. STOC 2008, pp. 197–206. ACM (2008)
9. Goldreich, O., Goldwasser, S., Halevi, S.: Public-Key Cryptosystems from Lattice Reduction Problems. In: Kaliski Jr., B.S. (ed.) CRYPTO 1997. LNCS, vol. 1294, pp. 112–131. Springer, Heidelberg (1997); full version vailable at ECCC as TR96-056
10. Hoffstein, J., Graham, N.A.H., Pipher, J., Silverman, J.H., Whyte, W.: NTRUSIGN: Digital signatures using the NTRU lattice. Full version of [11] Draft of April 2 (2002); Available on NTRU's website

11. Hoffstein, J., Howgrave-Graham, N., Pipher, J., Silverman, J.H., Whyte, W.: NTRUSIGN: Digital Signatures Using the NTRU Lattice. In: Joye, M. (ed.) CT-RSA 2003. LNCS, vol. 2612, pp. 122–140. Springer, Heidelberg (2003)
12. Hoffstein, J., Graham, N.A.H., Pipher, J., Silverman, J.H., Whyte, W.: Performances improvements and a baseline parameter generation algorithm for NTRUsign. In: Proc. of Workshop on Mathematical Problems and Techniques in Cryptology, pp. 99–126. CRM (2005)
13. Hoffstein, J., Howgrave-Graham, N., Pipher, J., Whyte, W.: Practical lattice-based cryptography: NTRUEncrypt and NTRUSign. In [25] (2010)
14. Hoffstein, J., Pipher, J., Silverman, J.H.: NTRU: A Ring-Based Public Key Cryptosystem. In: Buhler, J.P. (ed.) ANTS 1998. LNCS, vol. 1423, pp. 267–288. Springer, Heidelberg (1998); first presented at the rump session of Crypto 1996
15. Howgrave-Graham, N., Silverman, J.H., Whyte, W.: A meet-in-the-middle attack on an NTRU private key (2003),
    http://www.ntru.com/cryptolab/tech_notes.htm#004
16. Hu, Y., Wang, B., He, W.: NTRUSign with a new perturbation. IEEE Transactions on Information Theory 54(7), 3216–3221 (2008)
17. IEEE P1363.1. Public-key cryptographic techniques based on hard problems over lattices (June 2003),
    http://grouper.ieee.org/groups/1363/lattPK/index.html
18. Klein, P.: Finding the closest lattice vector when it's unusually close. In: Proc. of SODA 2000. ACM–SIAM (2000)
19. Lyubashevsky, V.: Lattice Signatures without Trapdoors. IACR Cryptology ePrint Archive, 2011:537 (2011); In: Pointcheval, D., Johansson, T. (eds.) EUROCRYPT 2012. LNCS, vol. 7237, pp. 738–755. Springer, Heidelberg (2012)
20. Malkin, T., Peikert, C., Servedio, R.A., Wan, A.: Learning an overcomplete basis: Analysis of lattice-based signatures with perturbations, 2009 manuscript cited in [26], available as [27, Chapter 6] (2009)
21. Micciancio, D.: Improving Lattice Based Cryptosystems Using the Hermite Normal Form. In: Silverman, J.H. (ed.) CaLC 2001. LNCS, vol. 2146, pp. 126–145. Springer, Heidelberg (2001)
22. Micciancio, D., Regev, O.: Lattice-based cryptography. In: Post-Quantum Cryptography, pp. 147–191. Springer, Berlin (2009)
23. Nguyen, P.Q., Regev, O.: Learning a Parallelepiped: Cryptanalysis of GGH and NTRU Signatures. J. Cryptology 22(2), 139–160 (2009); In: Vaudenay, S. (ed.) EUROCRYPT 2006. LNCS, vol. 4004, pp. 271–288. Springer, Heidelberg (2006)
24. Nguyen, P.Q., Stern, J.: The Two Faces of Lattices in Cryptology. In: Silverman, J.H. (ed.) CaLC 2001. LNCS, vol. 2146, pp. 146–180. Springer, Heidelberg (2001)
25. Nguyen, P.Q., Vallée, B. (eds.): The LLL Algorithm: Survey and Applications. Information Security and Cryptography. Springer (2010)
26. Peikert, C.: An Efficient and Parallel Gaussian Sampler for Lattices. In: Rabin, T. (ed.) CRYPTO 2010. LNCS, vol. 6223, pp. 80–97. Springer, Heidelberg (2010)
27. Wan, A.: Learning, cryptography, and the average case. PhD thesis, Columbia University (2010), http://itcs.tsinghua.edu.cn/~atw12/

On Polynomial Systems
Arising from a Weil Descent

Christophe Petit* and Jean-Jacques Quisquater**

UCL Crypto Group,
Université catholique de Louvain
Place du Levant 3
1348 Louvain-la-Neuve (Belgium)
{christophe.petit,jjq}@uclouvain.be

Abstract. In the last two decades, many computational problems arising in cryptography have been successfully reduced to various systems of polynomial equations. In this paper, we revisit a class of polynomial systems introduced by Faugère, Perret, Petit and Renault. Based on new experimental results and heuristic evidence, we conjecture that their degrees of regularity are only slightly larger than the original degrees of the equations, resulting in a very low complexity compared to generic systems. We then revisit the application of these systems to the elliptic curve discrete logarithm problem (ECDLP) for binary curves. Our heuristic analysis suggests that an index calculus variant due to Diem requires a *subexponential* number of bit operations $O(2^{c\,n^{2/3}\log n})$ over the binary field $\mathbb{F}_{2^n}$, where c is a constant smaller than 2. According to our estimations, generic discrete logarithm methods are outperformed for any $n > N$ where $N \approx 2000$, but elliptic curves of currently recommended key sizes ($n \approx 160$) are not immediately threatened. The analysis can be easily generalized to other extension fields.

1 Introduction

While linear systems of equations can be efficiently solved with Gaussian elimination, polynomial systems are much harder to solve in general. After their introduction by Buchberger [13], Gröbner bases have become the most popular way to solve polynomial systems of equations, in particular since the development of fast algorithms like F_4 [26] and F_5 [27]. Polynomial systems arising in cryptography tend to have a special structure that simplifies their resolution. In the last twenty years, many cryptographic challenges have been first reduced to polynomial systems of equations and then solved with fast and sometimes *dedicated* Gröbner basis algorithms [42,30,38,10,22,23,32,12,31].

* Supported by an F.R.S.-FNRS postdoctoral research fellowship at Université catholique de Louvain, Louvain-la-Neuve.
** This work was partly supported by the Belgian State's IAP program P6/26 BCRYPT.

X. Wang and K. Sako (Eds.): ASIACRYPT 2012, LNCS 7658, pp. 451–466, 2012.

Our Contribution

In this paper, we revisit a particular class of polynomial systems introduced by Faugère et al. [33,34]. These systems naturally arise by deploying a multivariate polynomial equation over an extension field into a system of polynomial equations over the ground prime field (a technique commonly called *Weil descent*).

We first observe that polynomial systems arising from a Weil descent are a natural generalization of a well-known family of polynomial systems appearing in the cryptanalysis of HFE [48,42,18,30,38,24,10,22,23]. Starting from this observation, we extend various experimental and theoretical results on HFE to the more general class of polynomial systems arising from a Weil descent. Our results suggest that the degrees of regularity of these systems are only sligthly larger than the degrees of their equations, essentially as small as they could be.

Following [34], we subsequently study an elliptic curve discrete logarithm algorithm of Diem [21] in the case of binary fields. Based on our heuristic analysis of polynomial systems arising from a Weil descent, we conjecture that the elliptic curve discrete logarithm problem can be solved over the binary field $\mathbb{F}_{2^n}$ in *subexponential time* $O(2^{c\,n^{2/3}\log n})$, where c is a constant smaller than 2. For n prime, this problem was previously thought to have complexity $O(2^{n/2})$.

Our analysis of polynomial systems arising from a Weil descent can also be applied to the factorization problem in $SL(2, \mathbb{F}_{2^n})$, to HFE and to other discrete logarithm problems. These applications will be discussed in an extended version of this paper [49]. Although we focus on characteristic 2 in this paper, most of our results can be easily extended to other characteristics.

Outline

The remaining of this paper is organized as follows. Section 2 contains most of the notations and definitions used in the paper. Section 3 provides general background on algebraic cryptanalysis with Gröbner bases. Section 4 contains our new analysis of polynomial systems arising from a Weil descent. The application to Diem's algorithm is detailed in Section 5 and Section 6 concludes the paper.

2   Definitions and Notations

We mostly follow the notations introduced in [33]. For any "small" prime p and any $n \in \mathbb{Z}$, we write $\mathbb{F}_{p^n}$ for the finite field with p^n elements. We see the field $\mathbb{F}_{p^n}$ as an n-dimensional vector space over $\mathbb{F}_p$ and we let $\{\theta_1, \ldots, \theta_n\}$ be a basis for $\mathbb{F}_{p^n}/\mathbb{F}_p$. With some abuse of notations, we use bold letters for all elements, variables and polynomials over $\mathbb{F}_{p^n}$ and normal letters for all elements, variables and polynomials over $\mathbb{F}_p$. If $x_1, \ldots, x_N$ are variables defined over a field $\mathbb{K}$, we write $R := \mathbb{K}[x_1, \ldots, x_N]$ for the ring of polynomials in these variables. Given a set of polynomials $f_1, \ldots, f_\ell \in R$, the *ideal* $I(f_1, \ldots, f_\ell) \subset R$ is the set of polynomials $\sum_{i=1}^{\ell} g_i f_i$, where, $g_1, \ldots, g_\ell \in R$. We write $\mathrm{Res}_{x_i}(f_1, f_2)$ for the *resultant* of $f_1, f_2 \in R$ with respect to the variable x_i. A *monomial* of R is a power product $\prod_{i=1}^{k} x_i^{e_i}$ where $e_i \in \mathbb{N}$. A *monomial ordering* for R is an ordering $>$ such that $m_1 > m_2 \Rightarrow m_1 m_3 > m_2 m_3$ for any monomials m_1, m_2, m_3 and $m > 1$ for any monomial m. The *leading monomial* $LM(f)$ of a polynomial $f \in R$ for a given ordering is equal to its largest monomial according to the ordering. Its *leading term* is the corresponding term. For any polynomial $f \in R$, we denote the set of monomials of f by $\mathtt{Mon}(f)$. We measure the memory and time complexities of algorithms by respectively the number of bits and bit operations required. Actual experimental results are given in megabytes and seconds. We write O for the "big O" notation: given two functions f and g of n, we say that $f = O(g)$ if there exist $N, c \in \mathbb{Z}^+$ such that $n > N \Rightarrow f(n) \le cg(n)$. Similarly, we write o for the "small o" notation: given two functions f and g of n, we say that $f = o(g)$ if for any $\epsilon > 0$, there exists $N \in \mathbb{Z}$ such that for any $n > N$, we have $|f(n)| \le \epsilon |g(n)|$. Finally, we write ω for the *linear algebra constant*. Depending on the algorithm used for linear algebra, we have $2.376 \le \omega \le 3$.

3 Background on Polynomial System Resolution

Let R be a polynomial ring and let $>$ be a fixed monomial ordering for this ring. A *Gröbner basis* [13,19] of an ideal $I(f_1, \ldots f_\ell) \subset R$ is a basis $\{f_1', \ldots, f_{\ell'}'\}$ of this ideal such that for any $f \in I(f_1, \ldots f_\ell)$, there exists $i \in \{1, \ldots, \ell'\}$ such that $LT(f_i')|LT(f)$. The first Gröbner basis algorithm was provided by Buchberger in his PhD thesis [13]. Lazard [44] later observed that computing a Gröbner basis is essentially equivalent to performing linear algebra on *Macaulay matrices* at a certain degree.

Definition 1 (Macaulay Matrix [45,46]). *Let R be a polynomial ring over a field K and let $\mathcal{B}_d := \{m_1 > m_2 > \cdots\}$ be the sorted set of all monomials of degree $\le d$ for a fixed monomial ordering. Let $F := \{f_1, \ldots, f_\ell\} \subset R$ be a set of polynomials of degrees $\le d$. For any $f_i \in F$ and $t_j \in \mathcal{B}_d$ such that $\deg(f_i) + \deg(t_j) \le d$, let $g_{i,j} := t_j f_i$ and let $c_{i,j}^k \in K$ be such that $g_{i,j} = \sum_{m_k \in \mathcal{B}} c_{i,j}^k m_k$. The Macaulay matrix $\mathcal{M}_d(F)$ of degree d is a matrix containing all the coefficients $c_{i,j}^k$, such that each row corresponds to one polynomial $g_{i,j}$ and each column to one monomial $m_k \in \mathcal{B}_d$.*

The idea behind Lazard's observation is *linearization*: new equations for the ideal are constructed by algebraic combinations of the original equations, every monomial term appearing in the new equations is treated as an independent new variable, and the system is solved with linear algebra. Gröbner basis algorithms like F_4 [26] and F_5 [27] successively construct Macaulay matrices of increasing sizes and remove linear dependencies in the rows until a Gröbner basis is found. Moreover, they optimize the computation by avoiding monomials t_j that would produce trivial linear combinations such as $f_1 f_2 - f_2 f_1 = 0$. The complexity of this strategy is determined by the cost of linear algebra on the largest Macaulay matrix occuring in the computation.

The degree of the largest Macaulay matrix appearing in a Gröbner basis computation with the algorithm F5 is called the *degree of regularity* D_{reg}. For a "generic" sequence of polynomials $f_1, \ldots, f_\ell \in R$ (with $\ell \leq n$), this degree is equal to $1 + \sum_{i=1}^{\ell}(\deg(f_i) - 1)$ [6]. The degree of regularity can be precisely estimated in the case of *regular* and *semi-regular* sequences [6,8] and (assuming a variant of Fröberg conjecture) in a few other cases [28,11]. However, precisely estimating this value for other classes of systems (in particular for the various structured systems appearing in cryptanalysis problems) may be a very difficult task. In practice, the degree of regularity may often be approximated by the first degree at which a non trivial *degree fall* occurs during a Gröbner basis computation.

Definition 2. *Let R be a polynomial ring over a field K and let $F := \{f_1, \ldots, f_\ell\} \subset R$. The* first fall degree *of F is the smallest degree $D_{firstfall}$ such that there exist polynomials $g_i \in R$ with $\max_i(\deg(f_i) + \deg(g_i)) = D_{firstfall}$, satisfying $\deg(\sum_{i=1}^{\ell} g_i f_i) < D_{firstfall}$ but $\sum_{i=1}^{\ell} g_i f_i \neq 0$.*

We have $D_{reg} \geq D_{firstfall}$. For many classes of polynomial systems, the two definitions lead to very close numbers. Although this is not true in general (counterexamples can be easily produced), it seems to be true for "random systems" and "most real-life systems of equations" [38, p. 350] including HFE and its variants [30,38,24,22,23,11]. This can intuitively be explained by the observation that an extremely large number of relations with a degree fall occur at the degree $D_{firstfall}$ or the degree $D_{firstfall}+1$ in these contexts, and these low degree relations can in turn be combined to produce lower degree relations [24, p. 561], until a Gröbner basis is finally found. In fact, the *first fall degree* has even sometimes been called *degree of regularity* in the cryptography community [24,22,23].

Many polynomial systems arising in cryptanalysis are very far from generic ones. In fact, their special structures often induce lower degrees of regularity, hence much better time complexities. Gröbner basis techniques have successfully attacked many cryptosystems, including HFE and its variants [48,42,30,38,10,22,23], the Isomorphism of Polynomials [32,12] and some McEliece variants [31]. In many cases, the resolution of these systems could be accelerated using *dedicated* Gröbner basis algorithms that exploited the particular structures. As was first pointed out in [33,34], this is also the case for polynomial systems arising from a Weil descent.

4   Polynomial Systems Arising from a Weil Descent

Let n, n', m be positive integers and let V be a vector subspace of $\mathbb{F}_{2^n}/\mathbb{F}_2$ with dimension n'. Let $\mathbf{f} \in \mathbb{F}_{2^n}[\mathbf{x_1}, \ldots, \mathbf{x_m}]$ be a multivariate polynomial with degrees bounded by $2^t - 1$ with respect to all variables. In [33,34], Faugère et al. considered the following problem:

$$\text{Find } \mathbf{x_i} \in V, i = 1, \ldots, m, \text{ such that } \mathbf{f}(\mathbf{x_1}, \ldots, \mathbf{x_m}) = \mathbf{0}. \qquad (1)$$

The constraints $\mathbf{x_i} \in V, i = 1, \ldots, m$ are called *linear constraints*. From now on, we assume that $mn' \approx n$ such that Problem (1) has about one solution on average. We also assume $n' \geq t$. The *multilinear* case $(t = 1)$ was first considered in [33] and later extended in [34].

Following [33,34], Problem (1) can be reduced to a system of polynomial equations. Let $\{\theta_1, \ldots, \theta_n\}$ be a basis of $\mathbb{F}_{2^n}$ over $\mathbb{F}_2$ and let $\{\mathbf{v_i}|i = 1, \ldots, n'\}$ be a basis of V over $\mathbb{F}_2$. We define $m \cdot n'$ variables x_{ij} over $\mathbb{F}_2$ such that $\mathbf{x_i} = \sum_{j=1}^{n'} x_{ij}\mathbf{v_j}$ and we group them into m *blocks of variables* $X_i := \{x_{ij}|j = 1, \ldots, n'\}$. By substituting each $\mathbf{x_i}$ in $\mathbf{f}$, decomposing in the basis $\{\theta_1, \ldots, \theta_n\}$ and reducing by the *field equations* $x_{ij}^2 - x_{ij} = 0$, we obtain $\mathbf{0} = \mathbf{f}(\mathbf{x_1}, \ldots, \mathbf{x_m}) = \mathbf{f}\left(\sum_{j=1}^{n'} x_{1j}\mathbf{v_j}, \ldots, \sum_{j=1}^{n'} x_{mj}\mathbf{v_j}\right) = [\mathbf{f}]_1^{\downarrow}\theta_1 + \ldots + [\mathbf{f}]_n^{\downarrow}\theta_n$ for some $[\mathbf{f}]_1^{\downarrow}, \ldots, [\mathbf{f}]_n^{\downarrow} \in \mathbb{F}_2[x_{11}, \ldots, x_{mn'}]$ that depend on $\mathbf{f}$ and on the vector subspace V. Problem (1) can therefore be reformulated as finding a solution to the (algebraic) system

$$[\mathbf{f}]_1^{\downarrow} = 0, \ldots, [\mathbf{f}]_n^{\downarrow} = 0. \tag{2}$$

Due to the bounds on the degrees of $\mathbf{f}$, this system has a *block structure*: the degrees of all polynomials $[\mathbf{f}]_k^{\downarrow}$ are bounded by t with respect to all blocks of variables. The resolution of System (2) can therefore be greatly accelerated using *block-structured* Gröbner basis algorithms [29,33,34].

Link to HFE. In this paper, we observe that a particular instance of Problem (1) had previously been studied in the cryptography literature. Indeed, the well-known problem of inverting HFE [48,30,38] leads to a particular instance of System (2), where the polynomial $\mathbf{f}$ is univariate $(m = 1)$ and the linear constraints are trivial $(V = \mathbb{F}_{2^n})$.[1] Interestingly, although the polynomial $\mathbf{f}$ used in HFE has a particular shape (it leads to quadratic equations over $\mathbb{F}_2$), we will see that this shape has generically little influence on the complexity of Problem (1).

Ten years of research on HFE systems have shown that their degrees of regularity are abnormally low compared to generic systems, resulting in very efficient attacks. Although no definitive proof of these results has been published yet, the experimental observations of [30] are now being supported by theoretical evidence such as the isolation of a subsystem with less variables [38], the existence of many low degree equations [17], first fall degree computations [22,24] and complexity results on the MinRank problem [11]. In this paper, we generalize some of these results to polynomial systems arising from a Weil descent.

Experimental Observations. We start our analysis of these systems with an experimental study of their degree of regularity for various parameters n, m, n', t. For each set of parameters, we generate a random vector space V of dimension n' and a random multivariate polynomial $\mathbf{f}(\mathbf{x_1}, \ldots, \mathbf{x_m})$ with degree bounded by $2^t - 1$ with respect to each variable. We then perform a Weil descent on this

---

[1] In HFE contexts, the attacker is not given $\mathbf{f}$ but only a "hidden" version of System (2). This can be ignored in the complexity analysis of Gröbner basis algorithms since the hiding transformation only consists of a linear combinations of the equations and a linear change of variables [48,38].

polynomial and we append the field equations to the system. Finally, we apply the Magma function *Groebner* to the result and we collect the maximal degree D reached during the computation, as given by the *Verbose* output of the Magma function. We repeat each experiment 100 times.

Table 1. Average maximal degree reached in Gröbner Basis experiments, average computation time (in seconds) and maximal memory requirements (in MB) for random polynomials

t	n	n'	m	$mt+1$	D_{av}	Time	Mem.	t	n	n'	m	$mt+1$	D_{av}	Time	Mem.
1	6	3	2	3	3.1	0	10	2	6	3	2	5	5.1	0	10
1	6	2	3	4	3.8	0	10	2	6	2	3	7	6.7	0	10
1	8	4	2	3	3.0	0	11	2	8	4	2	5	5.1	0	11
1	12	6	2	3	3.6	0	11	2	9	3	3	7	7.2	0	12
1	12	4	3	4	4.2	0	11	2	12	4	3	7	7.1	1	38
1	12	3	4	5	5.3	0	14	2	12	3	4	9	9.3	2	95
1	12	2	6	7	7.4	1	23	2	15	5	3	7	7.0	12	263
1	15	5	3	4	4.1	5	20	2	16	8	2	5	5.1	13	36
1	15	3	5	6	6.3	7	114	3	6	3	2	7	6.6	0	10
1	16	8	2	3	3.0	14	25	3	12	6	2	7	7.0	1	31
1	16	4	4	5	5.3	16	98	3	12	4	3	10	10.1	9	70
1	16	2	8	9	9.6	69	3388	3	12	3	4	13	12.6	70	113
1	18	9	2	3	3.0	85	74	3	15	5	3	10	10.0	118	2371
1	18	6	3	4	4.1	86	89	3	16	8	2	7	7.0	23	253
1	18	3	6	7	7.4	233	5398	3	16	4	4	13	13.2	1891	20135
1	20	10	2	3	3.0	487	291	4	8	4	2	9	8.7	1	11
1	20	5	4	5	6.2	515	733	4	12	4	3	13	12.6	199	116
1	20	4	5	6	6.2	669	3226	4	15	5	3	13	13.1	2904	6696

Table 1 reports the average value of D for these experiments, as well as the average computation time and the maximal memory used (all experiments were done on an Intel Xeon CPU X5500 processor running at 2.67 GHz, with 24 GB RAM). As is often the case in Gröbner basis computations, our experiments were limited more by the memory requirements than by the computation time.

For all parameter sets, the maximal degrees occuring during Gröbner basis computations were much smaller than the degrees of regularity of regular or semi-regular systems with the same degrees. In fact, our experiments suggest that the degree of regularity of System (2) is not much higher than the value $mt+1$. In other words since the original equations have degree mt, the degree of regularity is essentially as small as it could be. The even lower values obtained for all parameter sets such that $t = n'$ can be explained by a probable degeneracy in the degrees of the equations. Taking $m = 1$, we recover known experimental results on HFE [30].

Heuristic Upper Bound on D_{reg}. As a first step towards explaining these experimental results, we follow Granboulan et al. [38] and we bound of the degree of regularity of System (2) from above by the degree of regularity of a smaller system with a lower number of variables. We now suppose that $\{\theta_1, \ldots, \theta_n\}$ is a normal basis of $\mathbb{F}_{2^n}$ over $\mathbb{F}_2$, such that $\theta_i := \theta^{2^{i-1}}$ for some $\theta \in \mathbb{F}_{2^n}$. Let $v_{ij} \in \mathbb{F}_2$ such that $\mathbf{v_i} = \sum_{j=1}^n v_{ij}\theta_j$. We define nm auxiliary binary variables y_{ij} such that $\mathbf{x_i} = \sum_{j=1}^n y_{ij}\theta_j$. Proceeding to a Weil descent as above, we obtain a new system[2]

$$[\mathbf{f}]_1^{\downarrow y} = 0, \ldots, [\mathbf{f}]_n^{\downarrow y} = 0 \tag{3}$$

in the variables y_{ij}, to which we add $m(n + n')$ field equations $y_{ij}^2 - y_{ij} = 0$ and $x_{ij}^2 - x_{ij} = 0$, as well as mn linear equations $y_{ij} = \sum_{k=1}^n x_{ik}v_{kj}$ modeling the linear constraints. The resulting system of $m(n + n')$ variables and $n + m(n + n') + mn$ equations is equivalent to System (2) (with the field equations), hence they have the same degree of regularity.

Following Granboulan et al. [38], we perform additional modifications on this system to obtain a new system with less variables and higher or equal degree of regularity. First, we observe that linear equations do not contribute to the degree of regularity and can therefore be removed without affecting it. The resulting system is composed of $n + mn$ equations containing only the variables y_{ij} and mn' field equations $x_{ij}^2 - x_{ij} = 0$. Without decreasing the degree of regularity, we can focus on the first part containing Equations (3) and the field equations $y_{ij}^2 - y_{ij} = 0$.

In the next step, we observe that the degree of regularity of this system is not affected if we see the variables y_{ij} over $\mathbb{F}_{2^n}$ rather than over $\mathbb{F}_2$. Thanks to the field equations, the set of solutions is not affected by this change either. We then apply an invertible linear transformation on Equations (3), defined by $F_i := \sum_{j=1}^n \theta^{2^{i+j}} [\mathbf{f}]_j^{\downarrow y}$ for $i = 1, \cdots, n$. This transformation implies $F_i = F_1^{2^{i-1}}$. Finally, we perform a linear change of variables defined by $z_{ij} := \sum_{k=1}^n \theta^{2^{j+k-1}} y_{ik}$ for $i = 1, \ldots, m$, and $j = 1, \cdots, n$. Since this corresponds to setting $z_{i1} = \mathbf{x_i}$, $z_{i2} = \mathbf{x_i}^2$, ..., $z_{i,n} = \mathbf{x_i}^{2^{n-1}}$, each F_k only depends (linearly) on z_{ij}, $k \leq j \leq t+k-1$. A last linear transformation changes the field equations into $z_{ij}^2 = z_{i,j+1}$ and $z_{i,n}^2 = z_{i,1}$.

Since $F_2 = F_1 \cdot F_1$ modulo the field equations, the polynomial F_2 can be expressed at the degree $2mt$ as an algebraic combination of F_1 and the field equations. Similarly, all polynomials F_i, $i \geq 2$ can be recovered at degree $2mt$ from algebraic combination of F_1 and the field equations. Therefore, the degree of regularity of the original system is smaller than the maximum of $2mt$ and the degree of regularity of the system $\{F_1 = 0; z_{ij}^2 = z_{i,j+1}, i = 1, \ldots, m, j = 1, \ldots, n-1; z_{i,n}^2 = z_{i,1}, i = 1, \ldots, m\}$. Finally like [38], we bound this last degree by the degree of regularity of the subsystem $\{F_1 = 0; z_{ij}^2 = z_{i,j+1}, i = 1, \ldots, m, j = 1, \ldots, t-1\}$. Assuming that this system behaves like a generic system with the same degrees and the same number of variables[3], its degree of

---

[2] We add a subscript y to the arrows in System (3) to stress that the Weil descent is done on the y_{ij} variables and to distinguish this system from System (2).

[3] A similar assumption of semi-regularity is needed in [38] to apply Bardet's theorem.

regularity can be bounded by $m(2t - 1)$ using Macaulay's bound. Under this heuristic assumption, we conclude that the degree of regularity of System (3) is bounded by $2mt$.

We point out that the value $2mt$ is already much below the degree of regularity of a generic system of equations (or even a generic *binary* system of equations) with the same degrees [6,7]. Still, our experiments suggest that this bound is not even tight. A tighter bound can be obtained with a seemingly stronger (yet "classical") heuristic assumption.

First Fall Degree. An important characteristic of HFE systems is the existence of many algebraic combinations of the equations that have a degree lower than it would be expected for a generic system. Similar low degree equations were identified for System (2). More precisely, Faugère et al. [33,34] showed that for any monomial $\mathbf{m} \in \mathbb{F}_{2^n}[\mathbf{x_1}, \dots, \mathbf{x_{n'}}]$, the equations obtained by applying a Weil descent on the polynomial $\mathbf{m}f$ are algebraic combinations of the equations of System (2) that produce a degree fall. By the way they are constructed, the existence of these equations is very specific to polynomial systems arising from a Weil descent. For $\mathbf{m} := \mathbf{x_1}$, we immediately deduce:

Proposition 1. *The first fall degree of System* (2) *is at most* $mt + 1$.

This proposition provides a heuristic explanation for the degrees of regularity observed above since the first fall degree is often a good approximation of the degree of regularity. As recalled in Section 3, this heuristic assumption is "classical" in algebraic cryptanalysis, and it has in particular been verified for various HFE-like systems [38,24,22].

Assumption 1. *Let* $n, m, t, n' \in \mathbb{Z}$. *Let* $\mathbf{f}$ *be generated as in our experiments. For all but a negligible fraction of the resulting systems, we have* $D_{reg} = D_{firstfall} + o(D_{firstfall})$.

The assumption intuitively makes sense for System (2) since not only one but many degree falls are occuring at degree $D_{firstfall}$ and the next ones (each monomial $\mathbf{m}$ leads to new degree falls).

Heuristic Complexity Bounds for Problem (1). Given the degree of regularity, the complexity of Problem 1 simply follows from the cost of linear algebra.

Proposition 2. *If Assumption 1 holds, Problem 1 can be solved with* standard *Gröbner basis algorithms (like F4 or F5) in time* $O(n^{\omega D})$ *and memory* $O(n^{2D})$, *where* ω *is the linear algebra constant and* $D \approx mt$.

In the univariate case, this estimation reduces to $D \approx t$ which perfectly matches known cryptanalysis results on HFE algebraic systems [30,38]. Interestingly, the special shape of HFE polynomials (they deploy to *quadratic* equations over $\mathbb{F}_2$) seems to have no impact on the degree of regularity (although further restrictions on the shape may have an impact as pointed out in [22]). In the multilinear case, the estimation provided by Proposition 2 becomes $D \approx m$ which matches the experimental data of [33].

As observed in [33,34], the block structure of System (2) can be exploited to accelerate its resolution.

Proposition 3. *If Assumption 1 holds, Problem 1 can be solved with* block *Gröbner basis algorithms in time* $O((n')^{\omega D})$ *and memory* $O((n')^{2D})$, *where* ω *is the linear algebra constant and* $D \approx mt$.

Additional heuristic methods like hybrid approaches (consisting in mixing exhaustive search and polynomial system resolution [52,9]) may lead to substantial complexity improvements in practice, as was described in [33] for the multilinear case.

5   Index Calculus for Elliptic Curves

We now turn to the main application (so far) of Problem (1). As pointed out in [34], an instance of Problem (1) appears in the relation search step of an index calculus algorithm for elliptic curves proposed by Diem [21]. Given a cyclic (additive) group G, a generator P of this group and another element Q of G, the discrete logarithm problem asks for computing an integer k such that $Q = kP$. Groups typically used in cryptography include the multiplicative groups of finite fields, groups of points on elliptic curves and hyperelliptic curves and Jacobians of higher genus curves. Index calculus algorithms [43,25] with *subexponential* complexities have long been obtained for the multiplicative groups of finite fields [1,16,2,5,39] and more recently for the Jacobian groups of hyperelliptic curves [3,36,35].

In 2004, Semaev introduced his *summation polynomials* and identified their potential application to build index calculus algorithms on elliptic curves [51] over prime fields $\mathbb{F}_p$. These ideas were independently extended by Gaudry [37] and Diem [20] to elliptic curves over composite fields $\mathbb{F}_{p^n}$. Following this approach, Gaudry [37] and later Joux and Vitse [40,41] obtained index calculus algorithms running faster than generic algorithms for any p and any $n \geq 3$. On the other hand, Diem [20,21] identified some families of curves with a subexponential time index calculus algorithm by letting p and n grow simultaneously in an appropriate way. As far as was known at the moment, the two families of elliptic curves recommended by standards [47] (elliptic curves over prime fields $\mathbb{F}_p$ or over binary fields $\mathbb{F}_{2^n}$ with n prime) remained immune to these attacks. In 2012, Faugère et al. [34] observed that the computation of the relations in an algorithm of Diem for binary fields [21] could be reduced to special instances of Problem (1).

Diem's Variant of Index Calculus. Let K be a finite field and let E be an elliptic curve over K defined by the equation $E : y^2 + xy = x^3 + \mathbf{a_2}x^2 + \mathbf{a_6}$ for some $\mathbf{a_2}, \mathbf{a_6} \in \mathbb{F}_{2^n}$. Semaev's *summation polynomials* $\mathbf{S_r}$ are multivariate polynomials satisfying $\mathbf{S_r}(\mathbf{x_1}, \ldots, \mathbf{x_r}) = \mathbf{0}$ for some $\mathbf{x_1}, \ldots, \mathbf{x_r} \in \bar{K}$ if and only if there exist $y_1, \ldots, y_r \in \bar{K}$ such that $(x_i, y_i) \in E(\bar{K})$ and $(x_1, y_1) + \cdots + (x_r, y_r) = P_\infty$ [51]. The summation polynomials can be recursively computed as $S_2(x_1, x_2) := x_2 + x_1$, $S_3(x_1, x_2, x_3) := x_1^2 x_2^2 + x_1^2 x_3^2 + x_1 x_2 x_3 + x_2^2 x_3^2 + a_6$ and for any $r \geq 4$, any $k, 1 \leq k \leq r - 3$, $S_r(x_1, \ldots, x_r) := \mathrm{Res}_X (S_{r-k}(x_1, \ldots, x_{m-k-1}, X), S_{k+2}(x_{r-k}, \ldots, x_r, X))$. For $r \geq 2$, the polynomial S_r is symmetric and has degree 2^{r-2} in every variable x_i [51].

Summation polynomials were used by Gaudry [37], Joux and Vitse [40] and Diem [20,21] to compute relations in index calculus algorithms for elliptic curves over composite fields. The following variant is an adaptation of Diem [21].

1. **Factor Basis definition.** Fix two integers $m, n' < n$ with $mn' \approx n$ and a vector space $V \subset \mathbb{F}_{2^n} / \mathbb{F}_2$ of dimension n'. Let $\mathcal{F}_V := \{(x, y) \in E(K) | x \in V\}$ be the *factor basis*.
2. **Relation search.** Find about $2^{n'}$ relations $a_i P + b_i Q = \sum_{j=1}^m P_{ij}$ with $P_{ij} \in \mathcal{F}_V$. For each relation,
   (a) Compute $R_i := a_i P + b_i Q$ for random integers a_i, b_i.
   (b) Solve Semaev's polynomial $S_{m+1}(x_1, \ldots, x_m, (R_i)_x)$ with the constraints $x_i \in V$.
   (c) If there is no solution, go back to (a).
3. **Linear Algebra.** Perform linear algebra on the relations to recover the discrete logarithm value.

In previous works [37,20,21,40], a Weil descent was applied to Semaev's polynomials and the resulting systems were solved with resultants or Gröbner basis algorithms. In these works, the complexity of the relation search step was derived from the complexity of solving generic systems. However as pointed out in [33,34] and further demonstrated in Section 4 of the present paper, polynomial systems arising from a Weil descent are very far from generic ones.

A New Complexity Analysis. We now revisit Diem's algorithm [21] and its analysis by [34] in accordance with our new analysis of Problem (1). Let n, m, n' be integer numbers. Before starting Diem's algorithm, the $(m+1)$th summation polynomial must be computed. Using Collins' evaluation/interpolation method [15] for the resultant, this can be done in time approximately 2^{t_1} where[4] $t_1 \approx m(m + 1)$. We then compute about $2^{n'}$ relations. To obtain these relations, we solve special instances of Problem (1) where $f(x_1, \ldots, x_m) := S_{m+1}(x_1, \ldots, x_m, (a_i P + b_i Q)_x)$ has degree 2^{m-1} with respect to every variable. Since Semaev's polynomials are clearly not random ones, we perform additional experiments.

In our experiments, we apply Diem's algorithm to a randomly chosen binary curve $E : y^2 + xy = x^3 + a_2 x^2 + a_6$ defined over $\mathbb{F}_{2^n}$, where $n \in \{11, 17\}$. We first fix $m \in \{2, 3\}$ and $n' := \lceil n/m \rceil$. We then generate a random vector space V of dimension n' and a random point R on the curve such that f has solutions. As in Section 4, we finally use the *Groebner* function of Magma to solve Semaev's

---

[4] To compute S_{m+1}, we apply Collins' algorithm on S_k where $k = \lceil \frac{m+3}{2} \rceil$. This polynomial has degree $2^{\lceil (m-1)/2 \rceil}$ in each variable. Following Collins, Theorem 9, we have $t_1 \leq 2(m + 1)m/2 = m(m + 1)$.

Table 2. Average maximal degree reached in Gröbner Basis experiments, average computation time (in seconds) and maximal memory requirements (in MB) for Semaev polynomials. (R): Random curves. (K): Koblitz curves.

E	n	n'	m	t	$mt+1$	D_{av}	Time	Mem.	E	n	n'	m	t	$mt+1$	D_{av}	Time	Mem.
K	11	6	2	2	5	3.0	0	11	R	11	6	2	2	5	3.0	0	11
K	11	4	3	3	10	7.1	1	15	R	11	4	3	3	10	7.1	1	15
K	17	9	2	2	5	4.0	0	15	R	17	9	2	2	5	4.0	0	16
K	17	6	3	3	10	7.2	132	2133	R	17	6	3	3	10	7.1	130	2136

equation $\mathbf{S_{m+1}}(\mathbf{x_1}, \ldots, \mathbf{x_m}, R_x) = \mathbf{0}$ with the linear constraints. We repeat this experiment 100 times for each parameter set, then we repeat all our experiments with the Koblitz curve $E : y^2 + xy = x^3 + x^2 + 1$. The average value of the maximal degrees reached during the computation, the average computation time and the maximal memory requirements are reported in Table 2.

In all cases, the maximal degrees reached in the computations were even *below* the first fall degree bound given by Proposition (1). This phenomenon is probably due to the sparsity of Semaev's polynomials and will be exploited in future work (in particular, the degree of $\mathbf{S_{m+1}}$ with respect to every variable is 2^{m-1} but bounded by $2^m - 1$ in the analysis of Section 4). From now on in the analysis, we ignore this difference and analyze Semaev's polynomials as the random polynomials of Section 4.

Assumption 2. *Assumption 1 still holds if* $\mathbf{f}$ *is generated from Semaev's polynomials as in the experiments of this section.*

Under Assumption (2), Step 2(b) of Diem's algorithm can be solved using a dedicated Gröbner basis algorithm taking advantage of the block structure, in a time $(n')^{\omega D}$, where $D \approx (m^2 + 1)$ and ω is the linear algebra constant. Once the x components of a relation have been computed, the y components can be found by solving m quadratic equations and testing each possible combination of the solutions. This requires a time roughly 2^m, that can be neglected. On average, the probability that a point $R_i := a_i P + b_i Q$ can be written as a sum of m points from the factor basis can be heuristically approximated by $\frac{2^{mn'-n}}{m!}$ [21]. Assuming $mn' \approx n$, the total cost of the relation search step can therefore be approximated by 2^{t_2}, where $t_2 \approx m \log m + n' + \omega(m^2 + 1) \log n'$.

The last step of Diem's algorithm consists in (sparse) linear algebra on a matrix of rank about $2^{n'}$ with about m elements of size about n bits per row. This step takes a time approximately equal to $mn2^{\omega'n'} = 2^{t_3}$, where $t_3 \approx \log m + \log n + \omega'n'$ and ω' is the *sparse* linear algebra constant. If Assumption (2) holds and if $mn' \approx n$, the total time taken by Diem's algorithm can be estimated by $T := 2^{t_1} + 2^{t_2} + 2^{t_3}$, where t_1, t_2, t_3 are defined as above.

On the Hardness of ECDLP in Characteristic 2. We now evaluate the hardness of the elliptic curve discrete logarithm problem over the field $\mathbb{F}_{2^n}$ for "small" values of n. In our estimations, we use $\omega = \log(7)/\log(2)$ and $\omega' = 2$.

Table 3. Complexity estimates for Diem's algorithm in characteristic 2

n	m	n'	t_1	t_2	t_3	t_{max}	n	m	n'	t_1	t_2	t_3	t_{max}
50	2	25	6	92	57	92	2000	4	500	20	936	1013	1013
100	2	50	6	131	108	131	2500	5	500	30	1166	1014	1166
160	2	80	6	171	168	171	5000	6	833	42	1857	1682	1857
200	2	100	6	195	209	209	10000	7	1429	56	2919	2873	2919
500	3	167	12	379	344	379	20000	9	2222	90	4810	4462	4810
1000	4	250	20	638	512	638	50000	12	4167	156	9105	8353	9105

We consider $n \in \{50, 100, 160, 200, 500, 1000, 2000, 2500, 5000, 10^4, 2 \cdot 10^4, 5 \cdot 10^4, 10^5, 2 \cdot 10^5, 5 \cdot 10^5, 10^6\}$ and $m \in \{2, \ldots, n/2\}$. For every pair of values, we compute values t_1, t_2 and t_3 as above. Finally, we approximate the total running time of Diem's algorithm by $2^{t_{max}}$ where $t_{max} := \max(t_1, t_2, t_3)$. For every value of n, Table 3 presents the data corresponding to the value m for which t_{max} is minimal. We point out that the numbers obtained here have to be taken cautiously since they all rely on Assumption 2 and involve some approximations.

According to our estimations, Diem's version of index calculus (together with a sparse Gröbner basis algorithm) beats generic algorithms for any $n \geq N$, where N is an integer close to 2000. An actual attack for current cryptographically recommended parameters ($n \approx 160$) seems to be out of reach today, but the numbers in [34] suggest that medium-size parameters could be reachable with additional Gröbner basis heuristics like the hybrid method [9]. Large prime variations [35] of Diem's algorithm may also lead to substantial improvements in practice. This will be investigated in further work.

Letting n grow and fixing $n' := n^\alpha$ and $m := n^{1-\alpha}$ for a positive constant $\alpha < 1$, we obtain

$$t_1 \approx n^{2(1-\alpha)},$$
$$t_2 \approx (1-\alpha)n^{1-\alpha}\log n + n^\alpha + \alpha\omega n^{2(1-\alpha)}\log n,$$
$$t_3 \approx (2-\alpha)\log n + \omega' n^\alpha$$

Taking $\alpha := 2/3$, the relation search dominates the complexity of the index calculus algorithm and we deduce the following result.[5]

Proposition 4. *Under Assumption 2, the discrete logarithm problem over $\mathbb{F}_{2^n}$ can asymptotically be solved in time $O(2^{cn^{2/3}\log n})$, where $c := 2\omega/3$ and ω is the linear algebra constant.*

In particular if the Gaussian elimination algorithm is used for linear algebra, we have $\omega = 3$ and $c = 2$. We stress that Proposition 4 holds even when n is prime. Until now, the best complexity estimates obtained in that case corresponded to generic algorithms that run in time $2^{n/2}$.

---

[5] Note that the weaker bound $D_{reg} \leq 2mt$ derived in Section 4 with Macaulay's bound also leads to a subexponential complexity but with a constant $c = 4\omega/3$.

6   Conclusion and Perspectives

In this paper, we revisited the complexity of solving polynomial systems arising from a Weil descent, a class of polynomial systems previously introduced by Faugère et al. [33,34]. We observed that these systems can be seen as natural extensions of HFE systems and we generalized various results on HFE. Based on experimental results and heuristic arguments, we conjectured that the degree of regularity of these systems are only slightly larger than their original degrees, and we deduced new heuristic bounds on their resolution. Interestingly, our bounds nicely generalize previous bounds on HFE.

The most proeminent consequence of our analysis so far concerns the elliptic curve discrete logarithm problem (ECDLP) over binary fields. Indeed, our heuristic analysis suggests that ECDLP can be solved in *subexponential* time $O(2^{c\,n^{2/3}\log n})$ over the binary field $\mathbb{F}_{2^n}$, where c is a constant smaller than 2. This complexity is obtained with an index calculus algorithm due to Diem [20] and a block-structured Gröbner basis algorithm. In practice, our estimations predict that the resulting algorithm is faster than generic algorithms (previously thought to be the best algorithms for this problem) for any n larger than N, where N is an integer approximately equal to 2000. In particular, binary elliptic curves of currently recommended sizes ($n \approx 160$) are not immediately threatened.

Our complexity estimates are based on heuristic assumptions that differ from other index calculus algorithms, but are common in algebraic cryptanalysis. The polynomial systems appearing in the cryptanalysis of HFE have been intensively studied in the last 15 years, yet we have no definitive proof for their commonly admitted complexity. Our paper broadens the interest of these researches to all polynomial systems arising from a Weil descent and to their various applications. We leave further experimental and theoretical investigation of our heuristic assumptions to further work.

To conclude this paper, we point out that most of our results generalize quite easily to other fields, resulting in comparable asymptotic complexities.

Acknowledgements. We are indebted to Sylvie Baudine and the anonymous reviewers for their help in improving this paper. We also thank Jean-Charles Faugère, Ludovic Perret and Guénaël Renault for their useful comments on a preliminary version of this paper. Finally, Christophe Petit would like to thank Daniel Augot for his hospitality since this paper was partially written at LIX.

References

1. Leonard, M.A.: A subexponential algorithm for the discrete logarithm problem with applications to cryptography (abstract). In: FOCS, pp. 55–60. IEEE (1979)
2. Leonard, M.A.: The function field sieve. In: Adleman, Huang [4], pp. 108–121
3. Leonard, M.A., DeMarrais, J., Huang, M.-D.A.: A subexponential algorithm for discrete logarithms over the rational subgroup of the Jacobians of large genus hyperelliptic curves over finite fields. In: Adleman, Huang, [4] pp. 28–40

4. Huang, M.-D.A., Adleman, L.M. (eds.): ANTS 1994. LNCS, vol. 877. Springer, Heidelberg (1994)
5. Adleman, L.M., Huang, M.-D.A.: Function field sieve method for discrete logarithms over finite fields. Inf. Comput. 151(1-2), 5–16 (1999)
6. Bardet, M.: Etude des systèmes algébriques surdéterminés. Applications aux codes correcteurs et à la cryptographie. PhD thesis, Université Paris 6 (2004)
7. Bardet, M., Faugère, J.-C., Salvy, B.: On the complexity of Gröbner basis computation of semi-regular overdetermined algebraic equations. In: International Conference on Polynomial System Solving, ICPSS, pp. 71–75 (November 2004)
8. Bardet, M., Faugère, J.-C., Salvy, B.: Asymptotic expansion of the degree of regularity for semi-regular systems of equations. In: Gianni, P. (ed.) The Effective Methods in Algebraic Geometry Conference, Mega 2005, pp. 1–14 (May 2005)
9. Bettale, L., Faugère, J.-C., Perret, L.: Hybrid approach for solving multivariate systems over finite fields. Journal of Mathematical Cryptology 3(3), 177–197 (2010)
10. Bettale, L., Faugère, J.-C., Perret, L.: Cryptanalysis of multivariate and odd-characteristic hfe variants. In: Catalano, et al. [14], pp. 441–458
11. Bettale, L., Faugère, J.-C., Perret, L.: Cryptanalysis of HFE, Multi-HFE and Variants for Odd and Even Characteristic. Des. Codes Cryptography, 1–42 (accepted, 2012)
12. Bouillaguet, C., Faugère, J.-C., Fouque, P.-A., Perret, L.: Practical cryptanalysis of the identification scheme based on the isomorphism of polynomial with one secret problem. In: Catalano, et al. [14], pp. 473–493
13. Buchberger, B.: Ein Algorithmus zum Auffinden der Basiselemente des Restklassenringes nach einem nulldimensionalen Polynomideal. PhD thesis, Universität Innsbruck (1965)
14. Catalano, D., Fazio, N., Gennaro, R., Nicolosi, A. (eds.): PKC 2011. LNCS, vol. 6571, pp. 2011–2014. Springer, Heidelberg (2011)
15. Collins, G.: The calculation of multivariate polynomial resultants. Journal of the Association for Computing Machinery 18, 515–522 (1971)
16. Coppersmith, D.: Fast evaluation of logarithms in fields of characteristic two. IEEE Transactions on Information Theory 30(4), 587–593 (1984)
17. Courtois, N.T.: The Security of Hidden Field Equations (HFE). In: Naccache, D. (ed.) CT-RSA 2001. LNCS, vol. 2020, pp. 266–281. Springer, Heidelberg (2001)
18. Courtois, N., Klimov, A., Patarin, J., Shamir, A.: Efficient Algorithms for Solving Overdefined Systems of Multivariate Polynomial Equations. In: Preneel, B. (ed.) EUROCRYPT 2000. LNCS, vol. 1807, pp. 392–407. Springer, Heidelberg (2000)
19. Cox, D., Little, J., O'Shea, D.: Ideals, Varieties, and Algorithms, 1st edn. Springer, Heidelberg (1992)
20. Diem, C.: On the discrete logarithm problem in elliptic curves. Compositio Mathematica 147, 75–104 (2011)
21. Diem, C.: On the discrete logarithm problem in elliptic curves II (2011), http://www.math.uni-leipzig.de/~diem/preprints/dlp-ell-curves-II.pdf
22. Ding, J., Hodges, T.J.: Inverting HFE Systems Is Quasi-Polynomial for All Fields. In: Rogaway, P. (ed.) CRYPTO 2011. LNCS, vol. 6841, pp. 724–742. Springer, Heidelberg (2011)
23. Ding, J., Kleinjung, T.: Degree of regularity for HFE-. IACR Cryptology ePrint Archiv, 2011:570 (2011)
24. Dubois, V., Gama, N.: The Degree of Regularity of HFE Systems. In: Abe, M. (ed.) ASIACRYPT 2010. LNCS, vol. 6477, pp. 557–576. Springer, Heidelberg (2010)
25. Enge, A., Gaudry, P.: A general framework for subexponential discrete logarithm algorithms. Acta Arith. 102(1), 83–103 (2002)

26. Faugère, J.-C.: A new efficient algorithm for computing Gröbner bases (F4). Journal of Pure and Applied Algebra 139(1-3), 61–88 (1999)
27. Faugère, J.-C.: A new efficient algorithm for computing Gröbner bases without reduction to zero (F5). In: Proceedings of the 2002 International Symposium on Symbolic and Algebraic Computation, ISSAC 2002, pp. 75–83. ACM, New York (2002)
28. Faugère, J.-C., Din, M.S.E., Spaenlehauer, P.-J.: Computing loci of rank defects of linear matrices using gröbner bases and applications to cryptology. In: ISSAC, pp. 257–264 (2010)
29. Faugère, J.-C., Din, M.S.E., Spaenlehauer, P.-J.: Gröbner bases of bihomogeneous ideals generated by polynomials of bidegree (1, 1): Algorithms and complexity. J. Symb. Comput. 46(4), 406–437 (2011)
30. Faugère, J.-C., Joux, A.: Algebraic Cryptanalysis of Hidden Field Equation (HFE) Cryptosystems Using Gröbner Bases. In: Boneh, D. (ed.) CRYPTO 2003. LNCS, vol. 2729, pp. 44–60. Springer, Heidelberg (2003)
31. Faugère, J.-C., Otmani, A., Perret, L., Tillich, J.-P.: Algebraic Cryptanalysis of McEliece Variants with Compact Keys. In: Gilbert, H. (ed.) EUROCRYPT 2010. LNCS, vol. 6110, pp. 279–298. Springer, Heidelberg (2010)
32. Faugère, J.-C., Perret, L.: Polynomial Equivalence Problems: Algorithmic and Theoretical Aspects. In: Vaudenay, S. (ed.) EUROCRYPT 2006. LNCS, vol. 4004, pp. 30–47. Springer, Heidelberg (2006)
33. Faugère, J.-C., Perret, L., Petit, C., Renault, G.: New subexponential algorithms for factoring in $SL(2, \mathbb{F}_{2^n})$. Cryptology ePrint Archive, Report 2011/598 (2011), http://eprint.iacr.org/
34. Faugère, J.-C., Perret, L., Petit, C., Renault, G.: Improving the complexity of index calculus algorithms in elliptic curves over binary fields. In: Pointcheval, Johansson [50], pp. 27–44
35. Gaudry, P., Thomé, E., Thériault, N., Diem, C.: A double large prime variation for small genus hyperelliptic index calculus. Math. Comp. 76(257), 475–492 (electronic) (2007)
36. Gaudry, P.: An Algorithm for Solving the Discrete Log Problem on Hyperelliptic Curves. In: Preneel, B. (ed.) EUROCRYPT 2000. LNCS, vol. 1807, pp. 19–34. Springer, Heidelberg (2000)
37. Gaudry, P.: Index calculus for abelian varieties of small dimension and the elliptic curve discrete logarithm problem. J. Symb. Comput. 44(12), 1690–1702 (2009)
38. Granboulan, L., Joux, A., Stern, J.: Inverting HFE Is Quasipolynomial. In: Dwork, C. (ed.) CRYPTO 2006. LNCS, vol. 4117, pp. 345–356. Springer, Heidelberg (2006)
39. Joux, A., Lercier, R.: The Function Field Sieve in the Medium Prime Case. In: Vaudenay, S. (ed.) EUROCRYPT 2006. LNCS, vol. 4004, pp. 254–270. Springer, Heidelberg (2006)
40. Joux, A., Vitse, V.: Elliptic Curve Discrete Logarithm Problem over Small Degree Extension Fields. Application to the static Diffie-Hellman problem on $E(\mathbb{F}_{q^5})$. Cryptology ePrint Archive, Report 2010/157. Journal of Cryptology (2010), http://eprint.iacr.org/
41. Joux, A., Vitse, V.: Cover and decomposition index calculus on elliptic curves made practical - application to a previously unreachable curve over $\mathbb{F}_{p^6}$. In: Pointcheval, Johansson [50], pp. 9–26
42. Kipnis, A., Shamir, A.: Cryptanalysis of the HFE Public Key Cryptosystem by Relinearization. In: Wiener, M. (ed.) CRYPTO 1999. LNCS, vol. 1666, pp. 19–30. Springer, Heidelberg (1999)

43. Kraitchik, M.: Théorie des nombres. Gauthier-Villars (1922)
44. Lazard, D.: Gröbner-Bases, Gaussian Elimination and Resolution of Systems of Algebraic Equations. In: van Hulzen, J.A. (ed.) ISSAC 1983 and EUROCAL 1983. LNCS, vol. 162, pp. 146–156. Springer, Heidelberg (1983)
45. Macaulay, F.S.: The algebraic theory of modular systems. Cambridge Mathematical Library, vol. XXXI. Cambridge University Press (1916)
46. Macaulay, F.S.: Some properties of enumeration in the theory of modular systems. Proc. London Math. Soc. 26, 531–555 (1927)
47. National Institute of Standards and Technology. Digital Signature Standard (DSS). Federal Information Processing Standards Publication 186-3 (2009)
48. Patarin, J.: Hidden Fields Equations (HFE) and Isomorphisms of Polynomials (IP): Two New Families of Asymmetric Algorithms. In: Maurer, U.M. (ed.) EUROCRYPT 1996. LNCS, vol. 1070, pp. 33–48. Springer, Heidelberg (1996)
49. Petit, C., Quisquater, J.-J.: On polynomial systems arising from a weil descent. Cryptology ePrint Archive, Report 2012/146 (2012), http://eprint.iacr.org/
50. Pointcheval, D., Johansson, T. (eds.): EUROCRYPT 2012. LNCS, vol. 7237, pp. 2012–2031. Springer, Heidelberg (2012)
51. Semaev, I.: Summation polynomials and the discrete logarithm problem on elliptic curves (2004),
    http://www.isg.rhul.ac.uk/~ppai034/_pub/papers/Semaev%20%28Feb%29.pdf
52. Yang, B.-Y., Chen, J.-M., Courtois, N.T.: On Asymptotic Security Estimates in XL and Gröbner Bases-Related Algebraic Cryptanalysis. In: López, J., Qing, S., Okamoto, E. (eds.) ICICS 2004. LNCS, vol. 3269, pp. 401–413. Springer, Heidelberg (2004)

ECM at Work

Joppe W. Bos[1,*] and Thorsten Kleinjung[2]

[1] Microsoft Research, One Microsoft Way, Redmond, WA 98052, USA
[2] Laboratory for Cryptologic Algorithms, EPFL, Lausanne, Switzerland

Abstract. The performance of the elliptic curve method (ECM) for integer factorization plays an important role in the security assessment of RSA-based protocols as a cofactorization tool inside the number field sieve. The efficient arithmetic for Edwards curves found an application by speeding up ECM. We propose techniques based on generating and combining addition-subtracting chains to optimize Edwards ECM in terms of both performance and memory requirements. This makes our approach very suitable for memory-constrained devices such as graphics processing units (GPU). For commonly used ECM parameters we are able to lower the required memory up to a factor 55 compared to the state-of-the-art Edwards ECM approach. Our ECM implementation on a GTX 580 GPU sets a new throughput record, outperforming the best GPU, CPU and FPGA results reported in literature.

Keywords: Elliptic curve factorization, cofactorization, addition-subtraction chains, twisted Edwards curves, parallel architectures.

1 Introduction

Today, more than 25 years after its invention by Hendrik Lenstra Jr., the elliptic curve method [24] (ECM) remains the asymptotically fastest integer factorization method for finding relatively small prime factors of large integers. Although it is not the fastest general purpose integer factorization method, when factoring a composite integer $n = pq$ with $p \approx q \approx \sqrt{n}$ the number field sieve [32,23] (NFS) is asymptotically faster, it has recently received a renewed research interest due to the discovery of an interesting normal form for elliptic curves introduced by Edwards [13]. From a cryptologic point of view the practical performance of ECM is important since it is used to rapidly factor many small (up to one or two hundred bits) integers inside NFS. This is illustrated by the fact that it is estimated that five to twenty percent (cf. Section 2.2 why this is hard to estimate) of the total wall-clock time was spent in ECM in the current world-record factorization of a 768-bit RSA number [20] (and it is expected that this percentage will grow for larger factorizations). Using ECM as a tool to factor many small numbers inside NFS is an active research area by itself. Offloading this work to reconfigurable hardware such as field-programmable gate arrays is studied in [37,16,11,17,25,40] while [5,4]

---

* Part of this work was performed when the first author was working at the Laboratory for Cryptologic Algorithms, EPFL, Lausanne, Switzerland.

X. Wang and K. Sako (Eds.): ASIACRYPT 2012, LNCS 7658, pp. 467–484, 2012.

considers parallel architectures such as graphics processing units (GPUs) and the Cell broadband engine architecture. A comparison between software and hardware based solutions is presented in [21]. Traditionally, ECM is implemented using Montgomery curves [26] and uses the various techniques described in [39]. The most-widely used ECM implementation is GMP-ECM [41] and this implementation, or modifications to it, is responsible for setting all recent ECM record factorizations (see a description of some of these record factorizations in [8]). After the invention of Edwards curves Bernstein et al. explored the possibility to use these curves in the ECM setting [3]. Hisil et al. [19] published a coordinate system for Edwards curves which results in the fastest known realization of curve arithmetic. A follow-up paper by Bernstein et al. discusses the usage of these "$a = -1$" twisted Edwards curves [1] for ECM. The speedup from switching to Edwards curves comes at a price, addition chains [35] (or addition-subtraction chains [28]) equipped with large windowing sizes [9] are used (cf. [6] for a summary of these techniques). The memory requirement for Edwards ECM grows roughly linearly with the input parameters of ECM while a small constant number of residues modulo n are sufficient when using Montgomery curves.

In this paper we optimize ECM by exploiting the fact that the same scalar is often used when computing the elliptic curve scalar multiplication (ECSM), allowing one to prepare particularly good addition-subtraction chains for these fixed scalars. Our approach is inspired by the ideas used in the ECM implementation by Dixon and Lenstra [12] from 1992. In [12] the total cost to compute the ECSM, in terms of point doubling and point additions, is lowered by testing if the computation of the ECSM using batches of small prime products is cheaper (requires fewer point additions) than processing the primes one at a time (or all in one big batch). We generalize this idea: many billions of integers, which are constructed such that they can be computed using an addition-subtraction chain with a high doubling/addition ratio, are tested for smoothness and factored. By fixing different popular elliptic curve scalar values used in ECM inside NFS we are able to combine some of these integers using a greedy approach. This results in a *more efficient* ECSM algorithm with a *smaller* memory footprint. To illustrate, compared to the cofactorization setting considered by Bernstein et al. in [5,4] (using the parameter $B_1 = 2^{13}$) the techniques from this paper reduce the memory by a factor 55. This makes our approach particularly interesting for environments where the memory (per thread) is constrained; e.g. GPUs. We illustrate the practical benefits by implementing this approach for GPUs: setting a new throughput speed record compared to the current CPU, GPU and FPGA based results reported in literature. The best addition-subtraction chains found for the various popular B_1 values can be found online [7].

This paper is organized as follows. After recalling the preliminaries in Section 2 the notation and basic idea behind elliptic curve constant scalar multiplication is discussed in Section 3. Section 4 explains how to combine these chains such that they might result in a faster and more memory efficient ECM. Section 5 explains a side-effect why certain chains require more modular multiplications and Section 6 presents the obtained results. Section 7 concludes the paper.

2 Preliminaries

2.1 The Elliptic Curve Method

The elliptic curve method (ECM) for integer factorization [24] is analogous to the Pollard $p-1$ integer factorization method [33] and attempts to factor a composite integer n. The general idea behind ECM is as follows (we follow the description from [24]). First, pick a random point P and construct an elliptic curve E over $\mathbf{Z}/n\mathbf{Z}$ such that $P \in E(\mathbf{Z}/n\mathbf{Z})$ (cf. [22, Sec. 2.B]). Next, compute the elliptic curve scalar multiplication $Q = kP \in E(\mathbf{Z}/n\mathbf{Z})$. The positive integer k is selected such that it is divisible by many small prime powers: e.g. $k = \mathrm{lcm}(1, 2, \ldots, B_1)$ for some bound $B_1 \in \mathbf{Z}$. If for a prime p dividing n the order $\#E(\mathbf{F}_p)$ is B_1-powersmooth (an integer is defined to be B-powersmooth if none of the prime powers dividing this integer is greater than B) then $\#E(\mathbf{F}_p) \mid k$. In other words, $Q = kP$ and the neutral element of the curve become the same modulo p. In this event we have $p \mid \gcd(n, Q_z)$, where Q_z is the z-coordinate of the point Q when using projective Weierstrass coordinates. If $\gcd(n, Q_z) \neq n$ then we have split n.

Hasse proved (see e.g. [36, Theorem 1.1]) that the order $\#E(\mathbf{F}_p)$ is in the interval $[p + 1 - 2\sqrt{p}, p + 1 + 2\sqrt{p}]$. The advantage of ECM is that one can randomize the group order by trying different curves. It has been shown in [24] that the (heuristic) run-time of ECM depends mainly on p, the smallest non-trivial prime divisor of n, and can be expressed as

$$O(\exp((\sqrt{2} + o(1))(\sqrt{\log p \log \log p}))M(\log n))$$

where $M(\log n)$ represents the complexity of multiplication modulo n and the $o(1)$ is for $p \to \infty$. The approach described here is often referred to as "stage 1". There is a "stage 2" continuation for ECM which takes as input a bound $B_2 \in \mathbf{Z}$ and succeeds (in factoring n) if $Q = kP$ has prime order ℓ (for $B_1 < \ell < B_2$) in $E(\mathbf{F}_p)$. This means that $\#E(\mathbf{F}_p)$ is B_1-powersmooth except for one prime factor which is below B_2. There are several techniques [10,26,27] how to perform stage 2 efficiently. In the following we will focus on stage 1 only.

2.2 Cofactorization Using ECM

The relation collection phase, one of the two main phases of NFS, generates a lot of composite integers which need to be tested for powersmoothness. This is done using different factorization techniques and is denoted as the cofactorization phase. To illustrate, the total time spent in the cofactorization procedure was roughly one third of the sieving time when factoring the 768-bit RSA modulus in [20]. Note that this one third includes the time of pseudo primality tests and different factorization methods: quadratic sieve [34], Pollard $p - 1$ [33] and ECM. In this cofactorization phase only composites up to 140 bits were considered and ECM was used only for composites up to 109 bits. The parameters for ECM varied depending on the size of the composites and ranged from $B_1 = 150$ to $B_1 = 500$ where often only a single curve was tried with a maximum of around

Table 1. Performance comparison between GMP-ECM and EECM-MPFQ using the "$a = -1$" twisted Edwards curves in terms of modular multiplications (**M**) and squarings (**S**) together with the required number of residues modulo n (R) which needs to be kept in memory.

B1	GMP-ECM [41]				EECM-MPFQ [3]			
	#**S**	#**M**	#**S**+#**M**	#R	#**S**	#**M**	#**S**+#**M**	#R
256	1 066	2 025	3 091	14	1 436	1 638	3 074	38
512	2 200	4 210	6 410	14	2 952	3 183	6 135	62
1024	4 422	8 494	12 916	14	5 892	6 144	12 036	134
8192	35 508	68 920	104 428	14	47 156	45 884	93 040	550

eight curves. Observing the trend of past record factorizations, it is conceivable that cofactorization becomes more important in bigger factorizations (cf. [5] for more detailed arguments about the significance of ECM in NFS).

2.3   Montgomery versus Edwards Curves

The main motivation to use Edwards (over Montgomery) curves is performance. There is one implementation of ECM using Edwards curves available: EECM-MPFQ. This implementation includes the "$a = 1$" Edwards curves approach from [3] and the "$a = -1$" Edwards curves approach from [1]. The $a = -1$ Edwards ECM approach is the fastest in practice and we use this as the base setting to compare to. Table 1 compares the required number of multiplications and squarings required in GMP-ECM and EECM-MPFQ for different typical B_1 values used in ECM when used as a cofactorization method in NFS. These numbers show that using Edwards curves results in fewer modular multiplications and squarings. However, the required storage for GMP-ECM (Montgomery curves) is independent of B_1 while it grows almost linearly with the size of B_1 and is significantly higher, due to the use of windowing based methods, for EECM-MPFQ (Edwards curves, see [3, Table 4.1]).

3   Elliptic Curve Constant Scalar Multiplication

Most of the addition-subtraction chains based algorithms in practice use a w-bit windowing technique, for some (optimal) width w, to reduce the number of required elliptic curve additions. The total number of additions may be significantly reduced by using this approach but one also needs to store more points: 2^{w-1} when using sliding windows [38]. In environments where the available memory per thread is low, these methods cannot be used or one is forced to settle for a suboptimal window size. A prime example of such a platform are graphics processing units (GPUs); one of the latest GPU architectures [29] (Fermi) shares 64 kilobyte fast shared memory per 32 processors and each processor typically time-shares multiple threads (e.g., 16 to 32 corresponding to 128 to 64 bytes per thread).

We investigate two approaches to lower the number of elliptic curve additions *and* the storage required to compute the scalar product. Our approach is inspired by the results reported by Dixon and Lenstra [12]. Suppose we have a scalar $k = \mathrm{lcm}(1, \ldots, B_1) = \prod_{i=1}^{\ell} p_i$, where the p_i are primes which can occur multiple times. Typically, the ECSM is implemented processing one such p_i at a time [39]. In [12] it is suggested to process the p_i in *batches*; i.e. multiply a batch of p_i's at a time such that the weight of the product $w(\prod_i p_i)$, the number of ones in the binary representation of $\prod_i p_i$, is (much) lower than the sum of the individual weights $\sum_i w(p_i)$. If this is the case then the number of required EC-additions is reduced when using the straight forward double-and-add approach (which does not require to store any additional precomputed points). Such low-weight products can be constructed by greedily searching through b-tuples of the p_i where b is small. In [12] b was at most 3 which reduced the total weight by approximately a factor three. As an example the following triple is given

$$1028107 \cdot 1030639 \cdot 1097101 = 1162496086223388673$$
$$w(1028107) = 10, \quad w(1030639) = 16, \quad w(1097101) = 11,$$
$$w(1162496086223388673) = 8,$$

where the product of primes of weights 10, 16, and 11 results in a integer of weight eight. The resulting composite integer can be computed using an addition chain requiring only seven additions and 60 doublings using the naive double-and-add algorithm.

In this section we explore different methods to find numbers which can be constructed using even better (higher) doubling/addition ratios. These methods do not aim to construct sequences by combining the different p_i (as in [12]) but we propose an opposite approach by factoring many integers which are the result of addition-subtraction chains with high doubling/addition ratios and subsequently combining these integers such that all p_i's are used. These addition-subtraction chains are constructed such that they do not require any large lookup tables. Notice that the information encoding the sequence of arithmetic operations has to be stored (in all approaches). This does not pose a problem since this information is constant and can be shared among all the computational units (or streamed to the units or even hardcoded) and hence does not result in additional overhead in practice.

In the remainder of the paper we denote addition-subtraction chains simply as chains.

3.1   Chains with Restrictions

In order to generate integers which can be computed using a chain with a high doubling/addition ratio we need to construct and denote chains of a certain length m. A chain is a sequence of doublings, additions and subtractions denoted by D, A and S respectively. A doubling can always be assumed to apply to the previously generated element in the chain (instead of doubling any previous element), since one can reorder the symbols such that doubling always occurs on the last element without changing the result of the chain. In some cases this might result in a shorter (more efficient) sequence when the same element is doubled multiple times. Let us define the set of symbols $\mathscr{O}$ as

$$\mathscr{O} = \{D\} \cup \{A_{i,j} \mid i,j \in \mathbf{Z}, i > j\} \cup \{S_{i,j} \mid i,j \in \mathbf{Z}, i > j\},$$

where the subscripts indicate on which element in the chain we compute (this is made more precise later). The set of all m-tuples, ordered lists of m elements, of symbols in $\mathscr{O}$ with the restriction that no elements can be used which have not yet been generated is

$$\mathcal{O}_m = \{(o_{m-1}, \ldots, o_0) \in \mathscr{O}^m \mid o_k \in \{D\} \cup \{A_{i,j} \mid i \le k\} \cup \{S_{i,j} \mid i \le k\}, 0 \le k < m\}.$$

In order to construct a chain from such an m-tuple of symbols we define functions $\sigma_m : \mathscr{O} \times \mathbf{Z}^{m+1} \to \mathbf{Z}^{m+2}$ such that $(o, (t_m, \ldots, t_0)) \mapsto (t_{m+1}, t_m, \ldots, t_0)$ where

$$t_{m+1} = \begin{cases} 2t_m & \text{if } o = D, \\ t_i + t_j & \text{if } o = A_{i,j}, \\ t_i - t_j & \text{if } o = S_{i,j}. \end{cases}$$

Given an m-tuple of symbols $(o_{m-1}, \ldots, o_0) \in \mathcal{O}_m$ the $(m+1)$-tuple of integers associated to this chain is $\sigma_{m-1}(o_{m-1}, \sigma_{m-2}(o_{m-2}, \ldots, \sigma_0(o_0, 1) \ldots))$ and the *resulting integer* produced by this chain is t_m. As an example consider the 7-tuple of symbols $(S_{6,0}, D, D, A_{3,0}, D, D, D) \in \mathcal{O}_7$ which corresponds to the 8-tuple of integers in the chain $(35, 36, 18, 9, 8, 4, 2, 1)$ computed as

$$\sigma_6(S_{6,0}, \sigma_5(D, \sigma_4(D, \sigma_3(A_{3,0}, \sigma_2(D, \sigma_1(D, \sigma_0(D, 1))))))).$$

The function σ_m is the correspondence between a tuple of symbols and the actual chain. The example shows how to compute the resulting integer 35 using one subtraction, one addition and five doublings.

The set of tuples $\mathcal{O}_m$ consists of the most generic type of chains, a significant amount of tuples corresponds to chains which perform useless (unnecessary) computations. An example is computing the addition (or subtraction) of two previous values without using this result. To address this we define a more restricted set of tuples $\mathcal{P}_m \subset \mathcal{O}_m$ as

$$\mathcal{P}_m = \{(o_{m-1}, \ldots, o_0) \in \mathcal{O}_m \mid o_k \in \{D\} \cup \{A_{i,j} \mid i = k\} \cup \{S_{i,j} \mid i = k\}, 0 \le k < m\}.$$

These additional restrictions ensure that, just as for the doubling, we only add or subtract to the last integer in the sequence to obtain the next one. Such chains are known as *Brauer chains* or *star addition chains* [18, Section C6].

In this setting we write A_j and S_j for $A_{i,j}$ and $S_{i,j}$, respectively, and $k > 0$ subsequent instances of D are denoted by D^k. The previous example can now be written as $S_0 D^2 A_0 D^3 \in \mathcal{P}_7$ by abusing the notation: omitting the brackets and comma's. In practice we would generate sequences of symbols such that a number of elliptic curve additions $\mathbf{A}$ and doublings $\mathbf{D}$ are fixed and look at sequences of symbols of length $m = \mathbf{A} + \mathbf{D}$ which use $\mathbf{A}$ times A_j or S_j and

D times D. Different tuples might compute the same integer result. Using our example, the number 35 can be obtained with $\mathbf{D} = 5$ and $\mathbf{A} = 2$ in different ways

$$35 = (2^3 + 1) \cdot 2^2 - 1 \quad S_0 D^2 A_0 D^3 \in \mathcal{P}_7$$
$$= (2^4 + 1) \cdot 2 + 1 \quad A_0 D A_0 D^4 \in \mathcal{P}_7.$$

3.2 Generating Chains

We discuss how to efficiently generate the resulting integers t_m in a low-storage and no-storage setting.

The Low-Storage Setting. Let $\mathbf{A}$ be the number of elliptic curve additions and $\mathbf{D}$ the number of elliptic curve doublings (with $\mathbf{D} \geq \mathbf{A}$). The generation of *all* the tuples in $\mathcal{P}_m$, with $m = \mathbf{A} + \mathbf{D}$ results in many identical integers t_m. Removing these duplicate integers can be achieved by first generating and storing all the resulting integers and subsequently sorting and keeping exactly one of consecutive equal integers. To avoid storing all the resulting integers for a given pair $(\mathbf{A}, \mathbf{D})$, which requires a significant amount of storage as we will see later, and to avoid sorting this huge data set we define a more restricted set of rules $\mathcal{Q}_m \subset \mathcal{P}_m \subset \mathcal{O}_m$ as follows

$$\mathcal{Q}_m = \{(o_{m-1}, \ldots, o_0) \in \mathcal{P}_m \mid o_0 = D, o_{m-1} \in \{A_i, S_i\}, \text{and for } 0 < k < m - 1:$$
$$o_k \in \{D\} \cup \{A_i, S_i\}, o_k \in \{A_i, S_i\} \Rightarrow o_{k-1} = D$$
$$\wedge \ (i = 0 \ \vee \ o_{i-1} \in \{A_\ell, S_\ell\})\}.$$

The restrictions used in the definition of $\mathcal{Q}_m$ ensure that the resulting integer is odd and only addition (or subtraction) of an odd number to the current (even) number is allowed. This approach significantly reduces the amount of chains which produce the same resulting integer at the cost of slightly reducing the number of unique integers produced. To illustrate, for $\mathbf{D} = 50$ the total number of tuples generated by $\mathcal{P}_{53}$ is more than 140 times higher compared to $\mathcal{Q}_{53}$ while the number of unique odd resulting integers is only 1.09 times higher.

The list of $m + 1$ integers u_i corresponding to the m-tuple of symbols from $\mathcal{Q}_m$ can be efficiently generated recursively using

$$u_{i+1} = \begin{cases} 2u_i \\ u_i \pm u_j \text{ for } j < i \text{ and } 2 \mid u_i, 2 \nmid u_j \end{cases}$$

with $u_0 = 1$ and ensuring that the final operation is not a doubling (to make the resulting integer odd). Hence, the next integer in the sequence can always be obtained by doubling or adding a previous odd number u_j to the current even integer u_i. The required storage depends on which u_j are used in subsequent additions and at which indices they are used. In practice we generate all sequences using a fixed number of doublings $\mathbf{D}$ and additions $\mathbf{A}$ making sure that the resulting storage requirement is never too large.

A sequence of additions and doublings corresponding to the chains resulting from $\mathcal{Q}_m$ looks like

$$A_{i_{\mathbf{A}-1}} D^{d_{\mathbf{A}-1}} \ldots A_{i_1} D^{d_1} A_{i_0} D^{d_0} = (A_{i_{\mathbf{A}-1}} D) D^{d_{\mathbf{A}-1}-1} \ldots (A_{i_1} D) D^{d_1-1} (A_{i_0} D) D^{d_0-1}$$

$$(1)$$

with $\mathbf{D} = \sum_{i=0}^{\mathbf{A}-1} d_i$, $d_i > 0$, and indices i_j that satisfy the restrictions of $\mathcal{Q}_m$, i.e., i_j takes one of the values $\sum_{g=0}^{h}(d_g + 1)$ for $-1 \leq h < j$. Such a sequence starts with a doubling, ends with an addition and an addition is always preceded by a doubling. Hence, there are $\binom{\mathbf{D}-1}{\mathbf{A}-1}$ choices for the order of the $\mathbf{A} - 1$ pairs (A_{i_j}, D) and the $\mathbf{D} - \mathbf{A}$ doublings D. Since every addition can be substituted by a subtraction the number of possibilities is multiplied by a factor $2^{\mathbf{A}}$. The indices i_j can be chosen in $\mathbf{A}!$ ways, hence the total number of resulting integers produced by $\mathcal{Q}_m$ is

$$\binom{\mathbf{D}-1}{\mathbf{A}-1} \cdot \mathbf{A}! \cdot 2^{\mathbf{A}} = 2^{\mathbf{A}} \cdot \mathbf{A} \cdot \prod_{i=1}^{\mathbf{A}-1}(\mathbf{D} - \mathbf{A} + i).$$

The No-Storage Setting. The second setting we consider is constructing chains which do not require any additional stored points, besides the in- and output (and possibly some auxiliary variables required to calculate the elliptic curve group operation). This means we are looking for integers which can be computed using chains which only use doublings and add or subtract the input point. We can define the set of tuples $\mathcal{R}_m \subset \mathcal{Q}_m$ as $\mathcal{R}_m = \{(o_{m-1}, \ldots, o_0) \in \mathcal{Q}_m \mid o_k \in \{A_0, S_0, D\}, 0 \leq k < m\}$. All resulting integers of no-storage chains which can be constructed using $\mathbf{A}$ elliptic curve additions and $\mathbf{D}$ elliptic curve doublings are of the form

$$2^{\mathbf{D}} + \sum_{i=0}^{\mathbf{A}-1} \pm 2^{n_i}, \quad \text{with } 0 = n_0 < n_1 < \ldots < n_i < \ldots < n_{\mathbf{A}-1} < \mathbf{D}.$$

This follows from (1) by setting $i_j = 0$; we have $n_i = \sum_{g=1}^{i} d_{\mathbf{A}-g}$. Using the same argument as in the low-storage setting the number of resulting integers generated by $\mathcal{R}_m$ is $\binom{\mathbf{D}-1}{\mathbf{A}-1} \cdot 2^{\mathbf{A}}$. Compared to the low-storage setting the number is reduced by a factor of $\mathbf{A}!$, reflecting the missing choice of the indices i_j.

4   Combining Chains

Recall that, given a bound B_1, we want to perform an elliptic curve scalar multiplication with the integer $k = \prod_{i=1}^{\ell} p_i = \text{lcm}(1, \ldots, B_1)$ where the product ranges over ℓ (not necessarily distinct) primes. We can get rid of the problems posed by the primes 2 in this product by noticing that they can be handled by a sequence of doublings at the end of the ECSM and assuming in the following that all s_i are odd. The techniques from the previous section provide us with a lot of integers which can be constructed using a known number of additions (here we count subtractions as additions) and doublings. Since different chains can lead to the same integer we pick for each of these integers one chain (preferably the one with the lowest cost). In this way we get a list of distinct integers, each with an associated chain. We index this list by an index set I and call s_i the integer corresponding to $i \in I$. For $i \in I$ denote by $\text{add}(s_i)$ resp. $\text{dbl}(s_i)$ the number of additions resp. doublings in the chain and by $\{s_{i,1}, \ldots, s_{i,t_i}\}$ the multiset of the primes in the prime decomposition of s_i. Furthermore, let $\mathrm{cost}(s_i)$ be the cost of performing a scalar multiplication with s_i using the associated chain. A reasonable choice for Edwards curves is $\mathrm{cost}(s_i) = 7\mathrm{dbl}(s_i) + 8\mathrm{add}(s_i) + 1$ which will be discussed in the next section.

Ideally, we want to find a subset $I' \subset I$ such that $k \mid \prod_{i \in I'} s_i$ and $\sum_{i \in I'} \mathrm{cost}(s_i)$ is minimal. To facilitate our task we will modify this in two ways. If the product in the first condition is bigger than k we do more work than necessary. This can lead to a lower cost, but we assume that replacing the first condition by $k = \prod_{i \in I'} s_i$ will not increase the minimum of $\sum_{i \in I'} \mathrm{cost}(s_i)$ significantly. The second modification is the replacement of $\sum_{i \in I'} \mathrm{cost}(s_i)$ by $\sum_{i \in I'} \mathrm{add}(s_i)$. To explain why we think that this does not increase the minimum too much we consider subsets I' for which $\sum_{i \in I'} \mathrm{cost}(s_i)$ is close to the minimum. Then most s_i have a high ratio $\frac{\mathrm{dbl}(s_i)}{\mathrm{add}(s_i)}$ and therefore we have for most of them $s_i \approx 2^{\mathrm{dbl}(s_i)}$. Since $\prod_{i \in I'} s_i = k$ the sum $\sum_{i \in I'} \mathrm{dbl}(s_i) \approx \log_2(k)$ does not vary too much. Furthermore, the summand 1 in the cost function is the least significant term and the cardinality of I' does not vary much. We are aware that the second modification is more delicate than the first one, but, as explained below, we will generate many sets I' and will pick the best one amongst them using the more costly function $\mathrm{cost}(s_i)$.

The condition $k = \prod_{i \in I'} s_i$ implies that every s_i in this product is B_1-powersmooth which suggests the following two stage approach:

1. Restrict to $\hat{I} = \{i \in I \mid s_i \text{ is } B_1\text{-powersmooth}\}$.
2. Find a subset $I' \subset \hat{I}$ such that the multisets $\bigcup_{i \in I'} \{s_{i,1}, \ldots, s_{i,t_i}\} = \{p_1, \ldots, p_\ell\}$ coincide and that $\sum_{i \in I'} \mathrm{add}(s_i)$ is minimal.

Testing a large list of numbers for B_1-powersmoothness can be done using the method from [15, Section 4]. The main idea is to build a product tree from the list, replace the root node R (the product of all numbers of the list) by $k \bmod R$ (where $k = \mathrm{lcm}(1, \ldots, B_1)$ is precomputed) and then tree-wise replace each node by the residue of k modulo the node. The leaves resulting in 0 contained B_1-powersmooth numbers and their factorizations can be obtained by other means.

Finding an optimal set I' is in general a difficult problem and has been studied in [31]. We choose to use a greedy approach which produces satisfactory results. We start with an empty set I' and the multiset $M = \{p_1, \ldots, p_\ell\}$ of primes to be matched. As long as M is non-empty we select an integer $s_i = \prod_{j=1}^{t_i} s_{i,j}$ with $\{s_{i,1}, \ldots s_{i,t_i}\} \subset M$ such that the ratio $\frac{\mathrm{dbl}(s_i)}{\mathrm{add}(s_i)}$ is high and replace I' by $I' \cup \{i\}$ and M by $M \setminus \{s_{i,1}, \ldots s_{i,t_i}\}$. This may fail because we might not be able to satisfy the condition $\{s_{i,1}, \ldots s_{i,t_i}\} \subset M$ at a given point. There are several ways to overcome this problem, e.g., we could increase our supply of s_i by generating more chains. Another way consists in aborting the greedy search at this point, getting $k = c \cdot \prod_{i \in I'} s_i$ for some integer c. Using the method of Dixon/Lenstra, we can search for a decomposition of c into several factors, each having a good chain. For the sizes of B_1 considered in this paper, namely $B_1 \leq 8192$, c consisted of very few primes and was often 1. Therefore the usually lower doubling/addition ratio of the c-part does not pose a problem for small B_1.

A refinement to this approach is to also take the size of the prime factors $s_{i,j}$ into account. A strategy could be to prefer choosing integers s_i which have mostly large prime divisors, since the majority of the primes p_i is large. The idea is to attach a score to a B_1-powersmooth integer given its prime factorization with respect to the currently unmatched prime factors in k. For a multiset N of primes bounded by B_1 the ratio of j-bit primes is defined as

$$a_j(N) := \frac{\#\{p \in N \mid \lceil \log_2(p) \rceil = j\}}{\#M},$$

where $1 \leq j \leq \lceil \log_2(B_1) \rceil$. Given M, the multiset of currently unmatched primes, the *score* of s_i is defined as

$$\text{score}\left(s_i = \prod_{j=1}^{t_i} s_{i,j}, M\right) = \sum_{\substack{h=1: \\ a_h(M) \neq 0}}^{\lceil \log_2(B_1) \rceil} \frac{a_h(\{s_{i,1}, \ldots, s_{i,t_i}\})}{a_h(M)}$$

The higher the score the more small prime divisors are likely to be present. In general, for a given ratio, we select the integers which have a low score.

To illustrate, consider $B_1 = 1024$ where the initial a_i are

$$a_2 = 0.032, \ a_3 = 0.037, \ a_4 = 0.021, \ a_5 = 0.053, \ a_6 = 0.037,$$
$$a_7 = 0.069, \ a_8 = 0.122, \ a_9 = 0.229, \ a_{10} = 0.399$$

(with $\sum_{i=2}^{10} a_i = 1$). Almost 40 percent of all the primes fall in the largest (10-bit) category. An example of a low score-integer is $11529215054666795009 = 743 \cdot 719 \cdot 677 \cdot 461 \cdot 457 \cdot 449 \cdot 337$ where the size of the smallest prime is 9-bit, the score is 3.57 and this integer can be computed using 63 doublings and five additions as $A_0 D^{11} A_0 D^{12} A_0 D^{10} A_0 D^{28} A_0 D^2 \in \mathcal{R}_{68}$. On the other hand, an example of a high-score integer, consisting of mainly small primes, is $1048575 = 41 \cdot 31 \cdot 11 \cdot 5^2 \cdot 3$, its score is significant higher (29.62) and it can be computed with 20 doublings and a single subtraction as $S_0 D^{20} \in \mathcal{R}_{21}$.

This approach using scores is outlined in Algorithm 1. Note that the scores are recalculated each time an s_i is chosen. In practice one could reduce the amount of these costly recalculations by picking several s_i in lines 10-13 of the algorithm; in this case one has to check that the union of the prime factors of the chosen s_i is still a multisubset of M.

A Randomized Variant. In the current state, Algorithm 1 returns a single solution given a set of input parameters. To increase the amount of different subsets I', and thereby hopefully improving the results, we randomize the selection process of the index that is added in lines 10-13 of the algorithm. With probability $x \in \mathbf{R}$ $(0 < x < 1)$ select the s_i corresponding to score_1 or, with probability $1 - x$, skip it and repeat this procedure for score_2 and so on. If we have reached the end of the list (after j trials) one could apply a deterministic choice.

Algorithm 1. Given a bound B_1 and a set of B_1-powersmooth integers $\{s_i \mid i \in \hat{I}\}$, which can be computed with a chain using $\text{add}(s_i)$ resp. $\text{dbl}(s_i)$ elliptic curve additions resp. doublings, together with the prime factorization of these integers $(s_i = \prod_j s_{i,j})$ the algorithm attempts to output triples $(s_j, \text{add}(s_j), \text{dbl}(s_j))$ such that $\text{lcm}(1, \ldots, B_1) = c \cdot \prod_j s_j$ for a small integer c. This algorithm considers scores $\leq T$ only and combines integers s_i for which $\dfrac{\text{dbl}(s_i)}{\text{add}(s_i)} \geq r$ where r starts at r_h and is decreased until r_l.

Input: $\begin{cases} \text{Bound } B_1 \in \mathbf{Z}, \text{ we have } \text{lcm}(1, \ldots, B_1) = \prod_{i=1}^{\ell} p_i \text{ with } p_i \text{ prime.} \\ \text{Set of integers } \{s_i \mid i \in \hat{I}\} \text{ with } s_i = \prod_j s_{i,j} \text{ for } s_{i,j} \text{ prime and } i \in \hat{I}. \\ \text{Upper and lower bound on the doubling/addition ratio: } r_h \text{ and } r_l. \\ \text{A threshold value for the score: } T. \end{cases}$

Output: Triples $(s_i, \text{add}(s_i), \text{dbl}(s_i))$ and c such that $c \cdot \prod_i s_i = \text{lcm}(1, \ldots, B_1)$.

1. $M \leftarrow \{p_1, \ldots, p_\ell\}$, $I' \leftarrow \emptyset$
2. **for** $r = r_h$ to r_l **do**
3.   found $\leftarrow$ true
4.   **while** found=true **do**
5.   found $\leftarrow$ false, $j \leftarrow 0$
6.   **for** $i \in \hat{I}$ **do**
7.   **if** $\{s_{i,1}, \ldots, s_{i,t_i}\} \subset M$ and $\dfrac{\text{dbl}(s_i)}{\text{add}(s_i)} \geq r$ and $\text{score}(s_i, M) \leq T$ **then**
8. $j \leftarrow j + 1$, $\text{score}_j \leftarrow (\text{score}(s_i, M), i)$
9.   sort score_i for $1 \leq i \leq j$ with respect to $\text{score}(s_i, M)$
10.   **if** $j \geq 1$ **then**
11. $i \leftarrow$ index from score_1, output $(s_i, \text{add}(s_i), \text{dbl}(s_i))$
12. $I' \leftarrow I' \cup \{i\}$, $M \leftarrow M \setminus \{s_{i,1}, \ldots, s_{i,t_i}\}$
13.   found $\leftarrow$ true
14. output $\{(s_i, \text{add}(s_i), \text{dbl}(s_i)) \mid i \in I'\}$ and $c = \prod_{p \in M} p$

5 Additional Multiplications

The fastest arithmetic for Edwards curves is due to Hisil et al. [19]. They propose to use extended twisted Edwards coordinates, which are twisted Edwards coordinates plus an auxiliary coordinate. This allows faster addition but slower doubling. Using a mixing technique, by switching between extended twisted Edwards and regular twisted Edwards, the overall cost for scalar multiplication is reduced [19]. This is realized by performing the doublings using the cheaper regular twisted Edwards coordinates when a doubling is followed by a doubling. When an addition is required after a doubling one can use the doubling formula in the extended twisted Edwards coordinates (which does not need the auxiliary coordinate as input) at the cost of an extra multiplication to compute the auxiliary coordinate of the result. Next, the fast addition is performed in extended twisted Edwards coordinates; one multiplication (to compute the auxiliary coordinate of the output) can be saved, cancelling the extra multiplication used when doubling, since a doubling is always performed after an addition in

Table 2. The left table shows the number of integers (#int) generated with an addition-subtraction chain using **A** and **D** elliptic curve additions and doublings respectively. All these integers were tested for $2.9 \cdot 10^9$-powersmoothness and, if smooth, the prime divisors are stored. The **bold** ranges indicate that 2^{31} random integers per single **A**, **D** combination were tested for smoothness instead of the full range. The right table shows the number of unique B_1-powersmooth integers in the no-storage and low-storage setting for different values of B_1.

	No-storage setting			Low-storage setting	
A	**D**	#int	**A**	**D**	#int
1	$5 - 200$	$3.920 \cdot 10^2$	1	$5 - 250$	$4.920 \cdot 10^2$
2	$10 - 200$	$7.946 \cdot 10^4$	2	$10 - 250$	$2.487 \cdot 10^5$
3	$15 - 200$	$1.050 \cdot 10^7$	3	$15 - 250$	$1.235 \cdot 10^8$
4	$20 - 200$	$1.035 \cdot 10^9$	4	$20 - 250$	$6.101 \cdot 10^{10}$
5	$25 - 200$	$8.114 \cdot 10^{10}$	5	$25 - 158$	$2.956 \cdot 10^{12}$
			5	$\mathbf{159 - 220}$	$\mathbf{1.331 \cdot 10^{11}}$
6	$30 - 150$	$9.150 \cdot 10^{11}$	**6**	$\mathbf{60 - 176}$	$\mathbf{2.513 \cdot 10^{11}}$
7	$35 - 66$	$9.900 \cdot 10^{10}$			
Total		$1.096 \cdot 10^{12}$			$3.403 \cdot 10^{12}$

B_1	No-Storage	Low-Storage
256	$2.423 \cdot 10^5$	$9.210 \cdot 10^6$
512	$1.470 \cdot 10^6$	$3.159 \cdot 10^7$
1 024	$5.691 \cdot 10^6$	$7.861 \cdot 10^7$
8 192	$9.352 \cdot 10^7$	$4.400 \cdot 10^8$
$2.9 \cdot 10^9$	$2.274 \cdot 10^{10}$	$3.997 \cdot 10^{10}$

ECSM-algorithms. This approach assumes that both inputs of the elliptic curve addition are in extended twisted Edwards coordinates. This is the case for double-and-add algorithms and (signed) windowing algorithms where the computation of the auxiliary coordinates of the lookup table are a minor overhead.

In both our settings, where we consider low- and no-storage, this does not hold. The computation of the large elliptic curve scalar product is done by processing batches of prime products (the s_i) at a time. All the additions or subtractions required in the chain to compute s_i require that the points are in extended twisted Edwards coordinates. When required, the odd intermediate results are stored in extended twisted Edwards coordinates at a cost of a single additional multiplication. The cost of computing a low-storage chain $(o_{m-1}, \ldots, o_0) \in \mathcal{Q}_m$ resulting in s_i is increased by $x(s_i)$ multiplications, where $x(s_i) = \#\{j \mid \exists h : o_h \in \{A_j, S_j\}, 0 \leq h < m\}$; i.e. the unique number of indices used in the additions and subtractions. Therefore we get for the cost function from the previous section $\text{cost}(s_i) = 7\text{dbl}(s_i) + 8\text{add}(s_i) + x(s_i)$. In the no-storage setting we always have $x(s_i) = 1$ leading to the choice for $\text{cost}(s_i)$ given at the beginning of the previous section. In total we have #{chains used} additional multiplications in the no-storage setting and a potentially higher number in the low-storage setting. We can save one multiplication due to the sequence containing the power of 2 (which consists of doublings only) and another multiplication if we assume that the input point is already in extended twisted Edwards coordinates.

6   Results

Using the rules given in Section 3.2 for both the no-storage and the low-storage setting, we generated more than 10^{12} integers for many choices of the number of additions **A** and doublings **D**. Table 2 summarizes the ranges we have covered where bold ranges (in the low-storage setting) indicate that only 2^{31} random integers were generated instead of the full range. All these integers were subjected

Table 3. The table shows the number of modular multiplications (**M**) and squarings (**S**) required to calculate **A** elliptic curve additions and **D** doublings for various B_1 parameters when factoring an integer n with ECM. The memory required is expressed as the number of residues (R), integers modulo n, which are kept in memory. The performance speedup **PS** (in terms of #**M**+#**S**) and memory reduction **MR** compared to the ECM approach from [1] using "$a = -1$" twisted Edwards curves is given.

B_1		#**M**	#**S**	#**M** + #**S**	PS	A	D	#R	MR
256	[1]	1 638	1 436	3 074		69	359	38	
No-storage		1 400	1 444	2 844	1.08	38	361	10	3.80
Low-storage		1 383	1 448	2 831	1.09	35	362	14	2.71
512	[1]	3 183	2 952	6 135		120	738	62	
No-storage		2 842	2 964	5 806	1.06	75	741	10	6.20
Low-storage		2 776	2 964	5 740	1.07	65	741	18	3.44
1 024	[1]	6 144	5 892	12 036		215	1 473	134	
No-storage		5 596	5 912	11 508	1.05	141	1 478	10	13.40
Low-storage		5 471	5 904	11 375	1.06	123	1 476	18	7.44
8 192	[1]	45 884	47 156	93 040		1 314	11 789	550	
No-storage		43 914	47 160	91 074	1.02	1 043	11 790	10	55.00
Low-storage		42 855	47 136	89 991	1.03	878	11 784	18	30.56

to $2.9 \cdot 10^9$-powersmoothness tests which reduced the number of integers by about two orders of magnitude. This large powersmoothness-bound was chosen to facilitate searching for efficient chains for much larger B_1 parameters. From the reduced set of integers we extracted those that are B_1-powersmooth for the values of B_1 used in this paper (see right part in Table 2). These computations were done on five 8-core Intel Xeon E5430 (2.66GHz) and took more than a year, i.e., in total over 40 core years. The smoothness testing required most of the run-time and up to 4.6GB of memory. Using the approach outlined in Algorithm 1 one of these nodes was occasionally used for the combining experiments which consisted of thousands of runs of the randomized greedy approach, each of them taking only a couple of seconds for these low values of B_1.

Table 3 shows the results obtained using Algorithm 1 on our dataset (see Table 2). The memory required is expressed in the number of residues (R), integers modulo n, which need to be kept in memory. Here we assume that extended twisted Edwards coordinates are used, i.e., every point is represented by four coordinates. In the setting of EECM-MPFQ [3,1] we assume that an optimal window size is used and that besides the window table only the input point needs to be kept in memory while we assume that two points (the input point and the current active point) are required in the no- and low-storage setting. The implementation of the elliptic curve group operation is assumed to require at most two auxiliary variables (residues). Hence, the no-storage setting requires memory for $2 \times 4 + 2 = 10$ residues modulo n. The low-storage results presented in Table 3 require to store at most two additional points (8 more residues modulo n compared to the no-storage setting). This is still significantly less compared to the approach used in [3,1].

6.1   Application to GPUs

When running ECM on memory constrained devices, like GPUs, the large number of precomputed points required for the windowing methods cannot be stored in fast memory. Typically one is forced to settle for a (much) smaller window size reducing the advantage from using twisted Edwards curves. For example, in [5] no large window sizes are used at all, the authors remark: "Besides the base point, we cannot cache any other points". Memory is also a problem in [4], the faster curve arithmetic from Hisil et al. [19] is not used since this requires storing a fourth coordinate per point.

From the data given in Table 3 it becomes clear that our approach reduces the memory requirements significantly. For example, the memory required to run ECM in the cofactorization setting on GPUs using $B_1 = 8\,192$ can be reduced by a factor 55. This setting was already considered in [5,4] where the authors were forced to reduce memory requirements by using suboptimal window sizes. Hence, when using the methods described in this paper *less* memory is required allowing the usage of the *faster* curve arithmetic and *reducing* the number of elliptic curve additions required in the computation of the elliptic curve scalar multiplication.

6.2   Performance Comparison

In order to measure the practical speedup of the methods described in this paper we implemented the no-storage approach on GPUs. This implementation uses the Compute Unified Device Architecture (CUDA) which facilitates the development of massively-parallel general purpose applications for GPUs [30]. Our implementation is targeted at the third generation CUDA GPUs called "Fermi" [29]. Table 4 compares the performance results of different hardware platforms for $B_1 = 960$ and $B_1 = 8192$, numbers chosen such that we can directly compare to results reported in the literature on other (hardware) platforms. For $B_1 = 960$, which is used as the example B_1 value in [40,11] and not spending as much effort as for $B_1 = 1024$, we were able to construct a no-storage chain requiring 1 371 doublings and 135 additions. The FPGA and GTX295 results are quadratically scaled to 192-bit arithmetic to compare the different performance results. The other GPU results are from [4] and this implementation is optimized for the second generation CUDA GPUs. The pricing for this card is omitted since it is no longer sold (this card was launched January 2009). The results on the Intel i7-2600K CPUs have been obtained with the ECM implementation (using Montgomery curves) from the NFS software suite [14] which is responsible for all recent record NFS factorizations (e.g. [20]) and the EECM-MPFQ software package [2] which uses Edwards curves. The FPGA results are from [11,40] and the FPGA prices are taken from [40]. Note that the prices are for the GPU, CPU or FPGA devices only; in order to get a fully operational system more hardware is required. Note also that for all of the considered devices newer versions with better price performance ratio exist, but we do not expect that these will change this comparison significantly.

Table 4. Performance comparison of ECM on different platforms (using the "$a = -1$" twisted Edwards curves if available). The first table lists the different hardware properties. The second and third table state results for $B_1 = 960$ and $B_1 = 8192$ respectively. The scaled number of curves are when using 192-bit moduli. The performance ratio is the ratio between the GTX 580 no-storage row and the current row for the scaled number of curves per 100 USD.

properties	GPU		CPU	FPGA	
	GTX 295	GTX 580	Intel i7-2600K	V4SX35-10	V4SX25-10
#cores	480	512	4	24	1
clock (MHz)	1 242	1 544	3 400	200	220
price (USD)	-	400	300	468	298
#threads	46 080	8 192	4	24	1
#bits in moduli	210	192	192	202	135

	performance (#curves), $B_1 = 960$			performance
	(1/sec)	(1/sec, scaled)	(1/100 USD, scaled)	ratio
GTX 580, no-storage	171 486	171 486	42 872	1.00
GTX 580, windowing	79 170	79 170	19 793	2.17
Intel i7 [14]	13 661	13 661	4 554	9.41
Intel i7 [2]	8 677	8 677	2 892	14.82
V4SX35-10 [40]	3 240	3 586	766	55.97
V4SX25-10 [11]	16 000	7 910	2 654	16.15

	performance (#curves), $B_1 = 8192$			
GTX 295 [4]	4 928	5 895	-	-
GTX 580, no-storage	19 869	19 869	4 967	1.00
GTX 580, windowing	9 106	9 106	2 277	2.18
Intel i7 [14]	1 629	1 629	543	9.15
Intel i7 [2]	1 092	1 092	364	13.65

For the sake of comparison we also implemented Edwards ECM for GPUs using the same 192-bit arithmetic but using the windowing based approach. For $B_1 = 960$ ($B_1 = 8192$) we used a signed sliding window of size 2^6 (2^8), precomputing and storing 2^5 (2^7) extended twisted Edwards coordinates. These results are stated in Table 4 as well. On the GTX 580 the no-storage approach is more than twice as fast as the approach based on windowing techniques. This is significantly better than the theoretical numbers from Table 3. When running exactly the same experiment on 96-bit (three 32-bit limbs instead of six 32-bit limbs) moduli the number of curves per second for the no-storage and windowing approach is 76 665 and 75 584 for $B_1 = 8192$ and 649 904 and 618 111 for $B_1 = 960$, respectively. We think that this behaviour can be partially explained by an increased memory usage for the windowing approach and a better handling of the no-storage approach by the compiler since this approach uses fewer variables.

Another interesting observation is that the FPGA performance per 100 USD is lower than that of the CPU-based approaches. Furthermore, aided by the no-storage approach outlined in this paper, the GPU performance is almost an order of magnitude faster per 100 USD than the CPU and more than a order of magnitude faster compared to the fastest FPGA results. This suggests that GPUs are the best platform, i.e. give the best performance / price ratio, for integer cofactorization.

7   Conclusion

The relatively new Edwards curves combined with the fast arithmetic from extended twisted Edwards coordinates are faster compared to using Montgomery curves. This speed-up comes at a price, namely a larger memory requirement which, when optimizing for speed, grows roughly linearly in the size of B_1, whereas the memory requirement in the Montgomery curves setting is constant and small. Inspired by the approach from Dixon and Lenstra and using the fact that only a few popular B_1-values are used in practice in NFS, we have presented techniques to reduce the memory requirement significantly by doing precomputations for these B_1-values. In these precomputations we tested over 10^{12} integers coming from chains with a low addition/doubling ratio for smoothness and combined them using a greedy approach. Our results show that we require significantly less memory compared to the current state-of-the-art Edwards ECM approach, and are even slightly faster. This makes our approach extremely suitable for memory-constrained parallel architectures like GPUs. This is demonstrated by our GPU implementation which sets a new ECM cofactorization throughput speed record.

Acknowledgments. Much appreciated incisive comments by the Asiacrypt'12 reviewers helped improve the quality of this paper. This work was supported by the Swiss National Science Foundation under grant numbers 200020-132160 and 200021-119776.

References

1. Bernstein, D.J., Birkner, P., Lange, T.: Starfish on Strike. In: Abdalla, M., Barreto, P.S.L.M. (eds.) LATINCRYPT 2010. LNCS, vol. 6212, pp. 61–80. Springer, Heidelberg (2010)
2. Bernstein, D.J., Birkner, P., Lange, T., Peters, C.: EECM: ECM using Edwards curves (2010), Software, http://eecm.cr.yp.to/
3. Bernstein, D.J., Birkner, P., Lange, T., Peters, C.: ECM using Edwards curves. Mathematics of Computation (to appear, 2012)
4. Bernstein, D.J., Chen, H.-C., Chen, M.-S., Cheng, C.-M., Hsiao, C.-H., Lange, T., Lin, Z.-C., Yang, B.-Y.: The billion-mulmod-per-second PC. In: Special-purpose Hardware for Attacking Cryptographic Systems, SHARCS 2009, pp. 131–144 (2009)
5. Bernstein, D.J., Chen, T.-R., Cheng, C.-M., Lange, T., Yang, B.-Y.: ECM on Graphics Cards. In: Joux, A. (ed.) EUROCRYPT 2009. LNCS, vol. 5479, pp. 483–501. Springer, Heidelberg (2009)

6. Bernstein, D.J., Lange, T.: Analysis and Optimization of Elliptic-Curve Single-Scalar Multiplication. In: Mullen, G.L., Panario, D., Shparlinski, I.E. (eds.) Finite Fields and Applications. Contemporary Mathematics Series, vol. 461, pp. 1–19. American Mathematical Society (2008)

7. Bos, J.W., Kleinjung, T.: ECM at work, project page (2012), http://research.microsoft.com/ecmatwork/

8. Bos, J.W., Kleinjung, T., Lenstra, A.K., Montgomery, P.L.: Efficient SIMD arithmetic modulo a Mersenne number. In: IEEE Symposium on Computer Arithmetic, ARITH-20, pp. 213–221. IEEE Computer Society (2011)

9. Brauer, A.: On addition chains. Bulletin of the American Mathematical Society 45, 736–739 (1939)

10. Brent, R.P.: Some integer factorization algorithms using elliptic curves. Australian Computer Science Communications 8, 149–163 (1986)

11. de Meulenaer, G., Gosset, F., de Dormale, G.M., Quisquater, J.-J.: Integer factorization based on elliptic curve method: Towards better exploitation of reconfigurable hardware. In: Field-Programmable Custom Computing Machines, FCCM 2007, pp. 197–206. IEEE Computer Society (2007)

12. Dixon, B., Lenstra, A.K.: Massively Parallel Elliptic Curve Factoring. In: Rueppel, R.A. (ed.) EUROCRYPT 1992. LNCS, vol. 658, pp. 183–193. Springer, Heidelberg (1993)

13. Edwards, H.M.: A normal form for elliptic curves. Bulletin of the American Mathematical Society 44, 393–422 (2007)

14. Franke, J., Kleinjung, T.: GNFS for linux. Software (2012)

15. Franke, J., Kleinjung, T., Morain, F., Wirth, T.: Proving the Primality of Very Large Numbers with fastECPP. In: Buell, D.A. (ed.) ANTS 2004. LNCS, vol. 3076, pp. 194–207. Springer, Heidelberg (2004)

16. Gaj, K., Kwon, S., Baier, P., Kohlbrenner, P., Le, H., Khaleeluddin, M., Bachimanchi, R.: Implementing the Elliptic Curve Method of Factoring in Reconfigurable Hardware. In: Goubin, L., Matsui, M. (eds.) CHES 2006. LNCS, vol. 4249, pp. 119–133. Springer, Heidelberg (2006)

17. Güneysu, T., Kasper, T., Novotny, M., Paar, C., Rupp, A.: Cryptanalysis with COPACOBANA. IEEE Transactions on Computers 57, 1498–1513 (2008)

18. Guy, R.: Unsolved problems in number theory, 3rd edn., vol. 1. Springer (2004)

19. Hisil, H., Wong, K.K.-H., Carter, G., Dawson, E.: Twisted Edwards Curves Revisited. In: Pieprzyk, J. (ed.) ASIACRYPT 2008. LNCS, vol. 5350, pp. 326–343. Springer, Heidelberg (2008)

20. Kleinjung, T., Aoki, K., Franke, J., Lenstra, A.K., Thomé, E., Bos, J.W., Gaudry, P., Kruppa, A., Montgomery, P.L., Osvik, D.A., te Riele, H., Timofeev, A., Zimmermann, P.: Factorization of a 768-Bit RSA Modulus. In: Rabin, T. (ed.) CRYPTO 2010. LNCS, vol. 6223, pp. 333–350. Springer, Heidelberg (2010)

21. Kruppa, A.: A software implementation of ECM for NFS. Research Report RR-7041, INRIA (2009), http://hal.inria.fr/inria-00419094/PDF/RR-7041.pdf

22. Lenstra, A.K., Lenstra Jr., H.W.: Algorithms in number theory. In: van Leeuwen, J. (ed.) Handbook of Theoretical Computer Science (vol. A: Algorithms and Complexity), pp. 673–715. Elsevier and MIT Press (1990)

23. Lenstra, A.K., Lenstra Jr., H.W.: The Development of the Number Field Sieve. Lecture Notes in Mathematics, vol. 1554. Springer (1993)

24. Lenstra Jr., H.W.: Factoring integers with elliptic curves. Annals of Mathematics 126(3), 649–673 (1987)

25. Loebenberger, D., Putzka, J.: Optimization Strategies for Hardware-Based Cofactorization. In: Jacobson Jr., M.J., Rijmen, V., Safavi-Naini, R. (eds.) SAC 2009. LNCS, vol. 5867, pp. 170–181. Springer, Heidelberg (2009)
26. Montgomery, P.L.: Speeding the Pollard and elliptic curve methods of factorization. Mathematics of Computation 48(177), 243–264 (1987)
27. Montgomery, P.L.: An FFT extension of the elliptic curve method of factorization. PhD thesis, University of California (1992)
28. Morain, F., Olivos, J.: Speeding up the computations on an elliptic curve using addition-subtraction chains. Informatique Théorique et Applications/Theoretical Informatics and Applications 24, 531–544 (1990)
29. NVIDIA. NVIDIA's next generation CUDA compute architecture: Fermi (2009)
30. NVIDIA. NVIDIA CUDA Programming Guide 3.2 (2010)
31. Pisinger, D.: A minimal algorithm for the multiple-choice knapsack problem. European Journal of Operational Research 83(2), 394–410 (1995)
32. Pollard, J.M.: The lattice sieve. In: [23], pp. 43–49
33. Pollard, J.M.: Theorems on factorization and primality testing. Proceedings of the Cambridge Philosophical Society 76, 521–528 (1974)
34. Pomerance, C.: The Quadratic Sieve Factoring Algorithm. In: Beth, T., Cot, N., Ingemarsson, I. (eds.) EUROCRYPT 1984. LNCS, vol. 209, pp. 169–182. Springer, Heidelberg (1985)
35. Scholz, A.: Aufgabe 253. Jahresbericht der deutschen Mathematiker-Vereingung 47, 41–42 (1937)
36. Silverman, J.H.: *The Arithmetic of Elliptic Curves*. Gradute Texts in Mathematics, vol. 106. Springer (1986)
37. Šimka, M., Pelzl, J., Kleinjung, T., Franke, J., Priplata, C., Stahlke, C., Drutarovský, M., Fischer, V.: Hardware factorization based on elliptic curve method. In: Field-Programmable Custom Computing Machines, FCCM 2005, pp. 107–116. IEEE Computer Society (2005)
38. Thurber, E.G.: On addition chains $l(mn) \leq l(n) - b$ and lower bounds for $c(r)$. Duke Mathematical Journal 40, 907–913 (1973)
39. Zimmermann, P., Dodson, B.: 20 Years of ECM. In: Hess, F., Pauli, S., Pohst, M. (eds.) ANTS 2006. LNCS, vol. 4076, pp. 525–542. Springer, Heidelberg (2006)
40. Zimmermann, R., Güneysu, T., Paar, C.: High-performance integer factoring with reconfigurable devices. In: Field Programmable Logic and Applications, FPL 2010, pp. 83–88. IEEE (2010)
41. Zimmermann, P., et al.: GMP-ECM (elliptic curve method for integer factorization) (2012), Software, https://gforge.inria.fr/projects/ecm/

IND-CCA Secure Cryptography
Based on a Variant of the LPN Problem

Nico Döttling[1], Jörn Müller-Quade[1], and Anderson C.A. Nascimento[2]

[1] Karlsruhe Institute of Technology, Karlsruhe, Germany
{doettling,mueller-quade}@kit.edu
[2] University of Brasilia, Brasilia, Brazil
andclay@ene.unb.br

Abstract. In 2003 Michael Alekhnovich (FOCS 2003) introduced a novel variant of the learning parity with noise problem and showed that it implies IND-CPA secure public-key cryptography. In this paper we introduce the first public-key encryption-scheme based on this assumption which is IND-CCA secure in the standard model. Our main technical tool to achieve this is a novel all-but-one simulation technique based on the correlated products approach of Rosen and Segev (TCC 2009). Our IND-CCA1 secure scheme is asymptotically optimal with respect to ciphertext-expansion. To achieve IND-CCA2 security we use a technique of Dolev, Dwork and Naor (STOC 1991) based on one-time-signatures. For practical purposes, the efficiency of the IND-CCA2 scheme can be substantially improved by the use of additional assumptions to allow for more efficient signature schemes. Our results make Alekhnovich's variant of the learning parity with noise problem a promising candidate to achieve post quantum cryptography.

Keywords: IND-CCA2 Security, Learning Parity with Noise, All-But-One Decryption.

1 Introduction

This paper presents the first IND-CCA2 secure cryptosystem based on a computational assumption first introduced by Michael Alekhnovich in the year 2003 [Ale03]. This assumption essentially states that for a given random linear code C with a constant rate, a random code word with an inverse square root fraction of noise is indistinguishable from a random string. Alekhnovich [Ale03] was able to construct a semantically secure cryptosystem which was based solely on this assumption. It can be seen as an special case of the decisional learning parity with noise (LPN) problem. The decisional LPN problem (henceforth LPN problem), asks to distinguish noisy binary linear equations $Ax + e$ from uniformly random. The problem is parametrized by the number of samples provided (i.e the number of rows of A) and the amount of noise (i.e. the distribution of e). While most cryptographic constructions based on LPN (e.g. [HB01, JW05, KSS10]) use the standard parameter-choice of a polynomial number of samples and a constant

X. Wang and K. Sako (Eds.): ASIACRYPT 2012, LNCS 7658, pp. 485–503, 2012.

fraction of noise, Alekhnovich's LPN problem uses a *linear* number of samples and an inverse square root fraction of noise. These two parameter-choices are apparently incomparable. On one side, providing a larger amount of samples makes the problem apparently easier. On the other side, a larger amount of noise seems to make the problem harder. Nevertheless, Alekhnovich's parameter choice seems to yield the stronger assumption, as constructing a public key cryptosystem from LPN with a constant fraction of noise remains an important open problem.

LPN assumptions are of a more combinatorial nature and seem incomparable to the algebraic assumptions needed for the McEliece cryptosystem. For the security of the McEliece cryptosystem one has to additionally assume that scrambled Goppa-codes are computationally indistinguishable from random linear codes [McE78, BS08, NIKM08, DMQN09]. Moreover, though there is a syntactic similarity to the learning with errors (LWE) problem, LPN and LWE also seem rather incomparable. LWE asks to distinguish a polynomial number noisy linear equations over $\mathbb{Z}_q$ (for a polynomial sized q), where the error-distribution is euclidean, from uniformly random. IND-CCA2 encryption schemes based on LWE [PW08, Pei09, MP12] use properties that are very specific to LWE (e.g. short dual-lattice bases) and not available in the binary domain. It has been open for nine years if an IND-CCA2 secure scheme could be built from Alekhnovich's LPN problem. In this paper we present such a IND-CCA2 secure scheme which is based on the all-but-one approach [DDN00, PW08, RS09]. The new construction is asymptotically optimal for IND-CCA1 security. It has only a constant factor ciphertext-expansion and the ciphertexts are of size $O(k^{2/(1-2\epsilon)})$, where k is the security parameter and ϵ a small constant. To achieve IND-CCA2 security we use a generic transformation based on one-time-signatures [DDN00]. A more efficient construction is possible using additional assumptions yielding more efficient signature schemes. The trapdoor of our scheme is substantially different from Alekhnovich's original construction, but bears some similarities with the above-mentioned lattice-based constructions. It allows witness recovery and decryption with incomplete keys, which is necessary for applying the all-but-one approach. Different from [PW08, RS09, Pei09, DMQN09] we do not achieve the all-but-one property by repeatedly encrypting the same ciphertext or a correlated product. We employ a bitwise decryption and use error correction to cope with incomplete decryptions. The novel all-but-one simulation technique employed in this construction allows for a significant improvement in efficiency compared with previous constructions. While this new technique might be of interest in lattice-based cryptography, we see no obvious way to make use of our technique in McEliece-based constructions. Crucial to our technique is the ability to recover individual bits of a plaintext from the ciphertext using a partial secret key. This, however, seems out of reach for constructions based on the McEliece assumption.

Related Work. Ciphertext indistinguishability under chosen ciphertext attacks (IND-CCA2) security [RS91] is one of the strongest known notions of security for public key encryption schemes (PKE). Many computational assumptions have been used in the literature for obtaining cryptosystems meeting this security notion. Given one-way trapdoor permutations, CCA2 security can be obtained from any semantically secure public key cryptosystem [NY90, Sah99, Lin03]. Efficient constructions are also known based on number-theoretic assumptions [CS98, CS03, HK09], lattice-based assumptions [PW08, Pei09, MP12], the McEliece assumption [DMQN09] or identity based encryption schemes [CHK04].

2   Preliminaries

2.1   Coding-Theory

We need a few coding-theoretic facts and constructions for our schemes and proofs. We denote the finite field with q elements by $\mathbb{F}_q$. The hamming-weight $|x|$ of a vector $x \in \mathbb{F}_q^n$ is the number of its non-zero locations. The q-ary entropy function is defined as $H_q(\alpha) = \alpha \log_q(q - 1) - \alpha \log_q \alpha - (1 - \alpha) \log_q(1 - \alpha)$. It assumes its maximum at $\alpha = 1 - 1/q$ with $H_q(1 - 1/q) = 1$. The volume $\mathsf{Vol}_q(\alpha n, n)$ of the hamming-ball of radius αn in $\mathbb{F}_q^n$ can be bounded by $q^{H_q(\alpha) \cdot n - o(n)} \leq \mathsf{Vol}_q(\alpha n, n) \leq q^{H_q(\alpha) \cdot n}$.

Random Codes and the Gilbert-Varshamov bound. The Gilbert-Varshamov bound guarantees the existence of q-ary codes with almost maximal relative minimum-distance $1 - 1/q$. Moreover, with high probability, randomly chosen codes enjoy this property. Let $n, d, k \in \mathbb{N}$ and $\lambda > 0$. If it holds that $k \leq n - \log_q \mathsf{Vol}_q(d, n) - \lambda n$, then the code $\mathcal{C}(G)$ generated by a uniformly chosen matrix $G \in \mathbb{F}_q^{n \times k}$ has minimum-distance at least d, except with probability $q^{-\lambda n}$. Therefore, if $\delta < 1 - 1/q$ it holds that $n - \log_q \mathsf{Vol}_q(\delta n, n) \geq (1 - H_q(\delta))n =: \zeta n$. Thus, if $k \leq \zeta n/2$, a uniformly random chosen matrix $G \in \mathbb{F}_q^{n \times k}$ generates a code $\mathcal{C}(G)$ with minimum-distance at least δn, except with probability $q^{-\zeta n/2}$.

Asymptotically good codes with efficient error-correction. The decryption algorithm of our scheme will introduce errors in the plaintext when decrypting. We will therefore use asymptotically good error-correcting codes $\mathcal{C}$ with efficient error-correction algorithm $\mathsf{Decode}_{\mathcal{C}}$ to encode plaintexts. Prominent examples of such codes are binary expander-codes [SS96, Zém01]: There exists an explicit family of binary linear codes $\{\mathcal{C}_n\}$ of constant rate R arbitrarily close to 1 that can efficiently correct an α-fraction of errors, for a constant $\alpha > 0$.

2.2   Bernoulli Distributions and Bounds

In this section we will briefly gather some facts about low-noise Bernoulli distributions. While Alekhnovich's [Ale03] original proposal used a noise distribution that samples vectors of low-weight t uniformly at random, we will use Bernoulli-distributions where each bit of a vector is 1 with probability t/n and otherwise 0. The advantage of Bernoulli-distributions over the former distribution is that all components are independent of one another. We will take advantage of this fact when bounding the hamming-weight of matrix-vector products when the matrix is chosen from a Bernoulli distribution. The decryption-algorithm of Alekhnovich's and our encryption-scheme computes inner-products of Bernoulli-distributed vectors. To ensure that the inner-product of two Bernoulli-distributed vectors is 0 with high probability, we need to choose the bit-flip probability ρ below a $1/\sqrt{n}$ amount. If ρ is too big (e.g. constant), then the distribution of the inner-product would be statistically close to uniform and our decryption-approach would fail. Finally, we show that matrices X chosen from a component-wise low-noise Bernoulli distribution enjoy (with high probability) the property, that a product Xs has low-hamming-weight, for any vector s with sufficiently small hamming-weight. We will call such matrices *good*, and we will use this property for proving correctness of our schemes and in the proof of IND-CCA1 security.

Bernoulli distributions. For a noise-parameter ρ, we write χ_ρ for the Bernoulli-distribution that outputs 1 with probability ρ and 0 with probability $1 - \rho$. The distribution of the hamming-weight of a vector of n iid distributed Bernoulli-distributed random variables is the binomial distribution $B_{\rho,n}$. Throughout the paper, we frequently need to bound Binomial distributions. For this we require two different Chernoff bounds. Let x be distributed by χ_ρ^n.

1. It holds for any $R \geq 6\rho n$ that $\Pr[|x| > R] < 2^{-R}$.
2. It holds for any $0 < \delta < 1$ that $\Pr[||x| - \rho n| \geq \delta\rho n] < 2e^{-\delta^2 \rho n/3}$.

Distributions of inner products. For the decryption-algorithms of our schemes we require that the inner-product of a Bernoulli-distributed vector x and a vector s of small hamming weight is 0 with probability bounded away from $1/2$. We will thus show that the probability of the inner-product being 1 is sub-constant for a proper choice of ρ. Let $s \in \mathbb{F}_2^n$ be a fixed vector and x be distributed by χ_ρ^n. By a simple XOR-Lemma, it holds that

$$\Pr[x^T s = 1] = \frac{1}{2} \cdot (1 - (1 - 2\rho)^{|s|}),$$

i.e. the random variable $x^T s$ is distributed according to $\chi_{\rho'}$ with $\rho' = \frac{1}{2} \cdot (1 - (1 - 2\rho)^{|s|})$. If it holds that $\rho = \rho(n) = O(n^{-1/2-\epsilon})$ for some constant $\epsilon > 0$ and $|s| < \gamma\rho n$ for some constant $\gamma > 0$, we get the following estimate for ρ'. By the mean-value-theorem it holds for any p in the interval $(0, e^{-1})$ that $e^{-ep} \leq 1 - p$, therefore we get

$$\rho' = \frac{1}{2} \cdot (1 - (1 - 2\rho)^{|s|}) \leq \frac{1}{2} \cdot (1 - e^{-2e\rho|s|}) \leq \frac{1}{2} \cdot (1 - e^{-2e\gamma\rho^2 n}) = \frac{1}{2} \cdot (1 - e^{-O(n^{-2\epsilon})}).$$

The last term is sub-constant in n, i.e. $\rho'(n) = o(1)$. This means that for sufficiently large n ρ' is arbitrarily small.

Multiplication with random matrices. We will now give bounds for how much the hamming weight of a vector s increases when multiplied with a matrix $X \in \mathbb{F}_2^{l \times n}$ chosen from $\chi_\rho^{l \times n}$. Let x be distributed by χ_ρ^n and the hamming-weight of s be bounded by $|s| < \gamma \rho n$. Then by the above $\rho' = \Pr[x^T s = 1]$ can be made an arbitrarily small constant if $\rho = O(n^{-1/2-\epsilon})$. If $X \in \mathbb{F}_2^{l \times n}$ is distributed by $\chi_\rho^{l \times n}$, then $|Xs|$ is distributed by the Binomial-distribution $B_{\rho', l}$. The Chernoff-bound thus yields that for any $R \geq 6\rho' l$ it holds that $\Pr[|Xs| > R] < 2^{-R}$. The volume $\mathrm{Vol}_2(\gamma \rho n, n)$ of the hamming-ball of radius $\gamma \rho n$ in $\mathbb{F}_2^n$ is bounded by $2^{H_2(\gamma \rho)n}$. Thus, there are at most $2^{H_2(\gamma \rho)n}$ vectors s satisfying $|s| < \gamma \rho n$. A union-bound yields for any $R \geq 6\rho' l$

$$\Pr[\exists s \in \mathbb{F}_2^n : |s| < \gamma \rho n \text{ and } |Xs| > R] < 2^{H_2(\gamma \rho)n} \cdot 2^{-R}.$$

If $l = \Omega(n)$ and $\beta > 0$ it holds that

$$\Pr[\exists s \in \mathbb{F}_2^n : |s| < \gamma \rho n \text{ and } |Xs| > \beta l] < 2^{-\Omega(n)},$$

as $H_2(\gamma \rho)n$ is sub-linear in n (i.e. $o(n)$) since $\rho = O(n^{-1/2-\epsilon})$.

Definition 1. *Fix a constant β and $\epsilon = \epsilon(n)$. We shall call a matrix $X \in \mathbb{F}_2^{l \times n}$ (β, ϵ)-good, if for all $s \in \mathbb{F}_2^n$ with $|s| < \epsilon n$ it holds that $|Xs| \leq \beta l$.*

The above now implies that for $\rho = O(n^{-1/2-\epsilon})$, any fixed $\beta, \gamma > 0$ and sufficiently large n, a matrix X sampled from $\chi_\rho^{l \times n}$ is $(\beta, \gamma \rho)$-good with overwhelming probability in n.

2.3 Public Key Encryption

This Section is only meant to provide reference for the standard notions of security for encryption schemes and can be safely skipped. Let k be a security parameter.

Definition 2. *A public key encryption scheme* PKE *is a tuple* (KeyGen, Enc, Dec), *such that*

- KeyGen(1^k) *is a PPT-algorithm that takes a security-parameter k and outputs a pair of public and private keys* (pk, sk).
- Enc$_{pk}(m)$ *is a PPT-algorithm that takes a public key pk, a message m and outputs a ciphertext c.*
- Dec$_{sk}(c)$ *is an efficient deterministic algorithm taking as input a secret key sk and a ciphertext c and outputs a plaintext m.*

A standard-requirement for public key encryption is correctness.

Definition 3. *We say that* PKE $=$ (KeyGen, Enc, Dec) *is correct, if it holds for all plaintexts m that*

$$\Pr[\mathsf{Dec}_{sk}(\mathsf{Enc}_{pk}(m)) \neq m : (pk, sk) = \mathsf{KeyGen}(1^k)] < \mathsf{negl}(k).$$

The three security notions for public key encryption we are concerned with in this paper are IND-CPA, IND-CCA1 and IND-CCA2 security. Let $\mathcal{A}$ be an adversary.

Experiment: IND-CPA.

- Generate a pair of keys $(pk, sk) = \mathsf{KeyGen}(1^k)$. Run $\mathcal{A}$ on input pk.
- Once $\mathcal{A}$ outputs a pair (m_0, m_1), flip a coin b and compute $c^* = \mathsf{Enc}_{pk}(m_b)$. Give input c^* to $\mathcal{A}$ and continue its computation.
- Let b' be $\mathcal{A}$'s output. Output 1 if $b' = b$ an 0 otherwise.

Experiment: IND-CCA1.

- Generate a pair of keys $(pk, sk) = \mathsf{KeyGen}(1^k)$. Give $\mathcal{A}$ access to a decryption-oracle $\mathsf{Dec}_{sk}(\cdot)$ and run $\mathcal{A}$ on input pk.
- Once $\mathcal{A}$ outputs a pair (m_0, m_1), flip a coin b and compute $c^* = \mathsf{Enc}_{pk}(m_b)$. Give input c^* to $\mathcal{A}$ and continue its computation *without* access to the decryption-oracle.
- Let b' be $\mathcal{A}$'s output. Output 1 if $b' = b$ an 0 otherwise.

Experiment: IND-CCA2.

- Generate a pair of keys $(pk, sk) = \mathsf{KeyGen}(1^k)$. Give $\mathcal{A}$ access to a decryption-oracle $\mathsf{Dec}_{sk}(\cdot)$ and run $\mathcal{A}$ on input pk.
- Once $\mathcal{A}$ outputs a pair (m_0, m_1), flip a coin b and compute $c^* = \mathsf{Enc}_{pk}(m_b)$. Give input c^* to $\mathcal{A}$ and continue its computation *with* access to the decryption-oracle.
- Let b' be $\mathcal{A}$'s output. Output 1 if $b' = b$ an 0 otherwise.

Definition 4. *For $X \in \{CPA, CCA1, CCA2\}$, we say that the scheme* $\mathsf{PKE} = (\mathsf{KeyGen}, \mathsf{Enc}, \mathsf{Dec})$ *is IND-X secure, if it holds for every PPT-adversary $\mathcal{A}$ that* $\mathsf{Adv}_{IND\text{-}X}(\mathcal{A}) = |\Pr[IND\text{-}X(\mathcal{A}) = 1] - 1/2| \leq \mathsf{negl}(k)$.

2.4   One-Time Signatures

We also briefly recall the definition of one-time signatures [Lam79]. Let k be a security parameter.

Definition 5. *A one-time signature scheme* SIG *is a tuple* $(\mathsf{Gen}, \mathsf{Sign}, \mathsf{Verify})$, *such that*

- $\mathsf{Gen}(1^k)$ *is a PPT-algorithm that takes a security-parameter k and outputs a pair of verification and signature keys (vk, sgk).*
- $\mathsf{Sign}_{sgk}(m)$ *is a PPT-algorithm that takes a signature key sgk, a message m and outputs a signature σ.*
- $\mathsf{Verify}_{vk}(m, \sigma)$ *is a PPT-algorithm taking as input a verification key vk, a message m and a signature c and outputs a bit $b \in \{0, 1\}$.*

We require one-time signature schemes to be correct.

Definition 6. *We say that* $\mathsf{SIG} = (\mathsf{Gen}, \mathsf{Sign}, \mathsf{Verify})$ *is correct, if it holds for all messages m that*

$$\Pr[\mathsf{Verify}_{vk}(m, \mathsf{Sign}_{sgk}(m)) = 1 : (vk, sgk) = \mathsf{Gen}(1^k)] > 1 - \mathsf{negl}(k).$$

Moreover, we require existential unforgeability under one-time chosen message attacks (EUF-CMA security), specified by the following experiment. Let $\mathcal{A}$ be an adversary.

Experiment: EUF-CMA

- Generate a pair of keys $(vk, sgk) = \mathsf{Gen}(1^k)$. Give $\mathcal{A}$ a access to a signing-oracle $\mathsf{Sign}_{sgk}(\cdot)$ that signs one message m^* of $\mathcal{A}$'s choice and then outputs $\bot$ for any further signing-queries. Run $\mathcal{A}$ on input vk
- Once $\mathcal{A}$ outputs a pair (m, σ) with $m \neq m^*$, compute $b = \mathsf{Verify}_{vk}(m, \sigma)$ and output b. Otherwise output 0.

Definition 7. *We say that* $\mathsf{SIG} = (\mathsf{Gen}, \mathsf{Sign}, \mathsf{Verify})$ *is EUF-CMA secure, if it holds for every PPT-adversary* $\mathcal{A}$ *that* $\Pr[\text{EUF-CMA}(\mathcal{A}) = 1] \leq \mathsf{negl}(k)$.

EUF-CMA secure one-time signature schemes can be constructed from any one-way function [Lam79].

3   The Hardness-Assumption

The basic problem we will base the security of our scheme upon is a variant of the decisional learning parity with noise (LPN) problem. Roughly speaking, the LPN problem asks to distinguish a number of noisy samples of a linear function (specified by a secret vector x) from uniform random. The variant considered here differs from the standard LPN problem in two aspects. First, the distinguisher is provided only linear number of samples, rather than an arbitrary polynomial number. Second, the noise-level in this variant is significantly lower than in the standard LPN problem. While the standard LPN problem comes with an error-distribution that flips each output-bit with a small, but constant probability, for this variant the probability is sub-constant. More precisely, we will work with a bit-flip probability of the order $O(n^{-1/2-\epsilon})$ for some small constant ϵ. Here, n is the size of the secret x in bits.

Problem 1. Let $n \in \mathbb{N}$ be a problem parameter, $m = O(n)$ and $\epsilon > 0$ and $\rho = \rho(n) = O(n^{-1/2-\epsilon})$. Let $A \in \mathbb{F}_2^{m \times n}$ be chosen uniformly at random, $x \in \mathbb{F}_2^n$ be chosen uniformly at random and e according to χ_ρ^m. The problem is, given A and y, to decide whether y is distributed according to $Ax+e$ or chosen uniformly at random.

Currently, the best classical algorithms to attack Problem 1 require time of the order $2^{\Omega(n^{1/2-\epsilon})}$ [Ste88, CC98, MMT11, BLP11, BJMM12]. Moreover, there are no quantum algorithms known performing significantly better than the best classical algorithms. In our constructions we will choose n by $n = O(k^{2/(1-2\epsilon)})$, where k is the security parameter. This normalizes the hardness of Problem 1 to $2^{\Theta(k)}$. Thus, we choose ρ by $\rho(k) = O(k^{-(1+2\epsilon)/(1-2\epsilon)})$. In the full version of this paper, we provide a reduction establishing the hardness of problem 1 based on the hardness-assumption used in [Ale03], which uses a different error-distribution. It will be necessary to use a *normal-form* (as in [ACPS09]) of Problem 1 in our cryptographic constructions, which is stated in Problem 2. In this normal-form, the secret x is drawn from the noise-distribution χ_ρ^n.

Problem 2. Let $n \in \mathbb{N}$ be a problem parameter, $m = O(n)$, $\epsilon > 0$ and $\rho = O(n^{-1/2-\epsilon})$. Let $A \in \mathbb{F}_2^{m \times n}$ be chosen uniformly at random, x be distributed according to χ_ρ^n and e be distributed according to χ_ρ^m. The problem is, given A and y, to decide whether y is distributed according to $Ax+e$ or chosen uniformly at random.

The hardness of Problem 2 can be established by a simple reduction from Problem 1, given in the full version of this paper. By a simple hybrid-argument, it follows that that a matrix-version of problem 2 is also hard.

Problem 3. Let $n \in \mathbb{N}$ be a problem parameter, $m, k = \Theta(n)$, $\epsilon > 0$ and $\rho = O(n^{-1/2-\epsilon})$. Let $A \in \mathbb{F}_2^{n \times k}$ be chosen uniformly at random, $T \in \mathbb{F}_2^{m \times n}$ be distributed according to $\chi_\rho^{m \times n}$ and X be distributed according to $\chi_\rho^{m \times k}$. The problem is, given A and B, to decide whether B is distributed according to $TA + X$ or chosen uniformly at random in $\mathbb{F}_2^{m \times k}$.

In the security-proof for our schemes, we will use Problem 3 to establish pseudorandomness of the public keys, while we use Problem 2 to establish pseudorandomness of the ciphertexts.

4   Outline of the Techniques

In this Section, we will outline the techniques used to construct an IND-CCA1 secure scheme based on the hardness of Problem 2 and Problem 3. We will provide the full presentation in the subsequent sections. Let henceforth $\rho = O(n^{-1/2-\epsilon})$ for a small constant $\epsilon > 0$.

We will start with a rough outline of a scheme that encrypts single bits and has a substantial decryption-error. On a technical level, this first building block resembles the schemes of Regev [Reg05] and the Dual-Regev Scheme of Gentry et al. [GPV08] (which both live in the LWE realm). Public keys for our scheme are pairs (A, b^T), where $A \in \mathbb{F}_2^{l_1 \times n}$ is chosen uniformly at random and $b^T = t^T A + x^T$ with $t \in \mathbb{F}_2^{l_1}$ is distributed by $\chi_\rho^{l_1}$ and $x \in \mathbb{F}_2^n$ by χ_ρ^n. The secret key is t^T. To encrypt a message $m \in \mathbb{F}_2$, sample s according to χ_ρ^n, e_1 according to χ_ρ^l and e_2 according to χ_ρ. Compute $c = (As + e_1, b^T s + e_2 + m)$ and output c. To decrypt a ciphertext $c = (c_1, c_2)$, compute $y = c_2 - t^T c_1$ and output y. The output y is a *noisy* version of the plaintext m, since it holds that $y = c_2 - t^T c_1 = b^T s + e_2 + m - t^T (As + e_1) = m + t^T As + x^T s + e_2 - t^T As - t^T e_1 = m + x^T s + e_2 - t^T e_1$. By the properties of the distribution χ_ρ, the error-term $v = x^T s + e_2 - t^T e_1$ is 0 with probability bounded away from $1/2$, i.e. it holds $y = m$ with substantial probability.

This decryption-error can be dealt with by encoding m (which is now a bit-vector of length n) using an error-correcting code as follows. Let $G \in \mathbb{F}_2^{l_2 \times n}$ be the generator-matrix of a binary linear error-correcting code $\mathcal{C}$. The modified scheme works as follows. Public keys are of the form (A, B) with A as above and $B = TA + X$, where T is chosen from $\chi_\rho^{l_2 \times l_1}$ and X from $\chi_\rho^{l_2 \times n}$. The secret key is T. Messages $m \in \mathbb{F}_2^n$ are encrypted as $c = (As + e_1, Bs + e_2 + Gm)$, with s, e_1, e_2 sampled from the corresponding χ_ρ distributions. Decryption computes $y = c_2 - Tc_1 = Gm + Xs + e_2 - Te_1$. Since the matrices T and X were chosen from a χ_ρ distribution, they are good (as defined in Section 2.2) with overwhelming probability. Thus the error-term $v = Xs + e_2 - Te_1$ has a low hamming-weight and we can use the decoding-procedure of $\mathcal{C}$ to recover m. The IND-CPA security of this scheme follows easily by the hardness of Problem 2 and Problem 3. However, we will require a witness-recovering IND-CPA scheme for the construction of our IND-CCA scheme. A scheme is witness recovering if the decryption recovers the randomness used to encrypt. For the above scheme however, the vector s is "lost" during decryption. We circumvent this problem by using some sort of key-encapsulation. Instead of encrypting a plaintext-vector m using the above scheme, we encrypt the witness s (which has the same size as m). We will then use another instance of Problem 2 to encrypt the plaintext m (using s as symmetric key). Encrypting the witness s instead of m will not harm security. By Problem 3, the matrix B is pseudorandom. Therefore, the matrix $B + G$ is also pseudorandom. Thus, the second part of the ciphertext $c_2 = Bs + e_2 + Gs = (B + G)s + e_2$ is also pseudorandom by Problem 2. Observe that we do not need the entire secret key T to recover s from a ciphertext c. Let $y = c_2 - Tc_1 = Gs + Xs + e_2 - Te_1$. To recover the i-th component y_i of y, we merely need the i-th row t_i^T of the matrix T. If we posses a sufficient amount of the rows of T, yet not all of them, we can still recover s by computing y_i for all the i for which t_i^T is known and setting $y_i = \perp$ (erasure) otherwise. We can now recover s by performing a combined error- and erasure-correction on y using the decoding algorithm of $\mathcal{C}$. If it is guaranteed that the number of erasures is very low, we can simply set all erasures to random values (thereby introducing a few additional random errors) and use the standard decoding-algorithm $\mathsf{Decode}_{\mathcal{C}}$ of $\mathcal{C}$. Micciancio and Peikert [MP12] recently used a very similar witness-recovering mechanism in their construction of an improved LWE-based IND-CCA2 scheme. While our construction uses off-the-shelf binary error-correcting codes to encode the witness s, they needed to construct a special family of lattices for this purpose. These lattices have a short dual basis and an efficient decoding algorithm, thus they can be seen as a euclidean analogue to efficiently decodable error-correcting codes with large minimum distance. We can now give an outline of our IND-CCA1 construction. It is an adoption of the all-but-one simulation-paradigm [PW08, RS09] to the special structure of our CPA scheme. The key-generation samples not just one, but q (for a constant q) matrices $B_1, \ldots, B_q$ and $T_1, \ldots, T_q$. Encryption first samples a tag τ, then derives an instance-public-key B_τ from $B_1, \ldots, B_q$. It further proceeds as the INC-CPA variant using the matrix B_τ instead of B. The ciphertext is (τ, c). Decryption takes the tag τ, derives an instance secret-key T_τ and uses T_τ to decrypt c. After recovering the random coins it checks whether they suffice a certain hamming-weight criterium. If not, it aborts, otherwise it outputs the plaintext m. The instance-key derivation will assemble the matrix B_τ by picking certain rows from the matrices $B_1, \ldots, B_q$ depending on the tag τ. In the security proof, there will be a single tag τ^* for which the simulator is completely oblivious of the instance-secret key T_{τ^*} (this is the tag where the IND-CPA challenge will be embedded). For all other tags, the simulator needs to be able to simulate a decryption-oracle. This means that no other instance-secret-key T_τ should share too many rows with T_{τ^*}. If this is the case, the simulator will be able to use an incomplete secret key to answer decryption-queries by the above observation. To guarantee that the instance-secret-keys T_τ have small *overlap* with one another, we will use a q-ary error-correcting encoding for the tags τ. This simulation-strategy requires that the hamming-weight of the ciphertext-noise satisfies a certain bound, otherwise the simulator is unable to correct the additional erasure caused by the incomplete secret key. This is the reason why the decryption needs to check the hamming-weight of the witnesses. The IND-CCA2 construction is obtained by replacing the randomly chosen tags τ with the verification keys of a one-time signature scheme and appending an according signature to the ciphertext. This transformation has been used in several contexts to obtain CCA2 secure encryption from different primitives [DDN00, CHK04, PW08, RS09, DMQN09]. The encryption primitives admitting such a transformation can be generalized under the notion of tag-based encryption schemes [Kil06].

5   The IND-CPA Scheme

In this Section we will provide the full construction of an IND-CPA secure encryption scheme. We will use this scheme in the construction of our CCA1 secure scheme.

Let k be a security parameter, $n \in O(k^{2/(1-2\epsilon)})$, $l_1, l_2, l_3 \in O(k^{2/(1-2\epsilon)})$ and $\rho = O(k^{-(1+2\epsilon)/(1-2\epsilon)})$. Let $G \in \mathbb{F}_2^{l_2 \times n}$ be the generator-matrix of a binary linear error-correcting code $\mathcal{C}$ and $\mathsf{Decode}_\mathcal{C}$ an efficient decoding procedure for $\mathcal{C}$ that corrects up to αl_2 errors (for a constant α). Further let $\mathcal{D} \subseteq \mathbb{F}_2^{l_3}$ be a binary error-correcting code with efficient encoding $\mathsf{Encode}_\mathcal{D}$ and error-correction $\mathsf{Decode}_\mathcal{D}$ that corrects up to λl_3 errors.

Construction 1. *The scheme* $\mathsf{PKE}_1 = (\mathsf{KeyGen}, \mathsf{Enc}, \mathsf{Dec})$ *is specified by*

- $\mathsf{KeyGen}(1^k)$: *Sample matrices* $A \in \mathbb{F}_2^{l_1 \times n}$ *and* $C \in \mathbb{F}_2^{l_3 \times n}$ *uniformly at random, sample the matrix* T *from* $\chi_\rho^{l_2 \times l_1}$ *and the matrix* X *from* $\chi_\rho^{l_2 \times n}$. *Set* $B = G + T \cdot A + X$. *Set* $pk = (A, B, C)$ *and* $sk = T$. *Output* (pk, sk).
- $\mathsf{Enc}_{pk}(m)$: *Takes a public key* $pk = (A, B, C)$ *and a plaintext* $m \in \mathbb{F}_2^n$ *as input, samples* s *from* χ_ρ^n, e_1 *from* $\chi_\rho^{l_1}$, e_2 *from* $\chi_\rho^{l_2}$ *and* e_3 *from* $\chi_\rho^{l_3}$. *It sets* $c_1 = A \cdot s + e_1$, $c_2 = B \cdot s + e_2$ *and* $c_3 = C \cdot s + e_3 + \mathsf{Encode}_\mathcal{D}(m)$. *Output* $c = (c_1, c_2, c_3)$.
- $\mathsf{Dec}_{sk}(c)$: *Takes a secret key* $sk = T$ *and a ciphertext* $c = (c_1, c_2, c_3)$ *as input. Computes* $y = c_2 - T \cdot c_1$ *and* $s = \mathsf{Decode}_\mathcal{C}(y)$. *Outputs* $\perp$ *if decoding fails. Otherwise computes* $m = \mathsf{Decode}_\mathcal{D}(c_3 - C \cdot s)$ *and outputs* m.

We will now show that this scheme is correct, i.e. the probability that a decryption-error occurs is negligible in k.

Lemma 1. *The scheme* PKE_1 *is correct.*

Proof. Decryption only fails if one of the two decoding operations fails. We will thus bound the probability of failure for both decoding operations. It holds that

$$y = c_2 - T \cdot c_1 = B \cdot s + e_2 - T(A \cdot s + e_1) = G \cdot s + X \cdot s + e_2 - T \cdot e_1.$$

Thus, it is sufficient to bound the hamming-weight of the error-term $v = X \cdot s + e_2 - T \cdot e_1$. Fix constants $\beta, \gamma > 0$ such that $2\beta + \gamma\rho < \alpha$ and $\gamma\rho < \lambda$. By a Chernoff-bound, it holds that $|s| < \gamma\rho n$, $e_1 < \gamma\rho l_1$, $e_2 < \gamma\rho l_2$ and $e_3 < \gamma\rho l_3$ with overwhelming probability in k. The decoding procedure $\mathsf{Decode}_{\mathcal{C}}$ can correct up to αl_2 errors. With overwhelming probability in k, both matrices X and T are $(\beta, \gamma\rho)$-good (see Section 2.2). Thus it holds that $|Xs| < \beta l_2$ and $|Te_1| < \beta l_2$ (for sufficiently large k). All together, it holds that

$$|v| \leq |Xs| + |e_2| + |Te_1| \leq 2\beta l_2 + \gamma\rho l_2 < \alpha l_2.$$

Therefore, the decoding-procedure $\mathsf{Decode}_{\mathcal{C}}$ will successfully recover s. Moreover, $\mathsf{Decode}_{\mathcal{D}}$ will successfully recover m as $|e_3| < \gamma\rho \cdot l_3 < \lambda l_3$.

We now turn to proof IND-CPA security of the scheme PKE_1.

Theorem 1. *Assume that Problem 2 is hard. Then the scheme* PKE_1 *is IND-CPA secure.*

Proof. Let $\mathcal{A}$ be PPT-bounded IND-CPA adversary against PKE_1. Consider the following sequence of games.

- Game 1: This is the IND-CPA experiment.
- Game 2: This is the same as game 1, except that during key-generation, the matrix B is chosen uniformly at random by the experiment.
- Game 3: The same as game 2, except that during encryption of the challenge-ciphertext, $c^* = (c_1^*, c_2^*, c_3^*)$ is chosen uniformly at random.

Clearly, $\mathcal{A}$'s advantage of winning game 3 is zero, as the challenge-ciphertext c^* is statistically independent of the challenge bit b chosen by the experiment. It remains to show that the views of $\mathcal{A}$ are computationally indistinguishable in game 1, 2 and 3. For contradiction, assume that $\mathcal{A}$ distinguishes game 1 and game 2 with non-negligible advantage $\nu_1(n)$. We will construct a distinguisher $\mathcal{B}_1$ that distinguishes the distributions $(A, T \cdot A + X)$ and (A, U) with advantage $\nu_1(k)$, contradicting the hardness of Problem 3. The input of $\mathcal{B}_1$ is an instance $(A^\dagger, B^\dagger)$. $\mathcal{B}_1$ simulates the interaction with $\mathcal{A}$ in the same way as game 1 does, except for the key generation step. Instead of generating A and B as in game 1, it sets $A = A^\dagger$ and $B = G + B^\dagger$. After the simulation terminates, $\mathcal{B}_1$ outputs whatever $\mathcal{A}$ outputs. Clearly, if $(A^\dagger, B^\dagger)$ is chosen according to $(A, T \cdot A + X)$, then $\mathcal{A}$'s view in $\mathcal{B}_1$'s simulation is identically distributed as in game 1. On the other hand, if $(A^\dagger, B^\dagger)$ is distributed according to (A, U), then $\mathcal{A}$'s view in $\mathcal{B}_1$'s simulation is identical to game 2. Thus it holds that $| \Pr[\mathcal{B}_1(A, TA + X)] - \Pr[\mathcal{B}_1(A, U)]| = | \Pr[\mathsf{view}_\mathcal{A}(\mathsf{Game}_1)] - \Pr[\mathsf{view}_\mathcal{A}(\mathsf{Game}_2)]| \geq \nu_1(k)$, which contradicts the hardness of Problem 3. Now assume that $\mathcal{A}$ distinguishes between game 2 and game 3 with non-negligible advantage $\nu_2(k)$. We will construct a distinguisher $\mathcal{B}_2$ that distinguishes the distributions $(M, Ms + e)$ and (M, u) with advantage $\nu_2(k)$, contradicting the hardness of Problem 2. Let the input of $\mathcal{B}_2$ be (M, r), where $M \in \mathbb{F}_2^{(l_1+l_2+l_3) \times n}$ and $r \in \mathbb{F}_2^{l_1+l_2+l_3}$. $\mathcal{B}_2$ first partitions M in three matrices $M_1 \in \mathbb{F}_2^{l_1 \times n}$, $M_2 \in \mathbb{F}_2^{l_2 \times n}$ and $M_3 \in \mathbb{F}_2^{l_3 \times n}$. Likewise, it partitions r into $r_1 \in \mathbb{F}_2^{l_1}$, $r_2 \in \mathbb{F}_2^{l_2}$ and $r_3 \in \mathbb{F}_2^{l_3}$. $\mathcal{B}_2$ simulates the interaction with $\mathcal{A}$ exactly like game 2, except for two details. In the key-generation step, it sets $A = M_1$, $B = M_2$ and $C = M_3$. Moreover, the challenge-ciphertext $c^* = (c_1^*, c_2^*, c_3^*)$ by $c_1^* = r_1$, $c_2^* = r_2$ and $c_3^* = r_3 + \mathsf{Encode}_\mathcal{D}(m_b)$. After the simulation terminates, $\mathcal{B}_1$ outputs whatever $\mathcal{A}$ outputs. Clearly, if (M, r) is chosen according to $(M, Ms + e)$, then $\mathcal{A}$'s view is identically distributed to game 2. On the other hand, if (M, r) is distributed according to (M, u), then $\mathcal{A}$'s view is identically distributed to game 3. Therefore, it holds that $| \Pr[\mathcal{B}_2(M, Ms+e)] - \Pr[\mathcal{B}_2(M, u)]| = | \Pr[\mathsf{view}_\mathcal{A}(\mathsf{Game}_2)] - \Pr[\mathsf{view}_\mathcal{A}(\mathsf{Game}_3)]| \geq \nu_2(k)$, which contradicts the hardness of problem 2. This concludes the proof.

6   The IND-CCA1 Scheme

In this Section, we will construct an IND-CCA1 scheme based on the scheme PKE_1 constructed in the last section. We will extend the encryption and decryption algorithms with an instance-key derivation step, that assigns a tag to each ciphertext and derives an instance public or secret key for each tag. These instance-keys will be used as keys for PKE_1. Moreover, we need to ensure that decryption only outputs a plaintext if an incomplete key would have already been sufficient to decrypt. Decryption therefore checks if the hamming-weight of the randomness used to encrypt is small enough. When the scheme is used honestly, this is the case with overwhelming probability. As in the last section, let k be a security parameter, $n \in O(k^{2/(1-2\epsilon)})$, $l_1, l_2, l_3 \in O(k^{2/(1-2\epsilon)})$ and $\rho = O(k^{-(1+2\epsilon)/(1-2\epsilon)})$. Let $G \in \mathbb{F}_2^{l_2 \times n}$ be the generator-matrix of a binary linear error-correcting code $\mathcal{C}$ and $\mathsf{Decode}_\mathcal{C}$ an efficient decoding procedure that corrects up to αl_2 errors (for a constant α). Let $\mathcal{D} \subseteq \mathbb{F}_2^{l_3}$ be a binary error-correcting code with efficient encoding $\mathsf{Encode}_\mathcal{D}$ and error-correction $\mathsf{Decode}_\mathcal{D}$ as before. Let $\mathcal{E} \subseteq \Sigma^{l_2}$ be a q-ary code over the alphabet Σ (with $q = |\Sigma|$) with relative minimum-distance δ and dimension n. Such a code can be generated randomly (see Section 2.1). We will now explain how the parameters δ and q must be chosen. Recall that $\mathsf{Decode}_\mathcal{C}$ corrects up to αl_2 errors. As explained earlier, α must be big enough to correct the decryption-error, which has hamming-weight less than $(2\beta + \gamma\rho)l_2$ (for any constant $\beta > 0$). As the additional error induced by erasures will have hamming weight $\leq (1 - \delta)l_2$, it is sufficient to choose δ (which must be smaller than $1 - 1/q$) such that $2\beta + \gamma\rho + 1 - \delta < \alpha$. As we can choose β and γ arbitrarily small, we can always find q and δ such that the above is met. Therefore, fix β, γ, q and δ such that for sufficiently large n it holds that

$2\beta + \gamma\rho + 1 - \delta < \alpha$. We can choose the constant β arbitrarily small and it holds that $\gamma\rho \in o(1)$. There exist constructions of efficiently decodable linear codes $\mathcal{C}$ such that α is slightly larger than $1/400$ [Zém01]. Thus we can choose q as small as $q > 1/(\alpha - 2\beta - \gamma\rho) > 400$. We remark that this might be drastically improved if a more sophisticated joint error-and-erasure correction mechanism than ours was used. Our naive mechanism simply treats erasures as errors, but there might be much more efficient mechanism, maybe allowing to choose q as small as 2.

Construction 2. *The scheme* $\mathsf{PKE}_2 = (\mathsf{KeyGen}, \mathsf{Enc}, \mathsf{Dec})$ *is specified by*

- $\mathsf{KeyGen}(1^k)$: *Sample matrices* $A \in \mathbb{F}_2^{l_1 \times n}$ *and* $C \in \mathbb{F}_2^{l_3 \times n}$ *uniformly at random. For every* $j \in \Sigma$ *sample a matrix* T_j *from* $\chi_\rho^{l_2 \times l_1}$ *and a matrix* X_j *from* $\chi_\rho^{l_2 \times n}$. *Set* $B_j = G + T_j \cdot A + X_j$. *Set* $pk = (A, (B_j)_{j \in \Sigma}, C)$ *and* $sk = (T_j)_{j \in \Sigma}$. *Output* (pk, sk).
- $\mathsf{Enc}_{pk}(m)$: *Takes a public key* $pk = (A, (B_j)_{j \in \Sigma}, C)$ *and a plaintext* $m \in \mathbb{F}_2^n$ *as input. Write each* B_j *as* $B_j = (b_{j,1}, \ldots, b_{j,l_2})^T$ *(The* $b_{j,i}^T$ *are the rows of* B_j). *Sample a tag* $\tau \in \Sigma^n$ *uniformly at random and set* $\hat{\tau} = \mathsf{Encode}_{\mathcal{E}}(\tau)$. *It then sets* $B_{\hat{\tau}} = (b_{\hat{\tau}_1,1}, \ldots, b_{\hat{\tau}_{l_2},l_2})^T$, *i.e. the* i-*th row of* $B_{\hat{\tau}}$ *is* $b_{\hat{\tau}_i,i}$. *Encryption now samples* s *from* χ_ρ^n, e_1 *from* $\chi_\rho^{l_1}$, e_2 *from* $\chi_\rho^{l_2}$ *and* e_3 *from* $\chi_\rho^{l_3}$. *It sets* $c_1 = A \cdot s + e_1$, $c_2 = B_{\hat{\tau}} \cdot s + e_2$ *and* $c_3 = C \cdot s + e_3 + \mathsf{Encode}_{\mathcal{D}}(m)$. *Output* $c = (\tau, c_1, c_2, c_3)$.
- $\mathsf{Dec}_{sk}(c)$: *Takes a secret key* $sk = (T_j)_{j \in \Sigma}$ *and a ciphertext* $c = (\tau, c_1, c_2, c_3)$ *as input. Write each* T_j *as* $T_j = (t_{j,1}, \ldots, t_{j,l_2})^T$ *(The* $t_{j,i}^T$ *are the rows of* T_j). *Then it computes* $\hat{\tau} = \mathsf{Encode}_{\mathcal{E}}(\tau)$ *and* $T_{\hat{\tau}} = (t_{\hat{\tau}_1,1}, \ldots, t_{\hat{\tau}_{l_2},l_2})^T$. *Next it computes* $y = c_2 - T_{\hat{\tau}} \cdot c_1$ *and* $s = \mathsf{Decode}_{\mathcal{C}}(y)$. *Outputs* $\perp$ *if decoding fails. Otherwise compute* $m = \mathsf{Decode}_{\mathcal{D}}(c_3 - C \cdot s)$. *Now it computes* $e_1 = c_1 - A \cdot s$, $e_2 = c_2 - B_{\hat{\tau}} \cdot s$, $e_3 = c_3 - C \cdot s - \mathsf{Encode}_{\mathcal{D}}(m)$ *and checks whether* $|s| < \gamma\rho n$, $|e_1| < \gamma\rho l_1$, $|e_2| < \gamma\rho l_2$ *and* $|e_3| < \gamma\rho l_3$. *If yes it outputs* m, *otherwise* $\perp$.

Correctness of PKE_2 follows immediately from the correctness of PKE_1. The only additional step is the check of the hamming weights $|s|$, $|e_1|$, $|e_2|$ and $|e_3|$. However, this has been dealt with implicitly in Lemma 1. We will now prove IND-CCA1 security for the scheme PKE_2.

Theorem 2. *The scheme* PKE_2 *is IND-CCA1 secure, provided that the scheme* PKE_1 *is IND-CPA secure and the parameters* α, β, γ, q *and* δ *suffice* $\delta < 1 - 1/q$ *and* $2\beta + \gamma\rho + 1 - \delta < \alpha$.

Proof. Let $\mathcal{A}$ be PPT-bounded IND-CPA adversary against PKE_2. Consider the following sequence of games.

- Game 1: This is the IND-CCA1 experiment.
- Game 2: This is the same as game 1, except that the tag τ^* of the challenge-ciphertext $c^* = (\tau^*, c_1^*, c_2^*, c_3^*)$ is chosen before the experiment starts, and game 2 aborts if $\mathcal{A}$ sends a decryption-query with tag τ^*.

- Game 3 This is the same as game 2, except that the decryption-oracle is implemented differently. For a decryption-query $c = (\tau, c_1, c_2, c_3)$ the decryption-oracle proceeds as follows. Let $\hat{\tau} = \mathsf{Encode}_{\mathcal{E}}(\tau)$. For all $i \in \{1, \ldots, l_2\}$ with $\hat{\tau}_i \neq \hat{\tau}_i^*$, it computes $y_i = c_{2,i} - t_{\hat{\tau}_i, i}^T c_1$. For all remaining i it chooses y_i uniformly at random. The decryption-oracle then continues like in game 2, computing $s = \mathsf{Decode}_C(y)$ (and aborts if decoding fails) and $m = \mathsf{Decode}_D(c_3 - C \cdot s)$, setting $e_1 = c_1 - A \cdot s$, $e_2 = c_2 - B \cdot s$, $e_3 = c_3 - C \cdot s - \mathsf{Encode}_D(m)$ and checking whether $|s| < \gamma \rho n$, $|e_1| < \gamma \rho l_1$, $|e_2| < \gamma \rho l_2$ and $|e_3| < \gamma \rho l_3$. If yes it outputs m, otherwise $\perp$.

In game 2, the event that $\mathcal{A}$ sends a decryption-query with tag τ^* has probability at most $f(k)/q^n = \mathsf{negl}(k)$, where $f(k)$ is a polynomial upper bound for the number of decryption-queries $\mathcal{A}$ makes. If this event does not occur, game 1 and game 2 are identically distributed from $\mathcal{A}$'s view. Thus, from $\mathcal{A}$'s view game 1 and game 2 are statistically indistinguishable. We will now show that game 2 and game 3 are statistically indistinguishable from $\mathcal{A}$'s view. First, assume that for every tag τ the matrices $T_{\hat{\tau}}$ and $X_{\hat{\tau}}$ are $(\beta, \gamma \rho)$-good. If this is the case, we claim that the decryption oracles of game 2 and game 3 behave identical. We split the claim in two cases. The first case is simple: If either $|s| \geq \gamma \rho n$, $|e_1| \geq \gamma \rho l_1$, $|e_2| \geq \gamma \rho l_2$ or $|e_3| \geq \gamma \rho l_3$, then the decryption oracle will return $\perp$ in both games, regardless whether decoding fails or not. In the other case it holds that $|s| < \gamma \rho n$, $|e_1| < \gamma \rho l_1$, $|e_2| < \gamma \rho l_2$ and $|e_3| < \gamma \rho l_3$. Now it holds that the hamming-weight of the error-term $v = X_{\hat{\tau}} \cdot s + e_2 - T_{\hat{\tau}} \cdot e_1$ will be bounded by $2\beta l_2 + \gamma \rho l_2$. Thus, in game 2 the decoding-algorithm Decode_C has to correct at most $(2\beta + \gamma \rho) l_2 < \alpha l_2$ and will thus be successful and output the *unique* s. In game 3, there might be up to $(1 - \delta) l_2$ additional errors Decode_C has to deal with, as the decryption oracle chooses up to $(1 - \delta) l_2$ components of the codeword y at random. However, since $(2\beta + \gamma \rho + 1 - \delta) l_2 < \alpha l_2$ the decoding-algorithm Decode_C will also succeed in game 3 and output the unique s. This concludes the claim. What remains to show for this part of the proof is that, with overwhelming probability in k, it holds that for every tag τ the matrices $T_{\hat{\tau}}$ and $X_{\hat{\tau}}$ are $(\beta, \gamma \rho)$-good. We can think of each matrix $T_{\hat{\tau}}$ as a row-sub-matrix of a large matrix $T_{full} \in \mathbb{F}_2^{ql_2 \times n}$ that consists of all the rows of all T_i for $i \in \Sigma$ (i.e. T_{full} is just the vertical concatenation of all T_i). With overwhelming probability in k, T_{full} is $(\beta/q, \gamma \rho)$-good (since q is constant). This means that for each e_1 with $|e_1| < \gamma \rho l_1$ it holds that $|T_{full} e_1| < \beta/q \cdot (ql_2) = \beta l_2$. However, as each $T_{\hat{\tau}}$ is a row-sub-matrix of T_{full}, it also holds that $|T_{\hat{\tau}} e_1| < \beta l_2$. Showing that $|X_{\hat{\tau}} s| < \beta l_2$ works analogously, which concludes this part of the proof. Finally, $\mathcal{A}$'s advantage of winning game 3 is negligible in k, given that PKE_1 is IND-CPA secure. Assume for contradiction that $\mathcal{A}$ wins game 3 with non-negligible advantage $\nu(k)$. We will construct an IND-CPA adversary $\mathcal{B}$ against PKE_1 that wins the IND-CPA experiment with advantage ν. $\mathcal{B}$'s input from the IND-CPA experiment is a public key $pk' = (A', B', C')$ for the scheme PKE_1. $\mathcal{B}$ now runs the key-generation of game 3 with the following modifications. Instead of sampling the matrices A and C uniformly at random, it sets $A = A'$ and $C = C'$. Now it generates the B_j and T_j exactly like the key-generation in game 3. Then however, it replaces the public-key at the locations that constitute $B_{\hat{\tau}}$ with B', i.e. it sets $b_{\hat{\tau}_i,i}^T = b_i'^T$ for $i = 1, \ldots, l_2$. $\mathcal{B}$. Then it simulates the interaction between $\mathcal{A}$ and game 3, answering decryption-queries like game 3. This is possible, as game 3 never uses secret keys $t_{\hat{\tau}_i,i}$ (that correspond to public keys $b_{\hat{\tau}_i,i}$) to answer decryption queries. Once $\mathcal{A}$ sends challenge messages (m_0, m_1), $\mathcal{B}$ forwards (m_0, m_1) to the IND-CPA experiment and receives a challenge-ciphertext $c^\dagger = (c_1^\dagger, c_2^\dagger, c_2^\dagger)$. $\mathcal{B}$ sends $c^* = (\tau^*, c_1^\dagger, c_2^\dagger, c_2^\dagger)$ to $\mathcal{A}$ and continues the simulation. Once $\mathcal{A}$ terminates, $\mathcal{B}$ outputs whatever $\mathcal{A}$ outputs. From $\mathcal{A}$'s view, $\mathcal{B}$'s simulation and game 3 are perfectly indistinguishable, as the distributions of A and C are the same, as well as the distribution of the partial public keys $b_{j,i}^T$, which are independent of one another (only depending on the same A). Moreover, the decryption-oracle behaves identically in both experiments. Therefore, it holds that $\mathsf{Adv}_{\text{IND-CPA}}(\mathcal{B}) = \mathsf{Adv}_{\text{IND-CCA1}}(\mathcal{A}) = \nu(k)$ which contradicts the IND-CPA security of scheme PKE_1.

7 The IND-CCA2 Scheme

We will now provide details how the scheme PKE_2 can be transformed into an IND-CCA2 secure scheme PKE_3 using additional one-time signatures. We follow an approach by Dolev, Dwork and Naor [DDN00], which has been used in several other constructions [PW08, Pei09, RS09, DMQN09, MP12], especially in the world of lattice and coding assumptions, to achieve full CCA2 security. First observe that it is not necessary to choose the tag $\tau \in \Sigma^n$ uniformly at random in the encryption procedure of PKE_2. We only need to guarantee that a PPT-adversary $\mathcal{A}$ will have negligible probability guessing the secret tag τ^* correctly if it is granted a polynomial number of trials (this immediately yields the statistical indistinguishability of game 1 and game 2 in Theorem 2). Thus it is sufficient to sample the tags τ from a distribution with high min-entropy. Moreover, observe that the proof of Theorem 2 still holds if we allow $\mathcal{A}$ to make decryption-queries even after it has received the challenge-ciphertext c^*. This can be seen by noting that the decryption-oracle in game 3 can answer decryption-queries with $\tau \neq \tau^*$ regardless of whether the challenge-ciphertext has been given to $\mathcal{A}$ or not (decryption-queries with $\tau = \tau^*$ are rejected unconditionally). In fact, the decryption-oracle in game 3 is oblivious of whether the challenge-ciphertext has been given to $\mathcal{A}$ or not. Thus, the scheme PKE_2 can be recast as a tag-based encryption scheme [Kil06]. We will now outline PKE_3. Let $\mathsf{SIG} = (\mathsf{Gen}, \mathsf{Sign}, \mathsf{Verify})$ be an EUF-CMA secure one-time signature scheme. For simplicity, assume that the verification-keys vk of SIG are elements of Σ^n (this can always be accomplished by encoding vk in the q-ary alphabet Σ and choosing n large enough). The key-generation of PKE_3 is identical to the key-generation of PKE_2. The encryption procedure $\mathsf{PKE}_3.\mathsf{Enc}$ first computes a pair of verification and signature-keys $(vk, sgk) = \mathsf{SIG}.\mathsf{Gen}(1^k)$. Then it runs the encryption procedure $\mathsf{PKE}_2.\mathsf{Enc}$, with the difference that it sets $\tau = vk$ instead of choosing τ uniformly at random. Let c' be the output of $\mathsf{PKE}_2.\mathsf{Enc}$. $\mathsf{PKE}_3.\mathsf{Enc}$ then computes $\sigma = \mathsf{SIG}.\mathsf{Sign}_{sgk}(c')$ and outputs the ciphertext $c = (c', \sigma)$. The decryption procedure $\mathsf{PKE}_3.\mathsf{Dec}$ first checks if σ is a valid signature on c' using the verification-key $vk = \tau$ (where τ is the tag given in c'). If the check succeeds, it runs the decryption procedure $\mathsf{PKE}_2.\mathsf{Dec}$ on the ciphertext c' and outputs whatever $\mathsf{PKE}_2.\mathsf{Dec}$ outputs. We summarize this in the following construction. Let $\mathsf{Enc}'_{pk}(m, vk)$ be a procedure that does exactly the same as $\mathsf{PKE}_2.\mathsf{Enc}_{pk}(m)$, but sets $\tau = vk$ instead of choosing τ uniformly at random.

Construction 3. *The scheme* $\mathsf{PKE}_3 = (\mathsf{KeyGen}, \mathsf{Enc}, \mathsf{Dec})$ *is specified by*

- $\mathsf{KeyGen}(1^k)$: *Compute* $(pk, sk) = \mathsf{PKE}_2.\mathsf{KeyGen}(1^k)$ *and output* (pk, sk).
- $\mathsf{Enc}_{pk}(m)$: *Generate* $(vk, sgk) = \mathsf{SIG}.\mathsf{Gen}(1^k)$, *encrypt* $c' = \mathsf{Enc}'_{pk}(m, vk)$, *sign* $\sigma = \mathsf{SIG}.\mathsf{Sign}_{sgk}(c')$ *and output* $c = (c', \sigma)$.
- $\mathsf{Dec}_{sk}(c)$: *Let* $c = (c', \sigma)$ *and* $c' = (\tau, c_1, c_2, c_3)$. *Set* $vk = \tau$. *Check if* $\mathsf{SIG}.\mathsf{Verify}_{vk}(c', \sigma) = 1$, *if not abort. Otherwise compute* $m = \mathsf{PKE}_2.\mathsf{Dec}_{sk}(c')$ *and output* m.

Theorem 3. *The scheme* PKE_3 *is IND-CCA2 secure, provided that* SIG *is an EUF-CMA secure one-time signature scheme and the same requirements as in Theorem 2 are given.*

Proof. (Sketch) Let $\mathcal{A}$ be PPT-bounded IND-CCA2 adversary against PKE_3. It suffices to show that with overwhelming probability, every decryption-query by $\mathcal{A}$ tagged with τ^* (the tag of the challenge-ciphertext) is rejected. Thus, we can recycle the proof of Theorem 5 almost entirely, we only need to replace the indistinguishability of game 1 and game 2 in the proof of Theorem 2. The rest of the proof is identical. Consider the following two games.

- Game 1: This is the IND-CCA2 experiment.
- Game 2: This is the same as game 1, except that the tag τ^* of the challenge-ciphertext $c^* = (\tau^*, c_1^*, c_2^*, c_3^*, \sigma^*)$ is generated before the experiment starts, and game 2 aborts if $\mathcal{A}$ sends a decryption-query with tag τ^*.

Assume that $\mathcal{A}$ distinguishes between game 1 and game 2 with non-negligible advantage $\nu(k)$. Clearly, given that the decryption-oracle rejects every decryption-query tagged with τ^*, both games are identically distributed from $\mathcal{A}$'s view. Thus, to distinguish game 1 and game 2 $\mathcal{A}$ must generate a decryption-query tagged with τ^* that is accepted by the decryption-oracle. This implies that such a decryption-query $c = (c', \sigma)$ with $c' = (\tau^*, c_1, c_2, c_3)$ suffices the condition $\mathsf{SIG}.\mathsf{Verify}_{vk}(c', \sigma) = 1$, where $vk = \tau^*$. Thus we can assume that $\mathcal{A}$ generates such a decryption-query with probability $\nu(k)$. We construct an EUF-CMA adversary $\mathcal{B}$ that breaks the EUF-CMA security of SIG with probability $\nu(k)$. Let vk be the verification key provided to $\mathcal{B}$ by the EUF-CMA experiment. $\mathcal{B}$ simulates game 2 with $\mathcal{A}$, but makes the following changes. Instead of generating the tag τ^* itself, it sets $\tau^* = vk$. Moreover, $\mathcal{B}$ obtains the signature σ^* of the challenge-ciphertext c^* by querying its signature-oracle with c'^*, where $c'^* = \mathsf{Enc}'_{pk}(m_b, vk)$. Finally, once $\mathcal{A}$ sends a decryption-query $c = (c', \sigma)$ with $c' = (\tau^*, c_1, c_2, c_3)$ and $\mathsf{SIG}.\mathsf{Verify}_{vk}(c', \sigma) = 1$, $\mathcal{B}$ outputs (c', σ) an terminates.

Clearly, game 2 and the simulation of $\mathcal{B}$ are identically distributed from the view of $\mathcal{A}$. Thus, the event that $\mathcal{A}$ sends a decryption-query $c = (c', \sigma)$ with $c' = (\tau^*, c_1, c_2, c_3)$ and $\mathsf{SIG.Verify}_{vk}(c', \sigma) = 1$ happens with probability $\nu(k)$ in $\mathcal{B}$'s simulation. This means that $\mathcal{B}$ outputs a valid forged signature with probability $\nu(k)$, contradicting the EUF-CMA security of SIG.

8   Conclusion

In this work we constructed the first IND-CCA2 secure public key encryption scheme based solely on the hardness of a low-noise variant of the learning parity with noise problem. To achieve this, we introduced a novel all-but-one simulation technique. This new technique enabled the construction of a CCA1 secure scheme, which is more efficient than any previous such construction based on the correlated-products approach. The scheme enjoys a constant-factor ciphertext expansion as well as asymptotically efficient key-generation, encryption and decryption.

Acknowledgement. The authors would like to thank the anonymous reviewers of ASIACRYPT 2012 for providing helpful comments and pointing out insightful connections to related work. Nico Döttling was supported by IBM Research & Development Germany within the HomER-project.

References

[ACPS09]   Applebaum, B., Cash, D., Peikert, C., Sahai, A.: Fast Cryptographic Primitives and Circular-Secure Encryption Based on Hard Learning Problems. In: Halevi, S. (ed.) CRYPTO 2009. LNCS, vol. 5677, pp. 595–618. Springer, Heidelberg (2009)

[Ale03]   Alekhnovich, M.: More on average case vs approximation complexity. In: FOCS, pp. 298–307 (2003)

[BJMM12]   Becker, A., Joux, A., May, A., Meurer, A.: Decoding Random Binary Linear Codes in in 2 n/20: How $1 + 1 = 0$ Improves Information Set Decoding. In: Pointcheval, D., Johansson, T. (eds.) EUROCRYPT 2012. LNCS, vol. 7237, pp. 520–536. Springer, Heidelberg (2012)

[BLP11]   Bernstein, D.J., Lange, T., Peters, C.: Smaller Decoding Exponents: Ball-Collision Decoding. In: Rogaway, P. (ed.) CRYPTO 2011. LNCS, vol. 6841, pp. 743–760. Springer, Heidelberg (2011)

[BS08]   Biswas, B., Sendrier, N.: Mceliece cryptosystem implementation: Theory and practice. In: PQCrypto, pp. 47–62 (2008)

[CC98]   Canteaut, A., Chabaud, F.: A new algorithm for finding minimum-weight words in a linear code: Application to mceliece's cryptosystem and to narrow-sense bch codes of length 511. IEEE Transactions on Information Theory 44(1), 367–378 (1998)

[CHK04]   Canetti, R., Halevi, S., Katz, J.: Chosen-Ciphertext Security from Identity-Based Encryption. In: Cachin, C., Camenisch, J.L. (eds.) EUROCRYPT 2004. LNCS, vol. 3027, pp. 207–222. Springer, Heidelberg (2004)

[CS98]   Cramer, R., Shoup, V.: A Practical Public Key Cryptosystem Provably
           Secure against Adaptive Chosen Ciphertext Attack. In: Krawczyk, H.
           (ed.) CRYPTO 1998. LNCS, vol. 1462, pp. 13–25. Springer, Heidelberg
           (1998)
[CS03]   Camenisch, J., Shoup, V.: Practical Verifiable Encryption and Decryp-
           tion of Discrete Logarithms. In: Boneh, D. (ed.) CRYPTO 2003. LNCS,
           vol. 2729, pp. 126–144. Springer, Heidelberg (2003)
[DDN00]   Dolev, D., Dwork, C., Naor, M.: Nonmalleable cryptography. SIAM J.
           Comput. 30(2), 391–437 (2000)
[DMQN09]   Dowsley, R., Müller-Quade, J., Nascimento, A.C.A.: A CCA2 Secure Pub-
           lic Key Encryption Scheme Based on the McEliece Assumptions in the
           Standard Model. In: Fischlin, M. (ed.) CT-RSA 2009. LNCS, vol. 5473,
           pp. 240–251. Springer, Heidelberg (2009)
[GPV08]   Gentry, C., Peikert, C., Vaikuntanathan, V.: Trapdoors for hard lattices
           and new cryptographic constructions. In: STOC, pp. 197–206 (2008)
[HB01]   Hopper, N.J., Blum, M.: Secure Human Identification Protocols. In: Boyd,
           C. (ed.) ASIACRYPT 2001. LNCS, vol. 2248, pp. 52–66. Springer, Hei-
           delberg (2001)
[HK09]   Hofheinz, D., Kiltz, E.: Practical Chosen Ciphertext Secure Encryption
           from Factoring. In: Joux, A. (ed.) EUROCRYPT 2009. LNCS, vol. 5479,
           pp. 313–332. Springer, Heidelberg (2009)
[JW05]   Juels, A., Weis, S.A.: Authenticating Pervasive Devices with Human Pro-
           tocols. In: Shoup, V. (ed.) CRYPTO 2005. LNCS, vol. 3621, pp. 293–308.
           Springer, Heidelberg (2005)
[Kil06]   Kiltz, E.: Chosen-Ciphertext Security from Tag-Based Encryption. In:
           Halevi, S., Rabin, T. (eds.) TCC 2006. LNCS, vol. 3876, pp. 581–600.
           Springer, Heidelberg (2006)
[KSS10]   Katz, J., Shin, J.S., Smith, A.: Parallel and concurrent security of the hb
           and hb^+ protocols. J. Cryptology 23(3), 402–421 (2010)
[Lam79]   Lamport, L.: Constructing digital signatures from one-way functions. In:
           SRI Intl. CSL-98 (1979)
[Lin03]   Lindell, Y.: A Simpler Construction of CCA2-Secure Public-Key Encryp-
           tion Under General Assumptions. In: Biham, E. (ed.) EUROCRYPT
           2003. LNCS, vol. 2656, pp. 241–254. Springer, Heidelberg (2003)
[McE78]   McEliece, R.J.: A public-key cryptosystem based on algebraic coding the-
           ory. In: DSN Progress Report, Jet Propulsion Laboratory, California In-
           stitute of Technology, Pasadena, CA (1978)
[MMT11]   May, A., Meurer, A., Thomae, E.: Decoding Random Linear Codes in
           $\tilde{\mathcal{O}}(2^{0.054n})$. In: Lee, D.H. (ed.) ASIACRYPT 2011. LNCS, vol. 7073, pp.
           107–124. Springer, Heidelberg (2011)
[MP12]   Micciancio, D., Peikert, C.: Trapdoors for Lattices: Simpler, Tighter,
           Faster, Smaller. In: Pointcheval, D., Johansson, T. (eds.) EUROCRYPT
           2012. LNCS, vol. 7237, pp. 700–718. Springer, Heidelberg (2012)
[NIKM08]   Nojima, R., Imai, H., Kobara, K., Morozov, K.: Semantic security for the
           mceliece cryptosystem without random oracles. Des. Codes Cryptogra-
           phy 49(1-3), 289–305 (2008)
[NY90]   Naor, M., Yung, M.: Public-key cryptosystems provably secure against
           chosen ciphertext attacks. In: STOC, pp. 427–437 (1990)
[Pei09]   Peikert, C.: Public-key cryptosystems from the worst-case shortest vector
           problem: extended abstract. In: STOC, pp. 333–342 (2009)

[PW08]   Peikert, C., Waters, B.: Lossy trapdoor functions and their applications. In: STOC, pp. 187–196 (2008)

[Reg05] Regev, O.: On lattices, learning with errors, random linear codes, and cryptography. In: STOC, pp. 84–93 (2005)

[RS91]   Rackoff, C., Simon, D.R.: Non-interactive Zero-Knowledge Proof of Knowledge and Chosen Ciphertext Attack. In: Feigenbaum, J. (ed.) CRYPTO 1991. LNCS, vol. 576, pp. 433–444. Springer, Heidelberg (1992)

[RS09]   Rosen, A., Segev, G.: Chosen-Ciphertext Security via Correlated Products. In: Reingold, O. (ed.) TCC 2009. LNCS, vol. 5444, pp. 419–436. Springer, Heidelberg (2009)

[Sah99] Sahai, A.: Non-malleable non-interactive zero knowledge and adaptive chosen-ciphertext security. In: FOCS, pp. 543–553 (1999)

[SS96]   Sipser, M., Spielmanp, D.A.: Expander codes. IEEE Transactions on Information Theory 42(6), 1710–1722 (1996)

[Ste88] Stern, J.: A Method for Finding Codewords of Small Weight. In: Cohen, G., Godlewski, P. (eds.) Coding Theory 1986. LNCS, vol. 311, pp. 106–113. Springer, Heidelberg (1988)

[Zém01] Zémor, G.: On expander codes. IEEE Transactions on Information Theory 47(2), 835–837 (2001)

Provable Security of the Knudsen-Preneel Compression Functions

Jooyoung Lee[*]

Faculty of Mathematics and Statistics
Sejong University, Seoul, Korea 143-747
jlee05@sejong.ac.kr

Abstract. This paper discusses the provable security of the compression functions introduced by Knudsen and Preneel [11,12,13] that use linear error-correcting codes to build wide-pipe compression functions from underlying blockciphers operating in Davies-Meyer mode. In the information theoretic model, we prove that the Knudsen-Preneel compression function based on an $[r, k, d]_{2^e}$ code is collision resistant up to $2^{\frac{(r-d+1)n}{2r-3d+3}}$ query complexity if $2d \leq r + 1$ and collision resistant up to $2^{\frac{rn}{2r-2d+2}}$ query complexity if $2d > r + 1$. For MDS code based Knudsen-Preneel compression functions, this lower bound matches the upper bound recently given by Özen and Stam [23].

A preimage security proof of the Knudsen-Preneel compression functions has been first presented by Özen et al. (FSE '10). In this paper, we present two alternative proofs that the Knudsen-Preneel compression functions are preimage resistant up to $2^{\frac{rn}{k}}$ query complexity. While the first proof, using a wish list argument, is presented primarily to illustrate an idea behind our collision security proof, the second proof provides a tighter security bound compared to the original one.

1 Introduction

A cryptographic hash function takes a message of arbitrary length, and returns a bit string of fixed length. The most common way of hashing variable length messages is to iterate a fixed-size compression function (e.g. according to the Merkle-Damgård paradigm [7,20]). The underlying compression function can either be constructed from scratch, or be built upon off-the-shelf cryptographic primitives such as blockciphers. Recently, blockcipher-based constructions have attracted renewed interest as many dedicated hash functions, including those most common in practical applications, have started to exhibit serious security weaknesses [2,6,18,19,29,34,35,36]. By instantiating a blockcipher-based construction with an extensively studied (and fully trusted) blockcipher, one can conveniently transfer the trust in the existing blockcipher to the hash function.

---

[*] The work of J. Lee was supported by Basic Science Research Program through the National Research Foundation of Korea(NRF) funded by the Ministry of Education, Science and Technology(2012-0003157).

X. Wang and K. Sako (Eds.): ASIACRYPT 2012, LNCS 7658, pp. 504–525, 2012.

Compression functions based on blockciphers have been widely studied [3,4,9,10,14,22,25,26,27,28,30,31,32,33]. The most common approach is to construct a $2n$-to-n bit compression function using a single call to an n-bit blockcipher. However, such a function, called a *single-block-length* (SBL) compression function, might be vulnerable to collision attacks due to its short output length. For example, one could successfully mount a birthday attack on a compression function based on AES-128 using approximately 2^{64} queries. This observation motivated substantial research on constructions whose output size is larger than the block length of the underlying blockcipher(s). A typical approach has been to construct *double-block-length* (DBL) hash functions, where the output length is twice the block length of the underlying blockcipher(s). Since the 1990s various double-block-length constructions have been proposed mostly without formal security proofs. Those constructions were mainly focused on optimizing their efficiency in terms of the rate, while only recently have a few double-block-length constructions been supported by rigorous security proofs [8,15,17,24].

THE KNUDSEN-PRENEEL COMPRESSION FUNCTIONS. On the other hand, Knudsen and Preneel [11,12,13] adopted a different approach, aiming at achieving a particular level of security using a given number of ideal compression functions as building blocks. Specifically, they used r independent cn-to-n bit random functions to build the entire compression function producing rn-bit outputs. The parameter c is typically two or three so that the inner primitives can be constructed from n-bit key or $2n$-bit key blockciphers operating in Davies-Meyer mode. The main idea of Knudsen and Preneel's approach lies in the method of deriving the inputs to the inner primitives from the input to the entire compression function. They used an $[r, k, d]$ linear error-correcting code over a finite field in a way that its generator matrix extends a kcn-bit input to the entire compression function to an rcn-bit string. This string is parsed into r blocks of the same size, and the blocks go into the inner primitives in parallel. The output of the entire compression function is the concatenation of the n-bit outputs obtained from the r inner primitives. This Knudsen-Preneel (KP) compression function is fed to the Merkle-Damgård transform, producing the final output via a random finalization function whose output size might depend on the security target.

Due to the property of linear codes of minimum distance d, two different inputs to the KP compression function determine two sets of inputs to the inner primitives that are different at least at d positions. Based on this observation, Knudsen and Preneel made a certain plausible security assumption (see [11, Section 5]) which was used for their security proof that the KP compression function is collision resistant up to $2^{\frac{(d-1)n}{2}}$ query complexity. They also expected that the KP compression function would be preimage resistant up to $2^{(d-1)n}$ query complexity. In order to maximize the query complexity, Knudsen and Preneel suggested the use of MDS codes satisfying $d = r - k + 1$.

ATTACK HISTORY. For KP compression functions based on an MDS code, the designers described preimage attacks matching their security conjecture, while their collision attacks were far from tight for many of the parameter sets.

Afterwards Watanabe [37] proposed a collision attack beating the original conjecture for many cases. In particular, for $2k > r$ and $d \leq k$, one could find a collision with $k2^n$ query complexity.

Özen, Shrimpton and Stam [21] presented a preimage attack of $2^{\frac{rn}{k}}$ query complexity, far less than the bound of $2^{(d-1)n}$ that was originally conjectured by the designers. By giving a preimage security proof, they proved that their attack is tight. Their result also implies that one could expect a collision with about $2^{\frac{rn}{2k}}$ queries.

Subsequently, Özen and Stam [23] presented new collision attacks using the ideas of Watanabe and the preimage attack of Özen, Shrimpton and Stam. For $2k > r$ and $d \leq k$, their attacks require $2^{\frac{kn}{3k-r}}$ query complexity. This implies that the KP compression functions do not achieve the security level they were originally designed for. On the other hand, tightness of their attack remained an open question.

1.1   Our Contribution

In this paper, we prove that the KP compression function based on an $[r, k, d]_{2^e}$ code is collision resistant up to $2^{\frac{(r-d+1)n}{2r-3d+3}}$ query complexity if $2d \leq r+1$ and collision resistant up to $2^{\frac{rn}{2r-2d+2}}$ query complexity if $2d > r+1$. For KP compression functions based on an MDS code, this lower bound, simplified to $2^{\frac{kn}{3k-r}}$ for $2d \leq r+1$ and $2^{\frac{rn}{2k}}$ for $2d > r+1$ respectively, matches the upper bound given by [21,23]. For two parameter sets $[4, 2, 3]_8$ and $[5, 2, 4]_8$ such that $2d > r+1$, the collision security is proved up to the query complexity equal to or beyond the block-size of the underlying blockciphers.

Özen, Shrimpton and Stam [21] proved that the preimage finding advantage of a q-query adversary is not greater than

$$\epsilon_1(r, k) = \frac{q^{\frac{(r-k)k}{r}}}{2^{(r-k)n}} + \left(\frac{eq^{\frac{k}{r}}}{2^n} \right)^{kq^{\frac{(r-k)}{r}}},$$

where we set $\delta = \frac{r(k-1)-k^2}{r}$ in Theorem 10 of [21]. The upper bound $\epsilon_1(r, k)$ becomes negligible as q gets much smaller than $2^{\frac{rn}{k}}$. In this paper, we present two alternative preimage security proofs, where the second proof provides a tighter security bound compared to the original one. Specifically, the preimage finding advantage of a q-query adversary is upper bounded by

$$\epsilon_2(r, k) = \binom{r}{k} \frac{q^k}{2^{rn}}.$$

Our upper bound $\epsilon_2(r, k)$ is significantly smaller than $\epsilon_1(r, k)$ since $\epsilon_2(r, k) \leq \binom{r}{k}\epsilon_1(r, k)^{1+\frac{k}{r-k}}$. For example, for a $[5, 3, 3]_4$ code based KP compression function, we have $\epsilon_1(r, k) \geq \frac{q^{6/5}}{2^{2n}}$ while $\epsilon_2(r, k) = \frac{10q^3}{2^{5n}}$.

Our first preimage security proof, using a wish-list argument, is presented primarily to illustrate an idea behind our collision security proof. This proof is

Table 1. Provable security of Knudsen-Preneel constructions. Non-MDS parameters in italic. The parameter sets satisfying $r + 1 > 2k$ are $[4, 2, 3]_8$ and $[5, 2, 4]_8$.

$[r, k, d]_{2^e}$-Code	Basing Primitive	Compression Function	Collision Resistance			Preimage Resistance	
			Attack [23]	Security	Tightness	Attack [21]	Security
$[5, 3, 3]_4$		$(5+1)n \to 5n$	$2^{3n/4}$	$2^{3n/4}$	$\checkmark$	$2^{5n/3}$	$2^{5n/3}$
$[8, 5, 3]_4$		$(8+2)n \to 8n$	$2^{5n/7}$	$2^{3n/5}$		$2^{8n/5}$	$2^{8n/5}$
$[12, 9, 3]_4$	$2n \to n$	$(12+6)n \to 12n$	$2^{3n/5}$	$2^{5n/9}$		$2^{4n/3}$	$2^{4n/3}$
$[9, 5, 4]_4$		$(9+1)n \to 9n$	$2^{5n/6}$	$2^{2n/3}$		$2^{9n/5}$	$2^{9n/5}$
$[16, 12, 4]_4$		$(16+8)n \to 16n$	$2^{3n/5}$	$2^{13n/23}$		$2^{4n/3}$	$2^{4n/3}$
$[6, 4, 3]_{16}$		$(6+2)n \to 6n$	$2^{2n/3}$	$2^{2n/3}$	$\checkmark$	$2^{3n/2}$	$2^{3n/2}$
$[8, 6, 3]_{16}$		$(8+4)n \to 8n$	$2^{3n/5}$	$2^{3n/5}$	$\checkmark$	$2^{4n/3}$	$2^{4n/3}$
$[12, 10, 3]_{16}$	$2n \to n$	$(12+8)n \to 12n$	$2^{5n/9}$	$2^{5n/9}$	$\checkmark$	$2^{6n/5}$	$2^{6n/5}$
$[9, 6, 4]_{16}$		$(9+3)n \to 9n$	$2^{2n/3}$	$2^{2n/3}$	$\checkmark$	$2^{3n/2}$	$2^{3n/2}$
$[16, 13, 4]_{16}$		$(16+10)n \to 16n$	$2^{13n/23}$	$2^{13n/23}$	$\checkmark$	$2^{16n/13}$	$2^{16n/13}$
$[4, 2, 3]_8$		$(4+2)n \to 4n$	2^n [21]	2^n	$\checkmark$	2^{2n}	2^{2n}
$[6, 4, 3]_8$		$(6+6)n \to 6n$	$2^{2n/3}$	$2^{2n/3}$	$\checkmark$	$2^{3n/2}$	$2^{3n/2}$
$[9, 7, 3]_8$		$(9+12)n \to 9n$	$2^{7n/12}$	$2^{7n/12}$	$\checkmark$	$2^{9n/7}$	$2^{9n/7}$
$[5, 2, 4]_8$	$3n \to n$	$(5+1)n \to 5n$	$2^{5n/4}$ [21]	$2^{5n/4}$	$\checkmark$	$2^{5n/2}$	$2^{5n/2}$
$[7, 4, 4]_8$		$(7+5)n \to 7n$	$2^{4n/5}$	$2^{4n/5}$	$\checkmark$	$2^{7n/4}$	$2^{7n/4}$
$[10, 7, 4]_8$		$(10+11)n \to 10n$	$2^{7n/11}$	$2^{7n/11}$	$\checkmark$	$2^{10n/7}$	$2^{10n/7}$

tight only for the parameter sets of MDS codes. Table 1 summarizes these results for 16 parameter sets proposed by the original designers.

WISH LIST ARGUMENT. In the information-theoretic model, the most typical approach for a security proof has been upper bounding the probability that a single query of an adversary achieves a certain security goal (such as finding a collision or finding a preimage of a target image). The upper bound of the total adversarial advantage is obtained by multiplying this upper bound by the number of queries allowed to the adversary. Most single-block-length constructions can be analyzed in this way [25].

However, certain constructions might not allow an upper bound small enough to uniformly apply to all the queries. One of the techniques to address this difficulty is to define a certain bad event that happens with only small probability, and prove that it is hard for a single query to achieve an adversarial goal without the occurrence of the bad event. This approach was adopted in the collision security proof of MDC-2 and MJH hash functions [16,24] as well as the preimage security proof of the KP compression functions [21].

Another technique is to cleverly modify the adversary: the modified adversary, typically using the original adversary as a subroutine, is given slightly more power than the original one. So the success probability of the modified adversary is not reduced, while it becomes much easier to upper bound. With this approach, one can prove the security of Abreast-DM and Tandem-DM hash functions [8,15,17]. Our second alternative preimage security proof of the KP compression functions also follows this approach.

As yet another technique, one might use an observation that a security goal is usually achieved by a group of queries and the last query that achieves the goal is uniquely determined by the previous queries in the group. We assume, once a new query is obtained, the adversary computes a query that might become the last winning query along with a certain group of existing queries (including the new query). If this query has not been asked, the adversary includes it in a wish list expecting this wish is accomplished sometime later. If we have upper bounds on the size of the wish list (hopefully smaller than the total number of queries) and the probability that each wish in the list is accomplished, the total adversarial advantage can be obtained by a union bound. This technique, called a *wish list argument*, was first used in the preimage security proof of certain double-length blockcipher-based compression functions [1]. This work is the first application of a wish list argument to a collision security proof (combined with a bad event argument). In our extension, each wish is typically given as a set of unasked queries, rather than a single query.

EFFICIENCY. Unfortunately, for most of the parameter sets, the KP compression functions do not provide collision security beyond the block-size of the underlying blockcipher. However, from a practical point of view, some of the KP compression functions are still comparable to the existing blockcipher-based hash functions such as MDC-2, Abreast-DM and Tandem-DM in terms of efficiency and probable security.

In MDC-2, compression of a single n-bit message block requires two calls to the underlying n-bit key blockcipher, and it enjoys a $\frac{3n}{5}$-bit collision security proof. This construction is comparable to the KP compression functions using $[12,9,3]_4$, $[16,12,4]_4$ or $[8,6,3]_{16}$ codes: they are all of rate $\frac{1}{2}$ using $2n$-to-n bit primitives (or equivalently n-bit key blockciphers), and supported by a $\frac{3n}{5}$-bit security proof.

The compression function $H = \mathsf{KP}^1([6,4,3]_8)$ using $3n$-to-n bit primitives (or equivalently $2n$-bit key blockciphers) is supported by a $\frac{2n}{3}$-bit security proof. This construction has the same rate and the same provable security as MJH [16] using a $2n$-bit key blockcipher.

The compression function $H = \mathsf{KP}^1([4,2,3]_8)$ using $3n$-to-n bit primitives (or equivalently $2n$-bit key blockciphers) is supported by an n-bit security proof. This construction is comparable to Abreast-DM and Tandem-DM, both of which are of rate $\frac{1}{2}$ using a $2n$-bit key blockcipher. We also refer to [5] for comparison of this compression function with the other existing schemes in terms of AES driven implementations.

The compression function $H = \mathsf{KP}^1([5,2,4]_8)$ is relatively slow with rate $\frac{1}{5}$, while this is the first construction that enjoys the provable collision security beyond the block-size of the underlying blockciphers. However it remains open whether this KP compression function is still secure when the inner primitives are instantiated with $2n$-bit key n-bit blockciphers, since in general an n-bit blockcipher loses its randomness beyond 2^n queries (for a fixed key). The other open question raised here is the provable security of KP constructions where all the inner primitives are instantiated the same.

2 Preliminaries

2.1 The Knudsen-Preneel Compression Functions

An $[r, k, d]_{2^e}$ linear error-correcting code $\mathcal{C}$ is a k-dimensional subspace of $\mathbb{F}_{2^e}^r$, where $\mathbb{F}_{2^e}$ denotes a finite field of order 2^e. An $[r, k, d]_{2^e}$ code $\mathcal{C}$ can be represented by a $k \times r$ generator matrix G over $\mathbb{F}_{2^e}$ where every codeword of $\mathcal{C}$ is expressed as a linear combination of the row vectors of G, namely $w \cdot G$ for some $w \in \mathbb{F}_{2^e}^k$. Obviously, $k \leq r$, and the Singleton bound states that

$$d \leq r - k + 1.$$

When a code meets the equality of the Singleton bound, it is called *maximum distance separable* (MDS). As an important property of MDS codes, any k columns of a generator matrix of an MDS code are linearly independent.

Let $\mathbb{F}_{2^e} = \mathbb{F}(\omega)$ be an extension of $\mathbb{F}_2$ generated by the root ω of a primitive polynomial $p(x)$ of degree e, and let $\mathbb{F}_2^e$ be an e-dimensional vector space over $\mathbb{F}_2$. In order to clearly define the Knudsen-Preneel compression functions, we need to identify $\mathbb{F}_{2^e}$ and $\mathbb{F}_2^e$ by a group isomorphism $\psi : \mathbb{F}_{2^e} \to \mathbb{F}_2^e$ such that

$$\psi(a_{e-1}\omega^{e-1} + \cdots + a_1\omega + a_0) = (a_{e-1}, \ldots, a_1, a_0)^T.$$

For each $g \in \mathbb{F}_{2^e}$, consider a map

$$\Phi(g) : \mathbb{F}_2^e \longrightarrow \mathbb{F}_2^e$$
$$u \longmapsto \psi(g \cdot \psi^{-1}(u)),$$

where "$\cdot$" denotes the field multiplication of $\mathbb{F}_{2^e}$. This is a linear map, so it is associated with an $e \times e$ matrix over $\mathbb{F}_2$ with respect to the standard basis. We will denote this matrix as $\phi(g)$. Since for every $g, h \in \mathbb{F}_{2^e}$,

1. $\Phi(g + h) = \Phi(g) + \Phi(h)$,
2. $\Phi(gh) = \Phi(g) \circ \Phi(h)$,

we also have $\phi(g + h) = \phi(g) + \phi(h)$ and $\phi(gh) = \phi(g)\phi(h)$ for all $g, h \in \mathbb{F}_{2^e}$. This implies the map $\phi : \mathbb{F}_{2^e} \to \mathbb{F}_2^{e \times e}$ is a ring homomorphism.

Suppose that $\phi(g)$ is the identity matrix, or equivalently $\Phi(g)$ is the identity map. Since this implies $g \cdot \psi^{-1}(u) = \psi^{-1}(u)$ for every $u \in \mathbb{F}_2^e$, g should be the multiplicative identity of $\mathbb{F}_{2^e}$. This implies again that ϕ is injective.

This injective ring homomorphism naturally extends to $\bar{\phi} : \mathbb{F}_{2^e}^{r \times k} \to \mathbb{F}_2^{re \times ke}$ where ϕ is applied to each component and then $(\mathbb{F}_2^{e \times e})^{r \times k}$ is identified with $\mathbb{F}_2^{re \times ke}$. Now we are ready to define the Knudsen-Preneel compression functions.

Definition 1. *Let $\mathcal{C}$ be an $[r, k, d]_{2^e}$ linear code with a generator matrix $G \in \mathbb{F}_{2^e}^{k \times r}$ and let $\phi : \mathbb{F}_{2^e} \to \mathbb{F}_2^{e \times e}$ be the injective ring homomorphism defined above. Let $e = bc$ and $n = bn'$ for some positive integers b, c, n, n', and let $ek > rb$. Then the Knudsen-Preneel compression function*

$$H = \mathsf{KP}^b([r, k, d]_{2^e}) : \{0, 1\}^{kcn} \to \{0, 1\}^{rn}$$

making oracle queries to public random functions $f_l : \{0, 1\}^{cn} \to \{0, 1\}^n$, $l = 1, \ldots, r$, computes $H(W)$ for $W \in \{0, 1\}^{kcn}$ as follows.

1. *Compute $X \leftarrow (\bar{\phi}(G^T) \otimes I_{n'}) \cdot W$.*
2. *Parse $X = (x_1, \ldots, x_r)$, where $x_1, \ldots, x_r \in \{0, 1\}^{cn}$.*
3. *Make oracle queries $y_l = f_l(x_l)$ for $l = 1, \ldots, r$, and output the digest $Z = y_1 || \cdots || y_r$.*

Here $\otimes$ denotes the Kronecher product and $I_{n'}$ the identity matrix in $\mathbb{F}_2^{n' \times n'}$.

Example 1. The above mathematical description of Knudsen-Preneel constructions looks complicated, while the constructions themselves are very simple. For example, let $e = 2$ and let $\mathbb{F}_{2^2} = \mathbb{F}(\omega)$ for a root ω satisfying $\omega^2 + \omega + 1 = 0$. For $a_1 \omega + a_0 \in \mathbb{F}_{2^2}$,

$$\omega(a_1 \omega + a_0) = (a_0 + a_1)\omega + a_1.$$

This implies $\phi(\omega) = \begin{bmatrix} 1 & 1 \\ 1 & 0 \end{bmatrix}$. Since ϕ is an injective ring homomorphism,

$$\phi(0) = \begin{bmatrix} 0 & 0 \\ 0 & 0 \end{bmatrix}, \quad \phi(1) = \begin{bmatrix} 1 & 0 \\ 0 & 1 \end{bmatrix}, \quad \phi(\omega) = \begin{bmatrix} 1 & 1 \\ 1 & 0 \end{bmatrix}, \quad \phi(\omega+1) = \phi(\omega)+\phi(1) = \begin{bmatrix} 0 & 1 \\ 1 & 1 \end{bmatrix}.$$

Let $\mathcal{C}$ be a $[5, 3, 3]_4$ linear code with a generator matrix $G = \begin{bmatrix} 1 & 0 & 0 & 1 & 1 \\ 0 & 1 & 0 & 1 & \omega \\ 0 & 0 & 1 & 1 & \omega + 1 \end{bmatrix}$. If $c = 2$, then $b = 1$, $n = n'$ and

$$\bar{\phi}(G^T) = \begin{bmatrix} 1\,0\,|\,0\,0\,|\,0\,0\,|\,1\,0\,|\,1\,0 \\ 0\,1\,|\,0\,0\,|\,0\,0\,|\,0\,1\,|\,0\,1 \\ \hline 0\,0\,|\,1\,0\,|\,0\,0\,|\,1\,0\,|\,1\,1 \\ 0\,0\,|\,0\,1\,|\,0\,0\,|\,0\,1\,|\,1\,0 \\ \hline 0\,0\,|\,0\,0\,|\,1\,0\,|\,1\,0\,|\,0\,1 \\ 0\,0\,|\,0\,0\,|\,0\,1\,|\,0\,1\,|\,1\,1 \end{bmatrix}^T.$$

Let $H : \{0, 1\}^{6n} \to \{0, 1\}^{5n}$ be the resulting KP compression function using five public random functions $f_l : \{0, 1\}^{2n} \to \{0, 1\}^n$, $l = 1, \ldots, 5$. Then for $W = \omega_1 || \cdots || \omega_6$,

$$H(W) = f_1(x_1) || \cdots || f_5(x_5),$$

where $x_1 = (\omega_1||\omega_2)$, $x_2 = (\omega_3||\omega_4)$, $x_3 = (\omega_5||\omega_6)$, $x_4 = (\omega_1 \oplus \omega_3 \oplus \omega_5 || \omega_2 \oplus \omega_4 \oplus \omega_6)$, $x_5 = (\omega_1 \oplus \omega_3 \oplus \omega_4 \oplus \omega_6 || \omega_2 \oplus \omega_3 \oplus \omega_5 \oplus \omega_6)$.

Throughout this work, we will simply write $\mathcal{C}^{PRE}(W) = (\bar{\phi}(G^T) \otimes I_{n'}) \cdot W$. For the security analysis of H, we need to state some properties of $\mathcal{C}^{PRE}$.

Definition 2. *Let $\mathcal{I} \subset [1, r]$ and let $(x_l^*)_{l \in \mathcal{I}} \in \prod_{l \in \mathcal{I}} \{0, 1\}^{cn}$. $(x_1, \ldots, x_r) \in (\{0, 1\}^{cn})^r$ is called an* extension *of $(x_l^*)_{l \in \mathcal{I}}$ if there exists an input $W \in \{0, 1\}^{kcn}$ such that $\mathcal{C}^{PRE}(W) = (x_1, \ldots, x_r)$ and $x_l = x_l^*$ for $l \in \mathcal{I}$. We will say $(x_l^*)_{l \in \mathcal{I}}$ is* valid *if it has an extension.[1]*

For $\mathcal{I} = [1, r]$, valid tuples are exactly the images of $\mathcal{C}^{PRE}$. Due to the linearity of $\mathcal{C}^{PRE}$ (with respect to bitwise xor "$\oplus$"), we have the following property.

Property 1. *If $(x_l)_{l \in \mathcal{I}}$ and $(x_l')_{l \in \mathcal{I}}$ are valid, then $(x_l \oplus x_l')_{l \in \mathcal{I}}$ is also valid.*

Property 2. *Let $\mathcal{I}$ be a subset of $[1, r]$ such that $|\mathcal{I}| = r - d + 1$. If $(x_l^*)_{l \in \mathcal{I}} \in \prod_{l \in \mathcal{I}} \{0, 1\}^{cn}$ is valid, then it has a unique extension.*

Proof. Suppose that $(x_1, \ldots, x_r)$, and $(x_1', \ldots, x_r')$ are extensions of $(x_l^*)_{l \in \mathcal{I}}$. Then $(x_1 \oplus x_1', \ldots, x_r \oplus x_r')$ is also an extension of $(0)_{l \in \mathcal{I}}$. Since any nonzero codeword in $\mathcal{C}$ has at least d nonzero coordinates, we have $(x_1 \oplus x_1', \ldots, x_r \oplus x_r') = (0, \ldots, 0)$, and hence $(x_1, \ldots, x_r) = (x_1', \ldots, x_r')$. $\square$

2.2 Collision Resistance and Preimage Resistance

In this section, we review security notions of collision resistance and preimage resistance in an information theoretic sense. In the collision resistance experiment, a computationally unbounded adversary $\mathcal{A}$ makes oracle queries to public random functions f_l, $l = 1, \ldots, r$, and records a *query history* $\mathcal{Q}$, which is initialized as an empty set. When $\mathcal{A}$ makes a new query $f_l(x)$, a query-response pair $(l, x, f_l(x))$ is added to $\mathcal{Q}$.[2] We will loosely write $(l, x) \in \mathcal{Q}$ indicating that the value of $f_l(x)$ has been determined by $\mathcal{A}$'s query. Furthermore, we will denote $\mathcal{A}$'s i-th query as (l^i, x^i), $i = 1, \ldots, q$, indicating the i-th query is $f_{l^i}(x^i)$.

At the end of the collision-finding attack, $\mathcal{A}$ would like to find queries

$$(1, x^{i_1}), \ldots, (r, x^{i_r}), (1, x^{j_1}), \ldots, (r, x^{j_r}) \in \mathcal{Q}$$

satisfying the following two conditions.

1. $(x^{i_1}, \ldots, x^{i_r})$ and $(x^{j_1}, \ldots, x^{j_r})$ are *distinct* valid tuples.
2. $f_1(x^{i_1}) || \cdots || f_r(x^{i_r}) = f_1(x^{j_1}) || \cdots || f_r(x^{j_r})$.

---

[1] We regard $\prod_{l \in \mathcal{I}} \{0, 1\}^{cn}$ as the set of all functions from $\mathcal{I}$ to $\{0, 1\}^{cn}$. Thus, even in case $|\mathcal{I}| = |\mathcal{I}'|$, $\prod_{l \in \mathcal{I}} \{0, 1\}^{cn} \neq \prod_{l \in \mathcal{I}'} \{0, 1\}^{cn}$ as long as $\mathcal{I} \neq \mathcal{I}'$. We also naturally identify $(\{0, 1\}^{cn})^r$ with $\{0, 1\}^{crn}$.

[2] Unless stated otherwise, we will not allow any redundant query.

In this case, $(i_l, j_l)_{l \in [1,r]}$ is called an *index sequence of a collision*. The success probability of $\mathcal{A}$'s finding a collision is denoted by $\mathbf{Adv}_H^{\mathsf{col}}(\mathcal{A})$. The maximum of $\mathbf{Adv}_H^{\mathsf{col}}(\mathcal{A})$ over the adversaries making at most q queries is denoted by $\mathbf{Adv}_H^{\mathsf{col}}(q)$.

In the preimage resistance experiment, $\mathcal{A}$ chooses a target image $Z = z_1 || \cdots || z_r$ at the beginning of the attack, where $z_1, \ldots, z_r \in \{0,1\}^n$. After making a certain number of oracle queries to f_l, $l = 1, \ldots, r$, $\mathcal{A}$ would like to find queries

$$(1, x^{i_1}), \ldots, (r, x^{i_r}) \in \mathcal{Q}$$

such that $f_1\left(x^{i_1}\right) || \cdots || f_r\left(x^{i_r}\right) = z_1 || \cdots || z_r$. The success probability of $\mathcal{A}$'s finding a preimage is denoted by $\mathbf{Adv}_H^{\mathsf{pre}}(\mathcal{A})$, and $\mathbf{Adv}_H^{\mathsf{pre}}(q)$ is the maximum of $\mathbf{Adv}_H^{\mathsf{pre}}(\mathcal{A})$ over the adversaries making at most q queries. There might be several definitions of preimage resistance according to the distribution of a target image. The definition described here, called everywhere preimage resistance, is known as the strongest version in the sense that an adversary chooses its target image on its own.

3   Preimage Resistance Proofs

In this section, we will give two preimage resistance proofs of the KP compression functions. In both security proofs, we let $Z = z_1 || \cdots || z_r$ be the range point to be inverted where $z_1, \ldots, z_r \in \{0,1\}^n$. When an adversary $\mathcal{A}$ succeeds in finding a preimage of Z, predicate Pre is set to true by definition. So we need to upper bound the probability $\mathbf{Pr}[\mathsf{Pre}]$. Throughout this work, we will write $N = 2^n$.

3.1   The First Alternative Proof

Consider a subset $\mathcal{T} \subset [1, r]$ such that $|\mathcal{T}| = r - d + 1$. With respective to this subset, we define predicate $\mathsf{Pre}_\mathcal{T}$, where $\mathsf{Pre}_\mathcal{T}$ is true if $\mathcal{A}$ obtains an index sequence of a preimage $D = (i_l)_{l \in [1,r]}$ such that

1. $(l, x^{i_l}) \in \mathcal{Q}$ and $f_l(x^{i_l}) = z_l$ for $l = 1, \ldots, r$,
2. $\max_{l \in \mathcal{T}}\{i_l\} < \min_{l \in [1,r] \setminus \mathcal{T}}\{i_l\}$.

By the second condition, $\mathcal{T}$ specifies the function indices where the first $r - d + 1$ partial preimages are determined. More precisely, a partial preimage can be defined as follows.

Definition 3. *Let $\mathcal{T}$ be a subset of $[1, r]$ such that $|\mathcal{T}| = r - d + 1$. A sequence of indices*

$$P = (i_l)_{l \in \mathcal{T}}$$

is called a partial preimage at $\mathcal{T}$ if $(l, x^{i_l}) \in \mathcal{Q}$ and $f_l(x^{i_l}) = z_l$ for $l \in \mathcal{T}$.

We will upper bound $\mathbf{Pr}\,[\mathsf{Pre}]$ by using the following implication.

$$\mathsf{Pre} \Rightarrow \bigvee_{\substack{\mathcal{T}\subset[1,r]\\|\mathcal{T}|=r-d+1}} \mathsf{Pre}_{\mathcal{T}}$$

$$\Rightarrow \mathsf{Bad}(M) \vee \bigvee_{\substack{\mathcal{T}\subset[1,r]\\|\mathcal{T}|=r-d+1}} (\neg\mathsf{Bad}(M) \wedge \mathsf{Pre}_{\mathcal{T}}), \tag{1}$$

where the parameterized predicate $\mathsf{Bad}(M)$, $M > 0$, is true if there exists a subset $\mathcal{T} \subset [1,r]$ of size $r - d + 1$ such that the number of partial preimages at $\mathcal{T}$ is greater than M.

In order for a preimage finding adversary $\mathcal{A}$ to set $\mathsf{Pre}_{\mathcal{T}}$ to true, $\mathcal{A}$ has to first complete a partial preimage at $\mathcal{T}$. If $(x^{i_l})_{l\in\mathcal{T}}$ is valid at the point when a partial preimage $P = (i_l)_{l\in\mathcal{T}}$ is completed, then the remaining queries $(x_l)_{l\in[1,r]\backslash\mathcal{T}}$ that might complete a preimage of Z along with $(x^{i_l})_{l\in\mathcal{T}}$ are uniquely determined by Property 2. Specifically, it is required that $f_l(x_l) = z_l$ for $l \in [1,r]\backslash\mathcal{T}$. If any of these evaluations has not been determined, we include $(x_l, z_l)_{l\in[1,r]\backslash\mathcal{T}}$ into a wish list $\mathcal{L}$, expecting all of these evaluations to happen sometime later. A single query might include a multiple number of wishes into $\mathcal{L}$ by completing a multiple number of partial preimages at $\mathcal{T}$. However a single partial preimage at $\mathcal{T}$ is associated with a unique element in $\mathcal{L}$. Therefore the size of $\mathcal{L}$ would be at most M without the occurrence of $\mathsf{Bad}(M)$. Since each wish would be accomplished with probability $1/N^{|[1,r]\backslash\mathcal{T}|} = 1/N^{d-1}$, we have the following upper bound.

$$\mathbf{Pr}\,[\neg\mathsf{Bad}(M) \wedge \mathsf{Pre}_{\mathcal{T}}] \leq \sum_{i=1}^{M} \mathbf{Pr}\,[\text{the } i\text{-th wish is granted}] \leq \frac{M}{N^{d-1}}. \tag{2}$$

In order to address the remaining problem of upper bounding the probability of $\mathsf{Bad}(M)$, we will define a random variable X that counts the number of partial preimages at $\mathcal{T}$, and probabilistically upper bound the value of X using Markov's inequality.

Fix a subset $\mathcal{T} \subset [1,r]$ of size $r - d + 1$, and define a random variable X_P for each sequence $P = (i_l)_{l\in\mathcal{T}} \in \prod_{l\in\mathcal{T}}[1,q]$, where $X_P = 1$ if $(l, x^{i_l}) \in \mathcal{Q}$ and $f_l(x^{i_l}) = z_l$ for every $l \in \mathcal{T}$, and $X_P = 0$ otherwise. If we define

$$X = \sum_{P\in\prod_{l\in\mathcal{T}}[1,q]} X_P,$$

then X counts the number of partial preimages at $\mathcal{T}$. Since $\left|\prod_{l\in\mathcal{T}}[1,q]\right| = q^{r-d+1}$ and

$$\mathbf{Pr}[X_P = 1] = \mathsf{Ex}(X_P) \leq \frac{1}{N^{r-d+1}},$$

we have $\mathsf{Ex}(X) \leq \frac{q^{r-d+1}}{N^{r-d+1}}$. Using Markov's inequality, for $M > 0$ we have

$$\mathbf{Pr}\,[X \geq M] \leq \frac{q^{r-d+1}}{MN^{r-d+1}}.$$

Applying a union bound over subsets $\mathcal{T} \subset [1, r]$ of size $r - d + 1$, we have

$$\mathbf{Pr}[\mathsf{Bad}(M)] \leq \binom{r}{r-d+1}\frac{q^{r-d+1}}{MN^{r-d+1}} = \binom{r}{d-1}\frac{q^{r-d+1}}{MN^{r-d+1}}. \tag{3}$$

By (1), (2) and (3), we have

$$\mathbf{Pr}[\mathsf{Pre}] \leq \binom{r}{d-1}\frac{q^{r-d+1}}{MN^{r-d+1}} + \binom{r}{d-1}\frac{M}{N^{d-1}}.$$

Let

$$M = \frac{q^{\frac{r-d+1}{2}}}{N^{\frac{r-2d+2}{2}}}$$

by setting $q^{r-d+1}/(MN^{r-d+1}) = M/N^{d-1}$. Then we have

$$\mathbf{Pr}[\mathsf{Pre}] \leq 2\binom{r}{d-1}\frac{q^{\frac{r-d+1}{2}}}{N^{\frac{r}{2}}}.$$

The following theorem summarizes this result.

Theorem 1. *Let H be the Knudsen-Preneel compression function based on an $[r, k, d]_{2^e}$ code. Then we have*

$$\mathbf{Adv}_H^{\mathsf{pre}}(q) \leq 2\binom{r}{d-1}\frac{q^{\frac{r-d+1}{2}}}{N^{\frac{r}{2}}}.$$

For MDS codes, we have

$$\mathbf{Adv}_H^{\mathsf{pre}}(q) \leq 2\binom{r}{k}\frac{q^{\frac{k}{2}}}{N^{\frac{r}{2}}}.$$

Example 2. Let H be based on a $[5, 3, 3]_4$ MDS code. Then Theorem 1 implies

$$\mathbf{Adv}_H^{\mathsf{pre}}(q) \leq \frac{20q^{\frac{3}{2}}}{N^{\frac{5}{2}}}.$$

Therefore H is preimage resistant up to $N^{5/3}$ query complexity.

3.2   The Second Alternative Proof

The main idea of this proof is based on the observation that for any set of r queries to $f_1, \ldots, f_r$ that are in the range of $\mathcal{C}^{PRE}$, one can appoint k queries that expand the span. Whenever any of such queries is made by an adversary $\mathcal{A}$, we let the corresponding modified adversary $\mathcal{A}'$ immediately make any other queries that are added to the span. In this way, we can fix all the indices of queries at which $\mathcal{A}'$ obtains a full preimage of Z. This modification makes upper bounding the preimage finding advantage of $\mathcal{A}'$ much easier than $\mathcal{A}$.

To be precise, let $H = \mathsf{KP}^b([r, k, d]_{2^e})$ be given with a generator matrix

$$G = [G_1, G_2, \cdots, G_r]$$

where G_i is a $k \times 1$ column matrix for $i = 1, \ldots, r$. (G is not necessarily in standard form.) Fix a sequence

$$\mathcal{T} = (l_1, l_2, \ldots, l_k) \in [1, r]^k$$

such that column matrices $G_{l_1}, \ldots, G_{l_k}$ are linearly independent (which implies $l_1, l_2, \ldots, l_k$ are all different), and a sequence

$$P = (i_1, i_2, \ldots, i_k) \in [1, q]^k$$

such that $i_1 < i_2 < \cdots < i_k$. If partial preimages $f_{l_1}(x^{i_1}) = z_{l_1}, \cdots, f_{l_k}(x^{i_k}) = z_{l_k}$ are found,[3] then these queries uniquely determine the remaining $r - k$ queries x_l, $l \in [1, r] \backslash \mathcal{T}$, such that, setting $x_{l_j} = x^{i_j}$ for $l_j \in \mathcal{T}$, $(x_l)_{l \in [1,r]}$ is an image of $\mathcal{C}^{PRE}$. Specifically, each of the remaining queries is represented as a linear combination of $x^{i_1}, \ldots, x^{i_k}$. We define predicate $\mathsf{Pre}_{\mathcal{T},P}$ where $\mathsf{Pre}_{\mathcal{T},P}$ is true if the following two conditions are satisfied.

1. $(l_\alpha, x^{i_\alpha}) \in \mathcal{Q}$ and $f_{l_\alpha}(x^{i_\alpha}) = z_{l_\alpha}$ for $\alpha = 1, \ldots, k$.
2. For all $l \in [1, r] \backslash \mathcal{T}$, let α be the first index such that G_l is represented as a linear combination of $G_{l_1}, \ldots, G_{l_\alpha}$. $\mathcal{A}$ obtains $f_l(x_l) = z_l$ *after* $\mathcal{A}$ makes the i_α-th query. (Note that x_l is determined as a linear combination of $x^{i_1}, \ldots, x^{i_\alpha}$.)

Then we have the following implication.

$$\mathsf{Pre} \Rightarrow \bigvee_{(\mathcal{T}, P)} \mathsf{Pre}_{\mathcal{T},P}. \tag{4}$$

In order to prove the above implication, suppose that $\mathcal{A}$ sets Pre to true by obtaining $f_{l_1}(x^{i_1}) = z_1, \cdots, f_{l_r}(x^{i_r}) = z_r$ in an order of $i_1 < i_2 < \ldots < i_r$. From the sequence $(l_1, \ldots, l_r) \in [1, r]^r$, we can extract a subsequence $\mathcal{T} \in [1, r]^k$ using the following algorithm.

$\mathcal{T} \leftarrow \emptyset$
For $\alpha = 1, \ldots, r$,
   if G_{l_α} is not represented by a linear combination of G_l, $l \in \mathcal{T}$ **then**
      $\mathcal{T} \leftarrow l_\alpha$

Since G is of rank k, we have $|\mathcal{T}| = k$. We can also check that $\mathsf{Pre}_{\mathcal{T},P}$ is true with $P = (i_\alpha)$ where α satisfies $l_\alpha \in \mathcal{T}$.

Sequence P fixes the indices of queries when we need to obtain the partial preimages of z_l for $l \in \mathcal{T}$. In order to fix the indices of queries from which we obtain the remaining partial preimages, we construct a modified adversary $\mathcal{A}'$ that uses $\mathcal{A}$ as a subroutine. The behavior of $\mathcal{A}'$ can be illustrated as follows.

---

[3] Here we are using slightly different notations from Section 2.2 by assuming x^{i_α} is queried to f_{l_α} not f_α. This implies $l_\alpha = l^{i_\alpha}$ for $\alpha = 1, \ldots, k$.

1. Between $\mathcal{A}$ and the random function oracles, $\mathcal{A}'$ faithfully relays all the $\mathcal{A}$'s queries and the oracles' responses.
2. Once queries $f_{l_1}(x^{i_1}), \cdots, f_{l_\alpha}(x^{i_\alpha})$ are made for $\alpha = 1, \ldots, r$, $\mathcal{A}'$ searches for G_l that is represented as a linear combination of $G_{l_1}, \ldots, G_{l_\alpha}$ with a nonzero coefficient of G_{l_α}.
3. For such an index l, query x_l that is consistent with $x^{i_1}, \ldots, x^{i_\alpha}$ is determined as a linear combination of $x^{i_1}, \ldots, x^{i_\alpha}$. $\mathcal{A}'$ makes an additional query $f_l(x_l)$ without relaying the response to $\mathcal{A}$. When $\mathcal{A}$ makes a certain query, $\mathcal{A}'$ might need to make a multiple number of additional queries, while we fix an order between those queries.

In case $\mathcal{A}$ requests any of the additional queries later, $\mathcal{A}'$ would have to make a redundant query. Including the redundant queries, the number of queries made by $\mathcal{A}'$ is at most $q + r - k$. In this way, $(\mathcal{T}, P)$ induces new sequences

$$\mathcal{T}' = (l_1', l_2', \ldots, l_r') \in [1, r]^r,$$

$$P' = (i_1', i_2', \ldots, i_r') \in [1, q]^r$$

such that l_α' are all distinct, $i_1' < i_2' < \cdots < i_r'$, and $\mathcal{A}$ setting $\mathsf{Pre}_{\mathcal{T}, P}$ to true implies that $\mathcal{A}'$ obtains $f_{l_\alpha'}(x^{i_\alpha'}) = z_{l_\alpha'}$ as fresh queries for $\alpha = 1, \ldots, r$.[4]

Example 3. Let H be based on a $[5, 3, 3]_4$ MDS code with a generator matrix

$$G = [G_1, G_2, G_3, G_4, G_5].$$

Let $\mathcal{T} = (1, 5, 3)$ and $P = (i_1, i_2, i_3)$, and let $G_2 = \lambda G_1$ and $G_4 = \mu_1 G_1 + \mu_3 G_3 + \mu_5 G_5$ for some constants $\lambda, \mu_1, \mu_3, \mu_5$ where λ and μ_3 are nonzero. Then $(\mathcal{T}, P)$ induces $\mathcal{T}' = (1, \mathbf{2}, 5, 3, \mathbf{4})$ and $P' = (i_1, i_1 + 1, i_2 + 1, i_3 + 1, i_3 + 2)$. Note that i_2 and i_3 have been replaced by $i_2 + 1$ and $i_3 + 1$ respectively in P', since one additional query has been inserted right after the i_1-th query.

Since $(\mathcal{T}', P')$ fixes all query indices i_α' that determine a preimage of Z, we have

$$\mathbf{Pr}\left[\mathcal{A} \text{ sets } \mathsf{Pre}_{\mathcal{T}, P} \text{ to true}\right] \leq \mathbf{Pr}\left[\mathcal{A}' \text{ sets } \mathsf{Pre}_{\mathcal{T}', P'} \text{ to true}\right] \leq \frac{1}{N^r}. \qquad (5)$$

Since the number of possible choices for $(\mathcal{T}, P)$ is at most

$$\binom{r}{k} k! \cdot \binom{q}{k} \leq \binom{r}{k} q^k,$$

and by (4), (5) we conclude

$$\mathbf{Pr}\left[\mathsf{Pre}\right] \leq \binom{r}{k} \frac{q^k}{N^r}.$$

To summarize this result, we have the following theorem.

---

[4] Without allowing a redundant query, P' is not uniquely defined from $(\mathcal{T}, P)$. P' would be different according to the point of time when a redundant query is made.

Theorem 2. *Let H be the Knudsen-Preneel compression function based on an $[r, k, d]_{2^e}$ code. Then we have*

$$\mathbf{Adv}_H^{\mathrm{pre}}(q) \leq \binom{r}{k} \frac{q^k}{N^r}.$$

Example 4. Let H be based on a $[5, 3, 3]_4$ MDS code. Then Theorem 2 implies

$$\mathbf{Adv}_H^{\mathrm{pre}}(q) \leq \frac{10q^3}{N^5}.$$

4   Collision Resistance Proof

Consider two sets of evaluations $\big(f_l(x^{i_l})\big)_{l \in [1,r]}$ and $\big(f_l(x^{j_l})\big)_{l \in [1,r]}$ of the inner primitives for $H = \mathsf{KP}^b([r, k, d]_{2^e})$. Let $S \subset [1, r]$ and suppose that $i_l = j_l$ (and hence $x^{i_l} = x^{j_l}$) for $l \in S$. As long as $(x^{i_l})_{l \in [1,r]}$ and $(x^{j_l})_{l \in [1,r]}$ are valid, partial inner collisions $f_l(x^{i_l}) = f_l(x^{j_l})$ for $l \in [1, r]\backslash S$ suffice to guarantee an actual collision of H regardless of the evaluations of $f_l(x^{i_l})(= f_l(x^{j_l}))$ for $l \in S$. For this reason, we will call the indices in S *inactive* and the other indices *active*. The probability of finding a collision turns out to be closely related to the number of inactive indices that contribute a collision.

When a collision happens, let predicate Col be set to true by definition. Our security proof begins with decomposing this predicate into subcases according to the number of inactive indices. For $0 \leq s \leq r - d$, consider a subset $S \subset [1, r]$ such that $|S| = s$. With respective to this subset, we define predicate Col_S, where Col_S is true if $\mathcal{A}$ obtains an index sequence of a collision $C = (i_l, j_l)_{l \in [1,r]}$ such that

$$i_l = j_l \text{ if and only if } l \in S.$$

Note that more than $r - d$ inactive inner collisions enforce $(x^{i_1}, \ldots, x^{i_r}) = (x^{j_1}, \ldots, x^{j_r})$ since H is based on a code of minimum distance d. Therefore we have

$$\mathsf{Col} \Rightarrow \bigvee_{0 \leq s \leq r-d} \left(\bigvee_{\substack{S \subset [1,r] \\ |S|=s}} \mathsf{Col}_S \right). \tag{6}$$

4.1   Inner Collisions Compatible with Inactive Indices

For $s < d-1$, we will upper bound $\mathbf{Pr}\,[\mathsf{Col}_S]$ by a wish list argument. In order to upper bound the size of a certain wish list, we need a notion of partial collisions. Similar to partial preimages, each partial collision will uniquely determine a wish in the list, so the size of the wish list is upper bounded by the number of partial collisions.

Definition 4. *Let S and T be disjoint subsets of $[1, r]$. A sequence of indices*

$$P = (i_l, j_l)_{l \in T}$$

is called a partial collision at T compatible with inactive indices S *if*

1. $1 \le i_l, j_l \le q$ *are all distinct,*
2. $(l, x^{i_l}), (l, x^{j_l}) \in Q$ *and* $f_l(x^{i_l}) = f_l(x^{j_l})$ *for* $l \in T$,
3. $(\Delta_l)_{l \in S \cup T}$ *is valid where* $\Delta_l = 0$ *for* $l \in S$ *and* $\Delta_l = x^{i_l} \oplus x^{j_l}$ *for* $l \in T$.

Note that even in case of $S \cup T = [1, r]$, a partial collision need not correspond to an actual collision as $(x^{i_l})_{l \in T}$ and $(x^{j_l})_{l \in T}$ might not be valid. A partial collision also has the following property.

Property 3. *For disjoint subsets S and $T \subset [1, r]$, the number of partial collisions at T compatible with inactive indices S is a multiple of $2^{|T|}$.*

Proof. From a single partial collision $P = (i_l, j_l)_{l \in T}$, we can obtain $2^{|T|}$ different partial collisions by swapping i_l and j_l for each $l \in T$. Since we can define an equivalence relation between them, the total number of partial collisions is given as a multiple of $2^{|T|}$. □

By the following lemma, we can upper bound the number of partial collisions at T compatible with inactive indices S for a fixed subset T such that $S \cap T = \emptyset$ and $|S| + |T| \ge r - d + 1$. The proof, given in Appendix A in detail, is essentially based on the application of Markov's inequality.

Lemma 1. *Let S and T be disjoint subsets of $[1, r]$ such that $|S| \le r - d$ and $|S| + |T| \ge r - d + 1$, and let $|S| = s$ and $|T| = t$. Then for $M > 0$, the number of partial collisions at T compatible with inactive indices S is smaller than $2^{t-r+d+s-1} M$ except with probability*

$$\binom{t}{r - d - s + 1} \frac{q^{t+r-d-s+1}}{M N^t}.$$

4.2   Upper Bounding $\Pr[\text{Col}_S]$

According to the number of inactive indices, $s = |S|$, we distinguish two cases.

Case 1. $s < d - 1$: This case is analyzed by a wish list argument.
Note that $|[1, r] \backslash S| > r - d + 1$. For a subset $T \subset [1, r] \backslash S$ such that $|T| = r - d + 1$, we define predicate $\text{Col}_{S, T}$ where $\text{Col}_{S, T}$ is true if $\mathcal{A}$ obtains an index sequence of a collision $C = (i_l, j_l)_{l \in [1, r]}$ such that

1. $i_l = j_l$ if and only if $l \in S$,
2. $\max_{l \in T} \{i_l, j_l\} < \min_{l \in [1, r] \backslash (S \cup T)} \{\max\{i_l, j_l\}\}$.

Thus $\mathcal{T}$ specifies the indices where the first $r - d + 1$ *active* inner collisions are completed. For $M > 0$, we define predicate $\mathsf{Bad}(M)$ where $\mathsf{Bad}(M)$ is true if there exists a subset $\mathcal{T} \subset [1, r] \backslash \mathcal{S}$ of size $r - d + 1$ such that the number of partial collisions at $\mathcal{T}$ compatible with inactive indices $\mathcal{S}$ is greater than

$$L = 2^s M.$$

Then by Lemma 1 (with $t = r - d + 1$) and a union bound, we have

$$\mathbf{Pr}[\mathsf{Bad}(M)] \leq \binom{r - s}{r - d + 1} \binom{r - d + 1}{s} \frac{q^{2(r-d+1)-s}}{MN^{r-d+1}}. \tag{7}$$

In order to upper bound $\mathbf{Pr}[\mathsf{Col}_{\mathcal{S}}]$, we will use the following implication.

$$\mathsf{Col}_{\mathcal{S}} \Rightarrow \mathsf{Bad}(M) \vee \bigvee_{\substack{\mathcal{T} \subset [1,r] \backslash \mathcal{S} \\ |\mathcal{T}| = r - d + 1}} (\neg \mathsf{Bad}(M) \wedge \mathsf{Col}_{\mathcal{S}, \mathcal{T}}). \tag{8}$$

Now we will focus on upper bounding $\mathbf{Pr}[\neg \mathsf{Bad}(M) \wedge \mathsf{Col}_{\mathcal{S}, \mathcal{T}}]$ for fixed subsets $\mathcal{S}$ and $\mathcal{T}$. In order for $\mathcal{A}$ to set $\mathsf{Col}_{\mathcal{S}, \mathcal{T}}$ to true, $\mathcal{A}$ has to first complete a partial collision at $\mathcal{T}$ compatible with inactive indices $\mathcal{S}$. At the point when a partial collision $P = (i_l, j_l)_{l \in \mathcal{T}}$ is completed, the remaining queries $(x_l, x_l')_{l \in [1,r] \backslash (\mathcal{S} \cup \mathcal{T})}$ that could make a collision along with P are uniquely determined. (They *exist* only if $(x^{i_l})_{l \in \mathcal{T}}$ and $(x^{j_l})_{l \in \mathcal{T}}$ are valid.) If

1. $x_l \neq x_l'$ for $l \in [1, r] \backslash (\mathcal{S} \cup \mathcal{T})$,
2. any of collisions of $f_l(x_l)$ and $f_l(x_l')$ has not been determined for $l \in [1, r] \backslash (\mathcal{S} \cup \mathcal{T})$,

then we include $(x_l, x_l')_{l \in [1,r] \backslash (\mathcal{S} \cup \mathcal{T})}$ into a wish list $\mathcal{L}$, expecting all of the collisions to happen sometime later. A single query might include a multiple number of wishes into $\mathcal{L}$ by completing a multiple number of partial collisions. However a single partial collision is associated with a unique element in $\mathcal{L}$. Therefore without the occurrence of $\mathsf{Bad}(M)$, the size of $\mathcal{L}$ is at most L, and we have the following upper bound.

$$\mathbf{Pr}[\neg \mathsf{Bad}(M) \wedge \mathsf{Col}_{\mathcal{S}, \mathcal{T}}] \leq \sum_{i=1}^{L} \mathbf{Pr}[\text{the } i\text{-th wish is granted}]. \tag{9}$$

Since

$$\mathbf{Pr}[\text{the } i\text{-th wish is granted}] \leq \frac{1}{N^{|[1,r] \backslash (\mathcal{S} \cup \mathcal{T})|}} = \frac{1}{N^{d-s-1}},$$

for each $i = 1, \ldots, L$, and by (7), (8), (9), we have

$$\mathbf{Pr}[\mathsf{Col}_{\mathcal{S}}] \leq \binom{r - s}{r - d + 1} \binom{r - d + 1}{s} \frac{q^{2(r-d+1)-s}}{MN^{r-d+1}} + \binom{r - s}{r - d + 1} \frac{2^s M}{N^{d-s-1}}. \tag{10}$$

Case 2. $s \geq d-1$: This case might occur when $d-1 \leq r-d$. Let $\mathcal{T} = [1, r]\backslash\mathcal{S}$. In this case, $\mathsf{Col}_{\mathcal{S}}$ implies that there is a partial collision at $\mathcal{T}$ compatible with inactive indices $\mathcal{S}$. Here we can use Lemma 1 with $M = 1$ and $t = r - s$ since the number of partial collisions should be a multiple of $2^{|\mathcal{T}|} = 2^{r-s}$ but $2^{t-r+d+s-1}(= 2^{d-1})$ is smaller than 2^{r-s}. Therefore we have

$$\mathbf{Pr}[\mathsf{Col}_{\mathcal{S}}] \leq \mathbf{Pr}[\text{there is a partial collisions at } \mathcal{T} \text{ compatible with inactive indices } \mathcal{S}]$$

$$\leq \binom{r-s}{d-1}\frac{q^{2(r-s)-d+1}}{N^{r-s}}. \tag{11}$$

4.3   Putting the Pieces Together

By (6), (10) and (11), we obtain the following result.

$$\mathbf{Pr}[\mathsf{Col}] \leq \sum_{s=0}^{d-2}\binom{r}{s}\binom{r-s}{r-d+1}\left(\binom{r-d+1}{s}\frac{q^{2(r-d+1)-s}}{M(s)N^{r-d+1}} + \frac{2^s M(s)}{N^{d-s-1}}\right)$$

$$+ \sum_{s=d-1}^{r-d}\binom{r}{s}\binom{r-s}{d-1}\frac{q^{2(r-s)-d+1}}{N^{r-s}},$$

where the parameter $M(s)$ might depend on the size of $\mathcal{S}$ and the second term of the right hand side appears only when $d - 1 \leq r - d$. In order to optimize the right hand side of the inequality, set

$$M(s) = \binom{r-d+1}{s}^{\frac{1}{2}}\frac{q^{r-d+1-\frac{s}{2}}}{2^{\frac{s}{2}}N^{\frac{r+s}{2}-d+1}},$$

by solving

$$\binom{r-d+1}{s}\frac{q^{2(r-d+1)-s}}{M(s)N^{r-d+1}} = \frac{2^s M(s)}{N^{d-s-1}}.$$

Then we have the following theorem.

Theorem 3. *Let H be the Knudsen-Preneel compression function based on an $[r, k, d]_{2^e}$ code. Then we have*

$$\mathbf{Adv}_H^{\mathsf{col}}(q) \leq \sum_{s=0}^{d-2}\binom{r}{s}\binom{r-s}{r-d+1}\binom{r-d+1}{s}^{\frac{1}{2}}\frac{2^{\frac{s}{2}+1}q^{r-d+1-\frac{s}{2}}}{N^{\frac{r-s}{2}}}$$

$$+ \sum_{s=d-1}^{r-d}\binom{r}{s}\binom{r-s}{d-1}\frac{q^{2(r-s)-d+1}}{N^{r-s}}.$$

INTERPRETATION. Let $d - 1 \leq r - d$ or equivalently $2d \leq r + 1$. Assuming $N^{\frac{1}{2}} \leq q \leq N$, we have

$$\sum_{s=0}^{d-2}\binom{r}{s}\binom{r-s}{r-d+1}\binom{r-d+1}{s}^{\frac{1}{2}}\frac{2^{\frac{s}{2}+1}q^{r-d+1-\frac{s}{2}}}{N^{\frac{r-s}{2}}} = O\left(\frac{q^{r-\frac{3d}{2}+2}}{N^{\frac{r}{2}-\frac{d}{2}+1}}\right),$$

and

$$\sum_{s=d-1}^{r-d} \binom{r}{s}\binom{r-s}{d-1}\frac{q^{2(r-s)-d+1}}{N^{r-s}} = O\left(\frac{q^{2r-3d+3}}{N^{r-d+1}}\right).$$

In this case, H is collision resistant up to $N^{\frac{r-d+1}{2r-3d+3}}$ query complexity since

$$N^{\frac{1}{2}} \leq N^{\frac{r-d+1}{2r-3d+3}} \leq N^{\frac{r-d+2}{2r-3d+4}} \leq N.$$

Let $2d > r+1$. Assuming $q \geq N$, we have

$$\sum_{s=0}^{r-d} \binom{r}{s}\binom{r-s}{r-d+1}\binom{r-d+1}{s}^{\frac{1}{2}} \frac{2^{\frac{s}{2}+1}q^{r-d+1-\frac{s}{2}}}{N^{\frac{r-s}{2}}} = O\left(\frac{q^{r-d+1}}{N^{\frac{r}{2}}}\right).$$

In this case, H is collision resistant up to $N^{\frac{r}{2r-2d+2}}$ query complexity since

$$N \leq N^{\frac{r}{2r-2d+2}}.$$

We summarize this result as follows.

Corollary 1. *Let H be the Knudsen-Preneel compression function based on an $[r, k, d]_{2^e}$ code.*

(a) If $2d \leq r+1$, then H is collision resistant up to $N^{\frac{r-d+1}{2r-3d+3}}$ query complexity.
(b) If $2d > r+1$, then H is collision resistant up to $N^{\frac{r}{2r-2d+2}}$ query complexity.

Corollary 2. *Let H be the Knudsen-Preneel compression function based on an $[r, k, d]_{2^e}$ MDS code.*

(a) If $r+1 \leq 2k$, then H is collision resistant up to $N^{\frac{k}{3k-r}}$ query complexity.
(b) If $r+1 > 2k$, then H is collision resistant up to $N^{\frac{r}{2k}}$ query complexity.

Example 5. Let H be based on $[5, 3, 3]_4$ MDS code. Then

$$\mathbf{Adv}_H^{\mathrm{col}}(q) \leq \sum_{s=0}^{1} \binom{5}{s}\binom{5-s}{3}\binom{3}{s}^{\frac{1}{2}} \frac{2^{\frac{s}{2}+1}q^{3-\frac{s}{2}}}{N^{\frac{5-s}{2}}} + \binom{5}{2}\binom{3}{2}\frac{q^4}{N^3}$$

$$= \frac{20q^3}{N^{\frac{5}{2}}} + \frac{40\sqrt{6}q^{\frac{5}{2}}}{N^2} + \frac{30q^4}{N^3}.$$

Therefore H is collision resistant up to $N^{3/4}$ query complexity.

Acknowledgements. The author would like to thank Martijn Stam for many valuable comments.

References

1. Armknecht, F., Fleischmann, E., Krause, M., Lee, J., Stam, M., Steinberger, J.: The Preimage Security of Double-Block-Length Compression Functions. In: Lee, D.H. (ed.) ASIACRYPT 2011. LNCS, vol. 7073, pp. 233–251. Springer, Heidelberg (2011)
2. Biham, E., Chen, R., Joux, A., Carribault, P., Lemuet, C., Jalby, W.: Collisions of SHA-0 and Reduced SHA-1. In: Cramer, R. (ed.) EUROCRYPT 2005. LNCS, vol. 3494, pp. 36–57. Springer, Heidelberg (2005)
3. Black, J., Cochran, M., Shrimpton, T.: On the Impossibility of Highly-Efficient Blockcipher-Based Hash Functions. In: Cramer, R. (ed.) EUROCRYPT 2005. LNCS, vol. 3494, pp. 526–541. Springer, Heidelberg (2005)
4. Black, J., Rogaway, P., Shrimpton, T.: Black-Box Analysis of the Block-Cipher-Based Hash-Function Constructions from PGV. In: Yung, M. (ed.) CRYPTO 2002. LNCS, vol. 2442, pp. 320–325. Springer, Heidelberg (2002)
5. Bos, J.W., Özen, O., Stam, M.: Efficient Hashing Using the AES Instruction Set. In: Preneel, B., Takagi, T. (eds.) CHES 2011. LNCS, vol. 6917, pp. 507–522. Springer, Heidelberg (2011)
6. De Cannière, C., Rechberger, C.: Preimages for Reduced SHA-0 and SHA-1. In: Wagner, D. (ed.) CRYPTO 2008. LNCS, vol. 5157, pp. 179–202. Springer, Heidelberg (2008)
7. Damgård, I.B.: A Design Principle for Hash Functions. In: Brassard, G. (ed.) CRYPTO 1989. LNCS, vol. 435, pp. 416–427. Springer, Heidelberg (1990)
8. Fleischmann, E., Gorski, M., Lucks, S.: Security of Cyclic Double Block Length Hash Functions. In: Parker, M.G. (ed.) Cryptography and Coding 2009. LNCS, vol. 5921, pp. 153–175. Springer, Heidelberg (2009)
9. Hirose, S.: Provably Secure Double-Block-Length Hash Functions in a Black-Box Model. In: Park, C., Chee, S. (eds.) ICISC 2004. LNCS, vol. 3506, pp. 330–342. Springer, Heidelberg (2005)
10. Hirose, S.: Some Plausible Constructions of Double-Block-Length Hash Functions. In: Robshaw, M. (ed.) FSE 2006. LNCS, vol. 4047, pp. 210–225. Springer, Heidelberg (2006)
11. Knudsen, L.R., Preneel, B.: Construction of secure and fast hash functions using non binary error-correcting codes. IEEE Transactions on Information Theory 48(9), 2524–2539 (2002)
12. Knudsen, L.R., Preneel, B.: Fast and Secure Hashing Based on Codes. In: Kaliski Jr., B.S. (ed.) CRYPTO 1997. LNCS, vol. 1294, pp. 485–498. Springer, Heidelberg (1997)
13. Knudsen, L.R., Preneel, B.: Hash Functions Based on Block Ciphers and Quaternary Codes. In: Kim, K.-c., Matsumoto, T. (eds.) ASIACRYPT 1996. LNCS, vol. 1163, pp. 77–90. Springer, Heidelberg (1996)
14. Lai, X., Massey, J.L.: Hash Functions Based on Block Ciphers. In: Rueppel, R.A. (ed.) EUROCRYPT 1992. LNCS, vol. 658, pp. 55–70. Springer, Heidelberg (1993)
15. Lee, J., Kwon, D.: The security of Abreast-DM in the ideal cipher model. IACR ePrint Archive 2009/225 (2009)
16. Lee, J., Stam, M.: MJH: A Faster Alternative to MDC-2. In: Kiayias, A. (ed.) CT-RSA 2011. LNCS, vol. 6558, pp. 213–236. Springer, Heidelberg (2011)
17. Lee, J., Stam, M., Steinberger, J.: The Collision Security of Tandem-DM in the Ideal Cipher Model. In: Rogaway, P. (ed.) CRYPTO 2011. LNCS, vol. 6841, pp. 561–577. Springer, Heidelberg (2011)

18. Leurent, G.: MD4 is Not One-Way. In: Nyberg, K. (ed.) FSE 2008. LNCS, vol. 5086, pp. 412–428. Springer, Heidelberg (2008)
19. Mendel, F., Pramstaller, N., Rechberger, C., Rijmen, V.: Analysis of Step-Reduced SHA-256. In: Robshaw, M. (ed.) FSE 2006. LNCS, vol. 4047, pp. 126–143. Springer, Heidelberg (2006)
20. Merkle, R.C.: One Way Hash Functions and DES. In: Brassard, G. (ed.) CRYPTO 1989. LNCS, vol. 435, pp. 428–446. Springer, Heidelberg (1990)
21. Özen, O., Shrimpton, T., Stam, M.: Attacking the Knudsen-Preneel Compression Functions. In: Hong, S., Iwata, T. (eds.) FSE 2010. LNCS, vol. 6147, pp. 94–115. Springer, Heidelberg (2010)
22. Özen, O., Stam, M.: Another Glance at Double-Length Hashing. In: Parker, M.G. (ed.) Cryptography and Coding 2009. LNCS, vol. 5921, pp. 176–201. Springer, Heidelberg (2009)
23. Özen, O., Stam, M.: Collision Attacks against the Knudsen-Preneel Compression Functions. In: Abe, M. (ed.) ASIACRYPT 2010. LNCS, vol. 6477, pp. 76–93. Springer, Heidelberg (2010)
24. Steinberger, J.: The Collision Intractability of MDC-2 in the Ideal-Cipher Model. In: Naor, M. (ed.) EUROCRYPT 2007. LNCS, vol. 4515, pp. 34–51. Springer, Heidelberg (2007)
25. Preneel, B., Govaerts, R., Vandewalle, J.: Hash Functions Based on Block Ciphers: A Synthetic Approach. In: Stinson, D.R. (ed.) CRYPTO 1993. LNCS, vol. 773, pp. 368–378. Springer, Heidelberg (1994)
26. Ristenpart, T., Shrimpton, T.: How to Build a Hash Function from Any Collision-Resistant Function. In: Kurosawa, K. (ed.) ASIACRYPT 2007. LNCS, vol. 4833, pp. 147–163. Springer, Heidelberg (2007)
27. Rogaway, P., Steinberger, J.: Constructing Cryptographic Hash Functions from Fixed-Key Blockciphers. In: Wagner, D. (ed.) CRYPTO 2008. LNCS, vol. 5157, pp. 433–450. Springer, Heidelberg (2008)
28. Rogaway, P., Steinberger, J.: Security/Efficiency Tradeoffs for Permutation-Based Hashing. In: Smart, N.P. (ed.) EUROCRYPT 2008. LNCS, vol. 4965, pp. 220–236. Springer, Heidelberg (2008)
29. Sasaki, Y., Aoki, K.: Finding Preimages in Full MD5 Faster Than Exhaustive Search. In: Joux, A. (ed.) EUROCRYPT 2009. LNCS, vol. 5479, pp. 134–152. Springer, Heidelberg (2009)
30. Shrimpton, T., Stam, M.: Building a Collision-Resistant Compression Function from Non-compressing Primitives. In: Aceto, L., Damgård, I., Goldberg, L.A., Halldórsson, M.M., Ingólfsdóttir, A., Walukiewicz, I. (eds.) ICALP 2008, Part II. LNCS, vol. 5126, pp. 643–654. Springer, Heidelberg (2008)
31. Stam, M.: Beyond Uniformity: Better Security/Efficiency Tradeoffs for Compression Functions. In: Wagner, D. (ed.) CRYPTO 2008. LNCS, vol. 5157, pp. 397–412. Springer, Heidelberg (2008)
32. Stam, M.: Blockcipher-Based Hashing Revisited. In: Dunkelman, O. (ed.) FSE 2009. LNCS, vol. 5665, pp. 67–83. Springer, Heidelberg (2009)
33. Steinberger, J.: The Collision Intractability of MDC-2 in the Ideal-Cipher Model. In: Naor, M. (ed.) EUROCRYPT 2007. LNCS, vol. 4515, pp. 34–51. Springer, Heidelberg (2007)
34. Wang, X., Lai, X., Feng, D., Chen, H., Yu, X.: Cryptanalysis of the Hash Functions MD4 and RIPEMD. In: Cramer, R. (ed.) EUROCRYPT 2005. LNCS, vol. 3494, pp. 1–18. Springer, Heidelberg (2005)
35. Wang, X., Yin, Y.L., Yu, H.: Finding Collisions in the Full SHA-1. In: Shoup, V. (ed.) CRYPTO 2005. LNCS, vol. 3621, pp. 17–36. Springer, Heidelberg (2005)

36. Wang, X., Yu, H.: How to Break MD5 and Other Hash Functions. In: Cramer, R. (ed.) EUROCRYPT 2005. LNCS, vol. 3494, pp. 19–35. Springer, Heidelberg (2005)
37. Watanabe, D.: A note on the security proof of Knudsen-Preneel construction of a hash function (2006), http://csrc.nist.gov/pki/HashWorkshop/2006/UnacceptedPapers/WATANABE_kp_attack.pdf

A   Proof of Lemma 1

Let $\mathcal{T} = \mathcal{U} \cup \mathcal{V}$ be a disjoint decomposition of $\mathcal{T}$ such that $|\mathcal{S}| + |\mathcal{U}| = r - d + 1$. Let $\mathbb{D}[\mathcal{U}, \mathcal{V}]$ be the set of index sequences

$$D = ((i_l, j_l)_{l \in \mathcal{U}}, (h_l)_{l \in \mathcal{V}})$$

such that

1. $1 \le i_l, j_l, h_l \le q$ are all distinct,
2. $\max_{l \in \mathcal{U}}\{i_l, j_l\} < \min_{l \in \mathcal{V}}\{h_l\}$.

For a sequence $D = ((i_l, j_l)_{l \in \mathcal{U}}, (h_l)_{l \in \mathcal{V}}) \in \mathbb{D}[\mathcal{U}, \mathcal{V}]$, we define a random variable X_D where $X_D = 1$ if there is a sequence $(i_l, j_l)_{l \in \mathcal{V}}$ such that

1. $\max\{i_l, j_l\} = h_l$ for $l \in \mathcal{V}$,
2. $P = (i_l, j_l)_{l \in \mathcal{U} \cup \mathcal{V}}$ is a partial collision at $\mathcal{T}$ compatible with inactive indices $\mathcal{S}$,

and $X_D = 0$ otherwise. The condition

$$\max_{l \in \mathcal{U}}\{i_l, j_l\} < \min_{l \in \mathcal{V}}\{h_l\} = \min_{l \in \mathcal{V}}\{\max\{i_l, j_l\}\}$$

implies that the inner collisions at $\mathcal{V}$ are completed after the inner collisions at $\mathcal{U}$. Therefore for $D = ((i_l, j_l)_{l \in \mathcal{U}}, (h_l)_{l \in \mathcal{V}}) \in \mathbb{D}[\mathcal{U}, \mathcal{V}]$, $\mathbf{Pr}[X_D = 1]$ is the probability that

1. For $l \in \mathcal{U}$, $f_l(x^{i_l}) = f_l(x^{j_l})$,
2. For $l \in \mathcal{V}$, $f_l(x^{h_l}) = f_l(x^{h_l} \oplus \Delta_l^*)$, where
    (a) $(\Delta_l^*)_{l \in [1,r]}$ is a unique extension of $(\Delta_l)_{l \in \mathcal{S} \cup \mathcal{U}}$, where $\Delta_l = 0$ for $l \in \mathcal{S}$ and $\Delta_l = x^{i_l} \oplus x^{j_l}$ for $l \in \mathcal{U}$ (by Property 2),
    (b) $f_l(x^{h_l} \oplus \Delta_l^*)$ has been queried before the h_l-th query.

Since t inner collisions are necessary for $X_D = 1$, we have

$$\mathbf{Pr}[X_D = 1] = \mathsf{Ex}(X_D) \le \frac{1}{N^t}.^5$$

Let

$$X = \sum_{\substack{\mathcal{U} \cup \mathcal{V} = \mathcal{T} \\ \mathcal{U} \cap \mathcal{V} = \emptyset \\ |\mathcal{S}| + |\mathcal{U}| = r - d + 1}} \sum_{D \in \mathbb{D}[\mathcal{U}, \mathcal{V}]} X_D.$$

---

[5] If the extension $(\Delta_l^*)_{l \in [1,r]}$ does not exist, then $\mathbf{Pr}[X_D = 1] = 0$.

Since the number of possible decompositions of $\mathcal{T} = \mathcal{U} \cup \mathcal{V}$ such that $\mathcal{U} \cap \mathcal{V} = \emptyset$ and $|\mathcal{S}| + |\mathcal{U}| = r - d + 1$ is $\binom{t}{r-d-s+1}$ and

$$|\mathbb{D}[\mathcal{U}, \mathcal{V}]| \le q^{2|\mathcal{U}|+|\mathcal{V}|} = q^{|\mathcal{T}|+|\mathcal{U}|} = q^{t+(r-d+1)-s}$$

for each decomposition, we have

$$\mathsf{Ex}(X) = \binom{t}{r-d-s+1} q^{t+r-d-s+1} \mathsf{Ex}(X_D) \le \binom{t}{r-d-s+1} \frac{q^{t+r-d-s+1}}{N^t}.$$

Using Markov's inequality, for $M > 0$ we have

$$\mathbf{Pr}\,[X \ge M] \le \binom{t}{r-d-s+1} \frac{q^{t+r-d-s+1}}{MN^t}. \tag{12}$$

Let $P = (i_l, j_l)_{l \in \mathcal{T}}$ be a partial collision at $\mathcal{T}$ compatible with inactive indices $\mathcal{S}$. Then we always have a unique disjoint decomposition of $\mathcal{T} = \mathcal{U} \cup \mathcal{V}$ such that $|\mathcal{U}| = r - d - s + 1$ and

$$\max_{l \in \mathcal{U}}\{i_l, j_l\} < \min_{l \in \mathcal{V}}\{\max\{i_l, j_l\}\}.$$

In this case, we have $X_D = 1$ for $D = ((i_l, j_l)_{l \in \mathcal{U}}, (h_l)_{l \in \mathcal{V}})$ where $h_l = \max\{i_l, j_l\}$. If we regard this association of P with D as a mapping, then exactly $2^{|\mathcal{V}|}(= 2^{t-(r-d-s+1)})$ different partial collisions would be mapped to the same sequence D since (i_l, j_l) can be replaced by (j_l, i_l) for each index $l \in \mathcal{V}$ without changing the image of this mapping. Therefore the inequality (12) implies that the number of partial collisions at $\mathcal{T}$ compatible with inactive indices $\mathcal{S}$ is at most $2^{t-r+d+s-1}M$ except with probability $\binom{t}{r-d-s+1} q^{t+r-d-s+1}/(MN^t)$.

Optimal Collision Security in Double Block Length Hashing with Single Length Key

Bart Mennink

Dept. Electrical Engineering, ESAT/COSIC, KU Leuven, and IBBT, Belgium
bart.mennink@esat.kuleuven.be

Abstract. The idea of double block length hashing is to construct a compression function on $2n$ bits using a block cipher with an n-bit block size. All optimally secure double length hash functions known in the literature employ a cipher with a key space of double block size, $2n$-bit. On the other hand, no optimally secure compression functions built from a cipher with an n-bit key space are known. Our work deals with this problem. Firstly, we prove that for a wide class of compression functions with two calls to its underlying n-bit keyed block cipher collisions can be found in about $2^{n/2}$ queries. This attack applies, among others, to functions where the output is derived from the block cipher outputs in a linear way. This observation demonstrates that all security results of designs using a cipher with $2n$-bit key space crucially rely on the presence of these extra n key bits. The main contribution of this work is a proof that this issue can be resolved by allowing the compression function to make one extra call to the cipher. We propose a family of compression functions making three block cipher calls that asymptotically achieves optimal collision resistance up to $2^{n(1-\varepsilon)}$ queries and preimage resistance up to $2^{3n(1-\varepsilon)/2}$ queries, for any $\varepsilon > 0$. To our knowledge, this is the first optimally collision secure double block length construction using a block cipher with single length key space.

1 Introduction

Double (block) length hashing is a well-established method for constructing a compression function with $2n$-bit output based only on n-bit block ciphers. The idea of double length hashing dates back to the work of Meyer and Schilling [19], with the introduction of the MDC-2 and MDC-4 compression functions in 1988. In recent years, the design methodology got renewed attention in the works of [2,4,7,9,10,12,16,21,27]. Double length hash functions have an obvious advantage over classical block cipher based functions such as Davies-Meyer and Matyas-Meyer-Oseas [22,26]: the same type of underlying primitive allows for a larger compression function. Yet, for double length compression functions it is harder to achieve optimal n-bit collision and $2n$-bit preimage security.

We focus on the simplest and most-studied type of compression functions, namely functions that compress $3n$ to $2n$ bits. Those can be classified into two classes: compression functions that internally evaluate a $2n$-bit keyed block cipher $E : \{0,1\}^{2n} \times \{0,1\}^n \to \{0,1\}^n$ (which we will call the DBL^{2n} class), and

X. Wang and K. Sako (Eds.): ASIACRYPT 2012, LNCS 7658, pp. 526–543, 2012.

ones that employ an n-bit keyed block cipher $E : \{0,1\}^n \times \{0,1\}^n \to \{0,1\}^n$ (the DBL^n class). The DBL^{2n} class is well understood. It includes the classical compression functions Tandem-DM and Abreast-DM [8] and Hirose's function [6], as well as Stam's supercharged single call Type-I compression function design [25,26] (reconsidered in [14]) and the generalized designs by Hirose [5] and Özen and Stam [21]. As illustrated in Table 1, all of these functions provide optimal collision security guarantees (up to about 2^n queries), and Tandem-DM, Abreast-DM, and Hirose's function are additionally proven optimally preimage resistant (up to about 2^{2n} queries). These bounds also hold in the iteration, when a proper domain extender is applied [1]. Lucks [15] introduced a compression function that allows for collisions in about $2^{n/2}$ queries, but achieves optimal collision resistance in the iteration. Members of the DBL^n class are the MDC-2 and MDC-4 compression functions [19], the MJH construction [10], and a construction by Jetchev et al. [7]. For the MDC-2 and MJH compression functions, collisions and preimages can be found in about $2^{n/2}$ and 2^n queries, respectively[1]. The MDC-4 compression function achieves a higher level of collision and preimage resistance than MDC-2 [16], but contrary to the other functions it makes four block cipher calls. Jetchev et al.'s construction makes two block cipher calls and achieves $2^{2n/3}$ collision security. Stam also introduced a design based on two calls, and proved it optimally collision secure in a restricted security model where the adversary must fix its queries in advance. Therefore we did not include this design in the table. Further related results include the work of Nandi et al. [20], who presented a $3n$-to-$2n$-bit compression function making three calls to a $2n$-to-n-bit one-way function, achieving collision security up to $2^{2n/3}$ queries. They extended this result to a $4n$-to-$2n$-bit function using three $2n$-bit keyed block ciphers.

Unlike the DBL^{2n} class, for the DBL^n class no optimally secure compression function is known. The situation is the same for the iteration, where none of these designs has been proven to achieve optimal security. Determinative to this gap is the difference in the underlying primitive: in the DBL^{2n} class, the underlying primitive maps $3n$ bits to n bits and thus allows for more compression. In particular, if we consider Tandem-DM, Abreast-DM, and Hirose's function, the first cipher call already compresses the entire input to the compression function, and the second cipher call is simply used to assure a $2n$-bit output. In fact, these designs achieve their level of security merely due to this property, for their proofs crucially rely on this (see also Sect. 4).

Thus, from a theoretical point of view it is unreasonable to compare DBL^{2n} and DBL^n. But the gap between the two classes leaves us with an interesting open problem: starting from a single block cipher $E : \{0,1\}^n \times \{0,1\}^n \to \{0,1\}^n$, is it possible to construct a double length compression function that achieves optimal collision and preimage security? This is the central research question of this work. Note that Stam's bound [25] does not help us here: it claims that collisions can be found in at most $(2^n)^{(2r-1)/(r+1)}$ queries, where r denotes the

---

[1] In the iteration collision resistance is proven up to $2^{3n/5}$ queries for MDC-2 [27] and $2^{2n/3}$ queries for MJH [10].

Table 1. Asymptotic ideal cipher model security guarantees of known double length compression functions in the classes DBL^{2n} (first) and DBL^{n} (next). A more detailed comparison of some of these functions can be found in [3, App. A].

compression function	E-calls	collision security	preimage security	underlying cipher
Lucks'	1	$2^{n/2}$	2^n	
Stam's	1	2^n [26]	2^n [26]	
Tandem-DM	2	2^n [12]	2^{2n} [2,13]	
Abreast-DM	2	2^n [4,9]	2^{2n} [2,13]	
Hirose's	2	2^n [6]	2^{2n} [2,13]	
Hirose-class	2	2^n [5]	2^n [5]	
Özen-Stam-class	2	2^n [21]	2^n [21]	
MDC-2	2	$2^{n/2}$	2^n	
MJH	2	$2^{n/2}$	2^n	
Jetchev et al.'s	2	$2^{2n/3}$ [7]	2^n [7]	
MDC-4	4	$2^{5n/8}$ [16]	$2^{5n/4}$ [16]	
Our proposal	**3**	$\mathbf{2^n}$	$\mathbf{2^{3n/2}}$	

number of block cipher calls, which results in the trivial bound for $r \geq 2$. For $r \geq 2$, denote by $F^r : \{0,1\}^{3n} \to \{0,1\}^{2n}$ a compression function that makes r calls to its primitive E.

As a first contribution, we consider F^2, and prove that for a very large class of functions of this form one expects collisions in approximately $2^{n/2}$ queries. Covered by the attack are among others designs with linear finalization function (the function that produces the $2n$-bit output given the $3n$-bit input and the block cipher responses). We note that the compression function by Jetchev et al. [7] is not vulnerable to the attack due to its non-linear finalization function. Nevertheless, these results strengthen the claim that no practical optimally collision secure F^2 function exists. Motivated by this, we increase the number of calls to E, and consider F^3. In this setting, we derive a family of compression functions which we prove asymptotically optimal collision resistant up to $2^{n(1-\varepsilon)}$ queries and preimage resistant up to $2^{3n(1-\varepsilon)/2}$ queries, for any $\varepsilon > 0$. Our compression function family, thus, achieves the same level of collision security as the well-established Tandem-DM, Abreast-DM, and Hirose's function, albeit based on a much weaker assumption. In the DBL^n class, our design clearly compares favorably to MDC-4 that makes four block cipher evaluations, and from a provable security point of view it beats MDC-2 and MJH, still, an extra E evaluation has to be made which results in an efficiency loss. The introduced class of compression functions is simple and easy to understand: they are defined by 4×4 matrices over the field $GF(2^n)$ which are required to comply with easily satisfied conditions. Two example compression functions in this class are given in Fig. 1.

The security proofs of our compression function family rely on basic principles from previous proofs, but in order to accomplish optimal collision security (and

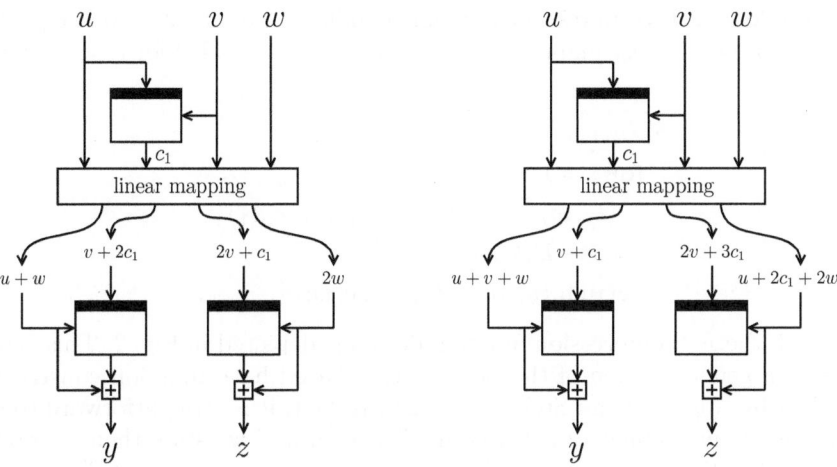

Fig. 1. Two example compression functions from the family of functions introduced and evaluated in this work. For these constructions, all wires carry $n = 128$ bits, and the arithmetic is done over $GF(2^{128})$. We further elaborate on these designs and their derivations in Sect. 4.

as our designs use n-bit keyed block ciphers) our proofs have become significantly more complex. The security proofs of all known DBL^{2n} functions (see Table 1) crucially rely on the property that one block cipher evaluation defines the input to the second one. For F^3 this cannot be achieved as each primitive call fixes at most $2n$ bits of the function input. Although one may expect this to cause an optimal proof to become unlikely, this is not the case. Using a new proof approach—we smartly apply the methodology of "wish lists" (by Armknecht et al. and Lee et al. [2,13]) to collision resistance—we manage to achieve asymptotically the close to 2^n collision security for our family of functions. Nonetheless, the bound on preimage resistance does not reach the optimal level of 2^{2n} queries. One can see this as the price we pay for using single key length rather than double key length block ciphers: a straightforward generalization of the pigeonhole-birthday attack of Rogaway and Steinberger [24] shows that, when the compression function behaves "sufficiently random", one may expect a preimage in approximately $2^{5n/3}$ queries (cf. Sect. 2). The asymptotic preimage bound of $2^{3n/2}$ found in this work closely approaches this generic bound.

Outline. We present and formalize the security model in Sect. 2. Then, in Sect. 3 we derive our impossibility result on F^2. We propose and analyze our family of compression functions in Sects. 4 and 5. This work is concluded in Sect. 6.

2   Security Model

For $n \geq 1$, we denote by $\mathrm{Bloc}(n)$ the set of all block ciphers with a key and message space of n bits. Let $E \in \mathrm{Bloc}(n)$. For $r \geq 1$, let $F^r : \{0,1\}^{3n} \to \{0,1\}^{2n}$

be a double length compression function making r calls to its block cipher E. We can represent F^r by mappings $f_i : \{0,1\}^{(i+2)n} \to \{0,1\}^{2n}$ for $i = 1, \ldots, r+1$ as follows:

$$
\begin{aligned}
&F^r(u,v,w) \\
&\quad \textbf{for } i = 1, \ldots, r\colon \\
&\qquad (k_i, m_i) \leftarrow f_i(u,v,w; c_1, \ldots, c_{i-1}), \\
&\qquad c_i \leftarrow E(k_i, m_i), \\
&\quad \textbf{return } (y,z) \leftarrow f_{r+1}(u,v,w; c_1, \ldots, c_r).
\end{aligned}
$$

For $r = 3$, the F^r compression function design is depicted in Fig. 2. This generic design is a generalization of the permutation based hash function construction described by Rogaway and Steinberger [24]. In fact, it is straightforward to generalize the main findings of [24] to our F^r design and we state them as preliminary results. If the collision- and preimage-degeneracies are sufficiently small (these values intuitively capture the degree of non-randomness of the design with respect to the occurrence of collisions and preimages), one can expect collisions after approximately $2^{n(2-2/r)}$ queries and preimages after approximately $2^{n(2-1/r)}$ queries. We refer to [24] for the details. First of all, these findings confirm that at least two cipher calls are required to get 2^n collision resistance. More importantly, from these results we can conclude that F^r can impossibly achieve optimal 2^{2n} preimage resistance. Yet, it may still be possible to construct a function that achieves optimal collision resistance and almost-optimal preimage resistance.

Throughout, we consider security in the ideal cipher model: we consider an adversary $\mathcal{A}$ that is a probabilistic algorithm with oracle access to a block cipher $E \xleftarrow{\$} \mathrm{Bloc}(n)$ randomly sampled from $\mathrm{Bloc}(n)$. $\mathcal{A}$ is information-theoretic: it has unbounded computational power, and its complexity is measured by the number of queries made to its oracles. The adversary can make forward queries and inverse queries to E, and these are stored in a query history Q as indexed tuples of the form (k_i, m_i, c_i), where k_i denotes the key input, and (m_i, c_i) the plaintext/ciphertext pair. For $q \geq 0$, by Q_q we define the query history after q queries. We assume that the adversary never makes queries to which it knows the answer in advance.

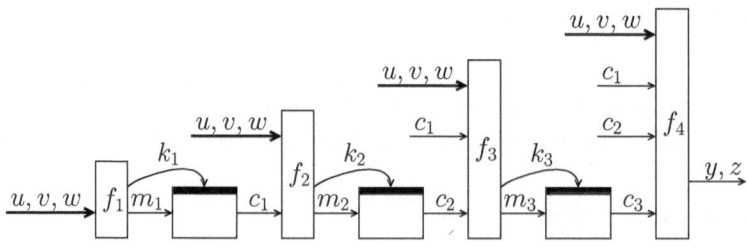

Fig. 2. $F^3 : \{0,1\}^{3n} \to \{0,1\}^{2n}$ making three block cipher evaluations

A collision-finding adversary $\mathcal{A}$ for F^r aims at finding two distinct inputs to F^r that compress to the same range value. In more detail, we say that $\mathcal{A}$ succeeds if it finds two distinct tuples $(u, v, w), (u', v', w')$ such that $F^r(u, v, w) = F^r(u', v', w')$ and Q contains all queries required for these evaluations of F^r. We define by

$$\mathbf{adv}_{F^r}^{\text{coll}}(\mathcal{A}) = \mathbf{Pr} \left(\begin{array}{c} E \xleftarrow{\$} \text{Bloc}(n), \; (u, v, w), (u', v', w') \leftarrow \mathcal{A}^{E, E^{-1}} : \\ (u, v, w) \neq (u', v', w') \; \wedge \; F^r(u, v, w) = F^r(u', v', w') \end{array} \right)$$

the probability that $\mathcal{A}$ succeeds in this. By $\mathbf{adv}_{F^r}^{\text{coll}}(q)$ we define the maximum collision advantage taken over all adversaries making q queries.

For preimage resistance, we focus on everywhere preimage resistance [23], which captures preimage security for every point of $\{0, 1\}^{2n}$. Before making any queries to its oracle, a preimage-finding adversary $\mathcal{A}$ first decides on a range point $(y, z) \in \{0, 1\}^{2n}$. Then, we say that $\mathcal{A}$ succeeds in finding a preimage if it obtains a tuple (u, v, w) such that $F^r(u, v, w) = (y, z)$ and Q contains all queries required for this evaluation of F^r. We define by

$$\mathbf{adv}_{F^r}^{\text{epre}}(\mathcal{A}) = \max_{(y,z) \in \{0,1\}^{2n}} \mathbf{Pr} \left(\begin{array}{c} E \xleftarrow{\$} \text{Bloc}(n), \; (u, v, w) \leftarrow \mathcal{A}^{E, E^{-1}}(y, z) : \\ F^r(u, v, w) = (y, z) \end{array} \right)$$

the probability that $\mathcal{A}$ succeeds, maximized over all possible choices for (y, z). By $\mathbf{adv}_{F^r}^{\text{epre}}(q)$ we define the maximum (everywhere) preimage advantage taken over all adversaries making q queries.

3   Impossibility Result for 2-Call Double Length Hashing

We present an attack on a wide class of double block length compression functions with two calls to their underlying block cipher $E : \{0, 1\}^n \times \{0, 1\}^n \to \{0, 1\}^n$. Let F^2 be a compression function of this form. We pose a condition on the finalization function f_3, such that if this condition is satisfied, collisions for F^2 can be found in about $2^{n/2}$ queries. Although we are not considering all possible compression functions, we cover the most interesting and intuitive ones, such as compression functions with linear finalization function f_3. Compression functions with non-linear f_3 are covered up to some degree (but we note that the attack does not apply to the compression function of [7], for which collision security up to $2^{2n/3}$ queries is proven).

We first state the attack. Then, by ways of examples, we illustrate its generality. For the purpose of the attack, we introduce the function left_n which on input of a bit string of length $2n$ bits outputs the leftmost n bits.

Proposition 1. Let $F^2 : \{0, 1\}^{3n} \to \{0, 1\}^{2n}$ be a compression function as described in Sect. 2. Suppose there exists a bijective function L such that for any $u, v, w, c_1, c_2 \in \{0, 1\}^n$ we have

$$\text{left}_n \circ L \circ f_3(u, v, w; c_1, c_2) = \text{left}_n \circ L \circ f_3(u, v, w; c_1, 0). \tag{1}$$

Then, one can expect collisions for F^2 after $2^{n/2}$ queries.

Proof. Let F^2 be a compression function and let L be a bijection such that (1) holds. First, we consider the case of L being the identity function, and next we show how this attack extends to the case L is an arbitrary bijection.

Suppose (1) holds with L the identity function. This means that the first n bits of $f_3(u, v, w; c_1, c_2)$ do not depend on c_2 and we can write f_3 as a concatenation of two functions $g_1 : \{0, 1\}^{4n} \rightarrow \{0, 1\}^n$ and $g_2 : \{0, 1\}^{5n} \rightarrow \{0, 1\}^n$ as $f_3(u, v, w; c_1, c_2) = g_1(u, v, w; c_1)\|g_2(u, v, w; c_1, c_2)$. Let $\alpha \in \mathbb{N}$. We present an adversary $\mathcal{A}$ for F^2. The first part of the attack is derived from [24].

- Make α queries $(k_1, m_1) \rightarrow c_1$ that maximize the number of tuples (u, v, w) with $f_1(u, v, w)$ hitting any of these values (k_1, m_1). By the balls-and-bins principle[2], the adversary obtains at least $\alpha \cdot 2^{3n}/2^{2n} = \alpha 2^n$ tuples $(u, v, w; c_1)$ for which it knows the first block cipher evaluation;
- Again by the balls-and-bins principle, there exists a value y such that at least α tuples satisfy $g_1(u, v, w; c_1) = y$;
- Varying over these α tuples, compute $(k_2, m_2) = f_2(u, v, w; c_1)$ and query (k_2, m_2) to the cipher to obtain a c_2. $\mathcal{A}$ finds a collision for F^2 if it obtains two tuples $(u, v, w; c_1, c_2)$, $(u', v', w'; c_1', c_2')$ that satisfy $g_2(u, v, w; c_1, c_2) = g_2(u', v', w'; c_1', c_2')$.

In the last round one expects to find a collision if $\alpha^2/2^n = 1$, or equivalently if $\alpha = 2^{n/2}$. In total, the attack is done in approximately $2 \cdot 2^{n/2}$ queries.

It remains to consider the case of L being an arbitrary bijection. Define $\overline{F}^2$ as F^2 with f_3 replaced by $\overline{f_3} = L \circ f_3$. Using the idea of equivalence classes on compression functions [18] we prove that F^2 and $\overline{F}^2$ are equally secure with respect to collisions. Let $\overline{\mathcal{A}}$ be a collision finding adversary for $\overline{F}^2$. We construct a collision finding adversary $\mathcal{A}$ for F^2, with oracle access to E, that uses $\overline{\mathcal{A}}$ to output a collision for F^2. Adversary $\mathcal{A}$ proceeds as follows. It forwards all queries made by $\overline{\mathcal{A}}$ to its own oracle. Eventually, $\overline{\mathcal{A}}$ outputs two tuples $(u, v, w), (u', v', w')$ such that $\overline{F}^2(u, v, w) = \overline{F}^2(u', v', w')$. Denote by c_1 the block cipher outcome on input of $f_1(u, v, w)$ and by c_2 the outcome on input of $f_2(u, v, w; c_1)$. Define c_1' and c_2' similarly. By construction, as (u, v, w) and (u', v', w') form a collision for $\overline{F}^2$, we have $L \circ f_3(u, v, w; c_1, c_2) = L \circ f_3(u', v', w'; c_1', c_2')$. Now, bijectivity of L implies that $f_3(u, v, w; c_1, c_2) = f_3(u', v', w'; c_1', c_2')$, and hence (u, v, w) and (u', v', w') form a collision for F^2. (Recall that F^2 and $\overline{F}^2$ only differ in the finalization function f_3, the functions f_1 and f_2 are the same.) We thus obtain $\mathbf{adv}_{\overline{F}^2}^{\text{coll}}(q) \leq \mathbf{adv}_{F^2}^{\text{coll}}(q)$. The derivation in reverse order is the same by symmetry. But $\overline{F}^2$ satisfies (1) for L the identity function. Therefore, the attack described in the first part of the proof applies to $\overline{F}^2$, and thus to F^2. □

We demonstrate the impact of the attack by giving several example functions that fall in the categorization. We stress that the requirement of Prop. 1 is in fact solely a requirement on f_3; f_1 and f_2 can be any function.

---

[2] If k balls are thrown in l bins, the α fullest bins in total contain at least $\alpha k/l$ balls.

Suppose F^2 uses a linear finalization function f_3. Say, f_3 is defined as follows:

$$\begin{pmatrix} a_{11}\ a_{12}\ a_{13}\ a_{14}\ a_{15} \\ a_{21}\ a_{22}\ a_{23}\ a_{24}\ a_{25} \end{pmatrix} (u, v, w, c_1, c_2)^\top = (y, z)^\top,$$

where addition and multiplication is done over the field $GF(2^n)$. Now, if $a_{25} = 0$ we set $L = \begin{pmatrix} 0\ 1 \\ 1\ 0 \end{pmatrix}$ which corresponds to swapping y and z. If $a_{25} \neq 0$, we set $L = \begin{pmatrix} 1 & -a_{15}a_{25}^{-1} \\ 0 & 1 \end{pmatrix}$, which corresponds to subtracting the second equation $a_{15}a_{25}^{-1}$ times from the first one. The attack also covers designs whose finalization function f_3 rotates or shuffles its inputs, such as MDC-2, where one defines L so that the rotation gets undone. We elaborate on this in the full version [17]. In general, if f_3 is a sufficiently simple add-rotate-xor function, it is possible to derive a bijective L that makes (1) satisfied. Up to a degree, the attack also covers general non-linear finalization functions. However, it clearly does not cover all functions and it remains an open problem to either close this gap or to come with a (possibly impractical) F^2 compression function that provable achieves optimal collision resistance. One direction may be to start from the compression function with non-linear finalization f_3 by Jetchev et al. [7], for which collision resistance up to $2^{2n/3}$ queries is proven.

4 Double Length Hashing with 3 E-calls

Motivated by the negative result of Sect. 3, we target the existence of double length hashing with three block cipher calls. We introduce a family of double length compression functions making three cipher calls that achieve asymptotically optimal 2^n collision resistance and preimage resistance significantly beyond the birthday bound (up to $2^{3n/2}$ queries). We note that, although the preimage bound is non-optimal, it closely approaches the generic bound dictated by the pigeonhole-birthday attack (Sect. 2).

Let $GF(2^n)$ be the field of order 2^n. We identify bit strings from $\{0,1\}^n$ and finite field elements in $GF(2^n)$ to define addition and scalar multiplication over $\{0,1\}^n$. In the family of double block length functions we propose in this section, the functions f_1, f_2, f_3, f_4 of Fig. 2 will be linear functions over $GF(2^n)$. For two tuples $x = (x_1, \ldots, x_l)$ and $y = (y_1, \ldots, y_l)$ of elements from $\{0,1\}^n$, we define by $x \cdot y$ their inner product $\sum_{i=1}^{l} x_i y_i \in \{0,1\}^n$.

Before introducing the design, we first explain the fundamental consideration upon which the family is based. The security proofs of all DBL^{2n} functions known in the literature (cf. Table 1) crucially rely on the property that one block cipher evaluation defines the input to the other one. For DBL^{2n} functions this can easily be achieved: any block cipher evaluation can take as input the full $3n$-bit input state (u, v, w). Considering the class of functions DBL^n, and F^r of Fig. 2 in particular, this can impossibly be achieved: one block cipher "processes" at most $2n$ out of $3n$ input bits. In our design, we slightly relax this requirement, by requiring that any *two* block cipher evaluations define the input to the third one. Although from a technical point of view one may expect that

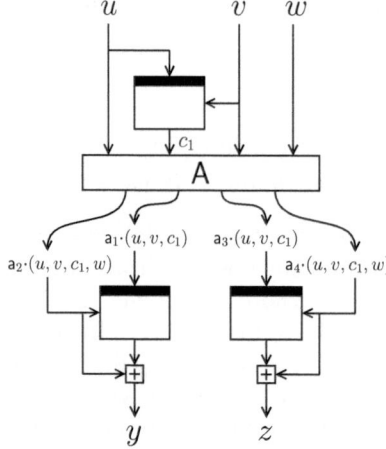

$$F_A^3(u, v, w) = (y, z), \text{ where:}$$
$$c_1 \leftarrow E(u, v),$$
$$k_2 \leftarrow a_1 \cdot (u, v, c_1),$$
$$m_2 \leftarrow a_2 \cdot (u, v, c_1, w),$$
$$y \leftarrow E(k_2, m_2) + m_2,$$
$$k_3 \leftarrow a_3 \cdot (u, v, c_1),$$
$$m_3 \leftarrow a_4 \cdot (u, v, c_1, w),$$
$$z \leftarrow E(k_3, m_3) + m_3.$$

Fig. 3. The family of compression functions F_A^3 where A is a 4×4 matrix as specified in the text. Arithmetics is done over $GF(2^n)$.

this change causes optimal collision resistance to be harder or even impossible to be achieved, we will demonstrate that this is not the case due to new proof techniques employed to analyze the collision resistance.

Based on this key observation we propose the compression function design F_A^3 of Fig. 3. Here,

$$A = \begin{pmatrix} a_1 \\ a_2 \\ a_3 \\ a_4 \end{pmatrix} = \begin{pmatrix} a_{11} & a_{12} & a_{13} & 0 \\ a_{21} & a_{22} & a_{23} & a_{24} \\ a_{31} & a_{32} & a_{33} & 0 \\ a_{41} & a_{42} & a_{43} & a_{44} \end{pmatrix} \tag{2}$$

is a 4×4 matrix over $GF(2^n)$. Note that, provided A is invertible and $a_{24}, a_{44} \neq 0$, any two block cipher evaluations of F_A^3 define (the inputs of) the third one. For instance, evaluations of the second and third block cipher fix the vector $A(u, v, c_1, w)^\top$, which by invertibility of A fixes (u, v, c_1, w) and thus the first block cipher evaluation. Evaluations of the first and second block cipher fix the inputs of the third block cipher as $a_{24} \neq 0$. For the proofs of collision and preimage resistance, however, we will need to posit additional requirements on A. As we will explain, these requirements are easily satisfied.

In the remainder of this section, we state our results on the collision resistance of F_A^3 in Sect. 4.1 and on the preimage resistance in Sect. 4.2.

4.1 Collision Resistance of F_A^3

We prove that, provided its underlying matrix A satisfies some simple conditions, F_A^3 satisfies optimal collision resistance. In more detail, we pose the following requirements on A:

- A is invertible;
- $a_{12}, a_{13}, a_{24}, a_{32}, a_{33}, a_{44} \neq 0$;
- $a_{12} \neq a_{32}$ and $a_{13} \neq a_{33}$.

We refer to the logical AND of these requirements as colreq.

Theorem 1. *Let $n \in \{0,1\}^n$. Suppose A satisfies colreq. Then, for any positive integral values t_1, t_2,*

$$\mathbf{adv}_{F_A^3}^{coll}(q) \leq \frac{2t_2^2 q + 3t_2 q + 11q + 3t_1 t_2^2 + 7t_1 t_2}{2^n - q} +$$

$$\frac{q^2}{t_1(2^n - q)} + 3 \cdot 2^n \left(\frac{eq}{t_2(2^n - q)} \right)^{t_2}. \tag{3}$$

The proof is given in Sect. 5. The basic proof idea is similar to existing proofs in the literature (e.g. [16,27]) and is based on the usage of thresholds t_1, t_2. For increasing values of t_1, t_2 the first term of the bound increases, while the second two terms decrease. Although the proof derives basic proof principles from literature, for the technical part we deviate from existing proof techniques in order to get a bound that is "as tight as possible". In particular, we introduce the usage of wish lists in the context of collisions, an approach that allows for significantly better bounds. Wish lists have been introduced by Armknecht et al. [2] and Lee et al. [11,13] for the preimage resistance analysis of DBL^{2n} functions, but they have never been used for collision resistance as there never was a need to do so. Our analysis relies on this proof methodology, but as for collisions more block cipher evaluations are involved (one collision needs six block cipher calls while a preimage requires three) this makes the analysis more technical and delicate.

The goal now is to find a good threshold between the first term and the latter two terms of (3). To this end, let $\varepsilon > 0$ be any parameter. We put $t_1 = q$ and $t_2 = 2^{n\varepsilon}$ (we can assume t_2 to be integral). Then, the bound simplifies to

$$\mathbf{adv}_{F_A^3}^{coll}(q) \leq \frac{5 \cdot 2^{2n\varepsilon} q + 10 \cdot 2^{n\varepsilon} q + 11q}{2^n - q} + \frac{q}{2^n - q} + 3 \cdot 2^n \left(\frac{eq}{2^{n\varepsilon}(2^n - q)} \right)^{2^{n\varepsilon}}.$$

From this, we find that for any $\varepsilon > 0$ we have $\mathbf{adv}_{F_A^3}^{coll}(2^n/2^{3n\varepsilon}) \to 0$ for $n \to \infty$. Hence, the F_A^3 compression function achieves close to optimal 2^n collision security for $n \to \infty$. For $n = 128$, we evaluate the bound in more detail in [17]. The advantage hits $1/2$ for $\log_2 q \approx 118.3$, relatively close to the threshold 127.5 for $q(q+1)/2^{2n}$. For larger values of n this gap approaches 0.

4.2   Preimage Resistance of F_A^3

In this section we consider the preimage resistance of F_A^3. Though we do not obtain optimal preimage resistance—which is impossible to achieve after all, due to the generic bounds of the pigeonhole-birthday attack (Sect. 2)—we achieve preimage resistance up to $2^{3n/2}$ queries, much better than the preimage bounds on MDC-2 and MDC-4 [16], relatively close to the generic bound. Yet, for the proof to hold we need to put slightly stronger requirements on A.

- $A - \begin{pmatrix} & 0\,0 \\ B_1 & 0\,0 \\ & 0\,0 \\ B_2 & 0\,0 \end{pmatrix}$ is invertible for any $B_1, B_2 \in \{\begin{pmatrix} 0\,0 \\ 0\,0 \end{pmatrix}, \begin{pmatrix} 1\,0 \\ 0\,0 \end{pmatrix}, \begin{pmatrix} 1\,0 \\ 0\,1 \end{pmatrix}\}$. In the

  remainder, we write $[B_1/B_2]$ to denote the subtracted matrix;
- $a_{12}, a_{13}, a_{24}, a_{32}, a_{33}, a_{44} \neq 0$;
- $a_{12} \neq a_{32}, a_{13} \neq a_{33}$, and $a_{24} \neq a_{44}$.

We refer to the logical AND of these requirements as $\mathtt{prereq}$. We remark that $\mathtt{prereq} \Rightarrow \mathtt{colreq}$, and that matrices satisfying $\mathtt{prereq}$ are easily found. Simple matrices complying with these conditions over the field $GF(2^{128})$ are

$$\begin{pmatrix} 0\,1\,2\,0 \\ 1\,0\,0\,1 \\ 0\,2\,1\,0 \\ 0\,0\,0\,2 \end{pmatrix}, \qquad \begin{pmatrix} 0\,1\,1\,0 \\ 1\,1\,0\,1 \\ 0\,2\,3\,0 \\ 1\,0\,2\,2 \end{pmatrix}. \tag{4}$$

These are the matrices corresponding to the compression functions of Fig. 1. Here, we use $x^{128} + x^{127} + x^{126} + x^{121} + 1$ as our irreducible polynomial and we represent bit strings as polynomials in the obvious way ($1 = 1$, $2 = x$, $3 = 1 + x$). Note that the choice of matrix A influences the efficiency of the construction. The first matrix of (4) has as minimal zeroes as possible, which reduces the amount of computation.

Theorem 2. *Let $n \in \{0,1\}^n$. Suppose A satisfies $\mathtt{prereq}$. Then, for any positive integral value t, provided $t \leq q$,*

$$\mathbf{adv}_{F_A^3}^{\mathrm{epre}}(q) \leq \frac{6t^2 + 18t + 26}{2^n - 2} + 4 \cdot 2^n \left(\frac{4eq}{t2^n}\right)^{t/2} + 8q \left(\frac{8eq}{t2^n}\right)^{\frac{t2^n}{4q}}. \tag{5}$$

The proof is given in the full version of this paper [17]. As for the bound on the collision resistance (Thm. 1), the idea is to make a smart choice of t to minimize this bound. Let $\varepsilon > 0$ be any parameter. Then, for $t = q^{1/3}$, the bound simplifies to

$$\mathbf{adv}_{F_A^3}^{\mathrm{epre}}(q) \leq \frac{6q^{2/3} + 18q^{1/3} + 26}{2^n - 2} + 4 \cdot 2^n \left(\frac{4eq^{2/3}}{2^n}\right)^{q^{1/3}/2} + 8q \left(\frac{8eq^{2/3}}{2^n}\right)^{\frac{2^n}{4q^{2/3}}}.$$

From this, we find that for any $\varepsilon > 0$ we have $\mathbf{adv}_{F_A^3}^{\mathrm{epre}}(2^{3n/2}/2^{n\varepsilon}) \to 0$ for $n \to \infty$. Hence, the F_A^3 compression function achieves close to $2^{3n/2}$ preimage security for $n \to \infty$. For $n = 128$, we evaluate the bound in more detail in [17]. The advantage hits $1/2$ for $\log_2 q \approx 180.3$, relatively close to the threshold 191.5 for $q^2/2^{3n}$. For larger values of n this gap approaches 0.

The result shows that F_A^3 with A compliant to $\mathtt{prereq}$ satisfies preimage resistance up to about $2^{3n/2}$ queries. We note that our proof is the best possible for this design, by demonstrating a preimage-finding adversary that with high probability succeeds in at most $O(2^{3n/2})$ queries. Let $\alpha \in \mathbb{N}$. The adversary proceeds as follows.

- Make $\alpha 2^n$ queries to the block cipher corresponding to the bottom-left position of Fig. 3. One expects to find α tuples (k_2, m_2, c_2) that satisfy $m_2 + c_2 = y$;
- Repeat the first step for the bottom-right position. One expects to find α tuples (k_3, m_3, c_3) satisfying $m_3 + c_3 = z$;
- By invertibility of A, any choice of (k_2, m_2, c_2) and (k_3, m_3, c_3) uniquely defines a tuple (u, v, c_1, w) for the F_A^3 evaluation. Likely, the emerged tuples (u, v, c_1) are all different, and we find about α^2 such tuples;
- Varying over all α^2 tuples (u, v, c_1), query (u, v) to the block cipher. If it responds c_1, we have obtained a preimage for F_A^3.

In the last round one expects to find a preimage if $\alpha^2/2^n = 1$, or equivalently if $\alpha = 2^{n/2}$. The first and second round both require approximately $2^{3n/2}$ queries, and the fourth round takes 2^n queries. In total, the attack is done in approximately $2 \cdot 2^{3n/2} + 2^n$ queries.

5   Proof of Thm. 1

The proof of collision resistance of F_A^3 follows the basic spirit of [16], but crucially differs in the way the probability bounds are computed. A new approach here is the usage of wish lists. While the idea of wish lists is not new—it has been introduced by Armknecht et al. [2] and Lee et al. [11,13] for double block length compression functions, and used by Mennink [16] for the analysis of MDC-4—in these works wish lists are solely used for the analysis of preimage resistance rather than collision resistance. Given that in a collision more block cipher evaluations are involved, the analysis becomes more complex. At a high level, wish lists rely on the idea that in order to find a collision, the adversary must at some point make a query that "completes this collision" together with some other queries already in the query history. Wish lists keep track of such query tuples, and the adversary's goal is to ever obtain a query tuple that is in such wish list. A more technical treatment can be found in the proof of Lem. 1.

We consider any adversary that has query access to its oracle E and makes q queries stored in a query history Q_q. Its goal is to find a collision for F_A^3, in which it by definition only succeeds if it obtains a query history Q_q that satisfies configuration $\mathsf{coll}(Q_q)$ of Fig. 4. This means,

$$\mathbf{adv}_{F_A^3}^{\mathrm{coll}}(q) = \mathbf{Pr}\left(\mathsf{coll}(Q_q)\right). \tag{6}$$

For the sake of readability of the proof, we label the block cipher positions in Fig. 4 as follows. In the left F_A^3 evaluation (on input (u, v, w)), the block ciphers are labeled $1L$ (the one on input (u, v)), $2L$ (the bottom left one), and $3L$ (the bottom right one). The block ciphers for the right F_A^3 evaluation are labeled $1R, 2R, 3R$ in a similar way. When we say "a query $1L$", we refer to a query that in a collision occurs at position $1L$.

For the analysis of $\mathbf{Pr}\left(\mathsf{coll}(Q_q)\right)$ we introduce an auxiliary event $\mathsf{aux}(Q_q)$. Let $t_1, t_2 > 0$ be any integral values. We define $\mathsf{aux}(Q_q) = \mathsf{aux}_1(Q_q) \vee \cdots \vee \mathsf{aux}_4(Q_q)$,

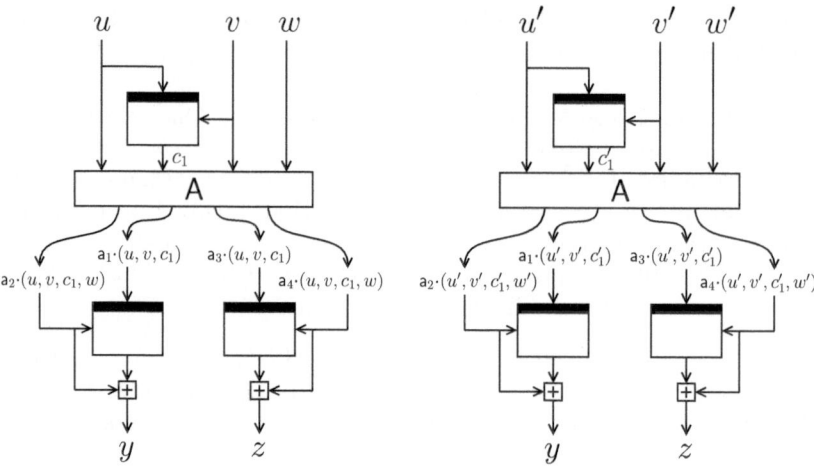

Fig. 4. Configuration $\mathsf{coll}(Q)$. The configuration is satisfied if Q contains six (possibly the same) queries that satisfy this setting. We require $(u, v, w) \neq (u', v', w')$.

where

$$\mathsf{aux}_1(Q_q): \left| \left\{ (k_i, m_i, c_i), (k_j, m_j, c_j) \in Q_q : i \neq j \wedge m_i + c_i = m_j + c_j \right\} \right| > t_1;$$

$$\mathsf{aux}_2(Q_q): \max_{z \in \{0,1\}^n} \left| \left\{ (k_i, m_i, c_i) \in Q_q : \mathsf{a}_1 \cdot (k_i, m_i, c_i) = z \right\} \right| > t_2;$$

$$\mathsf{aux}_3(Q_q): \max_{z \in \{0,1\}^n} \left| \left\{ (k_i, m_i, c_i) \in Q_q : \mathsf{a}_3 \cdot (k_i, m_i, c_i) = z \right\} \right| > t_2;$$

$$\mathsf{aux}_4(Q_q): \max_{z \in \{0,1\}^n} \left| \left\{ (k_i, m_i, c_i) \in Q_q : m_i + c_i = z \right\} \right| > t_2.$$

By basic probability theory, we obtain for (6):

$$\mathbf{Pr}\left(\mathsf{coll}(Q_q)\right) \leq \mathbf{Pr}\left(\mathsf{coll}(Q_q) \wedge \neg\mathsf{aux}(Q_q)\right) + \mathbf{Pr}\left(\mathsf{aux}(Q_q)\right). \tag{7}$$

We start with the analysis of $\mathbf{Pr}\left(\mathsf{coll}(Q_q) \wedge \neg\mathsf{aux}(Q_q)\right)$. For obtaining a query history that fulfills configuration $\mathsf{coll}(Q_q)$, it may be the case that a query appears at multiple positions. For instance, the queries at positions $1L$ and $2R$ are the same. We split the analysis of $\mathsf{coll}(Q_q)$ into essentially all different possible cases, but we do this in two steps. In the first step, we distinct among the cases a query occurs in both words at the same position. We define for binary $\alpha_1, \alpha_2, \alpha_3$ by $\mathsf{coll}_{\alpha_1\alpha_2\alpha_3}(Q)$ the configuration $\mathsf{coll}(Q)$ of Fig. 4 restricted to

$$1L = 1R \iff \alpha_1 = 1, \quad 2L = 2R \iff \alpha_2 = 1, \quad 3L = 3R \iff \alpha_3 = 1.$$

By construction, $\mathsf{coll}(Q_q) \Rightarrow \bigvee_{\alpha_1,\alpha_2,\alpha_3 \in \{0,1\}} \mathsf{coll}_{\alpha_1\alpha_2\alpha_3}(Q_q)$, and from (6-7) we obtain the following bound on $\mathbf{adv}_{F_\mathsf{A}^3}^{\mathsf{coll}}(q)$:

$$\mathbf{adv}_{F_\mathsf{A}^3}^{\mathsf{coll}}(q) \leq \sum_{\substack{\alpha_1,\alpha_2,\alpha_3 \\ \in \{0,1\}}} \mathbf{Pr}\left(\mathsf{coll}_{\alpha_1\alpha_2\alpha_3}(Q_q) \wedge \neg\mathsf{aux}(Q_q)\right) + \mathbf{Pr}\left(\mathsf{aux}(Q_q)\right). \tag{8}$$

Note that we did not make a distinction yet whether or not a query occurs at two "different" positions (e.g. at positions $1L$ and $2R$). These cases are analyzed for each of the sub-configurations separately, as becomes clear later. Probabilities $\mathbf{Pr}\left(\text{coll}_{\alpha_1\alpha_2\alpha_3}(Q_q) \wedge \neg\text{aux}(Q_q)\right)$ for the different choices of $\alpha_1, \alpha_2, \alpha_3$ are bounded in Lems. 1-4. The proofs are rather similar, and we only bound the probability on $\text{coll}_{000}(Q_q)$ in full detail (Lem. 1). A bound on $\mathbf{Pr}\left(\text{aux}(Q_q)\right)$ is given in Lem. 5. A part of the proof of Lem. 1, and the proofs of Lems. 2-5 are given in [17].

Lemma 1. $\mathbf{Pr}\left(\text{coll}_{000}(Q_q) \wedge \neg\text{aux}(Q_q)\right) \leq \frac{t_2 q + 7q + 3t_1 t_2^2 + 3t_1 t_2}{2^n - q}.$

Proof. Sub-configuration $\text{coll}_{000}(Q_q)$ is given in Fig. 5. The block cipher queries at positions a and $!a$ are required to be different, and so are the ones are positions $b, !b$ and $c, !c$.

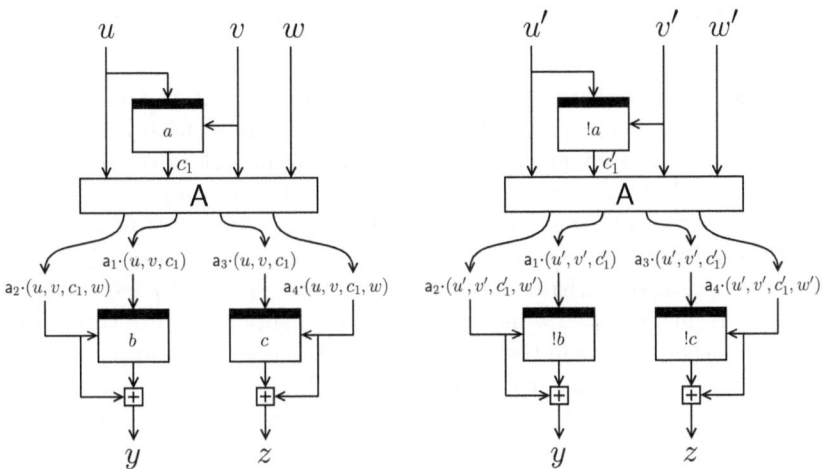

Fig. 5. Configuration $\text{coll}_{000}(Q)$. We require $(u, v, w) \neq (u', v', w')$.

We consider the probability of the adversary finding a solution to configuration $\text{coll}_{000}(Q_q)$ such that Q_q satisfies $\neg\text{aux}(Q_q)$. Consider the ith query, for $i \in \{1, \ldots, q\}$. We say this query is a winning query if it makes $\text{coll}_{000}(Q_i) \wedge \neg\text{aux}(Q_i)$ satisfied for any set of other queries in the query history Q_{i-1}. We can assume the ith query does not make $\text{aux}(Q_i)$ satisfied: if it would, by definition it cannot be a winning query.

Recall that, although we narrowed down the number of possible positions for a winning query to occur (in $\text{coll}_{000}(Q_q)$ it cannot occur at both $1L$ and $1R$, at both $2L$ and $2R$, or at both $3L$ and $3R$), it may still be the case that such a query contributes to multiple "different" positions, e.g. $1L$ and $2R$. Note that by construction, a winning query can contribute to at most three block cipher positions of Fig. 5. In total, there are 26 sets of positions at which the winning query can contribute at the same time. Discarding symmetric cases caused by swapping (u, v, w) and (u', v', w'), one identifies the following 13 sets of positions:

$$\begin{aligned}
&\mathcal{S}_1 = \{1L\}, &&\mathcal{S}_4 = \{1L, 2L\}, &&\mathcal{S}_7 = \{1L, 2R\}, &&\mathcal{S}_{10} = \{1L, 2L, 3L\}, \\
&\mathcal{S}_2 = \{2L\}, &&\mathcal{S}_5 = \{1L, 3L\}, &&\mathcal{S}_8 = \{1L, 3R\}, &&\mathcal{S}_{11} = \{1L, 2L, 3R\}, \\
&\mathcal{S}_3 = \{3L\}, &&\mathcal{S}_6 = \{2L, 3L\}, &&\mathcal{S}_9 = \{2L, 3R\}, &&\mathcal{S}_{12} = \{1L, 2R, 3L\}, \\
& && && &&\mathcal{S}_{13} = \{1L, 2R, 3R\}.
\end{aligned}$$

Note that there are many more symmetric cases among these, but we are not allowed to discard those as these may result in effectively different collisions. For $j = 1, \ldots, 13$ we denote by $\mathsf{coll}_{000:\mathcal{S}_j}(Q)$ configuration $\mathsf{coll}_{000}(Q)$ with the restriction that the winning query *must* appear at the positions in $\mathcal{S}_j$. By basic probability theory,

$$\mathbf{Pr}\left(\mathsf{coll}_{000}(Q_q) \wedge \neg\mathsf{aux}(Q_q)\right) \leq \sum_{j=1}^{13} \mathbf{Pr}\left(\mathsf{coll}_{000:\mathcal{S}_j}(Q_q) \wedge \neg\mathsf{aux}(Q_q)\right). \quad (9)$$

$\mathsf{coll}_{000:\mathcal{S}_1}(Q_q)$. Rather than considering the success probability of the ith query, and then sum over $i = 1, \ldots, q$ (as is done in the analysis of [4,5,6,7,9,12,16,21,26], hence all collision security proofs of Table 1), the approach in this proof is to focus on "wish lists". Intuitively, a wish list is a continuously updated sequence of query tuples that would make configuration $\mathsf{coll}_{000:\mathcal{S}_j}(Q_q)$ satisfied. During the attack of the adversary, we maintain an initially empty wish list $\mathcal{W}_{\mathcal{S}_1}$. Consider configuration $\mathsf{coll}_{000}(Q)$ with the query at position $\mathcal{S}_1 = \{1L\}$ left out (see [17] for a graphical intuition). If a new query is made, suppose it fits this configuration for some other queries in the query history (the new query appearing at least once), jointly representing queries at positions $\{2L, 3L, 1R, 2R, 3R\}$. Then the corresponding tuple (u, v, c_1) is added to $\mathcal{W}_{\mathcal{S}_1}$. Note that this tuple is uniquely determined by the queries at $2L$ and $3L$ by invertibility of A, but different combinations of queries may define the same wish. The latter does, however, not invalidate the analysis: this is covered by the upper bound on $\mathcal{W}_{\mathcal{S}_1}$ that will be computed later in the proof, and will simply render a slightly worse bound.

As we have restricted to the case the winning query only occurring at the position of $\mathcal{S}_1$, we can assume a query never adds itself to a wish list[3]. Clearly, in order to find a collision for F_{A}^3 in this sub-configuration, the adversary needs to wish for a query at least once. Suppose the adversary makes a query $E(k, m)$ where $(k, m, c) \in \mathcal{W}_{\mathcal{S}_1}$ for some c. We say that (k, m, c) is wished for, and the wish is granted if the query response equals c. As the adversary makes at most q queries, such wish is granted with probability at most $1/(2^n - q)$, and the same for inverse queries. By construction, each element from $\mathcal{W}_{\mathcal{S}_1}$ can be wished for only once, and we find that the adversary finds a collision with probability at most $\frac{|\mathcal{W}_{\mathcal{S}_1}|}{2^n - q}$.

---

[3] A winning query that would appear at multiple positions is counted in $\mathsf{coll}_{000:\mathcal{S}_j}(Q_q)$ for some other set $\mathcal{S}_j$.

Now, it suffices to upper bound the size of the wish list $\mathcal{W}_{\mathcal{S}_1}$ after q queries, and to this end we bound the number of solutions to configuration $\mathsf{coll}_{000:\mathcal{S}_j}(Q_q)$. By $\neg\mathsf{aux}_1(Q_q)$, the configuration has at most t_1 choices for $2L, 2R$. For any such choice, by $\neg\mathsf{aux}_2(Q_q)$ we have at most t_2 choices for $1R$. Any such choice fixes w' (as $\mathsf{a}_{24} \neq 0$), and thus the query at position $3R$, and consequently z. By $\neg\mathsf{aux}_4(Q_q)$, we have at most t_2 choices for $3L$. The queries at positions $2L$ and $3L$ uniquely fix (u, v, c_1) by invertibility of A. We find $|\mathcal{W}_{\mathcal{S}_1}| \leq t_1 t_2^2$, and hence in this setting a collision is found with probability at most $t_1 t_2^2/(2^n - q)$.

$\mathsf{coll}_{000:\mathcal{S}_j}(Q_q)$ **for** $j = 2, \ldots, 13$. In [17], $\mathbf{Pr}\left(\mathsf{coll}_{000:\mathcal{S}_j}(Q_q) \wedge \neg\mathsf{aux}(Q_q)\right)$ is bounded by $t_1 t_2^2/(2^n - q)$ for $j = 2, 3$, $q/(2^n - q)$ for $j = 4, 5, 6, 10, 11, 12, 13$, $t_1 t_2/(2^n - q)$ for $j = 7, 8$, and $(t_1 t_2 + t_2 q)/(2^n - q)$ for $j = 9$.

The proof is now completed by adding all bounds in accordance with (9).   □

Lemma 2. $\mathbf{Pr}\left(\mathsf{coll}_{100}(Q_q) \wedge \neg\mathsf{aux}(Q_q)\right) \leq \frac{2q + 2t_1 t_2}{2^n - q}$.

Lemma 3. $\mathbf{Pr}\left(\mathsf{coll}_{\alpha_1 \alpha_2 \alpha_3}(Q_q) \wedge \neg\mathsf{aux}(Q_q)\right) \leq \frac{t_2^2 q + t_2 q + q + t_1 t_2}{2^n - q}$ for $\alpha_1 \alpha_2 \alpha_3 \in \{010, 001\}$.

Lemma 4. $\mathbf{Pr}\left(\mathsf{coll}_{\alpha_1 \alpha_2 \alpha_3}(Q_q) \wedge \neg\mathsf{aux}(Q_q)\right) = 0$ when $\alpha_1 + \alpha_2 + \alpha_3 \geq 2$.

Lemma 5. $\mathbf{Pr}\left(\mathsf{aux}(Q_q)\right) \leq \frac{q^2}{t_1(2^n - q)} + 3 \cdot 2^n \left(\frac{eq}{t_2(2^n - q)}\right)^{t_2}$.

From (8) and the results of Lems. 1-5 we conclude the bound of (3). This completes the proof of Thm. 1.

6   Conclusions

In the area of double block length hashing, where a $3n$-to-$2n$-bit compression function is constructed from n-bit block ciphers, all optimally secure constructions known in the literature employ a block cipher with $2n$-bit key space. We have reconsidered the principle of double length hashing, focusing on double length hashing from a block cipher with *n-bit message and key space*. Unlike in the DBL^{2n} class, we demonstrate that there does not exist any optimally secure design with reasonably simple finalization function that makes two cipher calls. By allowing one extra call, optimal collision resistance can nevertheless be achieved, as we have proven by introducing our family of designs F_A^3.

In our quest for optimal collision secure compression function designs, we had to resort to designs with three block cipher calls rather than two, which moreover are not parallelizable. This entails an efficiency loss compared to MDC-2, MJH, and Jetchev et al.'s construction. On the other hand, our family of functions is based on simple arithmetic in the finite field: unlike constructions by Stam [25,26], Lee and Steinberger [14], and Jetchev et al. [7], our design does not make use of full field multiplications. The example matrices A given in (4) are designed to use a minimal amount of non-zero elements. We note that specific choices of A may be more suited for this construction to be used in an iterated design.

This work provides new insights in double length hashing, but also results in interesting research questions. Most importantly, is it possible to construct other collision secure F^3 constructions (beyond our family of functions F_A^3), that achieve optimal $2^{5n/3}$ preimage resistance? Given the negative collision resistance result for a wide class of compression functions F^2, is it possible to achieve optimal collision security *in the iteration* anyhow? This question is beyond the scope of this work. On the other hand, in line with ideas of [18], is it possible to achieve an impossibility result for F^3 restricted to the xor-only design (where $f_1, \ldots, f_4$ only xor their parameters)?

Acknowledgments. This work has been funded in part by the IAP Program P6/26 BCRYPT of the Belgian State (Belgian Science Policy), in part by the European Commission through the ICT program under contract ICT-2007-216676 ECRYPT II, and in part by the Research Council K.U.Leuven: GOA TENSE. The author is supported by a Ph.D. Fellowship from the Institute for the Promotion of Innovation through Science and Technology in Flanders (IWT-Vlaanderen). The author would like to thank Elena Andreeva and the anonymous ASIACRYPT 2012 reviewers for their valuable help and feedback.

References

1. Andreeva, E., Neven, G., Preneel, B., Shrimpton, T.: Seven-Property-Preserving Iterated Hashing: ROX. In: Kurosawa, K. (ed.) ASIACRYPT 2007. LNCS, vol. 4833, pp. 130–146. Springer, Heidelberg (2007)
2. Armknecht, F., Fleischmann, E., Krause, M., Lee, J., Stam, M., Steinberger, J.: The Preimage Security of Double-Block-Length Compression Functions. In: Lee, D.H. (ed.) ASIACRYPT 2011. LNCS, vol. 7073, pp. 233–251. Springer, Heidelberg (2011)
3. Bos, J.W., Özen, O., Stam, M.: Efficient Hashing Using the AES Instruction Set. In: Preneel, B., Takagi, T. (eds.) CHES 2011. LNCS, vol. 6917, pp. 507–522. Springer, Heidelberg (2011)
4. Fleischmann, E., Gorski, M., Lucks, S.: Security of Cyclic Double Block Length Hash Functions. In: Parker, M.G. (ed.) Cryptography and Coding 2009. LNCS, vol. 5921, pp. 153–175. Springer, Heidelberg (2009)
5. Hirose, S.: Provably Secure Double-Block-Length Hash Functions in a Black-Box Model. In: Park, C., Chee, S. (eds.) ICISC 2004. LNCS, vol. 3506, pp. 330–342. Springer, Heidelberg (2005)
6. Hirose, S.: Some Plausible Constructions of Double-Block-Length Hash Functions. In: Robshaw, M. (ed.) FSE 2006. LNCS, vol. 4047, pp. 210–225. Springer, Heidelberg (2006)
7. Jetchev, D., Özen, O., Stam, M.: Collisions Are Not Incidental: A Compression Function Exploiting Discrete Geometry. In: Cramer, R. (ed.) TCC 2012. LNCS, vol. 7194, pp. 303–320. Springer, Heidelberg (2012)
8. Lai, X., Massey, J.L.: Hash Functions Based on Block Ciphers. In: Rueppel, R.A. (ed.) EUROCRYPT 1992. LNCS, vol. 658, pp. 55–70. Springer, Heidelberg (1993)

9. Lee, J., Kwon, D.: The security of Abreast-DM in the ideal cipher model. Cryptology ePrint Archive, Report 2009/225 (2009)
10. Lee, J., Stam, M.: MJH: A Faster Alternative to MDC-2. In: Kiayias, A. (ed.) CT-RSA 2011. LNCS, vol. 6558, pp. 213–236. Springer, Heidelberg (2011)
11. Lee, J., Stam, M., Steinberger, J.: The collision security of Tandem-DM in the ideal cipher model. Cryptology ePrint Archive, Report 2010/409 (2010); full version of [12]
12. Lee, J., Stam, M., Steinberger, J.: The Collision Security of Tandem-DM in the Ideal Cipher Model. In: Rogaway, P. (ed.) CRYPTO 2011. LNCS, vol. 6841, pp. 561–577. Springer, Heidelberg (2011)
13. Lee, J., Stam, M., Steinberger, J.: The preimage security of double-block-length compression functions. Cryptology ePrint Archive, Report 2011/210 (2011)
14. Lee, J., Steinberger, J.: Multi-property-preserving Domain Extension Using Polynomial-Based Modes of Operation. In: Gilbert, H. (ed.) EUROCRYPT 2010. LNCS, vol. 6110, pp. 573–596. Springer, Heidelberg (2010)
15. Lucks, S.: A collision-resistant rate-1 double-block-length hash function. In: Symmetric Cryptography. Dagstuhl Seminar Proceedings, vol. (07021) (2007)
16. Mennink, B.: On the collision and preimage security of MDC-4 in the ideal cipher model. Cryptology ePrint Archive, Report 2012/113 (2012)
17. Mennink, B.: Optimal collision security in double block length hashing with single length key (2012); full version of this paper
18. Mennink, B., Preneel, B.: Hash Functions Based on Three Permutations: A Generic Security Analysis. In: Safavi-Naini, R., Canetti, R. (eds.) CRYPTO 2012. LNCS, vol. 7417, pp. 330–347. Springer, Heidelberg (2012)
19. Meyer, C., Schilling, M.: Secure program load with manipulation detection code. In: Proc. Securicom, pp. 111–130 (1988)
20. Nandi, M., Lee, W., Sakurai, K., Lee, S.: Security Analysis of a 2/3-Rate Double Length Compression Function in the Black-Box Model. In: Gilbert, H., Handschuh, H. (eds.) FSE 2005. LNCS, vol. 3557, pp. 243–254. Springer, Heidelberg (2005)
21. Özen, O., Stam, M.: Another Glance at Double-Length Hashing. In: Parker, M.G. (ed.) Cryptography and Coding 2009. LNCS, vol. 5921, pp. 176–201. Springer, Heidelberg (2009)
22. Preneel, B., Govaerts, R., Vandewalle, J.: Hash Functions Based on Block Ciphers: A Synthetic Approach. In: Stinson, D.R. (ed.) CRYPTO 1993. LNCS, vol. 773, pp. 368–378. Springer, Heidelberg (1994)
23. Rogaway, P., Shrimpton, T.: Cryptographic Hash-Function Basics: Definitions, Implications, and Separations for Preimage Resistance, Second-Preimage Resistance, and Collision Resistance. In: Roy, B., Meier, W. (eds.) FSE 2004. LNCS, vol. 3017, pp. 371–388. Springer, Heidelberg (2004)
24. Rogaway, P., Steinberger, J.: Security/Efficiency Tradeoffs for Permutation-Based Hashing. In: Smart, N.P. (ed.) EUROCRYPT 2008. LNCS, vol. 4965, pp. 220–236. Springer, Heidelberg (2008)
25. Stam, M.: Beyond Uniformity: Better Security/Efficiency Tradeoffs for Compression Functions. In: Wagner, D. (ed.) CRYPTO 2008. LNCS, vol. 5157, pp. 397–412. Springer, Heidelberg (2008)
26. Stam, M.: Blockcipher-Based Hashing Revisited. In: Dunkelman, O. (ed.) FSE 2009. LNCS, vol. 5665, pp. 67–83. Springer, Heidelberg (2009)
27. Steinberger, J.P.: The Collision Intractability of MDC-2 in the Ideal-Cipher Model. In: Naor, M. (ed.) EUROCRYPT 2007. LNCS, vol. 4515, pp. 34–51. Springer, Heidelberg (2007)

Bicliques for Permutations:
Collision and Preimage Attacks in Stronger Settings

Dmitry Khovratovich

Microsoft Research, USA, and Infotecs, Russia

Abstract. We extend and improve biclique attacks, which were recently introduced for the cryptanalysis of block ciphers and hash functions. While previous attacks required a primitive to have a key or a message schedule, we show how to mount attacks on the primitives with these parameters fixed, i.e. on permutations. We introduce the concept of sliced bicliques, which is a translation of regular bicliques to the framework with permutations.

The new framework allows to convert preimage attacks into collision attacks and derive the first collision attacks on the reduced SHA-3 finalist Skein in the hash function setting up to 11 rounds. We also demonstrate new preimage attacks on the reduced Skein and the output transformation of the reduced Grøstl. Finally, the sophisticated technique of message compensation gets a simple explanation with bicliques.

Keywords: Skein, SHA-3, hash function, collision attack, preimage attack, biclique, permutation, Grøstl.

1 Introduction

Meet-in-the-middle attacks have been known in cryptanalysis at least since the analysis of Double-DES [9], but got less attention in 90s and early 2000s because of more difficult key schedules in contemporary block ciphers. They regained prominence with the introduction of the splice-and-cut framework by Aoki and Sasaki for hash functions [2, 23]. Aoki and Sasaki considered various designs and demonstrated how to construct pseudo-preimages for compression functions based on block ciphers. Pseudo-preimages can be converted to regular preimages, though this reduces the advantage previously gained over brute force.

While the first splice-and-cut attacks were quite simple, they quickly became more sophisticated as cryptanalysts tried to increase the number of rounds broken [1, 24]. That number for the first attacks was determined by the length of *chunks* — two sections of a primitive each independent of its own set of key/message bits called *neutral bits*. For example, two DES calls in Double-DES are chunks each independent of half of the key. Later research showed how to start the attack with a sophisticated construction (so called *initial structure*) over several rounds to increase the total number of rounds in the attack [3, 24],

X. Wang and K. Sako (Eds.): ASIACRYPT 2012, LNCS 7658, pp. 544–561, 2012.

which culminated in the concept of bicliques [16]. While initial structures relied on slow diffusion, bicliques do not need that condition. In turn, they translated the condition on internal states being suitable for meet-in-the-middle attacks to the requirements on how these states map to each other under different sub-transformations.

Bicliques. The new biclique technique [8, 16] led to a few surprising attacks on AES, though many of them had only a constant factor improvement over exhaustive search. The attack has influenced those reducing the security level of the full Square [18], Kasumi [13], IDEA [15]. All these attacks need a small but noticeable number of operations to test a single key, and in our opinion they have smaller potential. Indeed, even a single operation for each key implies a lower bound on the complexity which is not far from exhaustive search. Also from the technical point of view, the use of bicliques in those settings is not much different from earlier use of initial structures.

From Parametrized Transformations to Permutations. The key/message schedule is a crucial element in the biclique attacks. In Section 2 we show how to enumerate N message candidates with only $2\sqrt{N}$ states.

However, there are several settings where an attacker can not manipulate a scheduled parameter, or there is no schedule at all. For example, preimage attacks on blockcipher-based hash functions first consider a compression function and produce pseudo-preimages, and then run a computationally expensive meet-in-the-middle attack to produce real preimages. If an attacker wants to reduce the cost by avoiding the second step, then he has to assign the chaining value (CV) with the original initial value (IV). If the compression function is based on the Matyas-Meyer-Oseas mode with E_K as a block cipher,

$$F(CV, M) = E_{CV}(M) \oplus M,$$

where M stands for a message block, then the attacker analyze the permutation $E_{IV}(\cdot)$.

Another example is the SHA-3 finalist Grøstl with output transformation $x \leftarrow$ Truncate$(x \oplus P(x))$, where P is a fixed permutation. Therefore, the translation of the biclique technique to permutations is quite promising.

Permutations have been subject to a few recent attacks [22, 30], which use a predecessor of biclique — initial structure. A natural question is whether the more general concept of bicliques can be carried out to this setting and even if so whether the advantages of long bicliques can be used similarly to AES.

Collisions for the MMO-Based Primitives. While the Matyas-Meyer-Oseas (MMO) and Davies-Meyer (DM) modes are equally resistant to generic attacks [7], they are way more different when dedicated methods are considered. Collision attacks typically fix the chaining value, so in the DM mode

$$F(CV, M) = E_M(CV) \oplus CV$$

an attacker is able to manipulate the round injections through the modification of M, while in the MMO mode he is able to choose only the input. From our point of view, famous collision attacks on the MD4/SHA family [5, 29] demonstrate that the first setting is much more friendly to the attacker. Indeed, the most powerful collision search method — differential cryptanalysis — works with related-key characteristics in the DM mode, and with regular characteristics in the MMO mode. Related-key attacks on the full AES [6] hint that the former setting is more suitable.

The hash function Skein follows the MMO mode and is an object of our analysis. The existing near-collision attacks on the compression function of Skein [4,26] are essentially free-start collisions, i.e. they inject the difference in the chaining value or the tweak. Therefore, we conclude that mounting a regular collision attack on the hash function based on MMO is quite difficult. The very recent pseudo-collision attack [17] on Skein is a great step forward, as we discuss in the further text.

Our Contributions

In Section 2 we introduce a new notion of *sliced biclique* as a translation of a regular biclique to permutations. The new concept helps to carry out the meet-in-the-middle attacks and the biclique technique to permutations without modifiable parameters. We call *parameters* both keys and messages.

We improve a very recent technique of finding pseudo-collisions with pseudo-preimages and show how to get regular collision attacks on the MMO-based primitives (Section 4). We obtain the first collision attacks on the reduced round Skein hash function (Section 5). The new attacks are also translated to new preimage attacks on Skein (see the extended version of this paper [14]).

Then we consider the output transformation of the SHA-3 finalist Grøstl-256 and derive the first shortcut 6-round attack (Section 6). Finally, we analyze a procedure from earlier meet-in-the-middle attacks called message compensation (Section 7). Previously ad-hoc, it gets a clear interpretation as a sliced biclique (see the details in the extended version) [14].

2   Splice-and-Cut Attacks and Bicliques

Splice-and-cut attacks [2, 23] were designed as a preimage search method. A simple splice-and-cut attack is applied to the Davies-Meyer-based compression function F:

$$F(CV, M) = E_M(CV) \oplus CV,$$

where CV is a chaining value, M is a message block, $E_K(\cdot)$ is a block cipher. An attacker is given an n-bit hash value H and has to find a preimage M. The preimage search is organized as follows. The attacker partitions the message space into sets, which are represented as two-dimensional array of messages $\{M[i,j]\}$, and process each set independently. Given $\{M[i,j]\}$, he selects an internal state S and an internal variable v such that v as a function of S in one direction does not depend on i, and in the other direction does not depend on j:

$$\text{CV} \cdots \boxed{\quad \xrightarrow{M[i,*]} S \xrightarrow{M[*,j]} v \xleftarrow{\quad} } \cdots \xrightarrow{M[i,*]} H$$

Then he assigns S with an arbitrary value and computes v in the forward direction for all possible j (denoted by $\overrightarrow{v}_j$) and in the other direction for all possible i (denoted by $\{\overleftarrow{v}_i\}$), computing CV and using H on the way. The overlap of the resulting two sets yields preimage candidates which are tested on the full state width. The indices i and j typically belong to $[0; 2^d - 1]$ for some d, which yields the matching probability 2^{2d-n} for a single set $\{M[i,j]\}$, and the complexity 2^{n-d} for the pseudo-preimage search. To find a full preimage the adversary generates $2^{d/2}$ pseudo-preimages, computes $2^{n-d/2}$ CVs out of the initial value, and checks for a matching pair. The total complexity is $2^{n-d/2+1}$ without optimizations (which are not always possible), so only $d \geq 3$ provides an advantage over brute force.

The basic attack was carried out to other modes and even block ciphers. For the latter, the encryption oracle plays the role of the feedforward to link the input and the output.

A biclique is an extension for the first step of the attack, which is based upon an earlier informal concept of *initial structure* [3, 24]. Instead of a single state S, a *biclique* is defined over a *sub-cipher* — a part of the primitive, typically several rounds long — and for a particular group of keys or messages that are subject to test. A biclique over f for parameters $\{M[i,j]\}$ is pair of state sets

$$\{Q_i\}, \{P_j\}$$

such that

$$Q_i \xrightarrow[f]{M[i,j]} P_j. \tag{1}$$

A biclique tests parameters $\{M[i,j]\}$ in the same way as in the basic attack. The matching variable v is computed in both directions:

$$CV \cdots \boxed{\quad \xrightarrow{M[i,*]} \overbrace{Q_i \qquad P_j}^{\text{biclique}} \xrightarrow{M[*,j]} v \xleftarrow{\quad} } \cdots \xrightarrow{M[i,*]} H$$

The condition (1) guarantees that if $M[i,j]$ is a preimage then the computations from P_j and Q_i meet in a biclique exactly as at the matching point.

The crucial property of a biclique is that it enumerates 2^{2d} parameters with only 2^{d+1} internal states. The value d is called *dimension* of a biclique, and the number of rounds in f — *length* of a biclique.

The computational advantage of a biclique attack is the same as in the basic attack, and hence is proportional to the dimension.

3   Bicliques for Permutations

The simplest way to turn a permutation into a preimage-resistant function is to xor the input to the output:

$$F(x) = E(x) \oplus x. \tag{2}$$

Our goal is to construct a preimage search algorithm, which recovers x from given $H = F(x)$. We proceed as follows.

Using a specific algorithm, we select a sub-permutation g within E and an internal state V in E but not in g. Denote the input state of g by Q, and the output state of g by P. We partition the space of all states into sets $\{Q_{i,j}\}$, which we represent as a two-dimensional array of states. Here i, j are d-bit values for some d. We test independently each set if it contains a state that correspond to a valid preimage x. Let us denote the g-image of $Q_{i,j}$ by $P_{i,j}$:

$$Q_{i,j} \xrightarrow{g} P_{i,j}.$$

We will explain how to choose g and partition of Q in a subsection "Construction algorithms", and it will also become clear why we use two indices to enumerate states Q.

The rest of this section is devoted to finding an improved way to test a single set of states. A straightforward way to check if one of $\{Q_{i,j}\}$ is a solution to (2) is to compute for each i, j the state V two times. First, as a function of P in the forward direction, let us denote this computation by $\overrightarrow{F}$. Second, compute V as a function of Q in the backward direction: computing x, then $E(x) = H \oplus x$, and then V; let us denote this computation by $\overleftarrow{F}$. Hence we check if

$$\exists i, j : \quad \overrightarrow{F}(P_{i,j}) = \overleftarrow{F}(Q_{i,j}). \tag{3}$$

This algorithm is equivalent to the exhaustive search and requires 2^{2d} computations of E.

The complexity can be reduced as follows. Let $v \subseteq V$ be an internal variable, and $\overrightarrow{f_v}$ and $\overleftarrow{f_v}$ be the projections of $\overrightarrow{F}$ and $\overleftarrow{F}$, resp., to v. We say that the states $Q_{i,j}$ and $P_{i,j}$ form a $sliced$ $biclique$, if the following conditions hold:

$$\forall i, j \quad \overleftarrow{f_v}(Q_{i,j}) = \overleftarrow{f_v}(Q_{i,0});$$
$$\forall i, j \quad \overrightarrow{f_v}(P_{i,j}) = \overrightarrow{f_v}(P_{0,j}).$$

Therefore, the necessary condition in Equation (3) can be reformulated as follows:

$$\exists i, j : \quad \overrightarrow{F}(P_{i,j}) = \overleftarrow{F}(Q_{i,j}) \implies \exists i, j : \quad \overrightarrow{f_v}(P_{i,j}) = \overleftarrow{f_v}(Q_{i,j}) \iff$$
$$\iff \exists i, j : \quad \overrightarrow{f_v}(P_{0,j}) = \overleftarrow{f_v}(Q_{i,0}). \tag{4}$$

Let us denote $\overrightarrow{f_v}(P_{0,j})$ by $\overrightarrow{v_j}$ and $\overleftarrow{f_v}(Q_{i,0})$ by $\overleftarrow{v_i}$. Hence one of $\{Q_{i,j}\}$ is a solution if

$$\exists i, j : \quad \overrightarrow{v_j} = \overleftarrow{v_i}. \tag{5}$$

To check it we need to call $\overrightarrow{f_v}$ and $\overleftarrow{f_v}$ 2^d times each, which is less than 2^d calls of E. The computations are depicted in Figure 1. The matching candidates yields a pair (i, j) and the state $Q_{i,j}$, which we retest as a preimage candidate. For the full attack we need to partition the full input domain into the groups of size 2^{2d} and construct bicliques for them.

If the complexity of constructing a biclique and retesting the false alarms is small compared to 2^d, then 2^{2d} states are tested with complexity 2^d, and the set of all n-bit states is tested with complexity 2^{n-d}. In the most of our attacks we test only a subset of states of cardinality 2^r with complexity 2^{r-d} The parameter d is called a *dimension* of sliced biclique.

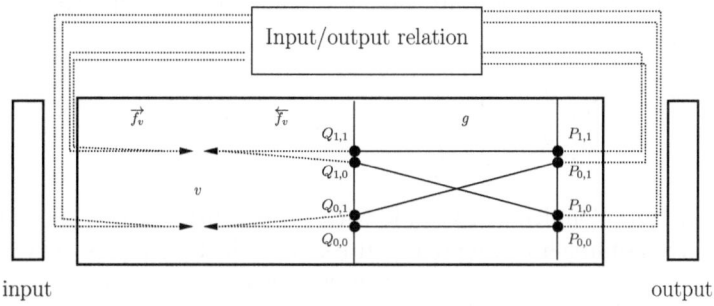

Fig. 1. Sliced biclique for a permutation

Construction Algorithms. Let us describe a construction algorithm for sliced bicliques, and then discuss its modifications. First we choose (below we will explain how) state $Q_{0,0}$ and two sets of differences $\{\Delta_i\}$ and $\{\nabla_j\}$, $i, j > 0$. We construct a biclique where

$$Q_{i,j} = Q_{i,0} \oplus \Delta_j; \tag{6}$$
$$P_{i,j} = P_{0,j} \oplus \nabla_i. \tag{7}$$

We proceed as follows

1. Compute $P_{0,0} \leftarrow g(Q_{0,0})$.
2. Set $Q_{0,j} \leftarrow Q_{0,0} \oplus \Delta_j$, compute $P_{0,j}$ for all $j > 0$.
3. Set $P_{i,j} \leftarrow P_{0,j} \oplus \nabla_i$, compute $Q_{i,j}$ for all i, j.

Hence Equation (7) is fulfilled by definition, and we need to prove Equation (6). We claim that it is fulfilled if the states $Q_{0,0}, Q_{i,0}, Q_{i,j}, Q_{0,j}$ form a boomerang quartet [28] over f with differences Δ_i and ∇_j, as demonstrated in Figure 2, a).

Indeed, $Q_{0,j} = Q_{0,0} \oplus \Delta_j$ by definition. We also have

$$P_{i,j} = g(Q_{0,j}) \oplus \nabla_i; \quad P_{i,0} = g(Q_{0,0}) \oplus \nabla_i.$$

Therefore, $g^{-1}(P_{i,j}) \oplus g^{-1}(P_{i,0}) = Q_{i,j} \oplus Q_{i,0} = \Delta_i$ if

$$(Q_{0,0}, Q_{i,0}, Q_{i,j}, Q_{0,j}) \text{ — boomerang quartet.} \tag{8}$$

In order to figure out sufficient conditions for the latter statement to hold, we consider two groups of differential trails. The trails in the first group are called Δ-trails and describe the evolution of differences Δ_j:

$$\begin{aligned} Q_{i,0} &\xrightarrow{g} P_{i,0}; \\ Q_{i,j} &\xrightarrow{g} P_{i,j}. \end{aligned} \quad \Longrightarrow \quad \Delta_j \to P_{i,0} \oplus P_{i,j}.$$

The trails in the second group are called ∇-trails and describe the evolution of differences ∇_i:

$$\begin{aligned} P_{0,j} &\xrightarrow{g^{-1}} Q_{0,j}; \\ P_{i,j} &\xrightarrow{g^{-1}} Q_{i,j}. \end{aligned} \quad \Longrightarrow \quad \nabla_i \to Q_{0,j} \oplus Q_{i,j}.$$

As proved in [16], Condition (8) holds with probability 1 if the Δ- and ∇-trails share no active nonlinear elements (Figure 2, b)). Such bicliques are called *based on non-interleaving trails*. A straightforward way to achieve this property is to select Δ_j and ∇_i so that their diffusion is minimum. A more sophisticated approach is to choose the state $Q_{0,0}$ so that the diffusion is minimum.

If Δ- and ∇-trail share nonlinear elements, we say that such bicliques are *based on interleaving trails*.

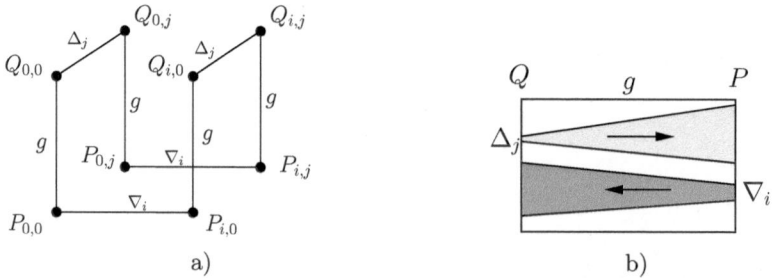

a)   b)

Biclique states as a boomerang quartet. Non-interleaving differential trails in g.

Fig. 2. Differential properties of sliced biclique

4   Framework of New Preimage and Collision Attacks on Skein

The SHA-3 finalist Skein [10] employs the Matyas-Meyer-Oseas mode to construct a compression function. It takes the block cipher Threefish (denoted by $E_K(\cdot)$) and computes:

$$F(CV, T, M) = E_{CV,T}(M) \oplus M,$$

where CV is the chaining value, and T is the tweak value. Due to difficulties in mounting collision attacks on the MMO mode, the only published attack on the Skein hash function is the preimage attack [16] based on regular bicliques. The parameter $M[i, j]$ in the biclique equation (1) is the chaining value. As a result, in the preimage attack on the compression function the attacker has to work with multiple CV's to get a pseudo-preimage. A full preimage requires another meet-in-the-middle procedure (Section 2). The first step must have complexity 2^{n-3} or smaller to yield an advantage over brute-force, which implies that only bicliques of dimension 3 or larger should be used.

Equipped with the concept of sliced bicliques, we can fix the chaining value and attack the permutation $E_{IV}()$. Hence we can generate full preimages without the pseudo-preimage step. The complexity drops to 2^{n-d} instead of $2^{n+1-d/2}$, and restrictions on the biclique dimension do not hold anymore. Meet-in-the-middle attacks on the first call of the MMO and similar modes exist [22,30], but do not use the long biclique approach yet, and were not applied to Skein.

Collision Attacks. A more interesting property of the MMO mode comes out if we consider a very recent pseudo-collision attack which uses regular bicliques [17]. The method produces pseudo-collisions out of biclique preimage attacks as follows. Assume we have a biclique of dimension d and are able to match deterministically on some l hash value bits. Then the adversary generates partial pseudo-preimages to a hash value with these l bits equal to an arbitrarily chosen constant h. Hence 2^{2d-l} l-bit partial pseudo-preimages to h can be generated with cost 2^d. Note that they collide on l output bits. The adversary generates $2^{n/2-l/2}$ such preimages and expect a pair of them to collide on the remaining $(n-l)$ bits by the birthday paradox. Since chaining values and schedule inputs are not fixed in the attack, this yields a pseudo-collision with the expected complexity $2^{(n/2-l/2)+(d)-(2d-l)} = 2^{n/2+l/2-d}$. The approach both for DM and MMO modes.

The optimal d satisfies the equation $d = 2d - l$, which implies $d = l$. The attack is optimal if all preimages are generated out of a single biclique, which implies

$$l = n/2 - l/2 \iff l = n/3.$$

Hence the minimum complexity of collision search is $2^{n/3}$.

Again, the chaining value can be fixed in the MMO mode if we apply the sliced biclique concept. Then we can generate real collisions instead of pseudo-collisions. However, we can break fewer rounds compared to the pseudo-collision attacks. The reason is the diffusion of differences Δ and ∇ to the whole state while computing v, whereas in the regular biclique attack the effect of those differences is postponed. Nevertheless, our approach is more interesting, since the real collision attacks are considered a much stronger setting as compared to pseudo-collisions (cf. collisions for MD5).

Memory. The straightforward version of the attack requires to store all the pseudo-preimages generated, which makes the memory complexity be of the same order as the time complexity. However, as the preimage step is non-deterministic, we can employ memoryless collision search methods [27], which multiply the time complexity by a small constant. Therefore, all the attacks described in the further text, except for the marginal ones, have memoryless equivalents.

5   Collision Attacks on Skein

Here we present the first collision attacks on the reduced Skein hash function. The MMO mode is considered to be difficult for collision search, since most methods require a fixed chaining value when attacking the compression function. Since the round injections in the MMO mode come from the chaining value, the cryptanalyst has no freedom there, and hence is unable to construct local collisions, apply message modification techniques, etc.. As a result, previous attacks on Skein [4, 26] dealt with the compression function only. The attacks are grouped according to the number of rounds covered by a biclique. Though we aim for the maximal dimension and the number of rounds attacked, for clarity we do not push the concept to the extreme and try to avoid complicated bicliques. Hence our attacks can be improved in the future.

Short Description of Skein. We consider three members of the Skein hash function family: Skein-512, Skein-256, and Skein-512-256. Skein-512 [10] operates on the internal state of eight 64-bit words, while Skein-256 works with a state of four words. We denote the state words by $S^0, S^1, \ldots, S^7$. All the versions have 72 rounds. Skein-512-256 merely truncates the output of Skein-512 to 256 bits. Each round of Skein-512 consists of four (two in Skein-256) simple transformations called MIX:

$$y_0 = x_0 + x_1 \pmod{2^{64}};$$
$$y_1 = (x_1 \lll_{R_{(d \bmod 8)+1,j}}) \oplus y_0.$$

where R is a constant depending on the round number d. The invocations of MIX are followed by a word permutation and, every four rounds, also by an addition of a linear function of the chaining value and the tweak (constants in our scenario).

The only published attack on the Skein hash function is a preimage attack [16] on 22 rounds of Skein-512.

5.1 Skein-512

As few as three rounds of Skein-512 are required to diffuse the contents of a single word to the full state. As a result, the bicliques based on non-interleaving trails are likely to cover two rounds only. We present bicliques of different kind that are capable to cover up to 4 rounds, and give some hints on how to construct longer bicliques.

2-Round Biclique. Our first examples deal with short bicliques of high dimension. As a result, the attacks have a significant advantage over brute-force. We use an additional enumeration of rounds in a biclique, starting with 0.

We use an algorithm from Section 3 with non-interleaving trails. We choose an arbitrary $Q_{0,0}$ and construct bicliques of dimension 64 out of the following differences Δ and ∇:

$$\Delta_j = \boxed{0}\,\boxed{0}\,\boxed{0}\,\boxed{0}\,\boxed{0}\,\boxed{j}\,\boxed{0}\,\boxed{0} \text{ after MIX in round 0 of the biclique. } j = 1\ldots 2^{64} - 1$$

$$\nabla_i = \boxed{0}\,\boxed{0}\,\boxed{0}\,\boxed{0}\,\boxed{0}\,\boxed{0}\,\boxed{i}\,\boxed{0} \text{ after MIX in round 2 of the biclique. } i = 1\ldots 2^{64} - 1$$

It is easy to check that the Δ- and ∇-trails do not share active non-linear components and hence produce a biclique (Figure 3).

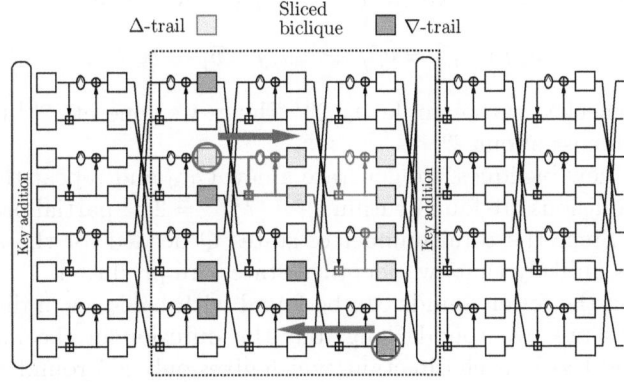

Fig. 3. Non-interleaving differential trails in a sliced biclique of dimension 64 in Skein-512

Only three rounds are required to diffuse a 64-bit word onto the full state. Hence we expect the matching part be two rounds long in both directions. A straightforward attack on 6 rounds uses a biclique in rounds 2-4 of Skein and word S^1 of the output of the 6-round transformation as the matching variable v (Figure 4).

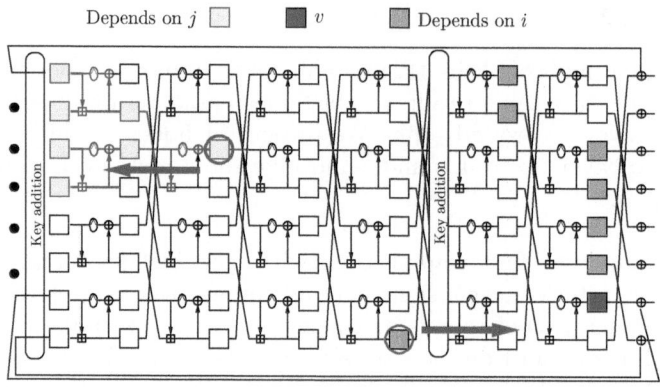

Fig. 4. Matching in 6-round Skein-512

However, we extend it by one round with the idea of the indirect partial matching [1]. Consider the state word S^0 after round 6 as a function of $P_{i,j}$. It is easy to check that

$$P_{i,j} = P_{0,j} \oplus i \implies S^0(P_{i,j}) = S^0(P_{0,j}) + i.$$

Therefore, if we set $v = S^0$, we get $\overrightarrow{v_{i,j}} = \overrightarrow{v_{0,j}} + i$. As a result,

$$\exists i, j : \overrightarrow{v_{i,j}} = \overleftarrow{v_{i,j}} \Leftrightarrow \exists i, j : \overrightarrow{v_j} = \overleftarrow{v_i} + i,$$

which can be checked with complexity 2^{64}. Hence we generate 2^{64} 64-bit partial preimages with cost about 2^{64}.

To produce new bicliques, we generate a new $Q_{0,0}$ and repeat the procedure. Full 7-round collisions are found within $2^{(512-64)/2} = 2^{224}$ partial preimages with the cost 2^{224}. Since the total number of states Q needed for the attack is less than 2^{256}, it is unlikely that two identical states are produced.

Collisions on the fewer rounds can be found with bicliques of dimension 128. These bicliques are two rounds long, but the diffusion in the matching part takes one round less in each direction, which gives only a 5-round collision. The complexity is 2^{192}.

3-Round Biclique. If we decrease dimension to 20 and lower, the diffusion takes more than three rounds. As a result, we can construct 3-round bicliques of dimension close to 20. We use an algorithm with non-interleaving trails with some modifications.

First, we carefully choose the position of the biclique in the compression function and bits where the difference is applied in Q and P. Since the rotation constants in each MIX function are distinct, the diffusion properties may change significantly when we shift the biclique over rounds and the active bits over the 64-bit word. The best configuration we have found places the biclique in rounds 5–7 (or $8k$ rounds further, because the rotation constants repeat every

8 rounds), where the states Q are defined before the MIX operation in round 5, and the states P are defined after the MIX operation in round 7. The Δ- and ∇-differences are defined for as follows:

$$\Delta_j = \boxed{0}\,\boxed{0}\,\boxed{0}\,\boxed{0}\,\boxed{0}\,\boxed{0}\,\boxed{j \ll 45}\,\boxed{0}\ \ j = 1 \ldots 2^{19} - 1$$
$$\nabla_i = \boxed{0}\,\boxed{i}\,\boxed{0}\,\boxed{0}\,\boxed{0}\,\boxed{0}\,\boxed{0}\,\boxed{0}\ \ i = 1 \ldots 2^{19} - 1$$

For $d < 19$ we simply set the most significant bits of Δ and ∇ to zero. We additionally require that the least 45 significant bits of the word S^6 in Q be equal to 0 in order to trails from interleaving. There is no other restriction on Q, so we can generate the states $Q_{0,0}$ in message sets simply by changing the words $S^0, \ldots, S^3$. Since we need less than 2^{256} states, it is unlikely that there would be a collision. This configuration produces a 3-round sliced biclique. Note that reducing dimension does not make the trails to interleave.

The length of the matching part decreases as the dimension grows. We have checked the diffusion on a PC and figured out that the matching part covers 7 rounds for $d = 17$. In this configuration we match on bits 30–33 of word S^2 and bits 20–32 of word S^3 of input to the compression function. The matching is not deterministic, as for some bits the difference is equal to zero with probability $p_i < 1$. We have calculated the type-I error probability as $\prod_i p_i = 0.6$ and conclude that probability 0.4 we miss a solution. Therefore, the total complexity is about two times larger compared to the deterministic case and is equal to about 2^{248}

For $d = 10$ the matching part takes 8 rounds. The matching variable consists of bits 17–21 of word S^0 and bits 24–31 of word S^2. The type-I error probability does not exceed 0.2, and the total complexity is 2^{251}. The other values are given in Table 1.

4-Round Biclique. A regular biclique in the preimage attack on Skein [16] covers 4 rounds with two key additions. If we consider these rounds without the key addition, we get exactly a sliced biclique of the same dimension. The diffusion in the matching part will be slightly different because of the rotation constants, but we still can bypass 10 rounds. Though the cost of the biclique construction is quite expensive — 2^{209} — there are 815 bit degrees of freedom left, of which 303 are in the internal state. We propose to use this freedom to amortize the biclique construction cost and generate new $Q_{0,0}$, so that a 14-round partial preimage is found with complexity 2^3. Full collisions are found with complexity $2^{254.5}$.

Longer Bicliques. Bicliques of dimension 1 can be constructed up to 8 rounds, but the advantage over brute-force attacks is really marginal. Another problem is that the construction cost of a single biclique is very high, and we are unaware of how to exploit the degrees of freedom over so many rounds given no freedom in the injections.

Table 1. Collision attacks on reduced Skein with large memory requirements (close to the computational complexity). Memoryless attacks add a small constant to the exponent.

Skein-256		Skein-512	
Rounds	Complexity	Rounds	Complexity
2	2^{85}	5	2^{192}
4	2^{96}	7	2^{224}
8	2^{120}	10	2^{248}
9	2^{124}	11	2^{251}
12	$2^{126.5}$	14	$2^{254.5}$

5.2 Skein-256

Diffusion in Skein-256 is generally faster, because the internal state consists of four words only. Typically it takes one round less to affect the whole state. As a result, non-interleaving biclique trails and the matching part are shorter. We figured out that collision attacks on Skein-256 with bicliques of the same dimension lag 2-3 rounds behind the attacks on Skein-512. For instance, bicliques of dimension 64 and 128 cover one round only, and the matching part is two rounds shorter. This results in 2-round collisions with complexity 2^{85} and 4-round collisions with complexity 2^{96}.

Bicliques of smaller dimension are found to be less sensitive to the smaller state size. The low-dimension attacks for Skein-512 lose two rounds when being translated to Skein-256 (Table 1).

The biclique construction, including trail details and partition of the state space, is very similar to that in Skein-512, so we do not give much details. The 2-round biclique yields 2- and 4-round attacks, which correspond to 5- and 7-round attacks on Skein-512. The 3-round biclique with dimension 17 yields an 8-round attack.

6 Certificational Preimage Attack on the Reduced Grøstl Output Transformation

In this section we present a certificational attack, i.e. it has only a small advantage over the exhaustive search, on Grøstl [11] — a SHA-3 finalist with a compression function not based on a block cipher. It invokes two permutations P and Q, both AES-based, and updates the chaining value CV as follows:

$$CV \leftarrow CV \oplus Q(M) \oplus P(M \oplus CV),$$

where M is a message block. The final call of the compression function is followed by the output transformation

$$F(x) = \text{Truncate}(x \oplus P(x)),$$

where the truncation operation takes half of the state to get 256- and 512-bit outputs. Hence Grøstl-256 operates on a 512-bit state and permutations P and Q, and Grøstl-512 operates on a 1024-bit state.

Permutations P and Q follow the AES design with very similar operations: SubBytes, ShiftBytes, MixBytes (8-byte analogue of MixColumns), and AddRoundConstant. The ShiftBytes operation in Grøstl-256 rotates i-th row by i positions to the left; details of the other operations are irrelevant for our attack. The sequence SubBytes–ShiftBytes–MixBytes–AddRoundConstant–SubBytes is equivalent to 8 (for Grøstl-256) parallel 64-bit Super S-boxes [12]. Due to the design simplicity, Grøstl has been the target of numerous cryptanalytic attacks [19, 21, 25], though only few of them violated collision or preimage resistance of the hash function [20, 30]. The paper [30] addresses virtually the same problem as we do, and obtains preimage attacks on the 5-round version of the compression function, including the preimage attack on the 5-round output transformation.

To run a preimage attack, and the first preimage attack in particular, it is desirable to invert the output transformation of Grøstl. As it is also claimed to be one-way, it serves as a natural target for sliced biclique attacks.

We adapt a differential view as it provides a simple explanation of the attack in differential trails, making it similar to both rebound attacks [19] and recent biclique attacks on AES. The main distinction is that there is no round without a difference because there is no schedule. However, the difference expansion in the outbound phase must be deterministic unless we have additional degrees of freedom in the inbound phase.

Attack. We denote the internal states within 6 rounds from #1 to #13, as depicted in Figure 5. We construct a sliced biclique of dimension 1 in states #4–#9, which contains the Super S-box layer in states #5–#8. The matching variable is a linear function of the variables in states #12 and #13 not affected by Δ- and ∇-differences. The Δ-difference has a single active byte, marked as lightblue. Its influence on the internal states within the matching part is also depicted as lightblue in Figure 5. The ∇-difference and its influence are depicted as green.

The matching condition is a linear function of the bytes not affected by the differences. Let us elaborate on this statement. Consider the rightmost columns of states #12 and #13 and denote them by A and B, respectively. Let us note that B is a linear function of A. In turn, 7 bytes of A do not depend on i, and 7 bytes of B do not depend on j. If the state $Q_{i,j}$ corresponds to a preimage, then a system of $8 - (8 - 7) - (8 - 7) = 6$ linear equations should hold. All the equations have form

$$A_j \oplus B_i = 0,$$

which is easily transformed to Equation (5).

We construct a sliced biclique based on interleaving trails. A biclique of dimension 1 is equivalent to a single boomerang quartet (Figure 2, left). In contrast to attacks on Skein, all the four relevant differences are distinct.

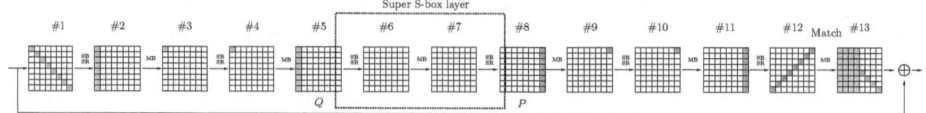

Fig. 5. Preimage attack on the reduced Grøstl-256 output transformation

Bicliques are constructed as follows. First, we arbitrarily choose $\Delta_{0,1} \neq \Delta_{1,1}$ and $\nabla_{1,0} \neq \nabla_{1,1}$, which are all active in one byte only, as specified earlier. We construct the states $\{Q_{i,j}\}_{i,j=0,1}$, which satisfy the following equations:

$$Q_{0,0} \oplus Q_{0,1} = \Delta_{0,1}; \quad Q_{1,0} \oplus Q_{1,1} = \Delta_{1,1}; \tag{9}$$
$$P_{0,0} \oplus P_{1,0} = \nabla_{1,0}; \quad P_{0,1} \oplus P_{1,1} = \nabla_{1,1}. \tag{10}$$

First, we derive the differences in #5 and in #8. Then we reformulate Equations (9) for each Super S-box, and solutions are found independently by exhaustive search with a total complexity around 2^{70}. The solutions are then concatenated. For the details, we refer to the long biclique attack on AES [8], which gives a description of an equivalent algorithm.

The complexity is amortized as follows. Each Super S-box has 7 inactive input S-boxes. There exist $2^{56-8} = 2^{48}$ alternative values for them which do not affect the active output S-box. Hence we can generate $2^{48 \cdot 8} = 2^{392}$ sliced bicliques out of a single one. As the hash value contains 256 bits only, we have enough freedom for the attack. For each biclique, i.e. 2^2 states, we recompute only a portion of the S-boxes in each round, with $2 \cdot (8 + 16 + 2 + 7 + 56 + 8) = 194$ S-boxes or 2^{-3} calls of the permutation. Hence the amortized cost of a single state test is 2^{-5}, and the total attack complexity is 2^{251}.

7 Message Compensation

The message compensation procedure [1, 16] instructs how to select message groups in the splice-and-cut attack in case of a strong, nonlinear message schedule. Existing applications are very ad-hoc and complicated. It is possible, however, to give a unified view on the message compensation problem and existing solutions with bicliques for permutations. The majority of this section is left for an extended version of this paper [14].

We propose the following algorithm a generic message schedule. Suppose you construct a biclique in rounds N_1–N_2, and want to describe a message set $\{M[i,j]\}$ such that

1. Injections $W^{N_0}, W^{N_0+1}, \ldots, W^{N_1-1}$ do not depend on j;
2. Injections $W^{N_2+1}, W^{N_2+2}, \ldots, W^{N_3}$ do not depend on i.

We propose to construct a sliced biclique without a matching point in rounds N_1–N_3. The difference Δ_j is defined before round N_1. To satisfy the first condition, we assign the words of Δ_j that correspond to $W^{N_0}, W^{N_0+1}, \ldots, W^{N_1-1}$ to zero. If some words are left undefined, then we get a freedom in these values and can use it to manipulate the difference propagation in the Δ-trails.

We define ∇_i after round N_3. To satisfy the second condition, we assign the words of ∇_i that correspond to $W^{N_2+1}, W^{N_2+2}, \ldots, W^{N_3}$ to zero. Again, if some words are undefined, we keep this freedom.

Finally, we construct a sliced biclique based on non-interleaving trails. We use undefined parts of Δ_j and ∇_i to control the diffusion on the word level, and select $M[0,0]$ to control the diffusion, if necessary, on the bit level. We may also choose other round indices for a biclique, if this makes the difference selection more clear. We may also have to deal with other constraints like padding, which further reduce the freedom in Δ and ∇. Finally, we may have to construct a biclique based on interleaving trails, if non-interleaving ones are impossible because of the diffusion.

8 Conclusions

We have introduced sliced bicliques as a new tool for the analysis of permutations in the context of preimage and collision attacks. We have demonstrated that the advantage in the number of rounds from the long biclique idea can be obtained also for permutations. The application of our concept to different design has interesting consequences.

First, our collision attacks on Skein demonstrate that the MMO mode may not be as resistant to collision attacks and the differential cryptanalysis in particular as it was considered. The fundament of our attacks is the new pseudo-collision search technique that has been recently introduced. Though we employ some elements of differential cryptanalysis, the details are completely different from the famous collision attacks on the SHA family. Hence we suppose that the potential of differential cryptanalysis for high-profile hash functions has not been exhausted.

Secondly, our preimage attacks on the Grøstl output transformation show that the concept of the Super S-box contributes not only to the biclique attacks on the designs with the key schedule (AES), but also on the ones without the schedule. We expect this type of attack to progress alongside with the future techniques for the Super S-box.

Finally, we explained the message compensation in the biclique terms. We expect that the designers of future meet-in-the-middle attacks on SHA-2 will be able to provide a compact two-step description of their results. First, a biclique in the schedule is constructed, and secondly, it is used to construct a biclique in the state. We are looking forward to new techniques that would combine these bicliques in an optimal way.

We leave a significant amount of targets for the future work. 7-round Grøstl-256, 9- and 10-round Grøstl-512, Whirlpool, BLAKE are natural targets. Construction of bicliques of high dimension out of interleaving trails remains an open problem.

Acknowledgements. The author thanks Maria Naya-Plasencia, Andrey Labunets, and Alexander Shalimov for fruitful discussions on the paper topic and possible improvements. The author also thanks reviewers of the paper, whose comments helped to improve the readability. Any further review or comment is warmly welcome.

References

1. Aoki, K., Guo, J., Matusiewicz, K., Sasaki, Y., Wang, L.: Preimages for Step-Reduced SHA-2. In: Matsui, M. (ed.) ASIACRYPT 2009. LNCS, vol. 5912, pp. 578–597. Springer, Heidelberg (2009)
2. Aoki, K., Sasaki, Y.: Preimage Attacks on One-Block MD4, 63-Step MD5 and More. In: Avanzi, R.M., Keliher, L., Sica, F. (eds.) SAC 2008. LNCS, vol. 5381, pp. 103–119. Springer, Heidelberg (2009)
3. Aoki, K., Sasaki, Y.: Meet-in-the-Middle Preimage Attacks Against Reduced SHA-0 and SHA-1. In: Halevi, S. (ed.) CRYPTO 2009. LNCS, vol. 5677, pp. 70–89. Springer, Heidelberg (2009)
4. Aumasson, J.-P., Çalık, Ç., Meier, W., Özen, O., Phan, R.C.-W., Varıcı, K.: Improved Cryptanalysis of Skein. In: Matsui, M. (ed.) ASIACRYPT 2009. LNCS, vol. 5912, pp. 542–559. Springer, Heidelberg (2009)
5. Biham, E., Chen, R.: Near-Collisions of SHA-0. In: Franklin, M. (ed.) CRYPTO 2004. LNCS, vol. 3152, pp. 290–305. Springer, Heidelberg (2004)
6. Biryukov, A., Khovratovich, D.: Related-Key Cryptanalysis of the Full AES-192 and AES-256. In: Matsui, M. (ed.) ASIACRYPT 2009. LNCS, vol. 5912, pp. 1–18. Springer, Heidelberg (2009)
7. Black, J., Rogaway, P., Shrimpton, T.: Black-Box Analysis of the Block-Cipher-Based Hash-Function Constructions from PGV. In: Yung, M. (ed.) CRYPTO 2002. LNCS, vol. 2442, pp. 320–335. Springer, Heidelberg (2002)
8. Bogdanov, A., Khovratovich, D., Rechberger, C.: Biclique Cryptanalysis of the Full AES. In: Lee, D.H. (ed.) ASIACRYPT 2011. LNCS, vol. 7073, pp. 344–371. Springer, Heidelberg (2011)
9. Diffie, W., Hellman, M.: Special feature exhaustive cryptanalysis of the NBS Data Encryption Standard. Computer 10, 74–84 (1977)
10. Ferguson, N., Lucks, S., Schneier, B., Whiting, D., Bellare, M., Kohno, T., Callas, J., Walker, J.: The Skein hash function family. Submission to NIST, Round 3 (2010), http://www.skein-hash.info/sites/default/files/skein1.3.pdf
11. Gauravaram, P., Knudsen, L.R., Matusiewicz, K., Mendel, F., Rechberger, C., Schläffer, M., Thomsen, S.S.: Grøstl – a SHA-3 candidate. Submission to NIST (2008), http://www.groestl.info/Groestl.pdf
12. Gilbert, H., Peyrin, T.: Super-Sbox Cryptanalysis: Improved Attacks for AES-Like Permutations. In: Hong, S., Iwata, T. (eds.) FSE 2010. LNCS, vol. 6147, pp. 365–383. Springer, Heidelberg (2010)
13. Jia, K., Yu, H., Wang, X.: A meet-in-the-middle attack on the full KASUMI. Cryptology ePrint Archive, Report 2011/466 (2011)
14. Khovratovich, D.: Bicliques for permutations: collision and preimage attacks in stronger settings. Cryptology ePrint Archive, Report 2012/141 (2012), http://eprint.iacr.org/2012/141
15. Khovratovich, D., Leurent, G., Rechberger, C.: Narrow-Bicliques: Cryptanalysis of Full IDEA. In: Pointcheval, D., Johansson, T. (eds.) EUROCRYPT 2012. LNCS, vol. 7237, pp. 392–410. Springer, Heidelberg (2012)

16. Khovratovich, D., Rechberger, C., Savelieva, A.: Bicliques for Preimages: Attacks on Skein-512 and the SHA-2 Family. In: Canteaut, A. (ed.) FSE 2012. LNCS, vol. 7549, pp. 244–263. Springer, Heidelberg (2012), http://eprint.iacr.org/2011/286.pdf

17. Li, J., Isobe, T., Shibutani, K.: Converting Meet-In-The-Middle Preimage Attack into Pseudo Collision Attack: Application to SHA-2. In: Canteaut, A. (ed.) FSE 2012. LNCS, vol. 7549, pp. 264–286. Springer, Heidelberg (2012)

18. Mala, H.: Biclique cryptanalysis of the block cipher Square. Cryptology ePrint Archive, Report 2011/500 (2011), http://eprint.iacr.org/

19. Mendel, F., Rechberger, C., Schläffer, M., Thomsen, S.S.: The Rebound Attack: Cryptanalysis of Reduced Whirlpool and Grøstl. In: Dunkelman, O. (ed.) FSE 2009. LNCS, vol. 5665, pp. 260–276. Springer, Heidelberg (2009)

20. Mendel, F., Rechberger, C., Schläffer, M., Thomsen, S.S.: Rebound Attacks on the Reduced Grøstl Hash Function. In: Pieprzyk, J. (ed.) CT-RSA 2010. LNCS, vol. 5985, pp. 350–365. Springer, Heidelberg (2010)

21. Peyrin, T.: Improved Differential Attacks for ECHO and Grøstl. In: Rabin, T. (ed.) CRYPTO 2010. LNCS, vol. 6223, pp. 370–392. Springer, Heidelberg (2010)

22. Sasaki, Y.: Meet-in-the-Middle Preimage Attacks on AES Hashing Modes and an Application to Whirlpool. In: Joux, A. (ed.) FSE 2011. LNCS, vol. 6733, pp. 378–396. Springer, Heidelberg (2011)

23. Sasaki, Y., Aoki, K.: Preimage Attacks on Step-Reduced MD5. In: Mu, Y., Susilo, W., Seberry, J. (eds.) ACISP 2008. LNCS, vol. 5107, pp. 282–296. Springer, Heidelberg (2008)

24. Sasaki, Y., Aoki, K.: Finding Preimages in Full MD5 Faster Than Exhaustive Search. In: Joux, A. (ed.) EUROCRYPT 2009. LNCS, vol. 5479, pp. 134–152. Springer, Heidelberg (2009)

25. Sasaki, Y., Li, Y., Wang, L., Sakiyama, K., Ohta, K.: Non-full-active Super-Sbox Analysis: Applications to ECHO and Grøstl. In: Abe, M. (ed.) ASIACRYPT 2010. LNCS, vol. 6477, pp. 38–55. Springer, Heidelberg (2010)

26. Su, B., Wu, W., Wu, S., Dong, L.: Near-Collisions on the Reduced-Round Compression Functions of Skein and BLAKE. In: Heng, S.-H., Wright, R.N., Goi, B.-M. (eds.) CANS 2010. LNCS, vol. 6467, pp. 124–139. Springer, Heidelberg (2010)

27. van Oorschot, P.C., Wiener, M.J.: Parallel collision search with cryptanalytic applications. J. Cryptology 12(1), 1–28 (1999)

28. Wagner, D.: The Boomerang Attack. In: Knudsen, L.R. (ed.) FSE 1999. LNCS, vol. 1636, pp. 156–170. Springer, Heidelberg (1999)

29. Wang, X., Yin, Y.L., Yu, H.: Finding Collisions in the Full SHA-1. In: Shoup, V. (ed.) CRYPTO 2005. LNCS, vol. 3621, pp. 17–36. Springer, Heidelberg (2005)

30. Wu, S., Feng, D., Wu, W., Guo, J., Dong, L., Zou, J.: (Pseudo) Preimage Attack on Reduced-Round Grøstl Hash Function and Others. In: Canteaut, A. (ed.) FSE 2012. LNCS, vol. 7549, pp. 127–145. Springer, Heidelberg (2012)

Investigating Fundamental Security Requirements on Whirlpool: Improved Preimage and Collision Attacks

Yu Sasaki[1], Lei Wang[2,3], Shuang Wu[4,*], and Wenling Wu[4]

[1] NTT Corporation
[2] The University of Electro-Communications
[3] Nanyang Technological University
wushuang@is.iscas.ac.cn
[4] Institute of Software, Chinese Academy of Sciences

Abstract. In this paper, improved cryptanalyses for the ISO standard hash function Whirlpool are presented with respect to the fundamental security notions. While a subspace distinguisher was presented on full version (10 rounds) of the compression function, its impact to the security of the hash function seems limited. In this paper, we discuss the (second) preimage and collision attacks for the hash function and the compression function of Whirlpool. Regarding the preimage attack, 6 rounds of the hash function are attacked with 2^{481} computations while the previous best attack is for 5 rounds with $2^{481.5}$ computations. Regarding the collision attack, 8 rounds of the compression function are attacked with 2^{120} computations, while the previous best attack is for 7 rounds with 2^{184} computations. To verify the correctness, especially for the rebound attack on the Sbox with an unbalanced Differential Distribution Table (DDT), the attack is partially implemented, and the differences from attacking the Sbox with balanced DDT are reported.

Keywords: Whirlpool, preimage, collision, meet-in-the-middle, guess-and-determine, local collision.

1 Introduction

Hash functions are taking important roles in various aspects of modern cryptography. Since the collision resistance of MD5 and SHA-1 has been broken by Wang *et al.* [1,2], cryptographers have looked for stronger hash function designs. While various new designs are discussed in the SHA-3 competition [3], some of existing hash functions seem to be much stronger than the MD4-family. Evaluating such hash functions is useful especially if they have been standardized internationally.

For hash functions, three security notions are classically considered: Collision Resistance, Second-Preimage Resistance, and Preimage Resistance. Besides,

---

* Corresponding author.

X. Wang and K. Sako (Eds.): ASIACRYPT 2012, LNCS 7658, pp. 562–579, 2012.

cryptographers recently have considered various non-ideal properties. Although considering such non-ideal properties is important especially for determining a new standard, focusing on vulnerabilities that can be exploited in practice is more important especially for evaluating hash functions in practice.

Whirlpool [4] is a 512-bit hash function proposed by Rijmen and Barreto in 2000. The compression function uses a 10-round AES based cipher with 8∗8-byte internal states, and the output is computed with the Miyaguchi-Preneel mode [5, Algorithm 9.43]. Whirlpool has been adopted by ISO [6] and NESSIE [7].

Regarding the collision attack, the rebound attack proposed by Mendel et al. [8] is very effective with respect to the differential attack against AES based structure. Indeed, Mendel et al. presented a 4-round collision attack on the hash function and a 5-round collision attack on the compression function of Whirlpool. Many improved techniques of the rebound attack have been devised such as start-from-the-middle technique [9], linearized match-in-the-middle technique [9], super-(S)box analysis [10,11], and multiple-inbound technique [11,12]. Besides, for the AES based structure with 8 ∗ 8 state including Whirlpool, more techniques have been proposed such as hyper-Sbox analysis [13], non-full-active super-Sbox analysis [14], efficient list-merging technique [15], and three inbound rounds [16]. Several practical results are given for round-reduced algorithms and intermediate rounds in [9,17,18]. This paper exploits the differences in both of data processing part and key schedule part. Some similarities can be seen in the analysis on AES-256 [19] and two analysis on Grøstl [20,21].

Regarding the preimage attack, meet-in-the-middle (MitM) attack with the splice-and-cut technique proposed by Aoki and Sasaki [22] has been actively discussed. Several papers proposed improved techniques [23,24]. For the preimage attack against the AES based structure, Sasaki showed a second preimage attack on 5 rounds of Whirlpool [25]. Later, Wu et al. improved its complexity and extended it to the preimage attack [26]. Note that Bogdanov et al. showed an attack on 10-round AES in hashing modes with the biclique technique [27]. Because this attack exploits the weakness in the AES key-schedule, the attack is specific to AES and cannot be directly applied to other AES based primitives.

Our Contributions. In this paper, we improve cryptanalyses on Whirlpool with respect to the fundamental security notions. The main results are a 6-round preimage attack on the hash function and an 8-round collision attack on the compression function. The results are summarized in Table 1.

Our preimage attack is based on the previous 5-round MitM attacks [25,26]. The number of attacked rounds is extended by applying the guess-and-determine approach during the MitM attack. Moreover, we increase the number of free bits for each chunk by exploiting the freedom degrees of the key, while previous attacks fix the key as a constant. More precisely, the key schedule function shares the same round function with the data process procedure, and thus we separate the key schedule function in the same way with the data process function.

Our collision attack is based on the rebound attack. We use the key difference to cancel the difference in the data part, while previous work avoided inserting differences to the key schedule. This leads to a differential path with a high

Table 1. Summary of attack results

Type		Target	#Rounds	Time	Mem.	Ref.	Remarks
Fundamental Properties	Preimage	Hash Function	5	$2^{481.5}$	2^{64}	[26]	
			5	2^{448}	2^{96}	Ours	
			5	2^{465}	$O(1)$	Ours	Memoryless MitM
			6	2^{481}	2^{256}	Ours	
			6	2^{504}	$O(1)$	Ours	Memoryless MitM
	Second Preimage	Hash Function	5	2^{504}	2^8	[25]	
			5	2^{448}	2^{64}	[26]	
			5	2^{464}	$O(1)$	Ours	Memoryless MitM
			6	2^{481}	2^{256}	Ours	
			6	2^{504}	$O(1)$	Ours	Memoryless MitM
	Collision	Hash Function	4	2^{64}	2^8	[9]	
			5	2^{120}	2^{64}	[10]	
		Compress. Function	7	2^{184}	2^8	[28]	semi-free-start
			7	2^{120}	2^{128}	[28]	semi-free-start
			4	2^8	2^8	Ours	free-start
			7	2^{64}	2^8	Ours	free-start
			8	2^{120}	2^8	Ours	free-start
Other Properties	Near-collision	Compress. Function	9	2^{176}	2^8	[28]	
			9	2^{112}	2^{128}	[28]	
	Distinguisher	Compress. Function	10	2^{188}	2^8	[28]	
			10	2^{121}	2^{128}	[28]	

probability. In this paper, we implement our 4-round collision attack which only requires 2^8 computations. Because all previous collision attacks require at least 2^{64} computations even for a small number of rounds, this is the first example of the collision for a reduced compression function. We also partially implement the 7-round collision attack. We show an example of the 40-byte near-collision.

2   Specification and Notations

Whirlpool [4] takes any message with less than 2^{256} bits as input, and outputs a 512-bit hash value. It adopts the Merkle-Damgård structure. The input message M is padded into a multiple of 512 bits. In details, the 256-bit binary expression of the bit length ℓ is padded according to the MD-strengthening, *i.e.* $M\|1\|0^*\|\ell$. The padded message is divided into 512-bit blocks $M_0\|M_1\|\cdots\|M_{N-1}$. Let H_n be a 512-bit chaining variable. First, an initial value IV is assigned to H_0. Then, $H_{n+1} \leftarrow \mathrm{CF}(H_n, M_n)$ is computed for $n = 0, 1, \ldots, N-1$, where CF is a compression function. H_N is produced as the hash value of M. CF uses an AES based block-cipher E_k, which takes a 512-bit chaining variable H_i as a key and a 512-bit message block M_i as a plaintext. The output of CF is computed by the Miyaguchi-Preneel mode, *i.e.* $E_{H_i}(M_i) \oplus M_i \oplus H_i$.

Inside the block cipher E_k, an internal state is represented by an $8 * 8$ byte array. At first, H_i is assigned to the key value k_0. Then, ten 512-bit subkeys $k_1, k_2, \ldots, k_{10}$ are generated by the key-schedule function defined as follows:

$$k_{n+1} \leftarrow \mathrm{AC} \circ \mathrm{MR} \circ \mathrm{SC} \circ \mathrm{SB}(k_n), \text{ for } n = 0, 1, \ldots, 9.$$

- SubBytes(SB): applies the Substitution-Box to each byte.
- ShiftColumns(SC): cyclically shift the j-th column downwards by j bytes.
- MixRows(MR): multiply each row of the state matrix by an MDS matrix.
- AddRoundConstant(AC): XOR a 512-bit constant defined in the specification.

For the data processing part, M_i is assigned to the plaintext p. Then, the whitening operation is performed and the result is stored into a variable s_0, i.e. $s_0 \leftarrow k_0 \oplus p$. The output s_{10} of the block cipher is computed as follows, where AddRoundKey(AK) takes the XOR with k_{n+1}.

$$s_{n+1} \leftarrow \text{AK} \circ \text{MR} \circ \text{SC} \circ \text{SB}(s_n), \text{ for } n = 0, 1, \ldots, 9.$$

Notations. Byte positions in a state S are denoted by integer numbers $0, 1, \ldots,$ 63, where the byte $8j + i$ corresponds to the byte in the i-th row and the j-th column of the state $\#S$, and is denoted by $\#S[8j+i]$. We denote the initial state for the data processing part in round x by $\#Dx^I$. Then, states immediately after SB, SC, MR, and AR in round x are denoted by $\#Dx^{SB}$, $\#Dx^{SC}$, $\#Dx^{MR}$, and $\#Dx^{AK}$, respectively. Obviously, $\#Dx^{AK}$ is identical with $\#D(x+1)^I$. Similarly, we use the notations $\#Kx^I, \#Kx^{SB}, \#Kx^{SC}, \#Kx^{MR}$, and $\#Kx^{AC}$ for the key schedule part. We often denote several bytes of state $\#S$ by $\#S[a, b, \ldots]$, e.g. 8 bytes in the right most column are denoted by $\#S[56, 57, \ldots, 63]$. We also use the following notations to denote specific byte positions.

- $\#S[row(i)]$: 8 byte-positions in the i-th row of state $\#S$
- $\#S[\text{SC}(row(i))]$: 8 byte-positions which SC is applied to $\#S[row(i)]$
- $\#S[\text{SC}^{-1}(row(i))]$: 8 byte-positions which SC^{-1} is applied to $\#S[row(i)]$

3   Related Work

3.1   Meet-in-the-Middle (Second) Preimage Attack on Whirlpool

In FSE 2011, Sasaki proposed the first MitM preimage attack on AES-like primitives [25]. Two main techniques were introduced: initial structure in an AES-like permutation and partial-matching across an MixColumn operation. As a direct application, a second preimage attack is found on 5-round Whirlpool hash function in [25]. In FSE 2012, Wu et al. improved the complexity of 5-round second preimage attack on Whirlpool [26] by exploiting more freedom degrees in the data state. They successfully represent the chunk separations by several essential integer parameters, and launched an automatic exhaustive search. Moreover, they also proposed a method to deal with the message padding and extended the attack into a first preimage attack.

3.2   Rebound Attack and Start-from-the-Middle Technique

The rebound attack was introduced by Mendel et al. [8]. If it is applied to Whirlpool, the 2-round path $8 \rightarrow 64 \rightarrow 8$ can be satisfied only with 2^8 computations. The path for rounds S and $S + 1$ is described in Fig. 1. First, an 8-byte

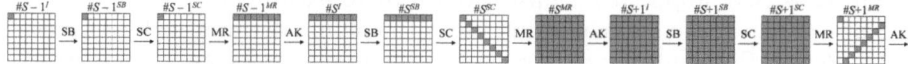

Fig. 1. Rebound and start-from-the-middle techniques

difference at $\#S + 1^{MR}$ is randomly chosen, and it is propagated to $\#S + 1^{SB}$. Then, a single-byte difference at one of the active bytes at $\#S^{SB}$ is randomly chosen, and it is propagated to 8 bytes of $\#S + 1^I$. For each S-box in round $S + 1$, randomly given input and output differences have solutions (paired values conforming the path) with probability about 2^{-1}, and the average number of solutions is 2. Hence, if we choose 2^8 differences for the single byte at $\#S^{SB}$, we obtain 2^8 solutions for the corresponding 8 S-boxes. By iterating it for 8 active bytes at $\#S^{SB}$, we obtain 2^8 solutions for each i of $\#S^{SB}[SC^{-1}(row(i))]$.

The start-from-the-middle technique is an improved procedure for the rebound attack, which was proposed by Mendel $et\ al.$ [9]. It satisfies a 3-round differential path with the same complexity as the rebound attack. After obtaining 2^8 solutions for each i of $\#S^{SB}[SC^{-1}(row(i))]$ with the rebound attack, each solution is computed until $\#S - 1^{MR}[SC^{-1}(row(i))]$. For each i, 127 kinds of differences are obtained at $\#S - 1^{MR}$. Then, a single-byte difference at $\#S - 1^{SB}$ is chosen. The attacker propagates it to $\#S - 1^{MR}$, and checks whether the 8-byte difference can be produced from the solutions of the rebound attack. Because there are 127 kinds of the differences for each i, the 8-byte differences can be produced with probability about 2^{-8}. Therefore, by choosing 2^8 differences at $\#S - 1^{SB}$, we expect to find the desired difference. In summary, the 3-round differential path $1 \rightarrow 8 \rightarrow 64 \rightarrow 8$ can be satisfied with a complexity of 2^8.

Note that the behavior of the S-box is explained based on the S-box of AES. Because the S-box of Whirlpool has a different property, the evaluation for AES cannot be applied to Whirlpool directly. We later discuss this issue in Sect. 5.4.

3.3   Distinguisher for the Full Whirlpool Compression Function

Lamberger $et\ al.$ proposed a distinguisher for the full Whirlpool compression function [11,28]. The distinguished property is called subspace distinguisher. The dimension of the input and output differences are defined before the analysis starts. The attacker aims to find paired values whose dimension of differences at input and output are lower than the defined ones. The core technique is running the rebound attack ($8 \rightarrow 64 \rightarrow 8$) at two parts independently without determining the key value. Then, two results are connected and a long differential path ($8 \rightarrow 64 \rightarrow 8 \rightarrow 8 \rightarrow 64 \rightarrow 8$) is satisfied by searching for an appropriate key value. Although the distinguisher beautifully breaks the full-round compression function, the impact is very limited. Nevertheless, collisions on compression function are generated with this technique for 7 rounds with (Time, Memory)$= (2^{184}, 2^8)$ or $(2^{120}, 2^{128})$.

3.4   Local Collision on AES-Like Primitives

For a distinguisher for AES-256, Biryukov *et al.* introduced differences to both the key and the data, and used the difference of round keys to cancel the difference of the internal states of the data process by the AddRoundKey operation, i.e. the local collision occurs [19]. The local collisions may help the attacker to build a high probability differential path on AES-like primitives.

4   Preimage Attack on 6-Round Whirlpool

4.1   Overview

Our first and main result is introducing the guess-and-determine approach to MitM preimage attack on Whirlpool hash function, and successfully increase one more attacked round. More specifically, during the independent chunk computation, even one unknown input byte of $MixRow$ makes all the 8 output bytes unknown, which is heavily unbalanced. So a chunk can guess a small number of unknown bytes in order to significantly increase the number of known bytes in the following rounds. Thus guess-and-determine approach is very effective for preimage attack on Whirlpool.

Our second result is exploiting the freedom degree in the key to increase the number of free bits in each chunk, and thus successfully reduce the complexity. Since the key schedule of Whirlpool is the same with the data process, we can separate the key schedule and the data process into two chunks in the same way, which doubles the number of free bits in both chunks.

Our third result is that we propose not only a first preimage attack on hash function with the lowest complexity, but also another memoryless preimage attack. Compared to the brute force attack, the second attack requires the same memory and a lower complexity. This is achieved by finding a last block attack first and then linking the chaining values with a fixed-key attack on the compression function. Since both the last block attack and the fixed-key attack can be implemented in a memoryless way [30], we obtain a memoryless first preimage attack.

4.2   Preimage Attack on 6-Round Compression Function

The chunk separation used in the 6-round attack is illustrated in Fig. 2. Five different colors are used to indicate the categories of the bytes. The gray bytes are constants which come from the hash/output value or the initial structure. The red/blue bytes belong to the backward/forward chunk, which can be determined by the red/blue byte in the initial structure. The white bytes are affected by both red and blue bytes and we can only determine their values after a partial match is found. The purple bytes are the guessed bytes.

Since that each row of the state $\#D1^{MR}$ has unknown bytes (in white color), if we went further back through MR^{-1}, all bytes would become unknown. The values of 24 white bytes in row 0 to row 5 are guessed. Thus we can maintain 6

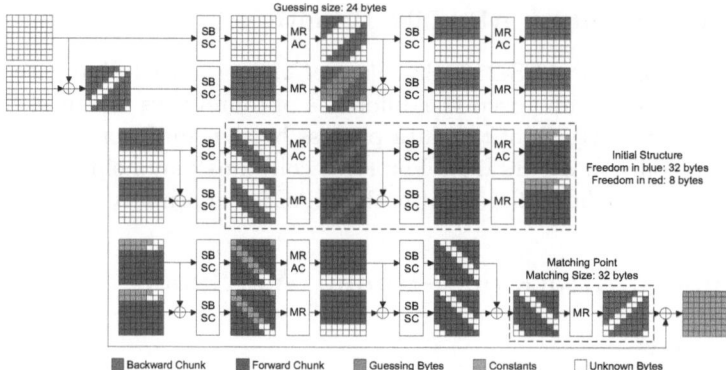

Fig. 2. Chunk separation for the preimage attack on 6-round compression function

red bytes in each of top 6 rows of the state $\#D1^{SC}$. In the key state, we do not guess the values, since the white key state $\#K1^{SC}$ does not affect the matching state through the feedforward operation.

All the possible values of the guessed bytes are used as extra freedom degrees to build the lookup table for the MitM. But after a partial match is found, we need to further check the correctness of the guessed values. More details about the guessing technique can be found in the following section.

The Attack Algorithm. In order to evaluate the attack complexity, we need to know the parameters: freedom degrees in red and blue bytes (D_r, D_b), size of the partial matching m and the number of guessed bits D_g. The explanation on calculating freedom degrees/size of matching point and how the partial matching works can be found in previous papers [25,26]. Here we omit these details due to the limited space.

To summarize, the parameters for MitM attack in Fig. 2 are as follows. Freedom degrees in red bytes: $D_r = 8$ bytes $= 64$ bits (4 bytes in the key and 4 bytes in the data). Freedom degrees in blue bytes: $D_b = 32$ bytes $= 256$ bits (16 bytes in the key and 16 bytes in the data). Size of the guessed value (purple bytes): $D_g = 24$ bytes $= 192$ bits. Size of the partial match: $m = 32$ bytes $= 256$ bits (only in the data). Size of the full match: $n = 512$ bits.

The attack algorithm is as follows:

Step 1. Randomly choose the values of the constants in the initial structure.
Step 2. For all the 2^{D_r} values $\{r_i\}$ of the red bytes in the initial structure and 2^{D_g} guessed values $\{g_j\}$, go backward to the matching point and store all $2^{D_r+D_g}$ partial matching values $F(r_i, g_j)$ in a look-up table L.
Step 3. For all the 2^{D_b} values $\{b_k\}$ of the blue bytes in the initial structure, go forward to obtain the partial matching value $G(b_k)$ and check if it is in L.

Step 4. Once a partial match (r_i, g_j, b_k) such that $F(r_i, g_j) = G(b_k)$ is found, use (r_i, b_k) to compute and check if the guessed value g_j is correct. If the guess is correct, check if it is a preimage.

Step 5. Repeat the above steps 1-4 to find a preimage.

The complexity is explained as follows:

Step 2. It takes $2^{D_r+D_g}$ computations and memory to build the look-up table.

Step 3. It takes 2^{D_b} computations to find all the $2^{D_b+D_r+D_g-m}$ partial matches.

Step 4. $2^{D_b+D_r+D_g-m}$ computations are needed to verify the correctness for all the partial matches. There would be $2^{D_b+D_r-m}$ valid partial matches that pass the correctness test, since the probability that g_j is correct is 2^{-D_g}.

Step 5. The probability that steps 1-4 succeed is $2^{D_b+D_r-m} \cdot 2^{-(n-m)} = 2^{D_b+D_r-n}$. The above steps are repeated for $2^{n-D_b-D_r}$ times to find a preimge.

Therefore, the complexity of the above algorithm is

$$2^{n-D_b-D_r} \cdot (2^{D_r+D_g} + 2^{D_b} + 2^{D_b+D_r+D_g-m}) = 2^n \cdot (2^{-D_r} + 2^{D_g-D_b} + 2^{D_g-m}) \quad (1)$$

With the given parameters, the complexity is about $2^{512} \cdot (2^{-64} + 2^{192-256} + 2^{192-256}) \approx 2^{448}$ compression function calls. Only step 2 requires $2^{64+192} = 2^{256}$ memory.

It is observed that the pattern for the chunk separation can be represented as several numbers: $b=$ the number of blue rows in $\#D2^{MR}$, $r=$ the number of red rows in $\#D2^I$, $w=$the number of white rows in $\#D5^{SC}$, $g=$the number of guessed rows in $\#D1^{MR}$. Then the parameters for the MitM attack can be calculated as: $D_b = 16(b - r)$ bytes, $D_r = 2w(8 - b)$ bytes, $D_g = g(8 - r)$ bytes and $m = 8(g + (8 - w) - 8) = 8(g - w)$ bytes. In the following sections, we will continue using the parameters of b, r, w and g to identify the pattern for chunk separations. We searched for all the possible patterns of the chunk separation by exhaustively enumerating the parameters b, r, w and g. Fig. 2 shows the optimal complexity case $(b, r, w, g = 6, 4, 2, 6)$. Note that the 6-round attack is also applicable without using freedom degrees of the key.

Memoryless MitM Attacks. In [30], Morita *et al.* proposed the memoryless MitM technique, which can be applied in our attack by designing the following three functions:

1) a mapping from the partial matching value to the blue value,
2) a mapping from the partial matching value to the red (and purple) value,
3) a pseudo-random boolean switching function taking the partial matching value as the input.

However, we found that the memoryless MitM has some limitations. The memoryless MitM is very efficient to find one match, its complexity is limited by half of the matching size m and increases linearly with the number of matches. Namely, at most $2^{max\{0,min\{D_b-D_g,D_r,m/2-D_g\}\}}$ computations can be saved using memoryless MitM. Using look-up tables, we can save at most $2^{max\{0,min\{D_b-D_g,D_r,m-D_g\}\}}$ computations. This difference results in different optimal chunk separations, which is considered in the following sections.

4.3   The First Preimage Attacks

A first preimage attack is the combination of a second preimage attack and an attack on the last compression function which produces message block with correct padding. In order to find optimal first preimage attacks, we need to consider a lot of different attacks.

Two Types of Last Block Attacks. The first preimage attack must fulfill the message length padding. In a fixed-key attack on the compression function, 10 padding bits can be chosen if the initial structure is placed at the beginning of the encryption. This technique was used in [26]. The probability that a random message block satisfies a constraint of the padding string is 2^{-9}. Details are explained in Appendix B.

In the chosen-key preimage attacks, the initial structure cannot be placed at the beginning of the compression function. So the chosen padding technique is not applicable. However, we can repeat the attack 2^9 times to obtain a valid last message block.

Since Whirlpool uses 256-bit length padding and we just satisfied a small part of it, the rest part of the length cannot be known before the attack. Therefore, we need the expandable messages [31] to fulfill it.

Two Types of Second Preimages. In previous attacks, the key (chaining value) is known before the attack. The preimage attack on the compression function is to find a message block that connect two chaining values. The fixed-key attack is equivalent to a second preimage attack if the input and output chaining values are chosen consecutively from the known ones.

If the key is chosen, the value of the key (chaining value) can only be determined after the attack. Then we need to connect it to one of the known chaining value. This is done using a MitM step on the chaining values.

Different Combinations for the First Preimage Attack. First, we analyzed all the 5/6-round fixed-/chosen-key attacks on compression functions and turn them into second preimage and last-block attacks. Second, we considered the fixed-key attacks with chosen padding and found more attacks on the last message block. At last, we combine the second preimage attacks and the last-block attacks to found the first preimage attacks with the lowest computations and the lowest memory respectively.

The detailed results of all preimage attacks are summarized in Table 2. Note that we can adjust the time-memory tradeoff by choosing different combinations of second preimages and the last-block attacks or changing the tradeoff of MitM on the chaining value for chosen key attacks.

Table 2. Detailed Results on all preimage attacks on CF and hash function

5-Round Attacks		b r w	D_b D_r m	Compression Function	Second Preimage	Last Block	Preimage
chosen-key	-	5 4 2	128 96 128	$2^{416}, 2^{96}$	$2^{465}, 2^{96}$	$2^{425}, 2^{96}$†	$2^{448}, 2^{96}$†
	ml	4 3 1	128 64 128	$2^{448}, O(1)$	$2^{481}, 2^{32}$	$2^{457}, 2^{32}$	
fixed-key	-	4 3 2	64 64 64	$2^{448}, 2^{64}$	$2^{448}, 2^{64}$†	$2^{457}, 2^{64}$	$2^{465}, O(1)$‡
	ml	5 4 2	64 48 128	$2^{464}, O(1)$	$2^{464}, O(1)$‡	$2^{473}, O(1)$	
fixed-key	-	4 3 2	55 63 64	-	-	$2^{457}, 2^{55}$	
chosen padding	ml	5 4 2	54 48 128	-	-	$2^{464}, O(1)$‡	

6-Round Attacks		b r w g	D_b D_r D_g m	Compression Function	Second Preimage	Last Block	Preimage
chosen-key	-	6 4 2 6	256 64 192 256	$2^{448}, 2^{256}$	$2^{481}, 2^{256}$†	$2^{457}, 2^{256}$†	$2^{481}, 2^{256}$†
	ml	7 6 2 6	128 32 96 256	$2^{480}, O(1)$	$2^{497}, 2^{16}$	$2^{489}, O(1)$‡	
fixed-key	-	6 5 1 2	64 16 48 64	$2^{496}, 2^{64}$	$2^{496}, 2^{64}$	$2^{505}, 2^{64}$	$2^{504}, O(1)$‡
	ml	7 5 1 5	128 8 120 256	$2^{504}, O(1)$	$2^{504}, O(1)$‡	$2^{513}, O(1)$	
fixed-key	-	6 4 1 3	118 16 96 128	-	-	$2^{496}, 2^{112}$	
chosen padding	ml	7 6 1 3	54 8 48 128	-	-	$2^{506}, O(1)$	

† : The attacks with the lowest computations.
‡ : The attacks with the lowest memory.
ml : The memoryless MitM attacks.

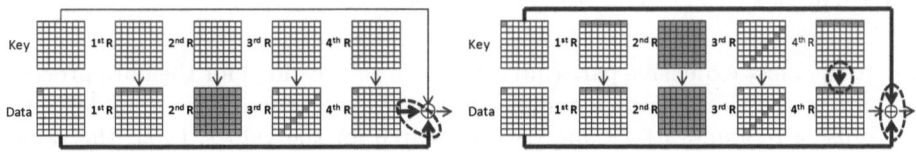

Fig. 3. Left: previous approach Right: our approach

5 Collision Attacks on the Compression Function

5.1 Overview

In order to generate collisions with previous rebound approaches, the state at the beginning and the end must have the same differential form so that they can cancel each other with the feed-forward operation. This is a strong constraint. We overcome this constraint by generating local collisions several times, *i.e.*, canceling differences of the data by using differences of the key. The idea is illustrated in Fig. 3. Because the diffusions for the data and key are identical, we can keep the same differential form. This makes possible to use the differential path with different differential forms between the beginning and the end.

The idea of using the key difference is advantageous not only for canceling the output difference but also constructing a high probability differential path by using the local collision. For example, we use the following differential path for an 8-round collision attack. Here, "WH" represents the whitening operation.

$$\text{Key: } 64 \xrightarrow{\text{WH}} 64 \xrightarrow{1^{st}R} 8 \xrightarrow{2^{nd}R} 1 \xrightarrow{3^{rd}R} 8 \xrightarrow{4^{th}R} 64 \xrightarrow{5^{th}R} 8 \xrightarrow{6^{th}R} 1 \xrightarrow{7^{th}R} 8 \xrightarrow{8^{th}R} 64,$$

$$\text{Data: } 64 \xrightarrow{\text{WH}} 0 \xrightarrow{1^{st}R} 8 \xrightarrow{2^{nd}R} 1 \xrightarrow{3^{rd}R} 8 \xrightarrow{4^{th}R} 0 \xrightarrow{5^{th}R} 8 \xrightarrow{6^{th}R} 1 \xrightarrow{7^{th}R} 8 \xrightarrow{8^{th}R} 0, \quad (2)$$

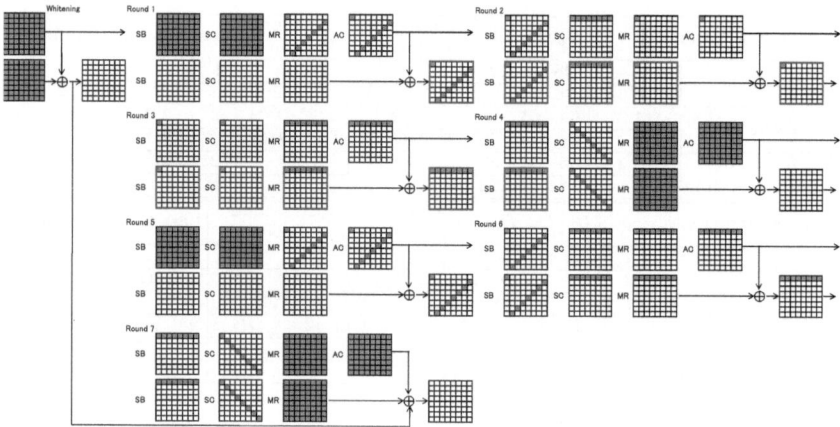

Fig. 4. Differential path for 7R attack. Grey bytes are active bytes. The inbound phase for the data processing part is stressed by red squares.

where, the most expensive part (full active state) is avoided for the data processing part to reduce the attack complexity and to keep enough freedom degrees.

We use a rebound-attack approach to search for the values. First, the values for the key are searched. Then, the values for the data are searched for the fixed key pairs. The complexity is a sum of two searching phases, not a product.

5.2   7-Round Collision Attack

We explain our 7-round collision attack, with 2^{64} computations and memory to store 2^8 state. The differential path is as follows. See its illustration in Fig. 4.

$$\text{Key: } 64 \xrightarrow{\text{WH}} 64 \xrightarrow{1^{st}R} 8 \xrightarrow{2^{nd}R} 1 \xrightarrow{3^{rd}R} 8 \xrightarrow{4^{th}R} 64 \xrightarrow{5^{th}R} 8 \xrightarrow{6^{th}R} 8 \xrightarrow{7^{th}R} 64,$$

$$\text{Data: } 64 \xrightarrow{\text{WH}} 0 \xrightarrow{1^{st}R} 8 \xrightarrow{2^{nd}R} 1 \xrightarrow{3^{rd}R} 8 \xrightarrow{4^{th}R} 0 \xrightarrow{5^{th}R} 8 \xrightarrow{6^{th}R} 8 \xrightarrow{7^{th}R} 0.$$

The key and the plaintext should have the same difference so that the plaintext difference can be canceled by the whitening operation. Then, we make a local collision after the 4th round, and another local collision after the 7th round.

Searching Procedure for Key Schedule Part. The goal is finding a single pair of key values satisfying the differential path for the key. The essential part of this procedure is finding two values satisfying the middle three rounds, $1 \rightarrow 8 \rightarrow 64 \rightarrow 8$. This can be done with the Start-from-the-Middle attack [9]. The complexity is only 2^8 computations and the amount of memory is 2^8 state. If the middle three rounds are satisfied, the entire path are also satisfied by simply extending the path by 2 rounds in backward and 2 rounds in forward. Because this transformation is deterministic, the complexity for 7 rounds is unchanged.

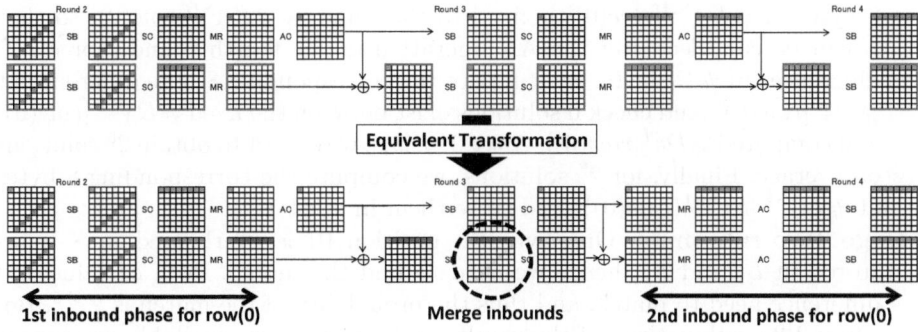

Fig. 5. Details of the inbound phase

Searching Procedure for Data Processing Part. This phase is performed after the key values are fixed. The goal is finding a pair of plaintexts which follow the differential path and generate a collision in the output. The procedure is divided into the inbound phase and the outbound phase.

Inbound Phase. The inbound phase is from state $\#D2^I$ to state $\#D4^{SB}$, which are stressed by red squares in Fig. 4. For the inbound phase, we search for the values with a similar approach to Mendel *et al.* [11]. The details of the inbound phase are described in Fig. 5. Note that the key values are already fixed. Hence, the differences for $\#2D^I$ and $\#D4^{SB}$ are uniquely fixed. First, we apply an equivalent transformation to the third round, *i.e.* AK is performed between SC and MR. Then, the inbound phase is further divided into three parts; first inbound phase, second inbound phase, and merge two inbounds.

First Inbound Phase for Row 0: We aim to find 2^8 paired values that satisfy the differential path between $\#D2^I[SC^{-1}(row(0))]$ and $\#D3^{SB}[row(0)]$ which are described by red in Fig. 5. We only compute a single row. The other rows remain unfixed. The difference for 8 bytes at $\#D2^I[SC^{-1}(row(0))]$ is fixed to the same as $\#K2^I[SC^{-1}(row(0))]$ so that the difference of $\#D2^I[SC^{-1}(row(0))]$ can be canceled by AK^{-1} in the first round. Then, for all 2^8 differences in $\#D2^{MR}[0]$, we compute the corresponding 8-byte difference at $\#D2^{SB}[SC^{-1}(row(0))]$. The average probability that the fixed difference at $\#D2^I[SC^{-1}(row(0))]$ and a computed one in $\#D2^{SB}[SC^{-1}(row(0))]$ have solutions for all 8 bytes is 2^{-8}. Because 2^8 differences are examined in $\#D2^{MR}[0]$, one pair is expected to have solutions and the number of obtained solutions is 2^8 on average. Finally, for all 2^8 solutions, we compute the corresponding 8 bytes at $\#D3^{SB}[row(0)]$ and store them in a list L_1.

Second Inbound Phase for Row 0: This part is similar to the first inbound phase. We aim to find 2^8 paired values that satisfy the differential path between $\#D3^{SB}[SC^{-1}(row(0))]$ and $\#D4^{SB}[row(0)]$ which are described by yellow in Fig. 5. Again we only compute a single row. The difference for

8 bytes at $\#D4^{SB}[row(0)]$ is fixed to the same as $\#K4^{SB}[row(0)]$ so that it can be canceled after the AK operation in the fourth round. For all 2^8 differences in $\#D3^{SC}[0]$, we compute the corresponding 8-byte difference at $\#D4^I[row(0)]$, and check if solutions exist between the fixed $\#D4^{SB}[row(0)]$ and computed $\#D4^I[row(0)]$. After 2^8 trials, we expect to obtain 2^8 solutions on average. Finally, for 2^8 solutions, we compute the corresponding 8 bytes at $\#D3^{SB}[SC^{-1}(row(0))]$ and store them in a list L_2.

Merge Two Inbounds: One byte (in position 0) is overlapped in 8 bytes stored in L_1 and L_2, hence we need to find the match. Both of value and difference need to match, and thus the probability of the match is 2^{-16}. Because 2^{16} combinations of the results in L_1 and L_2 are available, we expect to find a match. We use the other 49 unfixed bytes at $\#D3^{SB}$ as freedom degrees for the outbound phase. Because it can produce $2^{49*8} = 2^{392}$ values for the outbound phase, finding one match is enough for this phase.

The complexity for the inbound phase is 2^8 computations for both of the first and second inbound phases. A memory to store 2^8 state is required to generate L_1 and L_2. In summary, with 2^8 computations and a memory to store 2^8 state, up to 2^{392} solutions of the inbound phase can be produced.

Outbound Phase. Due to the inbound phase, the differential path is ensured to be satisfied up to the fourth round. The outbound phase is a brute force approach to satisfy the differential path after the fourth round by using solutions of the inbound phase. The only probabilistic event for the outbound phase is the cancelation of the difference at the final output. This occurs when the differences for $\#D7^{SB}[row(0)]$ is the same as $\#K7^{SB}[row(0)]$. Therefore, by examining 2^{64} solutions of the inbound phase, we can obtain a collision at the final output.

In summary, a collision is generated with 2^{64} in time and 2^8 in memory.

5.3   Extension to 8-Round Collision Attack and Other Variants

The 7-round attack in Sect. 5.2 can be extended to 8 rounds. The differential path up to the 4th round is exactly the same as the one for the 7-round attack. Therefore, the inbound part is unchanged. In the outbound phase, $8 \rightarrow 8 \rightarrow 64$ is replaced with $8 \rightarrow 1 \rightarrow 8 \rightarrow 64$. The entire path is given in Eq.(2).

Because the attack procedure is very similar, we only mention the difference from the 7-round attack. To search for the key values, we use the Start-from-the-Middle approach. In this time, the differential propagation $8 \xrightarrow{6^{th}R} 1$ needs to be satisfied probabilistically. Therefore, the complexity for the key schedule part is 2^{56} in time and 2^8 in memory. Note that the complexity can be improved to 2^{48} with the linearized match-in-the-middle technique [9]. Because this part is not the bottle-neck, we omit its detailed explanation. Also note that only 1 result is enough because the data processing part can produce many solutions.

For the data processing part, the inbound phase is exactly the same as the one for the 7-round attack, which requires 2^8 in time and 2^8 in memory, and can produce up to 2^{392} solutions. In the outbound phase, the probabilistic events are the differential propagation $8 \xrightarrow{6^{\text{th}} R} 1$ and the differential cancelation at the output state. Therefore, a collision for 8 rounds can be generated with a complexity of $2^{56+64} = 2^{120}$ computations and 2^8 state of memory.

It seems worth mentioning that our differential path is an iterative form;

$$\text{Key: } 64 \xrightarrow{x} 8 \xrightarrow{x+1} 1 \xrightarrow{x+2} 8 \xrightarrow{x+3} 64,$$

$$\text{Data: } 0 \xrightarrow{x} 8 \xrightarrow{x+1} 1 \xrightarrow{x+2} 8 \xrightarrow{x+3} 0.$$

Therefore, constructing a differential path for $4n$ rounds or $4(n-1)+3$ rounds is possible. However, we cannot find the attack for three iterations (12-rounds or 11-rounds) due to a too high complexity and too small freedom degrees.

Practical Near-Collision Attack on 7 Rounds. In some case, near-collisions can be a real threat because hash values are used after the truncation. Our 7-round attack in Sect. 5.2 can generate a 40-byte near-collision with a complexity of 2^{40} computations and 2^8 state of memory. For this attack, we only cancel the difference in 5-bytes between $\#K7^{SB}[row(0)]$ and $\#D7^{SB}[row(0)]$. Note that the brute force attack for 40-byte near-collision takes 2^{160} computations, and thus our attack is much faster. We also implemented the attack on a PC, and confirmed that the attack could work correctly. An example of the generated data is provided in Table 3 in Appendix C.

Practical Collisions on 4 Rounds. All previous attacks require at least 2^{64} computations to generate a collision even for a small number of rounds. Therefore, we investigate the practical collision attack on a small number of rounds.

Our differential path generates a local collision after the fourth round, and up to fourth round can be covered by the inbound phase. Therefore, we can generate collisions of the 4-round Whirlpool compression function only with 2^8 computations and 2^8 state of memory. No extra practical example is given here since the 7-round near-collision in Table 3 is also a 4-round collision.

5.4 Theory vs Practice: Implementation of Rebound Attacks

The DDT of the S-box is the core of the rebound attack, which provides an efficient method for satisfying the differential paths. The S-box of Whirlpool is not as balanced as the one in AES. For a non-zero difference pair, if there is a conforming value, we call it a match. The matching probability of Whirlpool S-box is lower than the one in AES.

The property of the Whirlpool S-box results in big differences between theory and practice. Theoretically, one valid key pair is enough to find a match of the MitM phase in the data processing part. But, practically, we tried 109 different valid key pairs to find a solution for the data part. In every matching step, we have to try more times to find a match. So the complexity to find one solution is increased. However, the expected number of solutions for a random difference pair does not depend on DDT. Hence, the total complexity is not increased if we need many solutions of the inbound phase. As a result, the complexity of our 7-round and 8-round attacks is not affected, since the complexity mainly depends on a lot of iterations in the outbound phase. The theoretical complexity of our inbound phase for both key and data (to find a 4-round collision) is 2^8. Because we only need one solution from the inbound phase, experiments show that the practical complexity for the inbound phase is increased by 2^4 to 2^7 times.

6   Concluding Remarks

In this paper, we improved the attacks on Whirlpool with respect to the fundamental security notions. For the preimage attack, the number of attacked rounds was extended by the guess-and-determine technique. Moreover, the complexity was improved by exploiting the freedom in the key value. For the collision attack, the difference was introduced in the key value, and a high probability differential path was constructed by canceling the difference in the data with the difference in the key. These results show several risks of using similar diffusions for the key and data. These also indicate that Whirlpool is still secure in practice.

References

1. Wang, X., Yu, H.: How to Break MD5 and Other Hash Functions. In: Cramer, R. (ed.) EUROCRYPT 2005. LNCS, vol. 3494, pp. 19–35. Springer, Heidelberg (2005)
2. Wang, X., Yin, Y.L., Yu, H.: Finding Collisions in the Full SHA-1. In: Shoup, V. (ed.) CRYPTO 2005. LNCS, vol. 3621, pp. 17–36. Springer, Heidelberg (2005)
3. U.S. Department of Commerce, National Institute of Standards and Technology: Federal Register, Notices, vol. 72(212), Friday, November 2, 2007/Notices (2007)
4. Rijmen, V., Barreto, P.S.L.M.: The WHIRLPOOL hashing function. Submitted to NISSIE (2000)
5. Menezes, A.J., van Oorschot, P.C., Vanstone, S.A.: Handbook of applied cryptography. CRC Press (1997)
6. International Organization for Standardization: ISO/IEC 10118-3:2004, Information technology – Security techniques – Hash-functions – Part 3: Dedicated hash-functions (2004)
7. New European Schemes for Signatures, Integrity, and Encryption(NESSIE): Nessie Project Announces Final Selection Of Crypto Algorithms (2003)
8. Mendel, F., Rechberger, C., Schläffer, M., Thomsen, S.S.: The Rebound Attack: Cryptanalysis of Reduced Whirlpool and Grøstl. In: Dunkelman, O. (ed.) FSE 2009. LNCS, vol. 5665, pp. 260–276. Springer, Heidelberg (2009)
9. Mendel, F., Peyrin, T., Rechberger, C., Schläffer, M.: Improved Cryptanalysis of the Reduced Grøstl Compression Function, ECHO Permutation and AES Block Cipher. In: Jacobson Jr., M.J., Rijmen, V., Safavi-Naini, R. (eds.) SAC 2009. LNCS, vol. 5867, pp. 16–35. Springer, Heidelberg (2009)
10. Gilbert, H., Peyrin, T.: Super-Sbox Cryptanalysis: Improved Attacks for AES-Like Permutations. In: Hong, S., Iwata, T. (eds.) FSE 2010. LNCS, vol. 6147, pp. 365–383. Springer, Heidelberg (2010)

11. Lamberger, M., Mendel, F., Rechberger, C., Rijmen, V., Schläffer, M.: Rebound Distinguishers: Results on the Full Whirlpool Compression Function. In: Matsui, M. (ed.) ASIACRYPT 2009. LNCS, vol. 5912, pp. 126–143. Springer, Heidelberg (2009)
12. Matusiewicz, K., Naya-Plasencia, M., Nikolić, I., Sasaki, Y., Schläffer, M.: Rebound Attack on the Full LANE Compression Function. In: Matsui, M. (ed.) ASIACRYPT 2009. LNCS, vol. 5912, pp. 106–125. Springer, Heidelberg (2009)
13. Wu, S., Feng, D., Wu, W., Su, B.: Hyper-Sbox View of AES-like Permutations: A Generalized Distinguisher. In: Lai, X., Yung, M., Lin, D. (eds.) Inscrypt 2010. LNCS, vol. 6584, pp. 155–168. Springer, Heidelberg (2011)
14. Sasaki, Y., Li, Y., Wang, L., Sakiyama, K., Ohta, K.: Non-full-active Super-Sbox Analysis: Applications to ECHO and Grøstl. In: Abe, M. (ed.) ASIACRYPT 2010. LNCS, vol. 6477, pp. 38–55. Springer, Heidelberg (2010)
15. Naya-Plasencia, M.: How to Improve Rebound Attacks. In: Rogaway, P. (ed.) CRYPTO 2011. LNCS, vol. 6841, pp. 188–205. Springer, Heidelberg (2011)
16. Jean, J., Naya-Plasencia, M., Peyrin, T.: Improved Rebound Attack on the Finalist Grøstl. In: Canteaut, A. (ed.) FSE 2012. LNCS, vol. 7549, pp. 110–126. Springer, Heidelberg (2012)
17. Jean, J., Fouque, P.-A.: Practical Near-Collisions and Collisions on Round-Reduced ECHO-256 Compression Function. In: Joux, A. (ed.) FSE 2011. LNCS, vol. 6733, pp. 107–127. Springer, Heidelberg (2011)
18. Wu, S., Feng, D., Wu, W.: Practical Rebound Attack on 12-Round Cheetah-256. In: Lee, D., Hong, S. (eds.) ICISC 2009. LNCS, vol. 5984, pp. 300–314. Springer, Heidelberg (2010)
19. Biryukov, A., Khovratovich, D., Nikolić, I.: Distinguisher and Related-Key Attack on the Full AES-256. In: Halevi, S. (ed.) CRYPTO 2009. LNCS, vol. 5677, pp. 231–249. Springer, Heidelberg (2009)
20. Mendel, F., Rechberger, C., Schläffer, M., Thomsen, S.S.: Rebound Attacks on the Reduced Grøstl Hash Function. In: Pieprzyk, J. (ed.) CT-RSA 2010. LNCS, vol. 5985, pp. 350–365. Springer, Heidelberg (2010)
21. Peyrin, T.: Improved Differential Attacks for ECHO and Grøstl. In: Rabin, T. (ed.) CRYPTO 2010. LNCS, vol. 6223, pp. 370–392. Springer, Heidelberg (2010)
22. Aoki, K., Sasaki, Y.: Preimage Attacks on One-Block MD4, 63-Step MD5 and More. In: Avanzi, R.M., Keliher, L., Sica, F. (eds.) SAC 2008. LNCS, vol. 5381, pp. 103–119. Springer, Heidelberg (2009)
23. Guo, J., Ling, S., Rechberger, C., Wang, H.: Advanced Meet-in-the-Middle Preimage Attacks: First Results on Full Tiger, and Improved Results on MD4 and SHA-2. In: Abe, M. (ed.) ASIACRYPT 2010. LNCS, vol. 6477, pp. 56–75. Springer, Heidelberg (2010)
24. Sasaki, Y., Aoki, K.: Finding Preimages in Full MD5 Faster Than Exhaustive Search. In: Joux, A. (ed.) EUROCRYPT 2009. LNCS, vol. 5479, pp. 134–152. Springer, Heidelberg (2009)
25. Sasaki, Y.: Meet-in-the-Middle Preimage Attacks on AES Hashing Modes and an Application to Whirlpool. In: Joux, A. (ed.) FSE 2011. LNCS, vol. 6733, pp. 378–396. Springer, Heidelberg (2011)
26. Wu, S., Feng, D., Wu, W., Guo, J., Dong, L., Zou, J.: (Pseudo) Preimage Attack on Reduced-Round Grøstl Hash Function and Others. In: Canteaut, A. (ed.) FSE 2012. LNCS, vol. 7549, pp. 127–145. Springer, Heidelberg (2012)
27. Bogdanov, A., Khovratovich, D., Rechberger, C.: Biclique Cryptanalysis of the Full AES. In: Lee, D.H., Wang, X. (eds.) ASIACRYPT 2011. LNCS, vol. 7073, pp. 344–371. Springer, Heidelberg (2011)

28. Lamberger, M., Mendel, F., Rechberger, C., Rijmen, V., Schläffer, M.: The rebound attack and subspace distinguishers: Application to Whirlpool. Cryptology ePrint Archive, Report 2010/198 (2010), http://eprint.iacr.org/2010/198

29. Aoki, K., Sasaki, Y.: Meet-in-the-Middle Preimage Attacks Against Reduced SHA-0 and SHA-1. In: Halevi, S. (ed.) CRYPTO 2009. LNCS, vol. 5677, pp. 70–89. Springer, Heidelberg (2009)

30. Morita, H., Ohta, K., Miyaguchi, S.: A Switching Closure Test to Analyze Cryptosystems. In: Feigenbaum, J. (ed.) CRYPTO 1991. LNCS, vol. 576, pp. 183–193. Springer, Heidelberg (1992)

31. Kelsey, J., Schneier, B.: Second Preimages on n-Bit Hash Functions for Much Less than 2^n Work. In: Cramer, R. (ed.) EUROCRYPT 2005. LNCS, vol. 3494, pp. 474–490. Springer, Heidelberg (2005)

A   Chunk Separation for Preimage Attack

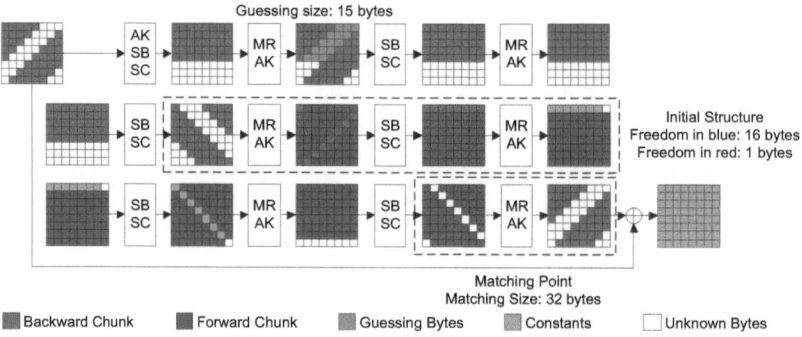

Fig. 6. Chunk separation $(b, r, w, g) = (7, 5, 1, 5)$ for the memoryless second preimage attack on 6-round hash function

B   On the Message Length Padding

In order to convert the attack on the compression function into an attack on the hash function, we need to deal with the message padding first. For the last message block, the lower half are the message length in binary expression. Here, we use L to denote the message length. If the last bit of the fourth row $\#M[row(3)]$ in the message block $\#M$ is 1, we can obtain that $L \equiv 255 \bmod 512$. So the last 9 bits of $\#M[row(7)]$ should be 011111111. If the last two bits of $\#M[row(3)]$ are 10, we know that $L \equiv 254 \bmod 512$. So the last 9 bits of $\#M[row(7)]$ should be 011111110. So, we can calculate the probability that a random message block is a valid block with correct padding by adding up all the probability for different suffix of the upper half of the message block:

$$\sum_{i=1}^{256} 2^{-(9+i)} \approx 2^{-9}.$$

C   Examples of Data Generated in the Experiments

Table 3. Collision for 4- and Near-Collision for 7-round Compression Function

Chaining Value	Message Block	Difference in Round 4	Difference in Round 7
7B A9 ED 44 E2 7A FE 2B	71 68 DA 09 4F B6 D0 B2		
FD 53 A5 EE 97 A6 72 F3	97 93 7B 9A FF 6C 41 BB		
FD 4E EF 3B F1 65 E8 64	C0 BD AF 12 72 FD A4 17		
B4 D0 84 01 F9 75 18 57	30 82 86 46 FF 83 47 D0	00 00 00 00 00 00 00 00	00 00 00 00 00 00 00 00
02 BB 5E 6F CA A3 E5 76	99 D8 0E 3C 03 C5 8E 06	00 00 00 00 00 00 00 00	00 00 00 00 00 00 00 00
86 F2 38 76 2B 9B 7F 58	EE 78 EF 01 74 65 7D AF	00 00 00 00 00 00 00 00	00 00 00 00 00 00 00 00
0E 80 06 67 58 65 90 0A	84 03 52 1B C3 F7 F2 BC	00 00 00 00 00 00 00 00	00 00 00 00 00 00 00 00
DD A7 64 C7 3A 6F ED AC	95 C9 BD 81 20 26 12 57	00 00 00 00 00 00 00 00	42 6E 9E 0D 4F 4F 21 4F
9A CB 57 95 CE 6B F7 17	90 0A 60 D8 63 A7 D9 8E	00 00 00 00 00 00 00 00	7D CF 94 FA B2 7D 7D E9
D9 35 99 D8 94 7D 35 F4	B3 F5 47 AC FC B7 06 BC	00 00 00 00 00 00 00 00	FA B0 E9 4A 7D 59 B0 B0
5E 1D 25 7F 45 10 E2 B2	63 EE 65 56 C6 88 AE C1	00 00 00 00 00 00 00 00	00 00 00 00 00 00 00 00
00 0F B4 6A A0 10 89 A0	84 5D B6 2D A6 E6 D6 27		
A6 16 D6 D4 6B 37 75 D4	3D 75 86 87 A2 51 1E A4		
B7 F8 03 25 F8 0D 9D 9D	DF 72 D4 52 A7 F3 9F 6A		
E8 1D 70 13 40 0E 47 94	62 9E 24 6F DB 9C 25 22		
52 58 53 E9 D0 C2 B5 0E	1A 36 8A AF CA 8B 4A F5		

Generic Related-Key Attacks for HMAC

Thomas Peyrin[1,*], Yu Sasaki[2], and Lei Wang[1,3]

[1] Division of Mathematical Sciences, School of Physical and Mathematical Sciences,
Nanyang Technological University, Singapore
`thomas.peyrin@gmail.com, wang.lei@ntu.edu.sg`
[2] NTT Secure Platform Laboratories, NTT Corporation
`sasaki.yu@lab.ntt.co.jp`
[3] The University of Electro-Communications

Abstract. In this article we describe new generic distinguishing and forgery attacks in the related-key scenario (using only a single related-key) for the HMAC construction. When HMAC uses a k-bit key, outputs an n-bit MAC, and is instantiated with an l-bit inner iterative hash function processing m-bit message blocks where $m = k$, our distinguishing-R attack requires about $2^{n/2}$ queries which improves over the currently best known generic attack complexity $2^{l/2}$ as soon as $l > n$. This means that contrary to the general belief, using wide-pipe hash functions as internal primitive will not increase the overall security of HMAC in the related-key model when the key size is equal to the message block size. We also present generic related-key distinguishing-H, internal state recovery and forgery attacks. Our method is new and elegant, and uses a simple cycle-size detection criterion. The issue in the HMAC construction (not present in the NMAC construction) comes from the non-independence of the two inner hash layers and we provide a simple patch in order to avoid this generic attack. Our work finally shows that the choice of the opad and ipad constants value in HMAC is important.

Keywords: HMAC, hash function, distinguisher, forgery, related-key.

1 Introduction

Hash functions are among the most important basic primitives in cryptography. Informally, a hash function H is a function that takes an arbitrarily long message M as input and outputs a fixed-length hash value of size n bits. Classical security requirements are collision resistance and (second)-preimage resistance. Namely, it should be impossible for an adversary to find a collision (two distinct messages that lead to the same hash value) in less than $2^{n/2}$ hash computations, or a (second)-preimage (a message hashing to a given challenge) in less than 2^n hash computations.

Hash functions are used in many applications such as digital signatures, message integrity check and message authentication codes (MAC). A MAC is a

---

* Supported by the Singapore National Research Foundation Fellowship 2012 (NRF-NRFF2012-06).

function that takes a k-bit secret key K and an arbitrarily long message M as inputs, and outputs a fixed-length tag of size n bits. A MAC algorithm should also meet some security requirements. It should be impossible to recover the secret key except by exhaustive search, and it should be computationally impossible to forge a valid MAC without knowing the secret key, the message being chosen by the attacker (existential forgery) or not (universal forgery).

MACs are crucial for many security systems and are often implemented with the HMAC [3] algorithm, in particular for banking protocols or protocols securing Internet connections (TLS and IPSEC). HMAC was designed by Bellare *et al.* in 1996 and is now widely standardized. It has the property to use an iterative hash function as internal component (thus composed of an iterative application of a compression function) and a proof of security is given in [2]: HMAC is a pseudo-random function under the assumption that the compression function is itself a pseudo-random function.

A trivial generic extension attack exists for HMAC: by asking for enough queries to obtain an internal collision, the attacker can then add extra message blocks to generate other colliding HMAC outputs, therefore breaking the existential forgery security criterion. In order to avoid this issue, many other MACs constructions have been proposed and analyzed [26,25,12], reaching a security beyond the $n/2$ birthday bound by using bigger hash function internal state sizes. For example, the extension attack applied to an n-bit hash function with a $2n$-bit internal state requires 2^n compression function calls.

In parallel to the recent impressive advances on standardized hash function cryptanalysis, the community studied the possible impact on the security of HMAC when instantiated with these standards (such as MD5 [19] or SHA-1 [21]). There have been also some related-key analysis of HMAC instantiated with real hash functions, but no generic attack is known in this model, i.e. without using any weakness from the internal hash function used. Note that the HMAC proof [2] only holds when considering a single-key scenario and says nothing in the related-key model.

The cryptanalysts also looked at other attacks such as distinguishing-R and distinguishing-H [15]. The aim of the former is to distinguish between a random function and the HMAC construction, while the latter aims at distinguishing if the compression function used inside a HMAC construction is a random function or a specific compression function instance. It is widely believed that for the ideal narrow-pipe hash function, the distinguishing-R should require about $2^{n/2}$ computations, while distinguishing-H should require about 2^n.

Our Contributions. In this article we introduce a new type of related-key distinguisher and forgery attacks for HMAC based on cycle length detection, requiring a birthday query complexity and only a single related-key. The attack complexities are summarized in Table 1 together with previous work that analyzed the HMAC instantiating a dedicated hash algorithm.

Our attacks work when the inner hash function is iterative (which is the case for almost all known hash functions, and is necessary for HMAC anyway) and when a special condition is met on the key input. This condition depends on the value of the HMAC constants opad and ipad (which shows for the first time the importance in the choice of their values) and it is always fulfilled when the key length k is equal to the message input length m of the compression function. HMAC is defined to even handle cases where $k > m$ and $k = m$ is likely to happen for example with lightweight hash functions for which the total internal state size has to remain rather small. One can cite DM-PRESENT or H-PRESENT [7] hash functions (PRESENT being already an ISO standard [6]), which have respectively 80 bits and 64 bits of message input for their compression function. Also, a block cipher-based hash function using a common mode such as DaviesMeyer or MatyasMeyerOseas [1] instantiated with the standardized AES [10] is also likely to meet the condition $k = m$.

We emphasize that this work is the first that exploits related-keys to attack HMAC when modeling the compression function as an ideal primitive. They are also the first attacks applying on HMAC and not on NMAC, which helps to understand the security loss when going from the latter to the former. Finally, our attacks are still applicable even when the internal hash function has a big l-bit internal state, unlike the known generic distinguishing or forgery attacks such as the extension attack. Note that many SHA-3 candidates are wide-pipe (like the finalists [14,5,24]) and it is the current trend in hash functions designs. Therefore, this work shows that a wide-pipe hash function used in HMAC can be weaker than the one used in simple MAC constructions such as a secret-prefix MAC and its strengthened version LPMAC [20]. In these schemes, the key (and the message length) is simply prepended to the input message, and the hash value is the MAC value. Due to the double size of the internal state, no attack is known with a smaller complexity than 2^n computations, while our attack on HMAC is more efficient, requiring only $2^{n/2+1}$ computations.

After a description of HMAC in Section 2, we introduce the generic distinguishing-R attack (requiring about $2^{n/2+1}$ computations) in Section 3, basis for the the internal state recovery attack in Section 4, the forgery attack in Section 5 and the distinguishing-H attack in Section 6. Finally, we discuss our results and propose a simple method to patch HMAC in Section 7.

2   Description of HMAC

A Hash Function H is a function that takes an arbitrary length input message M and outputs a fixed hash value of size n bits. When the hash function is iterative (for example see the classical Merkle-Damgård construction [17,11]), the message M is first padded and then divided into blocks m_i of m bits each. Then, the message blocks are successively used to update an l-bit internal state cv_i (where $l \geq n$) with a compression function h: $cv_{i+1} = h(cv_i, m_i)$, and cv_0 is initialized to a fixed public value $cv_0 = IV$. Once all the message blocks have been processed, an output function g is applied to the last internal state value cv_i so as to eventually obtain $hash = g(cv_i)$. The output function therefore transforms an l-bit value into an n-bit one.

Table 1. Summary of the attack complexities

Previous attacks on HMAC with dedicated hash algorithm

Attack	Key Setting	Target	Size	#Rounds	Attack	Ref.
Dist.-H	Single key	MD4	128	Full	$2^{121.5}$	[15]
Dist.-H	Single key	MD5	128	33/64	$2^{126.1}$	[15]
Dist.-H	Single key	3-pass HAVAL	128	Full	$2^{228.6}$	[15]
Dist.-H	Single key	4-pass HAVAL	128	102/128	$2^{253.9}$	[15]
Dist.-H	Single key	SHA0	128	Full	$2^{121.5}$	[15]
Dist.-H	Single key	SHA1	128	43/80	$2^{121.5}$	[15]
Dist.-H	Single key	SHA1	128	50/80	$2^{153.5}$	[18]
Inner key rec.	Single Key	MD4	128	Full	2^{63}	[9]
Inner key rec.	Single Key	SHA0	128	Full	2^{84}	[9]
Inner key rec.	Single Key	SHA1	128	34/80	2^{32}	[18]
Inner key rec.	Single Key	3-pass HAVAL	128	Full	2^{122}	[16]
Full key rec.	Single Key	MD4	128	Full	2^{95}	[13]
Full key rec.	Single Key	MD4	128	Full	2^{77}	[22]
Dist.-H	Single Key	MD5	128	Full	2^{97}	[23]
Dist.-H	Related Key	SHA1	128	58/80	$2^{158.74}$	[18]

New generic attacks on HMAC

Attack	Key Setting	Target	Old Generic Complexity	New Generic Complexity	Reference
Dist.-R	Related Key	Wide-pipe	$2^{l/2}$	$2^{n/2+1}$	This paper
Dist.-H	Related Key	Narrow-pipe†	2^n	$2^{n/2+1}$	This paper
Dist.-H	Related Key	Narrow or Wide†	2^n	$2^{n/2+2} + 2^{l-n+1}$	This paper
Inner state rec.	Related Key	Narrow or Wide†	2^n	$2^{n/2+2} + 2^{l-n+1}$	This paper
Ex. forgery	Related Key	Narrow or Wide†	2^n	$2^{n/2+2} + 2^{l-n+1}$	This paper

†: For a wide-pipe hash function with l-bit internal state, our attacks improve over the old generic complexity as long as $l < 2n - 1$.

The MAC Algorithm HMAC [3] is based on the NMAC construction that uses two k-bit keys K_{out} and K_{in}. NMAC replaces the public IV of a hash function $H(IV, M)$ by a secret key K to produce a keyed hash function $H(K, M)$. NMAC is defined by:

$$\text{NMAC}(K_{out}, K_{in}, M) = H(K_{out}, H(K_{in}, M)).$$

Since in practice a hash function is used as a black-box and has a fixed IV, HMAC simulates the keyed hash function $H(K, M)$ of NMAC by prepending a secret key block K to M, and computing $H(IV, K\|M)$, where $\|$ denotes the concatenation. Also, HMAC uses a single k-bit key K which is padded with zeros such that after padding the key length is equal to a multiple of m bits. For simplicity of the description and without loss of generality concerning our attacks, in the rest of this article we assume that the key can fit in one compression function message block $k \leq m$, and thus the length of the padded key is m bits (the notation of the keys therefore denotes the padded keys). K_{in} and K_{out} are defined by: $K_{in} = K \oplus \text{ipad} = K \oplus \text{0x3636} \cdots 36$ and $K_{out} = K \oplus \text{opad} = K \oplus \text{0x5C5C} \cdots 5\text{C}$, where ipad and opad have the same length than a padded key. HMAC is defined by:

$$\text{HMAC}(K, M) = H(IV, K \oplus \text{opad}\|H(IV, K \oplus \text{ipad}\|M)).$$

Since the key padding in HMAC enforces that the first compression function call(s) handles all and only the key material, we can rewrite

$$\text{HMAC}(K, M) = H_{K \oplus \text{opad}}(H_{K \oplus \text{ipad}}(M)) = H_{K_{out}}(H_{K_{in}}(M))$$

where $H_K(X)$ represents the iterative hash function H for which the initial value is changed to $h(IV, K)$.

3   Generic Related-Key Distinguisher for HMAC

3.1   General Description

Before describing our attacks, we first emphasize that for the rest of the section we will only use small n-bit messages M, such that after padding any message fit into one compression function message input. In other words, $|M||pad| = m$ and we will always compute a single compression function call in order to handle the whole message M. This is represented in Figure 1 and we have

$$\text{HMAC}(K, M) = g(h(h(IV, K \oplus \text{opad}), g(h(h(IV, K \oplus \text{ipad}), M||pad))||pad))$$
$$= f_{K_{out}}(f_{K_{in}}(M))$$

where $f_K(X) = g(h(h(IV, K), X||pad))$.

The general idea underlying our attacks came from the observation that, contrary to the case of NMAC, in HMAC the inner and outer functions are not fully independent. Indeed, both inner and outer hash functions are the same function H, and the inner and outer keys are related by the relation $K_{in} \oplus K_{out} = \text{ipad} \oplus \text{opad}$.

This is not an issue in the single key model, since when assuming the internal inner and outer compression functions as ideal, no information will leak on their

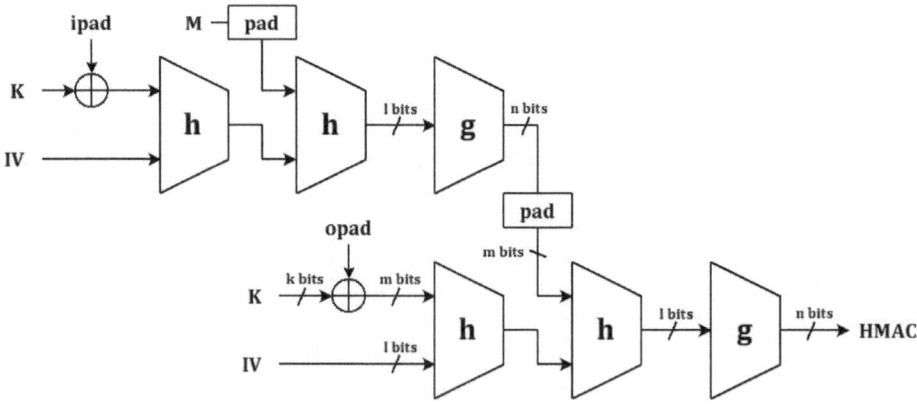

Fig. 1. The computation of HMAC with an iterated hash function when the padded message is small ($|M||pad| = m$)

output from this inner/outer key relation. However, in the related-key model the situation is different. When assuming that the key size k is equal to the padding size (thus one message block, i.e. $k = m$), then we can analyze what is happening when we query $\mathtt{HMAC}(K, M)$ and $\mathtt{HMAC}(K', M)$ with the related key $K' = K \oplus \mathtt{ipad} \oplus \mathtt{opad}$. For the first query the oracle will reply

$$\mathtt{HMAC}(K, M) = f_{K \oplus \mathtt{opad}}(f_{K \oplus \mathtt{ipad}}(M)) = f_{K_{out}}(f_{K_{in}}(M))$$

and for the second query the oracle will reply

$$\begin{aligned}
\mathtt{HMAC}(K', M) &= f_{K' \oplus \mathtt{opad}}(f_{K' \oplus \mathtt{ipad}}(M)) \\
&= f_{K \oplus \mathtt{ipad}}(f_{K \oplus \mathtt{opad}}(M)) \\
&= f_{K_{in}}(f_{K_{out}}(M))
\end{aligned}$$

One can easily see that the two oracles are doing the same computation, except that $\mathtt{ipad}$ and $\mathtt{opad}$ (or K_{in} and K_{out}) are inverted. In other words, we have two oracles, one that applies $f_{K_{in}}$ and then $f_{K_{out}}$ (top figure below), and one that does the opposite $f_{K_{out}}$ and then $f_{K_{in}}$ (bottom figure below).

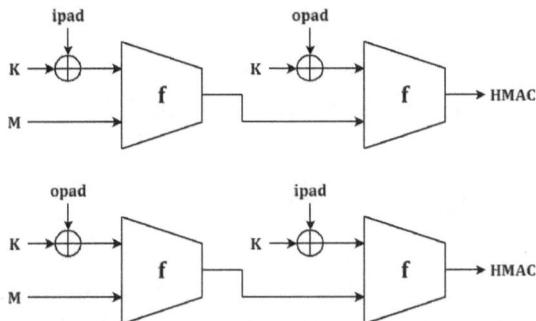

This non-random property seems not easy to detect since the functions $f_{K_{in}}$ and $f_{K_{out}}$ are parametrized with the secret key K, thus they are completely unknown to the attacker. However, it is possible to detect it using a cycle detection algorithm: the functions $f_{K_{in}} \circ f_{K_{out}}$ and $f_{K_{out}} \circ f_{K_{in}}$ have the same cycle structure. Indeed, it is easy to see that there is a one-to-one correspondence between each cycle from $f_{K_{in}} \circ f_{K_{out}}$ and $f_{K_{out}} \circ f_{K_{in}}$.

The attacker will start from an n-bit random input message, query the first oracle (with key K), and keep querying as new message the MAC he just received. He continues to do so for about $2^{n/2}$ queries until he gets a collision among the MACs received. This collision in fact represents a cycle in the successive computations of $f_{K_{in}} \circ f_{K_{out}}$ and this first phase defined a first walk that we denote walk A. In a second step the attacker finds also a cycle for the second oracle computations (with key $K' = K \oplus \mathtt{ipad} \oplus \mathtt{opad}$), i.e. for $f_{K_{out}} \circ f_{K_{in}}$ and that defines walk B. Finally, since the number of MACs obtained from the first and second oracle is big enough, there is a good chance that there is a collision between a MAC from walk A and an internal value of a MAC from walk B (the

internal value is the output of the first hash in HMAC). If so, then the cycle length
of the two cycles are necessarily the same since they follow exactly the same
computation path starting from the collision. This is depicted in Figure 2. An
attacker can use this criterion to distinguish between HMAC computations and
a randomly chosen function, since in the latter case there is only a very low
probability that the two cycles have the same length. We call the tail the part
of the walk that does not belong to the cycle and we denote Z_A (resp. Z_B) the
point where the tail enters the cycle for walk A (resp. walk B).

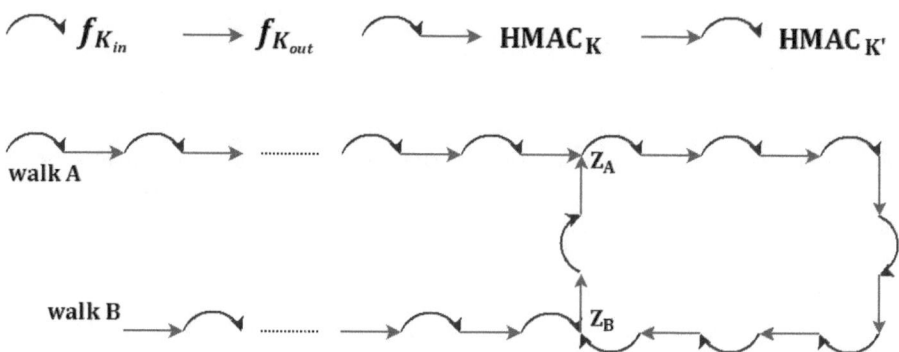

Fig. 2. The cycle structure built with access to oracles $f_{K_{out}} \circ f_{K_{in}}$ and $f_{K_{in}} \circ f_{K_{out}}$

3.2   The Distinguisher

Let $\mathcal{F}_n^n$ be the set of functions from n bits to n bits. We denote F_K and $F_{K'}$
the two oracles on which the adversary $\mathcal{A}$ can make queries. The oracles are
instantiated either with $F_K = \text{HMAC}_K$ and $F_{K'} = \text{HMAC}_{K'}$ (with K being a
randomly chosen k-bit key and $K' = K \oplus \text{ipad} \oplus \text{opad}$) or with two independent
randomly chosen functions R_K and $R_{K'}$ from $\mathcal{F}_n^n$. The goal of the adversary is
to distinguish between the two cases and its advantage is given by

$$Adv(\mathcal{A}) = |\Pr[\mathcal{A}(\text{HMAC}_K, \text{HMAC}_{K'}) = 1] - \Pr[\mathcal{A}(R_K, R_{K'}) = 1]|$$

1st Phase (Walk A). The attacker first chooses a random small message M_A
of size n bits and initializes $q_0^A = M_A$. Then, he will query $F_K(q_0^A)$ and store
the value obtained in q_1^A. He continues by querying $F_K(q_1^A)$ and by storing the
answer in q_2^A, etc. for $2^{n/2} + 2^{n/2-1}$ iterations. If he observes a collision among
the queries during the process, the attacker stops. If no collision is found or if
the collision occurred in the $2^{n/2}$ first queries, the attacker outputs 0.

2nd Phase (Walk B). This phase is identical to the first phase, except that
the attacker queries the oracle $F_{K'}$ instead of F_K. We denote q_i^B the queries
asked during this phase and M_B the starting message value.

3rd Phase (Cycle Detection). Since each query is obtained by applying the function F_K (or $F_{K'}$) on the previous query, a collision among the q_i^A (or among the q_i^B) naturally defines a cycle. If the cycle length of set A is equal to the cycle length of set B, the attacker outputs 1, otherwise he outputs 0.

3.3   Complexity and Success Probability

1st and 2nd Phases (Walk A and B). We first compute the probability that no collision is found when asking for the first $2^{n/2}$ queries in the first (or in the second) phase. In the case of randomly chosen functions:

$$P_{nc-rand} = \prod_{i=1}^{2^{n/2}} 1 - \frac{i}{2^n} \simeq \prod_{i=1}^{2^{n/2}} e^{-\frac{i}{2^n}} = e^{-2^{n/2}\cdot(2^{n/2}-1)/2^{n+1}} \simeq e^{-1/2}.$$

In the case of HMAC computations, a collision can occur either because of a collision on $f_{K_{in}}$ or because of a collision on $f_{K_{out}}$. Therefore, we have

$$P_{nc-hmac} = \left(\prod_{i=1}^{2^{n/2}} 1 - \frac{i}{2^n}\right)^2 \simeq \left(\prod_{i=1}^{2^{n/2}} e^{-\frac{i}{2^n}}\right)^2 = \left(e^{-2^{n/2}\cdot(2^{n/2}-1)/2^{n+1}}\right)^2 \simeq e^{-1}.$$

Then, we compute the probability that when querying the $2^{n/2-1}$ remaining elements, a collision will eventually be found in the first (or in the second) phase:

$$P_{c-rand} = 1 - \prod_{i=1}^{2^{n/2-1}} \left(1 - \frac{2^{n/2}+i}{2^n}\right) \simeq 1 - \prod_{i=1}^{2^{n/2-1}} e^{-\frac{2^{n/2}+i}{2^n}}$$

$$= 1 - e^{-2^{n/2-1}/2^{n/2}+2^{n/2-1}\cdot(2^{n/2-1}-1)/2^{n+1}} \simeq 1 - e^{-5/8}.$$

Again, in the case of HMAC computations, a collision can occur either because of a collision on $f_{K_{in}}$ or because of a collision on $f_{K_{out}}$. Therefore, we have

$$P_{c-hmac} = 1 - \left(\prod_{i=1}^{2^{n/2-1}} \left(1 - \frac{2^{n/2}+i}{2^n}\right)\right)^2 \simeq 1 - \left(\prod_{i=1}^{2^{n/2-1}} e^{-\frac{2^{n/2}+i}{2^n}}\right)^2$$

$$= 1 - (e^{-2^{n/2}/2^{n/2}+2^{n/2}\cdot(2^{n/2}-1)/2^{n+1}})^2 \simeq 1 - e^{-3}.$$

To summarize, the probability of the attacker to not output 0 during both the first and second phases is equal to $(P_{nc-rand} \cdot P_{c-rand})^2 \simeq 0.079$ with randomly chosen functions and to $(P_{nc-hmac} \cdot P_{c-hmac})^2 \simeq 0.122$ with HMAC.

3rd Phase (Cycle Detection).   We need to compute the probability that the cycle found in walk A and in walk B have the same length, for both the HMAC case and the randomly chosen functions case. We denote $P_{cl-hmac}$ the former and $P_{cl-rand}$ the latter.

When the oracles are instantiated with HMAC, we already explained that $HMAC_K$ and $HMAC_{K'}$ are related by their cycle structure. If there exists a collision between a member of walk A and an internal value of a member of walk B, then we are ensured that they will enter a cycle of the same length and the attacker will output 1. Thus $P_{cl-hmac}$ is the probability that such a collision occurs. Since the first phase (resp. second phase) ensured that a collision occurs after $2^{n/2}$ queries, we are ensured that at least $2^{n/2}$ distinct elements exist in walk A (resp. walk B). Therefore, the probability $P_{cl-hmac}$ is lower bounded by

$$P_{cl-hmac} \geq 1 - \prod_{i=1}^{2^{n/2}} (1 - \frac{2^{n/2}}{2^n}) = 1 - \prod_{i=1}^{2^{n/2}} e^{-\frac{1}{2^{n/2}}} = 1 - e^{-1}.$$

Now we need to evaluate the probability $P_{cl-rand}$ that the cycles in walk A and walk B have the same length for randomly chosen functions. Since we ensured that the collision happens in the last $2^{n/2-1}$ elements instead of the first $2^{n/2}$ elements for walk A, there must exist some value z_A, $1 \leq z_A \leq 2^{n/2-1}$, such that $q_{2^{n/2}+z_A}^A$ is the first query colliding with some previous query in walk A. So the cycle length of walk A is uniformly distributed between 1 and $2^{n/2} + z_A$. Similarly for walk B, there exists a value z_B, $1 \leq z_B \leq 2^{n/2-1}$, such that the cycle length of walk B is uniformly distributed between 1 and $2^{n/2} + z_B$. Without loss of generality, let z_A be smaller than or equal to z_B. Thus, the probability that the cycles in walk A and walk B have the same length is given by

$$P_{cl-rand} = \sum_{i=1}^{2^{n/2}+z_A} \frac{1}{2^{n/2} + z_A} \times \frac{1}{2^{n/2} + z_B} < \frac{1}{2^{n/2}} \times \sum_{i=1}^{2^{n/2}+z_A} \frac{1}{2^{n/2} + z_A} = 2^{-n/2}.$$

Overall the advantage of the adversary is

$$Adv(\mathcal{A}) = |\Pr[\mathcal{A}(HMAC_K, HMAC_{K'}) = 1] - \Pr[\mathcal{A}(R_K, R_{K'}) = 1]|$$
$$\geq |(P_{nc-hmac} \cdot P_{c-hmac})^2 \cdot P_{cl-hmac} - (P_{nc-rand} \cdot P_{c-rand})^2 \cdot P_{cl-rand}|$$
$$\simeq (e^{-1} \cdot (1 - e^{-1.5}))^2 \cdot (1 - e^{-1}) = 0.077$$

and it can be increased towards $(1-e^{-1}) = 0.63$ by allowing the attacker to spend a bit more computations in the first and second phases (instead of outputting 0, he just starts the phase over until he succeeds).

The complexity of the distinguisher is about $2^{n/2} + 2^{n/2-1}$ computations for each of the first and second phase, thus about $2^{n/2+1}$ computations in total.

As a proof of concept, we have implemented the attack for HMAC instantiated with SHA-2 truncated to 32 bits and the results can be found in the full version of the article.

4   Internal State Recovery Attack

In this section we extend the distinguisher from Section 3 and we present an internal-state-recovery attack that will be useful for the latter sections showing forgery and distinguishing-H attacks. These attacks are applicable to both narrow-pipe and wide-pipe hash functions under some conditions. As an example for a narrow-pipe hash function without finalization $g(\cdot)$, i.e. SHA-256 and SHA-512 [21], these attacks achieve a birthday-bound complexity $2^{n/2}$, thus significantly reducing the expected complexity of 2^n.

4.1   General Idea

We observe that if walk A and walk B follow the structure in Figure 2, then for any query in the cycle of walk A, denoted as q^A, the inner hash value $H_{K_{in}}(q^A)$ is necessarily equal to some query in the cycle of walk B, denoted as q^B. The goal is therefore to find this query among all $\#q^B$ candidate values (all the members of walk B that belong to the cycle). In other words, we would like to synchronize the two cycles from walk A and walk B, which we already know have the same length.

In general, even if we know that walk A and walk B have the same length and are actually doing the same computations, it seems hard to synchronize the two cycles because we do not know where the tail in walk A and in walk B is entering the cycle. However, in the special case where the collision between walk A and walk B happens in the tail (and not in the cycle), then we know that the tails are entering the cycle at the same position (see Figure 3). In that case, the cycles are directly synchronized and the attacker knows all the successive hash output values for every computation in the cycle (he knows the output values of all the $H_{K_{in}}$ and $H_{K_{out}}$ computed inside the cycle).

The first and second phases of the attack will be devoted to building a walk A and walk B with a rather long tail, such that during the third phase there is a good chance to get a collision between an element of the tail of walk A and an element of the tail of walk B. In order to recover an internal state, he will focus on one randomly chosen value belonging to the cycle, denoted q^A, and its next

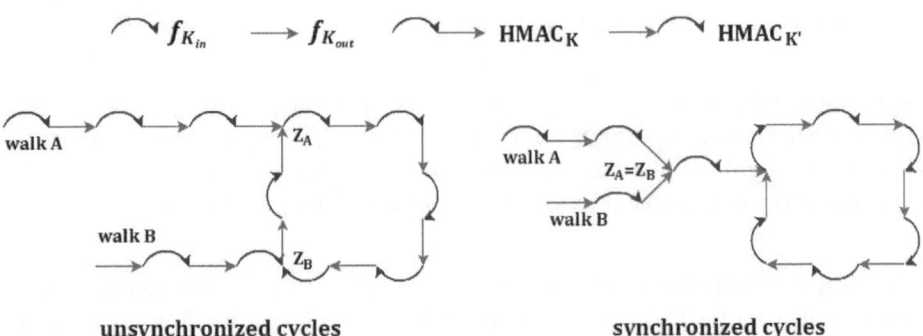

Fig. 3. Two walks A and B colliding and sharing a cycle. The left example shows unsynchronized cycles (the collision happens in the cycle, thus $Z_A \neq Z_B$), the right shows synchronized cycles (the collision happens before the cycle, in the tails, thus $Z_A = Z_B$).

hash output q^B, with $q^B = H(K_{in}, q^A)$. Then he will try to guess the internal hash value $X = h(h(IV, K_{in}), q^A || pad_1)$ that led to q^B, i.e. $g(X) = q^B$.

We assume that $g(\cdot)$ is easy to invert (given an output u, it is easy to find all preimages leading to u) and that it is balanced (given an output value, there exists 2^{l-n} corresponding input values through g). Inverting g provides 2^{l-n} candidates X_i such that $g(X_i) = q^B$. For each of these candidates, we will apply a filter to remove the bad guesses. The filter is based on an offline extension of the computation of $H_{K_{in}}$.

4.2 Detailed Procedure

1st Phase (Walk A). The attacker chooses a random small message M_A of size n bits and initializes $q_0^A = M_A$. Then he will query $\text{HMAC}_K(q_0^A)$ and store the value obtained in q_1^A. He continues by querying $\text{HMAC}_K(q_i^A)$, and by storing the answer in q_{i+1}^A for $i = 0, 1, \ldots, 2^{n/2}$. If no cycle is generated (no collision among the queries q_i^A) or if the walk A generated has a tail smaller than $2^{n/2-2}$, then the attacker chooses another random n-bit message as starting query q_0^A and repeats the search procedure until a walk A with a cycle and a tail of at least $2^{n/2-2}$ elements are found.

We evaluate the success probability of finding a proper walk A by trying one set of $2^{n/2}$ iterative queries. First we would like the first $2^{n/2-1}$ elements be distinct and the probability of this event is approximately $e^{-1/8}$ (the evaluation is similar to the one from Section 3, thus we omitted it here). Then the probability that the last $2^{n/2-1}$ queries produce a cycle is approximately $e^{-3/8}$. We evaluate the probability that the tail of walk A has at least $2^{n/2} - 2$ elements. Note that we have guaranteed that the query q_i^A causing the first collision happens during the i-th iteration, with $i > 2^{n/2} - 1$. Therefore, the probability that q_i^A does not collide with the first $2^{n/2} - 2$ elements is $1 - (2^{n/2-2}/i) \geq 1/2$. Finally, we conclude that by trying one set of $2^{n/2}$ iterative queries, the success probability of generating a proper walk A is at least $e^{-1/8} \times e^{-3/8} \times 1/2 \simeq 0.303$.

2nd Phase (Walk B). The procedure is identical to the first phase except that the attacker is querying $\text{HMAC}_{K'}$ with $K' = K \oplus \text{ipad} \oplus \text{opad}$ instead of HMAC_K. He obtains a walk B that has a cycle and whose tail contains at least $2^{n/2-2}$ elements with probability of about 0.303 (identical to 1st phase).

3rd Phase (Collision). The attacker checks that there is a collision between an element from walk A and one from walk B, which can be done by verifying that walk A and walk B have the same cycle length. He also wants this collision to happen more exactly between a member of the tail of walk A and a member of the tail of walk B. This event happens with probability $1 - e^{-1} \simeq 0.63$ and if such a collision occurs, then the cycles from walk A and walk B are synchronized. In other words, the attacker knows that the tail in walk A entered the cycle at the same position that the tail in walk B entered its own cycle and as a consequence he knows all the succesive internal values for the $\texttt{HMAC}_K$ and $\texttt{HMAC}_{K'}$ computations belonging to the cycle. We denote q^A, q^B and q^C three consecutive internal states, that is $q^B = H(K_{in}, q^A)$, $q^C = H(K_{out}, q^B)$ and $q^C = \texttt{HMAC}_K(q^A)$.

4th Phase (Recovery by Filtering). Given q^A, q^B and q^C, known by the attacker, the goal is now to recover the inner hash function internal state just before applying the output function g. In other words, the attacker is trying to recover $X = h(h(IV, K_{in}), q^A || pad_1)$, with $g(X) = q^B$. He first inverts the output function g from q^B and gets 2^{l-n} candidate values X_j.

The attacker chooses $2^{n/2}$ random distinct messages M_i, $0 \leq i < 2^{n/2}$, such that each $q^A || pad_1 || M_i || pad_2$ fits into exactly two message blocks. He queries the messages $q^A || pad_1 || M_i$ to $\texttt{HMAC}_K$ and look for collisions among the outputs. A collision happens in inner hash with a probability $1 - e^{-1/2}$. At the same time, we want to avoid faulty collision, i.e. collision in the outer hash instead of the inner hash, and this happens with probability $e^{-1/2}$. We denote (M, M') the pair of colliding message found and the success probability is $(1 - e^{-1/2}) \times e^{-1/2} \simeq 0.23$.

For each of the 2^{l-n} candidate values X_j, the attacker computes the values $g(h(X_j, M || pad_2))$ and $g(h(X_j, M' || pad_2))$, and checks whether they are equal. If it is the case, the attacker stores X_j as a very likely candidate for the yet unknown value of X. Since there are in total 2^{l-n} candidate values, and the filter is of n-bit, 2^{l-2n} candidates will be stored. The attacker repeats the colliding messages (M, M') search and the filtering process until only one candidate, namely the real value of X, is left.

Overall, the complexity of the attack is less than $2^{n/2+2}$ queries, and 2^{l-n+1} offline computations. The success probability is around $0.303 \times 0.303 \times 0.63 \times 0.23 = 0.013$. By repeating the phases from 2 and 4 several times, the success probability will be increased.

5   Forgery Attacks

This section describes the related-key forgery attacks on $\texttt{HMAC}$. The adversary is given access to two oracles $\texttt{HMAC}_K$ and $\texttt{HMAC}_{K' = (K \oplus \texttt{ipad} \oplus \texttt{opad})}$. After interacting with $\texttt{HMAC}_K$ and $\texttt{HMAC}_{K'}$, he outputs a message and MAC value (M, σ), such that the message has not be queried for $\texttt{HMAC}_K$. If σ is a valid MAC value for M through $\texttt{HMAC}$ with key K, the adversary is said to have successfully forged M for $\texttt{HMAC}_K$. More precisely, when the attacker is free to choose M it is an existential forgery, while if the message is fixed by the challenger beforehand it is a universal forgery.

A commonly known generic existential forgery attack on $\texttt{HMAC}$ (even in the single-key setting) is the so-called extension attack. The attacker first searches for a pair of messages (M, M') colliding on the last l-bit internal state of the inner hash (just before the application of the output function g in the inner hash function call), then appends each of them with the same additional message block X. Since the last internal state is the same for both messages (M, M'), the two computations of this extra message block X will also behave identically. Finally, by querying the HMAC value for one of the two message $M||X$, the attacker directly forge the other one $M'||X$ by outputting the same MAC value. The complexity of this existential forgery attack is around $2^{l/2}$ queries.

We extend the internal-state-recovery attack from Section 4 to an existential forgery attack. The method is simple. Following the procedure in Section 4, the attacker first recovers the internal state X during the $HMAC_K$ computation of one of the n-bit messages queried and we denote this message by M. Then, using about $2^{l/2}$ computations, he generates offline a pair of distinct messages M' and M'' of the same length satisfying $g(h(X, M'||pad_2)) = g(h(X, M''||pad_2))$, where pad_2 stands for the padding appended to the message $M||M'$ (or $M||M''$) when applying the hash function H. Finally, the attacker queries $M||pad_1||M'$ to the oracle $HMAC_K$ and receives a value T', where pad_1 stands for the padding added to the message M when applying the hash function H. He can forge the MAC value T'' for the message $M||pad_1||M''$ through $HMAC_K$ since $T'' = T'$. The overall complexity of this attack is $2^{n/2+2}$ queries and $2^{l-n} + 2^{l/2}$ computations. Note that in particular for the case $l < 2n$, our attack is faster than the commonly known existential forgery attack requiring $2^{l/2}$ computations.

One can trivially extend this existential forgery attack to an "almost-universal" forgery attack, where the attacker can only choose the first block and the $l/2$ first bits of the second block of the message to be forged. In practice, this would be very close to a universal forgery if one assumes that a few bytes of data in the header of the messages to be MACed can be controlled by the attacker.

6   Distinguishing-H Attacks

This section proposes two distinguishing-H attacks in the related-key setting. Let $\mathcal{F}_n^{m+n}$ be the set of functions from $m + n$ bits to n bits. The attacker is given access to two oracles $HMAC_K$ and $HMAC_{K'}$ with $K' = K \oplus \text{ipad} \oplus \text{opad}$. The compression function of the HMAC oracles is instantiated either with a known dedicated function h or with a random chosen function r from $\mathcal{F}_n^{m+n}$, which we denote $(HMAC_K^h, HMAC_{K'}^h)$ and $(HMAC_K^r, HMAC_{K'}^r)$ respectively. The goal of the adversary is to distinguish between the two cases and its advantage is given by

$$Adv(\mathcal{A}) = \left| \Pr[\mathcal{A}(HMAC_K^h, HMAC_{K'}^h) = 1] - \Pr[\mathcal{A}(HMAC_K^r, HMAC_{K'}^r) = 1] \right|.$$

6.1   Distinguishing-H Attack I: Comparing Cycles Lengths

The distinguisher in Section 3 can be extended to a distinguishing-H attack, as long as the finalization $g(\cdot)$ is bijective and invertible, for example the identity function. Without loss of generality, we omit the output function g. The only difference from the distinguisher in Section 3 will be that in order to produce walk A and walk B we will make full-block long iterative queries, namely m-bit queries, instead of n-bit queries. A graphical view of one iteration in a walk is given in Figure 4. Let pad_1 be the padding to an n-bit message and pad_2 the padding to an m-bit message. The attacker first chooses a small random n-bit value q_0^A. He then queries $q_0^A || pad_1$ to $\mathtt{HMAC}_K$ and receives X_0. He computes $h(X_0, pad_2)$ offline and stores the output as q_1^A. He continues to query $q_1^A || pad_1$, receive X_i and apply $h(X_i, pad_2)$ offline to produce q_{i+1}^A. With the same process, the attacker produces walk B, except that he queries $\mathtt{HMAC}_{K'}$ instead of $\mathtt{HMAC}_K$.

If $\mathtt{HMAC}$ oracles are instantiated with h, then $h(\mathtt{HMAC}_K(\cdot), pad_2)$ is $f_{K_{in}} \circ f_{K_{out}}$ and $h(\mathtt{HMAC}'_K(\cdot), pad_2)$ is $f_{K_{out}} \circ f_{K_{in}}$, where $f_{K_{in}}$ and $f_{K_{out}}$ are defined in Figure 4. So walk A and walk B have a good chance to have the structure explained in Section 3 and depicted in Figure 2, leading to cycles of equal length. On the other hand, if $\mathtt{HMAC}$ oracles are instantiated with r, walk A and walk B are independent. Thus by detecting the cycles lengths, the adversary can distinguish $(\mathtt{HMAC}_K^h, \mathtt{HMAC}_{K'}^h)$ from $(\mathtt{HMAC}_K^r, \mathtt{HMAC}_{K'}^r)$. The complexity and the success probability are identical to the ones for the distinguisher in Section 3.

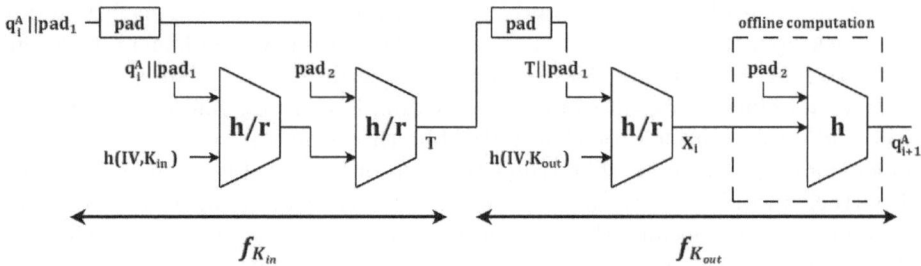

Fig. 4. Distinguisher-H attack I

6.2 Distinguishing-H Attack II: Recovering Internal State

The internal state recovery attack in Section 4 can be extended to a distinguishing-H attack as well. The adversary first regards the $\mathtt{HMAC}$ oracles as $(\mathtt{HMAC}_K^h, \mathtt{HMAC}_{K'}^h)$, and applies the internal state recovery procedure from Section 4 to obtain an internal state value X of some n-bit query q^A in a walk. Then he searches offline a pair of distinct messages (M, M') satisfying $g(h(X, M)) = g(h(X, M'))$, which costs $2^{n/2}$ computations. Finally, he queries $\mathtt{HMAC}_K$ with $q^A || pad_1 || M$ and $q^A || pad_2 || M'$ to check whether the two MAC values collide. If they do the attacker outputs 1, otherwise he outputs 0.

If the compression function is h, the probability that $\mathtt{HMAC}_K(q^A || pad_1 || M)$ collides with $\mathtt{HMAC}_K(q^A || pad_1 || M')$ is equal to the success probability of recovering X in the attack of Section 4. If the compression function is r, the probability that $\mathtt{HMAC}_K(q^A || pad_1 || M) = \mathtt{HMAC}_K(q^A || pad_1 || M')$ is negligible.

Overall, the complexity is $2^{n/2+2}$ queries, $2^{l-n+1} + 2^{n/2}$ offline computations and the success probability is 0.013.

7 Patching HMAC and Discussions

We emphasize again that the related-key issue depicted in this article only exists when the attacker can query $f_{K_{out}} \circ f_{K_{in}}$ and $f_{K_{in}} \circ f_{K_{out}}$ with related-key relations, and therefore keep the two computation chains synchronized if a collision happens. In the case of HMAC this is possible only when $k = m$ or $k = m - 1$ since the last bit of ipad and opad are equal (otherwise, for a smaller key the attacker can not build a proper related-key). This shows that **the choice of ipad and opad is not anecdotal**. For example, if ipad and opad were very similar, then our attacks would work for basically any key length. Also, we observe that our attacks are the first to apply to HMAC and not to NMAC, thus helping the community to understand what security we loose when going from NMAC to HMAC.

Even if our attack is only theoretical due to its high birthday complexity, it is interesting to study how one can patch the scheme and avoid this related-key issue. Since one of the best feature from HMAC is that it uses a hash function as a black box, without any need to change the primitive implementation, our goal is to find a patch that does not affect the hash function definition. Indeed, an easy and efficient tweak would be for example to force different IVs for the inner and outer instances of H in HMAC, but that would require modifying H's implementation. We note that truncating the output of HMAC would also work (the attacker would have to successively guess the truncated bits for each received query in order to continue the computation chain), but we do not consider this solution as satisfactory because reducing the output length will directly reduce the expected generic security of the MAC algorithm.

A first try could be to xor some distinct constants to the inner and/or outer hash message input in an attempt to separate the $f_{K_{out}}$ and $f_{K_{in}}$ computations. However, with such a patch, an attacker can adapt his query strategy and still perform a modified version of the attack from Section 3 to maintain the computation chains synchronized.

Our proposed solution is instead to force an extra fixed bit (or byte) before the input message M. This patch would not harm much the efficiency of the scheme since only one bit (or one byte) would be added to the message to hash for the inner hash function call (actually the efficiency will be the same if the message plus one bit still fit in the same number of message blocks). Also, this patch can even be applied on top of HMAC, as a preprocessing phase before calling the primitive, thus allowing to use existing HMAC libraries without having to modify them.

The related-key distinguishing-R attack from Section 3 is thwarted because now the inner and outer function are made distinct, even when querying with keys K and $K' = K \oplus \text{opad} \oplus \text{ipad}$. The attacker can no more adapt the queries to circumvent this countermeasure and keep the computation chains synchronized. The security proofs of HMAC still hold with this patch since it is trivial to see that any attack on this new proposal will also apply on HMAC.

Note that adding this extra bit (or byte) to the input of the outer hash function instead of the inner one, in an attempt to not reduce the efficiency (in most cases the hash function output size n is much smaller than its message input size m and fit in one block, thus the efficiency would actually be very likely to remain exactly the same), would not prevent the attack from Section 3 to be applicable, since the attacker could simply adapt his query strategy: instead of getting a value V from the HMAC oracle and then query this value V again etc., he could simply prepend a 0 to the received query $0||V$ before querying it again and eventually get the K and K' computations synchronized again.

We observed that appending or prepending the extra bit to the message have actually different impact on the security. For the former, the distinguishing-H attack (approach I) from Section 6 can still apply in the case of a narrow pipe internal hash function, while for the latter the attacker can no more play with pad_2 to absorb the prepended bit. Thus, **our final proposal is to simply prepend a 0 bit (or byte) to the input message of HMAC** . Namely, this new version HMAC' would be defined as

$$\mathtt{HMAC}'(K, M) = H_{K \oplus \mathtt{opad}}(H_{K \oplus \mathtt{ipad}}(0||M)) = H_{K_{out}}(H_{K_{in}}(0||M)) = \mathtt{HMAC}(K, 0||M)$$

Taking in account the fact that the related-key attacks described in this article only work for special key length, we propose to apply our patch to HMAC only when $k = m$ or $k = m - 1$.

We leave as an open problem to find a patch that has no impact on the efficiency (not even a single bit), without modifying the implementation of the hash function H (thus without using distinct IVs for the outer and inner hash calls).

As a final remark, we observe that for HMAC one should only consider related-keys of the same length than the original key. Indeed, for HMAC one can easily check that when the length of the key K is not a multiple of m, then the key $K' = K||0$ is equivalent to K in the sense that $\mathtt{HMAC}_K(M) = \mathtt{HMAC}_{K'}(M)$ for any message M (this related-key relation is even valid in the formalization of related-key attacks from Bellare and Kohno [4] since no two different keys have the same related-key). This is due to the fact that the padding of the key (so that its length becomes a multiple of m) is weak and do not distinguish between keys of different length. A possible patch in order to avoid any equivalent key would to simply pad the key with a 1 and as many zeros as needed (possibly none) such that $K||10\ldots0$ is a multiple of m, instead of the original $0\ldots0$ padding.

8   Conclusion

In this article we introduced a new type of distinguishing-R, distinguishing-H, internal state recovery and forgery attacks for HMAC in the related-key setting. While the applicability of this attack is only theoretical, it uses a novel attack angle, the cycle length. It is the first attack that applies on HMAC and not on NMAC and it provides a better understanding of the role of the constants ipad and opad. We also showed that our attacks can be avoided with a simple patch that only prepends 1 bit or 1 byte to the head of a message.

References

1. Menezes, A., van Oorschot, P., Vanstone, S.: CRC-Handbook of Applied Cryptography. CRC Press (1996)
2. Bellare, M.: New Proofs for NMAC and HMAC: Security Without Collision-Resistance. In: Dwork, C. (ed.) CRYPTO 2006. LNCS, vol. 4117, pp. 602–619. Springer, Heidelberg (2006)
3. Bellare, M., Canetti, R., Krawczyk, H.: Keying Hash Functions for Message Authentication. In: Koblitz, N. (ed.) CRYPTO 1996. LNCS, vol. 1109, pp. 1–15. Springer, Heidelberg (1996)
4. Bellare, M., Kohno, T.: A Theoretical Treatment of Related-Key Attacks: RKA-PRPs, RKA-PRFs, and Applications. In: Biham, E. (ed.) EUROCRYPT 2003. LNCS, vol. 2656, pp. 491–506. Springer, Heidelberg (2003)
5. Bertoni, G., Daemen, J., Peeters, M., Van Assche, G.: Keccak specifications. Submission to NIST (2008),
   http://keccak.noekeon.org/Keccak-specifications.pdf
6. Bogdanov, A., Knudsen, L.R., Leander, G., Paar, C., Poschmann, A., Robshaw, M.J.B., Seurin, Y., Vikkelsoe, C.: PRESENT: An Ultra-Lightweight Block Cipher. In: Paillier, P., Verbauwhede, I. (eds.) CHES 2007. LNCS, vol. 4727, pp. 450–466. Springer, Heidelberg (2007)
7. Bogdanov, A., Leander, G., Paar, C., Poschmann, A., Robshaw, M.J.B., Seurin, Y.: Hash Functions and RFID Tags: Mind the Gap. In: Oswald, E., Rohatgi, P. (eds.) CHES 2008. LNCS, vol. 5154, pp. 283–299. Springer, Heidelberg (2008)
8. Brassard, G. (ed.): CRYPTO 1989. LNCS, vol. 435. Springer, Heidelberg (1990)
9. Contini, S., Yin, Y.L.: Forgery and Partial Key-Recovery Attacks on HMAC and NMAC Using Hash Collisions. In: Lai, X., Chen, K. (eds.) ASIACRYPT 2006. LNCS, vol. 4284, pp. 37–53. Springer, Heidelberg (2006)
10. Daemen, J., Rijmen, V.: The Design of Rijndael: AES - The Advanced Encryption Standard. Springer (2002)
11. Damgård, I.: A Design Principle for Hash Functions. In: Brassard [8], pp. 416–427
12. Dodis, Y., Steinberger, J.: Domain Extension for MACs Beyond the Birthday Barrier. In: Paterson, K.G. (ed.) EUROCRYPT 2011. LNCS, vol. 6632, pp. 323–342. Springer, Heidelberg (2011)
13. Fouque, P.-A., Leurent, G., Nguyen, P.Q.: Full Key-Recovery Attacks on HMAC/NMAC-MD4 and NMAC-MD5. In: Menezes, A. (ed.) CRYPTO 2007. LNCS, vol. 4622, pp. 13–30. Springer, Heidelberg (2007)
14. Gauravaram, P., Knudsen, L.R., Matusiewicz, K., Mendel, F., Rechberger, C., Schläffer, M., Thomsen, S.S.: Grøstl- a SHA-3 candidate. Submitted to NIST (2008), http://www.groestl.info
15. Kim, J., Biryukov, A., Preneel, B., Hong, S.: On the Security of HMAC and NMAC Based on HAVAL, MD4, MD5, SHA-0 and SHA-1 (Extended Abstract). In: De Prisco, R., Yung, M. (eds.) SCN 2006. LNCS, vol. 4116, pp. 242–256. Springer, Heidelberg (2006)
16. Lee, E., Chang, D., Kim, J., Sung, J., Hong, S.: Second Preimage Attack on 3-Pass HAVAL and Partial Key-Recovery Attacks on HMAC/NMAC-3-Pass HAVAL. In: Nyberg, K. (ed.) FSE 2008. LNCS, vol. 5086, pp. 189–206. Springer, Heidelberg (2008)
17. Merkle, R.C.: One Way Hash Functions and DES. In: Brassard [8], pp. 428–446
18. Rechberger, C., Rijmen, V.: New Results on NMAC/HMAC when Instantiated with Popular Hash Functions. J. UCS 14, 347–376 (2008)

19. Rivest, R.L.: The MD5 message-digest algorithm. Request for Comments (RFC) 1320, Internet Activities Board, Internet Privacy Task Force (April 1992)
20. Tsudik, G.: Message Authentication with One-Way Hash Functions. ACM SIG-COMM Computer Communication Review 22(5), 29–38 (1992)
21. U.S. Department of Commerce, National Institute of Standards and Technology. Secure Hash Standard (SHS) (Federal Information Processing Standards Publication 180-3) (2008),
    http://csrc.nist.gov/publications/fips/fips180-3/fips180-3_final.pdf
22. Wang, L., Ohta, K., Kunihiro, N.: New Key-Recovery Attacks on HMAC/NMAC-MD4 and NMAC-MD5. In: Smart, N.P. (ed.) EUROCRYPT 2008. LNCS, vol. 4965, pp. 237–253. Springer, Heidelberg (2008)
23. Wang, X., Yu, H., Wang, W., Zhang, H., Zhan, T.: Cryptanalysis on HMAC/NMAC-MD5 and MD5-MAC. In: Joux, A. (ed.) EUROCRYPT 2009. LNCS, vol. 5479, pp. 121–133. Springer, Heidelberg (2009)
24. Wu, H.: The Hash Function JH. Submitted to NIST (2008),
    http://icsd.i2r.a-star.edu.sg/staff/hongjun/jh/jh.pdf
25. Yasuda, K.: Multilane HMAC— Security beyond the Birthday Limit. In: Srinathan, K., Pandu Rangan, C., Yung, M. (eds.) INDOCRYPT 2007. LNCS, vol. 4859, pp. 18–32. Springer, Heidelberg (2007)
26. Yasuda, K.: A Double-Piped Mode of Operation for MACs, PRFs and PROs: Security beyond the Birthday Barrier. In: Joux, A. (ed.) EUROCRYPT 2009. LNCS, vol. 5479, pp. 242–259. Springer, Heidelberg (2009)

The Five-Card Trick Can Be Done with Four Cards

Takaaki Mizuki, Michihito Kumamoto, and Hideaki Sone

Cyberscience Center, Tohoku University,
Aramaki-Aza-Aoba 6-3, Aoba-ku, Sendai 980-8578, Japan
tm-paper+ac2012@g-mail.tohoku-university.jp

Abstract. The "five-card trick" invented by Boer allows Alice and Bob to securely compute the AND function of their secret inputs using five cards—three black cards and two red cards—with identical backs. This paper shows that such a secure computation can be done with only four cards. Specifically, we give a protocol to achieve a secure computation of AND using only four cards—two black and two red. Our protocol is optimal in the sense that the number of required cards is minimum.

1 Introduction

Assume that two *honest-but-curious* players Alice and Bob, who hold secret bits $a \in \{0, 1\}$ and $b \in \{0, 1\}$, respectively, wish to *securely compute* the AND function, that is, they want to learn the value of $a \wedge b$ without revealing more of their own secret bits than necessary. The "five-card trick" invented in 1989 by Boer [2] achieves such a secure computation of AND using five cards ♣♣♣♡♡. Now, after over two decades since the invention of the five-card trick, this paper improves upon the result: we show that the same secure computation can be done using only four cards ♣♣♡♡.

This paper begins with an overview of the five-card trick.

1.1 The Five-Card Trick

The "five-card trick" by Boer [2] is an elegant secure AND computation protocol that uses three ♣s and two ♡s. Before going into the details of the protocol, we first mention the properties of cards appearing in this paper.

All cards of the same type (♣ or ♡) are assumed to be indistinguishable from one another. We use ? to denote a card lying face down. We also assume that the back ? of each card is identical. To deal with Boolean values, we use the following encoding:

$$\boxed{♣}\boxed{♡} = 0, \quad \boxed{♡}\boxed{♣} = 1. \tag{1}$$

Given a bit $x \in \{0, 1\}$, a pair of face-down cards ?? whose value is equal to x (according to the encoding rule (1) above) is called a *commitment to* x, and is expressed as

X. Wang and K. Sako (Eds.): ASIACRYPT 2012, LNCS 7658, pp. 598–606, 2012.

We now explain how to play the five cards in Boer's secure AND protocol. First, given two cards ♣♡ out of the five cards, Alice privately makes a commitment to her secret bit a (without Bob's knowing the order of the two cards); similarly, Bob makes a commitment to the negation $\bar{b}$ of his secret bit b. Then, with the remaining one card ♣, two commitments are put forth as follows:

It should be noted that the three cards in the middle would be ♣♣♣ only when $a = b = 1$ (if the second and fourth cards from the left were turned over).

Next, Alice and Bob turn the centered card ♣ face down, and apply a *random cut*, which is denoted by $\langle \cdot \rangle$:

$$\underbrace{?\,?}_{a}\underbrace{?\,?\,?}_{\bar{b}} \;\rightarrow\; \left\langle ?\,?\,?\,?\,? \right\rangle \;\rightarrow\; ?\,?\,?\,?\,?.$$

A random cut (also called a *random cyclic shuffling*) means that, as in the case of usual card games, a random number of leftmost cards are moved to the right without changing their order (of course, the random number must be unknown to Alice and Bob); to implement this, it suffices that Alice and Bob take turns cutting the deck until they are satisfied. Finally, Alice and Bob reveal all five cards. Then, the resulting sequence is either

$$\boxed{♣}\boxed{♣}\boxed{♡}\boxed{♡}\boxed{♣} \;\text{ or }\; \boxed{♡}\boxed{♣}\boxed{♡}\boxed{♣}\boxed{♣} \tag{2}$$

apart from cyclic rotations, where either the three ♣s are "cyclically" consecutive or not. One can easily verify that the former case implies $a \wedge b = 1$, and the latter case implies $a \wedge b = 0$.

This is the *five-card trick*, a simple and elegant secure AND protocol.

1.2 Our Result and Related Work

In this paper, we reduce the number of required cards by one, compared to the five-card trick, as listed in Table 1. That is, given commitments

$$\underbrace{?\,?}_{a}\underbrace{?\,?}_{b}$$

to Alice's bit a and Bob's bit b, our protocol needs no card other than the four cards constituting the two commitments, i.e., it can securely evaluate the value of $a \wedge b$ without the use of any additional card. Therefore, as long as one adopts

Table 1. The five-card trick and our protocol with their performance

○ *Secure AND in a non-committed format*

	# of card types	# of cards
Boer [2] (§1.1)	2	5
Ours (§2)	2	4

the encoding rule (1), our protocol is optimal in the sense that the number of required cards is minimum because at least four cards are necessary for the two inputs a and b.

Since the invention of the five-card trick, there have been several *card-based protocols* for secure computation, as listed in Table 2. All these protocols produce their output (say, $a \wedge b$) in a *committed format*, i.e., their output is described as a sequence like

$$a \wedge b$$

that follows the encoding rule (1) (and Alice and Bob have no knowledge about the value than their own secret bits). In contrast, the five-card trick and our protocol (given in Section 2) output the value of $a \wedge b$ in a *non-committed format*; the format of the output $a \wedge b$ differs from the format of inputs a and b, namely the encoding rule (1) (recall the resulting sequences (2), which are completely revealed to the public at the end of the protocol).

Table 2. The "committed format" protocols

○ *Secure AND in a committed format*

	# of card types	# of cards	avg. # of trials
Crépeau-Kilian [3]	4	10	6
Niemi-Renvall [7]	2	12	2.5
Stiglic [10]	2	8	2
Mizuki-Sone [4]	2	6	1

○ *Secure XOR in a committed format*

	# of card types	# of cards	avg. # of trials
Crépeau-Kilian [3]	4	14	6
Mizuki-Uchiike-Sone [5]	2	10	2
Mizuki-Sone [4]	2	4	1

Thus, all the card-based protocols are categorized into two types: "non-committed format" protocols (Table 1) and "committed format" protocols (Table 2); this paper addresses the former. Note that in Table 2 every protocol whose average number of trials is more than 1 is a Las Vegas algorithm.

While card-based protocols might fall within the area of *cryptography without computers* [7], *recreational cryptography* [1] or *human-centric cryptography* [6], we believe that this type of research will help professional cryptographers intuitively explain to nonspecialists the nature of their constructed cryptographic protocols (e.g. [9]). That is, card-based protocols would help ordinary people understand what secure computations are, or, more fundamentally, what cryptography is. Furthermore, it should be noted that some of the card-based protocols are implemented and used in online games [8].

The remainder of this paper is organized as follows. In Section 2, we give a description of our four-card secure AND protocol. In Section 3, we show the correctness of our protocol, that is, we prove that our protocol securely computes the AND function. This paper concludes in Section 4 with an open question.

2   Description of Our Protocol

In this section, we design a new card-based protocol that securely computes the AND function using only four cards ♣ ♣ ♡ ♡.

In Section 2.1, we first introduce the "random bisection cut" [4] used in our protocol. We then describe our protocol in Section 2.2.

2.1   Random Bisection Cuts

As seen in Section 1.1, applying a random cut to a sequence of face-down cards results in a sequence such that a random number of leftmost cards are moved to the right without changing their order. Whereas, a "random bisection cut" [4] works differently, as follows.

Given a deck of (an even number of) face-down cards, say

$$\underbrace{? \; ?}_{a} \; \underbrace{? \; ?}_{b},$$

bisect it and randomly switch the resulting two decks; such a card shuffling operation is called a *random bisection cut*. For the example above, a random bisection cut, denoted by $[\cdot \| \cdot]$, works as

$$\underbrace{? \; ? \; ? \; ?}_{a \quad b} \;\; \rightarrow \;\; \left[? \; ? \; \| \; ? \; ? \right] \;\; \rightarrow \;\; ? \; ? \; ? \; ?,$$

where the resulting deck of the four cards is either

$$\underbrace{? \; ?}_{a} \; \underbrace{? \; ?}_{b} \quad \text{or} \quad \underbrace{? \; ?}_{b} \; \underbrace{? \; ?}_{a},$$

and each case occurs with probability of exactly $1/2$.

Although at first glance a random bisection cut seems to be a little bit less natural operation compared to a (normal) random cut, we hope that people will feel a random bisection cut to be an easy-to-implement operation some day. If Alice and Bob are not familiar with playing cards, then they may hold each of the two bisected decks together using a clip before shuffling the two decks. Alternatively, they may put each of the two decks into an envelope (without changing the order of the cards), and shuffle the two envelopes.

2.2 The Protocol

Given a commitment to Alice's bit a and a commitment to Bob's bit b, our four-card AND protocol proceeds as follows.

1. Apply a random bisection cut:

$$\boxed{?}\boxed{?}\boxed{?}\boxed{?}_{\underbrace{}_{a}\underbrace{}_{b}} \rightarrow \left[\boxed{?}\boxed{?}\,\middle\|\,\boxed{?}\boxed{?}\right] \rightarrow \boxed{?}\boxed{?}\boxed{?}\boxed{?}.$$

2. Apply a random cut to the two cards in the middle, namely the second and third cards:

$$\boxed{?}\boxed{?}\boxed{?}\boxed{?} \rightarrow \boxed{?}\left\langle\boxed{?}\boxed{?}\right\rangle\boxed{?} \rightarrow \boxed{?}\boxed{?}\boxed{?}\boxed{?}.$$

3. Reveal the second card.

   (a) If the face-up second card is $\clubsuit$, then open the fourth card. We now have either

   $$\boxed{?}\boxed{\clubsuit}\boxed{?}\boxed{\clubsuit} \quad \text{or} \quad \boxed{?}\boxed{\clubsuit}\boxed{?}\boxed{\heartsuit}.$$

   The former case implies $a \wedge b = 1$, and the latter case implies $a \wedge b = 0$.

   (b) If the face-up second card is $\heartsuit$, then open the first card. We now have either

   $$\boxed{\clubsuit}\boxed{\heartsuit}\boxed{?}\boxed{?} \quad \text{or} \quad \boxed{\heartsuit}\boxed{\heartsuit}\boxed{?}\boxed{?}.$$

   The former case implies $a \wedge b = 0$, and the latter case implies $a \wedge b = 1$.

As described above, our protocol makes one random bisection cut (in step 1) and one random cut (in step 2). After those cuts, two cards are eventually revealed, namely either (a) the second and fourth cards, or (b) the first and second cards, depending on the result of revealing the second card in step 3. Note that if the two face-up cards are the same type ($\boxed{\clubsuit}\boxed{\clubsuit}$ or $\boxed{\heartsuit}\boxed{\heartsuit}$), then we have $a \wedge b = 1$; otherwise, we have $a \wedge b = 0$.

We show why our protocol works in the next section.

3 Correctness of Our Protocol

In this section, we prove that the protocol given in the previous section securely computes $a \wedge b$. First, in Section 3.1 we intuitively explain why our protocol works. Then, we verify the correctness of our protocol in Section 3.2.

3.1 An Intuitive Sketch

Given a commitment

$$\boxed{?}\boxed{?}_{\underbrace{}_{a}},$$

note that each individual card constituting the commitment inherently has the value of the bit a, that is, one can also write

$$\underbrace{\boxed{?}}_{a}\ \underbrace{\boxed{?}}_{\bar{a}}$$

where an encoding rule for a single card is taken: $\boxed{\clubsuit}$ expresses 0, and $\boxed{\heartsuit}$ expresses 1. Based on such a single-card encoding, commitments to a and b can be expressed as:

$$\underbrace{\boxed{?}}_{a}\ \underbrace{\boxed{?}}_{\bar{a}}\ \underbrace{\boxed{?}}_{b}\ \underbrace{\boxed{?}}_{\bar{b}}\ . \tag{3}$$

Now, for the expression (3), skip step 1 in our protocol and directly apply step 2. That is, apply a random cut to the second and third cards:

$$\underbrace{\boxed{?}}_{a}\ \Big\langle \underbrace{\boxed{?}}_{\bar{a}}\ \underbrace{\boxed{?}}_{b}\Big\rangle\ \underbrace{\boxed{?}}_{\bar{b}}$$

$$\rightarrow\ \text{(i)}\ \underbrace{\boxed{?}}_{a}\ \underbrace{\boxed{?}}_{\bar{a}}\ \underbrace{\boxed{?}}_{b}\ \underbrace{\boxed{?}}_{\bar{b}}\ \text{or (ii)}\ \underbrace{\boxed{?}}_{a}\ \underbrace{\boxed{?}}_{b}\ \underbrace{\boxed{?}}_{\bar{a}}\ \underbrace{\boxed{?}}_{\bar{b}}\ .$$

Next, applying step 3, reveal the second card. Assume that the face-up second card is $\boxed{\clubsuit}$ (as in step 3(a)), i.e.,

$$\text{(i)}\ \underbrace{\boxed{?}}_{a}\ \underbrace{\boxed{\clubsuit}}_{\bar{a}}\ \underbrace{\boxed{?}}_{b}\ \underbrace{\boxed{?}}_{\bar{b}}\ \text{or (ii)}\ \underbrace{\boxed{?}}_{a}\ \underbrace{\boxed{\clubsuit}}_{b}\ \underbrace{\boxed{?}}_{\bar{a}}\ \underbrace{\boxed{?}}_{\bar{b}}\ . \tag{4}$$

Then, it means that either (i) $\bar{a} = 0$ or (ii) $b = 0$ (because $\boxed{\clubsuit} = 0$). If (i) $\bar{a} = 0$, then $a = 1$ and hence $a \wedge b = b$; if (ii) $b = 0$, then $a \wedge b = 0 = b$. Therefore, in either case, we have $a \wedge b = b$, and hence one can notice that the value of $a \wedge b = b$ can be obtained by revealing the fourth card

$$\underbrace{\boxed{?}}_{\bar{b}}$$

in the sequence (4). Actually, in step 3(a), the fourth card is opened. If it is $\boxed{\clubsuit}$, then $\bar{b} = \boxed{\clubsuit} = 0$ and hence $a \wedge b = b = 1$; if it is $\boxed{\heartsuit}$, then $\bar{b} = \boxed{\heartsuit} = 1$ and hence $a \wedge b = b = 0$.

Thus, steps 2 and 3(a) surely compute the value of $a \wedge b$. One can similarly verify the claim for the case of step 3(b). Therefore, steps 2 and 3 can provide at least the value of $a \wedge b$. However, they also leak some secret information about a and b; indeed, for example, when the second card revealed in step 3 was $\boxed{\clubsuit}$, we have $\bar{a} = 0$ or $b = 0$ (as seen above), and hence the fact of $(a, b) \neq (0, 1)$ has been disclosed. Therefore, since executing only steps 2 and 3 is not secure, our protocol applies a random bisection cut in step 1 to guarantee secrecy, as intuitively explained below.

Note first that adding step 1 never affects the computation outcome of executing steps 2 and 3: after applying a random bisection cut in step 1, we have either

$$\underbrace{\boxed{?}\boxed{?}}_{a}\underbrace{\boxed{?}\boxed{?}}_{b} \quad \text{or} \quad \underbrace{\boxed{?}\boxed{?}}_{b}\underbrace{\boxed{?}\boxed{?}}_{a}, \tag{5}$$

and then applying steps 2 and 3 to the sequence (5) always provides the value of $a \wedge b$ as shown above (because $a \wedge b = b \wedge a$).

To see that the secrecy is preserved by introducing a random bisection cut in step 1, we enumerate all possibilities:

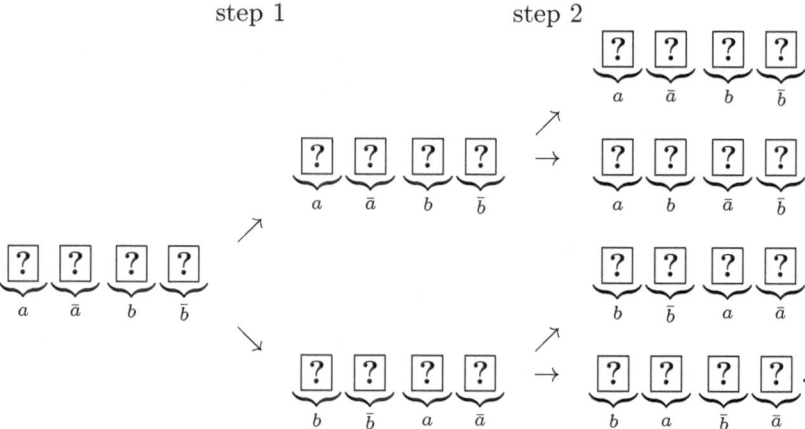

Therefore, opening the second card (in the rightmost sequence) means that one of $\bar{a}$, b, $\bar{b}$ and a is randomly revealed. Hence, opening the second card never leaks any information about each of a and b.

3.2 Proof of Correctness

In this subsection, we prove that our protocol works correctly.

Recall that a random bisection cut $[\boxed{?}\boxed{?} \| \boxed{?}\boxed{?}]$ in step 1 and a random cut $\boxed{?}\langle\boxed{?}\boxed{?}\rangle\boxed{?}$ in step 2 are applied to the two commitments; one can enumerate, as in Table 3, all possibilities of the four cards after each of steps 1 and 2. Note that the cases $(a, b) = (0, 1)$ and $(a, b) = (1, 0)$ both fall in the same status category after step 1 (and after step 2, of course).

Consider the actual execution of our protocol based on Table 3. After step 3, all possibilities can be enumerated as shown in Table 4. (Remember that the second card is opened in step 3, and that if it is $\boxed{\clubsuit}$, then the fourth card is opened; otherwise, the first one is opened.) Table 4 immediately implies that our protocol surely computes the value of $a \wedge b$—if the two revealed cards are the same type, then $a \wedge b = 1$; otherwise, $a \wedge b = 0$.

To verify the secrecy of the protocol, it suffices to show that

$$\Pr[(0,0) \| \boxed{?}\boxed{\clubsuit}\boxed{?}\boxed{\heartsuit}] = \Pr[(0,0) \| \boxed{\clubsuit}\boxed{\heartsuit}\boxed{?}\boxed{?}] = \Pr[(0,0) \mid a \wedge b = 0],$$

Table 3. All possibilities of the four cards after each of steps 1 and 2

(a,b)	initial	after step 1	after step 2
$(0,0)$	♣ ♡ ♣ ♡	♣ ♡ ♣ ♡	♣ ♣ ♡ ♡ or ♣ ♡ ♣ ♡
$(0,1)$	♣ ♡ ♡ ♣	♣ ♡ ♡ ♣ or ♡ ♣ ♣ ♡	♡ ♣ ♣ ♡ or ♣ ♡ ♡ ♣
$(1,0)$	♡ ♣ ♣ ♡	same as $(0,1)$	same as $(0,1)$
$(1,1)$	♡ ♣ ♡ ♣	♡ ♣ ♡ ♣	♡ ♣ ♡ ♣ or ♡ ♡ ♣ ♣

Table 4. All possibilities of the four cards after step 3

(a,b)	initial	after step 3
$(0,0)$	♣ ♡ ♣ ♡	? ♣ ? ♡ or ♣ ♡ ? ?
$(0,1)$	♣ ♡ ♡ ♣	? ♣ ? ♡ or ♣ ♡ ? ?
$(1,0)$	♡ ♣ ♣ ♡	? ♣ ? ♡ or ♣ ♡ ? ?
$(1,1)$	♡ ♣ ♡ ♣	? ♣ ? ♣ or ♡ ♡ ? ?

$$\Pr[(0,1) \,|\, \boxed{?}\,\boxed{♣}\,\boxed{?}\,\boxed{♡}] = \Pr[(0,1) \,|\, \boxed{♣}\,\boxed{♡}\,\boxed{?}\,\boxed{?}] = \Pr[(0,1) \,|\, a \wedge b = 0],$$

and

$$\Pr[(1,0) \,|\, \boxed{?}\,\boxed{♣}\,\boxed{?}\,\boxed{♡}] = \Pr[(1,0) \,|\, \boxed{♣}\,\boxed{♡}\,\boxed{?}\,\boxed{?}] = \Pr[(1,0) \,|\, a \wedge b = 0].$$

Let $\Pr[a = 0] = p$ and $\Pr[b = 0] = q$. Then, we have $\Pr[(0,0) \,|\, a \wedge b = 0] = pq/(p + q - pq)$. On the other hand,

$$\Pr[(0,0) \,|\, \boxed{?}\,\boxed{♣}\,\boxed{?}\,\boxed{♡}] = \frac{pq(1/2)}{pq(1/2) + p(1 - q)(1/2) + (1 - p)q(1/2)},$$

as desired. For all the remaining cases, one can easily check the equality.

4   Conclusions

In this paper, we presented a four-card secure AND protocol whose output is in a non-committed format. Since the existing protocol, namely the five-card trick, requires five cards, we have succeeded in reducing the number of required cards by one. Our protocol is optimal in the sense that at least four cards are required for commitments to the inputs a and b.

Note that the OR function can also be securely computed by using four cards, say, according to de Morgan's law $a \vee b = \overline{\overline{a} \wedge \overline{b}}$.

This paper showed a secure AND computation in a non-committed format using only four cards. For the committed format case, the best known protocol

[4] requires six cards as seen in Table 2. An intriguing open question is whether there exists a "committed format" AND protocol that requires fewer than six cards.

Acknowledgments. We thank the anonymous referees whose comments helped us to improve the presentation of the paper. This work was supported by JSPS KAKENHI Grant Number 23700007.

References

1. Balogh, J., Csirik, J.A., Ishai, Y., Kushilevitz, E.: Private computation using a PEZ dispenser. Theoretical Computer Science 306, 69–84 (2003)
2. den Boer, B.: More Efficient Match-Making and Satisfiability: The Five Card Trick. In: Quisquater, J.J., Vandewalle, J. (eds.) EUROCRYPT 1989. LNCS, vol. 434, pp. 208–217. Springer, Heidelberg (1990)
3. Crépeau, C., Kilian, J.: Discreet Solitary Games. In: Stinson, D.R. (ed.) CRYPTO 1993. LNCS, vol. 773, pp. 319–330. Springer, Heidelberg (1994)
4. Mizuki, T., Sone, H.: Six-Card Secure AND and Four-Card Secure XOR. In: Deng, X., Hopcroft, J.E., Xue, J. (eds.) FAW 2009. LNCS, vol. 5598, pp. 358–369. Springer, Heidelberg (2009)
5. Mizuki, T., Uchiike, F., Sone, H.: Securely computing XOR with 10 cards. Australasian Journal of Combinatorics 36, 279–293 (2006)
6. Moran, T., Naor, M.: Polling with Physical Envelopes: A Rigorous Analysis of a Human-Centric Protocol. In: Vaudenay, S. (ed.) EUROCRYPT 2006. LNCS, vol. 4004, pp. 88–108. Springer, Heidelberg (2006)
7. Niemi, V., Renvall, A.: Secure multiparty computations without computers. Theoretical Computer Science 191, 173–183 (1998)
8. Stamer, H.: Efficient electronic gambling: an extended implementation of the toolbox for mental card games. In: Proc. Western European Workshop on Research in Cryptology (WEWoRC 2005). LNI, vol. P-74, pp. 1–12 (2005)
9. Stamm, S., Jakobsson, M.: Privacy-preserving polling using playing cards. IACR Eprint archive (2005)
10. Stiglic, A.: Computations with a deck of cards. Theoretical Computer Science 259, 671–678 (2001)

A Mix-Net from Any CCA2 Secure Cryptosystem

Shahram Khazaei[1], Tal Moran[2], and Douglas Wikström[1]

[1] KTH Royal Institute of Technology
[2] IDC Herzliya

Abstract. We construct a provably secure mix-net from any CCA2 secure cryptosystem. The mix-net is secure against active adversaries that statically corrupt less than λ out of k mix-servers, where λ is a threshold parameter, and it is robust provided that at most $\min(\lambda - 1, k - \lambda)$ mix-servers are corrupted.

The main component of our construction is a mix-net that outputs the correct result if all mix-servers behaved honestly, and aborts with probability $1 - O(H^{-(t-1)})$ otherwise (without disclosing anything about the inputs), where t is an auxiliary security parameter and H is the number of honest parties. The running time of this protocol for long messages is roughly $3tc$, where c is the running time of Chaum's mix-net (1981).

1 Introduction

A *mix-net*, introduced by Chaum in 1981 [2], is a tool to provide anonymity for a group of senders. The main application is electronic voting, in which each sender submits an encrypted vote and the mix-net then outputs the votes in sorted order. Mix-nets have also found applications in other areas, e.g., anonymous web browsing [6], payment systems [13] and even as a building-block for secure multiparty computation [10].

A mix-net is constructed as a cryptographic protocol by invoking a set of *mix-servers*, arranged in a series. The original mix-net proposed by Chaum works as follows. To set up, each mix-server publishes a public key for an encryption system. Each sender then publishes a "wrapped" message with several layers of encryption: starting with the innermost layer—an encryption of her plaintext message using the last mix-server's public key—and ending with the outermost layer, encrypted using the first mix-server's public key. Once all senders have published their encrypted inputs, the mixing stage begins. In turn, each mix-server receives the encrypted values output from the previous server, "peels off" a layer of encryption, i.e., decrypts the values using his private key, sorts the decrypted values and passes them on to the next mix-server in the chain. The output of the final mix-server is the sorted list of the senders' original inputs.

Chaum's mix-net hides the correspondence between the input ciphertexts and the output plaintexts, but even a single mix-server can undetectably modify the output or refuse to take part in the protocol (forcing the protocol to abort without output). These drawbacks have been addressed in previous work. The most widely researched line of work is based on the idea of *re-encryption mixes* (originally proposed by Park, Itoh and Kurosawa [19]); these rely on homomorphic encryption schemes whose ciphertexts can be "re-randomized". Using the homomorphic properties of the encryption scheme, it is possible to generate very efficient zero-knowledge proofs that the mixing was performed correctly (e.g., Neff [16] or Furukawa and Sako [4]). While the state-of-the-art

X. Wang and K. Sako (Eds.): ASIACRYPT 2012, LNCS 7658, pp. 607–625, 2012.

re-encryption mixes are both provably secure and efficient for short inputs, their reliance on homomorphic properties limits them to a few specific encryption schemes.

1.1   Our Contribution

In this paper, we propose a new, efficient mix-net protocol that satisfies several highly-desirable properties:

- *Minimal Cryptographic Assumptions.* Our protocol can be based on any CCA2-secure cryptosystem, without requiring additional assumptions. In particular, we do not require the underlying encryption to have homomorphic properties.

    While interesting from a theoretical standpoint, this also has clear advantages in practice, as it gives greater flexibility in the choice of encryption scheme. For example, all currently practical homomorphic encryption schemes are susceptible to attacks from quantum computers. Although we do not currently know how to build quantum computers, it is important to take this vulnerability into account when using a mix-net as part of an electronic election scheme: ballot privacy is often required to be preserved for decades—these timeframes may be long enough for the development of a working quantum computer.

    Furthermore, the flexibility in the choice of encryption scheme makes it easy to deal efficiently with long inputs, while there do exist mix-nets that can deal with long inputs efficiently [11,17,5], these mixes require even more specialized encryption schemes tailored specifically to that purpose.
- *Provable Security.* Many of the existing mixing protocols do not have formal proofs of security. This may seem like a purely theoretical concern, but the history of cryptographic protocols, and mix-nets in particular, shows that there is good reason to distrust heuristic approaches. A notable example of this is the *Randomized Partial Checking* (RPC) scheme of Jakobsson, Juels and Rivest [12] (our main "competitor" in the field of generic CCA2-based mixes). The RPC scheme (and related constructions) have been around for over a decade, and have already been used in binding elections; however, recent work by Khazaei and Wikström [14] shows that RPC contains a subtle but serious security flaw, which was consistently missed in implementations. Other examples abound (see Section 1.2 for more).

    In contrast, our protocol is proven secure in the Universal Composability framework [1], a very strong notion of security that holds even when arbitrary additional protocols are run concurrently. (If a cryptosystem which allows recovering the randomness from a ciphertext using the secret key is used to implement the (non zero-knowledge) proof of correct decryption, then the result only holds in the stand-alone setting.)
- *Full Security.* The RPC scheme gains efficiency by relaxing slightly the security requirements. It prevents corrupt mix-servers from undetectably modifying *many* inputs of honest senders, but a malicious server can succeed in changing a constant number of inputs with non-negligible probability. For some uses, this may not be acceptable. RPC also relaxes the privacy guarantees: while the exact correspondence between senders and their inputs is hidden, some information may still be leaked. Our protocol, with comparable or better efficiency, provides full simulation-based security.

Our protocol is based on a new technique we call *Trip-Wire Tracing* (TWT). Our main idea is to do away with zero-knowledge proofs (that would be costly for a generic cryptosystem) used by existing protocols to guarantee correctness and replace them with a virtual "trip wire" system: we insert "fake" inputs into the mix to act as trip wires for catching misbehaving mix-servers (for more details, see Section 2).

Security Guarantees and Assumptions. The protocol preserves privacy and correctness against active adversaries that statically corrupt less than λ mix-servers, where λ is a threshold parameter, and it is robust provided that at most $\min(\lambda - 1, k - \lambda)$ mix-servers are corrupted, where k is the number of mix-servers. As for all other mix-nets in the literature, we assume the existence of an ideal bulletin board functionality (this is equivalent to a broadcast channel). We also need an ideal functionality for shared key generation. In the general case (when we can only assume a generic CCA2-secure cryptosystem without any additional structure), this functionality would have to be implemented using general MPC. However, if the chosen cryptosystem does have a more efficient shared-key-generation protocol, it can be used instead (in any case, the bulk of the work can always be carried out offline, in a preliminary key generation phase).

Finally, we need a functionality for proving that a ciphertext is correctly decrypted, but it suffices that this protocol hides the secret key. This functionality can be realized trivially if the cryptosystem allows recovery of the randomness (used to form the ciphertext) using the secret key. In any case this protocol is only used to identify corrupted parties and mix-servers, so during normal operation it is not used at all.

Limitations of Our Protocol. Our construction essentially uses privacy to ensure correctness (by hiding the "trip-wires" from malicious mix servers). Because a threshold coalition of malicious servers can always violate privacy, our protocol loses correctness as well in this case. This implies that our protocol cannot be "universally verifiable" (i.e., verifiable by third parties who do not trust any of the mix servers). In comparison, the state-of-the-art mix-nets based on homomorphic cryptosystems can provide integrity (but not privacy) even if all mix-servers are corrupt.

We remark that RPC only allows a restricted form of universal verifiability, i.e., its relaxed correctness degrades further and allows an adversary that controls all mix-servers to undetectably replace a notable number of ciphertexts.

1.2   Related Work

The literature on mix-nets and verifiable shuffling is extensive. Below, we mention a small sample of particularly relevant works. Park, Itoh and Kurosawa [19] introduced re-encryption mixes as a way to improve efficiency—the size of the ciphertexts and the amount of work performed by senders does not depend on the number of mix-servers. Sako and Kilian constructed the first universally-verifiable mix-net [22], where senders can verify that the entire shuffle was performed correctly (and not just that their own input was included in the output). Sako and Kilian's construction was based on cut-and-choose zero-knowledge proofs; Neff [16] and Furukawa and Sako [4] gave much more efficient zero-knowledge proofs of shuffle for homomorphic cryptosystems. Many of the works in the field aim to improve the efficiency of the mix-net. Our construction includes ideas that appear in several previous papers: Jakobsson used the idea of "dummy inputs" [9] and "repetition" [8] to increase correctness (although in a different way than we do). Golle, Zong, Boneh, Jakobsson and Juels [7] considered mix-nets that are "optimistic" (i.e., can be much more efficient in the case that no errors occur).

On the Importance of Formal Proofs. A recurring tale in the history of mix-net design is the proposal of a mix-net construction followed by discovery of security flaws. Following Chaum's seminal paper [2], Pfitzmann and Pfitzmann pointed out that Chaumian mixes are vulnerable to attack if the encryption scheme used is malleable [21]. The mix-net of Park et al. [19] was also shown to be vulnerable to similar attacks [20]. Jakobsson's scheme of [8] was broken in [3]. His other scheme [9], was broken by Mitomo and Kurosawa [15], who also suggested a fix; this in turn, in addition to the schemes of Jakobsson and Juels [11], of Golle, Zong, Boneh, Jakobsson and Juels [7] were all shown to be vulnerable (to various attacks) by Wikström [23].

While a formal proof of security is not an iron-clad guarantee that no vulnerabilities will ever be found (proofs may have subtle errors, and assumptions may be shown to be wrong), they do significantly improve the trust in the security of a cryptographic scheme. In fact, the need for some of the components of our protocol only became evident during the analysis of the protocol.

2 Informal Description of Our Protocol

We begin with an overview of our mix-net protocol and some intuition for why this protocol is secure. The main component of our construction is a mix-net that outputs the correct result if all mix-servers behave honestly, and aborts with overwhelming probability otherwise—without disclosing anything about the inputs. At a high level, our mix-net with abort protocol is a Chaumian mix-net with added verification. It is parametrized with an auxiliary security parameter t and uses two Chaumian mix-nets in sequence (one with "explicit verification" and one with "partial tracing") and three additional layers of encryption (labeled as "final", "repetition" and "outer"). Figure 1 presents a schematic of our protocol.

Each sender encodes her message as a bundle of t ciphertexts: First, she encrypts her plaintext message using the public key of the "final" layer of encryption and makes t identical copies of it. Next, each copy is further encrypted using the public key of the "repetition" encryption layer and then under the public keys of the mix-servers in the two Chaumian mix-nets. Finally, the t encryptions are concatenated and encrypted using the public key of the "outer" encryption layer. To generate the final list of inputs to the mix-net, each mix-server adds a "dummy" encryption of zero to the list of inputs submitted by the senders (the dummy input is constructed using the same operations as the real inputs).

Once all parties have submitted their bundles, the decryptions proceed in the reverse order. If all the parties are honest, there will be t identical copies of each innermost ciphertext before the final decryption takes place. In this case the dummies are traced and removed, the duplicates are ignored, and only one instance of each sender's innermost

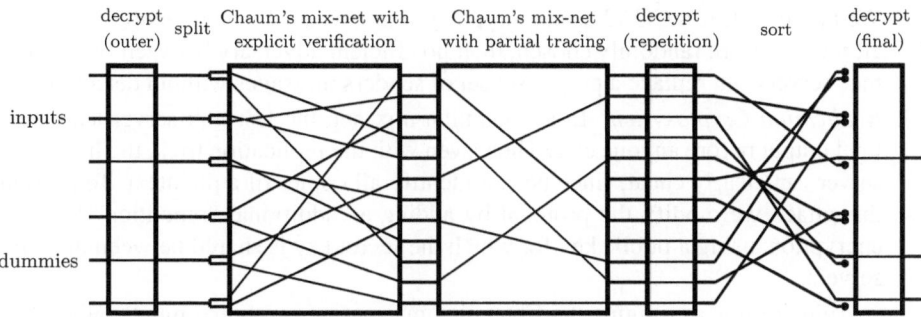

Fig. 1. Execution of Protocol 4 with $N = 3$ senders, $k = 3$ mix-servers, and $t = 2$ repetitions, where all parties are honest. Each party submits a bundle of two ciphertexts containing identical innermost ciphertexts. The bundle is decrypted and split into two ciphertexts. All ciphertexts are then individually shuffled in the two instances of Chaum's mix-net. Then the first is verified explicitly (revealing the permutation), the dummy ciphertexts are traced in the second (revealing the paths of the dummies) and the output is decrypted and verified to contain t copies of each ciphertext. If all tests passed, then a final round of decryption recovers the plaintexts.

ciphertext is decrypted. We stress that this is only an outline of the protocol. Additional measures are taken for ensuring correctness and privacy.

To help give the intuition for our construction, we will describe a sequence of attacks on the Chaumian mix-net and our corresponding modifications to the protocol that prevent them. The final protocol is a composition of all these modifications. We start with a "core" Chaumian mix, which ends up—after slight modifications—as the box labeled "Chaum's mix-net with partial tracing" in Figure 1. We call a set of ciphertexts containing identical innermost ciphertexts a *copyset*.

1. *Elementary Error Handling.* The first type of attack we consider is the introduction of "simple" errors that are publicly detectable. Invalid ciphertexts are simply ignored. If there are duplicates of a ciphertext in the input to a mix-server, then exactly one copy is considered part of the input and the rest is ignored.

2. *Replication.* In a Chaumian mix-net, any corrupt mix-server can change the output undetectably by replacing an output ciphertext with a new one generated by the malicious server (this new ciphertext can be completely valid, except for not being a decryption of any input ciphertext). To prevent this attack, each sender submits t independently formed ciphertexts of her message to the Chaumian mix-net.

   To see why this replication technique helps prevent replacement attacks, consider a corrupt mix-server that appears between two honest mix-servers in the mix-net chain. In this case, the corrupt mix-server cannot identify which of the ciphertexts encrypts the same messages due to the following two reasons.

   (a) He does not know the secret key of the succeeding honest mix-sever, and hence he can not fully decrypt the received ciphertexts and distinguish the copysets based on the final decrypted values.

   (b) The preceeding honest mix-server randomly permuted all of the ciphertexts and hence he does not know which ciphertext originated from which sender.

We prove that, if a CCA2 secure cryptosystem is used, t is sufficiently large, and all messages are randomly chosen, then no efficient adversary between two honest mix-servers can replace a proper subset of senders messages without detection.

3. *Replication Cryptosystem.* In a Chaumian mix-net, the last mix-server learns the final output before anyone else. Thus, even with the replication trick, the final mix-server can clearly cheat, since he can identify all copies of a plaintext. To prevent this attack we modify the protocol by adding an additional "repetition" layer of encryption, using a public key for which the secret key is shared between the mix-servers.

   Think of this as running the Chaumian mix-net on *encrypted* inputs rather than plaintexts, i.e., each sender makes t encryptions of her input with the shared public key of the "repetition" layer, and then uses the encrypted values as her "plaintexts". The output of the Chaumian mix-net is a list of ciphertexts encrypted with the shared public key, which prevents the last mix-server from identifying identical plaintexts and replacing all copies of a subset of the plaintexts. At the end of the mixing, the shared secret key is recovered and decryption is performed publicly. In Figure 1, this decryption step is the box labeled "decrypt (repetition)" right after the Chaum's mix-net with partial tracing.

4. *Additional Mix-net with Explicit Verification.* The first mix-server knows how to partition the input messages into copysets (since he receives the messages directly from the senders), hence he can replace all copies of a given plaintext undetectably. To prevent this attack, we add a new, unmodified Chaumian mix-net (the box labeled "Chaum's mix-net with explicit verification" in Figure 1) between the senders and the first mix-server in the "main" mix-net. Recall that the Chaumian mix-net does not give any correctness guarantees, but it does guarantee privacy if even a single mix-server is honest. This is exactly what we need to put the first mix-server in the Chaumian mix-net with partial tracing on an equal footing with the others in the chain.

   We rely on the privacy of the first Chaumian mix-net only to obtain *correctness* via replication. Therefore, once the second Chaumian mix-net finishes his process of mixing, the mix-servers can reveal the secret keys for the first Chaumian mix-net and verify its correctness completely (hence the name "mix-net with explicit verification"). If the verification fails, the guilty mix-server is publicly evident.

5. *Dummy Values.* If a corrupt mix-server in the second Chaumian mix-net wishes to replace a proper subset of senders' messages, he must guess the positions of the copysets, but he can still undetectably replace *all* of the inputs with his own values. To prevent this, we have every mix-server add a "dummy" value to the inputs of the mix-net. These dummy values are treated identically to the senders' inputs. Thus, any mix-server attempting to replace the entire list of inputs would also be replacing all dummy values. The mix-servers can "trace forward" the dummy values and remove them from the final decrypted list if the trace completes successfully. There is no privacy requirement for the dummy values; therefore, each mix-server can simply reveal all the randomness used in the encryption of the initial dummy values. This reveals all the internal layers of encryption in a verifiable way, allowing everyone to find the corresponding ciphertexts in each stage of the mix-net.

6. *Replication Verification and Error Tracing.* We need to handle the case where some of the final values, after recovering the shared secret key of the "repetition" encryption layer, do not have exactly t copies.

We now add another step to the protocol after decryption using the shared secret: replication verification (this occurs in the part labeled "sort" in Figure 1) and error tracing (this is not shown in Figure 1 since we assume all parties are honest). In the replication verification step, the (honest) mix-servers verify that there are exactly t duplicates of every output value. This clearly is the case if all servers and senders are honest. If the verification fails, however, we need to figure out who is to blame so that we can continue the protocol if it was just a corrupt sender. To do this, we need to trace errors through the system in two ways:

   (a) *Backwards Tracing.* After determining the messages with more or less than t duplicates, we trace them backwards to identify their original senders. Since each mix-server knows his own permutation, the backwards trace is easy to do: each mix-server in turn (starting from the last one and going backwards) publishes the "paths" taken by the traced messages along with a proof that the decryption was performed correctly. If a broken copyset being traced contains ciphertexts that were introduced by a cheating mix-server (i.e., ciphertexts that are not valid decryptions of the mix-server's inputs), the mix-server will not be able to provide a valid trace and will be identified as a cheater at this point.

   (b) *Forward Tracing.* If all the broken copysets were successfully traced back to their sender, there are still two remaining possibilities for casting blame:

      i. The mix-servers behaved honestly, and bad copysets were submitted by corrupt senders.

      ii. At least one ciphertext submitted by an honest party was replaced by a corrupted mix-server. (This could be the case even if no cheating was discovered during backwards tracing. To see this, consider the case that a corrupt mix-server arbitrarily chooses t ciphertexts from honest senders and replaces them with a valid copyset.)

   To distinguish these two cases, we identify the senders from which the broken copysets originated, and "trace forward" *all* the messages of these senders. This is done similarly to the backwards tracing, but in reverse: starting from the first mix-server and going forwards, each one in turn publishes the paths taken by the traced messages along with a proof of correct decryption. If a mix-server cheated, he will not be able to provide a valid trace—hence he will be fingered as the culprit. On the other hand, if only the identified senders were cheating (e.g., by not encrypting a valid copyset in the first place), we will be able to trace the messages all the way to the output.

   If the backwards and forward tracings complete successfully without identifying a mix-server as culprit, the ciphertexts of the corrupt senders are removed from the output (otherwise, the protocol outputs the identity of a guilty mix-server and aborts).

7. *Final Cryptosystem.* As we have described in Step 6, to catch a misbehaving mix-server we must sometimes trace messages of honest users through the system. Although we abort the protocol in this case, we must still preserve the honest senders' privacy. Therefore, we protect the senders' messages with an additional layer of

encryption (the last box labeled "decrypt (final)"). That is, a sender first encrypts her message under the "final" public key and uses this encrypted message as an input to the protocol as described so far. This innermost encryption layer is jointly decrypted only if the protocol does not abort. If the protocol does abort, only the encrypted values are revealed and privacy is protected by the final layer of encryption. The "final" layer of encryption also guarantees that the "plaintexts" of the protocol we have sketched so far (without the "final" layer) are distinct for all honest senders (and different from corrupt senders) with overwhelming probability.

8. *Outer Cryptosystem.* The protocol is still vulnerable to a subtle attack that uses the error-tracing mechanism itself to violate sender privacy. The problem is that tracing occurs in two additional indistinguishable cases:

   (a) Corrupt senders collude to create "colliding" ciphertexts (i.e., after removing some layers of encryption, the resulting ciphertexts are identical).

   (b) Corrupt mix-server(s) collude with corrupt sender(s) to copy some of an honest sender's ciphertexts.

In both cases tracing will complete successfully (since no inputs were replaced in the middle of the mix-net). Because in the first case the mix-servers are all honest, we cannot simply abort if this situation occurs. On the other hand, in the second case, we may be forced to trace an honest ciphertext from beginning to end (we trace a broken ciphertext back to a corrupt sender, then trace forward all of that sender's inputs, which include a copy of an honest ciphertext). Since the corrupt sender knows the identity of the sender from whom the ciphertext was copied, if we decrypt that value the honest sender's privacy is violated.

To prevent this, we add an "outer" layer of encryption (the box labeled "decrypt (outer)"): under a public-key whose secret key is shared by all the mix-servers, each sender formes a single "bundled" ciphertext. After all the ciphertext bundles are received, the mix-servers recover the secret key of the outer cryptosystem and the bundles are publicly decrypted and "split" into the separate copyset ciphertexts. This countermeasure works due to the CCA2 security of the cryptosystem: CCA2 security ensures that no corrupt coalition of mix-servers and senders can make partial copies of an honest sender's copyset: either they copy a bundle in its entirety (in which case they are removed due to being duplicates) or they create a bundle that is completely independent of the honest senders' bundles (in which case the probability of a collision is negligible).

3   Notation

For an integer e, we denote the set $\{1, \ldots, e\}$ by $[1, e]$. The security parameter, n, represented in unary, is an implicit input to all protocols and functionalities. Whenever we say a quantity ε is negligible, we mean that it is negligible in the security parameter, i.e., for every $c > 0$ we have $\varepsilon(n) < n^{-c}$ for all but finitely many n. We write $x \in a$ for a list $a = (a_1, \ldots, a_e)$ if and only if $x \in \{a_1, \ldots, a_e\}$. The length of a is denoted by $|a|$. For any index set $I \subset [1, e]$ of size ℓ, we write $(a_i)_{i \in I} = (a_{i_1}, \ldots, a_{i_\ell})$, where $I = \{i_1, \ldots, i_\ell\}$ with $i_1 < i_2 < \cdots < i_\ell$. We say that a list $b = (b_1, \ldots, b_\ell)$ is a subset of a and write $b \subset a$, if and only if $\{b_1, \ldots, b_\ell\} \subset \{a_1, \ldots, a_e\}$ (with multiplicity).

We use $\mathsf{Sort}(a)$ to denote the lexicographically sorted list of elements from a (with multiplicity). We write $a \setminus b$ for $\mathsf{Sort}(\{a_1, \ldots, a_e\} \setminus \{b_1, \ldots, b_\ell\})$ (with multiplicity in the set difference). We also write $a \circ b$ for the concatenated list $(a_1, \ldots, a_e, b_1, \ldots, b_\ell)$. We denote by $\mathsf{Unique}(a)$ the sorted list where each element of a appears exactly once.

We denote a cryptosystem by $\mathcal{CS} = (\mathsf{Gen}, \mathsf{Enc}, \mathsf{Dec})$, where Gen, Enc, and Dec denote the key generation algorithm, the encryption algorithm, and the decryption algorithm respectively. To deal with nested encryption as needed in a Chaumian mix-net, we simply assume that a plaintext of any length can be encrypted, but that indistinguishability only holds for plaintexts of the same length. We write $c = \mathsf{Enc}_{pk}(m, r)$ for the encryption of a plaintext m using randomness r, and $\mathsf{Dec}_{sk}(c) = m$ for the decryption of a ciphertext c. We often view Enc as a probabilistic algorithm and drop r from our notation. We assume that malformed ciphertexts are decrypted to a special symbol different from all normal plaintexts.

We extend our notation to lists of plaintexts, ciphertexts and keys as follows. For a plaintext $m = (m_1, \ldots, m_e)$ and a key pair (pk, sk) with $pk = (pk_1, \ldots, pk_\ell)$ and $sk = (sk_1, \ldots, sk_\ell)$ we write $c = \mathsf{Enc}_{pk}(m)$, where $c = (c_1, \ldots, c_e)$ with $c_i = \mathsf{Enc}_{pk_1}(\mathsf{Enc}_{pk_2}(\cdots \mathsf{Enc}_{pk_\ell}(m_i) \cdots))$. Similarly, $m = \mathsf{Dec}_{sk}(c)$ is defined by $m_i = \mathsf{Dec}_{sk_\ell}(\mathsf{Dec}_{sk_{\ell-1}}(\cdots \mathsf{Dec}_{sk_1}(c_i) \cdots))$. We stress that when we use Enc as a probabilistic algorithm with a list of messages or public keys, we assume that the random values used in each encryption are chosen randomly and independently. We use the notation $a \| b$ for the concatenation of two bitstrings. We define the function $\mathsf{Split}_t(a)$ to divide a bitstring a, whose length is a multiple of t, into t chunks of equals lengths and turn it into a list, i.e., $(a_1, \ldots, a_t) = \mathsf{Split}_t(a_1 \| \ldots \| a_t)$ when $|a_i|$s are equal.

4   Definitions and Conventions

We consider a mix-net employing k mix-servers $\mathcal{M}_1, \ldots, \mathcal{M}_k$ that provide anonymity for a group of N senders $\mathcal{P}_1, \ldots, \mathcal{P}_N$. Throughout, $\overline{\mathcal{M}}$ and $\overline{\mathcal{P}}$ denote the sets of all mix-servers and senders respectively. We let $J_\mathcal{M} \subset [1, k]$ and $I_\mathcal{P} \subset [1, N]$ denote the index sets of corrupted mix-servers and senders respectively. We let $J^* \subset J_\mathcal{M}$ denote the index set of mix-servers identified as corrupted so far. This set may grow throughout an execution.

We present and analyze the main components of our mix-net in the universal composability framework [1], with *non-blocking* adversaries, i.e., adversaries that do not block the delivery of messages indefinitely. We use superscripts to distinguish different functionalities and protocols, for example $\mathcal{F}^{bb}$ for a bulletin board and π^c for Chaum's mix-net. The ideal adversary (simulator) of the ideal model is denoted by $\mathcal{S}$. When there is no ambiguity, we use the same notation for dummy parties and real parties.

We use a number of conventions to simplify the exposition. Whenever we say a party "hands" a message to a functionality, we mean that the party sends the message to the corresponding dummy party who will then forward it to the functionality. All our functionalities capture distributed protocols where messages sent to more than one party can be delayed arbitrarily by the adversary, and all such messages are also given to the adversary. Thus, when we say that a functionality hands a message to more than one party, we mean that the message is passed to the adversary, who then schedules the delivery of the message to the parties. When a party "inputs" a message to a subprotocol, we mean that he executes the algorithm of the corresponding party with the same message. A party or protocol is said to "wait for" an input of a given form if any other input is immediately returned to the sender. Similarly, a party can wait for a message to appear on the bulletin board. In practice this would be implemented using a time-out, after which some default value is taken to be the message. Some of our functionalities give an output before receiving any input, which makes no sense in an event-driven model like the universal composability framework where execution starts by activating the environment. This is merely a useful convention, since we can easily fix this problem by allowing parties to request the given data.

In all of our protocols, security holds only as long as the adversary corrupts less than λ mix-servers, where $1 \leq \lambda \leq k$ is a parameter of the protocol. All our functionalities and protocols may fail to give an output if more than $\min(\lambda - 1, k - \lambda)$ mix-servers are corrupted. To capture the case $\lambda \geq k/2$ with minimal notational overhead, we simply assume that even a non-blocking simulator can block messages indefinitely in this case.

We use the subroutine Agree(Tag), parameterized by a label Tag, to simplify the description of some of our ideal functionalities. The subroutine waits until each mix-server $\mathcal{M}_j$ has submitted a pair (Tag, m_j) for some message m_j. If at least λ mix-servers submitted identical m_j, then the subroutine returns this value and otherwise it halts the complete ideal functionality, e.g., the functionality could hand $\perp$ to all parties and ignore inputs from then on. The message m_j can be an empty string in which case the subroutine is only used to capture the robustness property of the functionality. In Appendix A we give a formal definition of the subroutine Agree(Tag). We use the same convention for protocols, i.e., if an ideal functionality used by the protocol aborts, then the protocol aborts as well. These conventions allow us to capture the robustness of a protocol by requiring a non-blocking simulator for a non-blocking adversary.

4.1   Useful Functionalities

Our results are given in a hybrid model with distributed key generation functionalities of two types, a bulletin board functionality, and a proof of correct decryption functionality. In Appendix B, formal descriptions of these functionalities are presented. The first key generation functionality, $\mathcal{F}_j^{\mathrm{kg}}$, generates a public key pk_j such that only the jth mix-server knows the corresponding secret key sk_j. The second functionality, $\mathcal{F}^{\mathrm{dkg}}$, differs only in that no mix-server learns the secret key sk corresponding to the generated public key pk. In both functionalities, any subset of λ mix-servers can recover the secret key. The bulletin board functionality, denoted by $\mathcal{F}^{\mathrm{bb}}$, is used by parties to announce their messages. That is, a message can be posted by any party and read by any other one. To simplify the exposition, we simply say that a message is "published" when it appears on the public bulletin board. The published message can not be deleted or modified once posted. The proof of correct decryption functionality, $\mathcal{F}_j^{\mathrm{pd}}$, is used to prove that the jth mix-server has correctly decrypted a known ciphertext into a known plaintext. A subset of λ mix-servers must agree on the pair of plaintext and ciphertext. Assuming that the underlying cryptosystem allows the jth mix-server to recover the randomness used during encryption from the ciphertext itself, the realization of the functionality becomes trivial. In other words, the proof of correct decryption simply consists of revealing the randomness used for encryption. Our main result, Theorem 1, holds if this solution is employed, but only in a standalone model (see the full version of this paper for details). In any case, proofs of correct decryption are only used to trace ciphertexts to identify corrupted senders or mix-servers.

4.2 Mix-Nets

We use ideal mix-net functionalities similar to that in [24], but in a slightly simplified form in that we assume that each sender submits exactly one input. Functionality 1 presents a natural mix-net. Our results are easy to generalize to the case where senders can submit more than one input (this holds also for Functionality 2 and Functionality 3).

The protocol we construct does not quite implement the natural mix-net. Thus, we present a relaxed mix-net (Functionality 2) which we are able to securely realize and then argue that it still provides sufficient guarantees. The relaxed functionality first hands the adversary (simulator) a public key. Then it waits for inputs from all the senders, encrypts the messages of the honest senders, and then hands the resulting ciphertexts in sorted order to the adversary. The adversary is then asked to provide his own inputs in encrypted form on behalf of corrupted senders. The final output is the sorted decryption of the union of the ciphertexts computed by the functionality and those provided by the adversary (after duplicates are removed). For technical reasons the functionality uses several public keys and encrypts the messages under all keys.

This functionality provides *unconditional privacy* for honest senders. The relaxation lies in the ability of an *unbounded* adversary to *adaptively choose* the messages of the corrupted senders based on the *set of inputs* of the honest senders, but a CCA2-secure cryptosystem prevents this for efficient adversaries.

We define a mix-net with abort (Functionality 3) that either gives a proper output or aborts after identifying a mix-server as culprit (with no information about the submitted messages at all). A relaxed mix-net can be constructed using such a mix-net with abort. The mix-net with abort waits for inputs from all the senders and then outputs these messages in encrypted form (as in the relaxed mix-net). Then it allows the mix-servers to agree on a list of known corrupted mix-servers. Finally, the adversary decides if the mix-net should abort or not. In the former case, the adversary must provide the index of a *previously unknown* corrupted mix-server, and this is forwarded to all mix-servers. In the latter case, the mix-net outputs the result like in the relaxed mix-net.

In the full version of this paper we describe a protocol using Functionality 3 that securely realizes Functionality 2. The idea is to use λ instances of Functionality 3. Each sender submits a copy of his input to all functionalities. The mix-servers then run them sequentially until one produces an output without aborting. To ensure that this scheme eventually gives an output, the mix-servers jointly keep track of the identified corrupted mix-servers.

Functionality 1 (Natural Mix-Net). The *natural mix-net* functionality $\mathcal{F}^{\mathrm{mn}}$ executing with dummy senders $\overline{\mathcal{P}}$, dummy mix-servers $\overline{\mathcal{M}}$ and ideal adversary $\mathcal{S}$ proceeds as follows.

1. Let $I = [1, N]$. While $I \neq \emptyset$:
   a) Wait for a message (*Message*, m_i) with $m_i \in \{0,1\}^n$ from some dummy sender $\mathcal{P}_i$ with $i \in I$.
   b) Set $I \leftarrow I \setminus \{i\}$ and hand (*MessageReceived*, i) to $\mathcal{S}$.
2. Hand $(\textit{Mixed}, \mathsf{Sort}(m_1, \ldots, m_N))$ to $\mathcal{S}$ and $\overline{\mathcal{M}}$.

Functionality 2 (Relaxed Mix-Net). The *relaxed mix-net* functionality $\mathcal{F}^{\mathrm{mn}}$ executing with dummy senders $\overline{\mathcal{P}}$, dummy mix-servers $\overline{\mathcal{M}}$ and ideal adversary $\mathcal{S}$ proceeds as follows.

1. Hand $(\textit{PublicKeys}, (pk_\ell)_{\ell=1}^{\lambda})$ to $\mathcal{S}$, where $(pk_\ell, sk_\ell) = \mathsf{Gen}(1^n)$.
2. Let $I = [1, N]$. While $I \neq \emptyset$:
   a) Wait for a message (*Message*, m_i) with $m_i \in \{0,1\}^n$ from some dummy sender $\mathcal{P}_i$ with $i \in I$.
   b) Set $I \leftarrow I \setminus \{i\}$ and hand (*MessageReceived*, i) to $\mathcal{S}$.
3. Let $L_\ell = \mathsf{Sort}\big((\mathsf{Enc}_{pk_\ell}(m_i))_{i \in [1,N] \setminus I_\mathcal{P}}\big)$. Hand $\big(\textit{HonestCiphertexts}, (L_\ell)_{\ell=1}^{\lambda}\big)$ to $\mathcal{S}$ and wait to get back (*CorruptCiphertexts*, L', ℓ^*), where $|L'| \leq |I_\mathcal{P}|$ and $1 \leq \ell^* \leq \lambda$.
4. Hand (*SecretKey*, sk_{ℓ^*}) to $\mathcal{S}$ and $\big(\textit{Mixed}, \mathsf{Sort}\big(\mathsf{Dec}_{sk_{\ell^*}}(\mathsf{Unique}(L_{\ell^*} \circ L'))\big)\big)$ to $\overline{\mathcal{M}}$.

Functionality 3 (Mix-Net With Abort). The *mix-net with abort* functionality $\mathcal{F}^{\mathrm{mna}}$ executing with dummy senders $\overline{\mathcal{P}}$, dummy mix-servers $\overline{\mathcal{M}}$, and ideal adversary $\mathcal{S}$ proceeds as follows.

1. Generate $(pk, sk) = \mathsf{Gen}(1^n)$ and hand (*PublicKey*, pk) to $\mathcal{S}$.
2. Let $I = [1, N]$. Then while $I \neq \emptyset$:
   a) Wait for a message (*Message*, m_i) with $m_i \in \{0,1\}^n$ from some dummy sender $\mathcal{P}_i$ with $i \in I$.
   b) Set $I \leftarrow I \setminus \{i\}$, let $v_i = \mathsf{Enc}_{pk}(m_i)$, and hand (*MessageReceived*, i) to $\mathcal{S}$.
3. Wait for a common input $J^* \subset J_\mathcal{M}$ from dummy mix-servers, i.e., $J^* \leftarrow \mathsf{Agree}(\textit{Culprits})$.
4. Let $L = \mathsf{Sort}\big((v_i)_{i \in [1,N] \setminus I_\mathcal{P}}\big)$ and wait for a message *EncryptPlaintexts* from $\mathcal{S}$. Then hand (*HonestCiphertexts*, L) to $\mathcal{S}$ and wait to receive (*CorruptCiphertexts*, L') where $|L'| \leq |I_\mathcal{P}|$, or (*Culprit*, d) where $d \in J_\mathcal{M} \setminus J^*$. In the latter case, hand (*Culprit*, d) to $\overline{\mathcal{M}}$ and halt.
5. Hand (*SecretKey*, sk) to $\mathcal{S}$ and $\big(\textit{Mixed}, \mathsf{Sort}\big(\mathsf{Dec}_{sk}(\mathsf{Unique}(L \circ L'))\big)\big)$ to $\overline{\mathcal{M}}$.

---

5   Chaum's Mix-Net

Consider Chaum's original mix-net [2] with λ mix-servers in the chain. Each mix-server $\mathcal{M}_j$ generates a key pair (pk_j, sk_j) and a sender wraps her message m_i in λ layers

of encryptions and submits a ciphertext $c_i = \mathsf{Enc}_{pk_1}\big(\mathsf{Enc}_{pk_2}\big(\cdots\mathsf{Enc}_{pk_\lambda}(m_i)\cdots\big)\big)$.
Then the mix-servers form an initial list $L_0 = (c_i)_{i\in[1,N]}$, and sequentially peel off
layers of encryptions after removing the duplicates. That is, for $j = 1,\ldots,\lambda$, the jth
mix-server computes $L_j = \mathsf{Sort}\big(\mathsf{Dec}_{sk_j}(\mathsf{Unique}(L_{j-1}))\big)$. Thus, $\mathsf{Unique}(L_\lambda)$ is the
sorted list of plaintexts without duplicates. This mix-net is neither secure against active
adversaries nor robust, but it nevertheless forms the basis of our constructions. We formalize this in Protocol 1 below and later extend it in two different ways in Protocol 2
and Protocol 3. We assume that the main protocol (Protocol 4) keeps track of the set J^*
of indices of identified corrupted mix-server so far, see Step 3 of Protocol 1 below.

Protocol 1 (Chaum's Mix-Net, π^c).
Mix-servers. The jth mix-server $\mathcal{M}_j$ proceeds as follows when executing with
functionalities $\mathcal{F}^{\mathrm{bb}}$, and $\mathcal{F}_1^{\mathrm{kg}},\ldots,\mathcal{F}_\lambda^{\mathrm{kg}}$.

1. Wait for $(PublicKey, pk_\ell)$ from $\mathcal{F}_\ell^{\mathrm{kg}}$ for $\ell = 1,\ldots,\lambda$. Let $pk = (pk_1,\ldots,pk_\lambda)$
   and output $(PublicKey, pk)$. Wait for $(SecretKey, sk_j)$ from $\mathcal{F}_j^{\mathrm{kg}}$ if $j \in [1,\lambda]$.
2. Wait for an input $(Culprits, J^*)$. For $\ell = 1,\ldots,\lambda$: if $\ell \in J^*$, then hand $Recover$
   to $\mathcal{F}_\ell^{\mathrm{kg}}$ and wait for a response $(SecretKey, sk_\ell)$.
3. Wait for an input $(Ciphertexts, L_0)$. For $\ell = 1,\ldots,\lambda$ do the following and
   output $\big(Mixed, \mathsf{Unique}(L_\lambda)\big)$:
   (a) If $\ell \in J^*$ or $\ell = j$, then set $L_\ell = \mathsf{Sort}\big(\mathsf{Dec}_{sk_\ell}(\mathsf{Unique}(L_{\ell-1}))\big)$, and
       publish $(Decryption, L_\ell)$.
   (b) Otherwise, wait until $\mathcal{M}_\ell$ publishes $(Decryption, L_\ell)$ (or we published L_ℓ,
       since sk_ℓ was recovered), where $|L_\ell| = |\mathsf{Unique}(L_{\ell-1})|$.

Protocol 2 and Protocol 3 formalize the two nested mix-nets used in our main protocol. Recall from Section 2 that the first protocol is an *optimistic* execution of Chaum's
mix-net. The privacy of this mix-net is only required to *temporarily randomize* the input
to the second mix-net. This is needed to argue that it is hard to replace all ciphertexts
submitted by a non-empty proper subset of the honest senders without being identified as a cheater. When Protocol 3 has completed, the optimistic execution is verified
explicitly by simply recovering the secret keys of all mix-servers.

Protocol 2 (Chaum's Mix-Net with Explicit Verification, π^{cev}).
Mix-servers. The jth mix-server $\mathcal{M}_j$ when executing with functionalities $\mathcal{F}^{\mathrm{bb}}$, and
$\mathcal{F}_1^{\mathrm{kg}},\ldots,\mathcal{F}_\lambda^{\mathrm{kg}}$, first runs Chaum's mix-net (Protocol 1) and then proceeds as follows.

4. Wait for an input $Verify$. Then for $\ell = 1,\ldots,\lambda$, where $\ell \notin J^*$:
   (a) If $\ell = j$, then publish $(SecretKey, sk_j)$.
   (b) If $\ell \neq j$ and $\ell \notin J^*$, then wait until $\mathcal{M}_\ell$ publishes $(SecretKey, sk_\ell)$, and
       halt with output $(Culprit, \ell)$ if sk_ℓ does not correspond to pk_ℓ or if $L_\ell \neq$
       $\mathsf{Sort}\big(\mathsf{Dec}_{sk_\ell}(\mathsf{Unique}(L_{\ell-1}))\big)$.
5. Halt with output $(SecretKey, sk)$, where $sk = (sk_1,\ldots,sk_\lambda)$.

In our second variant of Chaum's mix-net (Protocol 3), the mix-servers proceed op-
timistically, but in contrast to Protocol 2 they do not later verify the complete execution

explicitly. Instead, they *trace* a subset of ciphertexts backwards and forwards through the mix-net and reveal how they are decrypted in the process. In the main protocol a small subset of *dummy ciphertexts* (submitted by the mix-servers) are always traced forward to show that these were processed correctly. As explained in Section 2, the idea is that starting from the randomly permuted output of Protocol 2, the adversary must avoid modifying the traced ciphertexts to avoid detection. In other words, to cheat without detection, a corrupted mix-server can not simply replace all ciphertexts. However, tracing starts by tracing any ciphertexts that do not have exactly t copies backwards to distinguish the case where a corrupted sender submits a malformed set of ciphertexts from the case where a corrupted mix-server processes his input incorrectly. Only then are the dummies, and possibly additional ciphertexts, traced forwards through the mix-net.

Protocol 3 (Chaum's Mix-Net with Partial Tracing, π^{cpt}).

Mix-servers. The jth mix-server $\mathcal{M}_j$, when executing with functionalities $\mathcal{F}^{\text{bb}}$, $\mathcal{F}_1^{\text{kg}}, \ldots, \mathcal{F}_\lambda^{\text{kg}}$, and $\mathcal{F}_1^{\text{pd}}, \ldots, \mathcal{F}_\lambda^{\text{pd}}$, runs Chaum's mix-net (Protocol 1), hands $(SecretKey, sk_j)$ to $\mathcal{F}_j^{\text{pd}}$ if $j \in [1, \lambda]$, and then proceeds as follows.

4. *Backward Tracing.* Wait for an input $(TraceB, B_\lambda)$, where B_λ is the list of ciphertexts to be traced backwards. For $\ell = \lambda, \ldots, 1$ do the following and then output $(Traced, B_0)$:
   (a) Expand B_ℓ to a list B_ℓ' by adding the removed duplicates, i.e., the expanded list B_ℓ' includes all copies in L_ℓ of every ciphertext occurring in B_ℓ.
   (b) If $\ell \in J^*$ or $\ell = j$, then identify $B_{\ell-1} \subset L_{\ell-1}$ such that $B_\ell' = \text{Dec}_{sk_\ell}(B_{\ell-1})$ and publish $(TracedB, B_{\ell-1})$. Otherwise, wait until $\mathcal{M}_\ell$ publishes $(TracedB, B_{\ell-1})$ with $B_{\ell-1} \subset L_{\ell-1}$.
   (c) If $\ell \notin J^*$, then hand $(Verify, B_\ell', B_{\ell-1})$ to $\mathcal{F}_\ell^{\text{pd}}$ and halt with $(Culprit, \ell)$ if it returns *False*.
5. *Forward Tracing.* Wait for an input $(TraceF, F_0)$, where F_0 is the ciphertexts to be traced forward. For $\ell = 1, \ldots, \lambda$ do the following and then halt with output $(Traced, F_\lambda)$:
   (a) Let $F_{\ell-1}' = \text{Unique}(F_{\ell-1})$.
   (b) If $\ell \in J^*$ or $\ell = j$, then let $F_\ell = \text{Dec}_{sk_\ell}(F_{\ell-1}')$ and publish $(TracedF, F_\ell)$. Otherwise, wait until $\mathcal{M}_\ell$ publishes $(TracedF, F_\ell)$ with $F_\ell \subset L_\ell$.
   (c) If $\ell \notin J^*$, then hand $(Verify, F_\ell, F_{\ell-1}')$ to $\mathcal{F}_\ell^{\text{pd}}$ and halt with $(Culprit, \ell)$ if it returns *False*.

Forward tracing of the dummy ciphertext list F_0, a subset of the input list L_0 submitted by the mix-servers, is done in the natural way. For $\ell = 1, \ldots, \lambda$, the ℓth mix-server computes $F_\ell = \text{Dec}_{sk_\ell}(\text{Unique}(F_{\ell-1}))$ and proves that he did so correctly. The other mix-servers verify the proof and that $F_\ell \subset L_\ell$.

Backward tracing of a list B_λ, a subset of the output list $\text{Unique}(L_\lambda)$, is more complicated in that we must invert the process of duplicate removal. For $\ell = \lambda, \ldots, 1$, all mix-servers first expand B_ℓ into a list B_ℓ' by including all copies in L_ℓ of each ciphertext in B_ℓ, and then the ℓth mix-server computes $B_{\ell-1} \subset L_{\ell-1}$ such that $B_\ell' = \text{Dec}_{sk_\ell}(B_{\ell-1})$

and proves that this relation holds. Thus, the expansion is the inversion of how Unique removed duplicates of traced ciphertexts during processing.

The correctness of the decryption for the jth mix-server is verified using the proof of correct decryption functionality $\mathcal{F}_j^{\mathrm{pd}}$. Notice that for the dummies, it suffices that each mix-server simply reveals the randomness used to encrypt his own dummy inputs. However, for senders' inputs this may not be possible for a general cryptosystem since the randomness is chosen by the corresponding sender and may not be known to the decrypting mix-server. Nevertheless, one possible incarnation of our protocol uses a cryptosystem that allows recovering the randomness used during encryption from the ciphertext itself during decryption. In this case, the proof of correct decryption used during tracing simply consists of revealing the randomness.

6   Constructing a Mix-Net with Abort

We are now ready to present the details of our mix-net with abort in Protocol 4. We use two nested instances of Chaum's mix-net: one with explicit verification (Protocol 2), and one with partial tracing (Protocol 3). The lists of public keys of these mix-nets are denoted by pk^{cev} and pk^{cpt}, each of which contains λ keys. Each sender encrypts her message m_i once using the additional joint "final" public key pk^{f} to form a ciphertext v_i. This layer of encryption hides the inputs of the honest senders if the execution aborts. The ciphertext v_i is then encrypted independently t times with the additional joint "replication" public key pk^{r}. Recall that this prevents the last mix-server in Chaum's mix-net with partial tracing (Protocol 3) from identifying all ciphertexts submitted by the same sender. The resulting ciphertexts are then encrypted using the lists pk^{cpt} and pk^{cev} of public keys of the two instances of Chaum's mix-net. Finally, the t encryptions are concatenated to form one plaintext chunk and then encrypted using the "outer" public key pk^{o}, which prevents a dishonest sender (with the collusion of some dishonest mix-servers) from partially copying an honest sender's submission to break his privacy. In addition to the ciphertexts submitted by senders, each mix-server submits a dummy encryption of the zero message computed like a sender's ciphertext. These ciphertexts prevent a corrupt mix-server from replacing all ciphertexts instead of guessing the positions of all ciphertexts submitted by a subset of the senders.

To process the ciphertexts, the mix-servers first remove the "outer" layer of encryption by jointly recovering the corresponding secret key sk^{o}. Then they execute the two instances of Chaum's mix-net in sequence. We stress that the t ciphertexts of each sender are processed independently at this stage. Then the secret keys in sk^{cev} (corresponding to pk^{cev}) are recovered and the mix-servers verify the execution of the first mix-net explicitly. The "replication" secret key sk^{r} corresponding to pk^{r} is then recovered and all ciphertexts are decrypted. Finally, the processing in the second mix-net is verified for: (1) all ciphertexts of which there are not exactly t copies (backward tracing), and (2) all dummy ciphertexts submitted by mix-servers and all ciphertexts intersecting with the ciphertexts traced backwards (forward tracing). If there is any inconsistency, the corrupted mix-server is identified and the execution aborts. If there is no inconsistency, then the "final" secret key sk^{f} corresponding to pk^{f} is recovered and the innermost layer of encryption is removed to reveal the plaintexts.

Theorem 1 captures security of Protocol 4. If we use a cryptosystem that allows recovering the randomness used for encryption, then our result still holds, but only in the standalone model where the simulator is allowed to rewind. The full version details this variation of the scheme.

Protocol 4 (Mix-Net with Abort π^{mna}). This protocol is executed with a bulletin board $\mathcal{F}^{\mathrm{bb}}$, a mix-net with explicit verification π^{cev}, a mix-net with partial tracing π^{cpt}, and distributed key generation functionalities $\mathcal{F}_{\mathrm{o}}^{\mathrm{dkg}}$, $\mathcal{F}_{\mathrm{r}}^{\mathrm{dkg}}$ and $\mathcal{F}_{\mathrm{f}}^{\mathrm{dkg}}$.

Senders. The ith sender $\mathcal{P}_i$ proceeds as follows on input $m_i \in \{0,1\}^n$.

1. Wait until λ of the mix-servers have published identical list $(PublicKeys, pk^{\mathrm{o}}, pk^{\mathrm{cev}}, pk^{\mathrm{cpt}}, pk^{\mathrm{r}}, pk^{\mathrm{f}})$. If no such list exists, then abort.
2. Let $v_i = \mathsf{Enc}_{pk^{\mathrm{f}}}(m_i)$.
3. Let $u_{i,s} = \mathsf{Enc}_{pk^{\mathrm{cev}}}(\mathsf{Enc}_{pk^{\mathrm{cpt}}}(\mathsf{Enc}_{pk^{\mathrm{r}}}(v_i)))$, for $s = 1, \ldots, t$.
4. Let $\overline{u}_i = \mathsf{Enc}_{pk^{\mathrm{o}}}(u_{i,1} \| \cdots \| u_{i,t})$ and publish $(Ciphertext, \overline{u}_i)$.

Mix-servers. The jth mix-server $\mathcal{M}_j$ proceeds as follows on input J_j^*.

1. *Public Keys.* Wait for public keys: $(PublicKey, pk^{\mathrm{o}})$ from $\mathcal{F}_{\mathrm{o}}^{\mathrm{dkg}}$, $(PublicKey, pk^{\mathrm{cev}})$ from π^{cev}, $(PublicKey, pk^{\mathrm{cpt}})$ from π^{cpt}, $(PublicKey, pk^{\mathrm{r}})$ from $\mathcal{F}_{\mathrm{r}}^{\mathrm{dkg}}$, and $(PublicKey, pk^{\mathrm{f}})$ from $\mathcal{F}_{\mathrm{f}}^{\mathrm{dkg}}$. Then publish $(PublicKeys, pk^{\mathrm{o}}, pk^{\mathrm{cev}}, pk^{\mathrm{cpt}}, pk^{\mathrm{r}}, pk^{\mathrm{f}})$. Wait until λ of the mix-servers have published the same list, or abort if no such list can be found.
2. *Input Ciphertexts.* Wait until every $\mathcal{P}_i$ has published her encrypted message $(Ciphertext, \overline{u}_i)$. Let $\overline{u}_{N+j}$ be an encryption of zero as computed by a sender and publish $(Ciphertext, \overline{u}_{N+j})$. Wait until every $\mathcal{M}_\ell$ has published $(Ciphertext, \overline{u}_{N+\ell})$ and let $L^{\mathrm{in}} = \mathsf{Unique}(\overline{u}_1, \ldots, \overline{u}_{N+k})$.
3. *Culprits Agreement.* Publish $(Culprits, J_j^*)$ and wait until λ of the mix-servers have published identical $(Culprits, J^*)$, or abort if no such set J^* can be found. Input $(Culprits, J^*)$ to π^{cev} and π^{cpt}.
4. *Decrypt and Split.* Hand *Recover* to $\mathcal{F}_{\mathrm{o}}^{\mathrm{dkg}}$ and wait for a response $(SecretKey, sk^{\mathrm{o}})$. Let $L^{\mathrm{o}} = \bigcirc_{\overline{u} \in L^{\mathrm{in}}} \mathsf{Split}_t(\mathsf{Dec}_{sk^{\mathrm{o}}}(\overline{u}))$.
5. *Chaum's Mix-Net.* Input $(Ciphertexts, L^{\mathrm{o}})$ to π^{cev} and wait for an output $(Mixed, L^{\mathrm{cev}})$.
6. *Chaum's Mix-Net.* Input $(Ciphertexts, L^{\mathrm{cev}})$ to π^{cpt}, and wait for an output $(Mixed, L^{\mathrm{cpt}})$.

This protocol is completed on the next page.

Theorem 1. *Let $\mathcal{CS}$ be a CCA2 secure cryptosystem. Then Protocol 4 securely realizes Functionality 3 with respect to static active adversaries that corrupt less than λ of the mix-servers and any number of senders, provided that t is chosen such that $H^{-(t-1)}$ is negligible, where $H > 1$ is the number of honest parties.*

Due to our conventions in Section 4 and the definition of Functionality 3, the theorem also captures the robustness of the protocol, i.e., it gives an output provided that at most $\min(\lambda - 1, k - \lambda)$ parties are corrupted.

Protocol 4 (Continued, including verifications.).

7. *Verifications.*

  (a) *Explicit Verification.* Input *Verify* to π^{cev}. If it outputs $(Culprit, d)$, then halt with this output, and otherwise let $(SecretKey, sk^{\text{cev}})$ be the output.

  (b) *Replication Check.* Hand *Recover* to $\mathcal{F}_{\text{r}}^{\text{dkg}}$ and wait for a response $(SecretKey, sk^{\text{r}})$. Compute $L^{\text{r}} = \text{Dec}_{sk^{\text{r}}}(L^{\text{cpt}})$ and let B be the ciphertexts in L^{cpt} that do not have exactly t copies after decryption with sk^{r}.

  (c) *Backwards Tracing.* Input $(TraceB, B)$ to π^{cpt}. If it outputs $(Culprit, d)$, then halt with this output, and otherwise let $(TracedB, B')$ be the output. Let L' be the list of all $\overline{u} \in L^{\text{in}}$ such that π^{cev} on input $\big(Ciphertexts, \text{Split}_t(\text{Dec}_{sk^{\circ}}(\overline{u}))\big)$ would output $(Mixed, B'')$ with $B' \cap B'' \neq \emptyset$.

  (d) *Forward Tracing.* Let F be the list such that π^{cev} on input $(Ciphertexts, L' \circ L'')$, where $L'' = \bigcirc_{\ell \in [1,k]} \text{Split}_t\big(\text{Dec}_{sk^{\circ}}(\overline{u}_{N+\ell})\big)$, would give an output $(Mixed, F)$. Input $(TraceF, F)$ to π^{cpt}. If it outputs $(Culprit, d)$, then halt with this output. Otherwise, let $(TracedF, F')$ be the output.

8. *Final Decryption.* Hand *Recover* to $\mathcal{F}_{\text{f}}^{\text{dkg}}$ and wait for a response $(SecretKey, sk^{\text{f}})$. Let $L^{\text{r}'} = \text{Unique}(\text{Dec}_{sk^{\text{r}}}(L^{\text{cpt}} \setminus F'))$ and halt with output $\big(Mixed, \text{Sort}(\text{Dec}_{sk^{\text{f}}}(L^{\text{r}'}))\big)$.

---

7 Conclusion

We construct a provably secure mix-net that unlike many other mix-nets in the literature do not require any homomorphic properties from the cryptosystem. This is a clear advantage for those concerned that quantum computers can be constructed in the future. In contrast to the only previous proposed mix-net based on any cryptosystem [12], our construction enjoys not only provable security but also full privacy and correctness. Our mix-net is fast there are many senders and plaintexts are large.

References

1. Canetti, R.: Universally composable security: A new paradigm for cryptographic protocols. In: FOCS, pp. 136–145. IEEE Computer Society (2001)
2. Chaum, D.: Untraceable electronic mail, return addresses, and digital pseudonyms. Commun. ACM 24(2), 84–88 (1981)
3. Desmedt, Y., Kurosawa, K.: How to Break a Practical MIX and Design a New One. In: Preneel, B. (ed.) EUROCRYPT 2000. LNCS, vol. 1807, pp. 557–572. Springer, Heidelberg (2000)
4. Furukawa, J., Sako, K.: An Efficient Scheme for Proving a Shuffle. In: Kilian, J. (ed.) CRYPTO 2001. LNCS, vol. 2139, pp. 368–387. Springer, Heidelberg (2001)
5. Furukawa, J., Sako, K.: An efficient publicly verifiable mix-net for long inputs. IEICE Transactions 90-A(1), 113–127 (2007)
6. Gabber, E., Gibbons, P.B., Matias, Y., Mayer, A.: How to Make Personalized Web Browsing Simple, Secure, and Anonymous. In: Hirschfeld, R. (ed.) FC 1997. LNCS, vol. 1318, pp. 17–32. Springer, Heidelberg (1997)

7. Golle, P., Zhong, S., Boneh, D., Jakobsson, M., Juels, A.: Optimistic Mixing for Exit-Polls. In: Zheng, Y. (ed.) ASIACRYPT 2002. LNCS, vol. 2501, pp. 451–465. Springer, Heidelberg (2002)
8. Jakobsson, M.: A Practical Mix. In: Nyberg, K. (ed.) EUROCRYPT 1998. LNCS, vol. 1403, pp. 448–461. Springer, Heidelberg (1998)
9. Jakobsson, M.: Flash mixing. In: PODC, pp. 83–89 (1999)
10. Jakobsson, M., Juels, A.: Mix and match: Secure function evaluation via ciphertexts. In: Okamoto [18], pp. 162–177
11. Jakobsson, M., Juels, A.: An optimally robust hybrid mix network. In: PODC, pp. 284–292. ACM Press, New York (2001)
12. Jakobsson, M., Juels, A., Rivest, R.L.: Making mix nets robust for electronic voting by randomized partial checking. In: Boneh, D. (ed.) USENIX Security Symposium, pp. 339–353. USENIX (2002)
13. Jakobsson, M., M'Raïhi, D.: Mix-Based Electronic Payments. In: Tavares, S., Meijer, H. (eds.) SAC 1998. LNCS, vol. 1556, pp. 157–173. Springer, Heidelberg (1999)
14. Khazaei, S., Wikström, D.: Randomized partial checking revisited. Cryptology ePrint Archive. Report 2012/063 (2012), http://eprint.iacr.org/2012/063
15. Mitomo, M., Kurosawa, K.: Attack for flash mix. In: Okamoto [18], pp. 192–204
16. Neff, C.A.: A verifiable secret shuffle and its application to e-voting. In: CCS 2001: Proc. of the 8th ACM Conference on Computer and Communications Security, pp. 116–125. ACM, New York (2001)
17. Ohkubo, M., Abe, M.: A length-invariant hybrid mix. In: Okamoto [18], pp. 178–191
18. Okamoto, T. (ed.): ASIACRYPT 2000. LNCS, vol. 1976. Springer, Heidelberg (2000)
19. Park, C., Itoh, K., Kurosawa, K.: Efficient Anonymous Channel and All/Nothing Election Scheme. In: Helleseth, T. (ed.) EUROCRYPT 1993. LNCS, vol. 765, pp. 248–259. Springer, Heidelberg (1994)
20. Pfitzmann, B.: Breaking an Efficient Anonymous Channel. In: De Santis, A. (ed.) EUROCRYPT 1994. LNCS, vol. 950, pp. 332–340. Springer, Heidelberg (1995)
21. Pfitzmann, B., Pfitzmann, A.: How to Break the Direct RSA-Implementation of MIXes. In: Quisquater, J.-J., Vandewalle, J. (eds.) EUROCRYPT 1989. LNCS, vol. 434, pp. 373–381. Springer, Heidelberg (1990)
22. Sako, K., Kilian, J.: Receipt-Free Mix-Type Voting Scheme — A Practical Solution to the Implementation of a Voting Booth. In: Guillou, L.C., Quisquater, J.-J. (eds.) EUROCRYPT 1995. LNCS, vol. 921, pp. 393–403. Springer, Heidelberg (1995)
23. Wikström, D.: Five Practical Attacks for "Optimistic Mixing for Exit-Polls". In: Matsui, M., Zuccherato, R.J. (eds.) SAC 2003. LNCS, vol. 3006, pp. 160–175. Springer, Heidelberg (2004)
24. Wikström, D.: A Universally Composable Mix-Net. In: Naor, M. (ed.) TCC 2004. LNCS, vol. 2951, pp. 317–335. Springer, Heidelberg (2004)

A   Agreement Subroutine

Subroutine 1 (Agree(Tag)).

1. Set $J \leftarrow \{1, \ldots, k\}$.
2. While $J \neq \emptyset$:
   a) Wait for a message (Tag, m_j) from the dummy mix-server $\mathcal{M}_j$ with $j \in J$.
   b) Set $J \leftarrow J \setminus \{j\}$ and hand $(TagReceived, j, m_j)$ to $\mathcal{S}$.
3. Return the value in $(m_j)_{j \in [1,k]}$ that has been submitted by λ of the mix-servers. If no such value exists, hand $\bot$ to $\overline{\mathcal{M}}$ and halt the main functionality.

B Functionalities Implemented by General MPC

Functionality 4 (Key Generation with VSS). The key generation with VSS functionality $\mathcal{F}_j^{\mathrm{kg}}$ executing with dummy mix-servers $\overline{\mathcal{M}}$ and ideal adversary $\mathcal{S}$ proceeds as follows.

1. Generate $(pk, sk) = \mathsf{Gen}(1^n)$, hand $(PublicKey, pk)$ to $\mathcal{S}$ and $\overline{\mathcal{M}}$, and $(SecretKey, sk)$ to $\mathcal{M}_j$, and wait until dummy mix-servers agree to recover, i.e., run $\mathsf{Agree}(Recover)$.
2. Hand $(SecretKey, sk)$ to $\mathcal{S}$ and $\overline{\mathcal{M}}$.

Functionality 5 (Distributed Key Generation with VSS). The distributed key generation with VSS functionality $\mathcal{F}^{\mathrm{dkg}}$ executing with dummy mix-servers $\overline{\mathcal{M}}$ and ideal adversary $\mathcal{S}$ proceeds as follows.

1. Generate $(pk, sk) = \mathsf{Gen}(1^n)$.
2. Hand $(PublicKey, pk)$ to $\mathcal{S}$ and $\overline{\mathcal{M}}$, and wait until dummy mix-servers agree to recover, i.e., run $\mathsf{Agree}(Recover)$.
3. Hand $(SecretKey, sk)$ to $\mathcal{S}$ and $\overline{\mathcal{M}}$.

Functionality 6 (Bulletin board). Executing with dummy senders $\overline{\mathcal{P}}$, dummy mix-servers $\overline{\mathcal{M}}$ and ideal adversary $\mathcal{S}$, the bulletin board functionality $\mathcal{F}^{\mathrm{bb}}$ keeps a private and a public[a] database and proceeds as follows.

1. Upon receiving a message (Tag, m) from a party $P \in \overline{\mathcal{P}} \cup \overline{\mathcal{M}}$, hand (P, Tag, m) to $\mathcal{S}$ and write (P, Tag, m) on the private database. Ignore any further message (Tag, m') from the party P.
2. Upon receiving a message (P, Tag, m) from $\mathcal{S}$, see if (P, Tag, m) already exists in the private database. If so, then write (P, Tag, m) on the public database. Ignore any further message (P, Tag, m) from $\mathcal{S}$.

---

[a] The contents of the public database is known to all parites. In our protocols, parties need to wait until a specific party P publishes (Tag, m) on the bulletin board. This means that, they wait until (P, Tag, m) appears on the public database.

Functionality 7 (Proof of Correct Decryption). The proof of correct decryption functionality $\mathcal{F}_j^{\mathrm{pd}}$ executing with dummy mix-servers $\overline{\mathcal{M}}$ and ideal adversary $\mathcal{S}$ proceeds as follows.

1. Wait for an input $(SecretKey, sk)$ from dummy mix-server $\mathcal{M}_j$ and then hand $(SecretKey, j)$ to $\mathcal{S}$.
2. Wait for a common input (m, c) from dummy mix-servers, i.e., $(m, c) = \mathsf{Agree}(Verify)$, and send *True* or *False* to $\mathcal{S}$ and $\overline{\mathcal{M}}$ depending on if $m = \mathsf{Dec}_{sk}(c)$ or not.

How Not to Prove Yourself: Pitfalls of the Fiat-Shamir Heuristic and Applications to Helios

David Bernhard[1], Olivier Pereira[2], and Bogdan Warinschi[1]

[1] University of Bristol
[2] Université Catholique de Louvain

Abstract. The Fiat-Shamir transformation is the most efficient construction of non-interactive zero-knowledge proofs.

This paper is concerned with two variants of the transformation that appear but have not been clearly delineated in existing literature. Both variants start with the prover making a commitment. The strong variant then hashes both the commitment and the statement to be proved, whereas the weak variant hashes only the commitment. This minor change yields dramatically different security guarantees: in situations where malicious provers can select their statements adaptively, the weak Fiat-Shamir transformation yields unsound/unextractable proofs. Yet such settings naturally occur in systems when zero-knowledge proofs are used to enforce honest behavior. We illustrate this point by showing that the use of the weak Fiat-Shamir transformation in the Helios cryptographic voting system leads to several possible security breaches: for some standard types of elections, under plausible circumstances, malicious parties can cause the tallying procedure to run indefinitely and even tamper with the result of the election.

On the positive side, we define a form of adaptive security for zero-knowledge proofs in the random oracle model (essentially simulation-sound extractability), and show that a variant which we call *strong Fiat-Shamir* yields secure non-interactive proofs.

This level of security was assumed in previous works on Helios and our results are then necessary for these analyses to be valid. Additionally, we show that strong proofs in Helios achieve non-malleable encryption and satisfy ballot privacy, improving on previous results that required CCA security.

1 Introduction

Zero-knowledge proofs of knowledge allow a prover to convince a verifier that she holds information satisfying some desirable properties without revealing anything else. To be useful, such proof systems should satisfy completeness (the prover can convince the verifier that a true statement is indeed true) and soundness (the prover cannot convince the verifier that a false statement is true). Zero-knowledge proofs can either be interactive or non-interactive; for the latter

X. Wang and K. Sako (Eds.): ASIACRYPT 2012, LNCS 7658, pp. 626–643, 2012.

the prover only sends his proof and the verifier decides to accept or reject the statement without any further interaction.

The focus of this paper is on the most common and efficient construction of non-interactive proofs, namely the Fiat-Shamir heuristic [1]. Here, one begins with an interactive sigma protocol, a special type of three-move protocol in which the prover sends a commitment, the verifier answers with a random challenge and the prover completes the protocol with a response. The idea behind the transformation is simple and appealing: have the prover compute the message of the verifier as the hash of the message sent by the prover — if the hash is modelled as a random oracle the message computed this way should look random as in an interactive execution, hence the properties of the original proof system should somehow be preserved.

The transformation appears in the literature in two different forms, depending on what is hashed. In the formalization of Bellare and Rogaway [2], which we refer to as the weak Fiat-Shamir transformation (wFS), the hash takes only the prover's first message as input. Other papers e.g. [3,4] suggest including the statement to be proved in the hash input. In the remainder of the paper we call this the strong Fiat-Shamir transformation (sFS).

CONTRIBUTIONS. The contributions of this paper fall into two main categories. First we identify weaknesses of the weak (sic!) Fiat-Shamir transformation and show that in applications it can be a serious source of insecurity. Secondly, we provide several positive results regarding the strong Fiat-Shamir transformation and its uses in applications.

Insecurity of wFS *and Attacks on Helios.* Our first results show that the security proofs commonly given for Fiat-Shamir proofs do not hold when applied to weak proofs and when the prover can chose his statement(s) to prove adaptively. This may or may not render a protocol using them insecure, as a protocol may have other means of dealing with adaptivity. For example, in the original application to identification protocols, weak proofs are sufficient.

As an example where weak proofs do not yield security, we consider Helios [5,6], a cryptographic voting protocol that has been used in practice. Versions of Helios have been employed, for example, for the election of the president of the Université catholique de Louvain [6], the Princeton University Undergraduate Student Government [7] and the board of the IACR [8]. We focus on the zero-knowledge protocols implemented since Helios 2.0 [6] for elections based on homomorphic tallying, which are still used in the latest version of Helios as documented on [9] at the time of writing. In brief, those elections work as follows. Trustees first jointly generate an election public key, using NIZK proofs to make sure that this key actually includes contributions from all trustees. Then, to cast a ballot, a voter encrypts a vote and attaches NIZK proofs that the vote is legal. All ballots are placed on a publicly readable bulletin board. Eventually, the election administrators homomorphically add all ballots, decrypt the result and use NIZK to prove the correctness of their actions. The encryption scheme is exponential ElGamal and the particular NIZKs involved are obtained by applying the weak Fiat-Shamir transformation to the Schnorr [10], Chaum-Pedersen [11] and disjunctive Chaum-Pedersen protocols (and variants thereof). These proofs are used to guarantee that the privacy of the votes rests on all trustees, to enforce that voters create ballots containing valid votes and to prevent dishonest administrators from claiming a wrong result. We show that the use of the wFS transformation is the source of three types of insecurity:

a) breaking verifiability by allowing colluding administrators to cast a single ballot that is not well-formed and contains any chosen number of votes for a specific candidate,
b) breaking liveness of the system by allowing colluding administrators to fail providing the election outcome while proving that they behave honestly, or by allowing voters to cast a random vote which leads to tallying taking superpolynomial time, and
c) breaking privacy by allowing the casting of related ballots that do not contain mere copies of previously submitted ciphertexts.

The first two of these attacks are undetectable under normal circumstances.

While our focus is on Helios which is our motivating application, in the full version of our paper we also show attacks against schemes constructed via the Naor-Yung paradigm and via the encrypt-then-prove construction: when using proofs derived through wFS these constructions may yield malleable encryption schemes.

Security of Strong Fiat-Shamir and Applications. The problems that we have identified in the use of the wFS do not apply to proofs obtained through the strong version of the transformation. It is then natural to ask what level of security does one get from these proofs. We provide several results. First, we formulate a security notion for non-interactive zero-knowledge proofs of knowledge which captures adversaries that can choose their statements adaptively. In essence, this notion is the analogue of simulation-sound extractability defined by Groth in the common reference string model [12]. Informally, a malicious prover is allowed to see simulated proofs (of potentially fake statements) and aims to provide valid looking proofs for adaptively chosen statements in such a way that an extractor cannot obtain witnesses. Interestingly, our definition is not simply a rehashing of the notion in [12]. In the random-oracle model, extraction requires the rewinding of the prover (as opposed to merely using a trapdoor) and in turn, this implies complex interaction between the adversary, the simulator and the extractor. We then show that applying sFS to Σ-protocols results in protocols that are simulation-sound extractable. Our result seems to be the first thorough investigation on the precise security guarantees offered by such proofs.

As a first application of this result, we investigate the security of non-malleable encryption schemes that are built by combining an IND-CPA encryption scheme with a proof of knowledge of the randomness used in the encryption process. We refer to this construction as the Enc+PoK approach. A well-known instantiation is the TDH0 scheme introduced and studied by Shoup and Gennaro [13]. Intuitively the construction should achieve IND-CCA security but so far, all attempts have failed to confirm or disprove this under natural assumptions (e.g., DDH in the random oracle model) [14,15]. As a consequence, the form of non-malleability ensured by Enc+PoK schemes is, surprisingly, still unknown. We provide a lower-bound on the answer to this question: if the proof of knowledge used in the encryption process is simulation-sound extractable, then the resulting scheme is NM-CPA secure. An immediate corollary is that the TDH0 scheme is NM-CPA secure in the random oracle model under the DDH assumption.

We then turn to the analysis of ballot privacy in Helios. Prior work shows that ballot privacy is guaranteed if the encryption scheme used in the construction is IND-CCA [16,17]. Since ballots in Helios use the Enc+PoK paradigm (which, as discussed above, is not known to be IND-CCA) a natural suggestion is then to replace it with something stronger. For example, Bernhard et al. suggested applying the Naor-Yung transformation to the underlying ElGamal encryption [17], while Bulens et al. used a variant of the TDH2 scheme [18]. These modifications both substantially increase the computational costs of the system and require major changes in the implementation.

Our final result is to show that although the NM-CPA notion is strictly weaker than IND-CCA [19], it is sufficient to ensure ballot privacy. In particular a minor tweak of the Enc+PoK construction currently used in Helios where we replace wFS with its strong counterpart and check for repeated ciphertexts is sufficient. The change that we require is easily accomplished by including additional elements in the inputs of the hash function and preserves the current level of efficiency.

2   The Fiat-Shamir/Blum Transformation

In this section we introduce the two variants of the Fiat-Shamir heuristic that we analyze. We start by fixing notation and recalling some standard notions. In the following we let $R \subseteq \mathcal{P}(\{0,1\}^* \times \{0,1\}^*)$ be an efficiently computable relation. R defines a language $\mathcal{L}_R = \{Y \in \{0,1\}^* | \exists w : R(w,Y)\}$ in NP. We further assume that there is a well-defined set $\Lambda \supseteq \mathcal{L}$ decidable in polynomial time.[1]

A non-interactive proof system for language $\mathcal{L}_R$ is a pair algorithms (Prove, Verify). Such a proof system is complete for $\mathcal{L}_R$ if for every $(w, Y) \in R$, with overwhelming probability if $\pi \leftarrow \mathsf{Prove}(w, Y)$ then $\mathsf{Verify}(Y, \pi) = 1$. We define soundness of such proof systems (the property that a cheating prover cannot make the verifier accept a false statement) later in the paper. Here we recall the notion of zero-knowledge in the random oracle model [20].

In this setting, a simulator $\mathcal{S}$ for a proof system is an algorithm in charge of answering random oracle queries and producing valid proofs for any statement $Y \in \Lambda$ with respect to this oracle. In particular, it can "patch" the oracle to create its simulated proofs. Such a simulator responds to the following queries:

---

[1] Suppose that $\mathcal{L}$ is the set of DDH triples (G^a, G^b, G^{ab}) over some group $\mathbb{G}$. Then Λ could be $\mathbb{G}^3$. The reason for defining this formally is that we will later expect our zero-knowledge simulator to produce valid "proofs" for some "false statements", but which ones? Can it produce a proof for the statement consisting of the empty string, for example? We use Λ as the class of statements on which the simulator can produce proofs.

$\mathcal{H}(s)$ $\mathcal{S}$ maintains a list of oracle query/response pairs. For repeated queries, $\mathcal{S}$ answers consistently; for fresh queries, $\mathcal{S}$ draws a random value r, adds (s, r) to its list and returns r.

Simulate(Y) For $Y \in \Lambda$, the simulator returns a proof π such that Verify$(Y, \pi) = 1$ if the verifier uses the simulator for its oracle queries. $\mathcal{S}$ can add query/response pairs to its oracle list to process a simulation query.

Definition 1 (Zero-Knowledge). *A proof system is zero-knowledge if there is a simulator $\mathcal{S}$ such that no adversary who can make queries to the random oracle and queries of the form* create-proof(w, Y) *can distinguish the following two settings with non-negligibly better than* $1/2$ *probability.*

1. Random oracle queries are answered by a random oracle. In response to create-proof(w, Y), the challenger checks that $R(w, Y)$. If not, he returns $\perp$. Otherwise, he returns Prove(w, Y).
2. The challenger runs a copy of the simulator $\mathcal{S}$. It forwards random oracle queries to $\mathcal{S}$ directly. For create-proof(w, Y), the challenger checks if $R(w, Y)$ holds: if not, the challenger returns $\perp$; if it holds, the challenger sends Simulate(Y) to $\mathcal{S}$ and returns the result to the adversary.

Sigma Protocols. A sigma protocol for a language $\mathcal{L}_R$ is a protocol for two parties, a prover and a verifier. Both share a statement $Y \in \mathcal{L}_R$ as input and the prover may additionally hold a witness w.

The prover begins by sending a value A known as the commitment. The verifier replies with a challenge c drawn uniformly from a fixed challenge set. The prover finishes the protocol with a response f whereupon the verifier applies a deterministic algorithm Verify to Y, A, c and f which can accept or reject this execution.

A sigma protocol is correct (w.r.t. $\mathcal{L}_R$) if the prover, on input a pair (w, Y) satisfying R and interacting with the verifier who has input Y, gets the verifier to accept with probability 1.

A sigma protocol has *special honest verifier zero knowledge* if there is an algorithm Simulate that takes as input a statement $Y \in \Lambda$, challenge c and response f and outputs a commitment A such that Verify$(Y, A, c, f) = 1$ and furthermore, if c and f where chosen uniformly at random from their respective domains then the triple (A, c, f) is distributed identically to that of an execution between the prover and the verifier. Notice that the verifier is supposed to work with statements that may be false.

A sigma protocol has *special soundness* if there is an algorithm Extract that takes as input a statement Y and any two triples (A, c, f) and (A', c', f') such that both verify w.r.t. Y, $A = A'$ and $c \neq c'$, and returns a witness w such that $R(w, Y)$.

The Fiat-Shamir Transformation. The Fiat-Shamir transformation [1] (which [2] attributes to Blum) is a technique to make sigma protocols non-interactive using a cryptographic hash function. There are two commonly used descriptions of this technique that we call weak and strong Fiat-Shamir and which we describe together in the following definition.

Definition 2 (Fiat-Shamir Transformation). *Let* $\Sigma = (\mathsf{Prove}_\Sigma, \mathsf{Verify}_\Sigma)$ *be a sigma protocol and* $\mathcal{H}$ *a hash function. The weak Fiat-Shamir transformation of* Σ *is the proof system* $\mathsf{wFS}_\mathcal{H}(\Sigma) = (\mathsf{Prove}, \mathsf{Verify})$ *defined as follows:*

$\mathsf{Prove}(w, Y)$ *Run* $\mathsf{Prove}_\Sigma(w, Y)$ *to obtain commitment* A. *Compute* $c \leftarrow \mathcal{H}(A)$. *Complete the run of* Prove_Σ *with* c *as input to get the response* f. *Output the pair* (c, f).

$\mathsf{Verify}(Y, c, f)$ *Compute* A *from* (Y, c, f), *then run* $\mathsf{Verify}_\Sigma(Y, A, c, f)$.

The strong Fiat-Shamir transformation of Σ, *i.e.,* $\mathsf{sFS}(\Sigma) = (\mathsf{Prove}, \mathsf{Verify})$ *is obtained as above with the difference that* c *is computed by* $c \leftarrow \mathcal{H}(Y, A)$.

3   Pitfalls of the Weak Fiat-Shamir Transformation

We now describe various standard protocols in which the use of the weak Fiat-Shamir transformation can have undesirable effects. We illustrate these effects through several new practical attacks on various components of the Helios voting system, which relies on these protocols.

SCHNORR PROOFS. The Schnorr [10] signature scheme is the weak Fiat-Shamir transformation of the Schnorr identification protocol. In a group $\mathbb{G}$ of order q generated by G, it proves knowledge of an exponent x satisfying the equation $X = G^x$ for a known X. Viewing (x, X) as a signing/verification key pair and including a message in the hash input yields a signature of knowledge.

To create a proof, the prover picks a random $a \leftarrow \mathbb{Z}_q$ and computes $A = G^a$. He then hashes A to create a challenge $c = \mathcal{H}(A)$. Finally he computes $f = a + cx$; the proof is the pair (c, f) and the verification procedure consists in checking the equation $c \stackrel{?}{=} \mathcal{H}(\frac{G^f}{X^c})$.

The weak Fiat-Shamir transformation can safely be used here, as discussed in previous analysis [10,21], since the public key X is selected first and given as input to the adversary who tries to produce a forgery.

However, if the goal of the adversary is to build a valid triple (X, c, f) for any X of his choice, then this protocol is not a proof of knowledge anymore unless the discrete logarithm problem is easy in $\mathbb{G}$. Suppose indeed that there is an extractor $\mathcal{K}$ that, by interacting with any prover $\mathcal{P}$ that provides a valid triple (X, c, f), extracts $x = \log_G(X)$. This extractor can be used to solve an instance Y of the discrete logarithm problem with respect to $(\mathbb{G}, G)$ as follows: use Y as the proof commitment, compute $c = \mathcal{H}(Y)$, choose $f \leftarrow \mathbb{Z}_q$ and set $X = (\frac{G^f}{Y})^{\frac{1}{c}}$. Since the proof (Y, c, f) passes the verification procedure for statement X, the extractor $\mathcal{K}$ should be able to compute $x = \log_G(X)$ by interacting with our prover. We now observe that, by taking the discrete logarithm in base G on both sides of the definition of X, we obtain the solution $\log_G(Y) = f - cx$ to the discrete logarithm challenge.

Application to Helios. Schnorr proofs are used during the key generation procedure of Helios as a way to prevent trustees from choosing their public key as a function of the public key of the other trustees, which could give them the possibility to select the election private key at will and to decrypt all individual votes [22]. While the scenario above shows that trustees who publish a public key together with a Schnorr proof for that public key do not necessarily know the corresponding private key, the fact that our scenario does not allow the prover to choose his statement (but just to compute it as a function of the elements of the proof) does not seem to give rise to any practical attack. These weak Schnorr proofs would, however, break the proof of ballot privacy that we give later in this paper (assuming strong proofs).

CHAUM-PEDERSEN PROOFS. Chaum and Pedersen [11] introduced a proof of discrete logarithm equality, which they make non-interactive using the strong form of the Fiat-Shamir transformation. More precisely, given two group elements (G, X), a prover who knows the discrete logarithm $x = \log_G(X)$ can prove that two group elements (R, S) satisfy the relation $\log_G(X) = \log_R(S)$ as follows. He picks a random $a \leftarrow \mathbb{Z}_q$, computes $A = G^a$, $B = R^a$, $c = \mathcal{H}(R, S, A, B)$ and $f = a + cx$. The proof is the pair (c, f) and the verification procedure consists in checking the equation $c \overset{?}{=} \mathcal{H}(R, S, \frac{G^f}{X^c}, \frac{R^f}{S^c})$.

We observe that this proof is not sound anymore if it is used as a proof that three elements (X, R, S) are such that $\log_G(X) = \log_R(S)$, that is, if the prover also has the possibility to choose X in the process of building his proof. Indeed, a prover could select $(a, b, r, s) \leftarrow \mathbb{Z}_q^4$ at random, compute $A = G^a$, $B = G^b$, $R = G^r$ and $S = G^s$ from which he can compute $c = \mathcal{H}(R, S, A, B)$ and $f = \frac{b+cs}{r}$. He now completes the proof by computing $x = (f - a)/c$ and setting $X = G^x$. Now, we observe that $\log_G(X) = \frac{s}{r} + \frac{b-ar}{rc}$ while $\log_R(S) = \frac{s}{r}$, which differ with overwhelming probability.

Application to Helios. Chaum-Pedersen proofs instantiated with the weak Fiat-Shamir transformation (that is, $c = \mathcal{H}(A, B)$) are used during the ElGamal decryption procedure of Helios, in order to demonstrate that the decryption of the product of the votes that is computed by the trustees is consistent with the public key. More precisely, given a public key X and a ciphertext (R, S) that encrypts the sum of all votes, a trustee is required to compute $T = R^x$ where $x = \log_G(X)$ and to publish it together with a Chaum-Pedersen proof that $\log_G(X) = \log_R(S)$. The ElGamal decryption is then computed as $\log_G(S/T)$.

In this proof, a malicious trustee does not have the possibility to choose his private key at decryption time, but has the possibility to select T as part of the proof computation process. He can do so as follows. Select $(a, b) \leftarrow \mathbb{Z}_q^2$ at random, compute the proof commitments $A = G^a$ and $B = G^b$, the challenge $c = \mathcal{H}(A, B)$ and the response $f = a + cx$. Eventually, compute the decryption factor $T = (\frac{R^f}{B})^{\frac{1}{c}}$. It is easy to verify that the proof (c, f) is valid for the tuple (G, X, R, S), but that $\log_R(T) = x + \frac{ar-b}{c}$, which will be different from x with overwhelming probability. As a result, the decryption procedure will provide an aberrant result: an essentially random element of $\mathbb{Z}_q$. This strategy provides a way to build a denial of service attack against a Helios election, without anyone being able to detect who was responsible.

A more dangerous attack can be mounted if we assume that the trustees have the possibility to passively eavesdrop on the randomness of all voters. Though demanding, such an attack is still easier to mount and harder to detect than a full active attack. We would expect the impact of such a scenario to be "only" a complete loss of privacy by the voters but we show that it actually provides a way for the trustees to announce any election outcome of their choice as soon as they can actively corrupt a single voter (which can happen simply if a trustee is a voter himself).

Consider an election with trustees who would like to announce the election outcome m. These trustees select a private key x and publish the public key X. They also select $(a, b) \leftarrow \mathbb{Z}_q^2$, compute $A = G^a$, $B = G^b$, $c = H(A, B)$ and $f = a + cx$. Then all the voters submit their votes, except the corrupted one who waits until the last minute of the election. At that time, the trustees compute the product of all encrypted votes that have been submitted and obtain a ciphertext $(R', S') = (G^{r'}, G^{m'} \cdot G^{xr'})$ for some values r' and m' that they can compute using the randomness of the voters. They now compute $r = \frac{b+c(m'-m)}{f-cx}$, as well as a ciphertext $(G^{r-r'}, G^{x(r-r')})$ which is an encryption of 0 for which they can compute a proof of validity since they know $r - r'$. This ciphertext and proof are submitted by the corrupted voter, with the effect that the product of all encrypted votes is $(R, S) = (G^r, G^{m'} \cdot G^{xr})$. It can now be verified that (c, f) form a valid proof that $\log_G(X) = \log_R(\frac{S}{G^m})$, which indicates that m is the outcome of the election.

DISJUNCTIVE CHAUM-PEDERSEN PROOFS. Disjunctive proofs allow proving that one of two statements holds without revealing which one is correct. These proofs have numerous applications. For instance, they can be used by a voter to demonstrate that a ciphertext he produced is an encryption of either 0 or 1 (but nothing else), expressing whether or not he supports a candidate.

Suppose that a voter builds an exponential ElGamal ciphertext (R, S) with respect to public key X and wants to prove that it encrypts 0 or 1. We consider the case where it is an encryption of 1 (the other case is similar). First, the voter simulates a proof that $\log_R(S) = x$ by selecting a random proof $(c_0, f_0) \leftarrow \mathbb{Z}_q^2$ and computing $A_0 = G^{f_0}/R^{c_0}$ and $B_0 = X^{f_0}/S^{c_0}$. Then he selects $a_1 \leftarrow \mathbb{Z}_q$, computes $A_1 = G^{a_1}$, $B_1 = X^{a_1}$, $c = \mathcal{H}(A_0, B_0, A_1, B_1)$, $c_1 = c - c_0$ and $f_1 = a_1 + c_1 r$. The proof consists of (c_0, c_1, f_0, f_1) and verification consists of verifying whether $c_0 + c_1 = \mathcal{H}(\frac{G^{f_0}}{R^{c_0}}, \frac{X^{f_0}}{S^{c_0}}, \frac{G^{f_1}}{R^{c_1}}, \frac{X^{f_1}}{(S/G)^{c_1}})$.

Application to Helios. The proof we just described is exactly the one used in Helios to guarantee that voters encode at most one vote for each candidate and it exhibits weaknesses that are similar to those described above, but with an even more dangerous effect. Consider an election organized by corrupted trustees who would like to influence the election outcome by adding (or removing) m approvals to a candidate of their choice. These trustees now have the freedom to choose any public key and ciphertext of their choice that would allow them to compute an encryption of m and to prove that it is an encryption of 0 or 1. They can achieve this as follows.

They first select $(a_0, b_0, a_1, b_1) \leftarrow \mathbb{Z}_q^4$, from which they compute the commitments $A_0 = G^{a_0}$, $B_0 = G^{b_0}$, $A_1 = G^{a_1}$, $B_1 = G^{b_1}$, the challenge $c = \mathcal{H}(A_0, B_0, A_1, B_1)$ and the private key $x = \frac{(b_0 + cm)(1-m) - b_1 m}{a_0(1-m) - a_1 m}$ and the public key $X = G^x$. Using this public key, they select a random encryption of m by selecting $r \leftarrow \mathbb{Z}_q$ and computing $(R, S) = (G^r, G^m X^r)$. Eventually, they compute the challenges $c_1 = \frac{b_1 - a_1 x}{1-m}$ and $c_0 = c - c_1$ and the responses $f_0 = a_0 + c_0 r$ and $f_1 = a_1 + c_1 r$. It can be verified that (c_0, c_1, f_0, f_1) form a proof that (R, S) encrypt 0 or 1, while it actually encrypts an arbitrary m.

Furthermore, it can be observed that this proof, like the others that we presented above, is indistinguishable from a regular one.

Other attack possibilities exist, based on the same techniques. For instance, a voter who does not know the election private key can build a ciphertext that encrypts a random value in $\mathbb{Z}_q$ and prove that it encrypts 0 or 1, which would make the decryption procedure fail. We do not know however whether it is possible to build such a proof in a way that is indistinguishable from a regular one.[2]

ENCRYPT + PoK. Adding a proof of knowledge of the plaintext/randomness to a ciphertext in an IND-CPA secure public key encryption scheme is a common way to yield a non-malleable encryption scheme.[3] We formalise this construction and show that using wFS does *not* yield non-malleable encryption.

Definition 3 (Encrypt+PoK). *Let* $\mathfrak{E} = (\mathsf{KeyGen}, \mathsf{Enc}, \mathsf{Dec})$ *be a public-key encryption scheme. Let* $R((m, r), (Y, pk)) := (Y = \mathsf{Enc}(pk, m; r))$ *be the relation that* Y *is an encryption of* m *with randomness* r *for public key* pk, *let* Λ *be the ciphertext space (or some suitable superset thereof) and let* $\mathfrak{P} = (\mathsf{Prove}, \mathsf{Verify})$ *be a NIZK-PoK for this relation.*

The Encrypt+PoK transformation $\mathfrak{E}_{\mathfrak{P}}$ *is the following encryption scheme.*

KeyGen' Run KeyGen.
Enc'(pk, m) Draw some randomness r and create a ciphertext $E = \mathsf{Enc}(pk, m; r)$.
    Create a proof $\pi \leftarrow \mathsf{Prove}(pk, E, m, r)$. The ciphertext is the pair (E, π).
Dec'(sk, E, π) First run $\mathsf{Verify}(pk, E, \pi)$. If this fails (returns 0), output $\bot$ and halt. Otherwise, return $\mathsf{Dec}(sk, E)$.

Consider the ElGamal encryption scheme with weak Schnorr proofs of the randomness used for encryption (which would allow one to extract the message), which would be a weak variant of the TDH0 scheme [13]. In other words, a ciphertext for a message M under public key X is $(G^r, M \cdot X^r, c, f)$ where $c = \mathcal{H}(G^f / (G^r)^c)$. We can rerandomise such a ciphertext (R, S, c, f) by picking a random u and setting the new ciphertext to be $(R \cdot G^u, S \cdot X^u, c, f + cu)$. The new plaintext is the same as the old one as $S/R^x = M \cdot X^{r+u}/(G^{r+u})^x = M$ and the proof still verifies as $c = \mathcal{H}\left(\frac{G^{f+cu}}{(G^{r+u})^c}\right) = \mathcal{H}\left(\frac{G^f}{(G^r)^c}\right)$. Clearly, this encryption scheme is malleable.

---

[2] Our current technique involves setting $c_0 = 0$. While such a ballot passes the current Helios verifier, this could be detected in an audit.

[3] Though the exact form of non-malleability that is provided is unclear [13].

Application to Helios. The same rerandomisation technique can be applied to current Helios ballots, giving a ballot privacy attack in the style of Cortier and Smyth [23] (based on the same principles as the attacks described in [24,25].) Helios ballots contain ElGamal ciphertexts (R, S) with disjunctive Chaum-Pedersen proofs (c_0, c_1, f_0, f_1). To rerandomise such a ciphertext, pick a random u and set $R' = R \cdot G^u$, $S' = S \cdot Y^u$, $f_0' = f_0 + c_0 u$ and $f_1' = f_1 + c_1 u$. Unlike previously known rerandomisation techniques, this one does not make use of a repeated ElGamal ciphertext or proof. It can be detected however by checking for repeated hash values, just as for the previous attacks.

FURTHER EXAMPLES. The various attacks that we described above focus on applications to the Helios voting system, which uses the weak Fiat-Shamir transformation in all proofs. We believe that these examples provide clear evidence that the weak Fiat-Shamir transformation should not be used in that context: in particular, we showed that malicious authorities can arbitrarily influence the outcome of an election, which is in clear contradiction with the universal verifiability properties expected from that system. In the next sections, we will focus on the properties of the strong Fiat-Shamir transformation and show the benefits that its adoption would provide for the Helios system.

We stress that there are various other contexts in which the weak Fiat-Shamir transformation should not be used. For instance, similarly to our observation for the weak variant of the TDH0 scheme, the scheme resulting from the Naor-Yung transformation [26] applied to ElGamal encryption may become malleable if the weak Fiat-Shamir transformation is used, contradicting the level of desired security. We provide attacks against a concrete instantiation of that transformation in the full version of our paper.

4   Simulation Sound Extractable Proofs

The examples discussed in the previous section show that the wFS transform fails to offer even the most basic soundness properties in many contexts. We now investigate the soundness properties of the sFS transform. More precisely, we formulate the notion of *simulation sound extractable proofs* in the random oracle model and show its applications to the sFS transformation. Our definition draws inspiration from that of witness-extended emulation [4] in which the existence of an extractor is demanded such that for any adversary returning a vector of statements and proofs, the extractor returns identically distributed elements along with the witnesses to the proven statements. However, the definition is perhaps more appropriately viewed as the analogue definition of simulation sound extractability defined by Groth [12] which combines the simulation soundess approach of Sahai [27], with proofs of knowledge [28].

We consider a malicious prover who may ask to see simulated proofs (as in simulation-soundness). The extractor that we consider gets the transcript of a run of the prover where the prover outputs several valid proofs together with the transcipt of random oracle queries. His goal is to extract witnesses of these proofs. In the process, we allow the extractor to invoke and communicate with copies of the prover that use the same randomness as the run it is trying to extract from. This ability is what permits the knowledge extractor to fork the prover's execution without giving the extractor access to the coins of the prover.[4] A bit more precisely, a malicious prover for a proof system $\mathfrak{P} = (\mathsf{Prove}, \mathsf{Verify})$ is an algorithm $\mathcal{A}$ that expects access to two oracles: a hashing oracle and a simulation oracle. Thus $\mathcal{A}$ may submit some string s and expects $\mathcal{H}(s)$ in return and it may also make simulation calls $\mathsf{Simulate}(Y)$ for any statement $Y \in \Lambda$ and expects to obtain a proof π such that $\mathsf{Verify}(Y, \pi) = 1$. The prover returns a pair of vectors $(\boldsymbol{Y}, \boldsymbol{\pi})$.

Definition 4 (Simulation Sound Extractability). *Let $\mathfrak{P}$ be a zero-knowledge proof system with simulator $\mathcal{S}$. We say that $\mathfrak{P}$ is simulation sound extractable (SSE) if there exists an extractor $\mathcal{K}$ such that for every prover $\mathcal{A}$, $\mathcal{K}$ wins the following game with non-negligible probability.*

1. *(Initial run.)* The game selects a random string ω for $\mathcal{A}$. It runs an instance of $\mathcal{A}$ with the simulator $\mathcal{S}$ until $\mathcal{A}$ makes his output and halts. If $\mathcal{A}$ does not output any proofs, any of the proofs do not verify (w.r.t. the instance of $\mathcal{S}$ used as the random oracle) or any of $\mathcal{A}$'s statement/proof pairs (Y, π) is such that π was the result of a $\mathsf{Simulate}(Y)$ query, then $\mathcal{K}$ wins the game directly.
2. *(Extraction.)* The game runs an instance of $\mathcal{K}$, giving it the transcript of all queries in the initial run and the produced $(\boldsymbol{Y}, \boldsymbol{\pi})$ as input. $\mathcal{K}$ may repeatedly make one type of query invoke in response to which the game runs a new invocation of $\mathcal{A}$ on the same randomness ω that it chose for the initial run. All queries made by these instances are forwarded to $\mathcal{K}$ who can reply to them.
3. $\mathcal{K}$ wins the game if it can output a vector of witnesses $\boldsymbol{w}$ that match the statements $\boldsymbol{Y}$ of the initial run, i.e. for all i we have $R(w_i, Y_i)$.

The following theorem confirms that the strong Fiat-Shamir transformation yields proof systems that satisfy the notion we described above.

Theorem 1. *Let Σ be a sigma protocol with a challenge space that is exponentially large in the security parameter, special soundness and special honest verifier zero-knowledge. Then $\mathsf{sFS}(\Sigma)$ is zero-knowledge and simulation sound extractable with respect to expected polynomial-time adversaries.*

Applications. The Schnorr and Chaum-Pedersen protocols are clearly both sigma protocols with special soundness and special honest verifier zero knowledge so Theorem 1 applies and the sFS versions of these protocols are SSE proofs. For disjunctive Chaum-Pedersen, the challenge is the actual c obtained from the verifier and the response is the tuple $f = (f_0, f_1)$. This is a sigma protocol with special soundness and almost special honest verifier zero knowledge — almost, because in our definition the simulator chooses c, f independently and uniformly

---

[4] Although not necessary for this paper, hiding the adversary's randomness from the extractor can be helpful in other contexts to prove separation results.

at random yet if $c \neq c_0 + c_1$ then the resulting proof will not verify, patched oracle or not. We could fix this by not sending c_1 in the response and having the verifier recompute $c_1 = c - c_0$. We will ignore this point as it is easy to see that, if the simulator chooses c, f at random and then adjusts c_0, all relevant theorems still hold. In particular, the sFS transformation of disjunctive Chaum-Pedersen is still a simulation-sound extractable proof.

Encrypt + PoK.

With this notion we can restore the folklore result that appending a PoK to an IND-CPA scheme gives a NM-CPA one, if the PoK is simulation-sound extractable. For space reasons the proof is only in the full version of our paper.

Theorem 2. *Let $\mathfrak{E}$ be an IND-CPA secure encryption scheme and $\mathfrak{P}$ be a simulation-sound extractable NIZK-PoK for the encryption relation. Then $\mathfrak{E}_{\mathfrak{P}}$ is non-malleable (NM-CPA) secure with respect to expected polynomial-time adversaries.*

5   Ballot Privacy in Helios

In this section we propose a modification to Helios and prove that it satisfies ballot privacy in the model of single-pass voting of Bernhard et al. [17].

SINGLE-PASS SCHEMES. A single-pass voting scheme is a protocol consisting of the following algorithms and execution protocol for a set $\mathcal{V}$ of voters, a set $\mathcal{T}$ of trustees and a bulletin board $\mathcal{B}$. The class of single-pass schemes includes not only Helios [5] but also several other cryptographic voting schemes [29,30,27,31]. Single-pass voting models two of the most popular approaches to cryptographic voting, homomorphic tallying and mix-nets. Voters need only read a single message off the board (the election specification and public keys) and post a single message (their ballot) in return. We assume some underlying voter authentication mechanism.[5]

Setup(1^λ) is an algorithm to create public parameters for the election and secret ones for all trustees.

Setup produces one public output Y known as the public key of the election and a secret output x_i for each trustee $\mathcal{T}_i$ in the set of trustees $\mathcal{T}$. The secret outputs of all trustees together are known as the secret key of the election.

Vote(id, v, Y) is a probabilistic algorithm run by voters to prepare their votes for submission. It takes as input a voter's identity id, a vote v and public information Y and outputs a ballot $s \leftarrow$ Vote(id, v, Y).

Validate(b, s) models the algorithm run by the bulletin board during voting. Its inputs are the current board state b and the submitted ballot s. It returns 1 if the submission is deemed valid (given the current state of the board) and 0 otherwise.

---

[5] In the election of the president of UC Louvain [6] using Helios, authentication was handled by the university's existing infrastructure. As such it escapes cryptographic modelling and we choose not to model authentication (in particular we do not wish to assume a PKI).

Tally(b) is a tallying protocol that is run by the trustees. Its inputs are the board so far and the private data kept by the trustees from the setup phase.

Result(b) is a deterministic algorithm that takes a bulletin board b of a completed election and returns the result of the election, or a special symbol $\perp$ if the board does not contain a valid result.

A single-pass scheme is executed as follows.

1. *Setup phase.* The trustees run the Setup algorithm and post the public key Y to the bulletin board.
2. *Voting phase.* Each voter may proceed as follows or abstain. He reads public key Y off the board and computes a ballot $s \leftarrow \mathsf{Vote}(id, v, Y)$ where id is his identity and v is his vote, and submits s to the board.
   The board runs Validate(b, s) on every submission it receives and appends valid ones to its state.
3. *Tallying phase.* The trustees run the Tally protocol and may post to the board.

A single-pass protocol is correct w.r.t. a result function ρ if as long as everyone follows the protocol, with overwhelming probability (in the security parameter) none of the algorithms abort, Result returns a result when executed on the board at the end of the tallying phase and this result corresponds to ρ evaluated on the votes cast by the voters.[6]

BALLOT PRIVACY. We base our definition of ballot privacy on previous work in this area by Bernhard et al. [17,32]. Ballot privacy is defined by means of a cryptographic indistinguishability game. The new feature of our definition is that it can deal with dishonest trustees; we introduce a simulator to handle tallying in this case.

Definition 5 (Ballot Privacy). *A single-pass protocol for n trustees and any number of voters has ballot privacy against up to $m < n$ dishonest trustees if there is a simulator S such that for any efficient adversary A, the advantage $\mathbf{Pr}[A\ wins\,] - 1/2$ against the following indistinguishability game is negligible (as a function of the security parameter). The simulator S is given black-box access to the adversary A and may invoke further copies of A using the same randomness as was used in the main run in the security game. We assume static corruption of trustees: the sets of honest and dishonest trustees are fixed in advance. The adversary can adaptively choose voters to be honest or dishonest, however.*

Setup Phase. The challenger picks a bit $\beta \leftarrow \{0, 1\}$ uniformly at random. He sets up two bulletin boards $\mathcal{L}$ and $\mathcal{R}$. The adversary is given access to either $\mathcal{L}$ if $\beta = 0$ or to $\mathcal{R}$ if $\beta = 1$.

The trustees jointly run the Setup protocol, the challenger playing the honest trustees and the adversary, the corrupt ones. This produces some output Y on the visible board. The challenger then copies Y from the visible board to the hidden one. If the setup phase fails to complete, the adversary loses the game.

---

[6] One may also want ρ to operate on (v, id) pairs: in the Helios election at UC Louvain, votes from students, faculty and staff were weighted differently.

Voting Phase. The adversary may make two types of queries.

>**Vote**$(id, v_\mathcal{L}, v_\mathcal{R})$ **Queries.** The adversary provides a voter identity id and two votes $(v_\mathcal{L}, v_\mathcal{R})$. The challenger runs $b_\mathcal{L} \leftarrow$ Vote$(id, v_\mathcal{L}, Y)$ and $b_\mathcal{R} \leftarrow$ Vote$(id, v_\mathcal{R}, Y)$, where Y is the public key of the election that can be computed using Keys on the public information on the board from the setup phase.
>
>The ballots $b_\mathcal{L}$ and $b_\mathcal{R}$ are submitted to the corresponding boards which process them normally (run Validate and append the ballot if it passes validation).
>
>**Ballot**(id, b) **Queries.** These are queries made on behalf of corrupt voters. Here the adversary provides a ballot b. The challenger first submits b to the board visible to the adversary, which validates it and appends it if validation is successful. If the ballot successfully validates on the visible board, the challenger also submits the ballot to the invisible board which again validates the ballot and appends it if successful.

Tallying Phase. If the adversary sees the $\mathcal{L}$ board ($\beta = 0$) then tallying can take place as normal. The trustees execute the Tally protocol, the challenger playing the honest ones and the adversary, the dishonest ones.

If the adversary sees the $\mathcal{R}$ board, the challenger starts up the simulator $\mathcal{S}$ and passes it both the $\mathcal{L}$ and $\mathcal{R}$ boards and the state of the honest trustees. In the random oracle model, the simulator is responsible for the random oracle from this point onwards (and gets a list of all previously queried input/output pairs). The simulator acts on behalf of the honest trustees from now onwards and may post to the board.

At the end of the game the adversary may make a guess of β and wins if his guess is correct.

We propose to fix Helios by changing all proofs to their strong counterparts. This allows us to state the following theorem. The proof along with a detailed description of modified Helios in the single-pass model can be found in the full version of our paper.

Theorem 3. *In the random oracle model, the modified Helios (using strong proofs) satisfies ballot privacy against up to $m = n - 1$ dishonest trustees, assuming that DDH is hard in the underlying group.*

6   Conclusion

The prominence of Helios (it has been used in several real elections, notably in the election of the IACR board of directors) justifies the level of attention it has recently received. Results are divided between finding attacks against ballot privacy (e.g. the method of casting related ballots [23,33] which we further refine in this paper) and proposing modifications that enable rigorous security proofs [23,17,34]. Our paper seems to be the natural convergence point. We identify the use of weak Fiat-Shamir proofs as a source of attacks much stronger than all those previously proposed: we have presented new and unforeseen consequences of these weak proofs and we have shown that switching to their strong counterpart allows for a proof of ballot secrecy for Helios, and provides a crucial assumption on which existing verifiability analyses of Helios rely [34]. In the process, we have made several conceptual contributions: we have defined simulation sound extractability in the random oracle model, proved that the strong Fiat-Shamir transformation yields secure non-interactive zero-knowledge proofs of knowledge, and justified the new notion through applications that include the Enc+PoK paradigm.

In the remainder of this section we discuss two points that naturally arise from our work.

Usability of wFS. Our results discourage the use of wFS proofs as they may lead to failures in the systems that employ them. Nonetheless, the transformation works well for its original application (and its generalizations [35]) in constructing signature schemes from identification schemes [1], since the statement (essentially the verification key) is fixed in advance. It is interesting to find other settings where wFS can actually be used safely. An intriguing possibility is to exploit malleability of wFS proofs as, for example, in the recent work of Chase et al. [36] that relies on controlled malleability of (standard model) non-interactive zero-knowledge proofs. A necessary first step in this direction is understanding precisely what is the level of malleability of wFS proofs, which we leave for further work.

Practical impact of our attacks. As Helios in its current form has been used in real elections, a discussion of the impact of our attacks in practice is in order. We note that our attacks have been tested and succeeded on the current version of the Helios system on http://vote.heliosvoting.org.

Our denial of service attacks may only have an impact on future elections: as far as we know, all Helios elections led to the successful computation of a tally. Regarding our attack on privacy, the scale and outcome of all known real-world elections based on Helios rule out the possibility of effectively violating the privacy of voters through ballot copying. We also checked the 2010 IACR bulletin board and verified that it does not contain any copied ballot.

Our most realistic new attack challenges the verifiability of elections: we showed that corrupted authorities colluding with a single voter can submit an encryption of an arbitrary (positive or negative) number of approvals for any candidate, and that this encryption is indistinguishable from a normal one. This attack could have a decisive impact on approval elections, where the addition of a reasonable number of votes for a single candidate can easily remain undetected.

Many important Helios elections did not use approval voting, though (e.g., the UCL president election and the IACR 2010 election): in those elections, voters were only allowed to select a limited number of candidates. The capability to submit a single malicious ciphertext has a much more limited effect in that case, due to the need to produce an overall proof of the validity of the ballot besides the individual 0/1 proofs. In this context, two possibilities are left to an attacker: either (1) cheat on an individual proof, that is, if allowed to choose up to n candidates, encrypt n votes for a single candidate, 0 for all others, and the overall proof could still be built normally; or (2) cheat on the overall proof, that is, select as many candidates as desired and fake the overall proof. The result of these limited manipulations could not have changed the outcome of the two particular elections mentioned above.

Extending our attack to more than a single ciphertext does not seem immediate. Indeed, our attack requires selecting the private key as a function of the hash of all the commitments in one proof. As a result, building two proofs based on different commitments would require using different election keys, which would not be possible in a single election.

Our second most damaging attack relies on authorities that gain access to the randomness that is used by all voters in order to encrypt their messages. This could possibly be achieved by hiding a function that sends this randomness in the JavaScript code sent by the Helios server to the voters for the ballot preparation, or by forcing server-side encryption. Though more demanding, the effect of this attack can also be more severe as the single actively corrupted voter now only needs to submit a regular ballot.

In all cases, including for approval elections (such as the 2011 IACR election) for which our first attack on verifiability applies, there remain possibilities to remove the concerns that our attacks may raise. For instance, a (possibly independent) set of trustees could be asked to run a mixnet on the ciphertexts posted on the bulletin board of the considered elections, which could then be followed by the individual decryption of all shuffled ballots. An invalid ballot would then be detected immediately, and the trustees would not be able to cheat on the decryption of a second ciphertext.

The existence of such a possibility shows that we still are in a better situation than the one obtained with postal voting. Here, the trustees still have a possibility to demonstrate that they did not manipulate the election. That would be much harder for postal voting, where there is no practical way for the tallying officers to demonstrate that the tally they announce actually corresponds to the authentic ballots.

Acknowledgement. We would like to thank Markulf Kohlweiss for helpful comments. The research leading to these results has received funding from the European Research Council under the European Union's Seventh Framework Programme (FP7/2007-2013) under the ICT-2 007-216676 European Network of Excellence in Cryptology II and under the HOME/2010/ISEC/AG/INT-011 B-CCENTRE project. Olivier Pereira is a Research Associate of the F.R.S.-FNRS.

References

1. Fiat, A., Shamir, A.: How to Prove Yourself: Practical Solutions to Identification and Signature Problems. In: Odlyzko, A.M. (ed.) CRYPTO 1986. LNCS, vol. 263, pp. 186–194. Springer, Heidelberg (1987)
2. Bellare, M., Rogaway, P.: Random oracles are practical: A paradigm for designing efficient protocols. In: Proceedings of the Annual Conference on Computer and Communications Security (CCS). ACM Press (1993)

3. Fouque, P.-A., Pointcheval, D.: Threshold Cryptosystems Secure against Chosen-Ciphertext Attacks. In: Boyd, C. (ed.) ASIACRYPT 2001. LNCS, vol. 2248, pp. 351–368. Springer, Heidelberg (2001)
4. Groth, J.: Evaluating Security of Voting Schemes in the Universal Composability Framework. In: Jakobsson, M., Yung, M., Zhou, J. (eds.) ACNS 2004. LNCS, vol. 3089, pp. 46–60. Springer, Heidelberg (2004)
5. Adida, B.: Helios: Web-based open-audit voting. In: Proceedings of the 17th USENIX Security Symposium (Security 2008), pp. 335–348 (2008)
6. Adida, B., de Marneffe, O., Pereira, O., Quisquater, J.J.: Electing a university president using open-audit voting: Analysis of real-world use of helios. In: Electronic Voting Technology Workshop/Workshop on Trustworthy Elections (2009)
7. Helios Headquarters, Princeton University Undergraduate Student Government (2010), http://usg.princeton.edu/officers/elections-center/helios-headquarters.html
8. International Association for Cryptologic Research, http://www.iacr.org/elections/2010
9. Helios Specification, http://documentation.heliosvoting.org/verification-specs
10. Schnorr, C.: Efficient signature generation by smart cards. Journal of Cryptology 4, 161–174 (1991)
11. Chaum, D., Pedersen, T.P.: Wallet Databases with Observers. In: Brickell, E.F. (ed.) CRYPTO 1992. LNCS, vol. 740, pp. 89–105. Springer, Heidelberg (1993)
12. Groth, J.: Simulation-Sound NIZK Proofs for a Practical Language and Constant Size Group Signatures. In: Lai, X., Chen, K. (eds.) ASIACRYPT 2006. LNCS, vol. 4284, pp. 444–459. Springer, Heidelberg (2006)
13. Shoup, V., Gennaro, R.: Securing threshold cryptosystems against chosen ciphertext attack. Journal of Cryptology 15(2), 75–96 (2002)
14. Tsiounis, Y., Yung, M.: On the Security of ElGamal Based Encryption. In: Imai, H., Zheng, Y. (eds.) PKC 1998. LNCS, vol. 1431, pp. 117–134. Springer, Heidelberg (1998)
15. Schnorr, C.-P., Jakobsson, M.: Security of Signed ElGamal Encryption. In: Okamoto, T. (ed.) ASIACRYPT 2000. LNCS, vol. 1976, pp. 73–89. Springer, Heidelberg (2000)
16. Wikström, D.: Simplified Submission of Inputs to Protocols. In: Ostrovsky, R., De Prisco, R., Visconti, I. (eds.) SCN 2008. LNCS, vol. 5229, pp. 293–308. Springer, Heidelberg (2008)
17. Bernhard, D., Cortier, V., Pereira, O., Smyth, B., Warinschi, B.: Adapting Helios for Provable Ballot Privacy. In: Atluri, V., Diaz, C. (eds.) ESORICS 2011. LNCS, vol. 6879, pp. 335–354. Springer, Heidelberg (2011)
18. Bulens, P., Giry, D., Pereira, O.: Running mixnet-based elections with helios. In: Shacham, H., Teague, V. (eds.) Electronic Voting Technology Workshop/Workshop on Trustworthy Elections, Usenix (2011)
19. Bellare, M., Desai, A., Pointcheval, D., Rogaway, P.: Relations among Notions of Security for Public-Key Encryption Schemes. In: Krawczyk, H. (ed.) CRYPTO 1998. LNCS, vol. 1462, pp. 26–45. Springer, Heidelberg (1998)
20. Bellare, M., Rogaway, P.: Entity Authentication and Key Distribution. In: Stinson, D.R. (ed.) CRYPTO 1993. LNCS, vol. 773, pp. 232–249. Springer, Heidelberg (1994)
21. Pointcheval, D., Stern, J.: Security arguments for digital signatures and blind signatures. Journal of Cryptology 13(3), 361–396 (2000)

22. Pedersen, T.P.: A Threshold Cryptosystem without a Trusted Party. In: Davies, D.W. (ed.) EUROCRYPT 1991. LNCS, vol. 547, pp. 522–526. Springer, Heidelberg (1991)
23. Cortier, V., Smyth, B.: Attacking and fixing helios: An analysis of ballot secrecy. In: CSF, pp. 297–311. IEEE Computer Society (2011)
24. Benaloh, J.: Verifiable Secret-Ballot Elections. PhD thesis, Yale University (January 1987)
25. Pfitzmann, B., Pfitzmann, A.: How to Break the Direct RSA-Implementation of MIXes. In: Quisquater, J.J., Vandewalle, J. (eds.) EUROCRYPT 1989. LNCS, vol. 434, pp. 373–381. Springer, Heidelberg (1990)
26. Naor, M., Yung, M.: Public key cryptosystem secure against chosen ciphertext attacks. In: Proceedings of the Annual Symposium on the Theory of Computing (STOC) 1990, pp. 33–43. ACM Press (1990)
27. Sahai, A.: Non-malleable non-interactive zero knowledge and adaptive chosen-ciphertext security. In: Proceedings of the Annual Symposium on Foundations of Computer Science (FOCS) 1999. IEEE Computer Society Press (1999)
28. De Santis, A., Persiano, G.: Zero-knowledge proofs of knowledge without interaction. In: Proceedings of the Annual Symposium on Foundations of Computer Science (FOCS) 1992, pp. 427–436. IEEE Computer Society Press (1992)
29. Cramer, R., Franklin, M.K., Schoenmakers, B., Yung, M.: Multi-authority Secret-Ballot Elections with Linear Work. In: Maurer, U.M. (ed.) EUROCRYPT 1996. LNCS, vol. 1070, pp. 72–83. Springer, Heidelberg (1996)
30. Cramer, R., Gennaro, R., Schoenmakers, B.: A Secure and Optimally Efficient Multi-authority Election Scheme. In: Fumy, W. (ed.) EUROCRYPT 1997. LNCS, vol. 1233, pp. 103–118. Springer, Heidelberg (1997)
31. Damgård, I., Groth, J., Salomonsen, G.: The theory and implementation of an electronic voting system. In: Gritzalis, D. (ed.) Secure Electronic Voting. Advances in Information Security, vol. 7, pp. 77–98. Springer (2003)
32. Bernhard, D., Pereira, O., Warinschi, B.: On necessary and sufficient conditions for private ballot submission. IACR Cryptology ePrint Archive 2012, 236 (2012)
33. Smyth, B.: Replay attacks that violate ballot secrecy in helios. IACR Cryptology ePrint Archive 2012, 185 (2012)
34. Küsters, R., Truderung, T., Vogt, A.: Clash Attacks on the Verifiability of E-Voting Systems. In: IEEE Symposium on Security and Privacy (S&P 2012), pp. 395–409. IEEE Computer Society (2012)
35. Camenisch, J.L., Stadler, M.A.: Efficient Group Signature Schemes for Large Groups. In: Kaliski, B.S. (ed.) CRYPTO 1997. LNCS, vol. 1294, pp. 410–424. Springer, Heidelberg (1997)
36. Chase, M., Kohlweiss, M., Lysyanskaya, A., Meiklejohn, S.: Malleable Proof Systems and Applications. In: Pointcheval, D., Johansson, T. (eds.) EUROCRYPT 2012. LNCS, vol. 7237, pp. 281–300. Springer, Heidelberg (2012)

Sequential Aggregate Signatures
with Lazy Verification
from Trapdoor Permutations
(Extended Abstract)

Kyle Brogle[1,*], Sharon Goldberg[2], and Leonid Reyzin[2]

[1] Stanford University Department of Computer Science
Stanford, CA 94305 USA
broglek@stanford.edu

[2] Boston University Department of Computer Science
Boston, MA 02215 USA
{goldbe,reyzin}@cs.bu.edu

Abstract. Sequential aggregate signature schemes allow n signers, in order, to sign a message each, at a lower total cost than the cost of n individual signatures. We present a sequential aggregate signature scheme based on trapdoor permutations (*e.g.,* RSA). Unlike prior such proposals, our scheme does not require a signer to retrieve the keys of other signers and verify the aggregate-so-far before adding its own signature. Indeed, we do not even require a signer to *know* the public keys of other signers!

Moreover, for applications that require signers to verify the aggregate anyway, our schemes support *lazy verification*: a signer can add its own signature to an unverified aggregate and forward it along immediately, postponing verification until load permits or the necessary public keys are obtained. This is especially important for applications where signers must access a large, secure, and current cache of public keys in order to verify messages. The price we pay is that our signature grows slightly with the number of signers.

We report a technical analysis of our scheme (which is provably secure in the random oracle model), a detailed implementation-level specification, and implementation results based on RSA and OpenSSL. To evaluate the performance of our scheme, we focus on the target application of BGPsec (formerly known as Secure BGP), a protocol designed for securing the global Internet routing system. There is a particular need for lazy verification with BGPsec, since it is run on routers that must process signatures extremely quickly, while being able to access tens of thousands of public keys. We compare our scheme to the algorithms currently proposed for use in BGPsec, and find that our signatures are considerably shorter nonaggregate RSA (with the same sign and verify times) and have an order of magnitude faster verification than nonaggregate ECDSA, although ECDSA has shorter signatures when the number of signers is small.

---

* Work done while at Boston University.

X. Wang and K. Sako (Eds.): ASIACRYPT 2012, LNCS 7658, pp. 644–662, 2012.
© International Association for Cryptologic Research 2012

1   Introduction

Aggregate signatures schemes allow n signers to produce a digital signature that authenticates n messages, one from each signer. This can be securely accomplished by simply concatenating together n ordinary digital signatures, individually produced by each signer. An aggregate signature is designed to maintain the security of this basic approach, while having length much shorter than n individual signatures. To achieve this, many prior schemes *e.g.,* [LMRS04,Nev08] relied on a seemingly innocuous assumption; namely, that each signer needs to *verify* the aggregate signature so far, before adding its own signature on a new message. In this paper, we argue that this can make existing schemes unviable for many practical applications, (in particular, for BGPsec [Lep12] / Secure BGP [KLS00]) and present a new scheme based on trapdoor permutations like RSA that avoids this assumption. In fact, our scheme remains secure even if a signer does not *know* the public keys of the other signers.

1.1   Aggregate Signatures from Trapdoor Permutations

Boneh, Gentry, Lynn, and Shacham [BGLS03] introduced the notion of aggregate signatures, in which individual signatures could be combined by *any third party* into a single constant-length aggregate. The [BGLS03] scheme is based on the bilinear Diffie-Hellman assumption in the random oracle model [BR93]. Subsequent schemes [LMRS04,Nev08] were designed for the more standard assumption of trapdoor permutations (*e.g.,* as RSA [RSA78]), but in a more restricted framework where third-party aggregation is not possible. Instead, the signers work *sequentially*; each signer receives the aggregate-so-far from the previous signer and adds its own signature.[1]

Lysyanskaya, Micali, Reyzin, and Shacham [LMRS04] constructed the first sequential aggregate signature scheme from trapdoor permutations, with a proof in the random oracle model.[2] However, their scheme has two drawbacks: the trapdoor permutation must be *certified* (when instantiating the trapdoor permutation with RSA, this means that each signer must either prove certain properties of the secret key or else use a long RSA verification exponent), and each signer needs to verify the aggregate-so-far before adding its own signature. Neven [Nev08] improved on [LMRS04] by removing the need for certified trapdoor permutations, but the need to verify before signing remained. Indeed, a signer who adds its own signature to an unverified aggregate in both [LMRS04] and [Nev08] (or, indeed, in any scheme that follows the same design paradigm) is exposed to a devastating attack: an adversary can issue a single malformed

---

[1] The need for the random oracle model was removed by Lu, Ostrovsky, Sahai, Shacham, and Waters [LOS+06], who constructed sequential aggregate signatures from the bilinear Diffie-Hellman assumption; however, it is argued in [CHKM10] that this improvement in security comes at a considerable efficiency cost. See also [RS09,CSC09] for other proposals based on less common assumptions.

[2] Bellare, Namprempre, and Neven [BNN07] showed how the schemes of [BGLS03] and [LMRS04] can be improved through better proofs and slight modifications.

aggregate to the signer, and use the signature on that malformed message to generate a *valid* signature on a message that the signer never intended to sign (we describe the attack in the full version of the paper [BGR11b]).

The nonsequential scheme of [BGLS03] does not, of course, require verification before signing. The only known sequential aggregate scheme to not require verification before signing is the history-free construction of Fischlin, Lehmann, and Schröder [FLS11] (concurrent with our work), but it, like [BGLS03], requires bilinear Diffie-Hellman.

Thus, the advantages of basing the schemes on trapdoor permutations (particularly a more standard security assumption and fast verification using low-exponent RSA) are offset by the disadvantage of requiring verification before signing. We argue below that this disadvantage is serious.

1.2   The Need for Lazy Verification

In applications with a large number of possible signers, the need to verify before signing can introduce a significant bottleneck, because each signer must retrieve the public keys of the previous signers before it can even begin to run its signing algorithm. Worse yet, signers need to keep their large caches of public keys secure and current: if a public key is revoked and a new one is issued, the signer must first obtain the new key and verify its certificate before adding its own signature to the aggregate.

A Key Application: BGPsec.   Sequential aggregate signatures are particularly well-suited for the BGPsec [Lep12] (formerly known as the Secure Border Gateway Protocol (S-BGP) [KLS00]), a protocol being developed to improve the security of the global Internet routing system. (This application was mentioned in several works, including [BGLS03,LOS+06,Nev08], and explored further in [ZSN05].) In BGPsec, autonomous systems (ASes) digitally sign routing announcements listing the ASes on the path to a particular destination. An announcement for a path that is n hops long will contain n digital signatures, added *in sequence* by each AS on the path. (Notice that the length of the BGPsec message *even without the signatures* increases at every hop, as each AS adds its name to the path, as well as extra information to the material in the routing message like its "subject key identifier" — a cryptographic fingerprint that is used to lookup its public key in the PKI [Lep12].) The BGPsec protocol is faced with two key performance challenges:

1. *Obtaining public keys.* BGPsec naturally requires routers to have access to a large number of public keys; indeed, a routing announcement can contain information from *any* of the 41,000 ASes in the Internet [COZ08] (this number is according to the dataset retrieved in 2012). Certificates for public keys are regularly rolled over to maintain freshness, and must be retrieved from a distributed PKI infrastructure [Hus12]. Caching more than 41,000 public keys is expensive for a memory-constrained device like a router (which often does not have a hard drive or other secondary storage [KR06]). Furthermore, whenever a router sees a BGPsec message containing a key that is not in

its cache, it incurs non-trivial delay on certificate retrieval (from a distant device that hosts the PKI) and verification.

2. *Dealing with routing table "dumps"*. When a link from a router to its neighboring router fails, the router receives a dump of the full routing table, often containing more than 300,000 routes [CID], from it neighbors. Because routers are CPU- and memory-constrained devices, dealing with these huge routing table dumps incurs long delays (up to a few minutes, even with plain, insecure BGP [BHMT09]!). The delays are exacerbated if cryptographic signing and verifying is added to the process, and even more so when a router comes online for the first time (or after failure) and needs to also retrieve and authenticate public keys for all the ASes on the Internet.

To deal with these issues, the BGPsec protocol gives a router the option to perform *lazy verification*: that is, to immediately sign the routing announcement with its *own* public key, and to delay verification until a later time, *e.g.*, when (a) it has time to retrieve the public keys of the other signers, or (b) when the router itself is less overloaded and can devote resources to verification [DHS]. It is important to note that lazy verification by one router need not hurt others: if a router has not verified a given announcement, routers further in the chain can verify it for themselves.

While there is legitimate concern that permitting lazy verification may cause routers to temporarily adopt unverified paths, the alternative may be worse: forbidding lazy verification can lead to problems with global protocol convergence (agreement on routes in the global Internet), because of routers that delay their announcements significantly until they can verify signatures (*e.g.*, during routing table dumps, or while waiting to retrieve a missing certificate). Such delays create their own security issues, enabling easier denial of service attacks and traffic hijacking during the long latency window. Thus, even though BGPsec recommends that every router *eventually* verifies BGPsec messages, requiring that routers always verify *before* signing and re-announcing BGPsec messages is considered a nonstarter by the BGPsec working group [Sri12, Section 8.2.1]. Lazy verification is written into the BGPsec protocol specification as follows [Lep12, Section 7]:

> ...it is important to note that when a BGPSEC speaker signs an outgoing update message, it is not attesting to a belief that all signatures prior to its are valid.

Requirement: No Public Keys in the Signing Algorithm!   Note that the primary obstacle here is *not* only verification time (which can perhaps be improved through batching and, anyway, can be considerably faster than signing time when using low-exponent RSA), but also the need to obtain public keys. Thus, lazy verification also requires that prior signers' public keys are *not* used in the signing algorithm (*e.g.,* hashed with the message as in [LMRS04,Nev08]).

Requirement: No Security Risk from Signing Unverified Aggregates!
As we already mentioned, a signer who adds its own signature to an unverified
aggregate in the schemes of [LMRS04] and [Nev08] is exposed to a devastating
attack. We already discussed how lazy verification may cause a signer to do so.
Moreover, even without lazy verification, BGPsec may sometimes require a signer
to add its own signature to an aggregate that is invalid. One such situation is
when a router knowingly adopts a path that fails verification—for example, if it
is the only path to a particular destination (the specification allows this [Lep12,
Section 5]). It will then add its own signature to the invalid one, because a
"BGPSEC router should sign and forward a signed update to upstream peers if
it selected the update as the best path, regardless of whether the update passed
or failed validation (at this router)" [Sri12, Section 8.2.1]. The need to sign a
possibly invalid aggregate also arises in the case each message is signed by two
different signature schemes (as will happen during transition times from one
signature algorithm to another), and "one set of signatures verifies correctly and
the other set of signatures fails to verify." In such a case the signer should still
"add its signature to each of the [chains] using both the corresponding algorithm
suite" [Lep12, Section 7]. Even if all BGPsec adopters avoid lazy verification
and always verify before signing, these guidelines make it impossible to adopt an
aggregate signature scheme that does not permit signing unverified aggregates,
because of the possibility of attack. In other words, lazy verification is still needed
for security even if no one uses it for efficiency!

Our Goal. We note that lazy verification is permitted by the trivial solution
of concatenating individual ordinary signatures, by aggregate signature schemes
defined in [BGLS03], and by history-free aggregate signature schemes defined
in [FLS11]. All of the above schemes do not require the current signer to know
anything about the previous signers: neither their public keys nor the messages
they signed. [3] Our goal is to obtain the same advantages, while relying on a more
basic security assumption than the bilinear Diffie-Hellman of [BGLS03,FLS11]
and saving space as compared to the trivial solution.

---

[3] Identity-based aggregate signatures [YCK04], [XZF05], [CLW05], [CLGW06],
[Her06], [GR06], [BGOY07], [HLY09], [SVSR10], [BJ10] also remove the need for
obtaining public keys and have been proposed for use in BGPsec. However, agreeing
on the secret-key-issuing authority for the global Internet seems politically infeasible. Moreover, on a technical level, the proposals either require interaction among
signers or are based on bilinear pairings. Interactive signatures would significantly
complicate the protocol. And if we are willing to rely on bilinear pairings, [BGLS03]
already gives us an excellent choice that allows for lazy verification.

Synchronized aggregate signatures (identity-based ones of [GR06] and regular ones
of [AGH10]) also allow for lazy verification, but require a common nonce for all
signers that, if repeated, breaks the security of the scheme. Implementing such a
nonce in BGPsec presents its own challenges, because each signer has to ensure it
never reuses a nonce, or else its secret key is at risk. The schemes are also pairing-
based.

1.3   Overview of Our Contributions

We present a sequential aggregate signature scheme that is secure even with lazy verification, based on any trapdoor permutation (such as RSA). Moreover, as in the nonsequential scheme of [BGLS03] and the history-free scheme of [FLS11], our signers do not need to know anything about each other—not even each other's public keys. To achieve this, we modify Neven's scheme [Nev08] by randomizing the H-hash function with a fresh random string per signer, which becomes a part of the signature, similarly to Coron's PFDH [Cor02] (Section 3). Our modification allows each signer to sign without verifying, and without even needing to know the public keys of all the signers that came before him, avoiding, in particular, the attack on [LMRS04,Nev08].

Although the ultimate goal in aggregate signatures is to produce schemes whose signature length is independent of the number of signers, signatures in our scheme grow slightly with the number of signers. However (as also pointed out by [Nev08]), while a constant-length aggregate signature is a theoretically interesting goal, what usually matters in practice is the *combined* length of signatures and messages, because that's what verifiers receive: signatures rarely live on their own, separately from the messages they sign. And the combined length of messages, if they are distinct, grows linearly with the number of signers, so the total growth of the amount of information received by the verifier is anyway linear. What matters, then, is *how fast* this linear growth is; below we derive parameters that show it to be much smaller than when ordinary trapdoor-permutation-based signatures are used as in the trivial solution.

We make the following contributions:

Generic Randomized Scheme. We present the basic version of our scheme, which requires each signer to append a *truly random* string to the aggregate (Section 3). Our scheme is as efficient for signing and verifying (per signer) as ordinary trapdoor-permutation based signatures, like the Full-Domain-Hash (FDH, [BR93, Section4]). We prove security (Section 4) in the random oracle model, based on the same assumption of trapdoor permutations (or claw-free permutations for a tighter security reduction) as in [Nev08]. Our security proof is more involved, because the reduction cannot know the public keys of other (adversarial) signers during the signature queries. We should note that our proof technique also shows that Neven's scheme need not hash other signer's public keys in the signing algorithm (however, Neven's scheme still fails under lazy verification).

Shortening the Randomness. We show that the per-signer random string can be shorter if it is made input-dependent (Section 5), ensuring that a given signer never produces two different signatures on the same input. The idea of input-dependent randomness has been used before in signature schemes (e.g., [KW03, Section 4]); however, our application requires a new combinatorial argument to show security.

Instantiating with RSA. In the full version of the paper [BGR11b] we show how to instantiate our schemes with practical trapdoor permutations like RSA, which have slightly different domains for different signers.

Detailed Specification. We provide a full, parameterized step-by-step specification of the truly-random and input-dependent-random versions of our signature when instantiated with RSA (see the full version of the paper [BGR11b], where we also provide guidelines on choosing parameters such as bit lengths).

Implementation, Benchmarking and Practical Considerations. We implement our specification as a module in OpenSSL (Section 6); the implementation is available from [BGR11a]. We compare our implementation's performance to other potential solutions that allow for lazy verification; namely, [BGLS03], and the "trivial" solution of using n RSA or ECDSA signatures (the two algorithms currently proposed for use in implementations of BGPsec [DHS]). When evaluating signatures schemes for use with BGPsec, we consider compute time as well as signature length. Thus, we show that our signature is shorter than trivial RSA when there are $n > 1$ signers and shorter than trivial ECDSA when there are $n > 6$ signers. (While our signature is longer than the constant-length [BGLS03] signature, it benefits from relying on the better-understood security assumption of RSA.) Moreover, our scheme enjoys the same extremely fast verify times as RSA.

2   Preliminaries

Sequential Aggregate Signature Security. The security definition for aggregate signatures (both original [BGLS03] and sequential [LMRS04]) is designed to capture the following intuition: each signer is individually secure against existential forgery following an adaptive chosen-message attack [GMR88] regardless of what all the other signers do. In fact, we will allow the adversary to give the attacked signer arbitrary—perhaps meaningless—aggregate-so-far signatures during the signature queries, thus making them adaptive "chosen-message-and-aggregate" queries. We also allow the adversary, which we call "the forger," to choose the public keys of all the other signers and to place the single signer who is under attack anywhere in the signature chain in the attempted forgery. This single attacked signer does not know any public keys other than its own and does not verify any aggregate-so-far given by the attacker.

Our formal definition, presented in the full version [BGR11b], is almost verbatim from [LMRS04], with one important difference needed to enable lazy verification: the public keys and messages of previous signers are not input to the signing algorithm. Therefore, each signer, by signing a message, is attesting only to that message, not to the prior signers' messages and public keys. At a technical level, this change implies that in security game the forger, in its query to ith signer, is required to supply only the aggregate-so-far signature allegedly produced by the first $i - 1$ signers, but not the messages or public keys with respect to which this aggregate was allegedly produced. And, of course, to be considered successful, the forger must use *a new message*—in other words, it is not enough to change a public key or message of someone else in the chain before the attacked signer (because such public keys and messages may not even be well defined during the attack). This definition is exactly the one that is satisfied by the trivial solution of concatenating n individual signatures (and therefore suffices, in particular, for BGPsec).

Fischlin, Lehmann, and Schröder [FLS11] propose a stronger security definition for their "history-free" signatures (building on history-free MACs of [EFG+10]), which prevents certain reordering and recombining of signatures. Their definition thus has a security property that the trivial solution of concatenating n individual signatures does not have. Although this security property is not needed in many applications (for example, in BGPsec reordering and recombining of signatures is prevented simply by the protocol message structure, where each message must, for the purposes of functionality, include all the signed information contained in previous messages), our signature scheme in fact also prevents reordering and recombining that are of concern to [FLS11]: see [BGR11b].

Cryptographic Primitives. We will use pseudorandom functions [GGM86]; the definition is omitted here because it is standard, but is presented in [BGR11b] for the sake of completeness. We will denote by $\varepsilon_{\mathsf{PRF}}(q, t)$ the maximum insecurity of PRF against any distinguisher who asks at most q queries and runs in time t.

We assume the reader is familiar with the trapdoor and claw-free permutations; we will denote by π the easy direction of the trapdoor permutation, by π^{-1} the hard direction, and by ρ the function such that it is hard to find a "claw" x, z with $\pi(x) = \rho(z)$.

3   Our Basic Signature Scheme

The intuition behind our construction is as follows. Like [Nev08], we use a random-oracle-based signature with message recovery, similar to PSS-R [BR96], as a basic building block. Signatures with message recovery embed a portion of the message into the signature, so it can be recovered on verification and does not need to be sent explicitly. In our case, the signature outputs two values: the output x of a trapdoor permutation and an additional hash value h. The ith signer receives (x_{i-1}, h_{i-1}) from the previous signer and wants to sign a message m_i. To enable aggregation, we view (x_{i-1}, m_i) together as a "message" to be signed with message recovery: we apply the signature with message recovery to this pair, so that x_{i-1} is embedded into the signature and does not have to be sent explicitly. The h portions of the signatures are exclusive-ored together for aggregation.

So far, what we described is a slightly simplified version of the scheme from [Nev08]. Note that verifying before signing is necessary in this scheme, because the transformation from (x_{i-1}, h_{i-1}) to (x_i, h_i) is deterministic, invertible, and can be performed by the adversary, except for the inversion of the trapdoor permutation performed at the last step. As we show in [BGR11b], no scheme constructed in this manner can permit lazy verification while protecting against a chosen message attack. Thus, to enable lazy verification, we require each signer to add a random string to the message, and concatenate and append these strings to the signature. Because the adversary lacks a priori knowledge about these random strings, the chosen message attack becomes useless and we can prove that this is sufficient to enable lazy verification.

Notation. We now describe the scheme precisely, using the following notation:

- Let m_i be the message signed by signer i.
- Let trapdoor permutation π_i be the public key of signer i and π_i^{-1} be the corresponding secret key. We assume all permutations operate on bit strings of length ℓ_π, *i.e.*, have domain and range $\{0,1\}^{\ell_\pi}$. (In the full version [BGR11b] we remove the assumption that all permutations operate on the same domain. Section 6 uses this to instantiate π from the RSA assumption, where π_i is the easy direction, and π_i^{-1} is the hard direction of the RSA permutation.)
- Let H (*resp.* G) be a cryptographic hash function (modeled as a random oracle) that outputs ℓ_H-bit (*resp.* ℓ_π-bit) strings.
- Let ℓ_r be a parameter denoting the length of the randomness appended by each signer.
- Let the notation $\boldsymbol{a_i}$ denote a vector of values $(a_1, a_2, ..., a_i)$.
- Let $\oplus$ to denote bitwise exclusive-or. Exclusive-or is not the only operation that can be used; any efficiently computable group operation with efficient inverse can be used here.
- ϵ is a special character denoting the empty string; we assume $\epsilon \oplus x = x$ for any x.

Sign: The ith Signer's algorithm

Require:
   $\pi_i, \pi_i^{-1}, m_i, x_{i-1}, h_{i-1}$
   (where $x_{i-1}, h_{i-1} = \epsilon, \epsilon$ if $i = 1$).

1: Draw $r_i \xleftarrow{R} \{0,1\}^{\ell_r}$
2: $\eta_i \leftarrow H(\pi_i, m_i, r_i, x_{i-1})$
3: $h_i \leftarrow h_{i-1} \oplus \eta_i$
4: $g_i \leftarrow G(h_i)$
5: $y_i = g_i \oplus x_{i-1}$
6: $x_i \leftarrow \pi_i^{-1}(y_i)$
7: **return** r_i, x_i, h_i {Note that x_i and h_i go to the next signer; *all* the r_i values go to the verifier, but only the last signer's x_i and h_i do.}

$\mathsf{Ver}^{H,G}$: The Verification Algorithm

Require: $\boldsymbol{\pi_n}, \boldsymbol{m_n}, \boldsymbol{r_n}, x_n, h_n$
1: **for** $i = n, n-1, ..., 2$ **do**
2:    $y_i \leftarrow \pi_i(x_i)$
3:    $g_i \leftarrow G(h_i)$
4:    $x_{i-1} \leftarrow g_i \oplus y_i$
5:    $\eta_i \leftarrow H(\pi_i, m_i, r_i, x_{i-1})$
6:    $h_{i-1} \leftarrow h_i \oplus \eta_i$
7: **if** $h_1 = H(\pi_1, r_1, m_1, \epsilon)$ and $\pi_1(x_1) = G(h_1)$ **then**
8:    **return** 1
9: **else**
10:    **return** 0

The ith signer's signing algorithm has no dependency on the number of signers; it takes in *only* the ith signers' own public key and message and the aggregated portion of the signature x_{i-1}, h_{i-1}. Moreover, the aggregated signature need not be verified before it is signed. For verification, only a single x_i and h_i—namely, the one from the last signer—is needed. However, every r_i, from the first signer to the last, is needed.

4   Security Proof

We prove our scheme secure if G and H are modeled as random oracles and π is a trapdoor permutation. The proof is easier to understand if π is additionally claw-free (in particular, any homorphic permutation, such as RSA, is claw-free if it is trapdoor). We therefore present the proof for the claw-free case. The more general case is addressed in the full version [BGR11b]. Our proof shows how a forger F on the aggregate signature scheme can be used to construct a reduction R that finds a claw in claw-free pair (π_*, ρ_*). R has F forge a signature for victim signer that uses permutation π_*, and then uses the resulting forgery to find the claw in the claw-free pair. The structure of our reduction is similar to [Nev08]; however, while [Nev08] constructs a "sequential forger" from forger F and then constructs reduction R from the sequential forger, our reduction must proceed in one step (since the notion of a sequential forger is undefined if hash queries do not include previous signers public keys).

F's Queries. We review what forger F expects to see on each one of its queries:

- **H-Query.** F asks query $Q = (\pi, m, r, x)$ (where x may be ϵ) and expects to see $H(Q) = \eta$.
- **G-query.** F asks query h, and expects to see $g = G(h)$.
- **Sign Query.** F asks query (m, h, x) to be signed by π_*, and expects to see r, h', x' back, where r looks uniform, $h' = h \oplus H(\pi_*, m, r, x)$, and $\pi_*(x') = G(h') \oplus x$.
- **Forgery.** Finally, F outputs a forgery, $\sigma = \boldsymbol{\pi_n}, \boldsymbol{m_n}, \boldsymbol{r_n}, x_n, h_n$ where $\pi_n = \pi_*$. (Value n is chosen by F).

Simplifying Assumptions about the Forger F.   The following simplifies our proof:

- We assume that the forger F forges the last signature in the signature chain; in other words, $\pi_n = \pi_*$ and m_n is a new message never queried by F to the signing oracle (whose public key is π_*). Indeed, any F can be easily modified to do so: if π_* and a new message $m_{n'}$ are present in $\boldsymbol{\pi_n}$ but at location $n' < n$, then we can run the verification algorithm loop for $n - n'$ iterations to obtain $x_{n'}, h_{n'}$ and output $\sigma' = \boldsymbol{\pi_{n'}}, \boldsymbol{m_{n'}}, \boldsymbol{r_{n'}}, x_{n'}, h_{n'}$ as the new forgery, which will be valid if an only if σ was valid. Note that we do <u>not</u> assume that π_* (or any other public key) is present in the signature chain only once.
- We assume that before forger F outputs its forgery and halts, it makes hash queries on all the hashes that will be computed during the verification of its forgery. Moreover, we assume that the forger does not output an invalid forgery; instead, it halts and outputs $\bot$. Indeed, any F can be modified to do so; simply run the verification algorithm upon producing the forgery, and check that m_n is different from every message asked in a sign query.

4.1   Description of the Reduction R

Data Structures Used by R. HT and GT Tables. The reduction R uses 'programmable random oracles', *i.e.*, it chooses answers for random oracle queries. R keeps track of queries whose answers have already been decided in two tables: HT for H and GT for G. We say $\mathsf{HT}(Q) = \eta$ if HT stores η as the answer to a query Q, and $\mathsf{HT}(Q) = \bot$ if HT has no answer for Q (similar for GT).

The HTree. The key challenge for the reduction is programming G, since G-queries are made on sums of H-query answers, rather than on individual H-query answers. Thus the reduction keeps an additional data structure, the HTree, that records responses to H-queries that may eventually be used as part of forger F's forgery. (HTree is inspired by the graph $\mathcal{G}$ in [Nev08, Lemma 5.3].)

The HTree is a tree of labeled nodes that stores a subset of the queries in HT. Each node in HTree (except the root) corresponds to an H-query that could potentially appear in the forger F's final forgery σ; the queries asked during verification of σ will appear on a path from one of the leaf nodes to the root (unless a very unlikely event occurs). The HTree has a designated *root node* that stores the value $h_0 = 0$. We consider the root to be at depth 0. A node N_i at depth $i > 0$ stores:

- a pointer to its parent node
- a query $Q_i = (\pi_i, m_i, r_i, x_{i-1})$ (where $x_{i-1} = \epsilon$ if and only if $i = 1$),
- the 'hash-response' values η_i and h_i (h_i is the XOR of the values $\eta_1, \dots, \eta_i$ on the path from the root to the node N_i; equivalently, $h_{i-1} \oplus \eta_i$, where h_{i-1} is stored in the parent node),
- an auxiliary value y_i that is used to determine how future queries are added to the HTree, computed as $G(h_i) \oplus x_{i-1}$ (note that y_i is the value to which the signer would apply π_i^{-1}),
- if $\pi_i = \pi_*$, an auxiliary value z that may be used to find a claw in (π_*, ρ_*).

Every node at depth $i = 2$ or deeper satisfies the relation $\pi_{i-1}(x_{i-1}) = y_{i-1}$, where π_{i-1} and y_{i-1} are stored at the node's parent. New H-queries Q are added as nodes to the HTree if they can satisfy this relation; we say that such a query can be *tethered* to an existing node in the HTree. Intuitively, a query tethered to N_i becomes a child of N_i in the HTree:

Definition 1 (Tethered queries). *An H-query Q containing $x \neq \epsilon$ is tethered to node N_i in the HTree if N_i stores π_i, y_i such that $\pi_i(x) = y_i$. If $x = \epsilon$, then Q is tethered to the root of the HTree.*

The HTree's Lookup function determines the HTree node to which query Q can be tethered. We can argue that Lookup finds at most *one node* with high probability.) The HTree is populated via the Sim-H algorithm. The reduction R adds an H-query Q to the HTree if and only if it is tethered to some node in the HTree *at the time that forger F makes the H-query*. It is possible that some query Q is not tethered at the time it is made, but becomes tethered at at *later* time (after some new nodes are added to the HTree). However, we show that this is highly unlikely.

Algorithms Used Used by R. The reduction R uses the following algorithms, which are formally specified in the full version [BGR11b].

G-Queries. R answers these queries using a simple algorithm Sim-G. Sim-G returns $GT(h)$ if it is already defined, or, if not, returns a fresh random value and records it in the GT.

Sign-Queries. The reduction R answers queries (m, h, x) to be signed by π_* using Sim-S. Since the reduction does not know the inverse of the challenge permutation π_*^{-1}, it 'fakes' a valid signature by carefully assigning certain entries in random oracle tables HT, GT, and ABORTS if these entries in HT, GT have been previously assigned. We are able to argue that Sim-S is unlikely to abort, since the entries added to HT, GT by Sim-S depend on a fresh random value r chosen as part of each signature query.

H-Queries. The reduction R answers these queries $Q = (\pi, m, r, x)$ using Sim-H. If there is an entry for Q in the HT, then Sim-H returns it. Otherwise, it assigns a fresh random value η as $HT(Q)$. Next, Sim-H needs to prepare for the event that Q could lead to a forgery by the forger F, and thus needs to be stored in the HTree. To do this, Sim-H uses the Lookup function to check if Q can be tethered and thus should be added to the HTree. If Q can be tethered, Sim-H adds a new node to the HTree containing Q, its hash response η, and an auxiliary value y that is used by the Lookup function to tether future H-queries. In order to ensure that HTree is a tree, it is important to ensure that y is a fresh random value; Sim-H aborts if that's not the case. Finally, if Q contains the challenge permutation π_*, Sim-H adds a value z to the HTree node that FindClaw will use to derive a claw from a valid forgery output by the forger F. To prepare these values, Sim-H behaves almost as if it is 'faking' the answer to a sign-query, except that instead of using the usual challenge permutation π_* (as in Sim-S), it uses the challenge permutation ρ_* applied to z (so as to benefit from forger F's forgery, which would invert π_* on the output of $\rho_*(z)$, thus producing a claw). As in Sim-S, this involves carefully assigning certain entries in GT, and aborting if these entries are already assigned. We are able to show that Sim-H is unlikely to abort.

Finding a Claw. Finally, forger F outputs a forgery $\boldsymbol{\pi_n, m_n, r_n}, x_n, h_n$, where $\pi_n = \pi_*$. Recall that our simplifying assumptions mean that the forgery is valid. The reduction R uses FindClaw to find a claw from the forgery. Because we assumed all the queries for verifying σ have already been asked, the query $(\pi_*, m_n, r_n, x_{n-1})$ is in HT. Moreover, if the forgery is valid, then with high probability it is in the HTree as a child of the node storing $(\pi_{n-1}, m_{n-1}, r_{n-1}, x_{n-2})$, which is in turn a child of the node storing $(\pi_{n-2}, m_{n-2}, r_{n-2}, x_{n-3})$, *etc.* This holds because in a valid forgery, each H-query made during verification is tethered to the next one, and all tethered queries are in the HTree with high probability. The value x_n (from the forgery σ) and value z_n (from HTree node of the query $Q = (\pi_*, m_n, r_n, x_{n-1})$) constitute a claw.

4.2 Analysis of the Reduction

Theorem 1. *If a forger F succeeds with probability ε, then the reduction R finds a claw for (π_*, ρ_*) in about the same running time as F with probability*

$$\varepsilon - (q_S + q_H)(q_S + q_G + q_H)2^{-\ell_H} - q_S(q_S + q_H)2^{-\ell_r} - q_H^2 2^{-\ell_\pi} \qquad (1)$$

where q_H is the number of H-hash queries, q_G is the number of G-hash queries, and q_S is the number of sign queries made by the forger F.

We prove this theorem in full version of the paper [BGR11b]. The proof hinges on two key statements about the HTree. First, the probability that $\mathsf{Lookup}(x)$ finds more than one HTree node is low (even though Lookup uses the functions π stored in the nodes of the HTree, which do not have to be permutations, because they are adversarially supplied and not certified like in [LMRS04]). Second, an H-query that was not added to HTree is unlikely to become tethered at *some later time*. Both statements rely on the fact that each time a query is placed on the HTree, its y value is random and independent of every other y value.

5 Shorter Signatures via Input-Dependent Randomness

To shorten our signature, we now show how to reduce ℓ_r (the length of the randomness appended by each signer). To do this, we replace the truly random r from our basic scheme with an r that is computed as a function of the inputs to the signer, and argue that it can be made shorter than the random r. Intuitively, we are able to maintain security with a shorter r because a given signer never produces two different signatures on the same input, thus limiting the information that an adversary can see and exploit. Of course, this input-dependent r need not be truly random; it suffices for a r to be a *pseudorandom* function of the input.

5.1 Modifying the Scheme

We now compute r as a pseudorandom function (PRF) over the input (m_i, h_{i-1}, x_{i-1}) received by that signer i. Let $\mathsf{PRF}_{\mathsf{seed}} : \{0,1\}^* \to \{0,1\}^{\ell_r}$ be a PRF with seed seed and insecurity $\varepsilon_{\mathsf{PRF}}(q, t)$ against adversaries asking q queries and running in time t. Add a uniformly chosen seed to the secret key of the signer and replace line 1 of the signing algorithm with $r \leftarrow \mathsf{PRF}_{\mathsf{seed}}(m, h, x)$.

In the previous section, we found that ℓ_r must be long enough to tolerate a security loss of $q_S(q_H + q_S)2^{-\ell_r}$ (Theorem 1). As we show below, ℓ_r in the modified scheme can be shorter, since it needs only to allow for a security loss of approximately $(q_G + q_H + q_S + \ell_H q_S^2)2^{-\ell_r}$. This is an improvement if we assume that $q_H \approx q_G$ (since both H and G are hash functions) and $q_S \ll q_H$ (since in practice hash queries can be made offline, while signing queries need access to an actual signer).

5.2   Key Insight for the Security Proof

Using the reduction of Section 4, we had to choose r long enough to make it unlikely that when a forger makes a sign query on $(\pi_*, m_i, x_{i-1}, h_{i-1})$, the algorithm Sim-S draws a random r_i that collides with a previously made H-query $Q_i = (\pi_*, m_i, r_i, x_{i-1})$. Indeed, if Q_i was answered by η_i and the forger chooses h_{i-1} so that h_i (which is computes as $h_{i-1} \oplus \eta_i$) has already been queried to G, then when r collides, the reduction would be prevented from programming the random oracle $G(h_i)$. Making r depend on the forger's input to the signer means that the forger gets only one chance (rather than q_S chances) to make this happen for a given Q_i, h_{i-1}, and h_i, because subsequent attemps by the forger will use the same r.

   We show in the full version [BGR11b] that the problem of proving this modified scheme secure hinges on the following combinatorial problem.

Combinatorial Problem. Suppose β values $\eta_1, \ldots, \eta_\beta$ are chosen uniformly at random as ℓ_H-bit strings and given to an adversary, who then chooses α distinct values $h'_1, \ldots, h'_\alpha$. The $\alpha \times \beta$-matrix ζ is constructed by XORing the η and the h' values. A *collision* in ζ is a set of entries that are all equal. What is the total number of entries in the γ biggest collisions?

Theorem 2. *With probability at least $1 - \beta^2 2^{\ell_H}$, the total size of the γ biggest collisions in ζ is at most $\alpha + (\ell_h + 2)\gamma^2$.*

The proof of this theorem, as well as the entire security analysis of the modified scheme, are found in [BGR11b].

6   Implementation and Evaluation

In the full version of the paper [BGR11b] we present details of instantiating our scheme with RSA (these include, in particular, dealing with the problem of slightly different domains for each signer's permutation). We implemented the input-dependent-r version as a module in OpenSSL [ope]. The code is available from [BGR11a].

Overview of Our Implementation. We instantiate the permutation π with 2048-bit RSA with public exponent 65537, hash H with SHA-256, full-domain hash G with the industry-standard Mask Generating Function (MGF) using SHA-256 [RSA02], and the pseudorandom function PRF with HMAC-SHA-256 [BCK96]. Instead of hashing the permutation π as-is inside the hash function H, we replace it with a short fingerprint of the RSA public key computed using SHA-256. Thus, we have parameters $\ell_\pi = 2048$, $\ell_h = 256$, and $\ell_r = 128$; the ℓ_r value is per signer, and each signer also adds one bit of information to deal with the problem that RSA gives each signer a slightly different domain. Therefore, the length of the aggregate signature for n signers is $2048 + 256 + 129n$ bits long (see Table 1). We justify this choice of parameters in [BGR11b].

Table 1. Benchmark results for n signers. Computed on a laptop with a Core i3 processor at 2.4GHz and 2GB RAM, running Ubuntu. The first three schemes were implemented using OpenSSL [ope] (with SHA-256 hashing and RSA public exponent of 65537); the BGLS scheme was implemented using MIRACL [Sco11] (with the curve BN-128 [BN05] and with precomputation on the curve generator but not on the public keys; further precomputation on the public keys seems to improve verification performance by up to 20% at the cost of additional storage). Results for specific values of n are not exactly in proportion due to rounding.

	2048-bit RSA	Our scheme	256-bit ECDSA	256-bit BGLS
Signature length (bits)	$2048n$	$2304 + 129n$	$512n$	257
Length for $n = 4.5$	9216	2885	2304	257
Length for $n = 7$	14336	3207	3584	257
Sign time (ms)	11.8	11.9	2.3	1.9
Verify time (ms)	$0.3n$	$0.3n$	$2.8n$	$\approx 18.9 + 6.6n$
Verify time for $n = 4.5$	1.3	1.3	12.5	47.6
Verify time for $n = 7$	2.1	2.1	19.4	64.8

Evaluation. We compare the implementation described in the previous paragraph to other signature schemes that allow for lazy verification. Table 1 contains data on our scheme as well as the "trivial" solution of using n RSA signatures, the solution of similarly using n ECDSA [Van92,IEE02] signatures (which are current contenders for adoption in BGPsec [Sri12, Section 4.1]), and the aggregate scheme of [BGLS03] (we do not compare against [FLS11], because it is a more complicated version of [BGLS03], so [BGLS03] performs better than [FLS11], anyway). In addition to providing formulas in terms of the number n of signers, we show results for specific values of $n = 4.5$ and $n = 7$. The value of 4.5 was chosen because it is roughly the average length of an AS path for a well-connected router on the Internet today (average length fluctuates with time and vantage point—see, e.g., here [Smi12]). We should note, however, that performance for higher than average values of n is particularly important: transition to BGPsec is expected to be particularly problematic for weaker routers, which are more likely to be located to in the less well-connected portions of the Internet, and that experience longer than average paths. We therefore also show results for $n = 7$.

The table shows that the [BGLS03] scheme is a clear winner in terms of signature length and signing time, but has considerably slower verification[4]. It should be noted, however, that it is not being considered for the BGPsec standard at this stage [Sri12, Section 4.1]: schemes relying on bilinear Diffie-Hellman are not considered ready for worldwide deployment on the internet backbone by the BGPsec working group, because a consensus has not emerged on which curves provide the right tradeoff between security and efficiency (for example, there is not a NIST-approved set of curves such as the one contained in [NIS09, Appendix D] for non-pairing-based elliptic-curve cryptography). It

---

[4] A more efficient pairing-based scheme of [WM08] with a constant total number of pairings was shown insecure by [SVS+09].

is also important to note that the time required to compute group operations and bilinear pairings depends very heavily on the curve used; improvements for various curves are produced frequently, and there is no generally accepted set of curves or algorithms at this point. We believe that, assuming continued progress to speed-up pairings on specific curves and sufficient confidence in the security of bilinear Diffie-Hellman on these curves, the scheme of [BGLS03] (as improved by [BNN07]) should be considered for real applications.

As far as the remaining three schemes are concerned, we observe that ECDSA provides the shortest signatures when $n < 6$, while our scheme dominates the three for $n > 6$ (as we already mentioned, performance for higher than average n is particularly important.) We also observe that our scheme has computation time almost identical to simple RSA while having much shorter signatures (RSA signature length is listed as a particular concern in [Sri12, Section 4.1.2]). While ECSDA has the fastest signing time, the *verification* times for RSA and our scheme are an order of magnitude faster than those of ECDSA. Note that, for a router, the time required to sign does not depend on n, but the time required to verify grows linearly with n, so verification times are also of particular importance to weaker routers at the edge of the network.

Thus, if one is interested in a scheme based on the standard assumption of trapdoor permutations (albeit in the random oracle model), then our proposal fits the bill. Moreover, even if one is willing to accept security of ECDSA (which is not known to follow from any standard assumptions), our scheme may be preferable based on fast verify times and comparable-length signatures. Our scheme also has much faster verifying that pairing-based BGLS.

Acknowledgements. We thank Anna Lysyanskaya for help with the early stages of this work, Hovav Shacham and Craig Costello for helpful pointers and explanations, and the DHS S&T CSD Secure Routing project for many useful discussions that informed the design of our schemes. We thank anonymous referees for helping us improve our presentation and for pointing out related work. This work was supported by NSF Grants 1017907, 0546614, 0831281, 1012910, 1012798, and a gift from Cisco.

References

AGH10.   Ahn, J.H., Green, M., Hohenberger, S.: Synchronized aggregate signatures: new definitions, constructions and applications. In: ACM Conference on Computer and Communications Security (2010)

BCK96.   Bellare, M., Canetti, R., Krawczyk, H.: Keying Hash Functions for Message Authentication. In: Koblitz, N. (ed.) CRYPTO 1996. LNCS, vol. 1109, pp. 1–15. Springer, Heidelberg (1996)

BGLS03.   Boneh, D., Gentry, C., Lynn, B., Shacham, H.: Aggregate and Verifiably Encrypted Signatures from Bilinear Maps. In: Biham, E. (ed.) EURO-CRYPT 2003. LNCS, vol. 2656, pp. 416–432. Springer, Heidelberg (2003)

BGOY07.   Boldyreva, A., Gentry, C., O'Neill, A., Yum, D.H.: Ordered multisigna-
            tures and identity-based sequential aggregate signatures, with applications
            to secure routing. In: ACM Conference on Computer and Communications
            Security, pp. 276–285. ACM (2007)
BGR11a.   Brogle, K., Goldberg, S., Reyzin, L.: Implementation of sequential aggre-
            gate signatures with lazy verification (2011),
            http://www.cs.bu.edu/fac/goldbe/papers/bgpsec-sigs.html
BGR11b.   Brogle, K., Goldberg, S., Reyzin, L.: Sequential aggregate signatures with
            lazy verification, Full version and implementation code (2011),
            http://www.cs.bu.edu/fac/goldbe/papers/bgpsec-sigs.html
BHMT09.   Houidi, Z.B., Meulle, M., Teixeira, R.: Understanding slow bgp routing
            table transfers. In: Proc. ACM SIGCOMM Internet Measurement Confer-
            ence, pp. 350–355. ACM, New York (2009)
BJ10.   Bagherzandi, A., Jarecki, S.: Identity-Based Aggregate and Multi-
            Signature Schemes Based on RSA. In: Nguyen, P.Q., Pointcheval, D. (eds.)
            PKC 2010. LNCS, vol. 6056, pp. 480–498. Springer, Heidelberg (2010)
BN05.   Barreto, P.S.L.M., Naehrig, M.: Pairing-Friendly Elliptic Curves of Prime
            Order. In: Preneel, B., Tavares, S. (eds.) SAC 2005. LNCS, vol. 3897, pp.
            319–331. Springer, Heidelberg (2006)
BNN07.   Bellare, M., Namprempre, C., Neven, G.: Unrestricted Aggregate Signa-
            tures. In: Arge, L., Cachin, C., Jurdziński, T., Tarlecki, A. (eds.) ICALP
            2007. LNCS, vol. 4596, pp. 411–422. Springer, Heidelberg (2007)
BR93.   Bellare, M., Rogaway, P.: Random oracles are practical: A paradigm for
            designing efficient protocols. In: ACM Conference on Computer and Com-
            munications Security, pp. 62–73 (1993)
BR96.   Bellare, M., Rogaway, P.: The Exact Security of Digital Signatures - How
            to Sign with RSA and Rabin. In: Maurer, U.M. (ed.) EUROCRYPT 1996.
            LNCS, vol. 1070, pp. 399–416. Springer, Heidelberg (1996)
CHKM10.   Chatterjee, S., Hankerson, D., Knapp, E., Menezes, A.: Comparing
            two pairing-based aggregate signature schemes. Des. Codes Cryptogra-
            phy 55(2-3), 141–167 (2010)
CID.   The CIDR report, http://www.cidr-report.org
CLGW06.   Cheng, X., Liu, J., Guo, L., Wang, X.: Identity-based multisignature and
            aggregate signature schemes from m-torsion groups. Journal of Electronics
            (China) 23(4) (July 2006)
CLW05.   Cheng, X., Liu, J., Wang, X.: Identity-Based Aggregate and Verifiably
            Encrypted Signatures from Bilinear Pairing. In: Gervasi, O., Gavrilova,
            M.L., Kumar, V., Laganá, A., Lee, H.P., Mun, Y., Taniar, D., Tan, C.J.K.
            (eds.) ICCSA 2005. LNCS, vol. 3483, pp. 1046–1054. Springer, Heidelberg
            (2005)
Cor02.   Coron, J.-S.: Optimal Security Proofs for PSS and Other Signature
            Schemes. In: Knudsen, L.R. (ed.) EUROCRYPT 2002. LNCS, vol. 2332,
            pp. 272–287. Springer, Heidelberg (2002)
COZ08.   Chi, Y.-J., Oliveira, R., Zhang, L.: Cyclops: The Internet AS-level obser-
            vatory. In: ACM SIGCOMM CCR (2008)
CSC09.   Chakrabarti, S., Chandrasekhar, S., Singhal, M., Calvert, K.L.: An ef-
            ficient and scalable quasi-aggregate signature scheme based on lfsr se-
            quences. IEEE Trans. Parallel Distrib. Syst. 20(7), 1059–1072 (2009)
DHS.   Department of Homeland Security, Science and Technology Directorate,
            Cyber Security Division, Secure Protocols for Routing Infrastructure
            project. Personal Communication

EFG+10.   Eikemeier, O., Fischlin, M., Götzmann, J.-F., Lehmann, A., Schröder, D.,
          Schröder, P., Wagner, D.: History-Free Aggregate Message Authentication
          Codes. In: Garay, J.A., De Prisco, R. (eds.) SCN 2010. LNCS, vol. 6280,
          pp. 309–328. Springer, Heidelberg (2010)
FLS11.   Fischlin, M., Lehmann, A., Schröder, D.: History-free sequential aggregate
          signatures. Technical Report 2011/231, Cryptology ePrint archive (2011),
          http://eprint.iacr.org
GGM86.   Goldreich, O., Goldwasser, S., Micali, S.: How to construct random func-
          tions. J. ACM 33(4), 792–807 (1986)
GMR88.   Goldwasser, S., Micali, S., Rivest, R.L.: A digital signature scheme secure
          against adaptive chosen-message attacks. SIAM J. Comput. 17(2), 281–
          308 (1988)
GR06.   Gentry, C., Ramzan, Z.: Identity-Based Aggregate Signatures. In: Yung,
          M., Dodis, Y., Kiayias, A., Malkin, T. (eds.) PKC 2006. LNCS, vol. 3958,
          pp. 257–273. Springer, Heidelberg (2006)
Her06.   Herranz, J.: Deterministic identity-based signatures for partial aggrega-
          tion. Comput. J. 49(3), 322–330 (2006)
HLY09.   Hwang, J.Y., Lee, D.H., Yung, M.: Universal forgery of the identity-based
          sequential aggregate signature scheme. In: Li, W., Susilo, W., Tupakula,
          U.K., Safavi-Naini, R., Varadharajan, V. (eds.) ASIACCS, pp. 157–160.
          ACM (2009)
Hus12.   Huston, G. (ed.): The Profile for Algorithms and Key Sizes for Use in the
          Resource Public Key Infrastructure (RPKI). IETF RFC 6485 (February
          2012), http://tools.ietf.org/html/rfc6485
IEE02.   IEEE Std 1363-2000. IEEE standard specifications for public-key cryptog-
          raphy (2002)
KLS00.   Kent, S., Lynn, C., Seo, K.: Secure border gateway protocol (S-BGP). J.
          Selected Areas in Communications 18(4), 582–592 (2000)
KR06.   Karpilovsky, E., Rexford, J.: Using forgetful routing to control bgp table
          size. In: Proceedings of the 2006 ACM CoNEXT Conference, CoNEXT
          2006, pp. 2:1–2:12. ACM, New York (2006)
KW03.   Katz, J., Wang, N.: Efficiency improvements for signature schemes with
          tight security reductions. In: Jajodia, S., Atluri, V., Jaeger, T. (eds.)
          ACM Conference on Computer and Communications Security, pp. 155–
          164. ACM (2003)
Lep12.   Lepinski, M. (ed.): BGPSEC Protocol Specification. IETF Network Work-
          ing Group, Internet-Draft (July 2012),
          http://tools.ietf.org/html/draft-ietf-sidr-bgpsec-protocol-04
LMRS04.   Lysyanskaya, A., Micali, S., Reyzin, L., Shacham, H.: Sequential Aggre-
          gate Signatures from Trapdoor Permutations. In: Cachin, C., Camenisch,
          J. (eds.) EUROCRYPT 2004. LNCS, vol. 3027, pp. 74–90. Springer, Hei-
          delberg (2004)
LOS+06.   Lu, S., Ostrovsky, R., Sahai, A., Shacham, H., Waters, B.: Sequential
          Aggregate Signatures and Multisignatures Without Random Oracles. In:
          Vaudenay, S. (ed.) EUROCRYPT 2006. LNCS, vol. 4004, pp. 465–485.
          Springer, Heidelberg (2006)
Nev08.   Neven, G.: Efficient Sequential Aggregate Signed Data. In: Smart, N.P.
          (ed.) EUROCRYPT 2008. LNCS, vol. 4965, pp. 52–69. Springer, Heidel-
          berg (2008)
NIS09.   FIPS publication 186-3: Digital signature standard (DSS) (June 2009),
          http://csrc.nist.gov/publications/PubsFIPS.html

ope.   OpenSSL toolkit, http://openssl.org/

RS09.   Rückert, M., Schröder, D.: Aggregate and Verifiably Encrypted Signatures
            from Multilinear Maps without Random Oracles. In: Park, J.H., Chen,
            H.-H., Atiquzzaman, M., Lee, C., Kim, T.-h., Yeo, S.-S. (eds.) ISA 2009.
            LNCS, vol. 5576, pp. 750–759. Springer, Heidelberg (2009)

RSA78.   Rivest, R.L., Shamir, A., Adleman, L.M.: A method for obtaining digital
            signatures and public-key cryptosystems. Commun. ACM 21(2), 120–126
            (1978)

RSA02.   PKCS #1: RSA Encryption Standard. Version 2.1. RSA Laboratories
            (June 2002),
            ftp://ftp.rsasecurity.com/pub/pkcs/pkcs-1/pkcs-1v2-1.pdf

Sco11.   Michael Scott. MIRACL library (2011), http://www.shamus.ie/

Smi12.   Smith, P.: BGP routing table analysis (2012),
            http://thyme.rand.apnic.net/. See historical data—e.g.,
            APNIC analysis summary for (September 7, 2012),
            http://thyme.apnic.net/ap-data/2012/09/07/0400/mail-global

Sri12.   Sriram, K. (ed.): BGPSEC Design Choices and Summary of Supporting
            Discussions. The Internet Engineering Task Force (IETF) Network Work-
            ing Group (July 2012), http://tools.ietf.org/html/
            draft-sriram-bgpsec-design-choices-02

SVS+09.   Sharmila Deva Selvi, S., Sree Vivek, S., Shriram, J., Kalaivani, S., Pandu
            Rangan, C.: Security analysis of aggregate signature and batch verification
            signature schemes. Technical Report 2009/290, Cryptology ePrint archive
            (2009), http://eprint.iacr.org

SVSR10.   Sharmila Deva Selvi, S., Sree Vivek, S., Shriram, J., Pandu Rangan, C.:
            Identity based partial aggregate signature scheme without pairing. Report
            2010/461, Cryptology ePrint archive (2010), http://eprint.iacr.org

Van92.   Vanstone, S.: Responses to NIST's proposal. Communications of the
            ACM 35, 50–52 (1992)

WM08.   Wen, Y., Ma, J.: An aggregate signature scheme with constant pairing
            operations. In: CSSE (3). IEEE Computer Society (2008)

XZF05.   Xu, J., Zhang, Z., Feng, D.: ID-Based Aggregate Signatures from Bilinear
            Pairings. In: Desmedt, Y.G., Wang, H., Mu, Y., Li, Y. (eds.) CANS 2005.
            LNCS, vol. 3810, pp. 110–119. Springer, Heidelberg (2005)

YCK04.   Yoon, H., Cheon, J.H., Kim, Y.: Batch Verifications with ID-Based Sig-
            natures. In: Park, C., Chee, S. (eds.) ICISC 2004. LNCS, vol. 3506, pp.
            233–248. Springer, Heidelberg (2005)

ZSN05.   Zhao, M., Smith, S.W., Nicol, D.M.: Aggregated path authentication for
            efficient BGP security. In: ACM Conference on Computer and Communi-
            cations Security, pp. 128–138. ACM (2005)

Commitments and Efficient Zero-Knowledge Proofs from Learning Parity with Noise*

Abhishek Jain[1], Stephan Krenn[2], Krzysztof Pietrzak[2], and Aris Tentes[3]

[1] Massachusetts Institute of Technology, USA, and
Boston University, USA
abhishek@csail.mit.edu
[2] Institute of Science and Technology Austria
{stephan.krenn,pietrzak}@ist.ac.at
[3] Department of Computer Science, New York University, USA
tentes@cs.nyu.edu

Abstract. We construct a perfectly binding string commitment scheme whose security is based on the learning parity with noise (LPN) assumption, or equivalently, the hardness of decoding random linear codes. Our scheme not only allows for a simple and efficient zero-knowledge proof of knowledge for committed values (essentially a Σ-protocol), but also for such proofs showing any kind of relation amongst committed values, i.e., proving that messages $m_0, \ldots, m_u$, are such that $m_0 = C(m_1, \ldots, m_u)$ for any circuit C.

To get soundness which is exponentially small in a security parameter t, and when the zero-knowledge property relies on the LPN problem with secrets of length ℓ, our 3 round protocol has communication complexity $\mathcal{O}(t|C|\ell \log(\ell))$ and computational complexity of $\mathcal{O}(t|C|\ell)$ bit operations. The hidden constants are small, and the computation consists mostly of computing inner products of bit-vectors.

1 Introduction

Commitment schemes and zero-knowledge proofs are fundamental cryptographic primitives. In this work we propose a simple string commitment scheme and show efficient zero-knowledge proofs for any relation amongst committed values. The security (more precisely, the computational hiding property) of our commitment scheme relies on the learning parity with noise (LPN) assumption, or equivalently, on the hardness of decoding random linear codes.

Commitment schemes. A commitment scheme allows a party to commit to a message m by publishing a commitment σ, and this commitment can be opened at a later point in time. The security properties required are called the hiding and binding property. Hiding means that one cannot learn anything about the

---

* This work was in part supported by the European Research Council under the European Unions Seventh Framework Programme (FP7/2007-2013) / ERC Starting Grant (259668-PSPC).

X. Wang and K. Sako (Eds.): ASIACRYPT 2012, LNCS 7658, pp. 663–680, 2012.

committed message m from the commitment σ, binding means that one cannot open a commitment σ to two different messages $m \neq m'$.

In our scheme, the commitment to a message m is simply the encoding of m using a random linear code, with some noise added to the codeword. Exploiting the linear structure of this scheme, we get simple and efficient zero-knowledge proofs for linear and multiplicative relations of committed values.

Zero-knowledge proofs of knowledge. Zero-knowledge proofs of knowledge are two party protocols, which allow a prover to convince a verifier that it knows some secret piece of information, without the verifier being able to learn anything about the secret value except for what is revealed by the claim itself.

The LPN assumption. The computationally hard problem underlying the security (i.e., the computational hiding property) of our commitment scheme is the *learning parity with noise* (LPN) assumption. This problem asks to distinguish "noisy" linear equations $\mathbf{A}.s \oplus e$ from uniformly random. Here $\mathbf{A}$ is a "skinny" public random binary $k \times \ell$ matrix, s is a uniformly random ℓ bit secret and e is a random vector of low weight (the exact distribution of e is discussed in §2.2). The LPN problem has found numerous applications as the assumption underlying provably secure cryptosystems, like secret-key [21,24,26,32] and public-key [36] authentication schemes or symmetric encryption [1,18].

LPN based cryptosystems are interesting for theoretical and practical reasons. On the one hand, the LPN problem is equivalent to the problem of decoding random linear codes, a problem that has been studied for over half a century [5,6,7,28,35]. The best known algorithms need $2^{\Theta(\ell/\log\ell)}$ time and samples (the number of samples is given by the number k of rows of $\mathbf{A}$) [7]. If $k = \Theta(\ell)$ is linear in ℓ, as it will be the case in this paper, the best algorithms need exponential $2^{\Theta(\ell)}$ time. Furthermore, unlike most number-theoretic problems used in cryptography, the LPN problem is not known to become insecure against quantum algorithms. On the practical side, LPN based cryptosystems tend to be extremely simple and efficient, and thus are good candidates for weak devices like RFID tags, where existing cryptographic algorithms cannot be implemented due to constraints on code-size, running-time or memory.

1.1   Our Contributions

Commitments from LPN. In our scheme the commitment to a message $m \in \mathcal{I}^v$ (where $\mathcal{I} \stackrel{\text{def}}{=} \{0,1\}$) is simply

$$\mathrm{Com}(m) = \mathbf{A}.(r\|m) \oplus e,$$

where $\mathbf{A} = \mathbf{A}'\|\mathbf{A}'' \in \mathcal{I}^{k\times(\ell+v)}$ is a public random binary matrix, $r \in \mathcal{I}^\ell$ is a uniformly random vector and $e \in \mathcal{I}^k$ is a random low-weight vector. To open a commitment σ, one reveals r, m, e and checks if $\sigma \stackrel{?}{=} \mathbf{A}.(r\|m) \oplus e$ and e is low weight. Here the length $\ell = |r|$ is chosen such that the LPN problem with secrets of length ℓ is hard. The length $v = |m|$ of the message can be arbitrary, but for efficiency reasons it is best to choose it roughly of the same size as ℓ.

Setting $k = \Theta(v + \ell)$ large enough, the commitment scheme becomes computationally hiding and perfectly binding (with overwhelming probability over the choice of $\mathbf{A}$). The binding property follows by the large distance of the code generated by the random matrix $\mathbf{A}$, the hiding property follows directly from the LPN assumption which implies that $\mathbf{A}'.r \oplus e$ is pseudorandom.

Zero-knowledge protocols for arbitrary circuits. We construct a zero-knowledge proof of knowledge, which is basically a so called Σ-protocol, that allows to prove knowledge of the message m hidden inside a commitment without revealing anything about it. Furthermore, we give a protocol for proving that committed messages m_0, m_1, m_2 satisfy a linear relation $m_0 = \mathbf{X}_1.m_1 \oplus \mathbf{X}_2.m_2$ (for any square matrices $\mathbf{X}_1, \mathbf{X}_2$). Based on this protocol, we construct proofs for any bitwise relations $m_0 = m_1 \circ m_2$, where $\circ$ can be any bitwise relation like AND, NAND, OR, NOR. As NAND is functionally complete, we can prove relations $m_0 = C(m_1, \ldots, m_t)$ for any boolean circuit C.

For $\mathbf{A} \in \mathcal{I}^{k \times m}$, the communication complexity of our proofs is $\Theta(k \log k)$. Setting $v = \ell$, we can set $k = \Theta(v + \ell) = \Theta(v)$, thus the proofs are quasilinear in the length of the committed messages. The soundness error of our protocol is $2/3$. To get soundness errors of 2^{-16} and 2^{-32} as specified by the ISO/IEC-9798-5 standard we would need 28 and 55 repetitions, respectively.

As one application (which we bring up to compare our scheme to existing schemes in the related work section below) consider an $\mathcal{NP}$ language $L = \{x : \exists w : \mathcal{R}(x, w) = 1\}$. Our scheme can be used to prove knowledge of a witness w for $x \in L$ as follows: commit to $m_0 = w$ and $m_1 = 1$ and prove that the committed values satisfy the relation $C_x(m_0) = m_1$ where $C_x(.)$ is the circuit computing the $\mathcal{NP}$ relation $\mathcal{R}(x, .)$. This proofs avoid expensive Karp reductions (to 3-coloring or Hamiltonian cycles) used in classical proofs.

1.2 Related Work

Our basic scheme for proving knowledge of a committed value is similar to Stern's [36] zero-knowledge proof of knowledge for the syndrome decoding problem, which can be seen as the "dual" of the LPN problem, and both are known to be $\mathcal{NP}$-complete [5]. Subsequent to Stern's work, Véron [37] proposed a Σ-protocol for proving knowledge of an LPN secret, but as we show in the full version of this paper [23], there is a gap in the proof of the ZK property of his protocol. Recently, several works have extended Stern's protocol to construct efficient identification schemes from various lattice-based and coding based assumptions (see [9,10,27] and references therein). In particular, Cayrel et al. [10] constructed an identification scheme with knowledge error $1/2$ based on the q-ary syndrome decoding problem. However, this improvement in the knowledge error adds two additional rounds to the protocol, and thus their construction does not decrease the total number of rounds required to reach a specified knowledge error. Very recently, Asharov et al. [3] constructed Σ-protocols for various learning with errors (LWE) related languages. We note that the ZK property of their protocols crucially relies on the ability to use large noise to "smudge out" small differences in distributions. Unfortunately, this technique does not extend to the setting of LPN (which is the focus of this work). We finally note that all the aforementioned works only construct ZK protocols for specific languages and, unlike our work, do not consider the general problem of constructing ZK proofs for circuit satisfiability.

There is a large body of work on efficient interactive or non-interactive zero-knowledge proofs and arguments, see, e.g., [8,13,14,15,20,22,25,30,31] and the references therein. For ZK *arguments* (as opposed to proofs), where the soundness property is only required to hold against computationally bounded malicious provers, one can construct schemes which asymptotically only require polylogarithmic communication (e.g., the interactive argument based on CRHFs [29] or the non-interactive argument in the random-oracle model [33]). These schemes rely on probabilistically checkable proofs (PCP), and are not really practical.

The beautiful work of Ishai et al. [22] on zero-knowledge proofs from secure multiparty computation aims at a similar goal as this work. They show how to construct ZK proofs from MPC; When instantiated with simple MPC protocols like GMW [19] they get ZK proofs for showing knowledge of a witness w such that $\mathcal{R}(x, w) = 1$ with communication complexity $O(ts)$, where 2^{-t} is the soundness error and s is the size of the circuit computing the relation $\mathcal{R}(x, .)$, which is the same asymptotic behavior we get (as explained in the previous section). Using protocols relying on sophisticated secret sharing schemes for constant-size fields based on algebraic-geometric codes [11] they even get an asymptotic communication complexity of $O(s) + poly(t, \log s)$, but due to the large hidden constants in such codes this scheme will only be more efficient than the simpler scheme for very large circuits.

A ZK proof for any $\mathcal{NP}$ relation can of course be used to prove any relation amongst *committed* values, but in general this would be rather expensive as the computation of the opening of the commitment must be part of the description of the relation. In contrast, our ZK proofs work directly on committed values, and we do not pay extra for this. Proving relations amongst *committed* values has been considered before, see [15] and references therein. These works give very efficient proofs for algebraic circuits over large fields, but are less suited for circuits over very small ones, in particular, for $\mathbb{Z}_2$ as in boolean circuits. As an application, consider the case where we need to prove that committed values satisfy $m_0 = \text{AES}(m_1, m_2)$, i.e., m_0 is the output of the AES block-cipher under key m_1 on input m_2.

1.3   Outline

We introduce some notation and recapitulate the basic definitions required for this paper in Section 2. In Section 3 we present a very simple commitment scheme based on the hardness of the LPN problem. Protocols allowing one to prove knowledge of the content of such commitments, and relations among them, are presented in Section 4. We finally conclude in Section 5.

2 Preliminaries

We use bold lower-case and upper-case letters like $\boldsymbol{a}, \mathbf{A}$ to denote vectors and matrices, respectively. Probabilistic polynomial time (PPT) algorithms are written by sans-serif letters like A. Calligraphic letters like $\mathcal{A}$ always denote sets. We write $a \overset{\text{R}}{\leftarrow} \mathcal{A}$ if a was drawn uniformly at random from set $\mathcal{A}$, $a \overset{\text{R}}{\leftarrow} \chi$ if a was drawn according to some probability distribution χ, and $a \overset{\text{R}}{\leftarrow} \mathsf{A}$ if a is the output of a randomized algorithm A.

We denote the set $\{0,1\}$ by $\mathcal{I}$, thus $\mathcal{I}^k$ denotes the set of strings of length k. The Hamming weight of $\boldsymbol{a} \in \mathcal{I}^k$ is denoted by $\|\boldsymbol{a}\|_1 = \sum_{i=1}^{k} \boldsymbol{a}[i]$. With $\mathcal{I}_w^k = \{\boldsymbol{a} \in \mathcal{I}^k : \|\boldsymbol{a}\|_1 = w\}$ we denote the set of all k-bit vectors of weight exactly w. The all-zeros and all-ones vectors of length k are denoted by $\boldsymbol{0}^k$ and $\boldsymbol{1}^k$, respectively. The concatenation of vectors $\boldsymbol{a}$ and $\boldsymbol{b}$ is written as $\boldsymbol{a}\|\boldsymbol{b}$. The symmetric group on k elements (i.e., the set of all permutations on k elements) is denoted $\mathcal{S}_k$. For $\pi \in \mathcal{S}_k$ and $\boldsymbol{a} \in \mathcal{I}^k$, $\pi(\boldsymbol{b})$ denotes the string $\boldsymbol{a}[i] = \boldsymbol{b}[\pi(i)]$.

2.1 Commitment Schemes

Definition 2.1. *A triple of algorithms* (KGen,Com,Ver) *is called a* commitment scheme *if it satisfies the following:*

- *On input 1^ℓ, the key generation algorithm* KGen *outputs a public commitment key pk.*
- *The commitment algorithm* Com *takes as inputs a message m from a message space $\mathcal{M}$ and a commitment key pk, and outputs a commitment/opening pair (c, d).*
- *The verification algorithm* Ver *takes a key pk, a message m, a commitment c and an opening d and outputs 1 or 0.*

The commitment scheme we construct satisfies the following security properties:

- *Correctness*: Ver evaluates to 1 whenever the inputs were computed by an honest party, i.e.,

$$\Pr[\mathsf{Ver}(pk, m, c, d) = 1; pk \overset{\text{R}}{\leftarrow} \mathsf{KGen}(1^\ell), m \in \mathcal{M}, (c, d) \overset{\text{R}}{\leftarrow} \mathsf{Com}(m, pk)] = 1$$

- *Perfect binding*: With overwhelming probability over the choice of the public key $pk \overset{\text{R}}{\leftarrow} \mathsf{KGen}(1^\ell)$, no commitment c can be opened in two different ways, i.e.,

$$(\mathsf{Ver}(pk, m, c, d) = 1) \wedge (\mathsf{Ver}(pk, m', c, d') = 1) \Rightarrow m = m'$$

- *Computational hiding*: A commitment c computationally hides the committed message: with overwhelming probability over the choice of $pk \overset{\text{R}}{\leftarrow} \mathsf{KGen}(1^\ell)$, for every $m, m' \in \mathcal{M}$ and $(c, d) \overset{\text{R}}{\leftarrow} \mathsf{Com}(m, pk), (c', d') \overset{\text{R}}{\leftarrow} \mathsf{Com}(m', pk)$ the distributions c and c' are computationally indistinguishable.

2.2   Learning Parity with Noise

The computational assumption underlying all our constructions is the *learning parity with noise* (LPN) assumption. Below we define the decisional version of LPN in a general form, not yet specifying the error distribution.

Definition 2.2. *For $k, \ell \in \mathbb{N}$, let χ be an error distribution over $\mathcal{I}^k$. The decisional (χ, ℓ, k)-LPN problem is (s, ϵ)-hard if for every distinguisher D of size s:*

$$\left| \Pr_{x, A, e}[D(\mathbf{A}, \mathbf{A}.x \oplus e) = 1] - \Pr_{r, A}[D(\mathbf{A}, r) = 1] \right| \leq \epsilon$$

where $\mathbf{A} \overset{R}{\leftarrow} \mathcal{I}^{k \times \ell}$, $e \overset{R}{\leftarrow} \chi_k$, $r \overset{R}{\leftarrow} \mathcal{I}^k$, and $x \overset{R}{\leftarrow} \mathcal{I}^{\ell}$ is fixed and secret. The search version is defined similarly, but we require that no D can find the secret x:

$$\left| \Pr_{x, A, e}[D(\mathbf{A}, \mathbf{A}.x \oplus e) = x] \right| \leq \epsilon$$

In the standard definition of the LPN problem, the error distribution χ is the Bernoulli distribution with some parameter $0 < \tau < \frac{1}{2}$, i.e., every bit $e[i]$ is chosen independently and identically distributed with $\Pr[e[i] = 1] = \tau$, we will refer to this version as LPN_τ. As mentioned in the introduction, for $k = \Theta(\ell)$ as used in this paper, the search version of LPN_τ is the same as the problem of decoding random linear codes, and is believed to be exponentially hard. The search and decision version of LPN_τ are known to be equivalent [6,26], but to show this search to decision reduction, the number of samples k in the decision version must be much larger than in the search version (by a factor of $\Omega(\ell/\epsilon)$). More recently, a sample preserving reduction has been shown [2, Lemma 4.4]. (cf. [34] for a more general treatment of sample preserving reductions).

Exact LPN. In this work we define a new version of the LPN problem, which we call *exact* LPN or xLPN for short. Similar to LPN_τ, xLPN is parameterized by some noise parameter $0 < \tau < \frac{1}{2}$, and the (search or decision) xLPN_τ problem is defined exactly like LPN_τ, except that the Hamming weight of the error vector is exactly $\lfloor k\tau \rceil$ (not of expected weight $k\tau$ as in LPN_τ). That is, e is sampled uniformly at random from the set $\mathcal{I}^k_{\lfloor k\tau \rceil}$.

In this work, we assume the hardness of *decisional* xLPN.[1] It is not hard to see that *search* xLPN_τ is hard iff search LPN_τ is hard.[2] Showing equivalence of *decisional* xLPN_τ and LPN_τ version is more tricky. The classical search to decision reduction for LPN_τ from [6,26] does not work for xLPN_τ, but the sample preserving reduction [2, Lemma 4.4] does. Summing up, we have

---

[1] The security of the basic commitment scheme can be based on decisional LPN, but our Σ-protocols to prove relations amongst committed values "leak" the weight of the error vectors. Thus, to be zero-knowledge, we need this value to be fixed.

[2] Any D who outputs x with advantage ϵ for xLPN_τ, will output the secret x with advantage at least $\epsilon/\sqrt{k}$ of LPN_τ, as the error vector sampled in LPN_τ has weight $\lfloor k\tau \rceil$ with probability $\geq 1/\sqrt{k}$, and conditioned on this being the case, the error distribution is exactly the same as in xLPN_τ.

Proposition 2.3. *The hardness of decisional* $\times\mathsf{LPN}_\tau$ *(used in this paper) is polynomially related to the hardness of search* LPN_τ.

The sample preserving reduction [2, Lemma 4.4] relies on the Goldreich-Levin theorem, and as a consequence is not very tight. Although we do not know of significantly more efficient attacks against $\times\mathsf{LPN}_\tau$ than against LPN_τ, if one insists on basing the security of our schemes on the standard LPN_τ assumption in a provable manner, one must take the loss in the reduction into account, which would result in rather large parameters. A protocol relying on the security of the standard decisional LPN_τ assumption can be found in the full version of this paper. The protocol given there can be extended to prove arbitrary relations amongst committed values, in the same manner as in the case of $\times\mathsf{LPN}_\tau$ assumption. However, this protocol is somewhat more complicated and has a worse soundness error (4/5 as compared to 2/3), and thus requires roughly twice the number of repetitions in order to achieve the same knowledge error.

As suggested in [26, Section 5], replacing the LPN assumption with an assumption where we have a fixed upped bound on the weight of the error vector (like it is the case in $\times\mathsf{LPN}$) would remove the completeness error (and thus allows for more efficient instantiations) also for other LPN based schemes, like HB type protocols. We thus think that investigating the exact hardness of the $\times\mathsf{LPN}$-problem is of interest beyond the realm of this work.

2.3 Zero-Knowledge Proofs of Knowledge and Σ-Protocols

Informally, a zero-knowledge proof of knowledge is a two party protocol between a prover P and a verifier V which allows the former to convince the latter that it knows some secret piece of information without revealing anything about it. A bit more precisely, in a zero-knowledge proof for a binary relation $\mathcal{R}$, the parties have common input y and the prover has private input w such that $(y, w) \in \mathcal{R}$. The protocol must then satisfy the following three properties: (i) For an honest prover, the verifier always accepts (*completeness*). (ii) For every potentially malicious verifier V^* there exists a PPT simulator only taking y as an input whose output is indistinguishable from conversations of V^* with an honest prover (*zero-knowledge*). (iii) From every prover P^* which can make the verifier accept with a probability larger than a threshold κ (the *knowledge error*), a w' satisfying $(y, w') \in \mathcal{R}$ can be extracted efficiently in a rewindable black-box way (*proof of knowledge*). For a formal definition we refer to Bellare and Goldreich [4].

The protocols we are going to design in the following are all instantiations of the following definition:

Definition 2.4 (Σ-Protocol). *Let* (P, V) *be a two-party protocol, where* V *is PPT, and let* $\mathcal{R}$ *be a binary relation. Then* (P, V) *is called a* Σ-*protocol for* $\mathcal{R}$ *with challenge set* $\mathcal{C}$*, public input* y *and private input* w*, if and only if it satisfies the following conditions:*

- **3-move form:** *The protocol is of the following form:*
  - *The prover* P *computes a commitment* t *and sends it to* V.

- *The verifier* V *draws a challenge* $c \xleftarrow{R} C$ *and sends it to* P.
- *The prover sends a response s to the verifier.*
- *Depending on the protocol transcript* (t, c, s), *the verifier accepts or rejects the proof.*

The protocol transcript (t, c, s) *is called* accepting, *if the verifier accepts the protocol run.*

- **Completeness:** *The verifier* V *accepts whenever* $(y, w) \in R$.
- **Special soundness:** *There exists a PPT algorithm* E *(the knowledge extractor) which takes a set* $\{(t, c, s_c) : c \in C\}$ *of accepting transcripts with the same commitment as inputs, and outputs* w' *such that* $(y, w') \in R$.
- **Special honest-verifier zero-knowledge:** *There exists a PPT algorithm* S *(the simulator) taking* y *and* $c \in C$ *as inputs, and which outputs triples* (t, c, s) *whose distribution is (computationally) indistinguishable from accepting protocol transcripts generated by real protocol runs.*

It is well known that every Σ-protocol is also a proof of knowledge for the same relation [16]. However, while in Σ-protocols the existence of a simulator is only required for the honest verifier, zero-knowledge proofs require this existence for arbitrary, potentially malicious, verifiers. This can be reached by applying generic standard techniques to Σ-protocols, e.g., Damgård et al. [17].

We note that our definition of Σ-protocols slightly differs from the standard definition found in the literature [12,16]. For the special soundness property, it is typically required that a valid witness can already be computed given any *two* accepting conversations with the same commitment but different challenges. We loosen this definition and only require that w' can be computed given valid responses to *all* challenges for a fixed commitment t. It can easily be seen that the aforementioned results showing that every Σ-protocol is also a proof of knowledge still hold true. However, while for the standard definition the knowledge error is given by $1/\#C$ it is only given by $1 - 1/\#C$ for Definition 2.4.

3   Perfectly Binding String Commitments from LPN

Our commitment scheme is parameterized by the main security parameter $\ell \in \mathbb{N}$, $0 < \tau < 0.25$, the message length $v \in \mathbb{N}$ and $k \in \mathcal{O}(\ell + v)$. Finally, we set $w = \lfloor \tau k \rfloor$. The algorithms of the commitment scheme are then given as follows:

- KGen: The public commitment key consists of the matrix $\mathbf{A} = \mathbf{A}' \| \mathbf{A}'' \in \mathcal{I}^{k \times (\ell + v)}$, where $\mathbf{A}' \xleftarrow{R} \mathcal{I}^{k \times \ell}$ and $\mathbf{A}'' \xleftarrow{R} \mathcal{I}^{k \times v}$.
- Com: The commitment to a message $\boldsymbol{m} \in \mathcal{I}^v$ is given by $\mathbf{A}.(\boldsymbol{r} \| \boldsymbol{m}) \oplus \boldsymbol{e}$, where $\boldsymbol{r} \xleftarrow{R} \mathcal{I}^\ell$ and $\boldsymbol{e} \xleftarrow{R} \mathcal{I}_w^k$. The opening of the commitment is given by $\boldsymbol{m}$ and $\boldsymbol{r}$.
- Ver: Given a commitment $\boldsymbol{c}$, a message $\boldsymbol{m}'$ and a randomness $\boldsymbol{r}'$, a verifier accepts if and only if $\boldsymbol{e}' = \boldsymbol{c} \oplus \mathbf{A}.(\boldsymbol{r}' \| \boldsymbol{m}')$ has weight w.

Theorem 3.1. *Let* $0 < \tau < 0.25$, *and* $\ell, k, v \in \mathbb{N}$ *be such that the decisional* xLPN$_\tau$ *problem (with secrets of length* ℓ *and* k *samples) is hard. Let* $k = \Theta(\ell + v)$ *be such that with overwhelming probability a randomly chosen generator matrix* of a linear code $\mathbf{A} \in \mathcal{I}^{k \times (\ell + v)}$ has distance larger than $2w$, i.e., $\|\mathbf{A}.\boldsymbol{x}\|_1 > 2w$ for all $\boldsymbol{x} \in \mathcal{I}^{\ell + v}$. Then the above commitment scheme is perfectly binding and computationally hiding.

Proof. The required security properties can be seen as follows:

Perfect binding. Assume, by contraposition, that $\boldsymbol{m}_i, \boldsymbol{r}_i, i = 1, 2$ are two different openings for a commitment $\boldsymbol{c}$. That is, we have that $\boldsymbol{e}_i = \boldsymbol{c} \oplus \mathbf{A}.(\boldsymbol{r}_i \| \boldsymbol{m}_i)$ has norm at most w for $i = 1, 2$. Thus we have that $\boldsymbol{e}_1 \oplus \boldsymbol{e}_2 = \mathbf{A}.(\boldsymbol{r}_1 \| \boldsymbol{m}_1 \oplus \boldsymbol{r}_2 \| \boldsymbol{m}_2)$ is a codeword of length $\|\boldsymbol{e}_1 \oplus \boldsymbol{e}_2\|_1 \leq \|\boldsymbol{e}_1\|_1 + \|\boldsymbol{e}_2\|_1 \leq 2w$, contradicting our assumption on the distance of the code generated by $\mathbf{A}$.

Computational hiding. We have that $\boldsymbol{c} = \mathbf{A}'.\boldsymbol{r} \oplus \boldsymbol{e} \oplus \mathbf{A}''.\boldsymbol{m}$. By the xLPN$_\tau$-assumption $\mathbf{A}'.\boldsymbol{r} \oplus \boldsymbol{e}$, and thus also $\boldsymbol{c}$, is pseudorandom. $\qquad\square$

4   Zero-Knowledge Proofs of Knowledge

In this section we first construct a Σ-protocol, which on common input $\mathbf{A}$ and $\boldsymbol{y}$ allows the prover to prove knowledge of a valid opening of $\boldsymbol{y}$ under the commitment scheme presented in Section 3. The protocol borrows some basic ideas from Stern [36], who gave a Σ-protocol for the syndrome decoding problem.

After presenting this basic protocol, we give two further Σ-protocols. The first can be used to prove that committed strings satisfy any linear relation. The second protocol can be used to show that committed strings satisfy any bitwise relation like bitwise AND, NAND, OR or NOR. As NAND is functionally complete, using this protocol we can construct Σ-protocols for any relation amongst committed messages.

4.1   Proving Knowledge of a Valid Opening

The following Σ-protocol proves knowledge of a valid opening for commitments of the form described in the previous section, i.e., it shows possession of $\boldsymbol{r}, \boldsymbol{m}, \boldsymbol{e}$ such that $\boldsymbol{y} = \mathbf{A}.(\boldsymbol{r} \| \boldsymbol{m}) \oplus \boldsymbol{e}$ for an error satisfying $\|\boldsymbol{e}\|_1 = w$. For notational convenience we will sometimes write $\boldsymbol{s}$ to denote the vector $\boldsymbol{r} \| \boldsymbol{m}$.

A first idea for such a protocol (which will not quite work) is to mimic Schnorr's protocol as follows: (1) the prover P commits to some value $\boldsymbol{t}_0 = \mathbf{A}.\boldsymbol{v} \oplus \boldsymbol{f}$, (2) the verifier V sends a challenge $c \xleftarrow{\text{R}} \{0, 1\}$, (3) the prover opens $\boldsymbol{t}_0$ (i.e., sends $\boldsymbol{v}, \boldsymbol{f}$) if $c = 0$ and opens $\boldsymbol{t}_0 \oplus \boldsymbol{y}$ (i.e., sends $\boldsymbol{v} \oplus \boldsymbol{s}, \boldsymbol{f} \oplus \boldsymbol{e}$) if $c = 1$.

If in this protocol $\boldsymbol{f}$ is sampled such that it has low weight, then $\boldsymbol{e} \oplus \boldsymbol{f}$ leaks information about $\boldsymbol{e}$, and the protocol is not zero-knowledge. On the other hand, if $\boldsymbol{f}$ is uniformly random (so $\boldsymbol{e} \oplus \boldsymbol{f}$ is independent of $\boldsymbol{e}$), the protocol is not sound (informally, all we can say is that from answers to both challenges we can extract $\boldsymbol{s}', \boldsymbol{e}'$ where $\boldsymbol{y} = \mathbf{A}.\boldsymbol{s}' \oplus \boldsymbol{e}'$, but $\boldsymbol{e}'$ can have arbitrary weight, and finding such a solution is trivial). In our protocol $\boldsymbol{f}$ is chosen uniformly at random, and to ensure soundness we use a trick from Stern [36]. We additionally commit to a random permutation $\pi \in \mathcal{S}_k$ and to $\pi(\boldsymbol{f})$, $\pi(\boldsymbol{f} \oplus \boldsymbol{e})$. On challenge $c = 0$ and $c = 1$ we now additionally make sure the openings are consistent with the committed errors by opening π and either $\pi(\boldsymbol{f})$ (if $c = 0$) or $\pi(\boldsymbol{f} \oplus \boldsymbol{e})$ (if $c = 1$). Moreover we extend the challenge space from two to three. The extra challenge $c = 2$ is used to verify that the weight of $\pi(\boldsymbol{f}) \oplus \pi(\boldsymbol{f} \oplus \boldsymbol{e}) = \pi(\boldsymbol{e})$ (and thus $\boldsymbol{e}$) is small, this will ensure soundness, as from valid answers to all three challenges we can extract $\boldsymbol{s}', \boldsymbol{e}'$ where $\boldsymbol{y} = \mathbf{A}.\boldsymbol{s}' \oplus \boldsymbol{e}'$ and $\boldsymbol{e}'$ has low weight. Opening the commitments to $\pi(\boldsymbol{f}), \pi(\boldsymbol{f} \oplus \boldsymbol{e})$ on $c = 2$ does not hurt the ZK property, as $\pi(\boldsymbol{f}), \pi(\boldsymbol{f} \oplus \boldsymbol{e})$ contains no information about $\boldsymbol{e}$ except its weight.

The common input to P, V is $\mathbf{A}$ and $\boldsymbol{y}$, P's secret input is $(\boldsymbol{e}, \boldsymbol{s})$. The protocol flow is then given as follows, where the commitment scheme $\mathsf{Com}(.)$ can be instantiated by an arbitrary perfectly binding string commitment scheme, potentially the scheme presented in Section 3 itself.

- P samples a permutation $\pi \xleftarrow{\text{R}} \mathcal{S}_k$ at random.
  It then draws $\boldsymbol{v} \xleftarrow{\text{R}} \mathcal{I}^{\ell+v}$, $\boldsymbol{f} \xleftarrow{\text{R}} \mathcal{I}^k$, and then sends the following commitments to the verifier V:
$$C_0 \leftarrow \mathsf{Com}(\pi' = \pi, \boldsymbol{t}_0 = \mathbf{A}.\boldsymbol{v} \oplus \boldsymbol{f})$$
$$C_1 \leftarrow \mathsf{Com}(\boldsymbol{t}_1 = \pi(\boldsymbol{f}))$$
$$C_2 \leftarrow \mathsf{Com}(\boldsymbol{t}_2 = \pi(\boldsymbol{f} \oplus \boldsymbol{e}))$$

- The verifier draws $c \xleftarrow{\text{R}} \mathbb{Z}_3$ and sends it to P.
- Depending on the value of c, P opens the following commitments:
  0. P opens C_0, C_1 by sending $\pi', \boldsymbol{t}_0, \boldsymbol{t}_1$ and the associated random coins.
  1. P opens C_0, C_2 by sending $\pi', \boldsymbol{t}_0, \boldsymbol{t}_2$ and the associated random coins.
  2. P opens C_1, C_2 by sending $\boldsymbol{t}_1, \boldsymbol{t}_2$ and the associated random coins.
- The verifier verifies the correctness of the openings received from the prover, and additionally performs the following checks depending on the challenge c:
  0. V accepts, iff $\boldsymbol{t}_0 \oplus \pi'^{-1}(\boldsymbol{t}_1) \overset{?}{\in} \mathrm{img}\,\mathbf{A}$ and $\pi' \overset{?}{\in} \mathcal{S}_k$.
  1. V accepts, iff $\boldsymbol{t}_0 \oplus \pi'^{-1}(\boldsymbol{t}_2) \oplus \boldsymbol{y} \overset{?}{\in} \mathrm{img}\,\mathbf{A}$.
  2. V accepts, iff $\|\boldsymbol{t}_1 \oplus \boldsymbol{t}_2\|_1 = w$.

Theorem 4.1. *The above protocol is a Σ-protocol for the following relation:*

$$\mathcal{R}_{LPN} = \{((\mathbf{A}, \boldsymbol{y}), (\boldsymbol{r}, \boldsymbol{m}, \boldsymbol{e})) : \boldsymbol{y} = \mathbf{A}.(\boldsymbol{r} \| \boldsymbol{m}) \oplus \boldsymbol{e} \quad \wedge \quad \|\boldsymbol{e}\|_1 = w\}$$

Proof. The 3-move form required for Definition 2.4 is clear. The remaining properties can be seen as follows.

Completeness. It is easy to see that an honest prover can always convince the verifier. Depending on the challenge c, we get:

0. $\boldsymbol{t}_0 \oplus \pi'^{-1}(\boldsymbol{t}_1) = (\mathbf{A}.\boldsymbol{v} \oplus \boldsymbol{f}) \oplus \pi^{-1}(\pi(\boldsymbol{f})) = \mathbf{A}.\boldsymbol{v} \in \mathrm{img}\,\mathbf{A}$ and π is a permutation.
1. $\boldsymbol{t}_0 \oplus \pi'^{-1}(\boldsymbol{t}_2) \oplus \boldsymbol{y} = (\mathbf{A}.\boldsymbol{v} \oplus \boldsymbol{f}) \oplus \pi^{-1}(\pi(\boldsymbol{f} \oplus \boldsymbol{e})) \oplus (\mathbf{A}.\boldsymbol{s} \oplus \boldsymbol{e}) = \mathbf{A}.(\boldsymbol{v} \oplus \boldsymbol{s}) \in \mathrm{img}\,\mathbf{A}$.
2. $\|\boldsymbol{t}_1 \oplus \boldsymbol{t}_2\|_1 = \|\pi(\boldsymbol{f}) \oplus \pi(\boldsymbol{f} \oplus \boldsymbol{e})\|_1 = \|\pi(\boldsymbol{f} \oplus \boldsymbol{f} \oplus \boldsymbol{e})\|_1 = \|\pi(\boldsymbol{e})\|_1 = \|\boldsymbol{e}\|_1 = w$.

Special Soundness. Assume that we have fixed values C_0, C_1, C_2 and openings for all challenges $c \in \mathbb{Z}_3$, such that the verifier accepts on all of them. Then, by the assumed perfect binding property of the underlying commitment scheme $\mathsf{Com}(.)$, we know that the openings to identical commitments must be identical across different challenges.

By adding the verification equations for $c = 0$ and $c = 1$ we get that $\pi'^{-1}(t_1 \oplus t_2) \oplus y \in \mathrm{img}\, \mathbf{A}$ and thus that $y = \mathbf{A}.s' \oplus \pi'^{-1}(t_1 \oplus t_2)$, where $s' = (r'\|m')$ is easy to compute. Now, using that $\|t_1 \oplus t_2\|_1 = w$, we have a valid witness of $(\mathbf{A}, y)$ is thus given by $(r', m', \pi'^{-1}(t_1 \oplus t_2))$.

Honest-Verifier Zero-Knowledge. In the following we describe an efficient simulator S, which for each challenge $c \in \mathbb{Z}_3$ outputs an accepting protocol transcript the distribution of which is computationally indistinguishable from real protocol transactions with an honest prover for challenge c.

0. The simulator S computes C_0 and C_1 like an honest prover, and computes C_2 as a commitment to 0. Then, clearly, the distribution of C_0, C_1, π', t_0, t_1 is identical to that in real protocol transcripts. Furthermore, by the computational hiding property of the commitment scheme $\mathsf{Com}(.)$, the distribution of C_2 is computationally indistinguishable from that in real protocol runs.

1. For $c = 1$, the simulator draws $\pi \xleftarrow{R} S_k$, $a \xleftarrow{R} \mathcal{I}^k$ and $b \xleftarrow{R} \mathcal{I}^{\ell+v}$. It sets $C_0 = \mathsf{Com}(\pi, \mathbf{A}.b \oplus y \oplus a)$ and $C_2 = \mathsf{Com}(\pi(a))$. The value of C_1 is computed as commitments to 0. It easy to see that the openings of C_0, C_2 pass the verification equations. To see the correctness of their distributions note that t_2 in the real protocol run and $\pi(a)$ in the simulated run are perfectly uniform in $\mathcal{I}^k$, and the permutations are also equally distributed both times. Concerning the opening of C_0, note the following: in the real protocol run, we have $t_0 = \mathbf{A}.v \oplus f$, where v is uniformly at random, and $f = \pi^{-1}(t_2 \oplus e)$; in the simulated transcript the content of C_0 is given by $\mathbf{A}.(b \oplus s) \oplus (a \oplus e)$. Now, v and $b \oplus s$ are both uniformly random, and the terms f and $a \oplus e$ are uniquely determined by the contents of C_0 and C_2. Thus, the distributions of C_0, C_2 and their openings are perfectly simulated. The distribution of C_1 is computationally indistinguishable by the assumed hiding property of $\mathsf{Com}(.)$.

2. Finally, for $c = 2$, the simulator draws $a \xleftarrow{R} \mathcal{I}^k$ and $b \leftarrow \mathcal{I}_w^k$ uniformly at random. It computed C_0 as a commitment to 0, $C_1 = \mathsf{Com}(a)$ and $C_5 = \mathsf{Com}(a \oplus b)$. As before, the distributions of C_0 is computationally indistinguishable from real protocol runs by the binding property of $\mathsf{Com}(.)$, and C_1 and C_2 as well as their openings can easily be seen to perfectly simulate the behavior of an honest prover. $\square$

4.2   Proving Linear Relations

We next describe a Σ-protocol which allows to prove that the messages hidden within commitments y_1, y_2, y_2 (where $y_i = \mathbf{A}.(r_i\|m_i) \oplus e_i$) satisfy arbitrary linear relations. That is, $\mathbf{X}_1.m_1 \oplus \mathbf{X}_2.m_2 = m_3$ for arbitrary matrices

$\mathbf{X}_1, \mathbf{X}_2 \in \mathcal{I}^{v \times v}$. The computational and communication complexity of the protocol is roughly the same as for proving the knowledge of the three committed messages using the protocol from the previous section, proving that they also satisfy the linear relation comes almost for free.

The high level idea of the protocol is as follows. P and V run the protocol from the previous section to prove knowledge of $\boldsymbol{m}_1, \boldsymbol{m}_2, \boldsymbol{m}_3$ for all the messages in parallel (but using the same challenge for all three). Recall that (oversimplifying a bit by ignoring the issue with the errors, i.e., the challenge $c = 2$) this protocol goes as follows: P commits to three random messages $\boldsymbol{v}_1, \boldsymbol{v}_2, \boldsymbol{v}_3$, and later opens the $\boldsymbol{v}_i$'s (if $c = 0$) or $\boldsymbol{v}_i \oplus \boldsymbol{m}_i$ (if $c = 1$). We change this protocol now a bit, and instead choosing $\boldsymbol{v}_3$ at random we compute it as $\boldsymbol{v}_3 = \mathbf{X}_1.\boldsymbol{v}_1 \oplus \mathbf{X}_2.\boldsymbol{v}_2$. Moreover the verifier now additionally checks if $\boldsymbol{v}_3 \stackrel{?}{=} \mathbf{X}_1.\boldsymbol{v}_1 \oplus \mathbf{X}_2.\boldsymbol{v}_2$ (if $c = 0$) and if $(\boldsymbol{v}_3 \oplus \boldsymbol{m}_3) \stackrel{?}{=} \mathbf{X}_1.(\boldsymbol{v}_1 \oplus \boldsymbol{m}_1) \oplus \mathbf{X}_2.(\boldsymbol{v}_2 \oplus \boldsymbol{m}_2)$ (if $c = 1$). With these changes, we get a stronger soundness property: not only can we extract the committed messages $\boldsymbol{m}_i$ from accepting answers to both challenges, but they will also satisfy $\boldsymbol{m}_3 = \mathbf{X}_1.\boldsymbol{m}_1 \oplus \mathbf{X}_2.\boldsymbol{m}_2$. At the same time the zero-knowledge property is not weakened, except of course for leaking the fact that the $\boldsymbol{m}_i$'s satisfy this linear relation.

The protocol flow is defined as follows:

- P samples permutations π_1, π_2, π_3 at random.
  It then draws $\boldsymbol{v}_1, \boldsymbol{v}_2 \stackrel{R}{\leftarrow} \mathcal{I}^v$, $\boldsymbol{u}_1, \boldsymbol{u}_2, \boldsymbol{u}_3 \stackrel{R}{\leftarrow} \mathcal{I}^\ell$, $\boldsymbol{f}_1, \boldsymbol{f}_2, \boldsymbol{f}_3 \stackrel{R}{\leftarrow} \mathcal{I}^k$, sets $\boldsymbol{v}_3 = \mathbf{X}_1.\boldsymbol{v}_1 \oplus \mathbf{X}_2.\boldsymbol{v}_2$ and then sends the following commitments for $i = 1, 2, 3$ to the verifier V:

$$C_{i0} \leftarrow \mathsf{Com}(\pi'_i = \pi_i, \boldsymbol{t}_{i0} = \mathbf{A}.(\boldsymbol{u}_i \| \boldsymbol{v}_i) \oplus \boldsymbol{f}_i)$$
$$C_{i1} \leftarrow \mathsf{Com}(\boldsymbol{t}_{i1} = \pi_i(\boldsymbol{f}_i))$$
$$C_{i2} \leftarrow \mathsf{Com}(\boldsymbol{t}_{i2} = \pi_i(\boldsymbol{f}_i \oplus \boldsymbol{e}_i))$$

- The verifier draws $c \stackrel{R}{\leftarrow} \mathbb{Z}_3$ and sends it to P.
- Depending on the value of c, P opens the following commitments:
  0. P opens C_{i0}, C_{i1} by sending $\pi'_i, \boldsymbol{t}_{i0}, \boldsymbol{t}_{i1}$ and the associated random coins.
  1. P opens C_{i0}, C_{i2} by sending $\pi'_i, \boldsymbol{t}_{i0}, \boldsymbol{t}_{i2}$ and the associated random coins.
  2. P opens C_{i1}, C_{i2} by sending $\boldsymbol{t}_{i1}, \boldsymbol{t}_{i2}$ and the associated random coins.
- The verifier verifies the correctness of the openings received from the prover, and additionally performs the following checks depending on the challenge c:
  0. V accepts, iff $\pi'_i \stackrel{?}{\in} \mathcal{S}_k$, there exist solutions $(\boldsymbol{a}_i, \boldsymbol{b}_i) \in \mathcal{I}^\ell \times \mathcal{I}^v$ to the equations $\boldsymbol{t}_{i0} \oplus \pi'^{-1}_i(\boldsymbol{t}_{i1}) = \mathbf{A}.(\boldsymbol{a}_i \| \boldsymbol{b}_i)$ and they satisfy $\boldsymbol{b}_3 = \mathbf{X}_1.\boldsymbol{b}_1 \oplus \mathbf{X}_2.\boldsymbol{b}_2$.
  1. V accepts, iff there exist solutions $(\boldsymbol{c}_i, \boldsymbol{d}_i) \in \mathcal{I}^\ell \times \mathcal{I}^v$ to the equations $\boldsymbol{t}_{i0} \oplus \pi'^{-1}_i(\boldsymbol{t}_{i2}) \oplus \boldsymbol{y}_i = \mathbf{A}.(\boldsymbol{c}_i \| \boldsymbol{d}_i)$ and they satisfy $\boldsymbol{d}_3 = \mathbf{X}_1.\boldsymbol{d}_1 \oplus \mathbf{X}_2.\boldsymbol{d}_2$.
  2. V accepts, iff $\|\boldsymbol{t}_{i1} \oplus \boldsymbol{t}_{i2}\|_1 \stackrel{?}{=} w$.

Theorem 4.2. *The above protocol is a Σ-protocol for the following relation:*

$$\mathcal{R}_{LLPN} = \Big\{ ((\mathbf{A}, \mathbf{X}_1, \mathbf{X}_2, \boldsymbol{y}_1, \boldsymbol{y}_2, \boldsymbol{y}_3), (\boldsymbol{r}_1, \boldsymbol{r}_2, \boldsymbol{r}_3, \boldsymbol{m}_1, \boldsymbol{m}_2, \boldsymbol{m}_3, \boldsymbol{e}_1, \boldsymbol{e}_2, \boldsymbol{e}_3)) :$$

$$\bigwedge_{i=1}^{3} \Big(\boldsymbol{y}_i = \mathbf{A}.(\boldsymbol{r}_i \| \boldsymbol{m}_i) \oplus \boldsymbol{e}_i \wedge \|\boldsymbol{e}_i\|_1 = w \Big) \wedge \boldsymbol{m}_3 = \mathbf{X}_1.\boldsymbol{m}_1 \oplus \mathbf{X}_2.\boldsymbol{m}_2 \Big\}.$$

As we can turn any commitment $\boldsymbol{y} = \mathbf{A}.(\boldsymbol{r} \| \boldsymbol{m}) \oplus \boldsymbol{e}$ for an (unknown) message $\boldsymbol{m}$ into a commitment for the message $\boldsymbol{m} \oplus \boldsymbol{x}$ as $\boldsymbol{y} \oplus \mathbf{A}.(\mathbf{0}^\ell \| \boldsymbol{x}) = \mathbf{A}.(\boldsymbol{r} \| (\boldsymbol{m} \oplus \boldsymbol{x})) \oplus \boldsymbol{e}$. Our protocol directly implies a protocol for affine relations

$$\mathcal{R}_{ALPN} = \Big\{ \Big((\mathbf{A}, \mathbf{X}_1, \mathbf{X}_2, \{\boldsymbol{x}_i, \boldsymbol{y}_i\}_{i=1}^3), (\{\boldsymbol{r}_i, \boldsymbol{m}_i, \boldsymbol{e}_i\}_{i=1}^3)\Big) :$$

$$\bigwedge_{i=1}^{3} \Big(\boldsymbol{y}_i = \mathbf{A}.(\boldsymbol{r}_i \| \boldsymbol{m}_i) \oplus \boldsymbol{e}_i \wedge \|\boldsymbol{e}_i\|_1 = w \Big)$$

$$\wedge (\boldsymbol{m}_3 \oplus \boldsymbol{x}_3) = \mathbf{X}_1.(\boldsymbol{m}_1 \oplus \boldsymbol{x}_1) \oplus \mathbf{X}_2.(\boldsymbol{m}_2 \oplus \boldsymbol{x}_2) \Big\}.$$

In particular, this allows to prove that $\boldsymbol{m}_1 = \mathbf{1}^v \oplus \boldsymbol{m}_2$, i.e., $\boldsymbol{m}_1$ is the bitwise negation of $\boldsymbol{m}_2$. Furthermore, the protocol can be seen to directly generalize to relations among more than 3 secret messages as well.

Proof. We do not give a full proof here, as it is very similar to that of Theorem 4.1. Besides technicalities, the only difference is to prove that the extracted witnesses indeed satisfy the required linear relation.

This can be seen as follows. From the verification equations of $c = 0$ and $c = 1$ we first get that $\boldsymbol{y}_i = \mathbf{A}.(\boldsymbol{a}_i \oplus \boldsymbol{c}_i \| \boldsymbol{b}_i \oplus \boldsymbol{d}_i) \oplus \pi_i'^{-1}(\boldsymbol{t}_{i1} \oplus \boldsymbol{t}_{i2})$, where the second addend has low weight by the same arguments as earlier. Using the linear relations among the $\boldsymbol{b}_i$ and the $\boldsymbol{d}_i$ we further get $(\boldsymbol{b}_3 \oplus \boldsymbol{d}_3) = \mathbf{X}_1.(\boldsymbol{b}_1 \oplus \boldsymbol{d}_1) \oplus \mathbf{X}_2.(\boldsymbol{b}_2 \oplus \boldsymbol{d}_2)$. Thus, a valid witness is given by $\boldsymbol{r}_i' = \boldsymbol{a}_i \oplus \boldsymbol{c}_i$, $\boldsymbol{m}_i' = \boldsymbol{b}_i \oplus \boldsymbol{d}_i$ and $\boldsymbol{e}_i' = \pi_i'^{-1}(\boldsymbol{t}_{i1} \oplus \boldsymbol{t}_{i2})$.

To see that the protocol is still honest-verifier zero-knowledge it suffices to note that the only additional information the verifier learns is that the random coins used in the protocol and the secret witnesses satisfy the linear relation which is already part of the description of the relation $\mathcal{R}_{LLPN}$. The rest of the protocol is just a parallel execution of independent instances of the protocol for $\mathcal{R}_{LPN}$. $\square$

4.3 Proving Multiplicative Relations

Finally, we present a protocol which can be used to prove a bitwise relation amongst commitments $\boldsymbol{y}_1, \boldsymbol{y}_2, \boldsymbol{y}_3$ (where $\boldsymbol{y}_i = \mathbf{A}.(\boldsymbol{r}_i \| \boldsymbol{m}_i) \oplus \boldsymbol{e}_i$). That is, it allows one to prove that the messages satisfy $\boldsymbol{m}_3 = \boldsymbol{m}_1 \circ \boldsymbol{m}_2$. The main idea of the protocol is to reduce the task of proving this multiplicative relation to a linear one, which we showed how to solve in the last section.

In the protocol, which is given in detail below, the prover P first samples vectors $\tilde{\boldsymbol{m}}_1, \tilde{\boldsymbol{m}}_2, \tilde{\boldsymbol{m}}_3 \xleftarrow{\text{R}} \mathcal{I}^{4v}$ such that (1) $\tilde{\boldsymbol{m}}_3 = \tilde{\boldsymbol{m}}_1 \circ \tilde{\boldsymbol{m}}_2$ and (2) for all $(a, b) \in \mathcal{I}^2$

the number of indices $j \in \{1, \ldots, 4v\}$ satisfying $(\tilde{m}_1[j], \tilde{m}_2[j]) = (a, b)$ is exactly v. Further, the prover draws a random matrix $\mathbf{R} \overset{\text{R}}{\leftarrow} \mathcal{I}^{v \times 4v}$ with full rank such that each row has Hamming weight exactly 1 and such that $\mathbf{R}.\tilde{m}_i = m_i$ for $i = 1, 2, 3$ (so $\mathbf{R}$ is a $v \times v$ permutation matrix with $3v$ additional zero columns).

Now P and V basically run the protocol from the previous section to prove the linear relation $\mathbf{R}.\tilde{m}_i = m_i$, with the crucial difference that the relation $\mathbf{R}$ is not know to V, instead the prover additionally sends a commitment to $\mathbf{R}$ with the first message. Moreover P sends commitments to the $\tilde{m}_i$'s to V.

The challenge space is extended from $\mathbb{Z}_3$ to $\mathbb{Z}_4$ (but will later merge $c = 2$ and $c = 3$ and get back to 3). If $c \in \{0, 1, 2\}$, the prover opens the commitment to $\mathbf{R}$ and sends the same answer as he would in the the protocol for proving the linear relation $\mathbf{R}.\tilde{m}_i = m_i$ for the given c. If $c = 3$, the prover opens the commitments to the $\tilde{m}_i$'s, and V checks if $\tilde{m}_3 \overset{?}{=} \tilde{m}_1 \circ \tilde{m}_2$.

The soundness of this protocol follows as $\mathbf{R}.\tilde{m}_i = m_i$ and $\tilde{m}_3 = \tilde{m}_1 \circ \tilde{m}_2$ together imply the claimed statement $m_3 = m_1 \circ m_2$.

The zero knowledge property holds as even though $\mathbf{R}$ together with the $\tilde{m}_i$'s determines the m_i's, each by itself is completely independent of the m_i's, and we never open both.

Finally, we observe that in our protocol for proving linear relations, the verifier does not need to know the linear relation $\mathbf{R}$ if $c = 2$. So we can collapse the challenges $c = 2$ and $c = 3$ as described above, but not open $\mathbf{R}$ in this case.

Formally, the protocol flow is defined by the following algorithms:

- P samples $\tilde{m}_1, \tilde{m}_2, \tilde{m}_3 \overset{\text{R}}{\leftarrow} \mathcal{I}^{4v}$ such that $\tilde{m}_3 = \tilde{m}_1 \circ \tilde{m}_2$ and such that for all $(a, b) \in \mathcal{I}^2$ the number of indices $j \in \{1, \ldots, 4v\}$ satisfying $(\tilde{m}_1[j], \tilde{m}_2[j]) = (a, b)$ is exactly v. Further, the prover draws a random matrix $\mathbf{R} \overset{\text{R}}{\leftarrow} \mathcal{I}^{v \times 4v}$ with full rank such that each row has at most Hamming weight 1 and such that $\mathbf{R}.\tilde{m}_i = m_i$ for $i = 1, 2, 3$.

  In the following we denote the jth v-bit block of $\tilde{m}_i$ by $\tilde{m}_i^j$, i.e., $\tilde{m}_i^j = (\tilde{m}_{iu})_{u=(j-1)v+1}^{jv}$. Similarly, $\mathbf{R}^j$ denotes the matrix given by columns $(j - 1)v + 1$ to jv of $\mathbf{R}$.

  In the remainder of this protocol description all computations are done for $i = 1, 2, 3$ and $j = 1, 2, 3, 4$, respectively.

  P draws $\tilde{r}_i^j \overset{\text{R}}{\leftarrow} \mathcal{I}^\ell$ and defines auxiliary images as $\tilde{y}_i^j = \mathbf{A}.(\tilde{r}_i^j \| \tilde{m}_i^j) \oplus \tilde{e}_i^j$ for $\tilde{e}_i^j \overset{\text{R}}{\leftarrow} \mathcal{I}_w^k$. It then samples permutations $\pi_i, \pi_i^j \leftarrow \mathcal{S}_k$ at random.

  It then draws $v_i^j \overset{\text{R}}{\leftarrow} \mathcal{I}^v$, $u_i, u_i^j \overset{\text{R}}{\leftarrow} \mathcal{I}^\ell$, $f_i, f_i^j \overset{\text{R}}{\leftarrow} \mathcal{I}^k$, sets $v_i = \sum_{j=1}^4 \mathbf{R}^j.v_i^j$ and then sends the following commitments to the verifier V:

$$\tilde{C} \leftarrow \mathsf{Com}(\tilde{y}_1'^1 = \tilde{y}_1^1, \ldots, \tilde{y}_3'^4 = \tilde{y}_3^4) \qquad C_\mathbf{R} \leftarrow \mathsf{Com}(\mathbf{R}' = \mathbf{R})$$

$$C_{i0} \leftarrow \mathsf{Com}(\pi_i' = \pi_i, t_{i0} = \mathbf{A}.(u_i \| v_i) \oplus f_i)$$

$$C_{i1} \leftarrow \mathsf{Com}(t_{i1} = \pi_i(f_i)) \qquad C_{i2} \leftarrow \mathsf{Com}(t_{i2} = \pi_i(f_i \oplus e_i))$$

$$C_{i0}^j \leftarrow \mathsf{Com}(\pi_i'^j = \pi_i^j, t_{i0}^j = \mathbf{A}.(u_i^j \| v_i^j) \oplus f_i^j)$$

$$C_{i1}^j \leftarrow \mathsf{Com}(t_{i1}^j = \pi_i^j(f_i^j)) \qquad C_{i2}^j \leftarrow \mathsf{Com}(t_{i2}^j = \pi_i^j(f_i^j \oplus \tilde{e}_i^j))$$

- The verifier draws $c \overset{\text{R}}{\leftarrow} \mathbb{Z}_3$ and sends it to P.

- Depending on the value of c, P opens the following commitments:
    0. P opens $C_{i0}, C_{i1}, C_{i0}^j, C_{i1}^j, C_{\mathbf{R}}$ by sending $\pi_i', t_{i0}, t_{i1}, \pi_i'^j, t_{i0}^j, t_{i1}^j, \mathbf{R}'$ and the associated random coins.
    1. P opens $C_{i0}, C_{i2}, C_{i0}^j, C_{i2}^j, \tilde{C}, C_{\mathbf{R}}$ by sending $\pi_i', t_{i0}, t_{i2}, \pi_i'^j, t_{i0}^j, t_{i2}^j, \tilde{\boldsymbol{y}}_i'^j, \mathbf{R}'$ and the associated random coins.
    2. P opens $C_{i1}, C_{i2}, C_{i1}^j, C_{i2}^j, \tilde{C}$ by sending $t_{i1}, t_{i2}, t_{i1}^j, t_{i2}^j, \tilde{\boldsymbol{y}}_i^j$ and the associated random coins.
- The verifier verifies the correctness of the openings received from the prover, and additionally performs the following checks depending on the challenge c:
    0. V accepts, iff $\pi_i', \pi_i'^j \overset{?}{\in} \mathcal{S}_k$, there exist solutions $(\boldsymbol{a}_i, \boldsymbol{b}_i), (\boldsymbol{a}_i^j, \boldsymbol{b}_i^j) \in \mathcal{I}^\ell \times \mathcal{I}^v$ to the equations $t_{i0} \oplus \pi_i'^{-1}(t_{i1}) = \mathbf{A}.(\boldsymbol{a}_i\|\boldsymbol{b}_i)$ and $t_{i0}^j \oplus (\pi_i'^j)^{-1}(t_{i1}^j) = \mathbf{A}.(\boldsymbol{a}_i^j\|\boldsymbol{b}_i^j)$, respectively, which satisfy $\boldsymbol{b}_i = \sum_{j=1}^4 \mathbf{R}'^j.\boldsymbol{b}_i^j$.
    1. V accepts, iff $\mathbf{R}'$ has full rank and each row has Hamming weight at most 1, and iff there exist solutions $(\boldsymbol{a}_i, \boldsymbol{b}_i), (\boldsymbol{a}_i^j, \boldsymbol{b}_i^j) \in \mathcal{I}^\ell \times \mathcal{I}^v$ to the equations $t_{i0} \oplus \pi_i'^{-1}(t_{i2}) \oplus \boldsymbol{y}_i = \mathbf{A}.(\boldsymbol{a}_i\|\boldsymbol{b}_i)$ and $t_{i0}^j \oplus (\pi_i'^j)^{-1}(t_{i2}^j) \oplus \tilde{\boldsymbol{y}}_i'^j = \mathbf{A}.(\boldsymbol{a}_i^j\|\boldsymbol{b}_i^j)$, respectively, which satisfy $\boldsymbol{b}_i = \sum_{j=1}^4 \mathbf{R}'^j.\boldsymbol{b}_i^j$.
    2. V accepts, iff $\|t_{i1} \oplus t_{i2}\|_1 = \|t_{i1}^j \oplus t_{i2}^j\|_1 \overset{?}{=} w$, $\tilde{\boldsymbol{y}}_i'^j \overset{?}{=} \mathbf{A}.(\tilde{\boldsymbol{r}}_i^j\|\tilde{\boldsymbol{m}}_i^j) \oplus \tilde{\boldsymbol{e}}_i^j$, $\|\tilde{\boldsymbol{e}}_i^j\|_1 \overset{?}{=} w$ and $\tilde{\boldsymbol{m}}_1^j \circ \tilde{\boldsymbol{m}}_2^j \overset{?}{=} \tilde{\boldsymbol{m}}_3^j$.

Theorem 4.3. *The above protocol is a Σ-protocol for the following relation:*

$$\mathcal{R}_{MLPN} = \Big\{ ((\mathbf{A}, \boldsymbol{y}_1, \boldsymbol{y}_2, \boldsymbol{y}_3), (\boldsymbol{r}_1, \boldsymbol{r}_2, \boldsymbol{r}_3, \boldsymbol{m}_1, \boldsymbol{m}_2, \boldsymbol{m}_3, \boldsymbol{e}_1, \boldsymbol{e}_2, \boldsymbol{e}_3)) :$$

$$\bigwedge_{i=1}^3 (\boldsymbol{y}_i = \mathbf{A}.(\boldsymbol{r}_i\|\boldsymbol{m}_i) \oplus \boldsymbol{e}_i \wedge \|\boldsymbol{e}_i\|_1 = w) \wedge \boldsymbol{m}_3 = \boldsymbol{m}_1 \circ \boldsymbol{m}_2 \Big\}.$$

Proof. The 3-move form of the protocol is easy to see. Furthermore, completeness directly follows from the construction and can easily be verified.

Special soundness. Concerning the special soundness of the protocol, note the following. Using the same arguments as in the proof of Theorem 4.1, we can extract openings $\boldsymbol{m}_i', \boldsymbol{r}_i'$ and $\boldsymbol{e}_i'$ of the $\boldsymbol{y}_i$, and similarly, we get $\tilde{\boldsymbol{m}}_i'^j, \tilde{\boldsymbol{r}}_i'^j$ and $\tilde{\boldsymbol{e}}_i'^j$ which are valid openings for the $\tilde{\boldsymbol{y}}_i'^j$. Now, by the same arguments as in Theorem 4.2 we can further infer that $\boldsymbol{m}_i' = \sum_{i=1}^4 \mathbf{R}'^j.\tilde{\boldsymbol{m}}_i'^j = \mathbf{R}'.\tilde{\boldsymbol{m}}_i'$. Furthermore, we know that $\tilde{\boldsymbol{m}}_1' \circ \tilde{\boldsymbol{m}}_2' = \tilde{\boldsymbol{m}}_3'$. Now, because of the special form of $\mathbf{R}'$, we can finally infer that the same relation must also be true for the $\boldsymbol{m}_i'$.

Honest-verifier zero-knowledge. We do not give a full simulator here, but only give the intuition why the protocol is zero-knowledge. Clearly, the $\tilde{\boldsymbol{m}}_i^j$ are uniformly random in their domain, and do not leak any information about the $\boldsymbol{m}_i$, as long as the matrix $\mathbf{R}$ is kept secret. Similarly, if the $\tilde{\boldsymbol{m}}_i^j$ are kept secret, the matrix $\mathbf{R}$ itself is a uniformly random matrix of full rank with the specified restriction on the weight of its rows. Computationally this still holds true even if the $\tilde{\boldsymbol{y}}_i^j$ are revealed, as they are pseudorandom by Theorem 3.1. The zero-knowledge property now follows from that of the protocol for $\mathcal{R}_{LLPN}$. $\square$

4.4   Proving Arbitrary Relations

We finally briefly explain how one can use the protocols presented in this section to prove that committed values m_0, m_1 satisfy $m_0 = C(m_1)$ for an arbitrary circuit C. Let $C_1, \ldots, C_d$ denote the layers of C, i.e., $C(m_1) = C_d(\ldots C_1(m_1) \ldots)$, where we assume that each C_i is either a linear function or a bitwise operation (e.g., bitwise NAND). For simplicity we assume the number of input and output wires to each C_i is ℓ, where ℓ is the length of the underlying LPN problem.

We use our string commitment scheme to commit to the values in the intermediate layers, i.e., to strings $x_1, \ldots, x_d$ where $x_1 = m_1, x_2 = C_1(m_1), \ldots, x_d = C(m_1)$ (note that we already have commitments to $x_1 = m_1$ and $x_d = m_0$). Now we use our Σ-protocols to prove that $x_{i+1} = C_i(x_i)$ for $i = 1 \ldots d - 1$.

The total communication complexity of this protocol is $\Theta(\sum |C_i| \ell \log \ell) = \Theta(|C| \ell \log \ell)$, the soundness error is $2/3$, and thus for most applications must be lowered by (parallel) repetition.

5   Conclusions and Open Problems

We presented a very simple and efficient string commitment scheme, whose security is based on the hardness of the LPN-problem, or, equivalently, on the hardness of decoding random linear codes. We further presented Σ-protocols which allow one to prove arbitrary relations among secret values m_i, i.e., $m_0 = C(m_1, \ldots, m_u)$ for any circuit C. The size of a proof is only quasi-linear in the length of the committed messages.

We introduced an "exact" version of the LPN-problem which is polynomially equivalent to the standard LPN problem. This new assumption might be of independent interest as basing existing LPN based schemes on this new assumptions removes the completeness error (cf. §2 for a discussion).

It would be interesting to find protocols which already achieve a small knowledge error in only run, and do not rely on repetitions. Furthermore, a tighter reduction for the hardness of the decisional xLPN problem, in particular not relying on the Goldreich-Levin theorem, would be desirable.

Acknowledgment. We are grateful to Petros Mol for helpful discussions on the reduction for the hardness of the xLPN problem.

References

1. Applebaum, B., Cash, D., Peikert, C., Sahai, A.: Fast Cryptographic Primitives and Circular-Secure Encryption Based on Hard Learning Problems. In: Halevi, S. (ed.) CRYPTO 2009. LNCS, vol. 5677, pp. 595–618. Springer, Heidelberg (2009)
2. Applebaum, B., Ishai, Y., Kushilevitz, E.: Cryptography with Constant Input Locality. Journal of Cryptology 22(4), 429–469 (2009)
3. Asharov, G., Jain, A., López-Alt, A., Tromer, E., Vaikuntanathan, V., Wichs, D.: Multiparty Computation with Low Communication, Computation and Interaction via Threshold FHE. In: Pointcheval, D., Johansson, T. (eds.) EUROCRYPT 2012. LNCS, vol. 7237, pp. 483–501. Springer, Heidelberg (2012)

4. Bellare, M., Goldreich, O.: On Defining Proofs of Knowledge. In: Brickell, E.F. (ed.) CRYPTO 1992. LNCS, vol. 740, pp. 390–420. Springer, Heidelberg (1993)
5. Berlekamp, E., McEliece, R., van Tilborg, H.: On the Inherent Intractability of Certain Coding Problems. IEEE Transactions on Information Theory 24(3), 384–386 (1978)
6. Blum, A., Furst, M.L., Kearns, M., Lipton, R.J.: Cryptographic Primitives Based on Hard Learning Problems. In: Stinson, D.R. (ed.) CRYPTO 1993. LNCS, vol. 773, pp. 278–291. Springer, Heidelberg (1994)
7. Blum, A., Kalai, A., Wasserman, H.: Noise-Tolerant Learning, the Parity Problem, and the Statistical Query Model. Journal of the ACM 50(4), 506–519 (2003)
8. Boyar, J., Damgård, I., Peralta, R.: Short Non-Interactive Cryptographic Proofs. Journal of Cryptology 13(4), 449–472 (2000)
9. Cayrel, P.-L., Lindner, R., Rückert, M., Silva, R.: Improved Zero-Knowledge Identification with Lattices. In: Heng, S.-H., Kurosawa, K. (eds.) ProvSec 2010. LNCS, vol. 6402, pp. 1–17. Springer, Heidelberg (2010)
10. Cayrel, P.-L., Véron, P., El Yousfi Alaoui, S.M.: A Zero-Knowledge Identification Scheme Based on the q-ary Syndrome Decoding Problem. In: Biryukov, A., Gong, G., Stinson, D.R. (eds.) SAC 2010. LNCS, vol. 6544, pp. 171–186. Springer, Heidelberg (2011)
11. Chen, H., Cramer, R.: Algebraic Geometric Secret Sharing Schemes and Secure Multi-Party Computations over Small Fields. In: Dwork, C. (ed.) CRYPTO 2006. LNCS, vol. 4117, pp. 521–536. Springer, Heidelberg (2006)
12. Cramer, R.: Modular Design of Secure yet Practical Cryptographic Protocols. PhD thesis, CWI and University of Amsterdam (1997)
13. Cramer, R., Damgård, I.: Linear Zero-Knowledge - A Note on Efficient Zero-Knowledge Proofs and Arguments. In: Leighton, F.T., Shor, P.W. (eds.) STOC 1997, pp. 436–445. ACM (1997)
14. Cramer, R., Damgård, I.: Zero-Knowledge Proofs for Finite Field Arithmetic or: Can Zero-Knowledge Be for Free? In: Krawczyk, H. (ed.) CRYPTO 1998. LNCS, vol. 1462, pp. 424–441. Springer, Heidelberg (1998)
15. Cramer, R., Damgård, I.: On the Amortized Complexity of Zero-Knowledge Protocols. In: Halevi, S. (ed.) CRYPTO 2009. LNCS, vol. 5677, pp. 177–191. Springer, Heidelberg (2009)
16. Damgård, I.: On Σ-Protocols. Lecture on Cryptologic Protocol Theory; Faculty of Science. University of Aarhus (2004)
17. Damgård, I., Goldreich, O., Okamoto, T., Wigderson, A.: Honest Verifier vs Dishonest Verifier in Public Coin Zero-Knowledge Proofs. In: Coppersmith, D. (ed.) CRYPTO 1995. LNCS, vol. 963, pp. 325–338. Springer, Heidelberg (1995)
18. Gilbert, H., Robshaw, M.J.B., Seurin, Y.: How to Encrypt with the LPN Problem. In: Aceto, L., Damgård, I., Goldberg, L.A., Halldórsson, M.M., Ingólfsdóttir, A., Walukiewicz, I. (eds.) ICALP 2008, Part II. LNCS, vol. 5126, pp. 679–690. Springer, Heidelberg (2008)
19. Goldreich, O., Micali, S., Wigderson, A.: How to Play Any Mental Game, or A Completeness Theorem for Protocols with Honest Majority. In: Aho, A.V. (ed.) STOC 1987, pp. 218–229. ACM (1987)
20. Groth, J., Sahai, A.: Efficient Non-interactive Proof Systems for Bilinear Groups. In: Smart, N.P. (ed.) EUROCRYPT 2008. LNCS, vol. 4965, pp. 415–432. Springer, Heidelberg (2008)
21. Hopper, N.J., Blum, M.: Secure Human Identification Protocols. In: Boyd, C. (ed.) ASIACRYPT 2001. LNCS, vol. 2248, pp. 52–66. Springer, Heidelberg (2001)

22. Ishai, Y., Kushilevitz, E., Ostrovsky, R., Sahai, A.: Zero-knowledge from secure multiparty computation. In: Johnson, D.S., Feige, U. (eds.) STOC 2007, pp. 21–30. ACM (2007)
23. Jain, A., Krenn, S., Pietrzak, K., Tentes, A.: Commitments and Efficient Zero-Knowledge Proofs from Hard Learning Problems. Cryptology ePrint Archive, Report 2012/513 (2012), http://eprint.iacr.org/
24. Juels, A., Weis, S.A.: Authenticating Pervasive Devices with Human Protocols. In: Shoup, V. (ed.) CRYPTO 2005. LNCS, vol. 3621, pp. 293–308. Springer, Heidelberg (2005)
25. Kalai, Y.T., Raz, R.: Succinct Non-Interactive Zero-Knowledge Proofs with Preprocessing for LOGSNP. In: FOCS 2006, pp. 355–366. IEEE Computer Society (2006)
26. Katz, J., Shin, J.S., Smith, A.: Parallel and Concurrent Security of the HB and HB$^+$ Protocols. Journal of Cryptology 23(3), 402–421 (2010)
27. Kawachi, A., Tanaka, K., Xagawa, K.: Concurrently Secure Identification Schemes Based on the Worst-Case Hardness of Lattice Problems. In: Pieprzyk, J. (ed.) ASIACRYPT 2008. LNCS, vol. 5350, pp. 372–389. Springer, Heidelberg (2008)
28. Kearns, M.J.: Efficient Noise-Tolerant Learning from Statistical Queries. Journal of the ACM 45(6), 983–1006 (1998)
29. Kilian, J.: A Note on Efficient Zero-Knowledge Proofs and Arguments (Extended Abstract). In: STOC 1992, pp. 723–732 (1992)
30. Kilian, J., Micali, S., Ostrovsky, R.: Minimum Resource Zero-Knowledge Proofs (Extended Abstract). In: Brassard, G. (ed.) CRYPTO 1989. LNCS, vol. 435, pp. 545–546. Springer, Heidelberg (1990)
31. Kilian, J., Petrank, E.: An Efficient Noninteractive Zero-Knowledge Proof System for NP with General Assumptions. Journal of Cryptology 11(1), 1–27 (1998)
32. Kiltz, E., Pietrzak, K., Cash, D., Jain, A., Venturi, D.: Efficient Authentication from Hard Learning Problems. In: Paterson, K.G. (ed.) EUROCRYPT 2011. LNCS, vol. 6632, pp. 7–26. Springer, Heidelberg (2011)
33. Micali, S.: Computationally Sound Proofs. SIAM Journal on Computing 30(4), 1253–1298 (2000)
34. Micciancio, D., Mol, P.: Pseudorandom Knapsacks and the Sample Complexity of LWE Search-to-Decision Reductions. In: Rogaway, P. (ed.) CRYPTO 2011. LNCS, vol. 6841, pp. 465–484. Springer, Heidelberg (2011)
35. Regev, O.: On Lattices, Learning with Errors, Random Linear Codes, and Cryptography. In: STOC 2005, pp. 84–93. ACM (2005)
36. Stern, J.: A New Identification Scheme Based on Syndrome Decoding. In: Stinson, D.R. (ed.) CRYPTO 1993. LNCS, vol. 773, pp. 13–21. Springer, Heidelberg (1994)
37. Véron, P.: Improved Identification Schemes Based on Error-Correcting Codes. Applicable Algebra in Engineering, Communication and Computing 8(1), 57–69 (1996)

Calling Out Cheaters:
Covert Security with Public Verifiability*

Gilad Asharov[1] and Claudio Orlandi[2,**]

[1] Department of Computer Science, Bar-Ilan University, Israel
[2] Department of Computer Science, Aarhus University, Denmark
asharog@cs.biu.ac.il, orlandi@cs.au.dk

Abstract. We introduce the notion of *covert security with public verifiability*, building on the covert security model introduced by Aumann and Lindell (TCC 2007). Protocols that satisfy covert security guarantee that the honest parties involved in the protocol will notice any cheating attempt with some constant probability ϵ. The idea behind the model is that the fear of being caught cheating will be enough of a deterrent to prevent any cheating attempt. However, in the basic covert security model, the honest parties are not able to persuade any third party (say, a judge) that a cheating occurred.

We propose (and formally define) an extension of the model where, when an honest party detects cheating, it also receives a *certificate* that can be published and used to persuade other parties, without revealing any information about the honest party's input. In addition, malicious parties cannot create fake certificates in the attempt of framing innocents.

Finally, we construct a secure two-party computation protocol for any functionality f that satisfies our definition, and our protocol is almost as efficient as the one of Aumann and Lindell. We believe that the fear of a public humiliation or even legal consequences vastly exceeds the deterrent given by standard covert security. Therefore, even a small value of the deterrent factor ϵ will suffice in discouraging any cheating attempt.

1 Introduction

One of the main goals of the theory of cryptographic protocols is to find security definitions that provide the participants with meaningful guarantees and that can, at the same time, be achieved by reasonably efficient protocols. Both standard security notions lack one of these two properties: the level of security offered by *semi-honest secure* protocols is unsatisfactory (as the only guarantee is that security is achieved if all parties follow the protocol specification) while *malicious secure* protocols (that offer security against arbitrarily behaving adversaries) are orders of magnitude slower than semi-honest ones.

* The research was supported by the European Research Council as part of the ERC project LAST.
** Research conducted while at Bar-Ilan University, Israel.

X. Wang and K. Sako (Eds.): ASIACRYPT 2012, LNCS 7658, pp. 681–698, 2012.

In *covert security*, introduced by Aumann and Lindell in 2007 [AL07], the honest parties have the guarantee that if the adversary tries to cheat in order to break some of the security properties of the protocol (correctness, confidentiality, input independence, etc.) then the honest parties will notice the cheating attempt with some constant probability ϵ. Here, unlike the malicious model where the adversary cannot cheat at all, the adversary can effectively cheat while taking the risk of being caught. This relaxation of the security model allows protocol designers to construct highly efficient protocols, essentially only a small factor away from the efficiency of semi-honest protocols.

The main justification for covert security is that, in many practical applications, the relationship between the participants of the protocol is such that the fear of being caught cheating is enough of a deterrent to avoid any cheating attempt. For example, two companies that decide to engage in a secure computation protocol might value their reputation and the possibility of future trading with the other company more than the possibility of learning a few bits of information about the other company's input, and therefore have no incentive in trying to cheat in the protocol at all.

However, a closer look at the covert model reveals that the repercussion of a cheating attempt is somewhat limited: Indeed, if Alice tries to cheat, the protocol guarantees that she will be caught by Bob with some predetermined probability, and so Bob will know that Alice is dishonest. Nevertheless, Bob will not be able to bring Alice in front of a judge or to persuade a third party Charlie that Alice cheated, and therefore Alice's reputation will only be hurt in Bob's eyes and no one else. This is due to the fact that Charlie has no way of telling apart the situation where Alice cheated from the situation where Bob is trying to frame Alice to hurt her reputation: Bob can always generate fake transcripts that will be indistinguishable from a real interaction between a cheating Alice and Bob.

This becomes a problem, as the fact that only Bob knows that Alice has tried to cheat may not be enough of a deterrent for Alice. In particular, consider the scenario where there is some social asymmetry between the parties, for instance if a very powerful company engages in a protocol with a smaller entity (i.e., a citizen). If the citizen does not have any clear evidence of the cheating she will not be able to get any compensation for the cheating attempt, as she will not be able to sue the company or persuade any other party of the misbehavior – who would believe her without any proof? This means that if we run a covert protocol between these parties, the fact that a party can detect the cheating may not be enough to prevent the more powerful one from attempting to cheat.

The scenario described above can be dramatically changed if, once a party is caught cheating, the other party receives some undeniable evidence of this fact, and this evidence can be independently verified by any third party. We therefore introduce the notion of *covert security with public verifiability* where if a party is caught cheating, then the honest parties receive a certificate – a small piece of evidence – that can be published and used to prove to all those who are interested that indeed there was a dishonest behavior during the interaction. Clearly, this provides a stronger deterrent than the one given by covert security.

Intuitively, we want cheating parties to be *accountable* for their actions i.e., if a party cheats then everyone can be persuaded of this fact. At the same time, we need also the system to be *defamation-free* in the sense that no honest parties can be framed i.e., no party can produce a fake cheating certificate.

Towards Better Efficiency: Choosing the Right ϵ. In order to fully understand the benefit of covert-security with public verifiability, consider the utilities of a rational Alice, running a cryptographic protocol with Bob for some task. Let $(U_h, U_c, U_f, U_f^{pub})$ be real numbers modeling Alice utilities: Alice's utility is U_h when she runs the protocol honestly, and so both parties learn the output and nothing else. If Alice attempts to cheat, she will receive utility U_c if the cheating attempt succeeds. If the cheating attempt fails (i.e., Alice gets caught), the utility received by Alice will be U_f in the standard covert security setting and U_f^{pub} in the setting with public verifiability. We assume that $U_c > U_h > U_f > U_f^{pub}$, namely, Alice prefers to succeed cheating over the outcome of an honest execution, prefers the latter over being caught cheating, and prefers losing her reputation in the eye of one parties over losing it publicly.

Remember that, since the protocol is ϵ-deterrent, whenever Alice attempts to cheat she will be caught with probability ϵ and succeed with probability $1 - \epsilon$. Therefore, assuming that Bob is honest, Alice's expected payoff is U_h when she plays the honest strategy and $\epsilon \cdot U_f' + (1 - \epsilon) \cdot U_c$ when she plays cheating, with $U_f' \in \{U_f, U_f^{pub}\}$ depending on whether the protocol satisfies public verifiability or not. Therefore if we set

$$\epsilon > \frac{U_c - U_h}{U_c - U_f'}$$

then Alice will maximize her expected utility by playing honest. This implies that the value of ϵ needed to discourage Alice from cheating is much higher in the standard covert security setting than in our framework.

As the value of the deterrent factor ϵ determines the replication factor and thus the efficiency of covert secure protocols, we believe that in practice using covert security with public verifiability will lead to an increase in efficiency, as the benefits obtained by the reduced replication factor will exceed the limited price to pay for achieving the public verifiability property on top of the covert secure protocol.

Main Ideas. It is clear that no solution to our problem exists in the plain model and that we need to be able to publicly identify parties. We therefore set our study in the public-key infrastructure (PKI) model, where the keys of all parties are registered in some public database. Note that in practice this is not really an additional assumption, as most cryptographic protocols already assume the existence of authenticated point-to-point channels, that can be essentially only implemented by having some kind of PKI and letting the parties sign all the messages they exchange to each other.

At this point it might seem that the problem we are trying to solve is trivial, and that the solution is simply to let all parties sign all the exchanged messages in a covert secure protocol. Here is why this naïve solution does not reach our goal: As a first problem, we need to make sure that the adversary cannot abort as a consequence of being caught cheating; think of a zero-knowledge (ZK) protocol with one bit challenge, where the prover only knows how to answer to a challenge $c = 0$. If the verifier asks for $c = 1$, the malicious prover has no reason to reply with an invalid proof and will abort instead. Surely, the honest party will suspect the prover of cheating but will have no certificate to show to a judge. The problem of an adversary aborting as an escape from being caught cheating was already raised in [AL07, Section 3.5], and the solution is to run all the *cut-and-choose* via an oblivious transfer (OT): here the prover (acting as a sender) inputs openings to all possible challenges and the verifier (acting as the receiver) inputs his random challenge. Due to the security of the OT, the prover now cannot choose whether to continue or abort the protocol as a function of the verifier's challenge. The prover needs to decide in advance whether to take the risk of being caught, or abort before the execution of the OT protocol.

Secondly, we need to ensure that the published certificate does not leak information about the honest party's input: when the honest party detects cheating, it computes a certificate as a function of its view i.e., the (signed) transcript of the protocol, his input and his random tape. Therefore, this certificate may (even indirectly) leak information about the input of the honest party. This is clearly unsatisfactory and leads us to the following unfortunate situation: a party knows that the other party has cheated, however, in order to prove this fact to the public he is required to reveal to the adversary his private information.

For the sake of concreteness, consider a protocol where Alice chooses a key pair (pk, sk) for a homomorphic encryption scheme E, and sends Bob $(pk, E_{pk}(x))$ where x is Alice's input. Later in the protocol, Alice and Bob use the homomorphic properties of E for a cut-and-choose; i.e., Bob sends the first message of a ZK proof, Alice sends an encrypted challenge $E_{pk}(c)$ and Bob obliviously computes the last message of the ZK proof for the challenge c, and signs all the transcripts of the protocol. Alice finally decrypts and checks the validity of the proof. Note that Bob cannot abort as a function of c (due to the semantic security of the encryption scheme). If Bob cheats and Alice detects it, she receives a proof, a signature on the (encrypted) incriminating messages. Alice can now publish the transcript and her secret key sk in order to enable the judge to verify that Bob cheated. However, once the certificate is made public, Bob will learn the secret, decrypt the first ciphertext and learn x.

Moreover, a malicious Alice might have a strategy to compute a different secret key sk' that makes the signed ciphertext decrypt to some "illegal" message that can be used to frame an innocent Bob. These examples show that things can easily go wrong, and motivates the need for a formal study of covert security with public verifiability.

Signed Oblivious Transfer. As a building block for our construction we introduce a new cryptographic primitive, that we shall call *signed oblivious-transfer*. In this primitive, the sender inputs two message (m_0, m_1) and a signature key sk, and the receiver inputs a bit b. At the end of the protocol, the receiver will learn the message m_b together with a signature on it, while the sender learns nothing. That is, the receiver learns: $(m_b, \mathsf{Sig}_{sk}(b, m_b))$.

To see the importance of this tool in constructing protocols that satisfy covert security with public verifiability it is useful to see how it can be used to fix the problems with the zero-knowledge protocols described before. A very high level description of the signed-OT based zero-knowledge protocol is: (1) First the prover prepares the first message of the zero-knowledge protocol and sends it to the verifier together with a valid signature on it; (2) Now the prover prepares the answers to both challenges $c = 0$ and $c = 1$ and inputs them, together with his secret key, to the signed OT; (3) The verifier inputs a random choice bit c to the signed OT and receives the last message of the zero-knowledge protocol together with a valid signature on it. The verifier checks this message and, if the proof passes the verification, it outputs accept. On the other hand, if the proof is invalid, the verifier can take the transcript of the protocol and send them to any third party as an undeniable proof that the prover attempted to cheat.

Note that this works only because b is included in the signature. Had b not be signed, the prover could input the simulated opening to both branches of the OT. This makes the (signed) transcript look always legit (in particular, it does not depend on the challenge bit b), and the verifier cannot persuade a third party that the prover did not properly answer to his challenge. Also, note that it is not enough to run a standard OT, where the prover inputs $(m_0, \mathsf{Sig}(0, m_0)), (m_1, \mathsf{Sig}(1, m_1))$, as in this case the prover could cheat by sending a valid signature on the valid opening, and no signature on the wrong opening – it is crucial for the security of the protocol that the verifier is persuaded that *both* signatures are valid, even if only one is received.

Our Model. Our security definition guarantees that when an honest party publishes the certificate, the adversary cannot gain any additional information from this certificate even when it is combined with the adversary's view, in a strong simulation sense. This, together with the fact that in the *strong explicit cheat formulation* of covert security a cheating party does not learn any information about the honest party's input and output, guarantees that the certificate does not leak any unintentional information to anyone seeing the certificate (i.e., the certificate can be simulated without the input/output of the honest party).

A covert secure protocol with public verifiability is composed of an "honest" protocol and two extra algorithms to deal with cheating situations: the first is used to produce a certificate when a cheating is detected, and the other to decide whether a certificate is authentic or not. The requirements for the two latter algorithms are the following: any time that an honest party outputs that the other party is corrupted, the evaluation of the verification algorithm on the produced certificate should output the identity of the corrupted party. In addition, no one should be able to produce incriminating certificates against honest parties.

Organization and Results. In Section 2, we define and justify the model of covert security with public verifiability. In Section 3 we show how to construct a signed-OT protocol: our starting point is the very efficient OT protocol due to Peikert, Vaikuntanathan and Waters [PVW08]. The resulting protocol is only slightly less efficient than the protocol of PVW.

Signed-OT will also be the main ingredient in our protocol for two-party secure computation using Yao's garbled circuit, described in Section 4. Here we show that for any two party functionality f, there exists an efficient covert secure protocol with ϵ-deterrent and public verifiability. Our protocol is roughly $1/\epsilon$ slower than a semi-honest secure protocol, and has essentially the same complexity as an ϵ-deterrent secure protocol without public verifiability.

Technically, our starting point is the protocol presented in [AL07, Section 6.3] (the variant where aborting is not considered cheating) the only differences with the original protocol are that every call to an OT is replaced by a call to a signed-OT, and that the circuit constructor will also send a few signatures in the right places. We believe that this is a very positive fact as the resulting protocol is only slightly less efficient than the original covert secure protocol, showing how covert security with public verifiability offers a much greater deterrent to cheating than standard covert security (as a cheater can face huge loss in reputation or even legal consequences), while only slightly decreasing the efficiency of the protocol.

Related Work. The idea of allowing malicious parties to cheat as long as this is detected with significant probability can be found in several works, e.g. [FY92, IKNP03, MNPS04], and it was first formally introduced under the name of covert security by Aumann and Lindell [AL07]. Since then, several protocols satisfying this definition have been constructed, for instance [HL08, GMS08, DGN10]. It is possible to add the public verifiability property to any of these protocols. Doing so in the most efficient way is left as a future work.

2   Definitions

Preliminaries. A function $\mu(\cdot)$ is negligible, if for every positive polynomial $p(\cdot)$ and all sufficiently large n's it holds that $\mu(n) < 1/p(n)$. A probability ensemble $X = \{X(a,n)\}_{a\in\{0,1\}^*;n\in\mathbb{N}}$ is an infinite sequence of random variables indexed by a and $n \in \mathbb{N}$. Usually, the value a represents the parties' inputs and n the security parameter. Two distributions ensembles $X = \{X(a,n)\}_{a\in\{0,1\}^*;n\in\mathbb{N}}$ and $Y = \{Y(a,n)\}_{a\in\{0,1\}^*,n\in\mathbb{N}}$ are said to be computationally indistinguishable, denoted $X \stackrel{c}{\equiv} Y$, if for every non-uniform polynomial-time algorithm D there exists a negligible function $\mu(\cdot)$ such that for every $a \in \{0,1\}^*$ and every $n \in \mathbb{N}$,

$$|\Pr[D(X(a,n)) = 1] - \Pr[D(Y(a,n)) = 1]| \leq \mu(n)$$

We assume the reader to be familiar with the standard definition for secure multiparty computation [Can00, Gol04].

Covert Security: Aumann and Lindell [AL07] present three possible definitions for this notion of security, where the three definitions constitute a strict hierarchy. We adopt the strongest definition that is presented, which is called "strong explicit cheat formulation" (Section 3.4 in [AL07]).

A protocol that is secure with respect to this definition is also secure with respect to the two other suggested definitions. Informally, in this stronger formulation, the adversary may choose to input a special input cheat to the ideal functionality. The ideal functionality will then flip a coin and with probability $(1 - \epsilon)$ will give to the adversary full control: the adversary will learn the honest party's input and instruct the functionality to deliver any output of its choice. However, with probability ϵ, the ideal functionality will inform the honest party of the cheating attempt by sending him a special symbol corrupted, and crucially, the adversary will not learn any information about the honest party's input.

2.1 Covert Security with Public Verifiability

For the sake of simplicity, we will present the definition and the motivation for the two-party case. The definition can be easily extended to the multi-party case.

Motivation: As discussed in the introduction, we work in the $\mathcal{F}^{\mathrm{PKI}}$-hybrid model where each party P_i registers a verification key vk_i for a signature scheme. This key will be used to uniquely identify a party. Note that we do not require parties to prove knowledge of their secret keys (i.e., the simulator will not know these secret keys), so this is the weakest $\mathcal{F}^{\mathrm{PKI}}$ formulation possible [BCNP04].

We extend the covert security model of Aumann and Lindell [AL07] and enhance it with the public verifiability property: As in covert security, if the adversary chooses to cheat it will be caught with probability ϵ, and the honest party outputs corrupted. However, in this latter case, the protocol in addition provides this party an algorithm Blame to distil a *certificate* from its view in the protocol. A third party who wants to verify the cheating ("the judge") should take the certificate and decide whether the certificate is authentic (i.e., some cheater has been caught) or it is a fake (i.e., someone is trying to frame an innocent). The verification is performed using an additional algorithm, which is called Judgement. We require the verification procedure to be *non-interactive*, which will enable the honest party to send the certificate to a judge or to publish it on a public "wall of shame".

In addition, as our interest is mainly to protect the interest of the honest party, we want to make sure that the certificate of cheating does not reveal any unnecessary information to the verifier. Therefore, we cannot simply publish the view (transcript and random tape) of the honest party, as those might reveal some information about the input or output of the honest party. In addition, we need to remember that the adversary sees the certificate once it is published and therefore we should take care that no one will be able to learn any meaningful information from this certificate, even when combining it with the adversary's view. To capture this fact, we use the convention that when a party detects a cheating, it creates the certificate and sends it to the adversary.

The fact that the certificate is part of the view of the adversary means that the simulator needs to include this certificate as a part of the view when it receives corrupted from the ideal functionality. Remember that in this case the simulator does not learn anything from the trusted party rather than the adversary got caught, and therefore this implies that our definition ensures that the certificate cannot reveal the private information of the honest party.

Regarding the Judgement algorithm, we require two security properties: whenever an honest party outputs corrupted, running the algorithm on the certificate will output the identity of the corrupted party. Moreover, no adversary (even interacting with polynomially many honest parties) can produce a certificate for which the verification algorithm outputs the identity of an honest party.

2.2   The Formal Definition

Let f be a two party functionality. We consider the triple $(\pi, \mathsf{Blame}, \mathsf{Judgement})$. The algorithm Blame gets as input the view of the honest party (in case of cheat detection) and outputs a certificate $Cert$. The verification algorithm, Judgement, takes as input a certificate $Cert$ and outputs the identity id (for instance, the verification key) of the party to blame or *none* in the case of an invalid certificate.

The Protocol: Let π be a two party protocol. If an honest party detects a cheating in π then the honest party is instructed to compute $Cert = \mathsf{Blame}(\text{view})$ and send it to the adversary.

Let $\mathrm{REAL}_{\pi,\mathcal{A}(z),i^*}(x_1, x_2; 1^n)$ denote the output of the honest party and the adversary on a real execution of the protocol π where P_1, P_2 are invoked with inputs x_1, x_2, the adversary is invoked with an auxiliary input z and corrupts party P_{i^*} for some $i^* \in \{1, 2\}$.

The Ideal World. The ideal world is exactly as [AL07, Definition 3.4]. Let $\mathrm{IDEAL}_{\pi,\mathcal{A}(z),i^*}(x_1, x_2)$ denote the output of the honest party, together with the output of the simulator, on an ideal execution with the functionality f, where P_1, P_2 are invoked with inputs x_1, x_2, respectively, the simulator $\mathcal{S}$ is invoked with an auxiliary input z and the corrupted party is P_{i^*}, for some $i^* \in \{1, 2\}$.

Notations. Let $\mathrm{EXEC}_{\pi,\mathcal{A}(z)}(x_1, x_2; r_1, r_2; 1^n)$ denote the messages and the outputs of the parties in an execution of the protocol π with adversary $\mathcal{A}$ on auxiliary input z, where the inputs of P_1, P_2 are x_1, x_2, respectively, and the random tapes are (r_1, r_2). Let $\mathrm{EXEC}_{\pi,\mathcal{A}(z)}(x_1, x_2; 1^n)$ denote the probability distribution of $\mathrm{EXEC}_{\pi,\mathcal{A}(z)}(x_1, x_2; r_1, r_2)$ where (r_1, r_2) are chosen uniformly at random. Let $\mathrm{OUTPUT}(\mathrm{EXEC}_{\pi,\mathcal{A}(z)}(x_1, x_2))$ denote the output of the honest party in the execution described above. We are now ready to define the security properties.

Definition 1 (covert security with ϵ-deterrent and public verifiability)
Let f, π, Blame and Judgement be as above. We say that $(\pi, \mathsf{Blame}, \mathsf{Judgement})$ securely computes f in the presence of a covert adversary with ϵ-deterrent and public verifiability if the following conditions hold:

1. **(Simulatability with ϵ-deterrent:)** *The protocol π (where the honest party broadcasts $Cert = \mathsf{Blame}(\text{view})$ if it detects cheating) is secure against a covert adversary according to the strong explicit cheat formulation with ϵ-deterrent (see [AL07, Definition 3.4]).*

2. **(Accountability:)** *For every PPT adversary $\mathcal{A}$ corrupting party P_{i^*} for $i^* \in \{1, 2\}$, there exists a negligible function $\mu(\cdot)$ such that for all sufficiently large $x_1, x_2, z \in (\{0, 1\}^*)^3$ the following holds:
If* $\text{OUTPUT}(\text{EXEC}_{\pi, \mathcal{A}(z), i^*}(x_1, x_2; 1^n)) = \text{corrupted}_{i^*}$ *then:*

$$\Pr[\text{Judgement}(Cert) = id_{i^*}] > 1 - \mu(n)$$

where Cert is the output certificate of the honest party in the execution.

3. **(Defamation-Free:)** *For every PPT adversary $\mathcal{A}$ controlling $i^* \in \{1, 2\}$ and interacting with the honest party, there exists a negligible function $\mu(\cdot)$ such that for all sufficiently large $x_1, x_2, z \in (\{0, 1\}^*)^3$:*

$$\Pr[Cert^* \leftarrow \mathcal{A}; \text{Judgement}(Cert^*) = id_{3-i^*}] < \mu(n)$$

Every Malicious Secure Protocol Is Also Covert Secure with Public Verifiability. As a sanity check, we note that any protocol that is secure against malicious adversaries satisfies all of the above requirements, with deterrence factor $\epsilon = 1 - \text{negl}(n)$: aborting is the only possible malicious behavior. Therefore the function Blame will never be invoked and the function Judgement outputs *none* on every input. In other words, given that no cheating strategy can succeed except with negligible probability, we have that by definition no one ever "cheats" and no one can be "framed".

3   Signed Oblivious Transfer

As discussed in the introduction, signed oblivious transfer (signed OT) is one of the main ingredient in our construction. For the sake of presentation, one can think of signed OT as a protocol implementing the following functionality:

$$(\perp; (m_b, \text{Sig}_{sk}(b, m_b))) \leftarrow \mathcal{F}((m_0, m_1, sk), (b, vk))$$

However it turns out that while this formulation certainly suffices for our goal, it is not necessary for our secure two-party computation protocol in Section 4. In particular, we don't need the signature to be computed by the ideal functionality. We therefore use a relaxed version of the signed OT functionality, that allows a malicious sender to choose any two strings (σ_0^*, σ_1^*) and input them to the functionality. If (σ_0^*, σ_1^*) are valid signatures on the messages $(0, m_0)$ and $(1, m_1)$ respectively, the functionality delivers (m_b, σ_b^*) to the receiver or abort otherwise. In other words, we allow a corrupted sender to influence the randomness involved in the generation of the signature, as long as it provides correct signatures for both messages. See Functionality 1 for the formal description.

3.1   A PVW Compatible Signature Scheme

As a first step, we will construct a (somewhat contrived) signature scheme, designed to combine efficiently with the OT protocol. Essentially, we are combining

---

FUNCTIONALITY 1 (The Signed OT Functionality – $\mathcal{F}_{\Pi}^{\mathrm{SignedOT}}$)

The functionality is parameterized by a signature Scheme $\Pi = (\mathsf{Gen}, \mathsf{Sig}, \mathsf{Ver})$.

Inputs: The receiver inputs (vk, b) – a verification key together with a bit $b \in \{0, 1\}$. The input of the sender is $(m_0, m_1, sk, \sigma_0^*, \sigma_1^*)$. An honest sender is restricted to input $(\sigma_0^*, \sigma_1^*) = (\bot, \bot)$.

Output: If $(\sigma_0^*, \sigma_1^*) = (\bot, \bot)$ the functionality computes $\sigma = \mathsf{Sig}_{sk}(b, m_b)$ and verifies that $\mathsf{Ver}_{vk}((b, m_b), \sigma) = 1$. It then outputs (m_b, σ) to the receiver or **abort** in case where the verification fails.

If $(\sigma_0^*, \sigma_1^*) \neq (\bot, \bot)$ the functionality outputs (m_b, σ_b^*) to the receiver if $\mathsf{Ver}_{vk}((0, m_0), \sigma_0^*) = 1$ and $\mathsf{Ver}_{vk}((0, m_0), \sigma_0^*) = 1$ or **abort** otherwise.

---

a signature scheme $\Pi' = (\mathsf{Gen}', \mathsf{Sig}', \mathsf{Ver}')$ with a computationally binding commitment $\mathsf{Com} = (\mathsf{Setup}, \mathsf{Com}, \mathsf{Open})$ (we do not need the commitment to be hiding). The verification key vk of the combined scheme is the same as the verification key of the original scheme vk'. On input a message m, the combined signature algorithm Sig chooses a random commitment key $ck = \mathsf{Setup}(1^n)$ and a string r, compute the commitment $(c, d) = \mathsf{Com}_{ck}(m; r)$ and outputs:

$$(ck, d, c, \mathsf{Sig}'_{sk}(ck, c)) \leftarrow \mathsf{Sig}_{sk}(m) . \tag{1}$$

On input $(m, (ck, d, c, \sigma))$, the verification algorithm Ver outputs 1 if and only if $\mathsf{Open}_{ck}(c, d) = m$ and $\mathsf{Ver}_{vk}((ck, c), \sigma) = 1$. Unforgeability of the combined scheme follows from the unforgeability of the original scheme together with the binding property of the commitment scheme. (Note that here is the signer creates both the commitment key and the commitment itself – differently from the standard game for computationally binding commitments, where the receiver needs to generate the key.) See the full version for details.

We present the commitment scheme that we use in the above template. Let $(\mathbb{G}, q)$ be a prime order group where the DDH assumption is believed to hold. Define the randomized function $RAND(g_0, h_0, g_1, h_1) = (u, v)$, where $u = (g_0)^s \cdot (h_0)^t$ and $v = (g_1)^s \cdot (h_1)^t$ and $s, t \in_R \mathbb{Z}_q$. Observe that if (g_0, h_0, g_1, h_1) is a DDH tuple for some x (i.e, there exists an x such that $g_1 = g_0^x$ and $h_1 = h_0^x$) then u is distributed at random in $\mathbb{G}$ and $v = u^x$. In case where (g_0, h_0, g_1, h_1) is not a DDH tuple (i.e, $\log_{g_0} g_1 \neq \log_{h_0} h_1$) then the pair (u, v) is distributed uniformly at random in $\mathbb{G}^2$. See [PVW08] for more details. The commitment scheme is as follows:

- **The Setup Algorithm** Setup: On input security parameter 1^n, the setup chooses a DDH tuple (g_0, h_0, g_1, h_1) in $\mathbb{G}$ and defines $ck = (g_0, h_0, g_1, h_1)$.
- **The Commitment Algorithm** Com_{ck}: On input message $(b, m) \in \{0, 1\} \times \mathbb{G}$, the Com algorithm chooses a random $r \in_R \mathbb{Z}_q$ and computes $(g, h) = (g_b, h_b)^r$ and $(u_b, v_b) = RAND(g_b, g, h_b, h)$, $w_b = m \cdot v_b$, $(u_{1-b}, w_{1-b}) \in_R \mathbb{G}^2$. Then, it defines $c = (g, h, u_0, w_0, u_1, w_1)$ and the decommitment value $d = (r; (b, m))$.

– **The Opening Algorithm** $\mathsf{Open}_{ck}(c, d)$: On input key $ck = (g_0, h_0, g_1, h_1)$, commitment $c = (g, h, u_0, w_0, u_1, w_1)$ and decommitment $d = (r; (b, m))$, the opening algorithm checks that $(g, h) = (g_b, h_b)^r$ and $w_b = m \cdot u_b^r$. If so it outputs (b, m), otherwise $\perp$.

Claim 1. *Assuming computing discrete logarithms is hard in $\mathbb{G}$, the scheme* $(\mathsf{Setup}, \mathsf{Com}, \mathsf{Open})$ *is computationally binding.*

Proof Sketch: To see that the scheme is binding, observe that there is a unique mapping between r and (b, m) in the following way: given a commitment $c = (g, h, u_0, w_0, u_1, w_1)$ and the decommitment r, we search for b for which $(g, h) = (g_b, h_b)^r$. Given (r, b), the message m is defined as: $w_b \cdot (u_b)^{-r}$. Therefore, the only way that an adversary can break the binding property of a given commitment c is by finding r' for which $(g, h) = (g_{1-b}^{r'}, h_{1-b}^{r'})$. But, to find such an r' the adversary needs to break the discrete logarithm assumption.   ∎

Our PVW compatible signature scheme $\Pi = (\mathsf{Gen}, \mathsf{Sig}, \mathsf{Ver})$ is the a combination of the signature scheme Π' and the commitment scheme Com as defined in Eq. (1). We conclude:

Corollary 1. *If $\Pi' = (\mathsf{Gen}', \mathsf{Sig}', \mathsf{Ver}')$ is an existentially unforgeable under an adaptive chosen-message attack signature scheme and the discrete logarithm problem is hard in $(\mathbb{G}, g_0, q)$, then $\Pi = (\mathsf{Gen}, \mathsf{Sig}, \mathsf{Ver})$ is also existentially unforgeable under an adaptive chosen-message attack.*

3.2 PVW-Based Signed OT

We present the protocol for signed OT in Protocol 1, combining the PVW OT protocol with the signature scheme described above. Like the original OT protocol [PVW08], our signed OT protocol can be extended in the straightforward way to an 1-out-of-ℓ signed OT (see the full version). Note that the overall protocol is just the DDH-based instantiation of the PVW OT framework with the following differences (clearly marked in the protocol description): (1) The sender chooses the "CRS" (g_0, h_0, g_1, h_1) and proves that it is a DDH tuple. (Remember that in this case the receiver's message hides his choice bit *statistically*). (2) The sender signs all the messages it sends to the receiver.

Note that the Com algorithm is *distributed*, in the sense that both parties contribute to the input and randomness: in particular the receiver chooses b while the sender specifies (m_0, m_1) without knowing which message is going to be chosen.

Lemma 1. *Let $\Pi = (\mathsf{Gen}, \mathsf{Sig}, \mathsf{Ver})$ be the PVW-compatible signature scheme defined above. Then, Protocol 1 securely implements the $\mathcal{F}_\Pi^{\mathrm{SignedOT}}$-functionality in the presence of a malicious adversary.*

PROTOCOL 1 (Signed One-out-of-Two OT Protocol)

Setup: This step can be done once and reused for multiple runs of the OT:
The sender S chooses $(g_0, h_0) \in_R \mathbb{G}^2$ and a random $\alpha \in_R \mathbb{Z}_q$ and compute
$g_1 = g_0^\alpha$ and $h_1 = h_0^\alpha$. The sender sends (g_0, h_0, g_1, h_1) to the receiver
R and gives a zero-knowledge proof-of-knowledge that (g_0, h_0, g_1, h_1) is a
DDH tuple.

Choose: R chooses random $r \in_R \mathbb{Z}_q$, computes $g = (g_b)^r$, $h = (h_b)^r$ and
sends (g, h) to S;

Transfer: The sender operates in the following way:
1. S computes $(u_0, v_0) = RAND(g_0, g, h_0, h)$ and $(u_1, v_1) = RAND(g_1, g, h_1, h)$;
2. S sends R the values (u_0, w_0) where $w_0 = v_0 \cdot m_0$, and (u_1, w_1) where
   $w_1 = v_1 \cdot m_1$;
3. **(diff)** S sends to the receiver
   $$\sigma' = \mathsf{Sig}'_{sk'}((g_0, h_0, g_1, h_1), (g, h, u_0, w_0, u_1, w_1));$$

Retrieve: **(diff)** Let $vk = vk'$. R checks that σ' is a valid signature on the
transcript of the protocol. If so, R outputs: $m_b = w_b \cdot (u_b)^{-r}$ and **(diff)**
$$\sigma = ((g_0, h_0, g_1, h_1), (r; (b, m_b)), (g, h, u_0, w_0, u_1, w_1), \sigma') \ .$$

Otherwise, it outputs **abort**.

Proof Sketch: As discussed in Corollary 1, σ is a proper signature on the
message (b, m_b), and therefore the correct functionality is implemented when
both parties are honest.

The proof of security of the underlying OT protocol is by now standard and
can be found in [PVW08, HL10]. When the receiver is corrupted, the simulator
plays as an honest sender except that it chooses instead a non-DDH tuple in
step "Setup" (i.e., some (g_0, g_0^x, g_1, g_1^y)) and then, given the pair (g, h) and using
the trapdoor (x, y), it can extract the receiver input's bit b by finding whether h
equals g^x or g^y. It then sends b to the functionality $\mathcal{F}_\Pi^{\mathrm{SignedOT}}$. Clearly, adding the
signature σ' does not break any security property of the original OT protocol (it
is easy to see that any attack to this protocol can be reduced to an attack to the
original protocol, where the reduction will simply produce this extra signature).

For the case of a corrupted sender, the simulator plays as an honest receiver
(with $b = 1$) except that it extracts α from the zero-knowledge proof in step
"Setup". Using this trapdoor, it can compute both messages m_0, m_1 (as in the
proof of the original protocol). Then, it computes the two signatures σ_0^*, σ_1^* as
follows:

$$\sigma_0^* = ((g_0, h_0, g_1, h_1), (\alpha \cdot r, (0, m_0)), (g, h, u_0, w_0, u_1, w_1), \sigma')$$
$$\sigma_1^* = ((g_0, h_0, g_1, h_1), (r, (1, m_1)), (g, h, u_0, w_0, u_1, w_1), \sigma')$$

In order to see that these are valid signatures on $(0, m_0), (1, m_1)$ respectively,
recall that $(g, h) = (g_1, h_1)^r = (g_0, h_0)^{\alpha \cdot r}$. This implies that $\alpha \cdot r$ is a valid
opening of c for $(0, m_0)$ whereas r is the opening of c for $(1, m_1)$. Finally, it is

easy to see that the distribution of the constructed signatures are the same as in the real execution. ∎

4   Two-Party Computation with Publicly Verifiable Covert Security

The protocol is an extension of the two party protocol of [AL07], which is based on Yao's garbled circuit protocol for secure two-party computation. We will start with an informal discussion of the ways that a malicious adversary can cheat in Yao's protocol[1] and we will present the (existing) countermeasures to make sure that such attacks will be detected with significant probability, thus leading to covert security. Finally we will describe how to add the public verifiability property on top of this. The ways that a malicious adversary can cheat in Yao's protocol are as follows:

1. **Constructing bad circuits:** To prevent P_1 from constructing a circuit that computes a function different than f, P_1 constructs ℓ independent garbled circuits and P_2 checks $\ell-1$ of them. Therefore if P_1 cheats in the construction of the circuits, P_2 will notice this with probability $> 1 - 1/\ell$. To make sure P_1 cannot abort if it is challenged on an incorrect circuit, we run the cut-and-choose through a 1-out-of-ℓ signed OT, so that P_2 will always receive some (signed) opening of the circuits that can be used to prove a cheating attempt to a third party.

2. **Selective failure attack on P_2's input values:** When P_2 retrieves its keys (using the OT protocol), P_1 may take a guess g at one of the inputs bits of P_2. Then, it may use some string r instead of the valid key k_{1-g}, as input to the OT protocol. Now, in case where that P_1 guesses correctly and indeed the input bit equals g, P_2 receives k_g and does not notice that there was anything wrong. However, in case the guess is incorrect, P_1 receives r instead of k_{1-g} which is an invalid key and thus it aborts. In both cases, the way P_2 reacts completely reveals this input bit. This problem can be fixed by computing a different circuit, where P_2's input is an m-out-of-m linear secret sharing of each one of the input bits of P_2. Now every $m-1$ input bits of P_2 to the protocol are uniformly random and therefore P_2 will get caught with probability $1 - 2^{-m+1}$ if it attempts to guess (the encoding of) an input bit. By using a signed OT we will ensure that P_2 receives a certificate on the wrong keys if P_1 cheats.

Let Com denote a perfectly-binding commitment scheme, where $\mathsf{Com}(x;r)$ denotes a commitment to x using randomness r. ($\mathsf{Gen_{ENC}}$, Enc, Dec) is a semantically secure symmetric encryption scheme. (Gen, Sig, Ver) is an existentially unforgeable signature scheme under an adaptive chosen-message attack. Note that it is crucial that every message is signed together with some extra-information about the role of this

---

[1] We assume the reader to be familiar with Yao's garbled circuit protocol. See [LP09] for more details and full proof of security.

message (i.e., with unique identifiers for the parties executing the protocols, the instance of the protocol, which type of message in the protocol, which gate/wire label is the message associated too etc.) but we will neglect these extra information in the description of our protocol for the sake of simplicity.

PROTOCOL 2 [Two-Party Secure Computation]

Inputs: *Party P_1 has input x_1 and Party P_2 has input x_2, where $|x_1| = |x_2|$. In addition, both parties have parameters ℓ and m, and a security parameter n. For simplicity, we will assume that the length of the inputs are n. (**diff**) Party P_1 knows a secret key sk for a signature scheme and P_2 received the corresponding verification key vk from the $\mathcal{F}^{\mathrm{PKI}}$.*

Auxiliary Input: *Both parties have the description of a circuit C for inputs of length n that computes the function f. The input wires associated with x_1 are $w_1, \ldots, w_n$ and the input wires associated with x_2 are $w_{n+1}, \ldots, w_{2n}$.*

The Protocol[2]

1. *Parties P_1 and P_2 define a new circuit C' that receives $m+1$ inputs x_1, x_2^1, $\ldots, x_2^m$ each of length n, and computes the function $f(x_1, \oplus_{i=1}^m x_2^i)$. Note that C' has $n + mn$ input wires. Denote the input wires associated with x_1 by $w_1, \ldots, w_n$, and the input wires associated with x_2^i by $w_{n+(i-1)m}, \ldots, w_{n+im}$ for $i = 1, \ldots, n$.*

2. *P_2 chooses $m-1$ strings $x_2^1, \ldots, x_2^{m-1}$ uniformly and independently at random form $\{0,1\}^n$, and defines $x_2^m = \left(\oplus_{i=1}^{m-1} x_2^i\right) \oplus x_2$, where x_2 is P_2's original input. Observe that $\oplus_{i=1}^m x_2^i = x_2$.*

3. *For each $i = 1, \ldots, mn$ and $\beta = 0, 1$, party P_1 chooses ℓ encryption keys by running $\mathsf{Gen_{ENC}}(1^n)$ for ℓ times. Denote the jth key associated with a given i and β by $k_{w_{n+i},\beta}^j$.*

4. *P_1 and P_2 invoke the mn times the (**diff**) $\mathcal{F}_\Pi^{\mathrm{SignedOT}}$ functionality with the following inputs: In the ith execution, party P_1 inputs the pair:*

$$\left(\left[k_{w_{n+i},0}^1, \ldots, k_{w_{n+i},0}^\ell \right], \left[k_{w_{n+i},1}^1, \ldots, k_{w_{n+i},1}^\ell \right] \right)$$

and party P_2 inputs the bit x_2^i (P_2 receives the keys $\left[k_{w_{n+i},x_2^i}^1, \ldots, k_{w_{n+i},x_2^i}^\ell \right]$ and a signature on this as output). If P_2 output in the OT is abort_i, then it outputs abort_i and halts.

5. *Party P_1 constructs ℓ garbled circuits $GC_1, \ldots, GC_\ell$ using independent randomness for the circuit C' described above. The keys for the input wires $w_{n+1}, \ldots, w_{n+mn}$ in the garbled circuits are taken from above (i.e., the keys associated with w_{n+i} are $k_{w_{n+i},0}^j$ and $k_{w_{n+i},1}^j$). The keys for the inputs wires $w_1, \ldots, w_n$ are chosen randomly, and are denoted in the same way. P_1 sends the ℓ garbled circuits to P_2 (**diff**) together with a signature on those.*

---

[2] The description of the protocol is almost verbatim from [AL07] to help the reader identify the few (clearly marked) differences between our protocol and the original protocol.

6. P_1 commits to the keys associated with its inputs. That is, for every $i = 1, \ldots, n$, $\beta = 0, 1$ and $j = 1, \ldots, \ell$, party P_1 computes (**diff**):

$$c^j_{w_i,\beta} = \mathsf{Com}\left(k^j_{w_i,\beta}; r^j_{i,\beta}\right), \sigma^j_{w_i,\beta} = \mathsf{Sig}_{sk}(c^j_{w_i,\beta})$$

The commitments and the signatures are sent as ℓ vectors of pairs (one vector for each circuit); in the jth vector the ith pair is $\{(c^j_{w_i,0}, \sigma^j_{w_i,0}), (c^j_{w_i,1}, \sigma^j_{w_i,0})\}$ in a random order (the order is randomly chosen independently for each pair). (**diff**) Party P_2 verifies that all the signatures are correct. If not, it halts and outputs abort$_1$.

7. P_2 chooses a random index $\gamma \in_R \{1, \ldots, \ell\}$.

8. (**diff**) P_1 and P_2 engage in a $\binom{\ell}{1}$-signed OT, where P_2 inputs γ and, for $i = 1, \ldots, \ell$, P_1 inputs as the ith message of the signed OT all of the keys for the inputs wires in all garbled circuits except for GC_i, together with the associated mappings and the decommitment values. P_1 sends also decommitments to the input keys associated with its input for the circuit GC_i.
   P_2 receives the openings for $\ell - 1$ circuits (all but GC_γ) together with a signature on them. P_2 receives also the decommitments and the keys associated with P_1's input for circuit GC_γ together with signatures on them. If any of the signatures are incorrect, it halts and outputs abort$_1$.

9. P_2 checks that:
   - That the keys it received for all GC_j, $j \neq \gamma$, indeed decrypt the circuits and the decrypted circuits are all C'. (**diff**) If not, add key = wrongCircuit to its view.
   - That the decommitment values correctly open all the commitments $c^j_{w_i,\beta}$ that were received, and these decommitments reveal the keys $k^j_{w_i,\beta}$ that were sent for P_1's wires. (**diff**) If not, add key = wrongDecommitment to its view.
   - That the keys received in the signed OT in Step 4 match the appropriate keys that it received in the opening. (**diff**) If not, add key = selectiveOTattack to its view.
   If all check pass, proceed to the next step, else (**diff**), P_2 computes Cert = Blame(view$_2$) (see the description of Blame for its output on different key values), it publishes Cert and output corrupted$_1$.

10. P_2 checks that the values received are valid decommitments to the commitments received above. If not, it outputs abort$_1$. If yes, it uses the keys to compute $C'(x_1, z_2) = C'(x_1, x^1_2, \ldots, x^m_2) = C(x_1, x_2)$, and outputs the result.

Theorem 1. Let ℓ and m be parameters in the protocol that are both upper-bound by $\mathrm{poly}(n)$, and set $\epsilon = (1 - 1/\ell)(1 - 2^{-m+1})$, and let f be a probabilistic polynomial-time function and let π denote Protocol 2. Then, assuming the DDH assumption, security of the commitment scheme, signature scheme and symmetric encryption scheme as described above, $(\pi, \mathsf{Blame}, \mathsf{Judgement})$ securely computes f in the presence of covert adversaries with ϵ-deterrent and public verifiability (i.e, satisfies Definition 1).

ALGORITHM 1 (The Blame Algorithm – Blame)

Input: The view of a honest party view, containing an error tag key.

Output: A certificate $Cert = (id, \mathsf{key}, \mathsf{message}, \sigma)$.

The Algorithm:

- **Case 1:** key = wrongCircuit: Let j be the smallest index s.t. the garbled circuit GC_j is not a garbling of C'. Let message be the commitment to GC_j concatenated with the opening obtained via the $\binom{\ell}{1}$-signed OT in Step 8, and σ the signature on these messages.
- **Case 2:** key = wrongDecommitment: Let message be (c, x, r) be a commitment where $c \neq \mathsf{Com}(x; r)$ and σ the signatures on c and (x, r).
- **Case 3:** key = selectiveOTattack: let message be a garbled circuit GC_i and two keys to one of its input gates. Let σ be the signature on the circuit and the signatures on the keys obtained in Step 8.

On any other case, output $\perp$.

ALGORITHM 2 (The Public Verification Algorithm – Judgement)

Input: A certificate $Cert = (id, \mathsf{key}, \mathsf{message}, \sigma)$.

Output: The identity id or *none*.

The Algorithm: If σ is not a valid signature on the message message according to verification key vk_{id} halt and output *none*. Else:

- **Case 1:** key = wrongCircuit: Parse message as a garbled circuit GC and the randomness r used to generate it. If GC is not an encryption of the circuit computing C' using randomness r output id or *none* otherwise.
- **Case 2:** key = wrongDecommitment: Parse message as (c, x, r). If $c \neq \mathsf{Com}(x; r)$ output id or *none* otherwise.
- **Case 3:** key = selectiveOTattack: Parse message as a circuit GC and two keys k^i, k^j for an input gate g of the circuit GC. If k^i, k^j do not decrypt the gate g output id or *none* otherwise.

Note that even for very small replication factors this construction gives reasonable level of deterrence factor e.g., $\ell = 3$ and $m = 3$ lead to $\epsilon = 50\%$. We can now proceed to the proof.

Proof Sketch: We show that our protocol satisfies each one of the properties as in Definition 1. We will use the similarity between our protocol and the one of [AL07] to argue for covert security with ϵ-deterrent.

Corrupted P_2. Our protocol achieves security in the presence of a *malicious* P_2. The security follows from the $\mathcal{F}_\Pi^{\mathsf{SignedOT}}$-functionality (that as we have seen, can be implemented efficiently with malicious security) and the same reasoning as in [AL07], with the exception that here we use a fully secure malicious OT instead of a a covert. We are therefore left with the case where P_1 is corrupted.

Simulatability with ϵ-deterrent. Our protocol is in fact the same protocol as in [AL07], with the following differences: (1) In Steps 5 and 6, P_1 sends its messages together with a signature on those. (2) In Steps 4 and 8, signed OT is used instead of standard OT. (3) In Step 9, if P_2 outputs corrupted$_i$, then it sends $Cert = \mathsf{Blame}(\mathrm{view}_2)$ to the adversary. Let π_0 be the protocol of [AL07, Section 6.3] and π_1, π_2, π_3 the protocols after the changes explained in bullets $1, 2, 3$ respectively.

Protocols π_1 and π_2 differ from π_0 only because P_1 signs the messages it sends to P_2. In the full version, we show that if π is a covert secure protocol with ϵ-deterrent and π' is the same protocol as π with the only change that parties sign on all the message they send, then π' is also a covert secure protocol with ϵ-deterrent. We therefore conclude that π_2 is also a covert secure protocol with ϵ-deterrent.

The only difference between π_3 and π_2 is that if P_2 outputs corrupted$_1$, then the adversary learns the certificate $Cert$. In the full version, we show that this extra information can be simulated as well and so the overall protocol is covert protocol with ϵ-deterrent.

Accountability. Accountability follows from the description of the protocol π and the $\mathsf{Blame}, \mathsf{Judgement}$ algorithms: an adversarial P_1 who constructs one faulty circuit must decide before the oblivious transfer in Step 9 if it wishes to abort (in which case there is no successful cheating) or if it wishes to proceed (in which case P_2 will receive an explicitly invalid opening and a signature on it). Note that due to the security of the oblivious transfer, P_1 cannot know what value γ party P_2 inputs, and so cannot avoid being detected.

Once the honest party outputs the certificate, it contains all the necessary information that caused the party to decide on the corruption. The verification algorithm $\mathsf{Judgement}$ performs exactly the same check as the honest party, and so accountability holds.

Defamation-Free. We need to show that for every PPT adversary $\mathcal{A}$ controlling $i^* \in \{1, 2\}$ and interacting with the honest party, there exists a negligible function $\mu(\cdot)$ such that for all sufficiently large $x_1, x_2, z \in (\{0, 1\}^*)^3$:

$$\Pr\left[Cert^* \leftarrow \mathcal{A}; \mathsf{Judgement}(Cert^*) = id_{3-i^*}\right] < \mu(n)$$

The above holds from the security of the signature scheme. Since $\mathsf{Judgement}$ never outputs the identity of P_2 and may just output the identity of P_1, the only interesting case is when the adversary controls P_2 and succeeds in creating a forged certificate $Cert^*$ for which $\mathsf{Judgement}(Cert^*) = id_1$. Since P_1 is honest, it follows the protocol specifications and creates all the circuits correctly, consistent and open the commitments correctly. Remember also that every signature the honest P_1 produces contains meta-information about the message (such as identity of the participating parties, protocol unique identifier, message identifier etc.) to ensure that a corrupted P_2^* cannot mix and match signatures obtained during different protocols to create a forged certificate. Therefore, if the adversary produces a certificate that passes the verification, it must have forged one of the messages. A more formal argument appears in the full version.   ■

Acknowledgements. The authors would like to thank Yehuda Lindell for many helpful discussions and his invaluable guidance.

References

AL07.   Aumann, Y., Lindell, Y.: Security Against Covert Adversaries: Efficient Protocols for Realistic Adversaries. In: Vadhan, S.P. (ed.) TCC 2007. LNCS, vol. 4392, pp. 137–156. Springer, Heidelberg (2007)

BCNP04.   Barak, B., Canetti, R., Nielsen, J.B., Pass, R.: Universally composable protocols with relaxed set-up assumptions. In: FOCS, pp. 186–195. IEEE Computer Society (2004)

Can00.   Canetti, R.: Security and composition of multiparty cryptographic protocols. J. Cryptology 13(1), 143–202 (2000)

DGN10.   Damgård, I., Geisler, M., Nielsen, J.B.: From Passive to Covert Security at Low Cost. In: Micciancio, D. (ed.) TCC 2010. LNCS, vol. 5978, pp. 128–145. Springer, Heidelberg (2010)

FY92.   Franklin, M.K., Yung, M.: Communication complexity of secure computation (extended abstract). In: STOC, pp. 699–710 (1992)

GMS08.   Goyal, V., Mohassel, P., Smith, A.: Efficient Two Party and Multi Party Computation Against Covert Adversaries. In: Smart, N.P. (ed.) EURO-CRYPT 2008. LNCS, vol. 4965, pp. 289–306. Springer, Heidelberg (2008)

Gol04.   Goldreich, O.: Foundations of Cryptography, Basic Applications, vol. 2. Cambridge University Press (2004)

HL08.   Hazay, C., Lindell, Y.: Efficient Protocols for Set Intersection and Pattern Matching with Security Against Malicious and Covert Adversaries. In: Canetti, R. (ed.) TCC 2008. LNCS, vol. 4948, pp. 155–175. Springer, Heidelberg (2008)

HL10.   Hazay, C., Lindell, Y.: Efficient secure two-party protocols: Techniques and constructions. Springer (2010)

IKNP03.   Ishai, Y., Kilian, J., Nissim, K., Petrank, E.: Extending Oblivious Transfers Efficiently. In: Boneh, D. (ed.) CRYPTO 2003. LNCS, vol. 2729, pp. 145–161. Springer, Heidelberg (2003)

LP09.   Lindell, Y., Pinkas, B.: A proof of security of yao's protocol for two-party computation. J. Cryptology 22(2), 161–188 (2009)

MNPS04.   Malkhi, D., Nisan, N., Pinkas, B., Sella, Y.: Fairplay - secure two-party computation system. In: USENIX Security Symposium, pp. 287–302 (2004)

PVW08.   Peikert, C., Vaikuntanathan, V., Waters, B.: A Framework for Efficient and Composable Oblivious Transfer. In: Wagner, D. (ed.) CRYPTO 2008. LNCS, vol. 5157, pp. 554–571. Springer, Heidelberg (2008)

A Unified Framework for UC from Only OT

Rafael Pass[1], Huijia Lin[2], and Muthuramakrishnan Venkitasubramaniam[3]

[1] Cornell University, Ithaca NY 14850, USA
[2] MIT and Boston University, Boston, MA, 02138, USA
[3] University of Rochester, Rochester, NY 14611, USA

Abstract. In [1], the authors presented a unified framework for constructing Universally Composable (UC) secure computation protocols, assuming only enhanced trapdoor permutations. In this work, we weaken the hardness assumption underlying the unified framework to only the existence of a stand-alone secure semi-honest Oblivious Transfer (OT) protocol. The new framwork directly implies new and improved UC feasibility results from only the existence of a semi-honest OT protocol in various models. Since in many models, the existence of UC-OT implies the existence of a semi-honest OT protocol.

Furthermore, we show that by relying on a more fine-grained analysis of the unified framework, we obtain concurrently secure computation protocols with super-polynomial-time simulation (SPS), based on the *necessary* assumption of the existence of a semi-honest OT protocol that can be simulated in super-polynomial times. When the underlying OT protocol has constant rounds, the SPS secure protocols constructed also have constant rounds. This yields the first construction of constant-round secure computation protocols that satisfy a meaningful notions of concurrent security (i.e., SPS security) based on tight assumptions.

A notable corollary following from our new unifed framwork is that stand-alone (or bounded-concurrent) password authenticated key-exchange protocols (PAKE) can be constructed from only semi-honest OT protocols; combined with the result of [2] that the existence of PAKE protocols implies that of OT, we derive a tight characterization of PAKE protocols.

1 Introduction

The notion of *secure multi-party computation* allows m mutually distrustful parties to securely compute a functionality $f(\bar{x}) = (f_1(\bar{x}), ..., f_m(\bar{x}))$ of their corresponding private inputs $\bar{x} = x_1, ..., x_m$, such that party P_i receives the value $f_i(\bar{x})$. Loosely speaking, the security requirements are that the parties learn nothing more from the protocol than their prescribed output, and that the output of each party is distributed according to the prescribed functionality. This should hold even in the case that an arbitrary subset of the parties maliciously deviates from the protocol.

Shortly after the notion was proposed, strong results were established for secure multi-party computation. Specifically, it was shown that any probabilistic polynomial-time computable multi-party functionality can be securely computed, assuming existence of enhanced trapdoor permutations [3, 4]. The original

X. Wang and K. Sako (Eds.): ASIACRYPT 2012, LNCS 7658, pp. 699–717, 2012.

setting in which secure multi-party protocols were investigated, however, only allowed the execution of a single instance of the protocol at a time; this is the so called *stand-alone setting*. A more realistic setting, is one which allows the concurrent execution of protocols. In the *concurrent setting*, many protocols are executed at the same time. This setting presents the new risk of a coordinated attack in which an adversary interleaves many different executions of a protocol and chooses its messages in each instance based on other partial executions of the protocol. The strongest (but also most realistic) setting for concurrent security— called *Universally Composable* (UC) security [5]—considers the execution of an unbounded number of concurrent protocols, in an arbitrary, and adversarially controlled, network environment. Unfortunately, security in the stand-alone setting does not imply security in the concurrent setting. In fact, without assuming some trusted set-up, the traditional simulation-based notion of concurrent security, and in particular UC security, cannot be achieved in general [6–8].

To circumvent the broad impossibility results, two distinct veins of research can be identified in the literature.

Trusted Set-Up Models: A first vein of work initiated by Canetti and Fischlin [6] and Canetti, Lindell, Ostrovsky and Sahai [9] (see also e.g., [10–13]) considers constructions of UC-secure protocol using various trusted set-up assumptions, where the parties have limited access to a trusted entity.

Relaxed Models of Security: Another vein of work considers relaxed models of security such as *quasi-polynomial simulation* [14–16] or *input-indistinguishability* [17]. These works circumvents the use of trusted set-ups, but, only provide weak guarantees about the computational advantages gained by an adversary in a concurrent execution of the protocol.

In [1], we provided a general unified framework to construct UC-secure protocols in both trusted set-up models and relaxed security models. In more detail, we showed that for any such model, the construction of UC protocols for realizing any multi-party functionality reduces to the construction of a so-called "UC-puzzle" and a so-called strongly non-malleable witness indistinguishable ($\mathcal{SNMWI}$) argument of knowledge. Intuitively, a "UC-puzzle" is a protocol that has the property that no adversary can successfully complete the puzzle and also obtain a trapdoor, but there exists a simulator who can generate (correctly distributed) puzzles together with trapdoors; and a $\mathcal{SNMWI}$ argument ensures that no man-in-the-middle adversary can correlate the witness it uses in a proof with the witness in the proof it receives[1]. They we showed that a $\mathcal{SNMWI}$ argument can be implemented using any non-malleable commitment scheme; therefore the task of realizing UC security in any model reduces to the task of constructing a "UC-puzzle" in that model, which can be easily achieved in almost all previously considered set-up and relaxed security models. Furthermore, in many models, we showed that the existence of a "UC-puzzle" is also necessary; in a sense, the notion of "UC-puzzle" characterizes the "minimal" set-up and relaxation of security needed for achieving UC security.

---

[1] A $\mathcal{SNMWI}$ argument can be viewed as an analogy of non-malleable commitments in the context of strongly $\mathcal{WI}$ proofs [18].

In this work, we focus on a different dimension: Namely, given the minimal set-up and relaxation of security need for UC, what is the "minimal" computational assumption additionally needed for constructing UC secure protocols. In [1], the construction of UC protocols from "UC-puzzles" is based on the existence of enhanced trapdoor-permutations (TDP's), whereas stand-alone secure multi-party computation protocols can be constructed based on the minimal assumption of the existence of stand-alone secure semi-honest OT protocols [19, 20], which clearly also is a necessary assumption. This immediately raises the following question.

Can we base UC security on the minimal assumption of the existence of a semi-honest OT protocol?

1.1 Previous Works

Immediately after the work of [1], there has been several works trying to address this problem in specific models.

In KRA and CRS model: Damgard, et al. [21] showed that UC security can be achieved assuming only semi-honest OT protocols in the key registration (KR), and common reference string (CRS), as well as uniform reference string (URS) models. Their constructions in the KR, and the more generalized arbitrary KR (A-KR), models achieve optimal round complexity, which have the same number of rounds as the underlying semi-hoest OT protocol (up to a constant factor). However, the round-complexity of their construction in the CRS and URS model grows linearly with the number of players in the protocol execution. Furthermore, their construction in the CRS and URS model only implements an ideal functionality $\mathcal{F}$ in a *single session*, meaning every execution of their protocol needs to invoke the CRS functionality to obtain an independently sampled reference string. In contrast, previous constructions of UC secure protocol in the CRS model directly implements the *multi session extension of $\mathcal{F}$* [1, 9] so that different protocol executions may share the same CRS.

In the $\mathcal{F}_{\mathsf{coin-toss}}$ hybrid model: In the context of characterizing functionalities that are complete for achiveing UC security, Maji, Prabhakaran and Rosulek [22] showed that the *ideal two-party coin-tossing functionality $F_{\mathsf{coin-toss}}$* is "complete", in the sense that, assuming the existence of semi-honest OT protocols, practically all functionalities[2] can be UC-securely realized when players have access to the $\mathcal{F}_{\mathsf{coin-toss}}$ functionality, with the same round complexity as the OT protocol.

In the tamper-proof hardware model: Goyal et. al. showed that in the model where players can generate and exchange tamper-proof hardware tokens, UC security can be achived assuming the weaker assumption of one-way functions or even unconditionally, in a constant-number of rounds.

The above mentioned previous works try to weaken the assumptions that UC security is based on using different techniques and exploiting different features

---

[2] More precisely, all well-formed functionalities can be UC-securely realized.

of the specific models under consideration. This immediately raises the question whether we can achieve UC security from semi-honest OT protocol in a generic way as in [1], independent of the specifics of different set-up or relaxed security models.

> *Can we base UC security only on the existence of semi-honest OT protocols,* generically?
> *Furthermore, can we achieve so with* optimal round complexity?

Such a generic construction would not only help us identify and undersand the key elements needed for achieving UC security, also allow us to obtain new UC-feasibility results in other models easily.

Furthermore, one common limitation of the previous results is that they all used the trusted set-ups in a strong way so that different protocol executions have *different and independent* "trapdoors", which makes UC security relatively easy to achieve. Let us explain the intuition. In order to construct a protocol secure in the concurrent setting, we need to establish two properties: *Concurrent simulation*, that is, the simulator can simulate messages from the honest players in many concurrent sessions for the adversary, and *concurrent simulation-soundness*, that is the adversary even when receiving simulated messages cannot break the security of the protocol against honest players. The concurrent simulation property can be established easily as long as there is a single trapdoor (or correlated trapdoors) shared by all protocol executions; the simulator can simply use that trapdoor to simulate. The concurrent simulation-soundness property, on the other hand, is much harder to establish, and often involves the use of *non-malleable* primitives to ensure independence of different protocol executions as in [1, 9, 23]. However, in the case where different sessions have independent trapdoors, concurrent simulation-soundness can be obtained "for free", as receiving simulated messages (containing information of one trapdoor) does not help the adversary obtain other trapdoors; hence, the security of the protocol w.r.t. the honest players remains.

Indeed, all previous works use the trusted set-up to generate independent trapdoors for different protocol executions. In the CRS (resp. URS) model, [21] constructed protocols that implement a general functionality $\mathcal{F}$ in a *single session*, meaning that each executions of their protocol invokes the CRS (resp. URS) functionality independently, which yields independent trapdoors (that is, independent secrets associated with different CRS's (resp. URS's)). In the KR and A-KR model of [21], every player is registered with a valid public key that has a corresponding secret key; furthermore, the secret key of any honest player is hidden even if the adversary obtains the secret keys of all other players. Naturally, the secret keys of players are used as independent trapdoors. The same happens in the tamper-proof hardware token model, where the freshly generated hardware tokens in each session yield independent trapdoors for different session. Finally, in the ideal coin-tossing hybrid model, the $F_{\text{coin-toss}}$ functionality is used to sample an independent URS in every session.

However, in many "weaker" models, there is only a single trapdoor (or correlated trapdoors) across many protocol executions. Then the techniques used in previous works no longer apply, and the protocol construction needs to explicitly "inject" independence to establish simulation soundness. Such set-up models include the CRS model, when the protocol construction directly implements the multi-session extension of functionalities, the single imperfect string (sunspot) model [13], the timing model [24] and the bounded concurrency model [25]. Furthermore, the super-polynomial time simultion model also share the same flavor: Though each protocol execution session may generate its own trapdoor (for instance, the pre-image of a randomly sampled image of a one-way function), receiving information of the trapdoor in one session, obtained via the super-polynomial time power of the simulator, does facilitate the adversary breaking the trapdoor in other sessions, as the adversary may create correlation between trapdoors in different sessions. Naturally, the question left open by previous works is,

Can we construct UC secure protocols when there are only correlated trapdoors, based on the existence of semi-honest OT protocols?

1.2   Our Results

In this work, we answer both questions above affirmatively. We improve upon the result in [1] to obtain a new unified framework for constructing UC secure protocols, assuming only the existence of semi-honest OT protocols. More precisely, the main theorem that we establish is:

Theorem 1 (Unified Framework from OT, Informal). *Assume the existence of a $t_1(\cdot)$-round UC-secure puzzle Σ using some set-up $\mathcal{T}$, and the existence of a $t_2(\cdot)$-round stand-alone secure semi-honest oblivious-transfer protocol. Then, for every m-ary functionality f, there exists a $O(t_1(\cdot) + t_2(\cdot))$-round protocol Π —using the same set-up $\mathcal{T}$—that UC-realizes the multi-session extension of f.*

We remark that since our main theorem is general and only requires the security model to admit a single UC-puzzle, the unified framework we provide encompasses both models where there are only correlated trapdoors, as detailed below.

Trusted Set-up Models: As shown in [1], many trusted set-up models admit *constant-round* UC-puzzles assuming the existence of one-way functions. Thus, our unified framework immediately yields UC feasibility results from only semi-honest OT, in a wide range of set-up models.

Corollary 1 (Trusted Set-up Models). *Assume the existence of a $t(\cdot)$-round stand-alone secure semi-honest oblivious-transfer protocol. Then, for every m-ary functionality f, there exists a $O(t(\cdot))$-round protocol Π that UC-realizes the multi-session extension of f in the following models:*

  − *Tamper proof hardware model [26],*
  − *Key registration (KR) model [10]*

- *Chosen common reference string (C-CRS) model [9], any common reference string (A-CRS) model [21], and uniform reference string (URS) model [9],*
- *Timing model [24],*
- *Multi-string model [27],*
- *Single imperfect string (sun-spot) model [13] (assuming additionally the existence of collision resistance hash functions).*

We compare our results with previous works. In the tamper-proof hardware model (line 1), our feasibility result is weaker than that of [28], which achieved UC unconditionally. In the key-registration models (line 2), we re-prove the result in [21]. In the CRS and URS models (line 3), we obtain new feasibility results that implement directly the multi-session extension of functionalities, instead of implementing only in single session as in [21]; furthermore, we improve the round complexity to that of the OT protocol, whereas in [21] the round-complexity grows linearly with the number of players in the protoccol execution. In the rest of set-up models (line 4 to 6) that only admit correlated trapdoors, we obtain new UC feasilibity results from only semi-honest OT.

Optimal Round-Complexity: We remark that round-complexity of our construction depends solely on and is at the same order as that of the underlying semi-honest OT protocol. Therefore, assuming the existence of a *constant-round* semi-honest OT protocol, we obtain *constant-round* UC secure protocols in all above mentioned models.

Sufficient and Necessary Assumption for UC Security: Our main theorem shows that t-round semi-honest OT protocols are sufficient for UC security in various models. In fact, it is also necessary in many models. As shown in [21, 29], t-round UC secure computation in the key registration, CRS and URS models (line 1 and 2) implies t-round semi-honest OT; since the single-CRS, and single-URS models are strictly weaker than their one-CRS-per-session and one-URS-per-session versions, the implication also holds in these two models. It is easy to see that the same is true in the timing model. Therefore, our result yields a *tight* characterization of the feasibility of t-round UC secure computation (from $\Omega(t)$-round semi-honest OT) in the key-registration, CRS, URS, single-CRS, single-URS and timing models.

Super-Polynomial Time Simulation Model. In a super-polynomial time simulation model with simulation time T—T can be, say, quasi-polynomial time (QPT) or sub-exponential time (subEXP)—assuming the existence of a one-way function that is hard to invert in polynomial time, but easy to invert (with probability 1) in T time, there exists a one-message UC-puzzle in T-time simulation model[3]. Note that when considering subEXP time simulation, the assumption of

---

[3] The UC puzzle simply consists of one message from the sender is the image of a random string through that one-way function. It is hard for polynomial time adversary to break the puzzle (i.e., obtain a pre-image), but easy for a T-time simulator.

one-way functions invertable in subEXP time is simply implied by the existence of any one-way functions[4]. Therefore, applying our main theorem[5], we have:

Corollary 2 (Super-Polynomial Time Simulation Models). *Assume the existence of a $t(\cdot)$-round stand-alone secure semi-honest oblivious-transfer protocols secure for subEXP-time. Then, for every m-ary functionality f, there exists a $O(t(\cdot))$-round protocol Π that realizes f with subEXP-time-simulation security. Furthermore, the real and ideal executions are indistinguishable to all subEXP-time distinguishers.*

This result weakens the assumptions that SPS secure protocols can be relied on: Previous constructions either requires strong complexity assumptions [15, 16] or the existence of enhanced trapdoor permutations secure against super-polynomial time [1].

Moreover, Our subEXP-secure protocols have optimal round-complexity. The construction relies on the existence of semi-honest OT protocols that are secure for subEXP time (i.e., semi-honest OT protocol that are simulatable by subEXP-time simulator and the simulation is indistinguishable to the real execution to subEXP-time distinguishers). This assumption is in fact *necessary*, in order to achieve the strong security guarantees provided by our unified framework: Protocols constructed through our unified framework admits simulation (i.e., the ideal world execution) that are indistinguishable from the real execution not only to all polynomial time distinguishers, but also to distinguishers with the same running time as the simulator; we call this *strong SPS-security*.

Constant-Round SPS Security from Poly-Time Secure OT. As discussed above, strong SPS security necessarily relies on super-polynomial time hard OT protocol. We show that, in fact, the use of super-polynomial time hardness assumption can be circumvented, when considering a weaker notion of security called *plain SPS-security*, where the simulator may take super-polynomial time, but the simulation produced is only indistinguishable w.r.t. polynomial time. (In fact, this is the security guarantee achieved in the first two positive results of SPS security in [15, 16], although they still requried super-polynomial time hardness assumptions.) Given a semi-honest OT protocol that is simulatable in subEXP-time but only indistinguishable to $\mathcal{PPT}$ distinguishers—call it a subEXP-simulatable semi-honest OT protocol—we have,

Theorem 2 (Plain SPS-Security from Polynomial-Time OT). *Assume the existence of a $t(\cdot)$-round stand-alone secure subEXP-simulatable semi-honest*

---

[4] Every one-way function can be inverted in exponential time using brute force. Therefore, by appropriately scale down the security parameter, we obtain one-way functions that can be inverted in sub-exponential time.

[5] The informal statement of our unified framework in Theorem 1 does not explicitly specify the complexity of the simulator and distinguisher, nor their relationship with the hardness of the OT in the assumption. More precisely, our unified framework holds for arbitrary classes $\mathcal{C}_{\mathsf{sim}}$ of simulators and distinguishers, assuming an OT protocol that is secure for $\mathcal{C}_{\mathsf{sim}}$. See Section 3 for a formal treatment of the security definition and statement of our unified framework in Theorem 3.

oblivious-transfer protocol. Then, for every m-ary functionality f, there exists a $O(t(\cdot))$-round protocol Π that realizes f with plain subEXP-time-simulation security.

Recently, Canetti, Lin, and Pass in [30] showed how to achieve plain SPS-security, assuming only enhanced trapdoor permutations; however, their construction requires polynomially many communication rounds, whereas our construction yields constant-round protocols assuming that the underlying OT protocol has constant rounds. In concurrent and independent work, Garg, Goyal, Jain and Sahai [31], also present a construction of constant-round SPS secure protocols; but they additionally assume the existence of collision resistant hash functions besides from that of semi-honest OT.[6] Finally, we remark that our assumption is again tight: secure protocols with plain subEXP-time-simulation security imply OT protocols that can be simulated using subEXP time.

Password-Key Exchange from OT. As another application of our unified framework, we consider another line of relaxation—bounded concurrency—that is, in the concurrent execution of protocols, there is *a priori* bound on the total number of sessions that may coexist at any time point. This line of relaxation has been previously considered in several works [8, 25, 32, 33]; they showed how to construct bounded-concurrent secure computation using non black-box techniques, based on the existence of collision resistant hash functions. We show that in fact, the model of bounded concurrency can be cast as a special case of our generalized model of UC security, by considering a restricted class of environment that respects the bound m_2 on the total number of concurrent executions, and additionally only exchanges a bounded number m_1 of messages with the the adversary. We call this the (m_1, m_2)-bounded concurrency model. Therefore, by constructing a $O(m_1 + m_2)$ UC-puzzle in this model, we immediately obtain the following feasibility result.

Corollary 3 (Bounded Concurrency Model). *Let m and m' be any polynomial. Assume the existence of constant-round stand-alone secure semi-honest oblivious-transfer protocol. Then, for every m-ary functionality f, there exists a $O(m_1 + m_2)$-round protocol Π that securely realizes f in the (m_1, m_2)-bounded concurrency model.*

Lindell [34] showed that $O(m)$ communication rounds are necessary for security in the $(m, 0)$-bounded concurrency model, when relying on black-box simulation techniques; therefore, our construction achieves the optimal round-complexity. Furthermore, it is shown in [35] that the existence of t-round two-party computation protocols in the $(2, 0)$-bounded concurrency model implies the existence of t Password-Authenticated Key-Exchange (PAKE) protocols. Therefore, we obtain $O(t)$-round PAKE protocols from any t-round semi-honest OT. Combined with the result of Nguyen [2] that t-round PAKE implies $O(t)$-round OT, this

---

[6] Their proof techniques, however, are significantly different, and it would seem that an advantage of their approach is that they not rely on non-uniform reductions to an as large extent as we do.

resolves the complexity of PAKE protocols. Previous constructions of PAKE protocols assume stronger assumptions, namely, the existence of enhanced trapdoor permutations and collision resistant hash functions [35]. Another related work due to Goyal, Jain and Ostrovsky [36] considered a weaker notion of security[7], and constructed PAKE protocols satisfying the weaker notion in the unbounded concurrent setting based on collision resistant hash functions.

1.3   Outline

We refer the reader to [1] for a formal definition of the generalized model of UC-security, and notions of UC-puzzle and $\mathcal{SNMWI}$ protocols. In Section 2 provide an overview of our techniques. In Section 3, we present our main result that general UC security can be based on sh-OT protocols, and provide a proof sketch. The remaining results and formal proofs will appear in the full version.

2   Techniques

2.1   The LPV Approach

By relying on previous results [4, 9, 25, 33, 37] the construction of a UC secure protocol for realizing any multi-party functionality reduces to the task of realizing the "ideal Zero-Knowledge functionality", which amounts to constructing a zero-knowledge protocol that is both *concurrently simulatable* and *concurrently simulation-extractable*—namely, we can concurrently extract a witness from every convincing proof given by the adversary, even if it receives multiple concurrent *simulated* proofs. The "simulation" part is usually easy to achieve; as shown in [1], it suffices to provide the simulator a single "trapdoor". This is formalized by the notion of a UC-puzzle in [1], which, intuitively, is a protocol that has the property that no adversary can successfully complete the puzzle and also obtain a trapdoor, but there is a simulator who can generate puzzle transcripts (distributed statistically close to real transcripts) together with trapdoors; the former is called the *soundness* property and the latter called the *statistical simulation* property. However, obtaining "simulation-soundness" it significantly harder. In [1], the authors achieve this in two steps: First construct a "special-purpose" zero-knowledge protocol that is *concurrently simulation-sound*—namely, even if an adversary receives multiple concurrent simulated proofs, it can not prove any false statements; then, strengthen security to get simulation-extractability.

The first step relies on a primitive called strong non-malleable witness-indistinguishable ($\mathcal{SNMWI}$) arguments, which captures the *non-malleability* property w.r.t. strongly witness indistinguishable proofs. Informally, a $\mathcal{SNMWI}$ argument ensures that no man-in-the-middle adversary can correlate the witness it uses in a

---

[7] More precisely, the security notion of [36] is defined through the simulation paradigm where the simulator may rewind the trusted functionality, for instance, the ideal PAKE functionality, for a limited number of times, whereas we achieve full security without rewinding. On the other hand, their protocols are secure in unbounded concurrent setting, however, ours are only secure in bounded concurrent setting.

proof with the witness in the proof it receives. It is shown in [1] that $\mathcal{SNMWI}$ arguments can be constructed from non-malleable commitments. At a high-level, the simulation-sound protocol follows the Feige-Shamir paradigm, in which the verifier first sends a UC-puzzle to establish a "trapdoor" (that is, the puzzle answer), and then the prover proves that either the statement is true or it knows a trapdoor, using a $\mathcal{SNMWI}$ argument[8]. In essence, the UC-puzzle enables concurrent simulation: A simulator can simulate the puzzle executions with the verifier to obtain corresponding answers, and then use them as trapdoors to successfully simulate the $\mathcal{SNMWI}$ arguments. On the other hand, the $\mathcal{SNMWI}$ property ensures simulation-soundness: Even if the adversary receives $\mathcal{SNMWI}$ proofs using the trapdoors as "fake witnesses", the adversary does not do the same.

The second step in [1] enhances the security by employing the compilation technique of [33, 37, 38], which transforms a concurrently simulation-sound protocol into one that is concurrently simulation-extractable, using enhanced trapdoor permutations (TDP).

2.2   UC-Security from Semi-honest OT

In this work, we weaken the assumption that UC security relies on, by providing a new compilation technique for transforming a simulation-sound protocol into a simulation-extractable one, relying only on stand-alone semi-honest oblivious transfer (sh-OT) protocols. Our compilation technique uses similar ideas as that in [21, 22] that achieves extractability using OT; furthermore, interestingly, though our compilation technique is non-black-box, it is inspired by the black-box compilation technique used in [39, 40] for transforming a sh-OT protocol into one secure against malicious adversaries (m-OT protocol). At a very high-level, we use the idea of having an OT execution with two random inputs at the prover's side (acting as the sender) and fixed input index 1 at the verifier's side (acting as the receiver), and later letting the prover use the second random input to hide the witness. This idea leads to a simple protocol as, even if the verifier deviates from the honest behavior in the OT execution, it learns no information of the witness; therefore, it suffices to require the verifier to prove of its honest behavior after the OT execution (instead of giving a proof after every message in the OT execution as the standard technique requires). Next we explain our compilation technique in more details.

First, it follows from standard techniques that the existence of a sh-OT protocols implies the existence of a full-fledged OT protocol against malicious adversaries (m-OT for short). Then given a simulation-sound ZK (ssZK) protocol, our compilation technique outputs a protocol $\langle P, V \rangle$ as follows: In the first stage, the prover and the receiver participates in an execution of a m-OT protocol where the prover acts as the OT sender using two random inputs r_1 and r_2 and the verifier acts as the OT receiver choosing the first input; in the second stage, the

---

[8] The actually protocol is more complicated, as the notion of $\mathcal{SNMWI}$ arguments are only defined with respect to languages with unique witness. But for an intuitive explanation of high-level ideas here, we omit the complication.

verifier proves that it has used input index 1 in the OT execution using the ssZK protocol; if the proof is accepting, the prover then sends the witness w padded with the second random input $w \oplus r_2$ in the third stage, followed by a proof in the fourth stage that this message XOR'ed with the second sender's input in the OT execution is indeed a valid witness of the statement being proved using again the ssZK protocol. The high level idea of the protocol $\langle P, V \rangle$ is simple. First of all, it is concurrently simulatable: To simulate a proof of statement x, a simulator can send a random string in the third stage in place of $w \oplus r_2$ and "cheats" in the proof in the last stage by relying on the concurrent-simulation property of the ssZK protocol; (it acts honestly in the first two stages). To see that $\langle P, V \rangle$ is further concurrently simulation-extractable, consider a man-in-the-middle adversary that receives many proofs, referred to as the left proofs, and gives many proofs, referred to as the right proofs, concurrently. We construct a simulator-extractor (which eventually corresponds to the simulator of our UC secure protocols) that concurrently simulates all the left proofs as described above and extracts a witness from every convincing right-proof as follows: In a right proof, the simulator-extractor (acting as the verifier) chooses the *second input* in the OT execution and "cheats" in the proof in the second stage relying again on the concurrent simulation property of the ssZK protocol; it then recovers the witness by simply XORing the third stage message with the second input it obtains in the OT execution. To show that simulator-extractor always extracts valid witnesses from the adversary, it boils down to show that the adversary is never able to prove a false statement using the ssZK protocol, even amid simulation, which essentially relies on the simulation-soundness property of the ssZK protocol.

However, some subtleties arise: The simulator-extractor simulates for the adversary both proofs of the ssZK protocol and OT executions. The simulation-soundness property only guarantees that the adversary cannot cheat when receiving simulated proofs of the ssZK protocols, but not simulated OT executions. (This problem is in the same spirit as the problems encountered in [41–43] when using non-malleable commitments as a sub-protocol in a larger protocol.) To solve this problem, we enhance the security of our ssZK protocol so that it is also simulation-sound w.r.t. the OT protocol—namely, even when the adversary receives many simulated executions of the OT protocol, it still cannot prove any false statement. In fact, we will design a protocol that is simulation-sound both w.r.t. itself and to any protocols with a fixed bounded number of rounds; this is achieved by relying on a notion of k-robust $\mathcal{SNMWI}$ protocol, which is a $\mathcal{SNMWI}$ protocol that additionally guarantees that no adversary can correlates the witness it uses in a proof with the "secret" in a k-round interaction it participates in, provided that messages in that interaction are indistinguishable (when generated with different secrets). This notion is in analogy to the notion of k-robust non-malleable commitments [42]; and as we show, can be realized using a k-robust non-malleable commitment scheme. Then since as shown in [42], k-robust non-malleable commitment can be constructed from the minimal assumption of OWF, so can k-robust $\mathcal{SNMWI}$ protocols. Finally, we remark that this problem of robustness is not present in [1]; there, the compilation technique of [25, 33, 37] only implicitly requires the ssZK protocol to be simulation-sound w.r.t. non-interactive protocols, which is satisfied by any ssZK protocol that is an argument of knowledge (as required by the compilation technique).

An additional issue that we encounter is that for the above argument to go through, we need the OT protocol to satisfy some additional properties. More precisely, recall that the proof of concurrently simulatability of $\langle P, V \rangle$ requires showing that as long as the adversary can prove that it has acted honestly in the OT execution with input 1, the sender's second random input is completely hidden. At a first glance, it seems that this follows directly from the security against malicious receiver of the OT protocol. However, it may be possible for a malicious receiver to obtain the second input in the OT execution, but later explain its behavior with input 1. Fortunately, the security property that we need is exactly captured by the notion of *defensible privacy for the receiver* introduced by [39], which, roughly speaking, ensures that as long as a malicious receiver can output a good "defense"—that is, explaining its behavior as an honest receiver with input b and random tape σ—at the end of the OT execution, then the honest sender's other input $b' \neq b$ must remain hidden. Furthermore, to show that $\langle P, V \rangle$ is simulation extractable, we need the OT protocol to satisfy that as long as a malicious sender can output a good "defense", with inputs r_1, r_2 and random tape σ', after an OT execution, the honest receiver with input b must obtain r_b. To formalize this security property, we adapt the notion of defensible privacy of [39] to consider the correctness requirement; we called it the *defensible correctness property*. Therefore, our compilation technique relies on a m-OT protocol that is defensibly private for the receiver and defensibly correct for the sender; we show that such a protocol is implied by the existence of sh-OT protocols.

Constant-round SPS-security from polynomial-time hard sh-OT: In [1], the authors constructed SPS-secure protocols with strong indistinguishability: Real-world executions of these protocols are indistinguishable to ideal-world simulations, against distinguishers of the same time complexity of the simulator, which is super-polynomial. To obtain a model of security that can be implemented in constant rounds from standard polynomial time harness assumptions (in the plain model), we weaken the generalized model of UC security in [1] to require only *plain indistinguishability* against $\mathcal{PPT}$ distinguishers. However, even with this weakening, at the first glance, it is still unclear how to achieve plain-SPS-security from only polynomial time hardness assumptions. Let us illustrate the difficulty using the above described protocol $\langle P, V \rangle$ that implements the ideal ZK functionality.

In order to simulate the view of and extract witnesses from a man-in-the-middle adversary, the simulator-extractor of $\langle P, V \rangle$ simulates all the ssZK proofs to the adversary, as well as all the OT executions it participates in. The latter can be simulated efficiently, but, the concurrent simulation of the ssZK arguments takes super-polynomial time in the SPS-model. Then it seems that in order to apply the security guarantees of the sh-OT protocol and the simulation soundness property of the ssZK protocol (to show that the view of the adversary is indistinguishable and it never proves any false statement), we need the security of the sh-OT and ssZK protocols to hold against super-polynomial machines, (since the adversary, though a $\mathcal{PPT}$ machine itself, receives many simulated proofs generated in super-polynomial time). Roughly, this is the technical reason why the LPV protocol relies on super-poly hardness assumption.

To get around this problem, we exploit the structure of the ssZK protocol constructed in [1]. Recall that it consists of a UC-puzzle execution where the verifier establishes a trapdoor, followed by a proof using the $\mathcal{SNMWI}$ argument that either the statement is true or a trapdoor is known. The key observation is that when simulating a proof of this protocol, only the simulation of the UC puzzle takes super-polynomial time; once a trapdoor is obtained, the rest of the simulation can be done efficiently. Therefore, if we modify the protocol $\langle P, V \rangle$ to have the puzzle executions in the two ssZK proofs sent at the beginning of the protocol—call it the preamble phase of the protocol—we obtain a protocol $\langle P', V' \rangle$ that has the same property: Only the preamble phase of the protocol takes super-polynomial time to simulate (the rest can be simulated efficiently given the puzzle answers). With this simple change, now we only need the sh-OT and $\mathcal{SNMWI}$ protocols to be secure for polynomial-time. To illustrate our idea, consider first the stand-alone setting. To show that $\langle P', V' \rangle$ is zero-knowledge, we rely on the "hiding" property of the sh-OT and the $\mathcal{SNMWI}$ protocols; since the simulation of the preamble phase happens before them, and thus the puzzle answers can be fixed non-uniformly, it suffices to rely on "hiding" against non-uniform $\mathcal{PPT}$ machines.

We use the same idea to prove the concurrent security of $\langle P', V' \rangle$: Establish the simulation-extractability property of $\langle P', V' \rangle$ in a sequence of hybrids that gradually simulate each session in two steps (the preamble phase first and then the rest) in a clever order. More precisely, consider a man-in-the-middle adversary that participates in m proofs; order all the proofs according to the sequence in which their preamble phases *completes*. Then consider a sequence of $2m + 1$ hybrids $H_0, \ldots, H_{2m+1}$'s, where in hybrid H_{2i} the first i sessions are simulated, and in hybrids H_{2i+1} and $H_{2(i+1)}$ (in addition to the first i sessions) the preamble phase and the rest of the $(i + 1)^{\text{th}}$ session are simulated respectively. To show that $\langle P', V' \rangle$ is simulation-extractable, it boils down to prove that every two subsequent hybrids are indistinguishable and the adversary never proves a false statement using the $\mathcal{SNMWI}$ argument in all hybrids. From hybrid H_{2i} to H_{2i+1} this follows directly from the statistical simulation property of the UC-puzzles. From hybrid H_{2i+1} to $H_{2(i+1)}$, this relies on the security of the sh-OT and $\mathcal{SNMWI}$ protocol executions in the $(i + 1)^{\text{th}}$ session; since in these two hybrids, only puzzles in the first $i + 1$ sessions are simulated, which happens before the OT and $\mathcal{SNMWI}$ executions in the $(i + 1)^{\text{th}}$ session and can be fixed non-uniformly, we only need the security of the OT and $\mathcal{SNMWI}$ protocols to hold against non-uniform $\mathcal{PPT}$ machines. Given that $\mathcal{SNMWI}$ arguments are implied by sh-OT protocols, $\langle P', V' \rangle$ implements the ideal ZK functionality with plain-SPS-security based on only polynomial-time hard sh-OT protocols.

Now, it seems that by simply combining $\langle P', V' \rangle$ with previous constructions of UC secure protocols Π that uses the ideal ZK functionality IdealZK [4, 9, 25, 33, 37], we can obtain constant-round plain-SPSsecure computation from sh-OT protocols. Unfortunately, previous constructions rely on the existence of sh-OT protocols; if composing them with $\langle P', V' \rangle$ in the straightforward way—replacing every IdealZK call in Π with an invocation of $\langle P', V' \rangle$—for the composed protocol $\Pi' = \Pi^{\mathsf{IdealZK}/\langle P',V' \rangle}$ to be secure in general, we need Π to be secure against super-poly time, which requires super-poly hard sh-OT! To address this, we modify the composed protocol Π' as we did to the protocol $\langle P, V \rangle$: Consider a protocol Π'' that is identical to Π' except that all the puzzle-executions in the invocations of $\langle P', V' \rangle$ are executed in parallel at the beginning of the protocol, call this again the preamble phase of the protocol; now Π'' has the property that only its preamble phase takes super-polynomial time to simulate, and the rest can be simulated efficiently with puzzle answers. Therefore, by considering a similar sequence of hybrids as in the proof of $\langle P', V' \rangle$, we prove the security of Π''.

UC-security with bounded concurrency: Let (m_1, m_2)-bounded concurrency denote a scenario where the UC environment communicates in at most m_1 rounds with the adversary, and when at most m_2 executions of some protocol take place. It follows from our unified framework that to construct secure computation protocols in this model, the key is to construct a UC-puzzle. Towards this, let us first examine a simple case where during the execution of any session, the total number of messages the adversary receives that *do not* belong to any UC-puzzle is bounded by a fixed number m. (These messages include ones from the environment and ones belonging to the *non-puzzle* part of the other sessions.) In this case, we can design the UC-puzzle as follows: The puzzle receiver sends the image $f(r)$ of a random value r through a OWF, followed by $m + 1$ witness hiding proof of knowledge (POK) of r; the answer to this puzzle is simply a pre-image of $f(r)$. It follows from the one-wayness of f and the witness-hiding property of the proofs that no adversary (acting as a puzzle receiver) can complete a puzzle and obtain an answer. But, there is a puzzle simulator that can simulate many concurrent puzzle executions with an adversary (acting as the puzzle sender) and extract an answer immediately after each accepting puzzle: The simulator emulates the puzzle receivers for the adversary honestly, and rewinds the adversary at one of the POK's to extract an answer. Since in this simple case, there are more, $m+1$, POK's than the number m of other non-puzzle messages, there must be one POK from which the simulator can rewind to extract an answer without needing to simulate any non-puzzle messages (messages belong to a puzzle can be simulated trivially by following the honest receiver strategy in the rewindings). Finally, we show that, in fact, this simple case always holds. As we will see later, secure computation protocols produced by our framework contains a constant number c of non-puzzle messages if the underlying sh-OT protocol has *constant* rounds. Therefore, in the (m_1, m_2)-bounded concurrent model, the total number of non-puzzle messages is bounded by $m_1 + cm_2$, yielding $O(m_1 + m_2)$-round bounded concurrent secure computation protocols.

3   UC Security Based on Stand-Alone Semi-honest OT

We consider the $(\mathcal{C}_{\mathsf{env}}, \mathcal{C}_{\mathsf{sim}})$-UC-model introduced in [1]. The model extends the framework of universal composability [5]. The key difference from UC lies in that in UC, the environment is modeled as a non-uniform $\mathcal{PPT}$ machine and the ideal-model adversary (or simulator) as a uniform $\mathcal{PPT}$ machines, whereas in the general model, the environment and the simulator are allowed to be from arbitrary complexity classes $\mathcal{C}_{\mathsf{env}}$ and $\mathcal{C}_{\mathsf{sim}}$ respectively. (Note, however, that the adversary is still uniform $\mathcal{PPT}$.) One important affect of this change is that the UC composition theorem [5] no longer holds; as a result, the stand-alone security of a protocol does not directly imply the concurrent security. In remedy, in the general model, an environment executing a protocol π can start many instances of the protocol, and thus implementing a functionality $\mathcal{F}$ in the general model means directly implementing the multi-session extension $\hat{\mathcal{F}}^9$ of $\mathcal{F}$. We focus only on static adversaries. Let $cl(\mathcal{C}_{\mathsf{env}}, \mathcal{C}_{\mathsf{sim}})$ represent the closure of $\mathcal{C}_{\mathsf{env}}$ and $\mathcal{C}_{\mathsf{sim}}$ that includes all computations by $\mathcal{PPT}$ oracle Turing machines M with oracle access to $\mathcal{C}_{\mathsf{env}}, \mathcal{C}_{\mathsf{sim}}$. In this section, we prove the following main technical theorem.

Theorem 3. *Assume the existence of a t_P-round $(\mathcal{C}_{\mathsf{env}}, \mathcal{C}_{\mathsf{sim}})$-secure UC-puzzle in a $\mathcal{G}$-hybrid model, and a t_{OT}-round stand-alone sh-OT protocol secure w.r.t $cl(\mathcal{C}_{\mathsf{sim}}, \mathcal{C}_{\mathsf{env}})$. Then, for every "well-formed" functionality $\mathcal{F}$, there exists a $O(t_P + t_{OT})$-round protocol Π in the $\mathcal{G}$-hybrid model that $(\mathcal{C}_{\mathsf{env}}, \mathcal{C}_{\mathsf{sim}})$-UC-realizes $\mathcal{F}$.*

3.1   Proof of Theorem 3

Recall that the IdealZK functionality parameterized with a language L implements the function $\mathsf{ZK}^L\left((x, w), x\right) = (\perp, b)$, where $b = 1$ if w is a valid witness for the membership of x in L and 0 otherwise. Then Theorem 3 follows from the following two lemmas.

Lemma 1 (IdealZK-Lemma). *Assume the existence of t-round stand-alone secure sh-OT secure w.r.t $cl(\mathcal{C}_{\mathsf{env}}, \mathcal{C}_{\mathsf{sim}})$. Then, for every well-formed functionality $\mathcal{F}$, there exists a $O(t)$-round protocol Π in the ZK-Hybrid model that $(\mathcal{C}_{\mathsf{env}}, \mathcal{C}_{\mathsf{sim}})$-UC-realizes $\mathcal{F}$.*

Lemma 2 (Puzzle-Lemma). *Let Π' be a protocol in the IdealZK model. Assume the existence of a $(\mathcal{C}_{\mathsf{env}}, \mathcal{C}_{\mathsf{sim}})$-secure t_P-round puzzle $\langle S, R \rangle$ in a $\mathcal{G}$-hybrid model, a t_{OT}-round stand-alone sh-OT protocol $\langle S_{OT}, R_{OT} \rangle$ that is secure w.r.t $cl(\mathcal{C}_{\mathsf{sim}}, \mathcal{C}_{\mathsf{env}})$, and a t_{WI}-round t_{OT}-robust $\mathcal{SNMWI}$ protocol $\langle P_s, V_s \rangle$ secure w.r.t $cl(\mathcal{C}_{\mathsf{sim}}, \mathcal{C}_{\mathsf{env}})$. Then, there exists a $O(t_P + t_{WI} + t_{OT})$-round protocol Π in the $\mathcal{G}$-hybrid that $(\mathcal{C}_{\mathsf{env}}, \mathcal{C}_{\mathsf{sim}})$-UC emulates Π'.*

The first lemma is implicit in previous works [4, 9, 25, 44] for normal UC-security (i.e., (n.u.$\mathcal{PPT}$, $\mathcal{PPT}$)-UC-security) and can be easily extended to the

---

[9] Informally speaking, $\hat{F}$ emulates many independent copies of $\mathcal{F}$ running concurrently; see [1, 9] for a formal definition.

general $(\mathcal{C}_{env}, \mathcal{C}_{sim})$-UC model assuming stand-alone sh-OT protocol secure w.r.t $cl(\mathcal{C}_{sim}, \mathcal{C}_{env})$; we omit the proof (see [1] for a similar proof assuming TDP's). Next, towards proving the puzzle lemma, we provide a general transformation that transforms any protocol Π in the ZK-Hybrid model into a protocol Π' in the real model using a special-purpose zero-knowledge protocol that is "concurrently simulatable" and "concurrently simulation-extractable".

Special-purpose ZK Protocol $\langle P, V \rangle$. The construction of $\langle P, V \rangle$ relies on the following three building blocks; all with security against class $cl(\mathcal{C}_{env}, \mathcal{C}_{sim})$. (1) A t'-round m-OT protocol $\langle S_{OT}, R_{OT} \rangle$ that is defensibly private for the receiver and defensibly correct for the sender; it follows from standard techniques [4, 18] that such a protocol exists assuming t_{OT}-round sh-OT protocols, and $t' = O(t_{OT})$. (2) A t'-robust $\mathcal{SNMWI}$ protocol $\langle P_s, V_s \rangle$; it follows from a similar proof as in [1] that such a protocol exists assuming OWF's and the round-complexity is of $O(t')$. (We defer the formal construction and proof of such m-OT and robust $\mathcal{SNMWI}$ to the full version.) (3) A $(\mathcal{C}_{env}, \mathcal{C}_{sim})$-secure puzzle $(\langle S, R \rangle, \mathcal{R})$ in a $\mathcal{G}$ hybrid model. For simplicity of exposition, our description below rely on a statistically binding commitment scheme com that has *unique decommitment*, that is the transcript of the commitment not only uniquely decides the value committed to inside but also the decommitment with overwhelming probability; but the protocol can be easily modified to work with any arbitrary statistically binding commitment (see the full version for more details). Then, the special-purpose ZK protocol $\langle P, V \rangle$ for a **NP** relation R_L proceeds as follows: To prove a statement x, the prover and verifier with identities id_P and id_V, and additional auxiliary input $w = R_L(x)$ for the prover, interacts in six stages.

Stage 1: The Prover and Verifier participate in a puzzle-interaction where the Verifier assumes the role of the sender and the Prover as the receiver. Let $\text{TRANS}_{V \to P}$ be the transcript of the messages exchanged.

Stage 2: The Prover and Verifier participate in a second puzzle-interaction with the roles reversed, i.e. the Prover is the sender and the Verifier is the receiver. Let $\text{TRANS}_{P \to V}$ be the transcript of the messages exchanged.

Stage 3: The Prover first selects two random string $r_1, r_2 \in \{0, 1\}^n$. Then the Prover and Verifier interact using $\langle S_{OT}, R_{OT} \rangle$, where the Prover is the sender with inputs (r_1, r_2) and the Verifier is the receiver with input 1. Let TRANS_{OT} be the transcript of the messages exchanged.

Stage 4: The Verifier commits to s using com. Then it proves using the protocol $\langle P_s, V_s \rangle$ and identity id_V, the statement that it either committed to a string s that contains a valid witness establishing the verifiers input as index 1 in TRANS_{OT} and the string output by the receiver at the end of the Stage 3 protocol or a string s such that $(s, \text{TRANS}_{P \to V}) \in \mathcal{R}$.

Stage 5: The Prover sends the string $r = r_2 \oplus w$ in the clear.

Stage 6: The Prover commits to s' using com. Then the prover proves using the protocol $\langle P_s, V_s \rangle$ and identity id_P, the statement that it either committed to a string s' that establishes that the inputs used by the prover in TRANS_{OT} is (r_1, r_2) such that $r_2 \oplus r \in R_L(x)$ or a string s' such that $(s', \text{TRANS}_{V \to P}) \in \mathcal{R}$.

Realizing the IdealZK-functionality: Given any protocol Π' in ZK-Hybrid model and the special-purpose zero-knowledge protocol $\langle P, V \rangle$, the protocol Π in the real model is constructed from Π' by instantiating the IdealZK functionality using $\langle P, V \rangle$. All invocations of the IdealZK functionality in which P_i provers to P_j a statement x using witness w is replaced with an subroutine call of $\langle P, V \rangle$ between P_i and P_j where P_i proves the statement x using witness w to P_j, using identities $\mathsf{id}_P = i$ and $\mathsf{id}_V = j$ respectively. The formal security proof that Π emulates Π' in the ZK-Hybrid will appear in the full version.

References

1. Lin, H., Pass, R., Venkitasubramaniam, M.: A unified framework for concurrent security: universal composability from stand-alone non-malleability. In: STOC, pp. 179–188 (2009)
2. Nguyen, M.-H.: The Relationship Between Password-Authenticated Key Exchange and Other Cryptographic Primitives. In: Kilian, J. (ed.) TCC 2005. LNCS, vol. 3378, pp. 457–475. Springer, Heidelberg (2005)
3. Yao, A.C.C.: How to generate and exchange secrets (extended abstract). In: FOCS, pp. 162–167 (1986)
4. Goldreich, O., Micali, S., Wigderson, A.: How to play any mental game or a completeness theorem for protocols with honest majority. In: STOC, pp. 218–229 (1987)
5. Canetti, R.: Universally composable security: A new paradigm for cryptographic protocols. In: FOCS, pp. 136–145 (2001)
6. Canetti, R., Fischlin, M.: Universally Composable Commitments. In: Kilian, J. (ed.) CRYPTO 2001. LNCS, vol. 2139, pp. 19–40. Springer, Heidelberg (2001)
7. Canetti, R., Kushilevitz, E., Lindell, Y.: On the Limitations of Universally Composable Two-party Computation without Set-up Assumptions. In: Biham, E. (ed.) EUROCRYPT 2003. LNCS, vol. 2656, pp. 68–86. Springer, Heidelberg (2003)
8. Lindell, Y.: General composition and universal composability in secure multi-party computation. In: FOCS, pp. 394–403 (2003)
9. Canetti, R., Lindell, Y., Ostrovsky, R., Sahai, A.: Universally composable two-party and multi-party secure computation. In: STOC, pp. 494–503 (2002)
10. Barak, B., Canetti, R., Nielsen, J.B., Pass, R.: Universally composable protocols with relaxed set-up assumptions. In: FOCS, pp. 186–195 (2004)
11. Canetti, R., Dodis, Y., Pass, R., Walfish, S.: Universally Composable Security with Global Setup. In: Vadhan, S.P. (ed.) TCC 2007. LNCS, vol. 4392, pp. 61–85. Springer, Heidelberg (2007)
12. Kalai, Y.T., Lindell, Y., Prabhakaran, M.: Concurrent general composition of secure protocols in the timing model. In: STOC, pp. 644–653 (2005)
13. Canetti, R., Pass, R., Shelat, A.: Cryptography from sunspots: How to use an imperfect reference string. In: FOCS, pp. 249–259 (2007)
14. Pass, R.: Simulation in Quasi-Polynomial Time, and Its Application to Protocol Composition. In: Biham, E. (ed.) EUROCRYPT 2003. LNCS, vol. 2656, pp. 160–176. Springer, Heidelberg (2003)
15. Prabhakaran, M., Sahai, A.: New notions of security: achieving universal composability without trusted setup. In: STOC, pp. 242–251 (2004)
16. Barak, B., Sahai, A.: How to play almost any mental game over the net - concurrent composition via super-polynomial simulation. In: FOCS, pp. 543–552 (2005)

17. Micali, S., Pass, R., Rosen, A.: Input-indistinguishable computation. In: FOCS, pp. 367–378 (2006)
18. Goldreich, O.: Foundations of Cryptography — Basic Tools. Cambridge University Press (2001)
19. Kilian, J.: Founding cryptography on oblivious transfer. In: STOC, pp. 20–31 (1988)
20. Ishai, Y., Prabhakaran, M., Sahai, A.: Founding Cryptography on Oblivious Transfer – Efficiently. In: Wagner, D. (ed.) CRYPTO 2008. LNCS, vol. 5157, pp. 572–591. Springer, Heidelberg (2008)
21. Damgård, I., Nielsen, J.B., Orlandi, C.: On the Necessary and Sufficient Assumptions for UC Computation. In: Micciancio, D. (ed.) TCC 2010. LNCS, vol. 5978, pp. 109–127. Springer, Heidelberg (2010)
22. Maji, H.K., Prabhakaran, M., Rosulek, M.: A Zero-One Law for Cryptographic Complexity with Respect to Computational UC Security. In: Rabin, T. (ed.) CRYPTO 2010. LNCS, vol. 6223, pp. 595–612. Springer, Heidelberg (2010)
23. Goyal, V.: Constant round non-malleable protocols using one way functions. In: STOC, pp. 695–704 (2011)
24. Dwork, C., Naor, M., Sahai, A.: Concurrent zero-knowledge. J. ACM 51(6), 851–898 (2004)
25. Pass, R.: Bounded-concurrent secure multi-party computation with a dishonest majority. In: STOC, pp. 232–241. ACM, New York (2004)
26. Katz, J.: Which Languages Have 4-Round Zero-Knowledge Proofs? In: Canetti, R. (ed.) TCC 2008. LNCS, vol. 4948, pp. 73–88. Springer, Heidelberg (2008)
27. Groth, J., Ostrovsky, R.: Cryptography in the Multi-string Model. In: Menezes, A. (ed.) CRYPTO 2007. LNCS, vol. 4622, pp. 323–341. Springer, Heidelberg (2007)
28. Goyal, V., Ishai, Y., Sahai, A., Venkatesan, R., Wadia, A.: Founding Cryptography on Tamper-Proof Hardware Tokens. In: Micciancio, D. (ed.) TCC 2010. LNCS, vol. 5978, pp. 308–326. Springer, Heidelberg (2010)
29. Damgård, I., Groth, J.: Non-interactive and reusable non-malleable commitment schemes. In: STOC, pp. 426–437 (2003)
30. Canetti, R., Lin, H., Pass, R.: Adaptive hardness and composable security in the plain model from standard assumptions. In: FOCS, pp. 541–550 (2010)
31. Garg, S., Goyal, V., Jain, A., Sahai, A.: Concurrently Secure Computation in Constant Rounds. In: Pointcheval, D., Johansson, T. (eds.) EUROCRYPT 2012. LNCS, vol. 7237, pp. 99–116. Springer, Heidelberg (2012)
32. Barak, B.: How to go beyond the black-box simulation barrier. In: FOCS, pp. 106–115 (2001)
33. Pass, R., Rosen, A.: Bounded-concurrent secure two-party computation in a constant number of rounds. In: FOCS. p. 404 (2003)
34. Lindell, Y.: Lower bounds and impossibility results for concurrent self composition. J. Cryptology 21(2), 200–249 (2008)
35. Barak, B., Canetti, R., Lindell, Y., Pass, R., Rabin, T.: Secure computation without authentication. J. Cryptology 24(4), 720–760 (2011)
36. Goyal, V., Jain, A., Ostrovsky, R.: Password-Authenticated Session-Key Generation on the Internet in the Plain Model. In: Rabin, T. (ed.) CRYPTO 2010. LNCS, vol. 6223, pp. 277–294. Springer, Heidelberg (2010)
37. Lindell, Y.: Bounded-concurrent secure two-party computation without setup assumptions. In: STOC, pp. 683–692 (2003)
38. Barak, B., Lindell, Y.: Strict polynomial-time in simulation and extraction. SIAM J. Comput. 33(4), 738–818 (2004)

39. Ishai, Y., Kushilevitz, E., Lindell, Y., Petrank, E.: Black-box constructions for secure computation. In: STOC, pp. 99–108 (2006)
40. Haitner, I.: Semi-honest to Malicious Oblivious Transfer—The Black-Box Way. In: Canetti, R. (ed.) TCC 2008. LNCS, vol. 4948, pp. 412–426. Springer, Heidelberg (2008)
41. Lin, H., Pass, R., Tseng, W.-L.D., Venkitasubramaniam, M.: Concurrent Non-Malleable Zero Knowledge Proofs. In: Rabin, T. (ed.) CRYPTO 2010. LNCS, vol. 6223, pp. 429–446. Springer, Heidelberg (2010)
42. Lin, H., Pass, R.: Non-malleability amplification. In: STOC, pp. 189–198 (2009)
43. Wee, H.: Black-box, round-efficient secure computation via non-malleability amplification. To appear in FOCS 2010 (2010)
44. Beaver, D., Micali, S., Rogaway, P.: The round complexity of secure protocols (extended abstract). In: STOC, pp. 503–513 (1990)

Four-Dimensional Gallant-Lambert-Vanstone Scalar Multiplication

Patrick Longa[1] and Francesco Sica[2]

[1] Microsoft Research, USA
plonga@microsoft.com
[2] Nazarbayev University, Kazakhstan
francesco.sica@nu.edu.kz

Abstract. The GLV method of Gallant, Lambert and Vanstone (CRYPTO 2001) computes any multiple kP of a point P of prime order n lying on an elliptic curve with a low-degree endomorphism Φ (called GLV curve) over $\mathbb{F}_p$ as $kP = k_1 P + k_2 \Phi(P)$, with $\max\{|k_1|, |k_2|\} \leq C_1 \sqrt{n}$, for some explicit constant $C_1 > 0$. Recently, Galbraith, Lin and Scott (EUROCRYPT 2009) extended this method to all curves over $\mathbb{F}_{p^2}$ which are twists of curves defined over $\mathbb{F}_p$. We show in this work how to merge the two approaches in order to get, for twists of any GLV curve over $\mathbb{F}_{p^2}$, a four-dimensional decomposition together with fast endomorphisms Φ, Ψ over $\mathbb{F}_{p^2}$ acting on the group generated by a point P of prime order n, resulting in a proven decomposition for any scalar $k \in [1, n]$ given by $kP = k_1 P + k_2 \Phi(P) + k_3 \Psi(P) + k_4 \Psi\Phi(P)$ with $\max_i(|k_i|) < C_2 n^{1/4}$, for some explicit $C_2 > 0$. Remarkably, taking the best C_1, C_2, we obtain $C_2/C_1 < 412$, independently of the curve, ensuring in theory an almost constant relative speedup. In practice, our experiments reveal that the use of the merged GLV-GLS approach supports a scalar multiplication that runs up to 50% times faster than the original GLV method. We then improve this performance even further by exploiting the Twisted Edwards model and show that curves originally slower may become extremely efficient on this model. In addition, we analyze the performance of the method on a multicore setting and describe how to efficiently protect GLV-based scalar multiplication against several side-channel attacks. Our implementations improve the state-of-the-art performance of point multiplication for a variety of scenarios including side-channel protected and unprotected cases with sequential and multicore execution.

Keywords: Elliptic curves, GLV-GLS method, scalar multiplication, Twisted Edwards curve, side-channel protection, multicore computation.

1 Introduction

The Gallant-Lambert-Vanstone (GLV) method is a generic approach to speed up the computation of scalar multiplication on some elliptic curves defined over fields of large prime characteristic. Given a curve with a point P of prime order n, it consists essentially in an algorithm that finds a decomposition of an

X. Wang and K. Sako (Eds.): ASIACRYPT 2012, LNCS 7658, pp. 718–739, 2012.

arbitrary scalar multiplication kP for $k \in [1, n]$ into two scalar multiplications, with the new scalars having only about half the bitlength of the original scalar. This immediately enables the elimination of half the doublings by employing the Straus-Shamir trick for simultaneous point multiplication.

Whereas the original GLV method as defined in [10] works on curves over $\mathbb{F}_p$ with an endomorphism of small degree (GLV curves), Galbraith-Lin-Scott (GLS) in [8] have shown that over $\mathbb{F}_{p^2}$ one can expect to find many more such curves by basically exploiting the action of the Frobenius endomorphism. One can therefore expect that on the particular GLV curves, this new insight will lead to improvements over $\mathbb{F}_{p^2}$. Indeed the GLS article itself considers four-dimensional decompositions on GLV curves with nontrivial automorphisms (corresponding to the degree one cases) but leaves the other cases open to investigation.

In this work, we generalize the GLS method to *all* GLV curves by exploiting fast endomorphisms Φ, Ψ over $\mathbb{F}_{p^2}$ acting on a cyclic group generated by a point P of prime order n to construct a proven decomposition with no heuristics involved for any scalar $k \in [1, n]$

$$kP = k_1 P + k_2 \Phi(P) + k_3 \Psi(P) + k_4 \Psi\Phi(P) \text{ with } \max_i(|k_i|) < Cn^{1/4}$$

for some explicitly computable C. In doing this we provide a reduction algorithm for the four-dimensional relevant lattice which runs in $O(\log^2 n)$ by implementing two Cornacchia-type algorithms [6,22], one in $\mathbb{Z}$, the other in $\mathbb{Z}[i]$. The algorithm is remarkably simple to implement and allows us to demonstrate an improved $C = O(\sqrt{s})$ (compared to the value obtained with LLL which is only $\Omega(s^{3/2})$). Thus, it guarantees a relative speedup independent of the curve when moving from a two-dimensional to a four-dimensional GLV method over the same underlying field. If parallel computation is available then the computation of kP can possibly be implemented (close to) four times faster in this case. When moving from two-dimensional GLV over $\mathbb{F}_p$ to the four-dimensional case over $\mathbb{F}_{p^2}$, our method still guarantees a relative speedup that is *quasi*-uniform among all GLV curves (see Section 7 for details). In fact, we present experimental results on different GLV curves that demonstrate that the relative speedup between the original GLV method and the proposed method (termed GLV-GLS in the remainder) is as high as 50%.

Twisted Edwards curves [2] are efficient generalizations of Edwards curves [7], which exhibit high-performance arithmetic. By exploiting this curve model, Galbraith, Lin and Scott showed in [9] that the GLS method can be improved in practice a further 10%, approximately (see also [19,18]). They also described how to write down j-invariant 0 and 1728 curves in Edwards form to combine a 4-dimensional decomposition with the fast arithmetic provided by this curve model. We exploit this approach and, most remarkably, lift the restriction to those special curves and show that in practice the GLV-GLS curves discussed in this work may achieve extremely high-performance and become virtually equivalent in terms of speed when written in Twisted Edwards form.

In the last years multiple works have incrementally shown the impact of using the GLS method for high performance [8,19,13]. However, it is still unclear how well the method behaves on settings where side-channel attacks are a threat. Since it is usually assumed that required countermeasures once in place degrade performance significantly, it is also unclear if the GLS method would retain its current superiority in the case of side-channel protected implementations. Here, we study this open problem and describe how to protect implementations based on the GLV-GLS method against timing attacks, cache attacks and similar ones and still achieve very high performance. The techniques discussed naturally apply to GLV-based implementations in general. Finally, we discuss different strategies to implement GLV-based scalar multiplication on modern multicore processors, and include the case in which countermeasures against side-channel attacks are required.

The presented implementations corresponding to the GLV-GLS method improve the state-of-the-art performance of point multiplication for all the cases under study: protected and unprotected versions with sequential and parallel execution. For instance, on one core of an Intel Core i7-2600 processor and at roughly 128 bits of security, we compute an *unprotected* scalar multiplication in only 91,000 cycles (which is 1.34 times faster than a previous result reported by Hu, Longa and Xu in [13]), and a *side-channel protected* scalar multiplication in only 137,000 cycles (which is 1.42 times faster than the protected implementation presented by Bernstein et al. in [3]).

Related Work. Recently, a paper by Zhou, Hu, Xu and Song [28] has shown that it is possible to combine the GLV and GLS approaches by introducing a three-dimensional version of the GLV method, which seems to work to a certain degree, with however no justification but through practical implementations. The first author together with Hu and Xu [13] studied the case of curves with j-invariant 0 and provided a bound for this particular case. Our analysis supplements [13] by considering all GLV curves and providing a unified treatment.

2   The GLV Method

In this section we briefly summarize the GLV method following [25]. Let E be an elliptic curve defined over a finite field $\mathbb{F}_q$ and P be a point on this curve with prime order n such that the cofactor $h = \#E(\mathbb{F}_q)/n$ is small, say $h \leq 4$. Let us consider Φ a non trivial endomorphism defined over $\mathbb{F}_q$ and $X^2 + rX + s$ its characteristic polynomial. In all the examples r and s are actually small fixed integers and q is varying in some family. By hypothesis there is only one subgroup of order n in $E(\mathbb{F}_q)$, implying that $\Phi(P) = \lambda P$ for some $\lambda \in [0, n-1]$, since $\Phi(P)$ has order dividing the prime n. In particular, λ is obtained as a root of $X^2 + rX + s$ modulo n.

Define the group homomorphism (the GLV reduction map)

$$\mathfrak{f} \colon \mathbb{Z} \times \mathbb{Z} \to \mathbb{Z}/n$$
$$(i, j) \mapsto i + \lambda j \pmod{n} .$$

Let $\mathcal{K} = \ker \mathfrak{f}$. It is a sublattice of $\mathbb{Z} \times \mathbb{Z}$ of rank 2 since the quotient is finite. Let $\Bbbk > 0$ be a constant (depending on the curve) such that we can find v_1, v_2 two linearly independent vectors of $\mathcal{K}$ satisfying $\max\{|v_1|, |v_2|\} < \Bbbk\sqrt{n}$, where $|\cdot|$

denotes the rectangle norm[1]. Express $(k, 0) = \beta_1 v_1 + \beta_2 v_2$, where $\beta_i \in \mathbb{Q}$. Then round β_i to the nearest integer $b_i = \lfloor \beta_i \rceil = \lfloor \beta_i + 1/2 \rfloor$ and let $v = b_1 v_1 + b_2 v_2$. Note that $v \in \mathcal{K}$ and that $u \stackrel{\text{def}}{=} (k, 0) - v$ is short. Indeed by the triangle inequality we have that

$$|u| \leq \frac{|v_1| + |v_2|}{2} < \Bbbk\sqrt{n} \ .$$

If we set $(k_1, k_2) = u$, then we get $k \equiv k_1 + k_2\lambda \pmod{n}$ or equivalently $kP = k_1 P + k_2 \Phi(P)$, with $\max(|k_1|, |k_2|) < \Bbbk\sqrt{n}$.

In [25], the optimal value of $\Bbbk$ (with respect to large values of n, i.e. large fields, keeping $X^2 + rX + s$ constant) is determined. Let $\Delta = r^2 - 4s$ be the discriminant of the characteristic polynomial of Φ. Then the optimal $\Bbbk$ is given by the following result[2] .

Theorem 1 ([25, Theorem 4]). *Assuming n is the norm of an element of $\mathbb{Z}[\Phi]$, then the optimal value of $\Bbbk$ is*

$$\Bbbk = \begin{cases} \dfrac{\sqrt{s}}{2}\left(1 + \dfrac{1}{|\Delta|}\right), & \text{if } r \text{ is odd,} \\[4mm] \dfrac{\sqrt{s}}{2}\sqrt{1 + \dfrac{4}{|\Delta|}}, & \text{if } r \text{ is even.} \end{cases}$$

3   The GLS Improvement

In 2009, Galbraith, Lin and Scott [8] realised that we do not need to have $\Phi^2 + r\Phi + s = 0$ in $\text{End}(E)$ but only in a subgroup of $E(\mathbb{F})$ for a specific finite field $\mathbb{F}$. In particular, considering $\Psi = \text{Frob}_p$ the p-Frobenius endomorphism of a curve E defined over $\mathbb{F}_p$, we know that $\Psi^m(P) = P$ for all $P \in E(\mathbb{F}_{p^m})$. While this tells nothing useful if $m = 1, 2$, it does offer new nontrivial relations for higher degree extensions. The case $m = 4$ is particularly useful here.

In this case if $P \in E(\mathbb{F}_{p^4}) \backslash E(\mathbb{F}_{p^2})$ then $\Psi^2(P) = -P$ and hence on the subgroup generated by P, Ψ satisfies the equation $X^2 + 1 = 0$. This implies that if $\Psi(P)$ is a multiple of P (which happens as soon as the order n of P is sufficiently large, say at least $2p$), we can apply the GLV approach and split again a scalar multiplication as $kP = k_1 P + k_2 \Psi(P)$, with $\max(|k_1|, |k_2|) = O(\sqrt{n})$. Contrast this with the characteristic polynomial of Ψ which is $X^2 - a_p X + p$ for some integer a_p, a non-constant polynomial to which we cannot apply as efficiently the GLV paradigm.

For efficiency reasons however one does not work with $E/\mathbb{F}_{p^4}$ directly but with $E'/\mathbb{F}_{p^2}$ isomorphic to E *over* $\mathbb{F}_{p^4}$ but not over $\mathbb{F}_{p^2}$, that is, a quadratic

---

[1] The rectangle norm of (x, y) is by definition $\max(|x|, |y|)$. As remarked in [25], we can replace it by any other metric norm. We will use the term "short" to denote smallness in the rectangle norm.

[2] There is a mistake in [25] in the derivation of $\Bbbk$ for odd values of r. This affects [25, Corollary 1] for curves E_2 and E_3, where the correct values of $\Bbbk$ are respectively $2/3$ and $4\sqrt{2}/7$.

twist over $\mathbb{F}_{p^2}$. In this case, it is possible that $\#E'(\mathbb{F}_{p^2}) = n \geq (p-1)^2$ be prime. Furthermore, if $\psi\colon E' \to E$ is an isomorphism defined over $\mathbb{F}_{p^4}$, then the endomorphism $\Psi = \psi\,\mathrm{Frob}_p\,\psi^{-1} \in \mathrm{End}(E')$ satisfies the equation $X^2 + 1 = 0$ and if $p \equiv 5 \pmod 8$ it can be defined over $\mathbb{F}_p$.

This idea is at the heart of the GLS approach, but it only works for curves over $\mathbb{F}_{p^m}$ with $m > 1$, therefore it does not generalise the original GLV method but rather complements it.

4   Combining GLV and GLS

Let $E/\mathbb{F}_p$ be a GLV curve. As in Section 3, we will denote by $E'/\mathbb{F}_{p^2}$ a quadratic twist $\mathbb{F}_{p^4}$-isomorphic to E via the isomorphism $\psi\colon E \to E'$. We also suppose that $\#E'(\mathbb{F}_{p^2}) = nh$ where n is prime and $h \leq 4$. We then have the two endomorphisms of E', $\Psi = \psi\,\mathrm{Frob}_p\,\psi^{-1}$ and $\Phi = \psi\phi\psi^{-1}$, with ϕ the GLV endomorphism coming with the definition of a GLV curve. They are both defined over $\mathbb{F}_{p^2}$, since if σ is the nontrivial Galois automorphism of $\mathbb{F}_{p^4}/\mathbb{F}_{p^2}$, then $\psi^\sigma = -\psi$, so that $\Psi^\sigma = \psi^\sigma\,\mathrm{Frob}_p^\sigma\,(\psi^{-1})^\sigma = (-\psi)\,\mathrm{Frob}_p(-\psi^{-1}) = \Psi$, meaning that $\Psi \in \mathrm{End}_{\mathbb{F}_{p^2}}(E')$. Similarly for Φ, where we are using the fact that $\phi \in \mathrm{End}_{\mathbb{F}_p}(E)$. Notice that $\Psi^2 + 1 = 0$ and that Φ has the same characteristic polynomial as ϕ. Furthermore, since we have a large subgroup $\langle P \rangle \subset E'(\mathbb{F}_{p^2})$ of prime order, $\Phi(P) = \lambda P$ and $\Psi(P) = \mu P$ for some $\lambda, \mu \in [1, n-1]$. We will assume that Φ and Ψ, when viewed as algebraic integers, generate disjoint quadratic extensions of $\mathbb{Q}$. In particular, we are not dealing with Example 1 from Appendix A, but this can be treated separately with a quartic twist as described in Appendix B of the full version of this article [21].

Consider the biquadratic (Galois of degree 4, with Galois group $\mathbb{Z}/2 \times \mathbb{Z}/2$) number field $K = \mathbb{Q}(\Phi, \Psi)$. Let $\mathfrak{o}_K$ be its ring of integers. The following analysis is inspired by [25, Section 8].

We have $\mathbb{Z}[\Phi, \Psi] \subseteq \mathfrak{o}_K$. Since the degrees of Φ and Ψ are much smaller when compared to n, the prime n is unramified in K and the existence of λ and μ above means that n splits in $\mathbb{Q}(\Phi)$ and $\mathbb{Q}(\Psi)$, namely that n splits completely in K. There exists therefore a prime ideal $\mathfrak{n}$ of $\mathfrak{o}_K$ dividing $n\mathfrak{o}_K$, such that its norm is n. We can also suppose that $\Phi \equiv \lambda \pmod{\mathfrak{n}}$ and $\Psi \equiv \mu \pmod{\mathfrak{n}}$. The four-dimensional GLV-GLS method works as follows.

Consider the GLV-GLS reduction map F defined by
$$F\colon \mathbb{Z}^4 \to \mathbb{Z}/n$$
$$(x_1, x_2, x_3, x_4) \mapsto x_1 + x_2\lambda + x_3\mu + x_4\lambda\mu \pmod{n} .$$
If we can find four linearly independent vectors $v_1, \ldots, v_4 \in \ker F$, with $\max_i |v_i| \leq Cn^{1/4}$ for some constant $C > 0$, then for any $k \in [1, n-1]$ we write
$$(k, 0, 0, 0) = \sum_{j=1}^{4} \beta_j v_j ,$$
with $\beta_j \in \mathbb{Q}$. As in the GLV method one sets $v = \sum_{j=1}^{4} \lfloor \beta_j \rceil v_j$ and
$$u = (k, 0, 0, 0) - v = (k_1, k_2, k_3, k_4) .$$

We then get

$$kP = k_1 P + k_2 \Phi(P) + k_3 \Psi(P) + k_4 \Psi\Phi(P) \quad \text{with } \max_i(|k_i|) \le 2Cn^{1/4} \ . \quad (1)$$

We focus next on the study of $\ker F$ in order to find a reduced basis v_1, v_2, v_3, v_4 with an explicit C. We can factor the GLV-GLS map F as

$$\mathbb{Z}^4 \xrightarrow{\ f\ } \mathbb{Z}[\Phi, \Psi] \xrightarrow[\text{mod } \mathfrak{n} \cap \mathbb{Z}[\Phi, \Psi]]{\text{reduction}} \mathbb{Z}/n$$

$$(x_1, x_2, x_3, x_4) \longmapsto x_1 + x_2\Phi + x_3\Psi + x_4\Phi\Psi \longmapsto x_1 + x_2\lambda + x_3\mu + x_4\lambda\mu \,(\text{mod } n).$$

Notice that the kernel of the second map (reduction mod $\mathfrak{n} \cap \mathbb{Z}[\Phi, \Psi]$) is exactly $\mathfrak{n} \cap \mathbb{Z}[\Phi, \Psi]$. This can be seen as follows. The reduction map factors as

$$\mathbb{Z}[\Phi, \Psi] \longrightarrow \mathfrak{o}_K \longrightarrow \mathfrak{o}_K/\mathfrak{n} \cong \mathbb{Z}/n$$

where the first arrow is inclusion, the second is reduction mod $\mathfrak{n}$, corresponding to reducing the x_i's mod $\mathfrak{n} \cap \mathbb{Z} = n\mathbb{Z}$ and using $\Phi \equiv \lambda, \Psi \equiv \mu \,(\text{mod } \mathfrak{n})$. But the kernel of this map consists precisely of elements of $\mathbb{Z}[\Phi, \Psi]$ which are in $\mathfrak{n}$, and that is what we want.

Moreover, since the reduction map is surjective, we obtain an isomorphism $\mathbb{Z}[\Phi, \Psi]/\mathfrak{n} \cap \mathbb{Z}[\Phi, \Psi] \cong \mathbb{Z}/n$ which says that the index of $\mathfrak{n} \cap \mathbb{Z}[\Phi, \Psi]$ inside $\mathbb{Z}[\Phi, \Psi]$ is n. Since the first map f is an isomorphism, we get that $\ker F = f^{-1}(\mathfrak{n} \cap \mathbb{Z}[\Phi, \Psi])$ and that $\ker F$ has index $[\mathbb{Z}^4 : \ker F] = n$ inside $\mathbb{Z}^4$.

We can also produce a basis of $\ker F$ by the following observation. Let $\Phi' = \Phi - \lambda$, $\Psi' = \Psi - \mu$, hence $\Phi'\Psi' = \Phi\Psi - \lambda\Psi - \mu\Phi + \lambda\mu$. In matrix form,

$$\begin{pmatrix} 1 \\ \Phi' \\ \Psi' \\ \Phi'\Psi' \end{pmatrix} = \begin{pmatrix} 1 & 0 & 0 & 0 \\ -\lambda & 1 & 0 & 0 \\ -\mu & 0 & 1 & 0 \\ \lambda\mu & -\mu & -\lambda & 1 \end{pmatrix} \begin{pmatrix} 1 \\ \Phi \\ \Psi \\ \Phi\Psi \end{pmatrix}$$

Since the determinant of the square matrix is 1, we deduce that $\mathbb{Z}[\Phi, \Psi] = \mathbb{Z}[\Phi', \Psi']$. But in this new basis, we claim that

$$\mathfrak{n} \cap \mathbb{Z}[\Phi', \Psi'] = n\mathbb{Z} + \mathbb{Z}\Phi' + \mathbb{Z}\Psi' + \mathbb{Z}\Phi'\Psi' \ .$$

Indeed, reverse inclusion ($\supseteq$) is easy since $\Phi', \Psi', \Phi'\Psi' \in \mathfrak{n}$ and so is n, because $\mathfrak{n}$ divides $n\mathfrak{o}_K$ is equivalent to $\mathfrak{n} \supseteq n\mathfrak{o}_K$. On the other hand, the index of both sides in $\mathbb{Z}[\Phi', \Psi']$ is n, which can only happen, once an inclusion is proved, if the two sides are equal. Using the isomorphism f, we see that a basis of $\ker F \subset \mathbb{Z}^4$ is therefore given by

$$w_1 = (n, 0, 0, 0), w_2 = (-\lambda, 1, 0, 0), w_3 = (-\mu, 0, 1, 0), w_4 = (\lambda\mu, -\mu, -\lambda, 1) \ .$$

The LLL algorithm [17] then finds, for a given basis $w_1, \ldots, w_4$ of $\ker F$, a reduced[3] basis $v_1, \ldots, v_4$ in polynomial time (in the logarithm of the norm of the w_i's) such that (cf. [5, Theorem 2.6.2 p.85])

---

[3] The estimates are usually given for the Euclidean norm of the vectors. But it is easy to see that the rectangle norm is upper bounded by the Euclidean norm.

$$\prod_{i=1}^{4} |v_i| \leq 8 \, [\mathbb{Z}^4 : \ker F] = 8n \ . \tag{2}$$

Lemma 1. *Let Φ and Ψ be as defined at the beginning of this section,*

$$\mathcal{N} \colon \mathbb{Z}^4 \to \mathbb{Z}$$
$$(x_1, x_2, x_3, x_4) \mapsto \sum_{\substack{i_1, i_2, i_3, i_4 \geq 0 \\ i_1 + i_2 + i_3 + i_4 = 4}} b_{i_1, i_2, i_3, i_4} x_1^{i_1} x_2^{i_2} x_3^{i_3} x_4^{i_4}$$

be the norm of an element $x_1 + x_2 \Phi + x_3 \Psi + x_4 \Phi \Psi \in \mathbb{Z}[\Phi, \Psi]$, where the b_{i_1, i_2, i_3, i_4}'s lie in $\mathbb{Z}$. Then, for any nonzero $v \in \ker F$, one has

$$|v| \geq \frac{n^{1/4}}{\left(\displaystyle\sum_{\substack{i_1, i_2, i_3, i_4 \\ i_1 + i_2 + i_3 + i_4 = 4}} |b_{i_1, i_2, i_3, i_4}| \right)^{1/4}} \ . \tag{3}$$

Proof. For $v \in \ker F$ we have $\mathcal{N}(v) \equiv 0 \pmod{n}$ and if $v \neq 0$ we must therefore have $|\mathcal{N}(v)| \geq n$. On the other hand, if we did not have (3), then every component of v would be strictly less than the right-hand side and plugging this upper bound in the definition of $|\mathcal{N}(v)|$ would yield a quantity $< n$, a contradiction. □

Let B be the denominator of the right-hand side of (3), then (2) and (3) imply that

$$|v_i| \leq 8B^3 \, n^{1/4} \quad i = 1, 2, 3, 4 \ . \tag{4}$$

Remark 1. In our case, where $\Psi^2 + 1 = 0$ and $\Phi^2 + r\Phi + s = 0$, we get as norm function

$$x_1^4 + s^2 x_2^4 + x_3^4 + s^2 x_4^4 - 2r x_1^3 x_2 - 2rs x_1 x_2^3 - 2r x_3^3 x_4 - 2rs x_3 x_4^3 +$$
$$(r^2 + 2s) x_1^2 x_2^2 + 2x_1^2 x_3^2 + (r^2 - 2s) x_1^2 x_4^2 + (r^2 - 2s) x_2^2 x_3^2 + 2s^2 x_2^2 x_4^2 + (r^2 + 2s) x_3^2 x_4^2$$
$$- 2r x_1^2 x_3 x_4 - 2rs x_2^2 x_3 x_4 - 2r x_1 x_2 x_3^2 - 2rs x_1 x_2 x_4^2 + 8s x_1 x_2 x_3 x_4 \ ,$$

and therefore

$$B = \left(4 + 4s^2 + 8s + 8|r| + 8|r|s + 2(r^2 + 2s) + 2|r^2 - 2s| \right)^{1/4} \ . \tag{5}$$

From (1) and (4) we have proved the following theorem.

Theorem 2. *Let $E/\mathbb{F}_p$ be a GLV curve and $E'/\mathbb{F}_{p^2}$ a twist, together with the two efficient endomorphisms Φ and Ψ, where everything is defined as at the start of this section. Suppose that the minimal polynomial of Φ is $X^2 + rX + s = 0$. Let $P \in E'(\mathbb{F}_{p^2})$ be a generator of the large subgroup of prime order n. There exists an efficient algorithm, which for any $k \in [1, n]$ finds integers k_1, k_2, k_3, k_4 such that*

$$kP = k_1 P + k_2 \Phi(P) + k_3 \Psi(P) + k_4 \Psi \Phi(P) \quad \text{with } \max_i(|k_i|) \leq 16B^3 n^{1/4}$$

and

$$B = \left(4 + 4s^2 + 8s + 8|r| + 8|r|s + 2(r^2 + 2s) + 2|r^2 - 2s| \right)^{1/4} \ .$$

4.1   Uniform Improvements

The previous analysis is only the first step of our work. It shows that the GLV-GLS method works as predicted in a four-way decomposition on twists of GLV curves over $\mathbb{F}_{p^2}$. However, the constant B^3 involved is rather large and, hence, does not guarantee a non-negligible gain when switching from 2 to 4 dimensions (especially on those GLV curves with more complicated endomorphism rings). A much deeper argument, which we develop in the full version of this article, allows us to prove the following result.

Theorem 3. *When performing an optimal lattice reduction on* $\ker F$, *it is possible to decompose any* $k \in [1, n]$ *into integers* k_1, k_2, k_3, k_4 *such that*

$$kP = k_1 P + k_2 \Phi(P) + k_3 \Psi(P) + k_4 \Psi\Phi(P) \ ,$$

with $\max_i(|k_i|) < 103(\sqrt{1 + |r| + s}) \, n^{1/4}$.

The significance of this theorem lies in the *uniform* improvement of the constant $16B^3$, which is $\Omega(s^{3/2})$ in Theorem 2, to a value that is an absolute constant times greater than the minimal bound for the 2-dimensional GLV method (Theorem 1). Hence, this guarantees in practice a quasi-uniform improvement when switching from 2-dimensional to 4-dimensional GLV independently of the curve.

To prove Theorem 3, first note that Lemma 1 gives a rather poor bound when applied to more than one vector, as is done three times for the proof of Theorem 2. A more direct treatment of the reduced vectors of $\ker F$ becomes necessary, and this is done via a modification of the original GLV approach. This results into a new, easy-to-implement lattice reduction algorithm which employs two Cornacchia-type algorithms [5, Section 1.5.2], one in $\mathbb{Z}$ (as in the original GLV method), the other one in $\mathbb{Z}[i]$ (Gaussian Cornacchia). The new algorithm is presented in Appendix B. The main difficulty lies in controlling arguments of complex numbers in the Gaussian Cornacchia algorithm and is technically rather delicate. This difficulty does not exist in the original GLV algorithm, as taking absolute values suffices to get the desired bounds. We refer to the full version [21] for details.

Remark 2. In the case of the LLL algorithm, we have not managed to demonstrate a bound as good as the one obtained with our lattice reduction algorithm.

Remark 3. Nguyen and Stehlé [23] have produced an efficient lattice reduction in four dimensions which finds successive minima and hence produces a decomposition with relatively good bounds. Our algorithm represents a very simple and easy-to-implement alternative that may be ideal for certain cryptographic libraries.

5   GLV-GLS Using the Twisted Edwards Model

The GLV-GLS method can be sped up in practice by writing down GLV-GLS curves in the Twisted Edwards model. Note that arithmetic on j-invariant 0

Weierstrass curves is already very efficient. However, some GLV curves do not exhibit such high-speed arithmetic. In particular, curves in Examples 3-6 from Appendix A have Weierstrass coefficients $a_4 \cdot a_6 \neq 0$ for curve parameters a_4 and a_6 and hence they have more expensive point doubling (even more if we consider the extra multiplication by the twisted parameter u when using the GLS method). So the impact of using Twisted Edwards is expected to be especially significant for these curves. In fact, if we consider that suitable parameters can be always chosen the use of Twisted Edwards curves isomorphic to the original Weierstrass GLV-GLS curves *uniformizes* the performance of all of them.

Let us illustrate how to produce a Twisted Edwards GLV-GLS curve with the GLV curve from Example 4, Appendix A. First, consider its quadratic twist over $\mathbb{F}_{p^2}$

$$E'/\mathbb{F}_{p^2} : \ x^3 - \frac{15}{2}u^2x - 7u^3 = (x + 2u) \cdot (x^2 - 2ux - \frac{7}{2}u^2)$$

The change of variables $x_1 = x + 2u$ transforms E' into

$$y^2 = x_1^3 - 6ux_1^2 + \frac{9u^2}{2}x_1 \ .$$

Let $\beta = 3u/\sqrt{2} \in \mathbb{F}_{p^2}$ and substitute $x_1 = \beta x'$ to get

$$\frac{1}{\beta^3}y^2 = x'^3 - \frac{6u}{\beta}x'^2 + x'$$

and this is a Montgomery curve $M_{A,B} : Bv^2 = u^3 + Au^2 + u$, where $A \neq \pm 2, B \neq 0$, with

$$B = \frac{1}{\beta^3} = \frac{2\sqrt{2}}{27u^3} \ , \quad A = -\frac{6u}{\beta} = -2\sqrt{2} \ .$$

The corresponding Twisted Edwards GLV-GLS curve is then $E_{a,d}: ax^2 + y^2 = 1 + dx^2y^2$ with

$$a = \frac{A+2}{B} = 27u^3 \left(\frac{\sqrt{2}}{2} - 1\right) \ , \quad d = \frac{A-2}{B} = -27u^3 \left(\frac{\sqrt{2}}{2} + 1\right) \ .$$

The map $E' \to E_{a,d}$ is

$$(x, y) \mapsto \left(\frac{x + 2u}{\beta y}, \frac{x + 2u - \beta}{x + 2u + \beta}\right) = (X, Y)$$

with inverse

$$(X, Y) \mapsto \left(\frac{\beta - 2u + (\beta + 2u)Y}{1 - Y}, \frac{1 + Y}{(1 - Y)X}\right) \ .$$

We now specify the formulas for Φ and Ψ, obtained by composing these endomorphisms on the Weierstrass model with the birational maps above. We found an extremely appealing expression in the case when $u = 1 + i$ and $i^2 = -1$. Then $\beta = 3u/\sqrt{2} = 3\zeta_8$ where ζ_8 is a primitive 8th root of unity. We have

$$\Phi(X,Y) = \left(-\frac{(\zeta_8^3 + 2\zeta_8^2 + \zeta_8)XY^2 + (\zeta_8^3 - 2\zeta_8^2 + \zeta_8)X}{2Y} , \frac{(\zeta_8^2 - 1)Y^2 + 2\zeta_8^3 - \zeta_8^2 + 1}{(2\zeta_8^3 + \zeta_8^2 - 1)Y^2 - \zeta_8^2 + 1} \right)$$

and

$$\Psi(X,Y) = \left(\zeta_8 X^p, \frac{1}{Y^p} \right) .$$

In this case

$$a = 54(\zeta_8^3 - \zeta_8^2 + 1) , \quad d = -54(\zeta_8^3 + \zeta_8^2 - 1) .$$

Finally, one would want to use the efficient formulas given in [12] for the case $a = -1$. After ensuring that $-a$ be a square in $\mathbb{F}_{p^2}$, we use the map $(x,y) \mapsto (x/\sqrt{-a}, y)$ to convert to the isomorphic curve $-x^2 + y^2 = 1 + d'x^2y^2$, where $d' = -d/a$.

6   Side-Channel Protection and Parallelization of the GLV-GLS Method

Given the potential threat posed by attacks that exploit timing information to deduce secret keys ([16,4]), many works have proposed countermeasures to minimize the risks and achieve the so-called constant-time execution during cryptographic computations. In general, to avoid leakage the execution flow should be independent of the secret key. This means that conditional branches and secret-dependent table lookup indices should be avoided [15]. There are *five* key points that are especially vulnerable during the computation of scalar multiplication: inversion, modular reduction in field operations, precomputation, scalar recoding and double-and-add execution.

A well-known technique that is secure and easy to implement for inverting any field element a consists of computing the exponentiation $a^{p-2} \mod p$ using a short addition chain for $p - 2$.

To protect field operations, one may exploit conditional move instructions typically found on modern x86 and x64 processors (a.k.a. cmove). Since conditional checks happen during operations such as addition and subtraction as part of the reduction step it is standard practice to replace conditional branches with the conditional move instruction. Luckily, these conditional branches are highly unpredictable and, hence, the substitution above does not only makes the execution constant-time but also more efficient in most cases. An exception happens when performing modular reduction during a field multiplication or squaring, where a final correction step could happen very rarely and hence a conditional branch may be more efficient.

In the case of precomputation, recent work by [15] and later by [3] showed how to enable the use of precomputed points by employing constant-time table lookups that mask the extraction of points. In our implementations (see Section 7), we exploit a similar approach based on cmove and conditional vector instructions instead, which is expected to achieve higher performance on some platforms than implementations based on logical instructions (see Listing 1 in [15]). Note that it is straightforward to enable the use of signed-digit representations that allow negative points by performing a second table lookup between the point selected in the first table lookup and its negated value.

To protect the scalar recoding and its corresponding double-and-add algorithm, one needs a regular pattern execution. Based on a method by [24], Joye and Tunstall [14] proposed a constant-time recoding that supports a regular execution double-and-add algorithm that exploits precomputations. The nonzero density of the method is $1/(w-1)$, where w is the window width. Therefore, there is certain loss in performance in comparison with an unprotected version with nonzero density $1/(w+1)$. In GLV-based implementations one has to deal with more than one scalar, and these scalars are scanned simultaneously during multi-exponentiation. So there are two issues that arise. First, how are the several scalars aligned with respect to their zero and nonzero digit representation?, and second, how do we guarantee the same representation length for all scalars so that no dummy operations are required? The first issue is inherently solved by the recoding algorithm itself. The input is always an odd number, which means that, from left to right, one obtains the execution pattern $(w-1)$ doublings, d additions, $(w-1)$ doublings, d additions, ... , $(w-1)$ doublings and d additions, for d-dimensional GLV. For dealing with even numbers, one may employ the technique described in [14] in a constant-time fashion, namely, scalars k_i that are even are replaced by k_i+1 and scalars that are odd are replaced by k_i+2 (the correction, also constant-time, is performed after the scalar multiplication computation using d point additions). Solution to the second issue was also hinted by [14]. The reader is referred to the full paper version for the modified recoding algorithm that outputs a regular pattern representation with fixed length. Note that in the case of Twisted Edwards one can alternatively use *unified* addition formulas that also work for doubling (see [2,12] for details). However, our analysis indicates that this approach is consistently slower because of the high cost of these unified formulas in comparison to doubling and the extra cost incurred by the increase in constant-time table lookup accesses.

6.1   Multicore Computation and Its Side-Channel Protection

Parallelization of scalar multiplication over prime fields is particularly difficult on modern multicore processors. This is due to the difficulty to perform point operations concurrently when executing the double-and-add algorithm from left to right. From right to left parallelization is easier but performance is hurt because the use of precomputations is cumbersome. Hence, parallelization should be ideally performed at the field arithmetic level. Unfortunately, current multicore processors still impose a severe overhead for thread creation/destruction. During our tests we observed overheads of a few thousands of cycles on modern 64-bit CPUs (that is, much more costly than a point addition or doubling). Given this limitation, for the GLV method it seems the ideal approach (from a speed perspective) to let each core manage a separate scalar multiplication with k_i. This is simple to implement, minimizes thread management overhead and also eases the task of protecting the implementation against side-channel attacks since each scalar can be recoded using Algorithm 4 [21, App. E]. Using d cores, the total cost of a protected d-dimensional GLV l-bit scalar multiplication (disregarding precomputation) is about l/d doublings and $l/((w-1)\cdot d)$ mixed additions. A somewhat slower approach (but more power efficient) would be to let one core manage all doublings and let one or two extra cores manage the additions corresponding to nonzero digits. For instance, for dimension four and three cores the total cost (disregarding precomputation) is about l/d doublings and $l/((w-1)\cdot d)$ *general* additions, always that the latency of $(w-1)$ doublings be equivalent or greater than the addition part (otherwise, the cost is dominated by non-mixed additions).

7   Performance Analysis and Experimental Results

For our analysis and experiments, we consider the *five* curves below: two GLV curves in Weierstrass form with and without nontrivial automorphisms, their corresponding GLV-GLS counterparts and one curve in Twisted Edwards form isomorphic to the GLV-GLS curve E_3' (see below).

- GLV-GLS curve with j-invariant 0 in Weierstrass form $E_1'/\mathbb{F}_{p_1^2} : y^2 = x^3 + 9u$, where $p_1 = 2^{127} - 58309$ and $\#E_1'(\mathbb{F}_{p_1^2}) = r$, where r is a 254-bit prime. We use $\mathbb{F}_{p_1^2} = \mathbb{F}_{p_1}[i]/(i^2+1)$ and $u = 1 + i \in \mathbb{F}_{p_1^2}$. E_1' is the quadratic twist of the curve in Example 2, Appendix A. $\Phi(x,y) = \lambda P = (\xi x, y)$ and $\Psi(x,y) = \mu P = (u^{(1-p)/3}x^p, u^{(1-p)/2}y^p)$, where $\xi^3 = 1 \mod p_1$. We have that $\Phi^2 + \Phi + 1 = 0$ and $\Psi^2 + 1 = 0$.
- GLV curve with j-invariant 0 in Weierstrass form $E_2/\mathbb{F}_{p_2} : y^2 = x^3 + 2$, where $p_2 = 2^{256} - 11733$ and $\#E_2(\mathbb{F}_{p_2})$ is a 256-bit prime. This curve corresponds to Example 2, Appendix A.
- GLV-GLS curve in Weierstrass form $E_3'/\mathbb{F}_{p_3^2} : y^2 = x^3 - 15/2\ u^2x - 7u^3$, where $p_3 = 2^{127} - 5997$ and $\#E_3'(\mathbb{F}_{p_3^2}) = 8r$, where r is a 251-bit prime. We use $\mathbb{F}_{p_3^2} = \mathbb{F}_{p_3}[i]/(i^2+1)$ and $u = 1 + i \in \mathbb{F}_{p_3^2}$. E_3' is the quadratic twist of a curve isomorphic to the one in Example 4, Appendix A. The formula for $\Phi(x,y) = \lambda P$ can be easily derived from $\psi(x,y)$, and $\Psi(x,y) = \mu P = (u^{(1-p)}x^p, u^{3(1-p)/2}y^p)$. It can be verified that $\Phi^2 + 2 = 0$ and $\Psi^2 + 1 = 0$.
- GLV-GLS curve in Twisted Edwards form $E_{T3}'/\mathbb{F}_{p_3^2} : -x^2 + y^2 = 1 + dx^2y^2$, where $d = 170141183460469231731687303715884099728 + 116829086847165810221872975542241037773i$, $p_3 = 2^{127} - 5997$ and $\#E_{T3}'(\mathbb{F}_{p_3^2}) = 8r$, where r is a 251-bit prime. We use again $\mathbb{F}_{p_3^2} = \mathbb{F}_{p_3}[i]/(i^2 + 1)$ and $u = 1 + i \in \mathbb{F}_{p_3^2}$. E_{T3}' is isomorphic to curve E_3' above and was obtained following the procedure in Section 5. The formulas for $\Phi(x,y)$ and $\Psi(x,y)$ are also given in Section 5. It can be verified that $\Phi^2 + 2 = 0$ and $\Psi^2 + 1 = 0$.
- GLV curve $E_4/\mathbb{F}_{p_4} : y^2 = x^3 - 15/2\ x - 7$, where $p_4 = 2^{256} - 45717$ and $\#E_4(\mathbb{F}_{p_4}) = 2r$, where r is a 256-bit prime. This curve is isomorphic to the curve in Example 4, Appendix A.

Let us first analyze the performance of the GLV-GLS method over $\mathbb{F}_{p^2}$ in comparison with the traditional 2-GLV case over $\mathbb{F}_p$. We assume the use of a pseudo-Mersenne prime with form $p = 2^m - c$, for small c (for our targeted curves, groups with (near) prime order cannot be constructed using the attractive Mersenne prime $p = 2^{127} - 1$). Given that we have a proven ratio $C_2/C_1 < 412$ that is independent of the curve, the only values left that could affect significantly a uniform speedup between GLV-GLS and 2-GLV are the quadratic non-residue β used to build $\mathbb{F}_{p^2}$ as $\mathbb{F}_p[i]/(i^2 - \beta)$, the value of the twisted parameter u and the cost of applying the endomorphisms Φ and Ψ. In particular, if $|\beta| > 1$ a few extra additions (or a multiplication by a small constant) are required per $\mathbb{F}_{p^2}$ multiplication and squaring. Luckily, for all the GLV curves listed in Appendix A one can always use a suitably chosen modulus p so that $|\beta|$ can be one or at least very close to it. Similar comments apply to the twisted parameter u. In this case, the extra cost (equivalent to a few additions) is added to the cost of point doubling always that the curve parameter a in the Weierstrass equation be different to zero (e.g., it does not affect j-invariant 0 curves). In the case of Twisted Edwards, we applied a better strategy, that is, we eliminated the twisted parameter u in the isomorphic curve. The cost of applying Φ and Ψ does depend on the chosen curve and it could be relatively expensive. If computing $\Phi(P)$, $\Psi(P)$ or $\Psi\Phi(P)$ is more expensive than point addition then its use can be limited to only one application (i.e., multiples of those values −if using precomputations− should be computed with point additions). Further, the extra cost can be minimized by choosing the optimal window width for each k_i.

To illustrate how the parameters above affect the performance gain we detail in Table 1 estimates for the cost of computing scalar multiplication with our representative curves. In the remainder, we use the following notation: M, S, A and I represent field multiplication, squaring, addition and inversion over $\mathbb{F}_p$, respect., and m, s, a and i represent the same operations over $\mathbb{F}_{p^2}$. Side-channel protected multiplication and squaring are denoted by m_s and s_s. We consider the cost of addition, substraction, negation, multiplication by 2 and division by 2 as equivalent. For the targeted curves in Weierstrass form, a mixed addition consists of 8 multiplications, 3 squarings and 7 additions, and a general addition consists of 12 multiplications, 4 squarings and 7 additions. For E_1' and E_2, a doubling consists of 3 multiplications, 4 squarings and 7 additions, and for E_3' and E_4, a doubling consists of 3 multiplications, 6 squarings and 12 additions. For Twisted Edwards we consider the use of mixed homogeneous/extended homogeneous projective coordinates [12]. In this case, a mixed addition consists of 7 multiplications and 7 additions, a general addition consists of 8 multiplications and 6/7 additions and a doubling consists of 4 multiplications, 3 squarings and 5 additions. We assume the use of interleaving [10] with width-w non-adjacent form (wNAF) and the use of the LM scheme for precomputing points on the Weierstrass curves [20] (see also [18, Ch. 3]).

According to our theoretical estimates, it is expected that the relative speedup when moving from 2-GLV to GLV-GLS be as high as 50%, approximately. To confirm our findings, we realized full implementations of the methods. Experimental

Table 1. Operation counts and performance for scalar multiplication at approximately 128 bits of security. To determine the total costs we consider 1i=66m, 1s=0.76m and 1a=0.18m for E_1', E_3' and E_{T3}'; and 1I=290M, 1S=0.85M and 1A=0.18M for E_2 and E_4. The cost ratio of multiplications over $\mathbb{F}_p$ and $\mathbb{F}_{p^2}$ is M/m=0.91. These values and the performance figures (in cycles) were obtained by benchmarking full implementations on a single core of a 3.4GHz Intel Core i7-2600 (Sandy Bridge) processor.

Curve	Method	Operation Count	Total Cost	Gain	Performance	Gain
$E_1'(\mathbb{F}_{p_1^2})$	4-GLV-GLS, 32pts.	2i + 617m + 404s + 847a	1209m	51%	99,000cc	53%
$E_2(\mathbb{F}_{p_2})$	2-GLV, 16pts.	1I + 904M + 690S + 1240A	2004M≈1824m	-	151,000cc	-
$E_{T3}'(\mathbb{F}_{p_3^2})$	4-GLV-GLS, 16pts.	1i + 742m + 225s + 767a	1117m	97%	91,000cc	102%
$E_3'(\mathbb{F}_{p_3^2})$	4-GLV-GLS, 16pts.	2i + 678m + 581s + 1200a	1468m	50%	121,000cc	52%
$E_4(\mathbb{F}_{p_4})$	2-GLV, 16pts.	1I + 950M + 970S + 1953A	2416M≈2199m	-	184,000cc	-

results, also displayed in Table 1, closely follow our estimates and confirm that speedups in practice are about 52%. Most remarkably, the use of the Twisted Edwards model pushes performance even further. In Table 1, the expected gains for E_{T3}' are 31% and 97% in comparison with 4-GLV-GLS and 2-GLV in Weierstrass form (respect.). In practice, we achieved similar speedups, namely, 33% and 102% (respect.). Likewise, a rough analysis indicates that a Twisted Edwards GLV-GLS curve for a j-invariant 0 curve would achieve roughly similar speed to E_{T3}', which means that in comparison to its corresponding Weierstrass counterpart the gains are 9% and 66% (respect.). This highlights the impact of using Twisted Edwards especially over those GLV-GLS curves relatively slower in the Weierstrass model. Timings were registered on a single core of a 3.4GHz Intel Core i7-2600 (Sandy Bridge) processor.

Let us now focus on curves E_1', E_2 and E_{T3}' to assess performance of implementations targeting *four* scenarios of interest: unprotected and side-channel protected versions with sequential and multicore execution. Operation counts for computing a scalar multiplication at approximately 128 bits of security for the different cases are displayed in Table 2. The techniques to protect and parallelize our implementations are described in Section 6. In particular, the execution flow and memory address access of side-channel protected versions are not secret and are fully independent of the scalar. For our versions running on several cores we used OpenMP. We use an implementation in which each core is in charge of one scalar multiplication with k_i. Given the high cost of thread creation/destruction this approach guarantees the fastest computation in our case (see Section 6 for a discussion). Note that these multicore figures are only relevant for scenarios in which latency rather than throughput is targeted. Finally, we consider the cost of constant-time table lookups (denoted by t) given its non-negligible cost in protected implementations.

Focusing on curve E_1', it can be noted a significant cost reduction when switching from non-GLV to a GLV-GLS implementation. The speedup is more than twofold for sequential, unprotected versions. Significant improvements are also expected when using multiple cores. A remarkable factor 3 speedup is expected when using GLV-GLS on four cores in comparison with a traditional execution (listed as non-GLV).

Table 2. Operation counts for scalar multiplication at approximately 128 bits of security using curves E_1', E_2 and E_{T3}' in up to four variants: unprotected and side-channel protected implementations with sequential and multicore execution. To determine the total costs we consider 1i=66m, 1s=0.76m and 1a=0.18m for unprotected versions of E_1' and E_{T3}'; 1i=79m$_s$, 1s$_s$=0.81m$_s$ and 1a=0.17m$_s$ for protected versions of E_1' and E_{T3}'; t=0.83m$_s$ for E_1' (32pts.); t=1.28m$_s$ for E_{T3}' (36pts.); t=0.78m$_s$ for E_{T3}' (20pts.); and 1I=290M, 1S=0.85M and 1A=0.18M for E_2. In our case, M/m=0.91 and m$_s$/m=1.11. These values were obtained by benchmarking full implementations on a 3.4GHz Intel Core i7-2600 (Sandy Bridge) processor.

Curve	Method	Protection	# Cores	Operation Count	Total Cost
$E_{T3}'(\mathbb{F}_{p_3^2})$	4-GLV-GLS, 16pts.	no	1	1i + 742m + 225s + 767a	1117m
$E_{T3}'(\mathbb{F}_{p_3^2})$	4-GLV-GLS, 36pts.	yes	1	1i + 1014m$_s$ + 217s$_s$ + 997a + 68t	1525m$_s$≈1693m
$E_{T3}'(\mathbb{F}_{p_3^2})$	4-GLV-GLS, 16pts.	no	4	1i + 420m + 198s + 484a	724m
$E_{T3}'(\mathbb{F}_{p_3^2})$	4-GLV-GLS, 20pts.	yes	4	1i + 503m$_s$ + 196s$_s$ + 532a + 22t	848m$_s$≈941m
$E_1'(\mathbb{F}_{p_1^2})$	4-GLV-GLS, 32pts.	no	1	2i + 617m + 404s + 847a	1209m
$E_1'(\mathbb{F}_{p_1^2})$	4-GLV-GLS, 36pts.	yes	1	2i + 849m$_s$ + 489s$_s$ + 1001a + 68t	1630m$_s$≈1809m
$E_1'(\mathbb{F}_{p_1^2})$	4-GLV-GLS, 32pts.	no	4	2i + 371m + 316s + 593a	850m
$E_1'(\mathbb{F}_{p_1^2})$	4-GLV-GLS, 36pts.	yes	4	2i + 425m$_s$ + 335s$_s$ + 637a + 17t	977m$_s$≈1084m
$E_1'(\mathbb{F}_{p_1^2})$	non-GLV, 8pts.	no	1	2i + 1169m + 1169s + 2141a	2575m
$E_2(\mathbb{F}_{p_2})$	2-GLV, 16pts.	no	1	1I + 904M + 690S + 1240A	2004M≈1824m
$E_2(\mathbb{F}_{p_2})$	2-GLV, 16pts.	no	2	1I + 681M + 615S + 1103A	1692M≈1540m

In general for our targeted GLV-GLS curves, the speedup obtained by using four cores is in between 42%-80%. Interestingly, the improvement is greater for protected implementations since the overhead of using a regular pattern execution is minimized when distributing computation among various cores. Remarkably, protecting implementations against timing attacks increases cost in between 28%-52%, approximately. On the other hand, in comparison with curve E_2, an optimal execution of GLV-GLS on four cores is expected to run 1.81 faster than an optimal execution of the standard 2-GLV on two cores.

To confirm our findings we implemented the different versions using curves E_1', E_2 and E_{T3}'. To achieve maximum performance and ease the task of parallelizing and protecting the implementations, we wrote our own standalone software without employing any external library. For our experiments we used a 3.4GHz Intel Core i7-2600 processor, which contains four cores. The timings in terms of clock cycles are displayed in Table 3. As can be seen, closely following our analysis GLV-GLS achieves a twofold speedup over a non-GLV implementation on a single core. Parallel execution improves performance by up to 76% for side-channel protected versions. In comparison with the non-GLV implementation, the four-core implementation runs 3 times faster. Our results also confirm the lower-than-expected cost of adding side-channel protection. Sequential versions lose about 50% in performance whereas parallel versions only lose about 28%. The relative speedup when moving from 2-GLV to GLV-GLS on j-invariant 0 curves is 53%, closely following the theoretical 50% estimated previously. Four-core GLV-GLS supports a computation that runs 81% faster than the standard 2-GLV on two cores. Finally, in practice our Twisted Edwards curve achieves up to 9% speedup on the sequential, non-protected scenario in comparison with the efficient j-invariant 0 curve based on Jacobian coordinates.

Table 3. Point multiplication timings (in clock cycles), 64-bit processor

Curve	Method	Protection	# Cores	Core i7
$E'_{T3}(\mathbb{F}_{p_3^2})$	4-GLV-GLS, 16pts.	no	1	**91,000**
$E'_{T3}(\mathbb{F}_{p_3^2})$	4-GLV-GLS, 36pts.	yes	1	137,000
$E'_{T3}(\mathbb{F}_{p_3^2})$	4-GLV-GLS, 16pts.	no	4	**61,000**
$E'_{T3}(\mathbb{F}_{p_3^2})$	4-GLV-GLS, 20pts.	yes	4	78,000
$E'_1(\mathbb{F}_{p_1^2})$	4-GLV-GLS, 32pts.	no	1	99,000
$E'_1(\mathbb{F}_{p_1^2})$	4-GLV-GLS, 36pts.	yes	1	145,000
$E'_1(\mathbb{F}_{p_1^2})$	4-GLV-GLS, 32pts.	no	4	70,000
$E'_1(\mathbb{F}_{p_1^2})$	4-GLV-GLS, 36pts.	yes	4	89,000
$E'_1(\mathbb{F}_{p_1^2})$	non-GLV, 8pts.	no	1	201,000
$E_2(\mathbb{F}_{p_2})$	2-GLV, 16pts.	no	1	151,000
$E_2(\mathbb{F}_{p_2})$	2-GLV, 16pts.	no	2	127,000

Let us now compare our best numbers with recent results in the literature. Focusing on one-core unprotected implementations, the first author together with Hu and Xu reported in [13] 122,000 cycles for a j-invariant 0 Weierstrass curve on an Intel Core i7-2600 processor. We report 91,000 cycles with the GLV-GLS Twisted Edwards curve E'_{T3}, improving that number in 34%. We benchmarked on the same processor the side-channel protected software recently presented by Bernstein et al. in [3], and obtained 194,000 cycles. Thus, our protected implementation, which runs in 137,000 cycles, improves that result in 42%. Our result is also 12% faster than the recent implementation by Hamburg [11]. Recent implementations on multiple cores are reported by Taverne et al. in [27]. However, they do not explore the 128-bit security level in their implementations and, hence, results are not directly comparable. They also report a protected implementation of a binary Edwards curve that runs in 225,000 cycles on a Core i7-2600 machine, which is 64% slower than our corresponding result. Since the advent of the carryless multiplier on recent Intel processors, it has been suspected that the only curves able to get performance as good as the GLV-GLS method over large prime characteristic fields are Koblitz curves over binary fields. In fact, Aranha et al. [1] very recently presented an implementation of the Koblitz curve K-283 that runs in 99,000 cycles on an Intel Core i7-2600, which is 9% slower than our GLV-GLS Twisted Edwards curve E'_{T3} (unprotected sequential execution). We remark that such performance for a binary elliptic curve can only be attained on very recent processors that possess the so-called carryless multiplier. Aranha et al. do not report timings for side-channel protected implementations. To the best of our knowledge, we have presented the first scalar multiplication implementation running on multiple cores that is protected against timing attacks, cache attacks and several others.

8   Conclusion

We have shown how to generalize the GLV scalar multiplication method by combining it with Galbraith-Lin-Scott's ideas to perform a proven almost fourfold speedup on GLV curves over $\mathbb{F}_{p^2}$. We have introduced a new and easy-to-implement reduction algorithm, consisting in two applications of the extended Euclidean algorithm, one in $\mathbb{Z}$ and the other in $\mathbb{Z}[i]$. The refined bound obtained from this algorithm has allowed us to get a relative improvement from 2-GLV to 4-GLV-GLS *quasi-independent* of the curve. Our analysis and experimental results on different GLV curves show that in practice one should expect speedups close to 50%. We improve performance even further by exploiting the Twisted Edwards model over a larger set of curves and show that this approach is especially significant to certain GLV curves with slow arithmetic in the Weierstrass model. This makes available to implementers new curves that achieve close to optimal performance. Moreover, we have shown how to protect GLV-based implementations against certain side-channel attacks with relatively low overhead and carried out a performance analysis on modern multicore processors. Our implementations of the GLV-GLS method improve the state-of-the-art performance of point multiplication for multiple scenarios: unprotected and side-channel protected versions with sequential and parallel execution.

Acknowledgements. We would like to thank the reviewers for their helpful comments and Mike Scott for advice on a first version of this work. Also, we would like to thank Diego F. Aranha for his advice on multicore programming and Joppe Bos for his help on looking for efficient chains for implementing modular inversion.

References

1. Aranha, D.F., Faz-Hernández, A., López, J., Rodríguez-Henríquez, F.: Faster Implementation of Scalar Multiplication on Koblitz Curves. In: Hevia, A., Neven, G. (eds.) LatinCrypt 2012. LNCS, vol. 7533, pp. 177–193. Springer, Heidelberg (2012)
2. Bernstein, D.J., Birkner, P., Joye, M., Lange, T., Peters, C.: Twisted Edwards Curves. In: Vaudenay, S. (ed.) AFRICACRYPT 2008. LNCS, vol. 5023, pp. 389–405. Springer, Heidelberg (2008)
3. Bernstein, D.J., Duif, N., Lange, T., Schwabe, P., Yang, B.-Y.: High-Speed High-Security Signatures. In: Preneel, B., Takagi, T. (eds.) CHES 2011. LNCS, vol. 6917, pp. 124–142. Springer, Heidelberg (2011)
4. Brumley, D., Boneh, D.: Remote timing attacks are practical. In: Mangard, S., Standaert, F.-X. (eds.) Proceedings of the 12th USENIX Security Symposium. LNCS, vol. 6225, pp. 80–94. Springer (2003)
5. Cohen, H.: A Course in Computational Algebraic Number Theory. Graduate Texts in Mathematics, vol. 138. Springer (1996)
6. Cornacchia, G.: Su di un metodo per la risoluzione in numeri interi dell'equazione $\sum_{h=0}^{n} C_h x^{n-h} y^h = P$. Giornale di Mathematiche di Battaglini 46, 33–90 (1908)

7. Edwards, H.: A normal form for elliptic curves. Bulletin of the American Mathematical Society 44, 393–422 (2007)

8. Galbraith, S.D., Lin, X., Scott, M.: Endomorphisms for Faster Elliptic Curve Cryptography on a Large Class of Curves. In: Joux, A. (ed.) EUROCRYPT 2009. LNCS, vol. 5479, pp. 518–535. Springer, Heidelberg (2009)

9. Galbraith, S.D., Lin, X., Scott, M.: Endomorphisms for faster elliptic curve cryptography on a large class of curves. J. Cryptology 24(3), 446–469 (2011)

10. Gallant, R.P., Lambert, R.J., Vanstone, S.A.: Faster Point Multiplication on Elliptic Curves with Efficient Endomorphisms. In: Kilian, J. (ed.) CRYPTO 2001. LNCS, vol. 2139, pp. 190–200. Springer, Heidelberg (2001)

11. Hamburg, M.: Fast and compact elliptic-curve cryptography. Cryptology ePrint Archive, Report 2012/309 (2012), http://eprint.iacr.org/2012/309

12. Hisil, H., Wong, K.K.-H., Carter, G., Dawson, E.: Twisted Edwards Curves Revisited. In: Pieprzyk, J. (ed.) ASIACRYPT 2008. LNCS, vol. 5350, pp. 326–343. Springer, Heidelberg (2008)

13. Hu, Z., Longa, P., Xu, M.: Implementing 4-dimensional GLV method on GLS elliptic curves with j-invariant 0. Designs, Codes and Cryptography 63(3), 331–343 (2012); also in Cryptology ePrint Archive, Report 2011/315, http://eprint.iacr.org/2011/315

14. Joye, M., Tunstall, M.: Exponent Recoding and Regular Exponentiation Algorithms. In: Preneel, B. (ed.) AFRICACRYPT 2009. LNCS, vol. 5580, pp. 334–349. Springer, Heidelberg (2009)

15. Kasper, E.: Fast elliptic curve cryptography in openssl. In: 2nd Workshop on Real-Life Cryptographic Protocols and Standardization (2011)

16. Kocher, P.C.: Timing Attacks on Implementations of Diffie-Hellman, RSA, DSS, and Other Systems. In: Koblitz, N. (ed.) CRYPTO 1996. LNCS, vol. 1109, pp. 104–113. Springer, Heidelberg (1996)

17. Lenstra, A.K., Lenstra Jr., H.W., Lovász, L.: Factoring polynomials with rational coefficients. Mathematische Ann. 261, 513–534 (1982)

18. Longa, P.: High-speed elliptic curve and pairing-based cryptography. PhD thesis, University of Waterloo (2011), http://hdl.handle.net/10012/5857

19. Longa, P., Gebotys, C.: Efficient Techniques for High-Speed Elliptic Curve Cryptography. In: Mangard, S., Standaert, F.-X. (eds.) CHES 2010. LNCS, vol. 6225, pp. 80–94. Springer, Heidelberg (2010)

20. Longa, P., Miri, A.: New Composite Operations and Precomputation Scheme for Elliptic Curve Cryptosystems over Prime Fields. In: Cramer, R. (ed.) PKC 2008. LNCS, vol. 4939, pp. 229–247. Springer, Heidelberg (2008)

21. Longa, P., Sica, F.: Four-dimensional Gallant-Lambert-Vanstone scalar multiplication (full version). Cryptology ePrint Archive, Report 2011/608 (2012), http://eprint.iacr.org/2011/608

22. Morain, F.: Courbes elliptiques et tests de primalité. PhD thesis, Université de Lyon I (1990), http://www.lix.polytechnique.fr/Articles/english.html; Chapter 2: On Cornacchia's algorithm (joint with J-L. Nicolas)

23. Nguyên, P.Q., Stehlé, D.: Low-Dimensional Lattice Basis Reduction Revisited. In: Buell, D.A. (ed.) ANTS 2004. LNCS, vol. 3076, pp. 338–357. Springer, Heidelberg (2004)

24. Okeya, K., Takagi, T.: The Width-w NAF Method Provides Small Memory and Fast Elliptic Scalar Multiplications Secure against Side Channel Attacks. In: Joye, M. (ed.) CT-RSA 2003. LNCS, vol. 2612, pp. 328–342. Springer, Heidelberg (2003)

25. Sica, F., Ciet, M., Quisquater, J.-J.: Analysis of the Gallant-Lambert-Vanstone Method Based on Efficient Endomorphisms: Elliptic and Hyperelliptic Curves. In: Nyberg, K., Heys, H. (eds.) SAC 2002. LNCS, vol. 2595, pp. 21–36. Springer, Heidelberg (2003)
26. Stark, H.M.: Class-numbers of Complex Quadratic Fields. In: Modular functions of one variable, I (Proc. Internat. Summer School, Univ. Antwerp, Antwerp, 1972). Lecture Notes in Mathematics, vol. 320, pp. 153–174. Springer, Berlin (1973)
27. Taverne, J., Faz-Hernández, A., Aranha, D.F., Rodríguez-Henríquez, F., Hankerson, D., López, J.: Software Implementation of Binary Elliptic Curves: Impact of the Carry-Less Multiplier on Scalar Multiplication. In: Preneel, B., Takagi, T. (eds.) CHES 2011. LNCS, vol. 6917, pp. 108–123. Springer, Heidelberg (2011)
28. Zhou, Z., Hu, Z., Xu, M., Song, W.: Efficient 3-dimensional GLV method for faster point multiplication on some GLS elliptic curves. Inf. Proc. Lett. 77(262), 1075–1104 (2010)

A   Examples

We give a few examples of GLV curves, which are curves defined over $\mathbb{C}$ with complex multiplication by a quadratic integer of small norm, corresponding to an endomorphism ϕ of small degree[4]. They make up an exhaustive list, up to isomorphism, in increasing order of endomorphism degree up to degree 3. While the first four examples appear in the previous literature, the next ones (degree 3) are new and have been computed with the Stark algorithm [26].

Example 1. Let $p \equiv 1 \pmod 4$ be a prime. Define an elliptic curve E over $\mathbb{F}_p$ by

$$y^2 = x^3 + ax \ .$$

If β is an element of order 4, then the map ϕ defined in the affine plane by

$$\phi(x, y) = (-x, \beta y)$$

is an endomorphism of E defined over $\mathbb{F}_p$ with $\mathrm{End}(E) = \mathbb{Z}[\phi] \cong \mathbb{Z}[\sqrt{-1}]$, since ϕ satisfies the equation[5]

$$\phi^2 + 1 = 0 \ .$$

Example 2. Let $p \equiv 1 \pmod 3$ be a prime. Define an elliptic curve E over $\mathbb{F}_p$ by

$$y^2 = x^3 + b \ .$$

If γ is an element of order 3, then we have an endomorphism ϕ defined over $\mathbb{F}_p$ by

$$\phi(x, y) = (\gamma x, y) \ ,$$

and $\mathrm{End}(E) = \mathbb{Z}[\phi] \cong \mathbb{Z}[\frac{1+\sqrt{-3}}{2}]$, since ϕ satisfies the equation

$$\phi^2 + \phi + 1 = 0 \ .$$

---

[4] By small we mean really small, usually less than 5. In particular, for cryptographic applications, the degree is much smaller than the field size.

[5] This is the only case when we cannot apply Lemma 1. It needs a separate treatment, given in [21], Appendix B.

Example 3. Let $p > 3$ be a prime such that -7 is a quadratic residue modulo p. Define an elliptic curve E over $\mathbb{F}_p$ by

$$y^2 = x^3 - \frac{3}{4}x^2 - 2x - 1 \ .$$

If $\xi = (1 + \sqrt{-7})/2$ and $a = (\xi - 3)/4$, then we get the $\mathbb{F}_p$-endomorphism ϕ defined by

$$\phi(x, y) = \left(\frac{x^2 - \xi}{\xi^2(x - a)}, \frac{y(x^2 - 2ax + \xi)}{\xi^3(x - a)^2} \right) \ ,$$

and $\mathrm{End}(E) = \mathbb{Z}[\phi] \cong \mathbb{Z}[\frac{1+\sqrt{-7}}{2}]$, since ϕ satisfies the equation

$$\phi^2 - \phi + 2 = 0 \ .$$

Example 4. Let $p > 3$ be a prime such that -2 is a quadratic residue modulo p. Define an elliptic curve E over $\mathbb{F}_p$ by

$$y^2 = 4x^3 - 30x - 28$$

together with the $\mathbb{F}_p$-endomorphism ϕ defined[6] by

$$\phi(x, y) = \left(-\frac{2x^2 + 4x + 9}{4(x + 2)}, y\frac{2x^2 + 8x - 1}{4\sqrt{-2}(x + 2)^2} \right) \ .$$

We have $\mathrm{End}(E) = \mathbb{Z}[\phi] \cong \mathbb{Z}[\sqrt{-2}]$ since ϕ satisfies the equation

$$\phi^2 + 2 = 0 \ .$$

Example 5. Let $p > 3$ be a prime such that -11 is a quadratic residue modulo p. We define the elliptic curve E over $\mathbb{F}_p$

$$y^2 = x^3 - \frac{13824}{539}x + \frac{27648}{539}$$

with $a = (1 + \sqrt{-11})/2$ and the endomorphism ϕ defined by

$\phi(x, y) =$

$$\left(\frac{\left(-\frac{539}{5184}a + \frac{539}{1728}\right)x^3 + \left(\frac{28}{27}a - \frac{35}{18}\right)x^2 + \left(-\frac{92}{9}a + \frac{8}{3}\right)x + \frac{1728}{77}a + \frac{192}{77}}{\left(\frac{2695}{5184}a - \frac{539}{864}\right)x^2 + \left(-\frac{217}{54}a + \frac{49}{18}\right)x + \frac{64}{9}a - \frac{4}{3}}, y \right.$$

$$\left. \frac{\left(\frac{3773}{373248}a - \frac{18865}{995328}\right)x^3 + \left(-\frac{2695}{20736}a + \frac{539}{3456}\right)x^2 + \left(\frac{7}{432}a - \frac{91}{144}\right)x + \frac{20}{27}a + \frac{1}{9}}{\left(-\frac{18865}{1492992}a + \frac{116963}{995328}\right)x^3 + \left(\frac{7007}{20736}a - \frac{539}{432}\right)x^2 + \left(-\frac{791}{432}a + \frac{581}{144}\right)x + \frac{74}{27}a - \frac{35}{9}} \right)$$

such that $\mathrm{End}(E) = \mathbb{Z}[\phi] \cong \mathbb{Z}[\frac{1+\sqrt{-11}}{2}]$. The characteristic polynomial of ϕ is

$$\phi^2 - \phi + 3 = 0 \ .$$

---

[6] We take the opportunity to correct a typo found and transmitted in many sources, where a y factor was absent in the second coordinate. Its sign is irrelevant.

Example 6. Let $p > 3$ be a prime such that -3 is a quadratic residue   mod p. We define the elliptic curve E over $\mathbb{F}_p$

$$y^2 = x^3 - \frac{3375}{121}x + \frac{6750}{121}$$

with the endomorphism ϕ defined by

$$\phi(x,y) = \left(-\frac{1331x^3 - 10890x^2 + 81675x - 189000}{33(11x - 45)^2}, y\,\frac{1331x^3 - 16335x^2 + 7425x + 43875}{3\sqrt{-3}(11x - 45)^3} \right)$$

such that[7] $\mathrm{End}(E) = \mathbb{Z}[\phi] \cong \mathbb{Z}[\sqrt{-3}]$. The characteristic polynomial of ϕ is

$$\phi^2 + 3 = 0 \ .$$

B   A New Four-Dimensional Lattice Reduction Algorithm

Algorithm 1 (Cornacchia's GCD algorithm in $\mathbb{Z}$)

Input: $n \equiv 1 \pmod{4}$ prime, $1 < \mu < n$ such that $\mu^2 \equiv -1 \pmod{n}$.
Output: $\nu = \nu_{(R)} + i\nu_{(I)}$ Gaussian prime dividing n, such that $\nu P = 0$.

---

1. **initialize:**
   $r_0 \leftarrow n$, $r_1 \leftarrow \mu$, $r_2 \leftarrow n$,
   $t_0 \leftarrow 0$, $t_1 \leftarrow 1$, $t_2 \leftarrow 0$,
   $q \leftarrow 0$.
2. **main loop:**
   while $r_2^2 \geq n$ do
       $q \leftarrow \lfloor r_0/r_1 \rfloor$,
       $r_2 \leftarrow r_0 - qr_1$, $r_0 \leftarrow r_1$, $r_1 \leftarrow r_2$,
       $t_2 \leftarrow t_0 - qt_1$, $t_0 \leftarrow t_1$, $t_1 \leftarrow t_2$.
3. **return:**
   $\nu = r_1 - it_1$, $\nu_{(R)} = r_1$, $\nu_{(I)} = -t_1$

---

[7] This is the first example where the endomorphism ring is not the maximal order of its field of fractions. It can be summarily seen as follows: $\mathrm{End}(E) \supseteq \mathbb{Z}[\sqrt{-3}]$. If not equal, then it must be the full ring of integers $\mathbb{Z}[\frac{1+\sqrt{-3}}{2}]$. This would imply that $j = 0$, as there is only $h(-3) = 1$ isomorphism class of elliptic curves with complex multiplication by $\mathbb{Z}[\frac{1+\sqrt{-3}}{2}]$, given in Example 2 (see [26] for an abridged description of the theory of complex multiplication). This is clearly not the case here. Alternatively, one can see that there would exist a nontrivial automorphism (a primitive cube root of unity) corresponding to $\frac{-1+\sqrt{-3}}{2}$. A direct computation then shows this is impossible.

Algorithm 2 (Cornacchia's algorithm in $\mathbb{Z}[i]$ - compact form)

Input: ν Gaussian prime dividing n rational prime, $1 < \lambda < n$ such that $\lambda^2 + r\lambda + s \equiv 0 \pmod{n}$.
Output: Two $\mathbb{Z}[i]$-linearly independent vectors v_1 & v_2 of $\ker F$ $\subset$ $\mathbb{Z}[i]^2$ of rectangle norms $< 51.5(\sqrt{1 + |r| + s})\,n^{1/4}$.

---

1. **initialize:**
   If $\lambda^2 \geq 2n$ then
       $r_0 \leftarrow \lambda$,
   else
       $r_0 \leftarrow \lambda + n$,
   $r_1 \leftarrow \nu$, $r_2 \leftarrow n$,
   $s_0 \leftarrow 1$, $s_1 \leftarrow 0$, $s_2 \leftarrow 0$,
   $q \leftarrow 0$.
2. **main loop:**
   while $|r_2|^4(1 + |r| + s)^2 \geq n$ do
       $q \leftarrow$ closest Gaussian integer to r_0/r_1,
       $r_2 \leftarrow r_0 - qr_1$, $r_0 \leftarrow r_1$, $r_1 \leftarrow r_2$,
       $s_2 \leftarrow s_0 - qs_1$, $s_0 \leftarrow s_1$, $s_1 \leftarrow s_2$.
3. **return:**
   $v_1 = (r_0, -s_0)$, $v_2 = (r_1, -s_1)$

Shuffling against Side-Channel Attacks: A Comprehensive Study with Cautionary Note

Nicolas Veyrat-Charvillon, Marcel Medwed,
Stéphanie Kerckhof, and François-Xavier Standaert

Université Catholique de Louvain, UCL Crypto Group,
B-1348 Louvain-la-Neuve, Belgium

Abstract. Together with masking, shuffling is one of the most frequently considered solutions to improve the security of small embedded devices against side-channel attacks. In this paper, we provide a comprehensive study of this countermeasure, including improved implementations and a careful information theoretic and security analysis of its different variants. Our analyses lead to important conclusions as they moderate the strong security improvements claimed in previous works. They suggest that simplified versions of shuffling (e.g. using random start indexes) can be significantly weaker than their counterpart using full permutations. We further show with an experimental case study that such simplified versions can be as easy to attack as unprotected implementations. We finally exhibit the existence of "indirect leakages" in shuffled implementations that can be exploited due to the different leakage models of the different resources used in cryptographic implementations. This suggests the design of fully shuffled (and efficient) implementations, were both the execution order of the instructions and the physical resources used are randomized, as an interesting scope for further research.

1 Introduction

Already in the first Differential Power Analysis (DPA) paper, Kocher et al. mentioned time randomization as a possible solution to improve security against side-channel attacks [15]. Following, different countermeasures have been proposed to exploit this idea, e.g. relying on the addition of random delays [7,31], shuffling the execution order of independent operations [13,26], or more generally, trying to build a non-deterministic processor [4,19]. As usual in side-channel attacks, the main question regarding these solutions is: "to what extent do they improve security and at which cost?". In this paper, we propose a comprehensive treatment of this question in the case of the shuffling countermeasure.

For this purpose, we start with the efficiency issue. In general, shuffling can be applied to any set of independent operations. The SubBytes layer of 16 S-boxes in the AES Rijndael is a typical example. Randomizing such operations can be done in different ways. Taking the extreme cases, either the S-boxes are executed according to a Random Permutation (RP) among 16! possible ones, or they are executed from a Random Start Index (RSI) among 16 possible ones, that is then incremented. This difference is nicely illustrated with previous works on shuffled

X. Wang and K. Sako (Eds.): ASIACRYPT 2012, LNCS 7658, pp. 740–757, 2012.

implementations of the AES. In a first paper from 2006 [13], the authors use partial shuffling and first-order masking based on S-box pre-computation. Whereas masking is applied to the whole cipher, shuffling is only applied to the first and last rounds. Furthermore, the RSI approach is pursued, for performance reasons. In a second work from Rivain et al., higher-order masking is implemented and shuffling is mainly based on the RP approach [26]. Yet, for the MixColumns operations, only the (first-order masked) columns are shuffled, accounting for 8! possible permutations. That is, thanks to the first-order masking, they have 8 positions that can be shuffled, vs. 4 if MixColumns was not masked, and 16 for the other AES transforms. Implementation details are not given in [26], but we assume that this choice is again motivated by performance reasons, with a MixColumn operation implemented with xtime tables [10]. Apart from those works, shuffling was also applied to hardware implementations with 8- or 32-bit datapaths, where RSI is usually preferred as it nearly comes for free [11,20,22].

Following this state-of-the art, our first contribution is to improve the performances of software implementations using the RP approach. In this respect, we start from the observation that in an unprotected block cipher implementation, one usually keeps as much data as possible in the processor registers, in order to minimize RAM access. By contrast, once random access to these registers is required (as in shuffled implementations), RAM usage is inevitable. This implies that any register access becomes a serial of load and store operations, resulting in major performance overheads. We mitigate these overheads by exploiting a different technique, which consists in manipulating the program flow. It allows us to operate on registers while at the same time randomizing the sequence of operations. In practice, we present two approaches: the first one changes the program flow "on-the-fly", while the other one re-writes the program memory prior to execution. The latter approach can be viewed as an adaptation of the self-modifying codes used in software engineering [2], also applied to counteract side-channel attacks in [1]. Our new solutions come with contrasted performance results. Namely, the on-the-fly proposal minimizes the overall cycle count, while the program memory manipulations allow very efficient online encryption at the cost of long (possibly offline) pre-computations. For illustration, we apply these proposals (and previously published ones) to the AES Furious implementation available from [23]. Besides, we also investigate the efficient generation of (small) random permutations in low-cost microcontrollers. That is, we take a well known optimal algorithm for permutation generation and modify it slightly, in order to improve its performances. As a result, we are able to generate close-to-uniform permutations, and obtain an efficient alternative to proposals such as [8].

Next, we investigate the security of shuffling against side-channel attacks. Here, we start from the observation that the existing literature usually evaluates the impact of shuffling based on a so-called "integrated DPA" (aka windowing attack), introduced in [7] and applied, e.g. in [26,30]. Intuitively, if the manipulation of a sensitive variable is spread over t time samples, its correlation with the actual leakages will be reduced by a factor $\sqrt{t}$ using such an attack, instead of t without integration. Integrating is a convenient tool for evaluation as it can be directly used to estimate the data complexity of a DPA using Mangard's formulas [17]. Yet, a possible limitation of this technique is that it treats the RSI and RP cases in the same way. Hence, a natural question is to determine whether these two approaches are indeed equivalent in general, or if advanced evaluation tools can be used to put forward additional weaknesses for RSI implementations. Our results regarding this question are summarized as follows.

First, we specialize the information theoretic and security analysis from [28] to the context of shuffled implementations. It allows us to confirm that integrated DPA is indeed suboptimal compared to a Bayesian exploitation of the leakages. While our worst-case evaluations rely on profiled attacks [6,27], we believe they are important to moderate claims of strong security improvements provided by shuffling (e.g. the data complexity increases by a factor 360 in [4]). In particular, these results complement the previous work of Asiacrypt 2010 [29], in which such an information theoretic and security analysis was performed for masking. As a result and for the first time, we obtain lower bounds for the data complexity of standard side-channel attacks against shuffled implementations.

Second, we notice that security evaluations for masking always combine the leakage corresponding to the masked data and its masks, e.g. [21,24]. Quite surprisingly, and to the best of our knowledge, the impact of such a scenario has not been investigated in the case of shuffling. Therefore, we include the possibility of a leakage on the permutation (or start index) manipulated when shuffling. We show that as soon as some information is leaked about them, attacks against RSI- and RP-based implementations become significantly different, the RSI case being much easier to attack, for computational reasons.

Finally, we observe that direct leakages about the start index or permutations naturally arise in practice and can be exploited. More surprisingly, we also show the existence of "indirect leakages", coming from the different power consumption models of the hardware resources manipulating the key bytes. For example, since the 16 registers used in our shuffled Furious implementations have (slightly) different models, marginalizing the distribution of the observed leakage over the 16 AES key bytes provides information about which S-box is computed.

Summarizing, we observe that *all* previous works on shuffling reduced the size of the permutation set for some of the operations in the protected algorithm. Hence, our results bring the important cautionary note that time complexity is critical in the security evaluation of this countermeasure, as permutations with a small size can be enumerated which leads to exploitable weaknesses. In this respect, an implementation protected with RSI-based shuffling can sometimes be as weak as an unprotected one. As for the RP-based solution, we recall that it can be used as a noise amplifier for leaking devices, but never as a noise generator.

2   Efficient Implementations

This section explores the software design space for shuffling the AES on an Atmel ATMega644P microcontroller [3]. We first describe an efficient way to obtain close-to-uniform permutations in this device. Next, we show how to obtain an AES implementation for which every transform can be shuffled according to such permutations, including MixColumns (and the key scheduling algorithm if needed). Afterwards, we describe different implementations: a basic one relying on a previously proposed "double indexing" method, and two optimized ones relying on randomized execution path and program memory. We finally provide precise performance evaluations and a comparison with previous works.

Permutation Generation. The first building block of a shuffled implementation is a permutation generator. From a sequence $S := \{1, \ldots, n\}$, a uniform permutation can be produced in linear time [14]. The original algorithm iterates over every element S_i (with i from 0 to $n - 2$), and swaps it with a random element from the remaining tail, i.e. $\{S_i, \ldots, S_{n-1}\}$. However, sampling from $\{i, \ldots, n - 1\}$ needs either a modulo operation and a random number greater than n to start from, or an approach with probabilistic run-time. We avoided this performance drawbacks by sampling from $\{0, \ldots, n - 1\}$. Permuting a sequence of 16 entries following this algorithm takes 362 cycles on our device, using 8 bytes of randomness. It still allows to generate all permutations, but with a slight bias that decreases with the size of the permuted set. To estimate the impact of this bias for different sizes of the permutation set N, we systematically sampled 10^8 permutations generated with this method, and built histograms with $N!$ bins. We then estimated the Euclidean distance between these biased histograms and a uniform distribution. In addition, we compared this situation with the one obtained with a quite minimum side-channel leakage. Namely, we assumed that the Hamming weight of the first entry of a (uniformly generated) permutation is known to be the least informative one (i.e. with half of the bits set to one). As can be observed in Table 1, the bias due to this small side-channel information is already significantly larger than the one due to the permutation generation algorithm. Furthermore, actual leakages in Sections 3 and 4 affect all the permutation entries, which further reduces the bias of the permutation generation algorithm compared to the one caused by physical information. Eventually, we will show in the next sections that exploiting these biases in a side-channel attack where we shuffle among 16! possible permutations is computationally hard. Therefore, we conclude that our performance optimized algorithm should not lead to a significant security reduction of the shuffling countermeasure.

Table 1. Bias of the optimized permutation algorithm vs. bias of a small SCA leakage

N	3	4	5	6	7	8	9
Perm. generation	0.04535	0.03522	0.02034	0.00993	0.00430	0.00170	0.00063
Small SCA Leak.	0.28868	0.20412	0.07454	0.03726	0.01627	0.00643	0.00234

Obtaining Independent Operations. Applying shuffling to an implementation requires finding sets of independent operations. In the AES case, sets of 16 independent operations naturally arise from the AddRoundKey and SubBytes transforms. By contrast, the situation is a little bit trickier for ShiftRows and MixColumns. For example, implementing ShiftRows requires one extra byte of storage in an unprotected implementation, and two in the case of RSI-based shuffling (i.e. when the permutation is "monotonous", which restricts the number of permutations to 16). But if 16 independent operations are desired, 16 bytes of temporary storage are required. As for MixColumns, four additional registers are sufficient if the state is processed column-wise, but this would then account for 4! permutations. Hence, having 16 independent operations again requires 16 bytes of temporary storage. Since our device has only 32 registers, some of which being already occupied, RAM usage becomes inevitable for shuffling. Besides, the key schedule has only four independent operations by default. This is because within one key schedule round, there are only four S-box executions. Thus, the smallest number of indistinguishable operations is four. Yet, applications requiring on-the-fly key expansion also need an appropriate SPA protection to prevent attacks such as [16]. In these cases, we interleaved the real key schedule with three dummy key schedules, in order to obtain 16 shuffleable operations.

Basic Implementation with Double Indexing. Direct shuffling requires an indirect indexing of the operands. That is, a counter is used to index a permutation vector, and the result is used to index the operand vector. Thus, instead of operating on registers directly, two RAM accesses are required for each (read or write) access to operands. This naturally leads to quite large cycle counts, as in AVR devices, load and store operations take two cycles (compared to one cycle for arithmetic and logic operations). Implementing a fully shuffled AES this way results in an execution time of 30 202 cycles, excluding the key schedule. In the following we propose two different strategies in order to improve on these figures. In both cases, instructions are shuffled rather than data location, in order to allow register usage. Precisely, we are still limited by the number of available registers when performing certain transforms. But contrary to the double indexing proposal we do not always access RAM when operating on intermediate data. The first solution changes the execution path on-the-fly while the second actually rewrites the program memory (i.e. assuming that this re-writing is pre-computed, this solution can be seen as a simplified one-time program [12]).

Optimized Implementation with Randomized Execution Path. For this implementation, the assembly code of every (compound of) round transform(s) is split into 16 independent blocks of instructions. Each of the 16 blocks is augmented with a label. This allows us to identify its address in ROM. Furthermore, every transform is associated with an array of 17 16-bit words, where the first 16 words hold the addresses of the 16 blocks, and the 17th holds the address of the return instruction. The array content is initialized with the addresses of the labels at compile time. Finally, we append a flow-control macro to each of the 16 blocks. This macro performs three things: fetch an address from the array, advance the pointer to the next array entry, and jump to the fetched address.

During the execution of the cipher, we first re-order the first 16 addresses in the array, according to a previously generated permutation. Then, when we enter a transform, we set a pointer to the beginning of the array and execute the flow-control macro. This causes the execution of the first block and sets a pointer to the address of the next block. The flow-control macro is executed 16 times, until it finally looks up the address of the return instruction. In practice, we defined several sets of transforms and therefore need an address array for each of them.

The first one is the compound of AddRoundKey, SubBytes, and ShiftRows. This transform reads the state from RAM and stores the result into the register file. The next one performs MixColumns and stores the result back to RAM. Afterwards, we perform one iteration of the key schedule. Similar to ShiftRows and MixColumns, this implies additional memory requirements, because of the RotWord operation. We finally need a standalone AddRoundKey layer for the last round. As each of the address arrays need 17x2 bytes, we have an additional RAM use of 170 bytes (for technical reasons, the key schedule uses two 17x2-byte address arrays). Permuting each of these arrays takes 205 cycles, implying an overhead of 1225 cycles. Eventually, for every set of transforms, we need to load an address and jump to this address 17 times, each of which takes 6 cycles. Together with the preamble to set up the array pointer, it leads to an additional overhead of 108 cycles for each of these compounds of transforms.

Optimized Implementation with Randomized Program Memory. For this implementation, we used the self-programming capabilities of the ATMega644p microcontrollers. As the shuffling applies to independent operations, and as for each operation, the state bytes are always stored in the same registers or RAM locations, the execution order of the operations can be permuted by modifying the data corresponding to these locations in program memory. In our target controller, the program memory has to be modified one page (i.e. 256 bytes) at a time. Hence, the shuffling can be prepared in five steps. First, the page is transferred from program memory to the RAM. Afterwards, the bytes of code corresponding to state-byte locations are modified according to the permutation vector. Then, the previous version of the page is erased from program memory, and the new page is loaded into a page buffer. Finally, this page buffer is written in program memory. This process is executed before each AES execution. The main advantage of this solution is that after pre-processing of a shuffled program memory, the execution time of the AES is nearly the same as for the unprotected implementation. Minor differences come from the fact that we need to have independent operations, which implies to use RAM for the storage of some intermediate results. Its main drawback is the long pre-computation time, which accounts for approximately 18 milliseconds independently of the clock frequency. This comes from the time-consuming instructions used to erase program memory and write page buffer in memory (4.5 millisecond per page writen or erased [9]), and the low granularity of these instructions (i.e. working at the page level) in the Atmel controllers. More flexible devices (e.g. devices with ARM architectures) would allow to improve this limitation. Note also that our target Atmel's EEPROM allows only for 10 000 re-write cycles, which could possibly lead to DoS attacks. If this is an issue, and depending on the actual available ROM, different areas can be used randomly and increase the number of possible encryptions by some factor. Again, alternative devices could be considered to relax this limitation. For example, the ARM LPC214x series allows already for 100 000 cycles. Note finally that, as this implementation mainly makes sense if pre-processing is allowed, it is naturally executed with a pre-computed key scheduling.

Implementation Results. The performance results of our implementations are compared with previous works in Table 2. Namely, we use the AES Furious as reference. As for protected implementations, we considered the basic one based on double indexing and the ones of Herbst et al. and Rivain et al. However, as mentioned above, they do not allow direct comparison. Herbst et al. only protect the outer rounds (one and ten) with RSI-based shuffling, but implement masking for all the rounds and the key schedule. Rivain et al. implement higher-order masking and use a "simplified" shuffling for the MixColumns operation (they also work on a different 8051-based architecture). The implementation for which we give cycle numbers is not masked except for MixColumns. By contrast, our implementations use log-table based polynomial multiplication, and are able to shuffle all bytes during MixColumns. Not surprisingly, our implementation based on double indexing is the slowest. Its performance is comparable to the one of Rivain et al. Manipulating the program counter allows us to get a performance improvement of almost a factor five and, excluding the key scheduling, leads to encryption time only twice as slow as Rijndael Furious. As previously mentioned, the larger overheads when executing the key scheduling come from the need to execute additional dummy schedulings, in order to keep a permutation among 16! for this part of the implementation. Finally, the randomized program memory allows the fastest online encryption (i.e. excluding program re-writing).

Table 2. Implementation result comparison

Implementation	Clock cycles	RAM [byte]
Furious [23]	2 739	176
Furious with KS [23]	3 546	176
Herbst et al. [13]	11 845	-
Rivain et al. [26]	29 400	-
Dbl. ind.	30 202	240
Dbl. ind. with KS	46 395	132
Rand. exec. path	6 934	394
Rand. exec. path with KS	14 834	302
Rand. prog. mem.	3299 ($+\simeq18$ msec)	480

3   Evaluation Framework

We now move to the security analysis of the shuffling countermeasures and its variants. For this purpose, we rely on the evaluation framework from [28] and adapt it to capture the specificities of shuffled implementations. In order to have a fair understanding of the strengths and weaknesses of the countermeasure, we pay a particular attention to worst-case (profiled) attacks. But for completeness, we also compare them with the integrated DPA used in previous works.

Notations. Variables are denoted with capital letters, sampled values with lowercase letters and functions with sans serif fonts. We consider the standard DPA attacks described in [18] and illustrate our notations with the case of the

AES Rijndael. In this context, the adversary tries to recover a 16-byte master key $\mathbf{k} = \{k_0, k_1, \ldots, k_{15}\}$, from a leakage corresponding to the first key addition and S-box layers. In attacks against unprotected implementations, each S-box is executed at a well defined time instant, giving rise to key leakages defined as:

$$L_0 \leadsto \mathsf{Sbox}(k_0 \oplus X_0), \qquad\qquad L_1 \leadsto \mathsf{Sbox}(k_1 \oplus X_1),$$
$$L_2 \leadsto \mathsf{Sbox}(k_2 \oplus X_2) \qquad\qquad \ldots$$

That is, we have 16 leakage points (or cycles) L_c (where c is the cycle index) and 16 subkeys k_s. If we denote the part of the master key that is manipulated at time c with a variable S_c, we straightforwardly have $S_c = c$ in this unprotected case. Note that the variable nature of the leakages comes both from possible noise in the measurements and the variable (known) inputs X_i. By contrast, in the case of a shuffled implementation, the execution order of the S-box computations is randomized according to a permutation P, leading to key leakages of the form:

$$L_0 \leadsto \mathsf{Sbox}(k_{\mathsf{P}(0)} \oplus X_{\mathsf{P}(0)}), \qquad L_1 \leadsto \mathsf{Sbox}(k_{\mathsf{P}(1)} \oplus X_{\mathsf{P}(1)}),$$
$$L_2 \leadsto \mathsf{Sbox}(k_{\mathsf{P}(2)} \oplus X_{\mathsf{P}(2)}) \qquad\qquad \ldots$$

That is, we have $S_c = \mathsf{P}(c)$ with P the secret permutation that is re-generated for every new input block, e.g. with the algorithm in Section 2. In this protected case, not only leakage about the S-box execution may be obtained, but also leakage on the permutation used in the shuffled implementation. In theory, an attack could exploit sixteen "direct" permutation leakages denoted as $L'_c \leadsto S_c$. Such notations allow us to reflect both the RSI- and RP-based shuffling methods. In the first case, we have $\mathsf{P}(c) = c + \tau \pmod{16}$, with $\tau \xleftarrow{R} [0:15]$. In the second case, P is directly picked up among the set of all 16! permutations, i.e. $\mathsf{P} \xleftarrow{R} \mathcal{P}_{16}$.

Information Theoretic Analysis. As a first step in our evaluation, we perform an information theoretic analysis that is aimed to capture the worst-case security of an implementation. In general, and for a fixed key byte K_s, we assume that the adversary can observe a leakage vector $\mathbf{L} = \{L_0, L_1, \ldots, L_{15}\}$. The goal of this evaluation is to obtain an accurate estimation of the mutual information[1]:

$$\mathrm{MI}(K_s; \mathbf{L}, X) = \mathrm{H}[K_s] - \sum_k \Pr[K_s = k] \sum_x \Pr[X = x]$$
$$\cdot \int_l \Pr[\mathbf{L} = \mathbf{l} | K_s = k, X = x] \cdot \log_2 \Pr[K_s = k | \mathbf{L} = \mathbf{l}, X = x] \; dl.$$

In this equation, the term $\Pr[K_s = k | \mathbf{L} = \mathbf{l}, X = x]$ is directly obtained from $\Pr[\mathbf{L} = \mathbf{l} | K_s = k, X = x]$ using Bayes' theorem. Hence, it is this last conditional leakage probability that is most critical to evaluate. For convenience, we will ignore the variable X in the rest of the paper, as it is assumed to be known for all computations. Next, we will consider two main evaluation scenarios.

---

[1] As discussed in [25], this mutual information can only be perfectly estimated when the evaluator knows the exact leakage model of his target device. This only happens in simulated analyses (e.g. as will be performed in the next section). Whenever a practical evaluation is carried out, it is formally a "perceived information" that is evaluated, with the goal to be as close as possible to the mutual information.

1. No Permutation Leakage, i.e. the adversary gets 16 leakage cycles, and each of them could correspond to the target subkey with probability $1/16$. That is:

$$\Pr[\mathbf{L} = 1|K_s = k] = \sum_c \frac{1}{16} \Pr[L_c = l_c|K_s = k].$$

We will refer to this attack as case $(1.a)$. Besides, and as mentioned in introduction, a usual trick to attack shuffled implementations is to integrate over the leakage cycles. In this case, the adversary defines a variable $\overline{L} = \sum_c L_c$, and performs the attack against this variable. It boils down to consider 15 cycles out of 16 as "algorithmic noise". We will refer to this attack as case $(1.b)$.

2. Leakage on the Permutation. In the same way as all the shares are assumed to leak in a masked implementation, it is natural to assume that the manipulation of a permutation may leak in a shuffled implementation. In practice, such leakages usually appear each time the permutation is manipulated in the microcode, e.g. when fetching the S_c'th part of the key, or when jumping to the S_c'th piece of code computing an S-box. We now show how to perform an information theoretic evaluation in these cases. As previously, the impact of different implementations of the countermeasure affects the term $\Pr[\mathbf{L} = 1|K_s = k]$. For this purpose, we start with the following general formulation:

$$\Pr[\mathbf{L} = 1|K_s = k] = \sum_c \frac{\mathsf{f}(c, s, \mathbf{l}')}{\sum_{c'} \mathsf{f}(c', s, \mathbf{l}')} \Pr[L_c = l_c|K_s = k],$$

with $\mathbf{l}'$ the vector of 16 leakages on the previously defined variable S_c (indicating the part of the master key used at time c). The function f essentially indicates how the knowledge available about this variable can be exploited by the adversary, as witnessed by the five examples that we now describe.

2.a. Unprotected implementation. In this case, we have $\mathsf{f}(c, s, \mathbf{l}') = 1$ if $c = s$ and 0 otherwise (i.e. the adversary knows exactly where each key byte is manipulated).

2.b. Direct template attack. In this case, we just add the permutation leakage in the conditional probabilities, yet without making any difference between the RSI and RP cases, by computing $\mathsf{f}(c, s, \mathbf{l}') = \Pr[L'_c = l'_c|S_c = s]$. Note that the case with no permutation leakage corresponds to $\mathsf{f}(c, s, \mathbf{l}') = 1/16$.

2.c. Taking advantage of RSI. Here, the the adversary exploits the fact that only 16 permutations are possible (out of the 16! ones), which can be enumerated. Hence, he can compute: $\mathsf{f}(c, s, \mathbf{l}') = \prod_{i=0}^{15} \Pr[L'_i = l'_i|S_i = (s - c + i) \bmod 16]$.

Contrary to the RSI case, using a RP implies that the permutation is picked up randomly among the $16! \simeq 2^{44}$, which is significantly harder to enumerate. Hence, our following experiments will additionally consider two heuristic solutions that can be used to mitigate this issue and attack more efficiently.

2.d. Restricted enumeration against RP. In this case, the function f is identical to the exhaustive one, i.e. $\mathsf{f}(c, s, \mathbf{l}') = \sum_{\mathsf{p}} \prod_{i=0}^{15} \Pr[L'_i = l'_i|S_i = \mathsf{p}(i)]$, but the sum only goes over an enumerable subset of most probable p's. A *beam search* is used for this purpose [32]. This is a breadth-first search that limits the number of nodes (i.e. permutations in the sum) by pruning the least probable ones, which is done by weighting permutations p's with $\prod_{i=0}^{15} \Pr[L'_i = l'_i | S_i = \mathsf{p}(i)]$.

2.e. Excluding heuristic. One alternative option to simplify the enumeration is to consider that whenever $S_c = s$, we have that $c \neq c'$ implies $S_{c'} \neq S_c$. This can be reflected with: $\mathsf{f}(c, s, \mathbf{l}') = \Pr[L'_c = l'_c | S_c = s] \cdot \prod_{c' \neq c} (1 - \Pr[L'_{c'} = l'_c | S_{c'} = s])$, which, up to normalization, is equivalent to:

$$\mathsf{f}(c, s, \mathbf{l}') = \frac{\Pr[L'_c = l'_c | S_c = s]}{1 - \Pr[L'_c = l'_c | S_c = s]}.$$

Overall, an intuition on the security of different implementations is obtained by quantifying the number of possible execution orders considered by the adversary (which may be more than the actual number of permutations, if attacks do not fully exploit their structure). In the unprotected case, only one order can occur. For the direct template attack, the adversary does not combine the different S_c informations and we implicitly have 16^{16} possible execution orders. In the RSI case, we exploit the fact that only 16 permutations are possible. The attack enumerating all possible permutations lists all 16! hypotheses. Finally, the excluding heuristic implicitly allows 16×15^{15} ones. This situation can be seen as an error correcting problem where 16 noisy values are transmitted, that can be integers from 0 to 15. The security of the countermeasure relies on a large probability of decoding error. In the RSI case, we only have 16 possible codewords, which gives us a very resilient code, lowering the probability of errors and thereby the strength of the countermeasure. For a RP, we have 16! codewords over a space of 16^{16} possible transmissions, hence increasing the probability of decoding errors.

As far as performing these attacks/evaluations in practice is concerned, case (a) is a classical template attack for which the computational complexity is usually neglected. Carrying out attacks/evaluations where L' is exploited naturally requires to build additional templates. Yet, the computational complexity of cases (b), (c) and (e) can also be neglected, as they only imply a few additional arithmetic operations. In fact, only case (d) may require intensive computations, if all permutations with non-negligible likelihood (with respect to L') are taken into account by the beam search. As will be shown in the next section, increasing the noise gradually implies that all permutation candidates have more similar likelihoods. Hence, this last attack is only applicable for low noise levels.

Security Analysis. The second step of our evaluation is to perform a security analysis. It allows measuring the extent to which the different strategies listed have a strong impact on the data complexity of successful side-channel attacks. For this purpose, we apply template attacks with the key selected as:

$$\tilde{k} = \underset{k^*}{\operatorname{argmax}} \prod_{j=1}^{q} \Pr[\mathbf{L}^j = \mathbf{l}^j, \mathbf{L}'^j = \mathbf{l}'^j | K_s = k^*],$$

and we compute their success rate, in function of the data complexity q.

4   Simulated Experiments

In order to gauge the impact of the proposed formulations and attacks, we first lead various experiments against simulated AES implementations. For this purpose, we re-use the notations introduced in the previous section and assume that the adversary is provided with a key leakage vector $\mathbf{L}$ with elements $L_c = \mathsf{HW}(\mathsf{Sbox}(k_{\mathsf{P}(c)} \oplus X_{\mathsf{P}(c)})) + \mathcal{N}(0, \sigma^2)$, and possibly a permutation leakage vector $\mathbf{L}'$ with elements of the form $L'_c = \mathsf{HW}(S_c) + \mathcal{N}(0, \sigma^2)$. In both cases, the second term is a Gaussian distributed random noise, with variance σ^2 that we will use as a parameter of our evaluations. Using these notations, there are various contexts that could be investigated. As illustrated in Table 3, we classify them among two axes: the target device and the adversary's means.

Table 3. Classification of the attacks

		Target devices		
		Unp.	RSI shuf.	RP shuf.
adversary's means	**L**	UNP-TA $(2.a)$	INT-TA $(1.b)$	
			UNI-TA $(1.a)$	
	L,L'		DPLEAK-TA $(2.b)$	
	L,L' + comp.		RSIENUM-TA $(2.c)$	RESENUM-TA $(2.d)$
				EXCLUDING-TA $(2.e)$

As far as the target device is concerned, we considered the case of an unprotected implementation for reference, an RSI-based shuffled implementation and a RP-based shuffled implementation. As far as the adversary's means are concerned, we first analyzed attacks where only the key leakage vector $\mathbf{L}$ is available. Next we evaluated attacks where the permutation leakage vector $\mathbf{L}'$ is additionally provided. Finally, we quantified the efficiency gains obtained when exploiting computational power, in order to enumerate (i.e. sum over) the possible permutations. Overall, this gives rise to seven attacks:

1. Template attack against the unprotected implementation (UNP-TA), i.e. the straightforward case where S-boxes are executed in deterministic order.
2. Template attack against integrated leakages (INT-TA), i.e. the attack against shuffled implementations previously used, e.g. in [7,26,30].

In these two first cases, template attacks and correlation DPA are essentially equivalent given that they exploit the same leakage model [18]. For coherence, we will keep on using template attacks everywhere. But as the experiments in Section 5 target a microcontroller with strong Hamming weight leakage dependencies, simpler (non-profiled) attacks would naturally apply as well. By contrast, the following attacks explicitly take advantage of a Bayesian description.

3. Template attack with uniform S_c (UNI-TA). In this case, the adversary follows the Bayesian strategy but does not exploit any information on the permutation (i.e. he assumes a uniform prior on the leakage cycles). Hence, the attacks still have identical efficiencies in the RSI and RP cases.

4. Template attack with direct permutation leakage (DPLEAK-TA). It corresponds to the attack (2.*b*) described in the previous section. Here, the leakage vector $\mathbf{L}'$ is simply added in the adversary's conditional probabilities. But again, it does not distinguish between the RSI and RP cases.

5. Template attack with permutation leakage enumerating the RSI permutations (RSIENUM-TA). It corresponds to the attack (2.*c*) in the previous section, where the adversary takes advantage of the 16 permutations that a RSI-based shuffling tolerates to combine its permutation leakages.

6. Template attacks with restricted enumeration (RESENUM-TA). It corresponds to the attack (2.*d*) described in the previous section. A beam search [32] is performed to enumerate the most likely permutations.

7. Template attacks with excluding heuristic (EXCLUDING-TA). It corresponds to the attack (2.*e*) in the previous section, where the likelihood of the permutations is weighted by simply excluding duplicates.

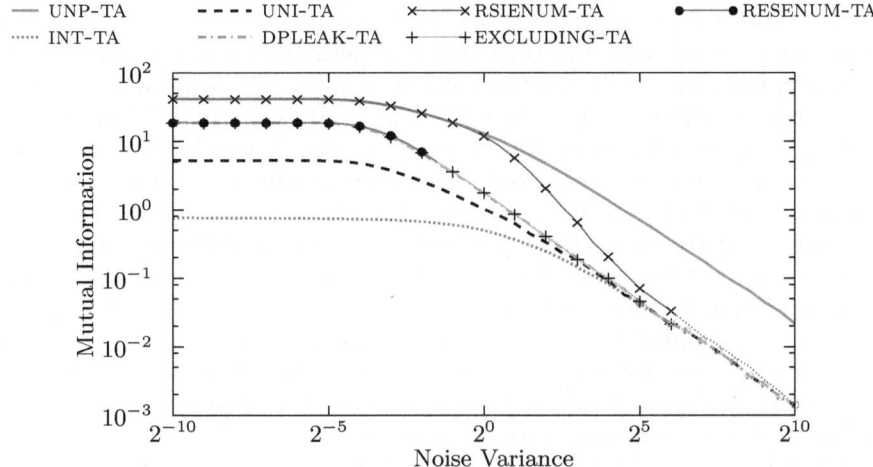

Fig. 1. Mutual information versus noise variance

The result of a simulated information theoretic analysis for these different attacks is given in Figure 1, in function of the noise variance. Several observations can be highlighted. First, and as usual in such worst-case evaluations, the asymptotic trend only appears for large noise levels. In this respect, the main conclusion is that (unlike masking [29]), the slope of the MI curves is the same for both the unprotected and all the shuffled implementations. Intuitively, it suggests that shuffling can (at best) be used to amplify the noise existing in side-channel measurements (i.e. imply a shift of the IT curves). Besides, one can observe that for lower noise levels, significant differences arise between the different scenarios of Table 3. For example, it is interesting to note that even without exploiting permutation leakage, the integrated attack is less efficient than the template attack with uniform prior. It confirms that this integrated attack is suboptimal in a profiled case, and is not suited to evaluate the worst-case security of an implementation in low-noise scenarios. Quite naturally, the distance between integrated and stronger attacks increases as permutation leakage becomes available. In this setting, the amount of information extracted is quite dependent on countermeasure implemented. If the RSI approach is chosen (and this information is exploited computationally), the implementation turns out to be as weak as an unprotected one until noise levels beyond $\sigma^2 = 2^0$. By contrast, in the RP case, the noise amplification happens earlier. In this respect, it is worth to notice the limited difference between the DPLEAK-, EXCLUDING-, and RESENUM-TAs for RP-shuffled implementations, the latter ones only bringing a small advantage. We also observe that as expected, the RESENUM-TA could only be launched until noise levels of approximately $\sigma^2 = 2^{-2}$: beyond this threshold, the large amount of permutations to enumerate with the beam search turned out to be hardly tractable. This last fact confirms the expectation in Section 2 that the small bias resulting from our efficient permutation generation algorithm should not lead to significantly improved side-channel attacks.

Note that the insecurity of RSI-based shuffling (and, to a lower extent, RP-based shuffling) for low noise levels has to be interpreted with care. What our analysis shows is *not* that the start index or permutation is trivially revealed with a template-based SPA (as the number of permutation candidates in the beam search already explodes when $\sigma^2 = 2^{-2}$). It is really the fact that the 16 leakage samples of the permutation can be exploited jointly that make these countermeasures weak. In other words, what these results show is the importance of computational power in the evaluation of shuffling: summing over 16 cases is easy, summing of 16! ones is harder, as highlighted by the different curves of the RSIENUM-TA and EXCLUDING-/RESENUM-TA information theoretic evaluations.

As a complement of information theoretic analyzes, we performed a security analysis, and computed the success rates of our different attacks, in function of the number of plaintexts measured by the adversary. This allows translating the IT curves of Figure 1 into data complexities. For illustration, we selected three different noise variances, corresponding to low (i.e. $\sigma^2 = 2^{-3}$), middle (i.e. $\sigma^2 = 2^0$) an large (i.e. $\sigma^2 = 2^3$) noise levels (where large refers to the fact that the IT curves are merging at this stage). The results of these simulated experiments are given in Figure 2 and confirm the previous observations. We again observe the weakness of the RSI-based shuffling in the low noise level case, and the lower efficiency of the integrated attack. The success rate curves also exhibit the slight advantage of the heuristic enumeration when exploiting the leakage of a RP for the smallest noise level, as well as the better behavior of the (computationally cheap) excluding heuristic when the noise increases (again, the RESENUM-TA evaluation could only be performed in the low noise case, i.e. upper figure).

5   Practical Experiments

The previous simulated attacks naturally raise the question whether our attacks similarly apply to real world implementations. In order to validate our conclusions, we also performed these attacks against shuffled implementations of the AES, based on the randomized execution path technique of Section 2.

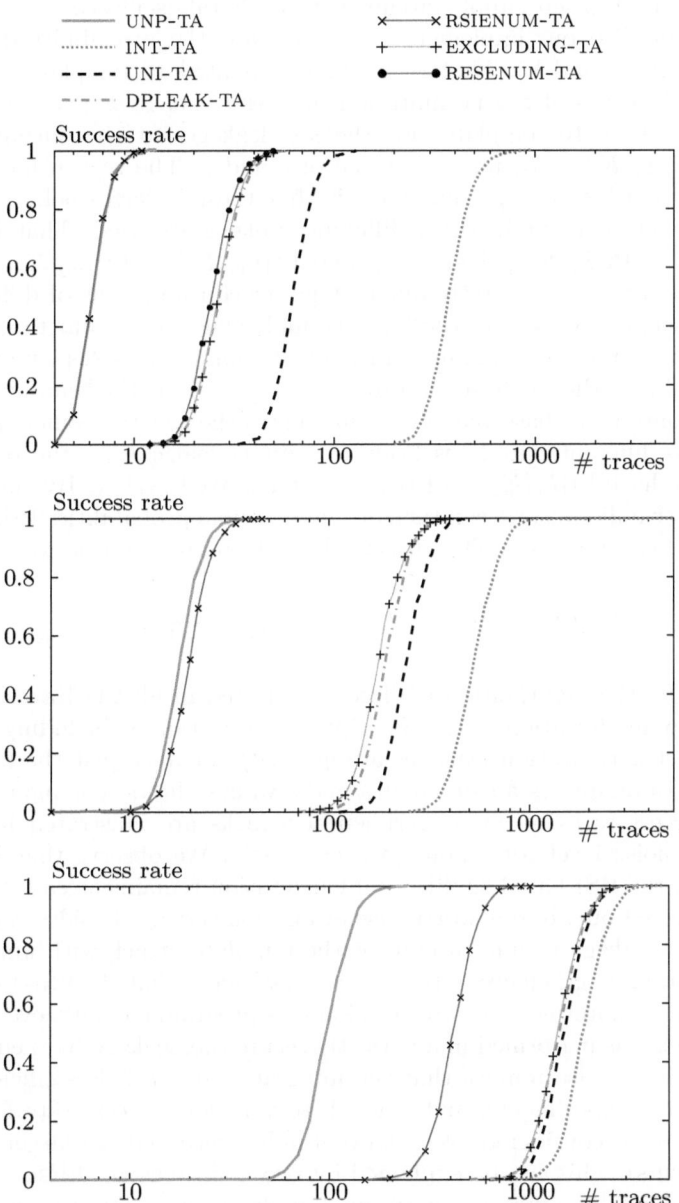

Fig. 2. Success rates of simulated attacks, $\sigma^2 = 2^{-3}$ (top), 2^0 (middle), 2^{+3} (bottom)

Our target device is an 8-bit Atmel microcontroller, and our measurement setup was monitoring the voltage variations of this target device over a small resistor inserted in our supply circuit, with a digital oscilloscope. Based on this setup, we profiled our implementation and built the probability distributions of the vectors $\mathbf{L}$ and $\mathbf{L'}$. That is, we first estimated 16 templates corresponding to the leakages of the permutation indexes c, i.e. $\Pr[L'_c|S_c = s]$. Next, we constructed 16×16 templates for the key leakages at the output of the S-box, i.e. $\Pr[L_c|K_s = k]$, for each value of c and s. The reason for having the 16×16 sets of key leakage templates is that these leakages behave differently when, at a given point in time, different subkeys are used. That is, we have $\Pr[L_c|K_{s_1}] \neq \Pr[L_c|K_{s_2}]$ if $s_1 \neq s_2$, and $\Pr[L_{c_1}|K_s] \neq \Pr[L_{c_2}|K_s]$ if $c_1 \neq c_2$. This fact is due to the slightly different power consumptions of different registers and memory accesses of the Furious implementation in our target device.

In order to limit the profiling efforts, our templates were kept univariate and constructed with the stochastic approach from [27], using the Hamming weight of the S-box outputs as base vectors. Interestingly, the fact that different key bytes give rise to different templates leads to indirect leakages on the permutation. That is, we have $\Pr[L_c|K_{s_1}] \neq \Pr[L_c|K_{s_2}]$ for a fixed cycle c. By summing over the 256 key candidates, we can then obtain marginal probabilities $\Pr[L_c = l_c|K_s]$ for all key byte indexes s. This directly leads to useful information of the type:

$$\Pr[S_c = s|L_c = l_c] = \frac{\Pr[L_c = l_c|K_s]}{\sum_{s'} \Pr[L_c = l_c|K_{s'}]}.$$

Furthermore, this information is directly reflected in all the Bayesian attacks, without any modification of the descriptions in Section 3 (including UNI-TA for which direct permutation leakages are ignored). That is, just the fact that we built 16×16 templates for different s and c values allows to exploit it.

The success rates of our experimental attacks are illustrated in Figure 3, where the noise level corresponds to $\sigma^2 = 3.25$. We observe that in this real case study, the RSI-based shuffled implementation remains as easy to attack as an unprotected one, in our worst-case evaluation setting. Besides, we note that the indirect leakage is quite useful for the template attack with uniform prior. One important consequence of this indirect leakage is that the UNI-TA could also apply to our countermeasure with randomized program memory, even if the precomputation was performed in a perfectly secure (i.e. leakage-free) environment. Interestingly, we also remark that the integrated attack is less efficient than in our simulated experiments, and is stuck to very low success rate for the data complexities we considered (yet, it eventually succeeded for larger number of measurements). This can be explained by two main reasons. First, the leakages on the permutation extracted with our templates (including the indirect ones) was larger than in our simulations, which naturally increases the gap between the integrating attack and the others. Second, the fact that different Atmel resources leak according to different models creates an additional noise for the integrating attack, due to a modeling error (i.e. these differences are lost after integration).

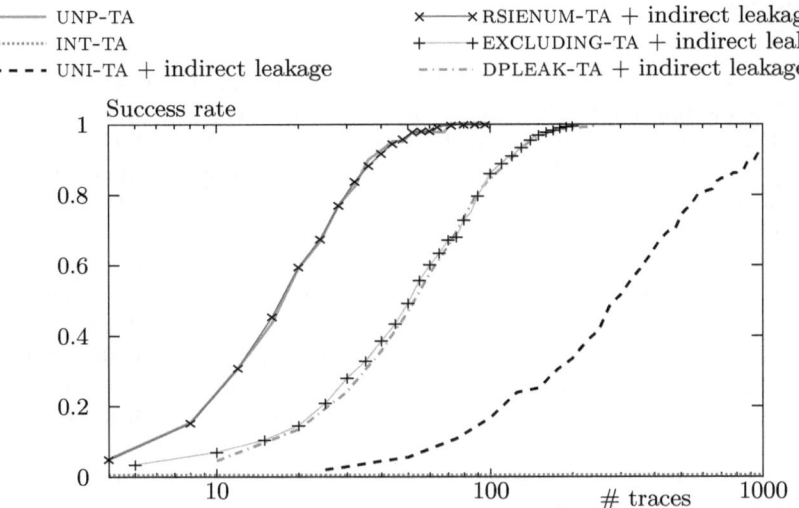

Fig. 3. Success rate of actual attacks on an ATMEL AVR implementation

6   Conclusions

In this paper, we first proposed two new implementations of the shuffling counter-measure in small (e.g. 8-bit) microcontrollers. They respectively allow improved performances in terms of overall cycle count and online cycle count. Next, we provided the first comprehensive evaluation of the shuffling countermeasure, in-cluding worst-case Bayesian attacks. For this purpose, we described intuitive formulas capturing the different variants of shuffling, and integrated them in a general evaluation framework from Eurocrypt 2009. These evaluation tools al-lowed us to show that previously used integrated attacks may not be enough for assessing the security of a shuffled implementations. We put forward that sim-plifying the permutation generation (e.g. by using RSI rather than RP) can lead to a complete breakdown of the countermeasure if not too noisy measurements are available (which turned out to be verified in a practical case study). We also explained the computational origin of these weaknesses (i.e. their relation with the total amount of permutations that are considered in the countermea-sure). Finally, we exhibited that indirect leakages may be available in shuffled implementations, due to the different leakage models of different resources. This suggest an interesting scope of further research. Namely, since our results show that randomizing the order of instructions in cryptographic implementations is not always sufficient, can we design efficient ways to randomize both the execu-tion order and the physical resources used in a cryptographic implementation?

Acknowledgements. This work has been funded in parts by the ERC project 280141 (acronym CRASH) and the 7th framework European project TAMPRES. S. Kerckhof is a PhD student funded by a FRIA grant. F.-X. Standaert is a Research Associate of the Belgian Fund for Scientific Research (FNRS-F.R.S).

References

1. Amarilli, A., Müller, S., Naccache, D., Page, D., Rauzy, P., Tunstall, M.: Can Code Polymorphism Limit Information Leakage? In: Ardagna, C.A., Zhou, J. (eds.) WISTP 2011. LNCS, vol. 6633, pp. 1–21. Springer, Heidelberg (2011)
2. Anckaert, B., Madou, M., De Bosschere, K.: A Model for Self-Modifying Code. In: Camenisch, J.L., Collberg, C.S., Johnson, N.F., Sallee, P. (eds.) IH 2006. LNCS, vol. 4437, pp. 232–248. Springer, Heidelberg (2007)
3. Atmel, http://www.atmel.com/products/microcontrollers/avr/
4. Bayrak, A.G., Velickovic, N., Ienne, P., Burleson, W.: An architecture-independent instruction shuffler to protect against side-channel attacks. TACO 8(4), 20 (2012)
5. Koç, Ç.K., Paar, C.: CHES 2000. LNCS, vol. 1965. Springer, Heidelberg (2000)
6. Chari, S., Rao, J.R., Rohatgi, P.: Template Attacks. In: Kaliski Jr., B.S., Koç, Ç.K., Paar, C. (eds.) CHES 2002. LNCS, vol. 2523, pp. 13–28. Springer, Heidelberg (2003)
7. Clavier, C., Coron, J.-S., Dabbous, N.: Differential power analysis in the presence of hardware countermeasures. In: Koç, Ç.K., Paar [5], pp. 252–263
8. Coron, J.-S.: A New DPA Countermeasure Based on Permutation Tables. In: Ostrovsky, R., De Prisco, R., Visconti, I. (eds.) SCN 2008. LNCS, vol. 5229, pp. 278–292. Springer, Heidelberg (2008)
9. Atmel Corporation. 8-bit Microcontroller with 16K/32K/64K Bytes In-System Programmable Flash - ATmega164P/V ATmega324P/V ATmega644P/V, Rev. 8011O- 07/10 (2010), http://www.atmel.com/images/8011s.pdf
10. Daemen, J., Rijmen, V.: The Design of Rijndael: AES - The Advanced Encryption Standard. Springer (2002)
11. Feldhofer, M., Popp, T.: Power Analysis Resistant AES Implementation for Passive RFID Tags. In: Lackner, C., Ostermann, T., Sams, M., Spilka, R. (eds.) Austrochip 2008, pp. 1–6 (2008)
12. Goldwasser, S., Kalai, Y.T., Rothblum, G.N.: One-Time Programs. In: Wagner, D. (ed.) CRYPTO 2008. LNCS, vol. 5157, pp. 39–56. Springer, Heidelberg (2008)
13. Herbst, C., Oswald, E., Mangard, S.: An AES Smart Card Implementation Resistant to Power Analysis Attacks. In: Zhou, J., Yung, M., Bao, F. (eds.) ACNS 2006. LNCS, vol. 3989, pp. 239–252. Springer, Heidelberg (2006)
14. Knuth, D.E.: The art of computer programming, 3rd edn. seminumerical algorithms, vol. 2. Addison-Wesley Publishing, Boston (1997)
15. Kocher, P.C., Jaffe, J., Jun, B.: Differential Power Analysis. In: Wiener, M. (ed.) CRYPTO 1999. LNCS, vol. 1666, pp. 388–397. Springer, Heidelberg (1999)
16. Mangard, S.: A Simple Power-Analysis (SPA) Attackon Implementations of the AES Key Expansion. In: Lee, P.J., Lim, C.H. (eds.) ICISC 2002. LNCS, vol. 2587, pp. 343–358. Springer, Heidelberg (2003)
17. Mangard, S.: Hardware Countermeasures against DPA – A Statistical Analysis of Their Effectiveness. In: Okamoto, T. (ed.) CT-RSA 2004. LNCS, vol. 2964, pp. 222–235. Springer, Heidelberg (2004)
18. Mangard, S., Oswald, E., Standaert, F.-X.: One for all – all for one: Unifying standard dpa attacks. IET Information Security 5(2), 100–110 (2011)
19. May, D., Muller, H.L., Smart, N.P.: Non-deterministic Processors. In: Varadharajan, V., Mu, Y. (eds.) ACISP 2001. LNCS, vol. 2119, pp. 115–129. Springer, Heidelberg (2001)

20. Medwed, M., Standaert, F.-X., Großschädl, J., Regazzoni, F.: Fresh Re-keying: Security against Side-Channel and Fault Attacks for Low-Cost Devices. In: Bernstein, D.J., Lange, T. (eds.) AFRICACRYPT 2010. LNCS, vol. 6055, pp. 279–296. Springer, Heidelberg (2010)

21. Messerges, T.S.: Using second-order power analysis to attack dpa resistant software. In: Koç, Ç.K., Paar [5], pp. 238–251

22. Moradi, A., Mischke, O., Paar, C.: Practical evaluation of dpa countermeasures on reconfigurable hardware. In: HOST, pp. 154–160. IEEE Computer Society (2011)

23. Poettering, B.: Rijndael Furious, http://point-at-infinity.org/avraes/

24. Prouff, E., Rivain, M., Bevan, R.: Statistical analysis of second order differential power analysis. IEEE Trans. Computers 58(6), 799–811 (2009)

25. Renauld, M., Standaert, F.-X., Veyrat-Charvillon, N., Kamel, D., Flandre, D.: A Formal Study of Power Variability Issues and Side-Channel Attacks for Nanoscale Devices. In: Paterson, K.G. (ed.) EUROCRYPT 2011. LNCS, vol. 6632, pp. 109–128. Springer, Heidelberg (2011)

26. Rivain, M., Prouff, E., Doget, J.: Higher-Order Masking and Shuffling for Software Implementations of Block Ciphers. In: Clavier, C., Gaj, K. (eds.) CHES 2009. LNCS, vol. 5747, pp. 171–188. Springer, Heidelberg (2009)

27. Schindler, W., Lemke, K., Paar, C.: A Stochastic Model for Differential Side Channel Cryptanalysis. In: Rao, J.R., Sunar, B. (eds.) CHES 2005. LNCS, vol. 3659, pp. 30–46. Springer, Heidelberg (2005)

28. Standaert, F.-X., Malkin, T.G., Yung, M.: A Unified Framework for the Analysis of Side-Channel Key Recovery Attacks. In: Joux, A. (ed.) EUROCRYPT 2009. LNCS, vol. 5479, pp. 443–461. Springer, Heidelberg (2009)

29. Standaert, F.-X., Veyrat-Charvillon, N., Oswald, E., Gierlichs, B., Medwed, M., Kasper, M., Mangard, S.: The World Is Not Enough: Another Look on Second-Order DPA. In: Abe, M. (ed.) ASIACRYPT 2010. LNCS, vol. 6477, pp. 112–129. Springer, Heidelberg (2010)

30. Tillich, S., Herbst, C.: Attacking State-of-the-Art Software Countermeasures—A Case Study for AES. In: Oswald, E., Rohatgi, P. (eds.) CHES 2008. LNCS, vol. 5154, pp. 228–243. Springer, Heidelberg (2008)

31. Tunstall, M., Benoit, O.: Efficient Use of Random Delays in Embedded Software. In: Sauveron, D., Markantonakis, K., Bilas, A., Quisquater, J.-J. (eds.) WISTP 2007. LNCS, vol. 4462, pp. 27–38. Springer, Heidelberg (2007)

32. Zhang, W.: State-space search - algorithms, complexity, extensions, and applications. Springer (1999)

Theory and Practice of a Leakage Resilient Masking Scheme

Josep Balasch[1], Sebastian Faust[2],
Benedikt Gierlichs[1], and Ingrid Verbauwhede[1]

[1] KU Leuven Dept. Electrical Engineering-ESAT/SCD-COSIC and IBBT
Kasteelpark Arenberg 10, B-3001 Leuven-Heverlee, Belgium
`firstname.lastname@esat.kuleuven.be`
[2] Aarhus University
Åbogade 34, DK-8200 Aarhus, Denmark
`sfaust@cs.au.dk`

Abstract. A recent trend in cryptography is to formally prove the *leakage resilience* of cryptographic implementations – that is, one formally shows that a scheme remains provably secure even in the presence of side channel leakage. Although many of the proposed schemes are secure in a surprisingly strong model, most of them are unfortunately rather inefficient and come without practical security evaluations nor implementation attempts. In this work, we take a further step towards closing the gap between theoretical leakage resilient cryptography and more practice-oriented research. In particular, we show that masking countermeasures based on the *inner product* do not only exhibit strong theoretical leakage resilience, but moreover provide better practical security or efficiency than earlier masking countermeasures. We demonstrate the feasibility of inner product masking by giving a secured implementation of the AES for an 8-bit processor.

Keywords: Inner product masking, AES, Leakage resilience.

1 Introduction

Side channel attacks (SCA) are among the most relevant threats for the security of implementations of cryptographic algorithms. Since the introduction of timing attacks to the research community in the late 1990s [22], more side channels have been discovered [13,23,25] and more powerful attacks have been developed [4,6,14]. It was soon clear that masking, i.e. concealing all sensitive intermediate values of a computation with random data, is an excellent way to prevent certain types of attacks [5,19]. As opposed to other countermeasures aiming at introducing noise in the side channel, e.g. random delays, random order execution, dummy operations, etc., one can formally argue the security masking provides.

The idea of dth order masking is to split every sensitive intermediate value in the implementation into $d + 1$ random shares, and to compute the algorithm on

X. Wang and K. Sako (Eds.): ASIACRYPT 2012, LNCS 7658, pp. 758–775, 2012.

these shares while maintaining that each tuple of d shares is independent of any sensitive value. The challenge is not to devise the masking scheme itself, i.e. to determine how a sensitive intermediate value is split, but rather to define the masked operations that process the independent shares, while still preserving the correctness of the computation. A dth order masked implementation can, in theory, always be broken by a $d + 1^{th}$ order side channel attack, i.e. an attack that exploits side channel leakage of $d + 1$ intermediate values in the masked implementation. However, given a sufficient amount of noise, attacking a dth order masked implementation becomes exponentially more difficult in d [5]. Motivated by this result, d^{th} order masking schemes (that can be implemented at any order d) based on *boolean* masking [27] and *polynomial* masking [18,24] have been recently proposed. Unfortunately, their security has so far been evaluated mainly by practice-oriented researchers, while a formal proof-driven analysis is either missing or is given only in a very weak security model.

In the theory community, masking-based countermeasures are analyzed within the framework of leakage resilient circuit compilers introduced by Ishai et al. [20]. A circuit compiler takes as input an arbitrary circuit C computing over some finite field and outputs a protected circuit C' that has the same functionality as C but comes with built-in security against certain classes of leakages. For the circuit compiler of [20] it can be shown that an adversary that learns up to d intermediate values during the computation of the transformed circuit C' does not learn anything beyond black-box access. That is, for instance, if C is an AES circuit then its implementation C' exhibits the standard black-box security even in the presence of side-channel leakage (in the given model).

The circuit compiler of Ishai et al. based on boolean masking with d masks has been recently extended, and a similar compiler (based on *any* linear secret sharing scheme) protecting against broader classes of leakages has been introduced [11]. Despite this progress, it has been suggested that masking schemes with greater algebraic complexity yield better resistance against side channel attacks. As boolean masking schemes only achieve weak provable security guarantees, attempts have been made to seek for alternatives. First examples are the compilers of Juma and Vahlis [21] and Goldwasser and Rothblum [15] which use as underlying masking a public key encryption scheme, i.e. every sensitive variable is encrypted with a suitable public key encryption scheme. While such compilers achieve strong security guarantees, namely, protection against *any polynomial-time computable* leakage function, they suffer from poor efficiency and rather provide theoretical feasibility results than a way towards a practical solution.

In two recent works [10,16], it was shown that such strong theoretical security guarantees can be achieved without relying on public-key encryption schemes. Instead, these works propose a purely information theoretic solution based on the inner product. While asymptotically these constructions are comparable to schemes based on public key encryption, they have the potential to achieve much better real-world efficiency as they only require simple algebraic operations. In this work, we show that this is indeed the case, if one is willing to accept a weaker security model. In a nutshell, our work shows that advances in leakage resilient cryptography can indeed have implications to real-world implementations and may even provide better practical security or efficiency than existing schemes.

Contributions. We rely on ideas of Dziembowski and Faust [10] for the inner product (IP) masking, and adjust the masked operations to improve their efficiency. As we are particularly interested in a secure implementation of the Advanced Encryption Standard (AES), we can exploit the linearity of the squaring operation in the underlying finite field $\mathbb{F}_{2^8}$. Moreover, we slightly simplify the masked multiplication operation of [10]. All these changes are done without affecting the theoretical security analysis. The bulk of our efficiency improvements, however, comes from using a simpler method to refresh a masked secret. Such a refreshing scheme takes as input a masked secret and outputs a masking of the same value with completely fresh randomness. The construction that we use in our implementation is essentially a simple variant of a scheme proposed in [9]. As such simple schemes only satisfy weaker security properties, we need to make additional restrictions to get a sound theoretical security analysis. We provide further details on how our changes affect the security and what additional assumptions are required in Section 4.

We also evaluate the security of the IP masking for practical parameters, i.e. when the number of shares is small. Our practical analysis reveals that the information leakage of IP masking is more than two orders of magnitude smaller than that of boolean masking for low levels of noise and the same number of shares. Finally, we detail how the AES can be implemented in a secure way using the IP masking scheme, and we provide an implementation and performance results to demonstrate its correctness and feasibility. We show that in particular non-linear operations in the IP masked domain, e.g. multiplication, clearly outperform polynomial-based masking solutions that enjoy similar algebraic complexity.

2   Inner Product Masking

In this section we introduce the circuit model assumed for the execution of the masked calculations, and we provide a detailed description of the masking scheme and its building blocks, including a complexity analysis and a comparison to other masking schemes.

Circuit Model. Following the model of Dziembowski and Faust [9,10], we consider that the target device running the masked computations contains two separate processors. Each of these processors, in the following referred to as *left processor* (P_L) and *right processor* (P_R), executes a part of the masked operations. Communication between processors is performed via a bidirectional data bus. Such a model is introduced in order to provide a framework to analyze the security of the masking scheme. As will be further explained in the following sections, its main purpose is to facilitate the assumption that P_L and P_R have completely independent side channel leakage, i.e. an adversary can only retrieve information specific to each physical processor. Notice that from a practical point of view, the required independent side-channel leakage can also be obtained by temporal (rather than physical) separation of the masked computations, e.g. in the context of sequential software implementations on a single processor.

Overview. The IP masking scheme can be instantiated to secure operations in any finite field $|\mathbb{F}| \geq 2$, such that all elements and operations in $\mathbb{F}$ can be mapped to and performed in the masked domain. This feature is extremely useful in the context of securing cryptographic applications, as the underlying field of the masking scheme can be adapted according to the characteristics of the cryptographic algorithm and/or the target platform. Without loss of generality, and driven by our goal to implement the AES, we provide in the following an efficient instantiation of the IP masking scheme for the field $\mathbb{F}_{2^8}$ of characteristic two.

Notation. We represent field elements with upper-case letters, e.g. $X \in \mathbb{F}_{2^8}$, and we use $\oplus$ to denote field addition and $\otimes$ to denote field multiplication. Vectors are represented with bold upper-case letters, e.g. $\boldsymbol{X} \in \mathbb{F}_{2^8}^n$ such that $\boldsymbol{X} = (X_1, \ldots, X_n)$. For two vectors $\boldsymbol{X}, \boldsymbol{Y} \in \mathbb{F}_{2^8}^n$ we denote by $\boldsymbol{X} \oplus \boldsymbol{Y}$ the vector addition in $\mathbb{F}_{2^8}^n$ calculated as $(X_1 \oplus Y_1, \ldots, X_n \oplus Y_n)$. The inner product $\langle \boldsymbol{X}, \boldsymbol{Y} \rangle \in \mathbb{F}_{2^8}$ is calculated as $\bigoplus_{i=1}^n X_i \otimes Y_i$.

Construction. In the IP masking scheme each sensitive variable $X \in \mathbb{F}_{2^8}$ is split into an even number of $2n$ shares such that:

$$X = L_1 \otimes R_1 \oplus \ldots \oplus L_n \otimes R_n. \tag{1}$$

We denote $\boldsymbol{L} = (L_1, \ldots, L_n)$ as *left vector* and $\boldsymbol{R} = (R_1, \ldots, R_n)$ as *right vector*. A variable X is represented in the masked domain as $(\boldsymbol{L}, \boldsymbol{R})$, and can be recovered by calculating the inner product of these two vectors, e.g. $X = \langle \boldsymbol{L}, \boldsymbol{R} \rangle$. In order to prevent a practically exploitable bias between the shares and the masked value, it is required that elements of $\boldsymbol{L}$ belong to $\mathbb{F}_{2^8} \setminus \{0\}$. We define $n \geq 2$ as the security parameter of our masking scheme.

Note that IP masking is a generalization of previously published masking schemes. Indeed, one trivially derives boolean masking from Eq. (1) by e.g. setting all elements in $\boldsymbol{L}$ (resp. $\boldsymbol{R}$) to one. Multiplicative masking [2] can be achieved by setting $n = 2$ and either of the shares L_2 and/or R_2 (resp. L_1 and/or R_1) to zero. Affine masking, described in [12] as $V = (A \otimes X) \oplus B$, can be obtained by fixing $n = 2$, $L_1 = L_2 = A^{-1}$, $R_1 = V$, and $R_2 = B$. Finally, as a secret variable in polynomial masking [18,24] is given by an interpolation polynomial in the Lagrange form, such masking scheme can be obtained by considering all elements in $\boldsymbol{L}$ to be public Lagrange coefficients.

Algorithm 1 depicts the procedure IPMask() to convert a variable X into the IP masked domain as two vectors $(\boldsymbol{L}, \boldsymbol{R})$ of size n. The function rand() returns a random element in $\mathbb{F}_{2^8}$, whereas the function randNonZero() returns a random element in $\mathbb{F}_{2^8} \setminus \{0\}$. The function IPUnmask() to convert a masked variable $(\boldsymbol{L}, \boldsymbol{R})$ of size n back to X consists in calculating the inner product $X = \langle \boldsymbol{L}, \boldsymbol{R} \rangle$.

Algorithm 1. Masking a variable: $(\boldsymbol{L}, \boldsymbol{R}) \leftarrow \texttt{IPMask}(X)$

Input: variable $X \in \mathbb{F}_{2^8}$
Output: masked variable $(\boldsymbol{L}, \boldsymbol{R})$
Ensure: $X = \langle \boldsymbol{L}, \boldsymbol{R} \rangle$

$\quad L_1 \leftarrow \texttt{randNonZero}()$
$\quad$ **for** $i = 2$ to n **do**
$\quad\quad L_i \leftarrow \texttt{randNonZero}(); R_i \leftarrow \texttt{rand}()$
$\quad$ **end for**
$\quad R_1 \leftarrow (X \oplus \bigoplus_{i=2}^n L_i \otimes R_i) \otimes L_1^{-1}$

2.1 Operations in the Masked Domain

After introducing how to convert variables between $\mathbb{F}_{2^8}$ and the IP masked domain, we need to provide a set of high-level functions that allows us to operate directly on the masked variables. In order to fulfill our security requirements, computations regarding the left vector $\boldsymbol{L}$ of masked variables should be executed in the left processor P_L, whereas calculations regarding $\boldsymbol{R}$ should be carried out in the right processor P_R. Moreover, the condition that elements of the vector $\boldsymbol{L}$ are different than zero must be inherited by all operations in order to avoid output masked values from being biased.

In the following we make use of a special operation called $\texttt{IPHalfMask}()$, which on input a variable X and a vector $\boldsymbol{L}$ calculates the corresponding vector $\boldsymbol{R}$ such that $X = \langle \boldsymbol{L}, \boldsymbol{R} \rangle$. It is thus a simplified version of Algorithm 1 for which the left vector $\boldsymbol{L}$ is already given and thus elements L_i do not need to be sampled.

Another operation that will be often used is $\texttt{IPRefresh}()$. This operation, depicted in Algorithm 2, takes as input a masked variable $(\boldsymbol{L}, \boldsymbol{R})$ and returns a new one $(\boldsymbol{L}', \boldsymbol{R}')$ such that $\langle \boldsymbol{L}, \boldsymbol{R} \rangle = \langle \boldsymbol{L}', \boldsymbol{R}' \rangle$. The purpose of the refreshing is to pump new randomness into the masking scheme. Algorithm 2 is tailored particular to work for the field $\mathbb{F}_{2^8}$. For a generalization we refer the reader to [9].

Algorithm 2. Refresh vector: $(\boldsymbol{L}', \boldsymbol{R}') \leftarrow \texttt{IPRefresh}(\boldsymbol{L}, \boldsymbol{R})$

Input: vector $\boldsymbol{L}$ in processor P_L, vector $\boldsymbol{R}$ in processor P_R
Output: vector $\boldsymbol{L}'$ in processor P_L, vector $\boldsymbol{R}'$ in processor P_R
Ensure: $\langle \boldsymbol{L}, \boldsymbol{R} \rangle = \langle \boldsymbol{L}', \boldsymbol{R}' \rangle$

P_L		P_R
$\boldsymbol{A} \in_R \mathbb{F}_{2^8}^n$		
$\boldsymbol{L}' = \boldsymbol{L} \oplus \boldsymbol{A}$	$\xrightarrow{\quad \boldsymbol{A} \quad}$	$X = \texttt{IPUnmask}(\boldsymbol{A}, \boldsymbol{R})$
	$\xleftarrow{\quad X \quad}$	
$\boldsymbol{B} = \texttt{IPHalfMask}(X, \boldsymbol{L}')$	$\xrightarrow{\quad \boldsymbol{B} \quad}$	$\boldsymbol{R}' = \boldsymbol{R} \oplus \boldsymbol{B}$

Although not clearly specified in Algorithm 2, it is necessary that the vector $\boldsymbol{A}$ sampled by P_L is such that the resulting elements of $\boldsymbol{L}'$ are non-zero. In other words, we need to ensure that $A_i \neq L_i$ for all $1 \leq i \leq n$. Details on how to implement this step efficiently, in constant time and flow are given in Section 5.

Addition. The procedure IPAdd() to calculate the addition of two masked variables is depicted in Algorithm 3. This algorithm requires a three vector additions, two joint executions of IPRefresh(), one of IPUnmask(), and one of IPHalfMask().

Algorithm 3. Masked addition: $(\boldsymbol{X}, \boldsymbol{Y}) \leftarrow$ IPAdd$((\boldsymbol{L}, \boldsymbol{R}), (\boldsymbol{K}, \boldsymbol{Q}))$

Input: vectors $\boldsymbol{L}$ and $\boldsymbol{K}$ in processor P_L, vectors $\boldsymbol{R}$ and $\boldsymbol{Q}$ in processor P_R
Output: vector $\boldsymbol{X}$ in processor P_L, vector $\boldsymbol{Y}$ in processor P_R
Ensure: $\langle \boldsymbol{X}, \boldsymbol{Y} \rangle = \langle \boldsymbol{L}, \boldsymbol{R} \rangle \oplus \langle \boldsymbol{K}, \boldsymbol{Q} \rangle$

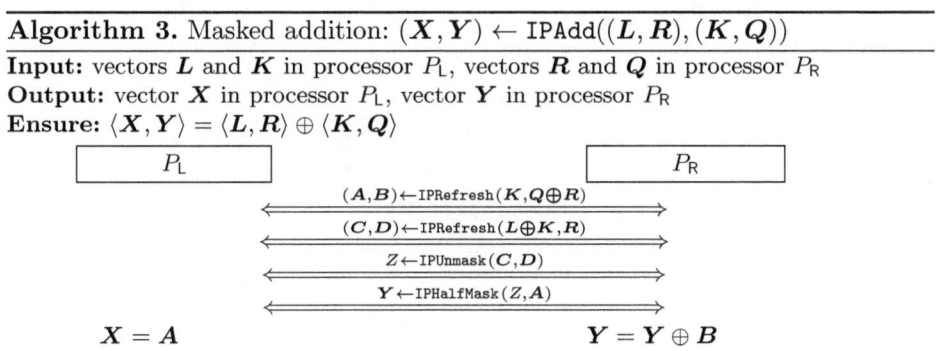

Notice that it might be the case that the component $\boldsymbol{L} \oplus \boldsymbol{K}$ in the second execution of IPRefresh() has elements equal to zero. While this is a source of first-order leakage in IP masking, i.e. the probability $\Pr(Z = 0 | (L_i \oplus K_i) = 0)$ is twice than that for any other value of Z, it is in this particular case not exploitable by an attacker. This is because $\Pr(\langle \boldsymbol{X}, \boldsymbol{Y} \rangle | Z = 0)$ is uniformly distributed, i.e. knowing that the intermediate value Z is zero does not give any information about the sensitive output value $(\boldsymbol{X}, \boldsymbol{Y})$.

Addition of a Constant. The procedure IPAddConst() to add a constant $Z \in \mathbb{F}_{2^s}$ to a masked variable $(\boldsymbol{L}, \boldsymbol{R})$ can be carried out more efficiently than Algorithm 3. Let $(\boldsymbol{L}, \boldsymbol{R})$ and Z be the input operands, and $(\boldsymbol{X}, \boldsymbol{Y})$ the output masked variable. Addition of a constant can be simply calculated by letting $\boldsymbol{X} = \boldsymbol{L}$ and $\boldsymbol{Y} = \boldsymbol{R}$, except for the first element $Y_1 = (R_1 \oplus Z) \otimes L_1^{-1}$.

Multiplication. The procedure IPMult() to calculate the multiplication of two masked variables is depicted in Algorithm 4. This algorithm requires $2n^2$ initial field multiplications, one execution of IPRefresh() with input/output vectors of size n^2, one execution of IPUnmask() with input vectors of size $n^2 - n$, one execution of IPHalfMask(), and one final vector addition.

Multiplication by a Constant. The procedure IPMulConst() to multiply a masked variable $(\boldsymbol{L}, \boldsymbol{R})$ by a constant $Z \in \mathbb{F}_{2^s}$ is efficiently computed in IP masking. Let $(\boldsymbol{L}, \boldsymbol{R})$ and Z be the input operands, and $(\boldsymbol{X}, \boldsymbol{Y})$ be the output masked variable. Multiplication by a constant can be performed by letting $\boldsymbol{X} = \boldsymbol{L}$ and calculating $\boldsymbol{Y} = (R_0 \otimes Z, \ldots, R_n \otimes Z)$. As will be further explained in Section 4, it is not necessary to execute IPRefresh() after IPMulConst().

Algorithm 4. Masked multiplication: $(\boldsymbol{X}, \boldsymbol{Y}) \leftarrow \texttt{IPMult}((\boldsymbol{L}, \boldsymbol{R}), (\boldsymbol{K}, \boldsymbol{Q}))$

Input: vectors $\boldsymbol{L}$ and $\boldsymbol{K}$ in processor P_L, vectors $\boldsymbol{R}$ and $\boldsymbol{Q}$ in processor P_R
Output: vector $\boldsymbol{X}$ in processor P_L, vector $\boldsymbol{Y}$ in processor P_R
Ensure: $\langle \boldsymbol{X}, \boldsymbol{Y} \rangle = \langle \boldsymbol{L}, \boldsymbol{R} \rangle \otimes \langle \boldsymbol{K}, \boldsymbol{Q} \rangle$

P_L		P_R
for $i = 1$ **to** n **do**		**for** $i = 1$ **to** n **do**
for $j = 1$ **to** n **do**		**for** $j = 1$ **to** n **do**
$\tilde{U}_{i*n+j} \leftarrow L_i \otimes K_j$		$\tilde{V}_{i*n+j} \leftarrow R_i \otimes Q_j$
	$\xleftrightarrow{(U,V) \leftarrow \texttt{IPRefresh}(\tilde{U}, \tilde{V})}$	
$\boldsymbol{A} = (U_1, \ldots, U_n)$		$\boldsymbol{B} = (V_1, \ldots, V_n)$
$\boldsymbol{C} = (U_{n+1}, \ldots, U_{n^2})$		$\boldsymbol{D} = (V_{n+1}, \ldots, V_{n^2})$
	$\xleftrightarrow{Z \leftarrow \texttt{IPUnmask}(C, D)}$	
	$\xleftrightarrow{Y \leftarrow \texttt{IPHalfMask}(Z, A)}$	
$\boldsymbol{X} = \boldsymbol{A}$		$\boldsymbol{Y} = \boldsymbol{Y} \oplus \boldsymbol{B}$

Squaring. The procedure `IPSquare()` can be carried out quite efficiently in the masked domain given that we work over a field of characteristic 2. Let the input masked variable be $(\boldsymbol{L}, \boldsymbol{R})$. The output masked variable $(\boldsymbol{X}, \boldsymbol{Y})$ can be calculated by squaring all elements of each vector independently, i.e. $X_i = (L_i)^2$ and $Y_i = (R_i)^2$. The masked squaring operation does not require refreshing the masks, and can be thus carried out with only $2n$ field squarings.

2.2 Complexity of Operations

The complexity of the main operations in the IP masked domain, namely addition and multiplication, is given in Table 1. We also provide a comparison with some masked operations that can be implemented at any order d, recently published in the literature for boolean and polynomial masking schemes, namely [18,24,27]. The complexity numbers are given in terms of d for all the schemes, where d indicates the number of random values in each masked variable. Recall that in IP masking, this number of random values is given by $d = 2n - 1$, with $n \geq 2$.

As shown in Table 1, the complexity of the addition operation in IP masking is slightly larger than in the other proposed methods. This is mainly due to the internal use of the `IPRefresh()` operation which, as opposed to the other masking schemes, involves several field multiplications. However, the results obtained for the multiplication operation are favourable for IP masking. In particular, both polynomial masked multiplications have complexity $O(d^3)$ while IP masked multiplications have complexity $O(d^2)$. The boolean masked multiplication has a similar complexity but, as we will show in the next sections, the masking scheme itself provides considerable less security from both practical and theoretical points of view.

Table 1. Complexity of IP masked operations and comparison to d^{th} order boolean masked operations and polynomial masked operations in the literature

Masked Operation	Scheme	Operations in $\mathbb{F}_{2^8}$			
		$\oplus$	$\otimes$	x^{-1}	Rand
ADDITION	Boolean [27]	$d+1$	-	-	-
	Polynomial [18]	$d+1$	-	-	-
	Polynomial [24]	$d+1$	-	-	-
	Inner Product	$(13d+1)/2$	$3d+3$	3	$(7d+3)/2$
MULTIPLICATION	Boolean [27]	d^2+d+1	$2d^2+2d$	-	$(d^2+d)/2$
	Polynomial [18]	$2d^3+7d^2+d$	$2d^3+5d^2+5d$	-	$2d^2+d$
	Polynomial [24]	$4d^3+8d^2+7d+2$	$4d^3+8d^2+3d$	-	$2d^2+d$
	Inner Product	$(5d^2+12d-9)/4$	$(5d^2+10d+5)/4$	2	$(3d^2+8d-3)/4$

3 Security Evaluation

In this section we evaluate the SCA resistance of IP masking and compare it to that of other masking schemes that can be implemented at any order, e.g. boolean masking and polynomial masking. We focus the analysis on the masking schemes themselves, i.e. we analyze the leakage of the shares of one masked value. We will show in the next section that the security relevant properties of IP masking carry over to the basic operations in the masked domain.

Attack Order. We begin the evaluation by deriving the minimum order for an attack against IP masking. For this we need the following definitions:

Definition 1: We say that a variable is sensitive, if it is an intermediate result in an implementation that leaks through side channels, and if it is a function of the input (resp. output), the key and possibly other constants that is not constant with respect to the key [27].

Definition 2: We say that a masking scheme is d^{th} order SCA secure, if every tuple of d or less shares is independent of the variable that is masked. Accordingly, a masked implementation of an algorithm is d^{th} order SCA secure, if every tuple of d or less intermediate variables is independent of any sensitive variable.

1st *order SCA resistance.* Clearly, IP masking with $n \geq 2$ is 1st order SCA secure. This is a simple consequence of the fact that, even if the value of one of the shares in $\mathbf{L}$ or $\mathbf{R}$ is known (in the worst case one R_i is known to be zero such that $L_i \otimes R_i = 0$), the value of the variable that is masked is still information theoretically hidden by the $\oplus$ with $n-1$ terms that are all uniformly distributed over $\mathbb{F}_{2^8}$.

2nd *order SCA resistance.* IP masking with $n = 2$ is not 2nd order SCA secure. This is because the product of two values is determined to be zero if one of the values is zero. Multiplicative masking [2] suffers from the same problem [17]. Suppose that the values of R_1 and R_2 are known to be zero. Then, $L_1 \otimes 0 \oplus L_2 \otimes 0 = s = 0$. This leads to a bias in the distribution $p(S = s | R_1 = r_1, R_2 = r_2)$, and the mutual information $I(s; (R_1, R_2))$ is non-zero.

d^{th} *order SCA resistance.* IP masking with $2n = d+1$ is SCA secure up to $n - 1^{th}$ (or $\frac{d+1}{2} - 1^{th}$) order, but not secure against n^{th} (or $\frac{d+1}{2}^{th}$) order SCA. Following the above examples, as long as the product of one pair $(L_i, R_i), i \in \{1, \ldots, n\}$ is unknown, the value of the variable s that is masked is still information theoretically hidden. On the other hand, if $\forall i \in \{1, \ldots, n\}$ the value of R_i is known to be zero, then the value of s is known to be zero. However, the probability that this case occurs is small and decreases rapidly with increasing n. More precisely, it is (2^{-8n}).

In summary, IP masking with $2n = d + 1$ can, in theory, be broken by a nth order SCA. On the other hand, similar to polynomial masking, it creates a much more complex relation between the shares than boolean masking, which is known to be more difficult to exploit. Hence, we expect IP masking with $2n = d + 1$ to provide much higher security in practice than boolean masking of order $d+1$, i.e. with the same number of random masks. Following this line, we opt to consider the leakage of all $2n$ or $d + 1$ shares in the following analysis, since an attack exploiting all shares is more powerful in an information theoretic sense, unless the noise levels are extremely high.

In polynomial masking half of the shares are non-zero public constants and the other half are random and secret masks. In particular, there is no direct correspondence to the notion of a masked variable. In the rest of the paper we refer only to the random and secret shares, and their number determines the masking order. For example, polynomial masking of order $d - 1$ uses d random and secret shares, and can theoretically be broken by a dth order SCA. We will compare polynomial masking of order $d - 1$ with boolean masking of order d ($d + 1$ shares, d masks) and with IP masking of order $2n = d+1$ ($d+1$ shares, d masks). One could expect IP masking with $2n = d + 1$ to provide a similar level of security as polynomial masking of order $d - 1$, i.e. both schemes should provide similar security when they use the same number of random and secret masks.

Information Leakage. As motivated and done in previous works [12,24,28,29], we use the mutual information between a variable and the leakage of all shares of its masked representation as a figure of merit. We estimate it using simulations. For IP masking, we set $n = 2$ and let $R_2 \in_R \mathbb{F}_{2^8}$ and $L_1, L_2 \in_R \mathbb{F}_{2^8} \setminus \{0\}$ such that $S = L_1 \otimes R_1 \oplus L_2 \otimes R_2$. Boolean masking uses $d+1$ shares $(M_1, \ldots, M_d, V)$ where the $M_i \in_R \mathbb{F}_{2^8}$ and V is computed such that $S = M_1 \oplus \ldots \oplus M_d \oplus V$ holds. We evaluate boolean masking for $d \in \{1, 2, 3\}$. Polynomial masking uses d shares $(Y_1, \ldots, Y_d)$ with $Y_i \in_R \mathbb{F}_{2^8}$ and d public Lagrange coefficients $(\beta_1, \ldots, \beta_d)$ with $\beta_i \in_R \mathbb{F}_{2^8} \setminus \{0\}$ and pairwise distinct [18]. We evaluate polynomial masking for $d \in \{2, 3\}$.

To quantify the amount of information leaked, we need to model the relation between the value of a variable and its physical leakage. We follow the approach that is usual in the literature [12,24,29]: we model that a variable leaks its Hamming weight, that each share leaks independently of all other shares, and that the leakage of each share is affected by independent Gaussian noise. The latter serves to mimic the noise effects that affect physical measurements. Putting this together, we model the leakage of IP masking as

$$\mathrm{Leak}(\mathbf{L}, \mathbf{R}) = (\mathrm{HW}(L_1) + n_1, \mathrm{HW}(R_1) + n_2, \mathrm{HW}(L_2) + n_3, \mathrm{HW}(R_2) + n_4),$$

the leakage of boolean masking as

$$\text{Leak}(M_1, \ldots, M_d, V) = (\text{HW}(M_1) + n_1, \ldots, \text{HW}(M_d) + n_d, \text{HW}(V) + n_{d+1})$$

and the leakage of polynomial masking as

$$\text{Leak}(Y_1, \ldots, Y_d) = (\text{HW}(Y_1) + n_1, \ldots, \text{HW}(Y_d) + n_d)$$

where the n_i are independent Gaussian variables with mean zero and standard deviation σ. The mutual information is then $I(S; \text{Leak}(\mathbf{L}, \mathbf{R}))$, $I(S; \text{Leak}(M_1, \ldots, M_d, V)$ resp. $I(S; \text{Leak}(Y_1, \ldots, Y_d))$. The number of measurements that a Template Attack [6], i.e. the worst case scenario of a profiled attack, requires to achieve a given success probability is directly related to this mutual information via $c \cdot I(\cdot; \cdot)^{-1}$, where the constant c is related to the success probability [29].

Figure 1 shows plots of the mutual information ($\log_{10}$) between S and the information leaked by all shares of its masked representation, over increasing noise levels σ, for all masking schemes considered[1] [2].

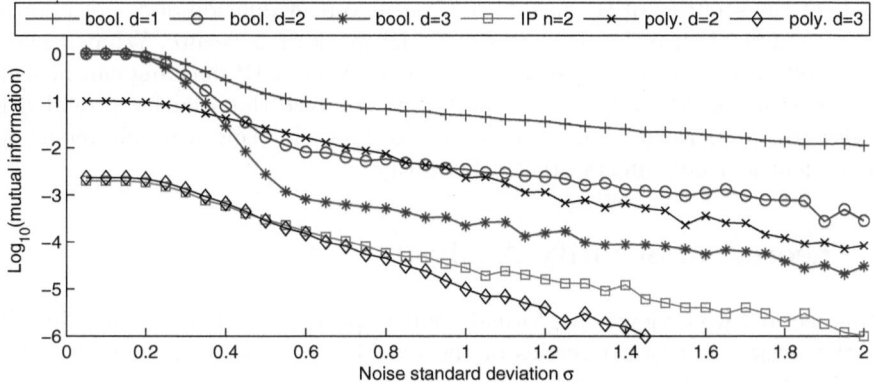

Fig. 1. Mutual information ($\log_{10}$) over increasing noise standard deviation σ for different masking schemes

The figure shows that IP masking with $n = 2$ leaks consistently less than boolean masking with $d \in \{1, 2, 3\}$ across the range of tested noise levels, which confirms our expectation. The advantage is more pronounced for low noise levels, where e.g. for $\sigma = 0.2$ the information leakage of IP masking is about 2.5 orders of magnitude(!) smaller than that of boolean masking. As expected, polynomial masking with $d = 2$ leaks consistently more than IP masking with $n = 2$. Polynomial masking with $d = 3$ provides a level of security very similar to IP masking

---

[1] Note that the mutual information values we computed for boolean masking are consistent with Figure 1 in [12] and Figure 3 in [24]. One has to take into account that the Y-axis in those figures is erroneously labeled $\log_{10}$ while it should be $\log_n$ [1].

[2] For polynomial masking with $d = 3$, reasonably accurate estimation of the mutual information values for high noise levels is beyond our computational budget.

with $n = 2$ for low noise levels. However, contrary to what one could expect, for high noise levels, polynomial masking with $d = 3$ leaks less than IP masking with $n = 2$. There are several possible explanations for this observation. For instance, IP masking with $n = 2$ involves two field multiplications while polynomial masking with $d = 3$ involves three field multiplications, i.e. the algebraic complexity of the masking is greater. Furthermore, IP masking with $n = 2$ is 1st order SCA secure while polynomial masking with $d = 3$ is 2nd order SCA secure. It is known that leakage of lower order is easier to exploit, in particular with increasing noise [29]. We leave the careful analysis of the observed difference in information leakage as an open question for future research.

Discussion. Our evaluation shows that IP masking with $n = 2$ provides high security even if there is little noise. However, although the simulated scenario (Hamming weight leakage, independent leakage of each share, Gaussian noise) is standard in the practice-oriented literature, it is synthetic and in particular meets the requirement of the masking schemes for independent leakage perfectly. It can be hard to achieve this for real-world implementations that are affected by effects such as coupling (we show in the extended version [3] that glitches do not affect the security of IP masking). Clearly, our evaluation does not allow to blindly assume that an implementation of IP masking is secure. What it shows is the level of security that a secure implementation of IP masking can provide. An interesting topic for future research is to analyze the security provided by a real-world implementation, and to analyze how violating a requirement, e.g. independent leakage, affects practical security.

4   Theoretical Security Analysis

In this section, we review some formal security properties of the IP masking. We give the basic security properties of the masking scheme itself, including very strong security guarantees with respect to non-adaptive leakages, and argue that these properties carry over to the basic operations in the masked domain. In the full version [3] we discuss further relaxations, and argue that our construction provides security against glitches similar to the results given in [24].

Notation. In the following we let $\mathbb{F}$ be a finite field, and we typically consider row vectors. We define the statistical distance between two random variables A, B over some set $\mathcal{X}$ as

$$\Delta(A; B) := \sum_{x \in \mathcal{X}} 1/2 \left| \Pr[A = x] - \Pr[B = x] \right|.$$

4.1   Security Properties of IP Masking

We have argued in Section 3 that even for small n, IP masking is robust to (noisy) Hamming weight leakage from the different shares of the masking. In this section, we back up these observations with a theoretical analysis showing strong security properties for IP masking that cannot be achieved, e.g. by linear masking schemes such as Boolean masking or masking schemes based on Shamir secret sharing. The analysis strongly relies on techniques and results from [10,11]. We repeat here part of the arguments where changes to the construction and model are required to get practical constructions. For a more formal analysis and full proof details the reader is referred to [10]. We emphasize that the theoretical analysis will typically require $n > 130$ to get meaningful security bounds.

As mentioned in Sect. 2, we assume that the device that runs the masked computation has two processors, P_L and P_R, leaking independently. Let S_L denote the state of processor P_L and S_R the state of processor P_R (resp.), then the adversary may interact with $\Omega(S_L, S_R)$ by sending functions f_L and f_R to the oracle and getting back $f_L(S_L)$ and $f_R(S_R)$. The only additional requirement that we make is that an adversary will not learn more than λ bits from each processor P_L and P_R. We call such an adversary λ-limited and denote the process of the adversary interacting with the leakage oracle by $(\mathcal{A} \leftrightarrows \Omega(\boldsymbol{L}, \boldsymbol{R}))$. For simplicity, we always assume that the output of $\mathcal{A}$ in the above leakage game is $f_L^1(S_L), f_L^2(S_L), \ldots, f_R^1(S_R), f_R^2(S_R), \ldots$. We emphasize that by modeling leakage in this way, we allow it to depend on any intermediate value that may be computed during the computation of the two processors.

To analyze the security of an IP masked value S from some finite field $\mathbb{F}$, we set $S_L := \boldsymbol{L}$ and $S_R := \boldsymbol{R}$, where $(\boldsymbol{L}, \boldsymbol{R}) \leftarrow \texttt{IPMask}_n(S)$, and let the adversary interact with the $\Omega(\boldsymbol{L}, \boldsymbol{R})$ leakage oracle. The following lemma was proven in [10].

Lemma 1. *Let $n \in \mathbb{N}$ and let $\mathbb{F}$ be such that $n \geq \log(|\mathbb{F}|)$. For any $1/2 > \delta > 0, \gamma > 0$, any two secrets $S, S' \in \mathbb{F}$ and any (unbounded) λ-limited adversary $\mathcal{A}$ we have*

$$\Delta((\mathcal{A} \leftrightarrows \Omega(\texttt{IPMask}_n(S))), (\mathcal{A} \leftrightarrows \Omega(\texttt{IPMask}_n(S')))) \leq \epsilon,$$

where $\lambda = (1/2 - \delta)n \log |\mathbb{F}| - \log \gamma^{-1} - 1$ and $\epsilon = 2(|\mathbb{F}|^{3/2-n\delta} + |\mathbb{F}|\gamma)$.

Informally, the lemma says that for any two (different) secrets S, S' no adversary can distinguish between leakage from a masking of S and a masking of S'.

As a special case, this gives us the following corollary when the underlying field is $\mathbb{F}_{2^8}$, namely:

Corollary 1. *Let $n \in \mathbb{N}$, then for any two secrets $S, S' \in \mathbb{F}_{2^8}$ and any λ-limited adversary $\mathcal{A}$, we have*

$$\Delta((\mathcal{A} \leftrightarrows \Omega(\texttt{IPMask}_n(S))), (\mathcal{A} \leftrightarrows \Omega(\texttt{IPMask}_n(S')))) \leq \epsilon,$$

where $\lambda = 3n$ and $\epsilon \leq 2^{13-0.1n} + 2^{-n}$.

Proof. Set $\delta := 0.1$ and $\gamma := 2^{-0.2n}$, then we get $\lambda = 3.2n - 0.2n - 1 = 3n$ and $\epsilon \leq 2^{13-0.1n} + 2^{-n}$. $\qquad\square$

Corollary 1 says that for sufficiently large n an adversary may learn up to $3n$ bits from each processor without being able to distinguish between a masking of S and S'. We notice that the bound on the statistical distance only gets meaningful, when $n > 130$, which, of course, is impractical.

One may ask if we can get stronger security guarantees for the masking scheme if we restrict our focus to certain special cases. To this end consider the case that the adversary cannot query adaptively the leakage oracle, i.e. he may learn only $f_L(L)$ and $f_R(R)$. In this case, it is easy to show that from the fact that the inner product is a strong randomness extractor [7,26], we can give to the adversary the entire L and up to $3n$ bits of R, and still it will be hard to decide whether (L, R) was sampled from $\text{IPMask}_n(S)$ or $\text{IPMask}_n(S')$.

Comparison with Linear Masking Schemes. We notice that linear masking schemes, such as the additive masking over finite fields [20,27], cannot achieve such strong security properties in our security model. Consider a secret $S \in \mathbb{F}$ that is masked by vectors (L, R) such that (L, R) are uniformly random in $\mathbb{F}^{2n}$ subject to the constraint that $S = \sum_i L_i + \sum_i R_i$. If we consider an adversary that can interact with $\Omega(L, R)$ then already a single field element of leakage entirely breaks the security: $f(L)$ may reveal $\sum_i L_i$, while $g(R)$ reveals $\sum_i R_i$, which together reveal S completely.

For fields of characteristic 2 such as $\mathbb{F}_{2^8}$ already a single bit of leakage suffices to learn information about the secret! Recall that in characteristic 2 fields addition works bit-wise. Similar arguments work for the polynomial masking based on Shamir secret sharing introduced in [18], as Lagrange polynomial interpolation is linear. Hence, such masking schemes can be broken in our model.

We emphasize that our leakage model includes certain classes of leakages that are very frequently used in practice, e.g. to model power consumption. One example is the Hamming weight leakage model. Of course, our theoretical analysis includes Hamming weight leakages as an adversary can learn the Hamming weight of a masked value and still the masked value remains information theoretically hidden. More precisely, as shown in Corollary 1 the IP masking remains provably secure even if an adversary learns L completely and $3n$ bits of R. As Hamming weight is a linear function we can compute the Hamming weight of (L, R) from just the Hamming weight of L *and* R separately. Notice that the Hamming weight of R can be compressed to $< 4 \log n$ bits, while according to Corollary 1 we are allowed to learn $3n$ bits of R. We emphasize that an adversary may even learn the individual Hamming weight of each share $R_1, \ldots R_n$ of the right vector and still the IP masking remains secure. This is easy to see as we can describe the Hamming weight of the n shares for sufficiently large n by at most $n \log(8) = 3n$ bits, which according to Corollary 1 an adversary may learn from R. We emphasize that for additive masking schemes, such as Boolean masking, it is not known whether such strong security guarantees hold.

4.2   Security of Masked Operations

So far, we looked at the robustness of the IP masking scheme in the presence of independent leakage, when we mask a secret value (or several secret values) and store the left part L on processor P_L, while R is stored on processor P_R. In the following, we "lift" the security analysis from just masking the secret state, e.g. the key of the AES, to arbitrary computation with masked values. More precisely, we describe why leakage from operations on masked values will not help to learn more about the masked value than just the leakage from a single masking. This can be viewed as a reduction from the security of "complicated" masked computation, to the security of a single masked value. The details of this analysis can be found in the full version.

In the security proof, we follow Dziembowski and Faust [10] and show two simple properties for the basic masked operations. These properties were introduced in [11] and are called *rerandomizing* and *reconstructability*. The first guarantees that for a masked operation the encoded output of the operation is distributed as a uniformly and independently sampled encoding. We show in the full version that all our masked operations satisfy this property. We notice that the algorithms for squaring and multiplication by a constant require only local computation, and hence do not require a refreshing.

To show reconstructability for a masked operation, we need to build a *reconstructor*. A reconstructor is a simulator that given the operations's masked inputs and outputs can reproduce the internal computation of the operation. The main requirement is that leakage from the reconstructor's output distribution (namely the internal computation) is indistinguishable from the leakage obtained from a real execution of the operation. At an intuitive level, this property guarantees that leakage from the internals of a masked operation will not reveal "more" information about the underlying secret than just the leakage from the masked inputs and outputs itself.

For practical reasons, we slightly adapt the construction of [10]. The three main differences are as follows: (1) the way in which we refresh masked secrets, (2) dedicated efficient masked operations for squaring and multiplication by a constant, and (3) a simplified masked multiplication operation (instead of a NAND we only build a simple multiplication). We discuss some details below. A more thorough discussion is deferred to the full version.

In the implementation, we use Algorithm 2 to refresh a masking of $(\boldsymbol{L}, \boldsymbol{R})$, which is a simple variant of the scheme given in [10]. To enable a security proof, we will in the following assume that the refreshing does not leak. This is required as Dziembowski and Faust show a theoretical attack on a similar refreshing scheme in [9]. Unfortunately, their attack also applies on the refreshing from Algorithm 2. The attack presented in [9] recovers the masked secret and requires an adversary to learn for n consecutive rounds the exact value of 3 field elements. While in theory such an attack completely breaks the masking scheme, we emphasize that for a real-world adversary it is very hard to learn the exact value of field elements. If learning the exact value of 3 field elements over n consecutive rounds is possible, then from a practical point of view it seems hard to argue why the adversary should not be able to learn the exact value of all $2n$ shares in one round of the refreshing. Notice also that practical SCA attacks typically require some knowledge of the inputs/outputs of the algorithm. For the refreshing algorithm this is not possible as both inputs and outputs are unknown and random. This makes attacking the refreshing a hard target. One may ask why we do not use alternative approaches of provably secure refreshing as presented in [9] and [16]. Our choice is motivated by practical limitations as existing refreshing schemes result in a quadratic blow-up.

5   Performance Evaluation

In this section we evaluate the performance and correctness of IP masking. We provide a general overview on how to implement the IP masking building blocks on an 8-bit embedded platform, and describe how to use them to protect an implementation of the AES.

5.1   Implementation of Masked Operations

The 8-bit Atmel AVR ATMega128 [8] is chosen as target platform. This device provides an advanced RISC architecture with 133 low-level instructions and it offers 128 kBytes of flash program memory and 4 kBytes of internal SRAM. The independent side channel leakage required by our model is in our implementation achieved by temporal separation, i.e. instead of using two physically separated processors P_L and P_R, we use a single 8-bit processor and we ensure independent leakage by not overlapping their respective operations.

For the sake of optimization, we have implemented all operations in assembly language. The ATMega128 does not provide an internal random number generator to implement the rand() and randNonZero() functionalities. Therefore, and only for the purposes of evaluating the implementation, the required random bytes are provided to the microcontroller externally previous to the encryption process. We note that modern devices with built-in TRNG or PRNG elements running in parallel would allow to generate such randomness internally.

Addition in $\mathbb{F}_{2^8}$ is carried out in a single clock cycle via the available XOR instruction, whereas the rest of field operations (multiplication, inversion, raisings to the power of 2) are implemented via lookup tables, requiring a total of 1,536 bytes in program memory. Besides the squaring, we have also implemented as lookup tables the rising to the powers of 4 and 16 required in the power function of the AES SubBytes step (see extended version for more details [3]). On devices with limited program memory these raisings can be alternatively carried out by consecutive squarings, effectively saving 512 bytes of program memory.

Special care has been taken in order to make the implementation not only time-constant, but flow-constant i.e. conditional execution paths, which can be a potential source of side channel leakage, have been avoided. A typical example of a function with conditional execution is the multiplication in $\mathbb{F}_{2^8}$ using log/alog tables. This method only works when both input operands are different than zero; otherwise, the result of the multiplication must be equal to zero. Implementing this routine in constant flow requires to calculate the potential outputs of *all* conditional paths, and thus it ends up requiring 22 clock cycles.

Worth mentioning is the implementation of the first part of Algorithm 2 for mask refreshing, namely sampling a vector $\boldsymbol{A}$ such that $A_i \neq L_i$ for $1 \leq i \leq n$.

This step is carried out as follows for each element A_i. First, we sample two elements $A_i' \in \mathbb{F}_{2^8}$ and $A_i'' \in \mathbb{F}_{2^8} \setminus \{0\}$. If $A_i' \neq L_i$ we simply set $A_i = A_i'$; otherwise, we assign $A_i = A_i' \oplus A_i''$. Independently of the sampled values A_i' and A_i'', this conditional statement ensures that i) the final value A_i is different than L_i, and ii) the final value of A_i is uniformly distributed over $\mathbb{F}_{2^8}$. Needless to say, such implementation is also performed in constant flow execution to prevent conditional execution branches.

5.2 Application to the AES

We have implemented and verified the correctness of a protected instance of AES using the IP masking scheme with $n = 2$. Due to space restrictions, we provide a high-level description about how to apply IP masking to the AES in the extended version of this work [3]. As shown in Table 2, our implementation requires around $1.9 \cdot 10^6$ clock cycles to perform a protected AES encryption (including on-the-fly key schedule calculation).

Table 2. Performance evaluation (in clock cycles) of AES round transformations and AES encryption with IP masking scheme with $n = 2$

AddRoundKey	SubBytes (Inverse)	SubBytes (Aff.Transf.)	ShiftRows	MixColumns	Full AES
8,796	45,632	72,128	200	27,468	1,912,000

We stress that these results should not be simply taken as an indicator to judge the practicality of IP masking, as they are obtained using a legacy general-purpose device without any type of hardware enhancements. If multiplication in $\mathbb{F}_{2^8}$ was available in the instruction set of the controller our timing for AES encryption would be instantly reduced to less than a million cycles. This could be achieved e.g. by providing instruction set extensions to the target device.

6 Conclusion

This work narrows the gap between the fields of theoretical leakage resilient cryptography and practice-oriented research, and it represents a first joint step towards the development and evaluation of common masking schemes. Although the levels of security required for each model differ considerably, we expect tighter bounds that allow to lower the value of the security parameters as the theory of leakage resilient cryptography advances. At the same time, technology advances steadily and what was impractical yesterday will be "normal" tomorrow. As a consequence one might expect that schemes such as IP masking can become practical for higher security levels.

Acknowledgment. This work was supported in part by the European Commission's ECRYPT II NoE (ICT-2007-216676), by the Flemish Government FWO G.0550.12N, by the Hercules Foundation AKUL/11/19, and by the Research Council KU Leuven: GOA TENSE (GOA/11/007). Benedikt Gierlichs is a Postdoctoral Fellow of the Fund for Scientific Research - Flanders (FWO). Sebastian Faust acknowledges support from the Danish National Research Foundation and The National Science Foundation of China (under the grant 61061130540) for the Sino-Danish Center for the Theory of Interactive Computation, within part of this work was performed; and from the CFEM research center, supported by the Danish Strategic Research Council. Josep Balasch is funded by a PhD grant within the covenant between KU Leuven and R.U. Nijmegen.

References

1. Personal communication with the respective authors (2012)
2. Akkar, M.-L., Giraud, C.: An Implementation of DES and AES, Secure against Some Attacks. In: Koç, Ç.K., Naccache, D., Paar, C. (eds.) CHES 2001. LNCS, vol. 2162, pp. 309–318. Springer, Heidelberg (2001)
3. Balasch, J., Faust, S., Gierlichs, B., Verbauwhede, I.: Theory and practice of a leakage resilient masking scheme – extended version –. Cryptology ePrint Archive (2012), http://eprint.iacr.org/
4. Brier, E., Clavier, C., Olivier, F.: Correlation Power Analysis with a Leakage Model. In: Joye, M., Quisquater, J.-J. (eds.) CHES 2004. LNCS, vol. 3156, pp. 16–29. Springer, Heidelberg (2004)
5. Chari, S., Jutla, C.S., Rao, J.R., Rohatgi, P.: Towards Sound Approaches to Counteract Power-Analysis Attacks. In: Wiener, M. (ed.) CRYPTO 1999. LNCS, vol. 1666, pp. 398–412. Springer, Heidelberg (1999)
6. Chari, S., Rao, J.R., Rohatgi, P.: Template Attacks. In: Kaliski Jr., B.S., Koç, Ç.K., Paar, C. (eds.) CHES 2002. LNCS, vol. 2523, pp. 13–28. Springer, Heidelberg (2003)
7. Chor, B., Goldreich, O.: Unbiased bits from sources of weak randomness and probabilistic communication complexity. SIAM J. Comput. 17(2), 230–261 (1988)
8. Atmel Corporation. ATmega128 data sheet
9. Dziembowski, S., Faust, S.: Leakage-Resilient Cryptography from the Inner-Product Extractor. In: Lee, D.H., Wang, X. (eds.) ASIACRYPT 2011. LNCS, vol. 7073, pp. 702–721. Springer, Heidelberg (2011)
10. Dziembowski, S., Faust, S.: Leakage-Resilient Circuits without Computational Assumptions. In: Cramer, R. (ed.) TCC 2012. LNCS, vol. 7194, pp. 230–247. Springer, Heidelberg (2012)
11. Faust, S., Rabin, T., Reyzin, L., Tromer, E., Vaikuntanathan, V.: Protecting Circuits from Leakage: the Computationally-Bounded and Noisy Cases. In: Gilbert, H. (ed.) EUROCRYPT 2010. LNCS, vol. 6110, pp. 135–156. Springer, Heidelberg (2010)
12. Fumaroli, G., Martinelli, A., Prouff, E., Rivain, M.: Affine Masking against Higher-Order Side Channel Analysis. In: Biryukov, A., Gong, G., Stinson, D.R. (eds.) SAC 2010. LNCS, vol. 6544, pp. 262–280. Springer, Heidelberg (2011)
13. Gandolfi, K., Mourtel, C., Olivier, F.: Electromagnetic Analysis: Concrete Results. In: Koç, Ç.K., Naccache, D., Paar, C. (eds.) CHES 2001. LNCS, vol. 2162, pp. 251–261. Springer, Heidelberg (2001)

14. Gierlichs, B., Batina, L., Tuyls, P., Preneel, B.: Mutual Information Analysis – A Generic Side-Channel Distinguisher. In: Oswald, E., Rohatgi, P. (eds.) CHES 2008. LNCS, vol. 5154, pp. 426–442. Springer, Heidelberg (2008)

15. Goldwasser, S., Rothblum, G.N.: Securing Computation against Continuous Leakage. In: Rabin, T. (ed.) CRYPTO 2010. LNCS, vol. 6223, pp. 59–79. Springer, Heidelberg (2010)

16. Goldwasser, S., Rothblum, G.N.: How to compute in the presence of leakage. Electronic Colloquium on Computational Complexity (ECCC) 19, 10 (2012)

17. Golic, J.D., Tymen, C.: Multiplicative Masking and Power Analysis of AES. In: Kaliski Jr., B.S., Koç, Ç.K., Paar, C. (eds.) CHES 2002. LNCS, vol. 2523, pp. 198–212. Springer, Heidelberg (2003)

18. Goubin, L., Martinelli, A.: Protecting AES with Shamir's Secret Sharing Scheme. In: Preneel, B., Takagi, T. (eds.) CHES 2011. LNCS, vol. 6917, pp. 79–94. Springer, Heidelberg (2011)

19. Goubin, L., Patarin, J.: DES and Differential Power Analysis The "Duplication" Method. In: Koç, Ç.K., Paar, C. (eds.) CHES 1999. LNCS, vol. 1717, pp. 158–172. Springer, Heidelberg (1999)

20. Ishai, Y., Sahai, A., Wagner, D.: Private Circuits: Securing Hardware against Probing Attacks. In: Boneh, D. (ed.) CRYPTO 2003. LNCS, vol. 2729, pp. 463–481. Springer, Heidelberg (2003)

21. Juma, A., Vahlis, Y.: Protecting Cryptographic Keys against Continual Leakage. In: Rabin, T. (ed.) CRYPTO 2010. LNCS, vol. 6223, pp. 41–58. Springer, Heidelberg (2010)

22. Kocher, P.C.: Timing Attacks on Implementations of Diffie-Hellman, RSA, DSS, and Other Systems. In: Koblitz, N. (ed.) CRYPTO 1996. LNCS, vol. 1109, pp. 104–113. Springer, Heidelberg (1996)

23. Kocher, P.C., Jaffe, J., Jun, B.: Differential Power Analysis. In: Wiener, M. (ed.) CRYPTO 1999. LNCS, vol. 1666, pp. 388–397. Springer, Heidelberg (1999)

24. Prouff, E., Roche, T.: Higher-Order Glitches Free Implementation of the AES Using Secure Multi-party Computation Protocols. In: Preneel, B., Takagi, T. (eds.) CHES 2011. LNCS, vol. 6917, pp. 63–78. Springer, Heidelberg (2011)

25. Quisquater, J.-J., Samyde, D.: ElectroMagnetic Analysis (EMA): Measures and Counter-Measures for Smart Cards. In: Attali, I., Jensen, T. (eds.) E-smart 2001. LNCS, vol. 2140, pp. 200–210. Springer, Heidelberg (2001)

26. Rao, A.: An exposition of bourgain's 2-source extractor. Electronic Colloquium on Computational Complexity (ECCC) 14(034) (2007)

27. Rivain, M., Prouff, E.: Provably Secure Higher-Order Masking of AES. In: Mangard, S., Standaert, F.-X. (eds.) CHES 2010. LNCS, vol. 6225, pp. 413–427. Springer, Heidelberg (2010)

28. Standaert, F.-X., Malkin, T.G., Yung, M.: A Unified Framework for the Analysis of Side-Channel Key Recovery Attacks. In: Joux, A. (ed.) EUROCRYPT 2009. LNCS, vol. 5479, pp. 443–461. Springer, Heidelberg (2009)

29. Standaert, F.-X., Veyrat-Charvillon, N., Oswald, E., Gierlichs, B., Medwed, M., Kasper, M., Mangard, S.: The World Is Not Enough: Another Look on Second-Order DPA. In: Abe, M. (ed.) ASIACRYPT 2010. LNCS, vol. 6477, pp. 112–129. Springer, Heidelberg (2010)

Erratum: Investigating Fundamental Security Requirements on Whirlpool: Improved Preimage and Collision Attacks

Yu Sasaki[1], Lei Wang[2,3], Shuang Wu[4], and Wenling Wu[4]

[1] NTT Corporation
[2] The University of Electro-Communications
[3] Nanyang Technological University
wushuang@is.iscas.ac.cn
[4] Institute of Software, Chinese Academy of Sciences

X. Wang and K. Sako (Eds.): ASIACRYPT 2012, LNCS 7658, pp. 562–579, 2012.
© International Association for Cryptologic Research 2012

DOI 10.1007/978-3-642-34961-4_46

The author missed to add the following acknowledgement to the paper:

Acknowledgement: Shuang Wu is supported by the National Natural Science Foundation of China Under Grant No.61202421.

The original online version for this chapter can be found at
http://dx.doi.org/10.1007/978-3-642-34961-4_34

Author Index